Handbook of Maize

Jeffrey L. Bennetzen • Sarah Hake

Editors

Handbook of Maize

Genetics and Genomics

 Springer

Editors
Jeffrey L. Bennetzen
Department of Genetics
University of Georgia
Athens, GA 30602-7223
USA
maize@uga.edu

Sarah Hake
USDA Plant Gene Expression Center
800 Buchanan Street
Albany, CA 94710
USA
maizesh@nature.berkeley.edu

ISBN: 978-0-387-77862-4   e-ISBN: 978-0-387-77863-1
DOI: 10.1007/978-0-387-77863-1

Library of Congress Control Number: 2008942387

Printed on acid-free paper

springer.com

Handbook of Maize

Jeff Bennetzen and Sarah Hake

Maize is one of the world's highest value crops, with a multibillion dollar annual contribution to agriculture. The great adaptability and high yield of maize as a food, feed and forage crop have led to its production on a massive scale, with acreage expanding at the expense of other crops. Maize has developed in its non-food usage, comprising a major source of ethanol for fuel in the United States. In addition, maize has occupied center stage in the transgenic plant controversy, serving as one of the first food crops with commercialized transgenic varieties. The release of a draft genome sequence of maize in 2009 will indicate the structure and gene content of the first average-size plant genome and will be the most complex genome sequenced from any organism to date.

Beyond its major agricultural and economic contributions, maize has been a model species for genetics since it was the first plant to have a genetic map, initially published by Emerson and colleagues in 1935. Such central genetic phenomena as transposable elements, nucleolar organizers, telomeres and epigenetic gene regulation were discovered first in maize, and later found to be universal eukaryotic genome properties. These key genetic contributions continue, including taking the lead in the characterization of the evolution of the highly unstable genomes so common in flowering plants.

Among plant science researchers, maize has the second largest basic science community, trailing only the Arabidopsis. Despite the size and scope of this community, a comprehensive book on the biology of maize – targeting genetics, genomics or overall biology - has not been published. Hence, a modern and comprehensive volume on the status (and future) of maize as a species for biological study is highly warranted.

Handbook of Maize: Genetics and Genomics centers on the past, present and future of maize as a model for plant genetics and crop improvement. The book includes chapters from the foremost maize experts on the role of maize in the origin of plant genetics, in modern crop improvement and in the study of genome structure, function and evolution.

Jeff Bennetzen, Ph.D. is the Norman and Doris Giles Professor of Molecular Biology and Functional Genomics at the University of Georgia, and is also a Georgia Research Alliance Eminent Scholar. He has studied the structure and evolution of the maize genome for the last 28 years.

Sarah Hake, Ph.D. is the Director of the Plant Gene Expression Center of the United States Department of Agriculture – Agricultural Research Service and University of California, Berkeley. She is also an adjunct Professor in the Department of Plant and Microbial Biology at U. C. Berkeley. She has worked on maize throughout her scientific career.

Preface

After the rediscovery of Mendel's research at the dawn of the 20th century, maize became the model genetic organism in the plant kingdom because of the simplicity of its analysis, particularly the ease of outcrossing and phenotyping. These facile technologies, rapidly joined by unmatched cytogenetics and mutagenesis, allowed the initiation of study into what has turned out to be a very complex genome. Although maize nuclear DNA content is less than half that of the average flowering plant, it is the most complicated genome in any species to be currently slated for full genome sequence analysis. Most of the maize genome, and most of the plant DNA on this planet, is comprised of transposable elements (TEs). Although first discovered in maize, TEs have been found to be ubiquitous across all the kingdoms of life and are the major motive force in the evolution of plant genome structure. Most of these TEs are kept silent, most of the time, by epigenetic regulation that itself may have arisen primarily as a mechanism to minimize the damage induced by parasite-like TEs and true parasites such as viruses. Perhaps the abundance of TEs in maize, and their close association with the great majority of genes, partly explains why the discovery that epigenetic phenomena existed and could regulate genes was also first made in maize.

In Volume II of *The Maize Handbook: Genetics and Genomics*, we have recruited chapters that describe part of the rich history of maize genetics, the history and modern practice of maize improvement, the nature and dynamics of the maize genome, the ever-strengthening toolkit of maize genetic technologies, the genetics of some important maize gene families, and a vison for the future of maize genetics. These chapters were provided by the leaders in their disciplines, both in maize and in the wider field of genetics, and they present state-of-the-art descriptions of the key areas in plant genetics and genomics.

The first two chapters describe the origins of maize genetics, featuring the exceptional contributions of East, Emerson and McClintock. Chapter 3 provides a very personal view, from Schwartz and Hannah, of the role of maize in the origin of the field called plant molecular biology. The development of one of the great glories of maize research, its outstanding collection of mutations with dramatic phenotypes, is described in Chapter 4. Chapters 5–7 describe modern maize improvement, from its US origins in the 19th Century, including its demonstration of the commercial value of hybrid vigor, through the focus on maize as a biotech target, to the current practice of maize breeding in the public and private sectors. Chapters 8–24 provide the most comprehensive description of the nature, biology and evolution of any plant genome yet published, with particular emphases on the great wealth of knowledge that continues to be amplified from TE studies. Chapters 8–11 emphasize maize genome structure and evolution, while chapter 12 by Dawe describes our current understanding of maize centromere structure and origin. Chapters 20–22 feature epigenetic characterization of maize, a field that is certain to continue to rocket towards insights of exceptional importance for understanding the evolution of complex regulatory circuits. Chapter 24 provides a comprehensive analysis of the organellar genomes in maize.

Chapters 25–34 supply a sampling of the major genetic and genomic technologies that are now in place in maize study, and indications of how these technologies are likely to improve in the foreseeable future. Chapter 28 on transposon tagging from McCarty and Meeley and Chapter 31 on transformation from Wang and coworkers describe components of vital reverse genetic tools that will soon lead maize into an era of unprecedented genetic study. Chapters 35–38 discuss some of the most important sets of genes in the maize genome, from regulatory loci to the genes responsible for cell wall synthesis (and for the possible future role of maize as a lignocellulosic source of liquid biofuels). Chapter 39 describes one person's prediction of the future of maize genetics, and its possible contributions to food, feed, energy and high-value industrial products.

In sum, Volume II of *The Maize Handbook: Genetics and Genomics* provides the most comprehensive analysis available of modern maize genetics and genomics, set in the historical context of how this important field of research arose. The visionaries who provided the chapters in *The Maize Handbook* also indicate where they see their disciplines headed, and how maize will contribute to this future. Maize has generated nearly a century of exceptional genetic discovery, and the rate of novel observations that this organism provides shows a continuation of its decades-long upward trajectory. We hope Volume II, and *The Maize Handbook* in its entirety, provides a valuable tool for understanding and further characterization of maize, other plants, eukaryotes, and the full spectrum of life on earth.

Contents

Part III   The Maize Genome

Part 1
Maize and the Origins of Plant Genetics

East, Emerson, and the Birth of Maize Genetics

Ed Coe

Abstract After the rediscovery of Mendel's work, acceptance and incorporation of the principles of heredity was natural to those whose interests were in agricultural advance. This chapter examines the circumstances and opportunities in which maize genetics research was founded, its role in establishing and spreading genetics widely, and its influence on cooperative sharing of information and data among investigators. Two scientists, E. M. East and R. A. Emerson, were largely responsible for the development of maize as a pivotal species in genetics.

1 Sources and Resources

The history of maize genetics has been offered in a number of previous reports, books, or reviews. Most appropriate it is, first, to feature for the reader some resources that provide more-extensive or more-intimate views, or offer selected perspectives. I have drawn heavily on each of these sources.

Wallace and Brown (1988) describe for novice and professional the history of major early contributions to the science of maize inheritance, to the development of a theoretical framework for modern genetics and breeding, and to hybrid corn. Every scientist who wishes to be informed about maize scientific history from Darwin and Mendel forward should have this book on (and off) the shelf.

Rhoades' (1984) personal review and memoir of the development of maize genetics and cytogenetics is central, authoritative, and captivating. He introduces his review of the early years of maize genetics with, first, an apology for any personal bias, then with "I accepted the invitation to write this account ... when I realized with something of a shock that I, now in my eighties, am one of the few geneticists living whose investigations with maize have extended from the 1920s to the

E. Coe
University of Missouri, Plant Sciences Unit
coee@missouri.edu

present." That modest statement perhaps identifies best the closeness of that review to his experiences of the times. My own claim can only be from the 1950s, overlapping with Rhoades' times, and numbers of my current colleagues have encompassed the same period. I ask their forbearance for any lack of balance or objectivity.

Peterson and Bianchi's (1999) sweeping narrative and background on genetics and breeding is focused especially on intellectual lineages and the research context of the 20th century. Their tree of scientists' pedigrees is an interesting visual map indicating some intellectual lineages. The book includes reprints of laudation articles in *Maydica* from 1977 to 1998, for maize scientists from W. A. Russell to J. D. Smith.

Crabb (1947, 1992) offers comprehensive and connected perspectives on the geneticists, botanists, agronomists, and plant breeders in public and commercial realms who led and contributed to hybrid corn and its advancements. The writing in the original and the supplemented update flows well and is difficult to lay down. Hearty helpings of information make the book an outstanding resource for both genetics and plant breeding.

Kass, Bonneuil and Coe (2005) and Coe (2001) relate lore and reality for the era in which maize genetics arose and spread. For the background of the Maize Genetics Cooperation and the News Letter, both can be recommended emphatically. More recently an extensive perspective of Emerson was prepared by Murphy and Kass (2007) on the occasion of the centennial of Cornell's Department of Plant Breeding and Genetics.

2   The Earliest Maize Genetic Studies

If you were an observant scientist in the late 1800's, consider what curiosity you might have had upon seeing purple and colorless kernels distributed over the ear: not graduated shades of color, but distinct, presence vs. absence of pigment. Similarly for plump vs. wrinkled (sugary) kernels. This would fly in the face of then-current wisdom that "like begets like" and "traits blend in progeny." Furthermore, pollen from a purple (or plump) strain, falling on a colorless (or sugary) strain, resulted in kernels with purple (or plump) endosperm, in the face of expectation that the endosperm ("albumen") was of maternal origin. The latter effect, termed xenia, is a particularly dramatic phenomenon in maize, and was under examination by a number of scientists, including Correns (1899) and deVries (1899), before the rediscovery of Mendel's work. Xenia is now understood to represent dominance in fertilized endosperms.

The re-discoverers' papers in 1900 were relatively brief acknowledgments of Mendel's work and insights. Credit belongs to Correns (1901) for a prompt, 161-page monograph, which must have required years of carefully planned, pedigreed research prior to 1900. This impressive monograph presented the first comprehensive genetic study of maize.

3 Emergence of Mendelian Research on Maize in the U.S.

Appreciation of the logic and general applicability of Mendel's laws was at least as rapid and widespread in the United States as it was in Europe. Edward Murray East (1879–1938) was in this very period a young graduate from the University of Illinois (B.S. 1900, M.S. 1904, Ph.D. 1907, the first Ph.D. in Agronomy from that institution). His first position was as an assistant chemist there in the laboratory of Cyril G. Hopkins, analyzing samples for the famous long-term selection experiments to improve protein and fat content in maize. East's first journal paper (Hopkins, Smith and East 1903) was on structure and composition of the corn kernel. In 1905, two years before receiving the Ph.D., East accepted a position at the Connecticut Agricultural Experiment Station/Yale University, then in 1909 took a position at the Bussey Institution, an endowed agriculture and horticulture institute at Harvard affiliated with the Arnold Arboretum. Many of East's early publications, including papers on maize and potato, were experiment station bulletins or reports, a regular route for dispensing scientific results and conclusions in that era (and for decades following), subject to review and approval by the Director of the Experiment Station. A seeding topic for our own century, with the title, "A mendelian interpretation of variation that is apparently continuous" (East 1910) was published as a journal paper.

Rollins Adams Emerson (1873–1947) in the same post-discovery period was a few years older than East and graduated earlier, from the University of Nebraska (B.S. 1897). Emerson's first positions were with the U.S. Dept. of Agriculture office of experiment stations (1897–1898) and assistant professor of horticulture to professor at Nebraska (1899–1914). His early publications were bulletins and reports on beans, orchards, apples, and potatoes. His first journal paper (Emerson 1910), in a later issue of the same journal as that of his future Ph.D. advisor and colleague (East 1910), was on the same seeding topic, "The inheritance of sizes and shapes in plants." During the year 1910–1911, Emerson pursued graduate study at Harvard with East, returned to Nebraska from study leave, and completed his Harvard Sc.D. in 1913. Emerson and East (1913) published together a classic paper on quantitative inheritance out of the Nebraska Agricultural Experiment Station entitled, "The inheritance of quantitative characters in maize." Cornell University in 1914 drew Emerson to the position of head of the Plant Breeding Department.

Between 1905 and 1911 maize genetics research at Yale, then at Harvard, advanced enough to provide the basis for a monumental treatment. East and Hayes (1911) published a 142-page bulletin, "Inheritance in maize", containing extensive observations, data, photographs, and analyses on Mendelian inheritance and on continuous variation. They clearly outlined technical requirements for an effective experimental object of genetics:

> "1. The genus or species under investigation should be variable ... [with] types which are differentiated by definite characters easy of determination. That is, the differences should be largely qualitative and not quantitative.
> 2. The different types should be freely fertile *inter se* ...

3. The flower structure should be such that the technique of crossing and selfing is simple and accurate.
4. … the subjects should return a large number of seed per operation …
5. The flowering branches should be numerous …
6. Seed should be viable for several years in order that different generations may be compared at the same time …
7. The subject material should be "workable" cytologically in order that it may be attacked from both standpoints."

The latter point was particularly prescient, and the advantages and disadvantages of maize were defined by the authors. With this paper East and Hayes made a signal contribution that instituted maize as an instrument for genetic study.

4   Institutions and Disciplines

Maize was, and is, both a biological curiosity and a major cereal crop. With the discovery of the laws of inheritance, maize served a central role in supporting the laws and in bridging between theory and application. Research investigators found a climate open to exploration of new territory and open to the establishment of science statesmanship and stature from agricultural roots.

4.1   Discipline Boundaries and Faculty Qualifications

In land-grant institutions during the first two dozen years of the 20th Century the classical boundaries of biological disciplines, Botany, Zoology, Chemistry, and Geology, began to be blurred by cross-discipline academic units. The growth of Agricultural Experiment Stations and Colleges of Agriculture emphasized agricultural development. Freshly minted disciplines included Agronomy, Plant Breeding, Soil Science, Plant Pathology, Entomology, and Biochemistry, which flowered by inter-fertilization among the classical disciplines and merging of theoretical with practical science.

Qualifications for professorial positions in teaching and research in this transitional period were often defined by letters of recommendation from respected scholars, and were at the discretion of department heads and deans. As indicated earlier, East was selected for a position at the Connecticut Agricultural Experiment Station/Yale University two years before completion of the Ph.D., and Emerson was on the faculty at Nebraska for 13 years before completing the Ph.D.

4.2   Birth from Agricultural Focus

Readers will have noted that both East and Emerson began their careers with an agricultural emphasis, extending from practice-oriented research and service to basic-knowledge research. As a matter of fact, the lifetime careers of both were at the interface between agricultural experiment stations and liberal arts institutions. Certain it is that this synergy stimulated the development of genetic research, at least on crop plants. Paul and Kimmelman (1986) highlight this synergy, for the innovation of hybrid corn as an example in the history of biology in America:

> "The development of hybrid corn was no simple matter of the transfer of theoretical science from an elite academic to an applied commercial context. Much of the theoretical work that made hybrids possible was pursued at institutions concerned with improving the efficiency and productivity of agriculture. In the 1880s, agricultural administrators began to promote hybridization as part of an effort to produce commercially viable new varieties. Ongoing interest in hybridization, in turn, underlay an enthusiastic response to Mendelism among researchers at the USDA and at agricultural colleges and experiment stations. ...
>
> "Mendelism between 1900 and 1910 was thus an applied science, in the literal sense of both; it was surely applied, and it was certainly science. The rapid development of genetics within an agricultural context, where breeding, selection techniques, hybridization, and even evolutionary issues had been addressed in the late nineteenth century, endowed Mendelism in the United States with a strongly practical and popular aspect. It also ensured that fundamental problems in genetics would be addressed within institutions oriented to practical ends – and that the subsequent development of genetic research would often reflect dominant social and economic interests in American agriculture."

Hybrid corn unquestionably is the most dramatic early example of the uniting of genetic theory with genetic application. The complementary phenomena of inbreeding depression and hybrid vigor for size traits were studied intensively and were interpreted as multifactorial Mendelian systems by East (1910), by Emerson (1910), by East and Hayes (1911) and by Emerson and East (1913). East and Hayes (1911) state tellingly:

> "... we do not attempt to analyze our results further than to say that they do show segregation in every case. And segregation is held to be the important and essential feature of Mendelism. Therefore we believe that size characters mendelize."

Rhoades (1984) briefly traces the genetic birth of hybrid corn, from Shull (1909) on pure line hybrids to Collins (1910) and Hayes and East (1911) on varietal hybrids to Jones (1918) on the double cross. The books by Wallace and Brown (1988) and by Crabb (1947, 1992) embrace the remarkable history of development of hybrid corn, influenced by the power of theory and by the dedication of enterprising individuals.

4.3 Freaks vs. Curiosities vs. Curiosity

Among the diverse races of maize, color variations were present in the aleurone layer of the endosperm, in the inner endosperm, in the pericarp, and in the tissues of the plant body. Variations also were present in endosperm texture (sugary, shrunken, or floury) and in the stature of plants and in the form of individual parts. Along with several heritable curiosities or 'freaks,' including anther-ear, dwarf, liguleless, striped leaves, chlorophyll reduction, tunicate, tassel seed, and ramosa, variations early-on offered a generous buffet for analysis of inheritance and genetic influences on development, anatomy, and phylogeny. East, Emerson, and G. N. Collins with J. H. Kempton each published extensively on inheritance of variations between 1910 and 1920. Notably the first demonstration of linkage in maize was for aleurone color, $c1$, with waxy, $wx1$, by Collins (1912).

A series of brief publications on inheritance of specific variants was instituted in the third post-rediscovery decade. Beginning in 1920 and continuing for over two decades, descriptive notes on particular variants, often with photographs, were published under a common header in the *Journal of Heredity*, the journal initiated ten years earlier by the American Breeders Association. The first maize note, by Collins and Kempton (1920), was assertively (and appropriately) subtitled, "Heritable Characters of Maize. I. Lineate Leaves: Description and Classification of Lineate Plant—Value of Maize as Material for Investigation, and Economic Importance of Discovering Latent Variations." The authors' intent in initiating this series was "… with the idea of facilitating reference, it is proposed to inaugurate a numbered series. … It is to be hoped that other workers with maize will find this a convenient place in which to publish illustrations and brief descriptions of their discoveries." Subsequent articles included ones by Emerson on tassel-seed, Kempton on brachytic culms, D. F. Jones on defective seeds, and Kempton on adherence, etc. By 1935, two dozen authors had described a total of 49 "classical" variants. Three more notes were published through 1953. Most of these variants, including several dominants, are still available for study from the collection at the Maize Genetics Cooperation Stock Center. Only a handful of the genes have been identified as to biochemical function or have been cloned to date, despite their many tantalizing developmental effects.

From where did these variants come in this early period? Some were displayed in farm shows – Emerson (1920) cites the interesting provenance of tassel-seed1: 'In the "freak" class at the Annual Corn Show held at Lincoln, Nebraska, in the winter of 1913–14, there was exhibited a corn tassel with a heavy setting of seeds.' On the other hand, other unusual strains that were grown as botanical curiosities, e.g., *Zea tunicata* (pod corn), and *Zea ramosa*, proved to be monogenic. The origins identified in descriptive notes often are from crosses between otherwise-normal races, or from variety collections. For example, waxy was found as a characteristic in a Chinese "glutinous" variety (Collins 1909). Few of the classical variants, including anthocyanin and chlorophyll as well as form variations, can be termed 'mutations' in the absence of molecular information because they were not discovered under defined experimental circumstances.

5   Emergence of Cooperation in Maize Genetics Research

Emerson posed the vision of maize workers cooperating "to such an extent that we can cover the field more quickly" in a letter to D. F. Jones in 1918 (Kass et al. 2005). Informal "Cornfest" meetings of colleagues, instituted by Emerson, occurred during the next 10 years and likely included discussions of common needs and interests. The "Heritable Characters" series was initiated in this period, with the same motivation.

A pace-setting tour de force was that by Emerson (1921a) on the complexities of the genetics of plant color, in which the remarkable variety of color expressions was codified to combinations involving a few genes (*a1, b1, pl1, r1*) that interact to determine the time, place, quality, intensity, and induction of anthocyanin pigmentation. A foundation for understanding the plant color interactions was set by prior analyses of aleurone color factor interactions (*a1, c1, C1-I, pr1, r1*) by East and Hayes (1911) and by Emerson (1912a, 1918). On these interactions and the regulation of the pathway rest a large part of the tools, regulation theory, imprinting, and paramutation systems for which maize is known.

Agreement on a symbolization system for genes was needed. Emerson raised Factor Notation issues in a 6-page mimeograph (mimeo) that was sent to maize colleagues, dated March 7, 1923 (reprinted in the Maize Genetics Cooperation Newsletter, *MNL*, 52:147–149). The letter posed alternatives for gene symbols and for linkage group numbering, and asked for preferences. A prelude to Maize Genetics Cooperation is evident in the purpose and collegiality of this mimeo. The end result is a simple, unambiguous system, which has been little modified subsequently, e.g., by adapting from subscripts and superscripts to linear symbols.

The first Maize Genetics Cooperation Newsletter (though not defined by number until the 1940 issue, number 14) was a mimeographed report from Emerson, dated April 12, 1929 (Kass et al. 2005). It began with often-quoted words, "You who attended the "cornfab" in my hotel room at the time of the winter science meetings in New York will recall that I promised to prepare a summary of the published data involving linkage groups in maize, to add my own unpublished data, and to send these records to each of you for criticism and the addition of such unpublished records as you may care to furnish me." With the two-page transmittal letter, which included a list of those who had responsibility for specific linkage groups, were 30 impressive pages of legal-size tables of shared compiled data and "rainbow" linkage maps. This mimeo was reprinted in *MNL* 53. During the next year, the shared information advanced greatly, shown in maps transmitted April 17, 1930 and data tables transmitted July 26, 1930. The influence of the Maize Newsletter spread: other newsletters for other species, patterned after that for maize, soon followed suit (Rhoades 1984). Coe and Kass (2005) tabulate *MNL* issues through 2004 with annotations about content. Distribution of *MNL* on an annual basis started with number 9, March 16, 1935, yet formal assignment of a volume number only came with volume 14, March 5, 1940. By then over 350 pages of data and descriptions had been shared among "Cooperators."

The landmark paper, "A Summary of Linkage Studies in Maize," (Emerson, Beadle and Fraser 1935), was much more significant than just tabulated data shared by Cooperators. It promoted the qualities and advantages of maize for genetic studies and celebrated the spirit of cooperation that provided for the compilation. It provided scientific descriptions of genes and their variants, tables of gene interactions, and linkage maps with documentation, prepared by Rhoades. The maps served as the skeleton on which subsequent linkage maps were built for 70 years, until molecular marker maps could be adequately interdigitated with gene-to-gene recombination data (Coe and Schaeffer 2005). Until the advent of the Maize Database, the 1935 summary was the core handbook for gene and linkage information.

The 1935 Summary was not held to be a theoretical tome, rather an aid to advancement. Emerson expressed appreciation to colleagues and a wish that the work would help investigators in their research: "The tables of data bear abundant evidence of the writers' indebtedness to the many investigators who have contributed previously unpublished records, an almost unique example of unselfish cooperation. It is the writers' hope that this summary presentation will prove to be sufficiently helpful to the contributors to compensate them in some measure for their aid in its preparation." His points epitomize the nature and intent of the Maize Genetics Cooperation, which was never structured as a membership organization but rather as an opportunity among investigators in maize genetics (Coe 2001; Kass et al. 2005).

6   Origins of East and Emerson "Schools" and Synergies

As in any field, in maize genetics and breeding healthy competition and independent corroboration have been mutually beneficial, just as cooperation has yielded hidden discoveries. East's Illinois/Yale/Harvard and Emerson's Nebraska/Harvard/Cornell platforms together produced a climate of effective training, collaboration, and interaction.

6.1   Teaching, Training, and Excitement

Figure 1 gives a representation of some of the academic and intellectual pedigrees of maize investigators. The drawing is limited to individuals who are no longer living. It attempts only to track primary academic origins and paths that could be identified readily, rather than to track influences (e.g., influences on postdoctorals, visiting scientists, or undergraduates). Influences are very significant indeed, and deserve scholarly treatment that is not attempted here.

Students of East, himself trained in applied botany and chemistry, developed a marked range of career emphases in maize biology, including its taxonomy, phylogeny, morphology, breeding, quantitative genetics, origin, chemistry, mutability, cytogenetics, heterosis with its applications, biochemistry, recombination, and genetic structure. Breeding, breeding training, and quantitative genetics theory stand out as part of a continuum that includes basic biological and biochemical research.

Emerson was one of East's students who had himself already established a research career in theoretical qualitative and quantitative genetics. Students of Emerson, himself trained in applied botany, developed also a marked range, including breeding, quantitative genetics, morphology, development, linkage, recombination, mutagenesis, cytogenetics, biochemistry, and cytoplasmic inheritance. The famous

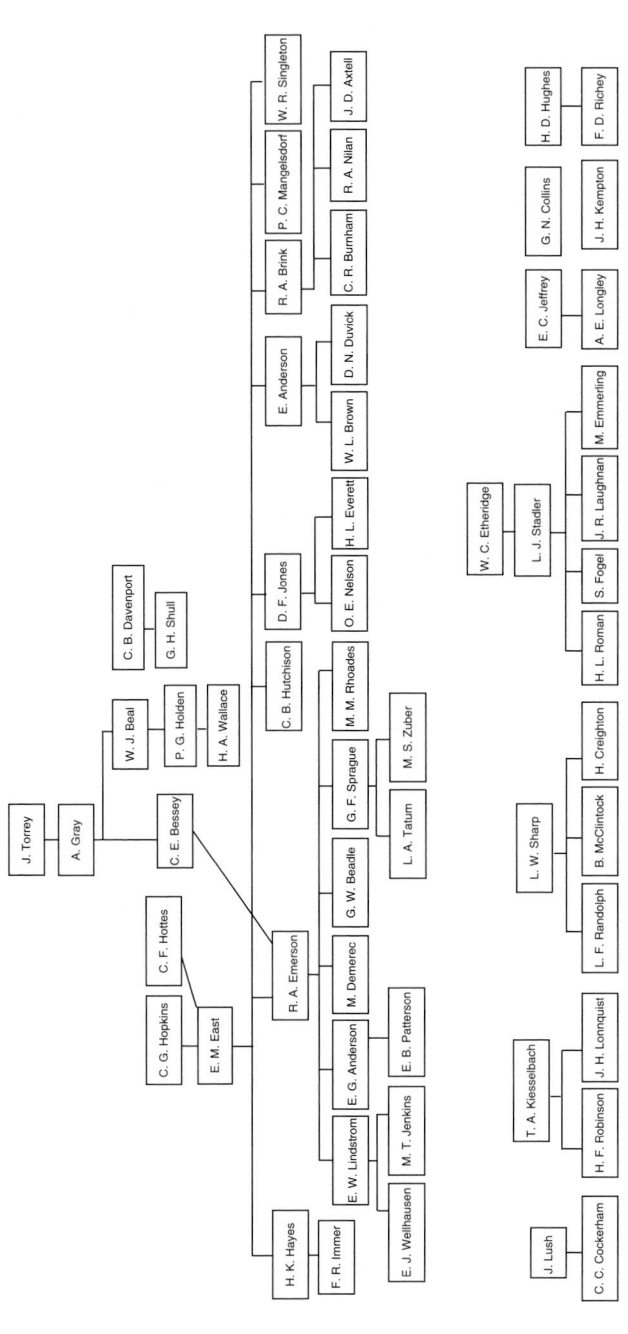

Fig. 1 Some academic and intellectual relationships in maize genetics. Students from the same mentor are shown chronologically on the horizontal

1928 photograph of Emerson with Graduate Students Beadle and Rhoades, National Research Council Postdoctoral Fellow Burnham, and Instructor McClintock, is but one sample. Training in the Emerson plots, of future international leaders in genetics and its applications in breeding, can be tasted in a 1927 photograph (see Kass et al. 2005) and in Rhoades (1984).

Lewis John Stadler, "One very prominent member of the first generation of maize geneticists who did not work for the PhD degree under either Emerson or East" (Rhoades 1984), went to Cornell in 1919 for graduate study, where his promise went unrecognized. It is noteworthy that, after returning to Missouri and completing the Ph.D. in 1922, Stadler developed a significant genetics program (Roman 1988) and an impressive "school" of students, postdoctorals, and associates (see 1938 photo in Kass 2003, and Redei 1971).

Many of the early scientists who carried out theoretical experiments in genetics, whether on single genes, interactions, or quantitative inheritance, also studied and conducted crop improvement per se (e.g., Jones, Hayes, Brink, Sprague, Stadler). For decades public breeders' common goal brought them together in close cooperation, planning and conducting research jointly and sharing data and germplasm. These interactions paralleled the Maize Genetics Cooperation and shared many of the same investigators, setting the knowledge foundation for quantitative genetics and hybrid vigor while contributing in a major way to advancing productivity during the early period of hybrid development. Collaborations with pathologists and entomologists likewise have contributed knowledge on inheritance and genes for resistance.

The excitement of discovery during the first 35 years of maize cytogenetics can be felt in 17 discoveries listed by Rhoades (1984). These include visualization of the individual chromosomes and the relations of linkage groups and chromosomal aberrations to them; the process of crossing over; chromosome behavior; breakage-fusion-bridge cycle; nucleolar organizer; radiation-induced mutations; and cytoplasmic inheritance. Emphasis can be added that Barbara McClintock was the central discoverer for many (Kass and Bonneuil 2004, Kass 2005). Some parallel exciting discoveries in genetics in the early period include dominance as the basis of xenia in the endosperm (Correns 1901, East and Hayes 1911); genetic modifications of endosperm properties (Correns 1901; East and Hayes 1911); glutinous (waxy) maize (Collins 1909); oil and protein content (Hopkins et al. 1903, East and Jones 1920); inbreeding depression (Shull 1909); hybrid vigor and heterosis (Shull 1910; Hayes and East 1911; Jones 1917, 1918); allelisms and interactions in anthocyanin formation (Emerson 1912a, 1918, 1921a; Anderson 1924; Emerson and Anderson 1932); linkage and recombination (Collins 1912); modifications of plant structure (East and Hayes 1911, Emerson 1912b, 1912c; Collins 1917); dosage effects on chemical content (East and Hayes 1911, Mangelsdorf and Fraps 1931); modifications in chlorophyll (Emerson 1912c, 1912d; Demerec 1924a); mutability (Emerson 1914, 1917; Anderson and Eyster 1928, Rhoades 1936); quantitative characters (Emerson and East 1913); compound loci (Emerson 1917); loss of chromosomes in sectors (Emerson 1921b); waxy expression in endosperm and pollen (Weatherwax 1922, Brink and MacGillivray 1924, Demerec 1924b); chromosome homology among races and relatives (Longley 1924); starch structure (Brink and Abegg

1926); gametophyte factors (Mangelsdorf and Jones 1926); supernumerary chromosomes (Longley 1927); hetero-fertilization (Sprague 1929); and spontaneous mutation (Stadler 1942).

Convergences and synergies with further disciplines such as cytology, statistics, biochemistry, physiology, and morphology, have contributed strongly to new insights.

6.2 Plant Scientists Burgeon and Branch Out

Because it emphasizes research on maize, Figure 1 does not adequately reflect the diverse careers of scientists trained in these schools. More than a few of those shown in the figure used additional objects of research, including floral species, beans, tobacco, cotton, sorghum, barley, oats, yeasts and other fungi, mammals, bacteria, and Drosophila, and broke ground in new theory directions, including population, biochemical, and molecular genetics. Not shown in the figure are several others of East's students, including N. H. Giles, C. Rick, K. Sax, E. R. Sears, H. H. Smith, and C. P. Swanson, whose careers developed notable impacts outside of maize. The contributions of the scientists trained in these schools, and those they trained, succeeded in branching out genetics, cytogenetics, and breeding among plants and other organisms.

7 Conclusions

Maize is a pivotal species in genetics, both as an object of theoretical study and as a crop that is subject to genetics-driven improvement. Its role naturally derived from its properties of trait variations, cytogenetic malleability, and reproductive generosity. Most important, however, are its investigators, with whom vigorous, collaborative, and cooperation-oriented spirit is a tradition. This tradition originated during the first few decades of the 20th Century, led largely by E. M. East and R. A. Emerson. Both trained students who became significant contributors to the field of genetics. Emerson sowed the seeds for sharing of maize linkage data in the 1920s, in informal meetings and in correspondence. The *Maize Genetics Cooperation Newsletter*, in which data were freely shared, and the Maize Genetics Cooperation, in which stocks were freely shared, became a paradigm for other species.

Acknowledgments I wish to express my appreciation to James Coors, John Dudley, Irwin Goldman, Major Goodman, Arnel Hallauer, Lee Kass, Jerry Kermicle, Royse Murphy, Ron Phillips, William Tracy and Forrest Troyer for their contributions to this historical account.

References

Anderson, E. G. (1924) Pericarp studies in maize. II. The allelomorphism of a series of factors for pericarp colors. Genetics 9, 442–453.

Anderson, E. G. and Eyster, W. H. (1928) Pericarp studies in maize. III. The frequency of mutation in variegated maize pericarp. Genetics 13, 111–120.

Brink, R. A. and Abegg, F. (1926) Dynamics of the waxy gene in maize. Genetics 11, 163–199.

Brink, R. A. and MacGillivray, J. (1924) Segregation for the waxy character in maize pollen and differential development of the male gametophyte. Amer. Jour. Bot. 11, 465–469.

Coe, E. (2001) The origins of maize genetics. Nat. Rev. Genet. 2, 898–905.

Coe, E. and Kass, L. B. (2005) Maize Genetics Cooperation News Letter files: Expanded chronological list of materials and related cooperation. Maize Genetics Cooperation Newsletter 79, 72–76.

Coe, E. and Schaeffer, M. L. (2005) Genetic, physical, maps, and database resources for maize. Maydica 50, 285–303.

Collins, G. N. (1909) A new type of Indian corn from China. USDA Bur. Plant Indust. Bull. 161, 1–30.

Collins, G. N. (1910) The value of first-generation hybrids in corn. US Bur. Plant Indus. Bull. 191, 1–45.

Collins, G. N. (1912) Gametic coupling as a cause of correlations. Am. Nat. 46, 559–590.

Collins, G. N. (1917) Hybrids of Zea tunicata and Zea ramosa. Proc. Natl. Acad. Sci., USA 3, 345–349.

Collins, G. N. and Kempton, J. H. (1920) Heritable characters of maize. I. Lineate leaves. J. Hered. 11, 3–6.

Correns, C. (1899) Untersuchungen uber die Xenien bei Zea Mays. Berichte Deut. Bot. Ges. 17, 410–417.

Correns, C. (1901) Bastarde zwischen Maisrassen, mit besonderer Berucksichtigung der Xenien. Bibliotheca Bot. 53, 1–161.

Crabb, A. (1947, 1992) *The Hybrid-Corn Makers: Prophets of Plenty.* Rutgers Univ. Press, New Brunswick, New Jersey.

Demerec, M. (1924a) Genetic relations of five factor pairs for virescent seedlings in maize. Cornell Univ. Agric. Exp. Stn. Memoir 84, 1–38.

Demerec, M. (1924b) A case of pollen dimorphism in maize. Amer Jour Bot 11, 461–464.

de Vries, H. (1899) Sur la fecondation hybride de l'albumen. Comptes Rendus Séances L'Acad. Sci. 129, 973–975.

East, E. M. (1910) A mendelian interpretation of variation that is apparently continuous. Am. Nat. 44:65–82.

East, E. M. and Hayes, H. K. (1911) Inheritance in maize. Conn. Agric. Exp. Stn. Bull. 167, 1–142.

East, E. M. and Jones, D. F. (1920) Genetic studies on the protein content of maize. Genetics 5, 543–610.

Emerson, R. A. (1910) The inheritance of sizes and shapes in plants. A preliminary note. Am. Nat. 44, 739–746.

Emerson, R. A. (1912a) The unexpected occurrence of aleurone colors in F2 of a cross between non-colored varieties of maize. Am. Nat. 46, 612–615.

Emerson, R. A. (1912b) The inheritance of the ligule and auricles of corn leaves. Nebr. Agric. Exp. Stn. Ann. Rep. 25, 81–88.

Emerson, R. A. (1912c) The inheritance of certain "abnormalities" in maize. Amer Breeders Assoc. Ann. Rpt. 7/8, 385–399.

Emerson, R. A. (1912d) The inheritance of certain forms of chlorophyll reduction in corn leaves. Nebr. Agric. Exp. Stn. Ann. Rep. 25, 89–105.

Emerson, R. A. (1914) The inheritance of a recurring somatic variation in variegated ears of maize. Am. Nat. 48, 87–115.

Emerson, R. A. (1917) Genetical studies of variegated pericarp in maize. Genetics 2, 1–35.

Emerson, R. A. (1918) A fifth pair of factors, Aa, for aleurone color in maize, and its relation to the C and Rr pairs. Cornell Univ. Agric. Exp. Stn. Memoir 16, 225–289.

Emerson, R. A. (1920) Heritable characters of maize. II. Pistillate flowered maize plants. J. Hered. 11, 65–76.

Emerson, R. A. (1921a) The genetic relations of plant colors in maize. Cornell Univ. Agric. Exp. Stn. Memoir 39, 1–156.

Emerson, R. A. (1921b) Genetic evidence of aberrant chromosome behavior in maize endosperm. Am. J. Bot. 8, 411–424.

Emerson, R. A. and Anderson, E. G. (1932) The A series of allelomorphs in relation to pigmentation in maize. Genetics 17, 503–509.

Emerson, R. A., Beadle, G. W. and Fraser, A. C. (1935) A summary of linkage studies in maize. Cornell Univ. Agric. Exp. Stn. Memoir 180, 1–83.

Emerson, R. A. and East, E. M. (1913) The inheritance of quantitative characters in maize. Nebr. Agric. Exp. Stn. Res. Bull. 2, 1–120.

Hayes, H. K. and East, E. M. (1911) Improvement in corn. Conn. Agric. Exp. Stn. Bull. 168, 3–21.

Hopkins, C. G., Smith, L. H., and East, E. M. (1903) The structure of the corn kernel and the composition of its different parts. Illinois Agric. Exp. Stn. Bull. 87, 79–112.

Jones, D. F. (1917) Increasing the yield of corn by crossing. Conn. State Sta. Ann. Rpt. 1915/16, 323–347.

Jones, D. F. (1918) The effects of inbreeding and cross-breeding upon development. Conn. Agric. Exp. Stn. Bull. 207, 5–100.

Kass, L. B. (2003) Records and Recollections: A new look at Barbara McClintock, Nobel-Prize-winning geneticist. Genetics 164, 1251–1260.

Kass, L. B. (2005) Missouri compromise: tenure or freedom? New evidence clarifies why Barbara McClintock left academe. Maize Genetics Cooperation Newsletter 79, 52–71.

Kass, L. B. and Bonneuil, C. (2004) Mapping and seeing: Barbara McClintock and the linking of genetics and cytology in maize, 1928–35. In: H. Rheinberger and J. P. Gaudilliere (Eds.), *Classical Genetic Research and its Legacy: The mapping cultures of twentieth-century genetics.* Rutledge, New York, pp. 91–118.

Kass, L. B., Bonneuil, C., and Coe, E. (2005) Cornfests, cornfabs and cooperation: the origins and beginnings of the maize genetics cooperation news letter. Genetics 169, 1787–1797.

Longley, A. E. (1924) Chromosomes in maize and maize relatives. J. Agric. Res. 28, 673–682.

Longley, A. E. (1927) Supernumerary chromosomes in Zea mays. J. Agric. Res. 35, 769–784.

Mangelsdorf, P. C. and Fraps, G. S. (1931) A direct quantitative relationship between vitamin A in corn and the number of genes for yellow pigmentation. Science 73, 241–242.

Mangelsdorf, P. C. and Jones, D. F. (1926) The expression of Mendelian factors in the gametophyte of maize. Genetics 11, 423–455.

Murphy, R. P. and Kass, L. B. (2007) *Evolution of Plant Breeding at Cornell University: A Centennial History, 1907–2006.* 98 pp., Appendices A1-A98, PhotoSection P1–P38. Department of Plant Breeding & Genetics, Cornell University, Ithaca, New York.

Paul, D. and Kimmelman, B. (1986) Mendel in America–theory and practice, 1900–1919. In: R. Rainger, K. R. Benson and J. Maienschein (Eds.), *American Development of Biology.* Rutgers Univ Press, New Brunswick, New Jersey, pp.281–310.

Peterson, P. A. and Bianchi, A. (1999). *Maize Genetics and Breeding in the 20th Century.* World Scientific Publishing Co., Inc., Singapore.

Redei, G. P. (1971) A portrait of Lewis John Stadler. Stadler Genet. Symp. 1, 5–20.

Rhoades, M. M. (1936) The effect of varying gene dosage on aleurone colour in maize. J. Genet. 33, 347–354.

Rhoades, M. M. (1984) The early years of maize genetics. Annu. Rev. Genet. 18, 1–29.

Roman, H. (1988) A diamond in a desert. Genetics 119, 739–742.

Shull, G. H. (1909) A pure-line method in corn breeding. Am. Breeders' Assoc. Ann. Rep. 5, 51–59.

Shull, G. H. (1910) Hybridization methods in corn breeding. Amer. Breeders Mag. 1, 98–107.

Sprague, G. F. (1929) Hetero-fertilization in maize. Science 69, 526–527.

Stadler, L. J. (1942). Some observations on gene variability and spontaneous mutation. The Spragg Memorial Lectures on Plant Breeding 1942, 1–15.

Wallace, H. A. and Brown, W. L. (1988). *Corn and Its Early Fathers (Revised Edition).* Iowa State Univ. Press, Ames, Iowa

Weatherwax, P. (1922) A rare carbohydrate in waxy maize. Genetics 7, 568–572.

Barbara McClintock

Lee B. Kass and Paul Chomet

Abstract Barbara McClintock, pioneering plant geneticist and winner of the Nobel Prize in Physiology or Medicine in 1983, is best known for her discovery of transposable genetic elements in corn. This chapter provides an overview of many of her key findings, some of which have been outlined and described elsewhere. We also provide a new look at McClintock's early contributions, based on our readings of her primary publications and documents found in archives. We expect the reader will gain insight and appreciation for Barbara McClintock's unique perspective, elegant experiments and unprecedented scientific achievements.

1 Introduction

This chapter is focused on the scientific contributions of Barbara McClintock, pioneering plant geneticist and winner of the Nobel Prize in Physiology or Medicine in 1983 for her discovery of transposable genetic elements in corn. Her enlightening experiments and discoveries have been outlined and described in a number of papers and books, so it is not the aim of this report to detail each step in her scientific career and personal life but rather highlight many of her key findings, then refer the reader to the original reports and more detailed reviews. We hope the reader will gain insight and appreciation for Barbara McClintock's unique perspective, elegant experiments and unprecedented scientific achievements.

Barbara McClintock (1902–1992) was born in Hartford Connecticut and raised in Brooklyn, New York (Keller 1983). She received her undergraduate and graduate education at the New York State College of Agriculture at Cornell University. In 1923, McClintock was awarded the B.S. in Agriculture, with a concentration in plant breeding and botany. She received both master's (1925) and doctoral degrees

L.B. Kass
Cornell University, Ithaca, NY

P. Chomet
Monsanto Co., Mystic, CT

(1927) from Cornell's Department of Botany, and held positions there as researcher, teaching assistant, and instructor from 1924–1931 (Kass 2000, Kass 2003). From 1931 to 1936, supported by the National Research Council, and the Guggenheim and Rockefeller Foundations, she continued the work that culminated in her discovery of transposable elements published in 1950. She held appointments at the University of Missouri (1936–1942) and then the Carnegie Institution of Washington's Department of Genetics, Cold Spring Harbor, New York, where she worked until her death in 1992 (Kass 2003).

2   The Early Years of Genetics

McClintock was awarded Cornell's "Graduate Scholarship in Botany" for 1923–1924, which provided support during her first year of graduate studies. She was elected to the graduate student's Honor Society, Sigma Xi (Kass 2000). Her Master's thesis was a literature review of cytological investigations in cereals with particular attention to wheat (McClintock 1925, Kass 1999). During this time, she did not study corn, the plant to which she would later devote her life's work, nor did she morphologically describe its chromosomes, although both stories have been widely circulated (i.e., Comfort 2001, Keller.1983).

In graduate school McClintock worked as a research assistant to L. F. Randolph, an Associate Cytologist of the United States Department of Agriculture. In the summer of 1925, she discovered a corn plant that had three complete sets of chromosomes (a triploid). Applying Belling's new chromosome staining technique they studied the meiotic behavior of the chromosomes in the pollen of this plant (Kass and Bonneuil 2004) and published their results the following February (Randolph and McClintock 1926). They described the repeated occurrence of tetraploid microsporocytes and suggested that the functioning of gametes derived from such cells might account for the origin of triploidy and conceivably of other forms of polyploidy in maize (Randolph 1927). It seems that McClintock was upset that her name appeared second on their article when she believed she had done most of the work, and soon after they ended their working relationship (Kass 2003, Kass and Bonneuil 2004).

The new focus of her Ph.D. dissertation was an investigation of the cytology and genetics of the unusual triploid corn plant, which, when crossed with normal diploid plants, produced offspring with extra chromosomes (trisomic plants). This permitted her to correlate an extra chromosome plant with a particular linkage group, although she had not yet determined the identity of the extra chromosome (McClintock 1927, 1929a, Kass and Bonneuil 2004). Her research was supported by an assistantship in the Botany Department with Lester Sharp, her Ph.D. advisor and one of the leading cytologists in the country. While McClintock was a graduate student in the Botany Department, several other women were awarded graduate degrees from Cornell's Department of Plant Breeding (Kass 2003, Murphy and Kass 2007), although it has been reported (i.e., Comfort 2001, Berg and Singer 2003) that Cornell's Department of Plant Breeding did not accept women graduate students.

3   The 10 Chromosomes of Maize

Upon completing her doctorate, McClintock became an instructor at Cornell, which afforded her faculty status with responsibilities for teaching and guiding graduate students. She continued to pursue her studies on the triploid corn plant and its trisomic offspring. Plants with extra chromosomes could be used for correlating genes with their chromosomes if one could cytologically distinguish the extra chromosome. McClintock's continued maize cytological investigations led her to devise a technique to distinguish all 10 chromosomes. She found that late prophase or metaphase in the first microspore *mitosis* (male gametophyte) provided clear chromosome observations and by June 1929 she published the first ideogram of maize chromosomes (McClintock 1929b, Kass and Bonneuil 2004).

4   Linking Genetics and Cytology

Having the ability to recognize each chromosome individually would now permit researchers to identify genes with their chromosomes. Using a technique of observing genetic ratios in her trisomic plants and comparing the ratios with plants having extra chromosomes, McClintock cooperated with and guided graduate students to determine the location of many genes grouped together (linkage groups) on six of the ten chromosomes in corn (summarized in Emerson et al. 1935; Rhoades and McClintock 1935; Kass et al. 2005). McClintock and graduate student Henry Hill were the first to identify the *R-G* linkage group with the smallest maize chromosome (10) (McClintock and Hill 1931).

McClintock was the first to observe pieces of one chromosome attached to another chromosome (interchange chromosomes) when pollen grains are produced during meiotic cell division (McClintock 1930). She used these translocation or interchange chromosomes, which she observed during the pachytene stage of meiosis, to locate the remaining four gene groups with their chromosomes (Kass 2000; Kass and Bonneuil 2004). In addition, the cooperators confirmed Belling's translocation hypothesis which offered an explanation of how translocations could confer semisterility (gametic sterility) in corn (Burnham 1930, see Rhoades and McClintock 1935).

5   Proof of Crossing Over and the Chromosome Theory of Inheritance

Also using interchange chromosomes, Instructor McClintock and graduate student Harriet Creighton (McClintock 1931, Creighton and McClintock 1931, Creighton and McClintock 1935) provided the first demonstration of cytological "crossing over," in which chromosomes break and recombine (exchange parts) to create genetic reassociations. It was the first cytological proof that demonstrated the genetic theory that linked genes on paired chromosomes (homologues) did exchange places from one homologue to another. For this study, Creighton and McClintock used a semisterile maize plant found to be associated with a reciprocal translocation (segmental interchange) between the second and third smallest chromosomes (chromosomes 9 and 8; McClintock 1930). By means of trisomic inheritance, McClintock (1931) first observed that chromosome 9 of the monoploid complement (n=10) was associated with the genes of the *c-sh-wx* linkage group. Using yet unpublished interchange data of C.R. Burnham, and trisomic inheritance for the genes *wx*, *sh*, and *c*, McClintock reported the order of the genes on chromosome 9 as *wx-sh-c*, beginning at the middle of the long arm and running toward the end of the short arm. Knowing the order of the genes was critical to the demonstration of a correlation of cytological and genetical crossing over (Creighton and McClintock 1931, Coe and Kass 2005).

Using a chromosome pair visibly heteromorphic in two regions, one chromosome of the homologue possessing an enlarged chromatic knob on the end of the short arm of the chromosome and a piece of a non-homologous chromosome attached to the same chromosome (a translocation), both pairs of contrasting features can be observed in meiotic prophases. When a heterozygous plant (in which an interchange chromosome includes the knob, and the non-interchange chromosome is knobless) is crossed to a standard knobless plant, the combinations knobless-interchange and knobbed-standard were found in the progeny. This indicated that chromosomal cross-overs had occurred in the region between the knob and the point of interchange. By the simultaneous use of genes known to be in the cytologically marked region of the chromosome, genetic crossing-over was correlated with the occurrence of chromosomal crossing-over (Creighton and McClintock 1931, Creighton 1934). The assumption that genetic crossing-over was due to an exchange of chromosome segments thus seemed to be justified (Creighton 1934). Creighton and McClintock's significant study gave further confirmation to Thomas Hunt Morgan's chromosome theory of inheritance, for which he won a Nobel Prize in 1933 (Kass, Bonneuil and Coe 2005; Coe and Kass 2005; Kass 2005).

In the *Dynamic Genome,* a gift to McClintock on her ninetieth birthday, Nina Fedoroff commended McClintock's early achievements: "The Influence of her early work is greater than that of any of her peers. ... Had she done no more, McClintock would have become a major figure in the history of genetics" (Fedoroff and Botstein 1992). McClintock hoped for a research appointment commensurate with her qualifications. By 1931, however, the country was deep into the Great Depression, which was spurred by the stock market crash in October 1929 (Kass 2000). Research jobs at most universities were not abundant and research appointments for educated women were even more limited (Kass 2005b).

6   Chromosomal Structural Changes and the Identification of Chromosome Mutations

From 1931–1934, sponsored by two National Research Council (NRC) Fellowships, and a prestigious Guggenheim Fellowship (resulting from excellent work and reputation), McClintock traveled to a series of important research institutions across the U.S., Germany, and back to Cornell, where she worked in the Department of Plant Breeding as an assistant to Rollins Adams Emerson, head of the department (Kass 2004, Kass and Bonneuil 2004). Invited by Lewis J. Stadler, University of Missouri, Columbia, McClintock studied the physical changes (mutations) in plants caused by X-rays and discovered that external phenotypic traits were caused by missing pieces of chromosomes in the cell (McClintock 1931). At the California Institute of Technology, as an NRC Fellow with Ernest Gustav Anderson, she employed interchange chromosomes to investigate the Nucleolar Organizer Region (NOR) in cells (McClintock 1934). Studying a translocation for chromosomes 9 and 6 (the satellite chromosome, which includes a "pycnotic body" –for the NOR) that had broken the NOR into distinct pieces, she could show that either large or small nucleoli would form in progeny cells that carried a piece of the NOR. This led her to conclude that the "pycnotic body" on chromosome 6 is a nucleolar-organizing-body, with one area more active than another. Pachytene configurations showed that the NOR was attached to the nucleolus. Microspores carrying chromosome 6 and a translocation 9^6 chromosome developed numerous small nucleolar bodies. The nucleolus, she concluded, develops from a definite nucleolar-organizing region of chromosome 6, the distal part of which, normally more closely associated with the nucleolus, is most active; but the proximal part of which may produce a full-sized nucleolus when not in competition with more active bodies. As summarized by Clausen (1936), McClintock suggested that the "function of the nucleolar-organizing body is to organize the nucleolar substance present in each chromosome into a definite body, the nucleolus, and that the nucleolar substance is either identical with or closely related to the matrix substance of the chromosomes, which may in turn be concerned with the distribution and dispersion of the chromatin within the chromosomes into the metabolic condition."

7   Variegated Phenotypes and Unstable Chromosomes

Returning to Missouri in 1932, she continued a research project on the cytology of X-rayed plants that she began there in 1931. This investigation of ring chromosomes in corn was influenced by a study of similar chromosomes first reported by M. Navashin (1930) in *Crepis*. McClintock had previously observed "ring fragments" in maize and hypothesized the mechanism for producing ring chromosomes (McClintock 1931; Creighton and McClintock 1932). In 1932, she correlated ring chromosomes with variegation occurring both spontaneously and in x-rayed maize plants (McClintock 1932, Anderson 1936). Since the ring chromosomes varied in size in different cells, and sometimes were totally eliminated, it was suggested that the variegation was caused by the somatic elimination of that part of the ring chromosome that carried the dominant allele. Studies of the synapsis and morphology of the different ring chromosomes and their normal homologues at the mid-prophase of meiosis [pachytene stage] gave the approximate location of the *b* and *pl* loci (Rhoades 1934). Gene *b* was found to be situated in the mid-portion of the short arm of the *B-lg* chromosome [chromosome 2] and *pl* was located towards the middle of the long arm of the satellite chromosome [chromosome 6] (McClintock 1932).

Later, McClintock (1938a) continued her studies of ring chromosomes and found that some plants harboring ring shaped chromosomes would produce dicentric chromosome bridges at mitotic anaphase, thereby losing genetic material and uncovering recessive traits in some of the cell descendents. Sister mitotic cells would gain the genetic material that was lost in the breakage of the chromosome bridge. The broken chromosome ends would fuse again to reform a ring, ready to undergo breakage again. The repeated losses resulted in variegated plant tissues due to the continual uncovering of recessive traits (McClintock 1938a).

McClintock returned to Cornell to complete her Guggenheim Fellowship after only four and one half months in Berlin and Freiburg, Germany (1933–1934). Her decision to return was due to uncomfortable circumstances caused by the rise of the Nazi Party, which forced many of her German-Jewish colleagues to leave academe, and to health problems she suffered while in Berlin. Upon completing her Guggenheim Fellowship, R. A. Emerson convinced the Rockefeller Foundation to grant him funds to employ McClintock as his research assistant. Her project provided insights to an understanding of variegation (Rhoades and McClintock 1935) and would eventually lead to her studies of the breakage-fusion-bridge cycle (see below).

In 1936, McClintock accepted an appointment as Assistant Professor of Botany at the University of Missouri to join L J. Stadler's genetics research group (Kass 2005, Kass 2007). Stadler (1928, 1930) was the first to use X-irradiation in plants to recover mutations. McClintock had previously performed extensive cytological examination of plants carrying Stadler's x-ray induced mutations (McClintock 1931, 1932, 1933). These studies revealed that many of these mutations were due to deficiencies or rearrangements of a chromosome segment carrying the dominant allele. This material was also the major source of chromosome rearrangements that later allowed McClintock (1938a, b) to study the behavior of broken chromosomes as well as the induction of transposable elements.

After years of understanding the consequences of mitotically unstable chromosome behavior in the form of ring chromosomes (McClintock 1931 through 1938a; see Appendix), McClintock sought to experimentally determine the process of chromosome breakage and fusion. Her 1938 studies (McClintock 1938b) of an inverted segment of chromosome 4L clearly showed that "fusion of broken meiotic chromatids does occur" resulting in anaphase bridge configurations and the formation of an acentric fragment. This curious but profound behavior of newly replicated chromosomes was termed the breakage-fusion-bridge (BFB) cycle (McClintock 1939). McClintock observed this chromosome behavior during the mitotic cell divisions in pollen cells following meiosis. The scientific question then followed, what happens to such chromosomes when they are passed into the next generation? (McClintock 1938b).

A definitive conclusion came only a year later (McClintock 1939). It was well established that deficient chromosomes would not transmit well through the haploid pollen or ovule, so a method was developed to transmit a newly broken chromosome that harbored a full complement of genes using rearrangements on chromosome 9.

These chromosome configurations were obtained from both Harriet Creighton and L. J. Stadler. One of the chromosomes was an inverted duplication of the short arm which carried a series of dominant color and starch conditioning genes that could be visualized in the kernel and plant when they were present or lost (including *Yg*, *C-I*, *Sh*, *Bz* and *Wx*). In a subsequent study, McClintock paired this chromosome with a 9S deficiency that did not pass through the male gamete. Each of these chromosomes did not transmit well (or at all) through the male gamete. Meiotic recombination of this chromosome with the 9S deficiency produced a dicentric chromosome, which ruptured after each centromere was pulled to opposite poles during anaphase. One of the ruptured chromosomes carried (at least) a full complement of 9S genes and was transmitted through the pollen. Since each of the parental chromosomes was not efficiently transmitted through the pollen, McClintock developed an exquisite method to select for a high percentage of broken chromosomes in the gametes. This method demonstrated the elegance of her experimentation, her deep understanding of the plant and the foresight she had in her studies.

Upon learning that Stadler's research unit might be eliminated, and preferring research over teaching, McClintock requested a leave of absence from Missouri in 1941 to seek employment elsewhere. In the summer of 1941, she was appointed a Visiting Professor at Columbia University (organized by M. M. Rhoades) and a Visiting Investigator at the Carnegie Institution of Washington (CIW), Department of Genetics (invited by A. F. Blakeslee), Cold Spring Harbor, Long Island, New York (Kass 2005). Soon after arriving at CIW, a former Cornell Plant Breeding student, Milislav Demerec, became Director of the Department of Genetics. He offered McClintock a permanent job commencing in 1943. The University of Missouri counter offered but she resigned effective August 31, 1942.

While with the CIW, Department of Genetics, McClintock elaborated on the behavior and characterization of chromosomes and genes in maize. She continued investigating the BFB cycle, but this time followed the fate of the broken chromosome into the progeny using the markers on the broken chromosome arm. By the patterned loss of markers in the endosperm, she definitively concluded that the BFB cycle continued in the cell divisions of the endosperm. Since the fusion of ends occurs between newly formed chromatids, McClintock referred to this type of chromosome behavior as the chromatid type BFB cycle (McClintock 1941a).

Cytological examination of the embryo and plant tissue revealed that the chromatid type BFB cycle did not continue in the somatic plant tissue or in the gametes of a plant that received such a broken chromosome. In fact, all sporophytic cells observed carried the same 9S chromosome complement suggesting that the chromosome had "healed" when passed to the embryo. Years later it was realized that the healing process is likely the formation of a telomere on the broken end of the chromosome entering the zygote (McClintock 1984a). This drastic difference in behavior of the chromosomes intrigued McClintock and further investigations on the behavior of broken chromosomes into the embryo followed. Elizabeth Blackburn (1992), who discovered the enzymes that add telomeres to chromosome ends, elaborated on the contribution McClintock made to the understanding of chromosome behavior and telomere formation.

Table 1 Description of loci and phenotypes described in the text. Note that the genes involved in anthocyanin accumulation require the presence of other genes in the pathway, including $A1$ $A2$ $C1$ $C2$ and R, to confer color to the plant or kernel

Gene Name	Chromosome-arm	Allele (historical*)	Phenotype
Activator		Ac	An autonomous transposable element capable of controlling the movement of itself and the movement of Ds; Increasing dosage of Ac delays transposition.
Anthocyaninless1	3L	A1	Red or purple anthocyanin accumulation in plant and kernel (aleurone).
		a1	Green, or brown plants, colorless aleurone, brown pericarp; recessive to A1.
		a1-m1 (a^{m-1})	Pale color in the kernel (aleurone) and plant. In the presence of a functional Spm element the pale pigmentation becomes colorless and red revertant sectors are detected. This allele responds to a fully functional Spm through a non-autonomous Spm present at the locus.
		a1-m2 (a^{m-2})	Pale pigmentation in the aleurone and plant with fully colored revertant sectors in the presence of Spm. The original allele contained a fully functional Spm at the a1 locus.
Anthocyaninless2	5S	A2	Red or purple plants and kernels (aleurone).
		a2	Green, or brown plants, colorless aleurone; recessive to A2.
		a2m-1 ($a2^{m-1}$)	Color in the kernel (aleurone) and plant in the absence of a fully functional Spm. In the presence of a functional Spm element the pale pigmentation becomes colorless and red revertant sectors are detected. This allele responds to a fully functional Spm through a non-autonomous Spm present at the locus.
Booster1	2S	B1 (B)	Plant color intensifier conferring a purple plant due to the accumulation of anthocyanin in major tissues.
		b1 (b)	Green plant; little or no accumulation of anthocyanins; recessive to B1.
Bronze1	9S	Bz1 (Bz)	Red or purple aleurone layer of the kernel due to the accumulation of anthocyanins.
		bz1 (bz)	Bronze or brownish colored aleurone layer; recessive to Bz1.
Colored aleurone1	9S	C1 (C)	Red or purple aleurone layer of the kernel due to the accumulation of anthocyanins.
		C-1 (I)	Color inhibitor, colorless aleurone layer; dominant to C1.
		c1 (c)	Colorless aleurone; recessive to C1.

Name	Symbol	Location	Description
	$c1$-$m1$ (c^{m-1})		Mutable color, red sectors on a colorless aleurone background in the presence of Ac. The red sectors are caused by restoration of the $c1$ locus after Ds transposition whereas Ds still inhibits the $c1$ locus in the colorless aleurone background. In the absence of Ac the kernel is colorless. A number following the "m" was used by McClintock to identify the next consecutively isolated mutable allele in her cultures; i.e., $c1$-$m1$ was the first mutable allele (caused by the insertion of Ds *at the C1* locus).
Dissociation	Ds		First named for the phenotype of dissociation (breakage) of chromosomes in the presence of Ac; First described on chromosome 9. Later shown to be a transposable element due to its ability to move (transpose) in the presence of Ac into and away from loci thereby disrupting or controlling their function. The presence of the element was designated Ds and the absence of the element was designated ds.
Dotted	Dt		An autonomous transposable element capable of controlling the transposition of the receptor element rDt at the $A1$ locus. In the presence of Dt, $a1$-dt responding alleles are unstable and express colored dots on a colorless aleurone background. In the absence of Dt, the $a1$-dt alleles are stable and exhibit a colorless aleurone.
Enhancer-Inhibitor	En-I		Discovered independently by P. Peterson (1953) yet structurally and functionally the same as Spm. The *Enhancer* function is equivalent to the *Suppressor* function (Sp) of Spm and the *Inhibitor* function is equivalent to the *mutator* function (m) of Spm.
Liguleless1	$Lg1$ (Lg) $lg1$ (lg)	2S	Leaves show normal ligule and auricle development and structure Leaf lacks ligule and auricles, leaves stand upright at base; recessive to $Lg1$.
Modulator of Pericarp	Mp		Discovered independently by Brink and Nilan (1952) yet structurally and functionally the same as Ac. The transposable element was studied in association with the P locus where it caused the red variegated or revertant tissue sectors in the P^{VV} allele.
Pericarp Color	$P1$ (P)	1S	Controls pigmentation accumulation in the pericarp and cob. All pericarp loci are designated P with superscript letters following the P, such as P^{RR} for red pericarp, red cob, P^{WW} for white pericarp, white cob. P^{VV} is variegated pericarp and cob color due to transposition of Mp and functional restoration of the P locus.

(continued)

Table 1 (continued)

Gene Name	Chromosome-arm	Allele (historical*)	Phenotype
Purple Plant1	6L	*Pl1 (Pl)*	Light independent accumulation of red or purple color (anthocyanin) in major tissue of the plant.
		pl1 (pl)	Sunlight dependent accumulation of pigmentation in the leaves and sheath.
Red (Colored1)	10L	*R1 (R.)*	Red or purple color in the plant and kernel. *R-G* designation refers to separate components of the locus that specifically confer color to the kernel or plant respectively.
		r1 (r)	Green plant, colorless aleurone in the kernel.
Shrunken1	9S	*Sh1 (Sh)*	Plump kernel, normal starch synthesis in endosperm.
		sh1 (sh)	Shrunken or collapsed endosperm due to reduced starch content, recessive to *Sh1*.
Suppressor-Mutator		*Spm*	An autonomous transposable element capable of controlling gene function as well as transposition of itself and nonautonomous *Spm* elements. The *Suppressor (Sp)* function enhances or suppresses gene expression of a locus containing a functional or defective *Spm* element, while the *Mutator* function (*m*) is necessary for transposition of the element.
Waxy1	9S	*Wx1 (Wx)*	Endosperm starch contains amylose and amylopectin; stains blue-black with iodine stain (I,KI).
		wx1 (wx)	Endosperm appears waxy, starch contains only amylopectin; stains red/brown with iodine stain (I,KI); recessive to *Wx1*.
Yellow Green2	9S	*Yg2 (Yg-2)*	A normal green plant.
		yg2 (yg-2)	Leaves of seedling and mature plant are yellow-green in color.

* Historical or alternative allelic designations for loci are indicated in parentheses.

The introduction of a broken chromosome from the male and the female gametes simultaneously into the zygote allowed McClintock to study the behavior of two newly broken chromosomes in the embryo and endosperm (McClintock 1942). From her ring chromosome studies she already realized that broken chromosomes, prior to replication into chromatids, should fuse together in embryo and plant tissue (McClintock 1938). Each of the chromosomes was distinctly marked so that she could follow the behavior of the chromosomes phenotypically in the endosperm. From such studies she realized that two broken chromosomes would fuse prior to replication in the embryo and undergo what she termed the chromosome type BFB cycle (McClintock 1942). This process was distinct from the chromatid type BFB cycle since it required two broken chromosomes. The fusion occurred prior to chromatid formation and the process occurred in sporophytic tissue. McClintock repeated experiments using her "improved method" of delivering a broken chromosome into the gametes (McClintock 1944b) to produce seed from plants that underwent either the chromatid or the chromosome type BFB cycle. Such plants were used as a system to induce new mutations by minute deletions and rearrangements of chromosome 9. Through careful cytological examination and subsequent genetic crosses, McClintock could associate new mutant "genes" to chromosome segments, a remarkable feat at the time and even today (McClintock 1945b, 1946).

In the winter of 1944 soon after her election to the National Academy of Sciences, George Beadle invited McClintock to Stanford University to study the chromosomes of the pink bread mold *Neurospora*. Beadle was a former student colleague and friend with whom she worked on corn genetics at Cornell from 1927–1930 and during her NRC years at Caltech. Within ten weeks, with the assistance of Beadle's staff, she was able to provide a preliminary study of the fungal chromosomes and demonstrate their movement during cell division (McClintock 1945a). This work was important to an understanding of the life history of the organism and was the basis for Jesse Singleton's Ph.D. dissertation research conducted at Caltech under McClintock's guidance in 1946 (Singleton 1948, Singleton 1953). Beadle and his colleagues would employ the fungus to elucidate how genes control cell metabolism (Berg and Singer 2003). In 1958 Beadle and Tatum shared a Nobel Prize with J. Lederberg.

8   Unstable Mutants and the Discovery of Transposable Elements

Returning to CIW in 1945, McClintock continued to investigate the behavior of chromosomes and the effects of deletions and rearrangements. Similar to her past (and future) studies, McClintock focused her attention on unique phenotypes. Such mutant or altered phenotypes were her key to unlocking the mechanisms of chromosome behavior, responses of the genome to stress and to gene regulation. In 1945, the same year she was elected the first woman President of the Genetics Society of America, she characterized four unusual unstable mutations (McClintock 1946). These were derived from selfed plants that underwent the chromatid or chromosome type BFB cycle (a white seedling variegated, a variegated light green seedling, a luteus seedling, and a chromosome breakage variegation pattern in the aleurone of the seed). They were all tied together by their unique, yet similar, frequency and timing of unstable phenotypes (mutability). In the plant, mutant tissue gave rise to wild type tissue in a very regular pattern. In the kernel mutant, loss of chromosome markers also occurred in the same "controlled" regular pattern. These patterns were unlike the random losses due to ring chromosomes or the usual BFB cycle and she suggested there was a different control mechanism at work.

What caught McClintock's eye was the unique pattern on the kernels showing a regular loss of all markers distal (towards the end of the chromosome) to the *Wx1* locus (see Table 1 for descriptions of genetic symbols) on chromosome 9S. In fact, crossover data and cytological observations indicated that there was a chromosome dissociating locus (designated *Ds*) just proximal to the *Wx1* locus. Due to the segregation of this trait, it was also apparent that a second locus was required, designated *Activator* (*Ac*), to cause the chromosome breakage. Many other mutable loci, like the ones described in 1946, were isolated from these cultures, including multiple alleles of a mutable color locus (*c-m1*, *c-m2*) and a mutable starch conditioning mutant (*wx-m1*, *wx-m2*). Each mutant was used to unify the theory of gene control while delineating and further defining the capabilities of what she later referred to as controlling elements (McClintock 1956).

A number of additional remarkable conclusions came from these early studies (McClintock 1946, 1947, 1948, 1949, 1950). The pattern of mutability of the reverting loci (*c1-m1*, *c1-m2*, *wx-m1*, see Table 1) was strikingly similar to the patterned loss of *Ds* and both types of loci were under the control of *Ac*. McClintock was convinced that a unifying mechanism was underlying these phenotypic similarities. In addition, somatic sectors and germinal changes in the plant and kernel were readily detected that altered the pattern of mutability in all descendent somatic cells. These heritable alterations were called "changes in state" of the locus, rather than simply mutations, since they occurred frequently and, at times, were twinned (one daughter cell gained what the other cell had lost). She also saw that the patterned loss of the distal third portion of chromosome 9S, including markers *C*, *Bz*, *Wx* could heritably change in a few kernels. The marker loss pattern in these derived kernels now clearly showed that marker losses were occurring at a new location on the chromosome. The *Ds* had moved (transposed) to a new position on chromosome 9.

McClintock hypothesized how the *Ds* element could move. Large chromosome rearrangements were often associated with the *Ds* locus since it was causing chromosome breaks. McClintock reasoned that in some instances, rather than a large, visible rearrangement, a submicroscopic chromatin segment carrying *Ds* was cut out and reinserted to the new position on the chromosome, not unlike larger rearrangements that she had observed in early studies. Similarly, she hypothesized that by the same *Ac* controlled mechanism, *Ds* could transpose to a position within or near a normal gene (the *C1* locus; McClintock 1948) thereby disrupting or inhibiting its function (*c1-m1*; McClintock 1949). McClintock also explained how the mutant *c1* locus could be restored in somatic and germinal tissues. She stated, "An event leading to removal of the inserted *Ds* segment from the *C* locus would give rise to two broken ends in the chromatid. Fusion of these broken ends would re-establish the former normal genic order, and remove the inhibitory action on the *C* locus induced by the inserted segment; and as a consequence a mutation from c to C would be evident." (McClintock 1949). This single hypothesis of Ds transposition unified many of the observations of Ds behavior, including the similarity in patterns of mutability of reverting loci (C and Wx) and the patterned loss and movement of Ds along the chromosome. It was for this discovery, 35 years later, that she was awarded the Nobel Prize in 1983.

9   Control of Gene Action

Transposition of the elements allowed McClintock to discover and study another key aspect of gene action, that of gene regulation. The observations that these transposable elements could influence the regulation of many genes was intensively studied over the next 15 years, particularly using an element designated *Suppressor-Mutator* (*Spm*).

By 1954 (McClintock 1954) the two-element mutation-controlling system designated *Ac/Ds* was well studied by McClintock and, like the *Dt* system described earlier by Rhoades (Rhoades 1938), investigators at other institutions were characterizing mutable systems, including *Mp* at the *P* locus (Brink and Nilan 1952). At this time McClintock switched much of her attention to a new and unique mutable system which she discovered was controlling the *A1* locus.

The new system, designated *Spm* for *Suppressor-Mutator* (McClintock 1954), had characteristics both similar to and distinct from the *Ds-Ac* system. Similarly, the *A1* somatic mutations were controlled by a two-element system that could transpose, although the instability was not controlled by the *Ac* or *Dt* loci. In addition, chromosome breaks were not detected with the *Spm* system. What was unique about *Spm* was the added layer of gene regulation that was not observed with *Ds-Ac*. The newly controlled allele of the *A1* gene (designated *a1-m1*) was lightly colored in the kernel or plant. Only upon introduction of a fully functional *Spm* element elsewhere in the genome in conjunction with the *a1-m1* allele, did the pale pigmentation turn to colorless. In addition, intensely colored sectors appeared much like the somatic reversion mutations recognized with other transposable elements. The function responsible for the trans-acting inhibition of pigmentation was termed the suppressor function (*Sp*). Likewise, the function responsible for the somatic reversion was designated the Mutator function (*m*). Independently, Peterson (1953) published on a transposable element system designated *En-I* (*Enhancer-Inhibitor*) which was later recognized as the same element system as Spm (Peterson 1965).

A second allele of *A1* under the control of *Spm* demonstrated a distinct suppressor regulation of gene function. In the case of the *a1-m2* allele a fully functional *Spm* element was originally identified at the locus (subsequently many defective derivatives were produced). When the element was functioning, the kernel exhibited pale pigmentation and fully colored sectors. The pale pigmentation in the presence of a functional *Spm* element was opposite of what was observed with the *a1-m1 (and a2-m1) Spm* controlled alleles. Utilizing other unique behaviors of *Spm*, McClintock showed that the *A1* locus expression was now dependent on the suppressor function of *Spm*; when *Sp* was active, the *A1* gene was expressed and when *Sp* was shut off, the *A1* gene was also inactivated giving rise to a colorless phenotype. Numerous cases were studied over the years to demonstrate and dissect the gene controlling action of the *Spm* system (McClintock 1965).

The fact that these elements could differentially regulate endogenous genes solidified McClintock's concept that they were truly "controlling elements". It was also clear that McClintock believed that by selecting for certain transpositions or other genic-altering events she could uncover gene regulation mechanisms that were not previously recognized but were an inherent component of the genome. Over the next 15 years or so, McClintock continued to uncover the complexities and intricacies of the *Spm* system and therefore the complexities of gene regulation in the nucleus of cells.

Further insights into gene regulation came with the description of reversible inactivation of controlling elements. After years of describing heritable and stable changes to genes and elements, McClintock realized that both *Spm* and *Ac* could undergo a cyclical change (designated phase change) in activity during plant and kernel development (McClintock 1957, 1958, 1964). She described cases in which both *Spm* and *Ac* activity could turn off and on during plant and kernel development. While this phase change seemed random, it was not; certain alleles were "programmed" to turn on and off with a given frequency and timing. Because this phenomenon could occur with different element families and did not involve movement of the element or irreversible changes to the locus, McClintock realized that phase change was a unique layer of gene regulation not previously accounted for. Since the advent of molecular biology it has been realized that elements and genes are regulated through multiple, complex mechanisms involving modifications to the DNA and chromatin structure. Controlling element phase change demonstrated a key process in this epigenetic gene regulation and is still under investigation today (Schwartz and Dennis 1986, Chomet et al. 1987, Banks et al. 1988, Lippman et al. 2003).

10   The Races and Varieties of Maize

Other areas of maize biology were of interest to McClintock. For over 20 years, she trained students and conducted research on the evolution and migration of varieties of corn in the Americas, which was initiated in the late 1950's (McClintock 1978, McClintock et al. 1981, Timothy 1984). Utilizing cytological knobs as polymorphic markers, she, and two of the students she trained at North Carolina State University in the 1960s, investigated and published a study of variation in chromosome constitutions for over 1,200 races, strains and varieties of maize. Far ahead of her time and predating the idea of restriction fragment length polymorphisms (1980), McClintock and associates utilized the shape and distribution of chromosomal knobs, the incidence of the abnormal 10 chromosome, and the number of B chromosomes to infer the makeup and predominance of certain maize varieties and relate them across geographical boundaries in the Americas. The distribution and frequency of knobs or chromosome components are interpreted in terms of where maize developed initially and when, where, and how it was introduced into other parts of the Americas, and its fate following introductions (McClintock et al. 1981, Timothy 1984). McClintock et al. (1981) develops a coherent theme on the evolution of maize races. For example, the distribution and frequency of knobs on chromosomes 4 and 5 link some Mexican races to one another and knobs on chromosome 7 were used to trace the migration routes by which maize from one area was carried to another. Additionally, groups of special knobs revealed the relationship of maize to teosinte, and geographical areas where some of the early diverse racial types developed. They also discuss the origins of knob complexes and their significance in association with migration pathways, introductions and introgression (for an excellent book review of McClintock et al. 1981, see Timothy 1984).

11   Recognition

Many persons have wondered why it took so long for McClintock's work to be recognized by the extensive scientific community. Although McClintock employed cytogenetic techniques to study corn chromosomes, other researchers studied simpler organisms (bacteria and their viruses) and used molecular techniques to investigate genes and inheritance. From 1944 through 1952, molecular genetic studies demonstrated that DNA was the hereditary material. This discovery launched the field of molecular biology to the forefront of scientific investigations. McClintock's experiments were complex, laborious, taking months or even years to explain results. Molecular studies in simpler organisms gave almost instant gratification and quickly answered questions about the hereditary material that had been unsolved for years. Additionally, McClintock's findings contradicted the prevailing view that all genes were permanently in a linear sequence on chromosomes.

McClintock's conclusion that genes could move from place to place in the maize genome was accepted, although the concept did not seem applicable to other organisms until transposable elements were found years later in prokaryotic (*E. coli*) and eukaryotic organisms (*Drosophila*, humans, etc.). Because McClintock was a very respected researcher (she received the Botanical Society of America Merit Award in 1957 and the National Academy of Science Kimber award in 1967), her work with corn was accepted but it was considered anomalous. In other words, even if genes did move around in corn the mechanism was probably not universally relevant to all organisms. Researchers in corn genetics however, immediately understood, explored, and expanded on her initial studies (Neuffer 1952, Brink and Nilan 1952, Dollinger 1954, Kreizinger 1960). During the 1970's, transposable elements were found in a number of other organisms; first in bacteria and then in most organisms studied by geneticists. This work has led to the revolution in modern recombinant DNA technology that has played a large role in medicine and agriculture, and which society now takes for granted. When McClintock's work was "rediscovered," she was recognized and rewarded for her great insights (Table 2, National Medal of Science 1970, Nobel Prize in Physiology or Medicine 1983, etc.). It is a lesson to be learned that new ideas in science, which contradict the

Table 2 Chronological list of Barbara McClintock's Major Awards and Recognitions

Undergraduate

1923	Phi Kappa Phi, Honorary Scholastic Society

Graduate Student

1923–1924	Cornell University Graduate Scholarship in Botany
1923–1924	Elected to Sigma Xi, Honorary Society

Postdoctoral

1931–1933	National Research Council, National Academy of Sciences, Fellowship
1933	First recognized in *American Men of Science*
1933–1934	Guggenheim Memorial Foundation Fellow

Academic, Professional and Public Recognition

1939–1940	Vice President, Genetics Society of America
1944	Elected to National Academy of Sciences
1944	Starred in Botany in *American Men of Science*
1945	President, Genetics Society of America
1946	Elected to American Philosophical Society
1947	Achievement Award, American Association of University Women
	Honorary Sc.D., University of Rochester
1949	Honorary Sc.D., Western College for Women, Oxford, OH
1953–1954	Visiting Professor, California Institute of Technology
1957	Merit Award, Botanical Society of America
1958	Honorary Sc.D. Smith College, Amherst, MA
1959	Elected Fellow American Academy of Arts and Sciences, MA
1965–1974	A.D. White Professor-at-Large, Cornell University
1967	Kimber Award, National Academy of Sciences
1967–1992	Distinguished Service Member, Carnegie Institution of Washington, Cold Spring Harbor, NY
1968	Honorary Sc.D., University of Missouri, Columbia, MO
1970	National Medal of Science, USA
1972	Honorary Sc.D., Williams College, Williamstown, MA
1973	McClintock Laboratory dedicated at Cold Spring Harbor Laboratory, Cold Spring Harbor, NY
1978	Lewis S. Rosensteil Award for Distinguished Work in Basic Medical Research, Waltham, MA
	Louis and Bert Friedman Foundation Award for Research in Biochemistry, New York Academy of Sciences

1979   Honorary Sc.D., The Rockefeller University, Bronx, NY
Honorary Sc.D., Harvard University, Cambridge, MA
Barbara McClintock Professorship of Genetics established, Rutgers, The State University of New
     Jersey, Rutgers, NJ

1980   Salute from the Genetics Society of America

1981   Thomas Hunt Morgan Medal, Genetics Society of America
Wolf Prize in Medicine, Wolf Foundation, Israel
MacArthur Prize Fellow Laureate, John D. and Catherine T. MacArthur Foundation
Honorary D.H.L., Georgetown University, Washington, DC
Albert Lasker Basic Medical Research Award, New York, NY

1982   Louisa Gross Horwitz Prize for Biology or Biochemistry, Columbia University, NY
Charles Leopold Mayer Prize, Academie des Sciences, Institut de France, Paris
Honorary Sc.D., Yale University, New Haven, CT
Honorary Sc.D., University of Cambridge, Cambridge, England

1983   Outstanding Alumni Award, Cornell University, Ithaca, NY
Honorary Sc.D., Bard College, Annandale-on-Hudson, NY
Honorary Sc.D., State University of New York, Stonybrook, NY
Honorary Sc.D., New York University, Manhattan, NY
Nobel Prize in Physiology or Medicine, Stockholm, Sweden

1984   Honorary Sc.D., Rutgers, The State University of New Jersey, Rutgers, NJ

1986   Elected to National Women's Hall of Fame, Seneca Falls, NY

1989   Elected Foreign Member, Royal Society UK, London, England

Posthumous Awards

1993   Benjamin Franklin Medal, Carnegie Institution of Washington, Wash. DC

2004   Barbara McClintock Professorships established at Cornell University, Ithaca, NY

2005   Barbara McClintock US Postal Service American Scientists Commemorative Postage Stamp

2007   Barbara McClintock Society [a philonthropic society], established by the Carnegie Institution of
     Washington, Wash, DC, 5 May 2007.

prevailing model (dogma or paradigm), are not easily accepted. Barbara McClintock died on September 2, 1992 in Huntington, Long Island, New York, yet her memorable words still resonate in the hearts and minds of the maize community:

> "Because I became actively involved in the subject of genetics only twenty-one years after the rediscovery, in 1900, of Mendel's principles of heredity, and at a stage when acceptance of these principles was not general among biologists, I have had the pleasure of witnessing and experiencing the excitement created by revolutionary changes in genetic concepts that have occurred over the past sixty-odd years. I believe we are again experiencing such a revolution. It is altering our concepts of the genome: its component parts, their organizations, mobilities, and their modes of operation. Also, we are now better able to integrate activities of nuclear genomes with those of other components of a cell. Unquestionably, we will emerge from this revolutionary period with modified views of components of cells and how they operate, but only, however, to await the emergence of the next revolutionary phase that again will bring startling changes in concepts" (McClintock 1984).

Acknowledgments LBK acknowledges the National Science Foundation (Grants SBR9511866 & SBR9710488), and the American Philosophical Society Library, Mellon Resident Research Fellowship and the Lilly Library, Helm Fellowship for support of archival research. And thanks the Department of Plant Biology and the Department of Plant Breeding and Genetics, Cornell University for logistical support. Special thanks to our colleagues Ed Coe, Royse P. Murphy, Kathleen Gale and Richard H. Whalen for helpful suggestions on revising the manuscript.

Appendix I
Current List of Barbara McClintock's Publications

(Modified from Kass, L. B. 1999. Current list of Barbara McClintock's publications. *Maize Genetics Cooperation Newsletter* 73: 42–48. Available online, 1998: http://www.agron.missouri.edu/mnl/73/110kass.html)
In 1987, Moore edited and reprinted a collection of Barbara McClintock's papers for the Great Books in Experimental Biology Series. McClintock's publications relevant to the discovery and characterization of transposable elements are reprinted in that work. The volume also includes a list of McClintock's published papers under the heading "Numbered List of Publications" (Moore 1987). I examined the journals, symposia, etc., where all of McClintock's papers appear. In the course of my research, I found 14 additional contributions. I subsequently compiled a chronological list of all known contributions published by McClintock, which are listed below.

My list updates and amends Moore's (1987) published list and brings the total number of publications to 93. I annotated citations to include, when available, dates when the papers were received, and the month they appeared in print. In some cases the publication date and inclusive pages were revised to reflect accurately these citations. This may be important for scholars who do not have direct access to the publications. I gratefully acknowledge Dr. Edward Coe for his support with this project.

Reference cited:

Moore, John A. (ed.) 1987. *The Discovery and Characterization of Transposable Elements. The Collected Papers of Barbara McClintock*. Great Books in Experimental Biology, Garland Publishing Co. New York.

Fig. 1 Gerry Neuffer, Om Sehgal, Barbara McClintock, Ed Coe at a party celebrating McClintock's honorary Sc.D. from University of Missouri (Columbia, MO), in 1968. (Used with permission of G. Neuffer, ID's Ed Coe)

Fig. 2 Barbara McClintock at Upland Farms, Cold Spring Harbor, NY, during the summer of 1986. Maize genetics was reinvigorated at Cold Spring Harbor Lab in the early 1980's by Steven Dellaporta and associates. McClintock was a mentor and teacher both in the field and in the lab for undergraduates, graduate students and post doctoral researchers. Shown here is (L to R): Scott Bernstein (undergraduate), Barbara McClintock, Brenda Lowe (Post doc), Jychien Chen (Post doc), and Paul Chomet (graduate student). (Used with permission of Paul Chomet)

ANNOTATED CHRONOLOGICAL LIST OF THE PUBLICATIONS
OF BARBARA MCCLINTOCK

Note: This list uses dates of publication as referenced by McClintock; i.e., the 1951 *Cold Spring Harbor Symposia on Quantitative Biology* is cited here as 1951, although the copyright date for the symposium is 1952. Additional pertinent information is enclosed in brackets.

*Appears in Moore's (1987) "Numbered list of publications," pgs. xiii–xv. I add month of publication and submission dates in brackets. I list publications chronologically and add letters following dates for more than one publication in the same year. I add subheadings following titles for *Carnegie Institution of Washington Year Book* reports, and complete titles for other publications. Inclusive years for *Carnegie Year Book* reports are in brackets.

**Appears in Moore (1987) and edited for accuracy; i.e., titles, page numbers, or dates corrected.

No Star(*) = additions to "Numbered list of publications" (Moore 1987).

> McClintock, Barbara. 1925. *A Resume of Cytological Investigations of the Cereals with Particular Reference to Wheat.* Ithaca, NY. 52 pgs. plus 25 unnumbered pgs. of tables and bibliographies. [Thesis, M. A. Cornell University. A literature review; no original research. Acknowledges Prof. L. W. Sharp.]

> *Randolph, L. F. and B. McClintock. 1926. Polyploidy in *Zea mays* L. *American Naturalist* LX (666) [Jan./Feb. 1926, received — no date given]: 99–102.

> McClintock, Barbara. 1927. *A Cytological and Genetical Study of Triploid Maize.* Cornell University, Ithaca, New York. 104 pgs. plus 39 unnumbered pgs. of tables, plates, and bibliographies. [Thesis, Ph.D. Acknowledges L.W. Sharp and A.C. Fraser.]

> *Beadle, G. W. and Barbara McClintock. 1928. A genic disturbance of meiosis in *Zea mays. Science* 68 (1766) [2 November 1928, received - no date given]: 433. [This became George Beadle's dissertation research project.]

> *McClintock, Barbara. 1929a. A cytological and genetical study of triploid maize. *Genetics* 14 (2) [11 March 1929, received 11 July 1928]: 180–222. [Publication of 1927 Ph.D. thesis. Genetics was issued bimonthly at this time.]

> *McClintock, Barbara. 1929b. A method for making aceto-carmin[e] smears permanent. *Stain Technology* IV (2) [April 1929, received - no date given]: 53–56. [In this publication carmine is incorrectly spelled in the title, throughout the text, and in the citation to Belling 1926.]

> *McClintock, Barbara. 1929c. A 2N-1 chromosomal chimera in maize. *Journal of Heredity* XX (5) [May 1929, received - no date given]: 218. [McClintock annotated the reprint she sent to T. H. Morgan indicating that only one photograph was intended to be published. She apparently submitted two exposures with the intent that the best one would be printed. The citation to Blakeslee and Belling *Science*, 55, is incorrect; the year, 1924, is missing, and the volume number should be 60 (LX) not 55.]

*McClintock, Barbara. 1929d. Chromosome morphology in *Zea mays. Science* 69 (1798) [14 June 1929, submitted - no date given]: 629. [The first published ideogram of Zea chromosomes. The chromosomes were identified in the "first division in the microspore" (Mitosis) not at pachytene of Meiosis I as described by some text book authors. The citation for McClintock *Genetics*, 14, is incomplete. The year, 1929, is missing.]

McClintock, Barbara and Henry E. Hill. 1929 [ABSTRACT]. The cytological identification of the chromosomes associated with the 'R-golden' and 'B-liguleless' linkage groups in *Zea mays. Anatomical Record* 44 (3) [25 December 1929]: 291. [The paper was "read by title" at the Joint Genetics Sections of the American Society of Zoologists and the Botanical Society of America, held with the AAAS, Des Moines, and Ames, Iowa, December 1929 - January 1930. Resulting manuscript submitted March 1930, and published one year later in *Genetics* 16: 175–190, March 1931. See McClintock 1933a (pg. 209) for correction of B-lg linkage group association with Chromosome 2 not Chromosome 4.]

*McClintock, Barbara. 1930a. A cytological demonstration of the location of an interchange between two non-homologous chromosomes of *Zea mays. Proceedings of the National Academy of Sciences* 16 (12) [15 December 1930, communicated 6 November 1930]: 791–796.

McClintock, Barbara. 1930b [ABSTRACT]. A cytological demonstration of the location of an interchange between two non-homologous chromosomes of *Zea mays. Anatomical Record* 47 (3) [25 December 1930]: 380. [Paper presented on 30 December 1930, at the Joint Genetics Sections of the American Society of Zoologists and the Botanical Society of America, held with the AAAS, Cleveland, Ohio, December 1930 - January 1931. Two weeks prior to these meetings, the results were published in *PNAS* 16: 791–796, December 1930.]

*McClintock, Barbara and Henry E. Hill. 1931. The cytological identification of the chromosome associated with the R-G linkage group in *Zea mays. Genetics* 16 (2) [16 March 1931, received 1 March 1930]: 175–190.

*McClintock, Barbara. 1931a. The order of the genes C, *Sh*, and *Wx* in *Zea mays* with reference to a cytologically known point in the chromosome. *Proceedings of the National Academy of Sciences* 17 (8) [15 August 1931, communicated 7 July 1931]: 485–491. [Communicated the same date and issued as one reprint with Creighton and McClintock 1931. The results reported in McClintock 1931a are necessary for an understanding of Creighton and McClintock 1931, which follows directly in the Journal. These papers were intended to be read together. McClintock 1931a ends with the following statement: "It was desired to present briefly the evidence at this time, since it lends valuable support to the argument in the paper which follows." Creighton & McClintock, 1931 state: "In the preceding paper it was shown that the knobbed chromosome carries the genes for colored aleurone" etc. Unfortunately the "preceding paper" (McClintock 1931a) is neither cited nor referenced.]

*Creighton, Harriet B. and Barbara McClintock. 1931. A correlation of cytological and genetical crossing-over in *Zea mays. Proceedings of the National Academy of Sciences* 17 (8) [15 August 1931, communicated 7 July 1931]: 492–497. [Communicated the same date and issued as one reprint with McClintock 1931a; see annotation for McClintock 1931a.]

*McClintock, Barbara. 1931b. Cytological observations of deficiencies involving known genes, translocations and an inversion in *Zea mays. Missouri Agricultural Experiment Station Research Bulletin* 163 [December, authorized 23 December 1931]: 1–30. [McClintock NRC Fellow at Missouri and Cal Tech, investigation conducted at Missouri beginning June 1, 1931; L. J. Stadler suggested the problem and furnished all the material in the growing state.]

McClintock, Barbara. 1932a [ABSTRACT]. Cytological observations in Zea on the intimate association of non-homologous parts of chromosomes in the mid-prophase of meiosis and its relation to diakinesis configurations. *Proceedings of the International Congress of Genetics II* [24–31 August 1932, preface dated 26 July 1932]: 126–128. [McClintock NRC Fellow at Cal Tech with E.G. Anderson. This paper was presented at the 6th International Congress of Genetics as a Sectional Paper in the session titled "Cytology I, Saturday August 27." McClintock presented paper number 6 of 11 papers. Resulting manuscript submitted in April 1933 and published in *ZZMA* 19:191–237, September 1933.]

Creighton, Harriet B. and Barbara McClintock. 1932 [EXHIBIT]. Cytological evidence for 4-strand crossing over in *Zea mays. Proceedings of the International Congress of Genetics II* [24–31 August 1932, preface dated 26 July 1932]: 392. [This was an exhibit that was part of the section on "General Cytology" in the "General Exhibits." The section was organized by Ralph E. Cleland.]

*McClintock, Barbara. 1932b. A correlation of ring-shaped chromosomes with variegation in *Zea mays. Proceedings of the National Academy of Sciences* 18 (12) [15 December 1932, communicated 2 November 1932]: 677–681. [McClintock NRC Fellow at Missouri with L. J. Stadler; her address is given as U of Missouri; Contribution from Dept of Field Crops, Missouri Agricultural Experiment Station Journal Series No. 355.]

**McClintock, Barbara. 1933a. The association of non-homologous parts of chromosomes in the mid-prophase of meiosis in *Zea mays*, with 51 figures in the text and plates VII–XII. *Zeitschrift fur Zellforschung und microskopische Anatomie* 19 (2) [22 September 1933, received 21 April 1933]: 191–237. [McClintock NRC Fellow in the biological Sciences, University of Missouri with L. J. Stadler and California Institute of Technology with E. G. Anderson; investigations conducted at Missouri and at Cal Tech.]

McClintock, Barbara. 1933b. News Items from Ithaca: 11. Brown midrib1 (bm1) ... . *Maize Genetics Cooperation News Letter* 4 [18 December 1933]: 2.

McClintock, Barbara. 1933c. News Items from Ithaca: 12. A new narrow leafed character is linked with a1. *Maize Genetics Cooperation News Letter* 4 [18 December 1933]: 2.

**McClintock, Barbara. 1934. The relation of a particular chromosomal element to the development of nucleoli in *Zea mays* with 21 figures in the text and plates VIII–XIV. *Zeitschrift fur Zellforschung und microskopische Anatomie* 21(2) [23 June 1934, received 2 March 1934]: 294–328. [McClintock NRC Fellow in the biological sciences, California Institute of Technology with E. G. Anderson; investigation conducted at Cal Tech. Paper written while McClintock was a Guggenheim Fellow in Berlin and Freiburg, Germany and submitted just prior to leaving Germany.]

*Creighton, Harriet B. and Barbara McClintock. 1935. The correlation of cytological and genetical crossing-over in *Zea mays*. A corroboration. *Proceedings of the National Academy of Sciences* 21 (3) [15 March 1935, communicated 9 February 1935]: 148–150. [Written while McClintock was a research assistant in the Department of Plant Breeding, Cornell University (address Botany Department).]

*Rhoades, Marcus M. and Barbara McClintock. 1935. The cytogenetics of maize. *Botanical Review*. 1 (8) [August 1935, received - no date given]: 292–325. [Written while McClintock was a research assistant in the Department of Plant Breeding, Cornell University.]

McClintock, Barbara. 1936a. Cornell University, Ithaca, N.Y. — 8. Mosaic plants in part heterozygous and in part homozygous for a chromosome 5 deficiency. *Maize Genetics Cooperation News Letter* 10 [4 March 1936]: 5–6.

McClintock, Barbara and Harriet Creighton. 1936. Cornell University, Ithaca, N.Y. — 9. Several inversions ... chromosome 9 ... and chromosome 4, ... detected and isolated by Creighton and [McClintock]. *Maize Genetics Cooperation News Letter* 10 [4 March 1936]: 6.

McClintock, Barbara. 1936b. Cornell University, Ithaca, N.Y. — 10. Disjunction studies on interchanges show that sister spindle fiber regions do not separate in I, ... *Maize Genetics Cooperation News Letter* 10 [4 March 1936]: 6.

McClintock, Barbara. 1936c. [ABSTRACT PREPRINT] The production of maize plants mosaic for homozygous deficiencies: Simulation of the bm1 phenotype through loss of the Bm1 locus. Abstracts of papers presented at the December 29–31, 1936 meetings of the Genetics Society of America, Atlantic City, New Jersey; Preprinted from *Genetics* 22:183–212, 1937, *in Records of the Genetics Society of America*, Number 5, 1936. [Paper delivered 29 December 1936.]

**McClintock, Barbara. 1937a. [ABSTRACT] The production of maize plants mosaic for homozygous deficiencies: Simulation of the bm1 phenotype through loss of the <u>Bm</u>1 locus. [In Abstracts of papers presented at the 1936 meetings of the Genetics Society of America, M. Demerec, Secretary.] *Genetics* 22 (1) [January 1937, presented 29 December 1936]: 200. [Investigations funded by the Rockefeller Foundation and conducted in Department of Plant Breeding, Cornell University; McClintock's address - Cornell University. In September 1936, McClintock left Cornell to begin her Assistant Professor appointment at U of Missouri. Results reported are part of a manuscript submitted February 1938 and published in *Genetics* 23: 315–376, July 1938. Note subheadings for sections V and VI in published paper are exactly the same as title of this abstract.]

McClintock, Barbara. 1937b [ABSTRACT PREPRINT] A method for detecting potential mutations of a specific chromosomal region. Abstracts of papers presented at the December 28–30, 1937 meetings of the Genetics Society of America, Indianapolis, Indiana. Preprinted from *Genetics* 23: 139–177, 1938 *in Records of the Genetics Society of America*, Number 6, 1937. [Demonstration Paper delivered 28 December 1937.]

**McClintock, Barbara. 1938a. [ABSTRACT] A method for detecting potential mutations of a specific chromosomal region. [In Abstracts of papers presented at the 1937 meetings of the Genetics Society of America] *Genetics* 23 (1) [January 1938, presented 28 December 1937]: 159. [McClintock Assistant Professor of Botany at U of Missouri; results reported here were based on investigations funded by the Rockefeller Foundation and previously conducted in Department of Plant Breeding, Cornell University.]

*McClintock, Barbara. 1938b. The production of homozygous deficient tissues with mutant characteristics by means of the aberrant mitotic behavior of ring-shaped chromosomes. *Genetics* 23 (4) [July 1938, received 25 February 1938]: 315–376. [Most of work undertaken at Cornell with aid of grant from the Rockefeller Foundation; original material supplied by L. J. Stadler.]

*McClintock, Barbara. 1938c. The fusion of broken ends of sister half-chromatids following breakage at meiotic anaphase. *Missouri Agricultural Experiment Station Research Bulletin* 290 [July 1938, authorized 12 July 1938]: 1–48. [Continuation of investigations begun at Cornell University between 1934–1936; cites McClintock 1938b.]

*McClintock, Barbara. 1939. The behavior in successive nuclear divisions of a chromosome broken at meiosis. *Proceedings of the National Academy of Sciences* 25 (8) [15 August 1939, communicated 7 July 1939]: 405–416.

[1940 NO PUBLICATIONS]

*McClintock, Barbara. 1941a. The stability of broken ends of chromosomes in *Zea mays*. *Genetics* 26 (2) [March 1941, received 27 November 1940]: 234–282. [Paper published just prior to McClintock's academic leave (1941–1942).]

*McClintock, Barbara. 1941b. The association of mutants with homozygous deficiencies in *Zea mays*. *Genetics* 26 (5) [September 1941, received 3 May 1941]: 542–571. [Both the journal article and reprints are dated inaccurately as September 1940.]

**McClintock, Barbara. 1941c [Issued December 1941, Symposium held June 1941]. Spontaneous alterations in chromosome size and form in *Zea mays*. pp. 72–80. In *Genes and Chromosomes - Structure and Organization. Cold Spring Harbor Symposia on Quantitative Biology* Volume IX [June 1941, Issued December 1941]. Katherine S. Brehme ed. The Biological Laboratory, Cold Spring Harbor, Long Island, New York. [McClintock was appointed guest investigator for academic year 1941–42, Department of Botany, Columbia University. During the summer of 1941, and from December 1941 through December 1942, McClintock was also guest investigator, Carnegie Institution of Washington, Department of Genetics, Cold Spring Harbor. McClintock resigned from University of Missouri effective August 1942.]

**McClintock, Barbara. 1942a. The fusion of broken ends of chromosomes following nuclear fusion. *Proceedings of the National Academy of Sciences* 28 (11) [15 November 1942, communicated 22 September 1942]: 458–463.

*McClintock, Barbara. 1942b [1 July 1941–30 June 1942]. Maize genetics: The behavior of "unsaturated" broken ends of chromosomes. Phenotypic effects of homozygous deficiencies of distal segments of the short arm of chromosome 9. *Carnegie Institution of Washington Year Book No. 41* [Issued 18 December 1942, submitted June 1942]: 181–186. [In the text McClintock cites her work as "McClintock 1941; see bibliography." The reprints do not include the bibliography, which lists three McClintock publications (1941a, b, & c, published in March, September, & December 1941, respectively.]

*McClintock, Barbara. 1943 [1 July 1942–30 June 1943]. Maize genetics: Studies with broken chromosomes. Tests of the amount of crossing over that may occur within small segments of a chromosome. Deficiency mutations: Progressive deficiency as a cause of allelic series. *Carnegie Institution of Washington Year Book No. 42* [Issued 7 December 1943, submitted June 1943]: 148–152. [McClintock was permanently appointed to the staff of Carnegie Institution of Washington, Department of Genetics, Cold Spring Harbor, in 1943.]

**McClintock, Barbara. 1944a. Carnegie Institution of Washington, Department of Genetics, Cold Spring Harber [sic], Long Island, N.Y. [This report is untitled in the *MGCNL*. This is a report on deficiencies in Chromosome 9]. *Maize Genetics Cooperation News Letter.* 18 [31 January 1944, submitted 1943]: 24–26. [The report concludes, " … the chromosomal breakage mechanism is a "mutation" inducing process which "induces" the same mutant time and again." Moore (1987) cites title as: "Breakage-fusion-bridge cycle induced deficiencies in the short arm of chromosome 9." However, the term "Breakage-fusion-bridge cycle" is not used in this report.]

*McClintock, Barbara. 1944b. The relation of homozygous deficiencies to mutations and allelic series in maize. *Genetics* 29 (5) [Sept. 1944, received 8 Feb. 1944]: 478–502.

*McClintock, Barbara. 1944c [1 July 1943–30 June 1944]. Maize genetics: Completion of the study of the allelic relations of deficient mutants. The chromosome-breakage mechanism as a means of producing directed mutations. Continuation of the chromatid type of breakage-fusion-bridge cycle in the sporophytic tissues. Homozygous deficiency as a cause of mutation in maize. *Carnegie Institution of Washington Year Book No. 43* [Issued 15 December 1944, submitted June 1944]: 127–135. [The text cites McClintock 1938, and McClintock 1941, but no references are listed.]

*McClintock, Barbara. 1945a. *Neurospora*. I. Preliminary observations of the chromosomes of *Neurospora crassa. American Journal of Botany* 32 (10) [December 1945, issued 14 January 1946, received 28 August 1945]: 671–678.

*McClintock, Barbara. 1945b [1 July 1944–30 June 1945]. Cytogenetic studies of maize and Neurospora: Induction of mutations in the short arm of chromosome 9 in maize. Preliminary studies of the chromosomes of the fungus *Neurospora crassa. Carnegie Institution of Washington Year Book No. 44* [Issued 14 December 1945, submitted June 1945]: 108–112.

*McClintock, Barbara. 1946 [1 July 1945–30 June 1946]. Maize genetics: Continuation of the study of the induction of new mutants in chromosome 9. Modification of mutant expression following chromosomal translocation. The unexpected appearance of a number of unstable mutants. *Carnegie Institution of Washington Year Book No. 45* [Issued 13 December 1946, submitted June 1946]: 176–186.]

*McClintock, Barbara. 1947 [1 July 1946–30 June 1947]. Cytogenetic studies of maize and *Neurospora*: The mutable *Ds* locus in maize. Continuation of studies of the chromosomes of *Neurospora crassa. Carnegie Institution of Washington Year Book No. 46* [Issued 12 December 1947, submitted June 1947]: 146–152.

*McClintock, Barbara. 1948 [1 July 1947–30 June 1948]. Mutable loci in maize: Nature of the *Ac* action. The mutable *c* loci. The mutable *wx* loci. Conclusions. *Carnegie Institution of Washington Year Book No. 47* [Issued 10 December 1948, submitted June 1948]: 155–169.

*McClintock, Barbara. 1949 [1 July 1948–30 June 1949]. Mutable loci in maize: The mechanism of transposition of the *Ds* Locus. The origin of *Ac*-controlled mutable loci. Transposition of the *Ac* locus. The action of *Ac* on the mutable loci it controls. Mutable loci *c m-2* and *wx m-1*. Conclusions. *Carnegie Institution of Washington Year Book No. 48* [Issued 9 December 1949, submitted June 1949]: 142–154.

*McClintock, Barbara. 1950a. The origin and behavior of mutable loci in maize. *Proceedings of the National Academy of Sciences*. 36 (6) [15 June 1950, communicated 8 April 1950]: 344–355.

*McClintock, Barbara. 1950b [1 July 1949–30 June 1950]. Mutable loci in maize: Mode of detection of transpositions of *Ds*. Events occurring at the *Ds* locus. The mechanism of transposition of *Ds*. Transposition and change in action of *Ac*. Consideration of the chromosome materials responsible for the origin and behavior of mutable loci. *Carnegie Institution of Washington Year Book No. 49* [Issued 15 December 1950, submitted June 1950]: 157–167.

*McClintock, Barbara. 1951a [1 July 1950–30 June 1951]. Mutable loci in maize. *Carnegie Institution of Washington Year Book No. 50* [Issued 14 December 1951, submitted June 1951]: 174–181.

**McClintock, Barbara. 1951b [C. 1952, Symposium held June 1951]. Chromosome organization and genic expression. Pgs. 13–47. In *Genes and Mutations, Cold Spring Harbor Symposia on Quantitative Biology*, Volume XVI [7–15 June 1951]. Katherine Brehme Warren ed. The Biological Laboratory, Cold Spring Harbor, Long Island, New York. [Copyright 1952. This reference has been cited as 1951 or 1952- see *Citation Index*. Moore (1987) lists it as 1951. McClintock cites it as 1951. *Carnegie Year Book No. 51*, Department of Genetics Bibliography, lists it as McClintock 1951.]

*McClintock, Barbara. 1952. [1 July 1951–30 June 1952]. Mutable loci in maize: Origins of instability at the *A*1 and *A*2 loci. Instability of *Sh*1 action induced by *Ds*. Summary. *Carnegie Institution of Washington Year Book No. 51* [Issued 12 December 1952, submitted June 1952]: 212–219.

*McClintock, Barbara. 1953a. Induction of instability at selected loci in maize. *Genetics* 38 (6) [November 1953, issued 20 January 1954, received 14 April 1953]: 579–599.

**McClintock, Barbara. 1953b [1 July 1952–30 June 1953]. Mutations in maize: Origin of the mutants. Change in action of genes located to the right of *Ds*. Comparison between *Sh1* mutants. Change in action of genes located to the left of *Ds*. Meiotic segregation and mutation. *Carnegie Institution of Washington Year Book No. 52* [Issued 11 December 1953, submitted June 1953]: 227–237. [Listed in Moore (1987) as a 1954 publication.]

*McClintock, Barbara. 1954 [1 July 1953–30 June 1954]. Mutations in maize and chromosomal aberrations in *Neurospora*: Mutations in maize. Chromosome aberrations in *Neurospora*. *Carnegie Institution of Washington Year Book No. 53* [Issued 10 December 1954, submitted June 1954]: 254–260.

*McClintock, Barbara. 1955a. Carnegie Institution of Washington, Department of Genetics, Cold Spring Harbor, Long Island, N.Y. 1. Spread of mutational change along the chromosome. 2. A case of *Ac*-induced instability at the bronze locus in chromosome 9. 3.

Transposition sequences of *Ac*. 4. A suppressor-mutator system of control of gene action and mutational change. 5. System responsible for mutations at *a1m-2. Maize Genetics Cooperation News Letter* 29 [17 March 1955]: 9–13.

*McClintock, Barbara. 1955b [1 July 1954–1955]. Controlled mutation in maize: The *a1m-1-Spm* system of control of gene action and mutation. Continued studies of the mode of operation of the controlling elements *Ds* and *Ac. Carnegie Institution of Washington Year Book No. 54* [Issued 9 December 1955, submitted June 1955]: 245–255.

**McClintock, Barbara. 1956a [Issued Feb. 1956, Symposium held 15–17 June 1955]. Intranuclear systems controlling gene action and mutation. pp. 58–74. In *Mutation, Brookhaven Symposia in Biology, No. 8*. Biology Department, Brookhaven National Laboratory, Upton, NY. [No editor listed for this volume. R.C. King, Symposium Chairman. Cited in Moore (1987) as "Issued 1956" but listed chronologically with the 1955 publications. Listed in *Carnegie Year Book No. 55* Bibliography as a 1956 publication.]

**McClintock, Barbara. 1956b. Carnegie Institution of Washington, Department of Genetics, Cold Spring Harbor, Long Island, N.Y. 1. Further study of the *a1m-1-Spm system*. 2. Further study of Ac control of mutation at the bronze locus in chromosome 9. 3. Degree of spread of mutation along the chromosome induced by *Ds*. 4. Studies of instability of chromosome behavior of components of a modified chromosome 9. *Maize Genetics Cooperation News Letter* 30 [15 March 1956]: 12–20. [In Moore (1987) the number 9 is missing following the last word of descriptive subtitle. This deletion is also transcribed in Buckner's (1997, *Women in the Biological Sciences*: Greenwood Press) bio-bibliography of McClintock.]

*McClintock, Barbara. 1956c [1 July 1955–1 June 1956]. Mutation in maize: *Ac* control of mutation at the bronze locus in chromosome 9. Control of gene action by a non-transposing *Ds* element. Continued examination of the *a1m-1-Spm* system of control of gene action. Changes in chromosome organization and gene expression produced by a structurally modified chromosome 9. *Carnegie Institution of Washington Year Book No. 55* [Issued 14 December 1956, submitted June 1956]: 323–332.

**McClintock, Barbara. 1956d [C.1957, Symposium held June 1956]. Controlling elements and the gene, pp. 197–216. In *Genetic Mechanisms: Structure and Function, Cold Spring Harbor Symposia on Quantitative Biology*, Volume XXI [4–12 June 1956]. K. B. Warren ed. The Biological Laboratory, Cold Spring Harbor, Long Island, New York. [Listed in *Carnegie Year Book 56* Bibliography as McClintock 1956.]

*McClintock, Barbara. 1957a. Carnegie Institution of Washington, Department of Genetics, Cold Spring Harbor, Long Island, N.Y. 1. Continued study of stability of location of *Spm*. 2. Continued study of a structurally modified chromosome 9. *Maize Genetics Cooperation News Letter* 31 [15 March 1957]: 31–39.

*McClintock, Barbara. 1957b [1 July 1956–30 June 1957]. Genetic and cyto-logical studies of maize: Types of *Spm* elements. A modifier element within the *Spm* system. The relation between *a1m*-1 and *a1m*-2. Aberrant behavior of a fragment chromosome. *Carnegie Institution of Washington Year Book 56* [Issued 9 December 1957, submitted June 1957]: 393–401.

*McClintock, Barbara. 1958 [1 July 1957–30 June 1958]. The suppressor-mutator system of control of gene action in maize: The mode of operation of the *Spm* element. A modifier element in the *a1m*-1-*Spm* system. Continued investigation of transposition of *Spm*. *Carnegie Institution of Washington Year Book 57* [Issued 19 December, submitted June 1958]: 415–429.

Moreno, Ulises, Alexander Grobman, and Barbara McClintock. 1959a. Escuela Nacional de Agricultura, La Molina, Lima, Peru: 5. Study of chromosome morphology of races of maize in Peru. *Maize Genetics Cooperation News Letter* 33 [1 April 1959]: 27–28.

*McClintock, Barbara. 1959b [1 July 1958–30 June 1959]. Genetic and cytologi-cal studies of maize: Further studies of the *Spm* system. Chromosome constitutions of some South American races of maize. *Carnegie Institution of Washington Year Book 58* [Issued 14 December 1959, submitted June 1959]: 452–456.

*McClintock, Barbara. 1960 [1 July 1959–30 June 1960]. Chromosome con-stitutions of Mexican and Guatemalan races of maize: General Conclusions. *Carnegie Institution of Washington Year Book 59* [Issued 12 December 1960, submitted June 1960]: 461–472. [Milislav Demerec, Director, Department of Genetics, retired 30 June 1960. He was succeeded by Berwind Kaufman.]

*McClintock, Barbara. 1961a. Some parallels between gene control systems in maize and in bacteria. *American Naturalist* XCV (884) [Sept.-Oct. 1961, received-no date given]: 265–277.

*McClintock, Barbara. 1961b [1 July 1960–30 June 1961]. Further studies of the suppressor-mutator system of control of gene action in maize: Control of *a1m*-2 by the *Spm* system. A third inception of control of gene action at the *A*1 locus by the *Spm* system. Control of gene action at the locus of *Wx* by the *Spm* system. Control of reversals in *Spm* activity phase. Nonrandom selection of genes coming under the control of the *Spm* system. *Carnegie Institution of Washington Year Book 60* [Issued 11 December 1961, submitted June 1961]: 469–476. [Berwind P. Kaufman, Acting Director, Department of Genetics.]

*McClintock, Barbara. 1962 [1 July 1961–30 June 1962]. Topographical rela-tions between elements of control systems in maize: Origin from *a1m*-5 of a two-element control system. Analysis of *a1m*-2. The derivatives of *bz* m-2. *Carnegie Institution of Washington Year Book 61* [Issued 10 December 1962, submitted June 1962]: 448–461. [Annual Report of the Director of the Department of Genetics: "as this report goes to press the

Department is being terminated" (pg. 438). Berwind P. Kaufman, Director, retired on 30 June 1962. Subsequently, McClintock's reports are published in the Annual Report of the Director (Alfred D. Hershey), Genetics Research Unit, Carnegie Institution of Washington. The Unit replaced the former Department of Genetics, active at Cold Spring Harbor from November 1, 1920 to June 30, 1962.]

*McClintock, Barbara. 1963 [1 July 1962–30 June 1963]. Further studies of gene-control systems in maize: Modified states of *a1m-2*. Extension of *Spm* control of gene action. Further studies of topographical relations of elements of a control system. *Carnegie Institution of Washington Year Book 62* [Issued 9 December 1963, submitted June 1963]: 486–493.

**McClintock, Barbara. 1964 [1 July 1963–30 June 1964]. Aspects of gene regulation in maize: Parameters of regulation of gene action by the *Spm* system. Cyclical change in phase of activity of *Ac* (Activator). *Carnegie Institution of Washington Year Book 63* [Issued December 1964, submitted June 1964]: 592–601, plus 2 plates and 2 plate legends. [Cited in Moore (1987) as 592–602.]

**McClintock, Barbara. 1965a. Carnegie Institution of Washington, Cold Spring Harbor, N.Y.: 1. Restoration of *A1* gene action by crossing over. *Maize Genetics Cooperation News Letter* 39 [15 April 1965]: 42–[45]. [Page 45 is unnumbered. This report and the one that follows are separate reports in the *MGCNL*. McClintock 1968b cites this report.]

**McClintock, Barbara. 1965b. Carnegie Institution of Washington, Cold Spring Harbor, N.Y.: 2. Attempts to separate *Ds* from neighboring gene loci. *Maize Genetics Cooperation News Letter* 39 [15 April 1965]:[45]–51. [Page 45 is unnumbered. This report and the one that precedes it are separate reports in the *MGCNL*; both are listed in Moore (1987) as one report.]

**McClintock, Barbara. 1965c [1 July 1964–30 June 1965]. Components of action of the regulators *Spm* and *Ac*: The component of *Spm* responsible for preset patterns of gene expression. Transmission of the preset pattern. Components of action of *Ac*. *Carnegie Institution of Washington Year Book 64* [Issued December 1965, submitted June 1965]: 527–534, plus 2 plates and 2 figure legends.

*McClintock, Barbara. 1965d [Issued December 1965, Symposium held 7–9 June 1965]. The control of gene action in maize. pp. 162–184. In *Genetic Control of Differentiation, Brookhaven Symposia in Biology: No. 18*. Biology Department, Brookhaven National Laboratory, Upton, N.Y. [No editor listed. H. H. Smith Chairman of the Symposium Committee.]

*McClintock, Barbara. 1967 [1 July 1965–30 June 1966]. Regulation of pattern of gene expression by controlling elements in maize: Pigment distribution in parts of the ear. Pigment distribution in the pericarp layer of the kernel. Presetting of the controlling element at the locus of *c2m-2*. Inheritance of modified pigmentation patterns. *Carnegie Institution of Washington Year Book 65* [Issued January 1967, submitted June 1966]: 568–576, plus 2 plates and 2 plate legends. [Cited in Moore as 568–578, which includes a non relevant page and one plate legend.]

**McClintock, Barbara. 1968a [1 June 1966–1 June 1967]. The states of a gene locus in maize: The states of *a*1*m*-1. The states of *a*1*m*-2. *Carnegie Institution of Washington Year Book 66* [Issued January 1968, submitted May 1967]: 20–28, plus 2 plates and 2 plate legends. [In 1967, McClintock was honored with the appointment of Distinguished Service Member of the Carnegie Institution of Washington. She held that position until her death in 1992.]

*McClintock, Barbara. 1968b [Symposium held June 1967]. Genetic systems regulating gene expression during development. In *Control Mechanisms in Developmental Processes, II*. The Role of the Nucleus. Michael Locke, ed. The 26th Symposium of the Society for Developmental Biology (June 1967) [La Jolla, CA, USA]. *Developmental Biology*, Supplement 1: 84–112. Academic Press. New York.

*McClintock, Barbara. 1971 [1 July 1970–30 June 1971]. The contribution of one component of a control system to versatility of gene expression: Relation of dose of *Spm* to pattern of pigmentation with the class II state of *a*2*m*-1. Distinctive phenotypes associated with activation of an inactive *Spm*. An example of versatility of control of gene expression associated with component-2 of *Spm. Carnegie Institution of Washington Year Book 70* [Issued December 1971, submitted June 1971]: 5–17. [This is the last report McClintock published in the *CIW Year Book*. The Genetics Research Unit closed 30 June 1971. McClintock was awarded the National Medal of Science that same year (Award year 1970, presented 21 May 1971).]

**McClintock, Barbara. 1978a [Symposium held September 1975]. Significance of chromosome constitutions in tracing the origin and migration of races of maize in the Americas. Chapter 11, pp. 159–184. In *Maize Breeding and Genetics*, David B. Walden ed., John Wiley and Sons, Inc., New York. [Moore (1987) cites editor as W. D. Walden. This volume is the Proceedings of the International Maize Symposium held September 1975, Champaign, Urbana, Illinois, USA. Note that page V incorrectly dates the Symposium as September 1977.]

**McClintock, Barbara. 1978b [Symposium held 13–15 June 1977]. Development of the maize endosperm as revealed by clones. pp. 217–237. In The *Clonal Basis of Development*. Stephen Subtelny, and Ian M. Sussex eds. The 36th Symposium of the Society for Developmental Biology (June 1977) [Raleigh, North Carolina, USA]. Academic Press, Inc., New York. [McClintock's paper appears in section IV. Nuclear and Genetic Events in Clone Initiation. Note correct spelling of endosperm in title.]

**McClintock, Barbara. 1978c. [Symposium held 7–8 April 1978] Mechanisms that rapidly reorganize the genome. *Stadler Genetics Symposia*, vol. 10, pp. 25–48. [Pg. 48 is a plate of photographs of

Symposium participants. Symposium held at Columbia, Missouri, USA. Proceedings published by University of Missouri, Agricultural Experiment Station.]

*McClintock, Barbara. 1980 [January 1980]. Modified gene expressions induced by transposable elements. In *Mobilization and Reassembly of Genetic Information. Proceedings of the Miami Winter Symposium.* W. A. Scott, R. Werner, D. R. Joseph, and Julius Schultz eds. *Miami Winter Symposium* 17 [January 1980]: 11–19. Academic Press, Inc. New York. [Lecture given on 7 January 1980. Symposium sponsored by Dept. of Biochemistry, University of Miami School of Medicine, Miami, Florida, and by The Papanicolaou Cancer Research Institute, Miami, Florida, USA].

*McClintock, Barbara, T. Angel Kato Y. and Almiro Blumenschein 1981. *Chromosome Constitution of Races of Maize. Its Significance in the Interpretation of Relationships Between Races and Varieties in the Americas.* Colegio de Postgraduados, Escuela National de Agricultura, Chapingo, Edo. Mexico, Mexico. xxxi, 517 pp.

McClintock, Barbara, 1983. [ABSTRACT]. Trauma as a means of initiating change in genome organization and expression. *In Vitro* 19 (3, Part II) [March 1983]: 283–284. [Paper presented at 34th annual meeting of the Tissue Culture Association, 12–16 June 1983, Orlando, Florida.]

McClintock, Barbara, 1984. The significance of responses of the genome to challenge. *Science* 226:792–801

*McClintock, Barbara, 1984a [Nobel lecture presented, 8 December 1983]. The significance of responses of the genome to challenge. *Science* 226 (4676) [16 November 1984]: 792–801. [Footnotes indicate that "Minor corrections have been made by the author," and that this lecture "will be included in the complete volume of *Les Prix Nobel en 1984* as well as in the series *Nobel Lectures* (in English) published by the Elsevier Publishing Company, Amsterdam and New York." The correct citations for *Les Prix Nobel* and *Nobel Lectures* are listed below (see McClintock 1984b and McClintock 1993).]

McClintock, Barbara, 1984b [Nobel lecture presented, 8 December 1983]. The significance of responses of the genome to challenge. pp. 174–193. *Les Prix Nobel 1983.* Almqvist & Wiksell International, Stockholm-Sweden. Copyright by the Nobel Foundation. [Note - some data bases incorrectly cite publisher as "Stockholm, Imprimeri Royal." In addition to McClintock's Nobel Prize lecture, this volume includes McClintock's photograph (pg. 170), biographical sketch (pgs. 171–173), and the short speech she delivered at the Nobel Banquet in Stockholm (pg. 42).]

McClintock, Barbara. 1987. Introduction. pp. vii-xi. In John A. Moore, ed. *The Discovery and Characterization of Transposable Elements. The Collected Papers of Barbara McClintock.* Garland Publishing Co. New York.

McClintock, Barbara, 1993. The significance of the genome to challenge. pp. 180–199. In Tore Frangsmyr and Jan Lindsten, eds. *Nobel Lectures*

Physiology or Medicine 1981–1990. World Scientific Pub. Co., Singapore, for the Nobel Foundation. [This volume also includes Presentation speeches (in English) and Laureates' photographs, and biographies. The complete section on McClintock's 1983 Nobel Prize is on pages 171–199. Note - Elsevier, who had published the *Nobel Lectures* from 1901 through 1970, discontinued the project.]

References

Anderson, E. G. 1936. Reduced chromosomal alterations in maize. Pgs 1297–1310 *in* Benjamin M. Duggar, ed., *Biological Effects of Radiation* volume II, McGraw-Hill Book Co. Inc., New York.

Banks, J.A. Masson, P. and N. Fedoroff 1988. Molecular mechanisms in the developmental regulation of the maize Suppressor-mutator transposable element. *Genes and Development* 2: 1364–1380.

Berg, P. and M. Singer 2003. *George Beadle, An Uncommon Farmer.* Cold Spring Harbor Lab Press, Cold Spring Harbor, NY.

Blackburn, E. 1992. Broken Chromosomes and Telomeres Pp 381–388, *in* Fedoroff, N. and D. Botstein, eds. *The Dynamic Genome, Barbara McClintock's Ideas in the Century of Genetics.* Cold Spring Harbor Press, Woodbury, NY.

Botstein, D., White, R. L., Skolnick, M. H. & Davis, R. W. 1980. Construction of a genetic linkage map in man using restriction fragment length polymorphisms. *Am. J. Hum. Genet.* 32 (3): 314–331.

Brink, R.A. and R.A. Nilan 1952. The relation between light variegated and medium variegated pericarp in maize. *Genetics* 37: 519–544.

Burnham, C. R. 1930.. Genetical and cytological studies of semisterility and related phenomena in maize. *Proceedings of the National Academy of Sciences* 16: 269–277.

Chomet, Paul S., Wessler, S. and S. Dellaporta 1987. Inactivation of the maize transposable element Activator (Ac) is associated with its DNA methylation. *EMBO J.* 6(2): 295–302.

Clausen, R. E. 1936. [Abstract #] 12585. McClintock, Barbara. The relation of a particular chromosomal element to the development of the nucleoli in *Zea mays. Zeitschr. Zellforsch. Sellforsch. u. Mikrosk. Anat.* 21(2): 294–328. 7 pl. 21 figs. 1934. *Biological Abstracts* 10(6): 1339–1340.

Coe, Edward and Lee B. Kass. 2005. Proof of physical exchange of genes on the chromosomes. *Proceedings of the National Academy of Sciences* 102: 6641–6656.

Comfort, Nathaniel C. 2001. *The Tangled Field.* Harvard U. Press, Cambridge, MA.

Creighton, Harriet B. and Barbara McClintock. 1931. A correlation of cytological and genetical crossing-over in *Zea mays. Proceedings of the National Academy of Sciences* 17: 492–497.

Creighton, H.B. 1934. [Abstract #] 57. Creighton, Harriet B. and Barbara McClintock. A correlation of cytological and genetical crossing-over in *Zea Mays. Proc. Nation. Acad. Sci. U.S.A.* 17(8): 492–497. 2 fig. 1931. *Biological Abstracts* 8(1, Jan.): 8.

Dollinger, E. J. 1954. Studies on induced mutations in maize. *Genetics* 39: 750–766.

Emerson, R.A. Beadle, G.W. and A.C. Fraser 1935. *A summary of linkage studies in maize.* Cornell University Agr. Exp. Sta. Memoir 180. Ithaca, NY.

Fedoroff, Nina V. and David Botstein eds. 1992. *The Dynamic Genome, Barbara McClintock's Ideas in the Century of Genetics.* Cold Spring Harbor Laboratory Press, Woodbury, NY.

Kass, Lee B. 2005a. "Harriet Creighton: proud botanist." *Plant Science Bulletin* 51: 118–125.

Kass, Lee B. 2005b. Missouri compromise: tenure or freedom? New evidence clarifies why Barbara McClintock left academe. *Maize Genetics Cooperation Newsletter* 79: 52–71. Available, online, April 2005: http://www.agron.missouri.edu/mnl/79/05kass.htm

Kass, Lee B. 2003. Records and recollections: A new look at Barbara McClintock, Nobel Prize-Winning geneticist. *Genetics* 164: 1251–1260.

Kass, Lee B. 2000. McClintock, Barbara, American botanical geneticist, 1902–1992. In *Plant Sciences,* edited by R. Robinson, pp. 66–69, New York: Macmillan Science Library, USA.

Kass, L. B. 1999. Current list of Barbara McClintock's publications. *Maize Genetics Cooperation Newsletter* 73: 42–48. Available online, 1998: http://www.agron.missouri.edu/mnl/73/110kass. html

Kass, Lee B. and Christophe Bonneuil. 2004. Mapping and seeing: Barbara McClintock and the linking of genetics and cytology in maize genetics, 1928–1935. In *Classical Genetic Research and its Legacy: The Mapping Cultures of 20th Century Genetics,* edited by Hans-Jörg Rheinberger and Jean-Paul Gaudilliere, pp. 91–118. Routledge, London.

Kass, Lee B., Christophe Bonneuil, and Edward H. Coe Jr. 2005. Cornfests, cornfabs and coopera-tion: The origins and beginnings of the *Maize Genetics Cooperation News Letter. Genetics* 169: 1787–1797.

Kass, L. B. and K. Gale. 2008. McClintock, Barbara, Pp. 200-201, in Bonnie Smith, Editor, *The Oxford Encyclopedia of Women in World History, Volume 3.* Oxford University Press.

Keller, Evelyn Fox. 1983. *A Feeling for the Organism. The Life and Work of Barbara McClintock.* W. H. Freeman & Co., San Francisco.

Kreizinger, J. D. 1960. Diepoxybutane as a chemical mutagen in *Zea mays. Genetics* 45(2): 143–154.

Lippman, Zachary and Rob Martienssen 2004. The role of RNA interference in heterochromatic silencing: *Nature* 431: 364–370.

McClintock, Barbara. 1987. *The Discovery and Characterization of Transposable Elements: The Collected Papers of Barbara McClintock.* Garland Publishing, New York.

McClintock, Barbara, 1984. The significance of responses of the genome to challenge. *Science* 226: 792–801.

McClintock, Barbara, T. Angel Kato Y. and Almiro Blumenschein 1981. *Chromosome Constitution of Races of Maize. Its Significance in the Interpretation of Relationships Between Races and Varieties in the Americas.* Colegio de Postgraduados, Escuela National de Agricultura, Chapingo, Edo. Mexico.

McClintock, Barbara. 1978. Significance of chromosome constitutions in tracing the origin and migration of races of maize in the Americas. Chapter 11, pp. 159–184. In *Maize Breeding and Genetics,* David B. Walden ed., John Wiley and Sons, Inc., New York.

McClintock, Barbara. 1965 The control of gene action in maize. pp. 162–184. In *Genetic Control of Differentiation, Brookhaven Symposia in Biology*: No. 18. Biology Department, Brookhaven National Laboratory, Upton, N.Y.

McClintock, Barbara. 1964. Aspects of gene regulation in maize. *Carnegie Institution of Washington Year Book 63*: 592–601.

McClintock, Barbara. 1958. The suppressor-mutator system of control of gene action in maize. *Carnegie Institution of Washington Year Book: No. 57*: 415–429.

McClintock, Barbara. 1957. Genetic and cytological studies of maize. *Carnegie Institution of Washington Year Book No. 56*: 393–401.

McClintock, Barbara. 1956. Intranuclear systems controlling gene action and mutation. pp. 58–74. In *Mutation, Brookhaven Symposia in Biology,* No. 8. Biology Department, Brookhaven National Laboratory, Upton, NY.

McClintock, Barbara. 1954. Mutations in maize and chromosomal aberrations in Neurospora. *Carnegie Institution of Washington Year Book No. 53*: 254–260.

McClintock, Barbara. 1950. The origin and behavior of mutable loci in maize. *Proceedings of the National Academy of Sciences.* 36 (6): 344–355.

McClintock, Barbara. 1951 [C. 1952, Symposium held June 1951]. Chromosome organization and genic expression. Pgs. 13–47. In *Genes and Mutations, Cold Spring Harbor Symposia on Quantitative Biology, Volume XVI* [7–15 June 1951].

McClintock, Barbara. 1946. Maize genetics: Continuation of the study of the induction of new mutants in chromosome 9. Modification of mutant expression following chromosomal translocation. The unexpected appearance of a number of unstable mutants. *Carnegie Institution of Washington Year Book No. 45*: 176–186.

McClintock, Barbara. 1945a. Neurospora. I. Preliminary observations of the chromosomes of *Neurospora crassa. American Journal of Botany* 32 (10): 671–678.

McClintock, Barbara. 1945b. Cytogenetic studies of maize and Neurospora: Induction of mutations in the short arm of chromosome 9 in maize. Preliminary studies of the chromosomes of the fungus *Neurospora crassa. Carnegie Institution of Washington Year Book No. 44*: 108–112.

McClintock, Barbara. 1944a. The relation of homozygous deficiencies to mutations and allelic series in maize. *Genetics* 29 (5): 478–502.

McClintock, Barbara. 1944b [1 July 1943–30 June 1944]. Maize genetics: Completion of the study of the allelic relations of deficient mutants. The chromosome-breakage mechanism as a means of producing directed mutations. Continuation of the chromatid type of breakage-fusion-bridge cycle in the sporophytic tissues. Homozygous deficiency as a cause of mutation in maize. *Carnegie Institution of Washington Year Book No. 43*.

McClintock, Barbara. 1942. The fusion of broken ends of chromosomes following nuclear fusion. *Proceedings of the National Academy of Sciences* 28 (11): 458–463.

McClintock, Barbara. 1941c [Issued December 1941, Symposium held June 1941]. Spontaneous alterations in chromosome size and form in *Zea mays*. pp. 72–80. In *Genes and Chromosomes - Structure and Organization. Cold Spring Harbor Symposia on Quantitative Biology* Volume IX.

McClintock, Barbara. 1941b. The association of mutants with homozygous deficiencies in Zea mays. *Genetics* 26 (5): 542–571.

McClintock, Barbara. 1941a. The stability of broken ends of chromosomes in *Zea mays. Genetics* 26 (2): 234–282.

McClintock, Barbara. 1939. The behavior in successive nuclear divisions of a chromosome broken at meiosis. *Proceedings of the National Academy of Sciences* 25 (8): 405–416.

McClintock, Barbara. 1938b. The fusion of broken ends of sister half-chromatids following breakage at meiotic anaphase. *Missouri Agricultural Experiment Station Research Bulletin* 290: 1–48.

McClintock, Barbara. 1938a. The production of homozygous deficient tissues with mutant characteristics by means of the aberrant mitotic behavior of ring-shaped chromosomes. *Genetics* 23 (4): 315–376.

McClintock, Barbara. 1934. The relation of a particular chromosomal element to the development of nucleoli in *Zea mays.. Zeitschrift fur Zellforschung und microskopische Anatomie* 21(2): 294–328.

McClintock, Barbara. 1933. The association of non-homologous parts of chromosomes in the mid-prophase of meiosis in *Zea mays*, with 51 figures in the text and plates VII–XII. *Zeitschrift fur Zellforschung und microskopische Anatomie* 19 (2): 191–237.

McClintock, B. 1932 A correlation of ring-shaped chromosomes with variegation in *Zea mays. Proceedings of the National Academy of Sciences*. 18(12): 677–681.

McClintock, Barbara and Henry E. Hill. 1931. The cytological identification of the chromosome associated with the R-G linkage group in *Zea mays. Genetics* 16 (2) 175–190.

McClintock. 1931. Cytological observations of deficiencies involving known genes, translocations, and an inversion in *Zea mays. Missouri Agric. Exp. Station Res. Bull.* 163:1–30.

McClintock, Barbara. 1930. A cytological demonstration of the location of an interchange between two non-homologous chromosomes of *Zea mays. Proceedings of the National Academy of Sciences* 16 (12): 791–796.

McClintock, Barbara. 1929a. A cytological and genetical study of triploid maize. *Genetics* 14 (2): 180–222.

McClintock, Barbara. 1929b. A 2N-1 chromosomal chimera in maize. *Journal of Heredity* XX (5): 218.

McClintock, Barbara. 1927. *A Cytological and Genetical Study of Triploid Maize*. Cornell University, Ithaca, New York. Ph.D Thesis.

McClintock, Barbara. 1925. *A Resume of Cytological Investigations of the Cereals with Particular Reference to Wheat*. Ithaca, NY. Thesis M. A. Cornell University.

Murphy, R. P. and L. B. Kass. 2007. *Evolution of Plant Breeding at Cornell University: A Centennial History, 1907–2006*. Department of Plant Breeding & Genetics, Cornell University, Ithaca, NY.

Navashin, M. 1930. Unbalanced somatic chromosomal variation. University of California Publications in Agricultural Sciences 6(3, March): 95–107.

Nuffer, M. G. 1952. *A study of a mutable allele of A1 and certain related modifiers of mutation in maize*. Ph.D. Dissertation. University of Missouri.

Peterson, P. A. 1953. A mutable pale-green locus in maize. Genetic*s* 38: 682–683.

Peterson, P.A. 1965. A relationship between the *Spm* and *En* control systems in maize. *Am. Nat.* 99:391–398.

Randolph, L.F. and B. McClintock. 1926. Polyploidy in *Zea mays* L. *Amer. Nat.* 60(666): 99–102.

Randolph, L. F. 1927. [Abstract #] 2050. Randolph, L.F.. and B. McClintock. Polyploidy in *Zea mays* L. *Amer. Nat.* 60(666): 99–102. 8 figs. 1926. *Biological Abstracts* 1(2–3, Apr.): 213.

Rhoades, M.M. 1934. [Abstract #] 5179. McClintock, Barbara. A correlation of ring-shaped chromosomes with variegation in *Zea mays*. *Proc. Nation. Acad. Sci. U. S. A.* 18(12): 677–681. 1932. *Biological Abstracts* 8(3): 571.

Rhoades, M. M. 1938. Effect of the Dt gene on the mutability of the *a1* allele in maize. *Genetics* 23(4): 377–397.

Rhoades, M. M. and B. McClintock. 1935. The cytogenetics of Maize. *Botanical Review*. 1(August): 292–325.

Schwartz, D. and E. Dennis 1986. Transposase activity of the Ac controlling element in maize is regulated by its degree of methylation. *Molecular and General Genetics* 205(3):476–482.

Singleton, J.R. 1953. Chromosome morphology and the chromosome cycle in the ascus of Neurospora crassa. *American Journal of Botany* 40(3): 124–144.

Singleton J. R. 1948. *Cytogenetic studies of Neurospora crassa*. Ph.D. Dissertation. California Institute of Technology, Pasadena, California. 134 pp.

Stadler, L.J. 1928. Genetic effects of X-rays in maize. *Proceedings of the National Academy of Sciences* 14(1): 69–75.

Stadler, L.J.1930. Some genetic effects of X-rays in plants. *Journal of Hererdity* 21 (1, Jan.): 3–20.

Timothy, D. H. 1984. Book Review, McClintock; Takeo Angel Kato Y.; Blumenschein, A.: *Chromosome Constitution of Races of Maize. Its Significance in the Interpretation of Relationships Between Races and Varieties in the Americas*. Chapingo, Mexico: Colegio de Postgraduados 1981. *Theoretical and Applied Genetics*. 67(2–3): 130.

The Birth of Maize Molecular Genetics

L. Curtis Hannah and Drew Schwartz

Abstract Long before recombinant DNA technology was invented, maize genetics was a vibrant and exciting science dominated by controlling elements, cytogenetics, gene mapping and heterosis. Genes were understood as mutationally-defined units of function that could be placed on chromosomes. Incorporation of the concept of DNA as genetic material and the central dogma of genetics (DNA $\Leftrightarrow$ RNA $\Rightarrow$ protein) into the thinking of maize geneticists occurred rapidly. But, the only way to propagate maize DNA was to plant a seed.

In this chapter, we provide a personal account of how maize molecular genetics came into existence. We focus on some of the original, urgent questions of maize genetics that required the tools of molecular biology for satisfying explanations. No pretense is implied concerning the completeness of the narrative below, and we emphasize that this is a personal account.

1 Introduction and Overview

Several aspects of maize genetics involved "getting as close as possible" to the gene and the technology available at the time usually directed people to the gene product. The ability to structurally analyze the gene at the DNA sequence level did not enter the realm of possibilities in these early times. Some of the pressing questions that quickened the application of recombinant DNA tools to maize genetics are outlined below:

L.C. Hannah
University of Florida, Department of Horticultural Sciences
lchannah@ufl.edu

D. Schwartz
Indiana University, Department of Biology
schwartz@indiana.edu

J.L. Bennetzen and S. Hake (eds.), *Maize Handbook - Volume II: Genetics and Genomics,*   53
© Springer Science+Business Media LLC 2009

1.1   The Nature of the Gene

One basic question that confronted the early maize geneticists was the nature of the gene. While Mendel's Law of Segregation clearly stated that alleles separate cleanly, definitive evidence for intragenic recombination became available in phage T4 of *E. coli* (Benzer 1955). This seemingly violated Mendel's first law. While recombination could occur within the A and within the B cistrons of the r_{II} region in this microbe, rates of recombination were quite low. Could recombination also occur in a gene of a eukaryote? Nelson (1968) exploited a large collection of recessive, loss-of-function *waxy (wx)* alleles and the fact that *wx* gene expression could be scored in the pollen. These factors allowed for the rapid examination of millions of meiotic products. Nelson found that, like phage T4, recombination could occur in plants heterallelic at the *wx* locus. Parallel work was ongoing in Drosophila and intragenic recombination was observed there as well. This and subsequent work with other systems showed that intragenic recombination is not limited to prokaryotes and is universal in nature.

While recombination could be observed in the *wx* locus, rates of recombination were dependent on the environment and genetic background. Recombination rates were not additive, and many mutants behaved as deletions/inversions. Nelson concluded his 1968 paper stating that fundamentally different types of analyses would be necessary to understand the nature of some of the mutations within the *wx* locus. Clearly a different set of experimental tools would be necessary to clarify the nature of mutations within *wx*.

1.2   Genetic Variability

Questions of genetic variability in maize were as important in the early days of maize genetics as they are today. Several investigators exploited the fact that the activity of some enzymes could be detected in gels following electrophoretic separation. Hence, the isoforms of an enzyme could be separated quickly without the need for lengthy purification procedures. What became obvious early in these experiments was the extensive amount of enzyme variation seen among tissues and among genotypes. Some of the earliest informative isoenzyme studies concerned esterase activity in various tissues of the maize plant (Schwartz 1960A, 1962, 1963, 1967). Initial studies revealed multiple bands of esterase activity and experiments were quickly designed to test whether these results were due to multiple alleles of a single locus or to multiple loci. Surprisingly, an activity band was found that was unique to heterozygotes. As judged by band intensities in plant (2N) and endosperm (3N) tissues, the most logical explanation was that the enzyme was a dimer composed of subunits encoded by a single locus. A hybrid enzyme, composed of subunits encoded by the two alleles of one gene, was found in the hybrid but not in either homozygous parent. Subsequent experiments with iodoacetamide and borohydride suggested that the various forms differed by a single charge that could be masked via protein folding in some cases.

Schwartz exploited a number of esterase allelic differences to infer important features of the enzyme. For example, enzyme activity did not require two functional subunits; a heterodimer composed of a functional and non-functional subunit had activity, albeit only one-half that of the wild-type enzyme. Some subunits conditioned enhanced enzyme stability, whereas others did not. He also isolated an allele that appeared identical to one of the other alleles by electrophoretic mobility, yet the timing of expression during development was strikingly different.

The Schwartz group at Indiana University also performed a series of seminal studies with isoenzymes of alcohol dehydrogenase (ADH) (Freeling, and Bennett 1985; Schwartz 1966, 1969, 1971; Schwartz, and Endo 1966). Two genes, *ADH1* and *ADH2,* encode this dimeric enzyme. Subunits encoded from both genes can dimerize to form active enzymes. Maize, again, proved to be highly polymorphic for alleles of both genes, and this variation proved important in uncovering a number of features of the enzyme. For example, one allele of *ADH1*, termed *Cm*, conditions a homodimer of low activity; however, the homodimer is quite stable. Stability is conferred to heterodimers containing the *Cm* subunit. One intriguing observation was that while levels of *ADH1* enzymatic activity were indistinguishable in pollen of plants homozygous for each of the two alleles, one allele exhibited higher levels of activity when heterozygotes were sampled. Furthermore, the hybrid band was missing in pollen extracts, showing that *ADH1* is expressed after separation of the two alleles by meiosis. This led to the concept of the "Competition Model". Before separation in meiosis, the two alleles compete for a limiting factor. One allele is a stronger competitor, explaining the higher level of activity of this isoform from heterozygotes. Competition occurs, but before gene expression. The implications of these observations and the model to explain it remain important today, since they point to allele-specific pre-programming of gene expression.

Analyses of esterase and alcohol dehydrogenase pointed to the prevalence of enzymes composed of identical or nearly identical subunits. The fact that heterozygotes contained unique, hybrid enzymes not found in either parent suggested parallels with hybrid vigor and specifically the "over-dominance" theory to account for it (Schwartz, and Laughner 1969).

1.3   Biochemical Genetics

A central and, to some, THE central question following mutant discovery is deciphering the underlying biochemical lesion. This was possible before transposon tagging; however, it required hard work and luck. These experiments usually began with some form of chemical analysis of the mutant to direct research to a particular pathway. Or the mutant phenotype itself would suggest a particular pathway. For example, mutants lacking purple or yellow pigments suggested alterations in anthocyanin or carotenoid biosynthesis, respectively. Shrunken mutants pointed to the starch biosynthetic pathway and opaque mutants directed investigators to the storage proteins. The second phase of investigations normally involved assays of known enzymological steps in the suspected pathway with the goal of finding an enzyme missing or greatly reduced in the mutant. As is now obvious, these approaches left acres of space to chase foul balls. For example, without invading the endosperm, it is easy to confuse *wx* mutants (starch biosynthetic enzyme) with *opaque* mutants (zein synthesis). Even if the correct pathway was initially identified, not all enzymatic steps in pathways were known, and even if they were known, some assay procedures were problematic at best. And perhaps most importantly, functional redundancy complicated interpretation of mutant phenotypes. In fact, the nature of the mutant phenotype may be a better indicator of partially overlapping levels of redundant functions than it is of a particular pathway.

Once a gene was associated with an enzyme, the next question concerned how the gene controlled the enzyme. Three options were usually considered: (i) the gene encoded the enzyme, (ii) the gene controlled the activity or amount of the enzyme or (iii) the effect of the gene on the enzyme was indirect. Of the three options, the first predicted that a structurally altered protein might be found in at least some of the leaky mutants of the locus. A second prediction of this model was that enzyme activity might exhibit a gene dosage effect if the gene encoded the enzyme. Accordingly, experiments were designed to detect allele-specific structural alterations within the enzyme and to determine whether the enzyme exhibited a gene dosage effect. While measurements of enzyme activity in parents and hybrids are quite straightforward, experimental searches for alterations in protein structure are complex and took many forms. Factors affecting experimental strategies included the abundance of the protein and ease of purification; availability of "in gel" stains or assays to detect enzyme activity after electrophoresis, availability of protein-specific antibodies and availability of leaky mutants.

Depending on the tools available, allele-specific alterations in enzyme electro-phoretic mobility, K_m values, heat stability and/or levels of protein recognized by enzyme specific-antibodies were usually measured. Allele-specific alteration in one or more of these parameters was usually taken as evidence that the mutant gene corresponded to the structural gene for the enzyme. While these were and are powerful approaches, post-translational modifications that alter enzyme parameters as well as the presence of other isoforms of the studied enzyme could complicate interpretation of enzyme/protein data. It was clear even then that direct examination of the wild-type and mutant genes would simplify experiments concerning the relationship between the gene and enzyme.

As an aside, we note one observation made repeatedly in the early days of biochemical genetics that is quite relevant today. While enzyme activity exhibited a dosage effect in hybrids involving a wild-type and mutant allele of the structural gene, heterozygotes are phenotypically indistinguishable from wild type. This is true of course because almost all loss-of-function mutants are recessive. Hence, in virtually all cases, much less than wild-type levels of gene product are needed for the wild-type phenotype. While an old observation, this fact should not be forgotten when interpreting expression experiments.

Despite all the caveats described above in deciphering the relationship between a gene and an enzyme, significant progress in biochemical genetics has been made in all the major biosynthetic pathways of maize. Obviously space constraints negate a complete description of these successes; however three classic examples are described.

Schwartz (1959, 1960B) exploited starch gel electrophoresis (Smithies 1955) to identify the protein encoded by the *shrunken1 (Sh1)* gene. Because starch comprises a major component of the endosperm and starch is deficient in this mutant, Schwartz reasoned that the SH1 protein might be a major protein of the endosperm. This reasoning proved correct. SH1 was shown to be a major soluble protein specific to the endosperm. The protein was completely missing in all alleles of the gene available at the time and was present in 17 other non-allelic genes that affected starch synthesis. Electrophoretic results were confirmed by immunochemical analyses using the Ouchterlony double diffusion technique, an extremely sensitive test for the presence of an antigenetic protein. Moving boundary electrophoresis patterns of crude extracts also clearly showed that the SH1 protein is a major soluble protein component in the endosperm.

The work with *Sh1* and its cognate protein was the first case reported in higher organisms where a protein product had been detected for a well-studied gene mapped to a specific position on a chromosome. The analyses of sickle cell anemia and the other abnormal hemoglobins were reported earlier but the genetic analyses had not been performed and it was not even known whether allelic or non-allelic genes were involved.

In 1962, Oliver Nelson and Howard Rines reported that the biochemical lesion associated with the *Wx* locus is a starch synthase. This enzyme prefers ADP-glucose, but can use UDP-glucose to elongate starch chains. This was the first documented association of a gene with an enzyme in the starch biosynthetic pathway of any plant. Not only was it a first, but this observation also led to a shift in the paradigm concerning our understanding of the starch biosynthetic pathway. Starch is composed of two polymers: the straight chain, amylose, (glucose polymers linked almost exclusively in alpha 1, 4 bonds) and the branched chain, amylopectin. Amylopectin contains straight-chained glucose linked through alpha 1, 4 bonds with branches composed of alpha 1, 6 bonds. What seemed obvious at the time was that amylose would be a precursor for amylopectin synthesis. However, *wx* mutants lack amylose and amylopectin levels are unaffected. The fact that *wx* affected a starch synthetic enzyme (and not a debranching activity) led to the inescapable conclusion that the syntheses of amylose and amylopectin occur through parallel pathways.

A second example concerns the high lysine mutants, *opaque-2 (o2)* and *floury-2 (fl2)* (Mertz, Bates, and Nelson 1964; Nelson, Mertz, and Bates 1965). Diets consisting exclusively or primarily of maize are not good for monogastric animals, because the level of an essential amino acid, lysine, is low relative to the amount needed by the animal. Consequently, a large germplasm screening program was initiated at Purdue University to identify high lysine lines. Oliver Nelson became aware of this program and also knew that high protein lines isolated elsewhere had two characteristics: the kernel phenotype was quite translucent and the increased protein was due almost exclusively to increased zein levels. Because zeins lack appreciable amounts of lysine, the enhanced zein lines were of limited value for animal feed. Nelson reasoned that mutants with the opposite phenotype – opaque – might have reduced zeins and perhaps elevated levels of proteins rich in lysine. Of the four opaque mutants originally analyzed (*o1, o2, fl1* and *fl2*), two had double the lysine content of wild-type kernels. Nelson was only half right, but that was good enough.

The association of the *o2* mutation with reduced zein synthesis led to a series of experiments from a number of laboratories aimed at deciphering this gene/protein relationship. "Zein" is a protein fraction classified by ethanol solubility, so initial experiments focused on resolving the number of different zein proteins, their levels of expression and the number of genes encoding them (reviewed in Kodrzycki, Boston, and Larkins 1989, also see chapter by Boston and Larkins, this volume). It became obvious early on that cloning of the various zein-encoding genes and of *o2* would be essential for any understanding of how *o2* controlled levels of zein proteins. Molecular studies of zein became tractable when it was discovered that polysomes synthesizing zeins could be isolated from the endosperm (Burr, and Burr, 1976; Larkins and Dalby, 1975). Southern blots probed with cDNA clones of the various zeins led to estimates of the number of genes encoding the various classes of zein proteins.

1.4 The Nature of Controlling Elements and Their Mechanism of Gene Inhibition

Before gene cloning, a number of properties were assumed about controlling or transposable elements; however, facts were lacking. First, it was assumed that these elements were gene insertions. This inference, of course, came from the high rates of instability or mutability of the affected genes. Subsequent molecular studies showed that this was indeed the case. In addition, parallels had been noted between controlling elements and the regulatory genes of the then recently-discovered *lac* operon of *E. coli* (McClintock 1965). A logical extension of this was that controlling elements acted like the *cis*-dominant *lac* regulatory genes. Controlling elements would affect transcription, and they would be located at one of the termini of the gene. That this might not be the case first came from the *wx* mapping work of Oliver Nelson mentioned above. Nelson noted that the transposable elements did not map at one end of the gene; rather, they were found throughout the *wx* gene as defined by mutants and fine structure analysis. Three reports (Dooner, and Nelson 1977; Hannah, and Nelson 1976; Schwartz 1960B) followed in which the protein product of the affected gene was examined. Mutants were found that completely lacked the protein product of the gene, while other transposable element-inhibited alleles showed altered developmental profiles and changes in enzyme kinetic parameters and heat stability. That structurally altered proteins were produced was not in keeping with the hypothesis that transposable elements acted simply to control the amount of protein produced by the affected gene.

The nature of the transposable element and the mechanism by which it affected gene function were questions of paramount importance within the maize and the biological sciences community. In our view, these questions provided the major impetus for gene cloning efforts in maize.

2   The Very Early Days of Gene Cloning in Maize

To some, life without Genbank or cloned transposable elements may be beyond belief. But it existed! In the beginning, the Holy Grail was isolation of a nucleic acid encoding a particular protein. Proteins produced in large amounts (for example, the product of the *shrunken-1* gene, sucrose synthase, as well as the anaerobically induced ADH proteins in maize roots) or proteins easily purified (for example the starch bound starch synthase encoded by the *wx* locus and zein proteins) made excellent first candidates. Purified protein was necessary for antibody production, and the identification of a tissue that highly expressed a protein usually reduced the amount of work needed to sort through clones.

The most direct approach to isolate an mRNA encoding a particular protein was through the use of columns containing covalently bound antibodies. These columns were used to fractionate polysomes. In theory, partially synthesized proteins on the ribosome would be recognized by the antibody and only polysomes synthesizing the protein of interest would stick to the column. Poly (A) RNA from the trapped polysomes could then be released and cDNA could be synthesized via use of poly-dT as the first primer. For technical reasons, however, hybrid-select translation became the method of choice in the early days to match cDNA clones with a particular protein. Mutants also provided valuable experimental power.

An example of one of the early cloning efforts is the isolation of the *ADH1* gene (Gerlach, Pryor, Dennis, Ferl, Sachs, and Peacock 1982). These investigators took advantage of the fact that *ADH* was induced by anaerobic treatments. Accordingly, a cDNA library was synthesized from poly (A) RNA isolated from anaerobic roots. The RNA used for cDNA synthesis was size selected for species that produced, via *in vitro* translation, proteins the size of ADH. This cDNA library was then hybridized in duplicate with cDNAs from RNA of anaerobic and aerobic roots. Colonies hybridizing to the former but not the latter cDNA were then examined by hybrid-select translation. Here, individual cDNA clones were hybridized to the RNA population and the transcript hybridizing to the clone was subsequently translated *in vitro*. The initial clone chosen gave rise to a protein having the same mobility as ADH on SDS gels. Gerlach et al (1982) then exploited a mutant *ADH* that altered the electrophoretic mobility of maize-synthesized ADH. When RNA from this mutant was hybrid selected, the protein produced exhibited the electrophoretic mobility specific to this mutant. Final proof that this was an ADH clone came with the demonstration that the protein produced with the clone cross-reacted to antibody made against ADH.

Many of the early maize genes targeted for cloning had alleles affected by transposable elements. The transposable elements could then be cloned via use of the wild-type gene as a probe to screen genomic DNA libraries.

What followed was a particularly exciting time in maize genetics (Bennetzen, Swanson, Taylor, and Freeling, 1984; Doring, and Starlinger 1986; Doring, Tillmann, and Starlinger 1984; Fedoroff, Wessler, and Shure 1983; Gierl, Saedler, and Peterson 1989; Mullerneumann, Yoder and Starlinger 1984; Pereira, Cuypers, Gierl, Schwarz-Sommer, and Saedler 1986; Pohlman, Fedoroff, and Messing 1984; Saedler, and Nevers 1985; Sutton, Gerlach, Schwartz, and Peacock 1984; Wessler, Baran, and Varagona 1987). A number of seminal facts emerged almost simultaneously. Transposable elements were indeed gene insertions and the sequence of the elements shed insight into family specificity and mechanisms of transposition. Autonomous, but not non-autonomous, elements contained open reading frames encoding proteins necessary for transposition. Autonomous and non-autonomous elements of the same family shared identical termini. Elements made duplications of host DNA sequences whose size was family specific. Transposition events did not usually restore the wild-type sequence. Elements altered the splicing pattern of the host gene, giving rise to multiple transcripts. Transcripts nearly identical, and in one case identical, to the wild-type gene could be produced by alternative splicing.

The utility of transposable elements as a tool for maize gene cloning was also demonstrated during this time. Fedoroff, Furtek, and Nelson (1984) took advantage of the fact that there are few copies of the element *activator* (*Ac*) in the maize genome and used a unique fragment of *Ac* to clone the maize *bronze1* gene. This opened the door to large scale cloning of maize genes via transposon tagging.

In many ways, the rest is history. With transposon tagging, genes can be cloned without any knowledge or guesswork of their underlying function or biochemistry. Through the power of PCR and large collections of transposon tagged lines, insertions into particular genes can now be identified in a matter of hours. And the insert need not condition an easily identifiable phenotype. In fact, with sequence-indexed tagged lines (McCarty, Settles, Suzuki, Tan, Latshaw, Porch, Robin, Baier, Avigne, Lai, Messing, Koch, and Hannah 2005), knockout mutants of a particular gene can simply be ordered from a supply source.

We hope that this chapter, written primarily for the young maize investigators provides some insight of how we got where we are. The maize genome should be completely sequenced soon and obviously maize genetics will change dramatically. With the tools now available, we clearly have entered the realm of systems biology. It is our sincere hope that as we move forward in this new world, we remember the spirit of discovery and the open and frank discussions that have made maize genetics what it is today.

Acknowledgments The study was supported by NSF grants IBN- 9982626 and 0444031 and USDA Competitive Grants. 2000-01488, 2006-03034 and 2007-03575 to LCH. We thank Rob Ferl and Brian Larkins for useful discussions, directions to seminal literature and very helpful comments on an earlier version of this paper. We thank Nick Georgelis and Brandon Futch for assistance in preparing this manuscript.

References

Bennetzen, J.L., Swanson, J., Taylor, W. C., and Freeling, M (1984) DNA insertion in the first intron of maize Adhl affects message levels: Cloning of progenitor and mutant Adhl alleles. Proc. Natl. Acad. Sci. USA 81, 4125–4128

Benzer, S. (1955) Fine structure of a genetic region in bacteriophage. Proc. Natl. Acad. Sci. USA 41, 344–354

Burr, B., and Burr F. A. (1976) Zein synthesis in maize endosperm by polyribosomes attached to protein bodies. Proc. Natl. Acad Sci USA 73, 515–519

Dooner, H. K., and Nelson, O. E. (1977) Controlling element-induced alterations in UDPglucose-flavonoid glucosyltransferase enzyme specified by the bronze locus in maize. Proc. Natl. Acad Sci. USA 74, 5623–5627

Doring, H. P., Tillmann, E., and Starlinger, P. (1984) DNA sequence of the maize transposable element *Dissociation*. Nature 307, 127–130

Doring, H. P., and Starlinger, P. (1986) Molecular genetics of transposable elements in plants. Ann. Rev. of Genet. 20, 175–200

Fedoroff, N., Wessler, S., and Shure, M. (1983) Isolation of the transposable maize controlling elements Ac and Ds. Cell 35, 235–242

Fedoroff, N. V., Furtek D. B., and Nelson. O. E. (1984) Cloning of the bronze locus in maize by a simple and generalizable procedure using the transposable controlling element activator (Ac). Proc. Natl Acad. Sci. USA 81, 3825–3829

Freeling, M., and Bennett, D. C. (1985) Maize Adhl. Ann. Rev. of Genet. 19, 297–323

Gerlach, W. L, Pryor, A. J., Dennis, E. S., Ferl, R. J., Sachs, M. M. and Peacock W. J. (1982) cDNA cloning and induction of the alcohol dehydrogenase gene (Adhl) of maize. Proc. Natl Acad. Sci. USA 79: 2981–2985

Gierl, A., Saedler, H., and Peterson P. A. (1989) Maize transposable elements. Ann. Rev. Genet. 23, 71–85

Hannah, L. C., and Nelson, O. E. (1976) Characterization of ADP-glucose pyrophosphorylase from *shrunken-2* and *brittle-2* mutants of maize. Biochem. Genet. 14:547–560

Kodrzycki, R., Boston, R. S., and Larkins B. A. (1989) The *opaque-2* Mutation of Maize Differentially Reduces Zein Gene Transcription. The Plant Cell 1, 105–114

Larkins, B. A., and Dalby, B. A. (1975) An *in vitro* synthesis of zein-like protein by maize polyribosomes. Biochem. Biophys. Res. Comm. 66, 1048–1054

McCarty, D. R., Settles, A. M., Suzuki, M., Tan, B. C., Latshaw, S., Porch, T., Robin, K., Baier, J., Avigne, W., Lai, J., Messing, J., Koch, K., and Hannah L.C. (2005) Steady-state transposon mutagenesis in inbred maize. Plant J. 44, 52–61

McClintock, B. (1965) The control of gene action in maize. Brookhaven Sym. in Biology 18, 162–184

Mertz, E. T., Bates, L. S., and Nelson, O. E. (1964) Mutant gene that changes protein composition and increases lysine content of maize endosperm. Science 17, 279–80

Mullerneumann, M., Yoder, J. I., and Starlinger, P. (1984) The DNA sequence of the transposable element Ac of *Zea mays*. Mol. Gen. Genet. 198, 19–24

Nelson, O. E. (1968) Waxy locus in maize. 2, Location of controlling element alleles. Genetics 60, 507–532

Nelson, O. E., and Rines, H.W. (1962) The enzymatic deficiency in the waxy mutant of maize. Biochem. Biophys. Res. Commun. 9, 297–300

Nelson, O. E., Mertz, E. T., and Bates, L. S. (1965) Second mutant gene affecting the amino acid pattern of maize endosperm proteins. Science 150, 1469–1470

Pereira, A., Cuypers, H., Gierl, A., Schwarz-Sommer, Z., and Saedler, H. (1986) Molecular analysis of the En/Spm transposable element system of *Zea mays*. EMBO J. 5, 835–841

Pohlman, R. R., Fedoroff, N. V., and Messing, J. (1984) The nucleotide sequence of the maize controlling element Activator. Cell 37, 635–643

Saedler, H., and Nevers, P. (1985) A molecular model of transposition in plants. EMBO J. 4, 585–590

Schwartz, D., (1959) Genetic studies on enzymes in maize and endosperm. Science 159, 1287

Schwartz, D., (1960A) Electrophoretic and immunochemical studies with endosperm proteins of maize mutants. Genetics 45, 1419–1427

Schwartz, D., (1960B) Genetic studies on mutant enzymes in maize: Synthesis of hybrid enzymes by heterozygotes. Genetics 16, 1210–1215

Schwartz, D., (1962) Genetic studies on mutant enzymes in maize. III. Control of gene action in the synthesis of the pH 7.5 esterase. Genetics 47,1609–1615

Schwartz, D., (1963) Genetic studies on mutant enzymes in maize. IV. Comparison of pH 7.5 esterases synthesized in seedling and endosperm. Genetics 49, 373–377

Schwartz, D., (1966) Genetic control of alcohol dehydrogenase in maize – gene duplication and repression. Proc. Natl. Acad. Sci. USA 56, 1431–1436

Schwartz, D., (1967) E1 esterase isozymes of maize: on the nature of the gene-controlled variation. Proc. Natl. Acad. Sci. 58, 568–575

Schwartz, D. (1969) Alcohol dehydrogenase in maize – genetic basis for multiple isozymes. Science. 164, 585–87

Schwartz, D. (1971) Genetic control of alcohol dehydrogenase – competition model for regulation of gene action. Genetics 67, 411–423

Schwartz, D., and Endo, T. (1966) Alcohol dehydrogenase polymorphism in maize – simple and compound loci. Genetics 53, 709–715

Schwartz, D., and Laughner, W. J. (1969) A molecular basis for heterosis. Science 166, 626–627

Smithies, O., (1955) Zone electrophoresis in starch gels. Biochem. J. 61, 629–641

Sutton, W.D., Gerlach, W.L., Schwartz, D., and Peacock, W. J. (1984) Molecular analysis of Ds controlling element mutations at the ADH1 locus of maize. Science 223, 1265–1268

Wessler, S. R., Baran, G., and Varagona, M. (1987) The maize transposable element *Ds* is spliced from RNA. Science 237, 916–918

Mutagenesis – the Key to Genetic Analysis

M.G. Neuffer, Guri Johal, M.T. Chang, and Sarah Hake

Abstract Mutagenesis is a major key to understanding gene function. Most chapters in this book take advantage of mutant alleles to advance the knowledge of maize traits. The chemical mutagen, EMS, has been particularly important because it has a very high efficiency and can be used in any genetic background. EMS also generates half-plant chimeras, which have interesting consequences for lethal dominant mutations. Although dominant mutants are often considered gain-of-function abnormalities, from analysis of thousands of mutants, it appears that most dominants mimic a set of recessive mutants. Examples in which the genes have been cloned demonstrate that a gene defined by a dominant mutation often functions in the same pathway as the gene defined by a recessive mutation with similar phenotype. We present an historical perspective of EMS mutagenesis and discuss frequencies of different types of mutations. Two types of dominant mutants that appear frequently and have recessive counterparts are described in more detail.

1 An Historical Perspective of EMS mutagenesis

Plant breeders and geneticists have long sought ways to increase mutation frequencies so as to acquire unique and useful mutant types. The pioneering work of L. J. Stadler at Missouri University established that ionizing radiation from X-rays and

M.G. Neuffer
Department of Agronomy, Curtis Hall, University of Missouri, Columbia MO 65211
gneuffer@aol.com

G. Johal
Dept of Botany and Plant Pathology, Purdue University, West Lafayette, IN 47907-2054
gjohal@purdue.edu

M.T. Chang
5333 Cervantes Dr, Ames IA 50014.

S. Hake
Plant Gene Expression Center, USDA-ARS, 800 Buchanan St. Albany, CA 94710
maizesh@nature.berkeley.edu

J.L. Bennetzen and S. Hake (eds.), *Maize Handbook - Volume II: Genetics and Genomics*,   63
© Springer Science+Business Media LLC 2009

atomic energy, applied to maize, was not a productive source of heritable changes but, instead, caused mostly re-arrangements or destructive deletions of genetic material. Some wave-lengths of UV light, on the other hand, did produce small changes in the gene which led to an early proof (13 years before Watson and Crick) that DNA was the basic genetic substance (Stadler and Uber, 1942). At the same time, McClintock showed that considerable variation could be produced by the transposable elements that she discovered and characterized (McClintock, 1950). Once DNA was established as the molecular basis for inheritance, the use of chemical agents that could change the nature of DNA was an appropriate strategy. The problem then became one of developing techniques for applying powerful and dangerous chemicals to the DNA of living germ cells without damaging them or the surrounding cells. Early efforts with radiation and harsh chemicals failed because they usually killed surrounding tissue before penetrating to the nuclei of the germ line. The paraffin oil technique for treating corn pollen (Neuffer and Coe, 1978) was ideal because it brought the chosen chemical (ethylmethane sulphonate, EMS) into close proximity with the chromosomes and, unlike seed treatment, reached the germ line at the one cell stage so that the consequences could be unambiguously identified in the progeny.

A large-scale experiment was initiated to determine the efficiency of different mutagens (Neuffer, 1966). Neuffer set out to test the stability of the colorless $a1$-m allele in the absence of Dt, knowing that, in the presence of Dt, hundreds of dots could be seen on each kernel (Nuffer, 1961). From looking at thousands of kernels with literally millions of aleurone cells, each carrying three $a1$-m alleles, it was estimated that the frequency of reversion from $a1$-m to $A1$ was less than 10^{-7}. Mutagenesis was carried out using X-rays, UV, and EMS on $a1$-m dt (lacking Dt) material. The subsequent 10,000 kernels produced in each treatment were screened for color phenotypes. No individual dots or colored kernels were found in any of the treatments. However, one or more sectors of dots were observed in each treatment (Figure 1). These can be interpreted as newly induced trans-acting Dt loci, not as excision of the suppressing rDt element from the $A1$ locus. The newly induced Dt loci could have been predicted as McClintock reported that new transposon activity often appears as a consequence of chromosome breakage. Thus, it was possible to show that while the three mutagenic agents were able to produce new transposon activator elements, none were able to dislodge or deactivate the rDt element at the $A1$ locus. Now that we know the molecular basis for transposable elements, the stability of $a1$-m dt is not surprising.

The M1 kernels from each treatment were planted to look for mutant seedlings as a comparative control measure of their mutagenic potential. M1 progeny from both X-rays and UV had a few small, weak, and abnormal aneuploid types. The UV treatment also produced two seedling mutants (one pale green and one dwarf). In contrast, the M1 from EMS treatment produced a large number of clear, mutant seedling phenotypes (such as white, yellow, yellow green, necrotic, dwarf, rolled leaf, virescent, adherent). Given that these mutants were dominant, it suggested that EMS-mutagenized progenies were a rich source of new mutants. More than half of the selfed ears also segregated for recessive heritable phenotypes. These spectacular

Fig. 1 Section of an *a1-m dt* ear crossed by *a1-m dt* pollen treated with EMS showing normal colorless kernels and one exceptional kernel with a sector of purple dots

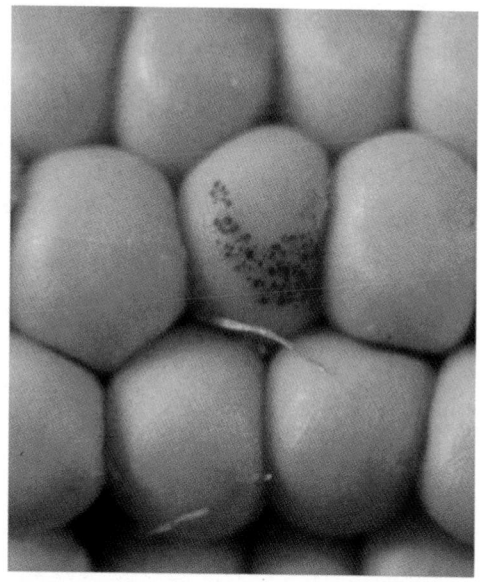

results led to a focus on EMS as a mutagen to produce the variation needed for crop improvement and for a better understanding of gene function.

2 Frequency and Types of Mutations with EMS

The protocol for EMS mutagenesis of maize pollen in paraffin oil is outlined in Mutants of Maize (Neuffer et al., 1997) and is elaborated in the chapter by Weil and Monde. One can optimize the variables for success of the mutagenesis prior to pollination by plating a subset of the pollen on media and checking for a reduction in germination frequency (Neuffer and Coe, 1978). Despite adherence to the protocol, considerable variability in results can occur. In some treatments, such as the one described below, the results have been truly startling while, in others, all the pollen died or very few mutants were found. Over the years, attempts have been made to understand the variables, one of which is the inbred background. For example, Mo17 has large pollen grains that survive the treatment very well but produce few mutants while B73 has small grains that are easily killed. A longer treatment for Mo17 and shorter for B73 improved the mutagenesis. Other variables, such as temperature and humidity on the day of treatment, have a big effect. Ideally, multiple treatments can be carried out to optimize the conditions in each genetic stock.

In addition to the very high mutation rates of EMS, it can be specifically applied to a single germ cell. Thus, when one sees multiple occurrences of a particular phenotype following pollen EMS treatment, it is clear that each one is a unique event. This is not the case for treatment of seeds in any plant and especially in plants where the male and female gametes are in separate flowers. With seed treatment, the germline is multicellular, thus leading to the formation of offspring with a mixture of mutant and non-mutant cells. The multiple copy mutant progeny that arise are often misread as multiple events leading to the conclusion of much higher frequencies and confusing results. The same is true for mutagenesis experiments with transposable elements where transposon insertions may occur at many stages of development.

In order to determine the frequency of dominant and recessive mutations, a particularly fruitful mutagenesis experiment was followed in detail (Figure 2).

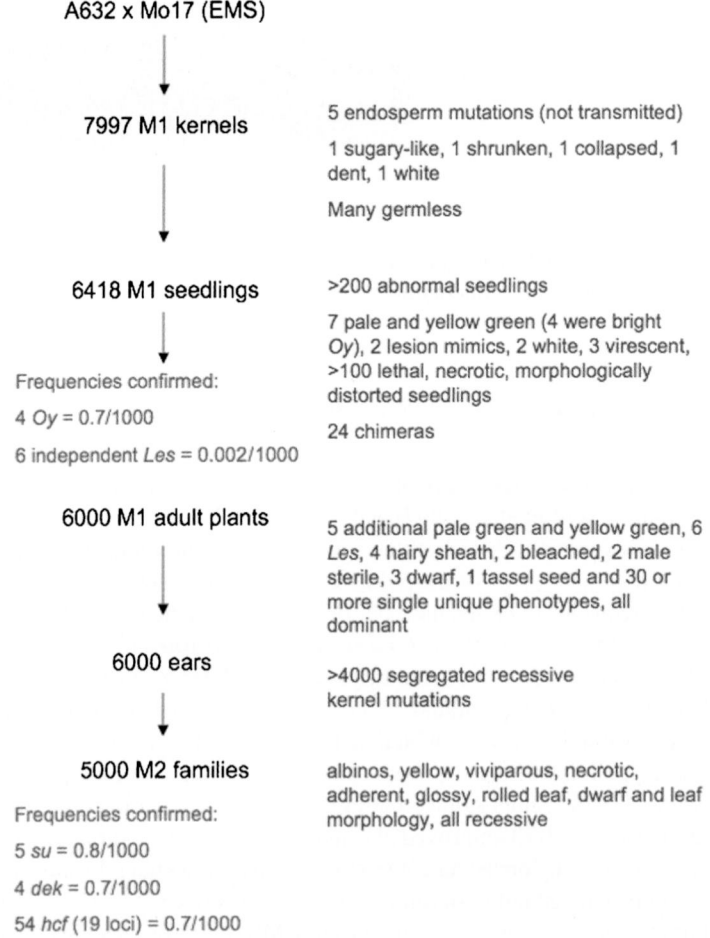

Fig. 2 Frequency and types of mutations in one particular treatment. The frequencies of mutations were confirmed by progeny testing

The progeny of a cross of A632 ears by EMS-mutagenized pollen of Mo17, including 7,997 M1 kernels, was screened for mutant phenotypes expressed in the triploid endosperm, which is derived from two untreated nuclei from the female parent and one EMS-treated nucleus from the male parent. As expected, kernels with mutant endosperm had normal embryos since it would be extremely rare for the same gene to be mutated in both sperm cells of a single pollen grain. In addition, a larger number of germless kernels with normal endosperm that failed to germinate were found. These potential mutants obviously had no progeny but provided an indication of the types of mutations that could affect the development of the embryo.

The M1 kernels were planted in the field and 6,418 M1 seedlings were screened for seedling mutant phenotypes. The germination frequency (80%) was significantly lower than that of the untreated control (96%). The kernels that failed to germinate were assumed to carry a significant but unknown number of lethal mutations, including the germless class described above. Treatment of pollen with EMS routinely causes this reduction in M1 seed viability. Over 200 dominant mutations were seen in the M1 seedlings, including 7 pale and yellow green (4 of which were bright *Oil yellow* (*Oy*) mutants), 2 lesion mimic, 2 white, 3 virescent, and more than 100 lethal, necrotic, morphologically distorted seedlings. The white, lethal and necrotic mutants died as seedlings, while the virescent mutants gradually became more normal and others persisted as mutant to maturity. Twenty-four seedlings with longitudinal stripes or chimeras of distinct mutant phenotypes, similar to those observed in whole seedling mutants, were also seen. Having normal adjoining tissue often sustained the mutant tissue, even in the case of lethal phenotypes, allowing these chimeras to persist till maturity.

Some phenotypes, such as tasselseed or male sterile, were seen only in adult plants (Figure 2). As with the seedling mutants, a corresponding number of half plant chimeras were also seen for these adult phenotypes. These chimeras occurred at approximately the same frequency as the whole plant mutants. More than thirty unique phenotypes were documented, as well as a considerable number of weak, slender or morphologically abnormal plants. At least that many more were seen but could not be confirmed either because no progeny were obtained or because they failed to transmit their mutant phenotype. Among those chimeras that were viable but failed to transmit the mutant phenotype, a large number of small slender plants that looked like aneuploids and haploids were found. The putative haploids were of two types; one of which looked like and had the glume bar allele (at the *b1* locus) of A632 and the other looked like and had the non-bar *b1* allele of Mo17. It was therefore inferred that they were maternal and paternal haploids, respectively. In total, 57 dominant mutants from the M1 seedling and plant screens were saved and assigned an identifying name and number.

Self-pollinated ears from 6,000 normal-appearing M1 plants were examined for kernel mutants segregating in a recessive 3:1 ratio. Two thirds segregated for at least one kernel mutant phenotype, essentially covering most of the known kernel phenotypes that could be expected to appear in a yellow dent background. Viviparous white kernel mutants were found with albino embryos that germinate precociously while still in the ear. Also included were various new types of defective kernel mutants; those with phenotypes in the endosperm, the embryo, or both. In addition, a wide range of semi-sterile ears were missing one quarter of the kernels or segregated tiny vestiges of what may have started out to be a kernel. Some ears had normal kernels on one side but less than one-quarter mutants of a particular type on the other side. These ears almost certainly corresponded to a half plant chimera for a recessive mutation. Given the recessive nature of these mutations, the chimeras had to include the ear and tassel of the M1 parent.

Twenty M2 seed samples of 5000 M1 ears were planted in sand benches and observed from emergence to the four-leaf stage. Most of the known mutant phenotypes were seen repeatedly along with some new types. The most frequent types were those relating to the absence of chlorophyll; the white, yellowish-white, and yellow seedlings that remained so until they died. This chlorophyll-less class appeared in approximately 10% of the progenies. The next largest group included those which showed variation in the type, quantity, and timing of chlorophyll production; the yellow green, pale green and virescent seedling mutants. Other common types were necrotic, adherent, glossy, rolled leaf, dwarf and leaf morphology, all with frequencies above 1%. Fifty-four *high chlorophyll fluorescence* (*hcf*) mutants were found from this population, which included 19 loci (Miles and Daniel, 1974).

To determine the prevalent recessive mutation frequency, two kernel mutants (*su1* and *dek1*) that were easily and unambiguously recognized were selected, as well as one group of seedling mutants (*hcf*) for which we were able to obtain precise data quickly (Figure 2). For these recessive phenotypes, we arrived at a frequency of 0.7 mutants per locus per 1,000 pollen grains. Thus, a 20 seed sample from 3,000 of our 5,000 ear collection would have a 95% chance to carry almost any desired mutant.

The dominant mutants fell into two classes. One class has a few unique loci with mutation rates as high as those seen for recessives. *Oy* is one such mutation (0.7/1000). The frequency for the majority of dominant mutants and a few unique recessive loci, is much lower than 0.1 per 1000 pollen grains per locus. For example, dominant *Lesion mimic* (*Les*) mutants are found at a frequency of 0.002/1000. Certainly, there are phenotypic classes that are recalcitrant to EMS. The dominant *Knotted1* (Hake et al., 2004) and recessive *tasselseed4* (Chuck et al., 2007b) mutations are two examples for which no EMS-induced alleles are known, all the mutant alleles result from chromosomal rearrangements or transposon insertions. In addition, some phenotypes are dependent on the genetic background, as will be discussed for the *Les* mutants. This dependence provides a very important reason for carrying out mutagenesis in multiple genetic backgrounds and with both EMS and transposons.

The frequency obtained from EMS mutagenesis is high enough that one could expect to find mutations in duplicate genes. For example, a 15:1 segregation ratio was seen on the self-fertilized M1 ear that led to the discovery of the *orange pericarp* (*orp*) loci (Figure 3A). The orange color is in the pericarp, which is the genetically identical maternal tissue covering all the kernels and should not differ in phenotype from kernel to kernel. When planted, these kernels produced very weak seedlings that survived only with ultimate care and grew into small morphologically defective plants smelling of indole (Figure 3C). Indole was identified as the substance accumulating in the endosperm that diffused into the

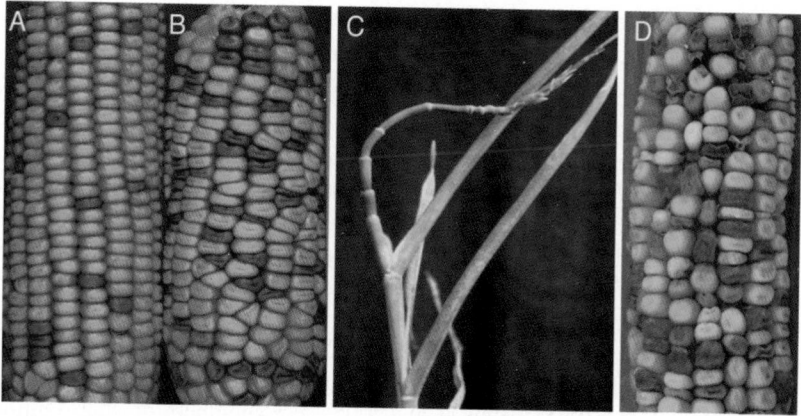

Fig. 3 Multiple mutations with EMS. A) A self-fertilized M1 ear segregated for two mutations, *orp1-1186A* and *orp2-1186B* at the ratio of 1:15. The pericarp is orange due to the secretion of indole from the underlying filial endosperm. B) A single factor ratio is obtained for an ear homozygous recessive for one factor and segregating for the other. C) Phenotype of the *orp1; orp2* homozygote showing orange color on green, narrow leaves and failure to develop properly. D) A selfed M1 ear segregating for four different recessive mutations (sugary, white, viviparous, and brown kernel)

maternal pericarp and turned it orange, explaining the unexpected phenotypic differences in the genetically identical pericarp (Wright and Neuffer, 1989). Analysis of *orp* led to an improved understanding of the tryptophan pathway (Wright et al., 1991; Wright et al., 1992).

Probably the most abundant class of recessive mutants was the *defective kernel* (*dek*) class, seen in selfed ears where 1/4 of the kernels failed to advance beyond the vestige of a kernel to produce a tiny empty shell of pericarp. If we consider only those mutants recognized as having a semblance of seed form, the *dek* class still constitutes a sizeable portion of the mutants produced (Neuffer and Sheridan, 1980; Sheridan and Clark, 1987; Clark and Sheridan, 1988, 1991). Chimeras have been useful to study the *dek* mutants as it allows one to see the recessive phenotype beyond the embryo lethal stage. A good example is the analysis of *colorless floury defective, dek1*. This mutant was found to be colorless because it lacked the aleurone layer (Cone et al., 1989), had a vigorous root but no shoot growth and was albino as determined in chimeral shoot tissue that was also morphologically altered (Neuffer, 1995; Becraft et al., 2002).

In total, the EMS Mutation Project has produced and observed more than 100,000 mutagenized M1 kernels. From 45,000 M1 plants, over 1,000 promising dominant mutants were found. Of these, 307 have proven heritable and were given name and number, 54 of which were located to chromosome arm. From 32,000 M2 ears, 52% had visible recessive variations segregating on the ear. Several ears had four visible kernel phenotypes (Figure 3D) and one of these proved to have three additional visible phenotypes in the seedling and the mature plant. A total of 5,737 putative recessive mutants of all types were observed, 706 of which have been assigned a chromosomal position.

3   The importance of Dominant Mutations

Dominant mutations exert a special attraction on the geneticist, as they are recognized in the F1, thereby simplifying pedigrees. Although rare, the high mutation rates of EMS provide the possibility of finding such alleles. Mutations that have been important for agriculture are often dominant, as is the case for the *Rht* dwarfing mutations of wheat (Peng et al., 1999) and some alleles that confer disease resistance. Dominant mutations have also had important consequences in evolution. Five major quantitative trait loci (QTL) account for most differences between maize and teosinte, two of which correspond to the *teosinte branched1* (*tb1*) and *teosinte glume architecture* (*tga*) loci (Doebley et al., 1997; Wang et al., 2005). The maize *tb1* and *tga* alleles are dominant over their respective teosinte alleles (Dorweiler et al., 1993; Doebley et al., 1995; Wang et al., 2005). For members of gene families, dominant mutations are often the only mutation that is visible.

3.1.   Dominant Morphological Mutants with Recessive Counterparts

Gibberellin mutants. A classic example of dominant and recessive mutations in a biological pathway comes from study of dwarf mutants in the maize gibberellin pathway (Phinney, 1956). The dominant dwarf, *D8*, and five recessive dwarfs all have short stature, dark green leaves, and a failure of stamen arrest in the ear. Recessive mutants can be rescued with exogenous GA and define genes that encode enzymes in the gibberellin (GA) biosynthetic pathway. In contrast, dominant *D8* mutants have high levels of GA and are not responsive to exogenous GA (Phinney, 1956). D8 is a member of the GRAS family of transcription factors that is unstable in the presence of GA. The dominant mutants have deletions in the DELLA domain, rendering the protein stable and thus unable to transduce the GA signal (Peng et al., 1999). This pattern of recessive mutations in biosynthetic genes and dominant mutations in receptor or signaling proteins is also seen with other hormones in Arabidopsis such as ethylene (Wang et al., 2002).

Leaf mutants. The *Knotted1* (*Kn1*) and related *knox* (*knotted1 homeobox*) mutants, *Roughsheath1*, *Gnarley1*, *Liguleless3* and *Liguleless4* mutants provide good examples of phenotypes that are due to misexpression. The genes encode a family of homeodomain transcription factors (Kerstetter et al., 1994) that are strongly expressed in vegetative and inflorescence meristems (Jackson et al., 1994). The dominant mutant phenotypes are due to misexpression in the leaf (Vollbrecht et al., 1991 Schneeberger et al., 1995; Foster et al., 1999; Muehlbauer et al., 1999; Bauer et al., 2004) and show defects in proximal-distal patterning, as discussed by Foster and Timmermans. Recessive mutants were found by screening for loss of the dominant phenotype. Because of functional redundancy, loss-of-function mutants may have no phenotype as in *Lg3* (Bauer et al., 2004) or may be background dependent (Vollbrecht et al., 2000).

Kn1 mutants have not been found in EMS screens although they appear frequently in *Mutator* lines (our observations). In twelve characterized *Kn1* mutations, two alleles have *Mutator* elements inserted 5′ of the transcription start site (Ramirez, 2007), and nine *Mutator* elements and one *Ds2* element are in a small region of the third intron (Greene et al., 1994; Vollbrecht et al., 2000). In two other mutants, an uncharacterized insertion is also in this third intron and an *rDt* element is found in the 4th intron. The original allele, *Kn1-O* (Bryan and Sass, 1941; Gelinas et al., 1969; Freeling and Hake, 1985), is a tandem duplication (Veit et al., 1990) and a new allele also appears to be a duplication (Ramirez, 2007). The position of the insertions in the intron suggests that intronic sequences are likely to be important for regulation. Indeed, several conserved non-coding sequences were found in this intron (Inada et al., 2003). The position of the insertions in the 5′ region and the break point between the copies of the tandem repeat in *Kn1-O* highlight promoter sequences that may be needed to keep the gene from being expressed in the leaf. Studies of homologous *kn1* genes in Arabidopsis have identified conserved sequences that are important for keeping *knox* expression out of the leaf (Uchida et al., 2007).

A number of recessive mutants show displaced sheath/blade boundary and misexpress *knox* genes, suggesting that their function is to negatively regulate *knox* genes in the leaf. The first studied was *roughsheath2* (*rs2*) (Timmermans et al., 1999; Tsiantis et al., 1999), which encodes a MYB transcription factor. KNOX proteins are misexpressed in *rs2* mutant leaf primordia. *indeterminate gametophyte* (*ig*) mutants have a leaf phenotype in addition to the gametophyte phenotype. Ectopic leaf flaps occur that are more reminiscent of abaxial/adaxial polarity defects. *ig* encodes a LOB domain protein, a homolog of which is implicated in negatively regulating Arabidopsis *knox* genes (Evans, 2007). Other genes that misexpress *knox* genes are not yet cloned. *corkscrew* mutants have displaced blade/sheath boundaries, altered phyllotaxy and show misexpression of *kn1*, *rs1* and *lg3* (Alexander et al., 2005). *semaphore* mutants misexpress *gn1* and *rs1* in the leaf and endosperm and show pleiotropic defects (Scanlon et al., 2002). These mutants also have reduced polar auxin transport. Whether this last defect is due to misexpression of *knox* or other genes is unknown.

Rolled leaf1 (*Rld1*) is another dominant leaf mutant for which recessive mutants with a related phenotype have been identified. *rld1* encodes a homeodomain-leucine zipper transcription factor that is normally expressed adaxially (Juarez et al., 2004b). In the dominant *Rld1* mutant, the gene is expressed throughout the leaf and the leaf is adaxialized (Figure 4B). Four *Rld1* alleles have been identified through *Mutator* and EMS screens and they all result in the same base pair substitution in the microRNA complementarity site of *mir166* (Juarez et al., 2004b). The recessive *milkweed pod1* (*mwp1*) mutant has similar patches of adaxialization (Figure 4A) and *rld1* is misexpressed in *mwp1* leaves (Candela et al., 2008). *mwp1* encodes a member of the KANADI family of transcription factors, which are known to promote abaxialization in Arabidopsis. The double mutant shows a more severe phenotype than either single mutant, suggesting that *mwp1* has additional functions besides the regulation of *rld1*. Other dominant mutants affecting leaf development are *Rough sheah4* and *Morph*, both identified in EMS screens (Figure 4C, D).

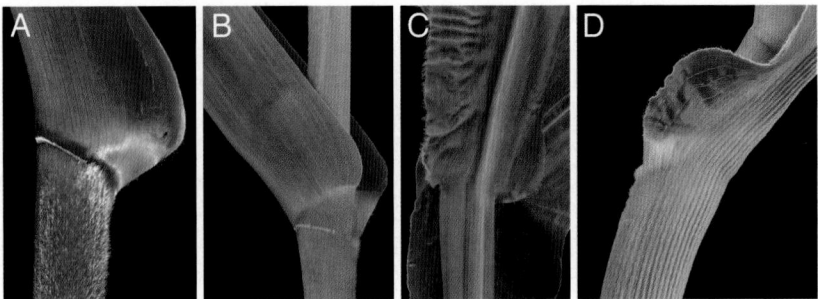

Fig. 4 The sheath and ligule region of leaf mutants. A) *milkweed pod*. B) *Rolled*. C) *Morph*
D) *Roughsheath4*. (photos courtesy of Hector Candela)

A related recessive phenotype is seen in *leafbladeless1* (*lbl1*) mutants. While *Rld1*
leaves are adaxialized (Nelson et al., 2002; Juarez et al., 2004a), *lbl1* leaves are
abaxialized (Timmermans et al., 1998; Nogueira et al., 2007). Double mutants of *lbl*
and *Rld1* show a suppressed phenotype. In *lbl1* mutants, *rld1* expression is decreased.
lbl1 encodes a protein involved in the small interfering RNA (siRNA) pathway and
mutants have an increase in *mir166* RNA levels (Nogueira et al., 2007).

Inflorescence mutants. The *tasselseed* mutants provide another example of
similar dominant and recessive mutants (Figure 5A, B). As mentioned by
Vollbrecht and Schmidt, *tasselseed6* encodes an *AP2* gene that was previously
identified by its recessive phenotype, *indeterminate spikelet* (*ids1*) (Chuck et al.,
1998). *ts4* is one of the maize *miR172* genes and regulates the expression of
ts6/ids1 posttranscriptionally (Chuck et al., 2007b). Like *Rld1*, the lesion in the
dominant *Ts6* allele is a base pair substitution in a microRNA complementarity site.
The presence of a mutant phenotype in *ts4* is impressive given the fact that there are
at least five *miR172* genes in the maize genome. Two other dominant *tasselseed*
mutants have been described in the literature. It will be interesting to determine if
they also encode targets of *ts4*.

thick tassel dwarf (*td1*) and *fasciated ear2* (*fea2*) are two recessive mutations
that have an enlarged tassel rachis and fasciated ear tips. They respectively encode
a leucine rich receptor kinase and a leucine rich receptor (Taguchi-Shiobara et al.,
2001; Bommert et al., 2005), whose Arabidopsis orthologs *CLAVATA1* and
CLAVATA2 are well studied (Clark et al., 1993; Kayes and Clark, 1998). Dominant
Fascicled (*Fas*) mutants have a similar phenotype (Haas and Orr, 1994; Orr et al.,
1997). *Fas* ears differ from *td1* and *fea2* in branching from the base of the ear. The main
rachis of the tassel also splits. It will be interesting to determine if the *Fas* gene
product encodes a component of the CLAVATA pathway.

The dominant *Barren inflorescence1* (*Bif1*) and recessive *bif2* mutations result
in similar phenotypes. *bif2* encodes a kinase with similarity to PINOID in
Arabidopsis (McSteen et al., 2007). The inflorescence is barren, although there are
a few spikelets in some inbred backgrounds. Leaf development is normal (McSteen

and Hake, 2001). Because PINOID is known to function in the regulation of auxin transport (Friml et al., 2004), we hypothesize that *Bif1* carries a dominant mutation that perturbs auxin transport regulation. Interestingly, the *bif* phenotypes are reminiscent of the *orange pericarp* double mutant phenotype (Figure 3). Indole is an intermediate in both the biosynthesis of tryptophan and auxin (indole-acetic acid), so both phenotypes are likely to be auxin-related.

Heterochronic mutants A group of mutants, referred to as heterochronic, shows delayed transition from the juvenile to the adult phase of vegetative development (Poethig, 1988a). The *Corngrass1* (*Cg1*) phenotype is most dramatic, the plant producing many tillers that continue to produce tillers (Figure 5C) (Singleton, 1951). In *Cg1* mutants, leaves are juvenile and roots are produced at all nodes. The defect extends into the inflorescence (Galinat, 1954a, b). In wild-type inflorescences, bract leaves are small and reduced. In contrast, *Cg1* bract leaves are large and vegetative in appearance (Figure 5D). Spikelet meristems and spikelet pair meristems are not apparent and floral meristems appear on the inflorescence in *Cg1* mutants. The tassel is also unbranched (Chuck et al., 2007a).

Cg1 was cloned and shown to encode *mir156*, a microRNA that targets transcripts of *Squamosa Promoter Binding Like* (*SPL*) genes. *Cg1* carries a transposon insertion in the 5′ region, which causes the misexpression of the microRNA (Chuck et al., 2007a). A second allele, identified by activation tagging, carries a T-DNA insertion that activates transcription of the microRNA gene. Twelve different *SPL* genes showed reduced expression in *Cg1* mutants. An analysis in Arabidopsis demonstrated that a parallel pathway of *mir156* regulation of *SPL* genes controls phase change in that species (Wu and Poethig, 2006). It is not likely that a single recessive mutation will mimic all of the *Cg1* phenotypes, however, there may be mutations that mimic one or two of the *Cg1* traits. One example is *tassel sheath*, which has elongated bract leaves, similar to those found in *Cg1* mutants. Another example is *unbranched* (see chapter 2). Given the nature of the lesion, it is not surprising that an EMS-induced *Cg1* mutation has never been identified.

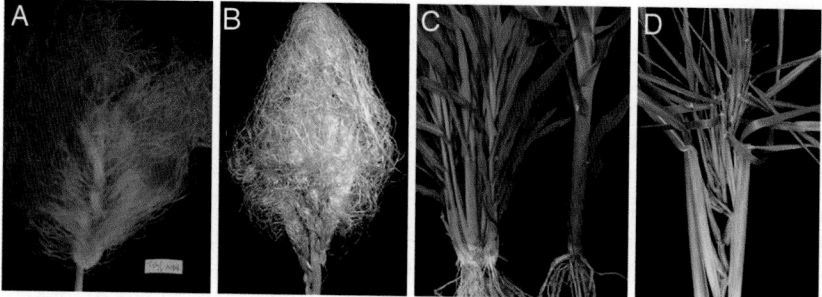

Fig 5 Mutations in microRNA regulated pathways. A) *tasselseed4*, B) *Tasselseed6*, C) *Corngrass* mutants (left) make multiple tillers that have juvenile phenotypes compared to wild type (right). D) In *Corngrass*, vegetative features continue into the inflorescence. (photos courtesy of George Chuck)

3.2   Disease Lesion Mimic Mutants

Disease lesion mimic mutants show symptoms mimicking disease or the resistance response in the absence of disease agents (Walbot et al., 1983). Both dominant and recessive lesion mimic mutants exist, which have been designated *Les* and *les* respectively. Although disease lesion mimic mutants are known to exist ubiquitously in plants (Dangl et al., 1996; Lorrain et al., 2003), they were initially recognized in maize as a unique class of mutants (Neuffer and Calvert, 1975). More than 50 loci have been identified in maize that cause *Les/les* phenotypes when defective (Johal, 2007) with a few represented by multiple alleles. Extrapolating from the general lack of confirmed allelic pairs at many of these loci, it has been suggested that more than 200 lesion mimic loci might exist in maize (Neuffer et al., 1983). Since more than half of these mutants are inherited in a partially- or completely-dominant fashion, *Les* loci constitute the largest class of gain-of-function mutations in maize (Johal, 2007).

Although every lesion mimic mutant is unique in some aspects, they fall into two general categories, determinative and propagative (Johal et al., 1995; Dangl et al., 1996). In determinative mutants, lesions are initiated frequently but their expansion is often curtailed. This gives the appearance of a massive hypersensitive response (HR), which is a programmed cell death reaction unleashed in resistant host cells in response to a diverse array of pathogens (Martin et al., 2003). In propagative mutants, lesions are initiated rarely, they tend to expand uncontrollably, covering large areas of the host tissue (Dangl et al., 1996; Lorrain et al., 2003). It is presumed that lesions in the determinative class arise from impairments that lower the threshold for cell death initiation (Walbot et al., 1983; Dangl et al., 1996). In contrast, propagative mutants are thought to represent defects in genes that encode negative regulators of cell death in plants (Walbot et al., 1983; Dangl et al., 1996).

The production of lesions in most maize *Les/les* mutants is developmentally programmed and influenced by genetic background (Neuffer et al., 1983; Johal, 2007) (also see MaizeGDB). Environmental factors, such as light and temperature, also have a significant effect on their etiology (Hoisington et al., 1982; Gray et al., 1997; Hu et al., 1998). Another unique aspect of most lesion mimic mutants is that they display their phenotype in a cell-autonomous manner (Fig. 6A). This characteristic, along with the fact that many are partially dominant, light-sensitive and developmentally programmed, suggests that there may be common factors contributing to the phenotypic manifestation of lesion mimic mutants.

Two features of lesion mimic mutants have triggered a great deal of interest. First, lesion mimic mutations often sensitize the host to pathogens, resulting in heightened defense responses (Dangl et al., 1996; Hu et al., 1996; Lorrain et al., 2003). This association has led to the belief that these mutants represent a valuable resource to study plant defense signaling and response in the absence of compounding effects from the pathogen. However, unlike the lesion mimic mutants of Arabidopsis and other dicots, maize *Les/les* mutants do not elicit a heightened systemic acquired response, even though some of the markers associated with such a response are upregulated in some of the maize lesion mimic mutants (Morris et al., 1998). A local resistance response in the immediate vicinity of lesions of some maize *Les*

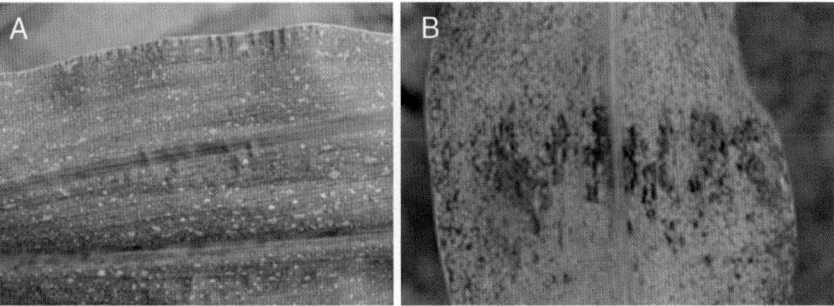

Fig. 6 A) A normal green somatic sector caused by insertion of a *Mutator* element in the *Les10* dominant mutant allele. B) Suppression of cell death underlying *Les17* lesions in the vicinity of common rust pustules

loci has been observed in a few cases (Johal, 2007). Curiously, the common rust pathogen, an obligate biotroph, can also suppress lesions associated with *Les17*, thereby producing areas on the leaf that are often referred to as 'green islands' in plant pathology literature (Fig. 6B). Second, tissue damage is a normal part of *Les/les* mutants. Cell death underlying this damage happens either precociously in these mutants or is not contained following normal onset (Johal, 2007). This has led many to suggest that *Les/les* mutants represent defects in genes and mechanisms that control programmed death of cells and tissues in plants (Johal et al., 1995; Dangl et al., 1996; Lorrain et al., 2003). In this regard, *Les/les* mutants appear to hold great promise because they may provide insights into mechanisms that control and signal cell death pathways in plants.

Why are there so many lesion mimic loci in plants? True to their name, one mechanism underlying some of these mimics involves defects in plant disease resistance genes (Johal et al., 1995). These R genes encode proteins that respond to pathogen ingress by triggering a rapid cell death response, HR (hypersensitive response) in affected host cells (Martin et al., 2003). Each R gene triggers HR only in response to a specific set of races of a single pathogen. An R gene can become defective such that it triggers an HR even in the absence of the pathogen (Johal et al., 1995; Martin et al., 2003), as first observed with maize *Rp1* that conditions resistance to common rust, caused by *Puccinia sorghi*. Occasionally, intragenic recombination within the *Rp1* locus, which is composed of tandemly duplicated copies of individual R gene paralogs, leads to the creation of novel genes, some of which confer a *Les* phenotype in which HR is triggered constitutively in the absence of pathogen ingress (Hu et al., 1996). Both dominant and recessive les mutants, differing in severity, have been identified at the *Rp1* locus (Hu et al., 1996). This suggests that weak alleles may behave as recessives and strong alleles may behave as dominants.

Notably, a majority of the maize *Les/les* mutants, however, do not seem to be involved in plant defense responses. Two other *Les/les* genes that have been cloned suggest errors or impairments in metabolism that lead to the lesion mimic phenotype. The maize mutant *Les22* is a key example of this (Hu et al., 1998). It is defective in a single copy of the *Urod* gene that encodes a tetrapyrrole biosynthetic enzyme required for the production of both heme and chlorophyll in plants. But why does *Les22* behave as a dominant mutant? The reason lies in the haplo-insufficient nature of the *urod* gene. When one copy of this gene is defective, the pathway runs into a bottleneck, causing the accumulation of a highly photodynamic intermediate, uroporphyrinogen. In the presence of light, this molecule leads to the production of singlet oxygen, which, in turn, leads to the *Les22* phenotype. However, if both copies of *urod* are defective, the mutants are albino due to lack of chlorophyll (Hu et al., 1998). Thus, the phenotype of a gene that leads to a *Les* phenotype as a heterozygote could be quite different from its homozygous phenotype.

Mutations in the chlorophyll degradative pathway, as well as in the biosynthetic pathway, also lead to a *les* phenotype (Johal, 2007). A good example is *lls1*, which controls the first committed step of the chlorophyll degradation pathway (Gray et al., 1997; Gray et al., 2002), and the maize ortholog of the Arabidopsis *acd2* mutant (G. Johal, unpublished results), which controls the next step following *lls1*. Again, cell death associated with both of these mutants is caused by the accumulation of phytotoxic intermediates that leads to cellular damage.

Among all the factors that impact the etiology of a maize *Les/les* mutant, the genetic background is perhaps the most important. A *Les/les* mutant may have a lethal pheno-type in one genetic background but a largely benign phenotype in another (Neuffer et al., 1983). Among the inbreds that tend to be suppressive is Mo20W, a 'stay-green' line that can withstand high heat and intense light (Neuffer et al., 1983). The W23 inbred, in contrast, enhances the severity of many mimics to the point that they become lethal when introgressed into its genome (Neuffer et al., 1983). Studies on *Les1* showed that the suppressible effect of Mo20W was dominant (over its enhanced expression in W23) and under the control of multiple factors (Neuffer et al., 1983).

A QTL approach involving an F_2 population between *les23*::Va35 and Mo20W was used to identify the modifiers responsible for the background dependence (Penning et al., 2004). A strong QTL, *slm1*, was identified which controlled more than 70% of the *les23* phenotypic variation in this population. *Slm1* has been mapped to the long arm of chromosome 2 (2L) in maize (Penning et al., 2004). A similar QTL capable of suppressing the phenotype of *Rp1-D21*, a constitutively active allele of *Rp1*, has been mapped to 10S (P. Balint-Kurti, personal communication). Suppressors of *Les/les* loci appear to be rather common in the maize genome, and can cause a lesion mimic mutant to become cryptic. Such is the case with Mo17, which fails to manifest the lesioned phenotype of the severe *les*-mo17* mutation because it carries two unlinked suppressors of *les*-mo17*. When these suppressors segregate away from *les*-mo17*, as happens in the IBM RILs or in the F2 populations of Mo17 with various inbreds, *les-Mo17* reveals itself (G. Johal, unpublished results).

3.3   Half Plant Chimeras

Half plant chimeras have been found for most of the dominant phenotypes observed, occurring at approximately the frequency of their whole plant equivalents in all the treatment progenies studied. Sometimes the difference is very subtle, such as a slightly different level of green, which can only be detected in side-by-side tissue comparisons of the chimera (Figure 71). Similarly, dominant mutants that grow

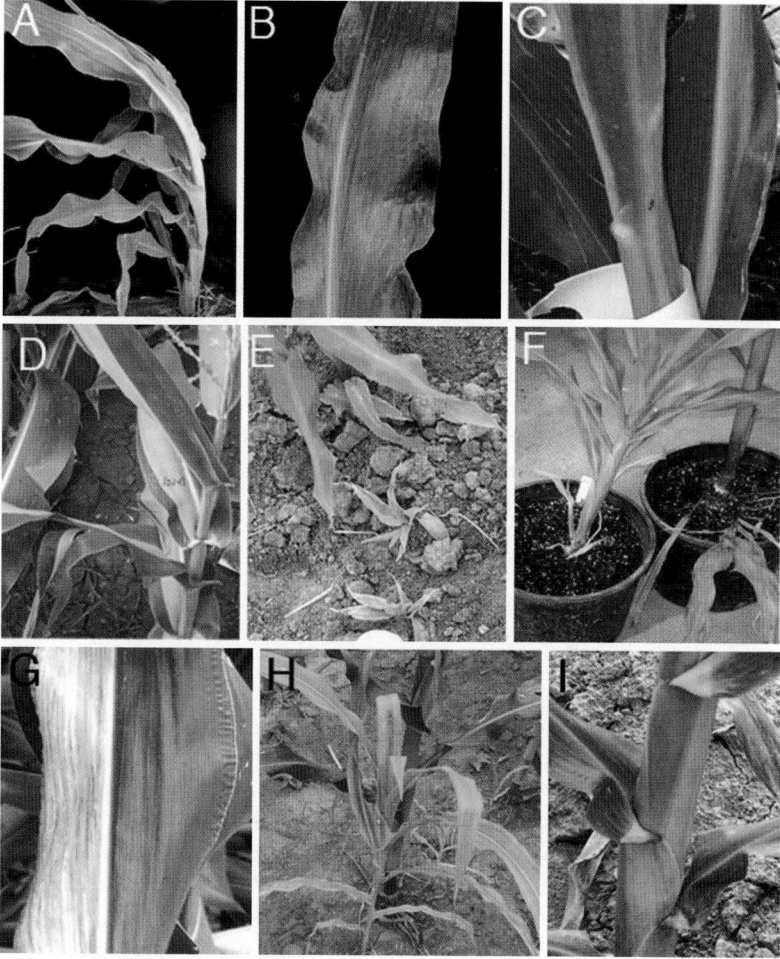

Fig. 7 Examples of half-plant chimeras. **A**) In this plant, half of each leaf is narrow, causing a bent posture (SH6842-141). **B**) Leaf from (A). **C**) Original *Liguleless narrow* chimera in which one half of the plant had narrower leaves and a displaced ligule. **D**) Half plant, pale green chimera (Ppg*Chi 2542). **E**) Progeny from the chimera in (D) segregated 1:1 for small pale green plants that made no tassels or ears. **F**) Progeny from (D) grown in the greenhouse show pale green plants that fall over because of very poor root growth but were able to make some pollen. **G**) Half plant chimera (Vsr*-2595) with yellowish-white tissue with tiny yellow green streaks that enlarge and merge to give a yellow green plant. **H**) Progeny from Vsr*-2595 that was pale green and infertile. Crosses to *R1-rsc* suggested a tight association with anthocyanin expression in the aleurone, but not necessarily with linkage to *r1*. This mutant may be allelic to one or more of the following similar mutants reported to be on chromosome 10L: *dek21, v29, Vsr1*, and *w2* (Neuffer et al., 1997). **I**) Pale sheath chimera (PlSh*-2562) showing the clear distinction between mutant and normal tissue

slower or faster than normal siblings are not easily recognized, but the mutation can be seen in a chimera. Such chimeras are often recognized as bent or curved plants depending on the nature of the gene product (Figure 7A, B). By examining the border between mutant and normal tissue, one may be able to determine if there is a sharp boundary, suggesting that the gene product acts autonomously. Alternatively, a blending gradient along the border may be seen indicating diffusion of gene product from one tissue into the other; such as is the case for *colorless floury defective (dek1)* and the *floury* endosperm (Neuffer, 1995). One may also fail to distinguish between mutant and wild-type tissue as in the case of *Cg* (Poethig, 1988b) and most sectors involving *D8* (Harberd and Freeling, 1989).

When the mutation occurs in an essential gene that is effectively lethal, the wild-type half often rescues the lethal mutant half. Some mutations turn out to be conditionally lethal, such that the chimera can be crossed to a different inbred and may survive in a vigorous hybrid background or survive when grown in the greenhouse. The *Liguleless narrow (Lgn)* chimera shown in Figure 7C was obvious at the ligule and auricle, which normally serves as a sharp boundary between blade and sheath. On one half of the leaf, the ligule and auricle were in their normal position and form, while the other half had a reduced auricle and displaced ligule. The chimera, which originated following B73 EMS pollen treatment onto B73, was crossed to A632 and produced vigorous plants with a mild ligule defect. Crosses of *Lgn* back to B73 produced very weak, liguleless plants that were nearly sterile. It is likely that this mutation would have been lost had it not been identified as a chimera.

PgV-2542* originated as a very light yellow green chimeric plant in the M1 of A619 × B73 treated pollen (Figure 7D). Pollen from this chimeric plant crossed on a standard stock produced progeny that segregated 1:1 for tiny, yellow green, and dwarf-like plants that failed to make a tassel or ear (Figure 7E). Replanting progeny under intensive care produced short, pale green mutant plants (Figure 7F), which fell over because they lacked normal root development. They also failed to make viable ears but did make one tassel with enough pollen for outcrossing and viable offspring for further analysis. A review of this mutant's history suggests that it would not have survived except as a chimera on a normal plant.

Vsr-2595*, originated as a chimeric plant with a large yellowish white half leaf sector on one side of the plant (Figure 7G), in a cross of Mo17 × A632 treated pollen. At the seedling stage it was almost white with tiny yellow green streaks, typical of those seen in recessive *v29* and dominant *Vsr1*. These streaks greened up to near normal green. Pollen from this chimeral plant crossed onto B73/A619 segregated 1:1 for yellowish white virescent seedlings that slowly greened up to produce small striped yellow green plants (Figure 7H), a few of which survived to maturity.

3.4   *Lethal Dominants*

Normally in genetic studies, mutants for which no progeny can be obtained are not described. However, with repeated occurrences obtained through chemical mutagenesis, such examples are worth noting. One example is the white or yellowish albino seedling (Figure 8A), the rare dominant phenotypic equivalent of the most frequent of all the recessive seedling mutants. Several of these have been seen both as chimeras and as whole seedling lethal cases. The same is true for the tannish necrotic seedling lethals. Another good example is a mutant that has small fleshy leaves, whose surface

Fig. 8 Dominant lethal mutations. **A)** Dominant yellowish, white lethal M1 seedling (W*-33:1022–58). **B)** Dwarf with fleshy sheen heart shaped leaf (DfShn*-33:1018–33). **C)** M1 plant, tangled midrib only (84:62–4). **D)** Putative DfShn type lethal chimera (Chi*79:116–4). **E)** Original Nl*-2598 mutant in Mo17 with narrow leaves and zig-zag culm. **F)** The progeny from the sib cross of heterozygous Nl-2598 plants segregated original mutant type and small, midrib only types (arrows) that were probably the homozygotes but looked just like the original heterozygote of another midrib only mutant (C above). **G)** Close-up of one of these small plants in F. **H)** *Leopard spot.* Unusual mutant with pale yellow background and green spots. The mutant arose in Mo17×A632 (81:ll108–1) but has no progeny

glistens like paint with metallic particles (Figure 8B). The leaf is broad and heart shaped. This phenotype has not been seen as a chimera but has been seen four times as a whole seedling, which does not usually grow beyond the four-leaf stage.

A third example that has been seen repeatedly is a narrow leaf mutant whose leaves are loosely tangled like cords of twine, and consists of mostly midrib (Figure 8C). These occur at a frequency as high as 1/1000 pollen grains for some treatments and not in others. They are similar to the recessive *leafbladeless* mutants discussed in Chapter 9. The presumed chimeras (Figure 8D) are distorted by the pulling of normal and mutant tissues against each other such that no normal tassels or ears are produced. We have seen the same phenotype in sib-crosses of NI*-2598 (Figure 8F, G). This mutant originated as a whole plant, with narrow leaf blades, hairy leaf margins and sheath, zigzag stalk and a few branched tassels with viable pollen, in an M1 from the cross of Mo17 by treated Mo17 pollen (Figure 8E). Pollen from this mutant plant crossed onto W22/W23 gave a 1:1 segregation for narrow leaf and normal plants. Subsequent sib progeny gave a wide range of phenotypes (Figure 8F, G) from bladeless tangled leaves to plants that looked like the original mutant parent. The extreme class (probably the homozygotes) appear to be identical with the heterozygotes of the dominant no progeny mutant, *Leafbladeless* (Figure 8C).

4 Conclusions

The ease of mutagenesis, mutant discovery and genetic analysis has kept maize at the forefront of plant genetics for decades. Many genes have been cloned thanks to transposable elements used as gene tags. The synteny of the maize genome with sequenced genomes of rice and sorghum has now made positional cloning also possible. In fact, the first maize gene cloned by position was a QTL and its identity was confirmed using EMS mutagenesis (Wang et al., 2005). Once the maize genome sequence is completed, positional cloning will become even more robust. To clone a gene defined by EMS mutagenesis one has only to develop a segregating population. The high frequency of mutation generated by EMS provides the chance of having multiple alleles. New alleles can also be obtained with reverse genetics resources described in this volume. The recent breakthroughs in high throughput sequencing technology suggest that it may even be possible to determine the mutated gene that results in lethal dominants, which precludes the creation of segregating populations. Half plant chimeras would be especially useful as the DNA from normal and mutant half of the leaf could be compared. Although there would be dozens of mutations, theoretically, there would be one that is only found in one half of the leaf. Future screens should keep careful phenotypic records of half-plant chimeras and lethal dominants along with a sample of the DNA for sequence analysis. In summary, EMS is an efficient, effective tool that differs from other mutagens in its production of valuable dominant lethals and half plant chimeras allowing for the study of genes that are recalcitrant to other forms of genetic analysis.

Acknowledgements We are very grateful to the many cooperators who have screened the EMS populations, found mutants and advanced our knowledge of genetics. Thanks also to Hector Candela-Anton and China Lunde for reading over the manuscript. The work was funded by NSF DBI 0604923.

References

Alexander, D.L., Mellor, E.A., and Langdale, J.A. (2005). CORKSCREW1 defines a novel mechanism of domain specification in the maize shoot. Plant Physiol 138, 1396–1408.

Bauer, P., Lubkowitz, M., Tyers, R., Nemoto, K., Meeley, R.B., Goff, S.A., and Freeling, M. (2004). Regulation and a conserved intron sequence of liguleless3/4 knox class-I homeobox genes in grasses. Planta 219, 359–368.

Becraft, P.W., Li, K., Dey, N., and Asuncion-Crabb, Y. (2002). The maize dek1 gene functions in embryonic pattern formation and cell fate specification. Development 129, 5217–5225.

Bommert, P.B., Lunde, C., Nardmann, J., Vollbrecht, E., Running, M.P., Jackson, D., Hake, S., and Werr, W. (2005). *Thick tassel dwarf1* encodes a putative maize orthologue of the Arabidopsis CLAVATA1 leucine-rich receptor-like kinase. Development 132, 1235–1245.

Bryan, A.A., and Sass, J.E. (1941). Heritable characters in maize. J. Hered. 32, 343–346.

Candela, H., Johnston, R., Gerhold, A., Foster, T., and Hake, S. (2008). The *milkweed pod1* gene encodes a KANADI protein that is required for abaxial-adaxial patterning in maize leaves. Plant cell 20, 2073–2087.

Chuck, G., Meeley, R., and Hake, S. (1998). The control of maize spikelet meristem fate by the APETALA2-like gene *indeterminate spikelet1*. Genes and Development 12, 1145–1154.

Chuck, G., Cigan, M., Saeteurn, K., and Hake, S. (2007a). The heterochronic maize mutant Corngrass1 results from overexpression of a tandem microRNA. Nat Genet. 39, 544–549.

Chuck, G., Meeley, R., Irish, E., Sakai, H., and Hake, S. (2007b). The maize tasselseed4 microRNA controls sex determination and meristem cell fate by targeting Tasselseed6/indeterminate spikelet1. Nat Genet. 12, 1517–1521.

Clark, J.K., and Sheridan, W.F. (1988). Characterization of the two maize embryo-lethal defective kernal mutants rgh*-1210 and fl*-1253B: Effects on embryo and gametophyte development. Genetics 120, 279–290.

Clark, J.K., and Sheridan, W.F. (1991). Isolation and characterization of 51 embryo-specific mutations of maize. Plant Cell 3, 935–951.

Clark, S.E., Running, M.P., and Meyerowitz, E.M. (1993). *CLAVATA1*, a regulator of meristem and flower development in *Arabidopsis*. Development 119, 397–418.

Cone, K.C., Frisch, E.B., and Phillips, T.E. (1989). dek1 interferes with aleurone differentiation. . Maize Genetics Cooperation Newsletter 63, 67–68.

Dangl, J.L., Dietrich, R.A., and Richberg, M.H. (1996). Death don't have no mercy: Cell death programs in plant-microbe interactions. Plant Cell 8, 1793–1807.

Doebley, J., Stec, A., and Gustus, C. (1995). *teosinte branched1* and the origin of maize: evidence for epistasis and the evolution of dominance. Genetics 141, 333–346.

Doebley, J., Stec, A., and Hubbard, L. (1997). The evolution of apical dominance in maize. Nature 386, 485–488.

Dorweiler, J., Stec, A., Kermicle, J., and Doebley, J. (1993). *Teosinte glume architecture1*: A genetic locus controlling a key step in maize evolution. Science 262, 233–235.

Evans, M.M. (2007). The indeterminate gametophyte1 gene of maize encodes a LOB domain protein required for embryo Sac and leaf development. Plant Cell 19, 46–62.

Foster, T., Yamaguchi, J., Wong, B.C., Veit, B., and Hake, S. (1999). *Gnarley* is a dominant mutation in the *knox4* homeobox gene affecting cell shape and identity. Plant Cell 11, 1239–1252.

Freeling, M., and Hake, S. (1985). Developmental genetics of mutants that specify Knotted leaves in maize. Genetics 111, 617–634.

Friml, J., Yang, S., Michniewicz, M., Weijers, D., Quint, A., Tietz, O., Benjamins, R., Ouwerkerk, P.B.F., Ljung, K., Sandberg, G., Hooykaas, P.J.J., Palme, K., and Offringa, R. (2004). A PINOID-dependent binary switch in apical-basal PIN polar targeting directs auxin efflux. Science 306, 862–865.

Galinat, W.C. (1954a). *Corn grass*. I. *Corn grass* as a prototype or a false progenitor of maize. Am. Nat. 88, 101–104.

Galinat, W.C. (1954b). *Corn grass*. II. Effect of the *Corn grass* gene on the development of the maize inflorescence. Am. J. Bot. 41, 803–806.

Gelinas, D., Postlethwait, S.N., and Nelson, O.E. (1969). Characterization of development in maize through the use of mutants. II. The abnormal growth conditioned by the Knotted mutant. Am. J. Bot. 56, 671–678.

Gray, J., Close, P.S., Briggs, S.P., and Johal, G.S. (1997). A novel suppressor of cell death in plants encoded by the Lls1 gene of maize. Cell 89, 25–31.

Gray, J., Janick-Buckner, D., Buckner, B., Close, P.S., and Johal, G.S. (2002). Light-dependent death of maize lls1 cells is mediated by mature chloroplasts. Plant Physiology 130, 1894–1907.

Greene, B., Walko, R., and Hake, S. (1994). *Mutator* insertions in an intron of the maize *knotted1* gene result in dominant suppressible mutations. Genetics 138, 1275–1285.

Haas, G., and Orr, A. (1994). Organogenesis of the maize mutant Fascicled ear (Fas). Maize Gen. Coop. Newsl. 68, 18–19.

Hake, S., Smith, H.M.S., Holtan, H., Magnani, E., Mele, G., and Ramirez, J. (2004). The role of *KNOX* genes in plant development. Annu. Rev. Cell Dev. Biol. 20, 125–151.

Harberd, N.P., and Freeling, M. (1989). Genetics of dominant gibberellin-insensitive dwarfism in maize. Genetics 121, 827–838.

Hoisington, D.A., Neuffer, M.G., and Walbot, V. (1982). Disease lesion mimics in maize. I. Effect of genetic background, temperature, developmental age, and wounding on necrotic spot formation with Les1. Dev Biol 93, 381–388.

Hu, G., Richter, T.E., Hulbert, S.H., and Pryor, T. (1996). Disease Lesion Mimicry Caused by Mutations in the Rust Resistance Gene rp1. Plant Cell 8, 1367–1376.

Hu, G., Yalpani, N., Briggs, S.P., and Johal, G.S. (1998). A porphyrin pathway impairment is responsible for the phenotype of a dominant disease lesion mimic mutant of maize. Plant Cell 10, 1095–1105.

Inada, D.C., Bashir, A., Lee, C., Thomas, B.C., Ko, C., Goff, S.A., and Freeling, M. (2003). Conserved noncoding sequences in the grasses. Genome Research 13, 2030–2041.

Jackson, D., Veit, B., and Hake, S. (1994). Expression of maize *KNOTTED1* related homeobox genes in the shoot apical meristem predicts patterns of morphogenesis in the vegetative shoot. Development 120, 405–413.

Johal, G.S. (2007). Disease lesion mimic mutants of maize: APSnet Feature Story July 2007, American Phytipathological Society. http://www.apsnet.org/online/feature/mimics/.

Johal, G.S., Hulbert, S.H., and Briggs, S.P. (1995). Disease Lesion Mimics of Maize - a Model for Cell-Death in Plants. Bioessays 17, 685–692.

Juarez, M.T., Twigg, R.W., and Timmermans, M.C. (2004a). Specification of adaxial cell fate during maize leaf development. Development 131, 4533–4544.

Juarez, M.T., Kui, J.S., Thomas, J., Heller, B.A., and Timmermans, M.C. (2004b). microRNA-mediated repression of rolled leaf1 specifies maize leaf polarity. Nature 428, 84–88.

Kayes, J.M., and Clark, S.E. (1998). *CLAVATA2*, a regulator of meristem and organ development in Arabidopsis. Development 125, 3843–3851.

Kerstetter, R., Vollbrecht, E., Lowe, B., Veit, B., Yamaguchi, J., and Hake, S. (1994). Sequence analysis and expression patterns divide the maize *knotted1*-like homeobox genes into two classes. Plant Cell 6, 1877–1887.

Lorrain, S., Vailleau, F., Balague, C., and Roby, D. (2003). Lesion mimic mutants: keys for deciphering cell death and defense pathways in plants? Trends Plant Sci 8, 263–271.

Martin, G.B., Bogdanove, A.J., and Sessa, G. (2003). Understanding the functions of plant disease resistance proteins. Annu Rev Plant Biol 54, 23–61.

McClintock, B. (1950). The origin and behavior of mutable loci in maize. Proc Natl Acad Sci U S A. 36, 344–355.

McSteen, P., and Hake, S. (2001). *barren inflorescence2* regulates axillary meristem development in the maize inflorescence. Development 128, 2881–2891.

McSteen, P., Malcomber, S., Skirpan, A., Lunde, C., Wu, X., Kellogg, E., and Hake, S. (2007). barren inflorescence2 Encodes a co-ortholog of the PINOID serine/threonine kinase and is required for organogenesis during inflorescence and vegetative development in maize. Plant Physiol 144, 1000–1011.

Miles, C.D., and Daniel, D.J. (1974). Chloroplast Reactions of Photosynthetic Mutants in Zea mays. Plant Physiol 53, 589–595.

Morris, S.W., Vernooij, B., Titatarn, S., Starrett, M., Thomas, S., Wiltse, C.C., Frederiksen, R.A., Bhandhufalck, A., Hulbert, S., and Uknes, S. (1998). Induced resistance responses in maize. Mol Plant Microbe Interact 11, 643–658.

Muehlbauer, G.J., Fowler, J.E., Girard, L., Tyers, R., Harper, L., and Freeling, M. (1999). Ectopic expression of the maize homeobox gene *liguleless3* alters cell fates in the leaf. Plant Physiology 119, 651–662.

Nelson, J.M., Lane, B., and Freeling, M. (2002). Expression of a mutant maize gene in the ventral leaf epidermis is sufficient to signal a switch of the leaf's dorsoventral axis. Development 129, 4581–4589.

Neuffer, M.G. (1966). Stability of the Suppressor Element in Two Mutator Systems at the A1 Locus in Maize. . Genetics 53, 541–549.

Neuffer, M.G. (1995). Chromosome breaking sites for genetic anlaysis in maize. Maydica 40, 99–116.

Neuffer, M.G., and Calvert, O.H. (1975). Dominant Disease Lesion Mimics in Maize. Journal of Heredity 66, 265–270.

Neuffer, M.G., and Coe, E.H. (1978). Paraffin oil technique for treating mature corn pollen with chemical mutagens. Maydica 23, 21–28.

Neuffer, M.G., and Sheridan, W.F. (1980). Defective Kernel Mutants of Maize. I. Genetic and Lethality Studies. Genetics 95, 929–944.

Neuffer, M.G., Coe, E.H., and Wessler, S.R. (1997). Mutants of maize. (Plainview, New York: Cold Spring Harbor Laboratory Press).

Neuffer, M.G., Hoisington, D.A., Walbot, V., and Pawar, S.E. (1983). The genetic control of disease symptoms. (Oxford and IBH Pub. Co, New Delhi, India).

Nogueira, F.T., Madi, S., Chitwood, D.H., Juarez, M.T., and C., T.M. (2007). Two small regulatory RNAs establish opposing fates of a developmental axis. Genes Dev. 21, 750–755.

Nuffer, M.G. (1961). Mutation Studies at the A1 Locus in Maize. I. A Mutable Allele Controlled by Dt. Genetics 46, 625–640.

Orr, A.R., Haas, G., and Sundberg, M.D. (1997). Organogenesis of *Fascicled* ear mutant inflorescences in maize (Poaceae). Am. J. Bot. 84, 723–734.

Peng, J., Richards, D.E., Hartley, N.M., Murphy, G.P., Devos, K.M., Flintham, J.E., Beales, J., Fish, L.J., Worland, A.J., Pelica, F., Sudhakar, D., Christou, P., Snape, J.W., Gale, M.D., and Harberd, N.P. (1999). Green revolution' genes encode mutant gibberellin response modulators. Nature 400, 256–261.

Penning, B.W., Johal, G.S., and McMullen, M.D. (2004). A major suppressor of cell death, slm1, modifies the expression of the maize (Zea mays L.) lesion mimic mutation les23. Genome 47, 961–969.

Phinney, B.O. (1956). Growth response of single-gene dwarf mutants in maize to gibberellic acid. Proc. Natl. Acad. Sci. USA 42, 185–189.

Poethig, R.S. (1988a). Heterochronic mutations affecting shoot development in maize. Genetics 119, 959–973.

Poethig, R.S. (1988b). A non-cell-autonomous mutation regulating juvenility in maize. Nature 336, 82–83.

Ramirez, J. (2007). thesis.

Scanlon, M.J., Henderson, D.C., and Bernstein, B. (2002). SEMAPHORE1 functions during the regulation of ancestrally duplicated knox genes and polar auxin transport in maize. Development 129, 2663–2673.

Schneeberger, R.G., Becraft, P.W., Hake, S., and Freeling, M. (1995). Ectopic expression of the knox homeo box gene rough sheath1 alters cell fate in the maize leaf. Genes and Development 9, 2292–2304.

Sheridan, W.F., and Clark, J.K. (1987). Maize enbryogeny: a promising experimental system. TIG 3, 3–6.

Singleton, W.R. (1951). Inheritance of Corn grass a macromutation in maize, and its possible significance as an ancestral type. Am Nat 305, 81–96.

Stadler, L.J., and Uber, F. (1942). Genetic effects of ultra-violet radiation in maize. IV. Comparison of monochromatic radiations. Genetics 27, 84–118.

Taguchi-Shiobara, F., Yuan, Z., Hake, S., and Jackson, D. (2001). The fasciated ear2 gene encodes a leucine-rich repeat receptor-like protein that regulates shoot meristem proliferation in maize. Genes Dev. 15, 2755–2766.

Timmermans, M.C., Schultes, N.P., Jankovsky, J.P., and Nelson, T. (1998). Leafbladeless1 is required for dorsoventrality of lateral organs in maize. Development 125, 2813–2823.

Timmermans, M.C., Hudson, A., Becraft, P.W., and Nelson, T. (1999). ROUGH SHEATH2: a Myb protein that represses knox homeobox genes in maize lateral organ primordia. Science 284, 151–153.

Tsiantis, M., Schneeberger, R., Golz, J.F., Freeling, M., and Langdale, J.A. (1999). The maize rough sheath2 gene and leaf development programs in monocot and dicot plants. Science 284, 154–156.

Uchida, N., Townsley, B., Chung, K.H., and Sinha, N. (2007). Regulation of SHOOT MERISTEMLESS genes via an upstream-conserved noncoding sequence coordinates leaf development. Proc Natl Acad Sci U S A. 104, 15953–15958.

Veit, B., Vollbrecht, E., Mathern, J., and Hake, S. (1990). A tandem duplication causes the *Kn1-O* allele of *Knotted*, a dominant morphological mutant of maize. Genetics 125, 623–631.

Vollbrecht, E., Veit, B., Sinha, N., and Hake, S. (1991). The developmental gene knotted is a member of a maize homeobox gene family. Nature 350, 241–243.

Vollbrecht, E., Reiser, L., and Hake, S. (2000). Shoot meristem size is dependent on inbred background and presence of the maize homeobox gene, *knotted1*. Development 127, 3161–3172.

Walbot, V., Hoisington, D.A., and Neuffer, M.G. (1983). Disease lesion mimic mutations. (New York: Plenum Publishing Corp.).

Wang, H., Nussbaum-Wagler, T., Li, B., Zhao, Q., Vigouroux, Y., Faller, M., Bomblies, K., Lukens, L., and Doebley, J.F. (2005). The origin of the naked grains of maize. Nature 436, 714–719.

Wang, K.L., Li, H., and Ecker, J.R. (2002). Ethylene biosynthesis and signaling networks. Plant Cell 14, S131–151.

Wright, A.D., and Neuffer, M.G. (1989). Orange pericarp in maize: filial expression in a maternal tissue. J. Hered. 80, 229–233.

Wright, A.D., Moehlenkamp, C.A., Perrot, G.H., Neuffer, M.G., and Cone, K.C. (1992). The maize auxotrophic mutant orange pericarp is defective in duplicate genes for tryptophan synthase beta. Plant Cell 4, 711–719.

Wright, A.D., Sampson, M.B., Neuffer, M.G., Michalczuk, L., Slovin, J.P., and Cohen, J.D. (1991). Indole-3-Acetic Acid Biosynthesis in the Mutant Maize orange pericarp, a Tryptophan Auxotroph. Science 254, 998–1000.

Wu, G., and Poethig, R.S. (2006). Temporal regulation of shoot development in Arabidopsis thaliana by miR156 and its target SPL3. Development 133, 3539–3547.

Part II
Maize Improvement

Development of Hybrid Corn and the Seed Corn Industry

A. Forrest Troyer, PhD

Plough deep while Sluggards sleep,
And you shall have Corn to sell and to keep.

–Benjamin Franklin (1976)

Abstract This is a history of the development of hybrid corn (Zea maize L.) and of the developing seed corn industry by review of the literature and by the personal testimony of colleagues. I identify, describe, and discuss pertinent background germplasm and provide a sampling of seed corn company histories. Some highlights of seed organizations and seed improvement associations are given. Charles Darwin's views in "The Effects of Cross- and Self-Fertilization in the Vegetable Kingdom" were instrumental in the development of commercial hybrid corn. I trace his hybrid vigor idea through Harvard University, Michigan Agricultural College, University of Illinois, and finally to Connecticut. Charles Darwin's views in "The Variation of Plants and Animals under Domestication" explain why only popular, widely adapted open-pollinated varieties persisted in the background of U.S. hybrid corn. Reid Yellow Dent contributed 56% of the germplasm in the documented background of current U.S. hybrid corn and other popular varieties, such as Lancaster Sure Crop and Minnesota 13, contributed the other 44%. These widely adapted varieties contributed to widely adapted hybrids. Corn hybrids were first commercially grown in the early 1930s when the annual U.S. corn yields averaged 1,518 kg per ha (24.2 bushels per acre), and corn production averaged 51 million Mg (2 billion bushels). In 2007, the average U.S. corn yield was estimated at 9,474 kg per ha (151.1 bushels per acre), and U.S. corn production was 332.7 million Mg (13.1 billion bushels). These increases were caused by better hybrids, improved cultural practices, and biotechnology. Corn has become the highest tonnage crop worldwide. Seed corn companies have grown larger, better, and fewer over time.

A.F. Troyer
Adjunct Professor, Department of Crop Sciences, University of Illinois

J.L. Bennetzen and S. Hake (eds.), *Maize Handbook - Volume II: Genetics and Genomics*,   87
© Springer Science+Business Media LLC 2009

1 Introduction

Corn, a New World crop, with the help of hybrid vigor and biotechnology surpassed rice and wheat, the Old World crops, in world production in the calendar year 2001 (UN/FAO. 2002; Fig. 1). Corn production has been increasing for 500 years since Columbus. This chapter is a history of the development of hybrid corn and the seed corn industry over the last 150 years by review of the literature and by the personal testimony of colleagues. I relate numerous happenings during the development of commercial hybrid corn, many of which occurred at or near the University of Illinois at Urbana, Illinois. I provide a sampling of hybrid seed corn cornpany histories including those that developed popular open-pollinated varieties whose germplasm is in the background of current U.S. hybrid corn. I give some highlights of seed organizations and seed improvement associations over time. The seed industry happenings are commingled with hybrid corn development in chronological order: Persistent, popular, widely grown, open-pollinated varieties have provided most of the seed corn grown in the USA; their histories explain how people and places improved corn by human and natural selection and acquaint the reader with

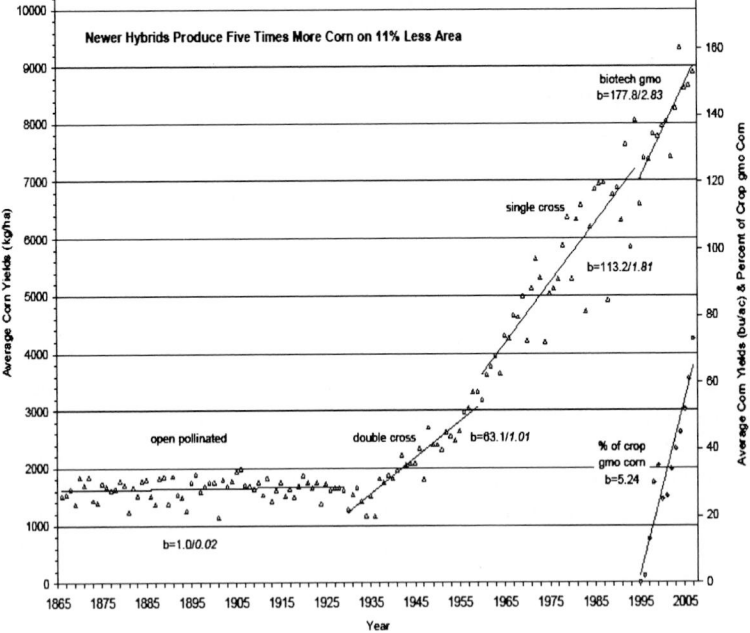

Fig. 1 Average U.S. Corn Yields & Kinds of Corn, Civil War to 2007; periods dominated by open pollinated varieties, four parent hybrids, two parent hybrids, and genetically modified hybrids are shown. "b" values (regressions) are yield gain per year (kg/bu). USDA data compiled by E. Wellin and F. Troyer

the background of today's corn. The business developers and innovators of the seed industry continue to make it all happen. Most of this corn history occurred during the last one and one half centuries.

2   1840

The idea for hybrid corn traces back to Charles Darwin, who married his first cousin, Emma Wedgwood, in 1839. Ten children were born, six boys and four girls. Two children died in infancy. Anne, everyone's favorite, died tragically at the age of ten, probably from tuberculosis. The relationship involved an aunt (Darwin's mother) and her niece (Darwin's wife); the resulting daughters generally were less healthy than their brothers. George, Francis, and Horace were knighted for contributions in astronomy, botany, and civil engineering, respectively. Leonard was a successful economist (Encyclopedia Britannica 1983). The retarded, youngest son, born when Emma was 48, probably had Down's syndrome, a trait unrelated to inbreeding; of course, Darwin didn't know that. Darwin's personal concern about and interest in inbreeding, eventually transmitted and enhanced through Asa Gray at Harvard University; and then William Beal at Michigan Agricultural College; followed by George Morrow, Eugene Davenport, Perry Holden, Cyril Hopkins, Harry Love, and Edward East at the University of Illinois, and finally by Edward East, Herbert Hayes, and Donald F. Jones in Connecticut led to commercial hybrid corn (Darwin 1875, Crabb 1948, Wallace and Brown 1956, Troyer 2004a).

Today's corn in the U.S. Corn Belt evolved during movement from the American Colonies in the east via American westward expansion across the North American continent. Natural and human selection during corn's movement was very important in forming new varieties. Evolution results from changes that occur in organisms over many generations and over long periods of time due to natural selection of mutations. Evolution for plant breeders involves selection, variation, and adaptation. Darwin (1859, 1868) explains them well. Dobzhansky (1947) found adaptive changes induced by natural selection in Drosophila that varied by season within years. The "long period of time" requirement for evolution to occur depends on the organism, the trait and the environment. Day length was particularly important in the movement of U.S. corn because of the north to south axis of the Americas. The 23.45° tilt of the earth's axis to the axis of its orbit is responsible for the seasons and the primary cause of the need for different corn maturities.

Jacob Leaming developed Leaming Corn from Cincinnati common corn selected near Wilmington, Ohio (56 km northeast of Cincinnati) from 1855 to 1885. Jake was an early user of legumes in his crop rotation, and he experimented with drilled spacing of corn back when all corn was check-wire planted to allow cross-cultivation. Jake produced several 6,271 kg per ha (100 bushels per acre) yields in favorable seasons, and became well known as a corn authority. He grew and selected the corn for 30 years. He selected for heavy, earlier ripening ears after removing weak and barren plants before pollination, and then he marked early ripening ears (with twine) on two-eared plants. These ears were more often tapered, and, over time, resulted in a noticeably tapered ear due to fewer kernel rows nearer the tip. Leaming Corn was the first popular corn variety. It was grown more than all other yellow, open-pollinated varieties together. About half of the corn grown was of white endosperm varieties. Leaming Corn was described as medium late maturity (120 Relative Maturity) for the central U.S. Corn Belt. Jake shipped seed corn to all parts of the United States. Leaming corn was recommended by seven state experiment stations in 1936 (DeKruif 1928, Troyer 1931, Jenkins 1936, Troyer 1999). About 2% of the documented background of current U.S. hybrid corn is derived from Leaming Corn per se (Troyer 1999, Fig. 2).

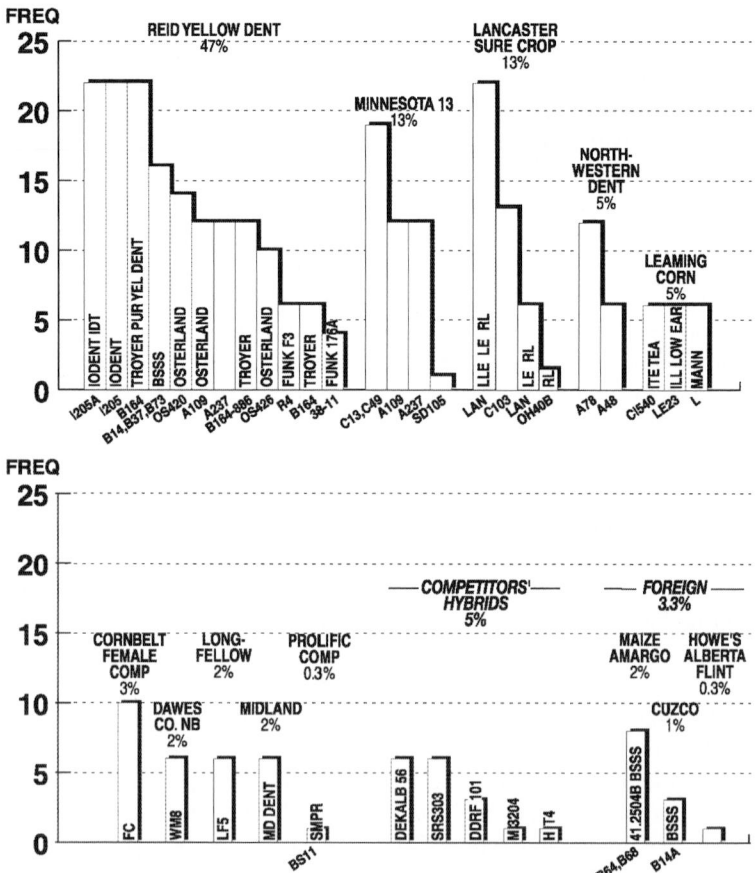

Fig. 2 Background of U.S. hybrid corn. Pedigree background frequencies for 33 Pioneer Hi-Bred Int'l, elite proprietary inbreds with parental background above and with publicly available representatives below. (Troyer 2004b)

The Chicago Board of Trade devised the standard grading of agricultural products that revolutionized commodities trading in the 1850s. The stabilized supply meant buyers could purchase goods before they were grown.

Robert Reid and his son James developed Reid Yellow Dent on the Delavan plains of Illinois about 40 km south of Peoria starting in 1846. They had moved from Russellville, Ohio, (56 km southeast of Cincinnati) in the late spring and planted some Gordon Hopkins seed corn, a semi-Gourdseed dent, which they had brought along. Robert harvested immature ears whose seed produced a poor stand that was, however, too good to destroy. He replanted missing hills with Little Yellow, an early native Indian flint corn, and the corns cross-pollinated. He planted this cross-pollinated seed and selected for the proper maturity over several years. Robert was an exacting father who pursued excellence as his guiding principle (Reid 1915, Crabb 1948, Troyer 1999).

James L. Reid worked on developing Reid Yellow Dent from 1866 until 1910. The oldest of three children, he received the brunt of his father's philosophy. Robert taught James to read at the age of four, to harness a team and to plow a straight furrow at the age of nine, and to practice the rudiments of corn selection soon after. James attended Tazewell College for two years and taught school a term before farming on his own in 1867. He was a quiet, thoughtful, ambitious person and a lover of books and of nature, according to his daughter, Olive. He was also described as withdrawn and deliberate by some of his envious competitors. He was introduced at the National Corn Exposition at Omaha in 1908 by Mr. J.W. Jones, exposition manager, as the man who had put more millions of dollars into the pockets of the farmers in the U.S. Corn Belt than any other man. He was also described by Country Gentleman Magazine as the man who put the Corn Belt on the map of America (Anon. 1894, Reid 1915, Crissey 1920, Everett 1935, Crabb 1948, Troyer 1999).

James selected for better agronomic traits and for beautiful ears. He selected for medium size ears, bright yellow kernel color with solid, deep, relatively smooth grain; a small red cob; and 18 to 22 kernel rows. James preferred ears being well filled over the tip and butt–leaving a small shank to ease hand husking. He saw vigorous plant growth and high shelling percentage as essential traits. James selected seed in the field at harvest time, thereby emphasizing mature, dry seed. Both Robert and James gave seed to their neighbors to keep Reid Yellow Dent pure. James won the World's Fair (Columbian Exposition) corn show in Chicago in 1893, and then he won countless local shows. Reid Yellow Dent became the most popular variety in the United States; 21 central states recommended it in the 1936 USDA Yearbook. Reid Yellow Dent underwent 50 years of evolution as the dominant corn variety. It has been estimated that up to 75% of then-current U.S. corn crop was Reid Yellow Dent variety (Shamel 1907, Anon. 1930, Jenkins 1936, Wallace and Brown 1956). Reid Yellow Dent was described as medium early relative maturity (115 RM) in the central U.S. Corn Belt.

About 56% of the documented background of current U.S. hybrid corn, including the prominent and persistent strains of Osterland Reid, Troyer Reid, Iodent Reid, and Funk Reid varieties and Iowa Stiff Stalk Synthetic is derived from Reid Yellow Dent (Fig. 2). These improved Reids benefited from human selection and from natural selection under newer and/or improved growing conditions. About 4% of the documented background of current U.S. hybrid corn is derived from Reid Yellow Dent per se (Troyer 1999, 2004b, Fig. 2).

3   1860

The Hershey Bros., then son Isaac, developed the Lancaster Sure Crop variety from a small, lavender colored, slender-eared, semi-flint variety in Lancaster County, Pennsylvania from 1860 until 1920. Isaac, one of eight children, was the seventh generation of the Hershey family in Lancaster County since 1717. In 1860, when Isaac was eight years old, Henry High, a neighbor, obtained the variety through the mail from the Patent Office (Hershey 1989). Isaac, helped by his brother Benjamin, selected among harvested ears in the seed house because corn in the area was cut and shocked to help dry the ears and to use the stalks for fodder. He preferred medium-length, well-matured, sound, firm ears with clean shanks and with neither moldy nor silk-cut kernels. He repeatedly selected smooth, flinty ears after crossing to later dent varieties. Later, Isaac's son Noah selected for stronger roots and larger ears in fields of standing corn. Over time, Lancaster Sure Crop was crossed to later dent cultivars then crossed back to Lancaster Sure Crop 8 to 12 times, by mixing seed in the field, followed by selection for flintier, smoother, and longer ears. The Hershey family started selling seed in 1910 when they quit crossing and blending their seed and started selecting for more uniformity (Anderson 1944, Hershey 1989, Troyer 1999). About 4% of the documented, background of U.S. hybrid corn is derived from Lancaster Sure Crop (Troyer 2004b, Fig. 2).

President Lincoln signed the Morrill Act for land grant colleges in 1862. The University of Illinois location in Champaign County was secured with 4 ha (10 acres) of Institute buildings and grounds, 405 ha (970 acres) of land, and $100,000 in county bonds in 1868. The state of Illinois grew about four million ha (10 million acres) of corn, which was about one third of current U.S. corn production. The University was soon to become (starting in about 1890) the world leader in corn breeding research. The University of Illinois in Champaign county had the right place (birthplace of,Burr-white and Chester Leaming corn varieties), the right people (Morrow, Davenport, Holden, Hopkins, Love, and East), and the right projects (hybridizing and inbreeding corn, improving corn quality, and improving soil fertility). Much of the research leading to hybrid corn occurred there (Troyer 2004a).

Mr. Ezra E. Chester of Champaign, Illinois, purchased ears of Leaming Corn from J. S. Leaming in 1885. Mr. Chester grew the corn in isolation away from other corn and selected for first ripening of the husks to obtain earlier ripening ears of corn. His resulting variety was called Chester Leaming. This was the parent variety of the first inbreds used commercially in the Burr-Leaming double cross in Connecticut and the Copper Cross hybrid in Iowa (Shamel 1901, Holden 1948, Troyer 2004a). About 3% of the documented background of current U.S. hybrid corn is derived from Chester Leaming (Troyer 2004b, Fig. 2). In addition to being a prominent seed man and land owner, Mr. Chester served admirably as the mayor of Champaign, Illinois and on the advisory board of the Illinois Agricultural Experiment Station (Cunningham 1905,Troyer 2004a).

David Richey, son Frank, and grandson Fredrick (F.D.) produced and selected Lancaster Sure Crop from 1888 to 1920 near LaSalle, Illinois. They selected for better germination, heavier test weight, and for longer ears. It was locally grown by neighbors

(Crabb 1948). F.D. Richey later self-pollinated Richey Lancaster at Arlington Farm, Virginia, where the Pentagon now stands, and he also gave 50 ears to M.T. Jenkins at Iowa State College. Glenn H. Stringfield made a synthetic of Richey Lancaster inbred lines for selection at Ohio State. About 9% of the documented background of current U.S. hybrid corn is derived from Richey Lancaster (Richey 1945, Crabb 1948, Troyer 2004b, Fig. 2).

Darwin was thorough in his concern about inbreeding. Over a 12 year period, he studied 57 species of plants and found in the majority of cases that the products of cross-fertilization were more numerous, larger, heavier, more vigorous, and more fertile than the products of self-fertilization–even in species that are normally self-fertilizing (Darwin 1875, DeBeer 1965). Darwin pointed out that silk delay is normal in corn, facilitating cross-fertilization. He described the advantage of cross-fertilized over self-fertilized plants under stress as follows: "To my surprise, the crossed plants when fully grown were plainly taller and more vigorous than the self-fertilized ones. The crossed plants had a decided advantage over the self-fertilized plants under this extremity of bad conditions" (Darwin 1875).

Heterosis is more pronounced under stress. Natural selection for adaptedness favored cross-pollinated plants in corn races and open-pollinated varieties. Silk delay was the mode of action to provide more vigorous, cross pollinated progeny. Because crossing was advantageous, heterozygosity became and still is an important cause of heterosis (hybrid vigor). Hybridization of hybrid corn causes a greater number of heterozygous loci just as cross fertilization in races and varieties did (Troyer 2006a).

Darwin's intimate friend and correspondent in America was Dr. Asa Gray at Harvard. They met at Kew in 1855. Darwin wrote a letter on divergence of species to Gray in 1857, which in mid-1858 was used by Lyle and Hooker to show Darwin's priority over A.R. Wallace on the concept of evolution and divergence through natural selection (Brackman 1980). The book "Origin of Species" was published in 1859. Gray wrote articles in Atlantic Monthly in 1861 where he stated that natural selection and natural theology were compatible. Darwin had Gray's articles printed and distributed as support for "The Origin of Species." Gray dined at Darwin's home and stayed overnight on October 24, 1868. Darwin dedicated his book "The Different Forms of Flowers" to Gray in 1876 (Freeman 1978).

Dr. William Beal trained under Dr. Gray at Harvard, graduating in 1865, taught two years at the University of Chicago, and then settled at Michigan Agricultural College, now Michigan State University. Dr. Beal wrote a letter to Charles Darwin after reading "Cross and Self Fertilization" and received a reply encouraging him to experiment with crossing. In 1877, Dr. William Beal became the first to cross-fertilize corn for the purpose of increasing yields through hybrid vigor. He asked his cooperators to obtain two seed lots that were essentially alike in all respects. One should have been grown at least five years (better if ten years or more) in one neighborhood and the other grown in another location about a hundred miles distant for five or ten years. He wanted different adaptedness through natural selection to have occurred. Beal ran yield tests of corn variety crosses in Michigan and across some central states; more often than not the crossed progeny yielded more than their parents (Beal 1876, 1881). Beal trained Eugene Davenport and later Perry Holden. Holden roomed in Davenport's home, and they became close friends (Holden 1948, Wallace and Brown 1956).

4   1890

Eugene D. Funk Sr. experimented with corn as early as 1892. His grandfather, Isaac Funk, came to central Illinois from the Scioto River Valley of Ohio and acquired more than 9,843 ha (25,000 acres) of land in central Illinois from 1824 to 1865. Gene was the fifth generation of Funks in America. They came to Pennsylvania in 1733 along with many of their fellow Mennonites (Anabaptists). Gene sought a method to utilize Funk farms to help agriculture. He had been impressed with the Vilmorin Seed Company in France during a trip to Europe. Perry Holden became the first manager of Funk Brothers Seed Co. in 1902. The stockholders were 13 members from the third generation of the Funk family in the USA., ranging in age from 26 to 35 years. They contacted leading corn growers in the U.S. Corn Belt to obtain about 2,500 bushels of seed corn that was sorted down to 3,000 of the finest ears for breeding. Seed from each ear was grown ear-to-row in isolated blocks of 80 to 100 ears each. They developed a three generation, ear-to-row, seed increase and selection scheme to produce open-pollinated variety seed. An ear-to-row planting is a row planted with the seed from a single ear, which enables observing the heritability of genetic traits. In addition to agronomic traits, they also bred varieties for high oil and for high protein content. They sold a pure line, double-cross hybrid (Funk 250) in 1922, but an appreciable amount of hybrid seed was not available to growers until 1933 (Cavanagh 1959, D. Sywassink pers.comm.).

Gene developed Funk Reid from Reid Yellow Dent near Shirley, Illinois in about 1900. Gene was active in state, national and seed trade affairs. He selected rough, deeply dented, show-type ears from disease resistant plants. Funk Reid was popular in Illinois and was also recommended by the Delaware Experiment Station in 1936 (Jenkins 1936). Funk Reid resembled Reid Yellow Dent with deeper, rougher dent. Some strains of Funk Reid were bred for higher content of protein and oil (Shoesmith 1910, Wallace and Bressman 1923). About 3% of the documented background of U.S. hybrid corn is derived from Funk Reid (Crabb 1948, Troyer 2004b, Fig. 2).

George McCluer (1892), assistant horticulturist of the College of Agriculture at the University of Illinois, studied hybrid vigor between dent, sweet, flour, and pop corn endosperm types and their second year progeny. The work was started in 1889 by T.F. Hunt. The dent varieties were Burr White and Chester Leaming. Sixteen of 18 crosses yielded more than the average of its parents. Only four crosses yielded more than its higher yielding parent. Only one cross, Chester Leaming x Mammoth Sweet, yielded more (4%) than Chester Leaming. Burr White was crossed to Chester Leaming for the purpose of observing seed and ears; but this crossed seed was not grown. All crosses of corn endosperm types were inbred by hand pollination to observe seed and ears; this inbred seed was not grown. The self-pollinated seed was made in 1890 and 1891.

McCluer (1892) observed that different corn endosperm types crossed readily, and crosses showed more uniform plants than varieties. Flintier (smoother) kernels and darker colors prevailed (dominance) in crosses as did heavier kernels in starchy by sweet kernel types. Color of seed coat was affected independently of endosperm color. Nearly all the corn grown a second year from the crosses (naturally pollinated seed harvested from each of the large plots) gave smaller plants than that grown from F1 seed in the first year. Progeny from the crosses of different types tended to run back (segregation and linkage) to the parent types.

George Morrow was trained as a lawyer, but made his living as an agricultural journalist for more than a decade. He became a professor of practical agriculture at Iowa State and six months later came to Illinois as the only agricultural professor. In 1878 he became the first Dean of the College of Agriculture. Agricultural student enrollment had dropped to 17 students. He visited Europe and was particularly impressed with the long-term crop trials at Rothamsted, England. He came home and started the Morrow Plots at Illinois. He and F.D. Gardner, agricultural assistant, used replicated test plots and multiple year averages to confirm that crosses of open pollinated varieties usually yield more than the average of their parents. Crossing varieties to develop newer varieties had been going on for about two centuries, but this was different. The new idea was to plant first generation hybrid seed each year. The hybrid seed would be produced by growing alternate rows of the two parents and detasseling the seed parent before pollination (Morrow and Hunt 1889, McCluer 1892, Morrow and Gardner 1893, 1894).

The American Seed Trade Association (ASTA) was founded by a small group of seeds men in early May of 1893, who called for a convention of seeds men to be held in the city of New York on June 12–14. The order of the day was to include: Protection of seeds men from unjust claims, reduction of postal rates on seeds, and tariffs on seeds. Thirty-three seeds men answered the first roll call. Officers were elected. The convention adopted an industry disclaimer clause. At its 50th meeting "A Half-Century of Progress" in Chicago in 1932, 181 members attended out of a total membership of 244. The ASTA has a policy of arranging for commodity-specific group meetings to accomplish more work in less time. The Hybrid Corn Group was established in 1939 (Cornell, 1958). The ASTA celebrate their 125th anniversary in 2008.

Mr. J.D. Whitesides selected Johnson County White from a cross between Boone County White and Forsythe's Favorite just south of Indianapolis, Indiana in 1893. It was improved later by several farmers. It replaced Boone County White in many areas to become the most popular white endosperm variety. Johnson County White matured in 120 to 125 days, and was adapted to the Southern Corn Belt. It was usually grown south of central Iowa (Wallace and Bressman 1923).

David Troyer selected Troyer Reid from Reid Yellow Dent near LaFontaine, IN from 1894 until 1936. Dave carefully removed husks, silks, and trash from ear corn before storage to reduce nesting materials for rodents. The corn dried better in the fall and germinated better the next spring. His good stands prompted neighbors to ask to buy seed from him. His elder son Chester resigned as the local high school principal and took over the seed corn business in 1908. They selected Troyer Reid in sub-irrigated river bottom fields that probably favored better, deeper, root growth (Troyer 2006b). Chester won corn shows repeatedly against tremendous odds by selecting the most beautiful ears of corn. Troyer Reid was popular at corn shows. The Indiana Experiment Station recommended it in 1936 (Jenkins 1936). Troyer Reid was described as adapted to northern Illinois, Indiana, and Ohio (110 RM) (Troyer Bros. 1916, Troyer 1999, 2006b). About 15% of the documented background of U.S. hybrid corn is derived from Troyer Reid (Troyer, 2004b, 2006b, Fig. 2).

Prof. Willet Hays developed Minnesota 13 in the 1890s. He recognized the individual plant as the unit of improvement saying: "There are Shakespeares among plants" (DeVries 1907, Boss 1929). He selected on a row basis and originated the centgener breeding method of growing 100 plants per generation from a single selected plant, selecting the best plant, and repeating the cycle. He selected for heavy, mature ears with high nitrogen (protein) in the grain. After three years of centgener selection, 9800 kg (300 bushels) were distributed as University 13 in 1896. In following years, a seed corn plot was planted, poor plants detasseled or entirely removed, and the best ears from the best plants selected for distribution as Minnesota 13 (Troyer 2007).

Minnesota 13 is a mid-early (95 RM) yellow dent corn. It is early maturing but heavy yielding, adapted from southern Minnesota northward. Ears are 15 to 21 cm long, 16.5 cm in circumference, slightly tapered, and have 12 to 16 kernel rows, medium size red cobs, and medium size shank diameter and length. Kernels are medium yellow, medium deep and medium broad with a medium smooth, dimpled dent. Kernels are compactly set on the cob. The butts are well rounded and the tips well filled. Plants are 2 to 2.1 m tall, and ear height is 71 to 79 cm high (Shoesmith 1910, Wallace and Bressman 1923). It was recommended by 15 state experiment stations within 40 years of its development (Jenkins 1936). About 13% of the documented background of current U.S. hybrid corn is derived from Minnesota 13 (Troyer 2004b, 2007, Fig. 2).

Eugene Davenport succeeded Morrow as Dean of the College of Agriculture at Illinois in January 1895. Davenport wanted to separate the study of plants and soils from animal studies in the College and contacted Holden, his friend and former tenant, to head crops and soils. Holden visited the University of Illinois in June 1895, and accepted Davenport's offer, effective in 1896. They spent several days planning corn improvement projects including the production of inbred seed. Holden instructed student W.J. Fraser (later Dairy Department Head) on how to self-pollinate corn with paper bags. Holden arrived in June 1896 and was able to harvest S2 (self-pollinated twice) inbred seed that fall. This inbreeding in 1895 began the development of the first inbreds grown commercially in corn hybrids. The result was the hybrid seed corn industry and higher yielding, more widely adapted corn. It was Davenport and Holden's idea to develop inbred seed, and East's idea to grow ears of inbred seed ear-to-row for purification, instead of growing them in bulk (Crabb 1948, Holden 1948, Troyer 2004a).

Prof. Holden became the first Head of the first Agronomy Department in America in 1899. From studies in Michigan, he was aware that crosses between varieties yielded more than crosses within varieties and that some crosses between varieties yielded more than others. At Illinois he learned more about natural cross pollination, measured progressively reduced plant and ear size due to successive inbreeding, and explained normal size plants growing in inbred materials as accidental crosses. Holden tested his explanation by making crosses (hybrids) between inbred plants in 1898, and planting the resulting seed to observe normal size plants in 1899 (Holden 1948). He reported that the normal (hybrid) plants greatly contrasted in size and vigor with those from inbred seed. He was assisted in these experiments by A.D. Shamel, who became a USDA plant breeder. Photographs of Holden in corn nurseries at Urbana, Illinois support his claim to early inbreeding (DeVries 1907). Wallace (1955) credits Holden as the first to cross corn inbreds in 1898, and that he did it at the University of Illinois.

Oscar Will selected popular and persistent Northwestern Dent from the Bloody Butcher variety near Bismarck, North Dakota from 1891 until 1894. Bloody Butcher came to North Dakota from south-central Indiana. He selected the 15 or 20 earliest plants whose ears showed more advanced development for three seasons in a severe, heat-unit limited, long day, droughty environment before growing a seed increase. Oscar first sold Northwestern Dent in 1896, and it soon became the most popular variety in the northwestern USA. (Olson et al. 1928). Nine state experiment stations recommended it in 1936 (Jenkins 1936). Northwestern Dent variety is mid-early to early (80 RM). It is distinctly earlier than Minnesota 13. Kernels are semi-dent, red in color (pericarp) with white or yellow (endosperm) indentations. (Atkinson and Wilson 1915). About 5% of the documented background of current U.S. hybrid corn is derived from Northwestern Dent (Troyer 2004b, Fig. 2).

Sturtevant (1899) listed 507 named corn varieties and 163 synonyms in 1898. He classified 323 as dent, 69 as flint, 63 as sweet, 27 as soft (floury), and 25 as pop corn. He discusses these endosperm type corns as groups and also provides morphological descriptions of each variety. He discusses the exceedingly different forms of corn: Height varied from 46 cm for Tom Thumb to 9.1 m for a West Indian variety. Some corns produced as many as 34 ears from a single seed (many tillers).

5   1900

The idea of pedigreed, certified grain for seed is believed to have originated in Sweden. The Canadian Seed Growers Association was organized in 1900 to encourage farmers to take better care of their seed grains. Quality and purity of grain was emphasized over quantity of grain and number of sales. The Illinois Corn Breeders Association was organized to improve corn varieties for Illinois farmers in 1900. It and the Illinois Corn Growers Association later joined the Illinois Crop Improvement Association founded and led by Prof. J.C, Hackelman in 1922. They incorporated in 1924 (Lang 1972). Professors Willet Hays and Coates Bull sponsored the Minnesota Field Crop Breeders Association in 1903. Its name was changed to Minnesota Crop Improvement Association in 1913 (H. Stoehr, pers. comm.). The Indiana Corn Growers Association, which was later superseded by the Indiana Crop Improvement Association, introduced seed certification in Indiana in 1920. Prof. Keller Beeson played a dominant role. Purdue Ag Alumni became the foundation seed organization to increase breeders' seed for sale to growers of certified seed (McFee and Ohm 2007). Other U.S. Corn Belt states organized similar state organizations with similar goals.

Prof. Holden accepted a high salary from the Illinois Sugar Refining Co. at Pekin, Illinois to work with sugar beet growers in March 1900. After about a year, the sugar beet operation was purchased by glucose (corn starch) interests soon-to-become Corn Products Co., and Holden was no longer needed. He then helped organize and headed Funk Bros. Seed Co., and later became a Professor at Iowa State College. (Moores 1970, Holden 1944, 1948). One of Holden's important achievements in Iowa was supplying Reid Yellow Dent seed corn from Illinois. In the fall of 1902, he purchased 19,600 kg (600 bushels) from Funk Bros. in Bloomington, Illinois that was distributed in Iowa by Wallaces' Farmer Magazine in 1903 and 1904 (Sizer and Silag 1981). In the fall of 1904, Holden purchased another 11,500 kg (350 bushels) selected by C.A. Peabody near Taylorville, Illinois. It was distributed in Iowa through grain elevators and Wallaces' Farmer for 25 cents a sack (Mosher 1962). Holden was very popular in Iowa. He was known as "The Great Corn Evangelist" and was persuaded to run for governor in 1912. He was defeated.

Edward M. East, an only child, was born and raised near DuQuoin, Illinois. A child prodigy, reputedly related to Sir Isaac Newton, he graduated from high school at age 15 and worked two years in a machine shop. He studied for a year at Case Western and then transferred to the University of Illinois, receiving a B.S. in 1901 and an M.S. in 1904. He was the first Agronomy Department graduate student to receive a Ph.D. (in chemistry) from the University of Illinois, in 1907. Previous Agronomy graduate students had gone overseas to finish. He learned genetics in Botany classes from Dr. Hottes, who was in Europe when Mendel's laws were rediscovered. After East earned the B.S. degree, his graduate adviser and supervisor was Dr. Hopkins, the new Agronomy Department Head. East did chemical analyses on corn protein and oil selection studies. He became interested in inbreeding after studying the ear numbers in the pedigree tails. He may have been the first to notice that all of the high protein samples traced back to a single ear (#21), and that yield was decreasing over time, though not directly inbred by self-fertilization. He suspected that intensive selection caused inbreeding. Would it have been better to self-fertilize ear 21 rather than cross-fertilize it? East decided he wanted to study inbreeding (Crabb 1948, Troyer 2004a).

After the M.S. degree, East was promoted to first assistant in plant breeding under Louie Smith. When Smith went to Halle, Germany to finish graduate work in 1904, East was put in charge of plant breeding, and Harry Love, who had been the chemist, was promoted to assistant in plant breeding (Crabb 1948). East and Love shared an office in room 328 of Davenport Hall. It is on the third floor across the hall from the present elevator that is near the main, west entrance. East got permission from Hopkins to grow the inbred material ear-to-row in 1905. When the American Breeders Association met in Urbana, Hopkins arranged a dinner conference for East and Dr. E.H. Jenkins, director of the Connecticut Experiment Station, who was looking for someone to head a corn breeding program for Connecticut. They hit it off well.

The inbred plot grown ear-to-row was very interesting. The inbreds were quite uniform and differed in size and shape. East arranged to have pictures taken, including Love and Craig, a technician, holding signs indicating LEAMING–HAND POLLINATED FOUR YEARS IN-BRED and LEAMING–HAND POLLINATED THREE YEARS IN-BRED. Finally the day arrived in mid-summer when East, Love, and Hopkins toured the plot. Hopkins rejected the idea of studying inbreeding and said that next year they would return to bulking the seed from all ears of inbred plants. East accepted Dr. E.H. Jenkins's offer at Connecticut effective September, 1 1905 (Crabb 1948).

We shouldn't judge Hopkins too harshly for East leaving Illinois. East came to Illinois to obtain a PhD. and to find gainful employment. Hopkins was instrumental in East achieving both goals. Hopkins had problems of his own: The bloom was off the chemical composition selection study because yield continued to decrease. Hopkins probably saw better soil fertility helping the farmers more than inbred corn. Hopkins was in a battle with the USDA Bureau of Soils about the efficacy of inorganic fertilizers. Hopkins recommended maintaining soil fertility by adding inorganic fertilizer. By sharp contrast, the USDA recommended only adding clover to the crop rotation for the grain farmer and only returning manure to the land for the livestock farmer to maintain fertility. This dispute got completely out of hand. The USDA Bureau of Soils withdrew from its agreement with the Illinois station in 1903 and did not resume cooperative soil-surveys until 1943 (Moores 1970). Hopkins won the fertility war in the 1910s with phosphorous supply-and-demand arithmetic for corn growing, and with phosphorous application demonstrations on "Poorland Farm" near Salem, Illinois (Hopkins 1903, 1913). Hopkins deserves much credit for perseverance and diligence in accomplishing and publishing results of the chemical composition selection study. He was a dedicated scientist.

Dr. E.H. Jenkins and East developed plans for Connecticut corn breeding. East related his ideas on a corn inbreeding study, and Jenkins encouraged him to do it. East contacted Love at Illinois to obtain seed of the Chester Leaming inbreds for use in the study (Crabb 1948). East went to the Bussey Institute at Harvard in 1909 but continued to interact with the Connecticut station, where he was replaced by H.K. Hayes. Hayes needed a thesis problem and decided to study the inheritance of protein content in corn. He requested and obtained high and low protein corn seeds from Dr. Louie Smith at Illinois. The materials were from the Burr White variety from Champaign County, Illinois and had been selected 14 years in the Illinois chemical composition selection study. Hayes further inbred and selected the Burr White lines in Connecticut (Hayes 1913, Troyer 2004a).

Dairyland Seed Co. Inc. originated in 1907 when Simon and Andrew Strachota began buying Dutch white clover seed, a usual component of lawn seed, from farmers near their small country store in St. Kilian, Wisconsin. They resold the seed to major seed firms in Milwaukee. Simon's son Orville, and his wife Marie, transitioned the company into a major marketer of seed and began developing hybrid corn, alfalfa, and soybean genetics in 1977. It is a family-owned business presently led by third-generation owner/managers Steve, John, and Tom Strachota, with Tom's son T.J. also active in the business. They have worked hard to establish high standards of ethics exemplified in many ways, including an industry-leading commitment of 15% of revenues and 40% of employees to research, maintaining a first-class physical plant, demonstrating employee pride in the organization, and the continued active involvement of the Strachota family. This effort has resulted in a positive and consistent image of Dairyland among customers and competitors alike. Dairyland is recognized for the development of the world's first herbicide tolerant soybeans (released in 1993) as well as the world's first hybrid alfalfa (released in 2001). Dairyland ranks among the top 15 seed companies within the corn division of the American Seed Trade Association of 800 members (Anonymous 1995; T. Strachota, pers. comm.).

Prof. Lyman Burnett, Perry Holden's brother-in-law, started selecting Reid Yellow Dent for earliness and yield using the ear-to-row method in 1909 and continued at least 13 years to develop Iodent Reid (Wallace 1923). Iodent is an abbreviation of Iowa Experiment Station Reid Yellow Dent. Prof. Burnett was one of the most successful small grain pure line breeders of the period. He developed several useful oat, winter wheat, and barley varieties. Prof. Burnett initiated the Standard Community Grain Trials in 1938 to obtain more data and to show newer varieties locally. This was the forerunner of corn strip trials that became popular in the 1960s (Iowa State University Archives, J. Pesek, pers. comm.). Wallace (1923) described Iodent Reid as an early, rather smooth eared strain of Reid Yellow Dent with horny, shiny kernels showing very little soft, white starch. About 13% of the documented background of current U.S. hybrid corn is derived from Iodent Reid (Troyer 2004b, Fig. 2).

Shull, a scientist at Cold Spring Harbor, spoke at three American Breeders Association meetings. These talks were published, in addition to an American Naturalist paper and a book chapter (Shull 1908, 1909, 1910, 1911, 1952). Shull quickly grasped the hybrid concept: "Hybrids will be grown; the question being whether they be known hybrids or unknown hybrids." He also grasped purification of traits by inbreeding. His papers were insightful. He effectively spread the idea of hybrid corn. None of his inbreds or hybrids was useful commercially. He did *not* inbreed popular varieties. East and Shull conferred and agreed to disagree on the practicality of hybrid corn; they apparently convinced each other: Dr. Shull, while in Germany in 1914, wrote letters to Gene Funk in Bloomington, Illinois, to H.K. Hayes in Connecticut, and to George Allee, president of the Iowa Agricultural Experiment Station to the effect that: "I consider the pure line method to be of theoretical rather than of practical interest" (Crabb 1948). East persevered with variety crosses, and his students, Hayes and Jones, advanced to the practical Burr-Leaming double-cross hybrid (Crabb 1948). Shull studied Luther Burbank's plant breeding methods in California during 11 visits in 5 years through a Carnegie Institute grant but never completed the project. Shull states: "Burbank deserved to be famous;" and "It was a psychological battle." Shull started and first edited the Genetics periodical (Kraft and Kraft 1967).

Henry Osterland developed Osterland Reid from Reid Yellow Dent near Faulkner, Iowa from 1910 until 1920. Henry had an eighth grade education and regularly read non-fiction books to improve his mind. His brother, brother-in-law, and nephew were ministers in the Evangelical Church. Henry twice refused the Iowa Master Farmer Award even after letters of encouragement from H.A. Wallace. Perhaps Henry's strong religious views were a factor. Twenty years later he answered a query from Dr. William L. Brown of Pioneer Hi-Bred to the effect that he doubted his corn breeding experiences would be useful to modern breeders. Henry Osterland was humble. He passed away in his 78th year in 1958; neither he nor his two brothers left a son to carry on the seed business. Henry grew seed from ears selected on standing plants with some bare cob at the tip to have room for more kernels in better years. He selected for earlier maturity, longer ears, and looser husks. He selected ears that extended beyond the husks to help the husks open up and hasten drying. He liked rather long ears with 18 or 20 kernel rows that were not too large in circumference, heavy with close set kernels of medium dimensions, medium to smooth indentation, and mostly free of soft starch (Ackley, Iowa. Historical Society; L. Osterland Bakley pers. comm.). About 11% of the documented background of current U.S. hybrid corn is derived from Osterland Reid (Troyer 1999, Fig. 2).

The DEKALB AgResearch Seed Co. traces back to the very first Farm Bureau county organization, which was founded in DeKalb, Illinois in 1912. Dr. Bill Eckhardt, with expertise in soils and seeds, was hired as manager. Charlie Gunn was hired in 1917 to purchase western forage seed for resale and to breed open-pollinated varieties of corn. Tom Roberts, Sr. became assistant manager in 1919 and manager from 1920 to 1932. Henry C. Wallace, Republican Secretary of Agriculture and father of Henry A. Wallace, visited the DeKalb Farm Bureau picnic and recommended hybrid corn in 1923. Tom Roberts Sr. accepted the challenge, realizing a 12 year lead time to market products. DEKALB AgResearch sold 14,000 bushels of hybrid seed corn in 1935. Roberts developed a seed grower's contract based on Chicago market price plus 25% with payment on demand between November 1 and May 1 the following year. He anticipated growers waiting until spring planting to be paid; thus money from customers preceded producer payments. He organized a farmer-dealer force with a 15% commission. He established an employee profit sharing plan of 15% before taxes. Their first popular hybrid was DEKALB 404A that sold 508,000 bags in 1947–its 10th year of sales and the first year that total DEKALB sales exceeded 2,000,000 bags of seed corn. DEKALB hybrids 805 and XL 45 were the first popular full- and short-season, large volume, single-cross hybrids. DEKALB was the leader in U.S. hybrid seed corn sales from the mid-1930s until the mid-1970s (Crabb 1948, Roberts 1999).

Henry A. Wallace started self-pollinating corn to develop inbreds in 1913 (Hayes 1956). He had written his first corn article in Wallaces' Farmer in 1907 and began plugging hybrids in a big way in 1919 (Wallace 1955). He and his friends, Jay Newlin and Simon Casady, formed the Hi-Bred Corn Co. in 1926. The name was changed to Pioneer Hi-Bred Corn Co. a few years later. Newlin was Wallace's farm manager. Casady had grown the first field of hybrid corn (Copper Cross) in Iowa on his farm south of Des Moines in 1924. Raymond Baker became a Wallace farm tenant and corn breeder in 1928 after winning the Banner Trophy in the Iowa Corn Yield Test for highest yielding entry (Hi-Bred B1) in the state. Garst and Thomas became associate producers and sellers in 1930. Bob Garst and Nelson Urban developed sales forces for hybrid seed corn. Casady designed their first seed corn plant built in 1931 at Johnston, Iowa. Perry Collins was the first full-time employee (corn breeder) hired in 1931. Fred Lehman, Wallace's attorney, became president when Wallace became U.S. Secretary of Agriculture and went to Washington in 1933. Pioneer Hi-Bred organized regional seed companies in Iowa, Illinois, Indiana, Ohio, and Canada. They started selling seed internationally in 1964. They went public in 1973. Pioneer Hi-Bred became first in U.S. seed corn sales in 1975 largely due to hybrids 3780 and 3369A. Two million bags of 3780 were grown in 1977 (Troyer 1990). Pioneer Hi-Bred bought Garst and Thomas inventory in 1982 and became a single U.S. company in 1986. DuPont purchased Pioneer Hi-Bred in 1999 (Crabb 1948; R. Baker, pers. comm.).

6   1915

Jim Holbert was born in a log cabin near Muncie, Indiana in 1890. He did undergraduate study at Indiana and Purdue Universities. He was influenced, particularly about hybrid corn, by Gene Funk and by Dr. H.K. Hayes of Minnesota. Jim started at Funk's in 1915. He bred corn for Funk Bros. and for the USDA (disease) on Research Acres at Funk Farms in Bloomington, Illinois. He completed a Ph.D. at Illinois in 1925. Dr. Holbert became Funk Brothers' research director in 1936 and general manager in 1943 (Crabb 1948).

Jim Holbert selected Strain 176A, a utility, smoother-eared variety, from Funk Reid. In 1915 he collected 4,200 sound, disease-free ears mostly from disease-free plants, but some were selected from cribs and from seed production plants. He discarded 3,000 ears based on germination tests, selected for ears with freedom from molds and from rotting that possessed unusually good seedling vigor and root development. He then grew 1,200 ears ear-to-row in isolation with 100 or more hills per row. He weighed each row and kept the best 20. After additional germination tests, the remnant seed from the best 12 ears became the parents of Strain 176A, which was recommended by the Illinois Experiment Station in 1936 (Crabb 1948). About 1% of the documented background of current U.S. hybrid corn is derived from Strain 176A (Troyer 1999, Fig. 2).

Dr. H.K. Hayes left Connecticut for Minnesota and was replaced by Donald F. Jones in 1915. Jones was Dr. East's graduate student. Jones (1917) published a theory of heterosis and made the first double-cross hybrid in 1917. A double cross is the hybrid progeny from a cross between two single-cross hybrids. Double-cross seed is produced on a vigorous single cross plant; this is the seed sold to the farmer. He planted and harvested the double-cross at Mt Carmel, Connecticut in 1918.

This feat is generally recognized as making hybrid corn available to farmers. The four-parent hybrid was called the Burr-Leaming double-cross hybrid. Two parents were from the Burr White variety and the other two were from the Chester Leaming variety (Jones 1927). Both varieties were developed in rural Champaign County, Illinois by local seeds men, E.E. Chester and F.E. Burr. Chester Leaming variety was selected in 1895 for the inbreeding study, and Burr White variety was selected in 1896 for the chemical composition selection probably by Eugene Davenport at the University of Illinois. They were selected several years at Illinois before being sent to Connecticut (Jones 1917, 1927, Troyer 2004a). East and Jones (1919) coauthored the book "Inbreeding and Outbreeding." East became a famous teacher and geneticist at Harvard. He provided Mendelian explanations for quantitative inheritance. East trained 20 Ph.D. students, who in turn, became research leaders and mentors for hundreds of the world's leading plant breeders. East died of stomach cancer at the age of 59 in 1938 (Troyer 2000).

The source of the Chester Leaming inbreds was assumed to be Connecticut until the book "The Hybrid Corn Makers–Prophets of Plenty" by Richard Crabb was published in 1948. Crabb's book was based on personal interviews with the principals in the development of the hybrid corn industry. Crabb scheduled a meeting with East's former Illinois, officemate Dr. Harry Love at Cornell. When Crabb arrived, he was told something had come up that conflicted with their meeting, and they would not be able to meet. At the last moment as Crabb was leaving, he was called back and asked if he could return the next morning. Love had been apprehensive about Crabb's visit. He had kept the secret about the origin of Leaming inbreds more than 40 years. The next morning, Love told Crabb he had sent seeds of the Leaming inbreds to East in 1905. Love said he was glad to get it off his chest (R. Crabb, pers. comm.).

Dr. D.F. Jones stayed in denial the rest of his life claiming the inbreds had only been self-fertilized once at Illinois–by East. This sentence ignores ten years of self-fertilization by other Illinois breeders. When H.A. Wallace's Copper Cross hybrid won the gold medal in the 1924 Iowa Corn Yield Tests, Jones wrote a note to Wallace suggesting the Connecticut station should have been awarded the medal because of the Connecticut Chester Leaming inbred female of Copper Cross. Wallace promptly sent him the medal, which Jones returned. Jones undoubtedly thought of this incident when he learned 24 years later that the Chester Leaming inbreds were developed at Illinois (Crabb 1948, Troyer 2004a). Hayes developed the Burr White inbreds. Dr. Jones remained the Dean of corn breeding for more than 50 years. He was active doing genetic research, developing inbred lines, working out the use of cytoplasmic male sterility and restorer for easier hybrid corn production, and authoring a newsletter periodical for corn breeders.

Montgomery (1916) stated that 1,000 named varieties of corn existed in the United States in 1916, 750 of which were developed since 1840 during American westward expansion. The newer varieties were earlier and more drought resistant. The area of cultivation for U.S. corn reached a high of 44.6 million ha (110 million acres) in 1917; this record was tied or slightly exceeded in 1932 (USDA/NASS 2006).

F.D. Richey began self-pollinating corn in 1916. He was put in charge of USDA corn investigations in 1922 and the hybrid corn model was adopted in 1925. He was followed by Dr. M.T. Jenkins in 1934 and then by Dr. G.F. Sprague in 1958, who recruited Dr. Arnel Hallauer to fill his vacant USDA position at Iowa State. Richey showed outstanding leadership in recruiting corn breeders and arranging summer meetings where federal and state corn breeders would gather and see actual field experiments. This led to informal cooperative exchange of materials, methods, and ideas under the Purnell Act (Wallace 1955, Hayes 1956).

Henry A. Wallace, early in 1920, suggested to his friend H.D. Hughes of the Iowa State faculty that the Agronomy Department initiate an Iowa Corn Yield Test made up of entries in the Iowa corn show exhibition. Wallace credits Hughes and F.D. Richey with working out the details of the test that had so much to do with bringing the superiority of hybrid corn forcibly to the attention of U.S. Corn Belt farmers. The design of the test was ahead of its time, dividing the state into 12 districts with two yield test locations per district (Crabb 1948, Wallace 1955).

7 1930

Ralph Crim, University of Minnesota agronomist, distributed 500 samples of double-cross hybrids in Minnesota in 1930. They had inbreds from Minnesota 13 in the female and inbreds from Rustler or from Northwestern Dent in the males. He directed the production of three one-acre plots of hybrids in cooperation with farmers that same year (Hayes 1956).

Dr. Sprague synthesized Stiff Stalk Synthetic from 16 inbreds plus four parent inbreds from 1932 to 1934 (Hallauer et al. 1983, Troyer 2000). The 20 inbreds had yellow endosperm and above average stalk quality. Fifteen of the twenty inbred-parent slots (75%) trace back to improved strains of Reid Yellow Dent (six from Troyer, three from Iodent, two from Funk Strain 176A, one each from Black, Krug, Osterland, and Walden Reid). Other varieties include one each from Illinois Two Ear (Chester Leaming), Illinois High Yield (from an indistinguishable variety), Illinois Low Ear (Chester Leaming), Eichenberger Clarage, and Kansas Sunflower). Dr. M.T. Jenkins developed seven of the inbreds, and Mr. Ben Duddleston developed four of the inbreds from Troyer Reid that fill six of the twenty inbred-parent slots (30%), plus an inbred from Strain 176A (Troyer 2004b). About 5% of the documented background of current U.S. hybrid corn is derived from Stiff Stalk Synthetic (Troyer 1999, Fig. 2).

In 1933, hybrid corn grew on 54,656 ha (135,000 acres) or more than 0.1% of U.S. corn acreage. This is generally accepted as the first year a significant amount of commercial hybrid corn was grown.

Beck's Superior Hybrids, founded in 1937 by Lawrence Beck and his son Francis Beck, has always been centered on customer service. The Atlanta, Indiana company started with Lawrence and Francis each growing a three acre allotment of hybrid parent seed from the Purdue Botany & Plant Pathology Department. From there, the company grew by servicing customers in several counties in central Indiana. Francis's motto "We're not selling you a bag of seed, we're selling you a stand of corn" continues yet today; the company still offers free replant seed (including trait fees) if there is any need to replant. In 1964, Lawrence (Sonny) Beck joined the company after graduation from Purdue University and led the company to its prominent status in the Eastern Corn Belt. Sonny currently serves as President. Beck's seed company is family owned and operated with all family members heavily involved in seed operations. Sonny and Glendia's oldest son, Scott, serves as Vice President and oversees the Practical Farm Research sites at Atlanta and Fort Branch, Indiana and near Bloomington, Illinois. Sonny and Glendia's son in-law, Todd Marschand, serves as Facilities Manager, and their youngest son, Tony, leads the production facility near Champaign, Illinois. Beck's Superior Hybrids currently markets in Indiana, Illinois, Ohio, Michigan and Kentucky. Beck's Superior Hybrids positions itself as an independent, family-owned company that provides its customers with innovative products, first-rate customer service, and outstanding seed quality (K.J. Cavanaugh, pers. comm.).

Illinois Foundation Seeds, Inc. (IFSI) was founded in 1937, signed an agreement with the University of Illinois to sell single-cross seed of their unreleased inbreds in 1942, hired a manager in 1943, and moved its principle office to the Champaign-Urbana area in 1944. They started advanced-contracting acreage in 1945, started a winter nursery in 1950–51 and started winter production in 1956–57. Sterile seed stocks were first available in 1952 and offered in quantity in 1954. Restorers were first offered in 1957. Substantial building was done in Illinois and in Florida from 1958–1961. Xtra-sweet production began in 1961. The Tolono facilities were approved in 1965. More building and expansion followed with research stations established in DeForest, Wisconsin; Arcanum, Ohio; and Seward, Nebraska.

IFSI incorporated in 1994. Dow AgroSciences made significant investments in IFSI in 1998 providing a path to transgenic traits and to research expansion. A continuous year-around research facility was added in Hawaii and additional research stations were added at Clarion, Iowa and Geneseo, Illinois. Long-term managers include Dale Cochran, Clarion Henderson, and Floyd Ingersoll. IFSI purchased SEED GENETICS, Inc. as their marketing representative in 2003. The IFSI/SGI corn research and marketing team is the largest, independently owned and operated developer and licenser of proprietary corn genetics in the U.S. today (D. Deutscher, pers. comm.).

Mr. H. Chris Hoegemeyer raised and sold seed of multiple crops locally near Hooper, Nebraska. In 1937 he was asked by the University of Nebraska to try raising hybrid seed corn for the farmers in the area. Leonard Hoegemeyer, who was then a student at the university, brought home two bags of single-cross seed, one each of the male and the female single-cross, enough to plant 11 acres of production. Leonard went on to get a master's degree in corn breeding under Dr. Lloyd Tatum at Kansas State, and worked on a Ph.D. program at the University of Missouri until WWII intervened. After the war, Leonard returned to the family farm, began breeding corn at Hooper, and expanded the seed business from a neighborhood to a regional enterprise. Tom Hoegemeyer, after completing a Ph.D. at Iowa State in 1974, returned to the family business, and expanded corn breeding and testing. Hoegemeyer Hybrids became one of the few family seed companies that developed a major share of their hybrids as the industry matured. Several inbreds developed at Hooper were licensed to the industry. In 2004, two of Leonard's grandsons, Stephan Becerra and Chris Hoegemeyer, took over management of the company, which has further expanded to serve farmers in the Western Corn Belt (Tom Hoegemeyer, pers. comm.).

Mr. R.R. St. John and C.L. Gunn for DEKALB, Lester Pfister for his company, Dr. Holbert for Funks Bros., and Raymond Baker and Perry Collins for Pioneer Hi-Bred not only developed inbreds of their own but also took the federal and state inbreds and greatly improved them by backcrossing and by recurrent selection (Wallace 1955).

Iowa grew 100%, U.S. Corn Belt grew 90%, and the U. S. overall grew 64% hybrid corn in 1946 (Wallace and Bressman 1949). The U.S. Corn Belt grew 99% hybrid corn in 1950.

8   1950

Walter Vandeventer of Pioneer Hi-Bred at Princeton, Indiana began work on Corn Belt Female Composite in 1951. He grew 22 proprietary inbreds used in female single-cross parents of Pioneer Hi-Bred commercial double crosses and 46 mostly S3 inbreds from the 22 inbreds in an isolated plot. Some entries were repeated up to eight times based on favor at the time. The background of the materials was predominantly Reid Yellow Dent, but five other varieties were represented. In 1952, each 1951 entry was bulked in equal amounts and planted in a 0.2 ha isolated plot. Vandeventer selected plants for favorable agronomic traits. Selected ears were shelled, selected for kernel appearance, and for cold test germination. In 1953, the selected ears from 1952 were yield tested per se, and remnant seed from the best performing ear-entries was used for inbred development by Vandeventer. About 3% of the documented background of current U.S. hybrid corn is derived from Corn Belt Female Composite (Troyer 2004b, Fig. 2).

Henry Wallace (1955) pointed out that the heart of a private company's corn improvement is the yield test that compares the producing power of new hybrids year after year under a variety of conditions. He stated that no experiment station begins to do the volume of careful yield testing that is done by some of the larger seed companies. Private companies at that time were spending more than $2 million (1955 dollars) annually in corn breeding and testing. The state Agricultural Experiment Stations probably spent less than half as much. The Federal Government in 1955 dedicated about $300 thousand on corn improvement, of which about $80 thousand was spent on basic research. Wallace added that anyone who breeds corn for a hybrid seed company should be able to produce some kind of practical result within 15 years (Wallace 1955). With present data-driven and turn-around winter nurseries, this time for evaluation can be cut to about 5 years (author).

Dr. I.J. Johnson (1957) responded to Wallace's comments on private companies versus public spending for seed corn research. Johnson stated that having both types of research was healthy, and that comparable 2:1 spending ratios in agricultural chemicals and the milling industry existed. He added that nearly all the basic research leading to the development of hybrid corn originated at public supported institutions. The USDA laid the background upon which this modern method of corn breeding reached the status of commercial application. The importance of the contributions of public stations in producing inbred lines should not be underestimated. He concluded that the public stations should not be placed in a position of contributing to the success or failure of companies without a breeding program

Mr. Ralph R. St. John developed DEKALB 805, the first popular (600 thousand bags of seed sold) single-cross hybrid in 1958. Dr. E.H. Rinke developed Minhybrid 4201 (perhaps 10 million bags) first commercially grown in 1967, and Glenn H. Stringfield and Henry Slade developed DEKALB XL45, the first popular, early single cross (8.5 million bags) in 1963. Corn row width narrowed and plant densities increased. Farmers increased nitrogen application to corn from 50 to 120 kg per ha from 1965 to 1976. Farmers narrowed row widths about 25% (102 to 76 cm) in the central U.S. Corn Belt. Corn combines provided harvest-time yield results (faster feedback). Tax credits helped farmers buy larger machinery for more timely operations (Troyer 2004c).

9   1980

Sprague (1980) addressed the changing role of the private and public sectors of corn breeding. He gave reasons why public hybrids were at a disadvantage in the market place. The same hybrid could be purchased from any one of several producers–no built-in repeat sales existed. Different producers had sizable price differentials for the same hybrid. The reputation of the hybrid was influenced by the poorest producer. He stated that all of the important developments in the theory and practice of corn breeding had come from the public sector: Jenkins and Brunson developed the top-cross test (Jenkins and Brunson 1932). Jenkins developed the double-cross prediction method (Jenkins 1934). Jenkins demonstrated the stability of inbred lines over generations of self-pollination (Jenkins 1935). Major efforts on disease and insect resistance, specialty types of corn, and quantitative genetic theory and practice efforts all came from the public sector. Major efforts in population improvement and use of exotic germplasm were also from the public sector. Dr. Sprague lamented the fact that funding for public plant breeding research was drying up because of the perception that all necessary plant breeding was being done by the private sector.

Sprague's message concluding that basic research would receive more and more of the public funding was heartfelt, and it was taken to heart by public corn breeders. They shifted their efforts away from inbred line development. Some changed with a sigh of relief because corn breeders are judged by the performance of their inbreds and hybrids. This has had a long term negative effect on better inbred development. As this is written, much concern exists about an impending great shortage of plant breeders. Without breeders' expertise, molecular-genetic approaches may never bear fruit. (Knight 2003, Fehr 2007). This has happened in corn because of the lack of role models in public inbred line development and because of a smaller U.S. farm population. In Illinois, only 1% of the human population presently lives on farms. Farm boys and girls learn to enjoy working outside with living plants during the growing season; like their parent role models. Plenty of students exist. They line up and wait in turn for medical school partly for future financial reward. An obvious answer to the shortage of plant breeders is to increase their salaries.

The Dow Chemical Company (Dow) entered the seed corn business with the purchase of United AgriSeeds, Inc in 1988. This leveraged Dow's agricultural biotechnology assets and diversified its portfolio beyond industrial and agricultural chemicals. United AgriSeeds owned Keltgen Seeds and Lynks Seeds. United AgriSeeds was rolled into Dow's agricultural business that became DowElanco in 1989 in a joint venture with Eli Lilly, and later became Dow AgroSciences when Dow bought Eli Lilly's interest in 2002. Mycogen Corporation, a maker of bioinsecticides, entered the seed corn business when it acquired Agrigenetics and its seven seed brands, AgriGene, Golden Acres, Growers, Jacques, McCurdy, ORO Hybrids and SIGCO, from Lubrizol Corporation in 1992. DowElanco acquired 46% of Mycogen and United AgriSeeds then became part of Mycogen in 1996. This strengthened Dow's seeds and agricultural biotechnology position. Mycogen expanded into South America acquiring Morgan Seeds of Argentina and Dinamilho, Colorado, HATA, and FT companies of Brazil in 1998.

Dow AgroSciences acquired complete ownership of Mycogen and added a corn breeding program in the Philippines in 1998. It acquired Cargill's U.S. and Canada seeds businesses and Zeneca seed in Brazil in 2000 and added a corn testing program in southern France in 2001. It added a corn breeding program in Mexico in 2006. Dow AgroSciences expanded its European seeds business by acquiring the corn assets of Maize Technologies International (MTI) and Duo Maize and further expanded its Brazilian business by acquiring Agromen Technologia in 2007 (D. Blackburn, pers. comm.).

10   1996

Monsanto entered the seed corn business by purchasing 40% of DEKALB Genetics in 1996. Monsanto purchased Holden's Foundation Seeds, Corn States Hybrid Service, and Asgrow Seeds in 1997. The foundation seed purchases complemented Monsanto's approach to widely licensing biotechnology traits. Monsanto purchased Cargill's international seed business, Agroceres in Brazil, and the complete purchase of DEKALB Genetics in 1998. (Prior to Sept.1, 2000, Monsanto was the agricultural business of Pharmacia Corp.) In total Monsanto assembled 36 germplasm bases into one unique global germplasm base. The introgression of YieldGard® Corn Borer, YieldGard® Rootworm, and Roundup Ready® Corn2 have further differentiated the global germplasm base. Additional investments in novel, automated seed analyses have enhanced screening and testing capabilities. Thus, Monsanto develops many more products than any other individual company. Today, Monsanto licenses germplasm and traits to more than 200 U.S. seed companies.

To better serve the seed industry, Monsanto formed American Seeds Inc. (ASI) in 2004. Channel Bio Corp. was followed by about 25 businesses to join ASI. Thus, Monsanto has three channels to market: The DEKALB® brand primarily distributes through the ag retail channel; ASI company brands primarily distribute direct to the grower; and 200+ licensee customers independently choose their own route to market. Monsanto's seed corn strategy claims to provide maximum genetic diversity offerings with industry-leading performance, ensuring all growers have access to the technology they need in the seed brands they want. Future traits under development, including drought tolerance and improved nitrogen use efficiency, are expected to help meet the growing grain demand for food, feed, and fuel (R. Fraley and C. Horner, pers. comm.).

DuPont purchased a 20 percent interest in Pioneer Hi-Bred on Sept. 18, 1997, for $1.7 billion, and embarked on a joint venture research alliance called Optimum Quality Grains, LLC. Two years of promising developments in new corn hybrids and soybean varieties led DuPont to buy the remaining 80 percent of Pioneer on Oct. 1, 1999, for $7.7 billion. Pioneer Hi-Bred, a DuPont business, is the world's leading source of customized solutions for farmers, livestock producers and grain and oilseed processors. With headquarters in Des Moines, Iowa, Pioneer provides access to advanced plant genetics in nearly 70 countries. DuPont is a science-based products and services company. Founded in 1802, DuPont puts science to work by creating sustainable solutions essential to a better, safer, healthier life for people everywhere. Operating in more than 70 countries, DuPont offers a wide range of innovative products and services for markets including agriculture and food; building and construction; communications; and transportation (D. Oestreich pers. comm).

BASF initiated seed research in 1998 with the formation of BASF Plant Science, specializing in three areas: (1) genetically enhanced crops that can better survive pests, disease and drought; (2) plants that contribute to healthier human and animal nutrition through optimized composition; and (3) plants that act as efficient "green factories" to produce valuable, complex molecules, thus reducing cost and environmental impact of select, traditional chemical processes. BASF Plant Science purchased ExSeed Genetics in 2001. ExSeed markets NutriDense® seed corn, a nutritionally-enhanced product for swine, poultry and dairy farmers, through seed partners. BASF also markets elite inbreds through Thurston Genetics, based in Olivia, MN. ExSeed and Thurston Genetics had merged in 1998, prior to the BASF acquisition. In 2001, BASF Plant Science initiated a corn breeding program near Sycamore, Illinois. The breeding group expanded within the next few years and now includes six research stations plus an all-season nursery in Hawaii, launched in 2006 (D. Gross, pers. comm.).

AgReliant Genetics is a joint venture of Groupe Limagrain and KWS SAAT AG that was formed on July 1, 2000. The formation of AgReliant was the most recent step in the business strategy of these two major European Seed companies in North America that began in the early 1980's. Both companies appreciated that, if they were to compete globally in the seed corn market, they must have a presence in the North

American market. Both Limagrain and KWS began by establishing a network of breeding stations and later adding commercial operations. Major acquisitions by Groupe Limagrain included AgriGold Hybrids (of St. Francisville, Illinois), BioTechnica International (which is today LG Seeds of Elmwood, Illinois), and Pride Seeds (of Chatham, Ontario, Canada). The major acquisition by KWS was Great Lakes Hybrids (of Ovid, Michigan). By the late 1990's both Limagrain and KWS agreed that the merger of their operations in North America would result in a sufficient critical mass to better compete and achieve operational synergies. Subsequent to the formation of AgReliant, the company acquired Wensman Seed Company (of Wadena, Minnesota in 2000) and Producers Hybrids (of Battle Creek, Nebraska in 2005). By pursuing a "multi-brand strategy" coupled with both organic growth and strategic acquisitions, AgReliant Genetics has grown to be the fourth largest corn and soybean seed company in North America (R. Journel, pers. comm.).

Syngenta was formed when Novartis and AstraZeneca merged in 2000 into today's largest global-company focused exclusively on agribusiness. Corn is central to this vision in developing leading corn genetics, crop protection chemicals and seed treatment products. Syngenta acquired Garst, Golden Harvest, and germplasm from CHS in 2004, and partnered in GreenLeaf Genetics in 2006. Syngenta biotech traits are branded as Agrisure. Novartis Seeds was created from the merger of Ciba Seeds and Northrup King Co., seed divisions of Ciba-Geigy and Sandoz AG, respectively, in 1997. Syngenta Seeds' legacy companies have a rich history of innovation: Funk Seeds originated Funks Yellow Dent in 1902 and C cytoplasm use. Ciba Seeds and Northrup King first developed and marketed Bt corn, the initial corn biotech product. Northrup King, since 1884, was first to develop a mail order seed business. NK460Bt was product of the year (1998) in North America. At Garst Seed Co., founded in 1930, Roswell Garst established the farmer dealer sales force and championed nitrogen fertilizer. Syngenta has a history of innovation in their acquired companies that continues to this day.

Golden Harvest's legacy companies were also innovators: The Garwood brothers were first to use a grain combine east of the Mississippi in 1924. Ted Sommer started his business in 1909 and was the first certified seed producer in Illinois. J.C. Robinson started his business in 1888 and was president of ASTA in 1909. Claude Thorp used a tractor when only a handful existed in Illinois before WW1 and experimented with hybrid corn in 1927. Claire Golden founded his business in 1920 and was a pioneer in selling hybrid corn in 1928.

Alas, hybrid seed corn companies are decreasing in number and increasing in size. High school age detasselers now get their check in the mail instead of the back door of the farm house. It's still a good summer job; the camaraderie and feeling of accomplishment remain. About 500 individuals each produced and sold hybrid seed corn in Iowa in 1940, decreasing to about 100 by 1957 (Johnson 1957), to 40 companies headquartered in Iowa in 1990, to 28 in 1997 (Troyer 2000), and to 10 companies approved for conditioning seed corn in Iowa in 2006 (Anon., 2006). Anderson (2007) estimated that approximately 70% of U.S. hybrid seed corn sales are controlled by four major seed companies. Seabrook (2007) stated that 55% of the world's seeds that are used to grow the world's food are sold by just ten global corporations. My lifelong experience in the hybrid seed corn business is that companies grow larger by better satisfying customers.

11 Summary

Darwin's views on cross-fertilization increasing vigor led to the development of hybrid corn (Darwin 1875, Troyer 2006a). Darwin's principles of natural and artificial (human) selection for adaptation during open-pollinated corn's growth in the U.S. Corn Belt followed by the hybrid seed companies' goal to provide more widely adapted hybrids (more weather-proof, more dependable), readily explain the preeminence (ca 73%) of Reid Yellow Dent, Lancaster Sure Crop and Minnesota 13 in the documented background of U.S. hybrid corn (Darwin 1859, 1868; Troyer 2004b,).

In the 1930s, when corn hybrids were first commercially grown, annual U.S. yields averaged 1518 kg per ha (24.2 bushels per acre), and corn production averaged 51 million Mg (2 billion bushels). Corn production grew to 76 million Mg (3 billion bushels) per year in the 1950s, to 150 million Mg (6 billion bushels) per year in the 1970s, to 219 million Mg (8.6 billion bushels) in the 1990s and to more than 254 million Mg (10 billion bushels) per year for the past five years. The final, 2007 U.S. corn estimate was 331 million Mg (13.1 billion bushels) from 34,582 ha (85,418 acres) for grain (USDA/NASS 2007). In 2001, corn became the highest tonnage crop worldwide: 557.6 million Mg of maize compared to 542.4 Mg of rice (*Oryza sativa* L.), and 535.6 Mg of wheat (*Triticum aestivum* L.) (UN/FAO 2002). It took more than 500 years for corn the New World cereal to surpass rice and wheat, the Old World cereals. In 2007, U.S. corn production was a record 331 million Mg (13.1 billion bushels) and the average yield in 2004 was estimated at a record 10,059 kg per ha (160.4 bushels per acre) (USDA/NASS 2004, 2007). These increases were caused by better hybrids, improved cultural practices and biotechnology (Fig. 1).

Is corn the answer to the biofuel quest? It's good that we are doing something substantive about the increasing price of oil. If corn doesn't work, we'll try something else. Corn's advantage is that it has experienced natural and human selection for increased yield for several thousand years. Corn will probably produce more renewable energy on less surface area than any other plant grown in the USA. This advantage is also its disadvantage because corn's efficiency has increased its use. A rise in the price of corn immediately signals a rise in the cost of feed and food. We need to look at the biofuel quest as an opportunity–a big opportunity.

References

Anderson, E. 1944. The sources of effective germplasm in hybrid maize. Ann. Mo. Bot. Gard. 31:355–361.
Anderson, R.M. 2007. A lesson learned. Seed World. June p.22–26. Grand Forks, ND.

Anonymous. 1894. Portrait and biographical record of Tazewell and Mason counties. Chicago Biographical Publishing Co. Chicago, IL.

Anonymous. 1930. James Reid. J. Heredity 21:402.

Anonymous. 1995. Dairyland Seed, 95 years of success. Dairyland Seed. West Bend, WI.

Anonymous. 2006. Iowa Seed Directory–for seed grown in 2006. Iowa Crop Improvement Association. Ames, IA.

Atkinson, A., and M.L. Wilson. 1915. Corn in Montana. Montana AES Bull. 107.

Beal, W.J. 1876. Rep. Mich. Board Agr. p, 212–213. Mich. Agr. Col. East Lansing, MI.

Beal, W.J. 1881. Rep. Mich. Board Agr. p, 98–153. Mich. Agr. Col. East Lansing, MI.

Boss, A. 1929. Willet Martin Hays. J. Heredity 20:496–509.

Brackman, A.C. 1980. A delicate arrangement. Times Books. Fitzhenry & Whiteside Ltd. Toronto.

Cavanagh, H.M. 1959. Seed, Soil, and Science. The Lakeside Press. Chicago.

Cornell, C. 1958. 1893–1958 Historical notes and anecdotes of seventy-five years. American Seed Trade Association. Washington, DC.

Crabb, A.R. 1948. The hybrid-corn makers–prophets of plenty. Rutgers Univ. Press. New Brunswick, NJ.

Crissey, F. 1920. James Reid, master of corn. Country Gentleman 85 (38). Sept 18.

Cunningham, J.O. 1905. History of Champaign County. Urbana Free Library. Urbana, IL

Darwin, C.R. 1859. On the origin of species by means of natural selection, or the preservation of favoured races in the struggle for life. John Murray. London.

Darwin, C.R. 1868. The variation of animals and plants under domestication. John Murray. London.

Darwin, C.R.1875. The effects of self- and cross-fertilization in the vegetable kingdom. John Murray. London.

De Beer, G. 1965. Charles Darwin: a Scientific Biography. Anchor Books Doubleday & Co. Inc. Garden City, NY.

DeKruif, P. 1928. Hunger fighters. Harcourt Brace. Rahway, NJ.

DeVries, H. 1907. Plant breeding. Open Court Publishing Co. Chicago.

Dobzhansky, Th. 1947. Adaptive changes induced by natural selection in Drosophila. Evolution 1:1–16.

East, E.M. and D.F. Jones. 1919. Inbreeding and outbreeding. J.B. Lippencott Co. Philadelphia and London.

Encyclopedia Britannica. 1983. Charles Darwin. 15th edition. Encyclopedia Britannica, Inc. Chicago.

Everett, E.E. 1935. James Reid. Dictionary of American Biography. Vol. 15:477–478 Chicago Biographical Publishing Co.

Fehr, S.2007. An endangered species? Seed World. May p.4–6. Issues Ink. Grand Forks, ND.

Franklin, B. 1976. Poor Richard: The almanacks for the years 1733–1758. Paddington Press LTD. New York and London.

Freeman, R.B.1978. Charles Darwin, a companion. Dawson Archon. W&J Mackay Ltd. Chatham, England.

Hallauer, A.R., W.A. Russell, and O.S. Smith. 1983. Quantitative analysis of Iowa Stiff Stalk Synthetic. p.359–360. In J.P. Gustafsen (ed). Proc. 15th Stadler Genetics Symposium. Washington University, St. Louis, MO. June 12–16. University of Missouri, AES. Columbia, MO.

Hayes, H.K. 1913. Report of the plant breeder. Conn. AES Rep. No.37..

Hayes, H.K. 1956. I saw hybrid corn develop. p.48–81. In Dolores Wilkinson (ed.) Eleventh hybrid corn industry-research conference. 28 29 Nov. Hyatt Regency Hotel Chicago, IL. Am. Seed Trade Assoc. 601 13th St. NW. Suite 570. Washington, DC.

Hershey, N.L. 1989. Descendents of John Eby Hershey and Anna Mellinger Hershey. Sutter House. Lititz, PA.

Holden, P.G. 1944. Records of the Holden, Wilson, and other related families. Belleville Publ. Belleville, MI.

Holden, P.G. 1948. Corn breeding at the University of Illinois 1895–1900. Archives. Michigan State Univ. East Lansing, MI.

Hopkins, C.G. 1903. Methods of maintaining the productive capacity of Illinois soils. Illinois AES Circ. 68.

Hopkins, C.G. 1913. The Illinois system of permanent fertility. Illinois AES Circ. 167.

Jenkins, M.T. 1934. Methods of estimating the performance of double-crosses in corn. Agron. J.26:199–204.

Jenkins, M.T. 1935. The effect of inbreeding within inbred lines of maize upon the hybrids made after successive generations of selfing. Iowa State College J. Science 9:429–450.

Jenkins, M.T. 1936. Corn Improvement. In E.S. Bressman (ed) Yearbook of Agriculture 1936. USDA. Washington, DC.

Jenkins, M.T. and A.M. Brunson. 1932.Methods of testing inbred lines of maize in crossbred combinations. Agron. J. 24:523–530.

Johnson, I.J. 1957. The role of the experiment stations in basic research relating to corn breeding. p.31–36. In: W. Heckendorn and B.H. Blankenship, Jr. (ed.) Twelfth hybrid corn industry research conference. 4 Dec. and 5 Dec. LaSalle Hotel Chicago, IL. Am. Seed Trade Assoc. Washington, DC.

Jones, D.F. 1917. Dominance of linked factors as a means of accounting for heterosis. Genetics 2: 466–497.

Jones, D.F. 1927. Double crossed Burr-Leaming seed corn. Conn. Ext. Bull. 108.

Kraft, K. and P. Kraft. 1967. Luther Burbank the wizard and the man. Meredith Press. New York, NY.

Knight, J. 2003. A dying breed. Nature 421:568–570.

Lang, A.L. 1972. 50 years of service–A history of seed certification in Illinois. Illinois Crop Improvement Association. Urbana, IL.

McCluer, G.W. 1892. Corn crossing. Illinois AES Bull. 21:83–101.

McFee, B. and H. Ohm. Small grain breeding at Purdue University. Purdue Agronomy–Alumni, Friends, News. Spring 2007 p.8, 9.

Montgomery, E.G. 1916. The Corn Crops. Macmillan. New York, NY.

Moores, R.G. 1970. Fields of rich toil. Univ. of Illinois Press. Urbana, IL.

Morrow, G.E. and F.D. Gardner. 1893. Field experiments with corn, 1892. Illinois AES Bull. 25:179–180.

Morrow, G.E. and F.D. Gardner. 1894. Field experiments with corn, 1893. Illinois AES Bull. 31: 359–360.

Morrow, G.E. and T.F. Hunt. 1889. Field experiments with corn, 1888. Illinois AES Bull. 4: 48–67.

Mosher, M.L. 1962. Early Iowa corn yield tests and related programs. Iowa State Univ. Press. Ames IA.

Olson, P.J., H.L. Walster, and T.H. Hopper. 1928. Corn for North Dakota. North Dakota Agric. Exp. Sta. Bull.207: 1–106.

Reid, O.G. 1915. One great accomplishment in corn breeding.. Breed. Gaz. 67:383–384.

Richey, F.D. 1945. Isolating better foundation inbreds for use in corn hybrids. Genetics 30:455–471.

Roberts, T.H. Jr. 1999. The story of the DEKALB "Ag." Carlith Printing, Inc. Carpentersville, IL.

Seabrook, J. 2007. Sowing for apocalypse. The New Yorker. Aug.27:61–71.

Shamel, A.D. 1901. Seed corn and some standard varieties for Illinois. Illinois AES. Bul.63:29–56.

Shamel, A.D. 1907. The art of seed selection and breeding. In G.W. Hill (ed). Yearbook of Agriculture, 1907. USDA. Washington, DC.

Shoesmith, V.M. 1910. The study of corn. Orange Judd. New York, NY.

Shull, G.H. 1908. The composition of a field of maize. Amer. Breed. Assoc. Rpt. 4:296–301.

Shull, G.H. 1909. A pure line method of corn breeding. Amer. Breed. Assoc. Rpt. 5:51–59

Shull, G.H. 1910. Hybridization methods in corn breeding. Amer. Breeders Mag. 1:98–107.

Shull, G.H. 1911. The genotypes of maize. Amer. Nat. 45:234–252.

Shull, G.H. 1952. Beginnings of the heterosis concept. p. 14–48 *In* J.W. Gowen (ed.) Heterosis. Iowa State College Press. Ames, IA.

Sizer, R. And W. Silag. 1981. Holden and the corn gospel trains. Palimpset 62 (3):66–71. Iowa State Cultural Affairs. Des Moines, IA.

Sprague, G.F. 1980. The changing role of the private and public sectors in corn breeding. p.1–9. *In* H. Loden and D. Wilkinson (ed.) Thirty-fifth annual corn and sorghum research conference. 9–11 Dec. Hyatt Regency Hotel Chicago, IL. Am. Seed Trade Assoc.1030 15th St., N.W.–Suite 964. Washington, DC.

Sturtevant, E. L. 1899. Varieties of corn. USDA Bul. 57. GPO. Washington, DC.

Troyer, A.F. 1990. A retrospective view of corn genetic resources. Jour. Heredity 17–24.

Troyer, A.F. 1999. Background of U.S. hybrid corn. Crop Sci.39:601–626.

Troyer, A.F. 2000. Temperate corn: background, behavior, and breeding. *In* Arnel Hallauer (ed). Specialty Corns, CRC Press. Boca Raton, London, New York, Washington, DC.

Troyer, A F. 2004a. Champaign County, Illinois, and the origin of hybrid corn. PBR 24:42–59.

Troyer, A.F. 2004b. Background of U.S. hybrid corn II: Breeding, climate, and food.. Crop Sci.44:370–380.

Troyer, A.F. 2004c. Persistent and popular germplasm in seventy centuries of corn evolution. p. 133–232. *In* C.W. Smith, J. Betran, and E.C.A. Runge (ed.). Corn; origin, history, technology, and production. John Wiley & Sons. Inc. Hoboken, NJ.

Troyer, A.F. 2006a. Adaptedness and heterosis in corn and mule hybrids. Crop Sci. 46:528–543.

Troyer, A.F. 2006b. Background and importance of Troyer Reid Corn. Crop Sci. 46:2460–2467.

Troyer, A.F. 2007. Background and importance of Minnesota 13 Corn. Crop Sci. 47:905–914.

Troyer Bros. 1916. Prize winning seed corn catalogue. Troyer Memorial Library. LaFontaine, IN.

Troyer, C.E. 1931. Origin and development of some modern corn cultivars. Ann. Rpt. 31. Ind. Corn Grow. Assoc. Purdue University. W.Lafayette, IN.

UN/FAO. 2002. United Nations. Food and Agricultural Organization. Dept. of Statistics. Rome, Italy.

USDA/NASS. 2006. United States Department of Agriculture. National. Agricultural Statistical Service. 1400 Independence Ave. S.W. Washington, DC.

Wallace, H.A. 1923. Burnett's Iodent. Wallaces Farmer.Vol.46, No.6. Des Moines, IA..

Wallace, H.A. 1955. Public and private contributions to hybrid corn—past and future. p. 107–115. *In*: W. Heckendorn and J. Gregory (ed.) Tenth hybrid corn industry research conference. 30 Nov. and 1 Dec. LaSalle Hotel Chicago, IL. Am. Seed Trade Assoc. Washington, DC.

Wallace, H.A. and W.L. Brown.1956. Corn and its early Fathers. The Michigan State University Press. The Lakeside Press, Chicago.

Wallace, H.A. and E.N. Bressman. 1923. Corn and corn growing. Wallace Publishing Co. DesMoines, IA.

Wallace, H.A. and E.N. Bressman. 1949. Corn and corn growing 5th Edition revised by J.J. Newlin, Edgar Anderson, and E.N. Bressman. John Wiley & Sons, Inc. New York.

Maize and the Biotech Industry

G. Richard Johnson and Zoe P. McCuddin

Abstract The story of maize and the biotech industry is predominantly a tale of technological innovations and scientific discoveries that have led to products which have added significant value to the crop and boosted farm income. The primary impact of biotechnology on maize has been through transformation. The initial development of plant transformation was enabled by the suitability of the Ti plasmid of *Agrobacterium tumefaciens* to serve as a natural transformation vector that could be engineered to introduce novel genes into dicots. Since maize is not naturally infected by *Agrobacterium tumefaciens*, transformation of maize was not accomplished until the invention of biolistic transformation, though with the later invention of super-binary vectors, efficient Agro-mediated transformation of maize was achieved. Advances in tissue culture and plant regeneration also played a significant role in enabling development of the maize biotech industry. Molecular markers have been enlisted to accelerate progress in conventional corn breeding. Herbicide tolerance derived from several sources and insect resistance through modification of gene constructs coding for *Bacillus thuringiensis* insecticidal proteins comprise the current product base of the industry. RNA interference technology is being harnessed and applied to gene regulation and insect pest control. Development of functional artificial chromosomes promises to extend the power of biotechnology in the modification of plants. Many patent applications have been filed and granted for a wide array of additional technological innovations and products, many of which are likely to be soon commercialized.

G.R. Johnson
University of Illinois, Department of Crop Sciences,
grjohnso@uiuc.edu

Z.P. McCuddin
Monsanto Company,
zoe.p.mccuddin@monsanto.com

J.L. Bennetzen and S. Hake (eds.), *Maize Handbook - Volume II: Genetics and Genomics,*
© Springer Science+Business Media LLC 2009

1   Introduction

In this chapter we shall attempt to cover the significant research and discoveries that led to the creation and further development of the crop plant biotech industry, highlighting as we do the role of maize. Biotechnology can be very broadly defined. The 4th edition of Webster's New World College Dictionary defines it as "the use of data and techniques of engineering and technology for the study and solutions of problems concerning living organisms." An expansive view of plant biotechnology relevant to maize genetics might include long-established disciplines such as Mendelian genetics and plant breeding. However, we will confine ourselves to discoveries, inventions, and technologies that have arisen in the past sixty years that have provided new tools for the heritable improvement of the value of maize to the benefit of society.

The history of the development of the plant biotech industry is largely a story of transformation and the regeneration of whole plants. The conception of the biotech industry came with the discovery that the causal agent of crown gall disease was a plasmid from *Agrobacterium tumefaciens* that integrated into the cells of the host plant, and with the realization that the plasmid could serve as a natural vector for transformation (Chilton, Drummond, Merlo, Sciaky, Montoya, Gordon, and Nester 1977; Matzke and Chilton 1981). Since monocotyledonous plants are not naturally infected by *A. tumefaciens*, efforts to transform the cereal grasses such as maize (*Zea mays*), rice (*Oryza sativa*), and wheat (*Triticum aestivum*) lagged until the invention in the mid-1980s of the 'gene gun' which facilitated direct transformation by means of microprojectile bombardment (Sanford, Wolf, and Allen 1990). With the advent of microprojectile bombardment, achievement of fertile, stably transformed maize plants followed quickly (Gordon-Kamm, Spencer, Mangano, Adams, Daines, Start, O'Brien, Chambers, Adams, Willets, Rice, Mackey, Krueger, Kausch, and Lemaux 1990; Fromm, Morrish, Armstrong, Williams, Thomas, and Klein 1990).

Intertwined with the story of transformation is the story of tissue culture and plant regeneration. Once cell cultures were transformed, there still remained the task of regenerating fertile plants, without which, transformation by itself would come to naught. Consequently, though transformation and tissue culture will be discussed in separate sections within the chapter, research and progress in both advanced hand-in-hand (Armstrong 1999). Transformation, tissue culture, and plant regeneration are the enabling technologies for the biotech industry. However, the new traits introduced via transformation were the elements that give added value to the crop by reducing reliance on chemical crop protection, increasing yield, and, more recently, improving nutritional quality. Hence, a section devoted to traits is included.

Independently of efforts in transformation and tissue culture, research on the application of molecular markers to plant breeding also arose with the consequence that marker-aided selection is now a major component of plant improvement within the biotech industry (Eathington, Crosbie, Edwards, Reiter, and Bull 2007; Graham 2007). Recently, the development of artificial chromosomes (Preuss, Copenhaver, and Keith 2005; Copenhaver, Keith, and Preuss 2007) promises to provide an entirely new technical level of genetic engineering for the biotech industry.

Likewise, new technologies such as RNA interference will provide new tools for control of gene expression and modification of plants.

Since hybrid corn is a high value cereal crop, successful transformation of maize assured the future of the plant biotech industry by enabling the development of highly beneficial products for, farmers and consumers. The price today in the US of a commercial conventional F1 hybrid is under $100 per unit while the same hybrid fortified with transgenes for corn borer, rootworm, and herbicide tolerance costs around $200 per unit (Hillyer 2005). This price differential suggests that the conventional hybrid corn industry has been transformed into a biotech industry and reflects the added benefit of transgenic corn to the farmer in terms of yield gain coupled with the reduced reliance on pesticides and herbicides. According to Brookes and Barfoot, the costs in 2005 of accessing transgenic insect resistance and herbicide tolerance relative to total farm income benefits derived were 44% and 38%, respectively (Brookes and Barfoot 2006). This means that if a $100 premium is paid per unit for the transgenic enhancements, the farmer grosses a profit of about $244 for a $144 net gain per unit purchased. The gain comes not only from increased yield but also in significant reductions in costs for weed and insect control. For example, Brookes and Barfoot estimated that, by reducing costs to farmer for weed management, average profitability for herbicide tolerant corn was approximately $25/hectare in 2005 in the US (Brookes et al. 2006). Globally, the farm income benefits in 2005 for herbicide tolerant and insect resistant corn was $212 million and $416 million, respectively (Brookes et al. 2006). In 2005, over 50% of the US corn acres were planted with transgenic hybrids, accompanied by a 10.8% decrease in insecticide application and an 8% reduction in herbicide use (Brookes et al. 2006).

In 2006, 41% of all genetically modified products involved maize herbicide tolerance or insect resistance traits (McDougall and Phillips 2007). Many of the products were comprised of multiple trait combinations (stacks). In 1997, 30.6 million acres of transgenic crops were planted globally. By 2005, the figure had reached 249.1 million total acres, with maize accounting for 67.5 million acres (McDougall et al. 2007). The past ten years have seen a corresponding decrease in conventional protection of crops against insect damage and competition from weeds. The increased use of transgenic products has been accompanied by consolidation of the biotech seed industry, blending plant breeding and biotech expertise into single functioning units. For example, the principally biotech corporations DuPont, Dow, Syngenta, and Monsanto have acquired and integrated large seed companies such as Pioneer Hi-Bred, Mycogen, Advanta, and DeKalb, respectively. Each applies a coordinated program of plant breeding and biotechnology to increase crop value.

2   Transformation

In 1947, Armin Braun achieved successful cultivation of crown gall tumors in periwinkle, free of the infecting agent, *Agrobacterium tumefaciens* (Braun 1947). Braun concluded that the normal cells of the host plant had somehow been transformed into tumor cells, undergoing uncontrolled growth, through introduction of a tumor inducing principle by the bacterium. Braun suggested that, among other possibilities, the tumor inducing principle might be DNA. Later, Braun established that the growth of tumor cell lines could be perpetuated free of the bacterium on media incapable of supporting normal cell lines. The implication of this discovery was that the normal cells had been transformed so as to produce growth substances necessary for the growth of tumors (Braun 1958). Georges Morel and colleagues subsequently discovered octopines and nopalines that were consumed exclusively by specific tumor-inducing *Agrobacteium* strains. The research team suggested that genetic information for the production of the specific opines by the tumor was transferred from the specific bacterial strain counterparts to the plant cell during tumorigenesis by a mechanism similar to bacterial transformation (Petit, Delhaye, Tempe, and Morel 1970). Further research led Zaenen et al. to propose that the tumor inducing principle, or transforming agent, was carried by one or more large plasmids in *A. tumefaciens* cells (Zaenen, Van Larebeke, Teuchy, Van Montagu, and Schell 1974).

The road to plant transformation was opened, and the conceptual basis of the plant biotech industry secured, when Mary-Dell Chilton and members of E.W. Nester's laboratory at the University of Washington demonstrated that stable integration of *A. tumefaciens* tumor inducing (Ti) plasmid DNA into host plant cells was the molecular basis of crown gall tumorigenesis (Chilton et al. 1977). This discovery immediately suggested the possibility that the naturally occurring Ti plasmid might be engineered to serve as a vehicle (or vector) for the introduction of novel gene constructs into plants. The Ti plasmid DNA (or T-DNA as it was called) was subsequently shown to be located in the nuclear fraction of the host plant DNA (Chilton, Saiki, Yadav, Gordon, and Quetier 1980; Willmitzer, DeBeuckeleer, Lemmers, Van Montague, and Schell 1980), indicating that an introduced gene construct could be heritably transmitted and sexually propagated. That *Agrobacterium*-mediated transformation for plant modification was feasible was suggested by Matzke and Chilton in a paper describing a site-specific insertion of a gene conferring resistance to the antibiotic gene kanamycin into the T-DNA of a Ti plasmid that became incorporated into the DNA of the transformed host cells (Matzke and Chilton 1980). In the paper, the authors outlined a procedure for introducing genes into higher plants using the Ti plasmid as a vector. In 1983, three research groups then reported successful transformation of higher plants with antibiotic resistance genes using the Ti plasmid as a vector (Bevan, Flavell, and Chilton 1983; Fraley, Rogers, Horsch, Sanders, Flick, Adams, Bittner, Brand, Fink, Fry, Galluppi, Goldberg, Hoffman, and Woo 1983; Herrara-Estrella, Depicker, Van Montague, and Schell 1983). At that time, it was noted that antibiotic resistance genes would be useful as selectable markers in future transformation experiments involving other genes.

The efficacy of *Agrobacterium*-mediated transformation was facilitated by two key developments. First was the construction of binary transformation vectors. The entire Ti plasmid was too large to be easily manipulated and cloned in *E. coli*. Hence, two plasmids were devised: one small containing just the portion of DNA actually transferred to the host (T-DNA), and another carrying the virulence genes necessary for infection (Hoekema, Hirsch, Hooykaas, and Schilperoort 1983).

The second key development was construction of plasmids in which all tumor inducing genes (oncogenes) had been removed permitting normal growth of the transformed plant cells (Zambryski, Joos, Genetello, Leemans, Van Montague, and Schell 1983). In their research, the authors also demonstrated that only the flanking repeat borders of the T-DNA are necessary for recombination of plasmid and host DNA and the stable integration of T-DNA into the host genome.

The *Agrobacterium*-mediated transformation and recovery of whole plants of petunia, tobacco, and tomato was achieved by Horsch et al. in a demonstration of the practicality of genetic engineering of higher plants (Horsch, Fry, Hoffmann, Eichholtz, Rogers, and Fraley 1985). In their paper, the authors described an integrated system of gene transfer, plant regeneration, and selection of transformants using selectable markers. A little earlier, direct (non-*Agrobacterium*-mediated) gene transfer into tobacco had been accomplished by Paszkowski et al. by incubating a cloned kanamycin resistance gene, under the control of a cauliflower mosaic virus (CaMV) promoter and terminator sequences, with totipotent protoplasts (Paszkowski, Shillito, Saul, Mandak, Hohn, Hohn, and Potrykus 1984). Whole plants were regenerated from transformed protoplasts. Shortly thereafter, Fromm et al. reported stable transformation of maize cells with a kanamycin resistance gene after electroporation of protoplast suspension cultures of Black Mexican sweet corn (Fromm, Taylor, and Walbot 1986). In 1984, Sanford et al. had submitted a patent for transformation via microprojectile bombardment (biolistic or gene gun transformation) (Sanford et al. 1990). Thus, by the mid-1980s three methods for transformation of higher plants were available: (1) *Agrobacterium*-mediated, (2) protoplast electroporation, and (3) microprojectile bombardment.

Rhodes et al. obtained maize plants regenerated from electroporated protoplast cultures co-cultivated with a plasmid vector containing a kanamycin resistance gene construct under the control of a CaMV promoter (Rhodes, Pierce, Mettler, Mascarenhas, and Detmer 1988). Mature regenerated transformed plants expressed the gene but were both male and female sterile. Cryptic mutations or epigenetic effects that reduced viability were thought to have occurred during time in suspension culture that compromised viability of the regenerated plants. Consequently, electroporation as a vehicle for transformation of maize was largely abandoned.

Though maize is not naturally infected by *A. tumefaciens,* Graves and Goldman presented results hinting that maize could be transformed using the bacterium (Graves and Goldman 1986). Later, Gould et al. reported Agrobacterium-mediated transformation of maize embryo and seedling ex-plants (Gould, Devey, Hasagawa, Ulian, Peterson, and Smith 1991). Though these results suggested that *Agrobacterium*-facilitated transformation of maize was possible, efficiency was poor. Consequently, transformation of maize by *A. tumefaciens* was not seriously considered until the development of the super binary vectors in the mid-1990s (Ishida, Saito, Ohta, Hiei, Komari, and Kumasharo 1996).

The practical breakthrough for maize transformation was provided by microprojectile bombardment. Klein et al. observed transient expression of a chloroamphenicol acetyltransferase (CAT) gene in intact cell suspension cultures of Black Mexican sweet (BMS) corn following microprojectile bombardment with tungsten beads coated with a plasmid construct containing the CAT gene that was driven by a CaMV promoter (Klein, Fromm, Weissinger, Tomes, Schaaf, Slatten, and Sanford 1988). The stable transformation of maize cells after particle bombardment of BMS suspension cultures was reported the following year (Klein, Kornstein, Sanford, and Fromm 1989). Then, in 1990, two groups reported regeneration of fertile transgenic plants following microprojectile bombardment of embryogenic maize suspension cultures (Gordon-Kamm et al. 1990; Fromm et al. 1990).

In one experiment, Gordon-Kamm et al. bombarded suspension cells derived from A188 type II callus with a construct containing the *Streptomyces hygrosopicus* bar gene that encodes the phosphinothricin acetyltransferase (PAT) enzyme (Gordon-Kamm et al. 1990). The PAT enzyme inactivates the herbicidal compound phosphinothricin (PPT) by acetylation. Calli from transformed cells of the suspension culture were selected using the PPT-containing herbicide bialaphos. All PAT-positive progeny of transformed plants contained the bar gene and were resistant to bialaphos, confirming that stable transformation had occurred. In another experiment by Gordon-Kamm et al., co-transformation with one plasmid containing the bar gene, and another containing the GUS gene, produced mature plants expressing both genes following selection of bialaphos-resistant calli and regeneration (Gordon-Kamm et al. 1990).

Fromm et al. obtained fertile transgenic maize plants expressing both a chlorsulferon-resistant form of the acetolactate synthase (ALS) gene and the firefly luciferase (LUC) gene following bombardment of an intact cell suspension culture with a construct containing the ALS and LUC genes (Fromm et al. 1990). Two cell lines produced transformed plants. One line was completely sterile. The other was male sterile but female fertile. Pollination of the female fertile line by a non-transformed plant resulted in progeny that were both male and female fertile. The progeny segregated 1:1 for luciferase activity. PCR analysis indicated that all luciferase expressing plants contained DNA of the transforming construct and were resistant to chlorsulferon, whereas all plants not expressing luciferase contained no construct DNA and were susceptible to chlorsulferon. These results suggested that the construct was integrated intact into the maize genome and inherited as a single unit. Walters et al. bombarded embryogenic callus tissue with a construct containing a hygromycin phosphotransferase (HPT) gene that confers resistance to hygromycin (Walters, Vetsch, Potts, and Lundquist 1992). Selection for hygromycin resistance followed by regeneration of whole plants produced three fertile transformed lines that contained one or more copies of the HPT gene. Transmission of the HPT gene was demonstrated through two generations. In one line, inheritance was consistent with that of a single dominant gene, suggesting a singe insertion event.

Compared to *Agrobacterium*-mediated transformation of dicots, maize biolistic transformation resulted in reduced transformation efficiency. Often, many transformation attempts needed to be screened before an adequate insertion event could be isolated. Nonetheless, biolistic transformation was the method of choice early-on. Though demanding in its application, through the hard work and persistence of its practitioners, biolistic transformation facilitated the incorporation of a number of very commercially important gene constructs into maize.

Other technologies for direct delivery transformation emerged during this period, including electroporation of plant tissue (D'Halluin, Bonne, Bossut, Beuckeleer, and Leemans 1992; Golovkin, Abraham, Morocz, Bottka, Feher, and Dudits 1993; Omirulleh, Ábrahám, Golovkin, Stefanov, Karabev, Mustárdy, Mórocz, and Dudits 1993; Krzyzek, Laursen, and Anderson. 1995) and "whisker"-mediated transformation which entailed silicon carbide fibers that mechanically disrupted cell walls (Kaeppler, Gu, Somers, Rines, and Cockburn 1990; Kaeppler, Somers, Rines, and Cockburn 1992). However, *Agrobacterium*-mediated transformation still held two attractive features. First, relatively large segments of DNA could be transferred with little rearrangement; and second, relatively few gene copies were inserted into the host plant chromosomes. Unfortunately, maize was not naturally infected with the bacterium. Thus the frequency with which transformed plants were obtained was quite low. Increased frequency of transformed plants was achieved by Ishida et al. through co-cultivation of immature embryos with *A. tumefaciens* super binary vectors containing extra copies of the virB, virC, and virG virulence genes (Ishida et al. 1996). In their experiments, the authors observed frequencies of transformation (independent transgenic plants/embryo) that ranged between 5% and 30%. Stable expression of transgenes was confirmed, and the morphology of nearly all transformants was normal.

The subsequent patenting of super binary vectors by Japan Tobacco, Inc. (Hiei and Komari 1997) imposed the added cost of royalty fees to users of the technology. Hence, efforts were directed at improvement of transformation efficiency using ordinary binary vectors. By including L-cysteine in the co-cultivation medium, Bronwyn et al. were able to demonstrate improved stable transformation efficiency with ordinary binary vectors (Bronwyn, Shou, Chikwamba, Zhang, Xiang, Fonger, Pegg, Li, Nettleton, Pei, and Wang 2002). Immature embryos were transformed with a bar selectable marker cassette and a GUS reporter gene cassette. The average stable transformation efficiency with L-cysteine co-cultivation was 5.3% while that for co-cultivation without L-cysteine was only 0.2%. Though addition of L-cysteine increased the overall rate of stable transformation, the rate of formation of embryogenic callus from embryos decreased. About 60% of embryos co-cultivated with L-cysteine produced embryogenic callus compared to 90% of the embryos from L-cysteine-free co-cultivation. Though the improvement in transformation efficiency was not dramatic, the increase is of practical significance.

Though *Agrobacterium*-mediated transformation is a widely used technology today, effective non-*Agrobacterium*-derived vectors and non-*Agrobacterium*-mediated transformation may be possible (Chung, Vaidya, and Tzfira 2006). Through suitable genetic engineering involving the Ti plasmid and T-DNA, vectors of *Sinorhizobium melliloti* have been shown to transform rice. Further, rice, tobacco, and *Arabidopsis* have been successfully transformed by multiple plant-associated symbiotic bacteria made competent for infection by the introduction of a disarmed Ti plasmid and a suitable binary vector (Broothaerts, Mitchell, Weir, Kaines, Smith, Yang, Mayer, Roa-Rodriguez, and Jefferson. 2005). Since maize is somewhat resistant to *Agrobacterium* infection, substitution of a more virulent bacterium might improve transformation. Conversely, use of a non-pathogenic bacterium might avoid the stimulation of a plant defense response that reduces transformation efficiency.

The ideal transformation event is comprised of a full-length, single-copy insertion of the gene construct without rearrangement of DNA sequence. The transgene should be expressed at its intended level, and the insertion event should cause no critical disturbance in the function of endogenous genes. Historically, the site of insertion has been beyond the control of the genetic engineer. Random insertion of transgenes into host chromosomes can disrupt the function of endogenous genes while causing variable and uncontrolled expression of the introduced genes. In addition, the odds of rearrangement during transformation rise as more genes are added to the transformation vector. The majority of the solutions to date have relied on recombination systems. For example, single-copy insertion can be enhanced by Cre-lox mediated recombination (Dale and Ow 1991; Russell, Hoopes, and Odell 1992; Ow 2007). Other developments for directed insertion include the FLP-*FRT* system. This system leverages the flippase recombinase (FLP) enzyme from yeast, which will recognize flippase recombinase target (*FRT*) sequence, wherein a pair of *FRT*s can be constructed to flank sequence of interest (Sadowski 1995; Sadowski 2003). Both Cre-lox and FLP-*FRT* have been described in conjunction with an engineered target site in a plant genome for site-specific, directed insertion of one or more expression cassettes (Baszczynski, Bowen, Peterson, and Tagliani 2002; Baszczynski, Bowen, Peterson, and Tagliani 2003a; Baszczynski, Bowen, Peterson, and Tagliani 2003b). More recently, endonuclease-mediated methods have been published (D'Halluin. Vanderstraeten, and Ruiter 2006) and meganucleases such as 1-CreI show potential as a more flexible platform for directed insertion (Arnould, Bruneau, Cabaniols, Chames, Choulika, Duchateau, Epinat, Gouble, Lacroix, Pagues, Smith, Perez-Michaut 2006; Duchateau and Paques 2006).

Artificial chromosomes may provide a single solution to the problems of multiple gene copies, rearrangement, and insertion site. Elements necessary for a functional artificial chromosome include autonomous replication sequences (ARS), centromeric sites of kinetochore assembly, and telomeres that stabilize the ends of the chromosome and facilitate complete replication of the DNA molecule. Luo et al. described a whole-genome fractionation technique that rapidly identifies bacterial artificial chromosome (BAC) clones derived from plant centromeric regions (Luo, Hall, Hall, and Preuss 2004). These regions of higher plants contain satellite sequences, transposons, and retroelements along with transcribed genes that mediate a variety of functions including nucleation of kinetochores, sister chromatid cohesion, and inhibition of recombination. Preuss et al., Copenhaver et al., and Mach et al. describe methods for isolation of a BAC clone harboring centromere DNA (Preuss et al. 2005; Copenhaver et al. 2007; Mach, Zieler, Jin, Keith, Copenhaver, and Preuss 2007a; Mach, Zieler, Jin, Keith, Copenhaver, and Preuss 2007b). Procedures are discussed for cloning structural genes expressing desired traits along with telomeric sequences selectable markers, and other necessary sequences, into the BAC to create a minichromosome. Processes are suggested for delivery of the minichromosome into a plant cell. Further research including construction of artificial chromosomes using the BAC clones identified with centromeric regions as substrate, followed by transformation and evaluation of inheritance pattern, could culminate in the development of autonomous vectors for maize providing for inserting exclusively single-copies genes under precise transcriptional control. Very recently, near-normal mitotic and meiotic transmission of an artificial chromosome through multiple generations has been demonstrated in maize (Carlson, Rudgers, Zieler, Mach, Luo, Gruden, Krol, Copenhaver, and Preuss 2007). Circular maize minichromosomes (MMC) containing maize centromeric elements and DsRed and nptII marker genes were constructed, and the constructs were introduced into maize embryonic tissue by particle bombardment. Transformed cells were selected, and plants were regenerated. Descendants of the regenerated plants were propagated for multiple generations. MMC meiotic behavior approached Mendelian expectations. Through four generations, the disomic propagation ratio was 93% (100% expected). The monosomic propagation ratio in crosses to wild type was 39% (50% expected) and 59% in self pollinations (75% expected). DsRed was nearly uniformly expressed in leaf tissue, and Southern blot analysis indicated that the reporter genes remained intact.

3 Tissue Culture

In the early 1960s, Murashige and Skoog formulated a tobacco tissue culture growth medium for use in bioassays of organic growth factors (Murashige and Skoog 1962). The medium provided for excellent growth of tobacco callus and excised pith tissue under ordinary environmental conditions. This growth medium became known as the MS media and was used extensively in plant tissue culture from then on. Chu et al. found that both ammonia and nitrate sources of nitrogen were required for abundant shoot induction from rice callus tissue (Chu, Wang, Sun, Hsu, Yin, Chu, and Bi 1975). They then devised a medium accordingly which became known as the N_6 medium. The MS and N_6 media have served as the basic growth media models in maize tissue culture up to the present day.

Using the MS growth medium, Green and Phillips assessed the percentage of embryos forming differentiated plantlets from callus cultures of inbred lines A188, A619, A632, B9A, and W64A, finding that A188 produced by far the highest percentage of differentiating cultures (Green and Phillips 1975). They then regenerated whole maize plants from embryo scutellular tissue of A188. Rapid callus growth was obtained through addition of the growth hormone 2,4-D to the medium. Callus growth was accompanied by localized chlorophyll development, the formation of very small leaves, the appearance of white compact structures resembling the organized scutellum of the original embryos, and the initiation of short roots. Differentiation to complete seedlings was accomplished by transfer of cultures with many small leaves to 2,4-D-free MS medium.

At about the same time, W.F. Sheridan created the BMS corn cell suspension cultures (Sheridan 1982). The BMS strain was the only line of 35 assayed by Sheridan that exhibited satisfactory callus growth in liquid medium. These cultures were fast growing and very easy to maintain. The cultures were non-regenerable (whole plants could not be grown from culture), but they were an invaluable tool for transient gene expression assays in evaluation and optimizing transformation techniques (Fromm et al. 1986; Klein et al. 1989; Spencer, Gordon-Kamm, Daines, Start, and Lemaux 1990).

Maize embryogenic callus is divided into two recognizable classes: type I and type II (Armstrong and Green 1985). Type I callus is easily derived from immature embryos and consists of a compact mass of cells that produce somatic embryos with complex, organized structure that grows very slowly in culture. On the other hand, Type II callus has a crumbly, friable texture, is embryogenic, and retains the capacity to rapidly regenerate plants over a relatively long period of time. Being friable, type II callus is also ideal for the establishment of cell suspension cultures. Unfortunately, type II callus forms at a lower rate than type I, but a significant step forward was made by Armstrong and Green by the discovery that the addition of L-proline to N_6 medium enabled routine initiation of type II callus from immature A188 embryos (Armstrong et al. 1985). Additional research provided evidence that incorporation of silver nitrate ($AgNO_3$) into the culture medium further enhanced the frequency of initiation of type II callus from immature embryos (Songstad, Duncan, and Widholm 1988; Vain, Flament, and Soudain 1989; Songstad, Armstrong, and Peterson 1991). Ethylene produced by cells in culture has an inhibitory effect on formation and growth of type II callus. Thus, the beneficial effect of $AgNO_3$ may be due to its antagonistic effect on ethylene in the cell culture.

The propensity to produce type II callus has been referred to as 'culturability' (Lowe, Way, Kumpf, Rout, Warner, Johnson, Armstrong, Spencer, and Chomet 2006). This propensity might more properly be called 'type II culturability', as culturability for one callus type does not imply equal culturability for others. Armstrong et al. demonstrated that type II culturability was a phenotypic trait under genetic control that could be improved by selection (Armstrong, Green, and Phillips 1991). Selection among partially inbred lines derived by selfing from the cross of A188×B73 isolated two lines with exceptionally high frequency of type II callus initiation. These two lines were crossed and the hybrid was released by the authors, under the name Hi-II, to the public for use in transformation projects.

A few maize inbred lines, notably A188, produce a high frequency of type II callus from immature embryos, but all are agronomically inferior (Lowe et al. 2006). Hence, the transfer to elite inbreds of genes introduced by transformation into culturable but agronomically poor lines is accomplished only through lengthy backcrossing and extensive evaluation to eliminate deleterious genomic background effects. The creation of agronomically suitable, highly culturable lines considerably improves transformation efficacy and speeds the trait integration process. Armstrong et al. conducted two culturability QTL mapping experiments (Armstrong, Romero-Severson, and Hodges 1992). First, a backcross selection experiment for culturability, in which B73 was the recurrent parent and A188 was the non-recurrent parent, was designed. A whole genome RFLP scan of selections from each generation identified five regions of the genome where A188 segments were retained. In a second experiment, culturability QTL were mapped in an F_2 population selfed out of Mo17×A188. Four of five significant QTL identified were mapped to the same segments that had been retained in the backcross selection experiment. In a marker-assisted selection experiment for culturability in maize, Lowe et al. used markers, linked to culturability in five regions of the genome, to derive a highly culturable inbred line while retaining an elite genomic background (Lowe et al. 2006). Selections retained at the end of the experiment exhibited agronomic performance nearly equal to the original elite line.

Although highly type II culturable elite lines can be developed, a problem common to all cell cultures remains. While in culture, cell lines accumulate cryptic mutations or epigenetic effects that increase in frequency over time (Lemaux 2007). These aberrations are passed on to the regenerated plants and, apparently, are heritably transmitted to the progeny of the regenerates through normal sexual propagation. The variability induced during culture is called somaclonal variation. The unwanted variability is an additional nuisance to contend with in transformation. Additional backcrossing and careful selection to ensure its complete removal is required if a gene construct from a transformed regenerate is being transferred by ordinary cross-pollination into a new line. Consequently, the extra effort to remove somaclonal variation negates to some extent the advantages of transforming into an elite background in the first place.

Since the degree of somaclonal variation is correlated with time in culture, methods that reduce culture time should be beneficial. Bregitzer et al. found that, in barley, reduction in somaclonal variation was achieved with cultures of highly differentiated meristematic shoot tissues derived from seedling axillary meristems (Bregitzer, Zhang, Cho, and. Lemaux 2002). Once established, the cultures grew rapidly, producing a relatively high frequency of transformed regenerates whose progeny displayed marked reduction in somaclonal variation. Earlier, similar preliminary results were obtained by Zhang et al. who successfully transformed barley using shoot meristematic cultures derived from germinating seedlings (Zhang, Cho, Koprek, Yun, Bregitzer, and Lemaux 1999). Cultures of a similar nature may prove useful in maize.

Improvements in cell culture have also been sought to improve the efficiency of *Agrobacterium*-mediated transformation. Zhao et al. and Armstrong and Rout found that suppression of bacterial growth during the inoculation and infection stages of transformation significantly improved transformation efficiency (Zhao, Gu, Cai, and Pierce 1999; Armstrong and Rout 2003). Bacterial-growth-suppressing agents may include silver nitrate, silver thiosulfate, penicillins, cephalosporins, and combinations of antibiotics with clavulanic acids. Armstrong and Rout also noted that the growth suppression procedures developed by them were efficient for ordinary as well as super binary vector (Armstrong et al. 2003).

4   Traits

Shortly after the development of efficient *Agrobacterium*-mediated transformation and plant regeneration systems in dicots (Horsch et al. 1985), tobacco and tomato were transformed by genes conferring economically valuable traits. Barton,

Whitley, and Yang (1987) transformed tobacco with an insecticidal protein coding sequence from *Bacillus thuringiensis* (Bt) that was toxic to lepidopteran insects feeding on host plant tissue. Bt transgenic constructs for lepidopteran insect resistance were also introduced into tomato (Fischhoff, Bowdish, Perlak, Marrone, McCormick, Niedermeyer, Dean, Kusano-Kretzmer, Mayer, Rochester, Rogers, and Fraley 1987), and insect resistant cotton was accomplished via transformation by Perlak et al. (Perlak, Deaton, Armstrong, Fuchs, Sims, Greenplate, and Fischhoff 1990). In each of these experiments, insect toxicity was evident in both transformed plants and progeny.

The efficacy of expression of Bt transgenic constructs in plants is compromised to some extent due to a lower overall content of guanine (G) + cytosine (C) (or higher adenine (A) + thymine (T)) content in the bacterium compared to plants. The bacterial gene sequence also displays a higher content of potential polyadenylation sequences, ATTTA sequences, potential transcription stop sequences, and A+T-rich regions. Each of these attributes can compromise gene expression in plants. Perlak et al. modulated these difficulties by exploiting the wobble in the genetic code to create modified Bt protein coding sequences by non-synonymous base substitutions that were effectively expressed in plants (Perlak, Fuchs, Dean, McPherson, and Fischhoff 1991). Similar modifications can be made regardless of the source of the transgene and with transgenes conferring various additional high value traits.

After the successful demonstration of transformation of maize in 1990, transformation with Bt transgenic constructs for resistance to the lepidopteran insect European corn borer quickly followed (Koziel, Beland, Bowman, Carozzi, Crenshaw, Crossland, Dawson, Desai, Hill, Kadwell, Launis, Lewis, Maddox, McPherson, Meghji, Merlin, Rhodes, Warren, Wright, and Evola 1993). In 1996, a patent was issued for glyphosate-resistant maize (Lundquist and Walters 1996). Also in 1996, three transformation events were released for unrestricted use in the United States: B16 (DLL25), conferring phosphinothricin (PPT) herbicide tolerance; MON809, for European corn borer and glyphosate resistance; and MON810, for European corn borer resistance (http://www.agbios.com/dbase). In 1998, a patent was granted for methods of producing insect resistant plants that included transformation to obtain resistance to coleopteran insects, primarily corn rootworm (Fischhoff, Fuchs, Lavrik, McPherson, and Perlak 1998). In 2003, the MON863 transformation event for control of corn rootworm was cleared for unrestricted use in the United States (http://www.agbios.com/dbase). The product array for Bt constructs has expanded from the single transgene events released in 1996–1997 by Novartis, Mycogen, and Monsanto to the first released stacks combining insect resistance with herbicide tolerance by AgrEvo and Monsanto (McDougall et al. 2007). The total acres of maize transformed with Bt constructs jumped from 0.5 to 46.2 million acres between 1996 and 2005.

Investigation of targets for herbicide tolerance was being pursued in parallel to the development of insect resistant crop plants (reviewed in Tan, Evans, and Singh 2006; CaJacob, Feng, Heck, Alibhai, Sammons, and Padgette 2004). Initial breakthroughs include the successful transformation of tomato with a glyphosate herbicide tolerance gene (Fillatti, Liser, Rose, and Comai 1987) and transformation of *Arabidopsis* with a chimeric ESP synthase construct from *Arabidopsis* and petunia (Klee, Muskopf, and Gasser 1987). The next decade saw tremendous activity in this area, culminating in the commercial releases of glyphosate tolerant soybean in 1996 by Monsanto, glufosinate tolerant corn in 1997 by AgrEvo, and glyphosate tolerant corn in 1998 by Monsanto. Currently, Monsanto's glyphosate tolerant corn products are the market leaders (McDougall et al. 2007) In 2010, Pioneer Hi-Bred plans to release a product developed through gene shuffling that confers both glyphosate and acetolactate synthase tolerance (www.pioneer.com).

In addition to second generation herbicide and insect resistance traits, recent years have seen a landslide of patents and patent applications relating to transgenes for many traits (Table 1). The list includes disease resistance, altered fatty acid, protein or carbohydrate metabolism, increased grain yield, increased oil, enhanced nutritional content, increased growth rates, enhanced stress tolerance, altered physiological maturity, enhanced organoleptic properties, altered morphological characteristics, male sterility, traits for industrial uses, and traits for improved consumer appeal. Traits such as stress tolerance, earlier maturity, or extended photoperiod may expand the climatic, and edaphic limits of corn production.

Quality traits, such as protein or oil concentration, may induce a shift away from corn being marketed strictly as a commodity toward an identity-preserved, value-added product. To date, transgenic quality traits have only penetrated the soy and canola markets, though Monsanto's LY038 event was recently released for unrestricted use in the United States (http://www.agbios.com/dbase). LY038 confers elevated levels of lysine in maize grain for livestock feed, primarily poultry and swine. The LY038 event consists of the incorporation of a *Corynebacterium glutamicum* cordapA gene that encodes endosperm-specific production of a lysine-insensitive dihydrodipicolinate synthase enzyme. Dihydrodipicolinate synthase regulates the first, and rate-limiting step, of lysine biosynthesis (Dizigan, Kelly, Voyles, Luethy, Malvar, and Malloy 2007). In non-transformed maize, expression of the enzyme is very sensitive to feedback inhibition. Transformation of maize with the bacterial protein coding sequence reduces feedback inhibition, permitting free lysine to accumulate to higher concentrations. The LY038 event is notable in that the kanamycin selectable marker gene NPT II was removed from transformed plants by site-specific recombination (Ow 2007).

An alternative approach to the engineering of gene expression has emerged via RNA interference (RNAi) technology. RNAi is a naturally occurring gene regulation process that suppresses the expression of target genes in a number of ways (reviewed in Mansoor, Amin, Hussain, Zafar, and Briddon 2006; Small 2007). RNAi technology seeks to harness the natural process through transformation with gene constructs that produce RNAi which suppresses expression of genes targeted by the biotechnologist or geneticist. For instance, Segal et al. found that maize transformed with RNAi constructs derived from a 22-kD zein encoding gene displayed a dominant high-lysine opaque phenotype in which 22-kD zeins were eliminated, but other zein proteins were unaffected (Segal, Song, and Messing 2003). Very recently, results suggesting that RNA interference may effectively control western corn rootworm and other coleopterans have been obtained (Baum,

Table 1 Examples of patent and patent application references for input and output transgenic traits

Trait	Gene/protein	Reference
Herbicide tolerance	5-enolpyruvylshikimate-3-phosphate synthases	U.S. Patents 5,094,945, 5,554,798, 5,627,061, 5,633,435, 6,040,497, 6,825,400; US Patent Application 20060143727; WO04009761
	glyphosate oxidoreductase (GOX)	U.S. Patent 5,463,175
	glyphosate decarboxylase	WO05003362; US Patent Application 20040177399
	glyphosate-N-acetyl transferase (GAT)	U.S. Patent Applications 20030083480, 20060200874
	dicamba monooxygenase	U.S. Patent Applications 20030115626, 20030135879
	phosphinothricin acetyltransferase (bar)	U.S. Patents 5,276,268, 5,273,894, 5,561,236, 5,637,489, 5,646,024; EP 275.957
	2,2- dichloropropionic acid dehalogenase	WO9927116
	acetohydroxyacid synthase or acetolactate synthase	U.S. Patents 4,761,373, 5,013,659, 5,141,870, 5,378,824, 5,605,011, 5,633,437, 6,225,105, 5,767,366, 6,613,963
	haloarylnitrilase (Bxn)	U.S. Patent 4,810,648
	acetyl-coenzyme A carboxylase (seq IDs)	U.S. Patent 6,414,222
	dihydropteroate synthase (sul I)	U.S. Patents 5,597,717, 5,633,444, 5,719,046
	32kD photosystem II polypeptide (psbA)	Hirschberg et al., 1983, Science, 222:1346–1349
	anthranilate synthase	U.S. Patent 4,581,847
	phytoene desaturase (crtI)	JP06343473
	hydroxy-phenyl pyruvate dioxygenase	U.S. Patent 6,268,549
	protoporphyrinogen oxidase I (protox)	U.S. Patent 5,939,602
	aryloxyalkanoate dioxygenase (AAD-1) (Seq IDs)	WO05107437
Male/female sterility system	Several	U.S. Patent Application 20050150013
	Glyphosate/EPSPS	U.S. Patent 6,762,344
	Male sterility gene linked to herbicide resistant gene	U.S. Patent 6,646,186
	Acetylated toxins/deacetylase	U.S. Patent 6,384,304
	Antisense to an essential gene in pollen formation	U.S. Patent 6,255,564

Intrinsic yield	DNAase or endonuclease/restorer protein	U.S. Patent 6,046,382
	Ribonuclease/barnase	U.S. Patent 5,633,441
	glycolate oxidase or glycolate dehydrogenase, glyoxylate carboligase, tartronic semialdehyde reductase	U.S. Patent Application 2006009598
	eukaryotic initiation Factor 5A; deoxyhypusine synthase	U.S. Patent Application 20050235378
	zinc finger protein	U.S. Patent Application 20060048239
	methionine aminopeptidase	U.S. Patent Application 20060037106
	2,4-D dioxygenase	U.S. Patent Application 20060030488
	serine carboxypeptidase	U.S. Patent Application 20060085872
	several	USRE38,446;U.S. Patents 6,716,474, 6,663,906, 6,476,295, 6,441,277, 6,423,828, 6,399,330, 6,372,211, 6,235,971, 6,222,098, 5,716,837, 6,723,897, 6,518,488; U.S. Patent Application 20060037106
Nitrogen use efficiency	fungal nitrate reductases, mutant nitrate reductases lacking post-translational regulation, glutamate synthetase-1, glutamate dehydrogenase, aminotransferases, nitrate transporters (high affinity and low affinities), ammonia transporters and amino acid transporters	U.S. Patent Application 20050044585
	glutamate dehydrogenase	U.S. Patent Application 20060090219
	cytosolic glutamine synthetase; root-specific glutamine synthetase.	EP0722494
	several	WO05103270; U.S. Patent Applications 20070044172, 20070107084
	glutamate 2-oxoglutarate aminotransferase	U.S. Patent 6,864,405
Abiotic stress tolerance including cold, heat, drought	succinate semialdehyde dehydrogenase	U.S. Patent Application 20060075522

(continued)

Table 1 (continued)

Trait	Gene/protein	Reference
	several	U.S. Patents 5,792,921, 6,051,755, 7,084,323, 6,229,069, 6,534,446, 6,951,971, 6,376,747, 6,624,139, 6,559,099, 6,455,468, 6,635,803, 6,515,202, 6,960,709, 6,706,866, 7,164,057, 7,141,720, 6,756,526, 6,677,504, 6,689,939, 6,710,229, 6,720,477, 6,818,805, 6,867,351, 7,074,985, 7,091,402, 7,101,828, 7,138,277, 7,154,025, 7,161,063, 7,166,767, 7,176,027, 7,179,962, 7,186,561, 7,186,563, 7,186,887, 7,193,130; U.S. Patent Applications 20030221224, 20040128712, 20040187175, 20050097640, 20050204431, 20050235382, 20050246795, 20050086718, 20060008874, 20060015972, 20060021082, 20060021091, 20060026716, 20060064775, 20060064784, 20060075523, 20060112454, 20060123516, 20060137043, 20060150285, 20060168692, 20060162027, 20060183137, 20060183137, 20060185038, 20060253938, 20070006344, 20070006348, 20070079400, 20070028333, 20070107084; WO06032708
Disease resistance	transcription factor	U.S. Patent Application 20060162027
	CYP93C (cytochrome P450)	U.S. Patent 7,038,113
	several	U.S. Patents 5,304,730, 5,516,671, 5,773,696, 5,850,023, 6,013,864, 6,015,940, 6,121,436, 6,215,048, 6,228,992, 6,316,407, 6,506,962, 6,573,361, 6,608,241, 6,617,496, 6,653,280, 7,038,113
Insect resistance	several	U.S. Patents 5,484,956, 5,763,241, 5,763,245, 5,880,275, 5,942,658, 5,942,664, 5,959,091, 6,002,068, 6,023,013, 6,063,597, 6,063,756, 6,093,695, 6,110,464, 6,153,814, 6,156,573, 6,177,615, 6,221,649, 6,242,241, 6,248,536, 6,281,016, 6,284,949, 6,313,378, 6,326,351, 6,468,523, 6,501,009, 6,521,442, 6,537,756, 6,538,109, 6,555,655, 6,593,293, 6,620,988, 6,639,054, 6,642,030, 6,645,497, 6,657,046, 6,686,452, 6,713,063, 6,713,259, 6,809,078, 7,049,491; U.S. Patent Applications 20050039226, 20060021087, 20060037095, 20060070139, 20060095986; WO05059103
Enhanced amino acid content	glutamate dehydrogenase	U.S. Patent 6,969,782
	threonine deaminase	U.S. Patent Application 20050289668

	dihydrodipicolinic acid synthase (dap A)	U.S. Patents 5,258,300, 6,329,574, 7,157,281
	chymotrypsin inhibitor	U.S. Patent 6,800,726
Enhanced protein content	several	U.S. Patent Application 20050055746
Modified fatty acids	several	U.S. Patents 6,380,462, 6,426,447, 6,444,876, 6,459,018, 6,489,461, 6,537,750, 6,589,767, 6,596,538, 6,660,849, 6,706,950, 6,770,465, 6,822,141, 6,828,475, 6,949,698
Enhanced oil content	several	U.S. Patents 5,608,149, 6,483,008, 6,476,295, 6,822,141, 6,495,739, 7,135,617
Carbohydrate production	raffinose saccharides	U.S. Patent 6,967,262
Starch production	several	U.S. Patent 5,750,876, 6,476,295, 6,538,178, 6,538,179, 6,538,181, 6,951,969
Phytic acid reduction	inositol polyphosphate 2-kinase	WO06029296
	inositol 1,3,4-triphosphate 5/6-kinases	U.S. Patent Application 20050202486
Processing enzymes production	several	WO05096804; U.S. Patent 5,543,576
Biopolymers	several	USRE37,543; U.S. Patents 5,958,745, 6,228,623; U.S. Patent Application 20030028917
Enhanced nutrition	several	US Patents 5,985,605, 6,171,640, 6,541,259, 6,653,530, 6,723,837
Pharmaceutical peptides and secretable peptides	several	U.S. Patents 6,080,560, 6,140,075, 6,774,283, 6,812,379
Improved processing trait	sucrose phosphorylase	U.S. Patent 6,476,295
Improved digestibility	thioredoxin and/or thioredoxin reductase	U.S. Patent 6,531,648

Bogaert, Clinton, Heck, Feldmann, Ilagan, Johnson, Plaetinck, Munyikawa, Pleau, Vaughn, and Roberts 2007). Maize plants were transformed by a construct containing a dsRNA sequence that targets an insect V-ATPase A gene. In a growth chamber experiment, transformed plants suffered far less larval root feeding damage than the non-transformed control. Presumably, silencing of the insect V-ATPase reduces larval growth and increases mortality. Protection from root feeding due to RNA interference was nearly as great as that provided by the MON863 BT event, a commercial standard. RNAi technology should be very useful, not only in the bio-engineering of gene regulation and pest control, but also for elucidation of gene networks.

5 Molecular Markers

The first indication that molecular markers might be useful selection tools was evident in the results of a maize selection experiment reported by Stuber and Moll (Stuber and Moll 1972). In the experiment, changes in allele frequencies of an acid phosphatase isozyme were correlated with artificial directional selection for grain yield. Stronger evidence was provided in an analysis of frequency changes at eight allozyme loci in four long-term yield selection experiments. The results of this analysis indicated that changes in allele frequency were likely due to directional selection, not drift (Stuber, Moll, Goodman, Schaffer, and Weir 1980). Direct evidence for yield improvement due to selection on molecular markers was demonstrated by Stuber et al. in an experiment in which selection for allozyme alleles previously shown to be associated with directional selection resulted in improvement of grain yield and number of ears per plant (Stuber, Goodman, and Moll 1982).

A few years previous to these experiments, H.O. Smith and co-workers in his laboratory had discovered that restriction endonucleases from *Hemophilus influenzae* cleaved at specific sites in phage DNA (Smith and Wilcox 1970; Kelley and Smith 1970). Cleavage of genomic DNA by restriction enzymes produced numerous fragments of varying size, which were dubbed restriction fragment length polymorphisms (RFLP). Danna and Nathans then suggested the use of restriction fragments identified by gel electrophoresis as a means to map genes of the simian SV 40 virus (Danna and Nathans 1971). Soon after, it became apparent that restriction fragment length polymorphisms (RFLP) could be utilized in the construction of genetic linkage maps of humans to aid in the identification of genes associated with phenotypic expression of complex human diseases (Botstein, White, Skolnick, and Davis 1980).

At the same time, Tanksley and Rick (1980) published an isozyme linkage map for tomato, followed quickly by publication of RFLP maps for both maize and tomato (Helentjaris, Slocum, Wright, Schaefer, and Nienhuis 1986). Almost immediately, RFLP were used to map quantitative trait loci in the two species (Stuber, Edwards, and Wendell 1987; Paterson, Lander, Hewitt, Peterson, Lincoln, and Tanksley 1988).

Lande and Thompson laid out the theoretical basis for marker-aided selection for improvement of quantitative traits (Lande and Thompson 1990). The authors showed that molecular markers could be treated as secondary correlated traits in a selection index for the improvement of a primary, economically valuable, trait. Efficiency of selection could be enhanced for quantitative traits with relatively low heritability and repeatability by selection for marker alleles favorably linked to alleles at loci controlling expression of the primary trait.

Practical applications of molecular markers in maize breeding soon followed. RFLP markers were used in recurrent selection for yield and quality traits in sweet corn (Edwards and Johnson 1994). Marker-trait associations identified in early generations of inbreeding have shown to be highly predictive of the breeding value of lines in advanced selfed generations, thus providing, if drawn upon, a substantial boost to the overall efficiency of a hybrid maize breeding program (Johnson and Mumm 1996; Eathington, Dudley, and Rufener 1997). Johnson and Mumm also presented results showing yield improvement in marker-aided reciprocal recurrent selection and related how molecular markers could be employed in selection for recurrent parent background to speed the introgression of transgenes into elite inbred lines. Particularly relevant to the biotech industry, as referenced above, marker-aided selection has been employed successfully to breed maize germplasm for improved culturability (Armstrong et al. 1992; Rosati, Landi, and Tuberosa 1994; Landi, Chiapetta, Salvi, Frascaroli, Lucchese, and Tuberosa 2002) as well as transformability (Lowe and Chomet 2004; Lowe et al. 2006). Initial studies indicated selection yielded improved regeneration in crosses including A188 as a parent (Armstrong et al. 1992; Rosati et al, 1994; Landi et al. 2002) though it was the introgression of regions of A188 into a more elite background that has shown the most promise for improved transformation and recovery of corn embryos (Lowe et al. 2006).

Invention of the polymerase chain reaction (PCR) by Mullis et al. enabled the development of new molecular markers that were less expensive and more amenable to high-throughput genotyping than RFLP (Mullis, Faloona, Scharf, Saiki, Horn, and Erlich 1986). Simple sequence repeats (SSR) were identified as a particularly valuable mapping tool in the late 1980s (Litt and Luty 1989; Weber and May 1989) and came into use during the 1990s. Advances in PCR-based technology (Livak, Flood, Marmaro, Guisti, and Deetz 1995) led to development of single nucleotide polymorphisms (SNPs) and their implementation after the year 2000. Each technological advance reduced the cost and increased the efficiency of marker-aided selection.

Today, marker-aided selection is firmly integrated into large-scale maize breeding programs within the plant biotech industry (Eathington et al. 2007; Graham 2007), marker-aided selection is firmly integrated into the large-scale maize breeding programs within the plant biotech industry (Eathington et al. 2007; Graham 2007). Monsanto estimates that marker-aided selection has doubled the rate of genetic gain in North American hybrid corn improvement in comparison to conventional breeding. The improvement comes principally from increased yield (Eathington et al. 2007). Marker-aided selection is also contributing to increased genetic gain over conventional selection in Brazil. At Pioneer, DNA markers promise to be very useful in cataloging genetic diversity within the breeding program and predicting superior commercial single cross candidates and breeding sources for new inbred line development (Graham 2007). Both companies have developed highly automated systems for high-throughput generation of SNP marker data and have implemented IT systems for sample tracking and data handling. Both organization use markers not only to speed and improve efficiency of biotech trait integration into elite lines, but directly in line and hybrid development breeding programs. The marker platforms used by these organizations require substantial capital and intellectual investment in high technology and may not be practical in some areas of the world.

The principles that inform marker-aided selection appertain also to the application of technologies such as transcript profiling to selection. Like marker genotype, the expression level of a gene can be treated as a secondary trait in index selection. Johnson found that expected progress from index selection for favorable response of photosynthetic rate to drought stress among inbred lines was more than doubled by inclusion of transcript expression of an NAD(H) dependent glutamate dehydrogenase (Johnson 2004). In response to drought stress, expression of the gene was highly repeatable with a strong genetic correlation to photosynthetic rate. Both markers and transcript profiles of genes can be added to an index so as to possibly add even further efficiency to selection.

6 Conclusions

Transgenic products have increased crop value and decreased overhead costs of maize production. From 1996 through 2005, farmers of transgenic insect resistant and herbicide tolerant corn have netted a $3.1 billion (USD) rise in farm income globally (Brookes et al. 2006) and have transformed the hybrid corn industry. To date, the productivity gain rests on a base of crop protection traits. Resistance to corn borer and corn rootworm are both due to transformation with modified constructs of the Bt protein coding sequence, while the dividends from herbicide resistance are mainly due to transformation with the *Agrobacterium* sp. CP4 protein coding sequence which confers tolerance to the herbicide glyphosate.

It may be worth noting that the products that have enjoyed success are due to transformation with genes derived from prokaryotes where a target phenotype is achieved via single-gene action. To date, modification of eukaryotic plant genes or engineering of endogenous maize genes has had little impact in the market. Glyphosate resistance is a case in point. Glyphosate acts by inhibiting the enzyme 5-enylpyruvylshikimate-3-phosphate synthase (EPSPS) which catalyzes the conversion of phosphoenolpyruvate (PEP) to shikimate-3-phosphate (S3P) in the shikimate pathway essential for the production of critical plant metabolites. Higher plants possess endogenous EPSPS genes. However, genetic engineering eliciting over-expression of the endogenous genes was ineffective (Shah, Horsch, Klee, Kishore, Winter, Turner, Hironaka, Sanders, Gasser, Aykent, Siegel, Rogers, and Fraley 1986). Success came only after transformation with the bacterial CP4 EPSPS gene, which is little affected by glyphosate (Dill 2005). The only commercialized product resulting from modification of an endogenous gene is the GA21 event for glyphosate resistance derived by directed mutagenesis of the EPSPS gene in a maize cell line (Monsanto; www.agbios.com). Other than this, nearly all glyphosate resistance products now sold contain a CP4 construct. The other dominant trait in the market is insect resistance.

The insecticidal properties of the Bt protein were well known for a considerable length of time, making the gene an obvious candidate for transformation.

As noted above, a large number of recent patent applications concerning plant biotech inventions have been filed or granted. Many of the phenotypic modifications dealt with are so-called 'output' traits. These modifications are intended to boost crop yield or quality directly, not just circuitously by reducing the impact of pests or pathogens. Others are aimed at reducing the effects of abiotic stress. The potential value of these traits is great, and some are likely to be commercialized in the near future.

As the number of traits increase, the opportunities to bundle various combinations into a single package concomitantly rises. The benefits of stacking multiple traits have already been realized by both the biotech industry and farmers. The influx of stacked traits in maize is substantial, and multiple companies are now offering more and more stacked products (McDougall et al. 2007). The capability to choose ideal combinations of traits for a given field, or growing season, provides increased flexibility for growers to manage risk and optimize return on operational inputs. In the future, farmers will have increasingly greater power to enhance yield potential by pyramiding combinations of output traits.

In the long run, continued growth of the maize biotech industry may rest on emerging scientific discoveries relating to gene expression and interaction on the whole genome level. Sequencing of the maize genome will likely draw all of maize genetics in the direction of whole-genome genetics. In the future, the disciplines of classical maize genetics and quantitative genetics may increasingly share a common ground. With the advent of systems biology, the next chapter written on biotechnology and maize should be fascinating indeed.

Acknowledgements The authors wish to thank Ravi Jain for support in compiling information for trait IP publications and Charles Armstrong, Sam Eathington, David Fischhoff, Cliff Lawson, Peggy Lemaux, Stephen Padgette, Eric Sachs, and David Songstad for guidance, discussion, and advice. The senior author also wishes to acknowledge the substantial benefits in understanding the history of plant transformation obtained from reading Paul Lurquin's book, the Green Phoenix (2001), and Mary-Dell Chiton's Scientific American essay (1983) on *Agrobacterium*-mediated transformation.

References

Armstrong, C.L. (1999) The first decade of maize transformation: A review and future perspective. Maydica 44, 101–109.

Armstrong, C.L. and Green, C.E. (1985) Establishment and maintenance of friable, embryogenic callus and the involvement of L-proline. Planta 164, 207–214.

Armstrong, C.L., Green, C.E. and Phillips, R.L. (1991) Development and availability of germplasm with high type-II culture formation response. Maize Genet. Coop. Newsletter 65, 92–93.

Armstrong, C.L., Romero-Severson, J. and Hodges, T.K. (1992) Improved tissue culture response of an elite maize inbred through backcross breeding, and identification of chromosomal regions important for regeneration by RFLP analysis. Theor. Appl. Genet. 84, 755–762.

Armstrong, C.L. and Rout, J.R. (2003) Agrobacterium-mediated transformation method. US Patent 6,603,061.

Arnould, S., Bruneau, S., Cabaniols, J.P., Chames, P., Choulika, A., Duchateau, P., Epinat, J.C., Gouble, A., Lacroix, E., Pagues, F., Smith, J. and Perez-Michaut, C. (2006) Custom-made meganuclease and use thereof. US Patent Application 20060206949.

Barton, K.A., Whitley, H.R. and Yang, N-S (1987) *Bacillus thuringiensis* δ-endotoxin expressed in transgenic *Nicotiana tabacum* provides resistance to lepidopteran insects. Plant Physiol. 85, 1103–1109.

Baszczynski, C.L., Bowen, B.A., Peterson, D.J. and Tagliani, L.A. (2002) Compositions and methods to stack multiple nucleotide sequences of interest in the genome of a plant. US Patent 6,455,315.

Baszczynski, C.L., Bowen, B.A., Peterson, D.J. and Tagliani, L.A. (2003a) Compositions and methods for locating preferred integration sites within a plant genome. US Patent 6,552,248.

Baszczynski, C.L., Bowen, B.A., Peterson, D.J. and Tagliani, L.A. (2003b) Compositions and methods to reduce the complexity of transgene integration in the genome of a plant. US Patent 6,573,425.

Baum, J.A., Bogaert, T., Clinton, W., Heck, G.R., Feldmann, P., Ilagan, O., Johnson, S., Plaetinck, G., Munyikawa, T., Pleau, M., Vaughn, T. and Roberts, J. 2007. Control of coleopteran insect pests through RNA interference. Nat. Biotech 25, 1322–1326.

Bevan, M.W., Flavell, R.B. and Chilton, M-D (1983) A chimaeric antibiotic resistance gene as a selectable marker for plant cell transformation. Nature 304, 184–187.

Botstein, D., White, R.L., Skolnick, M. and Davis, R.W. (1980) Construction of a genetic linkage map in man using restriction length polymorphisms. Am. J. Hum. Genet. 32, 314–331.

Braun, A.C. (1947) Thermal studies on the factors responsible for tumor induction in crown gall. Am. J. Bot. 34, 234–240.

Braun, A.C. (1958) A physiological basis for autonomous growth of the crown-gall tumor cell. Proc. Natl. Acad. Sci. USA 44, 344–349.

Bregitzer, P., Zhang, S., Cho, M-J and Lemaux, P.G. (2002) Reduced somaclonal variation in barley is associated with culturing highly differentiated meristematic tissues. Crop Sci. 42, 1303–1308.

Bronwyn, R.F., Shou, H., Chikwamba, R.K., Zhang, Z., Xiang, C., Fonger, T.M., Pegg, S.E.K., Li, B., Nettleton, D.S., Pei, D. and Wang, K. (2002) *Agrobacterium tumefaciens*-mediated transformation of maize embryos using a standard binary vector. Plant Physiol. 129, 13–22.

Brookes, G. and Barfoot, P. (2006) GM Crops: The First Ten Years – Global Socio-Economic and Environmental Impacts. *ISAAA Brief* No. 36. ISAAA, Ithaca, NY.

Broothaerts, W., Mitchell, H.J., Weir, B., Kaines, S., Smith, L.M.A., Yang, W., Mayer, J.E., Roa-Rodriguez, C. and Jefferson, R.A. (2005) Gene transfer to plants by diverse species of bacteria. Nature 433, 629–633.

CaJacob, C.A., Feng, P.C.C., Heck, G.R., Alibhai, M.F., Sammons, R.D. and Padgette, S.R. (2004) Engineering resistance to herbicides. In: P. Christou and H. Klee (Eds.), *Handbook of Plant Biotechnology*. John Wiley & Sons, New York, pp. 353–372.

Carlson, S.R., Rudgers, G.W., Zieler, H., Mach, J.M., Luo, S., Gruden, E., Krol, C., Copenhaver, G.P, and Preuss, D. 2007. Meiotic transmission of an in vitro-assembled maize minichromosome. PLOS Genet. 3, 1965–1974.

Chilton, M-D. (1983) A vector for introducing new genes into plants. Sci. Am. 248 No. 6, 50–59.

Chilton, M-D, Drummond, M.H., Merlo, D.J., Sciaky, D., Montoya, A.L., Gordon, M.P. and Nester, E.W. (1977) Stable incorporation of plasmid DNA into higher plant cells: The molecular basis of crown gall tumorigenesis. Cell 11, 263–271

Chilton, M-D, Saiki, R.K., Yadav, N., Gordon, M.D. and Quetier, F. (1980) T-DNA from *Agrobacterium* Ti plasmid is in the nuclear fraction of crown gall tumor cells. Proc. Natl. Acad. Sci. USA 77, 4060–4064.

Chu, C.C., Wang, C.C., Sun, C.S., Hsu, C., Yin, K.C., Chu, C.C. and Bi, F.Y. (1975) Establishment of an efficient medium for anther culture of rice through comparative experiments on the nitrogen sources. Scientia Sinica 18, 659–668.

Chung, S-M., Vaidya, M., and Tzfira, T. (2006) *Agrobacterium* is not alone: Gene transfer to plants by viruses and other bacteria. Trends Plant Sci. 11, 1–4.

Copenhaver, G.P., Keith, K. and Preuss, D. (2007) Methods for generating or increasing revenues from crops. US Patent 7,193,128.

Dale, E.C. and Ow, D.W. (1991) Gene transfer with the subsequent removal of the selection gene from the host genome. Proc. Natl. Acad. Sci. USA 88, 10558–10562.

Danna, K. and Nathans, D. (1971) Specific cleavage of simian virus 40 DNA by restriction endonucleases of *Hemophilus influenzae*. Proc. Natl. Acad. Sci. USA 68, 2913–2917.

D'Halluin, K., Bonne, E., Bossut, M., Beuckeleer, M.D., and Leemans, J. (1992) Transgenic maize plants by tissue electroporation. Plant Cell 4, 1495–1505.

D'Halluin, K., Vanderstraeten, C. and Ruiter, R. (2006) Targeted DNA insertion in plants. US Patent Application 20060282914.

Dill, G.M. (2005) Glyphosate resistant crops: History, status, and future. Pest Manag. Sci. 61, 219–224.

Dizigan, M.A., Kelly, R.A., Voyles, D.A., Luethy, M.H., Malvar, T.M. and Malloy, K.P. (2007) High lysine maize compositions and event LY038 maize plants. US Patent 7,157,281.

Duchateau, P. and Paques, F. (2006) 1-CreI meganuclease variants with modified specificity, methods of preparation and uses thereof. WO Patent Application 2006097853.

Eathington, S.R., Dudley, J.W. and Rufener II, G.R. (1997) Usefulness of marker-QTL associations in early generation selection. Crop Sci. 37, 1686–1693.

Eathington, S.R., Crosbie, T.M., Edwards, M.D., Reiter, R.S. and Bull, J.K. (2007) Molecular markers in a commercial plant breeding program. Proc. 43rd Ann. Ill. Corn Breed. School. March 5–6, Urbana, Illinois.

Edwards, M. and Johnson, L. (1994) RFLPs for rapid recurrent selection. In: *Proc. Joint Plant Breed. Symp. Series.* Am. Soc. Hort. And Crop Sci. Soc. Am. Corvallis, Oregon.

Environmental USDA/Aphis. 2006. Assessment of petition 04-229-01p. http://www.aphis.usda. gov/brs/aphisdocs/04_22901p_pea.pdf.

Fillatti, J.J., Kiser, J., Rose, R. and Comai, L. (1987) Efficient transfer of a glyphosate tolerance gene into tomato using a binary *Agrobacterium tumefaciens* vector. Bio/Technology 5, 726–730.

Fischhoff, D.A., Bowdish, K.S., Perlak, F.J., Marrone, P.G., McCormick, S.M., Niedermeyer, J.G., Dean, D.A., Kusano-Kretzmer, K., Mayer, E.J., Rochester, D.E., Rogers, S.G. and Fraley, R.T. (1987) Insect tolerant tomato plants. Bio/Technology 5, 807–813.

Fischhoff, D.A., Fuchs, R.L., Lavrik, P.B., McPherson, S.A. and Perlak, F.J. (1998) Insect resistant plants. US Patent 5,763,241.

Fraley, R.T., Rogers, S.G., Horsch, R.B., Sanders, P.R., Flick, J.S., Adams, S.P., Bittner, M.L., Brand, L.A., Fink, C.L., Fry, J.S., Galluppi, G.R., Goldberg, S.B., Hoffman, N.L. and Woo, S.C. (1983) Expression of bacterial genes in plant cells. Proc. Natl. Acad. Sci. USA 80, 4803–4807.

Fromm, M.E., Taylor, L.P. and Walbot, V. (1986) Stable transformation of maize after gene transfer by electroporation. Nature 319, 791–793.

Fromm, M.E., Morrish, F., Armstrong, C.A., Williams, R., Thomas, J. and Klein, T.M. (1990) Inheritance and expression of chimeric genes in the progeny of transgenic plants. Bio/Technology 8, 833–839.

Gianola, D., Perez-Enciso, M. and Toro, M.A. 2003. On marker-assisted prediction of genetic value: beyond the ridge. Genetics 163, 347–365.

Golovkin, M.V., Abraham, M., Morocz, S., Bottka, S., Feher, A. and Dudits, D. (1993) Production of transgenic maize plants by direct DNA uptake into embryogenic protoplasts. Plant Science 90, 41–52.

Gordon-Kamm, W.J., Spencer, T.M., Mangano, M.L., Adams, T.R., Daines, R.J., Start, W.G., O'Brien, J.V., Chambers, S.A., Adams, Jr., W.R., Willets, N.G., Rice, T.B., Mackey, C.J., Krueger, R.W., Kausch, A.P. and Lemaux, P.G. (1990) Transformation of maize cells and regeneration of fertile transgenic plants. Plant Cell 2, 603–618.

Gould, J., Devey, M., Hasagawa, O., Ulian, E.C., Peterson, G. and Smith, R.H. (1991) Transformation of *Zea mays* L. using *Agrobacterium tumefaciens* and the shoot apex. Plant Physiol. 95, 426–434.

Graham, G. (2007) The evolution of Pioneer's breeding program: Marker enhanced pedigree selection. Proc. 43rd Ann. Ill. Corn Breed. School. March 5–6, Urbana, Illinois.

Graves, A.C.F. and Goldman, S.L. (1986) The transformation of *Zea mays* seedlings with *Agrobacterium tumefaciens*. Plant Mol. Biol. 7, 43–50.

Green, C.E. and Phillips, R.L. (1975) Plant regeneration from tissue cultures of maize. Crop Sci. 15, 417–421.

Helentjaris, T., Slocum, M., Wright, S., Schaefer, A. and Nienhuis, J. 1986. Construction of genetic linkage maps in maize and tomato using restriction fragment length polymorphisms. Theor. Appl. Genet. 72, 761–769.

Herrara-Estrella, L., Depicker, A., Van Montague, M. and Schell, J. (1983) Expression of chimaeric genes transferred into plant cells using a Ti-plasmid-derived vector. Nature 303, 209–213.

Hiei, Y. and Komari, T. (1997) Method for transforming monocotyledons. US Patent 5,591,616.

Hillyer, G. (2005) Seed bank. Progressive Farmer Bus, 1-2.

Hoekema, A., Hirsch, P.R., Hooykaas, P.J.J. and Schilperoort, R.A. (1983) A binary plant vector strategy based on separation of the vir- and T-region of the *Agrobacterium tumefaciens* Ti-plasmid. Nature 303, 179–180.

Horsch, R.B., Fry, J.E., Hoffmann, N.L., Eichholtz, D., Rogers, S.G. and Fraley, R.T. (1985) A simple and general method for transferring genes into plants. Science 227, 1229–1231.

Illinois Agrinews. Sept. 15, 2006. Monsanto-Cargill venture takes root. http//www.renessen.com/news.release/ILagA3SEPT_15.pdf.

Ishida, Y., Saito, H., Ohta, S., Hiei, Y., Komari, T. and Kumasharo, T. (1996) High efficiency transformation of maize (*Zea* mays L.) mediated by *Agrobacterium tumefaciens*. Nat. Biotech. 14, 745–750.

Johnson, R. (2004) Marker-assisted selection. Plant Breed. Rev. 24 (Part 1), 293–309.

Johnson, G.R. and Mumm, R.H. (1996) Marker assisted maize breeding. Proc. Ann. Corn & Sorghum Ind. Res. Conf. 51, 75–84.

Kaeppler, H.F., Gu, W., Somers, D.A., Rines, H.W. and Cockburn, A.F. (1990) Silicon carbide fiber-mediated DNA delivery into plant cells. Plant Cell Rep. 9, 415–418.

Kaeppler, H.F., Somers, D.A., Rines, H.W., and Cockburn, A.F. (1992) Silicon carbide fiber-mediated stable transformation of plant cells. Theor. Appl. Genet. 84, 560–566.

Kelley, T.J. Jr., and Smith, H.O. (1970) A restriction enzyme from *Hemophilus influenzae*: II Base sequence of the recognition site. J. Mol. Bio. 51, 393–409.

Klee, H.J., Muskopf, Y.M. and Gasser, C.S. (1987) Cloning of an *Arabidopsis thaliana* gene encoding 5-enolpyruvylshikimate-3-phosphate synthase: sequence analysis and manipulation to obtain glyphosate-tolerant plants. Mol. Gen. Genet. 210, 437–442.

Klein, T.M., Fromm, M., Weissinger, A., Tomes, D., Schaaf, S., Slatten, M. and Sanford, J.C. (1988) Transfer of foreign genes into intact maize cells with high-velocity microprojectiles. Proc. Natl. Acad. Sci. USA 85, 4305–4309.

Klein, T.M., Kornstein, L., Sanford, J.C. and Fromm, M.E. (1989) Genetic transformation of maize cells by particle bombardment. Plant Physiol. 91, 440–444.

Koziel, M.G., Beland, G.L., Bowman, C., Carozzi, N.B., Crenshaw, R., Crossland, L., Dawson, J., Desai, N., Hill, M., Kadwell, S., Launis, K., Lewis, K., Maddox, D., McPherson, K., Meghji, M.R., Merlin, E., Rhodes, R., Warren, G.W., Wright, M. and Evola, S.V. (1993) Field performance of elite transgenic maize plants expressing an insecticidal protein derived from *Bacillus thuringiensis*. Bio/Technology 11, 194–199.

Krzyzek, R.A., Laursen, C.R.M. and Anderson, P.C. (1995) Genetic transformation of maize cells by electroporation of cells pretreated with pectin degrading enzymes. US Patent 5,384,253.

Lande, R. and Thompson, R. (1990) Efficiency of marker-assisted selection in the improvement of quantitative traits. Genetics 124, 743–746.

Landi, P., Chiapetta, L., Salvi, S., Frascaroli, E., Lucchese, C. and Tuberosa, R. (2002) Responses and allelic frequency changes associated with recurrent selection for plant regeneration from callus cultures in maize. Maydica 47, 21–32.

Lemaux, P.G. (2007) Personal communication.

Litt, M. and Luty, J.A. (1989) A hypervariable microsatellite revealed by in vitro amplification of a dinucleotide repeat within the cardiac muscle actin gene. Am. J. Hum. Genet. 44, 397–401.

Livak, K.J., Flood, S.J., Marmaro, J., Guisti, W. and Deetz, K. (1995) Oligonucleotides with fluorescent dyes at opposite ends provide a quenched probe system useful for detecting PCR product and nucleic acid hybridization. PCR Methods Appl. 4, 357–362.

Lowe, B. and Chomet, P. (2004) Methods and compositions for production of maize lines with increased transformability. US Patent Application 20040016030.

Lowe, B.A., Way, M.M., Kumpf, J.M., Rout, J., Warner, D., Johnson, R., Armstrong, C.L., Spencer, M.T. and Chomet, P.S. (2006) Marker assisted selection for transformability in maize. Mol. Breed. 18, 229–239.

Lundquist, R.C., and Walters, D.A. (1996) Fertile glyphosate-resistant transgenic corn plants. US Patent 5,554,798.

Luo, S., Hall, A.E., Hall, S.E. and Preuss, D. (2004) Whole-genome fractionation rapidly purifies DNA from centromeric regions. Nat. Meth. 1, 1–5.

Lurquin, P.F. (2001) *The green phoenix: A history of genetically modified plants*. Columbia Univ. Press, New York.

Mach, J., Zieler, H., Jin, J., Keith, K., Copenhaver, G. and Preuss, D. (2007a) Plant centromere compositions. US Patent 7,226,782.

Mach, J., Zieler, H., Jin, J., Keith, K., Copenhaver, G. and Preuss, D. (2007b) Plant centromere compositions. US Patent 7,227,057.

Mansoor, S., Amin, I., Hussain, M., Zafar, Y. and Briddon, R.W. (2006) Engineering novel traits in plants through RNA interference. Trends Plant Sci. 11, 559–565.

Matzke, A.J. and Chilton, M-D. (1981) Site-specific insertion of genes into T-DNA of the *Agrobacterium* tumor-inducing plasmid: An approach to genetic engineering of higher plant cells. J. Mol. Appl. Genet. 1, 39–49.

McDougall, J. and Phillips, M. (2007) Phillips McDougall AgriService, Edinburgh Scotland. 2007 Edition release 1.0.

Mullis, K., Faloona, F., Scharf, S., Saiki, R., Horn, G., and Erlich, H. (1986) Specific enzymatic amplification of DNA in vitro: The polymerase chain reaction. Cold Spring Harbor Symp. Quant. Biol. 51, 263–273.

Murashige, T. and Skoog, F. (1962) A revised medium for rapid growth and bioassays with tobacco tissue cultures. Physiol. Plant. 15, 473–497.

Omirulleh, S., Ábrahám, M., Golovkin, M., Stefanov, I., Karabev, M.K., Mustárdy, L., Mórocz, S. and Dudits, D. (1993) Activity of chimeric promoter with the doubled CaMV 35S enhancer element in protoplast-derived cells and transgenic plants in maize. Plant Mol. Biol. 21, 415–428.

Ow, D.W. (2007) GM maize from site-specific recombination technology, what next? Curr. Opinion Biotech. 18, 115–120.

Paszkowski, J., Shillito, R.D., Saul, M., Mandak, V., Hohn, T., Hohn, B. and Potrykus, I. (1984) Direct gene transfer to plants. EMBO J. 3, 2717–2722.

Paterson, A.H., Lander, E.S., Hewitt, J.D., Peterson, S., Lincoln, S.E. and Tanksley, S.D. (1988) Resolution of quantitative traits into Mendelian factors using a complete linkage map of restriction length polymorphisms. Nature 335, 721–726.

Perlak, F.J., Deaton, R.W., Armstrong, T.A., Fuchs, R.L., Sims, S.R., Greenplate, J.T.L. and Fischhoff, D.A. (1990) Insect resistant cotton plants. Bio/Technology 8, 939–943.

Perlak, F.J., Fuchs, R.L., Dean, D.A., McPherson, S.L. and Fischhoff, D.A. (1991) Modification of the coding sequence enhances plant expression of insect control protein genes. Proc. Natl. Acad. Sci. USA 88, 3324–3328.

Petit, A., Delhaye, S., Tempe, J. and Morel, G. (1970) Recherches sur les guanidines des tissus de crown gall. Mises en 'evidence d'une relation biochimique specifique entre les souches d'*Agrobacterium tumefaciens* et les tumeurs qu'elles induisent. Physiol. Veg. 8, 205–213.

Preuss, D., Copenhaver, G. and Keith, K.C. (2005) Plant chromosome compositions and methods. US Patent 6,972,197.

Rosati, C., Landi, P. and Tuberosa, R. (1994) Recurrent selection for regeneration capacity from immature embryo-derived calli in maize. Crop Sci. 34, 343–347.

Rhodes, C.A., Pierce, D.A., Mettler, I.J., Mascarenhas, D. and Detmer, J.J. (1988) Genetically transformed maize from protoplasts. Science 240, 204–207.

Russell, S.H., Hoopes, J.L. and Odell, J.T. (1992) Directed excision of a transgene from the plant genome. Mol. Gen. Genet. 234, 49–59.

Sadowski, P.D. (1995) The Flp recombinase of the 2-micron plasmid of *Saccharomyces cerevisiae*. Prog. Nucleic Acids Res. Mol. Biol. 51, 53–91.

Sadowski, P.D. (2003) The Flp double cross system a simple efficient procedure for cloning DNA fragments. BMC Biotechnol. 3, 9.

Sanford., J.C., Wolf, E.D. and Allen, E.D. (1990) Method for transporting substances into living cells and tissues and apparatus thereof. US Patent 4,945, 050.

Segal, G., Song, R. and Messing, J. (2003) A new opaque variant of maize by a single dominant RNA-interference-inducing transgene. Genetics 165, 387–397.

Shah, D.M., Horsch, R.B., Klee, H.J., Kishore, G.M., Winter, J.A., Turner, N.E., Hironaka, C.M., Sanders, P.R., Gasser, C.S., Aykent, S., Siegel, N.R., Rogers, S.G. and Fraley, R.T. (1986) Engineering herbicide tolerance in transgenic plants. Science 233, 478–481.

Sheridan, W.F. (1982) Black Mexican sweet corn: Its uses for tissue culture. In: W.F. Sheridan (Ed.) *Maize for Biological Research*. Plant Mol. Biol. Assoc. Charlottesville, Virginia, pp. 385–388.

Small, I. (2007) RNAi for revealing and engineering plant gene functions. Curr. Op. Biotechnol. 18,148–153.

Smith, H.O. and Wilcox, K.W. (1970) A restriction enzyme from *Hemophilus influenzae*: I. Purification and general properties. J. Mol. Biol. 51, 379–392.

Songstad, D.D., Duncan, D.R. and Widholm, J.M. (1988) Effect of l-aminocyclopropane-l-carboxylic acid, silver nitrate, and norbornadiene on plant regeneration from maize callus Plant Cell Rep. 7, 262–265.

Songstad, D.D., Armstrong, C.L. and Peterson, W.L. (1991) $AgNO_3$ increases type II callus production of maize inbred B73 and its derivatives. Plant Cell Rep. 9, 699–702.

Spencer, T.M., Gordon-Kamm, W.J., Daines, R.J., Start, W.G. and Lemaux, P.G. (1990) Bialaphos selection of stable transformants from maize cell culture. Theor. Appl. Genet. 79, 625–631.

Stuber, C.W. and Moll, R.H. (1972) Frequency changes of isozyme alleles in a selection experiment for grain yield in maize (*Zea mays* L.) Crop Sci. 12, 337–340.

Stuber, C.W., Moll, R.H., Goodman, M.M., Schaffer, H.E. and Weir, B.S. (1980) Allozyme frequency changes associated with selection for increased grain yield in maize (*Zea mays* L.). Genetics 95, 225–236.

Stuber, C.W., Goodman, M.M. and Moll, R.H. (1982) Improvement in yield and ear number resulting from selection at allozyme loci in a maize population. Crop Sci. 22, 737–740.

Stuber, C.W., Edwards, M.D. and Wendell, J.F. (1987) Molecular marker-facilitated investigations of quantitative trait loci in maize. II. Factors influencing yield and its component traits. Crop Sci. 27, 639–648.

Tan, S., Evans, R. and Singh, B. (2006) Herbicidal inhibitors of amino acid biosynthesis and herbicide-tolerant crops. Amino Acids 30, 195–204.

Tanksley, S.D. and Rick, C.M. 1980. Isozymic gene linkage map of the tomato.: applications to genetics and breeding. Theor. Appl. Genet. 57, 161–170.

Vaek, M., Reynaerts, A., Hofte, H., Jansens, S., De Beuckeleer, M., Dean, C., Zabeau, M., Van Montague, M. and Leemans, J. (1987). Transgenic plants protected from insect attack. Nature 328, 33–37.

Vain, P., Flament, P. and Soudain, P. (1989) Role of ethylene in embryogenic callus induction and initiation in *Zea mays* L. J. Plant Physiol. 135, 537–540.

Walters, D.A., Vetsch, C.S., Potts, D.E. and Lundquist R.C. (1992) Transformation and inheritance of a hygromycin phosphotransferase gene in maize plants. Plant Mol. Biol. 18, 189–200.

Weber, J.L. and May, P.E. (1989) Abundant class of human DNA polymorphisms which can be typed using the polymerase chain reaction. Am. J. Hum. Genet. 44, 388–396.

Willmitzer, L., DeBeuckeleer, M., Lemmers, M., Van Montague, M. and Schell, J. (1980) DNA from Ti plasmid present in nucleus and absent from plastids of crown gall plant cells. Nature 287, 359–361.

Zaenen, I., Van Larebeke, N., Teuchy, H., Van Montagu, M. and Schell, J. (1974) Supercoiled circular DNA in crown gall inducing *Agrobacterium* strains. J. Mol. Biol. 86, 109–127.

Zambryski, P., Joos, H., Genetello, C., Leemans, J., Van Montague, M. and Schell, J. (1983) Ti plasmid vector for the introduction of DNA into plant cells without alteration of their normal regeneration capacity. EMBO J. 2, 2143–2150.

Zhang, S., Cho, M-J, Koprek, T., Yun, R., Bregitzer, P. and Lemaux, P.G. (1999) Genetic transformation of commercial cultivars of oat (*Avena sativa* L.) and barley (*Hordeum vulgare* L.) using in vitro shoot meristematic cultures derived from germinated seedlings. Plant Cell Rep. 18, 959–966.

Zhao, Z.Y., Gu, W., Cai, T. and Pierce, D.A. (1999) Methods for *Agrobacterium*-mediated transformation. US Patent 5,981,840.

Modern Maize Breeding

Elizabeth A. Lee and William F. Tracy

1 Introduction

Maize breeders during the hybrid era (1939 to present) have been extremely successful in making continuous genetic improvement in commercial grain yield (Fig. 1). Commercial grain yield in the US increased from about 1,300 kg ha^{-1} in 1939 to 7,800 kg ha^{-1} in 2005, about 99 kg ha^{-1} year^{-1}, with similar gains observed in Canada during the hybrid era (80 kg ha^{-1} year^{-1}) (Lee and Tollenaar 2007). This 6-fold increase in grain yields over a 60 year period is unprecedented among cereals or oil seeds.

The USDA began collecting data on yield in 1862 and from that time maize yields did not change until the 1930s when they began an upward trend which has showed no signs of abating (Tracy et al. 2004). Prior to 1909, nearly all maize breeding was done by farmers or farmer/seedsmen, who used mass selection as their main breeding method (Hallauer et al. 1988). Mass selection is a method in which the best ears from the best plants would be selected from a population of maize plants. While this technique has been used since the domestication of plants, it is not very effective for traits with low heritability such as yield. Some farmer/breeders experimented with ear-to-row selection which entails growing and evaluating families derived from individual open-pollinated ears. While this method can be successful in improving traits of low heritability, the lack of knowledge of statistics and experimental design limited the success of this method and yields continued unchanged. Also contributing to static yields were the maize shows that were prevalent during the late 19th and early 20th centuries. In these shows, 10 ears were shown by each exhibitor and those groups that were most uniform and conformed to some ideal type in the mind of the judge were chosen as the winners, with no attention paid to yield or other economic traits. It is likely that the emphasis on uniformity in these competitions contributed to inbreeding and further suppressed any yield

E.A. Lee
University of Guelph, Department of Plant Agriculture, Guelph, ON N1G 2W1 Canada,
lizlee@uoguelph.ca

W.F. Tracy
wftracy@wisc.edu

J.L. Bennetzen and S. Hake (eds.), *Maize Handbook - Volume II: Genetics and Genomics,*   141
© Springer Science+Business Media LLC 2009

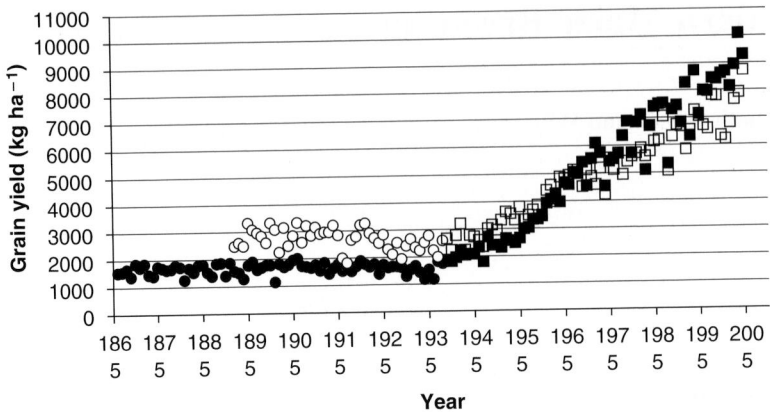

Fig. 1 Average US (1865 to 2005) and Canadian (1892 to 2005) maize yields in kg ha⁻¹ (15.5% moisture), 1866 to 1938 (pre-hybrid era; ● – US yields, ○ – Canadian yields), 1939 to 2005 (hybrid era; ■ – US yields, □ – Canadian yields). Data compiled by the USDA and OMAFRA. (From: Lee and Tollenaar 2007)

increases (Hallauer et al. 1988). A third breeding approach attempted during the 19th century was varietal hybridization in which two open-pollinated cultivars would be hybridized and the resulting hybrid would be grown for seed (Beal 1877). Some crosses did reveal reasonable amounts of heterosis (10–15%). But this yield increase was not enough to justify the extra energy and expense of producing the hybrid seed.

Hybrid maize traces its roots back to experiments on heterosis and inbreeding conducted by G.H. Shull (1908, 1909) at Cold Spring Harbor Laboratories in New York and E.M. East (1909) at Connecticut State College. Those observations, made nearly 100 years ago, and methodology outlined by Shull (1909), gave rise to the modern hybrid maize industry (c.f., Crow 1998). Because of the hybrid nature of the crop, modern temperate maize breeding in the US and Canada has evolved into two very distinct activities: inbred line development and hybrid commercialization (Duvick and Cassman 1999; Fig. 2). It is important to understand the distinctions between the two breeding activities, as genetic improvements first must occur within the inbred line development programs before they can be captured and realized in commercial hybrids. Inbred line development is the stage of maize breeding where the greatest amount of *de novo* genetic variation is present, created through recombination giving rise to novel alleles and new allelic combinations. In hybrid commercialization the genetic variation is potentially less, but represented by a far more refined germplasm pool; one that has been through extensive evaluation.

Maize breeding methodologies and philosophies have not remained static during the hybrid era. Breeding philosophies and practices have evolved; incorporating scientific advances in breeding and genetic theory such as early-generation testing (Jenkins 1935; Sprague 1946; El-Lakany and Russell 1971); rapidly adopting improvements in agronomic management practices such as increased plant population densities and modern herbicide chemistries; and recognizing how best to assess the

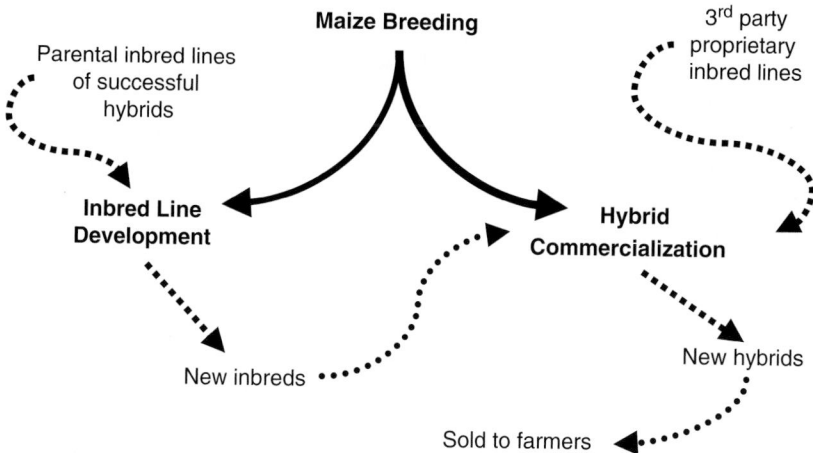

Fig. 2 Overview of a modern maize breeding program. (From: Lee and Tollenaar 2007)

genetic potential of a genotype, through improvements in experimental design and data analysis. Changes such as adoption of improved management practices impacted both inbred line development and hybrid commercialization, while other changes were geared toward one of the two maize breeding activities. Not all that surprising is that most of the innovations in breeding do not affect the generation of genetic material, but rather on how the genetic material is evaluated. The purpose of this chapter is to acquaint the reader with commercial maize breeding in North America circa 2008 – the methodology, the germplasm, and the philosophies.

2 Breeding Objectives

For the vast majority of maize breeding programs in the US and Canada, the primary breeding goals are high yield of machine-harvestable grain and wide adaptation within maturity zones (Troyer 1996). In a sense, these two traits encompass all the genes that determine the fitness of the organism. High yield of machine-harvestable grain indicates the need for alleles that positively affect all the pathways and functions affecting seed production. In addition, the term machine-harvestable indicates that the plant needs to be able to resist stresses that lead to root or stalk lodging or dropped ears. Wide adaptation indicates that the hybrid needs traits that protect yield from abiotic and biotic stresses. For a hybrid to succeed across the Corn Belt within its maturity zone, it must be able to resist the hot dry conditions of the western Corn Belt and the humid conditions which foster many fungal diseases in the east. Increased heterosis is not a breeding objective in the standard commercial maize breeding programs. Heterosis is a measure of the increased performance of a hybrid relative to its parents. Indeed because high seed yields of the inbred parents are so important to the profitability of a hybrid, the ideal hybrid would be very high yielding with low heterosis. In fact, heterosis of commercial hybrids calculated as a percent of inbred performance has actually decreased over time (Duvick 1984).

In addition to the performance of the hybrid, productivity of the inbreds is important. An inbred that will be used as a seed parent must produce high yields of high quality seed. That is, seed that has good germination and can tolerate mechanical handling and storage without losing vigor. Seed parents must be easy to detassel and have good stalk and root quality so that they can be easily harvested. Pollen parents should produce adequate quantities of viable pollen.

Traditionally, North American corn breeders have focused almost exclusively on harvestable yield and adaptation. However, breeding for specialty markets has increased. Specific goals include specialty starches for food and industrial uses (Boyer and Shannon 2003) and altered cell wall composition for silage and biofuel production (Frey et al. 2004). Sweet corn and pop corn have entirely different uses than corn grown for grain and, while yield remains important, greater attention is given to factors that affect table quality. In sweet corn, immature kernels are consumed, and they consist mainly of endosperm and ovary wall (immature pericarp). Sweet corn quality is determined by the flavor, aroma, and texture of the endosperm and tenderness of the pericarp. Genes affecting ear and kernel appearance are also important (Tracy 2000). Factors that confer high eating quality often result in poor germination and sweet corn breeders must strike a balance between table quality and germination (Marshall and Tracy 2003).

3   Inbreds, Hybrids, and Heterosis

Shull (1908, 1909) and East (1908) laid the theoretical groundwork for the inbred-hybrid breeding method. However, due to the poor performance of first generation inbreds, East and others did not believe this method could be commercially successful. D.F. Jones (1918) overcame this problem with the invention of the double cross hybrid. A double cross is created by making two single cross hybrids (A × B) and (C × D) and then crossing the two single crosses the following season. The seed sold to farmers was from this second cross. The male and female parents in a double cross are vigorous F_1 hybrids and the female parent produces large quantities of high quality seed. This made the production of hybrid seed possible and hybrid corn breeding programs sprang up around the country. However, for the first 30 years of the 20th century, the US agricultural economy was in recession and it was only when New Deal farm policies changed the economy of farming that farmers were willing to invest in increased inputs including hybrid seed. Cycles of inbred and hybrid development resulted in inbreds with improved performance per se. With better inbreds, double cross hybrids gave way to three ways and eventually single crosses in the 1970s. A three way cross uses three inbred lines, (A × B) × C, while single crosses only contain two, A × B (Hallauer et al. 1988). Single cross hybrids are the highest yielding and most uniform, and the most common type today in the Corn Belt.

Since Shull's initial papers, there has been debate about the underlying genetic mechanism of heterosis. The two main theories are the dominance theory and the overdominance theory. The dominance hypothesis attributes heterosis to the accumulation of favorable dominant genes or masking of deleterious recessive alleles in the hybrid (Davenport 1908; Bruce 1910; Keeble and Pellew 1910). In quantitative genetic terms, heterosis results when there is some degree of directional dominance (d) and the parents differ in gene frequency (Bruce 1910; Falconer 1981). The dominance hypothesis can be expressed in terms of a single-locus (B) with no epistasis as

$$\text{Heterosis} = d - \{a + (-a)\}/2 \tag{1}$$

where a and −a are the genotypic values of the parental genotypes (B_1B_1 and B_2B_2) and d is the genotypic value of the non-parental genotype (B_1B_2). The dominance theory is consistent with recent genomic evidence of differences in genic content between maize inbred lines (Fu and Dooner 2002; Song and Messing 2003; Brunner et al. 2005), and has been demonstrated as the underlying mechanism of a heterotic response for grain yield in a quantitative trait locus mapping study (Graham et al. 1997). The other hypothesis, over-dominance, argues that the heterozygous combination of the alleles at a single locus is superior to either of the homozygous combinations (Shull 1908; East 1908). The over-dominance hypothesis, unlike the dominance hypothesis, does not require the presence of either linkage or the involvement of multiple loci for heterosis to be expressed, nor is it necessarily based on classic Mendelian genetics. However, like the dominance hypothesis, it also requires that the parents differ in gene frequency. While there is no direct evidence in support of this hypothesis in the literature, it has not been completely rejected as an underlying genetic cause.

For heterosis to occur the parents must be genetically diverse and therefore the allele frequencies of effective heterotic groups must be divergent. However, a misperception persists on the absolute requirement for the groups to be based on geographical or phylogenetic diversity. Maize breeders in the 1940s created the basis for modern heterotic groups simply by splitting the available inbreds into two groups in an apparently arbitrary way with little attention to phylogeny (Tracy and Chandler 2006). This approach is supported by work by Cress (1967), who suggested that the way to make the most gain in a reciprocal recurrent selection program is to form one pool with the available germplasm and then arbitrarily split the pool into groups. Genetic drift will create an initial divergence of allele frequencies and the selection program will enhance those differences (Cress 1967). Molecular analysis of germplasm in the Pioneer Hi-bred breeding program supports Cress's approach (Duvick et al. 2004). It is clear that well-developed heterotic patterns of mature breeding programs are artificial constructs created by breeders and enhanced by the process of breeding hybrids (Tracy and Chandler 2006).

4 Methods of Inbred Development

While the primary trait of interest is grain yield in hybrid combinations, the inbred lines themselves need to be relatively productive. That is, inbred lines used as "females" in hybrid seed production must produce reasonable amounts of seed with good germination, and they also must resist lodging. "Male" inbred lines must produce adequate amounts of pollen. There are several methods that can be utilized for inbred line development, which will be outlined below. In general, the methods differ in one of two features, the amount of recombination that is occurring and/or the genome-wide allelic diversity (i.e., number of possible alleles present) of the starting material. Noticeably absent from the methods outlined below is recurrent selection. While recurrent selection has been widely used in the public sector for germplasm improvement, it is not a method that is routinely utilized by the commercial maize breeding sector for inbred line development.

4.1 Pedigree Breeding

Most if not all inbred line development activities in North America use the pedigree method of breeding (Duvick et al. 2004; Mikel and Dudley 2006; Lee and Tollenaar 2007; Fig. 3). Pedigree breeding, as it is structured in the commercial sector, is akin to reciprocal recurrent selection (RRS) (Hallauer and Miranda 1988; Duvick et al. 2004). Breeding crosses tend to be made by crossing inbred lines within a heterotic pattern. Inbred lines from the other heterotic patterns are used to improve the heterotic pattern represented by the breeding cross. This type of breeding scheme (i.e., akin to RRS) allows maize breeders to improve both additive and non-additive genetic

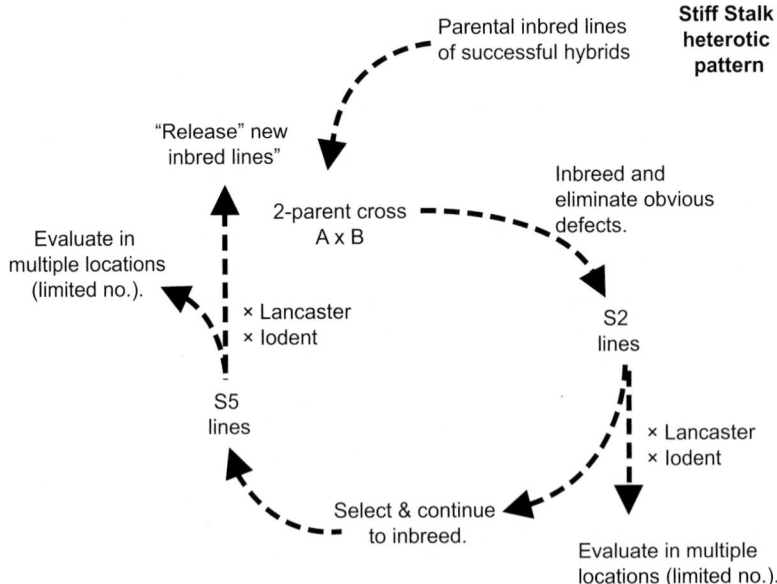

Fig. 3 An example of a typical inbred line development scheme, depicting a 2-parent breeding cross involving two inbred lines from the Stiff Stalk heterotic pattern. (From: Lee and Tollenaar 2007)

effects, resulting in greater overall genetic gains (Comstock et al. 1949). The typical pedigree breeding scheme generally consists of a 2-parent breeding cross within a heterotic pattern. Parent selection is based upon proven commercial utility of the inbred lines. An F_2 population is formed from the breeding cross. Inbreeding is performed for several generations (e.g., S_2) using ear-to-row with each family tracing back to different F_2 plants. During the inbreeding process, genotypes with obvious defects are eliminated. Early generation testing occurs around the S_2 generation, which involves forming topcross hybrids between the S_2 lines and an inbred line from each of the main heterotic patterns. The resulting topcross hybrids are evaluated in a limited number of environments, and selections based on agronomic performance are made. Only S_2 lines that correspond to selected topcross hybrids will be retained in the breeding program. Selected S_2 families are further inbred to the S_5 generation where a second round of topcross hybrid evaluation is performed. Again an inbred line from each of the main heterotic patterns is used to form the topcross hybrids. The hybrids are evaluated in a limited number of environments and selections are based on agronomic performance compared to commercial hybrids (i.e., checks). In general, all testing during inbred line development is done in hybrid combinations, involving a relatively limited number of hybrid combinations, in a relatively limited number of environments, and focused primarily on grain yield. At this stage, any superior inbred lines are then considered for release to the hybrid commercialization side of the breeding activities (Lee and Tollenaar 2007). Marker assisted breeding, a variation of the pedigree method, is used to varying degrees by many companies. Based on the phenotype data from early generation testing, molecular marker genotyping is used in conjunction with off-season nurseries to reduce inbred development time.

4.2   Dihaploids

The use of double haploids in breeding has been readily adopted for many homogenous, homozygous crops (e.g., canola, wheat, barley) (reviewed by Forster et al. 2007). The attraction for breeders is that the breeding material essentially moves directly from a segregating population of gametes to a homozygous, heterogeneous collection of plants (cultivars) without the time consuming process of inbreeding. The main characteristic that these crops share is that selection for performance is not practiced in the early generations of inbreeding, due to extremely limited seed quantities severely restricting the amount of field testing that can be carried out. Maize, unlike the crops mentioned above, does not suffer from restricted seed supply for testing early generations. All meaningful testing for grain yield is done in hybrid combinations, leading to essentially unlimited seed quantities from any generation. However, in recent years, there has been a renewed interest in using doubled haploids in maize inbred line development. The apparent attraction is probably a combination of several factors. (1) Reducing the time of the inbreeding process and therefore shortening the time between generations of inbred lines should lead to greater genetic gain per unit time. (2) In the large multinational breeding companies, resource utilization models probably favor separating the inbred line development process into two distinct activities - the inbreeding process which can be globally centralized in a single facility and the testing process which, due to maturity and environmental differences and genotype-by-environment interactions, still requires numerous regional testing facilities.

Maize haploid plants can be generated through several different means: *in vitro* via microspore (Pescitelli et al. 1989) or anther culture (Genovesi and Collins 1982; Barloy and Beckert 1993), or *in vivo* through the use of inducer lines. There are several inducer lines that have been used to generate maternal (gymnogenetic) haploids. Stock 6 (Coe 1959; Sharkar and Coe 1966), RWS (Röber et al. 2005), KEMS (Deimling et al. 1997), UH400 (MaizeGDB 2007), and KMS and ZMS (Chalyk et al. 1994; Chalyk and Chebotar 2000). Paternal (androgenetic) haploids can be generated using the *indeterminate gametophyte1* (*ig1*) mutation as the inducer (Kermicle 1969; U.S. Patent 5749169). Most double haploids in commercial maize breeding are generated using maternal inducer lines. While the original maternal inducer line, Stock 6, generated haploids at a frequency approaching 3%, the inducer lines used commercially have frequencies of haploid induction approaching 10% (Röber et al. 2005). To generate maternal haploids, the inducer line is used as the pollen donor and generally a visible recessive seedling mutant (e.g., *glossy1*, [*gl1*] or *liguleless2* [*lg2*]) is incorporated into the germplasm from which haploids are being derived. Seed from the cross between the donor germplasm (female parent) and the inducer line (male parent) is grown and screened for seedlings expressing the recessive seedling mutant. Those seedlings that express the seedling mutant are putative haploid plants. Alternatively, an anthocyanin marker system allows F_1 kernels to be screened for putative haploids. The inducer line may be homozygous for all of the functional alleles required for aleurone and embryo face color (e.g., *A1, A2, C1, C2,* and *R1-nj*). The F_1 seed will have colored aleurone and scutellum (i.e., embryo face) but, at a low frequency, colored aleurone and colorless scutellum kernels will appear. Those kernels represent putative haploids (reviewed in Röber et al. 2005). Regardless of the mechanism of haploid generation, chromosome doubling of haploid plants may rely upon spontaneous doubling, or the use of nitrous oxide (Kato 2002) or colchicine (Gayen et al. 1994).

4.3   Backcross Breeding

The goal of most backcrossing programs is to improve a particular strain (recurrent parent) for a specific characteristic, usually a single gene, obtained from a donor parent (Tracy 2004). In most backcross programs, the objective is to recover the recurrent parent essentially unchanged except for the introgression of the new characteristic. Backcrossing allows the plant breeder more precise control of allele frequencies than other traditional plant breeding methods. On average, with each backcross generation (BC) 50% of the recurrent parents alleles are recovered (e.g., F_1 – 50% [recurrent parent alleles]; BC_1 – 75%; BC_2 – 87.5 %; BC_3 – 93.25%; etc.). This is the theoretical average. Linkage and epistasis will result in the persistence of unwanted donor alleles and chromosome segments. Molecular markers can be used to reduce the size of donor linkage blocks and reduce the number of generations of backcrossing needed to recover the recurrent parent's genotype.

Compared to other crops, backcrossing has not been heavily used in maize breeding. Backcrossing has been used to incorporate disease resistance genes into elite inbreds or to transfer inbred nuclear genotypes into male sterile cytoplasms (Hallauer et al. 1988). However, with the widespread use of transgenes in maize breeding, backcrossing has become a standard tool of the maize breeder. Most commercial maize breeding programs have adopted the process of developing new inbreds from non-transgenic germplasm, and then, as new inbreds show promise, they will begin backcrossing in desired transgenes, at the same time maintaining the original non-transgenic inbred. Thus if a particular transgene is removed from the marketplace, or the use agreements are no longer valid, the new inbred is not lost. Also this permits the sale and production of non-transgenic versions of the new hybrids in locations that disallow the use of transgenes.

5 Hybrid Evaluation

As mentioned earlier, commercial maize grain yields during the hybrid era (1939 onward) have increased over 6-fold. This increase is not exclusively due to genetic improvement, as there have been substantial changes to agronomic practices during this 65–70 year period. Starting somewhere in the 1960s, commercial fertilizers were used, with increasing nitrogen levels occurring until the mid-1980s (cf., Troyer 2004; Crosbie et al. 2006). Better weed control was achieved through use of herbicides [e.g., 2,4-D was first used commercially in 1945 and Atrazine® was first used commercially in 1965 (cf., Troyer 2004)]. More uniform distribution of plants within a field was achieved by reducing row widths from 102 cm to 76 cm in the mid-1960s to early-1970s (cf., Troyer 2004). Earlier maize planting (Duvick 1989; Kucharik 2006) has effectively increased the duration of the growing season and, consequently, the period of time that plants can absorb incident solar radiation. Finally, plant population densities gradually increased from 30,000 plant ha^{-1} to 79,000 plants ha^{-1} (cf., Crosbie et al. 2006). In general, 60% of the increase in grain yield is attributed to genetic improvement (Duvick 1992) with 40% being attributed to improved agronomic practices (Cardwell 1982). As the agronomic practices changed, breeders incorporated these changes into their testing programs, meaning that, realistically, 100% of the increase in grain yield is actually due to the interaction between genetics and agronomic practices (Tollenaar and Lee 2002).

In contrast to inbred line development, hybrid commercialization generally involves a multi-tiered testing system (Duvick and Cassman 1999). More hybrid combinations are tested in fewer environments during the early testing phase while, in the later testing phases, fewer hybrid combinations are tested in more environments. Again, grain yield is the primary trait of interest. Testing for grain yield is always done in hybrid combinations, and testing is done using the most current agronomic practices. The specific traits that are assessed are: (1) machine harvestable grain yield adjusted to 15.5% grain moisture (bu ac^{-1} or Mg ha^{-1}); (2) grain moisture at harvest (%); (3) test weight, a measure of bulk density (lbs bu^{-1} or kg hl^{-1}); (4) broken stalks, dropped ears, root lodging; (5) days to 50% anthesis, days to 50% silking, and the interval between silking and anthesis; (6) germination under cold, wet conditions (i.e., cold germ test); (7) disease resistance (ear, stalk, and leaf diseases); and (8) general plant appearance.

Grain moisture at harvest is used by many breeders as the best assessment of the "maturity" of a hybrid. "Maturity" is a term that refers to whether a hybrid is adapted to a particular environment. Several systems have been developed to aid breeders and producers to place genotypes into the correct adaptation zones. In N. America, the Minnesota Relative Maturity (RM) system, Growing Degree Days (GDDs), and Ontario Corn Heat Units (OCHUs) are commonly used, while in Europe the Food and Agriculture Organization developed the FAO system (Troyer 2000a) (Table 1). The "ideal" hybrid is one that will maximize the full growing season available (i.e., flower late enough to produce maximal leaf area to intercept incident solar radiation), yet flower early enough that the grain reaches physiological maturity (i.e., black layer formation) before the first killing frost, and facilitate moisture loss from the kernel (i.e., dry down). At physiological maturity, grain moisture is >30% but <40%, while the target grain moisture for machine harvest is ~28% for "high moisture corn" and 15–22% for grain corn. Grain corn with moisture levels >15.5% is dried prior to storage and sale. If it is sold at grain moistures above 15.5%, the sale price will be reduced (i.e., docked), penalizing the producer. Several attributes influence maturity – flowering date, kernel type (flinty vs. floury), and husk characteristics (senescing, loose husks vs. green, tight husks). There is a strong positive linear relationship between flowering date and grain yield, flowering date and grain moisture, and grain moisture and grain yield (Fig. 4). Longer season

Table 1 Maize relative maturity rating systems: RM, GDUs, OCHUs, and FAO. (Adapted from Troyer 2000a)

Minnesota Relative Maturity (days)	U.S. Growing Degree Days (GDUs)	Ontario Corn Heat Units (OCHUs)	FAO (units)
70	1650	2100	100
75	1750	2300	
80	1850	2500	200
85	1950	2600	
90	2050	2700	300
95	2150	2800	
100	2250	2900	400
105	2350	3200	
110	2450	3400	500
115	2550	3500	
120	2650	3700	600
125	2750	3900	
130	2850	4100	700
135	2950	4300	
140	3050	4500	800

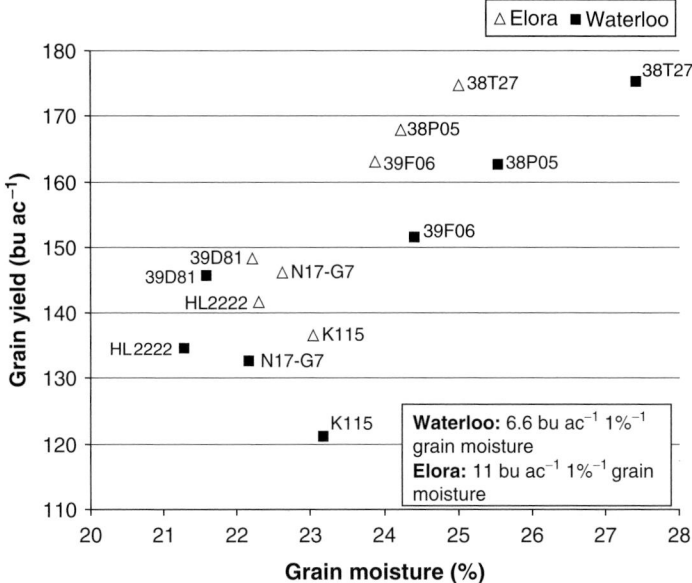

Fig. 4 The yield moisture paradigm in maize. Seven entries, 16 replications at each location, 2 locations, grown in 2002. Maturity ratings for the hybrids are: Pioneer hybrids 38T27 (2900 OCHUs), 38P05 (2850 OCHUs), 39F06 (2650 OCHUs), and 39D81 (2600 OCHUs); Hyland hybrid HL2222 (2450 OCHUs); Syngenta hybrid N17-G7 (2600 OCHUs); Pride hybrid K115 (2450 OCHUs). Accumulated heat units at Elora and Waterloo in 2002 were 2889 and 2750 OCHUs, respectively. Least significant differences (LSD) (0.05) for grain moisture and grain yield are 0.73% and 9.57 bu ac^{-1}, respectively.

hybrids are higher yielding and have higher grain moistures. Depending upon year and location, anywhere from a 6.6 to an 11 bu ac^{-1} decrease in grain yield for every 1% decrease in grain moisture is not uncommon (Fig. 4). To compensate for the yield-moisture paradigm in selection, many breeders assess hybrids based on their Y/M index, which is a selection index derived by dividing grain yield (adjusted to 15.5% grain moisture) by grain moisture at harvest.

Some of the experimental design and data analysis innovations, while also important for inbred line development, have substantially altered the hybrid commercialization aspect of maize breeding. In the early 1980s in North America, there was a shift in hybrid evaluation procedures. Instead of emphasizing relatively high precision per location at few locations (i.e., many replications, fewer locations), the new procedures emphasized relatively low precision at a large number of locations (i.e., one replication, but more locations) (Bradley et al. 1988). This shift encompassed an increase in the number of locations and years and a change in the type of location. The type of location shifted from high-yielding environments to environments that are most likely to occur in commercial maize production, including stress environments (Bradley et al. 1988). The traditional method of making hybrid comparisons, at that time, involved head-to-head comparisons of the experimental hybrid to various comparison hybrids grown in the same location and trial (e.g., Bradley et al. 1988; Crosbie et al. 2006). Another shift in evaluation procedures was facilitated by increased computational power and necessitated by the change in focus of the breeding companies' product lines. During the 1980s, several mergers and acquisitions occurred, creating a national or multi-national breeding focus rather than a regional breeding focus to their product lines. While genotype-by-environment interactions (G × E) have always been an impediment to identification of superior genotypes, shifting from breeding for a target population of environments (TPEs) encompassing large geographic areas increases the likelihood that G × E will confound identification of superior genotypes relative to breeding for regionally-defined TPEs. As an aid in handling G × E, geographic information system (GIS) approaches are now being utilized to define TPEs into environmental classes which, in turn, aids in predicting hybrid performance (Löffler et al. 2005).

6   Germplasm

Maize breeding in North America, unlike breeding activities for many other crop species, has been done primarily in the private sector for the past 30+ years. Movement of this activity out of the public domain has been accompanied by legal protection of the commercial germplasm through legally binding "use agreements", US patents, and the US Plant Variety Protection Act (PVP act). The goal of this section is to give an overview of the commercial North American germplasm circa 2008. The genetic base of the North American hybrid maize industry represents only a small portion of the entire *Zea mays* gene pool. There are 250 to 300 races of maize (Brown and Goodman 1977), of which only one, the Corn Belt Dent, is the predominant source of commercial germplasm (Goodman 1985). Of the hundreds of open-pollinated varieties (OPVs) of Corn Belt Dent that were grown up to the 1940s, only half a dozen or so can be considered as significant contributors to current inbred lines. But the overwhelming majority of the inbred lines trace their pedigree back to only two OPVs, Reid Yellow Dent and to a far lesser extent Lancaster Surecrop (Goodman and Holley 1988; Tracy and Chandler 2006; Troyer 2008).

One of the consequences of the hybrid concept was the development of heterotic patterns, also referred to interchangeably as heterotic groups. These groups were "created" by breeders as a means of maximizing the amount of hybrid vigor and ultimately grain yield in a more predictable manner (reviewed by Tracy and Chandler 2006). Modern hybrids, then, are the result of crossing an inbred line from one heterotic pattern with an inbred line from a different heterotic pattern. Today, inbred lines are classified into heterotic groups and are further sub-divided into families within a heterotic group. Classification of heterotic patterns is generally based upon several criteria such as pedigree, molecular marker-based associations, and performance in hybrid combinations (Smith et al. 1990), and has resulted in somewhere between two and seven distinct heterotic patterns being described (Smith and Smith 1989; Troyer 1999; Lu and Bernardo 2001; Gethi et al. 2002; Mikel and Dudley

2006). We have chosen to represent the germplasm as three main heterotic patterns "Stiff Stalk", "Lancaster" and "Iodent", and a miscellaneous category of lines that do not fit into the three primary heterotic patterns in the northern and central US and Canadian corn growing regions (Fig. 5). Most conventional inbred line development has focused on "recycling" (i.e., intermating) lines within a heterotic pattern and even within a family; however at least two novel heterotic patterns have arisen: "Maiz Amargo" and "Commercial Hybrid" derived germplasm (Mikel and Dudley 2006).

In the mid to late 1980s, virtually all commercial North American hybrids could be traced back to six public inbred lines or their close relatives: Lancaster-type inbreds C103, Mo17 and Oh43 and Reid-type lines B37, B73 and A632 (Goodman 1990). A recent US plant patent survey suggests that the current commercial North American maize germplasm pool is essentially based on seven inbred lines (B73, LH82, LH123, PH207, PH595, PHG39, and Mo17) with very little evidence of outside germplasm entering the commercial germplasm pool (Mikel and Dudley 2006). Two of these lines are the public inbreds B73 and Mo17 that were identified by Goodman (1990); however the remaining five are the product of commercial inbred line development activities. Inbred lines LH82 and LH123 were developed by Holden's Foundation Seeds (now a subsidiary of Monsanto), with LH82 representing a distinct family (referred to as "LH82-type") from the "Lancaster" heterotic pattern and LH123 representing a distinct family from the "commercial hybrid derived" heterotic pattern. Inbred lines PH207, PH595, and PHG39 were developed by Pioneer Hi-Bred International (now a subsidiary of DuPont). PH207 represents one of the distinct "Iodent" families. PH595 represents a distinct non-Stiff Stalk, non-Iodent

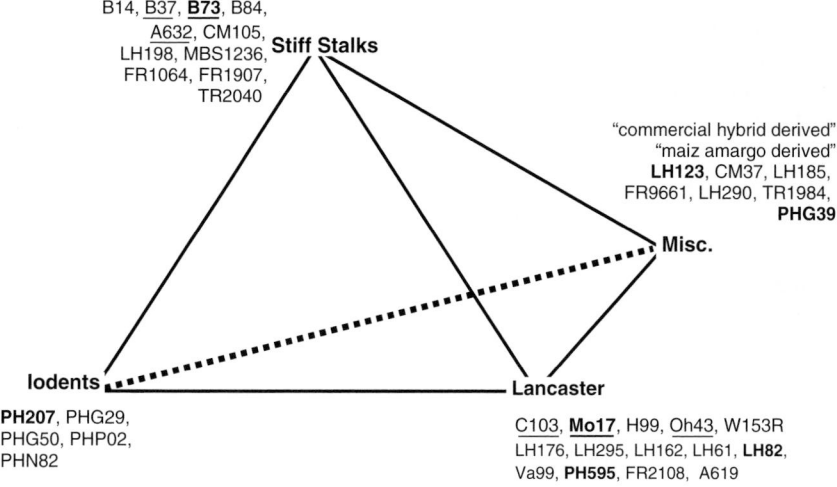

Fig. 5 Northern Corn Belt Dent heterotic patterns. Inbred lines in **bold** are the lines identified by Mikel and Dudley (2006) as those that formed the commercial germplasm pool. Lines under-scored are the public inbred lines identified by Goodman (1990) as being present in virtually all commercial hybrids. (Based on Troyer 1999; Mikel and Dudley 2006; Lee and Tollenaar 2007)

heterotic pattern from Pioneer that is probably best associated with the rather broad "Lancaster" heterotic pattern, through the Oh7-Midland and Oh43 families. The final line, PHG39, represents the "Maiz Amargo" heterotic pattern which traces its origins to the Argentine Maiz Amargo germplasm (Mikel and Dudley 2006).

The pedigree origin of the "Stiff Stalk" heterotic pattern traces back to a 16-line synthetic breeding population developed by George Sprague at Iowa State University in the 1930s, called Iowa Stiff Stalk Synthetic (BSSS) (Sprague 1946; Hallauer et al.1983). As a group, the 16 parental lines of BSSS were 75% Reid Yellow Dent (Troyer 2000b). The "Iodent" heterotic pattern traces its origins back to an OPV called Iodent and another called Minnesota 13, both of which were selections from the Reid Yellow Dent OPV (reviewed by Troyer 1999; Tracy and Chandler 2006; Mikel and Dudley 2006). The pedigree origin of the "Lancaster" heterotic pattern is probably the most misleading. The name implies that the pattern traces its origins to the OPV Lancaster Surecrop; however, the lines associated with this pattern are mostly Reid Yellow Dent in origin (Tracy and Chandler 2006). The "Lancaster" pattern, as we have chosen to represent it, is quite broad, being grouped into three main families of lines: LH82, C103, and Oh43. Inbred line C103, a parent of Mo17, is the only line that is directly from the Lancaster Surecrop OPV. Oh43 is 50% Lancaster Surecrop OPV by pedigree (Gerdes and Tracy 1993). We have chosen to associate LH82 with the "Lancaster" heterotic group based upon two pieces of evidence. First, even though LH82's pedigree (610 × LH7; 610 was a selection from W153R which was from the OPV Minnesota 13) does not indicate an association with the Lancaster OPV, molecular marker data suggests that W153R and Oh43 are more closely related to one another than they are to Mo17 and B73 (Gethi et al. 2002). Secondly, and perhaps more importantly, use of LH82 in further generations of inbred line development has involved crosses with other lines associated with Lancaster Surecrop, Oh43 and Oh07-Midland lines (see Figure 2 from Mikel and Dudley 2006).

The commercial germplasm base circa 2008 is almost exclusively from the Corn Belt Dent race of maize, with the Reid Yellow Dent OPV having the greatest contribution. The heterotic patterns that are used to group inbred lines are primarily a means of managing germplasm to exploit heterosis in a more predictable manner, they were created by breeders, and the names may no longer reflect the pedigree origins of the lines associated with the heterotic group. Finally and perhaps most importantly, the heterotic groups are dynamic. The "families" within a pattern evolve over time, the relative importance of a family changes over time, and "novel" heterotic patterns arise over time.

7   Germplasm Enhancement

Germplasm enhancement programs attempt to address the lack of racial diversity in the commercial germplasm base. The goal of most germplasm enhancement programs is to improve the adaptation and performance of non-commercial germplasm, so that it might be useful to hybrid development programs. In North America, germplasm improvement programs usually involve either introgression of non-Corn Belt Dent germplasm into the commercial germplasm pool, or employ one of a number of recurrent selection methods on Corn Belt Dent germplasm.

7.1 GEM Program

Concern about the narrow genetic base underlying the hybrid corn industry has led to programs designed to diversify breeding germplasm. The GEM[1] (Germplasm Enhancement of Maize) program represents one of these efforts. GEM represents a public-private sector collaboration in which elite tropical and sub-tropical germplasm (i.e., from non-Corn Belt Dent races of maize) is crossed to private sector inbred lines. Crosses for the southern US generally contain 50% Corn Belt Dent germplasm, while lines developed for the northern US generally contain 75% Corn Belt Dent germplasm. Inbreds are developed from these crosses following the pedigree method and early generation testing as described above (Carson et al. 2006).

7.2 Recurrent Selection

The general goal of recurrent selection programs is to increase the frequency of desirable alleles while maintaining genetic variation in the population (Hallauer and Miranda 1988). In maize breeding, a population is a heterogenous mixture of heterozygous genotypes derived by intermating inbred lines and/or OPVs. Populations are maintained by randomly intermating individuals in the population or allowing open-pollination under isolation. Recurrent selection programs are cyclic with a cycle consisting of a generation in which selection takes place followed by recombination of the selects. In academic research, the term recurrent selection generally indicates a closed population (i.e., no new germplasm being incorporated) unless otherwise specified. Hallauer and Miranda (1988) and Bernardo (2002) outline numerous recurrent selection programs. Recurrent selection programs are generally divided into either intra-population recurrent selection, in which selection is within one population, or inter-population, in which selection is based on test crosses to unrelated inbreds or populations. Intra-population systems target additive genetic effects, while inter-population programs attempt to select both additive and non-additive genetic effects. Recurrent selection programs can be very simple, such as mass selection, in which a large population of maize plants are grown and the best individuals are selected to supply seed to the following generation, or complex, such as some reciprocal recurrent selection schemes which may take as many as four growing seasons to complete a single cycle.

---

[1] Additional information on GEM can be found at www.public.iastate.edu/~usda-gem/homepage.html.

Maize is highly plastic and very responsive to selection. The generalized formula for response to selection is

$$R = h^2 S \qquad (2)$$

Where R equals the response to selection, h^2 is the narrow sense heritability, and S is the selection differential, which is the difference between the mean performance of the selected individuals and the mean of the parental population (Bernardo 2002). Narrow sense heritability is an estimate of the average effects of the alleles passed from parents to offspring and

$$h^2 = V_A/V_P \qquad (7.3)$$

V_A is the variance due to additive effects and V_P is the phenotypic variance.

If h^2 equals 1, then the trait is unaffected by the environment, and R=S. However, most quantitatively inherited traits are strongly affected by the environment and, for a traits such as grain yield, h^2 might be in the range of 0.1–0.4. For traits with low heritability, single plant selection or selection of families based on unreplicated evaluation is ineffective. Heritability can be increased by increasing the additive variance relative to the phenotypic variance. This can be done by replicating evaluation plots at a single location or increasing the number of environments in which plots are evaluated. Increasing the selection differential (S) will result in greater gain, but this means that a smaller proportion of the original population will be selected. If too few families are saved, inbreeding and genetic drift will result. A large selection differential and reduced inbreeding can be achieved if large numbers of families are evaluated and at least 15 families are saved for recombination.

Response to selection (R) can also be increased if selection can be made before pollination, such that both the male and female parents (biparental control) can be selected. In this case, gain will be twice as great as for a trait which can only be determined after pollination, such as grain yield, and in which only the female parents are selected and allowed to open-pollinate with the entire population (uniparental control). Schemes have been derived such that biparental control can be used for traits such as a grain yield, but these require an additional growing season that will increase gain per cycle but decrease the gain per season.

7.3 Divergent Recurrent Selection

Divergent recurrent selection is a form of recurrent selection in which, starting from an initial closed population, a trait is selected in opposite directions and populations divergent for that trait are developed. Perhaps the best known example is the Illinois high oil/ low oil and high protein/low protein long term selection experiment in which there have been more than 100 generations of selection (Dudley and Lambert 2004). There are numerous other examples including root and stalk strength populations at Missouri (Abedon et al. 1999; Flint-Garcia et al. 2003), Iowa ear length populations (Hallauer

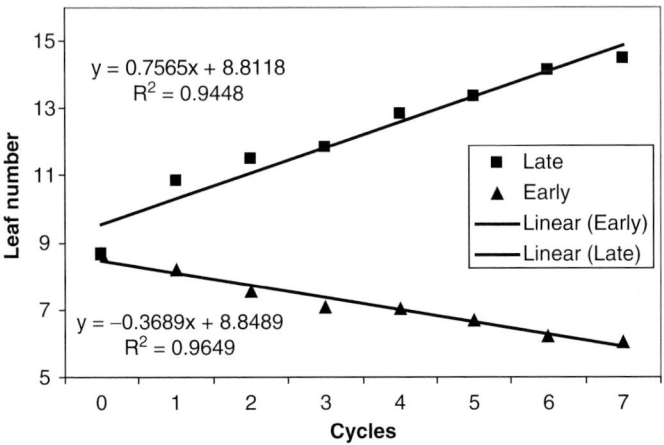

Fig. 6 Direct response to divergent recurrent selection for early and late last leaf with juvenile wax in Minn11 over seven cycles of selection including significant trend lines with corresponding equations and coefficients of determination (R^2). (From: Reideman et al. 2008)

et al. 2004) and endosperm and vegetative phase change populations at Wisconsin (Fig. 6) (Tracy and Chang 2007; Reideman et al. 2008). Divergent recurrent selection experiments can be useful in understanding the limits of selection and also the relationship between the selected trait and unselected traits that change in response to selection. Are such changes due to linkage, pleiotropy, physiological relationships (e.g. source – sink), or an artifact of the selection method? Genomics tools can be applied to divergent populations to determine the underlying genetic and physiological mechanisms between the selected trait and traits indirectly affected by selection.

8   Conclusions

The inbred-hybrid breeding method has been highly effective in increasing maize yields. Average grain yield in the US increased from about 1,300 kg ha^{-1} in 1939 to 7,800 kg ha^{-1} in 2005. Increased yields are in part due to improved agronomic practices and increased inputs, but increased yields could not have been realized without genetic improvements. Maize breeders created and continue to create genotypes that respond favorably to modern agronomic practices. During the first 70 to 80 years of the hybrid era, development and application of experimental design and biometry contributed greatly to genetic gain while genetic information was seldom used directly in maize breeding. Today, with genomics tools, plant breeders are able to apply genetic information directly. The combination of genomics, experimental design, biometry, and selection theory will allow maize breeders and maize growers to continue increasing maize productivity well into the 21st century.

References

Abedon BG, Darrah LL, Tracy WF (1999) Developmental changes associated with divergent selection for rind penetrometer resistance in the MoSCSS maize Synthetic. Crop Sci 39:108–114

Barloy D, Beckert M (1993) Improvement of regeneration ability of androgenetic embryos by early anther transfer in maize. Plant Cell Tissue Org Cult 33:45–50

Beal WJ (1877) Report of the professor of botany and horticulture. Michigan Board of Agric, Lansing, MI

Bernardo R (2002) Breeding for quantitative traits in plants. Stemma Press, Woodbury, MN

Boyer CD, Shannon JC (2003) Carbohydrates of the kernel. In: Ramstad PE, White P (eds) Corn chemistry and technology, 2nd edn. Am Assoc of Cereal Chemists, Minneapolis, pp 289–312

Bradley JP, Knittle KH, Troyer AF (1988) Statistical methods in seed corn product selection. J Prod Agric 1:34–38

Brown WL, Goodman MM (1977) Races of corn. In: Sprague GF (ed) Corn and corn improvement. Am Soc Agron, Madison, WI, pp 49–88

Bruce AB (1910) The Mendelian theory of heredity and the augmentation of vigor. Sci 32:627–628

Brunner S, Fengler K, Morgante M, Tingey S, Rafalski A (2005) Evolution of DNA sequence nonhomologies among maize inbreds. Plant Cell 17:343–360

Cardwell VB (1982) Fifty years of Minnesota corn production: Sources of yield increase. Agron J 74: 984–990

Carson ML, Balint-Kurti PJ, Blanco M, Millard M, Duvick S, Holley R, Hudyncia J, Goodman MM (2006) Registration of nine high-yielding tropical by temperate maize germplasm lines adapted for the southern USA. Crop Sci 46:1825–1826

Chalyk ST, Chebotar OD (2000) Regular segregation of four recessive marker genes among maternal haploids in maize. Plant Breeding 119:363–364

Chalyk ST, Bylich VG, Chebotar OD (1994) Transgressive segregation in the progeny of a cross between two inducers of maize maternal haploids. MNL 68:47

Coe EH (1959) A line of maize with high haploid frequency. Am Nat 93:381–382

Comstock R E, Robinson H F, Harvey PH (1949) A breeding procedure designed to make maximum use of both general and specific combining ability. J Am Soc Agron 41:360–367

Cress CE (1967) Reciprocal recurrent selection and modifications in simulated populations. Crop Sci 7:561–567

Crosbie TM, Eathington SR, Johnson GR, Edwards M, Reiter R, Stark S, Mohanty RG, Oyervides M, Buehler RE, Walker AK, Delannay R, Pershing JC, Hall MA, Lamkey KR (2006) Plant breeding: past, present, and future. In: Lamkey KR, Lee M (eds) Plant breeding: The Arnel R Hallauer international symposium. Blackwell Publishing, Oxford, pp 3–50

Crow JF (1998) 90 years ago: the beginning of hybrid maize. Genet 148:923–928

Davenport CB (1908) Degeneration, albinism and inbreeding. Sci 28:454–455

Deimling S, Röeber FK, Geiger HH (1997) Methodik und genetik der *in-vivo*-haploideninduktion bei mais. Vortr Pflanzenzuchtg 38:203–224

Dudley JW, Lambert RJ (2004) 100 Generations of selection for oil and protein in corn. Plant Breeding Rev 24:80–110

Duvick DN (1984) Genetic contributions to yield gains of US hybrid maize, 1930–1980. In: Fehr WR (ed) Genetic contributions to yield of five major crop plants. CSSA special publications number 7 Crop Sci Soc Amer, Madison, WI, pp 15–47

Duvick DN (1989) Possible genetic causes of increased variability in US maize yield. In: Anderson JR, Hazel PBR (eds) Variability in grain yields: Implications for agricultural research and policy in developing countries. John Hopkins Univ Press, Baltimore, pp 147–156

Duvick DN (1992) Genetic contributions to advances in yield of US maize. Maydica 37:69–79

Duvick DN, Cassman KG (1999) Post–green revolution trends in yield potential of temperate maize in the north-central United States. Crop Sci 39:1622–1630

Duvick DN, Smith JSC, Cooper M (2004) Long term selection in a commercial hybrid maize program. Plant Breed Rev 24:109–151

East EM (1909) Inbreeding in corn, 1907. In: Connecticut Agric Exp Stn Rep, pp 419–428

El-Lakany MA, Russell WA (1971) Effectiveness of selection in successive generations of maize inbred progenies for improvement of hybrid yield. Crop Sci 11:703–706

Falconer DS (1981) Introduction to quantitative genetics, 2nd edn. Longman, London

Flint-Garcia SA, Jampatong C, Darrah LL, McMullen MD (2003) Quantitative trait locus analysis of stalk strength in four maize populations. Crop Sci 43:13–22

Forster BP, Heberle-Bors E, Kasha KJ, Touraev A (2007) The resurgence of haploids in higher plants. Trends in Plant Sci 12:1360–1385

Frey TJ, Coors JG, Shaver RD, Lauer JG, Eilert DT, Flannery PJ (2004) Selection for silage quality in the Wisconsin quality synthetic and related maize populations. Crop Sci 44:1200–1208

Fu H, Dooner HK (2002) Intraspecific violation of genetic colinerity and its implications in maize. PNAS 99:9573–9578

Gayen P, Madan JK, Kumar R, Sarkar KR (1994) Chromosome doubling in haploids through colchicine. MNL 68:65

Genovesi AD, Collins GB (1982) In vitro production of haploid plants of corn via anther culture. Crop Sci 22:1137–1144

Gerdes JT, Tracy WF (1993) Pedigree diversity within the Lancaster surecrop heterotic group of maize. Crop Sci 33:334–337

Gethi JG, Labate JA, Lamkey KR, Smith ME, Kresovich S (2002) SSR variation in important US maize inbred lines. Crop Sci 42:951–957

Goodman MM (1985) Exotic maize germplasm: status, prospects, and remedies. Iowa State J of Res 59:497–527

Goodman MM (1990) Genetic and germplasm stocks worth conserving. J of Hered 81:11–16

Goodman MM, Holley RN (1988) US maize germplasm: origin, limitations and alternatives. In: Russell N, Listman GM (eds) Recent advances in the conservation and utilization of genetic resources, Proceedings of the global maize germplasm workshop, CIMMYT, Mexico, pp 130–148

Graham GI, Wolff DW, Stuber CW (1997) Characterization of a yield quantitative trait locus on chromosome five of maize by fine mapping. Crop Sci 37:1601–1010

Hallauer AR, Miranda Fo JB (1988) Maize breeding, 2nd edn. Iowa State Univ Press, Ames, IA

Hallauer AR, Russell WA, Lamkey KR (1988) Corn breeding. In: Sprague GF, Dudley JW (eds) Corn and corn improvement 3rd edn. Am Soc of Agron, Madison, WI, pp 463–564

Hallauer AR, Ross AJ, Lee M (2004) Long term divergent selection for ear length in maize. Plant Breeding Rev 24:153–168

Jenkins MT (1935) The effect of inbreeding and of selection with inbred lines of maize upon the hybrids after successive generations of selfing Iowa State Coll J Sci 3:429–450

Jones DF (1918) The effects of inbreeding and crossbreeding on development. In: Conn Agric Exp Stn Bull 207, pp 5–100

Keeble F, Pellew C (1910) The mode of inheritance of stature and of time of flowering in peas (*Pisum sativum*). J Genet 1:47–56

Kermicle JL (1969) Androgenesis conditioned by a mutation in maize. Sci 166:1422–1424

Kucharik CJ (2006) A multidecadal trend of earlier corn planting in the central USA. Agron J 98:1544–1550

Lee EA, Tollenaar M (2007) Physiological basis of sucessful breeeding strategies for maize grain yield. Crop Sci 47(S3):S202-S215

Löffler CM, Wei J, Fast T, Gogerty J, Langton S, Bergman M, Merrill B, Cooper M (2005) Classification of maize environments using crop simulation and geographic information systems. Crop Sci 45:1708–1716

Lu H, Bernardo R (2001) Molecular marker diversity among current and historical maize inbreds. Theor Appl Genet 103:613–617

MaizeGDB (2007) www.maizegdb.org verified Nov 17, 2007

Marshall SW, Tracy WF (2003) Sweet corn. In: Ramstad PE, White P (eds) Corn chemistry and technology. 2nd edn. Am Assoc Cereal Chemists, Minneapolis, MN, pp 537–569

Mikel MA, Dudley JW (2006) Evolution of North American dent corn from public to proprietary germplasm. Crop Sci 46:1193–1205

Peseitelli SM, Mitchell JC, Jones AM, Pareddy DR, Petolino JF (1989) High frequency androgenesis from isolated microspores of maize. Plant Cell Rep 7:673–676

Riedeman ES, Chandler MA, Tracy WF (2008) Seven cycles of divergent recurrent selection for vegetative phase change and indirect effects on resistance to common rust (*Puccinia sorghi*) and European corn borer (*Ostrinia nubilalis*). Submitted to Crop Sci

Röber FK, Gordillo GA, Geiger HH (2005) *In vivo* haploid production in maize – performance of new inducers and significance of double haploid lines in hybrid breeding. Maydica 50:275–283

Song R, Messing J (2003) Gene expression of a gene family in maize based on noncollinear haplotypes. PNAS 100:9055–9066

Sharkar KR, Coe EH (1966) A genetic analysis of the origin of maternal haploids in maize. Genet 54:453–464

Shull GH (1908) The composition of a field of maize. Amer Breeders' Assoc Rep 4:296–301

Shull GH (1909) A pureline method of corn breeding. Amer Breeders' Assoc Rep 5:51–59

Smith JSC, Smith OS (1989) The description and assessment of distances between inbred lines of maize. II. The utility of morphological, biochemical, and genetic descriptors and a scheme for the testing of distinctiveness between inbred lines. Maydica 34:141–150

Smith OS, Smith JSC, Bowen SL Tenborg RA, Wall SJ (1990) Similarities among a group of elite maize inbreds as measured by pedigree, F₁ grain yield, grain yield, heterosis, and RFLPs. Theor Appl Genet 80:833–840

Sprague GF (1946) Early testing of inbred lines of corn. J Am Soc Agron 38:108–117

Tollenaar M, Lee EA (2002) Yield, potential yield, yield stability and stress tolerance in maize. Field Crops Res 75:161–170

Tracy WF (2000) Sweet corn. In: Hallauer AR (ed) Specialty corns, 2nd edn. CRC, Boca Raton, FL, pp155–199

Tracy WF (2004) Breeding: the backcross method. In: Goodman RM (ed) Encyclopedia of crop science. Marcel Dekker, Inc, New York, pp 237–240

Tracy WF, Chandler MA (2006) The historical and biological basis of the concept of heterotic patterns in corn belt dent maize. In: Lamkey KR, Lee M (eds) Plant breeding: The Arnel R Hallauer international symposium. Blackwell Publishing, Ames, IA, pp 219–233

Tracy WF, Chang Y-M (2007) Effects of divergent selection for endosperm appearance in a *sugary1* maize population. Maydica 52:71–79

Tracy WF, Goldman IL, Tiefenthaler AE, Schaber MA (2004) Trends in productivity of US crops and long-term selection. Plant Breeding Rev 24:89–108

Troyer AF (1996) Breeding widely adapted, popular maize hybrids. Euphytica 92:163–174

Troyer AF (1999) Background of U.S. hybrid corn. Crop Sci 39:601–626

Troyer AF (2000a) Temperate corn: background, behavior, and breeding. In: Hallauer AR (ed) Specialty corns, CRC Press, Boca Raton, FL, pp 393–466

Troyer AF (2000b) Origins of modern corn hybrids. In: Wilkinson D (ed), Proceedings of the 55th annual corn and sorghum research conference. Am Seed Trade Assn, Washington DC, pp 27–42

Troyer AF (2004) Persistent and popular germplasm in seventy centuries of corn evolution. In: Smith CW, Betran J, Runge ECA (eds), CORN – Origin, history, technology, and production. John Wiley and Sons Inc, New Jersey, pp 133–231

Troyer AF (2008) Development of hybrid corn and the seed corn industry, this volume

Part III
The Maize Genome

Cytogenetics and Chromosomal Structural Diversity

James A. Birchler and Hank W. Bass

1 Introduction

The cytogenetics of maize can be traced to the pioneering work of Barbara McClintock, who first defined the ten chromosomes of maize at the pachytene stage of meiosis (McClintock 1929). The chromosomes at this stage are not as condensed as in somatic cells and have many distinguishing characteristics such as knob heterochromatin, chromomere patterns and arm ratios that permit each member of the karyotype to be identified. This seminal contribution allowed the genetic linkage groups to be associated with the respective chromosome and initiated a series of findings ranging from the demonstration of the cytological basis of crossing over to the nature of chromosomal aberrations such as inversions, translocations, deficiencies and ring chromosomes (see Birchler et al. 2004). This ability to identify each chromosome and its structure allowed the analysis of the Breakage-Fusion-Bridge (BFB) cycle that led to the concept of the need for a special structure at the ends of natural chromosomes (McClintock 1939; 1941), now referred to as the telomere. Cytological analysis was also important in the early recognition of transposable elements as the means to detect the fact that Disssociation changed its site for induction of chromosomal breakage (McClintock, 1950).

As alluded to above, the central fact of maize cytogenetics is that the gametic number of chromosomes is 10 and thus the sporophytic number is 20. The chromosomes were numbered from 1 to 10 in descending order of size, although variation exists in length due to the presence of extensive knob heterochromatin in different varieties. There is also variation for arm ratios due to small inversions surrounding the centromeres of some chromosomes (McClintock 1933; Lamb et al. 2007b). Monoploid as well as triploid to octoploid maize have been reported, although tetraploid varieties are the only other ploidy than diploidy that can be maintained

J.A. Birchler and H.W. Bass
Division of Biological Sciences, University of Missouri, Columbia, MO 65211
Biological Science, Florida State University,
Tallahassee, FL 32306-4370

from one generation to the next. This is due to the random assortment of chromosomes in monoploid and triploids or to sterility of the higher ploidies (Rhoades and Dempsey 1966).

In addition to the normal chromosomal complement (referred to as A chromosomes), there exists in some varieties a supernumerary chromosome called the B chromosome (Longley 1927; Randolph 1928). The B chromosome does not carry any genes that have been shown to complement any mutation in the A chromosomes so it is basically inert except for the functions required for its perpetuation (Carlson 1978). Being neither required nor detrimental at low copy numbers, the B chromosome is a selfish element that is maintained by an accumulation mechanism consisting of a high rate of nondisjunction at the second pollen mitosis (Roman 1947) coupled with the sperm with B chromosomes preferentially fertilizing the egg as opposed to the polar nuclei in the process of double fertilization (Roman 1948). Because of the property of nondisjunction at the mitosis that produces the two maize sperm, translocations between the B chromosome and various A chromosomes have served as a unique and valuable tool in maize genetics (Beckett 1978). Recently, they have been used to produce engineered minichromosomes via telomere mediated chromosomal truncation (Yu et al. 2007b). Using particle cobombardment of a particular sequence together with telomere repeats will place the desired construct onto a mini B chromosome that is truncated by the inclusion of the telomere sequences. This technique should allow the placement of desired genes or site-specific recombination cassettes for further additions onto an independent chromosome.

The genetic factors that have been identified on the B chromosome appear to be involved in affecting the nondisjunction property of the centromeric region (Roman 1947; Ward 1973; Lin 1978; Han et al. 2007). It has long been known that the very tip of the long arm is required in the same nucleus as the centromere, but not necessarily on the same chromosome, to cause the B centromeric region to undergo nondisjunction. Another site in the proximal euchromatin of the B chromosome is also required for conditioning nondisjunction of the centromeric region (Lin 1978).

The origin of the B chromosome is obscure. Because it has never been observed to complement any mutation in the A chromosomes, there is no evidence from genetics about any possible relationship to an A chromosome. However, in terms of the common retrotransposons known in maize (Theuri et al. 2005; Lamb et al. 2007c), the B chromosome carries some representation of all of them tracing back to Tekay, which is thought to have the oldest period of transposition of those known in maize. Thus, the B chromosome has likely been present in the same evolutionary lineage for a minimum of about 6 millions years (Ma et al. 2004), the half life for retrotransposons after insertion, either as an A or B chromosome. An alternative hypothesis that the B was donated from a relative of maize via a wide cross is unlikely in the noted time frame, given the presence of retroelements that have been active during the recent evolutionary history.

Despite the common DNA elements between A and B chromosomes, the B chromosome does possess several B specific DNA repeats. The first one discovered is present in and around the centromeric region and is referred to as the B specific repeat (ZmBs) (Alfenito and Birchler 1993). It is also present at some minor sites along the long arm of the B chromosome including a position near the end of the long arm that is detectable on somatic chromosomes (Lamb et al. 2005). A second B specific element is called the CL repeat, which is present at several sites in the long arm and near the centromeric region (Cheng and Lin 2004). Lastly, the Stark repeat is present in a block of the long arm heterochromatin (Lamb et al. 2007d). Moreover, the B chromosome is unique in containing numerous small arrays of the normally centromeric element, CentC, along the length of the chromosome, but without any apparent centromeric activity (Lamb et al. 2005).

2   Somatic Identification

Somatic chromosomes can be analyzed mainly by root tip mitotic metaphase spreads. When stained with traditional methods, for example with orcein, the chromosomes cannot be reliably identified other than chromosome 6, which possesses the secondary constriction at the site of the nucleolar organizer region (NOR). The B chromosome is also distinctive by virtue of its telocentric shape. Such traditional methods are useful for determining the number of chromosomes present in an individual, but they do not permit a definitive determination of each chromosome.

Two recent developments have allowed each chromosome to be identified in somatic chromosome preparations. The first such development was the application of pressurized nitrous oxide gas to root tips to yield a very high number of metaphase spreads (Kato 1999a). Root tips from young seedlings or from older plants are treated in 10 atmospheres of nitrous oxide. Comparisons of this technique to various previously used methods showed a far greater number of metaphase spreads (Kato 1999a).

The second development involved the assembly of a collection of probes of repeated gene arrays that could be used in fluorescent in situ hybridization (Kato et al. 2004). The presence of specific arrays at localized positions in the genome allows all the chromosomes to be identified when the probe collection is hybridized to metaphase spreads. The collection includes the 18S and 5S ribosomal RNA genes, the CentC centromere unit, the Cent4 repeat near the centromere of chromosome 4 (Page et al. 2001), a TAG microsatellite cluster of variable locations but usually on chromosomes 1, 2 and 4, two types of knob heterochromatin (180 bp (Peacock et al. 1981) and TR1 (Ananiev et al. 1998b) and lastly two types of subtelomere repeats (Gardiner et al. 1996; Kato et al. 2004) that are in greater abundance on some chromosomes compared to others. This collection of probes allows all ten pairs of homologues to be distinguished in virtually any inbred line. However, in some cases, an initial ambiguity of chromosomal assignment might occur, but these can be easily resolved by using single gene probes that are localized to chromosome (see below).

The advantages that somatic analyses have over meiotic studies include the fact that they can be conducted at any time without the need to grow the material to be analyzed to meiotic stages. The analyses can be completed from start to finish within a few days. Many individual seedlings can be analyzed in a short time frame and the examined individuals can be saved and grown to maturity for genetic studies or maintenance of novel chromosomal configurations.

The repeated array cocktail revealed that there is extensive variation for each type of sequence in the genome. Knob variation has long been known, but the karyotyping cocktail revealed that CentC, the microsatellite, both types of knobs and the rDNA clusters are highly variable in copy number across maize inbred lines. Extended exposures demonstrated that the knob repeat is represented at most if not all chromosomes ends to some degree, suggesting some function for this sequence involving chromosomal termini (Lamb et al. 2007c). In maize relatives, knobs are primarily terminal, but have been moved interstitially in maize due to their preferential segregation activity induced by abnormal chromosome 10 (Rhoades 1942; Buckler et al. 1999).

3   Single Gene Detection and Chromosome Identification

The development of more sensitive labeling techniques and procedures to crosslink the chromosomes to slides allowed the sensitivity of FISH on somatic chromosomes to detect a lower target limit on the order of 2–3 kb in length (Kato et al. 2006; Yu et al. 2007b; Lamb et al. 2007a). Because many maize genes are larger than this size, they can be visualized on root tip chromosomes. Thus, a collection of large genes and gene clusters was used to develop a single gene karyotyping cocktail (Lamb et al. 2007a). No variation has been found for the position of these genes among inbred lines, so such an approach can be used to identify each chromosome pair in a manner that bypasses the complication of the variability of the repeat arrays described above. This approach can obviously be customized to one's own interest or chromosomal region. In some cases, a combination of repeat arrays and single gene chromosome identification can be used for an inbred line to verify the karyotype. Thereafter, the repeat arrays can be used if each chromosome is distinct because they exhibit brighter signals and the labeling process for them is therefore not as expensive.

With this combination of chromosomal identification tools, it is possible to localize genes with unknown position in the genome. Moreover, the technique permits the visualization of transgenes to genomic position as well as the number of them in any one individual. The ability to detect transgene number and location greatly facilitates transformation efforts by eliminating any undesirable transformants early in the process. This method also can visualize individual transposable elements, such as *Activator*, *Suppressor-mutator* or *Mutator*, on the chromosomes (Yu et al. 2007b). All versions, active or inactive, will be detected, but transposition events can be documented by the absence of label at a progenitor site and the appearance of a new site elsewhere in the genome (Yu et al. 2007b).

Somatic chromosome analysis can also be used to locate bacterial artificial chromosomes (BACs) to genomic position. If a BAC is sequenced, the unique portions can be pooled and used as a probe onto somatic chromosomes. Despite the fact that the contributors to the pool represent different sites within about a 100 kb span, a signal can be detected on the chromosome. Depending on the location on the chromosome, the resolution of different ends of a BAC can be detected (Danilova and Birchler 2008). However, much greater distances of separation of the targets are needed for a definitive ordering of two probes.

4 Sorghum BACs as Locus-Specific FISH Probes

Another strategy for localizing sequences to maize chromosomes is to use syntenic sorghum genomic BAC clones as FISH probes (Zwick et al. 1998; Koumbaris and Bass 2003). This approach has been shown to be useful for pachtyene FISH mapping with maize chromosome addition lines of oat (Koumbaris and Bass 2003; Amarillo and Bass 2007). Maize marker-selected sorghum BACs overcome the probe-detection limit and provide an indirect way to define the cytogenetic location of sequences corresponding to targets such as maize RFLP probes, many of which are smaller than 1 kb. A detailed FISH map of maize chromosome 9 with the locations of more than 30 loci illustrates the use of this technique while uncovering regions of genome hyperexpansion (Amarillo and Bass 2007). A combination of cytogenetic approaches such as BAC FISH, single-gene FISH, and retroelement genome painting (described below) holds great promise for sorting out the structural diversity of the maize genome within and among different lines of maize and its relatives.

5 Cytological Maps of Meiotic Chromosomes

The analysis of mid-prophase chromosomes from pollen mother cells of maize has led to the production of cytological maps and a general meiotic karyotype (see Dempsey 1994). The foundation of maize cytogenetics comes from the early work of McClintock (1929) in which the meiotic karyotype of maize displayed the positions of the centromeres, knobs, and the nucleolus, but was otherwise devoid of genetic loci. This view of the normal complement of maize chromosomes remained unchanged for many decades, except for the mapping of translocation breakpoints and the more recent development of FISH and recombination nodule maps.

The cytological map coordinates are based on the relative position along the short or long arms, as widely used by Longley, Beckett, and colleagues, to characterize the cytological position of translocation breakpoints along pachytene fibers (Longley 1961; Beckett 1978). The relative arm position locus coordinate system is the basis for the modern meiotic karyotype. These cytological map units have recently been referred to as centiMcClintock, cMC, units (Lawrence et al. 2006) such that 1 cMC refers to 1% of the length of the chromosome arm upon which a given locus resides. The meiotic prophase karyotype serves to integrate several types of cytological maps including those derived from translocation breakpoint data (Sheridan and Auger, 2006), recombination nodule distribution data (Anderson et al. 2003; 2004), and FISH data (Chen et al. 2000; Sadder et al. 2000; Koumbaris and Bass 2003; Wang et al. 2006; Amarillo and Bass 2007).

6 Overview of Cytogenetic Tools

Compared to any other plant species, maize has an exceptional collection of cytogenetic tools available. These tools are described in other chapters in this volume and will only be summarized here. The complete set of whole chromosome trisomics was isolated some years ago (Rhoades 1955). The complementary collection of mono-somics for each chromosome must be produced anew from the r-X1 system that produces nondisjunction of chromosomes at the second gametophyte mitosis (Weber 1994; Zhao and Weber 1988). Monoploid maize can be produced by stock 6, which generates maternally derived monoploids (Coe 1959) or by the *indeterminate gametophyte* (*ig*) mutation that fosters paternal haploid production with the maternal cytoplasm (Kermicle 1969). Triploids can be routinely produced in any one inbred line or between inbred lines by treatment of developing pollen with trifluralin, which blocks the second pollen mitosis, thus producing a single diploid sperm (Kato 1997; 1999b; 1999c). This single sperm can function in fertilization to produce a triploid zygote and result in viable kernels can result if a follow-up pollination is conducted to fertilize the polar nuclei to form an accompanying triploid endosperm (Kato 1999b). Tetraploids can be produced using the *elongate* (*el*) mutation that generates unreduced female gametes (Rhoades and Dempsey 1966) or by nitrous oxide gas treatment of diploid zygotes shortly after fertilization followed by screen-ing the progeny for increased chromosome numbers (Kato and Birchler 2006). Translocations between the B chromosome and the various A chromosome arms confer onto the translocated segment the property of nondisjunction during the formation of the two sperm, thus creating deficient and duplicate male gametes (Roman 1947; Beckett 1978). These can be used to locate phenotypic mutations to chromosome arm, to conduct dosage series, to characterize mutations as to null, leaky or gain of function, to test for parental imprinting or for a myriad of other manipulations. In addition, there is a collection of several hundred A-A translocations that exchange portions of different nonhomologous chromosomes (Longley 1961). They can be used as additional mapping markers, as a means to bring transposable elements nearby a target locus (Auger and Sheridan 1999), as a method to produce segmental duplications (Gopinath and Burnham 1956; Birchler 1980; Vollbrecht and Hake 1995) or in combination with B-A translocations to produce compound B-A translocations that can expand the range of dosage manipulations (Rakha and Robertson 1970; Birchler 1981; Sheridan and Auger 2006) or to locate genes to chromosomal region as determined by the translocation breakpoints involved (Beckett 1978; Sheridan and Auger 2006). There is also a complete set of maize chromosomes, including the B chromosome, that have been transferred to an oat background to form a set of oat-maize addition lines (Ananiev et al. 1997; Kynast et al. 2001). These lines have been particularly useful in studying individual chromosomes of maize as separate entities from the remainder of the genome.

7 Interphase

Very little is known about interphase chromosomes in maize. As a byproduct of FISH on root tip chromosomes, interphase nuclei are routinely present and labeled by the same probes. From such analysis, it is clear that sites on homologous chromosomes do not appear to be routinely paired in interphase. Endoreduplication has been described to some degree from various differentiated maize tissues (Biradar and Rayburn et al. 1993; Biradar et al. 1993).

Endoreduplication occurs as a regular process in the differentiation of the endosperm. Following fertilization of the central cell of the female gameto-phyte, the primary endosperm nucleus begins several rounds of division ahead of the zygote (Mol et al. 1994). Subsequent cellularization occurs at about 4 days after pollination. At approximately 12 days after pollination, the chromosomes continue to replicate but cell division does not occur, resulting in endoreduplicated chromosomes (Kowles and Phillips 1985; Kowles et al. 1990). By using FISH probes onto endosperm cells at this stage, the replications can sometimes be determined by the number of hybridization signals detected on the endoduplicated chromosomes (Bauer and Birchler 2006). Interestingly, centromeres and knobs exhibit a single site of hybridization, suggesting a greater adhesion of the replicated sequences than those of single genes that often show a scatter shot pattern. Using various B specific repeats, the full length of the endoreduplicated B chromosome can be visualized (Bauer and Birchler 2006).

8 Pollen Cytogenetics

FISH probes can also be applied to mature pollen to visualize the behavior of specific chromosomes (Shi et al. 1996; Rusche et al. 1997; 2001). In pollen, the B chromosome has been the object of study because of its unusual behavior in pollen mitoses (Shi et al. 1996; Rusche et al. 1997). Using FISH, it was possible to establish that the B chromosome undergoes nondisjunction at a low frequency at the first male gametophyte mitosis and then at a very high frequency at the second division (Rusche et al. 1997). FISH was also used to establish that truncated B chromosomes fail to undergo this nondisjunction as predicted from the fact that they are missing the terminal portion of the B chromosome that is required for this process (Han et al. 2007).

It was also possible to visualize the behavior of an inactive B centromere translocated to the short arm of chromosome 9 (Han et al. 2007). In the absence of normal B chromosomes, this inactive centromere would disjoin normally. However, when normal B chromosomes were added to the genotype, nondisjunction of the B centromere sequences was induced despite the fact that the centromere was inactive. In some cases, this produced nondisjunction of the whole of chromosome 9, but more often the short arm of chromosome nine was fractured, resulting in the establishment of the breakage-fusion-bridge cycle. The failure of separation of the two chromosomes 9 with the terminal inactive B centromere could be visualized in young pollen (Han et al. 2007).

9   Organellar DNA Insertions into the Nucleus

The sequences of the Arabidopsis and rice genomes revealed the insertion of large fragments of the mitochondrial or plastid genome into the nucleus (Stupar et al. 2001; Yuan et al. 2002). Probing maize somatic chromosomes with individual cosmids of the mitochrondrial genome revealed numerous insertions across the genome (Lough et al. 2008). In the inbred line B73, there is most of the mitochrondrial genome inserted in a proximal position on the long arm of chromosome 9. Examination of ten inbred lines revealed extensive variation for the positions of mitochondrial sequences. Given that the FISH technique has a lower limit of sensitivity, there is likely to be further smaller insertions of organellar DNA. Within the span of a few generations within an inbred line, changes in insertions sites were documented, revealing the frequency with which changes can occur. Thus, organellar insertions contribute to the diversity of the maize genome and likely contribute to the spontaneous mutation frequency as well.

10   Retroelement Genome Painting

As noted above, there are numerous retrotransposons in maize that comprise a major fraction to the genome. By analyzing the divergence of the long terminal repeats (LTRs) that are identical upon insertion, the ages of the major bursts of transposition can be estimated (SanMiguel et al. 1996). In general, the transposon sequence will degenerate over evolutionary time with a half life on the order of 4–5 million years. In the maize lineage several retrotransposons have had major activity in recent evolutionary time (SanMiguel et al. 1998). Thus, the maize genome is populated by retroelements that are not major contributors to the related genus *Tripsacum* (Lamb and Birchler 2006). By using individual retroelements as probes, the chromosomes in maize-Tripsacum hybrids can be distinguished because the maize genome has a different copy number of specific elements, a procedure referred to as retroelement genome painting. By cloning retroelements from Tripsacum that have expanded in its lineage but not in Zea, contrasting probes accentuate the two genomes. In a tri-species hybrid of *Zea mays*, *Zea diploperennis* and *Tripsacum dactyloides*, contrasting probes from maize and Tripsacum can distinguish all three genomes. This technique can be used to detect introgressions of Tripsacum into maize and potentially the opposite. By combining retroelement painting with single gene probings, the syntenic relationships between the two genera could be examined in hybrid materials.

11   Retroelement Distributions

Careful examination of the distribution of individual retroelements painted onto maize pachytene chromosomes reveals that each as a distinct pattern along the chromosome (Lamb et al. 2007c). Some few are uniformly present throughout the length of the chromosomes, but most have either a more proximal or distal enrichment. This nonrandom distribution is likely due to the targeted insertion of specific retroelements to various chromatin configurations. The retroelements that are more concentrated around centromeric regions are typically found in greater abundance on the B chromosome than those elements that are enriched more distally (Lamb et al. 2005; 2007c).

12   Telomere Structure

The telomeres of maize consist of repeats of the sequence TTTAGGG, which is the common sequence in most plant species (Richards and Ausubel 1988). Genetic variation exists in maize for the control of telomere length (Burr et al. 1992). The telomeric complex proteins in maize are not yet characterized, but the single myb histone (SMH) proteins have biochemically defined double strand DNA-binding activity directed to telomere repeats (Marian et al. 2003). Another family of proteins, the initiator binding proteins (IBP1, IBP2) of maize, have a telomere DNA binding-like domain (Lugert and Werr 1994) and are predicted to be components of the telomeric complex.

Telomeres protect the ends of chromosomes from fusion as evidenced by the fact that broken chromosomes tend to fuse at the broken ends in the gametophyte generation and the endosperm (McClintock 1939; 1941). When chromosomes are broken in meiosis or subsequently in the gametophyte, a cycle of fusion of broken ends followed by dicentric bridge formation is established that continues. In sporophytic tissues, the broken ends are "healed" by the addition of telomeres to the ends of chromosomes (Chao et al. 1996). However, if two broken chromosomes are present in sporophytic tissues, then a "chromosome" type cycle is established that will continue throughout development until the developmental lineage is killed by the deficiencies produced or until dicentrics are no longer produced.

Using a foldback duplication recombined onto a B-A translocation involving chromosome arm 9S, minichromosomes produced from the BFB cycle can be recovered in the subsequent generation (Zheng et al. 1999). They are stabilized either by inactivation of one of the two centromeres in the dicentric or by becoming monocentric (Han et al. 2006). Some such minichromosomes consist of little more than the centromere of the B chromosome, but can be maintained from one generation to the next (Kato et al. 2005). The very small chromosomes typically have bipolar attachment in meiosis I and thus sister chromatid cohesion is dissolved in anaphase I even when two minichromosome "homologues" have paired in prophase of meiosis. Distribution of chromosomes in meiosis II is to one pole or the other, thus accounting for the reasonable levels of inheritance. The "tiny fragment" derived from an A chromosome behaves in the same fashion (Maguire 1987) indicating that this behavior is a property of small chromosomes.

13   Telomere Truncation

Transformation via Agrobacterium or biolistic bombardment of a plasmid containing telomere sequences at one end will often result in the fracture of chromosomes, apparently during the integration process (Yu et al. 2006). By designing the plasmids to different specifications, it is possible to introduce sequences of choice to the ends of chromosomes using this technique. Of particular note are site-specific recombination cassettes such as the Cre/lox or FLP/FRT systems. These have been introduced onto minichromosomes of maize and shown to recombine with sites on other chromosomes (Yu et al. 2007a).

14   Engineered Minichromosomes

Using telomere truncation, engineered minichromosomes were readily generated from the B chromosome (Yu et al. 2007a). By performing a cobombardment with telomere-containing plasmids, both DNAs tend to insert together and, with the telomere sequences being present, the terminal portion of the chromosome is cleaved off. The end product is a small chromosome with the desired genes and site-specific recombination cassettes at the terminus of a small chromosome. Reporter cassettes indicate that genes introduced in this fashion are faithfully expressed. Thus, genes of interest can be added to a mini B chromosome by cobombardment with telomere containing plasmids. The use of the B chromosome provided the opportunity to manipulate the dosage by the introduction of full length B chromosomes that supply the nondisjunction factors in the long arm.

Minichromosomes carrying the site-specific recombination cassettes were also recovered for A chromosomes in either spontaneous tetraploid events or by fission of a centromere such that no large deficiencies would cause selection against the transformation event (Yu et al. 2007a). Thus, it should be possible to produce minichromosomes in most plant species given the similarity of the telomere sequences. One could rely on spontaneous tetraploid events during the transformation procedure, which is not uncommon, or use polyploids as the starting material for transformation. Trisomic or addition lines should foster the truncation of a specific chromosome because there would be no selection against truncation events for that chromosome.

Engineered minichromosomes provide the opportunity to produce chromosomes to specification for studies of chromosome structure and function by the targeting of future additions to the site-specific recombination cassettes. In a practical sense, they provide the ability to stack genes on an independent chromosome that will overcome the problems of linkage drag present when multiple transgenes for various traits are combined. Desirable genes to be stacked would include herbicide and insect resistances together with stress tolerance genes. Eventually, it should become possible to add whole biochemical pathways to maize to confer new properties to the plant or to use maize as a factory for multiple foreign proteins or metabolites.

Acknowledgements This work was supported by the following grants from the National Science Foundation: MCB0091095, DBI 0321639, DBI 0421671, DBI 0423898, DBI 0501712 and DBI 0701297.

References

Alfenito, M.R. and J. A. Birchler, J.A. (1993) Molecular characterization of a maize B-chromosome centric sequence. Genetics 135, 589–597.

Amarillo, F.I.E. and Bass, H.W. (2007) A transgenomic cytogenetic sorghum (*Sorghum propinquum*) BAC FISH map of maize (*Zea mays* L.) pachytene chromosome 9, evidence for regions of genome hyperexpansion. Genetics (in press).

Ananiev, E., Riera-Lizarazu, O., Rines, H. and Phillips, R.L. (1997) Oat-maize chromosome addition lines: A new system for mapping the maize genome. Proc. Nat. Acad. Sci. USA 94, 3524–3529.

Ananiev, E., Phillips, R.L. and Rines, H. (1998a) Chromosome-specific molecular organization of maize (*Zea mays* L.) centromeric regions. Proc. Nat. Acad. Sci.USA 95, 13073–13078.

Ananiev, E., Phillips, R.L. and Rines, H. (1998b) A knob-associated tandem repeat in maize capable of forming fold-back DNA segments: Are chromosome knobs megatransposons? Proc. Nat. Acad. Sci. USA 95, 10785–10790.

Anderson L.K., Doyle, G.G., Brigham, B., Carter, J., Hooker, K.D., Lai, A., Rice, M. and Stack, S.M. (2003) High-resolution crossover maps for each bivalent of *Zea mays* using recombination nodules. Genetics 165, 849–865

Anderson, L.K., Salameh, H., Bass, H.W., Harper, L.C., Cande, W.Z., Weber, G. and Stack, S.M. (2004) Integrating genetic linkage maps with pachytene chromosome structure in maize. Genetics 166, 1923–1933.

Auger, D.L. and Sheridan, W. F. (1999) Maize stocks modified to enhance the recovery of Ac-induced mutations. J. Hered. 90, 453–459.

Bauer, M.J. and Birchler, J.A. (2006) Organization of endoreduplicated chromosomes in the endosperm of *Zea mays* L. Chromosoma 115, 383–394.

Beckett, J.B. (1978) B-A translocations in maize. J. Heredity 69, 27–36.

Biradar, D.P. and Rayburn, A.L. (1993) Intraplant nuclear DNA content variation in diploid nuclei of maize (*Zea mays* L.). J. Exp. Bot. 44, 1039–1044.

Biradar, D. P., Rayburn, A.L. and Bullock, D.G. (1993) Endopolyploidy in diploid and tetraploid maize (*Zea mays* L.). Annals of Botany 71, 417–421.

Birchler, J.A. (1980) The cytogenetic localization of the *alcohol dehydrogenase-1* locus in maize. Genetics 94, 687–700.

Birchler, J.A. (1981) The genetic basis of dosage compensation of *alcohol dehydrogenase-1* in maize. Genetics 97, 625–637.

Birchler, J.A., Auger, D.L. and Kato, A. (2004) Cytogenetics of corn. IN: *Corn: Origin, History, Technology and Production.* Edited by C. Wayne Smith, Javier Betran and Ed Runge. John Wiley and Sons, New York.

Buckler, E., Phelps-Durr, T.L., Buckler, C.S., Dawe, R.K., Doebley, J.F. and Holtsford, T.P. (1999) Meiotic drive of chromosomal knobs reshaped the maize genome. Genetics 153, 415–426.

Burr, B., Burr, F.A., Matz, E.C. and Romero-Severson, J. (1992) Pinning down loose ends: mapping telomeres and factors affecting their lengths. Plant Cell 4, 953–960.

Carlson, W.R. (1978). The B chromosome of corn. Annu Rev Genet 12, 5–23.

Chao, S., Gardiner, J.M., Melia-Hancock, S. and Coe, E.H., Jr. (1996) Physical and genetic mapping of chromosome 9S in maize using mutations with terminal deficiencies. Genetics 143, 1785–1794.

Chen C.C., Chen, C.M., Hsu, F.C., Wang, C.J., Yang, J.T. and Kao, Y.Y. (2000) The pachytene chromosomes of maize as revealed by fluorescence in situ hybridization with repetitive DNA sequences. Theor. Appl. Genet. 101, 30–36.

Cheng, Y.M. and Lin, B.Y. (2004) Molecular organization of large fragments in the maize B chromosome: indication of a novel repeat. Genetics 166, 1947–1961.

Coe, E.H. (1959) A line of maize with high haploid frequency. Am. Naturalist 93, 381–382.

Danilova, T.V. and Birchler, J.A. (2008) Integrated cytogenetic map of mitotic metaphase chromosome9 of maize, resolution, sensitivity and binding paint development. Chromosoma 117, 345–356.

Dempsey, E. (1994) Traditional analysis of maize pachytene chromosomes. Freeling, M. and V. Walbot (Ed.). *The Maize Handbook.* Springer-Verlag New York, Inc.: New York, New York, pp 432–441.

Gardiner, J.M., Coe, E.H. and Chao, S. (1996) Cloning maize telomeres by complementation in *Saccharomyces cerevisiae.* Genome 39, 736–748.

Gopinath, D. and Burnham, C.R. (1956) A cytogenetic study in maize of deficieny-duplication produced by crossing interchange involving the same chromosomes. Genetics 41, 382–395.

Han, F., Lamb, J.C. and Birchler, J.A. (2006) High frequency of centromere inactivation resulting in stable dicentric chromosomes of maize. Proc. Natl. Acad. Sci. USA 103, 3238–3243.

Han, F., Lamb, J.C., Yu, W., Gao, Z. and Birchler, J.A. (2007) Centromere function and nondisjunction are independent components of the maize B chromosome accumulation mechanism. The Plant Cell 19, 524–533.

Kato, A. (1997a) An improved method for chromosome counting in maize. Biotech. Histochem. 72, 249–252.

Kato, A. (1997b) Induced single fertilization in maize. Sex. Plant. Reprod. 10, 96–100.

Kato, A. (1999a) Air drying method using nitrous oxide for chromosome counting in maize. Biotech. Histochem. 74, 160–166.

Kato, A. (1999b) Induction of bicellular pollen by trifluralin treatment and occurrence of triploids and aneuploids after fertilization in maize. Genome 42, 154–157.

Kato, A. (1999c) Single fertilization in maize. J Hered. 90, 276–280.

Kato, A., Albert, P.S., Vega, J.M. and Birchler, J.A. (2006) Sensitive FISH signal detection in maize using directly labeled probes produced by high concentration DNA polymerase nick translation. Biotechnic & Histochemistry 81, 71–78.

Kato, A. and Birchler, J.A. (2006) Induction of tetraploid derivatives of maize inbred lines by nitrous oxide gas treatment. Journal of Heredity 97, 39–44.

Kato, A., Lamb, J.A. and Birchler, J.A. (2004) Chromosome painting in maize using repetitive DNA sequences as probes for somatic chromosome identification. Proc. Natl. Acad. Sci. USA 101, 13554–13559.

Kato, A., Zheng, Y.-Z., Auger, D.L., Phelps-Durr, T., Bauer, M.J., Lamb, J.C. and Birchler, J.A. (2005) Minichromosomes derived from the B chromosome of maize. Cytogenetic and Genome Research 109, 156–165.

Kermicle, J.L. (1969) Androgenesis conditioned by a mutation in maize. Science 166, 1422–1424.

Koumbaris, G.L. and Bass, H.W. (2003) A new single-locus cytogenetic mapping system for maize (*Zea mays* L.): overcoming FISH detection limits with marker-selected sorghum (*S. propinquum* L.) BAC clones. Plant J 35, 647–59.

Kowles, R. V. and Phillips, R.L. (1985) DNA amplification patterns in maize endosperm nuclei during kernel development. Proc. Natl. Acad. Sci. USA 82, 7010–7014.

Kowles, R.V., Srienc, F. and Phillips, R.L. (1990) Endoreduplication of nuclear DNA in the developing maize endosperm. Dev. Genet. 11, 125–132.

Kynast, R.G., Riera-Lizarazu, O., Vales, M.I., Okagaki, R.J., Maquieira, S.B., G. Chen, G., Ananiev, E.V., Odland, W.E., Russell, C.D., Stec, A.O., Zaia, S.M., Rines, H.W., and Phillips, R.L. (2001) A complete set of maize individual chromosome additions to the oat genome. Plant Physiol. 125, 1216–1227.

Lamb, J.C., Kato, A., and Birchler, J.A. (2005) Centromere associated sequences are present throughout the maize B chromosome. Chromosoma 113, 337–349.

Lamb, J.C. and Birchler, J.A. (2006) Retroelement Genome Painting: Cytological visualization of retroelement expansions in the genera Zea and Tripsacum. Genetics 173, 1007–1021.

Lamb, J.C., Danilova, T., Bauer, M.J., Meyer, J., Holland, J.J., Jensen, M.D. and Birchler, J.A. (2007a) Single gene detection and karyotyping using small target FISH on maize somatic chromosomes. Genetics 175, 1047–1058.

Lamb, J.C., Meyer, J.M., Corcoran, B., Kato, A., Han, F. and Birchler, J.A. (2007c) Distinct chromosomal distributions of highly repetitive sequences in maize. Chromosome Research 15, 33–49.

Lamb, J.C., Meyer, J.M. and Birchler, J.A. (2007b) A hemicentric inversion in the maize line knobless Tama flint created two sites of centromeric elements and moved the kinetochore-forming region. Chromosoma 116, 237–247.

Lamb, J.C., Riddle, N.C., Cheng, Y., Theuri, J. and Birchler, J.A. (2007d) Localization and transcription of a retrotransposon-derived element on the maize B chromosome. Chromosome Research 15, 33–49.

Lawrence, C.J., Seigfried, T.E., Bass, H.W. and Anderson, L.K. (2006) Predicting chromosomal locations of genetically mapped loci in maize using the Morgan2McClintock Translator. Genetics 172, 2007–2009.

Lin, B.-Y. (1978) Regional control of nondisjunction of the B chromosome in maize. Genetics 90, 613–627.

Longley, A. (1927) Supernumerary chromosomes in *Zea mays*. Journal of Ag. Research 35, 769–784.

Longley, A. (1961) Breakage points for four corn translocation series and other corn chromosome aberrations. *USDA-ARS Crops Research Bulletin* No. 34–16, 1–40.

Lough, A., Roark, L, Kato, A., Ream, T.S., Lamb, J.C., Birchler, J.A. and Newton, K.J. (2008) Mitochondrial DNA transfer to the nucleus generates extensive insertion site variation in maize. Genetics, in press.

Lugert, T. and Werr, W. (1994) A novel DNA-binding domain in the Shrunken initiator-binding protein (IBP1). Plant Mol Biol 25, 493–506.

Ma, J., Devos, K.M. and Bennetzen, J.L. (2004) Analyses of LTR-retrotransposon structures reveal recent and rapid genomic DNA loss in rice. Genome Res 14, 860–869.

Marian, C.O., Bordoli, S.J., Goltz, M., Santarella, R.A., Jackson, L.P., Danilevskaya, O., Beckstette, M., Meeley, R. and Bass, H.W. (2003) The maize Single myb histone 1 gene, Smh1, belongs to a novel gene family and encodes a protein that binds telomere DNA repeats in vitro. Plant Physiol 133, 1336–50.

McClintock, B. (1933) The association of non-homologous parts of chromosomes in the mid-prophase of meiosis in *Zea mays*. Z Zellforsch Mikrosk Anat 19, 191–237

Maguire, M.P. (1987) Meiotic behavior of a tiny fragment chromosome that carries a transposed centromere. Genome 29, 744–747.

McClintock, B. (1929) Chromosome morphology in *Zea mays*. Science 69, 629–630.

McClintock, B. (1939) The behavior in successive nuclear divisions of a chromosome broken at meiosis. Proc. Natl. Acad. Sci. USA 25, 405–416.

McClintock, B. (1941) The stability of broken ends of chromosomes in *Zea mays*. Genetics 26, 234–282.

McClintock, B. (1950) The origin and behavior of mutable loci in maize. Proc. Natl. Acad. Sci. USA 36, 344–355.

Mol, R., Matthys-Rochon, E. and Dumas, C. (1994) The kinetics of cytological events during double fertilizatioin in *Zea mays* L. The Plant Journal 5, 197–206.

Page, B.T., Wanous, M.K. and Birchler, J.A. (2001) Characterization of a maize chromosome 4 centromeric sequence: Evidence for an evolutionary relationship with the B chromosome centromere. Genetics 159, 291–302.

Peacock, W.J., Dennis, E.S., Rhoades, M.M. and Pryor, A.J. (1981) Highly repeated DNA sequence limited to knob heterochromatin in maize. Proc. Natl. Acad. Sci. USA 78, 4490–4494.

Rakha, F.A. and Robertson, D.S. (1970) A new technique for the productioin of A-B translocations and their use in genetic analysis. Genetics 65, 223–240.

Randolph, L.F. (1928) Types of supernumerary chromosomes in maize. Anatomical Record 41, 102.

Rhoades, M.M. (1955) The cytogenetics of corn. IN: G. F. Sprague (ed.), *Corn and Corn Improvement*. American Society of Agronomy, Crop Science Society of America, Soil Science Society of America, Madison, WI. pp. 123–219.

Rhoades, M.M. and Dempsey, E. (1966) Induction of chromosome doubling at meiosis by the *elongate* gene in maize. Genetics 54, 505–522.

Rhoades, M.M. (1942) Preferential segregation in maize. Genetics 27, 395–407.

Richards, E.J. and Ausubel, F.M. (1988) Isolation of a higher eukaryotic telomere from *Arabidopsis thaliana*. Cell 53, 127–136.

Roman, H.L. (1947) Mitotic nondisjunction in the case of interchanges involving the B- type chromosome in maize. Genetics 32, 391–409.

Roman, H.L. (1948) Directed fertilization in maize. Proc. Natl. Acad. Sci. USA 34, 36–42.

Rusche, M., Mogensen, H.L., Chaboud, A., Faure, J.E., Rougier, M., Kiem, P. and Dumas, C. 2001. B chromosomes of maize (*Zea mays* L.) are positioned nonrandomly within sperm nuclei. Sex Plant Reprod. 13, 231–234.

Rusche, M., Mogensen, H.L., Shi, L., Keim, P., Rougier, M., Chaboud, A. and Dumas, C. (1997) B chromosome behavior in maize pollen as determined by a molecular probe. Genetics 147, 1915–1921.

Sadder, M., Ponelies, N., Born, U., and Weber, G. (2000) Physical localization of single-copy sequences on pachytene chromosomes in maize (*Zea mays* L.) by chromosome in situ suppression hybridization. Genome 43, 1081–1083.

SanMiguel, P., Gaut, B.S., Tikhonov, A., Nakajima, Y. and Bennetzen, J.L. (1998) The paleontology of intergene retrotransposons of maize. Nat Genet 20, 43–45.

SanMiguel, P., Tikhonov, A., Jin, Y.K., Motchoulskaia, N., Zakharov, D., Melake-Berhan, A., Springer, P.S., Edwards, K.J., Lee, M., Avramova, Z. and Bennetzen, J.L. (1996) Nested retrotransposons in the intergenic regions of the maize genome. Science 274, 765–768.

Sheridan, W.F. and Auger, D.L. (2006) Construction and uses of new compound B-A-A maize chromosome translocations. Genetics 174, 1755–1765.

Shi, L., Zhu, T., Mogensen, H.L. and Keim, P. (1996) Sperm identification in maize by fluorescence in situ hybridization. Plant Cell 8, 815–821.

Stupar, R.M., Lilly, J.W., Town, C.D., Cheng, Z., Kaul, S., Buell, C.R. and Jiang, J. (2001) Complex mtDNA constitutes an approximate 620-kb insertion on *Arabidopsis thaliana* chromosome 2: implication of potential sequencing errors caused by large-unit repeats. Proc. Natl. Acad. Sci., USA 98, 5099–5103.

Theuri, J., Phelps-Durr, T., Mathews, S. and Birchler, J.A. (2005) A comparative study of retrotransposons in the centromeric regions of A and B chromosomes of maize. Cytogenetic and Genome Research 110, 203–208.

Vollbrecht, E., and Hake, S. (1995) Deficiency analysis of female gametogenesis in maize. Dev. Genet. 16, 44–63.

Wang, C.J., Harper, L. and Cande, W.Z. (2006) High-resolution single-copy gene fluorescence in situ hybridization and its use in the construction of a cytogenetic map of maize chromosome 9. Plant Cell 18, 529–544.

Ward, E. (1973) Nondisjunction: localization of the controlling site in the maize B chromosome. Genetics 73, 387–391.

Weber, D.F. (1994) Use of maize monosomics for gene localization and dosage studies. In: *The Maize Handbook*. M. Freeling and V. Walbot, Editors. Springer-Verlag, NY pp. 350–358.

Yu, W., Lamb, J.C., Han, F. and Birchler, J.A. (2006) Telomere-mediated chromosomal truncation in maize. Proc. Natl. Acad. Sci. USA 103, 17331–17336.

Yu, W., Lamb, J.C., Han, F. and Birchler, J.A. (2007) Cytological visualization of DNA transposons and their transposition pattern in somatic cells of maize. Genetics 175, 31–39.

Yu, W., Han, F., Gao, Z., Vega, J.M. and Birchler, J.A. (2007) Construction and behavior of engineered minichromosomes in maize. Proc. Natl. Acad. Sci., USA 104, 8924–8929.

Yuan, Q., Hill, J., Hsiao, J., Moffat, K., Ouyang, S., Cheng, Z., Jiang, J. and Buell, C.R. (2002) Genome sequencing of a 239-kb region of rice chromosome 10L reveals a high frequency of gene duplication and a large chloroplast DNA insertion. Molecular Genetics and Genomics 267, 713–720.

Zhao, Z. and Weber, D.F. (1988) Analysis of nondisjunction induced by the *r-X1* deficiency during microsporogenesis in *Zea mays* L. Genetics 119, 975–980.

Zheng, Y-Z., Roseman, R.R. and Carlson, W.R. (1999) Time course study of the chromosome-type breakage-fusion-bridge cycle in maize. Genetics 153, 1435–1444.

Zwick, M. S., Islam-Faridi, M.N., Czeschin, D.G.J., Wing, R.A., Hart, G.E., Stelly, D.M. and Price, H.J. (1998) Physical mapping of the *liguleless* linkage group in *Sorghum bicolor* using rice RFLP-selected sorghum BACs. Genetics 148, 1983–1992.

Maize Genome Structure and Evolution

Jeffrey L. Bennetzen

Abstract The nuclear genome of maize contains the most complex structure of any yet studied in depth, with small gene islands immersed in seas of nested transposable elements. The DNA between genes is exceptionally unstable in maize, such that the ancestral existence of most or all intergenic TE insertions is erased within a few million years. The genes appear to show a very high mobility that is partly an outcome of the mis-annotation of TEs as genes and the presence of ~10,000 pseudogenes, compared to ~35,000 true protein-encoding genes and ~210,000 TE genes. The primary mechanisms of genomic structural change, namely DNA breakage/repair, recombination and transposition, have been identified. All of these processes have been found to be exceptionally active for genome rearrangement in maize, compared to other angiosperms. Further research is needed on the specificities exhibited by these mechanisms, on the reasons for their very high rates of activity in maize, and on the biological outcomes of this continuous genomic fluidity.

1 Introduction

The nuclear genomes of flowering plants are exceptionally complex, with a mean size of about 6500 Mb (Zonneveld et al. 2005) and a very high content of repetitive DNA. The few plant genomes that have been sequenced to date, or are being sequenced, are all unusual in genome structure, as evidenced by genome size. These range from the very small sorghum genome (~750 Mb) to the exceedingly small rice genome (~390 Mb) to the outlandishly small Arabidopsis genome (~140 Mb) (IRGSP 2005; SGPWP 2005; Liu and Bennetzen 2008). The largest and most complex plant genome yet targeted for 'full' genome sequence analysis has been that of maize (http://maizesequence.org/index.html), with ~2400 Mb of DNA per haploid nucleus in the B73 inbred (Rayburn et al. 1993). Fortunately, the general

J.L. Bennetzen
Department of Genetics, University of Georgia, Athens, GA 30602-7223, USA
maize@uga.edu

J.L. Bennetzen and S. Hake (eds.), *Maize Handbook - Volume II: Genetics and Genomics,*   179
© Springer Science+Business Media LLC 2009

structure of the maize genome (SanMiguel et al. 1996; Tikhonov et al. 1999), most simply described as islands of genes in seas of repetitive DNA, has turned out to be quite similar to that seen in segmental analyses of larger (i.e., more average) genomes like those of barley and wheat (Dubcovsky et al. 2001; Feuillet et al. 2001; Ramakrishna et al. 2002; Rostoks et al. 2002; SanMiguel et al. 2002). Hence, it is likely that maize will serve as a good model for understanding flowering plant genome structures, although the concentration of global sequence analyses on grasses rather than a broader range of angiosperms may be problematic.

The first publications on the comprehensive sequence analysis of the maize genome are expected in early 2009, so one might conclude that this is precisely the wrong time to be writing a chapter reviewing what we know about maize genome structure and evolution. This conclusion would be incorrect, however. Because of the way that the maize genome is being sequenced, clone-by-clone across contiguous regions (contigs) of ordered clones, the sequence will not initially provide a great wealth of information regarding maize genome structure beyond what currently exists. By definition, the clone-by-clone approach will miss some regions (e.g., areas that are full of tandem and/or highly conserved repeats) and thus will provide a more biased description of the genome than already-published studies that employed a more random clone analysis (Meyers et al. 2001; Haberer et al. 2005; Liu et al. 2007).

The first maize sequence will be the foundation of additional "re-sequencing" studies of multiple orthologous regions or haplotypes, thus facilitating comparisons of several sequenced maize genomes to each other and to the sequenced rice and sorghum genomes. Such studies will be incredibly informative, especially regarding rearrangements at the level of one to a few centiMorgans (cM), where very little is currently known (Gale and Devos 1998; Wilson et al. 1999; Wei et al. 2007). On its own, though, prior to these additional genome structure and diversity studies, the first maize genome sequence will only incrementally improve our understanding of the subject of this chapter, although it will be a tremendous tool for gene discovery and future genetic investigation. Hence, this chapter will indicate our current level of knowledge regarding the genome of maize and will point out issues that need to be resolved in future investigations.

2   Maize Genome Structure

From the earliest cytogenetic investigations, it was clear that maize contained tremendous variety in its nuclear genome components, both within any single genome and between the genomes of different maize accessions (McClintock 1929; 1960; Cooper and Brink 1937). Many of the obvious cytogenetic features were eventually shown to consist of tandem repetitive arrays like the ribosomal DNA repeats of the nucleolus (Phillips 1978), variants of the selfish 180 bp repeat found in knob heterochromatin (Peacock et al. 1981; Ananiev et al. 1998a; Hsu et al. 2003) and the 156 bp CentC repeats that are believed to be necessary for centromere function (Ananiev et al. 1998b). Thus, it was not surprising that the first comprehensive renaturation analysis indicated that the majority of the maize genome was comprised of repetitive DNA (Hake and Walbot 1980). However, the major cytological features like centromeres and knobs were known from early days on to be largely free of protein-encoding genes, so genome structure in "standard" genic areas remained unknown well into the 1990s. In this chapter, I will concentrate on our understanding of the structure and evolution of these genic regions, and will refer the reader to the chapters in this book that discuss centromeres (Dawe, this volume) or other cytogenetic features (Birchler & Bass, this volume; Carlson, this volume) in much greater detail.

Most of the early DNA sequence analyses of the maize nuclear genome were targeted on single genes and the amount of nearby DNA that could be contained in a phage lambda or cosmid recombinant DNA vector. Many of these studies were targeted on exceptionally large tandem gene families like those encoding zeins (Kriz et al. 1987; Liu and Rubenstein 1992; Boston and Larkins, this volume), so they yielded results that could not be extrapolated to the overall maize genome structure. The first large insert clone to be comprehensively sequenced from maize, an ~225 kb insert in a yeast artificial chromosome vector containing an *adh1* allele from inbred LH82, uncovered an unexpected genome structure that has since proven to be standard across other grass genomes. Single genes were found to be separated by large blocks of repetitive DNA, and this repetitive DNA was shown to be primarily composed of long terminal repeat (LTR) retrotransposons (SanMiguel et al. 1996). These LTR retrotransposons, thought to be rare in plant genomes at that time, turned out to comprise about 70% of the *adh1* YAC and more than 70% of the overall maize genome (SanMiguel and Bennetzen 1998; Meyers et al. 2001). Subsequent analyses have shown that the abundance of LTR retrotransposons is significantly correlated with genome size across all flowering plants investigated, from the >15% LTR retrotransposons of the ~140 Mb Arabidopsis genome (Liu and Bennetzen 2008) up to the >70% LTR retrotransposon content of more average genomes like those of barley or wheat. Moreover, recent studies in *Oryza* (Piegu et al. 2006) and *Gossypium* (Hawkins et al. 2006) have shown that specific LTR retrotransposon amplification bursts have been responsible for dramatic genome size increases in these lineages. Thus, it is now accepted that differential LTR retrotransposon amplification is one of the three major factors (along with polyploidy and different rates of DNA removal, see below) responsible for the great variation in angiosperm genome size (Bennetzen et al. 2005: Vitte and Bennetzen 2006).

The genes of maize, like those of other plants, tend to be highly compact. Plant proteins on average are not unusually small compared to animal proteins, but the small average intron size in plants (for instance, a median of less than 150 bp in maize (Bruggman et al. 2006)) tends to make the genes quite small. Unlike some studied animals, intron size does not correlate with genome size in flowering plants (Wendel et al. 2002). Moreover, the promoters and other regulatory elements in plant genes have been observed to be very near to the structural portions of the genes. In most transgenics, for instance, a few hundred base pairs from the promoter of a studied gene are enough to drive appropriate expression that is tissue-, development- and signal-appropriate to a reporter cassette. The very few cases of gene regulatory elements that have been found to be somewhat (e.g., more than 5 kb) removed from a gene (Stam et al. 2002, Clarke et al. 2004) have been observed to be purely quantitative in nature. Hence, nothing with the orientation- or location-independence, or tissue/development/signal-specificity, seen for what was originally defined as a viral or cellular transcriptional enhancer in animal research (Banerji et al. 1981, Moreau et al. 1981), has been discovered yet in plants. If these exist at all in angiosperms, they may be rare and/or very subtle in effect. Still, too few genetic and transgenic studies have been performed in complex genomes like those of maize to fully resolve this issue.

Figure 1 shows a standard maize region, that around the *Orp1* gene on chromosome 4 (Ma et al. 2005). The four annotated genes are surrounded by blocks of repetitive DNA (dark shading). This region is somewhat unusual in containing two genes islands, each with two annotated genes. Over 60% of maize gene islands contain only one gene (Bennetzen et al. 2005; Liu et al. 2007), so most maize genes have evolved in regions where they are buried in non-genic DNA. From the sequence analysis of 74 randomly selected BACs, it appears that ~30% of ~100 kb segments of the maize genome contain no genes at all (Liu et al. 2007). Many of these gene-free regions are likely to be found in pericentromeric and other heterochromatic blocks.

At a quantitative level, the most prominent features in the maize genome are the transposable elements. Of these, the LTR retrotransposons comprise the greatest percentage of the genome, with a current estimate of >72% of the total nuclear genome based on the analysis of several thousand sequenced bacterial artificial chromosome (BAC) inserts (http://maizesequence.org/index.html) (Baucom and Bennetzen, unpub. obs.). The LTR retrotransposons in maize are mainly inserted into each other (i.e., nested) (Fig. 2). Like all other transposable elements (TEs) that have been studied, these insertions are not random in their distribution. No inactivational mutation in maize has ever been associated with a high-copy-number LTR retrotransposon insertion, although many mutations have been associated with the insertion of low-copy-number LTR retrotransposons and other types of TEs. Moreover, the LTR retrotransposons of maize show a marked (~5-fold) preference for inserting into LTRs rather than the internal portions of the elements (Bennetzen 2000). All other TEs are expected to have insertion site preferences as well, ranging from highly constrained (like the single preferred insertion site in the *E. coli* chromosome for Tn7; Craig 1997) to the relatively permissive, such as *Mutator* of maize (see chapters by Lisch and by McCarty & Meeley in this volume). For the cut-and-paste DNA elements like *Ac/Ds*, *En/Spm*, *Mutator* and *MITEs*, their general insertion preferences tend to be precisely opposite to those for the high-copy-number LTR retrotransposons;

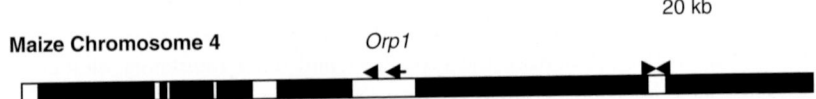

Fig. 1 The *Orp1* region on maize chromosome 4. Arrows indicate candidate genes, their approximate transcript sizes and their predicted direction of transcription. Black boxes indicate identified LTR retrotransposons

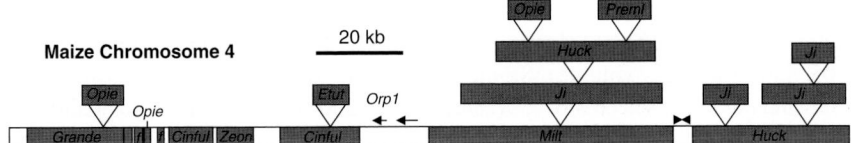

Fig. 2 LTR retrotransposon insertion patterns in the *Orp1* region on chromosome 4. Arrows indicate candidate genes as in Fig. 1. The LTR retrotransposons are shown as grey boxes with their names inside. Only the top grey boxes in the nested structure show the actual size of each element. The lower grey boxes in a nest include the size of the element plus the element indicated above that has inserted into it

namely an avoidance of repetitive/heterochromatic regions and a preference for insertion into or near genes (Bennetzen et al. 1988; Cone et al. 1988; Cresse et al. 1995; Tikhonov et al. 1999; 2000; Zhang et al. 2000; Kolkman et al. 2005).

3 The Genic Content of the Maize Genome

Precise gene annotation in plants is especially challenging because of the complexity of their genomes. For some classes of genes, particularly those that do not encode proteins, our annotation skills are quite primitive. Hence, as in this section, most gene number analyses are targeted only on protein-coding genes.

The TE genes within unrecognized transposable elements are commonly annotated as hypothetical nuclear genes, and this problem expands as genomes grow larger (more TEs) and if a large percentage of the TEs are present in a low-copy-number, meaning that they will be missed by repeat masking procedures (Bennetzen et al. 2004). Some investigators do not seem to trouble themselves with this issue, proceeding from the underlying conception that "genes are genes", so why worry about differentiating those inside TEs from those that are not inside TEs? This strategy, which is really an excuse for avoiding the effort of careful gene annotation, exhibits its absurdity when you appreciate that a smallish genome like maize will have over 200,000 genes inside TEs (e.g., *gag* and *pol* from LTR retrotransposons and *transposase* from DNA elements) and less than 50,000 genes that are not from TEs. Failing to differentiate TE genes would mean that TE content becomes the major determinant of gene number in a species, thereby missing the entire point of gene number determination. This is not just a theoretical issue: the approximate two-fold over-estimation of gene number in the original rice sequence releases, later shown to be an artifact primarily of mis-identifying TEs as genes (Bennetzen and Ma 2003; Bennetzen et al. 2004), initially led to much discussion and numerous studies to explain the biological significance of this mysteriously high copy number of genes in rice, and why so many were different from any genes found in Arabidopsis.

Maize has also been subject to inaccurate gene number predictions, for much the same reasons as rice and most other angiosperms, but with the additional problem created by the presence of numerous gene fragments inside a class of DNA elements called *Helitrons* (Kapitonov et al. 2001; Lal et al. 2003; Morgante et al. 2005; Lal et al., this volume). Studies using BAC end sequences (Messing et al. 2004) and randomly selected BACs (Haberer et al. 2005) for analysis led to predictions of a respective ~59,000 and 42,000–56,000 genes in the maize genome. These numbers are too high, mainly because many gene fragments and TE genes were annotated as candidate genes. Using a more conservative approach, with genes only considered real if they extended for >70% of the length of the protein-encoding gene model in other species, and if they were reasonably homologous to a gene in rice or a more distant relative of maize, a gene number of ~37,000 was determined for the haploid genome (Liu et al. 2007). The required homology to a species other than a Panicoid grass (like sorghum or pearl millet) was chosen because (a) lineage-specific genes are rare and (b) TEs evolve faster than standard nuclear genes. This study also provided a minimum estimate of ~5500 truncated pseudogenes in the maize genome, close to the ~4500 predicted for the number of gene fragments inside *Helitrons* in an inbred like B73 or Mo17 (Morgante et al. 2005).

As will be discussed below, the actual number of pseudogenes in maize is probably much greater than 5500. Not only *Helitrons*, but also LTR retrotransposons (Bureau et al. 1994; Jin and Bennetzen 1994; Wang et al. 2006) and *Mutator* elements (Talbert and Chandler 1988; Jiang et al. 2004) have been shown to acquire, transpose and sometimes express fragments from normal nuclear genes. Equally important, but often ignored, is the fact that many truncated or otherwise inactivated genes left over from the ancient polyploidization of maize (Swigonova et al. 2004) are still annotated as genes (Ilic et al. 2003). The polyploid origin of maize has now been well-documented (Swigonova et al. 2004) after numerous earlier predictions (Rhoades 1951; Goodman et al. 1980; Helentjaris et al. 1988; Gaut and Doebley 1997). This issue is discussed in detail by Messing (this volume), so it will not be further described herein. However, for gene number determinations, it should be noted that >50% of the genes originally duplicated by polyploidy have now been reduced back to a single copy state (Bruggmann et al. 2006), through many million years of small deletions (Ilic et al. 2003).

Of course, some of these truncated segments of genes may have acquired a function, and are thus no longer pseudogenes, but the rapid rate of DNA removal in maize and other higher plant genomes (see below) makes it very unlikely that a new protein-encoding function will be created from an initially non-functional mutated gene. Moreover, 30% truncation of protein-encoding domains is a severe and arbitrary criterion for pseudogene designation. A tiny deletion or missense substitution can turn a gene into a pseudogene, for example, but no sequence-based annotation will be sufficient by itself to come to such a pseudogene conclusion/assignment. Still, taking all of these factors into account, it is certain that many of the apparent protein-coding sequences that, even after passing the "this is not a TE" test, are annotated by standard gene-finder and homology criteria will turn out to be pseudogenes of one class or another. Hence, my current prediction for protein-encoding potential is that maize will have about 210,000 TE genes, about 35,000 functional genes and about 10,000 pseudogenes.

4   Maize Genome Change

Maize has a tremendous history in the study of chromosomal rearrangements, both induced and natural variants. Chromosome breakage, often associated with transposable element action, can be traced as the origin of many of these dramatic changes in gene order and/or centromere linkage. However, these major rearrangements are rare compared to the incredibly numerous local rearrangements that differentiate haplotypes within maize (Johns et al. 1983, Helentjaris et al. 1985, Fu and Dooner 2002; Wang and Dooner 2006). Further genome sequence assembly for maize, and its comparison to rice and sorghum genomes, will be needed to determine, in both haplotypic and interspecies comparisons, the frequency of rearrangements that are on the order of one or a few centiMorgans (Wilson et al. 1999; Wei et al. 2007). The ensuing review and discussion will focus on those genomic changes that we best understand in maize, the local rearrangement of genic and intergenic sequences.

4.1   Local Genome Rearrangement

When compared across species, local regions of the maize genome were found to differ dramatically in intergene DNA content and composition (Chen et al. 1997; Tikhonov et al. 1999; Song et al. 2002). The DNA in these intergenic regions, comprised mostly of TEs, did not hybridize to the orthologous regions even in closely related species like sorghum (Avramova et al. 1996), and this was later shown to be an outcome of the fact that these TEs were mostly LTR retrotransposons that had inserted into these sites within the last 3 million years or less (SanMiguel et al. 1996; 1998).

In stark contrast to TEs and other intergenic DNAs, the genes themselves tended to be conserved in content, order and orientation across long periods of evolutionary time, even the 50 million years or more since the rice and maize lineages shared a common ancestor (Chen et al. 1997; Tikhonov et al. 1999; Lai et al. 2004; Ma et al. 2005) (Fig. 3). However, significant exceptions were observed, as shown by the gene downstream of *Orp1* that has no homologue at that location in the homoeologous maize chromosome, in sorghum or in rice (Fig. 3). One could propose that this gene was lost three times independently (from rice, from sorghum and from one maize chromosome) and, given the high rates of sequence loss in grasses, this is not an impossible proposition. However, it is simpler to propose that either this gene downstream of *Orp1* moved to its location after the divergence of sorghum and the two diploid maize ancestors (Swigonova et al. 2004) or that it is actually a TE or gene fragment inside a TE that has been mis-annotated as a gene. Although the latter explanation seems most likely, cases of gene movement to new locations in specific lineages have been proven, including the movement of an *adh* gene in a common ancestor of maize and sorghum (Ilic et al. 2003).

How frequent is this plant gene movement, and how is it vectored? Both of these questions remain largely unanswered. A comprehensive analysis of two large segments of the maize genome, compared to rice, suggested that over 50% of genes have

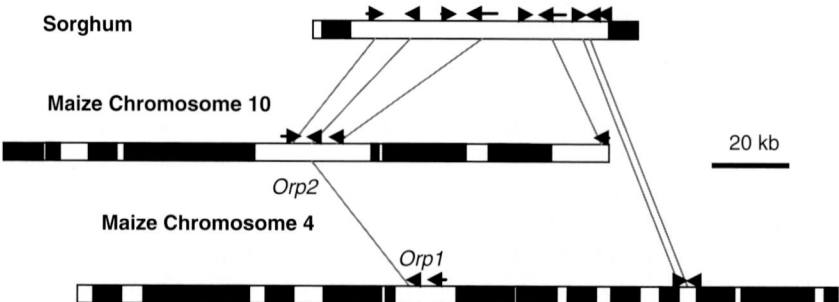

Fig. 3 *Orp1* orthologous regions in maize and sorghum. As in Figure 1, genes are indicated by arrows and LTR retrotransposons or other large TEs are indicated by black boxes. Slanted vertical lines connect orthologous genes between any two or all three genomic regions. The two orthologous regions for maize are due to its allotetraploid history, which was not shared with sorghum (Swigonova et al. 2004). Although only *Orp* genes appear to be conserved across all three regions, it is not possible to tell if the gene 3′ to *Orp2* or the two genes most 5′ to *Orp1* might be shared across all three, because insufficient sequence data are available across the maize regions to make this assessment

been lost or moved to new locations in one or the other of the lineages leading to rice and maize, and this incredibly high number does not include the common 'movements' associated with tandem gene duplication (Bruggmann et al. 2006). This is an amazing degree of genomic fluidity in a mere 50 million years. However, any TE sequence mis-identified as a gene would inflate these numbers, and this type of mis-identification is currently unavoidable. Still, even the most conservative interpretation of the data of Bruggman *et al.* (2006) and earlier studies (e.g., Lai et al. 2004) indicates that over 10% of genes are in new locations in rice/maize comparisons. Several mechanisms for gene fragment, or perhaps whole gene, movement are known in plants. These include TE acquisition of gene fragments (Talbert and Chandler 1988; Bureau et al. 1994; Jin and Bennetzen 1994; Lal et al. 2003; Jiang et al. 2004; Morgante et al. 2005; Wang et al. 2006), TE mobilization of flanking or intervening genomic segments (see chapter by Zhang, Peterson and Peterson in this volume for several examples with *Ac/Ds*), unequal homologous recombination (often with TEs as the sites of unequal homology) and gap filling during double-strand break repair (Kirik et al. 2000). However, few individual cases of functional gene movement have been "caught in the act" to date, so we do not know the relative contributions of any of these mechanisms, or whether other unidentified processes are also involved.

4.2   TE Amplification

Transposable element activity is unusually high in maize, both historically and currently, relative to all other studied angiosperms. This is part of the reason, along with the exceptional genetic skills of Dr. Barbara McClintock, that TEs were first discovered in maize (McClintock 1948). Several cut-and-paste DNA transposons, such as the classically defined elements *Ac/Ds* and *En/Spm* (see Zhang et al., this volume) and *Mutator* (see Lisch, this volume), are currently hopping around the nuclei in different maize populations, making mutants and otherwise rearranging genomes. Although the most abundant elements, the LTR retrotransposons, do not have a confirmed active member in maize, their recent mutant induction at *adh1* (Johns et al. 1985) and the frequent high degree of sequence identity across multiple members indicate recent activity. Similarly, *Helitrons* have been found as insertions in recently-derived mutants (Lal et al. 2003; Gupta et al. 2005). In short, at least a few families of all classes of TEs probably have active members in some maize population or other, and many more probably have quiescent TEs that can be activated by tissue culture or other "genomic shock" (McClintock 1984).

Although broad TE activity is a current property of the maize genome, this does not necessarily mean *per se* that TE activity has always been high in the *Zea* lineage. Moreover, high current levels of activity also does not imply any sort of continuous process. In fact, the dating of LTR retrotransposon activity in maize suggests some bursts of activity, and that the timing of these bursts was different for different LTR retrotransposon families (Liu et al. 2007; Baucom and Bennetzen, unpub. obs.). Similar bursts of TE activity, specifically by LTR retrotransposons, have been uncovered in many plant species and some have been shown to be responsible for rapid increases in overall genome size (Hawkins et al. 2006; Piegu et al. 2006; Wicker and Keller 2007).

In maize, the amazing result that intergenic TEs are mostly or totally different between different genome haplotypes (Fu and Dooner 2002; Song and Messing 2003; Wang and Dooner 2006) provides independent confirmation that TE activity has been enormous over the last two million years (SanMiguel et al. 1998). Has this been an anomalous time, where TE activity was activated in this lineage to a level not seen previously? Probably not. DNA removal has been so rapid in the *Zea* lineage (see below), that any TE that inserted more than a few million years ago is probably mostly deleted and largely unrecognizable. Certainly, TEs are currently rearranging the maize genome through transposition, as sites for unequal homologous recombination, by breaking chromosomes, and by mobilizing gene fragments and genes. There is every reason to believe that this process has been going on at similar rates for many millions of years in this lineage, and also (albeit, usually at slower rates) in the genomes of all other flowering plants.

4.3   DNA Loss

When TE amplification, primarily by LTR retrotransposons, was shown to be the major cause of genome growth in higher plants (SanMiguel et al. 1996; SanMiguel and Bennetzen 1998; Tikhonov et al. 1999; Bennetzen 2000; Bennetzen et al. 2005; Hawkins et al. 2006; Piegu et al. 2006), it was initially unclear what mechanism (if any) might significantly compete with this progressive "genomic obesity". Hence, a call was put out to search for a process that could, at the very least, attenuate or reverse genome expansion (Bennetzen and Kellogg 1997). Because this DNA loss process could not remove any large number of genes and retain the observed gene content conservation across species, and because the genes were interspersed with LTR retrotransposons in maize, the mechanism could not be simple events that removed hundreds of kb of DNA in single steps. Moreover, because the vast majority of LTR retrotransposons in the seas between gene islands had no obvious effects on nearby gene expression, it seemed unlikely that natural selection could be a major player.

Many in the "C-value-paradox" field (Gregory 2005) felt that selection on genome size *per se* could be a major factor, but this seemed an unlikely possibility to me in a "day-to-day" sense. That is, if one genome is 2000 Mb and the other (due to a new TE insertion of 5 kb) is now 2000.005 Mb, it is hard to see how the tiny difference needed in ATP to replicate or express this extra DNA could generate enough selective value to compete with all of the gene-based selections or the chance founder/bottleneck events that drive evolution in real-world populations. I believe that genome size selection, on processes like the length of S phase and subsequent time from seed to seed (Bennett 1972), is probably a major factor affecting fitness in some environments once populations within a plant differ by 1% or more in nuclear DNA content (Rayburn et al. 1993), but it seems most likely that the processes that initially generate this variation are selfish (Rhoades 1942; Roman 1948) or indirect outcomes of variation in other processes.

Research in Arabidopsis and rice identified illegitimate recombination as the major process that removes genomic DNA (Devos et al. 2002). Unequal homologous recombination, mostly converting LTR retrotransposons into solo LTRs, can also be a significant factor (Bennetzen and Kellogg 1997; Shirasu et al. 2000; Ma et al. 2004). In rice, these mechanisms were shown to have removed at least 190 Mb of LTR retrotransposon DNA in the last 8 million years (Ma et al. 2004). However, because the major DNA removal mechanism, illegitimate recombination, was shown to be active on all other sequences in the rice genome, and because the original clock used to date the sequence loss was exceedingly conservative (Ma and Bennetzen 2004), a more realistic DNA loss schedule for rice indicates about 140 Mb of un-selected DNA removed per million years over the last few million years. Hence, any LTR retrotransposon insertion in rice has a half-life of about 2 million years; that is, a 10 kb TE will be missing 5 kb of its sequence on average within 2 million years.

The mechanism of illegitimate recombination responsible for most of this DNA removal is not absolutely clear. The key hallmark of illegitimate recombination, short stretches (as low as 1 bp!) of variable-length homology flanking a deletion or insertion site, could be associated with a number of biological phenomena. I believe that the most likely mechanism is the manner in which double-strands breaks are repaired in plant, usually by non-homologous end joining, thereby leading to common tiny deletions and rarer insertions (Kirik et al. 2000). Recent work that directly generated double-strand breaks in rice and maize has shown that the repair events are commonly associated with small deletions (Lu and Bennetzen, unpub. obs.) like those observed for natural indel variation between rice haplotypes (Ma and Bennetzen 2004). The lion's share of these indels are tiny (median 1–2 bp), but most of the DNA removal is contributed by the fairly rare deletions that are a few hundred to a few thousand bp in size (Ma and Bennetzen 2004).

An astounding outcome of studies measuring the structure of LTR retrotransposons in maize and several other plant species has been the observation that the relative rates of DNA removal are quite different even between closely related plant lineages (Vitte and Bennetzen 2006; Wicker and Keller 2007). Maize, for instance, appears to have a faster DNA removal rate than rice. The much larger (~6X) genome size of maize compared to rice must then be an outcome of a much greater rate of TE amplification (and an extra round of polyploidy; Swigonova et al. 2004) in the maize lineage that has overwhelmed this rapid rate of DNA removal. Hence, we expect that the half-lives of TE sequences and other non-selected DNA in maize are less than two million years. This observation, along with a very high level of recent TE activity, explains why the intergenic regions in different maize haplotypes can be so very different (Fu and Dooner 2002; Song and Messing 2003; Wang and Dooner 2006). The non-functional DNAs (mostly TEs) in these regions even two million years ago are completely or nearly completely removed, and new TEs have arrived to take their place.

5 Biological Outcomes of Rapid Genome Change

The rapid and dramatic changes in genome composition and structure that have been ongoing in the *Zea* lineage have apparently led to surprisingly little change in genetic function. For instance, the movement of *Adh1* from its location in other grasses to its current location on a new chromosome in maize and its close relatives (Ilic et al. 2003) does not appear to have altered the tissue specificity or developmental timing of expression of this gene. The expression of *Bz1* has also not been noticeably affected by the fact that the *Bz1* gene has very different flanking sequences in different maize haplotypes (Wang and Dooner 2006). In fact, we do not yet have a case in maize where natural movement of a gene into a new chromosomal location has altered its subsequent expression properties. It seems likely that such cases will be found, but they must be relatively rare in this species. Some of this conservation of function should be due to selection and/or drift (a gene moved to a silenced location, if therefore nonfunctional, would be rapidly lost by accumulated small deletions generated by illegitimate recombination). However, it also seems likely that maize genes are unusually well-insulated from the heterochromatic effects of nearby silenced TEs. This is a likely to be an obligate property of most maize genes because they are generally flanked by inactivated TEs in the repetitive DNA seas that surround their tiny gene islands (Liu et al. 2007). The nature of these "insulation" factors remains unclear, although apparent matrix/scaffold attachment regions are routinely associated with the boundaries between gene islands and the adjacent heterochromatic seas of nested TEs (Avramova et al. 1995; Tikhonov et al. 2000).

Numerous studies have shown that TE insertions, for instance, can alter the regulation of adjacent genes (reviewed in Girard and Freeling 1999; also see the chapters by Lisch and by Zhang et al. in this volume). However, as with any genetic mutation, the great majority of these genetic changes will be deleterious and thus lost due to natural selection. Moreover, when these TE-associated regulatory changes are observed, they are usually insertions within a few hundred bp of the structural gene. Over time, it has been observed that many (probably most or all) plant promoters have acquired TE fragments that are often involved in gene regulation as sites for the binding of cis-acting regulatory factors (White et al. 1994), and this phenomenon is not unique to plants (Mariño-Ramírez et al. 2005). Because most changes are deleterious, it is expected that the majority of the "regulatory" TE fragments that persist are likely to be those that do not change gene expression, but are conservative in nature. This situation should arise when a TE and its regulatory components insert near a gene and thereby duplicate the same regulatory properties that the gene already exhibited, thus allowing either the old regulatory domain or the new regulatory domain (i.e., the TE component) to be subsequently lost. A good example of this phenomenon is the propensity of MITEs (a class of tiny TEs) to insert near matrix attachment regions (MARs) (Tikhonov et al. 2000). Because most MITEs appear to have a molecular structure that allows them to act as MARs, their insertion at a MAR site creates a duplicated function that (a) might be a stronger boundary/insulating element and/or (b) could be reduced back to a single component by loss either of the MITE acting as a MAR or by the original MAR at that site (Tikhonov et al. 2000).

Are plant genomes generating new genes? Despite early reports that different flowering plant genomes, for instance rice and Arabidopsis, differed dramatically in gene content, comparative mapping studies in the grasses suggest that gene content is very highly conserved (Bennetzen et al. 2004; Bennetzen 2005). Although gene family numbers can vary quickly, and perhaps even directionally in specific lineages, the core gene content (that is, number of gene families) of any two compared grasses is unlikely to differ by even as much as 10%. However, having said this, it is amazing to see the rate at which plants are generating new gene candidates.

Studies in rice and maize have now shown that three different classes of TEs, the cut-and-paste TEs of the *Mutator* family (Jiang et al. 2004), the apparent rolling circle TEs called *Helitrons* (Morgante et al. 2005) and the RNA-vectored LTR retrotransposons (Wang et al. 2006), are all very active in the acquisition and transpositional amplification of gene fragments. Multiple independent gene fragments are commonly acquired by these elements, especially *Helitrons*, and these are often associated with identified transcripts that fuse exons from different plant genes. Gilbert (1978) initially proposed this process, exon shuffling, as the reason that introns currently exist in many eukaryotic genes. In the early days of life on earth, small peptide-encoding sequences with one functional domain could be brought together into longer sequences with two or more functional domains by ligation, and intron processing would be employed to make an appropriate transcript and subsequent peptide without the exceedingly limiting requirements that the sequence fusion be in frame and without any intervening nucleotides.

Exon shuffling, however, was believed to be an ancient process with little likely relevance to gene creation in the last billion years or so, with a few fascinating examples of recent events (e.g., Babushok et al. 2007). Contemporary generation of novel genes, in contrast, is thought to be primarily an outcome of gene duplication followed by divergence (Ohno 1970; Long et al. 2003). However, in maize, for instance, many hundreds of *Helitron* and other TE-driven gene fragment fusions are present in every nucleus, many in different locations in each haplotype (Morgante et al. 2005). Hence, one could argue that maize and other angiosperms are generating new genes by exon shuffling at prodigious rates. However, despite this breakneck rate of candidate gene creation, no newly shuffled 'gene' inside any class of TE has yet been associated with a "whole plant" phenotype. That is, when the genes associated with mutations that change a plant's phenotype, development or response to the environment are identified, none has yet been associated with a mutation inside a TE that contains a gene fragment or fragments. One expects that some new gene creation by TE-mediated exon shuffling will eventually be found in plants, but this outcome is expected to be associated with a tiny minority of the candidate exon-shuffled genes that are observed. Given the very rapid rate of removal of non-functional sequences in angiosperms, and especially in maize (Vitte and Bennetzen 2006), a shuffled set of genes will rapidly drift into an irretrievably inactive state by the accumulation of indels (mostly deletions), followed by eventual complete loss (Ma et al. 2004). Still, even if only 1 in one million such TE-mediate exon shufflings leads to a new functional gene, this would provide an astoundingly high rate for the creation of genes with truly novel characteristics.

Even if TE-driven exon shuffling in maize and other angiosperms is only rarely, if ever, sufficient to create new protein-encoding genes, it is impossible to imagine that the expressed gene fragments inside Pack-MULEs, *Helitrons* and other TEs are not influencing genetic function at an RNA level. Antisense gene fragments inside expressed TEs should lead to some decrease in transcript levels for the wild type 'donor' gene (Ecker and Davis 1986) and some sense fragments should also create RNAi-based inhibition through siRNA generation (Hamilton and Baulcombe 1999). If translated, gene fragments are also likely to poison some multi-enzyme complexes. The recent proposal that some microRNAs are derived from TEs (Priyapongsa and Jordan 2008) could be the beginning of our understanding of the long term and immediate implications of the interactions between the gene-capture properties of TEs and the "RNA world", a still-mysterious part of the coding genome that may contain as many genes as the protein-encoding part (see chapter by Kaeppler in this volume). In maize, the haplotypic differences in TE-derived gene fragments (Morgante et al. 2005) suggest that each maize line will have a somewhat unique contribution of antisense and other gene-fragment-derived RNAs, suggesting a truly novel explanation for some aspects of heterosis.

The mechanisms that affect genome complexity in angiosperms, TE amplification, ectopic homologous recombination and illegitimate recombination (primarily double-strand break repair), all have internal rationales for action that are unrelated to genome instability *per se*. TE amplification can be viewed as a fully selfish action by a highly adapted parasite, while ectopic homologous recombination may be an unavoidable outcome of meiotic recombination and/or DNA repair, or it may also be programmed for the specific generation of tandem gene family rearrangements (Nagy and Bennetzen, 2008). Similarly, illegitimate recombination may be an obligate outcome of the ways in which chromosomal DNAs are replicated and repaired. From this perspective, the complexity of plant genomes could be generated without any need for a specific purpose to this complexity, or any significant selection for or against this complexity. However, any of these processes or components could theoretically evolve into a useful (i.e., domesticated) version. The best examples of this process involve multiple independent TE domestications (Volff 2006). Outside of plants, there are a few dramatic cases, like the replacement of telomeres by telomere-specific retroelements in some *Drosophila* species (Levis et al. 1993) [and/or vice versa (Eickbush 1997)] and the neofunctionalization of a retroviral envelope gene to become an essential gene for placental development in hominoids (Mallet et al. 2004). The first cases of TE gene domestication have now been found in plants, with encoded proteins derived from the *Mutator* transposase (Hudson et al. 2003; Cowan et al. 2005). Plants will probably have more of these genes, and their discovery will be a major benefit of the comprehensive reverse genetic technologies that are beginning to be applied across the plant kingdom.

Will there be regions, or biological traits, where selection acts to influence genome structure? Almost certainly, but it is surprising how few concrete examples we have at this time. Perhaps the most obvious case is related to intron size and the overall compactness of genes. It seems likely, although the evidence is scattershot, that most plant genes will not tolerate large insertions or deletions in either their exons or introns. Unusually large introns in plants may often be an indication of the presence of a regulatory function. Centromeres provide a likely site where selfish natural selection may dramatically influence genome structure and, in this case, lead to greater diversity that allows centromeric competition in the developing egg (Henikoff et al. 2001; Dawe and Henikoff 2006; see the chapter by Dawe in this volume). The recent observation that the core region of a centromere in rice is a hotspot for recombination, rather than the dogmatic attribution as a coldspot, provides a mechanism by which this variability can be generated (Ma and Bennetzen 2006). Another place to look for selection that acts on genome structure and structural rearrangement will be in the study of unequal recombination that is promoted in some tandem gene families (e.g., Nagy et al. 2008) and may be inhibited in others. We know little about the three dimensional structures of plant nuclear genomes, and how it changes across the cell cycle, in different tissues, and during development, so it is guaranteed that variable aspects of these processes that are subject to natural selection will be uncovered in future studies (see the chapter by Cande et al. in this volume).

6 Future Directions

For the foreseeable future, maize will maintain its status as the premiere organism for investigations into genome rearrangement, and there is no shortage of additional questions that need to be answered. Although the level of genomic instability in maize is high even compared to other angiosperms, we have no information suggesting that it is exceptional in either the types or outcomes of these genomic changes, relative to other flowering plants. Analysis of genetic change is more feasible in lineages where such changes are routine, hence the great history of maize as the first site of discovery of the existence of telomeres (in contrast to broken chromosome ends) (McClintock 1941), of transposable elements (McClintock 1948) and epigenetics (Brink 1958). Interestingly, all three of these maize-initiated subfields of genetics are still major topics of investigation in eukaryotic genetics, and maize will continue its contributions to the next generation of advances.

An understanding of the 'epigenome' is one of the great missions for the next decade of eukaryotic biology. Plants share many aspects of a universal epigenetic program in eukaryotes, but also have several unique specializations. From a genome structure and evolution perspective, it will be especially important to determine the nature of the processes that insulate the genes in normal plants (those with genomes >2000 Mb) from adjacent heterochromatin. On a related issue, we need to know much more about the higher order structure of chromatin, especially in the interphase nucleus. Are there specific patterns that are maintained over evolutionary time and, if so, do they exhibit tissue-, developmental- or environmental-specificity?

We now are aware of most, perhaps all, of the mechanisms responsible for genome rearrangement. We also know that the relative aggressiveness of these mechanisms varies between species and across different chromosomal domains. It is less clear why, especially between species, processes like illegitimate recombination or TE amplification might be more or less tightly constrained. The genes involved in regulating the general or positional activities of these mechanisms of rearrangement are also unknown. It will be challenging to identify these genes, given that their natural phenotype is the direction and rate at which genome structure evolves. Because this evolution is moving so quickly in maize, it will be a good organism in which to search for the genes involved.

Although evolutionary analysis indicates that most genome rearrangement in the grasses, including maize, has not led to altered gene function, it will be fascinating to discover how often and in what cases genome rearrangement has led to new phenotypes. This question will be most effectively investigated with *de novo* events, before selection can winnow the general phenomena down to those that are acceptable for overall organismal fitness.

With the arrival of a 'benchmark' genome sequence for maize inbred B73 (http://maizesequence.org/index.html), the foundation will be laid to begin to describe the full molecular diversity in maize. High throughput sequencing of multiple maize genomes will allow facile description of the variability that resides within genes, but more clever approaches will be needed to extract information that can describe changes in genome structure within the genus *Zea* and beyond. When 'single molecule' sequencing reaches the whole chromosome level, whenever that might be, the great genomic diversity within maize will become the resource to describe the full array of patterns, processes, preferences and possibilities in maize genome structure and evolution.

Acknowledgements The author wishes to thank Vicki Chandler, Erich Grotewald and Maike Stam for helpful discussions and to acknowledge the continuing support of NSF and the USDA in pursuing much of the research described in this chapter. The writing of this chapter was supported by NSF grant DBI 0607123.

References

Ananiev, E.V., R.L. Phillips and H.W. Rines (1998a) A knob-associated tandem repeat in maize capable of forming fold-back DNA segments: Are chromosome knobs megatransposons? *Proc. Natl. Acad. Sci. USA* **95**: 10785–10790.

Ananiev, E.V., R.L. Phillips and H.W. Rines (1998b) Chromosome-specific molecular organization of maize (*Zea mays* L.) centromeric regions. *Proc. Natl. Acad. Sci. USA* **95**: 13073–13078.

Avramova, Z., P. SanMiguel, E. Georgieva and J.L. Bennetzen (1995) Matrix attachment regions and transcribed sequences within a long chromosomal continuum containing maize *Adh1*. *Plant Cell* **7**: 1667–1680.

Avramova, Z., A. Tikhonov, P. SanMiguel, Y.-K. Jin, C. Liu, S.-S. Woo, R.A. Wing and J.L. Bennetzen (1996) Gene identification in a complex chromosomal continuum by local genomic cross-referencing. *Plant J.* **10**: 1163–1168.

Babushok, D.V., K. Ohshima, E.M. Ostertag, X. Chen, Y. Wang, P.K. Mandal, N. Okada, C.S. Abrams and H.H. Kazazian Jr. (2007) A novel testis ubiquitin-binding protein gene arose by exon shuffling in hominoids. *Genome Res.* **17**: 1129–1138.

Banerji, J., S. Rusconi and W. Schaffner (1981) Expression of a β-globin gene is enhanced by remote SV40 DNA sequences. *Cell* **27**: 299–308.

Bennett, M.D. (1972) Nuclear DNA content and minimum generation time. *Proc. Royal Soc. London, Series B* **181**: 109–135.

Bennetzen, J. L. (2000) Transposable element contributions to plant gene and genome evolution. *Plant Mol. Biol.* **42**: 251–269.

Bennetzen, J. L. (2005) Transposable elements, gene creation and genome rearrangement in flowering plants. *Curr. Opin. Gen. Dev.* 15:1–7.

Bennetzen, J. L., W. E. Brown and P. S. Springer (1988) DNA modification within and flanking maize transposable elements. In O.E. Nelson, Jr., ed., *Plant Transposable Elements*. Plenum Press, New York, pp. 237–250.

Bennetzen, J.L., C. Coleman, J. Ma, R. Liu and W. Ramakrishna (2004) Consistent over-estimation of gene number in complex plant genomes. *Curr. Opin. Plant Biol.* **7**: 732–736.

Bennetzen, J.L. and E.A. Kellogg (1997) Do plants have a one way ticket to genomic obesity? *Plant Cell* **9**: 1509–1514.

Bennetzen, J. L. and J. Ma (2003) The genetic colinearity of rice and other cereals based on genomic sequence analysis. *Curr. Opin. Plant Biol.* 6:128–133.

Bennetzen, J.L., J. Ma and K.M. Devos (2005) Mechanisms of recent genome size variation in flowering plants. *Annals Bot.* **95**: 127–132.

Bennetzen, J.L., R. Liu, J. Ma and A. Pontaroli (2005) Maize genome structure and rearrangement. *Maydica* **50**: 387–392.

Brink, R.A. (1958) Paramutation at the *R* locus in maize. *Cold Spring Harbor Symp. Quant. Biol.* 23: 379–391.

Bruggmann, R., A.K. Bharti, H. Gundlach, J. Lai, S. Young, A.C. Pontaroli, F. Wei, G. Haberer, G. Fuks, C. Du, C. Raymond, M.C. Estep, R. Liu, J.L. Bennetzen, A.P. Chan, P.D. Rabinowicz, J. Quackenbush, W.B. Barbazuk, R.A. Wing, B. Birren, C. Nusbaum, S. Rounsley, K.F.X. Mayer and J. Messing (2006) Uneven chromosome contraction and expansion in the maize genome. *Genome Res.* **16**: 1241–1251.

Bureau, T.E., S.E. White and S.R. Wessler (1994) Transduction of a cellular gene by a plant retroelement. *Cell* **77**: 479–480.

Chen, M., P. SanMiguel, A.C. de Oliveira, S.-S. Woo, H. Zhang, R.A. Wing and J.L. Bennetzen (1997) Microcolinearity in the *sh2*-homologous regions of the maize, rice and sorghum genomes. *Proc. Natl. Acad. Sci. USA* **94**: 3431–3435.

Clark, R.M., E. Linton, J. Messing and J.F. Doebley (2004) Pattern of diversity in the genomic region near the maize domestication gene *tb1*. *Proc. Natl. Acad. Sci. USA* **101**: 700–707.

Cone, K.C, R.J. Schmidt, B. Burr and F. Burr (1988) Advantages and limitations of using *Spm* as a transposon tag. In O.E. Nelson, Jr., ed., *Plant Transposable Elements*. Plenum Press, New York, pp. 149–159.

Cooper, D.C. and R.A. Brink (1937) Chromosome homology in races of maize from different geographical regions. *Am. Nat.* **71**: 582–587.

Cowan, R.K., D.R. Hoen, D.J. Schoen and T.E. Bureau (2005) *MUSTANG* is a novel family of domesticated transposase genes found in diverse angiosperms. *Mol. Biol. Evol.* 22: 2084–2089.

Craig, N.L. (1996) Transposon Tn7. *Curr. Top. Microbiol. Immunol.* **204**: 27–48.

Cresse, A. D., S. H. Hulbert, W. E. Brown, J. R. Lucas and J. L. Bennetzen (1995) *Mu1*-related transposable elements of maize preferentially insert into low copy number DNA. *Genetics* 140:315–324.

Dawe, K. and S. Henikoff (2006) Centromeres put epigenetics in the driver's seat. *Trends Biochem. Sci.* **31:** 662–669.

Devos, K.M., J.K.M. Brown and J.L. Bennetzen (2002) Genome size reduction through illegitimate recombination counteracts genome expansion in *Arabidopsis*. *Genome Res.* **12:** 1075–1079.

Dubcovsky, J., W. Ramakrishna, P. SanMiguel, C.S. Busso, L. Yan, B. A. Shiloff and J.L. Bennetzen (2001) Comparative sequence analysis of colinear barley and rice BACs. *Plant Physiol.* **125:** 1342–1353.

Ecker, J.R. and R.W. Davis (1986) Inhibition of gene expression in plant cells by expression of antisense RNA. *Proc. Natl. Acad. Sci. USA* **83:** 5372–5376.

Eickbush, T.H. (1997) Telomerase and retrotransposons: Which came first? *Science* **277:** 911–912.

Feuillet, C., A. Penger, K. Gellner, A. Mast and B. Keller (2001) Molecular evolution of receptorlike kinase genes in hexaploid wheat. Independent evolution of orthologs after polyploidization and mechanisms of local rearrangements at paralogous loci. *Plant Physiol.* **125:** 1304–1313.

Fu, H. and H.K. Dooner (2002) Intraspecific violation of genetic colinearity and its implications in maize. *Proc. Natl. Acad. Sci. USA* **99:** 9573–9578.

Gale, M.D., and K.M. Devos (1998) Plant comparative genetics after 10 years. *Science* **282:** 656–659.

Gaut, B.S. and J.F. Doebley (1997) DNA sequence evidence for the segmental allotetraploid origin of maize. *Proc. Natl. Acad. Sci. USA* **94:** 6809–6814.

Gilbert, W. (1978) Why genes in pieces? *Nature* **271:** 501.

Girard, L. and M. Freeling (1999) Regulatory changes as a consequence of transposon insertion. *Dev. Genet.* **25:** 291–296.

Goodman, M.M., C.W. Stuber, K. Newton and H.H. Weissinger (1980) Linkage relationships of 19 enzyme loci in maize. *Genetics* **96:** 697–710.

Gregory, T.R. (2005) The C-value enigma in plants and animals: a review of parallels and an appeal for partnership. *Ann. Bot.* **95:** 133–146.

Gupta, S., A. Gallavotti, G.A. Stryker, R.J. Schmidt and S.K. Lal (2005) A novel class of *Helitron*-related transposable elements in maize contain portions of multiple pseudogenes. *Plant Mol. Biol.* **57:** 115–27.

Haberer, G., S. Young, A.K. Bharti, H. Gundlach, C. Raymond, G. Fuks, E. Butler, R.A. Wing, S. Rounsley, B. Birren, B. Nusbaum, K.F.X. Mayer and J. Messing (2005) Structure and architecture of the maize genome. *Plant Phys.* **139:** 1612–1624.

Hake, S. and V. Walbot (1980) The genome of *Zea mays*, its organization and homology to related species. *Chromosoma* **79:** 251–270.

Hamilton, A.J. and D.C. Baulcombe (1999) A novel species of small antisense RNA in post-transcriptional gene silencing. *Science* **286:** 950–952.

Hawkins, J.S., HR. Kim, J.D. Nason, R.A. Wing and J. F. Wendel (2006) Differential lineage-specific amplification of transposable elements is responsible for genome size variation in *Gossypium*. *Genome Res.* **16:** 1252–1261.

Helentjaris, T., G. King, M. Slocum, C. Siedenstrang and S. Wedman (1985) Restriction fragment polymorphisms as probes for plant diversity and their development as tools for applied plant breeding. *Plant. Mol. Biol.* **5:** 109–118.

Helentjaris, T., D. Weber, and S. Wright (1988). Identification of the genomic locations of duplicate nucleotide sequences in maize by analysis of restriction fragment length polymorphism. *Genetics* **118:** 353–363.

Henikoff, S., K. Ahmad and H.S. Malik (2001) The centromere paradox: stable inheritance with rapidly evolving DNA. *Science* **293:** 1098–1102.

Hsu, F. C., C.J. Wang, C.M. Chen, H.Y. Hu and C.C. Chen (2003) Molecular characterization of a family of tandemly repeated DNA sequences, TR-1, in heterochromatic knobs of maize and its relatives. *Genetics* **164:** 1087–1097.

Hudson, M.E., D.R. Lisch and P.H. Quail (2003) The *FHY3* and *FAR1* genes encode transposase-related proteins involved in regulation of gene expression by the phytochrome A-signaling pathway. *Plant J.* **34:** 453–471.

Ilic, K., P.J. SanMiguel and J.L. Bennetzen (2003) A complex history of rearrangement in an orthologous region of the maize, sorghum and rice genomes. *Proc. Natl. Acad. Sci. USA* **100:** 12265–12270.

IRGSP [International Rice Genome Sequencing Project] (2005) The map-based sequence of the rice genome. *Nature* **436:** 793–800.

Jiang, N., Z. Bao, X. Zhang, S.R. Eddy and S.R. Wessler (2004) Pack-MULE transposable elements mediate gene evolution in plants. *Nature* **431:** 569–573.

Jin, Y.-K. and J.L. Bennetzen (1994) Integration and nonrandom mutation of a plasma membrane proton ATPase gene fragment within the *Bs1* retroelement of maize. *Plant Cell* **6:** 1177–1186.

Johns, M.A., J. Mottinger and M. Freeling (1985) A low copy number, copia-like transposon in maize. *EMBO J.* **4:** 1093–1101.

Johns, M.A., J.N. Strommer and M. Freeling (1983) Exceptionally high levels of restriction site polymorphism in DNA near the maize *adh1* gene. *Genetics* **105:** 733–743.

Kapitonov, V.V. and J. Jurka (2001) Rolling-circle transposons in eukaryotes. *Proc. Natl. Acad. Sci. USA* **98:** 8714–9.

Kirik, A., S. Salomon and H. Puchta (2000) Species-specific double-strand break repair and genome evolution in plants. *EMBO J.* **19:** 5562–5566

Kolkman, J.M., L.J. Conrad, P.R. Farmer, K. Hardeman, K.R. Ahern, P.E. Lewis, R.J.H. Sawers, S. Lebejko, P. Chomet and T.P. Brutnell (2005) Distribution of *Activator* (*Ac*) throughout the maize genome for use in regional mutagenesis. *Genetics* 169: 981–995.

Kriz, A.L., R.S. Boston and B.A. Larkins (1987) Structural and transcriptional analysis of DNA sequences flanking genes that encode 19 kilodalton zeins. *Mol. Gen. Genet.* **207:** 90–98.

Lai, J., J. Ma, Z. Swigonova, W. Ramakrishna, E. Linton, V. Llaca, B. Tanyolac, Y.-J. Park, O.-Y. Jeong, J.L. Bennetzen, and J. Messing (2004) Gene loss and movement in the maize genome. *Genome Res.* **14:** 1924–1931.

Lal, S.K., M.J. Giroux, V. Brendel, C.E. Vallejos and L.C. Hannah (2003) The maize genome contains a *Helitron* insertion. *Plant Cell* **15:** 381–91.

Levis, R.W., R. Ganesan, K. Houtchens, L.A. Tolar and F.M. Sheen (1993) Transposons in place of telomeric repeats at a Drosophila telomere. *Cell* **75:** 1083–1093.

Liu, C.-N., and I. Rubenstein (1992) Genomic organization of an alpha-zein gene cluster in maize. *Mol. Gen. Genet.* **321:** 304–312.

Liu, R., C. Vitte, J. Ma, A.A. Mahama, T. Dhliwayo, M. Lee and J.L. Bennetzen (2007) A GeneTrek analysis of the maize genome. *Proc. Natl. Acad. Sci. USA* **104:** 11844–11849.

Liu, R. and J.L. Bennetzen (2008) ENCHILADA REDUX: How complete is your genome sequence? *New Phytol.* **179:** 249–250.

Long, M., E. Betran, K. Thornton and W. Wang (2003) The origin of new genes: Glimpses from the young and old. *Nat. Rev. Genet.* **4:** 865–875.

Ma, J., J. Lai, J. Messing and J.L. Bennetzen (2005) DNA rearrangement in orthologous *orp* regions of the maize, rice and sorghum genomes. *Genetics* **170:** 1209–1220.

Ma, J., and J.L. Bennetzen (2004) Rapid recent growth and divergence of rice nuclear genomes. *Proc. Natl. Acad. Sci. USA* **101:** 12404–12410.

Ma, J. and J.L. Bennetzen (2006) Recombination, rearrangement, reshuffling and divergence in a centromeric region of rice. *Proc. Natl. Acad. Sci. USA* **103:** 383–388.

Ma, J., K.M. Devos and J.L. Bennetzen (2004) Analyses of LTR-retrotransposon structures reveal recent and rapid genomic DNA loss in rice. *Genome Res.* **14:** 860–869.

Mallet, F., O. Bouton, S. Prudhomme, V. Cheynet, G. Oriol, B. Bonnaud, G. Lucotte, L. Duret and B. Mandrand (2004) The endogenous retroviral locus ERVWE1 is a bona fide gene involved in hominoid placental physiology. *Proc. Natl. Acad. Sci. USA* **101:** 1731–1736.

Mariño-Ramírez, L., K.C. Lewis, D. Landsman and I.K. Jordan (2005) Transposable elements donate lineage-specific regulatory sequences to host genomes. *Cytogenet. Genome Res.* **110:** 333–341.

McClintock, B. (1929) Chromosome morphology in *Zea mays. Science* **69:** 629.

McClintock, B. (1941) The stability of broken ends of chromosomes in *Zea mays. Genetics* **26:** 234–282.

McClintock, B. (1948) Mutable loci in maize. *Carnegie Inst. Wash. Yearbook* **47:** 155–169.

McClintock, B. (1960) Chromosome constitutions of Mexican and Guatelmalan races of maize. *Carnegie Inst. Wash. Yearbook* **59:** 461–472.

McClintock, B. (1984) The significance of responses of the genome to challenge. *Science* **226:** 792–801.

Messing J., A.K. Bharti, W.M. Karlowski, H. Gundlach, H.R. Kim, Y. Yu, F. Wei, G. Fuks, C.A. Soderlund, K.F.X. Mayer and R.A. Wing (2004) Sequence composition and genome organization of maize. *Proc Natl Acad Sci USA* **101:** 14349–14354.

Meyers, B.C., S.V. Tingey and M. Morgante (2001) Abundance, distribution, and transcriptional activity of repetitive elements in the maize genome. *Genome Res.* **11:** 1660–1676.

Moreau, P., R. Hen, B. Wasylyk, R. Everett, M.P. Gaub and P. Chambon (1981) The SV40 72 base repair repeat has a striking effect on gene expression both in SV40 and other chimeric recombinants. *Nucl. Acids Res.* **9:** 6047–6068.

Morgante, M., S. Brunner, G. Pea, K. Fengler, A. Zuccolo and A. Rafalski (2005) Gene duplication and exon shuffling by *Helitron*-like transposons generate intraspecies diversity in maize. *Nat. Genet.* **37:** 997–1002.

Nagy, E.D. and J.L. Bennetzen (2008) Pathogen corruption and site-directed recombination at a plant disease resistance gene cluster. *Genome Res.*, in press.

Ohno, S. (1970) *Evolution by Gene Duplication.* Springer, Berlin.

Peacock, W.J., E.S. Dennis, M.M. Rhoades and A.J. Pryor (1981) Highly repeated DNA sequence limited to knob heterochromatin in maize. *Proc. Natl. Acad. Sci. USA* **78:** 4490–4494.

Phillips, R.L. (1978) Molecular cytogenetics of the nucleolus organizer region. In: Walden, D.B. (Ed.) *Maize Breeding and Genetics,* chapter 43. John Wiley and Sons, New York.

Piegu, B., R. Guyot, N. Picault, A. Roulin, A. Saniyal, H. Kim, K. Collura, D.S. Brar, S. Jackson, R.A. Wing and O. Panaud. (2006) Doubling genome size without polyploidization: Dynamics of retrotransposition-driven genomic expansions in *Oryza australiensis*, a wild relative of rice. *Genome Res.* **16:** 1262–1269.

Priyapongsa, J. and I.K. Jordan (2008) Dual coding of siRNAs and miRNAs by plant transposable elements. *RNA* **14:** 814–821.

Ramakrishna, W., J. Dubcovsky, Y.-J. Park, C. Busso, J. Emberton, P. SanMiguel and J.L. Bennetzen (2002) Different types and rates of genome evolution detected by comparative sequence analysis of orthologous segments from four cereal genomes. *Genetics* **162:** 1389–1400.

Rayburn, A.L., D.P. Biradar, D.G. Bullock and L.M. McMurphy (1993) Nuclear DNA content in F1 hybrids of maize. *Heredity* **70:** 294–300.

Rhoades, M.M. (1942) Preferential segregation in maize. *Genetics* **27:** 395–407.

Rhoades, M.M. (1951) Duplicate genes in maize. *Amer. Nat.* **85:** 105–110.

Roman, H. (1948) Directed fertilization in maize. *Proc. Natl. Acad. Sci. USA* **34:** 36–42.

Rostoks, N., Y.-J. Park, W. Ramakrishna, J. Ma, A. Druka, B.A. Shiloff, P.J. SanMiguel, Z. Jiang, R. Brueggeman, D. Sandhu, K. Gill, J.L. Bennetzen and A. Kleinhofs (2002) Genomic sequencing reveals gene content, genomic organization, and recombination relationships in barley. *Functional and Integrative Genomics* **2:** 70–80.

SanMiguel, P., A. Tikhonov, Y.-K. Jin, N. Motchoulskaia, D. Zakharov, A. Melake-Berhan, P.S. Springer, K.J. Edwards, M. Lee, Z. Avramova and J.L. Bennetzen (1996) Nested retrotransposons in the intergenic regions of the maize genome. *Science* **274:** 765–768.

SanMiguel, P., B.S. Gaut, A. Tikhonov, Y. Nakajima and J.L. Bennetzen (1998) The paleontology of intergene retrotransposons of maize. *Nat. Genet.* **20:** 43–45.

SanMiguel, P., and J.L. Bennetzen (1998) Evidence that a recent increase in maize genome size was caused by the massive amplification of intergene retrotransposons. *Annals Bot.* **82:** 37–44.

SanMiguel, P.J., W. Ramakrishna, J.L. Bennetzen, C.S. Busso and J. Dubcovsky (2002) Transposable elements, genes and recombination in a 215-kb contig from wheat chromosome 5A^m. *Functional and Integrative Genomics* **2:** 51–59.

SGPWP [Sorghum Genomics Planning Workshop Participants] (2005) Toward sequencing the sorghum genome. A U.S. National Science Foundation-sponsored workshop report. *Plant Physiol.* **138:** 1898–1902.

Shirasu, K., A.H. Schulman, T. Lahaye and P. Schulze-Lefert (2000) A contiguous 66-kb barley DNA sequence provides evidence for reversible genome expansion. *Genome Res.* **10:** 908–915.

Song, R., and J. Messing (2003) Gene expression of a gene family in maize based on noncollinear haplotypes. *Proc. Natl. Acad. Sci. USA* **100:** 9055–9060.

Song, R., V. Llaca and J. Messing (2002) Mosaic organization of orthologous sequences in grass genomes. *Genome Res.* **12:** 1549–1555.

Stam, M., C. Belele, W. Ramakrishna, J. Dorweiler, J.L. Bennetzen and V.L. Chandler (2002) The regulatory regions required for *B'* paramutation and expression are located far upstream of the maize *b1* transcribed sequences. *Genetics* **162:** 917–930.

Swigonova, Z., J. Lai, J. Ma, W. Ramakrishna, V. Llaca, J.L. Bennetzen and J. Messing (2004) Close split of sorghum and maize genome progenitors. *Genome Res.* **14:** 1916–1923.

Talbert, L.E. and V.L. Chandler (1988) Characterization of a highly conserved sequence related to mutator transposable elements in maize. *Mol. Biol. Evol.* **5:** 519–529.

Tikhonov, A. P., J.L. Bennetzen and Z. Avramova (2000) Structural domains and matrix attachment regions along colinear chromosomal segments of maize and sorghum. *Plant Cell* **12:** 249–264.

Tikhonov, A.P., P.J. SanMiguel, Y. Nakajima, N.D. Gorenstein, J.L. Bennetzen and Z. Avramova (1999) Colinearity and its exceptions in orthologous *adh* regions of maize and sorghum. *Proc. Natl. Acad. Sci. USA* **96:** 7409–7414.

Vitte, C. and J.L. Bennetzen (2006) Analysis of retrotransposon diversity uncovers properties and propensities in angiosperm genome evolution. *Proc. Natl. Acad. Sci. USA* **103:** 17638–17643.

Volff, J.-N. (2006) Turning junk into gold: domestication of transposable elements and the creation of new genes in eukaryotes. *BioEssays* **28:** 913–922.

Wang, Q. and H.K. Dooner (2006) Remarkable variation in maize genome structure inferred from haplotype diversity at the *bz* locus. *Proc. Natl. Acad. Sci. USA* **103:**17644–17649.

Wang, W., H. Zheng, C. Fan, J. Li, J. Shi, Z. Cai, G. Zhang, D. Liu, J. Zhang, S. Vang, Z. Lu, G.K. Wong, M. Long and J. Wang (2006) High rate of chimeric gene origination by retroposition in plant genomes. *Plant Cell* **18:** 1791–1802.

Wei, F., E. Coe, W. Nelson, A.K. Bharti, F. Engler, E. Butler, H. Kim, J.L. Goicoechea, M. Chen, S. Lee, G. Fuks, H. Sanchez-Villeda, S. Schroeder, Z. Fang, M. McMullen, G. Davis, J.E. Bowers, A.H. Paterson, M. Schaeffer, J. Gardiner, K. Cone, J. Messing, C. Soderlund and R.A. Wing (2007) Physical and genetic structure of the maize genome reflects its complex evolutionary history. *PLoS Genet.* **3:** e123.

Wendel, J.F., R.C. Cronn, I. Alverez, B. Liu, R.L. Small and D.S. Senchina (2002) Intron size and genome size in plants. *Mol. Biol. Evol.* **19:** 2346–2352.

Wicker, T., B. Keller (2007) Genome-wide comparative analysis of copia retrotransposons in Triticeae, rice, and Arabidopsis reveals conserved ancient evolutionary lineages and distinct dynamics of individual copia families. *Genome Res.* **17:** 1072–1081.

White, S.E., L. F. Habera and S.R. Wessler (1994) Retrotransposons in the flanking regions of normal plant genes: a role for copia-like elements in the evolution of the gene structure and expression. *Proc. Natl. Acad. Sci. USA* **91:** 11792–11796.

Wilson, W.A., S.E. Harrington, W.L. Woodman, M. Lee, M.E. Sorrells and S.R. McCouch (1999) Inferences on the genomes structure of progenitor maize through comparative analysis of rice, maize and domesticated panicoids. *Genetics* **153:** 453–473.

Zhang, Q., J. Arbuckle and S.R. Wessler (2000) Recent, extensive, and preferential insertion of members of the miniature inverted-repeat transposable element family *Heartbreaker* into genic regions of maize. *Proc. Natl. Acad. Sci. USA* **97:** 1160–1165.

Zonneveld, B.J.M., I.J. Leitch and M.D. Bennett (2005) First nuclear DNA amounts in more than 300 angiosperms. *Ann. Bot.* **96:** 229–244.

Genetic Diversity, Linkage Disequilibrium and Association Mapping

Antoni Rafalski and Evgueni Ananiev

Abstract Maize, at all levels of resolution, is one of the most diverse crop species. Large insertions and deletions are common between maize inbreds, and include tandem repeat clusters, abundant retroelement and transposons. At the gene level, single nucleotide polymorphisms are common, especially in introns and untranslated regions of genes. Depending on choice of experimental population and region in the genome, linkage disequilibrium between polymorphic sites could decay very rapidly, or persist for hundreds of Kb. Appropriately chosen germplasm collections may be used for genetic association mapping (also called linkage disequilibrium mapping), either with candidate genes, or by scanning the whole genome with thousands of markers at high density. This approach, in favorite circumstances, could provide high resolution. The power of association mapping is variable, and has not been thoroughly investigated. Rapid advances in genome sequencing and high density genotyping are making this approach to relating genotype with phenotype increasingly attractive.

1 Introduction

Maize is known as one of the most phenotypically diverse crop species. Detailed analyses of maize diversity at the molecular level, especially in genic and other single copy regions, have been available for some time (Doebley et al., 1984; Gaut and Clegg, 1993; Tenaillon et al., 2001; Vigouroux et al., 2005; Yamasaki et al.,

A. Rafalski

DuPont Crop Genetics Research, DuPont Experimental Station E353, Route 141 and Henry Clay Road, Wilmington, DE19880-0353

j-antoni.rafalski@cgr.dupont.com

E. Ananiev (deceased)

Pioneer Hi-Bred, A DuPont Business, Johnston, IA 50131-1004, USA

J.L. Bennetzen and S. Hake (eds.), *Maize Handbook - Volume II: Genetics and Genomics*, 201
© Springer Science+Business Media LLC 2009

2005; Buckler et al., 2006). However, in the last few years, a new understanding of maize diversity at the whole genome level has been developed. At the same time, studies of genetic diversity in diverse germplasm collections were combined with the analysis of phenotype in what is known as genetic association mapping, or linkage disequilibrium mapping. These studies, coupled with technological developments which allow many thousands of polymorphic loci to be genotyped simultaneously, promise to significantly enhance our understanding of complex traits in maize.

Completion of the maize genome sequence (expected in 2008) and rapid developments in high throughput sequencing technologies will increase the resolving power of diversity studies and further enhance our understanding of genome evolution in grasses.

For technical reasons, most analyses of genetic diversity were, until recently, limited to single copy or low copy number regions, especially open reading frames of genes. RFLP and PCR based methods are less effective in multi-copy rather than in low-copy regions. This was thought not to be a significant issue because much of functionally relevant variation is usually considered to be genic in nature. Yet, regulatory regions, sometimes quite distant from the coding region of genes, play a very important role in maize evolution (Doebley and Lukens, 1998; Stam et al., 2002; Clark et al., 2006). Realization that multi-copy genetic elements may play a regulatory role occurred relatively recently, at the time when genomic sequencing became more efficient. In a pioneering study, Fu and Dooner (Fu and Dooner, 2002) compared genic and intergenic sequences over a more than 100 kilobase region in two inbreds belonging to different heterotic groups, B73 and Mo17. While genic regions were largely conserved, the intergenic segments lacked similarity. Studies from several laboratories (Song and Messing, 2003; Brunner et al., 2005b; Lai et al., 2005; Wang and Dooner, 2006) later confirmed and expanded on these findings.

It has become clear that in addition to transposons and ubiquituous retrotransposons (SanMiguel et al., 1996; SanMiguel et al., 1998; Kumar and Bennetzen, 2000; Bruggmann et al., 2006), some recently identified families of elements such as helitrons (Lal et al., 2003; Brunner et al., 2005a; Gupta et al., 2005; Lai et al., 2005; Morgante et al., 2005; Xu and Messing, 2006) play a role in generating intraspecific diversity in maize. This phenomenon occurs faster than maize populations can become homogenized by fixation or elimination by stochastic or selective forces. Consequently, these genomic elements must have been established relatively recently ($<<1\times10^6$ years), and may be continually colonizing the genome (Brunner et al., 2005b; Liu et al., 2007).

A rapidly developing understanding of the abundance and biological role of diverse RNA transcripts, such as miRNAs and siRNAs lends additional arguments to the importance of the study of genetic diversity outside of classically defined genes (Nobuta et al., 2007; Rafalski, 2007). In this chapter, we will review recent developments in the analysis of maize genome diversity and applications of these results to the understanding of complex traits.

2   Gene Molecular Diversity in Maize and its Ancestors

Many different molecular mechanisms contribute to the evolution of maize genetic diversity (Table 1).

Protein coding genes constitute only 5%-6% of the maize genome, but much of the knowledge of genetic diversity in maize has been gained by analysis of such genes and DNA sequences in their immediate vicinity, and of simple sequence repeats (SSRs, microsatelites) which evolve by a different mechanism and more rapidly than single nucleotide polymorphisms (SNPs) (Vigouroux et al., 2002)

Pairwise genetic diversity (Tajima, 1983) have been estimated to be π=0.0063 in cultivated maize (Ching et al., 2002), on the basis of SNP analysis in genes, including untranslated regions. A nearly identical value, π=0.0067 was obtained by Yamasaki in a survey of 1095 genes (Yamasaki et al., 2005) in a set of inbreds selected to maximize genetic diversity. Our sampling of SNPs in a large collection of maize inbreds (excluding landraces) at an unbiased set of genes resulted in lower mean diversity estimate of π=0.0047, median π=0.0035 (A. Rafalski et al., unpublished observations). The difference in diversity estimates between the two studies most likely reflects different germplasm sets, as well as an ascertainment bias (non-random choice of markers). In reference to other species, maize is about seven to tenfold more diverse than humans (π=0.0007, http://egp.gs.washington.edu/summary_stats. html) and 2.7× more diverse than soybean (Zhu et al., 2003). While comparative SNP-based surveys are still lacking for many crops, grape (*Vitis vinifera*) has been recently reported to have diversity levels comparable to those of maize (π=0.0066) (Lijavetzky et al., 2007).

Most recent genic SNP-based estimates indicate that maize retains about 57% of the diversity of its progenitor, teosinte (Wright et al., 2005). Relative to landraces, maize inbreds retained 77% of the diversity, according to an analysis of 21 genes located on chromosome 1 (Tenaillon et al., 2001). A similar estimate was obtained using microsatellites (Liu et al., 2003). However, SSR-based estimates of the diversity

Table 1   Contributions to genome diversity in maize

◦ Genome or large genome segment duplications followed by structural and functional diversification
◦ Gene duplications, followed by diversification of DNA sequence and gene function
◦ Mutational processes including those associated with recombination and DNA replication, including gene conversion events
◦ Insertion and loss of DNA transposons
◦ Insertion and partial loss of retroelements
◦ Capture and translocation of gene segments by specialized classes of transposons (Pack-Mules and Helitrons)
◦ Expansion and contraction of simple sequence repeats (SSRs)
◦ Expansion and contraction of tandemly repeated sequences
◦ Possible gene flow between maize and teosinte

change relative to teosinte results in a less drastic loss of alleles (Vigouroux et al., 2005). This is especially pronounced for rapidly-mutating dinucleotide repeats (Vigouroux et al., 2002; Vigouroux et al., 2005) suggesting that the microsatellite-based number is an underestimate resulting from rapid recovery after the domestication bottleneck. As expected, distribution of diversity values among genes is very broad (standard deviation of the $\pi=0.0047$ estimate is 0.045, our unpublished observations), with a substantial number of loci with very low diversity ($\pi <0.0005$), and loci with unusually high diversity ($\pi >0.03$, (Yamasaki et al., 2005)). Although a systematic survey of high diversity genes has not been reported in maize, it is well established that certain classes of genes, such as those involved in defense against pathogens and pests, evolve rapidly and frequently exhibit high allelic diversity (Rose et al., 2004; Tiffin et al., 2004).

A genetic bottleneck during the domestication of maize as well as ongoing selection by breeders is expected to result in diversity reduction across the whole genome as well as at selected loci. Reduced diversity was observed at a major domestication locus, *tb1* (Wang et al., 1999). The diversity loss was especially strong in the upstream regulatory region while the open reading frame was almost unaffected (Wang et al., 1999; Clark et al., 2006). A recent SSR study indicated a strong bottleneck effect and a less pronounced loss of diversity at linked loci due to selection (Vigouroux et al., 2005). Such effects are quite pronounced at genes known to be under strong breeders' selection. At the maize *Y1* gene, which encodes a phytoene synthase gene (Buckner et al., 1990; Buckner et al., 1998) selection for the yellow endosperm phenotype preferred in some markets resulted in >10-fold reduction of diversity (Palaisa et al., 2003). The effect of this selective sweep extends to several hundred Kb (Palaisa et al., 2004). Reduction of diversity and other effects indicative of selection has also been demonstrated in the starch pathway (Whitt et al., 2002) and in the anthocyanin regulatory locus *c1* (Hanson et al., 1996). More recent studies attempt a more comprehensive survey of the maize genome in search of evidence for selection. Yamasaki (Yamasaki et al., 2005) surveyed over 1,000 genes and identified eight which show evidence of selection with strongly reduced diversity. In the near future, scanning of all 50,000 or so maize genes, using high throughput sequencing or oligonucleotide array technology, will allow identification of a more complete set of genes under natural or artificial selection. Extended selective sweeps will be indicative of more recent selection by breeders and may point to regions where very selective diversity enhancement by introgressions from landraces or even teosinte may be beneficial.

3   Genetic Diversity Outside of Genes

The analysis of intraspecific diversity has to include the understanding of genetic diversity in intergenic regions. In maize, these regions differ dramatically between inbreds and are filled with large insertions and deletions frequently resulting from activities of retrotransposons and DNA transposons (Banks et al., 1985; Bureau and

Wessler, 1994; SanMiguel et al., 1996; Bennetzen, 2000; Fedoroff, 2000; Kumar and Bennetzen, 2000; Zhang et al., 2000; Wessler, 2006). About 77% of the maize genome is repetitive and composed primarily of retroelements which constitute ~60% of the genome (Meyers et al., 2001). The first direct comparison of extended intergenic DNA of two maize inbreds, B73 and Mo17, at the *Bz* locus revealed numerous large insertion / deletion polymorphisms (Fu and Dooner, 2002) including pseudogenes translocated by transposons. This extreme intergenic diversity was found to be common throughout the genomes of maize inbreds (Wang and Dooner, 2006) and not limited to the *Bz* locus (Song and Messing, 2003; Brunner et al., 2005b; Brunner et al., 2005a; Morgante et al., 2005).

Analysis of the distribution of retrotransposon insertion times by evaluating sequence divergence between LTRs (SanMiguel et al., 1998) of retrotransposon copies not shared between maize inbreds indicates that the youngest elements are most common (Brunner et al., 2005b). This finding agrees with the expectation of population genetics that the older polymorphisms will tend to be either fixed in the population or eliminated over time. Commonly found co-existence of transposons and their unfilled insertion sites in maize germplasm indicates that the insertion events occurred in the time frame comparable to the half-life of a polymorphism expected from population genetic considerations; on the order of a few million years (Brunner et al., 2005b).

Among the unexpected results of the comparative sequencing of maize inbreds was the identification of numerous pseudogenes consisting of imperfectly concatenated introns and exons originating from genes present elsewhere in the genome (Brunner et al., 2005a; Gupta et al., 2005; Lai et al., 2005; Morgante et al., 2005). Helitron elements, DNA transposons replicating by a rolling-circle mechanism (Kapitonov and Jurka, 2001) have been implicated in these events. Although the mechanism is not fully characterized, it is thought helitrons create and move these pseudogenes (Lal et al., 2003; Brunner et al., 2005a). Some of the pseudogene-carrying helitrons are transcribed, presumably having inserted near a promoter, and thus may have phenotypic effects. Anti-sense expression may lead to silencing of genes from which the helitron-carried sequences originated, possibly affecting phenotype. It could be hypothesized that, over evolutionary time, novel activities may have been selected by accidental concatenation and expression of exons from different genes (Brunner et al., 2005a). Given short sequence signatures of non-autonomous helitrons, such events would be very difficult to demonstrate conclusively. Common in rice, but also detected in maize and Arabidopsis, Pack-Mule transposable elements have been found also to carry gene fragments and possibly contribute to genome re-modelling (Jiang et al., 2004).

4 Genetic Recombination and the Distribution of Diversity Along Maize Chromosomes

The positive correlation between frequency of genetic recombination and genetic diversity is both predicted theoretically and well-documented experimentally (Hudson and Kaplan, 1995). In maize, Tenaillon et al. (Tenaillon et al., 2001) studied the distribution of genetic diversity in 21 loci distributed along chromosome 1. However, no obvious pattern of diversity was identified, perhaps because the sampling density was insufficient and only two of the sampled loci were located in centromeric regions, where recombination is expected to be very low. With the availability of high density genetic and physical maps, it is now possible to address this issue with better resolution. The relationship between genetic and physical distance along the chromosomes has been precisely described (Anderson et al., 2004; Fengler et al., 2007), so genetic diversity measures can now be placed in the same framework. A pairwise genetic diversity measure π (Tajima, 1983), calculated as a running average over 50 loci, is non-uniformly distributed along maize chromosome 2 (Figure 1). The relationship with recombination frequency is complex: the expected pericentromeric reduction of diversity is pronounced, but high diversity regions appear to directly flank the region of low gene density (data not shown). Thus, chromosome 2 exhibits reduced recombination and reduced genetic diversity encompassing approximately one third of its physical length. It remains to be determined if this pattern can be generalized throughout the genome.

The non-uniform distribution of cytogentically identifiable features such as knobs (McClintock, 1931) along maize chromosomes and their polymorphisms between inbreds has been the subject of extensive cytogenetic studies (Longley, 1939; McClintock, 1978). In addition to knobs (composed of long tracts of 180-bp tandem repeat units) (Dennis and Peacock, 1984) and knob-specific TR1 repeat

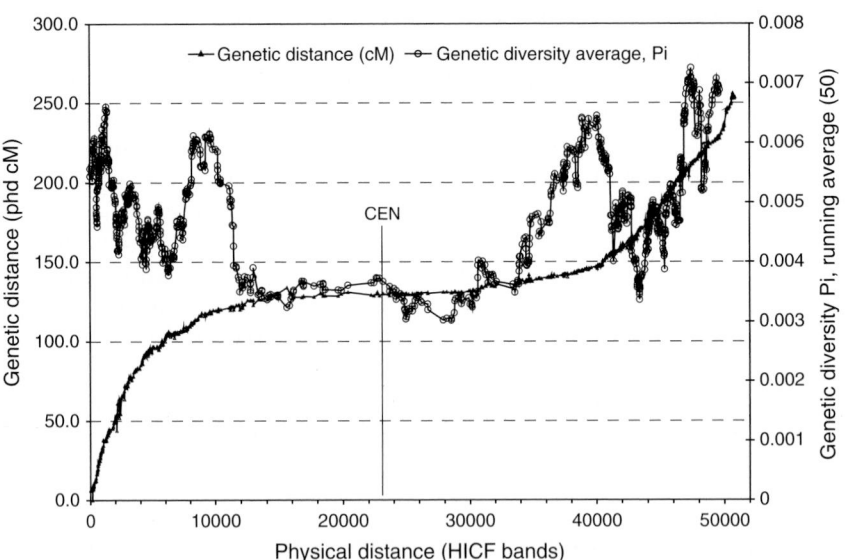

Fig. 1 Genetic diversity (π) of maize chromosome 2 in an elite inbred collection, as a function of physical distance (expressed as high information content fingerprinting (HICF) band count along the physical map of the chromosome) and recombination frequency (expressed as genetic distance, cM). (A. Rafalski and Scott Tingey., unpublished results). Location of the centromere is indicated (E. Ananiev, unpublished results). In the broad region surrounding the centromere, recombination is suppressed and diversity is relatively low

units (Ananiev et al., 1998c), numerous classes of tandemly repeated sequences have been characterized. These include 18-28S rDNA (Zimmer et al., 1988; Li and Arumuganathan, 2001), 5S rDNA (Mascia et al., 1981; Zimmer et al., 1988; Li and Arumuganathan, 2001), centromeric CentC tandem repeats (Ananiev et al., 1998a), centromeric CRM (Nagaki et al., 2003) and CentA (Ananiev et al., 1998a) clustered repeats, megatracts of AGT and AGC microsatellites (Ananiev et al., 2005b), sub-centromeric Cent4 tandem repeat (Page et al., 2001), B-centromere-specific repeats (Alfenito and Birchler, 1993), CL clustered repeats on the B chromosome (Cheng and Lin, 2003), subtelomeric tandem repeats TR1350 (Chen et al., 2000) and pBF266 (Burr et al., 1992; Gardiner et al., 1996). While these features will not be reviewed here, they remain an important tool in the analysis of diversity from a chromosomal perspective (see below).

The application of the fluorescent in situ hybridization (FISH) chromosome painting technique revealed that many of these repetitive sequences exhibit large copy number variation and even presence/absence polymorphsms between different maize inbreds (Kato et al., 2004). It is unknown how frequently and by what mechanisms these repeat clusters generate diversity. Do the clustered/tandem repeats move across the corn genome, or are there "seed" copies of repeated units which can be locally amplified or reduced in copy number?

Ananiev (Ananiev et al., 2005a) documented large difference in the length and presence of extended microsatellite-like megatracts which may amplify by slippage mechanisms postulated for microsatellite (SSR) sequence with contributions from a recombinational mechanism.

Tandem repeat polymorphisms may be conveniently visualized in diplotene chromosomes derived from a heterozygous individual (Figure 2). Recently, to increase

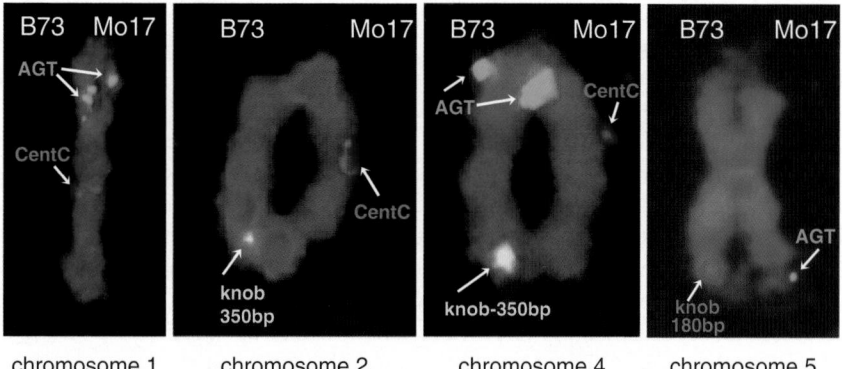

Fig. 2 An example of multi-color FISH on the diplotene chromosomes of maize F1 hybrid B17 × Mo17 (E. Ananiev, M. Chamberlin and S. Svitashev, unpublished results). Four probes each labeled with a different fluorophore were hybridized simultaneously to meiotic diplotene stage preparations of the maize hybrid. The CentC repeat probe (red) is centromere-specific and allows positioning of markers on either chromosome arm. The AGT microsatellite probe (green) marks chromosomes 1, 2, 4 and 5. The 350 bp knob specific probe (yellow) marks chromosomes 2, 4 and 5. The 180 bp knob probe (also red) marks chromosome 5. Numerous polymorphisms are visible

resolution, FISH mapping of tandem repeat polymorphisms was performed on pachytene chromosomes from two maize inbred lines, Mo17 and B73. Copy number or presence / absence polymorphisms of at least nine types of tandem repeats were characterized. Molecular analysis indicated that tandem arrays or loose clusters of repeats are frequently interrupted by retroelements (Ananiev et al., 1998a, b; Mroczek and Dawe, 2003; Nagaki et al., 2003; Ananiev et al., 2005b). Nucleotide sequence comparison of one particular site from different inbred lines indicated that an entire small cluster of 180 bp knob tandem repeats might be completely absent in one inbred but present in another. Such insertions/deletions or transpositions of entire repeat clusters constitute another mechanism of disruption of the linear gene order in different maize inbreds, in addition to the mechanisms discussed earlier (Fu and Dooner, 2002; Song and Messing, 2003; Brunner et al., 2005b).

Insertion/deletion polymorphisms are likely to affect local recombination frequency beyond the immediate vicinity of the repeat by interfering with chromosome pairing or by other mechanisms, as has been demonstrated at the *Bz* locus (Dooner and Martinez-Ferez, 1997). The existence of these large-scale polymorphisms involving tandemly repeated DNA is likely to bias the distribution of recombinants in a manner dependent upon the choice of parents of the genetic cross.

5 Linkage Disequilibrium and Population Structure in Maize

The concept of linkage disequilibrium (LD), or non-random association of alleles (Hartl, 2000), is defined at the population level. Very close linkage between genetic loci may result in strong disequilibrium, however the amount of disequilibrium, measured by the LD parameter r^2 or D' (Flint-Garcia et al., 2003) is expected to decline with distance between markers, due to recombinations occurring over generations. In the absence of other factors, LD should be close to 0 at unlinked loci. The distribution of alleles at these loci, relative to each other, should be random. Factors affecting the rate of decline of LD with distance have been reviewed recently (Flint-Garcia et al., 2003; Gaut and Long, 2003). In maize, an outcrossing species with large effective population size, the decay of LD is expected to be rapid. However, the observed LD depends very strongly on the population being analyzed. Indeed, in highly diverse collections of germplasm, LD has been shown to decay rapidly. For example, Tenaillon and co-workers (Tenaillon et al., 2001) found that LD decays to half of its starting value within a few hundred bp in maize. In collections of breeding lines, which do not resemble natural populations at all, LD could decay more slowly (Remington et al., 2001; Ching et al., 2002). Such populations usually consist of more closely related sub-groups, that is, are highly structured, increasing observed LD. At the *Adh1* locus, which is not known to be under selection, (Jung et al., 2004) found that LD extends on average over 500 kb in a collection of elite maize inbreds. Moreover, LD could also decay more slowly in the presence of selection (Remington et al., 2001; Whitt et al., 2002). Long-range LD has also been found at the *Y1* locus, which is under strong breeding selection for endosperm color (Palaisa et al., 2004). The availability of germplasm collections with varying degree of kinship, and thus varying amount of ancestral recombination and LD, allows the researcher to choose the most appropriate germplasm collections for association mapping (see below).

6   Genetic Diversity and Selection by Breeders

In the past 50 to 100 years, maize has been under intense selection by breeders aiming to maximize yield and reduce detrimental effects of biotic and abiotic stresses. The expected result of selection is a reduction of diversity at loci under selection. However, in a manifestation of LD, diversity at adjacent loci, which may be beneficial, is reduced as well (a phenomenon known as selective sweep (Hartl, 2000)). It has been suggested that the prevalence of diversity reduction at loci adjacent to those directly selected for may be low (Wright et al., 2005), but the previously cited *Y1* locus may be a counterexample (Palaisa et al., 2004).

Since the advent of maize F1 hybrids, breeders have worked to increase heterotic yield. As a consequence, genetic distance between two heterotic groups increased (Cooper et al., 2004). Allele frequencies at many loci have changed in opposite directions in different heterotic groups such as Stiff Stalk Synthetic (SSS) and Non-Stiff Stalk (NSS). This effect deepens the non-random population structure of maize germplasm collection, which includes contemporary cultivated lines. Efforts are being made to expand the range of genetic diversity available to breeders by producing introgression populations derived from exotic or unadapted donor parents (Tanksley and McCouch, 1997; Xiao et al., 1998; Hoisington et al., 1999). Such efforts have been productive in other species such as tomato (Lippman and Tanksley, 2001; Frary et al., 2004). While cultivated maize is significantly more genetically diverse than cultivated tomato, it is expected that such introductions may further expand allelic diversity for disease and pest resistance.

7   Genetic Association Mapping: Principles and the Connection to Linkage Disequilibrium

Historically, genetic mapping of traits has been done in purpose-developed segregating biparental populations. More recently, advances in rapid genotyping of hundreds of individuals at multiple SNP loci increased interest in the use of diverse collections of germplasm, usually representing a substantial fraction of breeding germplasm, for genetic mapping. This population-based approach was initially applied in human genetics (Cardon and Bell, 2001; Cardon and Abecasis, 2003), but was rapidly adopted in plants, under the name "genetic association mapping" or "linkage disequilibrium mapping" (Remington et al., 2001; Thornsberry et al., 2001; Rafalski, 2002;

Whitt et al., 2002; Mackay and Powell, 2007). This approach relied on the realization that the distribution of phenotypic scores (such as plant height) between collections of individuals differing in the functionally relevant alleles (such as in a plant height gene) should also be different, even in the presence of other segregating genes (Figure 3). This simple principle can be confounded by many factors, such as population structure. This is intuitively obvious, as two groups of individuals differing in their ancestry may not only differ in trait means but also in allele frequencies at many genetic loci. Association mapping methods can be roughly divided into two categories: candidate gene association analysis and whole genome association mapping.

7.1 Candidate Gene Analysis

One or several genes suspected or known to be involved in the determination of the trait of interest may be tested for association with the phenotype. For example, genes of carbohydrate and starch biosynthesis are reasonable candidates for an endosperm starch content trait (Whitt et al., 2002). It is essential to remember that

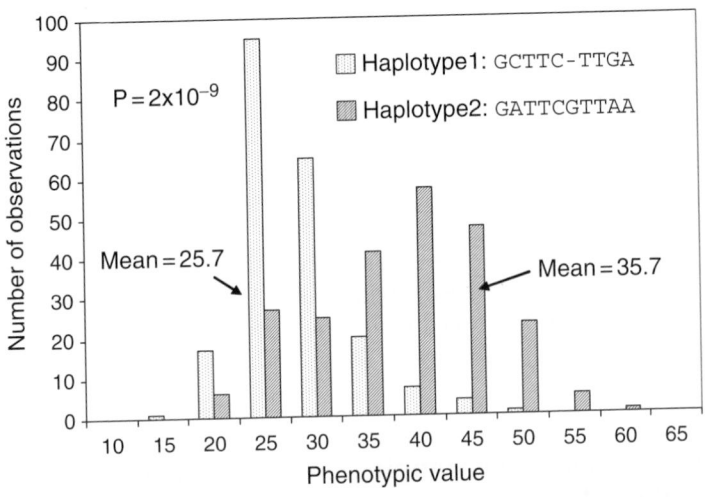

Fig. 3 Principles of candidate gene–based association mapping – a hypothetical example. A 550-bp PCR amplicon derived from a candidate gene has been sequenced in 442 inbred lines and two haplotypes have been identified at four SNP loci. The same individuals have also been phenotyped for a quantitative trait. The graph represents the histogram of distribution of phenotypic values for individuals containing haplotype 1 and for those containing haplotype 2. The difference between the two distributions is significant, as determined by a permutation test. In absence of additional information, it is not possible to ascertain weather the significant p value results from true association, or from a population structure, that is high frequency of haplotype 1 in a subpopulation characterized by low phenotypic values, while another subpopulation has high frequency of haplotype 2 and high average phenotypic values, without causal relationship. Moreover, distribution of phenotypic values for haplotype 2 appears may be bimodal, possibly indicating that this haplotype may be further subdivided into two groups by additional genotyping

a positive result may also mean that the gene being assayed is a marker for a nearby gene if these two genes are in LD. In turn, a negative result does not eliminate the gene from consideration – as lack of proof of association may be due to insufficient diversity in the population in question or diversity associated with a rare allele. Low allele frequencies combined with small to medium phenotypic effects usually would not lead to statistically significant results given phenotypic noise or population size. Interpretation of probability values resulting from association testing is non-trivial and p values frequently accepted as significant ($p<0.05$) may be found to be insufficient. Although it has been suggested (Curtis et al., 2007) that p values resulting from candidate gene approach should be treated differently from those from whole genome scans (see below) to account for other supporting evidenc. We find such an arbitrary "fix" difficult to justify.

In an early study of candidate gene associations in maize, the role of the *dwarf8* gene on flowering time was evaluated (Pritchard, 2001; Thornsberry et al., 2001). Realizing that the complex population structure of maize might cause statistical errors in an association study, the method of Pritchard (Pritchard and Rosenberg, 1999) was modified to extend its applicability to quantitative traits. The population structure was estimated on the basis of SSR genotypes of the individuals in the population under study and then used to correct association probability values. The *dwarf8* gene was found to be significantly associated with flowering time. A similar approach was later used to confirm the association of a *su1* mutant allele with a sweet phenotype of corn (Whitt et al., 2002). In their subsequent work, the same group developed a more advanced method for the correction of population structure effects (Yu et al., 2006).

Palaisa (Palaisa et al., 2003; Palaisa et al., 2004) identified a strong association between certain polymorphisms in the maize phytoene synthase gene *Y1*, and yellow endosperm phenotype, while another phytoene synthase gene, *PSY2*, showed no association. The *Y1* gene had been independently shown to be involved in the yellow endosperm trait (Buckner et al., 1990; Buckner et al., 1998; Tochtrop and Buckner, 2000), so this pilot study convincingly demonstrates the power of candidate gene association analysis. On the other hand, the same work also demonstrates that, in elite maize, LD may extend to several hundred Kbp, making definitive identification of the causal gene elusive.

As these studies show, the candidate gene association method could be successful, provided several conditions are met:
- Candidate genes are well-chosen based on substantial prior knowledge. In many cases, some genes affecting a trait of interest are known, such as structural genes for enzymes in well-defined pathways, but frequently other genes such as transcriptional regulators may not be known, making the choice of candidates a hit-or-miss proposition. This issue, especially relevant for agronomic traits such as drought tolerance for which biochemical pathways are not well known, makes whole genome scan association mapping, described below, attractive.
- There is an appropriate amount of LD in the germplasm collection available for investigation (Gaut and Long, 2003). Strong LD, extending over tens or hundreds of Kbp, results in poor resolution and a candidate gene may be just

a marker for a gene in the vicinity. On the other hand, very rapidly declining LD, such as has been demonstrated in some maize populations (Tenaillon et al., 2001) may also be refractory to candidate gene association analysis. In that case, surveying a few markers in a gene of interest may be insufficient, and full sequencing of the whole gene and surrounding region may be necessary to identify an association. The resolution, however, is likely to be sufficiently high to allow discovery of the causal polymorphism. In maize, LD is very dependent on the choice of germplasm, and LD extending the length of the gene (Remington et al., 2001) and beyond (Jung et al., 2004; Palaisa et al., 2004) is common in elite germplasm.

- Alleles with contrasting phenotypic effects are present in the germplasm collection in sufficient frequency to allow the differences in the distribution of phenotypes to yield statistical significance. Rare alleles, such as some disease resistance alleles, cannot be mapped by association and mapping, such that mapping in a biparental population is a better choice.

A detailed study of the effects of various factors on the power of association mapping is still lacking. In cases where meaningful candidate genes are not available, whole genome scan association mapping may be an attractive alternative.

7.2   Whole genome scan

Rather than focusing on imperfectly understood candidate genes, it is more powerful to test loci across the genome (or, perhaps, soon, all genes) for genetic association with a traits. This approach removes some of the caveats of the candidate gene approach, but it is costly because of the number of loci that are required. Unlike trait mapping in biparental populations, where a few hundred markers are usually adequate, the number of markers in a genome scan can only be determined by first understanding the extent of LD in the genome (Remington et al., 2001); i.e., how distant a marker can be from the causal gene and still allow detection of the association. In diverse maize populations, LD was found to decay very rapidly, within a few hundred bases as discussed earlier. In such circumstances, the number of markers required for a genome scan will exceed by several fold the number of genes (estimated at 50,000). Today, this is at the limit of technical feasibility. However in populations of elite maize breeding lines, the decay of LD (while variable) was found to be much slower, in some cases exceeding few hundred Kb.

To evaluate the feasibility of genome scans in maize with ca. 10,000 markers, we chose to use red cob trait as a best-case test. The gene responsible for red coloration of cobs, pl, located on chromosome 1 bin 1.03 is well-characterized (Grotewold et al., 1994) and phenotyping of cob color is unambiguous. The results of a gene scan performed with 1200 genetic loci on chromosome 1 are shown in Fig 4. A sharp peak of probability was found to coincide with the known location of the pl gene. The genetic marker with the best p value ($p = 10^{-17}$) is located on the same BAC clone as the p1 gene (K. Fengler, personal communication). The breadth of

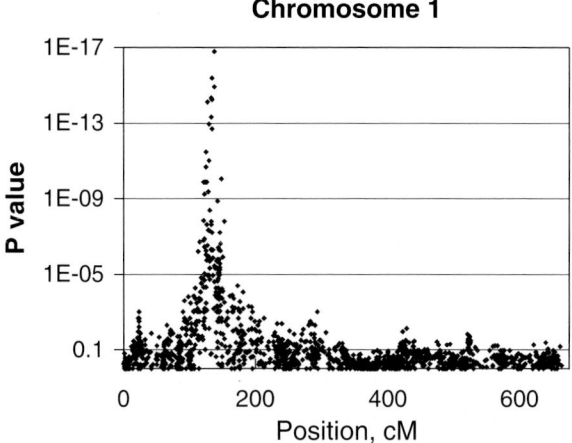

Fig. 4 Genome scan mapping of the *red cob* trait. Probability of associations between the phenotype (red or white cob) and each of the SNP haplotypes at 1200 chromosome 1 markers was computed in a collection of non-stiffstalk maize elite germplasm, using a permutation test. The horizontal axis represents a genetic map of maize chromosome 1, the vertical axis is the probability of association with the *red cob* trait. The marker at the probability peak (p=10^{-17}) is located within a BAC distance (<150 kb) of the red cob gene, *P1* (S. Luck and O. Smith, unpublished data)

the peak indicates a large interval of linkage disequilibrium in this region of the genome. Indeed, LD initially declined quite rapidly, but the width of the LD peak at $r^2=0.5$ is still ca. 25 cM (data not shown)

Beló et al. (Beló et al., 2007) have conducted a maize genome scan for genes affecting oleic acid content in maize grain. A number of peaks of association were identified. The phenotypic effect of three of the strongest associations was independently tested by following segregation of the allele putatively associated with high oleic acid content in an F_2 population. One of the tested associations was confirmed and completely explained the observed phenotypic effect. The markers with the strongest associations to oleic acid content were located at 380–384 cM (Intermated B73/Mo17 maize genetic map with neighbors (IBM2+neighbors; http://www.maizemap.org) on chromosome 4, and the marker with the best p value was located within 2 kbp of fatty acid desaturase-2 (*fad2*), known to catalyze the 18:1 > 18:2 desaturation (Mikkilineni and Rocheford, 2003). Two other peaks of association could not be validated by this approach. It was hypothesized that the unconfirmed associations may be false positives resulting from population structure effects. This conclusion was strengthened by the finding that all three associations are in LD with each other. This finding alone could alternatively be explained by co-selection for loci increasing oleic acid content, but this explanation cannot be reconciled with the lack of effect of the two putative false positives individually on the oleic acid phenotype.

Whole genome scan mapping of traits has been validated in maize, at least in the simplest cases, but we do not have detailed understanding of the power of this approach, given highly variable LD, noisy phenotypic data and complex population structure. Appropriate methods of correcting probability values for multiple testing, when the tests are not fully independent (loci in LD with each other, analysis of both individual SNPs as well as SNP haplotypes), also requires further study.

8 Future Developments

As we look forward to developments in the next few years, several technological advances promise to change the depth of understanding of maize genetic diversity and of the association between naturally occurring alleles and phenotypes. High throughput sequencing technology currently allows the generation of one giga-base (1×10^9 bp) or more of DNA sequence per machine run, and this limit is expected to increase rapidly. Combined with a maize genome sequence of the B73 inbred line, which will be essentially complete in the next couple of years, these methods will allow re-sequencing of at least the single-copy regions of many maize inbreds. Some form of reduced representation sequencing may be needed initially, but in the near future it may be possible to fully describe genetic diversity at single copy regions of a maize inbred in a single experiment, and even to identify with high resolution all recombinations which occurred during a meiosis.

Very high density microarrays, such as those from Affymetrix for the human genome, allow genotyping of ~900,000 SNP loci per "chip" (www.affymetrix.com). Given the expected ~50,000 maize genes, a haplotype composed of 5–10 SNPs per gene is in principle within reach in a single experiment. A data set of this resolution produced for a large population will allow precise association mapping even if LD is low, and will dramatically accelerate gene identification assuming bioinformatic capabilities follow suit. The availability of high quality phenotypic data is, however, already limiting. This limitation is unlikely to be removed by a quick technological fix.

In contrast, a full description of the diversity of intergenic, largely repetitive sequences, in a collection of germplasm will be much more challenging. Recent developments in the understanding of the transcriptome, including discovery of diverse classes of non-coding RNAs, such as miRNAs and siRNAs underscore the need for detailed description of diversity beyond protein-coding genes. Together, these advances will enable very fine-grained descriptions of genetic diversity and identification of recombination events in collections of germplasm and in defined populations.

Acknowledgements We thank Scott Tingey for continuing support, frequent sharing of ideas and for comments on the manuscript.

The authors are grateful to Mark Chamberlin, Sergei Svitashev, André Beló, Howie Smith and Stan Luck for sharing unpublished results and to Beth Holloway and Petra Wolters for helping to improve the manuscript.

References

Alfenito, M.R., and Birchler, J.A. (1993). Molecular characterization of a maize B chromosome centric sequence. Genetics 135, 589–597.

Ananiev, E.V., Phillips, R.L., and Rines, H.W. (1998a). Chromosome-specific molecular organization of maize (Zea mays L.) centromeric regions. Proc Natl Acad Sci U S A. 95, 13073–13078.

Ananiev, E.V., Phillips, R.L., and Rines, H.W. (1998b). Complex structure of knob DNA on maize chromosome 9. Retrotransposon invasion into heterochromatin. Genetics 149, 2025–2037.

Ananiev, E.V., Phillips, R.L., and Rines, H.W. (1998c). A knob-associated tandem repeat in maize capable of forming fold-back DNA segments: are chromosome knobs megatransposons?. Proc Natl Acad Sci U S A 95, 10785–10790.

Ananiev, E.V., Chamberlin, M.A., Klaiber, J., and Svitashev, S. (2005a). Microsatellite megatracts in the maize (Zea mays L.) genome. Genome 48, 1061–1069.

Ananiev, E.V., Chamberlin, M.A., Klaiber, J., and Svitashev, S. (2005b). Microsatellite megatracts in the maize (Zea mays L.) genome. Genome 48, 1061–1069.

Anderson, L.K., Salameh, N., Bass, H.W., Harper, L.C., Cande, W.Z., Weber, G., and Stack, S.M. (2004). Integrating genetic linkage maps with pachytene chromosome structure in maize. Genetics 166, 1923–1933.

Banks, J., Kingsbury, J., Raboy, V., Schiefelbein, J.W., Nelson, O., and Fedoroff, N. (1985). The Ac and Spm controlling element families in maize. Cold Spring Harb Symp Quant Biol. 50, 307–311.

Beló, A., Zheng, P., Luck, S., Shen, B., Meyer, D.J., Li, B., Tingey, S., and Rafalski, A. (2007). Whole genome scan detects an allelic variant of fad2 associated with increased oleic acid levels in maize. Mol Genet Genomics 279, 1–10.

Bennetzen, J.L. (2000). Transposable element contributions to plant gene and genome evolution. Plant Mol Biol. 42, 251–269.

Bruggmann, R., Bharti, A., Gundlach, H., Lai, J., Young, S., Pontaroli, A., Wei, F., Haberer, G., Fuks, G., Du, C., Raymond, C., Estep, M., Liu, R., Bennetzen, J., Chan, A., Rabinowicz, P., Quackenbush, J., Barbazuk, W., Wing, R., Birren, B., Nusbaum, C., Rounsley, S., Mayer, K., and Messing, J. (2006). Uneven chromosome contraction and expansion in the maize genome. Genome Res. 16, 1241–1251.

Brunner, S., Pea, G., and Rafalski, A. (2005a). Origins, genetic organization and transcription of a family of non-autonomous helitron elements in maize. The Plant Journal 43, 799–810.

Brunner, S., Fengler, K., Morgante, M., Tingey, S., and Rafalski, A. (2005b). Evolution of DNA sequence nonhomologies among maize inbreds. Plant Cell 17, 343–360.

Buckler, E.S., Gaut, B.S., and McMullen, M.D. (2006). Molecular and functional diversity of maize. Curr Opin Plant Biol. 9, 172–176.

Buckner, B., Kelson, T.L., and Robertson, D.S. (1990). Cloning of the y1 Locus of Maize, a Gene involved in the Biosynthesis of Carotenoids. The Plant Cell 2, 867–876.

Buckner, B., San Miguel, P., Janick-Buckner, D., and Bennetzen, J.L. (1998). The y1 Gene of Maize codes for Phytoene Synthase. Genetics 143, 479–488.

Bureau, T.E., and Wessler, S.R. (1994). Mobile inverted-repeat elements of the Tourist family are associated with the genes of many cereal grasses. Proc. Natl. Acad. Sci. USA 91, 1411–1415.

Burr, B., Burr, F.A., Matz, E.C., and Romero-Severson, J. (1992). Pinning down loose ends: mapping telomeres and factors affecting their length. Plant Cell 4, 953–960.

Cardon, L.R., and Bell, J.I. (2001). Association Study Designs for Complex Diseases. Nature Reviews Genetics 2, 91–99.

Cardon, L.R., and Abecasis, G.R. (2003). Using haplotype blocks to map human complex trait loci. Trends Genet 19, 135–140.

Chen, C., Yan, H., Zhai, W., Zhu, L., and Sun, J. (2000). Identification and chromosomal location of a new tandemly repeated DNA in maize. Genome 43, 181–184.

Cheng, Y.M., and Lin, B.Y. (2003). Cloning and characterization of maize B chromosome sequences derived from microdissection. Genetics 164, 299–310.

Ching, A., Caldwell, K.S., Jung, M., Dolan, M., Smith, O.S., Tingey, S., Morgante, M., and Rafalski, A.J. (2002). SNP frequency, haplotype structure and linkage disequilibrium in elite maize inbred lines. BMC Genet 3, 19.

Clark, R.M., Wagler, T.N., Quijada, P., and Doebley, J. (2006). A distant upstream enhancer at the maize domestication gene tb1 has pleiotropic effects on plant and inflorescent architecture. Nat Genet. 38, 594–597.

Cooper, M., Smith, O.S., Graham, G., Arthur, L., Feng, L., and Podlich, D.W. (2004). Genomics, Genetics and Plant Breeding: A Private Sector Perspective. Crop Sci. 44, 1907–1913.

Curtis, D., Vine, A., and Knight, J. (2007). A pragmatic suggestion for dealing with results for candidate genes obtained from genome wide association studies BMC Genetics 8, 20.

Dennis, E.S., and Peacock, W.J. (1984). Knob heterochromatin homology in maize and its relatives. J. Mol. Evol. 20, 341–350.

Doebley, J., and Lukens, L. (1998). Transcriptional regulators and the evolution of plant form. Plant Cell 10, 1075–1082.

Doebley, J., Goodman, M.M., and Stuber, C.W. (1984). Isoenzymatic variation in Zea (Gramineae). Systematic Botany 9, 203–218.

Dooner, H.K., and Martinez-Ferez, I.M. (1997). Recombination occurs uniformly within the bronze gene, a meiotic recombination hotspot in the maize genome. Plant Cell 9, 1633–1646.

Fedoroff, N. (2000). Transposons and genome evolution in plants. Proc Natl Acad Sci U S A 97, 7002–7007.

Fengler, K., Allen, S.M., Li, B., and Rafalski, A. (2007). Distribution of genes, recombination, and repetitive elements in the maize genome. The Plant Genome [A Supplement to Crop Science], 83–95.

Flint-Garcia, S.A., Thornsberry, J.M., and Buckler IV, E.S. (2003). Structure of linkage disequilibrium in plants. Annu. Rev. Plant Biol. 54, 357–374.

Frary, A., Fulton, T.M., Zamir, D., and Tanksley, S.D. (2004). Advanced backcross QTL analysis of a Lycopersicon esculentum x L. pennellii cross and identification of possible orthologs in the Solanaceae. Theor Appl Genet. 108, 485–496.

Fu, H., and Dooner, H.K. (2002). Intraspecific violation of genetic colinearity and its implications in maize. Proc Natl Acad Sci U S A 99, 9573–9578.

Gardiner, J.M., Coe, E.H., and Chao, S. (1996). Cloning maize telomeres by complementation in Saccharomyces cerevisiae. Genome 39, 736–748.

Gaut, B.S., and Clegg, M.T. (1993). Molecular evolution of the Adh1 locus in the genus Zea. Proc. Natl. Acad. Sci. USA 90, 5095–5099.

Gaut, B.S., and Long, A.D. (2003). The Lowdown on Linkage Disequilibrium. The Plant Cell 15, 1502–1506.

Grotewold, E., Drummond, B.J., Bowen, B., and Peterson, T. (1994). The myb-homologous P gene controls phlobaphene pigmentation in maize floral organs by directly activating a flavonoid biosynthetic gene subset. Cell 76, 543–553.

Gupta, S., Gallavotti, A., Stryker, G.A., Schmidt, R.J., and Lal, S.K. (2005). A novel class of Helitron- related transposable elements in maize contain portions of multiple pseudogenes. Plant Mol Biol 57, 115–127.

Hanson, M.A., Gaut, B.S., Stec, A.O., Fuerstenberg, S.I., Goodman, M.M., Coe, E.H., and Doebley, J.F. (1996). Evolution of anthocyanin biosynthesis in maize kernels: the role of regulatory and enzymatic loci. Genetics 143, 1395–1407.

Hartl, D. (2000). A Primer of Population Genetics. (Sunderland, MA, USA.: Sinauer Associates).

Hoisington, D., Khairallah, M., Reeves, T., Ribaut, J.M., Skovmand, B., Taba, S., and Warburton, M. (1999). Plant genetic resources: what can they contribute toward increased crop productivity? Proc. Natl. Acad. Sci. U.S.A. 96, 5937–5943.

Hudson, R.R., and Kaplan, N.L. (1995). Deleterious background selection with recombination. Genetics 141, 1605–1617.

Jiang, N., Bao, Z., Zhang, X., Eddy, S.R., and Wessler, S.R. (2004). Pack-MULE transposable elements mediate gene evolution in plants. Nature 431, 569–573.

Jung, M., Ching, A., Bhattramakki, D., Dolan, M., Tingey, S., Morgante, M., and Rafalski, A. (2004). Linkage disequilibrium and sequence diversity in a 500-kbp region around the adh1 locus in elite maize germplasm. Theor Appl Genet 109, 681–689.

Kapitonov, V.V., and Jurka, J. (2001). Rolling-circle transposons in eukaryotes. Proc Natl Acad Sci U S A 98, 8714–8719.

Kato, A., Lamb, J.C., and Birchler, J.A. (2004). Chromosome painting using repetitive DNA sequences as probes for somatic chromosome identification in maize. Proc Natl Acad Sci U S A. 101, 13554–13559.

Kumar, A., and Bennetzen, J.L. (2000). Retrotransposons: central players in the structure, evolution and function of plant genomes. Trends Plant Sci 5, 509–510.

Lai, J., Li, Y., Messing, J., and Dooner, H.K. (2005). Gene movement by Helitron transposons contributes to the haplotype variability of maize. Proc Natl Acad Sci U S A 102, 9068–9073.

Lal, S.K., Giroux, M.J., Brendel, V., Vallejos, C.E., and Hannah, L.C. (2003). The maize genome contains a helitron insertion. Plant Cell 15, 381–391.

Li, L., and Arumuganathan, K. (2001). Physical mapping of 45S and 5S rDNA on maize metaphase and sorted chromosomes by FISH. Hereditas 134, 141–145.

Lijavetzky, D., Cabezas, J.A., Ibanez, A., Rodriguez, V., and Martinez-Zapater, J.M. (2007). High throughput SNP discovery and genotyping in grapevine (Vitis vinifera L.) by combining a re-sequencing approach and SNPlex technology. BMC Genomics 8, 424.

Lippman, Z., and Tanksley, S.D. (2001). Dissecting the genetic pathway to extreme fruit size in tomato using a cross between the small-fruited wild species Lycopersicon pimpinellifolium and L. esculentum var. Giant Heirloom. Genetics. 158, 413–422.

Liu, K., Goodman, M., Muse, S., Smith, J.S., Buckler, E., and Doebley, J. (2003). Genetic structure and diversity among maize inbred lines as inferred from DNA microsatellites. Genetics 165, 2117–2128.

Liu, R., Vitte, C., Ma, J., Mahama, A.A., Dhliwayo, T., Lee, M., and Bennetzen, J.L. (2007). A GeneTrek analysis of the maize genome. Proc Natl Acad Sci U S A 104, 11844–11849.

Longley, A.E. (1939). Knob positions on corn chromosomes. J. Agric. Res. 59, 475–490.

Mackay, I., and Powell, W. (2007). Methods for linkage disequilibrium mapping in crops. Trends Plant Sci. 12, 57–63.

Mascia, P.N., Rubenstein, I., Phillips, R.L., Wang, A.S., and Xiang, L.Z. (1981). Localization of the 5S rRNA genes and evidence for diversity in the 5S rDNA region of maize. Gene 15, 7–20.

McClintock, B. (1931). The Order of the Genes C, Sh, and Wx in Zea Mays with Reference to a Cytologically Known Point in the Chromosome. Proc Natl Acad Sci U S A 17, 485–491.

McClintock, B. (1978). Significance of chromosome constitutions in tracing the origin and migration of races of maize in the Americas. In: Maize Breeding and Genetics. Walden, D.B., ed., 159–184. In in: Maize Breeding and Genetics, D.B. Walden, ed (New York: Wiley), pp. 159–184.

Meyers, B.C., Tingey, S.V., and Morgante, M. (2001). Abundance, distribution and transcriptional activity of repetitive elements in the maize genome. Genome Res. 11, 1660–1676.

Mikkilineni, V., and Rocheford, T. (2003). Sequence variation and genomic organization of fatty acid desaturase-2 (fad2 and fatty acid desaturase-6 (fad6) cDNAs in maize. Theor Appl Genet 106, 1326–1332.

Morgante, M., Brunner, S., Pea, G., Fengler, K., Zuccolo, A., and Rafalski, A. (2005). Gene duplication and exon shuffling by helitron-like transposons generate intraspecies diversity in maize. Nat Genet 37, 997–1002.

Mroczek, R.J., and Dawe, R.K. (2003). Distribution of retroelements in centromeres and neocentromeres of maize. Genetics 165, 809–819.

Nagaki, K., Song, J., Stupar, R.M., Parokonny, A.S., Yuan, Q., Ouyang, S., Liu, J., Hsiao, J., Jones, K.M., Dawe, R.K., Buell, C.R., and Jiang, J. (2003). Molecular and cytological analyses of large tracks of centromeric DNA reveal the structure and evolutionary dynamics of maize centromeres. Genetics 163, 759–770.

Nobuta, K., Venu, R.C., Lu, C., Belo, A., Vemaraju, K., Kulkarni, K., Wang, W., Pillay, M., Green, P.J., Wang, G.L., and Meyers, B.C. (2007). An expression atlas of rice mRNAs and small RNAs. Nat Biotechnol. 25, 473–477.

Page, B.T., Wanous, M.K., and Birchler, J.A. (2001). Characterization of a maize chromosome 4 centromeric sequence: evidence for an evolutionary relationship with the B chromosome centromere. Genetics 159, 291–302.

Palaisa, K., Morgante, M., Williams, M., and Rafalski, A. (2003). Contrasting effects of selection on sequence diversity and linkage disequilibrium at two phytoene synthase loci. The Plant Cell 15, 1795–1806.

Palaisa, K., Morgante, M., Tingey, S., and Rafalski, A. (2004). Long-range patterns of diversity and linkage disequilibrium surrounding the maize Y1 gene are indicative of an assymetric selective sweep. Proc Natl Acad Sci U S A 101, 9885–9890.

Pritchard, J.K. (2001). Denconstructing maize population structure. Nature Genetics 28, 203–204.

Pritchard, J.K., and Rosenberg, N.A. (1999). Use of unlinked genetic markers to detect population stratification in association studies. Am J Hum Genet 65, 220–228.

Rafalski, A. (2002). Applications of single nucleotide polymorphisms in crop genetics. Current Opinion in Plant Biology 5, 94–100.

Rafalski, A. (2007). Tagging the rice transcriptome. Nat Biotechnol. 25, 430–431.

Remington, D.L., Thornsberry, J.M., Matsuoka, Y., Wilson, L.M., Whitt, S.R., Doebley, J., Kresovich, S., Goodman, M.M., and Buckler, E.S.t. (2001). Structure of linkage disequilibrium and phenotypic associations in the maize genome. Proc Natl Acad Sci U S A 98, 11479–11484.

Rose, L.E., Bittner-Eddy, P.D., Langley, C.H., Holub, E.B., Michelmore, R.W., and Beynon, J.L. (2004). The maintenance of extreme amino acid diversity at the disease resistance gene, RPP13, in Arabidopsis thaliana. Genetics 166, 1517–1527.

SanMiguel, P., Gaut, B.S., Tikhonov, A., Nakajima, Y., and Bennetzen, J.L. (1998). The paleontology of intergene retrotransposons of maize. Nat.Genet. 20, 43–45.

SanMiguel, P., Tikhonov, A., Jin, Y.K., Motchoulskaia, N., Zakharov, D., Melake-Berhan, A., Springer, P.S., Edwards, K.J., Lee, M., Avramova, Z., and Bennetzen, J.L. (1996). Nested retrotransposons in the intergenic regions of the maize genome. Science 274, 765–768.

Song, R., and Messing, J. (2003). Gene expression of a gene family in maize based on noncollinear haplotypes. Proc Natl Acad Sci U S A 100, 9055–9060.

Stam, M., Belele, C., Ramakrishna, W., Dorweiler, J.E., Bennetzen, J.L., and Chandler, V.L. (2002). The regulatory regions required for B' paramutation and expression are located far upstream of the maize b1 transcribed sequences. Genetics 162, 917–930.

Tajima, F. (1983). Evolutionary relationship of DNA sequences in finite populations. Genetics 105, 437–460.

Tanksley, S.D., and McCouch, S.R. (1997). Seed banks and molecular maps: unlocking genetic potential from the wild. Science 277, 1063–1066.

Tenaillon, M.I., Sawkins, M.C., Long, A.D., Gaut, R.L., Doebley, J.F., and Gaut, B.S. (2001). Patterns of DNA sequence polym orphism along chromosome 1 of maize (Zea mays ssp. mays L.). Proc. Natl. Acad. Sci. USA 98, 9161–9166.

Thornsberry, J.M., Goodman, M.M., Doebley, J., Kresovich, S., Nielsen, D., and Buckler, E.S.I. (2001). Dwarf8 polymorphisms associate with variation in flowering time. Nature Genetics 28, 286–289.

Tiffin, P., Hacker, R., and Gaut, B.S. (2004). Population genetic evidence for rapid changes in intraspecific diversity and allelic cycling of a specialist defense gene in Zea. Genetics 168, 425–434.

Tochtrop, C., and Buckner, B. (2000). Sequence analysis of a recessive allele of the y1 gene of maize. In Maize Genetics Conference Abstracts, pp. P96.

Vigouroux, Y., Jaqueth, J.S., Matsuoka, Y., Smith, O.S., Beavis, W.D., Smith, J.S., and Doebley, J. (2002). Rate and pattern of mutation at microsatellite loci in maize. Mol Biol Evol. 19, 1251–1260.

Vigouroux, Y., Mitchell, S., Matsuoka, Y., Hamblin, M., Kresovich, S., Smith, J.S., Jaqueth, J., Smith, O.S., and Doebley, J. (2005). An analysis of genetic diversity across the maize genome using microsatellites. Genetics 169, 1617–1630.

Wang, Q., and Dooner, H.K. (2006). Remarkable variation in maize genome structure inferred from haplotype diversity at the bz locus Proc Natl Acad Sci U S A 103, 17644–17649.

Wang, R.L., Stec, A., Hey, J., Lukens, L., and Doebley, J. (1999). The limits of selection during maize domestication. Nature 398, 236–239.

Wessler, S.R. (2006). Transposable elements and the evolution of eukaryotic genomes Proc Natl Acad Sci U S A 103, 17600–17601.

Whitt, S.R., Wilson, L.M., Tenaillon, M.I., Gaut, B.S., and Buckler, E.S.t. (2002). Genetic diversity and selection in the maize starch pathway. Proc Natl Acad Sci U S A 99, 12959–12962.

Wright, S.I., Bi, I.V., Schroeder, S.G., Yamasaki, M., Doebley, J.F., McMullen, M.D., and Gaut, B.S. (2005). The Effects of Artificial Selection on the Maize Genome. Science 308, 1310–1314.

Xiao, J., Li, J., Grandillo, S., Ahn S, N., Yuan, L., Tanksley, S.D., and McCouch, S.R. (1998). Identification of trait-improving quantitative trait loci alleles from a wild rice relative, Oryza rufipogon. Genetics 150, 899–909.

Xu, J.H., and Messing, J. (2006). Maize haplotype with a helitron-amplified cytidine deaminase gene copy. BMC Genet 7, 52.

Yamasaki, M., Tenaillon, M.I., Bi, I.V., Schroeder, S.G., Sanchez-Villeda, H., Doebley, J.F., Gaut, B.S., and McMullen, M.D. (2005). A large-scale screen for artificial selection in maize identifies candidate agronomic loci for domestication and crop improvement. Plant Cell 17, 2859–2872.

Yu, J., Pressoir, G., Briggs, W.H., Vroh, B.I., Yamasaki, M., Doebley, J.F., McMullen, M.D., Gaut, B.S., Nielsen, D.M., Holland, J.B., Kresovich, S., and Buckler, E.S. (2006). A unified mixed-model method for association mapping that accounts for multiple levels of relatedness. Nat Genet. 38, 203–208.

Zhang, Q., Arbuckle, J., and Wessler, S.R. (2000). Recent, extensive, and preferential insertion of memebers of the miniature inverted-repeat transposable element family Heartbreaker into genic regions of maize. PNAS 97, 1160–1165.

Zhu, Y.L., Song, Q.J., Hyten, D.L., Van Tassell, C.P., Matukumalli, L.K., Grimm, D.R., Hyatt, S.M., Fickus, E.W., Young, N.D., and Cregan, P.B. (2003). Single-nucleotide polymorphisms in soybean. Genetics 163, 1123–1134.

Zimmer, E.A., Jupe, E.R., and Walbot, V. (1988). Ribosomal gene structure, variation and inheritance in maize and its ancestors. Genetics 120, 1125–1136.

The Polyploid Origin of Maize

Joachim Messing

Abstract Maize has exhibited a remarkable level of genetic variation leading to the assumption that continuous breeding of a rich repertoire of haplotypes facilitated its success as a major crop worldwide. Now, with genomic sequences of two close relatives, rice and sorghum, and a gene-dense physical map of maize chromosomes at hand, we can use DNA sequence alignments to further our understanding of the molecular basis of its genetic variability and the origin of its chromosomes. There are two striking features emerging from such studies, one is polyploidy, the other is recent chromosome expansion and contraction. Based on synteny, collinear arrangement of chromosomal segments, rice and sorghum match maize at a ratio of 1:2, which is typical for a whole-genome duplication event. Because meiosis offers a strong selection against polyploidy, different pathways have evolved to stabilize chromosome structure. It appears that, in case of maize, polyploidy has triggered chromosome breakage and fusion events reshaping today's maize chromosomes relative to its predecessors. Diploidization, a process to transition a genome from polyploid to diploid status, seems to have benefited from the uneven expansion of maize chromosomes by retrotransposition, thereby preventing pairing of homoeologous chromosomal segments during meiosis. In addition, loss of orthologous gene copies was followed by "copy and paste" of paralogous gene copies enhancing non-collinearity in syntenic regions.

1 Introduction

Polyploidy is quite common among crop plants (Wolfe, 2001). For instance, soybean, cotton, sugarcane, and wheat are polyploid. In all these cases, their genome underwent at least one round of whole-genome duplication (WGD). Besides WGD, segmental chromosomal duplications are also quite common and occur before and

J. Messing
Waksman Institute of Microbiology, Rutgers University, Piscataway, NJ, USA

J.L. Bennetzen and S. Hake (eds.), *Maize Handbook - Volume II: Genetics and Genomics*,   221
© Springer Science+Business Media LLC 2009

after speciation. For example, segmental duplications in rice occurred multiple times over a span of 50 million years (Yu et al., 2005). The largest of about 3 Mb occurred on the tip of the short arms of rice chromosomes 11 and 12 about 7.7 million years ago (mya) (Rice-Chromosomes-11-and-12-Sequencing-Consortia, 2005). Because these segmental duplications occurred at different times to only single chromosomal segments, they arose by mechanisms different than WGD (Gaut and Doebley, 1997). The presence of many small chromosomal duplications spread throughout a genome mimics polyploidization. However, if some of these occur prior to speciation, they would be duplicated again in closely related species that are true polyploids. Polyploids in plants are usually allopolyploids rather than autopolyploids, because the hybridization of two closely, but diverged progenitors has a better ability to maintain two sets of chromosomes. Therefore, the prominence of allotetraploidy indicates that meiosis selects against having two homologous chromosomes, and that dissimilarities between homoeologous chromosomes, chromosomes derived from the same ancestral chromosome, are more stable.

To better understand the dynamic processes that follow polyploidization, synthetic polyploids have been produced and examined. The outcome of these investigations is that the synthetic polyploids trigger a range of responses, including epigenetic modifications, activation of transposition, chromosome breakage and rearrangements. However, the responses seem to differ widely among synthetics. While the genome of synthetic polyploid cotton seems to be rather unaffected (Liu et al., 2001), neopolyploid wheat underwent major changes, including methylation and loss of sequences (Ozkan et al., 2001; Shaked et al., 2001). Besides an impact on meiosis, that on gene expression was clearly evident. When all the genes are duplicated simultaneously, the number of regulatory *cis*-acting elements needs to be reduced and the level of gene expression needs to be adjusted. Therefore, in neopolyploids of Arabidopsis, rapeseed, cotton and wheat, global changes in gene expression occurred (Soltis and Soltis, 1995; Comai et al., 2000; Lee and Chen, 2001; Adams et al., 2003; Kashkush et al., 2003; Osborn et al., 2003). Deletion of entire genes might be the most drastic mechanism to silence regulatory elements and genes, but epigenetic silencing, and changes of expression from adjacent transposable elements also occur (Kashkush et al., 2003).

While the responses to the formation of neopolyploids can serve as a model for what happened in ancient polyploids, they do not explain the chronology of events and the specific genetic constitution of the progenitors of ancient polyploids. The long-term effects of WGD can only be studied in non-synthetic polyploids. What has hampered these studies in the past is the lack of reference genomes. For instance, in the case of *Saccharomyces cerevisiae*, its polyploid origin only became apparent after sequencing a related yeast species, *Kluyveromyces waltii*. Aligning the two genomic sequences showed a 2:1 genetic relationship that had deteriorated overtime by extensively eliminating the second copy of duplicated genes in the *Saccharomyces cerevisiae* genome (Kellis et al., 2004). However, these losses made it impossible to reconstruct the WGD nature of *Saccharomyces cerevisiae* without the reference genome. The same is true for many polyploid plants. For instance, classifying Arabidopsis as an ancient polyploid is probably premature without a reference genome that has maintained a diploid state. On the other hand, with the two reference-genome sequences of rice and sorghum at hand, maize will be the first plant genome where we can study a historical polyploid at the DNA sequence level.

2 Duplicated Regions of the Maize Genome

Although the maize genome is nearly the size of mammalian genomes, it has only 10 chromosomes. Sizes of chromosomes, however, were critical in selecting maize and fruitfly as models in early genetic studies and facilitated the microscopic examination of chromosome architecture and their behavior during meiosis. Cytological studies also showed that chromosome size can vary among races and wild species due largely to structures called knobs. Therefore, the total DNA content in the haploid maize nucleus ranges between 2.3 and 3.3 Gb (Bennett and Leitch, 2005). Today, we know that knobs consist of tandemly repeated sequences. These cytological landmarks also served as a tool to classify maize in races and characterize their geographical spread (Anderson and Cutler, 1942; Goodman and Brown, 1988). In addition to knob regions, a larger DNA content is proportional to extranumerary B chromosomes. Cytological studies have shown that different maize chromosomes have regions of homology. For instance, semisterility in maize was associated with an interchange of two different chromosomes during meiosis (McClintock, 1930). For such an interchange, pairing has to occur between these chromosomes and presumably requires long stretches of homology. Furthermore, one can envision that such pairing could result in convergent evolution of duplicated genes. Subsequent cytogenetic studies further advanced the assumption that maize is an ancient polyploid, and it was believed for many years that the maize lineage arose from the hybridization of two species, each with five basic chromosomes (Molina and Naranjo, 1987).

Besides cytological studies, mutation research was an early indicator of duplicated regions of the maize genome. Segregation of duplicate and triplicate orthologous genes in tetraploid oats and hexaploid wheat also provided examples. Orthologous chromosomal regions are those where gene order is derived by ancestry. Paralogous regions arise from duplications within the same genome. Still, the sets of known duplicated genes in maize orthologous chromosomal regions was rather small, so it was difficult to decide whether they reflected that maize is an ancient polyploid or had later-occurring segmental duplications. Today, the limitations of these studies are clear. Most genes that became duplicated were found to have differentiated and evolved in their phenotype, so they behave as single rather than duplicated factors. For instance, genes that became duplicated during polyploidization include the *R1* and *B1* transcription factors (Swigonova et al., 2005). Although their function was conserved, their tissue specificity differentiated, and therefore they behave genetically as diploid genes. With the advent of biochemical tools, like the separation of isozymes, mapping studies were successful in identifying duplicated regions in the maize genome (Goodman et al., 1980; Wendel et al., 1986).

3   Comparative Mapping with DNA Markers

A breakthrough in understanding the extent of duplicated regions in maize came with the use of restriction enzyme fragment length polymorphisms (RFLPs) (Helentjaris et al., 1988). Out of 217 DNA probes used in maize mapping populations, 29% hybridized to DNA fragments that mapped to more than one locus. Moreover, duplicated loci mapped to different chromosomes. Interestingly, duplicated loci were often arrayed in the same order in both chromosomal locations, therefore providing evidence of chromosomal duplications rather than simple gene duplications, and being consistent with the earlier cytological studies. The first evidence that these duplications were the result of polyploidy came from comparative analysis of the sorghum and maize genome maps using 85 DNA probes (Whitkus et al., 1992). Because one location in sorghum corresponded to two in maize, it was possible to distinguish between orthologous and paralogous regions in maize. A similar 2:1 mapping relationship was also found between maize and rice, although rice is more distantly related to maize than sorghum (Ahn and Tanksley, 1993).

Use of a larger marker set for maize (250) and rice (422) led to several additional insights (Wilson et al., 1999). It confirmed the ancient polyploidy of maize, but it also indicated that polyploidization resulted in the loss of centromeres and 20 chromosomal rearrangements. In contrast to earlier cytogenetic studies, these low-density marker maps suggested that the two progenitors of maize had eight chromosomes instead of five. These maps were also sufficient to indicate that, over 60 million years, the genomes of several species of the grass family retained conserved genetic maps (Gale and Devos, 1998; Kellogg, 2001). To align these maps, it was necessary to dissect the chromosomes into fragments based on their collinearity with the rice chromosomes and display the different segments of the larger genomes in concentric circles around the rice genome as the smallest circle with orthologous genes along the radii (Moore et al., 1995). This analysis indicated that the C-value paradox (Gregory, 2005) is largely due to the presence of non-genic sequences and the genetic size of genomes is constant, despite the enormous expansion of intergenic space. It also showed that polyploid genomes, like wheat, consist of intact homoeologous chromosomes, while maize has one set of chromosomes with two sets of homoeologous regions (Fig. 1).

Fig. 1 Homoeologous regions in maize do not match progenitor chromosomes 1:1, but are rather composites of different progenitor chromosomes

4 Pattern of Sequence Divergence of Duplicated Genes

Because the sequences of genes that mapped to homoeologous regions of the maize genome were known, it became possible to use sequence divergence to examine the origins of the duplications (Gaut and Doebley, 1997). As discussed above, one possibility is that multiple segmental duplications could mimic polyploidy. In such a case, divergence of collinear genes of the same duplicated regions would be the same, while they would differ for other duplicated regions, as was shown for the rice genome (Yu et al., 2005). Autopolyploidy would result from the doubling of unreduced gametes of a diploid progenitor. There would be little divergence during the tetraploid state, but since maize is diploid, one would need to assume that a switch from tetrasomic to disomic inheritance occurred. In such a case, the onset of divergence of gene pairs would begin after this switch. If such a switch were genome-wide, one would expect a uniform distribution of divergence. Another possibility would be allotetraploidy, where two closely related species would hybridize. In such a case, the two sets of chromosomes would have already diverged. Here, one could find two groups of gene pairs, one reflecting the split of the two progenitors and the other consisting of those that diverged since the polyploidization event. There are also more complex models, where hybridization occurs between two progenitors that have only partially diverged chromosome sets, so called segmental allotetraploidy. In such a case, one would expect divergence rates of gene pairs to also fall into two groups, one reflecting autotetraploidy and one allotetraploidy. The difference between these alternatives is that, in the case of allotetraploidy, examples of the two groups are physically linked within the same collinear region, whereas, in the case of segmental allotetraploidy, collinear gene pairs are either derived from progenitor chromosomes that have already diverged or ones that have not diverged because segmental allotetraploidy requires chromosome sets that are only partially differentiated.

Using a set of 14 gene pairs that had been mapped to duplicated regions of various maize chromosomes, nucleotide substitution rates could be calculated (Gaut and Doebley, 1997). They differ from 3-fold, for synonymous, to 10-fold for non-synonymous nucleotide substitutions between gene pairs. However, variations appear to fall into two groups of gene pairs; four into group A, the most diverged, and 10 into group B. Curiously, the distance analysis of a single conserved gene pair of sorghum and maize falls right between group A and B. Furthermore, out of the 14 pairs only 6 outgroups from the same reference taxon were available for this study. However, orthology of these outgroups could not be verified within the limits of this study. Therefore, dating the split of the progenitors and the polyploidization at 20.5 and 11.4 mya, respectively, could be considered a preliminary estimate. Given the limited data set, this study concluded that maize arose by segmental allotetraploidization rather than allotetraploidization. It also concluded that one of the progenitors was closer to the progenitor of sorghum because the group A set appeared to be older than the maize-sorghum gene pair in this study. However, if one would find other gene pairs in maize that fall between groups A and B, providing a straight line of variations in Ks values, the bivalent model would not be validated.

Furthermore, if genes in sorghum orthologous to the maize gene pairs would also show a straight line of ranges in Ks variation, one could not conclude that one of the progenitors was closer to the progenitor of sorghum.

5 Contiguous Sequences Linking Duplicated Genes

To resolve the above issues between allotetraploidy and segmental allotetraploidy, it became necessary to obtain contiguous sequences that were orthologous between the two duplicated regions of maize and the single regions of sorghum and rice. Contiguous sequences had to be long enough to have at least two genes in the same order in all four taxa to be sure that they were orthologous and would provide reliable nucleotide substitution rates. Because the whole rice genome was sequenced based on a physical map of overlapping clones (International-Rice-Genome-Sequencing-Project, 2005), one could easily use the sequences of duplicated loci in maize to select orthologous regions of rice *in silico*. For sorghum and maize it was necessary to isolate clones from bacterial artificial chromosome (BAC) libraries of maize B73 and sorghum Btx623. Consistent with the previous 2:1 mapping results, two genomic regions of maize could be identified for each sorghum region that would correspond to the sequences of the two duplicated loci. Furthermore, the sequences of sorghum clones showed that, in addition, other genes in each region were conserved in the same order in rice, confirming that the selection had resulted in the isolation of chromosomal regions that were related by ancestry. The same was true for the sequences of two duplicated regions of the maize genome, establishing that they represent homoeologous chromosomal fragments. However, a striking difference was that the number of collinear genes was much higher between sorghum and rice than with either of the two homoeologous regions of maize, indicating that, at least in half of the cases, one of the homoeologous regions of maize has lost an orthologous gene copy. As a consequence, the number of genes collinear in all four regions was significantly lower than expected (Lai et al., 2004). Furthermore, in one of the cases, a gene present in all four regions appeared to be orthologous, but orthology was only valid for rice and sorghum, but not for the homoeologous regions in maize. In this case rice and sorghum had two tandem copies of the gene. One would have expected to have two copies in each maize region, but these regions had retained only one of the two tandem copies. By determining the nucleotide substitution rates of each copy, it became clear that the 5′ copy in rice and sorghum was present on one maize chromosome and the 3′ copy on the other maize chromosome. This exception demonstrates that validation of orthologous gene pairs for the study of the ancestry of chromosomes requires not only the gene sequences themselves but also the flanking gene sequences and their chromosomal organization (Swigonova et al., 2004).

Given the validation of the data set, 11 genes derived from the same ancestral chromosomes present in all four descending chromosomal regions could be used to determine distances between sorghum and each of maize homoeologous regions. Again, like in the earlier study, nucleotide substitution rates varied between different gene pairs. Between maize orthologs, synonymous rates varied 2.8-fold and 3.2-fold in maize-sorghum comparisons, close to the 3-fold variation found earlier. However, statistical evaluation of these rates did not show any difference in the divergence of the two maize homoeologs and sorghum, except for the *tbp1/tbp2* gene pair. Therefore, it appears that the three progenitors, two for maize and one for sorghum, have about the same distance to the progenitor of rice (Swigonova et al., 2004). Because this is the case, then the earlier conclusion that one of the maize progenitors is closer to sorghum than the other one was not confirmed. The *tbp1/tbp2* gene pair had a much lower rate of synonymous and, in particular, non-synonymous nucleotide substitutions than the other gene pairs. This is especially significant because three of the remaining 10 gene pairs are in close proximity of the *tbp1/tbp2* gene pair on the long arm of chromosome 1, less than 10 cM apart. Having two such distinct groups of gene pairs within close chromosomal proximity would make a segmental allotetraploid origin of this chromosome unlikely. Rather, one can assume that, after polyploidization, two homoeologous chromosomes could still pair even after the two progenitors had diverged, consistent with McClintock's cytological studies. In such a scenario, gene conversion could reset the clock for the divergence rate for the *tbp1/tbp2* gene pair compared to the other ten gene pairs. Consequently, the 10 pairs would give us the divergence of the progenitors and the *tbp1/tbp2* gene pair a minimal time since polyploidization. This conclusion is based on the assumption that the frequency of such conversions between homoeologous chromosomes was greatly reduced after the WGD event because of the formation of 10 new chromosomes with a mosaic of homoeologous regions. From this analysis, it appears that maize originated from allotetraploidization (Swigonova et al., 2004). Its two progenitors split from the progenitor of sorghum at the same time. Indeed, sorghum provides an excellent reference genome for studying the events after allotetraploidization, in particular because it preserved a similar gene organization as rice. The divergence time for the split and the WGD event have been placed at 11.9 and 4.8 mya, respectively. Given the sample size of gene pairs, these numbers still might change, but one would still expect two distinct values, because the progenitors had already diverged significantly before polyploidization.

6   Progenitor Chromosomes from a High-Density Gene Map

Two post-allotetraploidization features of the maize genome are emerging from the studies of the 2:1 mapping of maize and sorghum/rice references and the distance analysis of orthologous gene pairs: one is the chromosomal rearrangements and loss of centromeres, the other is the loss of genes. A deeper look at these aspects necessitated the construction of a high-density gene map of maize. To achieve this in a short time frame and cost-effective manner with a genome the size of maize required the development of high-throughput experimentation. Large-insert libraries of a reference inbred had to be generated: inbred B73 was selected because of its significance in breeding. Because of the size and the complexity of the maize genome, it was desirable to create libraries with high redundancy and different restriction enzymes. Using three different restriction enzymes, *Hind*III, *Eco*RI, and *Mbo*I, the genome was covered about 30 times based on a total size of 2.3 Gb for inbred B73 with bacterial artificial chromosome (BAC) clones with average insert sizes of 154 kb (Yim et al., 2002). To construct a map from this resource, it was necessary to process nearly half a million clones for DNA fingerprinting reactions and sequence information from the ends of clones (BES). The restriction digest pattern of each clone can be compared to other ones by common sizes of fragments (DNA fingerprints), which makes it possible to stack clones based on overlaps. A high redundancy of contiguous sequences is critical for overcoming ambiguities. To permit DNA fingerprinting and BAC end sequencing to occur with the same template preparations and their analysis by capillary DNA sequencers, fingerprints had to have sizes under 1 kb, thereby creating a high information content (High-Information-Content-Fingerprinting). The maize genome was the first to have a physical map by HICF (Nelson et al., 2005). Overlapping clones resulting in contiguous chromosomal segments are referred to as fingerprinting contigs (FPCs).

Because clones are assigned sequence identities (IDs) through their BESs, the sequence IDs are also anchored to the map. Given the number of sequences, sequence IDs were placed on average every 6.2 kb along the maize chromosomes (Messing et al., 2004). BESs also provided a first glimpse at the sequence organization of the maize genome. Such an analysis of BESs, however, required first the construction of a repeat database of maize. Because at that time only a few BAC clones were completely sequenced, knowledge of intact repeat sequences was limited. Still, a repeat database could be constructed from a variety of sources. The main conclusions were that about 60% of BESs were derived from repeats and 7.5% from genes, with the rest from unassigned sequences. Actually, the latter could be composed of medium- and low-copy retrotransposons, which have been estimated to amount up to 17% of the genome based on hybridization experiments (Sanmiguel and Bennetzen, 1998). The majority of repeats were LTR-retrotransposons, as suggested previously (Meyers et al., 2001). However, an unexpected result was that nearly half of the *copia*-like elements were hypomethylated. This conclusion was facilitated because of sequences derived from methylation-filtered (MF) clone libraries (Whitelaw et al., 2003). MF libraries are genomic libraries that are cloned in bacterial strains that cleave methylated DNA. Comparison of BESs with MF sequences can be used to quantify what proportion of sequence families are methylated. The repeat libraries can also be used to examine how efficient different restriction sites sample BESs with repeats. Clearly, restriction enzymes counter select cleavage of repetitive DNA, in particular *Eco*RI (50%) compared to *Hind*III (60%).

The small set of BESs that were derived from genes could be further enhanced with EST resources to obtain a set of 9,129 gene sequences that were anchored to the physical map (Messing et al., 2004). Two interesting features of the maize genome could therefore be observed. One is the occurrence of tandem gene amplification, as was observed in the sequenced plant genomes of Arabidopsis and rice (Arabidopsis-Genome-Initiative, 2000; International-Rice-Genome-Sequencing-Project, 2005). The other is that the maize genome has been duplicated, but that after duplication a dramatic loss of one copy of the duplicated genes had occurred, providing evidence on a genome-wide level of what had been proposed based on a few orthologous contiguous chromosomal regions.

Furthermore, the gene set of BESs also aligned with the rice genome sequence. Because of gene collinearity in local chromosomal regions between rice and maize, this information then could be used to merge FPCs by manual editing and order them along the maize chromosomes (Wei et al., 2007). After hybridizing 1,902 genetic markers to filters containing all the BAC clones, it became possible to align FPCs along the genetic map. Out of a total of 721 FPCs that covered 93.5% of the genome, 421 could be anchored to about 86.1% of the genetic map. The unanchored 300 FPCs were rather small, indicating they were difficult to merge and probably represented heterochromatic regions of the genome. The BACs that became ordered along the genetic map were hybridized with 14,877 gene sequences (mostly overgos; short overlapping oligonucleotides designed from exons, whose cohesive ends could be end-labeled by a fill-in reaction), resulting in a total of 25,908 gene sequences, probably more than half of the gene content of the maize genome, that were distributed over the physical map. With such a density of gene sequences placed on the maize map, one could renew the earlier analysis of the fragmentation of the progenitor chromosomes after polyploidization, but at a higher resolution. Again the anchored sequences could be aligned with the sequenced rice genome and one could ask where breakpoints in synteny occurred. This analysis revealed that 62 major rearrangements involving chromosome breakages and fusions had to occur to reconstruct rice from two sets of maize chromosomes. One also could detect quadrupled segments of ancient segmental duplications that preceded the split of the progenitors of rice and maize, confirming segmental duplications in rice

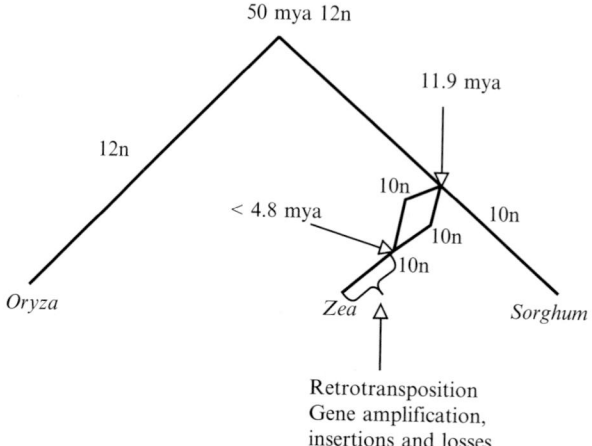

Fig. 2 If we assume that the ancestral genome for rice and maize had 12 chromosomes, the progenitors for sorghum and maize changed to a chromosome number of 10. While the progenitor of sorghum maintained this chromosome number, ancient allotetraploid maize went from 20 to 10 chromosomes. Maize chromosomes increased further in size mainly by retro-transposition

that did not arise from a WGD event. From this reconstruction, it appeared that the maize progenitor genomes possessed 10 chromosomes. Therefore, contrary to earlier beliefs, one now has to assume that the allotetraploidization of maize led to a loss of 10 centromeres and a doubling of chromosome size. Furthermore, the sorghum chromosomes increased in size without loss of centromeres during the same period (Fig. 2).

7 Chromosome Expansion and Contraction

In addition to the WGD event, the maize genome increased by about 65% even further than expected from the hybridization of two sorghum-like progenitors. To gain a first glimpse at the additional chromosomal expansion relative to sorghum and rice, it was necessary to obtain contiguous sequence information beyond the small intervals described earlier from the divergence studies (Swigonova et al., 2004). The enabling resource for such a study came from the physical map. Because the maize physical map permitted the selection of BAC clones with minimal overlap, contiguous sequences could be determined from sequencing these clones (megacontig). This approach was actually used to obtain the entire sequence of the B73 genome. One could therefore consider sequencing these megacontigs as a pilot project for sequencing the entire genome. To maximize insight from such a pilot project, it made sense to sequence two megacontigs that permitted the sampling of chromosomal changes that occurred in the two progenitor genomes following allotetraploidization. While the earlier studies provided knowledge of duplicated regions from different locations of the genome, the two megacontigs would give a better overview on what happened in larger intervals. Selecting such intervals had to be based on anchored gene sequences. It is not surprising that the selection resulted in alignment of two telomeric segments, because it was observed that gene density tends to be higher in these as opposed to pericentromeric regions. Sequence analysis of the two regions confirmed this (Bruggmann et al., 2006). One region is from chromosome 1S, bin 1.00–1.01, and spans about 17.4 cM in 7.8 Mb. The other is from chromosome 9 L, bin 9.07, and spans 25.6 cM in only 6.6 Mb. These frequencies amount to an average of 450 and 256 Kb/cM, respectively, and are below the genome-wide average of 1,200 Kb/cM. To assess an average gene density, 100 random regions of the genome, represented by single BACs, had been sequenced to provide an unbiased organization of contiguous sequences throughout the genome rather than regions selected by known gene loci (Haberer et al., 2005). Using this reference, the recombination frequencies of the megacontigs are consistent with the proposition that meiotic crossover occurs primarily in genes (Thuriaux, 1977). While gene density in the 100 random regions is 2.3 genes/100 Kb, it is 3.0 and 3.7 genes/100 Kb for chromosome 1 S and 9 L regions, respectively.

Based on genes in both regions, a 2:1 ratio can be found relative to a single chromosomal fragment on 3S of the rice genome that is collinear with both maize regions. The most striking feature of the alignment is that collinearity is split into synteny blocks. The junction regions consist of non-conserved sequences of variable length and sites of inversions. Previously, comparisons of orthologous sequences have been pointed out as chromosomal sites that might have been preserved through speciation as hot spots for chromosomal rearrangements, leading to a mosaic structure of recently formed chromosomes (Song et al., 2002). Alignment of chromosomal regions in intervals of synteny blocks revealed that chromosomal expansions in maize were very uneven. As described above, genome-wide alignment of the 26K gene map with rice, suggested an expansion factor of 3.2; for every 1 Kb in rice there would be on average two regions in maize with about 3.2 Kb (Wei et al., 2007). However, some blocks of maize chromosome 1S expanded by a factor of over four, whereas blocks on chromosome 9L contracted by 0.8 relative to the orthologous regions in rice. Interestingly, the predominant sequences contributing to expansion appeared to be *copia*-like elements. As already described above, there was an indication that *copia*-like elements are heavily hypomethylated (Messing et al., 2004). One wonders whether there is a correlation between their accumulation in gene-dense regions and their general methylation status. One could also envision contraction to occur by unequal crossing over between LTRs, particularly if not heterochromatized. In such a scenario, sequences between retroelements would also be lost. In this respect, it is interesting to note that the contracted regions on chromosome 9L appear to be the preferential site where the second copy of the two duplicated genes from the WGD event were lost. In this case, retroelements could contribute two-fold to the disomic state of maize, one way could facilitate divergence between homoeologous regions through sequence insertions and the second could remove points of crossing over by deleting gene sequences.

While the dynamic features of the maize genome resemble those for synthetic polyploids as described above, one has to note that the switch to a disomic state must have occurred immediately after polyploidization. There is a single locus in wheat, the *Ph1* locus that is needed to prevent pairing of non-homologuous chromosomes (Griffiths et al., 2006). Maize does not have the equivalent activity and without it genetic stability and fertility is impaired. Still, the dynamic changes of the maize genome appear to have been a continuous process. Therefore, one can hardly say that they all were caused by the WGD event. Probably, mosaic chromosomes were formed immediately, but the process of change continued, leading to the accumulation of diverged homoeologous regions (Fig. 1). Still, it appears that the original WGD event could be considered the highest orderly state of the genome. Therefore, one is reminded of the second law of thermodynamics according to the German physicist Rudolf Clausius, who postulated that entropy forces matter from a higher to a lower orderly state (Clausius, 1868). The dissimilarities between homoeologous regions could therefore be considered a lower orderly state.

8   Gene Mobility in Diploids and Polyploids

Comparison of the orthologous regions of the maize megacontigs with sorghum and rice indicated that rice and sorghum, evolving from a single progenitor, have a greater degree of conserved gene order than any of the two maize homoeologous regions (Messing, manuscript in preparation). Interestingly, if one compares the AA and the BB genomes of the *Oryza* genus, which are also derived from a common progenitor like the two progenitors of maize, the degree of divergence is largely confined to non-genic elements (Kim et al., 2007), suggesting that gene collinearity is more stable if closely related genomes stay as separate species, as opposed to undergoing hybridization. Therefore, the diploidization of the maize genome might have triggered a greater degree of genome dynamics than had the two progenitors been maintained as separate species. It appears that in the megacontigs only a third of the genes are orthologous, less than 50% of the previous studies from different loci, but close to the number already predicted from the genome-wide analysis of anchored BESs (34%). Even the first sampling of duplicated regions in maize indicated that 29% of the gene probes hybridized to two locations (Helentjaris et al., 1988). Genes that remain duplicated are not necessarily redundant or produce a higher level of a gene product. The examples that were used in the previous orthologous study showed differentiation at the genetic level, giving rise to distinct phenotypes for each of the two copies. For instance, although the orthologous *R1* and *B1* genes act at the same step in anthocyanin biosynthesis, their transcriptional control differentiated and allowed gene expression to become tissue-specific. Interestingly, there is also a category where both orthologous genes were lost in maize. This classification is based on orthologous genes absent in both homoeologous regions of maize, but conserved in sorghum and rice. Genome-wide analysis of genes missing in orthologous regions has been possible both in rice and maize. Although maize had not been completely sequenced like rice, genome survey sequences (GSSs) and EST tags cover a large percentage of the gene repertoire of maize. Genes that would have been expected based on rice contigs appear to be present somewhere else in the maize genome. The same is true for rice. Interestingly, one can find that such repositioning of genes seems to occur in more than one genomic position. Indeed, gene movement appears to have been common in rice and maize. Although gene movement might contribute to the transition from tetrasomic to disomic inheritance, because it makes homoeologous chromosomal regions more divergent, it seems to occur also in the evolution of diploid species. The same is true for transposition. While transposition can make homoeologous chromosomal regions more dissimilar, they are not restricted to polyploids. Therefore, it appears that mechanisms of epigenetic gene silencing, 'copy and paste', and 'cut and paste' of sequences can be triggered in different ways, but the degree to which it occurred in maize is unusually high.

9 Dosage Compensation of Transgenics and Aneuploids

The emerging picture from the analysis of genomic and genetic data from maize and close relatives is that maize arose by allotetraploidization of two progenitors that diverged from a third progenitor that became sorghum. Distance analysis of LTRs of retrotransposons from orthologous regions indicated that genome expansion occurred in recent times and might not have played a major role in the structure of the three progenitors before the speciation of maize (Du et al., 2006). When retroelements insert into the genome, the LTRs are identical. Therefore, nucleotide substitutions in LTRs can be used as a measure to determine the age of elements (SanMiguel et al., 1998). While retroelements in maize are found throughout genic regions, this does not seem to be the case in sorghum. It is more likely that sorghum like tomato expanded predominantly in pericentromeric regions. Given the unusual path that allotetraploidization has taken in generating today's maize chromosomes, one wonders about the impact polyploidization might have had in different species on the organization of genes and gene expression patterns and which dynamic features of chromosomal architecture are common among diploid and polyploid plant genomes.

Studies of synthetic polyploids indicated that duplication of the entire gene set has adverse effects on plant development. The most common compensation effect for gene duplication seems to be gene silencing (Matzke et al., 1989; Napoli et al., 1990). Interestingly, there is a good correlation between gene silencing and DNA methylation, which serves as a hallmark for epigenetic remodeling of chromatin structure (Matzke et al., 1994; Lund et al., 1995). Mechanisms of gene silencing through epigenetic changes can occur at different levels, including transcriptional regulation and RNA turnover (Matzke and Birchler, 2005). There could be additional mechanisms not related to epigenetic changes that can alter gene dosage effects. In maize, extranumerary B chromosomes can increase in copy number by non-disjunction (Roman, 1947). Therefore, B-A translocations can be used to investigate the effect of changes in ploidy for chromosome arms. Although B chromosomes can increase significantly in copy number, there is a limitation in the increase of B-A chromosomes because of the increase in gene copies located on the respective chromosome arms. Still, analysis of individual gene products has shown that gene expression does not necessarily increase with copy number (Guo and Birchler, 1994). Furthermore, increase in gene copy number through B-A translocations might not necessarily implicate the gene encoding a regulatory factor and its target gene at the same time, because they could be on different chromosome arms. Therefore, one would expect that such aneuploid states would be better tolerated than a tetraploid state.

10   Accelerated Rate of Variability after Polyploidization

An interesting example of the extent of gene mobility in maize is illustrated by the zein storage protein genes. A detailed description of zein genes is presented in a different chapter of this book. From a polyploidy point of view, they exemplify orthologous and paralogous gene amplification in maize. In brief, there are four different storage protein genes in maize called zeins, α, β, γ, and δ zeins. The α zeins can be divided into four subfamilies, *z1A*, *z1B*, *z1C*, and *z1D*. A different classification groups α zeins into zein-1, hence z1 in their designations, and β, γ, and δ zeins into zein-2, because of their difference in sulfur-rich amino acids and therefore solubility (Heidecker and Messing, 1986). Based on comparisons of orthologous regions of maize and sorghum, loci of γ and δ zein genes remained duplicated after tetraploidization; they are present in a single orthologous location in sorghum. In the case of the β zein gene, the second duplicated locus in maize was lost. Because flanking regions contain genes that have been retained as duplicated copies, the absence of the second β zein gene in the other homoeologous region can be easily identified. The situation for α zein genes is more complicated. Although the α zein genes originated before the progenitors of maize and sorghum split, no duplicated loci were retained after polyploidization. The four subfamilies in maize are spread over four chromosomal locations; however, part of *z1A*, *z1A1*, occurs in one location with *z1C1*, and the other part, *z1A2*, with *z1C2* in a second location. Furthermore, only two of the four loci have orthologous regions in sorghum, the *z1D* and *z1A1/z1C1* loci (Xu and Messing, 2008). The *z1A2/z1C2* and the *z1B* loci are unique to maize and absent in sorghum; they are clearly paralogous. Divergence rates confirm that *z1A2/z1C2* arose after polyplodization and *z1B* in one of the two progenitors of maize. Previous studies had shown that the *z1A1/z1C1* locus, which is present both in maize and sorghum, is absent in rice. In this respect, this locus is not orthologous between the progenitors of sorghum and maize and the progenitor of rice. Rather, it belongs to a category of new genes; for instance, 10% of all genes found in the alignment of the megacontigs were unique (Bruggmann et al., 2006). Interestingly, insertion of the first copy of α zein genes in maize and sorghum led to tandem gene amplification independently in each lineage after the time maize underwent polyploidization. In comparison, it appears that gene movement and amplification was more extensive in maize than in sorghum.

It is also interesting to note that paralogous gene copies, like their orthologous counterparts, do not result in increased levels of gene expression. In contrast to the α zein genes, the β, γ, and δ zein genes remained low-copy number genes. All their copies are expressed in most inbred lines studied so far. However, nearly half of the α zein genes are not expressed (Song and Messing, 2002). Early mapping studies of paralogous gene copies indicated that paralogous α zein gene copies might be inbred-specific (Wilson, 1989). Indeed, it has been shown that haplotypes of zein loci differ in their collinearity (Song and Messing, 2003). Furthermore, allelic and non-allelic combinations of different inbred lines deviate from additive effects in their expression pattern. In one case, transcriptional regulation changed with gene amplification (Song et al., 2001). Such drastic effects on amplified genes seem to be absent in sorghum. Although we are still at the beginnings of our understanding of genomic organization of genes, it appears that the degree of gene mobility is proportional to the variability in gene expression patterns. Because the extent of gene mobility that seems to have occurred after the hybridization of two of the three progenitors of sorghum and maize appears to be greater in maize, maize might possess greater variability in gene expression than other crop plants. An interesting scenario of the variability in the expression of storage protein genes comes from studies of the interaction of regulatory factors controlling zein gene expression. It was shown that transcription factors like O2 that activate expression of a subset of storage protein genes interact with a regulatory factor known as the prolamin-box binding factor (PBF) (Ueda et al., 1994; Vicente-Carbajosa et al., 1997; Wang et al., 1998), which turns out to be one of the domestication loci of maize (Jaenicke-Despres et al., 2003). Given the fact that PBF is critical for the morphology of maize, one would assume that the fine-tuning of different functions is afforded through the amplification of target genes and cofactors rather than master switches.

There are other examples of extensive gene paralogy in maize. Comparisons of orthologous regions between maize inbreds not only revealed non-allelic zein gene copies, but also gene copies present in one haplotype but absent in the other (Fu and Dooner, 2002; Brunner et al., 2005). It was shown that these arose after the polyploidization of maize from a 'copy and paste' mechanism of the orthologous gene copy (Xu and Messing, 2006). However, in most cases, only gene fragments were copied, which is different from zein genes (Lai et al., 2005; Morgante et al., 2005). The copying mechanism is also different. We do not yet know by which mechanism zein genes were copied, but many other cases appear to be the result of a rolling circle replication (Kapitonov and Jurka, 2001). Again insertions of genes and gene fragments into intergenic chromosomal regions provide yet another mechanism to make homoeologous regions dissimilar. In particular, since amplification of gene fragments does not necessarily result in new functional gene products, they could be one of the features contributing to the diploidization of the maize genome.

What might be the purpose of such gene mobility in plant genomes and is there any precedence in nature that could give us an indication of their role in the evolution of these species? One interesting system that comes to mind is the role of extrachromosomal DNA in bacteria. In bacteria, we frequently find extrachromomal DNA, like F-factors, transducing phage, and plasmids, some in low copy and some in high copy number, that carry genes and spread to other bacteria by conjugation or transfection, making them meridiploid and providing unique expression properties (Falkow, 2004). For instance, plasmids harbor genes that encode enzymes that inactivate antibiotics, produce toxins, or even plant growth factors. The effect of having these genes on plasmids rather than on the bacterial chromosome is that they provide a form of gene amplification and mobility. Because they are only required under certain environmental conditions, they represent a class of genes that are dispensable for normal growth of the organism and can be cured from the bacterial cell. Integration of the T-DNA of *Agrobacterium* into plant chromosomes also exemplifies a mechanism by which extrachromosomal bacterial DNA can be transferred to plants, where it becomes integrated in the plant genome rather than remaining in an extrachromosomal state; it also exemplifies non-collinear, new genes acquired by the plant genome (Zupan et al., 2000). Therefore, one is tempted to hypothesize that gene mobility is caused in part by changes in environmental conditions, which McClintock referred to as the response of the genome to challenge (McClintock, 1984).

Acknowledgements Work reported here was supported in part by National Science Foundation Plant Genome Grants #9975618, #0211851, and Department of Energy grant # DE-FG05-95ER20194. I also would like to thank Brian Larkins for critical reading of the manuscript.

References

Ahn, S., and Tanksley, S.D. (1993). Comparative linkage maps of the rice and maize genomes. Proc Natl Acad Sci USA 90, 7980–7984.

Anderson, E., and Cutler, H.C. (1942). Races of Zea Mays: I. Their Recognition and Classification. Annals of the Missouri Botanical Garden 29, 69–88.

Arabidopsis-Genome-Initiative. (2000). Analysis of the genome sequence of the flowering plant Arabidopsis thaliana. Nature 408, 796–815.

Bennett, M.D., and Leitch, I.J. (2005). Nuclear DNA amounts in angiosperms: progress, problems and prospects. Ann Bot (Lond) 95, 45–90.

Clausius, R. (1868). On the Mechanical Theory of Heat. Philos Mag 40, 122.

Du, C., Swigonova, Z., and Messing, J. (2006). Retrotranspositions in orthologous regions of closely related grass species. BMC Evol Biol 6, 62.

Falkow, S. (2004). Molecular Koch's postulates applied to bacterial pathogenicity–a personal recollection 15 years later. Nat Rev Microbiol 2, 67–72.

Gaut, B.S., and Doebley, J.F. (1997). DNA sequence evidence for the segmental allotetraploid origin of maize. Proc Natl Acad Sci U S A 94, 6809–6814.

Goodman, M.M., and Brown, W.L. (1988). Races of corn. In Corn and corn improvement, G.F. Sprague and J.W. Dudley, eds (Madison: Amer. Soc. Agron), pp. pp. 33–79.

Goodman, M.M., Stuber, C.W., Newton, K., and Weissinger, H.H. (1980). Linkage Relationships of 19 Enzyme Loci in Maize. Genetics 96, 697–710.

Gregory, T.R. (2005). The C-value enigma in plants and animals: a review of parallels and an appeal for partnership. Ann Bot (Lond) 95, 133–146.

Griffiths, S., Sharp, R., Foote, T.N., Bertin, I., Wanous, M., Reader, S., Colas, I., and Moore, G. (2006). Molecular characterization of Ph1 as a major chromosome pairing locus in polyploid wheat. Nature 439, 749–752.

Guo, M., and Birchler, J.A. (1994). Trans-Acting Dosage Effects on the Expression of Model Gene Systems in Maize Aneuploids. Science 1999–2002.

Haberer, G., Young, S., Bharti, A.K., Gundlach, H., Raymond, C., Fuks, G., Butler, E., Wing, R.A., Rounsley, S., Birren, B., Nusbaum, C., Mayer, K.F., and Messing, J. (2005). Structure and architecture of the maize genome. Plant Physiol 139, 1612–1624.

Heidecker, G., and Messing, J. (1986). Structural Analysis of Plant Genes. Annual Review of Plant Physiology 37, 439–466.

Helentjaris, T., Weber, D., and Wright, S. (1988). Identification of the genomic locations of duplicate nucleotide sequences in maize by analysis of restriction fragment length polymorphism. Genetics 118, 353–363.

International-Rice-Genome-Sequencing-Project. (2005). The map-based sequence of the rice genome. Nature 436, 793–800.

Jaenicke-Despres, V., Buckler, E.S., Smith, B.D., Gilbert, M.T., Cooper, A., Doebley, J., and Paabo, S. (2003). Early allelic selection in maize as revealed by ancient DNA. Science 302, 1206–1208.

Kapitonov, V.V., and Jurka, J. (2001). Rolling-circle transposons in eukaryotes. Proc Natl Acad Sci U S A 98, 8714–8719.

Kashkush, K., Feldman, M., and Levy, A.A. (2003). Transcriptional activation of retrotransposons alters the expression of adjacent genes in wheat. Nat Genet 33, 102–106.

Kellis, M., Birren, B.W., and Lander, E.S. (2004). Proof and evolutionary analysis of ancient genome duplication in the yeast Saccharomyces cerevisiae. Nature 428, 617–624.

Kim, H., San Miguel, P., Nelson, W., Collura, K., Wissotski, M., Walling, J.G., Kim, J.P., Jackson, S.A., Soderlund, C., and Wing, R.A. (2007). Comparative physical mapping between Oryza sativa (AA genome type) and O. punctata (BB genome type). Genetics 176, 379–390.

Lai, J., Ma, J., Swigonova, Z., Ramakrishna, W., Linton, E., Llaca, V., Tanyolac, B., Park, Y.J., Jeong, O.Y., Bennetzen, J.L., and Messing, J. (2004). Gene loss and movement in the maize genome. Genome Res 14, 1924–1931.

Liu, B., Brubaker, C.L., Mergeai, G., Cronn, R.C., and Wendel, J.F. (2001). Polyploid formation in cotton is not accompanied by rapid genomic changes. Genome 44, 321–330.

Matzke, M.A., and Birchler, J.A. (2005). RNAi-mediated pathways in the nucleus. Nat Rev Genet 6, 24–35.

McClintock, B. (1930). A cytological demonstration of the location of an interchange between two non-homologous chromosomes of Zea mays. Proc Natl Acad Sci U S A 16, 791–796.

McClintock, B. (1984). The significance of responses of the genome to challenge. Science 226, 792–801.

Messing, J., Bharti, A.K., Karlowski, W.M., Gundlach, H., Kim, H.R., Yu, Y., Wei, F., Fuks, G., Soderlund, C.A., Mayer, K.F., and Wing, R.A. (2004). Sequence composition and genome organization of maize. Proc Natl Acad Sci U S A 101, 14349–14354.

Meyers, B.C., Tingey, S.V., and Morgante, M. (2001). Abundance, distribution, and transcriptional activity of repetitive elements in the maize genome. Genome Res 11, 1660–1676.

Molina, M.d.C., and Naranjo, C.A. (1987). Cytogenetic studies in the genus Zea. TAG Theoretical and Applied Genetics 73, 542–550.

Moore, G., Devos, K., Wang, Z., and Gale, M.D. (1995). Grasses, line up and form a circle. Curr Biol 5, 737–739.

Nelson, W.M., Bharti, A.K., Butler, E., Wei, F., Fuks, G., Kim, H.-R., Wing, R.A., Messing, J., and Soderlund, C. (2005). Whole-Genome Validation of High-Information-Content Finger-printing. Plant Physiol. 139, 27–38.

Rice-Chromosomes-11-and-12-Sequencing-Consortia. (2005). The sequence of rice chromosomes 11 and 12, rich in disease resistance genes and recent gene duplications. BMC Biol 3, 20.

Roman, H. (1947). Mitotic Nondisjunction in the Case of Interchanges Involving the B-Type Chromosome in Maize. Genetics 32, 391–409.

Sanmiguel, P., and Bennetzen, J.L. (1998). Evidence that a Recent Increase in Maize Genome Size was Caused by the Massive Amplification of Intergene Retrotransposons. Annals of Botany 82, 37.

SanMiguel, P., Gaut, B.S., Tikhonov, A., Nakajima, Y., and Bennetzen, J.L. (1998). The paleontology of intergene retrotransposons of maize. Nat Genet 20, 43–45.

Song, R., and Messing, J. (2002). Contiguous genomic DNA sequence comprising the 19-kD zein gene family from maize. Plant Physiol 130, 1626–1635.

Song, R., and Messing, J. (2003). Gene expression of a gene family in maize based on noncol-linear haplotypes. Proc Natl Acad Sci U S A 100, 9055–9060.

Song, R., Llaca, V., and Messing, J. (2002). Mosaic organization of orthologous sequences in grass genomes. Genome Res 12, 1549–1555.

Song, R., Llaca, V., Linton, E., and Messing, J. (2001). Sequence, regulation, and evolution of the maize 22-kD alpha zein gene family. Genome Res 11, 1817–1825.

Swigonova, Z., Bennetzen, J.L., and Messing, J. (2005). Structure and evolution of the r/b chromosomal regions in rice, maize and sorghum. Genetics 169, 891–906.

Swigonova, Z., Lai, J., Ma, J., Ramakrishna, W., Llaca, V., Bennetzen, J.L., and Messing, J. (2004). Close split of sorghum and maize genome progenitors. Genome Res 14, 1916–1923.

Thuriaux, P. (1977). Is recombination confined to structural genes on the eukaryotic genome? Nature 268, 460–462.

Wei, F., Coe, E., Nelson, W., Bharti, A.K., Engler, F., Butler, E., Kim, H., Goicoechea, J.L., Chen, M., Lee, S., Fuks, G., Sanchez-Villeda, H., Schroeder, S., Fang, Z., McMullen, M., Davis, G., Bowers, J.E., Paterson, A.H., Schaeffer, M., Gardiner, J., Cone, K., Messing, J., Soderlund, C., and Wing, R.A. (2007). Physical and Genetic Structure of the Maize Genome Reflects Its Complex Evolutionary History. PLoS Genet 3, e123.

Wendel, J.F., Stuber, C.W., Edwards, M.D., and Goodman, M.M. (1986). Duplicated chromosomal segments in Zea mays L.: Further evidence from Hexokinase isozymes. Theor Appl Genet 72, 178–185.

Whitkus, R., Doebley, J., and Lee, M. (1992). Comparative genome mapping of Sorghum and maize. Genetics 132, 1119–1130.

Wilson, C.M., Spraque, G.F., and Nelsen, T.C. (1989). Linkages among zein genes determined by isoelectric focusing. Theor. Appl. Genet. 77, 217–226.

Wolfe, K.H. (2001). Yesterday's polyploids and the mystery of diploidization. Nat Rev Genet 2, 333–341.

Xu, J.H., and Messing, J. (2006). Maize haplotype with a helitron-amplified cytidine deaminase gene copy. BMC Genet 7, 52.

Xu, J.H., and Messing, J. (2008). Organization of the prolamin gene family provides insight into the evolution of the maize genome and gene duplications in grass species. Proc Natl Acad Sci U S A 105, 14330–14335.

Zupan, J., Muth, T.R., Draper, O., and Zambryski, P. (2000). The transfer of DNA from agrobacterium tumefaciens into plants: a feast of fundamental insights. Plant J 23, 11–28.

Maize Centromeres and Knobs (neocentromeres)

R. Kelly Dawe

Abstract In most species, the only chromosomal domains that interact with the cytoskeleton are the centromeres. However in maize there are two motile domains: the centromeres and knobs/neocentromeres. Intensive research has been conducted on both domains. The intent of this review is to provide a broad overall perspective on centromere and knob structure, to compare and contrast their behavior, and to summarize current interpretations of their evolutionary past.

1 Centromeres

1.1 Centromeric DNA

The spindle interacts primarily with kinetochores, which mark the centromeric DNA. Contrary to popular interpretations (Pennisi 2001) the DNA sequence of most centromeres has very little or no impact on kinetochore location. Current views suggest that kinetochores can 'move' under selection and that new kinetochores can attach in regions that have no sequence similarity to centromeres elsewhere in the genome. For instance, in humans, functional kinetochores have formed over apparently random gene-containing regions (Warburton 2004). Similarly, a barley centromere can move laterally to a new position that lacks any sequences found at other centromeres (Nasuda et al. 2005). Even on stable centromeres there is no obvious delineation in sequence between known centromeric DNA and flanking heterochromatic (pericentromeric) DNA (Nagaki et al. 2004).

Although particular sequence motifs are probably not required for centromere function, the overall repetitive structure of centromeric DNA may have a strong impact on centromere stability over evolutionary time (Dawe and Henikoff 2006). Most centromeres are characterized by some type of simple tandem repeat array.

R.K. Dawe
Departments of Plant Biology and Genetics, University of Georgia, Athens, GA 30602

The most common repeat unit is 150–180 bp, which roughly correlates with the size of a nucleosome. Centromeres evolve very quickly and show astonishing variation even among the grasses. For instance the tandem repeats of maize and barley centromeres have no homology (Cheng and Murata 2003). However the same repeats tend to be found at each chromosome within a species, indicating that centromeres do not evolve independently. In maize, the major tandem repeat is a 156 bp sequence known as CentC (Ananiev et al. 1998c). We presume the arrays can extend continuously for many kilobases. Analysis of stretched DNA fibers suggests that the total length of CentC arrays varies among centromeres from as little as <100 kb to as much as several thousand kb (Jin et al. 2004).

Grass centromeres contain a novel class of retroelements known as Centromeric Retroelements (Jiang et al. 2003). The term 'CR element' is used to describe Centromeric Retroelements generally, and more specific notations are used to describe the elements in individual species. For instance maize CR elements are called CRM and rice CR elements are called CRR. CR elements were first discovered as conserved small centromere-specific sequences (Aragon-Alcaide et al. 1996; Jiang et al. 1996) and later shown to be portions of full length Ty3/Gypsy retroelements (Presting et al. 1998). CR elements are among the most conserved known retroelements, having overall identities as high as 85% across cereal species that diverged 60 million years ago (Zhong et al. 2002).

Within the centromere proper, defined by the presence of kinetochore proteins, CR elements are particularly abundant. However CR elements are also found in the pericentromeric regions – the heterochromatic domains that surround all centromeres [pericentromeres and centromeres are strikingly different at the cytological chromatin level, with centromeres staining weakly for DNA and pericentromeres staining very brightly]. The copy numbers of CR elements range in the thousands per genome and are found in virtually all grasses with the possible exception of *Oryza brachyantha* (Lee et al. 2005). CRM elements are active on an evolutionary time scale and are known to insert within CentC arrays and other CRM elements (Nagaki et al. 2003). The unique centromere-specific nature of CR elements, their broad conservation within a dynamic context, and the polymorphism they provide for sequence analysis (below) have made the grasses the preeminent models for plant centromere research.

1.2   Centromeric Chromatin

Given the epigenetic nature of centromere specification and their striking sequence polymorphism, how do we know that CentC and CRM are centromeric DNAs? Centromeres are defined by the presence of kinetochore proteins, and the most fundamental kinetochore protein is a histone variant known as CENH3. CENH3 is similar to H3 in the core domain that binds to other histones, but differs in the key N-terminal domain that interacts with the outside chromatin environment. As a rule the N-termini of CENH3s are entirely different from those of H3 (Henikoff et al. 2000) and appear to exclude most euchromatic and heterochromatic binding proteins. In their place a third class of chromatin (centrochromatin) is formed that organizes kinetochores. The unique N-terminus is also useful in the laboratory since antibodies can be prepared that differentiate CENH3 from H3.

With specific antisera, chromatin immunoprecipitation (ChIP) can be employed to identify the DNAs that interact with the kinetochore. In this technique, antisera are incubated with fragmented chromatin, precipitated, and subjected to analysis. By this method both CentC and CRM were shown to interact with CENH3 (Zhong et al. 2002). It appears that only a fraction of the CentC and CRM are bound to CENH3 at any given time. This interpretation is supported by experiments where DNA and CENH3 were visualized simultaneously (Jin et al. 2004). Thus, CentC and CRM are not sufficient and probably not necessary to organize the overlying kinetochore.

Efforts to assay individual loci relative to their CENH3 association are still in their infancy. As yet only two sequenced CentC-containing BACs are published (Nagaki et al. 2003). Analysis of the sequence revealed no classical single-copy domains (as expected), but it was discovered that CRM insertion points are often unique (Luce et al. 2006). Using primers directed against such insertion points, single copy polymorphic markers were developed. This made it possible to map the BACs, as well as assay the markers in ChIP samples (Luce et al. 2006). One BAC mapped to a central position on chromosome 8 and was shown to interact with CENH3, providing the first precise mapping of a maize centromere. It should be possible to use similar methods to map all ten maize centromeres as new sequence data are released and annotated.

The available data suggest that CentC and CRM may be particularly effective centromeric DNA sequences: they are not required, but are 'better' than most sequences at recruiting the kinetochore. So far this is only a correlative argument. To prove this, we will need to transform centromeric BACs into maize and show that the introduced DNA can independently organize kinetochores. These experiments have yet to be completed for maize, but have been done in rice (Phan et al. 2007). Centromeric BACs were introduced by biolistic transformation and the resulting plants studied. The insertions appeared to be as large or larger than the true centromeres, however there was no evidence of secondary kinetochore activity associated with them. These data support the view that centromeres are defined epigenetically, and that even the most common centromere repeats are not sufficient to organize kinetochores.

2 Knobs and Neocentromeres

2.1 Knob Structure

Maize is known for the 'knobs' that McClintock and others used to identify and track chromosomes. Unlike centromeres, which appear as weakly stained constrictions on chromosomes, knobs appear as darkly staining, large, bulbous structures. Knobs are found at 23 known positions near the ends or in mid-arm (interstitial) positions

(Kato 1976). They are highly polymorphic, with some strains having no visible knobs and others with as many as 14 (Kato 1976). They also vary widely in size, with nearly invisible knobs in some strains and massive knobs in others (Kato 1976; Adawy et al. 2004).

One of the first cloned maize DNAs was the primary knob unit, now called the 180 bp repeat (Peacock et al. 1981). More recently a second knob repeat called TR1 was identified (Ananiev et al. 1998a) which is present on many knobs, though in minor proportion relative to the 180 bp repeat. Both repeats, like CentC, are arrayed in tandem for many kilobases/knob (Adawy et al. 2004). Knobs also contain a smattering of transposable elements, though the overall frequency/bp is much lower than in any other segment of the genome (Ananiev et al. 1998b; Mroczek and Dawe 2003). Simple selfish DNA scenarios do not adequately explain the existence and polymorphism of knobs – classic selfish elements should be more evenly distributed and generally located in low recombination regions (Charlesworth et al. 1994).

2.2   Abnormal Chromosome 10 and Meiotic Drive of Knobs

The largest knob in maize occurs on a rare chromosomal variant known as Abnormal chromosome 10 (Ab10) (Longley 1938). While observing meiosis in strains that contained Ab10, Marcus Rhoades made an important discovery. In strains that carry Ab10, all knobs (not just the Ab10 knob) move rapidly forward and arrive at spindle poles far in advance of the centromeres during anaphase I and II (Rhoades and Vilkomerson 1942). Rhoades referred to them as neocentromeres since they showed centromere-like behavior. Neocentromeres can be very dramatic – often stretching chromosome arms the entire length of the spindle (Fig. 1A). At the time Rhoades suspected that neocentromeres were knobs in a new role. Peacock et al. (1981) later demonstrated by in situ hybridization that knobs and neocentromeres were indeed the same.

In further studies, Rhoades discovered that Ab10 is preferentially segregated to progeny. Instead of the expected 50/50 ratio in testcrosses, Ab10 showed roughly 75/25 (Rhoades 1942). He ruled out all known and trivial explanations and concluded that Ab10 must be somehow promoting its own transmission (Rhoades 1942). Rhoades and others went on to show that at least three other knobs show the same levels of preferential segregation when Ab10 is present (Longley 1945; Rhoades and Dempsey 1985). These data suggest that all 23 knobs are preferentially segregated, but only when Ab10 is present.

The general phenomenon of preferential segregation is now referred to as meiotic drive (Sandler and Novitski 1957; Burt and Strivers 2006). In a classic model for meiotic drive in maize (Rhoades 1952), Rhoades proposed that Ab10 shows preferential segregation because of neocentromere activity (Fig. 2). Only plants that are heterozygous for a knob show drive, since those without knobs have no means to, and those homozygous for a knob show a standard 50/50 ratio (since they compete against themselves). Preferential segregation proceeds as follows:

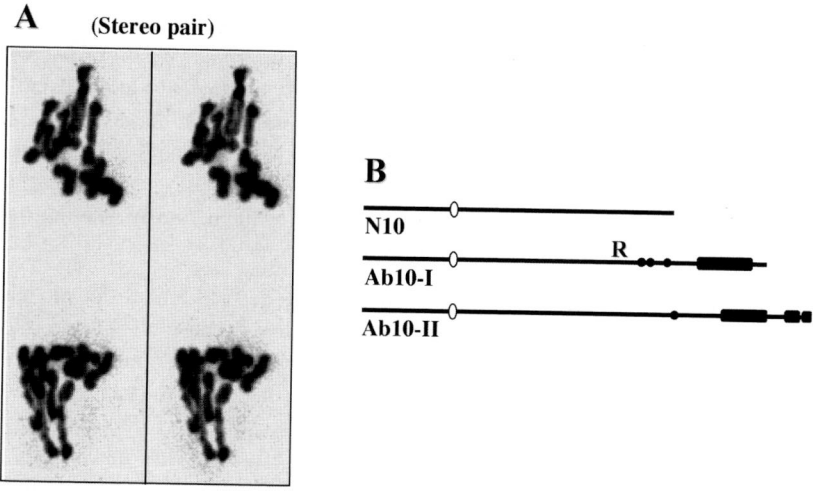

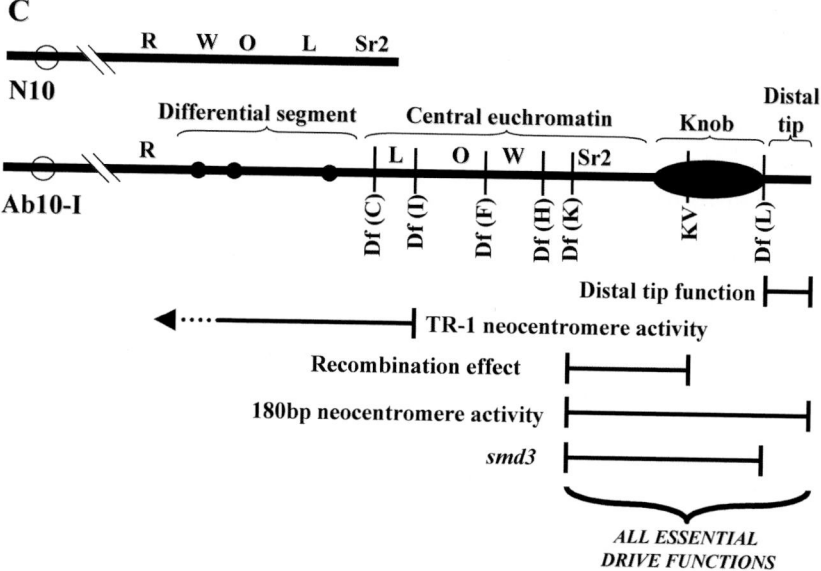

Fig. 1 Ab10 and neocentromeres. **A)** Neocentromeres. Anaphase II is shown as a stereo pair with knobs pulled towards poles. To see the image in pseudo-3D, cross your eyes or use a pair of appropriate viewing glasses. **B)** The major variants of chromosome 10. Ab10-I differs from Ab10-II by the number of TR1-rich chromomeres (small dots) and the presence of an additional knob on the tip of the long arm. **C)** The Ab10 haplotype. Only the non-recombining segment of Ab10 is shown (for point of reference note the R gene, which is also shown in 'B'). Rhoades discovered that W2, O7, and L13 were inverted on Ab10. This inversion, as well a second inversion, has been confirmed by RFLP mapping. Also shown are the locations of known meiotic functions. For additional detail see (Hiatt and Dawe 2003b; Mroczek et al. 2006)

1) Recombination occurs between the centromere and knob. This occurs frequently since all 23 knobs are far enough from centromeres to be genetically unlinked.
2) Once exchanged, the chromatids (known as heteromorphic dyads) orient with the knob towards the pole as neocentromere activity commences at meiosis I.
3) The knobs retain a polar position through meiotic interphase, and again line up with the knobs facing towards the outside of what will become the linear tetrad in meiosis II.
4) At the completion of meiosis, the upper three spores die naturally, leaving only the bottom cell (containing Ab10) to become the egg.

It is important to note that the Rhoades model was derived and tested in male meiocytes, when in fact male parents do not show drive because all four products of meiosis produce gametophytes. Unfortunately empirical tests on female meiocytes are exceedingly difficult. Only a few female meiocytes have been observed in Ab10 strains, just enough to show that neocentromeres are present (I. Golubovskaya, personal communication). The maximum possible drive by this mechanism is 83.3%, which is the maximum number of heteromorphic dyads that can be produced at a genetic distance >50 cM from the centromere (allowing for multiple crossovers among the four chromatids; see (Buckler et al. 1999).

2.3   Cell Biology of Neocentromeres

Several forms of data suggest that neocentromeres move on the spindle by a mechanism that is quite different from kinetochore-mediated chromosome movement. Perhaps the most straightforward evidence is that neocentromeres are known to lack three key kinetochore proteins: CENH3, CENP-C, and MAD2 (Dawe et al. 1999; Yu 2000; Dawe and Hiatt 2004). However, there are also other compelling arguments. Live-cell imaging revealed that neocentromeres move 50% faster than centromeres on the meiotic spindle, and that they begin poleward movement in prometaphase, which is earlier than the centromeres (Yu et al. 1997). Neocentromeres move along the sides of microtubules, whereas the kinetochores interact with microtubules in an end-on fashion (Yu et al. 1997). The ends of microtubules are known to regulate kinetochore movement. Thus, while centromeres/kinetochores are moving (relatively) slowly towards the poles in procession, neocentromeres rapidly slide along microtubule sidewalls by an unknown mechanism.

Other data show that the two major knob repeats differ in their neocentromere behavior. When TR1 arrays are present along with the 180 bp repeat in the same knob, they occupy separate domains. Both TR1 and 180 bp repeats are active in neocentromeres, but TR1 is much more effective. It appears to bind very tightly to microtubules, often spreading out along a fiber for a long distance. Further, TR1 arrays always precede 180 bp arrays on the spindle. The result is that TR1 appears as 'beaks' extending well ahead of the ball-shaped masses of 180 bp, with the entire complex of knob repeats moving faster than centromeres.

A frequently asked question is why neocentromeres never cause chromosome breakage. How does the neocentromere know to move in the same direction as the kinetochore? The answer probably lies in the fact that neocentromeres begin to move poleward as soon as there is a semblance of a spindle to move on, which is prior to the movement of the kinetochores towards the poles. The early movement could swing the attached kinetochores towards the same pole as the neocentromere. Under this view a neocentromere chooses a pole and the kinetochore follows (Yu et al. 1997).

2.4 Ab10 Structure and Trans-Acting Factors

Genetic analyses indicate that only the last third of the long arm of Ab10 is responsible for meiotic drive. The terminal segment is surprisingly large; probably close to 55 map units and containing up to 75 megabases of DNA (Hiatt and Dawe 2003a; Mroczek et al. 2006). The functional portion of Ab10 is referred to as the Ab10 haplotype. There are at least two major forms of the haplotype: Ab10-I, which occurs in maize and teosinte, and Ab10-II, which is only found in teosinte. The two chromosomes differ in external appearance but have many of the same genes and functions (Fig. 1B; (Rhoades and Dempsey 1988). In this review only a brief treatment of Ab10 structure is given. Further information can be found in (Hiatt and Dawe 2003b; Dawe and Hiatt 2004; Burt and Strivers 2006; Mroczek et al. 2006).

Ab10 can be subdivided into four major domains: the distal tip, the knob, the central euchromatin, and the differential segment. The large knob is composed almost entirely of the 180 bp knob repeats, while the differential segment contains three small knobs (called chromomeres) composed of TR1. Rhoades believed that the large knob itself was causing preferential segregation and referred to Ab10 as 'K10' (K being a general nomenclature for knobs). However, late in his life he showed that removal of the major Ab10 knob abolished drive but not neocentromere activity at other knobs (Rhoades and Dempsey 1986). These data indicate that neocentromere activity is not caused by the special structure of Ab10 but by its encoded trans-acting factors.

Building on prior studies, a mutant screen was developed to identify genes that controlled meiotic drive (Dawe and Cande 1996). Mutants and deletions derived from this screen, as well as a known set of terminal deficiencies, were used to construct a map of functions within the haplotype (Fig. 1C; (Hiatt and Dawe 2003a; Hiatt and Dawe 2003b). The map has also been integrated with an RFLP map of N10-derived genes (Mroczek et al. 2006). The combined data show that the central euchromatin contains a large inversion of N10-derived genes as well as a smaller inversion within it (the inversion of *L13 O7* and *W2* was first described by Rhoades; (Rhoades and Dempsey 1985)). The inverted regions appear to have little role in meiotic drive, but serve to block recombination with N10 and stabilize the haplotype.

On the proximal side of the central euchromatin is the differential segment that contains three TR1-rich chromomeres. Interestingly, the differential segment contains at least one gene that is specialized for TR1 neocentromere activity (Hiatt et al. 2002). The spatial separation of the TR1 region and its linkage to a sequence-specific neocentromere-activating gene suggests that a TR1 'cassette' evolved independently of the primary 180-bp system. Supporting this interpretation is the fact that the Ab10-II variant has very little TR1 repeat and lacks the TR1-specific activating phenotype (Mroczek et al. 2006).

With the exception of the TR1 region, all known drive functions map distal to the last N10-homologous gene (*sr2*). Nothing is known of the single copy sequence there, and the available data suggest that it was obtained from another species several million years ago (Mroczek et al. 2006). Among the meiotic drive functions that lie distal to *sr2* are: 1) the 180 bp neocentromere-activating gene(s); 2) a gene that increases recombination in structural heterozygotes (presumably to ensure recombination between the Ab10 haplotype and centromere is >50 cM); 3) two unknown genes that are required for meiotic drive but have no role in neocentromere activity or the recombination effect. The unknown genes may be involved in stabilizing knob position between meiosis I and II (Fig. 2) (Hiatt and Dawe 2003b; Mroczek et al. 2006).

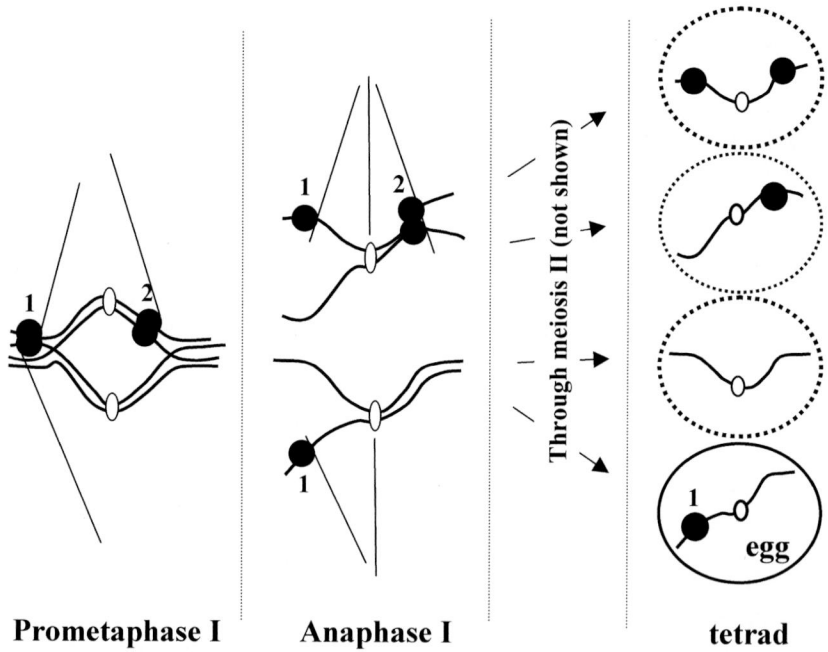

Prometaphase I   **Anaphase I**   **tetrad**

Fig. 2 The Rhoades model for meiotic drive. A chromosome that contains two knobs is used for the purposes of illustration (in fact it is rare for a chromosome to have two knobs). In the meiosis shown, recombination occurred proximal to knob 1, but distal to knob 2. Note that the crossover places knob 1 on both chromosomes, whereas without a crossover knob 2 remains on one chromosome. All knobs form neocentromeres at meiosis I and II. The orientation established in meiosis I is maintained through meiosis II by an unknown mechanism. The result is that knob 1 is segregated to the bottom, functional megaspore (egg). Given this recombination pattern, knob 1 is assured a position in the egg cell, but knob 2 can only segregate to the egg cell half the time

The presence of Ab10 explains one of the 23 knobs. The other 22 knobs appear to have evolved as a consequence of the presence of Ab10, by taking advantage of both the repeats and transacting factors provided by Ab10 (Buckler et al. 1999). The size and frequency of knobs in maize races correlates well with the presence of Ab10 (Buckler et al. 1999). Rare transposition events presumably brought samples of knob repeats to other chromosome arms at appropriate positions >50cM from centromeres. Such arrays are expected to expand and contract by occasional unequal recombination (Smith 1976). This would in turn set the stage for rapid escalations in knob size, since large knobs are more effective neocentromeres (Yu et al. 1997) and preferentially segregated over small knobs (Kikudome 1959).

Given the mechanism it employs, it would appear that Ab10 (and knobs) have an overwhelming advantage and should rapidly go to fixation. However Ab10 remains a rare chromosomal variant. There are at least three reasons why this is the expected outcome. The first is that Ab10 only shows drive when it is heterozygous; once it is common enough to be frequently present as homozygous, it loses its edge. The second is that the Ab10 haplotype contains a long section of required maize genes in the central euchromatin. These do not readily exchange with N10 and as a result are expected to accumulate deleterious alleles at a high frequency. Homozygous Ab10 plants do appear to be 'sicker', but it is difficult to separate this effect from inbreeding. No systematic studies have been carried out to test whether Ab10 is a deleterious chromosome in the homozygous state. The third is that any successful meiotic drive system is inherently bad for a species since selfish components can control the organism's evolutionary path. The expectation, which has been demonstrated in Drosophila (see Ardlie 1998), is that host modifiers will evolve to suppress drive. We assume similar modifiers are present in maize and that these have helped limit the spread of Ab10.

3   Using Meiotic Drive Logic to Understand Centromere Evolution

An important aspect of meiotic drive is that it has the capacity to evolve without regard to host fitness (Ardlie 1998). This produces a 'genomic conflict' (Burt and Strivers 2006), where the selfish interests of the DNA are at odds with the interests of the organism. Any allele linked to a knob is constrained in evolutionary terms, since it is fated to increase in the population whether or not it is a fit allele. As the majority of the maize genome is linked to a knob, meiotic drive is presumed to have had a major impact on the makeup of maize (Buckler et al. 1999). At least eight other species have neocentromeres (Dawe and Hiatt 2004). It is possible that many of the interstitial heterochromatic blocks in eukaryotes have a similar history of meiotic drive.

If an occasional neocentromere can have a major affect on the genetic makeup of a species, then centromeres gone awry could have a debilitating impact (Henikoff et al. 2001). With even a small segregation advantage, selfish centromere repeats could rapidly sweep through a population. Given this potential, it is likely that there is strong (selfish) selection for centromere repeats to increase their capacity to 'attract' the kinetochore (Henikoff et al. 2001; Dawe and Henikoff 2006). This probably occurs through mutation events that confer sequence-specific binding interactions between repeats and inner kinetochore proteins. Once a group of repeats have acquired sequence-specific roles, they could further increase their transmission by increasing the size of repeat arrays, the size of the kinetochore, and the likelihood they will interact with the spindle (Henikoff et al. 2001; Dawe and Henikoff 2006). A driven centromere would drag linked genes with it, almost certainly to the detriment of the organism.

In principle, centromere drive can be thwarted with epigenetics (Dawe and Henikoff 2006). If the DNA's grip on the kinetochore were loosened so that sequence has little or no consistent role, then the effect of meiotic drive would be lessened or eliminated. It has been proposed that the respective roles of genetics and epigenetics have cycled over time (Dawe and Henikoff 2006). When centromeric DNA acquires the capacity to bind tightly to inner kinetochores, the organism responds by changing the structure of the inner kinetochore proteins and restoring sequence independent (epigenetic) inheritance. Supporting this view is the fact that two fundamental DNA binding proteins of the kinetochore, CENH3 and CENP-C, show strong evidence of adaptive evolution (Talbert et al. 2004). A cycling pattern also explains the rapid evolution of centromeric DNA. Whenever the inner kinetochores change, the DNA sequences start anew, reinventing themselves to adapt to the new binding interface (Dawe and Henikoff 2006). In addition, a cycling pattern helps to explain why centromeres do not expand uncontrollably to encompass larger and larger regions.

It is noteworthy that the Ab10 meiotic drive system – the first discovered and most thoroughly understood meiotic drive system – provided the framework on which the centromere drive hypothesis was built. Thus, although maize neocentromere and centromeres have very different roles and mechanisms of movement, their modes of evolution may have much in common.

Acknowledgements I thank Lisa Kanizay and Evelyn Hiatt for critically reading the manuscript.

References

Adawy, S. S., R. M. Stupar and J. Jiang, 2004 Fluorescence in situ hybridization analysis reveals multiple loci of knob-associated DNA elements in one-knob and knobless maize lines. J Histochem Cytochem **52:** 1113–1116.

Ananiev, E. V., R. L. Phillips and H. W. Rines, 1998a A knob-associated tandem repeat in maize capable of forming fold-back DNA segments: Are chromosome knobs megatransposons? Proc. Natl. Acad. Sci. USA **95:** 10785–10790.

Ananiev, E. V., R. L. Phillips and H. W. Rines, 1998b Complex structure of knob DNA on maize chromosome 9: Retrotransposon invasion into heterochromatin. Genetics **149:** 2025–2037.

Ananiev, E. V., R. L. Phillips and H. W. Rines, 1998c Chromosome-specific molecular organization of maize (*Zea mays* L.) centromeric regions. Proc. Natl. Acad. Sci. USA **95:** 13073–13078.

Aragon-Alcaide, L., T. Miller, T. Schwarzacher, S. Reader and G. Moore, 1996 A cereal centromeric sequence. Chromosoma **105:** 261–268.

Ardlie, K. G., 1998 Putting the brake on drive: meiotic drive of *t* haplotypes in natural populations of mice. Trends Genet. **14:** 189–193.

Buckler, E. S. I., T. L. Phelps-Durr, C. S. K. Buckler, R. K. Dawe, J. F. Doebley et al., 1999 Meiotic drive of chromosomal knobs reshaped the maize genome. Genetics **153:** 415–426.

Burt, A., and R. Strivers, 2006 *Genes in Conflict: The Biology of Selfish Genetic Elements.* Harvard University Press, Cambridge, Massachusetts.

Charlesworth, B., P. Sneglowski and W. Stephan, 1994 The evolutionary dynamics of repetitive DNA in eukaryotes. Nature **371:** 215–220.

Cheng, Z. J., and M. Murata, 2003 A centromeric tandem repeat family originating from a part of Ty3/gypsy-retroelement in wheat and its relatives. Genetics **164:** 665–672.

Dawe, R. K., and W. Z. Cande, 1996 Induction of centromeric activity in maize by suppressor of meiotic drive 1. Proc. Natl. Acad. Sci. USA **93:** 8512–8517.

Dawe, R. K., and S. Henikoff, 2006 Centromeres put epigenetics in the driver's seat. Trends Biochem Sci **31:** 662–669.

Dawe, R. K., and E. N. Hiatt, 2004 Plant neocentromeres: fast, focused, and driven. Chromosome Res **12:** 655–669.

Dawe, R. K., L. Reed, H.-G. Yu, M. G. Muszynski and E. N. Hiatt, 1999 A maize homolog of mammalian CENPC is a constitutive component of the inner kinetochore. Plant Cell **11:** 1227–1238.

Henikoff, S., K. Ahmad and H. S. Malik, 2001 The centromere paradox: stable inheritance with rapidly evolving DNA. Science **293:** 1098–1102.

Henikoff, S., K. Ahmad, J. S. Platero and B. V. Steensel, 2000 Heterochromatic deposition of centromeric histone H3-like proteins. Proc. Natl. Acad. Sci. USA **97:** 716–721.

Hiatt, E. N., and R. K. Dawe, 2003a The meiotic drive system on maize abnormal chromosome 10 contains few essential genes. Genetica **117:** 67–76.

Hiatt, E. N., and R. K. Dawe, 2003b Four loci on Abnormal chromosome 10 contribute to meiotic drive in maize. Genetics **164:** 699–709.

Hiatt, E. N., E. K. Kentner and R. K. Dawe, 2002 Independently-regulated neocentromere activity of two classes of satellite sequences in maize. Plant Cell **14:** 407–420.

Jiang, J., J. A. Birchler, W. A. Parrott and R. K. Dawe, 2003 A molecular view of plant centromeres. Trends Plant Sci. **8:** 570–575.

Jiang, J., A. Nasuda, F. Dong, C. W. Scherrer, S.-S. Woo et al., 1996 A conserved repetitive DNA element located in the centromeres of cereal chromosomes. Proc. Natl. Acad. Sci. USA **93:** 14210–14213.

Jin, W., J. R. Melo, K. Nagaki, P. B. Talbert, S. Henikoff et al., 2004 Maize centromeres: organization and functional adaptation in the genetic background of oat. Plant Cell **16:** 571–581.

Kato, Y. T. A., 1976 Cytological studies of maize (*Zea mays* L.) and teosinte (*Zea mexicana* Shrader Kuntze) in relation to their origin and evolution. Mass. Agric. Exp. Stn. Bull. **635:** 1–185.

Kikudome, G. Y., 1959 Studies on the phenomenon of preferential segregation in maize. Genetics **44:** 815–831.

Lee, H. R., W. Zhang, T. Langdon, W. Jin, H. Yan et al., 2005 Chromatin immunoprecipitation cloning reveals rapid evolutionary patterns of centromeric DNA in Oryza species. Proc Natl Acad Sci U S A **102:** 11793–11798.

Longley, A. E., 1938 Chromosomes of maize from North American Indians. J. Agric. Res. **56.**

Longley, A. E., 1945 Abnormal segregation during megasporogenesis in maize. Genetics **30:** 100–113.

Luce, A. C., A. Sharma, O. S. Mollere, T. K. Wolfgruber, K. Nagaki et al., 2006 Precise centromere mapping using a combination of repeat junction markers and chromatin immunoprecipitation-polymerase chain reaction. Genetics **174:** 1057–1061.

Mroczek, R. J., and R. K. Dawe, 2003 Distribution of retroelements in centromeres and neocentromeres of maize. Genetics **165:** 809–819.

Mroczek, R. J., J. R. Melo, A. C. Luce, E. N. Hiatt and R. K. Dawe, 2006 The maize Ab10 meiotic drive system maps to supernumerary sequences in a large complex haplotype. Genetics **174:** 145–154.

Nagaki, K., Z. Cheng, S. Ouyang, P. B. Talbert, M. Kim *et al.*, 2004 Sequencing of a rice centromere reveals active genes. Nature Genet. **36:** 138–145.

Nagaki, K., J. Song, R. Stupar, A. S. Parokonny, Q. Yuan *et al.*, 2003 Molecular and cytological analyses of large tracks of centromeric DNA reveal the structure and evolutionary dynamics of maize centromeres. Genetics **163:** 759–770.

Nasuda, S., S. Hudakova, I. Schubert, A. Houben and T. R. Endo, 2005 Stable barley chromosomes without centromeric repeats. Proc Natl Acad Sci U S A **102:** 9842–9847.

Peacock, W. J., E. S. Dennis, M. M. Rhoades and A. J. Pryor, 1981 Highly repeated DNA sequence limited to knob heterochromatin in maize. Proc. Natl. Acad. Sci. USA **78:** 4490–4494.

Pennisi, E., 2001 Genetics. Closing in on the centromere. Science **294:** 30–31.

Phan, B. H., W. Jin, C. N. Topp, C. X. Zhong, J. Jiang *et al.*, 2007 Transformation of rice with long DNA-segments consisting of random genomic DNA or centromere-specific DNA. Transgenic Res **16:** 341–351.

Presting, G., L. Malysheva, J. Fuchs and I. Schubert, 1998 A TY3/GYPSY retrotransposon-like sequence localizes to the centromeric regions of cereal chromosomes. Plant J. **16:** 721–728.

Rhoades, M., and E. Dempsey, 1986 Evidence that the K10 knob is not responsible for preferential segregation and neocentromere activity. Maize Genet. Coop. Newslett. **60:** 26–27.

Rhoades, M. M., 1942 Preferential segregation in maize. Genetics **27:** 395–407.

Rhoades, M. M., 1952 Preferential segregation in maize, pp. 66–80 in *Heterosis*, edited by J. W. Gowen. Iowa State College Press, Ames, Iowa.

Rhoades, M. M., and E. Dempsey, 1985 Structural heterogeneity of chromosome 10 in races of maize and teosinte, pp. 1–18 in *Plant Genetics*, edited by M. Freeling. Alan R. Liss, New York.

Rhoades, M. M., and E. Dempsey, 1988 Structure of K10-II chromosome and comparison with K10-I. Maize Genet. Coop. Newslett. **62:** 33.

Rhoades, M. M., and H. Vilkomerson, 1942 On the anaphase movement of chromosomes. Proc. Natl. Acad. Sci. USA **28:** 433–443.

Sandler, L., and E. Novitski, 1957 Meiotic drive as an evolutionary force. Am Nat. **91:** 105–110.

Smith, G. P., 1976 Evolution of repeated DNA sequences by unequal crossover. Science **191:** 528–535.

Talbert, P. B., T. D. Bryson and S. Henikoff, 2004 Adaptive evolution of centromere proteins in plants and animals. J Biol **3:** 18.

Warburton, P. E., 2004 Chromosomal dynamics of human neocentromere formation. Chromosome Res **12:** 617–626.

Yu, H.-G., 2000 The maize kinetochore: composition, structure and roles in meiotic chromosome segregation, pp. 180 in *Department of Botany*. University of Georgia, Athens.

Yu, H.-G., E. N. Hiatt, A. Chan, M. Sweeney and R. K. Dawe, 1997 Neocentromere-mediated chromosome movement in maize. J. Cell Biol. **139:** 831–840.

Zhong, C. X., J. B. Marshall, C. Topp, R. Mroczek, A. Kato *et al.*, 2002 Centromeric retroelements and satellites interact with maize kinetochore protein CENH3. Plant Cell **14:** 2825–2836.

Transposons *Ac/Ds*, *En/Spm* and their Relatives in Maize

Jianbo Zhang, Thomas Peterson, and Peter A. Peterson

Abstract Many wild and cultivated species of plants exhibit unstable traits and variegated phenotypes, hallmarks of transposable element activity. However, the unique advantages of maize for genetic and cytogenetic research greatly facilitated the discovery and characterization of transposable element systems by McClintock. While initially dismissed by some scientists as being peculiar to maize, transposable elements are now recognized as major components of all eukaryotic genomes. Transposition of single elements can exert powerful effects on the structure and expression of individual genes. Recent research is showing how alternative transposition reactions involving multiple elements can have a major impact on the evolution of the genome as a whole.

1 Introduction

In her efforts to study gene structure through analysis of chromosomal changes, McClintock (McClintock, 1944) developed an inverse-duplicate segment that included the terminal segment of the short arm of chromosome nine in maize (Peterson, 1987). Crossovers that occurred between homologous segments of this inverted duplication would result in the formation of chromatid bridges that connected centromeres. Using appropriate genetic markers, she could detect losses of chromatid segments resulting from the events of the Bridge-Breakage-Fusion (B-B-F) cycle (Figure 1; also see Figures 3-5 in Peterson 1999). However, the breaking chromatid bridges produced an unexpected result in the subsequent progenies: an abundance of mutants, many of which were unstable (McClintock, 1946). Concurrent with McClintock's investigations, on the other side of the continental United States, Dr. E. G. Anderson (1947) found a pale green mutable seedling. This mutant was produced from seed exposed to radiation from the 1946 US Navy atomic bomb test at Eniwetok Atoll in the South Pacific (Anderson, 1948; Anderson et al., 1949; Peterson, 1991). Thus began one of the most significant scientific advances of 20th century biology: the discovery of the *Ac/Ds* and *En/Spm* transposable element systems.

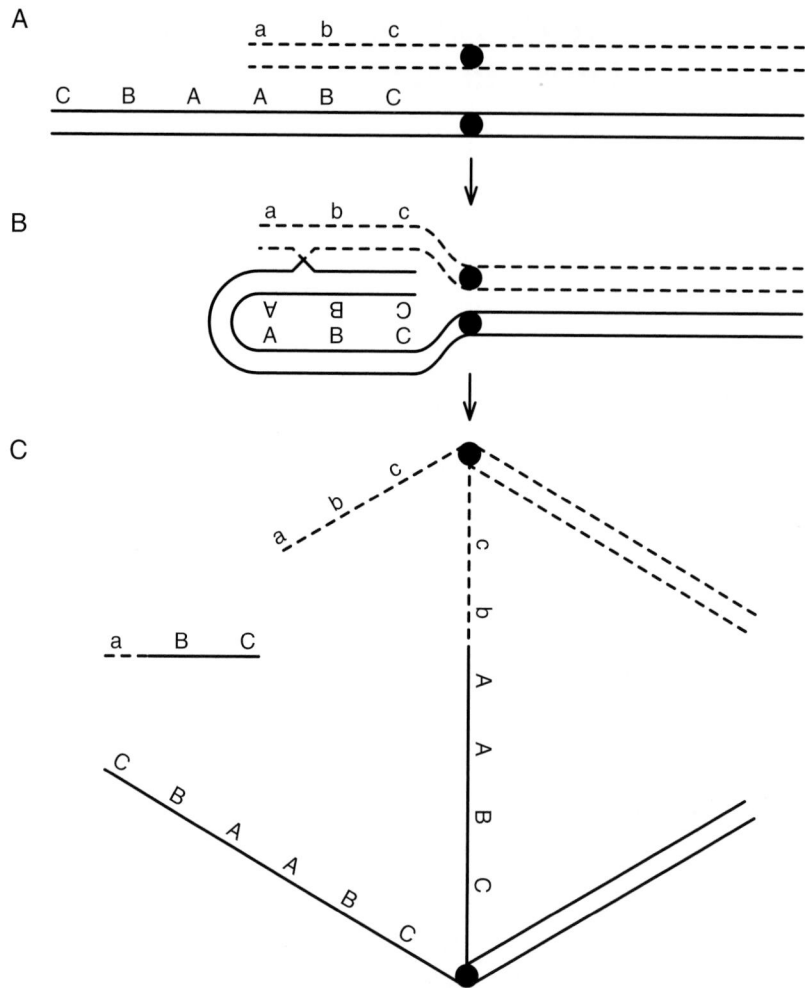

Fig. 1 Bridge-Breakage-Fusion cycle. (**A**) Two chromosomes 9, each with identical sister chromatids; upper chromosome (dashed lines) is normal, and lower chromosome (solid lines) has a terminal inverted duplication of the interval *ABC*. (**B**) The distal copy of the inverted duplication pairs with the corresponding region of the normal homologous chromosome. A crossover occurs in the *a, b* interval, generating a chromatid bridge and an acentric fragment. (**C**) Movement of centromeres to opposite poles at anaphase will break the chromatid bridge. Fusion of broken ends can occur following subsequent replication of the broken chromatids (not shown)

In this chapter, we describe briefly the initial discovery and characterization of the classical maize transposable element systems. For a detailed account of McClintock's early investigations, see the chapter in this volume by Kass and Chomet. A number of excellent recent reviews are available: R.N. Jones (Jones, 2005) provides a detailed description of the genetic characterization of *Ac/Ds* elements; Kunze and Weil (Kunze and Weil, 2002) provide a thorough molecular characterization of both *Ac/Ds* and *En/Spm*; and Dooner and Weil (2007) describe the interactions of transposons and host genomes.

2   Discovery and Genetic Characterization of *Ac/Ds* and *En/Spm*

2.1   *Ac/Ds*

When studying mutants derived from the BBF cycle, McClintock observed unexpected variegation in both endosperm and leaf traits. Genetic and cytological analyses revealed that the variegation phenotype was associated with chromosome breakage at a particular site in the short arm of chromosome 9 designated *Dissociation* (*Ds*) (McClintock, 1946; see Figures 4 - 7 in McClintock 1951a for photographs of chromosome breakage at *Ds*.). Chromosome breakage occurred at the *Ds* locus only when the dominant factor *Activator* (*Ac*) was also present in the genome. McClintock further observed that increased *Ac* copy number delayed the timing of *Ds*-induced chromosome breakage (McClintock, 1947; McClintock, 1948). This phenomenon, termed the *Ac* negative dosage effect, was also observed as a characteristic feature of the variegated kernel pericarp pigmentation conditioned by *Ac* inserted in the maize *p1-vv* allele (Brink and Nilan, 1952; Greenblatt and Brink, 1962). Unlike all other genetic factors known at that time, both *Ac* and *Ds* were observed to move from one location to another in the genome. The presence of *Ac* in the genome was also required for both *Ds*-induced chromosome breakage and for *Ds* transposition. To distinguish different *Ds* elements, McClintock designated the original *Ds* proximal to *wx1* as "*Ds* in standard position". In one case, *Ds* transposed from the standard position to a new site in or near the *c1* gene; this new *c1* mutant was designated *c1-m1*. In *c1-m1* and related alleles, McClintock observed that *Ds* elements could exist in two distinct states: **state I** *Ds* elements cause a high frequency of chromosome breakage, but exhibit a low frequency of transposition; whereas, **state II** *Ds* elements show a low frequency of chromosome breakage and a high frequency of transposition (McClintock, 1949).

While McClintock was growing up in Brooklyn NY, her future advisor, Rollins Emerson, began to study the striking variegated maize kernel pericarp and cob phenotype specified by the *p1-vv* allele. By carefully analyzing the pericarp variegation pattern of *p1-vv*, Emerson concluded that the red stripes in the *p1-vv* kernels were the result of somatic modification of a factor that otherwise behaved according to Mendelian laws (Emerson, 1914). Emerson further observed that the variegated pericarp phenotype exhibited two predominant patterns of instability: a heavily-sectored pattern termed medium-variegated, and a less-sectored pattern termed light variegated. These variegation patterns were dependent on the genetic constitution of the *p1-vv* allele (homozygotes being light variegated, and heterozygotes usually medium variegated), and could also be affected by the presence of a factor outside the *pl* locus. We now know that Emerson's observations reflect the negative dosage effect of *Ac*: increased dosage of *Ac* delays the developmental timing of *Ac/Ds* excision and reduces the frequency of revertant sectors. Interestingly, maize ears with variegated pericarp often exhibit twinned sectors in which an area of light variegated pericarp lies adjacent to an area of red pericarp (Figure 2A). By analysis of these twinned sectors, Brink and Nilan (Brink and Nilan, 1952) determined that the *p1-vv* allele contains an insertion of an element they termed *Mp* (*Modulator of pericarp*), which was later shown to be functionally and molecularly (Chen et al., 1987; Chen et al., 1992; Lechelt et al., 1989) equivalent to *Ac*. Moreover, the light-variegated area contains two *Mp/Ac* elements: the original element at the *p1-vv* locus, and a second element that was lost from the red sector and gained by the light variegated sector (Figure 3; Greenblatt, 1966; Greenblatt and Brink, 1962). These and other genetic results indicated that *Modulator* (*Ac*) preferentially transposes during DNA

A   B

Fig. 2A Maize ear with medium-variegated kernel pericarp and sector of light-variegated pericarp (upper right) twinned with red pericarp sector (lower left). Purple aleurone sectors in some kernels indicate excision of *Ds* from *r1-sc:m3* induced by *Ac*. **B** Maize ear showing colored kernel aleurone phenotypes resulting from *En/Spm* interactions with the *a1-m1* allele. Ear 07 06×1 × *a1-m1/ a1-m1*; colored kernels are *a1-m1*, no *En/Spm*; spotted kernels are *a1-m1* with *En/Spm*

replication, an interpretation that was subsequently confirmed by molecular analysis (Chen et al., 1992; Wirtz et al., 1997).

2.2   En/Spm

Peterson identified the *En/I* transposable element system in the *pg* (*pale green*) locus (Peterson, 1953), while McClintock independently identified *Spm/dSpm* in the *a1* locus (McClintock, 1954). Later it was shown that *En* is genetically identical to *Spm* (Peterson, 1965). *En/Spm* is the autonomous component of the two-element system; transposition of the nonautonomous *I/dSpm* elements requires the presence of *En/Spm* in the genome.

2.2.1   The Origin of the *En/I* Transposable Element System

The pale green seedling mutant originally identified by Anderson was sent to Professor Marcus M. Rhoades (U. Illinois) who provided the *pg* seed progeny to

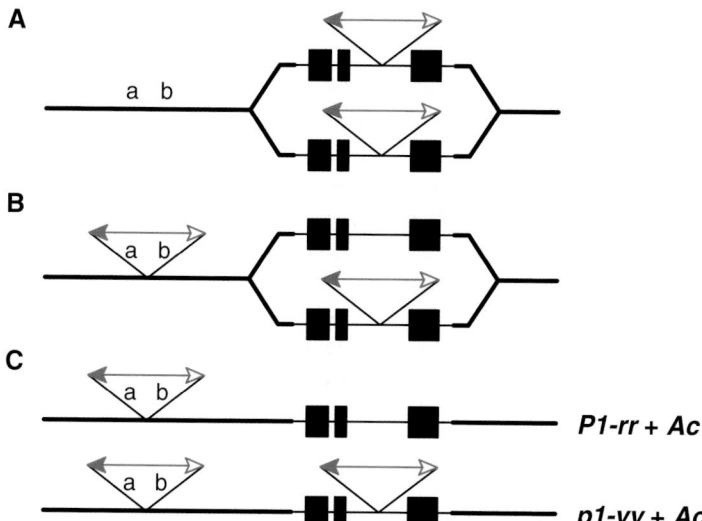

Fig. 3 Model for twin sector formation by *Ac* transposition. The black boxes represent maize *p1* gene exons. The *p1-vv* allele contains an *Ac* element (red line with two arrowheads) inserted in intron 2 of the *p1* gene; the solid and open arrowheads indicate the 5' end and 3' end, respectively, of *Ac*. (**A**) Replication of the *p1-vv* allele generates two daughter chromatids, each carrying an *Ac* insertion in the *p1* gene. (**B**) *Ac* in one daughter chromatid excised from *p1-vv* and inserted into an unreplicated site between *a* and *b*. (**C**) After DNA replication is complete, the upper chromatid carries a functional P1-rr gene and a transposed Ac element, while the lower chromatid carries the p1-vv allele and an Ac element. Segregation of these two chromatids into adjacent daughter cells generates an incipient red/light variegated twinned sector

this author (PAP) for his graduate research project. In outcrosses of the plants from the mutant seedlings, two mutant seedling phenotypes appeared: pale green stable *(pg-s)*, and pale green mutable *(pg-m)* (see photos in Peterson, 1959). The latter had a pale green background (similar to the stable), with darker green stripes. The factor that inhibited normal green color resulting in the pale green phenotype was identified as *Inhibitor (I)*. Further analysis indicated the presence of an independently-segregating factor that was named *Enhancer (En)* because it enhanced mutability in the *pg-m* class. Taking a cue from the *Ac-Ds* elements (McClintock, 1951a) and the *Dt-dt* (Rhoades, 1938) mutable systems, Peterson (Peterson, 1953) (1952, 1953, 1959) identified the factors responsible for the pale green mutable alleles as the *En-I* system of transposable elements. Thus was discovered the third transposon system in maize; additional systems would be identified as more research was done on other genetically unstable loci (see below). The *pg* locus was later isolated and defined molecularly as *golden 2 (g2)* by Hall *et al.* (1998).

2.2.2 The *Spm* Transposon and its Relationship to *En*

In 1954, McClintock reported her characterization of the *Spm* element at the *a1-m1 5719A1* allele (hereafter referred as *a1-m1*). This unstable allele was found as a single variegated kernel in the progeny of the cross: *A/a1 × a1/a1* (Figure 2B; McClintock, 1951b). Because the patterns of kernel mutability of many of the unstable alleles isolated independently by McClintock and Peterson appeared similar and often involved the same loci, the relationships of the various mutability systems were tested genetically by crossing stocks containing autonomous elements *(En, Spm, Ac, Dt)* with a series of stocks containing reporters (defective elements such as *I, Ds,* and *rDt*). In these tests, only the *Spm* reporter was activated by the introduced *En*. This indicated that *En* and *Spm* functioned equally in activating a common reporter (Peterson, 1965). This genetic equivalence was confirmed at the molecular level when *En* and *Spm* were independently cloned (Masson et al., 1987; Pereira et al., 1986; Pereira et al., 1985). On the initiative of the Saedler laboratory (Cologne), these elements were collectively designated as the *En/Spm* transposable element system.

2.2.3 *En* Transposition to and from the *a1* Locus

Crosses of a *pg-m* stock with a purple plant type *(aa × AA)* yielded one exceptional progeny plant containing a sectored tassel. The tassel was colored except for a colorless sector, which in turn contained smaller colored sectors (i.e., $A \rightarrow a \rightarrow A$). Pollen from the colorless sector of the tassel was crossed to a *a1* tester line, leading to the isolation of an autonomous mutable *a1 (a1-m)* allele controlled by *En* (Peterson, 1961). Further tests with this material uncovered various mutable kernel pattern types including the following:

a1-m = autonomous *a1* mutable; i.e. ***A1::En.***

a1-(mr) = non-autonomous *a1* mutable; i.e. ***A1::I.*** (became a reporter of *En*.)

a1-m(nr) = non-responsive colorless derivatives from *a1-m* (see following)

In further crosses with the highly sectored *a1-m* allele, a number of colorless kernels appeared in the progeny at frequencies similar to the appearance of *pg-s* from *pg-m*. These colorless derivatives of *a1-m* were termed *non-responsive* [*a1-m (nr)*], and they were derived by defective excisions of *En* from the *A1* gene. Many *a1-m (nr)* alleles retained *En* in the genome, and it was possible to determine the chromosome location of the transposed *En* using the *a1-(mr)* reporter. It was determined that many of the transposed *En*'s inserted into chromosome 3 on both proximal and distal sides of the *a1* locus (Nowick and Peterson, 1981; Peterson, 1956). This tendency for local transposition of *En* resembled a similar, but more pronounced, local transposition preference of the *Mp (Ac)* element reported by the Brink laboratory (Brink and Nilan, 1952; Greenblatt and Brink, 1963).

3   Molecular Analyses

3.1   Structure and Gene Products of Ac

The maize *Ac* element was initially cloned from the *waxy1* locus (Fedoroff et al., 1983; Fedoroff et al., 1984) and is 4565 bp in length. *Ac* is bounded by 11 bp imperfect terminal inverted repeat sequences (TIRs), and also contains multiple copies of hexamer motifs (AAACGG or similar) located within 250 bp of the element termini (Kunze and Weil, 2002). Both the TIR sequences and the hexamer motifs appear to be critical for element transposition.

Transcription of *Ac* starts at several sites, with a major site located at position 334 (referring to GenBank sequence X05424). The poly (A) site is at position 4302. Splicing of the four introns yields a mature mRNA of ~3.5 kb in size. The first ATG is at position 988, hence the 5' untranslated region (5' UTR) of the major transcript is 654 bp in length (Figure 4; Kunze et al., 1987). Deletion of a 537 bp segment of the GC-rich 5' UTR increases *Ds* transposition frequency in Arabidopsis, but does not affect *Ac* transcript levels (Bancroft et al., 1992).

The *Ac* mRNA encodes an 807 amino acid transposase protein. The N-terminal 102 amino acids are not required for *Ac/Ds* transposition. *Ac* transposase binds most efficiently to A/TCGG, part of the AAACGG motifs that occur frequently in the *Ac/Ds* 5' and 3' subterminal regions. Transposase also binds to the TIRs, but at much lower affinity (Becker and Kunze, 1997; Kunze and Starlinger, 1989). Transposase contains a basic domain (residues 159–206) that is involved in DNA binding; its N-terminal subdomain (residues 159–176) binds to the TIR, while its C-terminal subdomain binds to both the TIR and the AAACGG motif (Feldmar and Kunze, 1991). *Ac* transposase contains three nuclear localization signals, two of which are located in the basic domain (Boehm et al., 1995). A C-terminal dimerization

Fig. 4 Structures of *Ac* and a representative *Ds* element. The black boxes indicate exons of *Ac*, and the gray boxes indicate sequences essential for *Ac/Ds* transposition. The gap in *Ds* represents a deletion

domain (aa 674–754) is highly conserved among the *hAT* transposon superfamily, of which *Ac* is a founding member. The active form of the transposase is probably a dimer, because transposase mutants that lack a functional DNA binding domain can suppress the activity of wild type transposase (Essers et al., 2000). *Ac* transposase aggregates into rod-shape structures in nuclei of maize endosperm containing three copies of *Ac,* and in petunia protoplasts in which *Ac* is expressed by a strong promoter. In petunia protoplasts, transposase aggregation increases with stronger promoters, but the frequency of *Ds* transposition does not increase proportionally, suggesting that the transposase aggregates are not active in transposition (Heinlein et al., 1994).

3.2 Structures of Ds Elements

Several *Ds* elements have been cloned and sequenced. Although these vary in length and internal sequence, the state II (non-chromosome-breaking) *Ds* elements generally have a simple structure with 11 bp terminal inverted repeats and subterminal sequences containing at least one AAACGG motif (Figure 4). In contrast, state I (chromosome-breaking) *Ds* elements typically have a more complex structure; for example, *doubleDs* contains an ~2 kb state II *Ds* inserted into another copy of the same element in the opposite orientation (Doring et al., 1984). The structural basis for the chromosome-breaking activity of state I *Ds* elements will be discussed below.

It has been proposed that McClintock's original state I *Ds* in standard position is a *doubleDs* element, because all of the mutants directly or indirectly derived from *Ds* in standard position and analyzed at the molecular level carry *doubleDs* or a *doubleDs*-derived structure. Recent PCR analysis (JZ and TP, in preparation) indicates that maize lines with *Ds* in standard position contain a double *Ds* located approximately ~1 Mb proximal to *waxy1* on the short arm of chromosome 9, which is consistent with McClintock's cytogenetic localization of *Ds* in the standard position.

3.3 Formation of Ds from Ac

Ac can spontaneously convert into *Ds*. Studies in transgenic tobacco indicate that conversion of *Ac* to *Ds* is transposase-dependent, and that *Ds* alone is stable (Rubin and Levy, 1997). Many *Ds* elements are internally-deleted versions of *Ac*, while some *Ds* elements contain additional sequences ("filler DNA") that may or may not be related to *Ac* sequences. The internal deletion junctions often exhibit short direct repeats (1–5 bp). These features support a model for *Ds* formation by abortive gap repair that is often accompanied by template slippage and mispairing. *Ac* excision results in a double-strand break (DSB) at the donor site; most often, the DSB ends would be rejoined by non homologous end joining (NHEJ). Rarely, the DSB may be repaired by synthesis-dependent strand-annealing (SDSA) using the sister chromatid as template. SDSA is an error prone process, in which template slippage may occur frequently. The 3′ end of newly synthesized DNA may pair with nearby sequences at regions of microhomology (1–5 bp), followed by resumption of DNA replication. If slip-mispairing occurs only once, a *Ds* with an internal deletion will be formed; if it occurs more than once, the resulting *Ds* could carry filler DNA. *Ds* elements that lack microhomologies at the deletion junctions would likely be generated by both SDSA and NHEJ (Conrad et al., 2007; Yan et al., 1999).

3.4   Molecular Structure of En/Spm

3.4.1   Isolation of a *Wx* Clone

When molecular biological tools became available in the 1970's (Morrow *et al* 1974), it was opportune to examine the molecular structure of *En/Spm*. This required an insertion of *En/Spm* in a gene for which a molecular probe could be obtained. One approach was to target a highly active gene with abundant kernel transcripts. The prime candidate for this was the *Waxy* gene, which was known to be involved in starch metabolism and whose protein product could be verified. Immature kernel cDNAs were tested for the production of the *Wx* product in an *in vitro* transcription-translation system; among 800 + cDNAs tested, one clone was identified as a *Waxy* product and hence could be used as a molecular probe for the isolation of *En/Spm* elements in *Waxy*. The first insertion isolated was a 2 kb, non-autonomous *I* element inserted in the *wx-m8* allele (Schwarz-Sommer et al., 1984). However, isolation of a full-length *En/Spm* element still required the identification of an autonomous *En/Spm* insertion in the *Waxy* gene, which turned out to be more difficult than expected. What was needed were the following genetic stocks: *Wx* colorless that contained *En* to be used as a female (ear) parent, and colorless *wx* to be used as the male parent. Because the *En* source material also contained genes for purple kernel color, it was decided to cross this stock with a color inhibitor allele at the *C1* locus, *C1-I*, in order to eliminate kernel color formation. In the progeny of the cross of *Waxy En C1-I* by *waxy*, a single *wx*-mutable kernel was identified among nearly four million kernels screened. Using the *Waxy* probe described above, *En* was isolated from this *wx*-mutable and identified as *wx-844* by Pereira et al (Pereira et al., 1986; Pereira et al., 1985). (*Subsequent analysis indicated that many wx-mutable kernels were likely misclassified as* Waxy *because of their heavy* wx→ Wx *variegation. Interestingly, the single* wx-m *kernel that was isolated possessed a modifier that reduced the frequency of* wx→ Wx *mutability, allowing for detection. This modifier will be discussed in the section on Suppression.)*

3.4.2   *En/Spm* Structure

En/Spm elements belong to the CACTA transposon superfamily, so-named because of the conserved first 5 bp sequences of the TIRs. Like other CACTA elements, *En/ Spm* insertion results in a 3 bp Target Site Duplication (TSD). The *En* element in the *wx-844* allele is 8287 bp long, and contains 13 bp terminal inverted repeats. Additionally, *En/Spm* elements contain GC-rich Subterminal Repetitive Regions (SRRs) that play an important role in element and target gene expression (Masson et al., 1987, 1991). With the available *En* probe, other *En*-containing genes were isolated, including *a1* (O'Reilly et al., 1985), *c1* (Paz-Ares et al., 1986), and *c2* (Wienand et al., 1986).

3.4.3   Transcription Products of *En*

RNA gel blot analysis of polyA+ RNA has identified two *En/Spm*-specific RNAs: a predominant 2.5 kb RNA and a less-pronounced 6 kb RNA (Figure 5). The 2.5-kb transcript is translated to yield the TNPA protein, which binds to 12-bp DNA motifs that are present in the *En/Spm* SRR sequences, resulting in the suppression function of the element (Grant et al., 1990). The 6 kb transcript encodes the TNPD protein which is required for element excision.

In order to determine which of the 2.5 and 6 kb RNAs correspond to the genetically-identified M (Mutator) and S (Suppressor) functions of *En/Spm*, a mutant was needed that removed the M function but maintained the S function. Such a mutant was isolated and identified as *En2*; the structure of *En2* included a deletion of most of ORF1 and almost all of ORF2 (Gierl et al., 1988). Tests with the suppressible

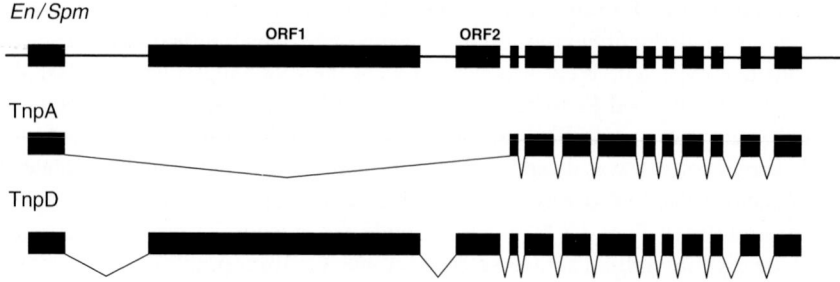

Fig. 5 Structures of *En/Spm* and its major transcripts *TnpA* (2.5 kb) and *TnpD* (6 kb). The black boxes indicate exons

a1m-1 allele indicated that *En2* retained the Suppressor function of *En* and severely reduced but did not entirely eliminate the Mutator function. These results assign the M function to the 6 kb RNA and the S function to the 2.5 kb RNA. Interestingly, the *En2* allele exhibits substantial *En* activity when crossed with the *a1m(Au)* derivative suppressible allele (*a1-m (pale-mr)*; (Peterson, 1995)). Research to localize the residual *En* acitivity within *En2* is ongoing.

To verify the authenticity of the proposed functions of the TNPA and TNPD proteins, Frey *et al.* (Cardon et al., 1991; Frey et al., 1990; Frey et al., 1989) in the Saedler laboratory under the direction of A. Gierl reconstructed the *En/Spm* excision system in transgenic tobacco using three components: (1) a reporter gene containing a *dSpm* insertion into a GUS coding sequence; (2) a construct to express the TNPA protein; and (3) a construct to express the TNPD protein. Excision of *dSpm,* visualized as GUS-expressing clones of cells, occurred only when TNPA was expressed together with TNPD; TNPA or TNPD alone were ineffective in promoting excision. These molecular data matched the genetic results that first showed that *En/Spm* excision requires both S and M functions (McClintock, 1954). Based on these results, Frey et al. (1990) proposed that associations of the TNPA protein bring the *En/Spm* termini together, which then allows the TNPD protein to excise the element. Masson et al. (1991) later confirmed these roles of the *tnpA* and *tnpD* gene products in tobacco.

4 Suppressors and Modifiers of *En/Spm* Activity

Transposon expression can be suppressed by modifiers in the genome. For example, the *wx-844* allele is sensitive to a modifier that segregates independently and suppresses *En/Spm* activity. The modifier was isolated and found to be a defective *En* (*En-I102*) (Cuypers et al., 1988; Gierl et al., 1988), which is hypothesized to produce aberrant *En* proteins that inhibit the wild type *En/Spm* functions. Other modifiers termed *Rst* (Restrainers of mutability) have been identified that act upon other *En* transposons and other transposon systems (P.A. Peterson, unpublished). In contrast, some *I/dSpm* elements appear to require a "helper" to facilitate excisions. In one case, excision of the *I/dSpm* insertion in the *c2-m88 1058Y* allele requires, in addition to *En/Spm*, a third factor which has been termed *Mediator* (Muszynski et al., 1993). *Mediator* did not influence excision of other *I/dSpm* elements tested, indicating that the effects of accessory factors such as *Mediator* may be specific to either particular loci or classes of *I/dSpm* elements.

The *En*-controlled *a2-(mr)* allele exhibits tissue-specific mutability patterns. The kernels from the main stalk of the plant show a low frequency and late pattern of excision, while kernels on the tiller of the same plant show early excision events (Fowler and Peterson, 1978). This differential expression was ascribed to *En-v,* a derivative of *En* that shows a lower rate of expression in kernels of the main plant vs. those from the tiller. Interestingly, the frequent and early excision pattern of the tillers is heritable and appears on the main stalk ears of the next generation. These results illustrate the plasticity of *En/Spm* expression and its exquisite sensitivity to developmental and possibly environmental influences.

5   Other *hAT* and CACTA Elements

Ac/Ds and *En/Spm* were the first transposons discovered, and they have been studied extensively at both genetic and molecular levels. Many other transposons share similarity to either *Ac/Ds* or *En/Spm* in certain respects. An ~50 amino acid segment of *Ac* transposase is highly conserved among the *hAT* transposon superfamily, named after the founding members *hobo, Ac,* and *Tam3,* from Drosophila, maize, and snapdragon, respectively (Calvi et al., 1991). In addition to similarity in amino acid sequence, all of the *hAT* superfamily members have short Terminal Inverted Repeats (TIRs) and create an 8 bp Target Site Duplication (TSD) upon insertion (Rubin et al., 2001). Additional members of the *hAT* superfamily include *rDt, Bg-rBg,* and *Mx-rMx* from maize (Hartings et al., 1991; Xiao et al., 2006; Xu and Dooner, 2005), *Tag1* and *Tag2* from Arabidopsis (Henk et al., 1999; Tsay et al., 1993), *Slide* from tobacco (Grappin et al., 1996), and *hermes* from Drosophila (Warren et al., 1994).

Maize *En/Spm*, snapdragon *Tam1* (Saedler et al., 1984), and soybean *Tgm* (Rhodes and Vodkin, 1988) are members of the CACTA transposon superfamily. CACTA elements generate a 3 bp target site duplication upon insertion. Other CACTA elements include sorghum *Candystripe1* (Chopra et al., 1999), carrot *Tdc1* (Ozeki et al., 1997), petunia *Ps1* (Snowden and Napoli, 1998), and Arabidopsis *CAC1* (Miura et al., 2001). Unlike the *hAT* elements that are widespread in the plant, animal and fungal kingdoms, CACTA elements are found mainly in plants.

6   Transposons in Corn Populations

6.1   *Uq and Mrh*

In several surveys of corn breeding populations, two active transposons (*Uq* and *Mrh*) were found in stocks of the BSSS type (Iowa Stiff Stalk Synthetic) (Cormack et al., 1988; Peterson, 1986; Peterson and Salamini, 1986). Interestingly, tests showed that the *Uq* element has been present in the BSSS stocks throughout the 90-year period in which BSSS types were a critical component of commercial inbred line development (Lamkey et al., 1991). The observed persistence of active transposons in corn breeding lines supports the notion that transposable elements continue to generate important genetic variation in corn breeding populations. This idea is further supported by molecular investigations which indicate that both RNA and DNA transposons have been active in the recent evolutionary history of maize (SanMiguel et al., 1998; Fu and Dooner, 2002; Song and Messing, 2003).

7 The impact of *Ac/Ds* and *En/Spm* on Gene Expression

7.1 *Position vs. Composition*

Prior to the advent of molecular biology, researchers noted that TE-containing alleles of the same gene could have strikingly different levels of residual expression and patterns of mutability (see Figure 2 in (Peterson, 1961)). These observations sparked a vigorous debate among maize geneticists as to whether the variations in gene expression were caused by differences in the **position** or the **composition** of the insert. To address this question, Peterson (Peterson, 1978) used large plantings to obtain new independent cases of TE insertions in target genes such as *c1* and *a2*. The new independent alleles were then tested for their effects on a common reporter (e.g. *a1-mr*). If the patterns of the targeted genes varied, but the mutability patterns on the common reporter were *identical*, it would mean that the alleles differed in the **position** of the insert. If however both the targeted gene and reporter gene expression varied, it would mean that the **composition** of the element differed (Peterson, 1977; Peterson, 1978). When the results of molecular studies became available, it became clear that both models were possible: i.e., some differences were due to position, while others were due to composition (Cuypers et al., 1988; Tacke et al., 1986).

7.2 *Position and Orientation-Dependent Effects on Gene Expression*

Ac/Ds elements usually inhibit expression of the genes into which they insert; insertions into exonic open reading frames cause a severe loss of function, while insertion into introns or untranslated regions can have a lesser effect. In the *p1* locus, *Ac* insertion into the ORF results in very light variegated pericarp, whereas insertion into introns or untranslated regions results in medium variegated or orange variegated pericarp (Athma et al., 1992). The red stripes in these kernels result from somatic *Ac* excision; for *Ac* insertion in exons, only the rare excision events that restore an open reading frame can restore *p1* gene function, hence the frequency of red revertant stripes is very low (Grotewold et al., 1991).

The orientation of *Ac/Ds* insertions can also affect gene expression through orientation-dependent effects on transcription and RNA splicing. In the maize *p1* gene, *Ac* insertions in intronic sequences result in two distinct phenotypes that correlate with the orientation of the *Ac* insertion. Alleles that give a standard *p1-vv* phenotype (red stripes on a colorless pericarp background) have *Ac* insertions in which the transcriptional orientation of *Ac* is the same as that of the *p1* gene. In contrast, insertions of *Ac* in the opposite orientation lead to *P1-ovov* alleles which exhibit *orange-variegated* pericarp and cob. Transcript analysis shows that the *p1-vv* allele generates hybrid *p1::Ac* transcripts that lack *p1* exon 3 and hence are non-functional.

In contrast, the *P1-ovov* allele generates significant quantities of normal-sized transcripts (Peterson, 1990). These results suggest that, in the *p1-vv* alleles, *p1* transcription is prematurely terminated by the transcription terminator signal in *Ac*; whereas, in the *P1-ovov* alleles, *p1* transcription can proceed through the *Ac* element, and the *Ac* sequences in the primary RNA transcripts can be spliced out as a part of intron 2 (Figure 6). Because *Ds* cannot self-excise, *Ds* insertions produce stable mutations without *Ac* in the genome; whereas in the presence of *Ac*, *Ds* insertion results in phenotypes similar to that caused by *Ac* insertion.

Ds insertion does not always eliminate the expression of a gene even when inserted into an exon. Wessler et al. (Wessler et al., 1987) showed that *Ds* element sequences could be efficiently spliced out of RNA transcripts to allow significant expression of *waxy* alleles containing *Ds* insertions in exonic sequences. There are several RNA splicing donor sites in the 5′ end of *Ds*, and differential splicing can generate multiple transcripts. Most of the *Ds* sequence can be spliced in some transcripts; if splicing restores the open reading frame, significant gene function can be restored (Wessler et al., 1987). *Ds* insertion can also create a donor site and/ or an acceptor site that may enable *Ds* sequences to be spliced from the primary transcript. One such case is *sh2-m1,* which contains a 1.68 kb *Ds* inserted into exon 16 of the *shrunken2* gene. The sequence immediately upstream of the *Ds* insertion contains a splice donor site, and the insertion of *Ds* created a splice acceptor site immediately downstream of the *Ds* insertion; in some cases the entire *Ds* element was spliced from the RNA and the open reading frame was precisely restored (Giroux et al., 1994).

Finally, transposon sequences can become permanently fixed as new introns in a gene. An excellent example of this phenomena is the *a2-m1 (Class II)* allele, which contains an *I/dSpm* insertion in the *a2* gene. The 5′ end of this element has a deletion of the TNPA binding motif, which renders the element immobile (Thatiparthi et al., 1995). Interestingly, the *a2* gene containing this insertion is still expressed at levels sufficient to specify colored kernels. Thus, the transposon-derived sequences now

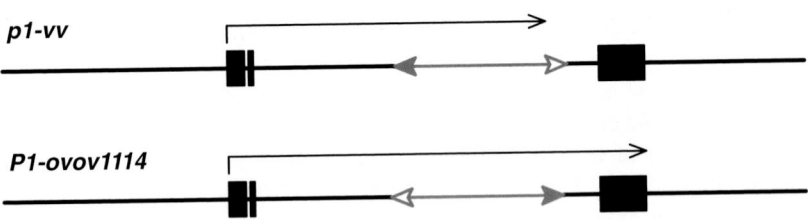

Fig. 6 Structures of *p1-vv* and *P1-ovov1114* alleles and model for differential expression dependent upon orientation of *Ac*. The black boxes indicate exons of the *p1* gene. The double-head arrow indicates the *Ac* element, which is in opposite orientations in the two alleles. The black lines with arrowheads indicate the primary RNA transcripts generated by each allele. In *p1-vv,* transcripts begin normally but terminate within the *Ac* element, prior to reaching *p1* exon 3. In *P1-ovov1114,* transcription proceeds through *Ac* and includes *p1* exon 3. The *Ac* sequences are spliced out of the *P1-ovov1114* transcript, yielding functional *p1* mRNA

behave as an intron in the otherwise intron-less *a2* gene. Similar results have been described for the *bronze-mutable13* allele (Schiefelbein et al., 1988).

7.3   Suppression and Dependence

Insertion of *I/dSpm* elements can have either positive or negative effects on gene expression. In some cases, genes with *I/dSpm* insertions exhibit significant residual expression in the absence of an *En/Spm* element in the genome, while in the presence of *En/Spm* expression of the mutant gene is completely eliminated; these alleles are designated *Spm*-suppressible. In other cases, insertion of an *I/dSpm* element abolishes expression of the gene; if an *En/Spm* is present in the genome, expression of the mutant gene can be restored to relatively high levels; these alleles are designated *Spm*-dependent (Masson et al., 1987).

7.3.1   *En/Spm*-Suppressible Allele: *a1-m1*

In the absence of *En/Spm,* the *a1-m1* allele specifies a stable pale colored phenotype. The addition of *En/Spm* has two distinct effects: 1) suppression of the colored phenotype and 2) mutation in some cells to a fully colored state (See Figure 2B). Similar results led McClintock to conclude that *Spm* contained two distinct functions she termed suppressor (Sp) and mutator (M). Molecular analyses showed that the insertion in *a1-m1* is quite short – less than 900 bp (Tacke et al., 1986). In the absence of *En/Spm,* this and other suppressible alleles are expressed because the *I/dSpm* insert is removed by splicing at the element termini (Menssen et al., 1990; Schiefelbein et al., 1988). In the presence of *En/Spm,* suppression of the *a1-m1* allele is caused by binding of the TNPA protein to the *I/dSpm* insertion which interferes with transcription, presumably by blocking the progression of RNA polymerase.

7.3.2   *En/Spm*-Dependent Allele: *a1-m2 8004*

The expression of the *a1-m2 8004* allele is nearly opposite to that of the *a1-m1* allele. In the absence of *En/Spm,* the *a1-m2 8004* allele specifies a colorless kernel phenotype. In contrast, the presence of *En/Spm* with *a1-m2 8004* both activates *a1* expression resulting in colored kernels, and induces excision of the *I/dSpm* insert resulting in darker colored revertant sectors. The unusual behavior of this allele was analyzed by Schwarz-Sommer (Schwarz-Sommer et al., 1987), who found that the *I* insert in this allele is located in the promoter of the *A1* gene. In this position, binding of *En/Spm* protein to the *I/dSpm* insert induces *A1* transcription. In this and similar cases analyzed in the Fedoroff group (Masson et al., 1987), *En/Spm* activates gene expression.

8   The Impact of *Ac/Ds* and *En/Spm* on the Maize Genome

8.1   Single Element Insertion and Excision

Insertion of *Ac/Ds* elements generates an 8 bp target site duplication (TSD), and excision creates a double strand break (DSB) at the donor site. These DSBs are most
frequently repaired by non-homologous end joining (NHEJ). Because repair by
NHEJ is usually not precise, considerable variation occurs in the footprint sequences
that remain at the transposon excision sites. Only rarely (~2%) is the TSD precisely
removed and the original target sequence restored; the vast majority of *Ac/Ds* excision
events leave part or all of the TSD behind as a "footprint" of the transposon insertion.
Footprints usually consist of a several bp deletion, insertion, or base substitution. The
footprint sequences are not completely random: most common are transversions or
deletion of the base adjacent to the *Ac/Ds* element (Rinehart et al., 1997). Interestingly,
the sequences flanking the *Ac/Ds* element affect footprint formation, so that the same
element produces different footprint patterns at different donor sites (Scott et al.,
1996). These footprint sequences may have little or no effect on untranslated sequences,
but can cause a frameshift mutation in an open reading frame. In some cases, the
footprint remaining after excision of *Ac/Ds* may leave a multiple of three nucleotides
which would not disrupt the ORF, but which would result in insertion of one or more amino
acids into the protein. These small insertions can offer a means to generate functional
variation at specific sites in a protein. Thus, excision of *Ac/Ds* can produce an allelic
series including deletions, frameshifts, and small insertions with a wide range of
activities (Grotewold et al., 1991; Wessler et al., 1986).

Although the predominant *Ac/Ds* footprints are very similar in both maize and
Arabidopsis, a broader range of minor footprints are produced in Arabidopsis
(Rinehart et al., 1997). In yeast, *Ds* excision results in footprints distinctly different
from that observed in plants: some contain palindromic repeats centered around a
base apparently derived from the transposon terminus. These footprint sequences
suggest that, in yeast, *Ac* transposase makes a staggered cut one base in from the
transposon end, and hairpins are formed at the cut end (Weil and Kunze, 2000).

8.2   Macrotransposition

A pair of *Ac/Ds* elements in direct orientation flanking some length of host sequence
may behave as a macrotransposon. Several macrotransposons containing *Ac/Ds* have
been reported (Ralston et al., 1989). The *bz1-m4 D6856* allele carries one partial and
three intact *Ds* elements, all of which are in direct orientation; macrotransposons
composed of the first and second *Ds* elements, or the first and third *Ds* elements can
excise from the *bz1* locus and reinsert elsewhere in the genome (Dowe et al., 1990;
Klein et al., 1988). The *sh1-m5933* allele contains a 30 kb insertion with a *doubleDs*
at one end, and a one and a half *Ds* at the other end; the entire 30 kb insertion can be

excised as a macrotransposon to restore *sh1* gene function at high frequency (Courage-Tebbe et al., 1983). Recently, Dooner and Weil (2007) reported that a *Ds* element in *bz1* and an *Ac* element in the nearby *stc1* gene, together with the contained 6.5 kb host sequence, can excise and reinsert as a macrotransposon. We (J.Z. and T.P.) identified an approximately 15 kb macrotransposon composed of a *fractured Ac* element, an intact *Ac* element, and genomic sequences from the vicinity of the *p1* locus. Together, these cases demonstrate that transposition of macrotransposons could result in segmental deletions, duplications, and the shuffling of regulatory and coding sequences to generate new genes. Because *Ac/Ds* elements tend to transpose locally, macrotransposon formation should be relatively frequent and thus may have played an important role in genome evolution.

8.3   Chromosome Breakage

McClintock discovered transposable elements while studying state I *Ds*-induced chromosome breakage. In addition to state I *Ds* elements, pairs of closely-linked *Ac/Ds* elements can also cause chromosome breakage (Dooner and Belachew, 1991; Ralston et al., 1989). Molecular cloning and sequencing revealed that state II *Ds* elements have relatively simple structures derived for example by internal deletions of *Ac*; in contrast, chromosome-breaking state I *Ds* elements have more complex structures such as that found in *double Ds, double Ds* derivatives, and other composite *Ds* elements (Doring et al., 1984). PCR and sequencing analysis indicate that *doubleDs* and other complex elements can undergo transposition reactions that involve the ends of different *Ds* elements; this alternative (or aberrant) transposition reaction generates a chromatid bridge that breaks at anaphase (English et al., 1993; Weil and Wessler, 1993). Dissection of the double *Ds* element and analysis in transgenic plants revealed that chromosome breakage can be induced by a pair of *Ds* 5′ and 3′ ends in direct orientation (the double *Ds* carries two pairs of such ends) (English et al., 1993; English et al., 1995).

8.4   Complex Elements: Alternative Transposition Events and Genome Rearrangements

Transposition of *Ac* and state II *Ds* elements changes only the position of the element in the genome. In contrast, state I *Ds* elements can elicit not only chromosome breakage, but can also induce gross chromosome rearrangements such as deletions, duplications, inversions, reciprocal translocations, and ring chromosomes (McClintock, 1950). Recent results have shown that these chromosomal rearrangements can be generated when closely-linked *Ac/Ds* termini undergo two different alternative transposition reactions:

8.4.1 Sister-Chromatid Transposition

In a standard *Ac* or *Ds* element (state II), the 5′ and 3′ termini at each end of the element are said to be in opposite orientation. In contrast, *doubleDs* and certain configurations of *Ac* and a nearby *fAc* element have their 5′ and 3′ termini arranged in direct orientation with respect to each other. When in direct orientation, transposition may involve an *Ac/Ds* 5′ end from one sister chromatid and an *Ac/Ds* 3′ end from the other sister chromatid in a reaction termed Sister Chromatid Transposition (SCT; Figure 7). Although the control of standard transposition vs. SCT is still unknown, the frequent occurrence of SCT with certain transposon configurations is consistent with a model for *Ac* transposase interaction with *Ac/Ds* termini that are hemimethylated on specific DNA strands (Wang et al., 1996). In SCT, excision of the *Ac/Ds* termini and subsequent joining of the sequences flanking the *Ac/Ds* termini can generate a chromatid bridge. Reinsertion of the excised transposon termini into the chromatid bridge can generate a deletion and a reciprocal inverted duplication (Zhang and Peterson, 1999), and independent events can produce a series of nested deletions and inverted duplications (Zhang and Peterson, 2005). If, during SCT, the excised *Ac/Ds* ends reinserted into a site in another chromosome, a translocation

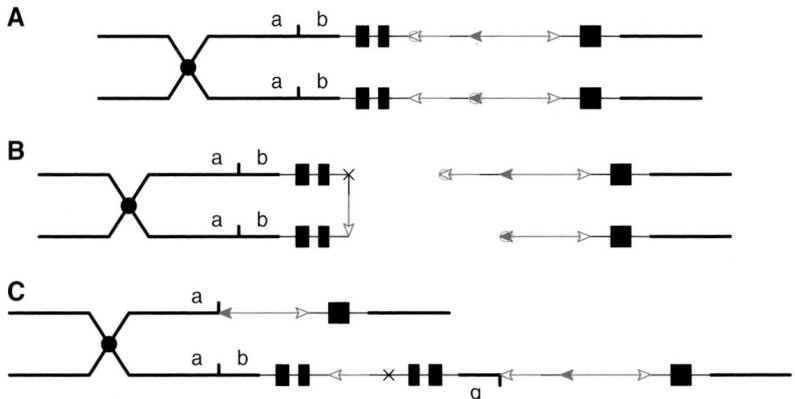

Fig. 7 Sister chromatid transposition model (modified from Figure 1 in Zhang and Peterson, 2005). The two lines indicate sister chromatids joined at the centromere (small black circle). Black boxes indicate *p1* exons; open and solid arrowheads indicate 3′ and 5′ *Ac* termini, respectively. Double-headed arrows indicate *Ac,* and single-headed arrows indicate *fracturedAc (fAc)*, a 2 kb 3′ part of *Ac*. **(A)** *Ac* transposase (small ovals) binds to the 5′ terminus of *Ac* in one sister chromatid and the 3′ terminus of *fAc*, a terminally-deleted copy of *Ac*, in the other sister chromatid. **(B)** Cuts are made at the *Ac* and *fAc* termini. The two nontransposon ends join together to generate a chromatid bridge, and minor sequence changes occur at the junction to form a transposon footprint indicated by *x*. **(C)** The excised transposon termini insert at the target site between *a* and *b* to generate one sister chromatid with an inverted duplication, and a second sister chromatid with a corresponding deficiency. To view an animation of the alternative transposition model, see http://jzhang.public.iastate.edu/Transposition.html

would be generated; such rearrangements were reported by McClintock from *Ds5245* and *Ds4864A* (McClintock, 1953).

8.4.2   Reversed-Ends Transposition

If two *Ac* or *Ds* elements are inserted near each other in the same orientation, the juxtaposed 5′ and 3′ ends of the different elements would be in reversed orientation. These reverse-oriented *Ac/Ds* termini on the same sister chromatid can serve as substrates in a reaction termed reversed-ends transposition (Figure 8).

After excision of the transposon termini, the sequences flanking the transposon ends may join together to form a circle. Reinsertion of the excised *Ac/Ds* termini into the circle generates a permutation of the original sequence termed a local rearrangement (Zhang and Peterson, 2004). Insertion of the excised *Ac/Ds* termini into other sites in the genome is predicted to generate a variety of rearrangements, including deletions, duplications, inversions, reciprocal translocations, and ring chromosomes. So far, we have identified local rearrangements, deletions, inversions (both paracentric and pericentric), and reciprocal translocations produced by reversed-ends transposition reactions at the maize *p1* locus (Zhang and Peterson, 2004). Similarly, Dooner and Weil (2007) describe the formation of deletions, inversions and local rearrangements at the *bz1* locus. These types of chromosomal rearrangements may shuffle coding and regulatory sequences to generate new genes or patterns of gene expression; we indeed isolated four new functional chimeric genes formed by fusion of the *p1* gene and a nearby paralog through reversed *Ac* ends transposition (Zhang et al., 2006).

8.4.3   Alternative Transposition During DNA Replication: Formation of Complex Alleles

Various studies have shown that standard *Ac/Ds* transposition occurs during DNA replication, and that excised *Ac/Ds* elements can reinsert into either replicated or unreplicated sites. If alternative transposition events (SCT or reversed-end transposition) also occur during DNA replication, a number of complex chromosomal rearrangements may be generated. Recent results from our lab suggest that alternative transposition during DNA replication does indeed generate complex structures. For example, we isolated a number of maize *p1* alleles, each of which has a large inverted duplication and a composite *Ac/Ds* element flanking one of the duplications (Zhang and Peterson, in preparation). The structures of these *p1* alleles resemble that of the *sh1-m5933* allele, which contains a complex 30kb insertion (Doring et al., 1989; Doring et al., 1990; Doring and Starlinger, 1984). These results reveal another possible route by which transposable elements may have contributed to the formation of the maize genome. Further research is required to determine the precise mechanism by which these complex alleles were generated.

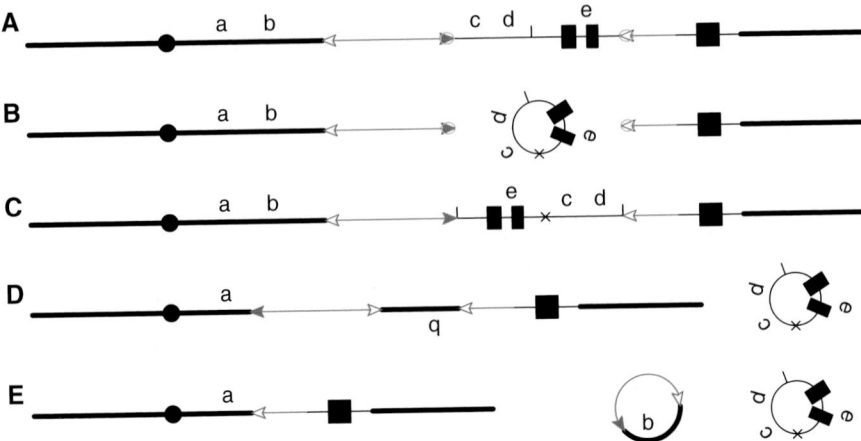

Fig. 8 Reversed-*Ac*-ends transposition model (modified from Figure 3 in Zhang and Peterson, 2004). Symbols have the same meaning as in Figure 6. (A) *Ac* transposase cleaves at the 5′ end of *Ac* and the 3′ end of *fAc*. (B) Following transposase cleavage at the junctions of *Ac/p1* and *fAc/p1*, the internal *p1* genomic sequences are joined to form a 13-kb circle. The "x" on the circle indicates the site where the joining occurred, marked by a transposon footprint. The *Ac* 5′ and *fAc* 3′ ends are competent for insertion anywhere in the genome. (C–E) The structures expected from insertion into two possible target sites. (C) The transposon ends insert into the 13-kb circle. The *Ac* 5′ end joins to the end adjacent to exon 1 and *fAc* 3′ end joins to the other end. The 13-kb sequence is rearranged (segment c–d and segment e exchanged positions). (D) The transposon ends insert into a site between *a* and *b*: the *Ac* 5′ end joins to the end adjacent to *a*, and the *fAc* 3′ end joins to the end adjacent to *b*. Segment *b* is inverted, and the 13-kb circle is lost. The resulting chromosome contains an inversion of sequences from the *Ac* 5′ end to the insertion site proximal to *b*. (E) The transposon ends insert into a site between *a* and *b*: the *fAc* 3′ end joins to the end adjacent to *a*, and the *Ac* 5′ end joins to the end adjacent to *b*. Segment *b* is circularized and presumed lost; the 13-kb circle is also lost. The resulting chromosome contains an interstitial deletion from the *Ac* 5′ end to the insertion site distal to *a*. To view an animation of the alternative transposition model, see http://jzhang.public.iastate.edu/Transposition.html

9   Summary: Remaining Questions and Opportunities for Future Research

Although nearly 60 years have elapsed since the discovery of *Ac/Ds* and *En/Spm*, we still have much to learn about the biochemistry of transposition of these elements. For example, how is the transposition machinery assembled? How do transposase and host factors interact at the transposon termini to excise the element? How does the excised transposon/transposase complex insert at a target site? For other transposon systems, these types of questions have been best answered using in vitro transposition assays (Kaufman and Rio, 1992; Zhou et al., 2004); unfortunately no such in vitro system has been demonstrated for *Ac/Ds* or *En/Spm* elements.

Transposable elements have been extremely valuable for applications such as gene tagging. Natural and modified *Ac/Ds* and *En/Spm* elements have been shown to transpose in maize, rice, Arabidopsis, and other plants. Recently, it was shown that *Ds* can transpose in zebrafish and human cells (Emelyanov et al., 2006). The potential use of these elements for gene tagging in additional plant and animal species could be improved if the elements were modified to achieve higher levels of reinsertion following element excision. In addition, if the *Ac* negative dosage effect and the tendency of *Ac* transposase to aggregate into non-functional complexes could be reduced or eliminated, transposition frequencies and gene tagging efficiencies may be significantly increased.

Recent results have shown that alternative *Ac/Ds* transposition can generate major genome rearrangements, such as deletions, duplications (both inverted and direct), inversions, and reciprocal translocations. Evidence suggests that other transposable elements, such as the *P* element in Drosophila (Gray et al., 1996; Preston et al., 1996), *impala* in fungus (Hua-Van et al., 2002), and *SleepingBeauty* in vertebrates (Geurts et al., 2006), can also undergo alternative transposition to generate chromosomal rearrangements. These types of gross chromosomal rearrangements are often associated with speciation. Because alternative transposition reactions produce characteristic footprint and target site duplication sequences, it may be possible to compare genome sequences of closely related species to identify chromosomal rearrangements generated by alternative transposition. If confirmed, these results would further demonstrate the major impact that transposable elements have in shaping eukaryotic genomes.

Acknowledgements The research and preparation of this manuscript was supported by NSF award 0450243 to T. Peterson and J. Zhang.

References

Anderson, E. G. (1948). On the Frequency and Transmitted Chromosome Alterations and Gene Mutations Induced by Atomic Bomb Radiations in Maize. Proceedings of the National Academy of Sciences of the United States of America *34*, 387–390.

Anderson, E. G., Longley, A. E., Li, C. H., and Retherford, K. L. (1949). Hereditary effects produced in maize by radiations from the bikini atomic bomb. I. Studies on seedlings and pollen of the exposed generation. Genetics *34*, 639–646.

Athma, P., Grotewold, E., and Peterson, T. (1992). Insertional mutagenesis of the maize P gene by intragenic transposition of Ac. Genetics *131*, 199–209.

Bancroft, I., Bhatt, A. M., Sjodin, C., Scofield, S., Jones, J. D., and Dean, C. (1992). Development of an efficient two-element transposon tagging system in Arabidopsis thaliana. Mol Gen Genet *233*, 449–461.

Becker, H. A., and Kunze, R. (1997). Maize Activator transposase has a bipartite DNA binding domain that recognizes subterminal sequences and the terminal inverted repeats. Mol Gen Genet *254*, 219–230.

Boehm, U., Heinlein, M., Behrens, U., and Kunze, R. (1995). One of three nuclear localization signals of maize Activator (Ac) transposase overlaps the DNA-binding domain. Plant J *7*, 441–451.

Brink, R. A., and Nilan, R. A. (1952). The relation between light variegated and medium variegated pericarp in maize. Genetics *37*, 519–544.

Calvi, B. R., Hong, T. J., Findley, S. D., and Gelbart, W. M. (1991). Evidence for a common evolutionary origin of inverted repeat transposons in Drosophila and plants: hobo, Activator, and Tam3. Cell *66*, 465–471.

Cardon, G. H., Frey, M., Saedler, H., and Gierl, A. (1991). Transposition of En/Spm in transgenic tobacco. Maydica *36*, 305–308.

Chen, J., Greenblatt, I. M., and Dellaporta, S. L. (1987). Transposition of Ac from the P locus of maize into unreplicated chromosomal sites. Genetics *117*, 109–116.

Chen, J., Greenblatt, I. M., and Dellaporta, S. L. (1992). Molecular analysis of Ac transposition and DNA replication. Genetics *130*, 665–676.

Chopra, S., Brendel, V., Zhang, J., Axtell, J. D., and Peterson, T. (1999). Molecular characterization of a mutable pigmentation phenotype and isolation of the first active transposable element from Sorghum bicolor. Proc Natl Acad Sci U S A *96*, 15330–15335.

Conrad, L. J., Bai, L., Ahern, K., Dusinberre, K., Kane, D. P., and Brutnell, T. P. (2007). State II Dissociation (Ds) Element Formation Following Activator (Ac) Excision in Maize. Genetics.

Cormack, J. B., Cox, D. F., and Peterson, P. A. (1988). Presence of the transposable element Uq . in maize breeding material. Crop science *28*, 941–944.

Courage-Tebbe, U., Doring, H. P., Fedoroff, N., and Starlinger, P. (1983). The controlling element Ds at the Shrunken locus in Zea mays: structure of the unstable sh-m5933 allele and several revertants. Cell *34*, 383–393.

Cuypers, H., Dash, S., Peterson, P. A., Saedler, H., and Gierl, A. (1988). The defective En-I102 element encodes a product reducing the mutability of the En/Spm transposable element system of Zea mays. EMBO J *7*, 2953–2960.

Dooner, H. K., and Belachew, A. (1991). Chromosome breakage by pairs of closely linked transposable elements of the Ac-Ds family in maize. Genetics *129*, 855–862.

Doring, H. P., Nelsen-Salz, B., Garber, R., and Tillmann, E. (1989). Double Ds elements are involved in specific chromosome breakage. Mol Gen Genet *219*, 299–305.

Doring, H. P., Pahl, I., and Durany, M. (1990). Chromosomal rearrangements caused by the aberrant transposition of double Ds elements are formed by Ds and adjacent non-Ds sequences. Mol Gen Genet *224*, 40–48.

Doring, H. P., and Starlinger, P. (1984). Barbara McClintock's controlling elements: now at the DNA level. Cell *39*, 253–259.

Doring, H. P., Tillmann, E., and Starlinger, P. (1984). DNA sequence of the maize transposable element Dissociation. Nature *307*, 127–130.

Dowe, M. F., Jr., Roman, G. W., and Klein, A. S. (1990). Excision and transposition of two Ds transposons from the bronze mutable 4 derivative 6856 allele of Zea mays L. Mol Gen Genet *221*, 475–485.

Emelyanov, A., Gao, Y., Naqvi, N. I., and Parinov, S. (2006). Trans-kingdom transposition of the maize dissociation element. Genetics *174*, 1095–1104.

Emerson, R. A. (1914). The Inheritance of a Recurring Somatic Variation in Variegated Ears of Maize. The American Naturalist *48*, 87–115.

English, J., Harrison, K., and Jones, J. D. (1993). A genetic analysis of DNA sequence requirements for Dissociation state I activity in tobacco. Plant Cell *5*, 501–514.

English, J. J., Harrison, K., and Jones, J. (1995). Aberrant Transpositions of Maize Double Ds-Like Elements Usually Involve Ds Ends on Sister Chromatids. Plant Cell *7*, 1235–1247.

Essers, L., Adolphs, R. H., and Kunze, R. (2000). A highly conserved domain of the maize activator transposase is involved in dimerization. Plant Cell *12*, 211–224.

Fedoroff, N., Wessler, S., and Shure, M. (1983). Isolation of the transposable maize controlling elements Ac and Ds. Cell *35*, 235–242.

Fedoroff, N. V., Furtek, D. B., and Nelson, O. E. (1984). Cloning of the bronze locus in maize by a simple and generalizable procedure using the transposable controlling element Activator (Ac). Proc Natl Acad Sci U S A *81*, 3825–3829.

Feldmar, S., and Kunze, R. (1991). The ORFa protein, the putative transposase of maize transposable element Ac, has a basic DNA binding domain. Embo J *10*, 4003–4010.

Fowler, R. G., and Peterson, P. A. (1978). An Altered State of a Specific En Regulatory Element Induced in a Maize Tiller. Genetics *90*, 761–782.

Frey, M., Reinecke, J., Grant, S., Saedler, H., and Gierl, A. (1990). Excision of the En/Spm transposable element of Zea mays requires two element-encoded proteins. Embo J *9*, 4037–4044.

Frey, M., Tavantzis, S. M., and Saedler, H. (1989). The maize En-1/Spm element transposes in potato. Mol Gen Genet *217*, 172–177.

Fu, H., and Dooner, H. K. (2002). Intraspecific violation of genetic colinearity and its implications in maize. Proc Natl Acad Sci U S A *99*, 9573–9578.

Geurts, A. M., Collier, L. S., Geurts, J. L., Oseth, L. L., Bell, M. L., Mu, D., Lucito, R., Godbout, S. A., Green, L. E., Lowe, S. W., *et al.* (2006). Gene mutations and genomic rearrangements in the mouse as a result of transposon mobilization from chromosomal concatemers. PLoS Genet *2*, e156.

Gierl, A., Lütticke, S., and Saedler, H. (1988). TnpA product encoded by the transposable element En-1 of Zea mays is a DNA binding protein. Embo J *7*, 4045–4053.

Giroux, M. J., Clancy, M., Baier, J., Ingham, L., McCarty, D., and Hannah, L. C. (1994). De novo synthesis of an intron by the maize transposable element Dissociation. Proc Natl Acad Sci U S A *91*, 12150–12154.

Grant, S. R., Gierl, A., and Saedler, H. (1990). En/Spm encoded tnpA protein requires a specific target sequence for suppression. EMBO J *9*, 2029–2035.

Grappin, P., Audeon, C., Chupeau, M. C., and Grandbastien, M. A. (1996). Molecular and functional characterization of Slide, an Ac-like autonomous transposable element from tobacco. Mol Gen Genet *252*, 386–397.

Gray, Y. H., Tanaka, M. M., and Sved, J. A. (1996). P-element-induced recombination in Drosophila melanogaster: hybrid element insertion. Genetics *144*, 1601–1610.

Greenblatt, I. M. (1966). Transposition and replication of modulator in maize. Genetics *53*, 361–369.

Greenblatt, I. M., and Brink, R. A. (1962). Twin Mutations in Medium Variegated Pericarp Maize. Genetics *47*, 489–501.

Greenblatt, I. M., and Brink, R. A. (1963). Transpositions of Modulator in maize into divided and undivided chromosome segments. Nature *197*, 412–413.

Grotewold, E., Athma, P., and Peterson, T. (1991). A possible hot spot for Ac insertion in the maize P gene. Mol Gen Genet *230*, 329–331.

Hall, L. N., Rossini, L., Cribb, L., and Langdale, J. A. (1998). GOLDEN 2: a novel transcriptional regulator of cellular differentiation in the maize leaf. Plant Cell *10*, 925–936.

Hartings, H., Spilmont, C., Lazzaroni, N., Rossi, V., Salamini, F., Thompson, R. D., and Motto, M. (1991). Molecular analysis of the Bg-rbg transposable element system of Zea mays L. Mol Gen Genet *227*, 91–96.

Heinlein, M., Brattig, T., and Kunze, R. (1994). In vivo aggregation of maize Activator (Ac) transposase in nuclei of maize endosperm and Petunia protoplasts. Plant J *5*, 705–714.

Henk, A. D., Warren, R. F., and Innes, R. W. (1999). A new Ac-like transposon of Arabidopsis is associated with a deletion of the RPS5 disease resistance gene. Genetics *151*, 1581–1589.

Hua-Van, A., Langin, T., and Daboussi, M. J. (2002). Aberrant transposition of a Tc1-mariner element, impala, in the fungus Fusarium oxysporum. Mol Genet Genomics *267*, 79–87.

Jones, R. N. (2005). McClintock's controlling elements: the full story. Cytogenet Genome Res *109*, 90–103.

Kaufman, P. D., and Rio, D. C. (1992). P element transposition in vitro proceeds by a cut-and-paste mechanism and uses GTP as a cofactor. Cell *69*, 27–39.

Klein, A. S., Clancy, M., Paje-Manalo, L., Furtek, D. B., Hannah, L. C., and Nelson, O. E., Jr. (1988). The mutation bronze-mutable 4 derivative 6856 in maize is caused by the insertion of a novel 6.7-kilobase pair transposon in the untranslated leader region of the bronze-1 gene. Genetics *120*, 779–790.

Kunze, R., and Starlinger, P. (1989). The putative transposase of transposable element Ac from Zea mays L. interacts with subterminal sequences of Ac. Embo J *8*, 3177–3185.

Kunze, R., Stochaj, U., Laufs, J., and Starlinger, P. (1987). Transcription of transposable element Activator (Ac) of Zea mays L. Embo J *6*, 1555–1563.

Kunze, R., and Weil, C. F. (2002). The hAT and CACTA superfamilies of plant transposons, In Mobile DNA II, N. L. Craig, R. Craigie, M. Gellert, and A. Lambowitz, eds. (ASM Press), pp. 565–610.

Lamkey, K. R., Peterson, P. A., and Hallauer, A. R. (1991). Frequency of the transposable element uq in iowa stiff stalk synthetic maize populations. Genetical research *57*, 1–9.

Lechelt, C., Peterson, T., Laird, A., Chen, J., Dellaporta, S. L., Dennis, E., Peacock, W. J., and Starlinger, P. (1989). Isolation and molecular analysis of the maize P locus. Mol Gen Genet *219*, 225–234.

Masson, P., Banks, J. A., and Fedoroff, N. (1991). Structure and function of the maize Spm transposable element. Biochimie *73*, 5–8.

Masson, P., Surosky, R., Kingsbury, J. A., and Fedoroff, N. V. (1987). Genetic and molecular analysis of the Spm-dependent a-m2 alleles of the maize a locus. Genetics *117*, 117–137.

McClintock, B. (1944). The relation of homozygous deficiencies to mutations and allelic series in maize. Genetics *29*, 478–502.

McClintock, B. (1946). Maize genetics. Carnegie Institution of Washington Year Book *45*, 176–186.

McClintock, B. (1947). Cytogenetic studies of maize and Neurospora. Carnegie Institution of Washington Year Book *46*, 146–152.

McClintock, B. (1948). Mutable loci in maize. Carnegie Institution of Washington Year Book *47*, 155–169.

McClintock, B. (1949). Mutable Loci in Maize. Carnegie Institution of Washington Year Book *48*, 142–154.

McClintock, B. (1950). The Origin and Behavior of Mutable Loci in Maize. Proceedings of the National Academy of Sciences of the United States of America *36*, 344–355.

McClintock, B. (1951a). Chromosome organization and genic expression. Cold Spring Harb Symp Quant Biol *16*, 13–47.

McClintock, B. (1951b). Mutable Loci in Maize. Carnegie Institute of Washington Yearbook *50*, 174–181.

McClintock, B. (1953). Induction of Instability at Selected Loci in Maize. Genetics *38*, 579–599.

McClintock, B. (1954). Mutations in maize and chromosomal aberrations in Neurospora. Carnegie Inst Wash Year Book *53*, 254–260.

Menssen, A., Hohmann, S., Martin, W., Schnable, P. S., Peterson, P. A., Saedler, H., and Gierl, A. (1990). The En/Spm transposable element of Zea mays contains splice sites at the termini generating a novel intron from a dSpm element in the A2 gene. Embo J *9*, 3051–3057.

Miura, A., Yonebayashi, S., Watanabe, K., Toyama, T., Shimada, H., and Kakutani, T. (2001). Mobilization of transposons by a mutation abolishing full DNA methylation in Arabidopsis. Nature *411*, 212–214.

Muszynski, M. G., Gierl, A., and Peterson, P. A. (1993). Genetic and molecular analysis of a three-component transposable-element system in maize. Mol Gen Genet *237*, 105–112.

Nowick, E. M., and Peterson, P. A. (1981). Transposition of the enhancer controlling element system in maize. Molecular Genetics and Genomics *183*, 440–448.

O'Reilly, C., Shepherd, N. S., Pereira, A., Schwarz-Sommer, Z., Bertram, I., Robertson, D. S., Peterson, P. A., and Saedler, H. (1985). Molecular cloning of the a1 locus of Zea mays using the transposable elements En and Mu1. The EMBO Journal *4*, 877–877.

Ozeki, Y., Davies, E., and Takeda, J. (1997). Somatic variation during long-term subculturing of plant cells caused by insertion of a transposable element in a phenylalanine ammonia-lyase (PAL) gene. Mol Gen Genet *254*, 407–416.

Paz-Ares, J., Wienand, U., Peterson, P. A., and Saedler, H. (1986). Molecular cloning of the c locus of Zea mays: a locus regulating the anthocyanin pathway. The EMBO Journal *5*, 829–829.

Pereira, A., Cuypers, H., Gierl, A., Schwarz-Sommer, Z., and Saedler, H. (1986). Molecular analysis of the En/Spm transposable element system of Zea mays. Embo J *5*, 835–841.

Pereira, A., Schwarz-Sommer, Z., Gierl, A., Bertram, I., Peterson, P. A., and Saedler, H. (1985). Genetic and molecular analysis of the Enhancer (En) transposable element system of Zea mays. Embo J *4*, 17–23.

Peterson, P. A. (1953). A mutable pale green locus in maize. Genetics *38*, 682–683.

Peterson, P. A. (1956). An a1 mutable arising in pgm stocks. Maize Co-op News L *30*, 82.

Peterson, P. A. (1961). Mutable a1 of the En system in maize. Genetics *46*, 759–771.

Peterson, P. A. (1965). A Relationship between the Spm and En Control Systems in Maize. The American Naturalist *99*, 391–398.

Peterson, P. A. (1977). The position hypothesis for controlling elements in maize, In DNA Insertion Elements, Plasmids and Episomes. (Cold Spring Harbor Laboratory Press), pp. 429–435.

Peterson, P. A. (1978). Controlling elements: the induction of mutability at the A2 and C loci in maize, In Maize breeding and genetics, D. B. Walden, ed. (New York: Wiley and Sons), pp. 601–631.

Peterson, P. A. (1986). Mobile elements in maize. Plant Breed Rev *4*, 122–122.

Peterson, P.A. (1987). Mobile Elements in Plants. CRC Critical Reviews in Plant Sciences *6*, 105–208.

Peterson, P. A. (1991). The transposable element-En-four decades after Bikini. Genetica *84*, 63–72.

Peterson, P. A. (1995). Genetic Analysis of the Functions of the Transposable Element En in Zea mays: Limited Transposase Elicits a Differential Response on Reporter Alleles. Genetics *141*, 1135–1145.

Peterson, P. A., and Salamini, F. (1986). A search for active mobile elements in the Iowa Stiff Stalk Synthetic maize population and some derivatives. Maydica *31*, 163–172.

Peterson, T. (1990). Intragenic transposition of Ac generates a new allele of the maize P gene. Genetics *126*, 469–476.

Preston, C. R., Sved, J. A., and Engels, W. R. (1996). Flanking duplications and deletions associated with P-induced male recombination in Drosophila. Genetics *144*, 1623–1638.

Ralston, E., English, J., and Dooner, H. K. (1989). Chromosome-breaking structure in maize involving a fractured Ac element. Proc Natl Acad Sci U S A *86*, 9451–9455.

Rhoades, M. M. (1938). Effect of the Dt gene on the mutability of the a1 allele in maize. Genetics *23*, 377–397.

Rhodes, P. R., and Vodkin, L. O. (1988). Organization of the Tgm family of transposable elements in soybean. Genetics *120*, 597–604.

Rinehart, T. A., Dean, C., and Weil, C. F. (1997). Comparative analysis of non-random DNA repair following Ac transposon excision in maize and Arabidopsis. Plant J *12*, 1419–1427.

Rubin, E., and Levy, A. A. (1997). Abortive gap repair: underlying mechanism for Ds element formation. Mol Cell Biol *17*, 6294–6302.

Rubin, E., Lithwick, G., and Levy, A. A. (2001). Structure and evolution of the hAT transposon superfamily. Genetics *158*, 949–957.

Saedler, H., Bonas, U., Gierl, A., Harrison, B. J., Klosgen, R. B., Krebbers, E., Nevers, P., and Peterson, P. A. (1984). The Plant Transposable Elements TarnI, Tarn2 and Spm-18. The Impact of Gene Transfer Techniques in Eucaryotic Cell Biology.

SanMiguel P., Gaut B.S., Tikhonov A., Nakajima Y., Bennetzen J.L. (1998) The paleontology of intergene retrotransposons of maize. Nat Genet. *20*, 43–45.

Schiefelbein, J. W., Raboy, V., Kim, H. Y., and Nelson, O. E. (1988). Molecular characterization of suppressor-mutator (Spm)-induced mutations at the bronze-1 locus in maize: the bz-m13 alleles. Basic Life Sci *47*, 261–278.

Schwarz-Sommer, Z., Gierl, A., Klösgen, R.B, Wienand, U., Peterson, P. A., and Saedler, H. (1984). The Spm (En) transposable element controls the excision of a 2-kb DNA insert at the wxm-8 allele of Zea mays. EMBO J *3*, 1021–1028.

Schwarz-Sommer, Z., Shepherd, N., Tacke, E., Gierl, A., Rohde, W., Leclercq, L., Mattes, M., Berndtgen, R., Peterson, P. A., and Saedler, H. (1987). Influence of transposable elements on the structure and function of the A1 gene of Zea mays. EMBO J *6*, 287–294.

Scott, L., LaFoe, D., and Weil, C. F. (1996). Adjacent sequences influence DNA repair accompanying transposon excision in maize. Genetics *142*, 237–246.

Snowden, K. C., and Napoli, C. A. (1998). Psl: a novel Spm-like transposable element from Petunia hybrida. Plant J *14*, 43–54.

Song, R., and Messing, J. (2003). Gene expression of a gene family in maize based on noncollinear haplotypes. Proc Natl Acad Sci U S A *100*, 9055–9060.

Tacke, E., Schwarz-Sommer, Z., Peterson P., and Saedler, H. (1986). Molecular analysis of states of the A locus of Zea mays. Maydica *31*, 83–91.

Thatiparthi, V. R., Dinesh-Kumar, S. P., and Peterson, P. A. (1995). Permanent Fixation of a Transposable Element Insert in the A2 Gene of Maize (Zea mays L.). Journal of Heredity *86*, 167–167.

Tsay, Y. F., Frank, M. J., Page, T., Dean, C., and Crawford, N. M. (1993). Identification of a mobile endogenous transposon in Arabidopsis thaliana. Science *260*, 342–344.

Wang, L., Heinlein, M., and Kunze, R. (1996). Methylation pattern of Activator transposase binding sites in maize endosperm. Plant Cell *8*, 747–758.

Warren, W. D., Atkinson, P. W., and O'Brochta, D. A. (1994). The Hermes transposable element from the house fly, Musca domestica, is a short inverted repeat-type element of the hobo, Ac, and Tam3 (hAT) element family. Genet Res *64*, 87–97.

Weil, C. F., and Kunze, R. (2000). Transposition of maize Ac/Ds transposable elements in the yeast Saccharomyces cerevisiae. Nat Genet *26*, 187–190.

Weil, C. F., and Wessler, S. R. (1993). Molecular evidence that chromosome breakage by Ds elements is caused by aberrant transposition. Plant Cell *5*, 515–522.

Wessler, S. R., Baran, G., and Varagona, M. (1987). The maize transposable element Ds is spliced from RNA. Science *237*, 916–918.

Wienand, U., Weydemann, U., Niesbach-Kl<chi>?</chi>gen, U., Peterson, P. A., and Saedler, H. (1986). Molecular cloning of the c2 locus of Zea mays, the gene coding for chalcone synthase. Molecular Genetics and Genomics *203*, 202–207.

Wirtz, U., Osborne, B., and Baker, B. (1997). Ds excision from extrachromosomal geminivirus vector DNA is coupled to vector DNA replication in maize. Plant J *11*, 125–135.

Xiao, W., Brown, R. C., Lemmon, B. E., Harada, J. J., Goldberg, R. B., and Fischer, R. L. (2006). Regulation of Seed Size by Hypomethylation of Maternal and Paternal Genomes. Plant Physiol.

Xu, Z., and Dooner, H. K. (2005). Mx-rMx, a family of interacting transposons in the growing hAT superfamily of maize. Plant Cell *17*, 375–388.

Yan, X., Martinez-Ferez, I. M., Kavchok, S., and Dooner, H. K. (1999). Origination of Ds elements from Ac elements in maize: evidence for rare repair synthesis at the site of Ac excision. Genetics *152*, 1733–1740.

Zhang, J., and Peterson, T. (1999). Genome rearrangements by nonlinear transposons in maize. Genetics *153*, 1403–1410.

Zhang, J., and Peterson, T. (2004). Transposition of reversed Ac element ends generates chromosome rearrangements in maize. Genetics *167*, 1929–1937.

Zhang, J., and Peterson, T. (2005). A segmental deletion series generated by sister-chromatid transposition of Ac transposable elements in maize. Genetics *171*, 333–344.

Zhang, J., Zhang, F., and Peterson, T. (2006). Transposition of reversed Ac element ends generates novel chimeric genes in maize. PLoS Genet *2*, e164.

Zhou, L., Mitra, R., Atkinson, P. W., Hickman, A. B., Dyda, F., and Craig, N. L. (2004). Transposition of hAT elements links transposable elements and V(D)J recombination. Nature *432*, 995–1001.

Mutator and MULE transposons

Damon Lisch and Ning Jiang

Abstract Because it is highly mutagenic and relatively non-specific, the *Mutator* system of transposons has proved to be an extraordinarily useful tool for maize geneticists. It has also proved to be a valuable model system for understanding the basic biology of transposons in higher eukaryotes, particularly the means by which transposons are epigenetically silenced by their hosts. Further, the wide distribution and remarkable variety of *Mu*-like elements (MULEs) among plant species has illuminated the role that transposons can play in the evolution of genomes. This chapter will provide an overview of the biology and evolution of this highly active and diverse family of transposable elements.

1   Introduction

In 1978 Donald Robertson described a line of maize with an extraordinarily high forward mutation rate, as much as 100 times that of background (Robertson, 1978, 1983). Robertson found that when plants from this line were self fertilized, they gave rise to a wide variety of mutants, many of which exhibited characteristic late somatic reversion patterns suggestive of transposon excision. However, unlike the classically defined two element systems such as *Ac/Ds* or *Spm/dSpm*, what he called "mutator" activity did not transmit in any obvious Mendelian pattern. On outcrossing, 90% of the progeny of a mutator plant exhibited the mutator trait (Robertson, 1985). He also found clear evidence of epigenetic silencing of the entire system; when a mutator line was self fertilized for several generations, the mutator activity was lost, suggesting that although the regulatory, or autonomous, elements were there, they must have

D. Lisch
Department of Plant and Microbial Biology, 111 Koshland Hall, U.C. Berkeley,
Berkeley, California 94720
dlisch@berkeley.edu

N. Jiang
Department of Horticulture, Michigan State University, East Lansing, Michigan 48824
jiangn@msu.edu

J.L. Bennetzen and S. Hake (eds.), *Maize Handbook - Volume II: Genetics and Genomics*,   277
© Springer Science+Business Media LLC 2009

become inactive (Robertson, 1986). Non-Mendelian genetics and spontaneous epigenetic silencing were for many years characteristic features of mutator lines.

The mutations in Robertsons lines are caused by a novel class of transposons, called *Mutator,* or *Mu* elements (Strommer et al., 1982; Bennetzen, 1984; Taylor and Walbot, 1987) and lines carrying large numbers of active copies of these element are now referred to as "*Mutator* lines". These elements can reach much higher copy numbers than classically defined maize elements because they can transpose at frequencies averaging 100% (Alleman and Freeling, 1986; Walbot and Warren, 1988), making them among the most virulent transposons yet described. This has made the *Mutator* system an extraordinarily useful tool for gene tagging; the expected frequency of insertion into any given gene can range from roughly 10^{-4} to 10^{-5} per locus per generation (Bennetzen et al., 1993). *Mutator* has also proved to be a particularly useful model for understanding epigenetic regulation in maize, and for understanding the evolution of new transposon families. This review will provide an overview of the *Mutator* system in maize: it's genetics, evolution and regulation.

2   Classes of *Mu* Elements in Maize

2.1   The Non-Autonomous Elements

Transposons that encode the proteins necessary for their own transposition are referred to as autonomous; those that require the presence of an autonomous element to transpose are non-autonomous. Non-autonomous *Mu* elements all carry similar, ~220 bp terminal inverted repeats (TIRs), but each class of element carries a distinct captured, or transduplicated, maize genic region between those TIRs (Chandler et al., 1986; Lisch, 2002). Transduplication is a characteristic feature of non-autonomous *Mu-like* elements (MULEs) in a number of other species in addition to maize and it suggests a generic propensity for MULEs to capture and mobilize fragments of host genes (Jiang et al., 2004; Holligan et al., 2006) (See section 6). There are currently six classes of these elements fully characterized in maize (Figure 1). Those that are known to be transpositionally active are *Mu1* (Strommer et al., 1982; Bennetzen, 1984), *Mu1.7* (Taylor and Walbot, 1987), *Mu3* (Oishi and Freeling, 1983), *Mu4* (Talbert et al., 1989; Dietrich et al., 2002), *Mu7/rcy* (Schnable et al., 1989) and *Mu8* (Fleenor et al., 1990). Each of these has been shown to transpose and is fully sequenced. In addition, there are at least four other active elements that have only been partially sequenced: *Mu10, Mu11, Mu12* (Dietrich et al., 2002) and *Mu13* (M. Robbins, personal communication). Finally, an artificial *Mu* element was constructed, which is composed of a complete *Mu1* element into which was added a bacterial plasmid that carries a selectable marker, an origin of replication and a unique sequence tag. This element, called *RescueMu*, has facilitated the sequencing of large portions of maize gene space as well as detailed analysis of *Mu* element transposition behavior (Raizada et al., 2001a). None of the non-autonomous elements

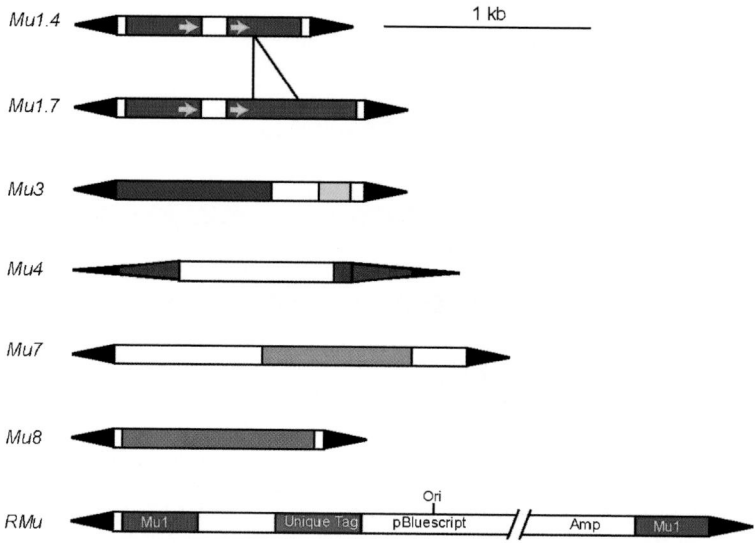

Fig. 1 The structural organization of various non-autonomous elements in maize. TIRs are depicted as black arrowheads. The colored boxes represent various fragments of host genes captured by these elements (Lisch, 2002). *RMu* refers to the artificially constructed *RescueMu* element, which includes all of *Mu1.4*, along with a plasmid that contains a unique tag, a selectable marker and an origin of replication (Raizada et al., 2001a)

are present in the sorghum genome (Lisch, unpublished data), and some are as much as 98% identical in portions to their cognate host genes, suggesting that non-autonomous elements in maize are likely to have arisen in the relatively recent past.

2.2 *MuDR*

The *Mutator* system is regulated by the *MuDR* class of elements (Chomet et al., 1991; Hershberger et al., 1991; Qin et al., 1991; James et al., 1993). *MuDR* elements encode two genes, *mudrA,* the putative transposase, and *mudrB,* a helper gene of that is required for new insertions (Hershberger et al., 1995; Lisch et al., 1999; Raizada and Walbot, 2000; Woodhouse et al., 2006a) (Figure 2). These genes encode two proteins, MURA and MURB respectively. Homologs of the *mudrA* gene are present in a wide variety of plants (Lisch, 2002), animals (Makarova et al., 2002), fungi (Chalvet et al., 2003) and bacteria (Eisen et al., 1994), suggesting an ancient origin for this transposase. In contrast, *mudrB* appears to be restricted to the genus *Zea* (Lisch et al., 2001). MULEs in other organisms carry *mudrA* ortholog but all lack *mudrB*. Also present in maize are *hMuDR* elements, orthlogs of *MuDR* that lack some diagnostic restriction sites but that can be as much as 98.9 % identical in segments to *MuDR* (Rudenko and Walbot, 2001). Although some of these elements have the potential to express both *mudrA* and *mudrB,* they do so only at very low levels

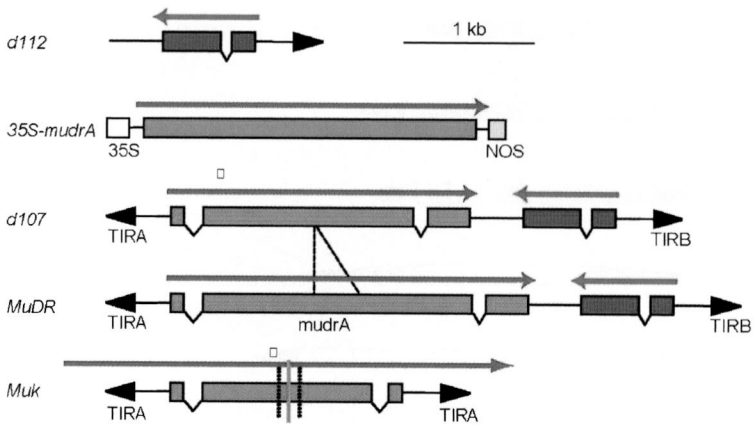

Fig. 2 *MuDR* and various derivatives of that element. TIRs are depicted as black arrowheads. *mudrA* and *mudrB* ORFs are depicted as blue boxes and green boxes respectively. The arrows indicate the direction of transcription. Deletion derivatives *d112* and *d107* are missing the indicated regions of *MuDR*. *Muk* represents a derivative of *MuDR* in which roughly half of the element (the 5' end) has been duplicated and inverted relative to itself. 35S-*mudrA* contains the *mudrA* cDNA flanked by the CaMV 35S promoter and an nopaline synthetase (nos) terminator (Raizada and Walbot, 2000)

(Hershberger et al., 1995), and the transcript that they do produce appears to be localized primarily in the nucleus (Rudenko and Walbot, 2001). As *Mutator* activity appears to be exclusively associated with *MuDR*, *hMuDR* elements do not appear to have a role in activity, however, as we will see, they may play a role in reinforcing epigenetic silencing of *MuDR* elements.

The maize genome also hosts at least three other classes of MULEs: *Jittery*, which has been shown to excise from a reporter (Xu et al., 2004), TRAP (Comelli et al., 1999) and TAFT (Wang and Dooner, 2006), a member of the KI-MULE family (Hoen et al., 2006) that specifically targets TA satellite sequences in maize (Figure 3). All of these elements have homologs in rice that are more similar to them than is *MuDR*, and none of these element encodes a *mudrB* gene (Lisch, 2002). Thus, the maize genome is host to a number of active or potentially active *MuDR* relatives that diverged before the divergence of maize and rice. The elements do not appear to cross-mobilize, as *Mu* element activity in maize is strictly dependent on *MuDR*.

3   *Mu* Element Behavior

3.1   *Mechanism of Transposition*

Excisions of *Mu* elements in maize requires the presence of an intact *mudrA* gene (Lisch et al., 1999; Raizada et al., 2001a). Isolation of the MURA protein made it possible to demonstrate that this protein interacts with a specific 32 bp conserved

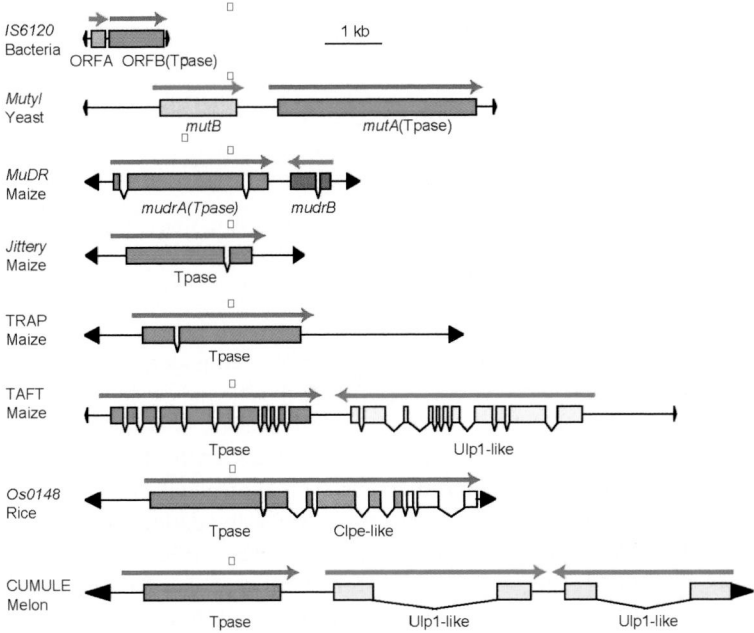

Fig. 3 Structural diversity of autonomous MULEs from bacteria, fungi, and plants. MULE TIRs are shown as black arrowheads. Exons are depicted as colored boxes and introns as the lines connecting exons; other sequences are shown as horizontal lines. Long coloured arrows indicate the orientation of transcription. Gene predictions were based on the annotation provided with the relevant submission. The accession numbers for each element or ORF are: M69182 (IS6120), AJ621543 (*Mutyl*), U14597 and U14598 (*MuDR*), AF247646 (*Jittery*), AJ238507 (TRAP), AF466931 (TAFT), AC090683 (Os0148), AY524004 (CUMULE). "Ulp" stands for "ubiquitin-like protease" and "Clpe" for "calcineurin-like phosphoesterase"

region of *Mu* TIRs (Benito and Walbot, 1997) (Figure 4). This region is distinct from two previously identified binding sites within the *Mu1* TIRs, one of which is bound only in *Mu* active plants and the other of which is bound in both inactive and active plants (Zhao and Sundaresan, 1991). MURA binding is to both methylated and unmethylated forms of this sequence, and may bind as a multimer of MURA. Presumably, MURA binds to *Mu* TIRs, brings them together and catalyzes a double-stranded break. In maize, the available data suggest that new insertions require MURB (Lisch et al., 1999; Raizada et al., 2001a; Woodhouse et al., 2006b). It is not known how this protein facilitates reintegration, but deletions, transgenes, or epigenetic variants that lack MURB lack new insertions as well. Interestingly, unlike MURA, MURB lacks a nuclear localization signal (Ono et al., 2002). Although immunolocalization experiments suggest that it may be localized to the nucleus (Donlin et al., 1995), localization of this protein in transgenic onion epidermal cells is primarily in the cytoplasm, and MURB does not appear to interact with MURA (Ono et al., 2002). Thus, the role of MURB in the transposition reaction remains largely mysterious.

The consequences of *Mutator* activity vary dramatically depending on whether or not that activity is in a germinal or terminally differentiated tissue. Although the

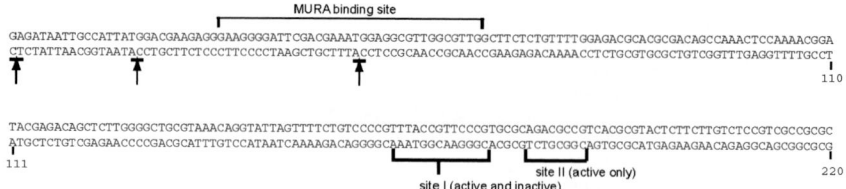

Fig. 4 The sequence of the *Mu1* left TIR. Binding sites for MURA, as well as two other previously identified binding sites are as indicated. Transcript initiation sites identified in the *hcf106::Mu1* allele are indicated by arrows. Adapted from (Benito and Walbot, 1997)

Mutator system shows a very high frequency of new germinal insertions, germinal reversion events are exceedingly rare, less than 10^{-4} (Schnable and Peterson, 1988; Brown et al., 1989; Levy et al., 1989; Levy and Walbot, 1991; Walbot, 1991; Bennetzen, 1996). In contrast, somatic reversions are frequent, and characteristically late (Figure 5). Although germinally transmitted insertions can occur throughout development (Lisch et al., 1995), they most frequently occur just prior to, during or after meiosis (Robertson, 1980; Robertson, 1981; Robertson and Stinard, 1993). Analyses of lines carrying a single *MuDR* element and a single *Mu1* element reveal that germinal insertions are not associated with the loss of the donor element, and they are most often into genetically unlinked sites (Lisch et al., 1995), a propensity that distinguished *Mu* from elements such as *Ac* and *Spm*. Insertion of *Mu* elements into new sites is associated with the production of a 9 bp direct repeat, a diagnostic repeat length for MULEs in a number of organisms (Le et al., 2000; Turcotte et al., 2001).

Somatic events are associated with excision of the element and reversion of the target allele. In some cases these reversion events can be frequent enough such that most of the tissue appears revertant (Figure 5). Examination of somatic excision of *Mu* elements from cell autonomous markers suggest that the majority (90%) of reversions occur in single cells, and nearly all reversion events occur in the last three or four cell divisions (McCarty et al., 1989; Raizada et al., 2001a), although it should be noted that there appears to be variation in the size of revertant sectors depending on the allele being examined (Schnable et al., 1989). Analyses of somatic excisions are consistent with a "cut and paste" mechanism that results in a variety of footprints, which includes small deletions and the insertion of "filler" DNA (Figure 6). In contrast with germinal insertions, direct observation of *RescueMu* insertions in somatic cells using *in situ* hybridization reveals that new insertions are associated with the loss of the donor element (Yu et al., 2007).

Two models have been put forward to explain the available data (Raizada et al., 2001a; Lisch, 2002). One model posits a shift in the mode of transposition, from a "cut and paste" mechanism in somatic cells, to a duplicative mechanism in germinal tissues. In support of this model is the observation that the alteration of a single amino acid in the *Tn7* transposon can cause a shift from a "cut and paste" mode of transposition to true replicative transposition (May and Craig, 1996). This model would predict an alteration of either MURA or MURB in germinal versus somatic tissues, but no such alteration has been observed. It also would predict that efficient double-stranded gap repair would not be expected to occur in the germ line.

Fig. 5 Representative phenotypes associated with *Mutator* activity. In each case *MuDR* is driving excision of a *Mu1* element from *a1-mum2*, resulting in spots of color. A) The kernel on the left is exhibiting a typical pattern of late and frequent somatic reversion events. A similar frequency of events can be observed in any tissue with the correct color factors. The kernel on the right has the same genetic constitution as the one on the left, but it has a duplicate copy of the same *MuDR* element at a weak position (Lisch et al., 1995). B) A somatic sector in which the *MuDR* element at the weak position has transposed during the development of an ear to a more typical, high activity position. This sector reveals that although duplications of *Mu* elements are often late, they can also occur earlier. C) Kernels in which multiple *MuDR* elements are becoming epigenetically silenced during development. Note the patches within the kernels displaying various levels of activity

An alternative model suggests that the difference has to do with variations in the means by which double-stranded gaps introduced by excision of *Mu* elements is resolved (Donlin et al., 1995; Hsia and Schnable, 1996). In the germ line, these gaps are repaired primarily using the sister chromatid. Since the sister chromatid contains an un-excised copy of the element, complete repair using it as a template would result in restoration of the element at its original position. Incomplete, or interrupted repair would be expected to result in deletions whose break-points would be at short direct repeats, and this is what is observed (Hsia and Schnable, 1996). In a single-copy *MuDR* line with a reporter that can monitor the loss of functional transposase, these deletion events can be observed at various points during plant development (Lisch et al., 1995) (Figure 6). In somatic lineages, particularly late during development, it is hypothesized that the repair pathway switches to one in which double-stranded breaks are resolved primarily via non-homologous end joining (NHEJ). According to this model, the characteristically late and relatively uniform reversions characteristic of *Mutator* are a function of a shift in repair pathways, an hypothesis that is supported by the observation that 90% of reversions can occur during or after the last cell division when the sister chromatid may no longer be available for use as a template (Raizada et al., 2001a). This model is also supported by analysis of mutations in the redundant maize genes *Rad51a1* and *Rad51a2*, which play a central role in homologous recombination. The frequency of deletions of a *MuDR* element or sequences flanking it in the *A1* gene in *rad51a/b* double mutant is at least forty times as high as wild type, and the breakpoints of these deletions are consistent with NHEJ (P. Schnable, personal communication). These data strongly suggest that *Mu* elements are making breaks in the germline, and in the absence of RAD51a/b, these breaks are repaired using NHEJ instead of homologous repair using the sister chromatid. This hypothesis is also supported by the observation that *Mutator* activity increases the rate of homologous recombination at the site of a *Mu* insertion in the *A1* gene (Yandeau-Nelson et al., 2005). It is also interesting to note that *Rad51* expression is higher in rapidly dividing mitotic and meiotic tissues than in differentiated tissues (Franklin et al., 1999). To summarize, the available data supports a model for developmentally regulated shifts in repair of double-stranded gaps introduced by *Mu* excision.

3.2   Target Site Preferences

Mutator has been used to mutagenize large numbers of genes and to clone genic regions by cloning and sequencing flanking sequences in an undirected manner. Thus, a considerable body of data has been accumulated with respect to target site specificity. The most obvious preference *Mu* elements use is for genic, or low copy sequences. The maize genome is roughly 70% retrotransposon (SanMiguel et al., 1996a) and other repetitive sequences, and yet a large-scale analysis of *RescueMu* insertions revealed that only 8% of the 14,265 *RescueMu* insertions isolated were into repetitive sequences, and roughly half were into previously identified expressed sequences (Fernandes et al., 2004). This data is in agreement with more limited analysis of natural *Mu* insertions (Cresse et al., 1995; Hanley et al., 2000). The insertions were into all ten maize chromosomes and into a wide variety of genes. It is not known how it is that *Mu* insertions specifically target genic regions; these regions are distinct with respect to chromatin configuration, DNA methylation and recombinational activity (Bennetzen, 2000), each of which could play a role in *Mu* element targeting. The *RescueMu* elements also showed

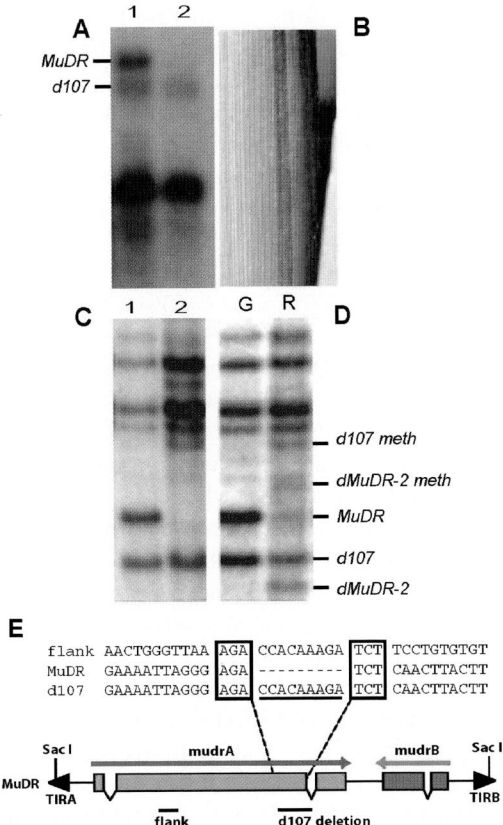

Fig. 6 Dynamic changes in *MuDR* structure and *Mu* element methylation during plant development. The *a1-mum2* allele is suppressible, so the loss of *Mutator* activity leads to expression of the *a1-mum2* allele, resulting in red color in plant tissues (Chomet et al., 1991). **A)** Northern blot analysis of a plant that carried both *d107* and a full length *MuDR* element (lane 1), and a sibling that carried only a deletion derivative of *MuDR*, *d107* (lane 2). Note that the *d107* element by itself or in combination with the full length element expresses *mudrB* and a truncated *mudrA* transcript. **B)** A red sector of sheath tissue in which *MuDR* activity has been lost and the *Mu1* element at *a1-mum2* is methylated (Lisch et al., 1995). This plant carried a full length *MuDR* element and *d107*. **C)** *Sac*I digestion of DNA from a plant with both *MuDR* and *d107* (lane 1) and with only *d107* (lane 2). *Sac*I is a methyl-sensitive enzyme that recognizes sites in the ends of *MuDR* elements (Martienssen and Baron, 1994). The expected sizes for fully digested *MuDR* and *d107* elements are as indicated. Methylation of sites within the ends of *d107* in the absence of *MuDR* (lane 2), results in a larger fragment, as indicated. Note that the partial methylation of *d107* does not result in the loss of *d107* transcript (panel A). **D)** *Sac*I digestion of DNA from a green *MuDR* active part of the sheath and an adjacent inactive red sector. Note that in the sector there is a new, smaller derivative of *MuDR* in place of the full length element, and that both *d107* and the new deletion derivative, *dMuDR-2* are both methylated. **E)** Sequence analysis of *d107*, which is missing the sequences in *MuDR* indicated by the dotted lines. In place of those sequences is a short sequence from a flanking region of *MuDR* (flank). The inserted sequence has short, three base-pair sequences that are present on either side of the deletion in *d107*. This arrangement of sequences suggest an aberrant gap repair pathway in which short stretches of homology mediated replacement of the missing sequences with a short stretch of nearby sequences. Experimental protocols are those as described in (Lisch et al., 1995)

a bias against genes transcribed using polymerase I and III, which suggests that MURA or MURB may specifically interact with some transcription factors and not others. There is certainly precedent for transposons targeting genes transcribed by specific polymerases. In yeast, *Ty3* is known to target genes transcribed by polymerase III due to a specific interaction with TFIIIB and TFIIIC. Conversely, *Ty5* is specifically targeted to heterochromatin due to an interaction between a targeting domain in the *Ty5* integrase and structural components of silenced heterochromatin (reviewed by (Lesage and Todeschini, 2005)).

Surveys of both *RescueMu* and natural *Mu* element 9 bp target site duplications show a very weak consensus of CTC(T/CG)(G/C)(A/C)(G/A)(A/G)C (Cresse et al., 1995; Hanley et al., 2000; Dietrich et al., 2002; Fernandes et al., 2004). Within sequences immediately flanking the 9 bp TSD, there is also a consensus of CCT-(TSD)-AGG. Given the degree of variation between individual insertion sites, the preference appears to be for structural features, rather than for specific DNA sequences. This is supported by the observation that the TSDs tend to occur at transitions from high to low GC content, and have distinct profiles with respect to GC content, bendability, B-DNA twist, α-philicity and protein-induced deformability relative to flanking sequences (Dietrich et al., 2002). There also hotspots for *Mu* element insertion. In a site-selected mutagenesis experiment, three independent insertions of *Mu1* were into the same site in the *hcf106* gene (Das and Martienssen, 1995). Similarly, in a survey of 79 independent *Mu* insertions into the *glossy 8* gene, roughly 80% were into the 5′ UTR, and 69% were into a roughly 60 nucleotide interval. A single position accounted for 19% of the insertions (Dietrich et al., 2002). These results were consistent with early analysis of insertions into other genes (reviewed by (Bennetzen et al., 1993)). Finally, there is some evidence that different *Mu* elements may have different preferences. In one experiment, although *Bz1* was exclusively targeted by *Mu1*, the closely linked *Sh1* gene was primarily targeted by *MuDR* (Hardeman and Chandler, 1989, 1993). Since *Mu1* and *MuDR* are mobilized by the same transposase, this would imply the target specificity is determined by both the transposase and element, which possibly form a unique complex during transposition.

3.3   Suppressible Alleles

Suppressibility is a feature of several *Mu*-induced mutations. The phenotype of suppressible alleles is dependent on the activity state of the transposon system. Examples of *Mu*-suppressible alleles include *Les28* (Martienssen and Baron, 1994), *hcf106::Mu1* (Martienssen et al., 1990), and *a1-mum2* (Chomet et al., 1991; Pooma et al., 2002), as well as mutant alleles of *rs1*, *lg3* (Girard and Freeling, 2000), *kn1* (Lowe et al., 1992; Greene et al., 1994) and *rf2a* (Cui et al., 2003). In each case the phenotype of the mutant allele returns to that of the progenitor allele when *Mu* activity is lost. Thus ectopic expression of dominant *Mu*-suppressible revertant alleles such as *Lg-0r422* and *Rs1-Or11* is restored when *Mu* activity is lost, as is normal expression of recessive alleles such as *hsf106::Mu1* and *a1-mum2*. Suppressible alleles can arise from insertions into promoters, 5′ UTRs, 3′ UTRs or introns. It is not known what the mechanism of suppression is, but a parsimonious general explanation for this phenomenon is that the MURA transposase binds to the insertion and sterically blocks binding of proteins in or near *Mu* TIRs. For example, it is known that expression of the *Hcf106::Mu1* allele is a consequence of an outward-reading promoter within the *Mu1* element inserted into the normal promoter of the *hcf106* gene (Barkan and Martienssen, 1991). Binding of MURA to the *Mu1* TIR, where the outward–reading promoter is located (Figure 4), would be expected to block the binding of transcription factors. In contrast, the evidence suggests that the large second intron of *kn1* contains a negative regulatory element that prevents expression of *kn1* in the leaves. An insertion of a *Mu* element into this region does not by itself affect this negative regulation, but the presence of MURA does, presumably because it binds to the *Mu* TIRs. *Mu* suppression can also affect the ectopic polyadenylation introduced by a *Mu7* insertion into the 3′ UTR of an *rf2a* mutant allele (Cui et al., 2003). The *Mu7* element provides a novel polyadenylation signal, which results in a functional transcript. In this case, binding of MURA to the *Mu7* TIR could simply prevent transcription to this polyadenylation signal.

Although sometimes difficult to work with, suppressible alleles can be quite useful because they can be used for clonal analysis. In single copy *MuDR* lines, *MuDR* elements often delete at various times during development, resulting in clonal sectors of lost activity (Figure 6). If a single *MuDR* element is combined with a suppressible allele, the result will be clonal sectors in which that allele has reverted to its progenitor function (Fowler et al., 1996). Similar (if less elegant or predictable) experiments can be done by silencing high copy *Mutator* lines using *Mu killer* (see section 5).

4   Regulation of *Mutator* Activity

4.1   *MuDR Regulates the Mutator System*

All aspects of *Mutator* activity in maize are dependent on the presence of active *MuDR* elements (Chomet et al., 1991; Hershberger et al., 1991; Qin et al., 1991). This was demonstrated conclusively in families segregating for a single *MuDR* elements (Schnable and Peterson, 1988; Chomet et al., 1991; Stinard et al., 1993). Such lines, which contain as few as one *MuDR* element and one non-autonomous element, are referred to as "minimal lines" (Lisch et al., 1995). In the presence of *MuDR*, methyl-sensitive sites within non-autonomous element TIRs are unmethylated, the elements excise from reporter genes and duplicate germinally (Lisch et al., 1995; Lisch et al., 1999). In the absence of *MuDR*, none of these events occur. Duplications of the single *MuDR* element give rise to families that segregate for two regulatory loci, and deletions within single *MuDR* elements result in the loss of activity (Figure 6). Although there has been confusion concerning the distinction between classically defined "mutator activity" (large numbers of new mutations) and somatic reversion of various reporters, this is almost certainly a result of the distinction between the activity of a single element at a single reporter and the mass action of large number of active *Mu* transposons in the generation of new mutations.

4.2   *MuDR Expression*

The two genes encoded by *MuDR* are transcribed convergently from very similar TIRs, suggesting that they are regulated in similar ways (Hershberger et al., 1995). Expression of both *mudrA* and *mudrB* is correlated with the presence of active *MuDR* elements and is highest in actively dividing cells, particularly in floral tissues and embryos (Joanin et al., 1997). Similar results were obtained using a transgenic construct containing TIRB driving expression of GUS or luciferase (Raizada et al., 2001b). That analysis also revealed particularly high levels of expression of these constructs in pollen, and both cell cycle and pollen-specific motifs were identified within *MuDR* TIRs. Similar results have been obtained from immulocalization of MURA and MURB, although these results can be difficult to interpret due to a lack of specificity, as some antibodies raised against these proteins detect equivalent amounts of them regardless of the level of expression of the genes or the level of activity (Rudenko and Walbot, 2001). One polyclonal antibody raised against MURB did show specificity, and its localization was in good agreement with *in situ* and Northern blot data. Interestingly, although the level of MURB was highest in developing floral structures, it dropped dramatically in pre-pollen mother cells, the cells that are to undergo meiosis. This observation lead to the hypothesis that *Mutator* activity may be specifically down-regulated during meiosis, perhaps to limit the damage that the breaks caused by *Mu* activity could cause (Donlin et al., 1995). However, more recent data suggests that *Mutator* activity enhances recombination (Yandeau-Nelson et al., 2005), a process that occurs during meiosis, and so the loss of MURB in pre-pollen mother cells remains an enigmatic observation. Overall, the pattern of expression of *MuDR* genes is rather typical for transposons as a whole, with the highest expression levels in tissues most likely to become part of the germ line. At least 8% of transcripts in maize sperm cells, for example, are derived from transposons (Engel et al., 2003), and the same is true of apical meristem tissue in maize, where 14% of the expressed sequences were retrotransposon-related (Ohtsu et al., 2007). This is likely to be the result of selection at the level of transposons to maximize the chances of having duplicates find their way into the next generation; duplication events in determinant tissues would only result in damage to the host genome without benefit to the transposon.

In addition to the sense-strand versions of *mudrA* and *mudrB*, antisense transcripts have also been detected (Kim and Walbot, 2003). This is hardly surprising given that the two genes are convergently transcribed from nearly identical promoters (Hershberger et al., 1995). One deletion derivative that removes the 3′ end of *mudrA* along with some of the intergenic region exhibits enhanced production of antisense *mudrA* transcript (Lisch et al., 1999). It has been tempting to speculate that variations in the production of antisense transcript may be involved in regulation of *MuDR*, but there is no obvious correlation between the level of antisense transcript and the level or form of *MuDR* activity. Most strikingly, transgenes specifically engineered to produce large quantities of antisense *mudrA* have no effect on activity (Kim and Walbot, 2003). It appears that *MuDR* elements may in fact be particularly resistant to antisense down-regulation, although it should be noted that several decades of research into antisense transgenes (prior to the discovery of RNAi) suggests that only some forms of antisense transcripts are sufficient to induce silencing.

Alternative forms of MURA and MURB can be produced via alternative splicing, and differentially spliced forms of *mudrA* and *mudrB* transcripts have been detected (Hershberger et al., 1995). Depending on splicing of the third intron, *mudrA* can produce an 823 amino acid protein or a 736 amino acid protein, both of which retain conserved transposase and nuclear localization domains (Ono et al., 2002). It has been suggested that somatic excision versus germinal duplication may related to the production of differentially spliced versions of these genes (Rudenko and Walbot, 2001). However, to date, no correlations have been found between splicing variation or transcriptional start site use and transposition outcome. However, if it were found that an artificially spliced version of one of these genes specifically promoted duplication or excision, the hypothesis would be supported.

5 Epigenetic Regulation of the *Mutator* System

5.1 *Spontaneous Epigenetic Silencing*

A great deal of progress has been made in the past few years with respect to epigenetic regulation of *Mu* transposons. When *Mutator* activity is lost *Mu* elements are invariably methylated (Chandler and Walbot, 1986; Bennetzen, 1987; Chomet et al., 1991; Martienssen and Baron, 1994; Slotkin et al., 2005). DNA methylation is a hallmark of epigenetic silencing that is often associated with transcriptional silencing (Bender, 2004). Somatic sectors in which a single *MuDR* element becomes spontaneously deleted contain methylated *Mu1* elements and *MuDR* deletion derivatives, demonstrating that the machinery for *de novo* DNA methylation of non-autonomous elements is present throughout development (Lisch et al., 1995) and Figure 6). The default methylation of non-autonomous elements that occurs in the absence of transposase is fully reversible once the transposase is re-introduced. This could be due to a passive process, in which binding of the transposase to the TIRs blocks methyl-transferase activity, or it could be the result of active demethylation, as has been suggested for *Spm* (Raina et al., 1998). In any event, default methylation of non-autonomous elements does not appear to be associated with transcriptional silencing. For instance, the outward-reading *Mu1* promoter in *hcf106::Mu1* is specifically active when *Mu1* is methylated (Martienssen et al., 1990; Barkan and Martienssen, 1991), and deletion derivatives of *MuDR* become methylated at at least one site, and yet they remain fully transcriptionally active (Figure 6 and (Lisch et al., 1999)).

In contrast to default methylation of non-autonomous elements and deletion derivatives, methylation of otherwise active *MuDR* elements is invariably associated with transcriptional gene silencing (Martienssen and Baron, 1994; Slotkin et al., 2003; Takumi and Walbot, 2007). This process is often progressive (Brown et al., 1994) and occurs preferentially in the tassel (Walbot, 1986). Changes in activity can be observed throughout the life of a single plant (Bennetzen, 1994; Brown et al., 1994; Martienssen and Baron, 1994), or over several generations (Robertson, 1983; Walbot, 1986). This spontaneous silencing occurs preferentially in the tassel. In general however, once *MuDR* silencing begins in a given plant or lineage rarely reverses itself, and it is associated with the loss of *MuDR* expression (Hershberger et al., 1991; Takumi and Walbot, 2007).

5.2   *Mu Killer*

The cause of spontaneous silencing of complex lines remains unknown. However, analysis of a silencing in a low copy number minimal line revealed the presence of a locus, *Mu killer* (*Muk*), that can reliably trigger *Mutator* silencing (Slotkin et al., 2003). *Muk* arose spontaneously in the minimal line. When crossed to a line carrying one or many *MuDR* elements, these elements become methylated and transcriptionally inactive. Further, even when *Muk* is segregated away, *MuDR* elements remain heritably silenced, a process that is quite reminiscent of paramutation. This is the first locus in any organisms that can reliably and heritably silence a transposon, and it has proved to be a valuable tool for understanding the initiation of transposon silencing.

Muk was cloned and was found to be a deletion derivative of *MuDR* that has a portion of the 5′ end of the element duplicated and inverted relative to itself (Slotkin et al., 2005). The rearrangement that produced *Muk* also resulted in a deletion of flanking sequences, which turns out to be important for *Muk* function. Portions of two genes as well as complete copy of a third gene are deleted, and the resulting locus has four potentially functional promoters: two *MuDR* TIRA promoters, and two promoters reading into the *MuDR* derivative from the remaining portions of the two flanking genes. Neither TIRA promoter appears to be active, presumably because they, like *MuDR* promoters silenced by *Muk*, are targeted for silencing by the *Muk* transcript. The promoter from one of the two flanking genes, *Accomplice1* (*Acc1*), is very active and promotes expression through the entire *MuDR* derivative and into the other flanking gene, which contains an ectopic polyA signal. The pattern of expression of the *Muk* locus mirrors that of the progenitor *Acc1* gene (Slotkin and Lisch, unpublished data). The *Muk* transcript forms a long perfect hairpin RNA. This transcript is processed into small interfering RNAs (siRNAs) that trigger heritable *MuDR* silencing. Using various deletion derivatives, it was possible to demonstrate that *Muk* silencing specifically targets the *mudrA* gene for transcriptional gene silencing.

The siRNAs associated with *Muk* are present in relatively small quantities, and they correspond to the sense strand of *mudrA* (Slotkin et al., 2005). However, in plants carrying both *Muk* and *MuDR,* the quantity of siRNA corresponding to both sense and antisense *mudrA* is dramatically increased. It should be noted that the siRNAs corresponding to *mudrA* and *mudrB* are present in germinal tissues in all maize plants and are likely to be the result of processing of silenced *hMuDR* elements (Rudenko et al., 2003; Woodhouse et al., 2006b). The presence of siRNAs corresponding to silenced transposons in these tissues has been well documented in *Arabidopsis* (Lu et al., 2005) and rice (Chen et al., 2006). What is novel about the siRNAs associated with *Muk*-induced silencing of *MuDR* is that they are present at high levels in young leaves, where siRNAs from *hMuDRs* are absent. Further, unlike *hMuDR* siRNAs, *Muk*-induced siRNAs are not dependent on a maize homolog of RNA-dependent RNA polymerase, MOP1/*Zm*RDR2 (Woodhouse et al., 2006b). Thus, the initiation of silencing via *Muk* is a distinct process from the maintenance of silencing.

Although *mudrA* transcriptional activity is lost in *Muk;MuDR* F1 plants, *mudrB* remains transcriptionally active. Further, in the same tissue in which *mudrA* siRNAs are readily apparent (leaf 2), there are no detectable *mudrB* siRNAs (Slotkin et al., 2005). Nevertheless, *mudrB* is affected in these plants; although the overall quantity of *mudrB* transcript remains constant in F1 plants, it is no longer polyadenylated (Slotkin et al., 2003). By the next generation, this transcript is no longer detectable either by RT-PCR or Northern blots. Interestingly, when a derivative of *MuDR* that expresses only *mudrB* is put *in trans* with a full length *MuDR* element in the presence of *Muk*, the derivative does not lose polyadenylation. Thus, it appears that the silencing of *mudrB* is a process that only occurs in *cis*, and likely is the result of "spreading" of a silencing signal from *mudrA*. It will be interesting to see what factors mediate this process, as opposed to the initial silencing of *mudrA*.

It is not known how high copy *Mutator* lines are silenced, although it clearly happens at varying frequencies, particularly when active lines are inter-crossed or self fertilized (Robertson, 1983 and reviewed by Bennetzen 1996). We do know that a rearranged version of *MuDR* (*Muk*) can trigger heritable silencing, and that *MuDR* elements are subject to frequent rearrangement (Lisch and Freeling, 1994; Hsia and Schnable, 1996), at least at some positions. Thus, it is certainly possible that in a given high copy number *Mutator* line at a given frequency, derivatives will occasionally form that trigger silencing. However, it should be noted that there is conflicting evidence for the appearance of dominant factors in *Mutator* lines that cause epigenetic silencing (Brown and Sundaresan, 1992), and the frequency of production of new deletion derivatives in high copy lines is relatively low. Thus, it is also possible that the production of antisense or otherwise aberrant transcript reaches a threshold level once the copy number of elements gets high enough. Perhaps in combination with the *hMuDR* derived siRNAs, this could trigger heritable silencing (Rudenko et al., 2003). If the elements that contributed to the aberrant RNAs themselves became transcriptionally silenced (unlike *Muk*, which uses an ectopic promoter), then they may not be competent to silence new *MuDR* elements that are crossed back into the inactive lines. In this scenario, the cumulative effect of large numbers of transcripts could produce a transient trigger for silencing that would not necessarily propagate as a single dominant locus. However, depending on a variety of factors, it could certainly propagate as a dominant quantitative trait, as has been observed (Bennetzen, 1987, 1996).

5.3 Factors that Influence Epigenetic Silencing

Little is known about variables that affect the process of *Mutator* silencing. In order to look for such factors, RNAi knockdown transgenic lines targeting a number of chromatin remodeling genes has been screened for its effects on *MuDR* silencing in a minimal line (McGinnis et al., 2007). Two transgenes, targeting two distantly related maize homologs of a gene encoding Nuclear Assembly Protein 1 (NAP1) each affect the establishment of a heritably silenced *MuDR* element (Woodhouse et al., 2006b). When *MuDR(p1)* was combined with *Muk* in the presence of these transgenes, although *Muk* appeared to eliminate *MuDR* activity in the F1 plants, once *Muk* was segregated away, the *MuDR* element became reactivated, suggesting that these NAP1 homologs are required to establish the heritable component of *Muk*-induced silencing of *MuDR*. Interestingly, *MuDR* elements at certain chromosomal positions have been identified that mimic this effect, but in a non-mutant background. *MuDR* elements at these positions are silenced by *Muk*, but they reactivate once *Muk* is lost (Singh and Lisch, 2008). *MuDR* elements have also been identified that exhibit a reversible position effect with respect to somatic activity (Figure 5). It will be interesting to see what *cis* factors are responsible for these position effects.

Once an otherwise active *MuDR* element has been silenced by *Muk*, it can be maintained in a silenced state indefinitely. Because of its relative stability, it is possible to then examine the effects of various mutations on the silenced state. One such mutation that has proved to be particularly informative is *mop1* (Dorweiler et al., 2000). This mutation was originally isolated due to its effects on paramutable alleles, which are alleles of genes that can express at high or low levels depending on their epigenetic state (Chandler et al., 2000). When a high-expressing version of such a gene is crossed to a low expression version of this gene, its expression level is heritably reduced. In the absence of functional *MuDR* elements, *mop1* mutants lack methylation at *Mu* element TIRs, suggesting that the default methylation observed in the absence of MURA is mediated by this gene (Lisch et al., 2002). When a silenced *MuDR* element is maintained in a *mop1* mutant background for multiple generations, *mudrA* becomes transcriptionally reactivated, which results in somatic excisions of a reporter element. At first, the reactivated state is dependent on the mutant background, but in subsequent generations the element becomes heritably active even in a wild type background (Woodhouse et al., 2006a). Thus, the loss of MOP1 results in a progressively less stable silenced state. Interestingly, even when *mudrA* has become heritably active, *mudrB* remains inactive, suggesting that although *mudrA* and *mudrB* are both silenced by *Muk*, they are maintained in their silenced state via distinct mechanisms.

Mop1 is an RNA-dependent RNA polymerase (RDR) whose closest homolog in *Arabidopsis* is RDR2 (Alleman et al., 2006; Woodhouse et al., 2006b). RDRs are required for the production of double-stranded RNA from RNA templates. RDR2 in *Arabidopsis* is specifically required for the production of siRNAs required to maintain transposon silencing (Xie et al., 2004; Lu et al., 2006; Pikaard, 2006; Pontes et al., 2006; Zhang et al., 2007). In maize, ZmRDR2 is required for the production of *MuDR* and *hMuDR* siRNAs that accumulate in reproductive structures and embryos, but not for the production of the siRNAs associated with *Muk*-induced silencing in young leaves. Not surprisingly, then, *Muk* is competent to heritably silence *MuDR* in a *mop1* mutant background (Woodhouse et al., 2006b). This observation, along with the fact that *MuDR* elements are only gradually reactivated in a *mop1* mutant background suggests that the pathway that includes ZmRDR2/MOP1 acts to reinforce, rather than to initiate previously established silent chromatin states at transposons such as *MuDR*.

6 Evolution of the MULEs

Early in the analysis of the *MuDR* element from maize, it became apparent that the transposase encoded by this element is an ancient one, as homologous sequences were detected in bacterial transposases (harbored in insertion sequence, IS) (Eisen et al., 1994). Since then, MULEs have been detected in many species of plants, including both monocots (Lisch et al., 2001) and dicots (Le et al., 2000), as well as fungi (Chalvet et al., 2003; Neuveglise et al., 2005), and possibly animals (Makarova et al., 2002; Pace and Feschotte, 2007). Although there are many structural variations, most potentially autonomous eukaryotic MULEs share *mudrA* homologs and nine base-pair target site duplications. The length of terminal inverted repeats (TIRs), however, varies greatly among different families. All the IS elements with *mudrA* transposase have relatively short (<50 bp), often imperfect, TIRs (Eisen et al., 1994). In plants and fungi most MULEs are associated with long TIRs (100 to 600 bp). Recently, the presence of MULEs with short TIRs (<50 bp), the so-called "non-TIR MULEs", have been reported in *Arabidopsis*, *Lotus japonicus*, maize, and yeast (Yu et al., 2000; Neuveglise et al., 2005; Holligan et al., 2006; Wang and Dooner, 2006). This suggests a widespread distribution of MULEs with short TIRs. Among all the major families of DNA transposons in plants (*Ac/Ds*, *Spm*, *Helitron*, MULE, *PIF/Pong*, *Tc1/Mariner*), the association with long TIR is largely limited to MULEs, except a subset of MITEs (*PIF/Pong*, *Tc1/Mariner*). At present it is not known whether the long TIRs are responsible for some specific features of MULEs, such as the frequent acquisition of genomic sequences (see below). If we consider the short-TIR MULEs as the more ancient form (such as those in bacteria), it is intriguing to consider how the long TIR evolved in plants and fungi and what the selective advantages are to having them.

In addition to their TIRs, dramatic variation is also observed in the coding region of MULEs. Some autonomous MULEs contain a single open reading frame (ORF) that encodes a transposase, while others harbor two or more ORFs. Even the small autonomous MULEs (IS elements, < 2 kb) in bacteria can contain as many as three ORFs (Yates et al., 1988). When there are two or more ORFs, transcription orientation can be either the same or convergent. As mentioned above, the maize *MuDR* element encodes *mudrA* and *mudrB*. Although no MULEs other than *MuDR* contain *mudrB*, it appears that other MULEs have captured additional coding regions that may contribute to function. The best studied example of this are CUMULEs, also referred to as KI-MULEs (Hoen et al., 2006; van Leeuwen et al., 2007) (Figure 3). This class of elements encodes both *mudrA* homologs as well as an ORF or ORFs that encode a peptidase C48 domain derived from a cellular small ubiquitin-like protease. KI-MULEs in *Arabidopsis* are non-TIR MULEs, but those in melon have extended TIRs. The *Ulp1* protease can be present in one or two copies, depending on the element, and can be present in various orientations. Interestingly, other transposons, such as *Spm*-like elements, can also carry the same protease domain, suggesting that there might be a generic advantage for transposons to capture this sequence (van Leeuwen et al., 2007). Together, these data suggest that DNA transposons, like some retroviruses, are competent to capture fragments of host sequence that can assist in their proliferation.

Single genomes often contain distantly related families of MULEs, which share more similarity with the same family in other species than they do with each other. In most instances it appears that MULEs are inherited vertically, as is the case for all other plant transposons examined to date. However, there is evidence that MULEs can be horizontally transferred as well. The best evidence comes from analysis of one class of MULEs present in the genus *Setaria* and in rice. Analysis of the sequence of these elements revealed a remarkably high degree of sequence similarity in non-coding sequences, strongly suggestive of horizontal transfer (Diao et al., 2006). Additional analysis of "garden blots" using various MULE probes suggests that horizontal transfer may be a common feature of at least some MULEs (Lisch, unpublished data). What is surprising is how infrequent horizontal transfer of plant transposons is considering how often it has been observed in animals.

6.1 Pack-MULEs

Like their autonomous partners, the non-autonomous MULEs are very diverse. As mentioned above, non-autonomous *Mutator* elements from a single family have similar TIRs but the internal sequence (between TIRs) varies from element to element. In many cases, non-autonomous DNA elements are simply deletion derivatives of their cognate autonomous elements (Figure 6). In contrast, the internal regions of non-autonomous MULEs seem to derive from other genomic sequences including fragments of genes. This phenomenon was first described about two decades ago, when the *Mu1/Mu1.7* elements were shown to contain part of a gene of unknown function called *MRS-A* (Talbert and Chandler, 1988). Subsequently, a few *Arabidopsis* and rice MULEs harboring fragments of host genes were reported (Yu et al., 2000; Turcotte et al., 2001). Because of the diverse sequences they carry, this type of element is now referred to as Pack-MULEs (Jiang et al., 2004).

The scale of gene capturing by MULEs was first demonstrated in rice, which possesses one of the smallest of the grass genomes. In the rice genome, there are about 3,000 Pack-MULEs which carry fragments from over 1000 distinct genes (Jiang et al., 2004), with an average element size of 1.9 kb, similar to the 1.7 kb average size of maize non-autonomous elements. Some of the Pack-MULEs are amplified to multiple copies (elements with similar TIRs and internal regions) through transposition, while others are present as single copy elements. The average copy number of Pack-MULE is three, suggesting a selective pressure against amplification of individual elements or an intrinsic mechanism for fast shuffling or exchange of internal sequences. The fact that in an active *Mu* line, the copy number of *Mu1* can be fifty or more suggests that Pack-MULEs are able to amplify to a relatively high copy number (Bennetzen, 1996), and the low copy number of individual Pack-MULEs observed in rice is likely to be attributed to the subsequent loss or dilution in the population of elements after MULE activity is lost (Walbot and Warren, 1988). In addition to rice, Pack-MULEs are also abundant in *Lotus japonicus* (Holligan et al., 2006), suggesting that the process of transduplication by MULEs is of ancient origin. However, it should be noted that in both rice and *L. japonicus*, not all MULEs are associated with the formation of Pack-MULEs.

Recently, the availability of maize genomic sequences allowed a more comprehensive study about the number and families of Pack-MULEs in maize. An initial search using 177 Mb (about 7% of the genome) of maize genomic sequences indicated that the *MuDR* and *Jittery* classes of elements, represents MULEs with lowest copy numbers in maize. In contrast, there are at least fifty distinct MULE families, including TRAP and TAFT, that are more abundant than *MuDR* and *Jittery* elements, and some of the non-autonomous versions of these elements appear to carry gene fragments. It is estimated there are at least 1000 copies of Pack-MULEs in the maize genome, with diverse TIRs and internal regions (Jiang, unpublished data). This provides additional evidence for the general abundance of Pack-MULEs in plants, even in genomes such as that of maize that are dominated by LTR retrotransposons (SanMiguel et al., 1996b; SanMiguel et al., 1998; International-Rice-Sequencing-Project, 2005; Holligan et al., 2006).

A comparison between Pack-MULEs in rice and their genomic copies (from which the transduplicated sequences are derived) revealed many interesting features of sequence acquisition by Pack-MULEs. First, Pack-MULE transduplication is error prone; there are frequent deletion and rearrangements (Figure 7), and most of the captured parts are gene fragments instead of complete genes. Second, introns are retained in the captured gene fragments, suggesting that transduplication occurs at the DNA, not the RNA level (Figure 7B). Third, about one fourth of the rice Pack-MULEs contain fragments from more than one gene, suggesting a mechanism for exon-shuffling. Fourth, the sequence similarities between the internal sequence carried by Pack-MULEs and their cognate genomic copies demonstrate a continuous distribution, ranging from 78% to 100%. Assuming that the captured sequence and the genomic homologue were initially identical, sequence acquisition must have occurred over a long time frame and might be still occurring. Finally, at least 5% of the Pack-MULEs in rice are transcribed and most transcripts seem to be initialized from within TIRs. Interestingly, some transcripts are in sense orientation and others are in anti-sense orientation (Figure 7), with a small portion of Pack-MULEs that are transcribed bi-directionally, although the transcripts with different orientations are detected in different tissues. The presence of anti-sense transcripts raises the possibility that the expression of Pack-MULEs may negatively regulate the expression of the relevant host genes at the post-transcriptional level. Even transcripts expressed in sense orientation could be expressed in the wrong quantities, or at the

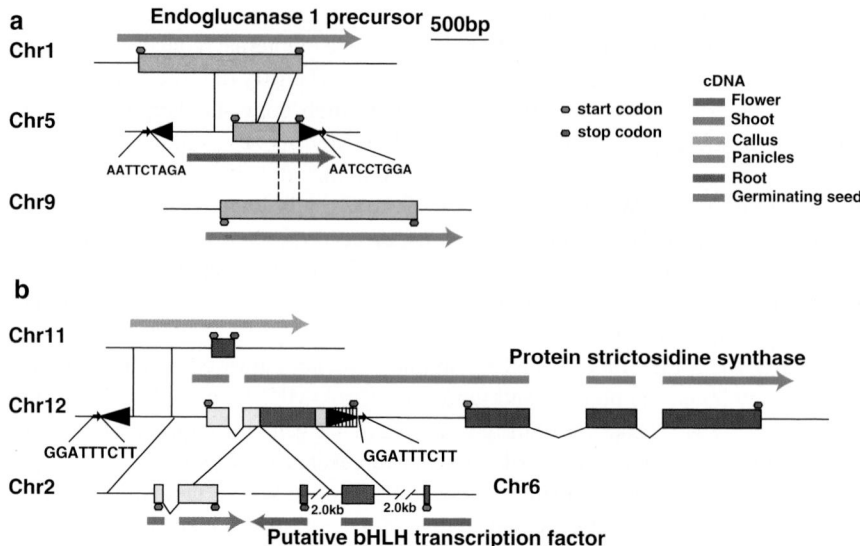

Fig. 7 Pack-MULEs rearrange genomic sequence and may alter gene expression patterns. (**A**) A Pack-MULE containing fragments from two genes that are expressed in shoot and panicles, respectively. One acquisition was accompanied by a deletion and that Pack-MULE is transcribed in roots. B) A Pack-MULE with gene fragments from three genomic loci including an intron and the regulatory region from one of these genes. A transcript appears to be initiated from the relevant regulatory region and extends into the gene immediately downstream of the Pack-MULE. Pack-MULE TIRs are shown as black arrowheads, and black horizontal arrows indicate target site duplications (TSDs) with their sequences shown underneath. Exons are depicted as colored boxes and introns as the lines connecting exons; other sequences are shown as horizontal lines. Homologous regions are associated with solid or dashed lines. The light-blue box in B represents part of an exon where the origin of the sequence is not clear. The striped box in B indicates that the TIR overlaps with the putative exon. Long colored arrows indicate sequences matching cDNAs from the designated tissues. The gene name is given for putative genes; all other genes encode 'unknown proteins'. Gene predictions were based on the annotation provided by the rice full-length cDNA consortium (Kikuchi et al., 2003). In B, the ORF inside the Pack-MULE was defined by ORFfinder (http://www.ncbi.nlm.nih.gov) based on the corresponding cDNA sequence (Kikuchi et al., 2003)

wrong time (Figure 7), resulting in feedback regulation of the progenitor gene (Lisch, 2005). Further, Pack-MULEs can mobilize regulatory regions as well as coding regions, and they can combine promoter or enhancer sequences from one location and coding regions from another (Figure 7).

Given the fact that transposon sequence, such as TIRs, becomes unrecognizable in a few million years (through mutation and deletion) (Ma et al., 2004), the time frame of transduplication by Pack-MULEs may be beyond what the sequence similarity of Pack-MULE and their genomic copies would suggest. The further implication is that some of the "normal genes" in the genome might be ancient Pack-MULEs that lost recognizable TIRs. If we assume that the acquisition process has been occurring at a steady rate after the divergence of rice and maize, then it is quite possible that a large proportion of genes in those two species (and perhaps other organisms as well) have been transduced by MULEs or some other transposable elements (e.g, *Helitrons*) (Lisch, 2005; Morgante et al., 2005; Wang and Dooner, 2006). The fact that many, even most genes within the grasses are syntenous (collinear between related species) suggests that the majority of those duplicated gene sequences were simply lost to recognition due to gradual sequence degradation, as may happen to any other type of gene duplication. However, there are many exceptions to synteny, and an intriguing question will be how many of those have arisen from recent transduction and subsequent transposition (Lai et al., 2004). In regard to this issue, a key question is whether any of the Pack-MULEs are able to produce functional proteins, in addition to their potential to regulate the expression of pre-existing genes (see above). In one study, Jiang et al (2004) showed that six Pack-MULEs have perfect matches with entries in a rice peptide database, and a small subset of Pack-MULEs demonstrated functional constraint based on the ratios of the synonymous (Ks) to non-synonymous (Ka) substitution rates. This suggested that proteins were made from the relevant elements and some of them might have retained the function of their genomic copies. In another study, Juretic et al (2005) examined a total of 66 MULEs with gene fragments and found that all of them contain one or more pseudogenic features including fragmented conserved domains, frameshifts, and premature stop codons. Based on this fact as well as the comparison of the 66 rice Pack-MULEs with human peudogenes, Juretic et al. concluded that all rice Pack-MULEs represent pseudogenes (Juretic et al., 2005). Recently, an expression analysis in rice indicated that the transcription activity of Pack-MULEs fall in between the levels of TE-related and non-TE-related gene models, suggesting some of the Pack-MULEs may represent new evolving functional genes (Jiao and Deng, 2007). Thus, the coding capacity of Pack-MULEs is still an open question. The ultimate resolution to this question is to test whether any of the Pack-MULEs, when mutagnized, will lead to a detectable phenotype.

Despite the abundance of Pack-MULEs in the genomes of many plants, little is known about the mechanism involved in the acquisition of genomic sequences. In 1988, Talbert and Chandler (Talbert and Chandler, 1988) proposed that a sequence closely related to *MRS-A* was encompassed by *Mu* termini, followed by the deletion of a portion of the gene. This would explain the presence of sequence homologous to *MRS-A* in *Mu1.7*. This model requires the mobility of an individual *Mu* terminus, as is seen in IS elements in bacteria. However, there is no evidence to support this hypothesis. A second model suggested by Bennetzen and Springer (Bennetzen and Springer, 1994) raised the possibility of sequence acquisition through ectopic gene-conversion across a nicked cruciform structure. According to this model, a stem-loop structure will form (the TIR will be the double-stranded stem) if the element is somehow single-stranded. A nick in the loop will be generated due to the action of endonuclease. If the sequence around the nick is homologous to some sequence elsewhere in the genome, the relevant genomic sequence will be used as a template to repair the nick (Bennetzen and Springer, 1994). If the nick in the loop is the primary cause for the introduction of new sequences, the formation of a cruciform structure will promote the exchange of internal sequences. The fact that most of the Pack-MULEs reported so far are associated with long TIRs seems to favor this model in that the stem-loop structure will be more readily formed with long TIRs. The question is, in which circumstances would a nearly 2 kb element be able to form a cruciform, without interference from other proteins, such as single strand DNA binding proteins? A third model, as mentioned earlier, explains the sequence acquisition by an aberrant gap repair event using ectopic sequence as template. This would predict that the sequence acquisition happens at the donor site after an element is excised and following the repair of the TIRs but not internal sequences, which would be replaced with the ectopic sequences. Understanding of the mechanism(s) of sequence acquisition by Pack-MULEs awaits further biochemical, genetic, and informatics analyses.

In addition to Pack-MULEs, other DNA transposable elements are involved in the duplication and mobilization of gene fragments. For instance, a mutation in soybean that caused the alternation of flower color and seed weight was due to an insertion of a non-autonomous *Spm* element that contains five gene fragments (Zabala and Vodkin, 2005). In addition, the heterogeneity of internal sequences demonstrated by *PIF* elements may suggest their potential to capture genomic sequences (Zhang et al., 2001; Zhang et al., 2004). Nevertheless, *Helitron* elements seem to be the only other DNA transposon that carries and mobilizes gene fragments on a large scale, with possibly 10,000 copies in the maize genome (Morgante et al., 2005). Like Pack-MULEs, non-autonomous *Helitrons* frequently carry gene fragments, and introns are retained. Some of the captured fragments are transcribed. The gene fragments in *Helitrons* are usually in the same orientation with respect to the direction of transcription (Brunner et al., 2005), although this is not observed for Pack-MULEs (Jiang, unpublished). This difference might reflect the distinct mechanisms of sequence acquisition by the two transposon families or perhaps selective pressures following capture.

6.3 MULE Domestication

Transposons are in large measure parasites. They proliferate to the extent that they can, and their effects are largely neutral or even deleterious (Kidwell et al., 1988; Le Rouzic and Capy, 2005). However, there are some examples of cases in which individual transposons provide selective benefit to their hosts. Examples include the RAG genes in vertebrates which are responsible for programmed DNA rearrangement of immunoglobulin genes (Agrawal et al., 1998), the CENP-B binding protein (Kipling and Warburton, 1997), involved in centromere function, telomerases (Malik et al., 1999) and the JERKEY gene in mice (Toth et al., 1995), which is involved in proper brain function. In each case, some aspect of the transposase function has been co-opted by the host, a process which has been termed "transposon domestication" (Miller et al., 1999). In plants there are relatively few examples of domestication, but several of them involve MULEs. The first evidence for domestication of any plant transposon came from analysis of mutations in the *FAR1* and related *FHY3* genes in *Arabidopsis*. Mutations in these genes exhibit an inability to respond to far-red light. Sequence analysis revealed that the *FAR1* gene exhibits a high degree of similarity to the transposase encoded by the *bona fide* transposon in maize, *Jittery* (Hudson et al., 2003), and recent work has demonstrated that each of these genes acts as a *bona fide* transcription factor (Lin et al., 2007). The fact that mutations in *FAR1* and *FHY3* genes have a clear phenotypic effect unambiguously demonstrates that this gene has a well-defined cellular function in plants, and members of the *FAR1* subclade are present in multiple plant species, including both monocots and dicots, suggesting that the domestication event occurred prior to the monocot-dicot split. Interestingly it appears that MULEs may have been domesticated more than once. The MUSTANG family of genes share extensive homology with MULEs, but, like the *FAR1* family, lack TIRs and are unlikely to be mobile, a fact supported by their presence at syntenous sites within different plant species (Cowan et al., 2005). Orthologs of both MUSTANG and *FAR1* in rice are distinct from most other *mudrA*-related sequences in that they are each transcribed in a broad range of tissues (Jiao and Deng, 2007). Thus, transcriptional profiling and phylogenetic analysis may serve as a powerful means to identify domesticated transposases. Finally, analysis of a zinc finger encoded by both MULE transposases and some transcription factors has led to the hypothesis that MULEs have been domesticated many times over the course of evolution of these transcription factors (Babu et al., 2006). Theoretically, most transposases are DNA binding proteins with the ability to bind sequences such as TIRs, so it is not surprising that they have the potential to evolve into transcription factors. Indeed, suppressible *Mu* insertions provide excellent examples of how transposon insertions could evolve into a two-component mechanism for either negative or positive gene regulation.

7 Conclusion

The *Mutator* family of elements has proved to be a remarkably useful tool for gene discovery in maize (see chapter 10, MaCarty et al). If, as has been proposed, *Mutator* is used to generate large numbers of sequence-indexed insertion lines, *Mutator* promises to become an even more important resource for maize geneticists. The development of minimal lines in maize has also made *Mutator* an excellent model for understanding epigenetic silencing. The ability to silence *MuDR* elements in a controlled and heritable fashion using *Mu killer*, and to test the effects of various mutations and chromosomal positions on the process of silencing, should continue to provide valuable insights into the means by which epigenetic information is introduced and then propagated across generations. Finally, analysis of the MULE superfamily of transposons continues to provide a wealth of data relevant to important evolutionary processes. The process by which MULEs can capture, amplify, and utilize host genes and regulatory information (i.e. *Ulp1* and Pack-MULEs), and that by which MULE transposases can be domesticated, illustrates the lack of clear lines between host genes and transposons. Given the fact that what defines a "transposon" can be as little as the appropriate transposase binding sites flanking a given genomic sequence, all genes are potentially mobile. And, given the vast array of regulatory and enzymatic information carried by transposons, all transposons are potentially useful to the host. What is particularly intriguing about this blurry line is the potential it provides for innovation and variation; MULEs and other transposons can undergo radical changes "under the radar" of selection at the level of the host, driven in large measure by selection for the capacity to replicate themselves. In doing so, they produce a large reservoir of potentially useful biochemical and regulatory information for the host. Thus, rapidly evolving transposons such as MULEs may act as an engine to drive evolutionary change.

Acknowledgements DL was supported by grant DBI 031726 from the National Science Foundation, Plant Genome Research Initiative. NJ was supported by grant DBI 0607123 from the National Science Foundation, Plant Genome Research Initiative and grant 2006-35604-16631 from USDA NRI-CSREE.

References

Agrawal, A., Eastman, Q.M., and Schatz, D.G. (1998). Transposition mediated by RAG1 and RAG2 and its implications for the evolution of the immune system. Nature 394, 744–751.

Alleman, M., and Freeling, M. (1986). The *Mu* transposable elements of maize: evidence for transposition and copy number regulation during development. Genetics 112, 107–119.

Alleman, M., Sidorenko, L., McGinnis, K., Seshadri, V., Dorweiler, J.E., White, J., Sikkink, K., and Chandler, V.L. (2006). An RNA-dependent RNA polymerase is required for paramutation in maize. Nature 442, 295–298.

Babu, M.M., Iyer, L.M., Balaji, S., and Aravind, L. (2006). The natural history of the WRKY-GCM1 zinc fingers and the relationship between transcription factors and transposons. Nucleic Acids Res 34, 6505–6520.

Barkan, A., and Martienssen, R.A. (1991). Inactivation of maize transposon *Mu* suppresses a mutant phenotype by activating an outward-reading promoter near the end of Mu1. Proc Natl Acad Sci U S A 88, 3502–3506.

Bender, J. (2004). DNA methylation and epigenetics. Annu Rev Plant Biol 55, 41–68.

Benito, M.I., and Walbot, V. (1997). Characterization of the maize *Mutator* transposable element MURA transposase as a DNA-binding protein. Mol Cell Biol 17, 5165–5175.

Bennetzen, J.L. (1984). Transposable element *Mu*1 is found in multiple copies only in Robertson's Mutator maize lines. J Mol Appl Genet 2, 519–524.

Bennetzen, J.L. (1987). Covalent DNA modification and the regulation of *Mutator* element transposition in maize. Mol. Gen. Genet. 208, 45–51.

Bennetzen, J.L. (1994). Inactivation and reactivation of mutability at a *Mutator*-derived *bronze-1* allele in maize. Maydica 39, 309–317.

Bennetzen, J.L. (1996). The *Mutator* transposable element system of maize. Curr Top Microbiol Immunol 204, 195–229.

Bennetzen, J.L. (2000). The many hues of plant heterochromatin. Genome Biol 1, REVIEWS107.

Bennetzen, J.L., and Springer, P.S. (1994). The generation of *Mutator* transposable element subfamilies in maize. Theor Appl Genet 87, 657–667.

Bennetzen, J.L., Springer, P.S., Cresse, A.D., and Hendrickx, M. (1993). Specificity and regulation of the *Mutator* transposable element system in maize. Critical Reviews in Plant Sciences 12, 57–95.

Brown, J., and Sundaresan, V. (1992). Genetic study of the loss and restoration of *Mutator* transposon activity in maize: evidence against dominant-negative regulator associated with loss of activity. Genetics 130, 889–898.

Brown, W.E., Robertson, D.S., and Bennetzen, J.L. (1989). Molecular analysis of multiple *Mutator*-derived alleles of the *bronze* locus of maize. Genetics 122, 439–445.

Brown, W.E., Springer, P.S., and Bennetzen, J.L. (1994). Progressive modification of *Mu* transposable elements during development. Maydica 39, 119–126.

Brunner, S., Pea, G., and Rafalski, A. (2005). Origins, genetic organization and transcription of a family of non-autonomous helitron elements in maize. Plant J 43, 799–810.

Chalvet, F., Grimaldi, C., Kaper, F., Langin, T., and Daboussi, M.J. (2003). Hop, an active *Mutator*-like element in the genome of the fungus Fusarium oxysporum. Mol Biol Evol 20, 1362–1375.

Chandler, V., Rivin, C., and Walbot, V. (1986). Stable non-mutator stocks of maize have sequences homologous to the *Mu1* transposable element. Genetics 114, 1007–1021.

Chandler, V.L., and Walbot, V. (1986). DNA modification of a maize transposable element correlates with loss of activity. Proc Natl Acad Sci U S A 83, 1767–1771.

Chandler, V.L., Eggleston, W.B., and Dorweiler, J.E. (2000). Paramutation in maize. Plant Mol Biol 43, 121–145.

Chen, Z., Zhang, J., Kong, J., Li, S., Fu, Y., Li, S., Zhang, H., Li, Y., and Zhu, Y. (2006). Diversity of endogenous small non-coding RNAs in Oryza sativa. Genetica 128, 21–31.

Chomet, P., Lisch, D., Hardeman, K.J., Chandler, V.L., and Freeling, M. (1991). Identification of a regulatory transposon that controls the *Mutator* transposable element system in maize. Genetics 129, 261–270.

Comelli, P., Konig, J., and Werr, W. (1999). Alternative splicing of two leading exons partitions promoter activity between the coding regions of the maize homeobox gene *Zmhox1a* and *Trap* (transposon-associated protein). Plant Mol Biol 41, 615–625.

Cowan, R.K., Hoen, D.R., Schoen, D.J., and Bureau, T.E. (2005). MUSTANG is a novel family of domesticated transposase genes found in diverse angiosperms. Mol Biol Evol 22, 2084–2089.

Cresse, A.D., Hulbert, S.H., Brown, W.E., Lucas, J.R., and Bennetzen, J.L. (1995). *Mu1*-related transposable elements of maize preferentially insert into low copy number DNA. Genetics 140, 315–324.

Cui, X., Hsia, A.P., Liu, F., Ashlock, D.A., Wise, R.P., and Schnable, P.S. (2003). Alternative transcription initiation sites and polyadenylation sites are recruited during *Mu* suppression at the *rf2a* locus of maize. Genetics 163, 685–698.

Das, L., and Martienssen, R. (1995). Site-selected transposon mutagenesis at the *hcf106* locus in maize. Plant Cell 7, 287–294.

Diao, X., Freeling, M., and Lisch, D. (2006). Horizontal transfer of a plant transposon. PLoS Biol 4, e5.

Dietrich, C.R., Cui, F., Packila, M.L., Li, J., Ashlock, D.A., Nikolau, B.J., and Schnable, P.S. (2002). Maize *Mu* transposons are targeted to the 5′ untranslated region of the gl8 gene and sequences flanking *Mu* target-site duplications exhibit nonrandom nucleotide composition throughout the genome. Genetics 160, 697–716.

Donlin, M.J., Lisch, D., and Freeling, M. (1995). Tissue-specific accumulation of MURB, a protein encoded by *MuDR*, the autonomous regulator of the *Mutator* transposable element family. Plant Cell 7, 1989–2000.

Dorweiler, J.E., Carey, C.C., Kubo, K.M., Hollick, J.B., Kermicle, J.L., and Chandler, V.L. (2000). m*ediator of paramutation1* is required for establishment and maintenance of paramutation at multiple maize loci. Plant Cell 12, 2101–2118.

Eisen, J.A., Benito, M.I., and Walbot, V. (1994). Sequence similarity of putative transposases links the maize *Mutator* autonomous element and a group of bacterial insertion sequences. Nucleic Acids Res 22, 2634–2636.

Engel, M.L., Chaboud, A., Dumas, C., and McCormick, S. (2003). Sperm cells of *Zea mays* have a complex complement of mRNAs. Plant J 34, 697–707.

Fernandes, J., Dong, Q., Schneider, B., Morrow, D.J., Nan, G.L., Brendel, V., and Walbot, V. (2004). Genome-wide mutagenesis of *Zea mays* L. using *RescueMu* transposons. Genome Biol 5, R82.

Fleenor, D., Spell, M., Robertson, D., and Wessler, S. (1990). Nucleotide sequence of the maize *Mutator* element, *Mu8*. Nucleic Acids Res 18, 6725.

Fowler, J.E., Muehlbauer, G.J., and Freeling, M. (1996). Mosaic analysis of the *liguleless3* mutant phenotype in maize by coordinate suppression of *Mutator*-insertion alleles. Genetics 143, 489–503.

Franklin, A.E., McElver, J., Sunjevaric, I., Rothstein, R., Bowen, B., and Cande, W.Z. (1999). Three-dimensional microscopy of the Rad51 recombination protein during meiotic prophase. Plant Cell 11, 809–824.

Girard, L., and Freeling, M. (2000). *Mutator*-suppressible alleles of *rough sheath1* and *liguleless3* in maize reveal multiple mechanisms for suppression. Genetics 154, 437–446.

Greene, B., Walko, R., and Hake, S. (1994). Mutator insertions in an intron of the maize *knotted1* gene result in dominant suppressible mutations. Genetics 138, 1275–1285.

Hanley, S., Edwards, D., Stevenson, D., Haines, S., Hegarty, M., Schuch, W., and Edwards, K.J. (2000). Identification of transposon-tagged genes by the random sequencing of *Mutator*-tagged DNA fragments from *Zea mays*. Plant J 23, 557–566.

Hardeman, K.J., and Chandler, V.L. (1989). Characterization of *bz1* mutants isolated from mutator stocks with high and low numbers of *Mu1* elements. Dev Genet 10, 460–472.

Hardeman, K.J., and Chandler, V.L. (1993). Two maize genes are each targeted predominantly by distinct classes of *Mu* elements. Genetics 135, 1141–1150.

Hershberger, R.J., Warren, C.A., and Walbot, V. (1991). *Mutator* activity in maize correlates with the presence and expression of the *Mu* transposable element *Mu9*. Proc Natl Acad Sci U S A 88, 10198–10202.

Hershberger, R.J., Benito, M.I., Hardeman, K.J., Warren, C., Chandler, V.L., and Walbot, V. (1995). Characterization of the major transcripts encoded by the regulatory *MuDR* transposable element of maize. Genetics 140, 1087–1098.

Hoen, D.R., Park, K.C., Elrouby, N., Yu, Z., Mohabir, N., Cowan, R.K., and Bureau, T.E. (2006). Transposon-mediated expansion and diversification of a family of ULP-like genes. Mol Biol Evol 23, 1254–1268.

Holligan, D., Zhang, X., Jiang, N., Pritham, E.J., and Wessler, S.R. (2006). The transposable element landscape of the model legume *Lotus japonicus*. Genetics 174, 2215–2228.

Hsia, A.P., and Schnable, P.S. (1996). DNA sequence analyses support the role of interrupted gap repair in the origin of internal deletions of the maize transposon, *MuDR*. Genetics 142, 603–618.

Hudson, M.E., Lisch, D.R., and Quail, P.H. (2003). The FHY3 and FAR1 genes encode transposase-related proteins involved in regulation of gene expression by the phytochrome A-signaling pathway. Plant J 34, 453–471.

International-Rice-Sequencing-Project. (2005). The map-based sequence of the rice genome. Nature 436, 793–800.

James, M.G., Scanlon, M.J., Qin, M., Robertson, D.S., and Myers, A.M. (1993). DNA sequence and transcript analysis of transposon *MuA2*, a regulator of *Mutator* transposable element activity in maize. Plant Mol Biol 21, 1181–1185.

Jiang, N., Bao, Z., Zhang, X., Eddy, S.R., and Wessler, S.R. (2004). Pack-MULE transposable elements mediate gene evolution in plants. Nature 431, 569–573.

Jiao, Y., and Deng, X.W. (2007). A genome-wide transcriptional activity survey of rice transposable element-related genes. Genome Biol 8, R28.

Joanin, P., Hershberger, R.J., Benito, M.I., and Walbot, V. (1997). Sense and antisense transcripts of the maize *MuDR* regulatory transposon localized by in situ hybridization. Plant Mol Biol 33, 23–36.

Juretic, N., Hoen, D.R., Huynh, M.L., Harrison, P.M., and Bureau, T.E. (2005). The evolutionary fate of MULE-mediated duplications of host gene fragments in rice. Genome Res 15, 1292–1297.

Kidwell, M.G., Kimura, K., and Black, D.M. (1988). Evolution of hybrid dysgenesis potential following P element contamination in *Drosophila melanogaster*. Genetics 119, 815–828.

Kikuchi, S., Satoh, K., Nagata, T., Kawagashira, N., Doi, K., Kishimoto, N., Yazaki, J., Ishikawa, M., Yamada, H., Ooka, H., Hotta, I., Kojima, K., Namiki, T., Ohneda, E., Yahagi, W., Suzuki, K., Li, C.J., Ohtsuki, K., Shishiki, T., Otomo, Y., Murakami, K., Iida, Y., Sugano, S., Fujimura, T., Suzuki, Y., Tsunoda, Y., Kurosaki, T., Kodama, T., Masuda, H., Kobayashi, M., Xie, Q., Lu, M., Narikawa, R., Sugiyama, A., Mizuno, K., Yokomizo, S., Niikura, J., Ikeda, R., Ishibiki, J., Kawamata, M., Yoshimura, A., Miura, J., Kusumegi, T., Oka, M., Ryu, R., Ueda, M., Matsubara, K., Kawai, J., Carninci, P., Adachi, J., Aizawa, K., Arakawa, T., Fukuda, S., Hara, A., Hashizume, W., Hayatsu, N., Imotani, K., Ishii, Y., Itoh, M., Kagawa, I., Kondo, S., Konno, H., Miyazaki, A., Osato, N., Ota, Y., Saito, R., Sasaki, D., Sato, K., Shibata, K., Shinagawa, A., Shiraki, T., Yoshino, M., Hayashizaki, Y., and Yasunishi, A. (2003). Collection, mapping, and annotation of over 28,000 cDNA clones from japonica rice. Science 301, 376–379.

Kim, S.H., and Walbot, V. (2003). Deletion derivatives of the *MuDR* regulatory transposon of maize encode antisense transcripts but are not dominant-negative regulators of mutator activities. Plant Cell 15, 2430–2447.

Kipling, D., and Warburton, P.E. (1997). Centromeres, CENP-B and Tigger too. Trends Genet 13, 141–145.

Lai, J., Ma, J., Swigonova, Z., Ramakrishna, W., Linton, E., Llaca, V., Tanyolac, B., Park, Y.J., Jeong, O.Y., Bennetzen, J.L., and Messing, J. (2004). Gene loss and movement in the maize genome. Genome Res. 14, 1924–1931.

Le, Q.H., Wright, S., Yu, Z., and Bureau, T. (2000). Transposon diversity in *Arabidopsis thaliana*. Proc Natl Acad Sci U S A 97, 7376–7381.

Le Rouzic, A., and Capy, P. (2005). The first steps of transposable elements invasion: parasitic strategy vs. genetic drift. Genetics 169, 1033–1043.

Lesage, P., and Todeschini, A.L. (2005). Happy together: the life and times of *Ty* retrotransposons and their hosts. Cytogenet Genome Res 110, 70–90.

Levy, A.A., and Walbot, V. (1991). Molecular analysis of the loss of somatic instability in the *bz2::mu1* allele of maize. Mol Gen Genet 229, 147–151.

Levy, A.A., Britt, A.B., Luehrsen, K.R., Chandler, V.L., Warren, C., and Walbot, V. (1989). Developmental and genetic aspects of *Mutator* excision in maize. Dev Genet 10, 520–531.

Lin, R., Ding, L., Casola, C., Ripoll, D.R., Feschotte, C., and Wang, H. (2007). Transposase-derived transcription factors regulate light signaling in Arabidopsis. Science 318, 1302–1305.

Lisch, D. (2002). *Mutator* transposons. Trends Plant Sci 7, 498–504.

Lisch, D. (2005). Pack-MULEs: theft on a massive scale. Bioessays 27, 353–355.

Lisch, D., and Freeling, M. (1994). Loss of *Mutator* activity in a minimal line. Maydica 39, 289–300.

Lisch, D., Chomet, P., and Freeling, M. (1995). Genetic characterization of the *Mutator* system in maize: behavior and regulation of *Mu* transposons in a minimal line. Genetics 139, 1777–1796.

Lisch, D., Girard, L., Donlin, M., and Freeling, M. (1999). Functional analysis of deletion derivatives of the maize transposon *MuDR* delineates roles for the MURA and MURB proteins. Genetics 151, 331–341.

Lisch, D., Carey, C.C., Dorweiler, J.E., and Chandler, V.L. (2002). A mutation that prevents paramutation in maize also reverses *Mutator* transposon methylation and silencing. Proc Natl Acad Sci U S A 99, 6130–6135.

Lisch, D.R., Freeling, M., Langham, R.J., and Choy, M.Y. (2001). *Mutator* transposase is widespread in the grasses. Plant Physiol 125, 1293–1303.

Lowe, B., Mathern, J., and Hake, S. (1992). *Active Mutator* elements suppress the knotted phenotype and increase recombination at the *Kn1-O* tandem duplication. Genetics 132, 813–822.

Lu, C., Tej, S.S., Luo, S., Haudenschild, C.D., Meyers, B.C., and Green, P.J. (2005). Elucidation of the small RNA component of the transcriptome. Science 309, 1567–1569.

Lu, C., Kulkarni, K., Souret, F.F., MuthuValliappan, R., Tej, S.S., Poethig, R.S., Henderson, I.R., Jacobsen, S.E., Wang, W., Green, P.J., and Meyers, B.C. (2006). MicroRNAs and other small RNAs enriched in the Arabidopsis RNA-dependent RNA polymerase-2 mutant. Genome Res 16, 1276–1288.

Ma, J., Devos, K.M., and Bennetzen, J.L. (2004). Analyses of LTR-retrotransposon structures reveal recent and rapid genomic DNA loss in rice. Genome Res 14, 860–869.

Makarova, K.S., Aravind, L., and Koonin, E.V. (2002). SWIM, a novel Zn-chelating domain present in bacteria, archaea and eukaryotes. Trends Biochem Sci 27, 384–386.

Malik, H.S., Burke, W.D., and Eickbush, T.H. (1999). The age and evolution of non-LTR retrotransposable elements. Mol Biol Evol 16, 793–805.

Martienssen, R., and Baron, A. (1994). Coordinate suppression of mutations caused by Robertson's *Mutator* transposons in maize. Genetics 136, 1157–1170.

Martienssen, R., Barkan, A., Taylor, W.C., and Freeling, M. (1990). Somatically heritable switches in the DNA modification of *Mu* transposable elements monitored with a suppressible mutant in maize. Genes Dev 4, 331–343.

May, E.W., and Craig, N.L. (1996). Switching from cut-and-paste to replicative Tn7 transposition. Science 272, 401–404.

McCarty, D.R., Carson, C.B., Stinard, P.S., and Robertson, D.S. (1989). Molecular analysis of *viviparous-1:* an abscisic acid-insensitive mutant of maize. Plant Cell 1, 523–532.

McGinnis, K., Murphy, N., Carlson, A.R., Akula, A., Akula, C., Basinger, H., Carlson, M., Hermanson, P., Kovacevic, N., McGill, M.A., Seshadri, V., Yoyokie, J., Cone, K., Kaeppler, H.F., Kaeppler, S.M., and Springer, N.M. (2007). Assessing the efficiency of RNA interference for maize functional genomics. Plant Physiol 143, 1441–1451.

Miller, W.J., McDonald, J.F., Nouaud, D., and Anxolabehere, D. (1999). Molecular domestication–more than a sporadic episode in evolution. Genetica 107, 197–207.

Morgante, M., Brunner, S., Pea, G., Fengler, K., Zuccolo, A., and Rafalski, A. (2005). Gene duplication and exon shuffling by helitron-like transposons generate intraspecies diversity in maize. Nat Genet 37, 997–1002.

Neuveglise, C., Chalvet, F., Wincker, P., Gaillardin, C., and Casaregola, S. (2005). *Mutator*-like element in the yeast *Yarrowia lipolytica* displays multiple alternative splicings. Eukaryot Cell 4, 615–624.

Ohtsu, K., Smith, M.B., Emrich, S.J., Borsuk, L.A., Zhou, R., Chen, T., Zhang, X., Timmermans, M.C., Beck, J., Buckner, B., Janick-Buckner, D., Nettleton, D., Scanlon, M.J., and Schnable, P.S. (2007). Global gene expression analysis of the shoot apical meristem of maize *(Zea mays L.)*. Plant J.

Oishi, K., and Freeling, M. (1983). The Mu3 transposon in maize. In Plant Transposable Elements, O.N.e. al, ed (New York: Plenum Press), pp. 289–292.

Ono, A., Kim, S.H., and Walbot, V. (2002). Subcellular localization of MURA and MURB proteins encoded by the maize *MuDR* transposon. Plant Mol Biol 50, 599–611.

Pace, J.K., 2nd, and Feschotte, C. (2007). The evolutionary history of human DNA transposons: evidence for intense activity in the primate lineage. Genome Res 17, 422–432.

Pikaard, C.S. (2006). Cell biology of the Arabidopsis nuclear siRNA pathway for RNA-directed chromatin modification. Cold Spring Harb Symp Quant Biol 71, 473–480.

Pontes, O., Li, C.F., Nunes, P.C., Haag, J., Ream, T., Vitins, A., Jacobsen, S.E., and Pikaard, C.S. (2006). The Arabidopsis chromatin-modifying nuclear siRNA pathway involves a nucleolar RNA processing center. Cell 126, 79–92.

Pooma, W., Gersos, C., and Grotewold, E. (2002). Transposon insertions in the promoter of the *Zea mays a1* gene differentially affect transcription by the Myb factors P and C1. Genetics 161, 793–801.

Qin, M.M., Robertson, D.S., and Ellingboe, A.H. (1991). Cloning of the Mutator transposable element *MuA2*, a putative regulator of somatic mutability of the *a1-mum2* allele in maize. Genetics 129, 845–854.

Raina, R., Schlappi, M., and Fedoroff, N. (1998). Epigenetic mechanisms in the regulation of the maize Suppressor-*mutator* transposon. Novartis Found Symp 214, 133–140; discussion 140–133, 163–137.

Raizada, M.N., and Walbot, V. (2000). The late developmental pattern of *Mu* transposon excision is conferred by a cauliflower mosaic virus 35S -driven MURA cDNA in transgenic maize. Plant Cell 12, 5–21.

Raizada, M.N., Nan, G.L., and Walbot, V. (2001a). Somatic and germinal mobility of the *RescueMu* transposon in transgenic maize. Plant Cell 13, 1587–1608.

Raizada, M.N., Benito, M.I., and Walbot, V. (2001b). The *MuDR* transposon terminal inverted repeat contains a complex plant promoter directing distinct somatic and germinal programs. Plant J 25, 79–91.

Robertson, D.S. (1978). Characterization of a mutator system in maize. Mutat. Res. 51, 21–28.

Robertson, D.S. (1980). The Timing of *Mu* Activity in Maize. Genetics 94, 969–978.

Robertson, D.S. (1981). Mutator Activity in maize: timing of its activation in ontogeny. Science 213, 1515–1517.

Robertson, D.S. (1983). A possible dose-dependent inactivation of mutator (*Mu*) in maize. Mol Gen Genet 191, 86–90.

Robertson, D.S. (1985). Differential activity of the maize mutator *Mu* at different loci and in different cell lineages. Mol. Gen. Gent. 200, 9–13.

Robertson, D.S. (1986). Genetic Studies on the Loss of *Mu* mutator activity in Maize. Genetics 113, 765–773.

Robertson, D.S., and Stinard, P.S. (1993). Evidence for *Mu* activity in the male and female gametophytes of maize. Maydica 38, 145–150.

Rudenko, G.N., and Walbot, V. (2001). Expression and post-transcriptional regulation of maize transposable element *MuDR* and its derivatives. Plant Cell 13, 553–570.

Rudenko, G.N., Ono, A., and Walbot, V. (2003). Initiation of silencing of maize *MuDR/Mu* transposable elements. Plant J 33, 1013–1025.

SanMiguel, P., Gaut, B.S., Tikhonov, A., Nakajima, Y., and Bennetzen, J.L. (1998). The paleontology of intergene retrotransposons of maize. Nat. Genet. 20, 43–45.

SanMiguel, P., Tikhonov, A., Jin, Y.K., Motchoulskaia, N., Zakharov, D., Melake-Berhan, A., Springer, P.S., Edwards, K.J., Lee, M., Avramova, Z., and Bennetzen, J.L. (1996a). Nested retrotransposons in the intergenic regions of the maize genome. Science 274, 765–768.

Schnable, P.S., and Peterson, P.A. (1988). The *Mutator*-related Cy transposable element of *Zea mays* L. behaves as a near-mendelian factor. Genetics 120, 587–596.

Schnable, P.S., Peterson, P.A., and Saedler, H. (1989). The *bz-rcy* allele of the *Cy* transposable element system of *Zea mays* contains a *Mu-like* element insertion. Mol Gen Genet 217, 459–463.

Singh, J., Freeling, M., and D. Lisch, (2008). A position effect on the heritability of silencing. PLoS Genetics. 4:e1000216. PMID: 18846225.

Slotkin, R.K., Freeling, M., and Lisch, D. (2003). *Mu killer* causes the heritable inactivation of the *Mutator* family of transposable elements in *Zea mays*. Genetics 165, 781–797.

Slotkin, R.K., Freeling, M., and Lisch, D. (2005). Heritable transposon silencing initiated by a naturally occurring transposon inverted duplication. Nat Genet 37, 641–644.

Stinard, P.S., Robertson, D.S., and Schnable, P.S. (1993). Genetic isolation, cloning, and analysis of a mutator-induced, dominant antimorph of the maize amylose extender1 Locus. Plant Cell 5, 1555–1566.

Strommer, J.N., Hake, S., Bennetzen, J.L., Taylor, W.C., and Freeling, M. (1982). Regulatory mutants of the maize *Adh1* gene caused by DNA insertions. Nature 300, 542–544.

Takumi, S., and Walbot, V. (2007). Epigenetic silencing and unstable inheritance of MuDR activity monitored at four *bz2-mu* alleles in maize (*Zea mays L.*). Genes Genet Syst 82, 387–401.

Talbert, L.E., and Chandler, V.L. (1988). Characterization of a highly conserved sequence related to mutator transposable elements in maize. Mol. Biol. Evol. 5, 519–529.

Talbert, L.E., Patterson, G.I., and Chandler, V.L. (1989). *Mu* transposable elements are structurally diverse and distributed throughout the genus *Zea*. J Mol Evol 29, 28–39.

Taylor, L.P., and Walbot, V. (1987). Isolation and characterization of a 1.7-kb transposable element from a mutator line of maize. Genetics 117, 297–307.

Toth, M., Grimsby, J., Buzsaki, G., and Donovan, G.P. (1995). Epileptic seizures caused by inactivation of a novel gene, *jerky*, related to centromere binding protein-B in transgenic mice. Nat Genet 11, 71–75.

Turcotte, K., Srinivasan, S., and Bureau, T. (2001). Survey of transposable elements from rice genomic sequences. Plant J 25, 169–179.

van Leeuwen, H., Monfort, A., and Puigdomenech, P. (2007). *Mutator*-like elements identified in melon, Arabidopsis and rice contain ULP1 protease domains. Mol. Genet. Genomics 277, 357–364.

Walbot, V. (1986). Inheritance of mutator activity in *Zea mays* as assayed by somatic instability of the *bz2-mu1* allele. Genetics 114, 1293–1312.

Walbot, V. (1991). The *Mutator* transposable element family of maize. Genet Eng (N Y) 13, 1–37.

Walbot, V., and Warren, C. (1988). Regulation of *Mu* element copy number in maize lines with an active or inactive *Mutator* transposable element system. Mol Gen Genet 211, 27–34.

Wang, Q., and Dooner, H.K. (2006). Remarkable variation in maize genome structure inferred from haplotype diversity at the *bz* locus. Proc Natl Acad Sci U S A 103, 17644–17649.

Woodhouse, M.R., Freeling, M., and Lisch, D. (2006a). The *mop1 (mediator of paramutation1)* mutant progressively reactivates one of the two genes encoded by the *MuDR* transposon in maize. Genetics 172, 579–592.

Woodhouse, M.R., Freeling, M., and Lisch, D. (2006b). Initiation, establishment, and mainte-nance of heritable *MuDR* transposon silencing in maize are mediated by distinct factors. PLoS Biol 4, e339.

Xie, Z., Johansen, L.K., Gustafson, A.M., Kasschau, K.D., Lellis, A.D., Zilberman, D., Jacobsen, S.E., and Carrington, J.C. (2004). Genetic and functional diversification of small RNA path-ways in plants. PLoS Biol 2, E104.

Xu, Z., Yan, X., Maurais, S., Fu, H., O'Brien, D.G., Mottinger, J., and Dooner, H.K. (2004). *Jittery*, a *Mutator* distant relative with a paradoxical mobile behavior: excision without reinser-tion. Plant Cell 16, 1105–1114.

Yandeau-Nelson, M.D., Zhou, Q., Yao, H., Xu, X., Nikolau, B.J., and Schnable, P.S. (2005). *MuDR* transposase increases the frequency of meiotic crossovers in the vicinity of a *Mu* inser-tion in the maize a1 gene. Genetics 169, 917–929.

Yates, J.R., Cunningham, R.P., and Holmes, D.S. (1988). IST2: an insertion sequence from Thiobacillus ferrooxidans. Proc Natl Acad Sci U S A 85, 7284–7287.

Yu, W., Lamb, J.C., Han, F., and Birchler, J.A. (2007). Cytological visualization of DNA trans-posons and their transposition pattern in somatic cells of maize. Genetics 175, 31–39.

Yu, Z., Wright, S.I., and Bureau, T.E. (2000). *Mutator*-like elements in Arabidopsis thaliana. Structure, diversity and evolution. Genetics 156, 2019–2031.

Zabala, G., and Vodkin, L.O. (2005). The *wp* mutation of *Glycine max* carries a gene-fragment-rich transposon of the CACTA superfamily. Plant Cell 17, 2619–2632.

Zhang, X., Jiang, N., Feschotte, C., and Wessler, S.R. (2004). PIF- and Pong-like transposable elements: distribution, evolution and relationship with Tourist-like miniature inverted-repeat transposable elements. Genetics 166, 971–986.

Zhang, X., Henderson, I.R., Lu, C., Green, P.J., and Jacobsen, S.E. (2007). Role of RNA polymer-ase IV in plant small RNA metabolism. Proc Natl Acad Sci U S A 104, 4536–4541.

Zhang, X., Feschotte, C., Zhang, Q., Jiang, N., Eggleston, W.B., and Wessler, S.R. (2001). P instability factor: an active maize transposon system associated with the amplification of Tourist-like MITEs and a new superfamily of transposases. Proc Natl Acad Sci U S A 98, 12572–12577.

Zhao, Z.Y., and Sundaresan, V. (1991). Binding sites for maize nuclear proteins in the terminal inverted repeats of the *Mu1* transposable element. Mol Gen Genet 229, 17–26.

The LTR-Retrotransposons of Maize

Phillip SanMiguel and Clémentine Vitte

Abstract The maize genome comprises 150,000–250,000 long terminal repeat (LTR)-retrotransposons, mostly in nested clusters, intermingled with other transposable elements and, more rarely, genes. All told, the genomic landscape of maize is 50–80% retrotransposons. Myriad families exist but >80% of maize retrotransposons belong to the five largest: *Opie-Ji*, *Cinful-Zeon*, *Huck*, *Prem1* and *Grande*. Closely related to animal retroviruses, retrotransposons utilize an RNA intermediate to initiate their transposition. Despite extensive proliferation they are nevertheless suppressed by a variety of mechanisms, including DNA methylation, conversion to heterochromatin and various types of recombinational deletion. Retrotransposons play a large role in the size, structure, gene function and haplotype variation of the maize genome.

1 Introduction

Long terminal repeat (LTR)-retrotransposons are mobile genetic elements that are related to retroviruses, with which they share many properties. They represent 50–80% of maize nuclear DNA and are therefore major players in this genome. Among these many elements, a few high-copy families (Table 1), *Opie-Ji*, *Cinful-Zeon*, *Huck*, *Prem1* and *Grande*, compose a large fraction of the genome (SanMiguel and Bennetzen 1998). Retrotransposons are distributed throughout the maize genome. These large accumulations of retrotransposons may have resulted from higher transposition rates, a low rate of removal, or both in maize when compared to smaller genome grasses such as sorghum or rice.

Molecular dating of maize retrotransposons indicates that no single "amplification event", where large increases in the numbers of all retrotransposons occurred

P. SanMiguel
Purdue University, Department of Horticulture and Landscape Architecture,
pmiguel@purdue.edu

C. Vitte
UMR de Génétique Végétale, Equipe Génétique Evolutive : Adaptation et Redondance, vitte@moulon.inra.fr

J.L. Bennetzen and S. Hake (eds.), *Maize Handbook - Volume II: Genetics and Genomics*,   307
© Springer Science+Business Media LLC 2009

Table 1 The major retrotransposon families of maize

Family	Type	Size (kb)	LTR (kb)	Inner (kb)	GC %	Reference
Opie	RLC	8.7	1.2	6.3	46	SanMiguel et al. 1996
Ji	RLC	9.5	1.3	6.9	45	Turcich et al. 1996
Cinful	RLG	9.5	0.6	8.3	43	Sanz-Alferez et al. 2003
Zeon	RLG	7.5	0.7	6.1	44	Hu, Das et al. 1995
Huck	RLG	13.3	1.6	10.1	62	SanMiguel et al. 1996
Prem1	RLG	9.4	3.3	2.8	43	Turcich and Mascarenhas 1994
Grande	RLG	13.4	0.6	12.2	52	Garcia-Martinez and Martinez-Izquierdo 2003

over a short span of time, can account for the numbers of elements currently seen in the maize genome. However, it is possible that a large proportion of them appeared during just a few major amplification events (SanMiguel, Gaut et al. 1998). Their origin through successive waves of transposition is evident in the characteristic nested clusters of insertion (Fig. 1) that allows the insertion order and even the timing of their insertion to be determined. This review describes maize retrotransposons, and gives insights on their structure, behavior, genomic organization, and regulation. It also describes their possible fates after insertion in the genome and discusses their many contributions to maize genome structure and evolution.

1.1   Classification

Transposable elements compose a large boundary class of genetic entities that are neither gene nor virus. Like cellular genes, they share the fate of the cell and organism in which they reside. But, like viruses, they encode the means to produce additional copies of themselves. In other words, a transposable element family, though fully a component of a normal genome, also exists in an evolutionary space separate from that of the genic components of that genome. Probably as a result of this, vast numbers of transposable element families, possessing various mechanisms of transposition, inhabit any genome.

Recently, an attempt to formalize the existing classification and nomenclature system for eukaryotic transposable elements has been undertaken (Wicker, Sabot et al. 2007). Using this scheme retrotransposons would be classified under a "class, order, superfamily" nomenclature as either "RLG" or "RLC". That is, they are class retroelements (R), order LTRs (L) and they are either of superfamily *Copia* (C), or *Gypsy* (G). For example: "*RLC_Opie*" or "*RLG_Huck*". However, to simplify the text, we will largely refer to element families without prefixing them with this field. Neither will we prefix "retrotransposon" with "LTR-". Frequently the terms "retroelement" and "retrotransposon" are used interchangeably, but here we will only use the word "retrotransposon" to refer to LTR-retrotransposons.

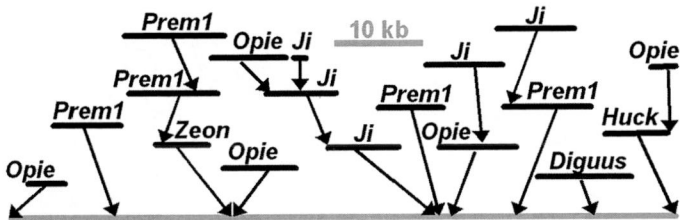

Fig. 1 Diagram of insertion orders of retrotransposons in a 230 kb segment of the maize genome (Ilic et al. 2003)

1.2   Structure and Mechanism of Transposition

Retrotransposons transpose via an elaborate mechanism outlined in Fig. 2. Briefly, a "genomic RNA", the full-length mRNA of an element, is reverse-transcribed into double-stranded DNA and inserted back into the host cell's genome in a different location. This inserted DNA may then be transcribed, starting a new transposition cycle.

The full details of the regeneration of a DNA retrotransposon from its genomic RNA likely vary among families and, in any case, have not been comprehensively studied in plants. But the process is thought to proceed similarly to that of retroviruses, as described in Fig. 2. The genomic RNA is transported to the cytoplasm, translated and bundled together with its gene products into a structure called a virus-like particle (VLP). There, its full-length DNA form is recreated from the RNA and imported back into the nucleus, complexed with INTEGRASE (IN). At the site of integration, IN cuts both strands of a chromosome, staggering the cuts to create 5 base overhangs. The retrotransposon is integrated and the overhangs are filled-in, thus creating characteristic 5 bp target site duplications (TSD).

This process affords the opportunity for recombination to occur, especially where synthesis of the full double-stranded DNA is forced into "template switches" after polymerization halts at the ends of the RNA template, creating "strong-stop" DNA intermediates. Note that although only one RNA molecule is represented in Fig. 2, multiple genomic RNAs are co-packaged during VLP formation (Feng, Moore et al. 2000). This allows strand-transfer to occur both intra-molecularly (as depicted in Fig. 2) and/or inter-molecularly.

1.3   Major Maize Retrotransposon Families

Classically, transposable element families were defined genetically. If trans-acting factors of one element could catalyze the transposition of another element both were considered members of a single family. However, the exponentially falling price of sequence data has resulted in a surfeit of structural data and little functional

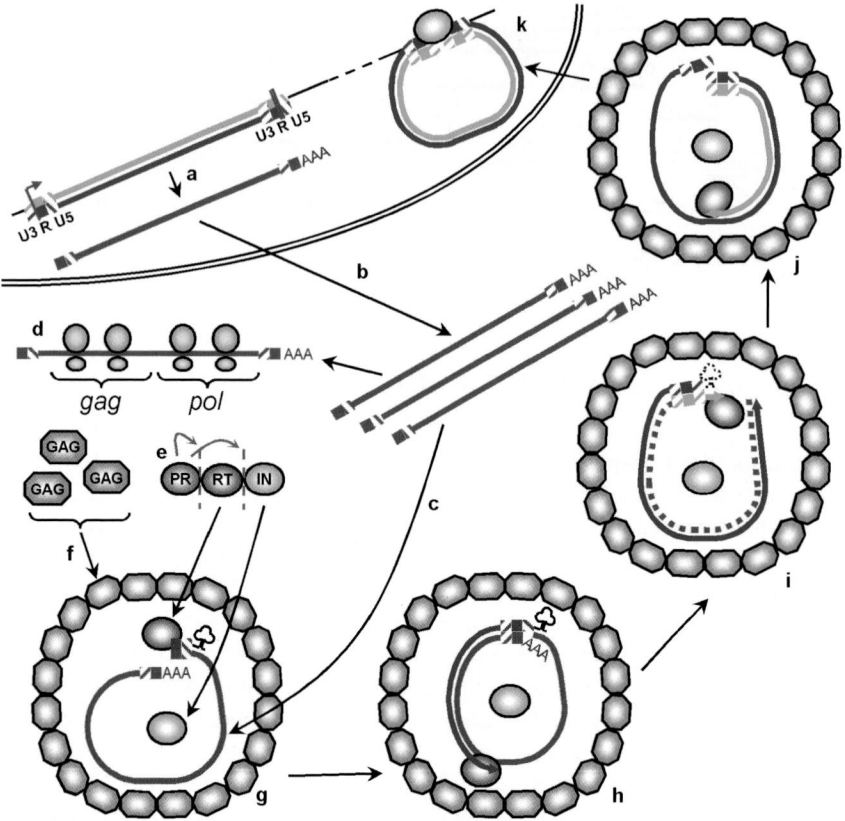

Fig. 2 Model life-cycle of retrotransposons. **a.** Transcription produces a genomic RNA (green) plus-strand form of the provirus. **b.** Export of the RNA to the cytoplasm. **c.** Incorporation of the RNA into a Virus-Like Particle (VLP) **d.** Translation of *gag* and *pol* mRNAs. **e.** Cleavage of POL by PR (protease) and incorporation of RT (reverse transcriptase) and IN (integrase) proteins into the VLP **f.** GAG assembly to form the VLP **g.** Synthesis of the minus-strand strong-stop DNA by RT, initiated by annealing of a tRNA (clover shape). **h.** After "strand switch" to the R-U3 region, minus strand (dark blue) synthesis continues. **i.** Start of the plus strand (light blue) strong-stop DNA synthesis primed from the RNAse-resistant PolyPurine Tract (PPT) just upstream of the U3 region. Once the PBS (primer binding site) is synthesized, degradation of the tRNA j.Annealing of the PBS regions of the two DNA strands and end of synthesis of plus strand DNA k. Integration of the nascent retrotransposon at a new location by IN. Symbols and abbreviations: GAG: structural protein, PR: protease, RT: reverse-transcriptase, IN: integrase. Symbols are drawn at arbitrary scale and stoichiometry

data. Thus sequence-based definitions of "families" have and are being devised to allow family classification in the absence of any functional data. The actual overlap of these two methods of classification is uncertain.

How sequence-stringent is the transposition machinery of these families? At minimum, IN must bind to and insert the termini of a retrotransposon into host DNA for the nascent element to successfully insert into the genome. In yeast, Ty1

IN has been shown to integrate segments of DNA containing as little as 12 bp of each Ty1 LTR terminus at 50% efficiency (Eichinger and Boeke 1990) suggesting that even meager stretches of terminal homology may be sufficient to bind fairly distantly related elements into functional families. Table 2 shows the termini of fourteen commonly observed maize retrotransposon families and groups them into putative meta-families. Continuing this theme, Table 3 shows the similarity of sequence adjacent to the internal boundaries of the LTRs of meta-family members.

1.3.1 Opie-Ji

The most numerous family in maize, *Opie-Ji*, is of the *Copia* superfamily. *Opie* and *Ji* LTRs generally diverge by >50%. Thus, by at least one definition (SanMiguel, Tikhonov et al. 1996) they compose separate families. However, given the similarity of their LTR termini (Table 2) it is possible that *Opie*, *Ji*, *Ruda* and perhaps even *Giepum* form a single genetic family.

One component of these LTRs, the "maize palindromic unit" (MPU) (Quayle, Brown et al. 1989), is a short sequence (consensus: CACCGGACANTGTCCGGTG) that exists in multiple copies in every LTR of this family. As a result the MPU is among the most common palindromes in the maize genome, with tens of thousands of copies. Presumably these sequences serve some sort of regulatory function. *Ji* (also known as "PREM-2") is heavily transcribed during early microspore development (Turcich, Bokhari-Riza et al. 1996) but is also transcribed in other tissues (Avramova, SanMiguel et al. 1995).

1.3.2 Cinful-Zeon

Cinful is a *Gypsy* superfamily member. *Cinful* and *Zeon* both appear to be defective, the former lacks *gag* and the latter lacks *pol*. Sequence similarity between their LTRs, especially at their termini (Table 2), is sufficient to suggest that *Cinful* and *Zeon* share gene products in order to successfully transpose (Sanz-Alferez, SanMiguel et al. 2003).

1.3.3 Huck

Huck is a large, GC-rich, *Gypsy* superfamily retrotransposon. Several sub-families exist, including one that encodes POL and another that does not. Some estimates (Meyers, Tingley et al. 2001) place this family, rather than *Opie-Ji*, as the most numerous in the maize genome. But even if lower (SanMiguel and Bennetzen 1998) estimates of the numbers of *Huck* insertions in the maize genome are correct, at least 10% of the maize genome consists of this element. Given the GC-content of *Huck*, which averages 62%, the amplification of this family has increased the maize genome GC content by at least a few percent.

Table 2 LTR Termini of several common maize retrotransposon families

Family	Type	5′ End	3′ End
Opie	RLC	TGAAAGGGAAATGTGCCCTT	CCCCCCTCTAGGTGCTCTCA
Ji	RLC	TGAAAGGGAATTAGGCTTAC	CCCCCTCTAGGCGACTWTCA
Giepum	RLC	TGAAAGCATCTAGGSCCCTG	TGGGCATCGTGATCCTTTCA
Ruda	RLC	TGTAAGGAAAATGGACCCCG	GGCGTCACCCGTTTCCTACA
Cinful	RLG	TGTTGGGACCATGCTTCGTC	TTGAGAACAAGTCCCCAACA
Zeon	RLG	TGTTGGGGGCCTTCGGCTTC	TTGAGAACAAGTCCCCAACA
Prem1	RLG	TGTAATACCCAMNGTTGAAN	CCTTAAAGCCGGGTGTGACA
Xilon	RLG	TGTAACGCCCCGAATTTTGC	TCTAAAAACCGGGTGTGACA
Diguus	RLG	TGTAACACCCTGAATTTTGG	CTTCAAAACCGGGTGTGACA
Tekay	RLG	TGTAACACCATAAAAGCCAT	CCCTGGAACCGGGTGTGACA
Huck	RLG	TGTCGGGGACCATAATTAGG	CCKGTCTCGAAACGCCGACA
Grande	RLG	TGTGGGGGATAGATATCCCC	TCGGACCCAAAACACCGACA
Fourf	RLC	TGTTGGATCTTTTATGGGCT	AAATGCCTATATTTCTAACA
Machiavelli	RLC	TGTTAGGATTTATGGGCTTG	ATTTGTCTAATTATTCAACA

Table 3 The ends of the internal domains (between the LTRs) of several common maize retrotransposon families

Family	Type	(Primer Binding Site) 5′ End	(PolyPurine Tract) 3′ End
Opie	RLC	ATTGGTATCRGAGCCGTTCT	ATCACCAAAAAGGGGGAGAT
Ji	RLC	ATTGGTATCAGAGCCCGGTG	ATCACCAAAAAGGGGGAGAT
Giepum	RLC	ATTGGTATCAAAGCCTTGTT	ATTACCAAAAAGGGGGAGAT
Ruda	RLC	AGTGGTATCAAAGCCCGGTT	ATCACCAAAAAGGGGGAGAT
Cinful	RLG	TTGGCGCCCACCTCCGGTGA	TAGCACCGCGAAGGGGCTAC
Zeon	RLG	TTGGCGCCCACCTCCGGTGA	TTAATATTGCGAGGGGCTAC
Prem1	RLG	GAAGTGGTATCAGAGGAAAT	TTTTTAAGGGGGGTAGGATT
Xilon	RLG	TAAGTGGTATCAAAGCCGTG	TCTTTTAAGGGGGATAGGTT
Diguus	RLG	TAAGTGGTATCAAAGTCGTG	CCTGTTAAGGGGGTTAGATT
Tekay	RLG	GAAGTGGTATCAAAGCTATG	CTTTTTAAGAGGGGTAGGTT
Huck	RLG	GTTGGCGCGCCAGGTAGGGG	CTTCGAGGCTCGGGGGCTAC
Grande	RLG	GCTGGCGCGCCAGGTAGGGG	CAACCCAAAGATCGGGGGCT
Fourf	RLG	ATCCAAAAACCTAATGTTAG	CCCTAGAGTTTGGTGGGGAT
Machiavelli	RLG	ATCCAAAAACCTTATTGTAG	TTTGAGATCTGGTGGGGGAT

1.3.4   Prem1

Prem1 is a *Gypsy* superfamily element with unusually large LTRs (Table 1). While termini-intact full insertions of this family have been found (Fu, Zheng et al. 2002), the degree of deletion and sequence divergence seen among its members suggests that the majority inserted long ago. Sequence similarity, especially at the LTR termini (Table 2) suggest that *Xilon* (Fu and Dooner 2002), *Diguus* (Ilic, SanMiguel et al. 2003) and *Tekay* (SanMiguel and Bennetzen 1998) are members of an extended *Prem1* family.

Prem1 appears to be highly expressed in early to mid-stage uninucleate microspores and, to a lesser extent in root and cob tissues (Turcich and Mascarenhas 1994).

1.3.5 Grande

The *Gypsy* superfamily element, *Grande*, contains the largest internal domain (Table 1) of all high copy maize retrotransposons. These elements also feature multiple clusters of 50–150 base tandem repeats that appear to have stable secondary structures (Monfort, Vicient et al. 1995). An unnamed element in tobacco and *Cinful-Zeon* of maize share this feature (Sanz-Alferez et al. 2003). The conservation of these tandem repeats across such distant families of retrotransposons suggests some structural purpose—perhaps a binding site for GAG or a genomic RNA dimerization site.

2   The Retroviral Nature of Retrotransposons

Retrotransposons display a number of behaviors in common with the behaviors of retroviruses. As they are closely related this is to be expected. However the structure and extent of their relatedness is unclear.

2.1   *Transduction of Cellular Genes*

Like their retroviral cousins, retrotransposons have been shown to incorporate host cellular genes into their own sequence. Note that the process of reverse transcription occurring in a VLP containing multiple RNAs combined with RT's proclivity for template switching provides an obvious mechanism for how this might have occurred.

Many or all of the *Gypsy* superfamily retrotransposons in Table 1 contain segments of DNA of uncertain origin that may have been obtained from functional maize genes. However, the evidence of a retrotransposon transducing a cellular gene is strongest for the maize *Bs1* retrotransposon (Jin and Bennetzen 1989). *Bs1* is a defective element, lacking a functional *pol* ORF, but does possess 654 bp of a spliced proton ATPase gene (Bureau, White et al. 1994; Jin and Bennetzen 1994; Palmgren 1994). A later study noted that portions of two other cellular genes were also present in *Bs1* (Elrouby and Bureau 2001).

The functional significance of these transduction events is uncertain. It remains possible that host genes carried in this manner somehow confer a special resistance to host repression of transposition or even are intermediates on the path to acquisition of a functional *env* that would allow the element to "go viral" and exit the host cell. But while plant retroviruses may or may not exist (see below), plant retrotransposons do. So it should also be considered that these gene fragments may simply serve as "stuffers" that bring what would otherwise be a completely non-mobile element up to a size and composition that allows favorable competition for incorporation into VLPs of a given retrotransposon family.

2.2   The Case for Plant Retroviruses

Retrotransposons are closely related to retroviruses (Xiong and Eickbush 1990). In principle, a retrotransposon differs from a retrovirus only in its ability to exit the cell (and perhaps the organism) of its origin and enter another cell and integrate into that genome. The gene conferring this ability is located downstream of *pol* in retroviruses and is called *env* for the "envelope protein" it encodes. Unlike *pol* and, to a lesser extent, *gag*, the sequence of *env* is not well conserved.

Discovery that a *Drosophila* retrotransposon, *gypsy* (Kim, Terzian et al. 1994; Song, Gerasimova et al. 1994), can form particles that can infect some *Drosophila* lines provided the first example of a non-vertebrate retrovirus.

While no direct evidence for the infection of a plant by a retrovirus has yet been revealed, various indirect evidence suggests plant retroviruses might exist. The discovery of ORFs downstream of *pol* in a number of plant *Gypsy* superfamily retrotransposons (Wright and Voytas 1998; Wright and Voytas 2002) as well as one case in a *Copia* superfamily retrotransposon (Laten, Majumdar et al. 1998) seemed likely to quickly result in a discovery paralleling that made in *Drosophila*. Disappointingly, strong evidence is still lacking.

The structure of plant cells may make a retroviral lifestyle less effective than it is in animals. Given the failure, as of yet, to demonstrate their ability to produce infectious plant retroviral particles, it may be that these downstream ORFs have an alternate, and completely unrelated, function.

3   Host Control and Domestication of Retrotransposons

Once inserted in the genome, a new copy of a retrotransposon can be affected by several processes, including DNA methylation, unequal recombination and the accumulation of indels. Methylation is expected to generally lead to the transcriptional inactivation of the retrotransposon and a progressive decay of its sequence through the accumulation of substitutions (particularly transitions, as we discuss below), whereas recombination and the accumulation of small deletions will eventually lead to the elimination of the inserted sequence. On the other hand, the inserted copy may sometimes provide a selective advantage to the host plant, leading to the "domestication" of the corresponding element by the plant genome.

3.1   DNA Methylation and Heterochromatinization

Retrotransposons are targeted, and likely silenced, by RNA-directed DNA methylation (Hamilton, Voinnet et al. 2002; Huettel, Kanno et al. 2007; Matzke, Kanno et al. 2007). Cytosine methylation has been associated with histone modifications that characterize a heterochromatic state: lack of transcription, and inactivation of transposition (Lippman, Gendrel et al. 2004). Because transposable elements are insertional mutations and can cause chromosomal rearrangements, this methylation process has been proposed to modify and reorganize chromatin to inactivate retrotransposons, therefore reducing the impact of these elements on the "host" genome (Martienssen 1998; Bender 2004).

In contrast to genic sequences, maize repetitive DNAs, largely retrotransposons, show a high degree of cytosine 5-methylation at 5'-CG-3' and 5'-CNG-3' sites (Bennetzen, Schrick et al. 1994). So prevalent is this modification to repetitive DNA that it has been used to build gene-enriched libraries of maize (Palmer, Rabinowicz et al. 2003; Emberton, Ma et al. 2005). Sequence analysis of these libraries confirmed that most DNA methylation in maize is restricted to transposable elements. However, it showed that not all retrotransposon sequences are methylated, suggesting that some of them escape this type of regulation. This is further supported by retrotransposon sequences present in EST databases (Meyers et al. 2001), although some of these ESTs may derive from transcripts driven by flanking, rather than retrotransposon, promoters.

DNA methylation is also associated with a higher C to T transition rate. For instance, transition to transversion ratios in retrotransposon sequences from maize and four other angiosperm species were found to be ~1.5:1, and therefore higher than the 1:1 rate generally observed for genic sequences, including introns (Vitte and Bennetzen 2006). 5-Methylation both drastically increases cytosine's spontaneous deamination rate (Coulondre, Miller et al. 1978) and prevents uracil N-glycosylase-mediated repair mechanisms from reversing the transition. Hence, retrotransposons should mutate faster than non-methylated DNA and gradually grow depleted in cytosines at CG and CNG sites.

3.2   Unequal Recombination

The observation of rearranged retrotransposon copies deriving from unequal recombination in several plant genomes has revealed the importance of this process in shaping the structure of retrotransposons. Different types of rearrangements are described below. They suggest that the extent of unequal recombination is low in maize.

3.2.1   Solo LTRs

As shown in yeast (Roeder and Fink 1980), unequal intra-strand homologous recombination between the two LTRs of a given retrotransposon leads to the elimination of the internal region and one recombined LTR, and the formation of a solo LTR flanked by the original TSD of the complete element (Fig. 3B).

Solo LTRs are seen in any plant genome that contains retrotransposons. Therefore, intra-strand unequal homologous recombination would seem common in plants.

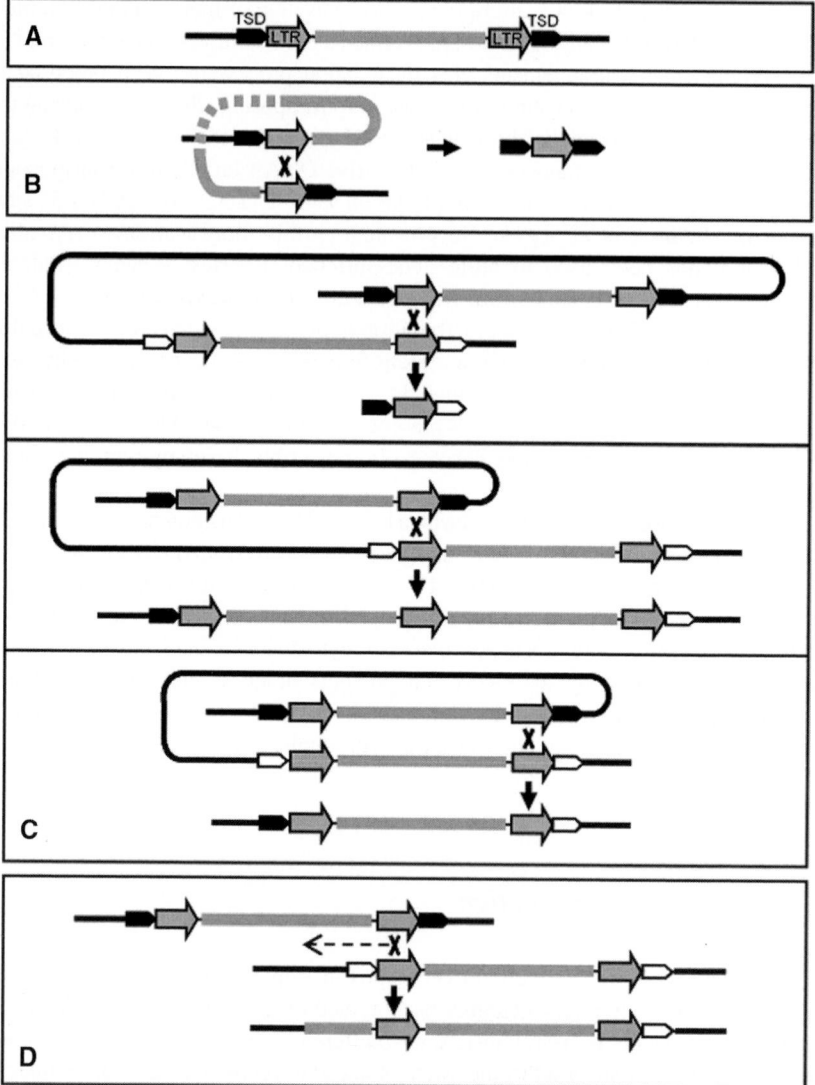

Fig. 3 Unequal recombination of retrotransposons. A. Complete retrotransposon, flanked by 5 bp TSD (not to scale). B. Solo LTR with TSD resulting from intra-element recombination. C. Three outcomes of intra-strand inter-element unequal recombination. D. Branch migration (conversion) during inter-element unequal recombination without exchange of flanking markers. "X" indicates the site of initiation and the dashed arrow shows the direction of branch migration

However, the extent of this process seems to differ among species (Vitte and Bennetzen 2006). For instance, Arabidopsis and rice, two species for which extensive analyses of solo LTRs have been performed, show different rates of intact elements to solo LTRs (<0.6:1 in rice, compared to <1:1 in Arabidopsis). In maize, intact retrotransposons outnumber solo LTRs by a ratio of 5:1 to 25:1 in several genomic regions investigated (SanMiguel et al. 1996; Fu and Dooner 2002; Brunner, Fengler et al. 2005; Ma, SanMiguel et al. 2005; Liu, Vitte et al. 2007).

3.2.2   Inter-Element Recombination

Intact elements and solo LTRs lacking TSDs are common (Devos, Brown et al. 2002; Ma, Devos et al. 2004; Vitte and Bennetzen 2006). These likely derive from unequal homologous recombination between two different elements of the same family (Fig. 3C). The frequency of inter-element recombination also varies among species. For instance, they represent 5.6% of the copies in rice (Ma et al. 2004), and <1% in Arabidopsis (Devos et al. 2002). Other types of inter-element recombination have been described in Arabidopsis. For example, a rare event led to a conjoined element with 3 LTRs and lacking TSD (Fig. 3C), (Devos et al. 2002).

3.2.3   Conversion through Branch Migration During Recombination

In some cases, the absence of TSD has been attributed to gene conversion. Unequal recombination need not, in principle, lead to an exchange that results in rearrangement of the retrotransposon structure. A large proportion of recombination events "abort" and do not induce any exchange of sequence around the initiation site. Nevertheless, a "conversion tract" is formed, resulting from branch migration during the initiation of recombination. This feature has been observed for two retrotransposon families in rice (Vitte and Panaud 2003). Recently, a similar case involving a duplication of at least 2,145 bases of a *Grande* element has been documented in maize (Ma et al. 2005). Presumably, after an unequal recombination event between LTRs, branch migration continued outside one LTR overwriting the flanking sequence with the internal sequence of the element (Fig. 3D). The process could lead to more complex rearrangements if the repair is not continuous over the conversion tract, suggesting that unequal recombination may explain a larger group of rearrangements (Ma et al. 2005).

3.3   *Sequence Erosion Through Accumulation of Deletions*

In addition to the discovery of recombined elements, sequence analysis of retrotransposons has revealed frequent deletions in retrotransposon sequences in Arabidopsis (Devos et al. 2002), and cereals including maize (Jin and Bennetzen 1989; Ramakrishna, Emberton et al. 2002a). Therefore, deletion serves as a common mechanism for elimination of retrotransposon sequences in plants, as it does in other organisms like yeast (Asami, Jia et al. 2002), or insects (Petrov, Sangster et al. 2000). Detailed analysis of these deletions in the retrotransposons of several plant genomes has revealed that they range in size from 1 to 3766 bp with a mean size around 10–300 bp (Devos et al. 2002; Ma et al. 2004; Vitte and Bennetzen 2006). In one multi-locus study in maize, deletions from 1 to 116 bp, with a mean size of ~19 bp (Vitte and Bennetzen 2006) were observed while deletions of up to 2.5 kb are evident in retrotransposons in the maize *adh1* region (SanMiguel and Bennetzen 1998). Most sequences deleted in this manner are flanked by direct duplications (Devos et al. 2002; Wicker, Yahiaoui et al. 2003; Ma et al. 2004; Vitte and Bennetzen 2006), suggesting that they are legacies of illegitimate recombination. These duplications are usually small, from 2 to 53 bp, with an average size around 9 bp, although possibly larger in maize than in other species (mean size of 11.3 bp, compared to a mean size of ~5 bp in other species) (Vitte and Bennetzen 2006).

The accumulation of small deletions in retrotransposon sequences from Arabidopsis (Devos et al. 2002) and rice (Ma et al. 2004; Vitte, Panaud et al. 2007) account for large amounts of DNA loss. Thus, it appears that gradual erosion of retrotransposons in these two genomes leads to the highly fragmented status of most of their retrotransposons. The elimination of retrotransposons in maize appears less aggressive (Vitte and Bennetzen 2006).

3.3.1 Domestication of Retrotransposons

Complete or partial retrotransposon sequences are expected to occasionally contribute to plant fitness. This process is commonly referred to as "molecular domestication" of the element by the host genome (Miller, McDonald et al. 1997).

Several cases of such domestication of retrotransposons have been described in plants, including maize. For instance, retrotransposons contribute to the promoters of many "wild-type" plant genes including many in maize (White, Habera et al. 1994), or can provide new regulatory properties to a gene (Wessler, Bureau et al. 1995). In maize and other cereals, retrotransposons are also suspected of contributing to centromere function, as suggested by reports of a retrotransposon conserved at specific centromere regions in distantly related grass species (Miller, Dong et al. 1998; Presting, Malysheva et al. 1998).

4   Retrotransposons and Maize Genome Evolution

4.1   *Retrotransposons Are the Major Constituents of the Maize Genome*

Maize has a 1C genome size of ~2,700 Mbp. Sequence analyses of subsets of the genome from an unfiltered sheared library (~200,000 sequences representing ~132 Mb), (Whitelaw, Barbazuk et al. 2003), BAC ends (~475,000 sequences representing ~307 Mb, (Messing, Bharti et al. 2004) or random BACs (100 BACs representing 14.38 Mb, (Haberer, Young et al. 2005; Liu et al. 2007) have led to the estimation that at least 58–66% of the genome is repetitive. However, these numbers are underestimates, as a perusal of on-going sequence of maize BAC sequences (such as that available at http://www.agcol.arizona.edu/cgi-bin/msll/mini.cgi) clearly shows. Retrotransposons represent over 78% of this repetitive fraction, and therefore dominate any other class of repeat in the genome (Messing et al. 2004). Clearly, retrotransposons are the major constituent of the maize genome.

Some families of retrotransposons are highly repeated (those listed in Table 1) whereas some are middle- (less than 1,000 copies, e.g. *Fourf, Kake, Victim*) or low-copy families (~10 copies, e.g. *Reina*). Altogether, the five most abundant families (*i.e. Opie-Ji, Cinful-Zeon, Huck, Prem1* and *Grande*) are estimated to represent 33–62% of the total nuclear maize genome (SanMiguel and Bennetzen 1998).

4.2   The Location of Retrotransposons in the Genome

High copy retrotransposons are mostly located between genes, a feature that seems characteristic, since *Ac/Ds*, *En/Spm* and *Mu* DNA transposons are known to insert preferentially into genes and low-copy DNA (Kumar and Bennetzen 1999), and MITEs and LINEs are often found in introns (Tikhonov, SanMiguel et al. 1999; Haberer et al. 2005).

Differences in location among retrotransposons families are shown by cytogenetics studies. *Huck* is dispersed along all chromosomes but is more abundant in centromeric regions (Ananiev, Phillips et al. 1998b) and absent from the knob regions, whereas *Grande, Cinful-Zeon* and *Opie-Ji* are abundant in the knob regions and poorly represented around centromeres (Ananiev, Phillips et al. 1998a; Ananiev et al. 1998b; Mroczek and Dawe 2003; Lamb, Meyer et al. 2007). Moreover, *Opie-Ji* is significantly more abundant in gene-containing than gene-free regions, whereas the opposite is true for the *Gyma* family (Liu et al. 2007).

4.3   Retrotransposon Age

The insertion date of a retrotransposon can be estimated from the sequence divergence between its two LTRs (SanMiguel et al. 1998). One presumes that its LTRs are initially identical in sequence (Fig. 2) and that a constant rate of nucleotide substitution occurs over the time since insertion. Knowing an approximate substitution rate, the LTR divergence can then be converted into an insertion date.

Although estimating the divergence between the two LTRs of a retrotransposon is fairly easy, approximating a substitution rate is less so. A common substitution rate used for maize retrotransposons is 1.3×10^{-8} substitutions per site per year. It derives from the substitution rate of synonymous sites of the *adh1* and *adh2* gene sequences combined with dated fossil evidence of the origin of rice (Gaut, Morton et al. 1996) that has been calibrated for the observation that rice retrotransposons sequences mutate 2 to 3-fold faster than genes (Ma and Bennetzen 2004).

Using this method, it appears that all "datable" maize retrotransposons have inserted within the past 4 My, with a mean insertion time around 1 My (SanMiguel et al. 1998; Brunner et al. 2005; Ma et al. 2005; Du, Swigonova et al. 2006; Vitte and Bennetzen 2006; Liu et al. 2007). This does not imply that retrotransposons have only been active in the last 4 My. Older elements are frequently missing at least one LTR, which prevents insertion date estimates. Where the insertion date of highly truncated copies could be estimated in rice, some old copies (e.g., up to 10 My) were found (Ma et al. 2004; Vitte et al. 2007).

4.4 Contribution to Genome Evolution

4.4.1 Impact on Maize Genome Size

Analyses of orthologous regions from maize and closely related species have yielded considerable information on the contribution of retrotransposons to angiosperm genome size variations and maize genome size evolution. Comparisons of *adh1* (Tikhonov et al. 1999; Ilic et al. 2003), *rp1* (Ramakrishna, Emberton et al. 2002b), *z1C1* (Song and Messing 2003), *lg2* and *lrs1* (Langham, Walsh et al. 2004) and *orp1* regions (Ma et al. 2005) showed that genes are mainly conserved, but that the intergenic regions, predominantly retrotransposons, are not. In particular, although maize and sorghum diverged from a common ancestor only 12 million years ago (Swigonova, Lai et al. 2004), the maize genome is more than three-fold larger than the sorghum genome (Arumuganathan and Earle 1991). The presence of large retrotransposon blocks in maize that are not present in orthologous loci of sorghum (Tikhonov et al. 1999) suggests that the difference in genome size observed between maize and sorghum is mainly due to an extensive proliferation of retrotransposons in maize.

In contrast to this amplification process, retrotransposons are gradually eliminated from the genome, through mechanisms detailed above and in Fig. 3. Comparison of five angiosperms including maize revealed that their different sizes, contents, and structures derive from the same mechanistic processes, but acting with different relative efficiencies in different lineages (Vitte and Bennetzen 2006). With the growing amount of sequence data for maize, the timing and extent of these processes will soon be more precisely described.

4.4.2 Impact on Maize Genome Structure

Due to their predominance in the maize genome, retrotransposons play a major role in shaping its structure. Contrary to early expectations of a relatively unorganized mass of intermixed repetitive DNA between genes (Bennetzen et al. 1994; Springer, Edwards et al. 1994) detailed sequence analysis of maize BACs revealed that retrotransposons are largely intact and mostly organized into nested structures (Fig. 1). Comparison of paired loci derived from whole-genome duplication (Swigonova, Bennetzen et al. 2005), for example, shows that the variation of gene density observed in the maize genome is largely the result of retrotransposon insertions.

Due to their high copy numbers and their dispersed nature throughout the genome, retrotransposons are also potential sites of unequal (ectopic) recombination. Ectopic recombination between different genomic locations can lead to net deletion or duplication of nuclear DNA between elements, or inversions and reciprocal translocations, depending on the chromosomal location and orientation of the participating elements (Garfinkel 2005). Intra-strand inter-element recombination leading to genomic deletions and the formation of elements lacking matching TSD has been described in maize (see above), but the size of the corresponding deletions is, for the moment, unknown. More complex inversions or translocations mediated by retrotransposon have, to our knowledge, not yet been clearly established in maize, although they have been shown to occur in other eukaryotes (Lim and Simmons 1994; Mieczkowski, Lemoine et al. 2006).

4.4.3 Impact on Maize Gene Function

Retrotransposons can generate mutation through insertion within or near genes. In particular, many retrotransposon insertions within or near genes modify their expression, for instance through the production of aberrant transcripts (Kashkush, Feldman et al. 2003) or DNA methylation (Lippman et al. 2004; Huettel et al. 2007). In maize, retrotransposon insertions have been shown to induce inactivation of a gene, alteration of size and amount of the transcript, or reduction of protein level (Kumar and Bennetzen 1999). Tissue-specific alternative splicing has also been shown in maize (Marillonnet and Wessler 1997). Moreover, a paralogous gene (*Fie2*) nested within a retrotransposon had a different expression pattern than its non-nested paralog (*Fie1*) (Du et al. 2006), which suggests that the regulatory elements of the retrotransposon flanking the *Fie2* gene modified its expression.

Altogether, these studies confirm the various processes by which retrotransposons can mediate extensive sequence rearrangements and yield diversification of the genome. In the case of maize, the extent of the functional diversification mediated by retrotransposons could have been more extensive than in other plant genomes due to its ancient polyploidy. That is, because of gene redundancy, a new expression pattern would not at first be negatively selected, allowing subsequent sequence modifications to lead to an advantageous change in expression.

4.5 Differences Among Maize Haplotypes

The B73 cultivar has been chosen for "full-genome" sequencing. Only a few studies of local genome structure have been made on other maize inbreds such as McC (Fu, Park et al. 2001) or BSSS53 (Song, Llaca et al. 2001). However, estimates of maize genome size (2C) vary from 4.9 pg to 6.1 pg (Rayburn, Price et al. 1985) and early comparative cytogenetic studies showed high chromosomal variation between distinct maize races (McClintock, Kato et al. 1981). These findings suggest possible sequence variation among maize races, and possibly among maize inbreds.

A sequence comparison of two inbreds, McC and B73, at the *bz* locus revealed extensive non-colinearity between them (Fu and Dooner 2002). Only the genes they shared could be aligned, whereas the make-up and sizes of the flanking retrotransposon blocks were totally different. Similar results have then been reported in a comparison of the *z1C* zein genomic regions of B73 and BSSS53 (Song and Messing 2003), and for three random regions in a comparison between B73 and Mo17 (Brunner et al. 2005; Morgante, Brunner et al. 2005). Subsequent comparison of the *bz* genomic region among eight maize cultivars confirmed the existence of many polymorphic insertions of retrotransposons (Wang and Dooner 2006). Haplotypes were highly variable, with only 25% to 84% of sequence shared between any two. Some haplotypes were found to be chimeric, suggesting that the large diversity of genome organization observed in modern maize cultivars derived from the structural diversity created ancestrally by intervals of intense transposition, followed by shuffling of retrotransposon blocks through recombination (Wang and Dooner 2006).

Estimates of insertion dates in orthologous regions reveals that non-shared retrotransposons are significantly more recent than the shared ones; most non-shared elements being younger than 0.5 Myr (Brunner et al. 2005). Nevertheless, almost all non-shared insertions presumably occurred much earlier than the domestication of maize from its wild progenitor (teosinte), around 9,000 years ago (Matsuoka, Vigouroux et al. 2002). The structural diversity observed in today's maize germplasm was therefore already present in the wild progenitor of maize at the time of domestication. This is in accordance with the estimation that maize still possesses ~75% of the allelic diversity that was present in teosinte (Eyre-Walker, Gaut et al. 1998), even though it has gone through a domestication bottleneck that decreased sequence diversity at domestication genes (Wang, Stec et al. 1999).

5   Conclusions and Perspectives

To a first approximation, the maize genome is retrotransposons. So, as a whole, the structure and composition of most of the genome derive from the retrotransposons of which it is formed. Yet the bulk of these, despite their vast numbers, so well segment themselves from the genic portion of the genome, that they went largely unnoticed until studies of large-insert clones commenced. Most of their characteristics are in direct counterpoint to those of genes. They increase their copy number in the genome, but any given copy has a fleeting existence, eroding away in a few million years. Rather than contributing to the survival of the host organism, for the most part they inhabit a selective grey area, follow an ecology of their own. Further, while their fate appears to be tied to that of their host, it is conceivable that some fraction of them are actually viruses—capable of infecting the genomes of other organisms.

Further study is warranted. Compared to detailed knowledge present for animal retroviruses, only the scantest outline of the structure and function of plant retrotransposons has been sketched. In this context, the release of a draft sequence of the maize genome is welcome and will provide a wealth of data. But, ultimately, biochemical studies of the structure and function of plant retrotransposons need to be much expanded to provide any comprehensive understanding of these ubiquitous components of the maize genome.

References

Ananiev, E. V., R. L. Phillips and H. W. Rines (1998a) Chromosome-specific molecular organization of maize (Zea mays L.) centromeric regions. Proc Natl Acad Sci USA 95: 13073–13078.

Ananiev, E. V., R. L. Phillips and H. W. Rines (1998b) Complex structure of knob DNA on maize chromosome 9: Retrotransposon invasion into heterochromatin. Genetics 149: 2025–2037.

Arumuganathan, K. and E. Earle (1991) Nuclear DNA content of some important plant species. Plant Molecular Biology Reporter 9: 208–218.

Asami, Y., D. W. Jia, K. Tatebayashi, K. Yamagata, M. Tanokura and H. Ikeda (2002) Effect of the DNA topoisomerase II inhibitor VP-16 on illegitimate recombination in yeast chromosomes. Gene 291: 251–257.

Avramova, Z., P. SanMiguel, E. Georgieva and J. L. Bennetzen (1995) Matrix Attachment Regions and Transcribed Sequences within a Long Chromosomal Continuum Containing Maize Adh1. Plant Cell 7: 1667–1680.

Bender, J. (2004) DNA methylation and epigenetics. Annu Rev Plant Biol 55: 41–68.

Bennetzen, J. L., K. Schrick, P. S. Springer, W. E. Brown and P. SanMiguel (1994) Active Maize Genes Are Unmodified and Flanked by Diverse Classes of Modified, Highly Repetitive DNA. Genome 37: 565–576.

Brunner, S., K. Fengler, M. Morgante, S. Tingey and A. Rafalski (2005) Evolution of DNA sequence nonhomologies among maize inbreds. Plant Cell 17: 343–360.

Bureau, T. E., S. E. White and S. R. Wessler (1994) Transduction of a Cellular Gene by a Plant Retroelement. Cell 77: 479–480.

Coulondre, C., J. H. Miller, P. J. Farabaugh and W. Gilbert (1978) Molecular-Basis of Base Substitution Hotspots in Escherichia-Coli. Nature 274: 775–780.

Devos, K. M., J. K. M. Brown and J. L. Bennetzen (2002) Genome size reduction through illegitimate recombination counteracts genome expansion in Arabidopsis. Genome Res 12: 1075–1079.

Du, C. G., Z. Swigonova and J. Messing (2006) Retrotranspositions in orthologous regions of closely related grass species. BMC Evol Biol 6: 62.

Eichinger, D. J. and J. D. Boeke (1990) A Specific Terminal Structure Is Required for Ty1 Transposition. Gene Dev 4: 324–330.

Elrouby, N. and T. E. Bureau (2001) A novel hybrid open reading frame formed by multiple cellular gene transductions by a plant long terminal repeat retroelement. J Biol Chem 276: 41963–41968.

Emberton, J., J. X. Ma, Y. N. Yuan, P. SanMiguel and J. L. Bennetzen (2005) Gene enrichment in maize with hypomethylated partial restriction (HMPR) libraries. Genome Res 15: 1441–1446.

Eyre-Walker, A., R. L. Gaut, H. Hilton, D. L. Feldman and B. S. Gaut (1998) Investigation of the bottleneck leading to the domestication of maize. Proc Natl Acad Sci USA 95: 4441–4446.

Feng, Y.-X., S. P. Moore, D. J. Garfinkel and A. Rein (2000) The Genomic RNA in Ty1 Virus-Like Particles Is Dimeric. J Virol 74: 10819–10821.

Fu, H. H. and H. K. Dooner (2002) Intraspecific violation of genetic colinearity and its implications in maize. Proc Natl Acad Sci USA 99: 9573–9578.

Fu, H. H., W. K. Park, X. H. Yan, Z. W. Zheng, B. Z. Shen and H. K. Dooner (2001) The highly recombinogenic bz locus lies in an unusually gene-rich region of the maize genome. Proc Natl Acad Sci USA 98: 8903–8908.

Fu, H. H., Z. W. Zheng and H. K. Dooner (2002) Recombination rates between adjacent genic and retrotransposon regions in maize vary by 2 orders of magnitude. Proc Natl Acad Sci USA 99: 1082–1087.

Garcia-Martinez, J. and J. A. Martinez-Izquierdo (2003) Study on the evolution of the Grande retrotransposon in the Zea genus. Mol Biol Evol 20: 831–841.

Garfinkel, D. J. (2005) Genome evolution mediated by Ty elements in Saccharomyces. Cytogenet Genome Res 110: 63–69.

Gaut, B. S., B. R. Morton, B. C. McCaig and M. T. Clegg (1996) Substitution rate comparisons between grasses and palms: Synonymous rate differences at the nuclear gene Adh parallel rate differences at the plastid gene rbcL. Proc Natl Acad Sci USA 93: 10274–10279.

Haberer, G., S. Young, A. K. Bharti, H. Gundlach, C. Raymond, G. Fuks, E. Butler, R. A. Wing, S. Rounsley, B. Birren, C. Nusbaum, K. F. X. Mayer and J. Messing (2005) Structure and architecture of the maize genome. Plant Physiol 139: 1612–1624.

Hamilton, A., O. Voinnet, L. Chappell and D. Baulcombe (2002) Two classes of short interfering RNA in RNA silencing. EMBO J 21: 4671–4679.

Hu, W. M., O. P. Das and J. Messing (1995) Zeon-1, a Member of a New Maize Retrotransposon Family. Molecular & General Genetics 248: 471–480.

Huettel, B., T. Kanno, L. Daxinger, E. Bucher, J. van der Winden, A. J. M. Matzke and M. Matzke (2007) RNA-directed DNA methylation mediated by DRD1 and Pol IVb: A versatile pathway for transcriptional gene silencing in plants. Biochimica Et Biophysica Acta-Gene Structure and Expression 1769: 358–374.

Ilic, K., P. J. SanMiguel and J. L. Bennetzen (2003) A complex history of rearrangement in an orthologous region of the maize, sorghum, and rice genomes. Proc Natl Acad Sci USA 100: 12265–12270.

Jin, Y. K. and J. L. Bennetzen (1989) Structure and Coding Properties of Bs1, a Maize Retrovirus-Like Transposon. Proc Natl Acad Sci USA 86: 6235–6239.

Jin, Y. K. and J. L. Bennetzen (1994) Integration and Nonrandom Mutation of a Plasma-Membrane Proton Atpase Gene Fragment within the Bs1 Retroelement of Maize. Plant Cell 6: 1177–1186.

Kashkush, K., M. Feldman and A. A. Levy (2003) Transcriptional activation of retrotransposons alters the expression of adjacent genes in wheat. *Nat Genet* **33**: 102–106.

Kim, A., C. Terzian, P. Santamaria, A. Pelisson, N. Prudhomme and A. Bucheton (1994) Retroviruses in Invertebrates - the Gypsy Retrotransposon Is Apparently an Infectious Retrovirus of Drosophila-Melanogaster. Proc Natl Acad Sci USA 91: 1285–1289.

Kumar, A. and J. L. Bennetzen (1999) Plant retrotransposons. Annu Rev Genet 33: 479–532.

Lamb, J. C., J. M. Meyer, B. Corcoran, A. Kato, F. P. Han and J. A. Birchler (2007) Distinct chromosomal distributions of highly repetitive sequences in maize. Chromosome Research 15: 33–49.

Langham, R. J., J. Walsh, M. Dunn, C. Ko, S. A. Goff and M. Freeling (2004) Genomic duplication, fractionation and the origin of regulatory novelty. Genetics 166: 935–945.

Laten, H. M., A. Majumdar and E. A. Gaucher (1998) SIRE-1, a copia/Ty1-like retroelement from soybean, encodes a retroviral envelope-like protein. Proc Natl Acad Sci USA 95: 6897–6902.

Lim, J. K. and M. J. Simmons (1994) Gross Chromosome Rearrangements Mediated by Transposable Elements in Drosophila-Melanogaster. Bioessays 16: 269–275.

Lippman, Z., A. V. Gendrel, M. Black, M. W. Vaughn, N. Dedhia, W. R. McCombie, K. Lavine, V. Mittal, B. May, K. D. Kasschau, J. C. Carrington, R. W. Doerge, V. Colot and R. Martienssen (2004) Role of transposable elements in heterochromatin and epigenetic control. Nature 430: 471–476.

Liu, R., C. Vitte, J. Ma, A. A. Mahama, T. Dhliwayo, M. Lee and J. L. Bennetzen (2007) A GeneTrek analysis of the maize genome. Proc Natl Acad Sci USA 104: 11844–11849.

Ma, J. X. and J. L. Bennetzen (2004) Rapid recent growth and divergence of rice nuclear genomes. Proc Natl Acad Sci USA 101: 12404–12410.

Ma, J. X., K. M. Devos and J. L. Bennetzen (2004) Analyses of LTR-retrotransposon structures reveal recent and rapid genomic DNA loss in rice. Genome Res 14: 860–869.

Ma, J. X., P. SanMiguel, J. S. Lai, J. Messing and J. L. Bennetzen (2005) DNA rearrangement in orthologous Orp regions of the maize, rice and sorghum genomes. Genetics 170: 1209–1220.

Marillonnet, S. and S. R. Wessler (1997) Retrotransposon insertion into the maize waxy gene results in tissue-specific RNA processing. Plant Cell 9: 967–978.

Martienssen, R. (1998) Transposons, DNA methylation and gene control. Trends Genet 14: 263–264.

Matsuoka, Y., Y. Vigouroux, M. M. Goodman, G. J. Sanchez, E. Buckler and J. Doebley (2002) A single domestication for maize shown by multilocus microsatellite genotyping. Proc Natl Acad Sci USA 99: 6080–6084.

Matzke, M., T. Kanno, B. Huettel, L. Daxinger and A. J. M. Matzke (2007) Targets of RNA-directed DNA methylation. Curr Opin Plant Biol 10: 512–519.

McClintock, B., Y. T. A. Kato and A. Blumenshein (1981) In: Chromosome Constitution of Races of Maize. Colegio do Postgraduados, Chapingo, Mexico.

Messing, J., A. K. Bharti, W. M. Karlowski, H. Gundlach, H. R. Kim, Y. Yu, F. S. Wei, G. Fuks, C. A. Soderlund, K. F. X. Mayer and R. A. Wing (2004) Sequence composition and genome organization of maize. Proc Natl Acad Sci USA 101: 14349–14354.

Meyers, B. C., S. V. Tingley and M. Morgante (2001) Abundance, distribution, and transcriptional activity of repetitive elements in the maize genome. Genome Res 11: 1660–1676.

Mieczkowski, P. A., F. J. Lemoine and T. D. Petes (2006) Recombination between retrotransposons as a source of chromosome rearrangements in the yeast Saccharomyces cerevisiae. DNA Repair 5: 1010–1020.

Miller, J. T., F. G. Dong, S. A. Jackson, J. Song and J. M. Jiang (1998) Retrotransposon-related DNA sequences in the centromeres of grass chromosomes. Genetics 150: 1615–1623.

Miller, W., J. McDonald and W. Pinsker (1997) Molecular domestication of mobile elements. Genetica 100: 261–270.

Monfort, A., C. M. Vicient, R. Raz, P. Puigdomenech and J. A. Martinez-Izquierdo (1995) Molecular Analysis of a Putative Transposable Retroelement from the Zea Genus with Internal Clusters of Tandem Repeats. DNA Res 2: 255–261.

Morgante, M., S. Brunner, G. Pea, K. Fengler, A. Zuccolo and A. Rafalski (2005) Gene duplication and exon shuffling by helitron-like transposons generate intraspecies diversity in maize. Nat Genet 37: 997–1002.

Mroczek, R. J. and R. K. Dawe (2003) Distribution of retroelements in centromeres and neocentromeres of maize. Genetics 165: 809–819.

Palmer, L. E., P. D. Rabinowicz, A. L. O'Shaughnessy, V. S. Balija, L. U. Nascimento, S. Dike, M. De la Bastide, R. A. Martienssen and W. R. McCombie (2003) Maize genome sequencing by methylation filtration. Science 302: 2115–2117.

Palmgren, M. G. (1994) Capturing of Host DNA by a Plant Retroelement - Bs1 Encodes Plasma-Membrane H+-Atpase Domains. Plant Mol Biol 25: 137–140.

Petrov, D. A., T. A. Sangster, J. S. Johnston, D. L. Hartl and K. L. Shaw (2000) Evidence for DNA loss as a determinant of genome size. Science 287: 1060–1062.

Presting, G. G., L. Malysheva, J. Fuchs and I. Z. Schubert (1998) A TY3/GYPSY retrotransposon-like sequence localizes to the centromeric regions of cereal chromosomes. Plant J 16: 721–728.

Quayle, T. J. A., J. W. S. Brown and G. Feix (1989) Analysis of Distal Flanking Regions of Maize 19-Kda Zein Genes. Gene 80: 249–257.

Ramakrishna, W., J. Emberton, M. Ogden, P. SanMiguel and J. L. Bennetzen (2002a) Structural analysis of the maize Rp1 complex reveals numerous sites and unexpected mechanisms of local rearrangement. Plant Cell 14: 3213–3223.

Ramakrishna, W., J. Emberton, P. SanMiguel, M. Ogden, V. Llaca, J. Messing and J. L. Bennetzen (2002b) Comparative sequence analysis of the sorghum Rph region and the maize Rp1 resistance gene complex. Plant Physiol 130: 1728–1738.

Rayburn, A. L., H. J. Price, J. D. Smith and J. R. Gold (1985) C-Band Heterochromatin and DNA Content in Zea mays. Am J Bot 72: 1610–1617.

Roeder, G. S. and G. R. Fink (1980) DNA Rearrangements Associated with a Transposable Element in Yeast. Cell 21: 239–249.

SanMiguel, P. and J. L. Bennetzen (1998) Evidence that a recent increase in maize genome size was caused by the massive amplification of intergene retrotransposons. Ann Bot 82: 37–44.

SanMiguel, P., B. S. Gaut, A. Tikhonov, Y. Nakajima and J. L. Bennetzen (1998) The paleontology of intergene retrotransposons of maize. Nat Genet 20: 43–45.

SanMiguel, P., A. Tikhonov, Y. K. Jin, N. Motchoulskaia, D. Zakharov, A. MelakeBerhan, P. S. Springer, K. J. Edwards, M. Lee, Z. Avramova and J. L. Bennetzen (1996) Nested retrotransposons in the intergenic regions of the maize genome. Science 274: 765–768.

Sanz-Alferez, S., P. SanMiguel, Y. K. Jin, P. S. Springer and J. L. Bennetzen (2003) Structure and evolution of the Cinful retrotransposon family of maize. Genome 46: 745–752.

Song, R. T., V. Llaca, E. Linton and J. Messing (2001) Sequence, regulation, and evolution of the maize 22-kD alpha zein in gene family. Genome Res 11: 1817–1825.

Song, R. T. and J. Messing (2003) Gene expression of a gene family in maize based on noncollinear haplotypes. Proc Natl Acad Sci USA 100: 9055–9060.

Song, S. U., T. Gerasimova, M. Kurkulos, J. D. Boeke and V. G. Corces (1994) An Env-Like Protein Encoded by a Drosophila Retroelement - Evidence That Gypsy Is an Infectious Retrovirus. Genes Dev 8: 2046–2057.

Springer, P. S., K. J. Edwards and J. L. Bennetzen (1994) DNA Class Organization on Maize Adh1 Yeast Artificial Chromosomes. Proc Natl Acad Sci USA 91: 863–867.

Swigonova, Z., J. L. Bennetzen and J. Messing (2005) Structure and evolution of the r/b chromosomal regions in rice, maize and sorghum. Genetics 169: 891–906.

Swigonova, Z., J. S. Lai, J. X. Ma, W. Ramakrishna, M. Llaca, J. L. Bennetzen and J. Messing (2004) On the tetraploid origin of the maize genome. Comp Funct Genomics 5: 281–284.

Tikhonov, A. P., P. J. SanMiguel, Y. Nakajima, N. M. Gorenstein, J. L. Bennetzen and Z. Avramova (1999) Colinearity and its exceptions in orthologous adh regions of maize and sorghum. Proc Natl Acad Sci USA 96: 7409–7414.

Turcich, M. P., A. Bokhari-Riza, D. A. Hamilton, C. P. He, W. Messier, C. B. Stewart and J. P. Mascarenhas (1996) PREM-2, a copia-type retroelement in maize is expressed preferentially in early microspores. Sexual Plant Reproduction 9: 65–74.

Turcich, M. P. and J. P. Mascarenhas (1994) PREM-1, a Putative Maize Retroelement Has LTR (Long Terminal Repeat) Sequences That Are Preferentially Transcribed in Pollen. Sexual Plant Reproduction 7: 2–11.

Vitte, C. and J. L. Bennetzen (2006) Analysis of retrotransposon structural diversity uncovers properties and propensities in angiosperm genome evolution. Proc Natl Acad Sci USA 103: 17638–17643.

Vitte, C. and O. Panaud (2003) Formation of solo-LTRs through unequal homologous recombination counterbalances amplifications of LTR retrotransposons in rice Oryza sativa L. Mol Biol Evol 20: 528–540.

Vitte, C., O. Panaud and H. Quesneville (2007) LTR retrotransposons in rice (Oryza sativa, L.): recent burst amplifications followed by rapid DNA loss. BMC Genomics 8: 218.

Wang, Q. H. and H. K. Dooner (2006) Remarkable variation in maize genome structure inferred from haplotype diversity at the bz locus. Proc Natl Acad Sci USA 103: 17644–17649.

Wang, R. L., A. Stec, J. Hey, L. Lukens and J. Doebley (1999) The limits of selection during maize domestication. Nature 398: 236–239.

Wessler, S. R., T. E. Bureau and S. E. White (1995) LTR-Retrotransposons and Mites - Important Players in the Evolution of Plant Genomes. Curr Opin Genet Dev 5: 814–821.

White, S. E., L. F. Habera and S. R. Wessler (1994) Retrotransposons in the Flanking Regions of Normal Plant Genes - a Role for Copia-Like Elements in the Evolution of Gene Structure and Expression. Proc Natl Acad Sci USA 91: 11792–11796.

Whitelaw, C. A., W. B. Barbazuk, G. Pertea, A. P. Chan, F. Cheung, Y. Lee, L. Zheng, S. van Heeringen, S. Karamycheva, J. L. Bennetzen, P. SanMiguel, N. Lakey, J. Bedell, Y. Yuan, M. A. Budiman, A. Resnick, S. Van Aken, T. Utterback, S. Riedmuller, M. Williams, T. Feldblyum, K. Schubert, R. Beachy, C. M. Fraser and J. Quackenbush (2003) Enrichment of gene-coding sequences in maize by genome filtration. Science 302: 2118–2120.

Wicker, T., F. Sabot, A. Hua-Van, J. L. Bennetzen, P. Capy, B. Chalhoub, A. Flavell, P. Leroy, M. Morgante, O. Panaud, E. Paux, P. SanMiguel and A. H. Schulman (2007) A unified classification system for eukaryotic transposable elements. Nat Rev Genet 8: 973–982.

Wicker, T., N. Yahiaoui, R. Guyot, E. Schlagenhauf, Z. D. Liu, J. Dubcovsky and B. Keller (2003) Rapid genome divergence at orthologous low molecular weight glutenin loci of the A and A(m) genomes of wheat. Plant Cell 15: 1186–1197.

Wright, D. A. and D. F. Voytas (1998) Potential retroviruses in plants: Tat1 is related to a group of Arabidopsis thaliana Ty3/gypsy retrotransposons that encode envelope-like proteins. Genetics 149: 703–715.

Wright, D. A. and D. F. Voytas (2002) Athila4 of Arabidopsis and Calypso of soybean define a lineage of endogenous plant retroviruses. Genome Res 12: 122–131.

Xiong, Y. and T. H. Eickbush (1990) Origin and Evolution of Retroelements Based Upon Their Reverse-Transcriptase Sequences. EMBO J 9: 3353–3362.

Helitrons: Their Impact on Maize Genome Evolution and Diversity

Shailesh K. Lal, Nikolaos Georgelis, and L. Curtis Hannah

Abstract Gene piece movement by the newly-described *Helitron* family of transposable elements has significantly impacted the evolution of the maize genome. *Helitrons* have been implicated in causing gene non-colinearity between different maize inbred lines, however capture and movement of functional genes to different regions of the genome by *Helitrons* remains to be demonstrated. The abundance of these elements and the extent of diversity among them remain largely undetermined. Several hypotheses have been proposed to explain their transposition and mechanism by which these elements prolifically capture and mobilize gene sequences, but each lacks supporting experimental evidence. A more complete understanding of this process requires molecular and genetic evidence of *Helitron* activity in modern maize genome.

1 Introduction and Overview of *Helitrons*

Helitron refers to a recently-described class of transposable elements that has a number of important and unique features. *Helitrons* appear to underlie much of the +/− polymorphisms or "intraspecific violation of genetic colinearity" recently observed in maize genome sequencing projects. This chapter reviews the short documented history of *Helitrons* and highlights many of the unresolved questions concerning this intriguing class of elements

Helitrons were first hypothesized from computer analysis of the fully sequenced genomes of *Caenorhabditis elegans* (*C. elegans*), rice and Arabidopsis

S.K. Lal
Oakland University, Department of Biological Sciences
lal@oakland.edu

N. Gerogelis and L.C. Hannah
University of Florida, Department of Horticultural Sciences
lchannah@ufl.edu

J.L. Bennetzen and S. Hake (eds.), *Maize Handbook - Volume II: Genetics and Genomics*,   329
© Springer Science+Business Media LLC 2009

(Kapitonov and Jurka 2001). These investigators proposed that *Helitrons* comprise a repetitive class of DNA representing ~2% of the total genome. An autonomous *Helitron* was predicted by computer simulation of the consensus sequence of non-autonomous elements. The autonomous element was proposed to contain coding information for a HEL protein composed of a rolling-circle replication initiator and DNA helicase domains needed for transposition. It also contains a replication protein A (RPA)-like protein with single-stranded DNA-binding activity. These proteins are associated with bacterial transposons which transpose via rolling circle (RC) replication (Khan 2000; Tavakoli, Comanducci, Dodd, Lett, Albiger, and Bennet 2000).

Helitrons are novel because they bear unique characteristics heretofore not shared by known class 1 (transposition through RNA) and class 2 (transposition through DNA) eukaryotic transposable elements. Despite their abundance, *Helitrons* have remained elusive because of the small size of invariant terminal sequences and the polymorphic nature of internal sequences. The latter characteristic is uniquely attributable to the presence of different gene sequences apparently captured by these elements.

Helitrons lack terminal repeats and do not duplicate host sequence upon insertion. The 5′ and 3′ terminal ends of *Helitrons* characterized by Kapitonov and Jurka (2001) contain sequences TC and CTRR, respectively. They consistently insert between the bases of dinucleotide AT and contain a non-conserved short palindrome sequence located 15-20 nucleotides upstream of their 3′ end capable of forming a hairpin structure (Kapitonov and Jurka 2001).

Because an autonomous *Helitron* was not found from computer searches of the *C. elegans,* rice and Arabidopsis genomes, there exists the possibility that *Helitrons* are no longer active in these organisms and the isolated *Helitron* sequences are simply evolutionary remnants of once-active elements. That *Helitrons* are active in modern genomes came from three recent reports of mutants caused by *Helitron* insertions. Two maize mutants, *sh2-7527* and *ba1-ref* are caused by insertion of non-autonomous *Helitrons* (Gupta, Gallavotti, Stryker, Schmidt, and Lal 2005; Lal, Giroux, Brendel, Vallejos, and Hannah 2003). The mutants likely were caused by relatively recent *Helitron* insertions since (1) the insertion mutants do not contain any other alteration inactivating the gene (a signature of the evolutionary decay of functional genes) and (2) loss of *Sh2* or *Ba1* leads to germination and seed production deficiencies, respectively. Natural selection pressure would significantly limit propagation of these mutants in the absence of human selection. These studies then provided the first evidence that the present-day maize genome harbors an active *Helitron*. Furthermore, the reports of nearly identical *Helitrons* inserted in different regions of the maize genome indicate that these elements were active and have moved very recently and may still be active (Gupta et al. 2005; Lai, Li, Messing, and Dooner 2005).

A third plant mutant was recently reported in morning glory, *Ipomoea tricolor*. Isolated in the 1940's, the white flower mutant termed pearly-s-mutant was recently shown to be caused by a *Helitron* insertion into the dihydroflavonol 4-reductase B gene involved in anthocyanin biosynthesis (Choi, Hoshino, Park, Park, and Iada

2007). Intriguingly and unlike maize mutants, this *Helitron* contains non-functional remnants of sequences encoding helicase and nuclease/ligase.

Helitrons have now been found via sequence analysis in many organisms (reviewed in Kapitonov and Jurka 2007). These include other plants, invertebrates, fungi and fish (Choi et al. 2007; Kapitonov and Jurka 2001; Lal et al. 2003; Poulter, Goodwin, and Butler 2003). Massive abundance of *Helitrons* has been recently reported in a mammalian genome, the brown bat *Myotis lucifugus*, where they constitute ~3% of the genome (Pritham and Feschotte 2007).

2 Mass Movement of *Helitron*-Mediated Gene Sequences Bestow Apparent Intra-specific Violation of Gene Colinearity in Maize

Compared to *Helitrons* in other organisms, maize *Helitrons* appear unique in their ability to incorporate portions of host gene sequences into the transposable element (reviewed in Kapitonov and Jurka 2007). In doing so, *Helitrons* contribute to the lack of gene colinearity between different maize inbred lines. The maize genome exhibits tremendous diversity that has played a vital role in its domestication. Plant breeders have successfully exploited this diversity for crop improvement. Intriguingly, data emerging from analysis of multiple regions of the genome indicated that significant differences in gene composition and colinearity are widely prevalent among different maize inbred lines. (Brunner, Pea, and Rafalski 2005; Fu and Dooner 2002; Song and Messing 2003). This so-called, "intraspecific violation of gene colinearity" or "+/− polymorphism"" has been shown to be primarily due to the presence/absence of gene piece-rich fragments between different inbred lines. This perceived lack of gene colinearity between maize lines and the possible contribution of *Helitrons* to this variation were first reported by Fu and Dooner (2002) who sequenced the *bronze* (*bz*) region of maize inbreds McC and B73 and discovered the presence of several annotated "genes" in McC and their absence in B73. They later observed that these genes were actually pieces of genes inside two nested *Helitrons* (Lai et al. 2005). This accounted for the haplotype variability at the *bz* locus between these two inbred lines. In a parallel study, the extent of lack of apparent gene colinearity between inbred lines B73 and Mo17 was investigated by using oligonucleotide probes (Morgante, Brunner, Pea, Fengler, Zuccolo, and Rafalski 2005). Here, 20% of the detected 20,656 genes were found in only one of the inbred lines. These differences were primarily caused by segments of gene piece-rich insertions rather than deletions among the two lines. Detailed sequence analysis of nine insertions indicated that eight were caused by *Helitrons*. If this limited data set has any bearing at the genomic level, one would assume that *Helitrons* may have captured and mobilized thousands of gene pieces across the maize genome. The complete sequencing of maize inbred B73 genome will not provide the complete picture of the extent and diversity of gene piece movement by *Helitrons*, since *Helitron* identification and the demarcation of their precise boundaries often entails "+/– polymorphism" studies between different inbred lines. Nevertheless, precise identification and annotation of *Helitrons* is important because the majority of the elements identified thus far represents non-autonomous *Helitrons* bearing segments of multiple genes. This, no doubt, will complicate the proper annotation of wild type genes in the sequenced maize genome.

3 Isolated Maize *Helitrons* Represent a Sub-class of *Helitrons*

Twenty two maize *Helitrons* displaying +/– polymorphisms have been reported to date (Table 1). Fourteen were found initially through characterization of +/– DNA polymorphisms between maize lines while eight were discovered through sequence similarity to the original 14 elements. Subsequent analysis showed that all eight "derived" elements are associated with +/– polymorphism at their allelic sites or in paralogous genes (references in Table 1). The characterized *Helitrons* clearly define two closely related families as judged by the similarity of their 3′ terminal sequence (Table 1). And, placement of gaps in the two families of 3′ sequences allows for the detection of relatedness between the two families.

The major sub-family consists of 19 members while the minor group is represented by three *Helitrons*. As noted above, sequence similarity can be found in the 3′ termini and, importantly, relatedness of these maize *Helitrons* can be seen by the presence of conserved sequences in a 14 bp motif in the 5′ termini (data not shown).

Table. 1 Classification of maize *Helitrons* based on the conservation of 3′ terminal sequence

	HELITRON NAMES	3′ TERMINUS	FIRST REFERENCE
Hei1	sh2-7527*	CTCCGTAGC-AACGCACGGACATTCAC-----CTAG	Lal et al., 2003
	ba1-Ref*	TTCCGTAGC-AACGCACGGCTATATAC-----CTAG	Gupta et al., 2005
	Rp1B73**	GCCCGTTGC-AACGCACGAGCACTGAC-----CTAG	Gupta et al., 2005
	ZeinBSSS53**	-TTCGTGGCTAACGCACGGCACATAC-----CTAG	Gupta et al., 2005
	Hei1-19**	TCCCGTGGC-AACGCACGGGCACGAAC-----CTAG	Lal et al., unpublished
	HeiA-1*	TCCCGTCGC-AACGCACGGGCACTCAC-----CTAG	Lai et al., 2005
	HeiA-2**	TCTCGTCGC-AACGCACGGGCACTCAC-----CTAG	Lai et al., 2005
	GHIJKLM9002*	AACCGTAGC-AACGCACGGGCATTCAA-----CTAG	Morgante et al., 2005
	NOPQ9002*	CCCCGTTGC-AACGCACGGGCACTCAC-----CTAG	Morgante et al., 2005
	NOPQB73_14578**	CCCCGTTGC-AACGCACGGGCACTCAC-----CTAG	Brunner et al., 2005
	NOPQMo17_14594**	CCCCATTGC-AACGCACGGGCACTCAC-----CTAG	Brunner et al., 2005
	NOPQB73_9002**	CCCCGTTGC-AACGCACGAGCACTGGC-----CTAG	Brunner et al., 2005
	Mo17NOPQ_14577**	GCCCGTTGC-AACGCACGAGCACTGGC-----CTAG	Brunner et al., 2005
	RST9002*	TCCCGTCGC-AACGCACGTGCACTCAC-----CTAG	Morgante et al., 2005
	U9002*	TTACGTAGCGAAGGCACGGGCACATAC-----CTAG	Morgante et al., 2005
	HI9002*	TCCCGTTGC-AACGCACGGGCACTCAC-----CTAG	Morgante et al., 2005
	Hei-BSSS53-zICI*	TTCCGTGGC-ATCGCACGGGCACATAC-----CTAG	Xu and Messing, 2006
	Hei1-4*	TTCCGTGGC-ATCGCACGGGCACCTAA-----CTAG	Wang and Dooner, 2006
	Hei1-5*	AGTCGTCGC-AACGCACGGGCAACGGA-----CTAG	Wang and Dooner, 2006
Hei2	TUVW9002*	-AGCGCCCG-AT-----AGGGCGCTTTCCTATTCTAG	Brunner et al., 2005
	HeiB*	GGGCGCCCG-TA-----TGGGCGCCC-CATGTTCTAG	Lai et al., 2005
	Hei1-6*	GGGCGCACG-AA-----GTGG-GCCT-CCTGATCTAG	Wang and Dooner, 2006

*discovered initially by +/– polymorphism

**discovered by similarity of either terminal or internal sequence to other *Helitrons*

While termini of maize *Helitrons* are strikingly similar, internal sequences exhibit little sequence similarity. Like other transposable elements, the conserved terminal ends of these non-autonomous *Helitrons* suggest that their movement may be mediated by the transposition protein provided in trans by a master or autonomous member of the family. Searches for candidate autonomous *Helitrons* have identified several instances in which sequences bearing similarity to DNA helicases are localized in close proximity to Replication protein A, a characteristic of the proposed autonomous *Helitron* (Gupta et al. 2005; Morgante et al. 2005). However, the presence of premature in-frame stop-codons or large insertions shows that the isolated sequences represent inactive elements. In addition, these putative, mutationally-silenced autonomous elements do not exhibit sequence similarity to the 3′ conserved terminal ends of isolated maize *Helitrons*. This suggests that these inactive remnants of proposed autonomous *Helitrons* may not be closely related to the *Helitrons* in Table 1. The completion of the B73 genome sequence will determine if this inbred harbors a functional autonomous *Helitron*. However, a functionally autonomous *Helitron* may not be necessary for *Helitron* transposition if the helicase and replicase activity are provided by host genes. To date there is no experimental evidence that links the coding sequences of the proposed autonomous *Helitrons* to transposition activity. In fact there is no direct experimental evidence that helicase and replicase activities are even needed for transposition.

The complete divergence in internal sequence of the characterized maize non-autonomous *Helitrons* makes it difficult to search for other members in the genome. The highly repetitive nature of genomic DNA hybridizing to *Helitron* termini (Lal et al. 2003) suggests that there are hundreds to thousands of other family members. Accordingly, the terminal ends of the *Helitrons* have been used to effectively search for other family members in the maize genome database (Gupta et al. 2005). The scheme used to discover *Helitrons* is outlined in Fig. 1.

The initial search entails detection in a single clone of sequences highly similar to the *Helitron* 5′ and 3′ termini. The intervening sequence then is viewed as a candidate *Helitron*. Intervening sequences are then perused for the presence of gene piece-rich sequences since characterized non-autonomous maize *Helitrons* are rich in gene piece sequences. However, maize *Helitrons* that do not ferry gene sequences have recently been reported (Wang and Dooner 2006) so the presence of gene sequences within a *Helitron* is not invariant. Definitive validation that a sequence fulfilling the criteria outlined above is, in fact, a *Helitron* requires demonstration of +/− polymorphism. One approach, accounting for the isolation of the majority of *Helitrons*, involves parallel sequencing and comparison of homologous regions in different inbreds. As an alternative, more cost effective strategy, Gupta et al. (2005) searched maize GSS, EST and HTGS clones for sequences aligning with sequences flanking the putative *Helitron*. Primers were then designed to scan different inbred lines for "empty" sites. This approach has identified 336 putative *Helitrons* in B73. Of these, 22 exhibit +/− polymorphisms just within the B73 inbred (Lal, unpublished). Since B73 is an inbred, the +/− polymorphism here likely reflects comparison of paralogous loci.

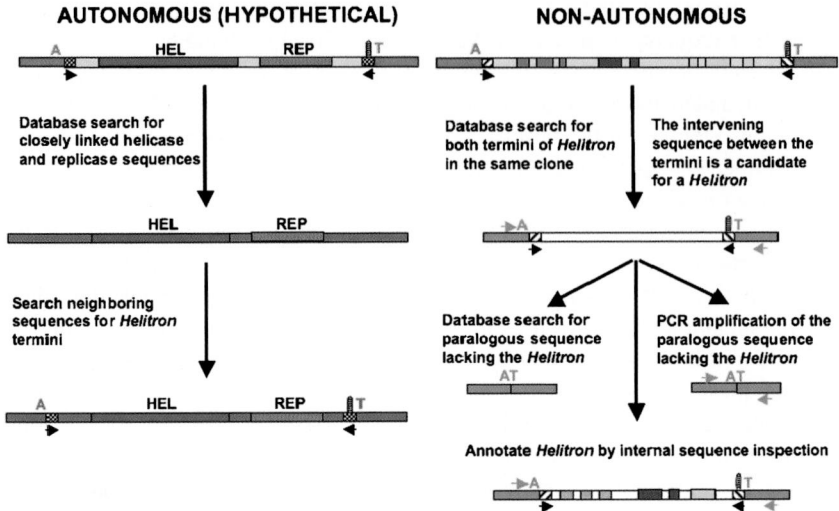

Fig. 1 Strategy for Discovering Maize *Helitrons*: The upper left and right panels display the structure of autonomous and non-autonomous maize *Helitrons*. An autonomous *Helitron* is predicted by computational analysis of related repeat sequences in the genomes of Arabidopsis, rice and *Caenorhabditis elegans* (Kapitonov and Jurka, 2001). The coding regions of DNA helicase (HEL) and single stranded DNA binding protein (REP) of a theoretical autonomous *Helitron* are marked. The exons captured by the non-autonomous *Helitron* are displayed by colored blocks. The terminal ends of the *Helitron* are displayed by pattern-filled boxes and the loop near the 3′ terminus represents the palindrome sequence. The A and T nucleotide immediately flanking the insertion site of the *Helitron* are indicated

4 The Mechanism by Which *Helitrons* Capture and Mobilize Gene Sequences Remains Unknown

While the occurrence of host gene fragments embedded within transposable elements has been reported in the past, it was generally regarded as a rare incidence. However, a recent report of mass movement of gene sequences by a *Mutator*-like family of transposable elements in rice termed MULEs indicates that this process is far more prevalent than initially anticipated (Jiang, Bao, Zhang, Eddy, and Wessler 2005). This may play a major role in genome evolution. The MULE family of elements has captured and mobilized more than 1000 gene fragments in the rice genome (Jiang et al. 2005). However, the mechanism by which MULE and *Helitron* super families of transposable elements acquire gene fragments may differ. For example, 20% of the gene fragments captured by MULEs bear no apparent wild-type counterpart in the rice genome but display strong similarity to intact genes from other plants. This suggests that the host gene was destroyed during the acquisition process. In contrast, Morgante et al. (2005) demonstrated that the wild type progenitor of four different gene fragments captured by a *Helitron* exist in the maize genome.

These investigators speculate that *Helitron* transposition involves replication and strand replacement that preserves the integrity of the donor host sequence. On the other hand, the mechanism of gene capture may not differ for the two types of elements. Given that only one of the two alleles of a gene participates in the events of gene capture, alteration in one allele (and not in the other) creates a heterozygous condition. Given heterozygosity and assuming no selection, it would simply be random chance whether the mutant, non-functional allele or the wildtype allele becomes fixed in the population following gene capture.

The similarity of *Helitron* transposition proteins, helicase and nuclease/ligase, to host genes led Kapitonov and Jurka (2001) to postulate that these genes also were recruited by the *Helitron* from the host genome. Further understanding of this process will require understanding of the mechanism by which these elements capture and mobilize host gene sequences.

While *Helitron* gene piece capture has been reported in several organisms, *Helitrons* of maize are clearly the most prolific at incorporating gene pieces. Several hypotheses have been forward to explain the mechanism of gene piece capture by *Helitrons* (reviewed in Kapitonov and Jurka 2007). For example, it was proposed that transposition is initiated at the 5′ end of the element and the inefficient recognition of the 3′ palindrome sequence as a termination signal by the transposition machinery results in the capture of 3′ flanking host sequences (Feschotte and Wessler 2001). The strong conservation of the 3′ end and not the 5′end of a family of non-autonomous *Helitrons* led Morgante et al. (2005) to propose that transposition is initiated at the 3′ end of the *Helitrons* and the capture of 5′ host flanking sequence occurs during transposition via an unknown mechanism. Kapitonov and Jurka (2007) have proposed that gene piece incorporation occurs through double stranded breaks within the transposing *Helitron* and the random incorporation of "filler" DNA coming from the host genome.

4.1 Helitrons and Bacterial Integrons Share a Number of Common Features

A number of parallels between *Helitrons* and bacterial integrons suggests that integrons may be the bacterial counterpart of *Helitrons* (Fig. 2). Like *Helitrons*, integrons are mobile, lack terminal repeats and do not make host duplications upon insertion. Like *Helitrons*, integrons capture genes (Hall and Callis 1995). Integrons have the capability to acquire gene sequences from mobile elements called gene cassettes via site specific recombination involving circular intermediates. Analogous to *Helitrons*, bacterial integrons consist of conserved terminal sequences encompassing a variable region primarily consisting of captured gene cassettes. The captured gene cassettes, like gene pieces in *Helitrons,* are oriented in the same direction. Approximately 86% of all captured maize exons are in the direct orientation (Kapitonov and Jurka 2007). This allows captured exons of both *Helitrons* and integrons to be transcribed by the same promoter located in

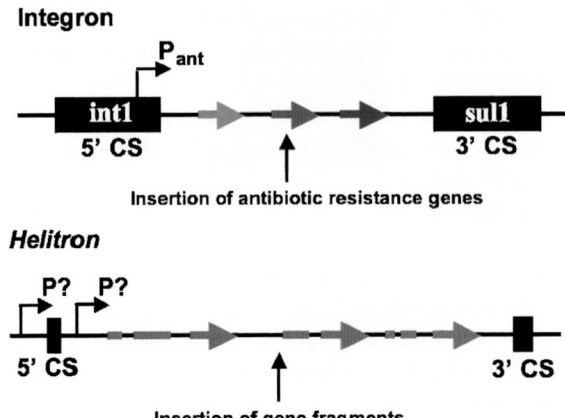

Fig. 2 General structural similarity between Class 1 integrons and gene-carrying non-autonomous maize *Helitrons*: The upper and lower panels display the basic structure of a Class 1 integron and a maize non-autonomous *Helitron,* respectively. The letter P marks the position of the promoter. The 5′ and 3′ conserved terminal sequences (CS) are indicated. The integrase (int1) and sulfonamide resistance (sul1) are also indicated in the integron structure. The color coded arrows display the coding sequences and direction of the captured genes

the 5′ conserved sequence. Also like *Helitrons*, different gene cassettes captured by integrons are often co-transcribed as a single unit. In integrons, the site specific recombination during gene capture is mediated by an integrase gene located in the conserved 5′terminus. Whether *Helitrons* capture gene sequences using mechanisms similar to integrons needs further investigation; however sequences bearing similarity to an integron integrase have not been reported to date in maize *Helitrons* (Gupta et al. 2005).

4.2   *Potential Role of* Helitrons *in Maize and Gene Evolution and their Impact on Host Gene Expression*

Integron capture of genes for antibiotic resistance in bacteria has been noted in a number of cases. Expression of the captured gene then provides an obvious advantage to bacteria exposed to the toxin. In contrast, no clear cut example of *Helitron* gene capture conditioning advantages to the *Helitron* or to the host organism is available. In fact, all characterized *Helitron*-captured maize genes are truncated and likely nonfunctional (Brunner et al. 2005; Gupta et al. 2005; Lai et al. 2005; Lal et al. 2003; Morgante et al. 2005). *Helitron* capture of an almost-intact copy of the maize gene encoding cytidine deaminase was recently reported (Xu and Messing 2006).

The captured gene lacks its promoter, exon 1 and a major portion of intron 1. While the truncated open reading frame and the introns have been conserved, the lack of detectable transcript of this gene suggests that the gene may not be active. While no *a priori* reason exists for the absence of intact gene capture in maize, their nonexistence in extant databases is curious. Intact gene capture is possible, at least in *Aspergillus nidulans* (Cultrone, Dominguez, Drevet, Scazzocchio, and Fernandez-Martin 2007).

The expression of captured genes by *Helitrons* can potentially have significant impact on host genome expression. For example, the *wp* mutation in soybean is caused by an insertion of the *Tgm-Express1* transposon in the flavanone-3- hydroxylase gene. Similar to *Helitrons*, *Tgm-Express1* contains fragments of different genes which are transcribed, giving rise to multiple, alternatively-spliced, chimeric transcripts. Intriguingly, the *wp* mutant also displays a significant increase in both seed weight and protein content. Since the genes pieces captured by *Tgm-Express1* represent enzymes involved in diverse metabolic pathways, it was postulated that their transcript may somehow modulate the expression of their wild type genes causing pleiotropic effects of the *wp* mutant (Zabala and Vodkin 2005; Zabala and Vodkin 2007).

A potentially significant role for *Helitrons* is in the synthesis of new genes. The gene pieces captured by *Helitrons* are often transcribed into eclectic transcripts conjoining coding regions of several different genes. These chimeric transcripts are sometimes alternatively spliced, generating several different transcripts via differential utilization of splice sites during pre-mRNA processing (Brunner et al. 2005; Lal et al. 2003; Morgante et al. 2005). For example, the *Helitron* insertion in maize mutant, *sh2-7527* generates several, alternatively-spliced isoforms, at least two of which are caused by alternative retention of a wild type *Sh2* intron (Lal et al. 2003). Since mutations acting from a distance are known to influence splice site recognition, it is possible that change in the context of captured gene fragments induces alternatively splicing events (Lal, Choi, Shaw, and Hannah 1999).

Promoters located inside or outside *Helitrons* can initiate transcription. Expression of the *sh2-7527* gene uses the *Sh2* promoter to produce a chimeric transcript containing sequences of several genes. In contrast, transcription initiation of captured genes within a *Helitron* has been suggested (Brunner et al. 2005; Morgante et al. 2005). Similarly, transcription of a maize cytochrome P450 monoxygenase captured by a *Helitron* seems to be initiated inside the element (Lal and Hannah, unpublished results). In this regard, *Helitrons* parallel pack-MULEs since transcription initiation within many pack-MULEs has been documented (Jiang et al. 2004). The possibility that genes captured by *Helitrons* may lead to evolution of new genes by "promoter shuffling" was recently reported for the xanthine α-ketoglutarate-dependent dioxygenase (xanA) gene captured by *Helitron-N1_AN* in *Aspergillus nidulans* (Cultrone et al. 2007). Apparently, the 5' region of this gene, including its tightly regulated promoter, is composed of a *Helitron*. It is proposed that the 3' portion of a *Helitron* inserted upstream to the ancestral was deleted and the rolling circle replication initiating from the intact 5' end of the *Helitron* provided read-through, which terminated in the coding sequence of the downstream xanA gene.

4.3   Future Prospects and Overview

The newly discovered *Helitrons* constitute a vastly unexplored family of novel transposable elements. Observations to date suggest that maize contains a particular subfamily of *Helitrons*. In contrast to a "generic" *Helitron*, the maize members bear much sequence similarity in termini. This feature will greatly aid in the quick identification of *Helitrons* in large-scale sequencing projects. Maize *Helitrons* also appear to be unique in their ability to prolifically capture gene pieces.

Existing data point to two fundamentally different ways that *Helitrons* could significantly alter the maize genome and gene expression. First, the ability to capture genes provides a simple mechanism to alter gene dosage. The presence of a gene in one inbred and its absence at the homologous site in another inbred provides an uncomplicated explanation for heterosis. So far however, no case of complete gene capture has been documented in maize so the gene dosage/heterosis model awaits validation. The availability of complete maize genome sequence in the near future will provide a valuable resource to investigate whether *Helitrons* mediate the mobilization of active genes, and if they do, whether it has occurred to an extent that might impact heterosis.

Second, the ability to place segments of different genes into a common transcribed region provides the opportunity to create new genes and functions. While this aspect of *Helitron* function awaits validation, it does suggest that "hybrid" transcripts will be found in large EST collections. Studies are in progress.

Despite strong evidence that these elements have moved recently, there is no direct genetic proof of their activity in the present day maize genome. Analysis of *Helitron* movement in the two maize mutants caused by *Helitron* insertion may provide clues concerning mechanisms of transposition. Also, a more complete understanding of the process by which two short conserved terminal ends of maize non-autonomous elements mediate movement and integration of large and diverse segments of DNA may aid in designing novel strategies for transformation and transgenic expression for crop improvement.

Acknowledgments The study was supported by NSF grant 0514759 to SL and NSF grants IBN-9982626 and 0444031 and IOS 0815104 to LCH and USDA Competitive Grants. 2000-01488, 2006-03034 and 2007-03575 to LCH. We thank Hugo Dooner for many interesting and fruitful discussions.

References

Brunner, S., Fengler, K., Morgante, M., Tingey, S. and Rafalski, A. (2005) Evolution of DNA sequence nonhomologies among maize inbreds. Plant Cell. 17, 343–360.
Brunner, S., Pea. G. and Rafalski, A. (2005) Origins, genetic organization and transcription of a family of non-autonomous *Helitron* elements in maize. Plant J. 43, 799–810.
Choi, J.D., Hoshino, A., Park, K.I., Park, I.S. and Iida, S. (2007) Spontaneous mutations caused by a *Helitron* transposon, Hel-It1, in morning glory, Ipomoea tricolor. Plant J. 49, 924–34.

Cultrone, A., Dominguez, Y.R., Drevet, C., Scazzocchio, C. and Fernandez-Martin, R. (2007) The tightly regulated promoter of the *xanA* gene of *Aspergillus nidulans* is included in a *Helitron*. Mol. Microbiol.63, 1577–87.

Feschotte, C. and Wessler, S.R.(2001) Treasures in the attic: Rolling circle transposons discovered in eukaryotic genomes. Proc. Natl. Acad. Sci. USA. 98, 8923–8924.

Fu, H. and Dooner, H.K. (2002) Intraspecific violation of genetic colinearity and its implication in maize. Proc. Natl. Acad. Sci. USA. 99, 9573–9578.

Gupta, S., Gallavotti, A., Stryker, GA., Schmidt, R.J. and Lal, S.K. (2005) A novel class of *Helitron*-related transposable elements in maize contain portions of multiple pseudogenes. Plant Mol. Biol. 57, 115–27.

Hall, R.and Callis, C. (1995) Mobile gene cassettes and integrons: capture and spread of genes by site specific recombination. Mol. Micro. 15, 593–600.

Jiang, N., Bao, Z., Zhang, X., Eddy, S.R. and Wessler S.R. (2004) Pack-MULE transposable elements mediate gene evolution in plants. Nature. 431:569–73.

Kapitonov, V.V. and Jurka, J. (2001) Rolling-circle transposons in eukaryotes. Proc. Natl. Acad. Sci. USA. 98, 8714–9.

Kapitonov, V.V. and Jurka, J. (2007) *Helitrons* on a roll: eukaryotic rolling-circle transposons. Trends Genet. Trends Genet 10, 521–529.

Khan, S.A. (2000) Plasmid rolling circle replication: recent development. Mol Microl. 37, 477–484.

Lai, J., Li, Y., Messing, J. and Dooner, H.K. (2005) Gene movement by *Helitron* transposons contributes to the haplotype variability of maize. Proc. Natl. Acad. Sci. U S A. 102, 9068–73.

Lal, S.K., Giroux, M.J., Brendel, V., Vallejos, C.E. and Hannah, L.C. (2003) The maize genome contains a *Helitron* insertion. Plant Cell. 15, 381–91.

Lal, S.K., Choi, J.H., Shaw, J. and Hannah, L.C. (1999) A splice site mutant of maize activates cryptic splice sites, elicits intron inclusion and exon exclusion, and permits branch point elucidation. Plant Physiol. 121, 411 –418.

Lal, S.K. and Hannah, L.C. (2005) *Helitrons* contribute to the lack of gene colinearity observed in modern maize inbreds. Proc. Natl. Acad. Sci. U S A. 102, 9993–4.

Lal, S.K. and Hannah, L.C. (2005) Plant genomes: massive changes of the maize genome are caused by *Helitrons*. Heredity. 95, 421–2.

Morgante, M., Brunner, S., Pea, G., Fengler, K., Zuccolo, A. and Rafalski, A. (2005) Gene duplication and exon shuffling by *Helitron*-like transposons generate intraspecies diversity in maize. Nat. Genet. 37, 997–1002.

Poulter, R.T., Goodwin, T.J. and Butler, M.I. (2003) Vertebrate helentrons and other novel *Helitrons*. Gene. 313, 201–12.

Pritham, E.J. and Feschotte, C. (2007) Massive amplification of rolling-circle transposons in the lineage of the bat Myotis lucifugus. Proc Natl Acad Sci USA.104, 1895–900.

Song, R. and Messing, J. (2003) Gene expresiion of a gene family in maize based on noncollinear haplotypes. Proc. Natl. Acad. Sci. USA. 100, 9055–9060

Tavakoli, N., Comanducci, A., Dodd, H.M., Lett, M.C. and Albiger Bennett, P. (2000) IS1294, a DNA element that transposes by RC transposition. Plasmid. 44, 66–84.

Wang, Q. and Dooner, H.K. (2006) Remarkable variation in maize genome structure inferred from haplotype diversity at the bz locus. Proc. Natl. Acad. Sci. U S A. 103, 17644–9.

Xu, J.H. and Messing, J. (2006) Maize haplotype with a *Helitron*-amplified cytidine deaminase gene copy. BMC. Genet. 7, 52.

Zabala, G. and Vodkin, L.O. (2005) The wp mutation of Glycine max carries a gene-fragment-rich transposon of the CACTA superfamily. Plant Cell. 17, 2619–32.

Zalaba, G. and Vodkin, L.O. (2007) Novel exon combinations generated by alternative splicing of gene fragments mobilized by a CACTA transposon in Glycine max. BMC Plant Biol.7, 38

Maize GEvo: A Comparative DNA Sequence Alignment Visualization and Research Tool

Eric Lyons, Sara Castelletti, Brent Pedersen, Damon Lisch, and Michael Freeling

Abstract Comparing the DNA sequence of maize genes and chromosomal regions to their maize homeologs and to other grass orthologs provides a quick check for many annotation errors and may identify useful regions of conservation and divergence. Fractionation and subfunctionalization, two processes that shape the evolution of genes and genomes, may be evaluated by comparing syntenic chromosomal regions within and among genomes. We provide an overview of maize's genome evolution, emphasizing the processes of fractionation and subfunctionalization. We then provide a tutorial of our software for multiple DNA sequence alignment display and analysis. As an example, we begin with one maize gene sequence, find a pair of maize homeologous gene regions, and add an orthologous segment from sorghum and rice; we then analyze this 4-way DNA alignment.

1 Introduction

BAC-by-BAC sequencing of the genome of maize inbred B73 is in progress. As these sequences are assembled and gene models annotated, the evidence on which models are based may be evaluated on a number of displays where frozen data conforms to community standards. Maize genes are displayed at Gramene (http://www.gramene.org/), The Maize Genome Browser (http://www.maizesequence. org/index.html, PlantGDB (http://www.plantgdb.org/prj/GSSAssembly/) and perhaps will be at the official maize information clearing-house, MaizeGDB (www.maizegdb.org/). Seeing each maize gene model is important, but the accuracy

E. Lyons, S. Castelletti, B. Pedersen, D. Lisch, and M. Freeling
Department of Plant and Microbial Biology,
University of California at Berkeley, CA, USA
freeling@nature.berkeley.edu

S. Castelletti
Department of Agroenvironmental Science and Technology,
University of Bologna, Viale Fanin, 44, 40127 Bologna, Italy, EU

J.L. Bennetzen and S. Hake (eds.), *Maize Handbook - Volume II: Genetics and Genomics*,   341
© Springer Science+Business Media LLC 2009

of any maize gene model and the "wholeness" of any single maize gene DNA sequence are very much in doubt. By "gene" we refer to both coding sequence and cis-regulatory elements. By "wholeness" we are referring to the fact that a gene with a history of retention post-tetraploidy is expected to subfunctionalize those regulatory elements, as will be discussed. Understanding the evolutionary relationships within and among genomes of related species, and using comparative genomics to compare syntenous genomic regions, vastly improves our ability to validate models and estimate wholeness of a gene's cis-acting sequence. Here, we briefly review the evolution of the maize genome with regards to its two sequential, ancient tetraploidies– how the resulting duplications of genomic regions and subsequent fractionation rip chromosomes apart, discard centromeric DNA, and often distribute dispensable parts of an ancestral gene's cis-acting units of function over more than one extant gene. Comparative genomics can help validate gene models and infer ancestral genes.

That any one maize gene encodes the full set of functions encoded by a pre-grass ancestral gene could happen, but is not an expectation. This argument (involving fractionation and subfunctionalization) is a primary topic of the next few paragraphs. We envision a need to begin maize gene research by using DNA sequence comparisons of homeologous chromosomal regions within maize, and among orthologous regions in other sequenced grasses. Such alignments identify regions of conservation and permit a quick evaluation of model annotation error and address gene "wholeness" issues as well. The comparative genomics system we introduce here, CoGe, allows researchers to easily specify, compare and visualize regions of similarity among any set of genomic regions. CoGe makes available the most recently annotated versions of published Angiosperm genomes (*Arabidopsis*, rice, poplar, grape), and some other genomes if assembled sequence has been submitted to GenBank. For maize, CoGe obtains its primary maize sequences and annotations from the community information support websites mentioned above. As with the maize genome, sorghum (same tribe as maize: Phytozome *Sorghum bicolor*) browser (http://www.phyto-zome.net/cgi-bin/gbrowse/sorghum/), *Brachypodium distachyon* (same subfamily as rice, but distantly related) and *Setaria italica* (foxtail millet; sister tribe to maize) are in progress, and will be in CoGe. Our fully documented web application is available at http://synteny.cnr.berkeley.edu/CoGe/.

Although sequence alignment algorithms are at the core of comparative genomics, the ability to easily visualize and interpret alignment results is equally important. We recently reviewed and evaluated various alignment algorithms and visualization tools useful for plant comparative genomics (Lyons and Freeling 2008). In that review, we introduced one of CoGe's tools called GEvo (for Genome Evolution Analysis). Although GEvo makes use of our custom genome visualization package, the Vista display, popular with animal biologists, (http://genome.lbl.gov/vista/index. shtml) can also be used with excellent results. Please refer to this publication (Lyons and Freeling 2008) for citations, a list of comparative genomics definitions, ways to import DNA sequences into GEvo for sequence comparisons, and a description on how to detect synteny, conserved noncoding sequence (CNS), fractionation, subfunctionalization, inversions, local duplications and annotation errors. As of yet, there is no pre-computed list of maize genes matched to their homeologous or orthologous partners for use in comparative genomics. (Such a list of sorghum – rice GEvo links is available at CoGe.) Such lists of GEvo links removes much of the tedium associated with "looking" at alignment outputs. In the meanwhile, researchers will need to identify their own useful homeologs and orthologs for maize comparative genomics. Following this Introduction, we provide a brief tutorial to emphasize the challenges confronting a maize molecular geneticist and how to surmount them. We start with a single maize sequence carrying a gene of interest, identify its homeolog in the in-progress maize genome, identify its orthologs in rice and sorghum, and use comparative genomics to detect synteny among these four homologous chromosomal regions (to validate their evolutionary history), annotation errors, and CNSs.

Simply observing the official model for a maize gene and evaluating the textual data underlying that model are not the best ways to quickly assess the accuracy of a gene model or to understand the extent to which any individual maize gene is a self-contained unit of function. As mentioned previously, any individual maize gene is not likely to do everything that its ancestor gene did 100 million years ago because the extant maize genome has undergone two tetraploidies since the divergence of the monocots from the rest of the angiosperms. The first tetraploidy occurred before the major radiation of common grass subfamilies as deduced from syntenic chromosomal regions within rice (Paterson et al. 2004, Tian 2005, Yu et al. 2005). This tetraploidy is present in all grasses. The second maize lineage tetraploidy (Gaut and Doebley 1997) occurred just after the sorghum and maize lineages diverged (Swigonova et al. 2004). Almost all flowering plants are derived from multiple, sequential tetraploidies over 10 million years old [called "ancient" or "paleo" (Adams and Wendel 2005); (Jaillon et al. 2007)], and many plant species have polyploidies so recent that they may be deduced from multiple chromosomal sets (Adams and Wendel 2005). The maize tetraploidy is ancient enough so that the process of fractionation—a mechanism to be explained in detail—is well underway, but recent enough so that functionless DNA has not fully randomized. This obfuscates identifying true CNSs in maize-maize homeologous sequence alignments because neutral sequence is simply carried-over from the ancestor. The sorghum sequence is an ideal outgroup to maize because the sorghum genome has not undergone a tetraploidy event since their divergence. However, pairwise comparisons between maize and sorghum are also too close for CNS discovery. Maize-rice comparisons are ideal for CNS discovery (Inada et al. 2003). For this reason, comparisons of any maize chromosomal region with its homeologous region (most recent tetraploidy), and with its orthologous sequences in sorghum and rice, provides important data for understanding the evolutionary history of any gene or genomic region. An analysis of this type requires the comparison of four genomic regions: two from maize, one from sorghum, and one from rice. If we extend our analysis to include the pre-grass tetraploidy as well, any ancestral pre-grass genomic region (60-100 MYA) is expected to be represented by two syntenic regions in both rice and sorghum and four such regions in maize. *Therefore, to access the fate of a single gene or region from the pre-grass ancestor would require the simultaneous alignment and evaluation of four maize genomic regions, and two each from sorghum and rice; an 8-way comparison.* Our alignment and visualization software, GEvo, permits finding, retrieving and comparing multiple DNA sequences of this sort on-the-fly.

2   Fractionation and subfunctionalization

Why, if a researcher's interest is with a single maize gene, are multiple evolution-arily-related genomic regions important? The answer to this question involves an understanding of the related terms *"fractionation"* and *"subfunctionalization."* The next two sections define these conceptually similar terms. In short, any one maize gene is not expected to encode an entire ancestral gene function. If one thinks "one gene-one function" in the context of repeated ancient tetraploidies, then serious confusions of parts for wholes will occur, and "novelty" will be impossible to recognize. GEvo is built to explore and quantify *fractionation* of genomic regions and *subfunctionalization* of genes.

Definitions

2.1 *Fractionation* is defined as the mechanism by which a duplicated gene, chromosomal segment or genome tends to return to pre-duplication information content, but not necessarily its pre-duplication gene order (review: (Lockton and Gaut 2005). Fractionation is the loss of one or the other of the initial homeologs, but not both. The process of fractionation may be associated with chromosomal rearrangements and "transcriptome shock" (Wang et al. 2006), and may help cluster dose-sensitive genes via an epigenetic mutational targeting mechanism (Thomas et al. 2006).

The fractionation mechanism is particularly important for genome evolution because particular types of genes are susceptible or resistant to fractionation. The latter genes accumulate in the genome following tetraploidy. In *Arabidopsis*, where this matter of retention bias following tetraploidy has been studied most extensively, gene families encoding transcription factors, protein kinases, ribosomal proteins, proteasome core proteins and other components of complexes or networks have expanded. (Blanc and Wolfe 2004, Maere et al. 2005, Seoighe and Gehring 2004, Thomas, Pedersen and Freeling 2006). The fractionation process for maize is still ongoing (Ilic et al. 2003).

2.2 *Subfunctionalization* is defined as the natural tendency of a duplicated cis-acting unit of function (gene) to lose dispensable sequences (functions) on one but not both duplicates, such that the ancestral function is spread over both duplicates (Force *et al.* 1999). This idea was originally used to explain the over-retention of pairs following duplication (Lynch and Force 2000), but gene content change data following tandem versus whole-genome duplications supports a dosage explanation and *not* subfunctionalization as the duplicate retention mechanism (Freeling 2008, Freeling and Thomas 2006). In any case, subfunctionalize.

If DNA sequence alignments are to be used to measure fractionation and subfunctionalization, outgroup genomes and appropriate visualization software are essential in order to define the ancestral "whole" gene. For a detailed step-by-step version of this tutorial, please visit http://tinyurl.com/59x3nb.

3 Maize GEvo Tutorial: From one Maize Gene to Useful Comparative Genomic Information.

This tutorial begins with a putative coding sequence from maize identified in an unordered contig sequenced by The Maize Sequencing Consortium and deposited in NCBI under the accession AC210314.1 (http://tinyurl.com/2vyl99). This contig is typical of in-progress sequencing projects as it has 21 unordered and unannotated pieces. The specific subsequence from this contig can be downloaded from http://synteny.cnr.berkeley.edu/CoGe/data/distrib/example_maize.faa. The tutorial that follows uses this maize sequence, but feel free to follow along with your own favorite sequence. From now on, the reader will be referred to as "you". For work with grasses, you are advised to use genomes that have been masked for sequences present in over 50 copies. We have found that use of these masked genomes, supplied by CoGe, greatly reduces the noise derived from unannotated transposons masquerading as CNSs.

3.1 Finding Homologs

CoGe has a genome-oriented Blast interface that allows you to search a sequence against any number of genomes in CoGe and display the results in an interactive webpage. This makes it especially useful for identifying putative homologs. Figure 1A. shows a blast analysis configured to search against the genomes of maize, sorghum, and rice. Once the blast analyses are completed, the results are shown on the top of the screen while your analysis configuration remains in place to allow you to modify and rerun the analysis quickly. On the left of the results is an overview of the genomic segments (e.g. chromosomes, contigs) with blast high-scoring sequence pairs (Blast hits; HSPs) identified by triangles and lines. On the right is a list of HSPs and any genomic feature (e.g. gene) that they overlap. To get additional information about an HSP, you can click on an HSP in the genomic overview graphic or the HSP number in the HSP list. This will give you a graphical overview of the query sequence coverage and the matched genomic region. Any other HSPs in that genomic region will also be displayed in both graphics (albeit in a different color), allowing you to quickly evaluate the coverage of your entire sequence to a genomic region. Also, HSP-specific information such as percent identity and e-value will be displayed in a table below the HSP graphics. By evaluating the HSPs in this manner, it is easy to identify which overlapping genomic features are of interest. Figures 1B-1D show HSP graphics

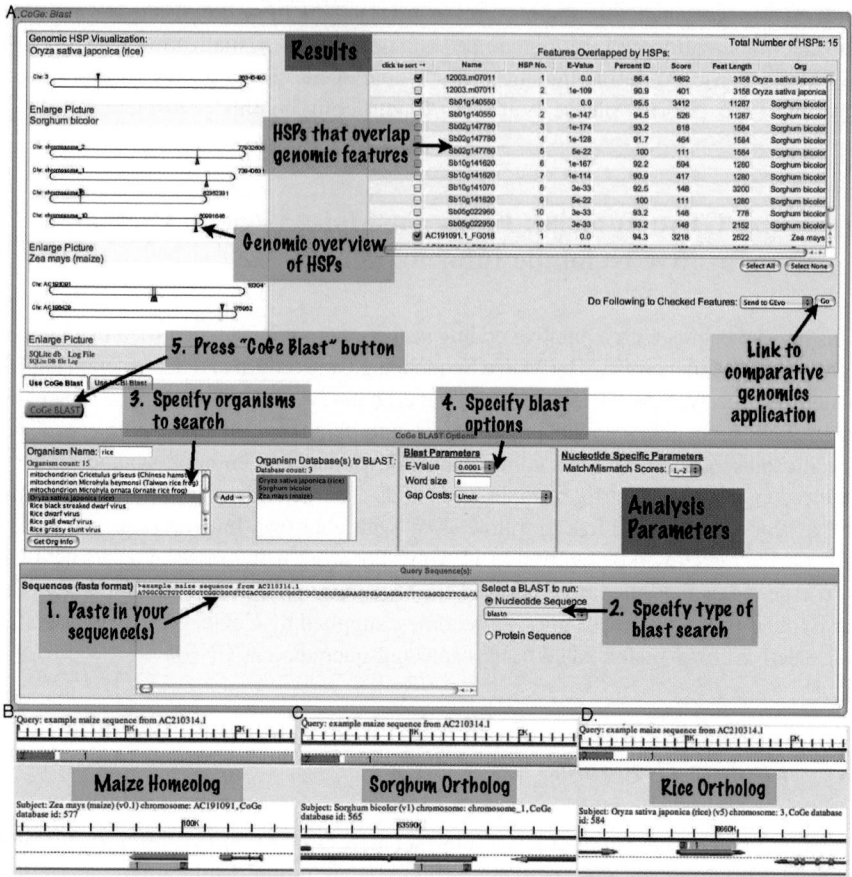

Fig. 1 Output from CoGeBlast (http://synteny.cnr.berkeley.edu/CoGe/CoGeBlast.pl) using the example maize sequence (http://tinyurl.com/2ua942) as the query sequence, Blastn for the Blast algorithm, and the genomes of maize, sorghum, and rice as search databases. (A.) Results are visualized on the top of the screen after the analysis is complete showing a graphical overview of chromosomes or contigs hit by Blast, and a table of overlapping annotated genomic features (e.g. genes) on the right. The raw Blast results are obtained by clicking on the organism name, and detailed information about an HSP is obtained by clicking on the colored arrow in the graphical overview or on an HSP number in the table. (B-D.) Graphical overviews of top scoring HSPs in maize, sorghum, and rice (respectively) showing nearly full coverage to the query sequence. The query sequence and genomic region are shown in the top and bottom graphic respectively. Note the incongruence of the sorghum gene model to the query sequence; this is an annotation error

for the potential maize homeolog, and rice and sorghum orthologs. In all of these, you can see that there is nearly full coverage of the query sequence, which, for DNA sequence alignments, is indicative of a high degree of sequence relatedness. Interestingly, there is full coverage of the gene models shown in the genomic regions of maize and rice, but only partial coverage of sorghum's gene model. The sorghum model is likely in error. Although you would expect to find two regions in maize that are very similar to the query sequence due to maize's recent tetraploidy, only one was identified. Since the percent identity of this HSP is 94.3%, this is probably, but not certainly, the homeolog to the query sequence. As the maize genome is not finished, we do not expect complete sequence or complete assembly. For this example, the query sequence was not found in the maize genome. Once a candidate list of homologs has been made, we can use another of CoGe's tools—GEvo– to compare these genomic regions in order to identify synteny, evaluate fractionation and discover CNSs.

3.2   Detecting Synteny

CoGe's Blast tool has the ability to send your selected genomic features to CoGe's comparative genomics tool, GEvo (Fig. 1A). To do this, simply click on the button marked "send to GEvo" in the lower right-hand side of the results panel. GEvo (Fig. 2) lets you specify genomic regions for comparison, choose and configure an alignment algorithm, and modify the results visualization. The Blast analysis identified three homologs which, when sent to GEvo, are used to identify genomic regions from CoGe's genome database that can be extended by specifying additional sequence to the left and right of these genes. In addition to these genomic regions, you will want to add your original sequence. You can do this by adding an additional sequence submission form, selecting the type of sequence submission (NCBI retrieval in this case), and entering in the GenBank accession. By default, GEvo will use Blastz (Schwartz et al. 2003, Schwartz et al. 2000) for its alignment algorithm. Blastz is a variant of Blast that has been optimized for finding large blocks of conserved sequence and works well for aligning genomic sequence and for interferring synteny.

Figure 2 shows the results from a GEvo analysis comparing (pair-wise) these four genomic regions. Each region is drawn horizontally with gene models in the middle and HSPs drawn as numbered colored blocks with those on the top half representing HSPs in the (++) orientation and those on the bottom in the (+–) orientation. Note that the maize sequence retrieved from NCBI does not have annotated gene models. When models are added to GenBank records, we will import them into CoGe. Each set of HSPs for a pair-wise sequence comparison is drawn in its own track. Using GEvo's results visualization interface, lines have been drawn connecting HSPs between rice, sorghum, and the maize sequence from CoGe's database. These help infer synteny by identifying collinear HSPs. Since all of the HSPs (except one) are collinear and overlap coding regions, these regions are syntenic. Although the maize sequence retrieved from NCBI is unordered and unannotated, there are several conserved regions that match the other regions which shows that it is likely syntenic as well. Also, note the discordance of gene models among regions with sequence similarity.

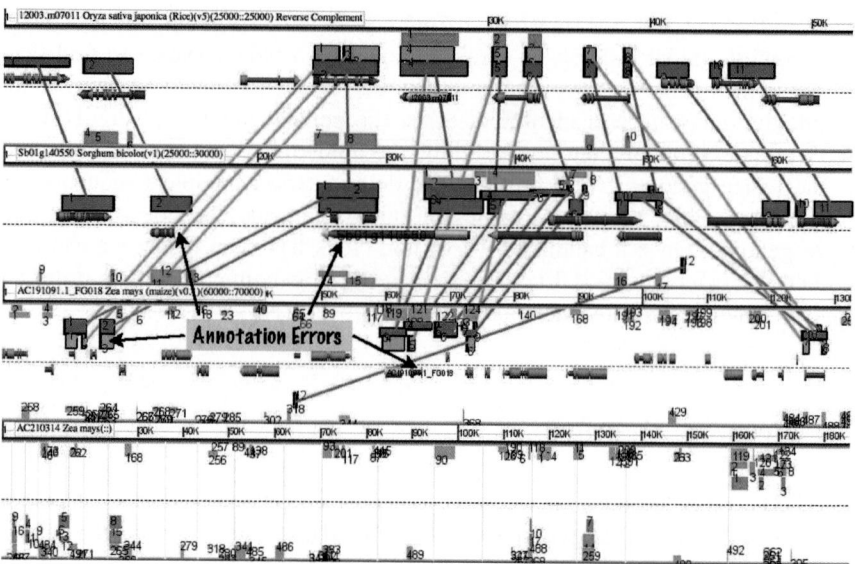

Fig. 2 Output from GEvo comparing the genomic regions around homologous genes identified in Fig. 1 (highlighted with yellow gene models) and the complete maize contig from NCBI from which the example query sequence originated. Drawn from top to bottom are rice, sorghum, maize homeolog 1 and maize homeolog 2. Colored numbered blocks are pairwise regions of sequence similarity identified by Blastz. Colored lines show synteny among rice, sorghum, and maize homeolog 1. Four annotation errors are indicated on the figure where gene models are not congruent between the genomic regions and their respective regions of sequence similarity

As the maize sequence is assembled and finished, longer and better annotated contigs will emerge. Eventually, chromosomes with canonical gene loci names (e.g. *ZmXgXXXXXX*) will be available. CoGe will periodically update its maize genome sequence. When the maize genome reaches a "finished" state, we will prepare a menu of *Zm1-Zm2-Sb-Os* GEvo links. Currently, the *Sb-Os* GEvo links list is available online (http://tinyurl.com/2dckrf).

By knowing that these regions of Fig. 2 are all syntenic and thus derived from a common ancestral genomic region, we are able to perform a higher-resolution analysis of the sequences neighboring our gene and its homologs to find conserved noncoding sequences (CNSs).

3.3   Detecting CNSs

GEvo's interface is much like CoGe's Blast interface where the results are shown above the analysis configuration. This makes it easy to quickly reconfigure and rerun an analysis. For detecting CNSs, we'll want to examine smaller genomic regions at a resolution just above the noise level. Blastn (Altschul et al. 1990) works well, but GEvo supports alternative alignment algorithms Lagan (Brudno et al. 2003b), Chaos (Brudno et al. 2003a) and DiAlign (Morgenstern 1999). Plant CNSs are smaller than animal CNSs, and tend to disappear more quickly over evolutionary time (Lyons and Freeling 2008); operationally plant CNSs are defined as having a Blast e-value less than that obtained for a 15 nucleotide exact match (Inada et al. 2003). GEvo makes use of this CNS "noise" filter when using Blastn by default, but it can be switched "off."

Figure 3 shows the results of examining the region around the homologs of one gene. Note that the gene models are drawn with the 5′ end to the right (as displayed in Fig. 2). The lowermost maize sequence of Fig. 3 has no model, and notice the approximately 1.3 kb insertion between *Zm1-Zm2* HSPs 7 and 8 (purple). Also note that there are several collinear HSPs in noncoding sequence. This is particularly evident if you examine the 5′ (right side) noncoding hits plotted in the *Os* track, below the "Fractionated CNSs" label. From upper to lower for the *Os* track: coral is *Os-Zm2*, forest green is *Os-Zm1*, and blue is *Os-Sb*. If you were interested in conserved gene regulation, the 5′ and 3′ fully conserved CNSs would be a good place to start because they are present in all syntenic regions. As you can see from the various comparisons, using the rice gene as an outgroup is essential due to the "carry-over" of sequences when comparing the more closely related sorghum and maize genes.

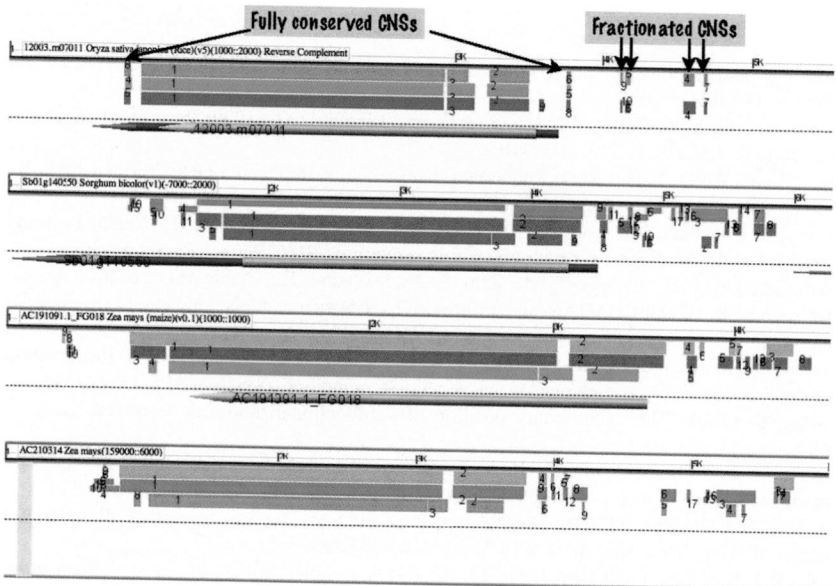

Fig. 3 Output from GEvo comparing the same homologs as in Fig. 2, but using less surrounding genomic sequence and Blastn (with a noise filter for remove HSPs less significant than a 15/15 exact nucleotide match) for identifying conserved noncoding sequences. Blastz is useful for finding large blocks of conserved sequence while Blastn is useful for finding small blocks of highly conserved sequence

You may have noticed that there are several CNSs that are only partially conserved among all these syntenic regions. This is especially apparent for the two maize homeologs where one region has a CNS shared in rice and sorghum, while the other does not. Since the two maize regions are more closely related to one another than either is to rice, you can infer that the ancestral region in the pre-tetraploid maize lineage contained this region and it was lost (subfunctionalized) in one of the maize homeologs following duplication. Further research on these CNSs would test this subfunctionalization hypothesis.

4   Conclusion

Comparing the DNA sequence of maize genes and chromosomal segments to their maize homeologs and to other grass orthologs provides a quick check for many annotation errors and may identify useful regions of conservation and divergence. CoGe– with its Blast tools– and CoGe's comparative genomics tool, GEvo, facilitate such comparisons.

References

Adams, K.L. and Wendel, J.F. (2005) Polyploidy and genome evolution in plants. *Curr Opin Plant Biol*, 8, 135–141.

Altschul, S.F., Gish, W., Miller, W., Myers, E.W. and Lipman, D.J. (1990) Basic local alignment search tool. *J Mol Biol*, 215, 403–410.

Blanc, G. and Wolfe, K.H. (2004) Functional divergence of duplicated genes formed by polyploidy during Arabidopsis evolution. *Plant Cell*, 16, 1679–1691.

Brundo, M., Chapman, M., Gottgens, B., Batzoglou, S. and Morgenstern, B. (2003a) Fast and sensitive multiple alignment of large genomic sequences. *BMC Bioinformatics*, 4.

Brudno, M., Do, C.B., Cooper, G.M., Kim, M.F., Davydov, E., Green, E.D., Sidow, A. and Batzoglou, S. (2003b) LAGAN and Multi-LAGAN: efficient tools for large-scale multiple alignment of genomic DNA. *Genome Res*, 13, 721–731.

Force, A., Lynch, M., Pickett, F.B., Amores, A., Yan, Y.L. and Postlethwait, J. (1999) Preservation of duplicate genes by complementary, degenerative mutations. *Genetics*, 151, 1531–1545.

Freeling, M. (2008) The evolutionary position of subfunctionalization, downgraded. *Genome Dynamics*, 4, 26–40.

Freeling, M. and Thomas, B.C. (2006) Gene-balanced duplications, like tetraploidy, provide predictable drive to increase morphological complexity. *Genome Research*, 16, 805–814.

Gaut, B.S. and Doebley, J.F. (1997) DNA sequence evidence for the segmental allotetraploid origin of maize. *Proc Natl Acad Sci U S A*, 94, 6809–6814.

Ilic, K., P. J. SanMiguel and Bennetzen, J.L. (2003) A complex history of rearrangement in an orthologous region of the maize, sorghum and rice genomes. Proc. Natl. Acad. Sci. USA 100, 12265–12270.

Inada, D.C., Bashir, A., Lee, C., Thomas, B.C., Ko, C., Goff, S.A. and Freeling, M. (2003) Conserved noncoding sequences in the grasses. *Genome Res*, 13, 2030–2041.

Jaillon, O., Aury, J.M., Noel, B., Policriti, A., Clepet, C., Casagrande, A., Choisne, N., Aubourg, S., Vitulo, N., Jubin, C., Vezzi, A., Legeai, F., Hugueney, P., Dasilva, C., Horner, D., Mica, E.,

Jublot, D., Poulain, J., Bruyere, C., Billault, A., Segurens, B., Gouyvenoux, M., Ugarte, E., Cattonaro, F., Anthouard, V., Vico, V., Del Fabbro, C., Alaux, M., Di Gaspero, G., Dumas, V., Felice, N., Paillard, S., Juman, I., Moroldo, M., Scalabrin, S., Canaguier, A., Le Clainche, I., Malacrida, G., Durand, E., Pesole, G., Laucou, V., Chatelet, P., Merdinoglu, D., Delledonne, M., Pezzotti, M., Lecharny, A., Scarpelli, C., Artiguenave, F., Pe, M.E., Valle, G., Morgante, M., Caboche, M., Adam-Blondon, A.F., Weissenbach, J., Quetier, F. and Wincker, P. (2007) The grapevine genome sequence suggests ancestral hexaploidization in major angiosperm phyla. *Nature*, 449, 463–467.

Lockton, S. and Gaut, B.S. (2005) Plant conserved non-coding sequences and paralogue evolution. *Trends Genet*, 21, 60–65.

Lynch, M. and Force, A. (2000) The probability of duplicate gene preservation by subfunctionalization. *Genetics*, 154, 459–473.

Lyons, E. and Freeling, M. (2008) How to usefully compare homologous plant genes and chromosomes as DNA sequence. *The Plant Journal*, 53: 661–673.

Maere, S., De Bodt, S., Raes, J., Casneuf, T., Van Montagu, M., Kuiper, M. and Van de Peer, Y. (2005) Modeling gene and genome duplications in eukaryotes. *Proc Natl Acad Sci U S A*, 102, 5454–5459.

Morgenstern, B. (1999) DIALIGN 2: improvement of the segment-to-segment approach to multiple sequence alignment. *Bioinformatics*, 15, 211–218.

Paterson, A.H., Bowers, J.E. and Chapman, B.A. (2004) Ancient polyploidization predating divergence of the cereals, and its consequences for comparative genomics. *Proc Natl Acad Sci U S A*, 101, 9903–9908.

Schwartz, S., Kent, W.J., Smit, A., Zhang, Z., Baertsch, R., Hardison, R.C., Haussler, D. and Miller, W. (2003) Human-mouse alignments with BLASTZ. *Genome Res*, 13, 103–107.

Schwartz, S., Zhang, Z., Frazer, K.A., Smit, A., Riemer, C., Bouck, J., Gibbs, R., Hardison, R. and Miller, W. (2000) PipMaker–a web server for aligning two genomic DNA sequences. *Genome Res*, 10, 577–586.

Seoighe, C. and Gehring, C. (2004) Genome duplication led to highly selective expansion of the Arabidopsis thaliana proteome. *Trends Genet*, 20, 461–464.

Swigonova, Z., Lai, J., Ma, J., Ramakrishna, W., Llaca, V., Bennetzen, J.L. and Messing, J. (2004) Close split of sorghum and maize genome progenitors. *Genome Res*, 14, 1916–1923.

Thomas, B.C., Pedersen, B. and Freeling, M. (2006) Following tetraploidy in an Arabidopsis ancestor, genes were removed preferentially from one homeolog leaving clusters enriched in dose-sensitive genes. *Genome Research*, 16, 934–946.

Tian, C., Y. Xiong, T. Liu, S. Sun, L. Chan and M. Chen. (2005) Evidence for ancient whole genome duplication event in rice and other cerials. *Acta Genetica Sinica*, (in press).

Wang, J., Tian, L., Lee, H.S., Wei, N.E., Jiang, H., Watson, B., Madlung, A., Osborn, T.C., Doerge, R.W., Comai, L. and Chen, Z.J. (2006) Genomewide nonadditive gene regulation in Arabidopsis allotetraploids. *Genetics*, 172, 507–517.

Yu, J., Wang, J., Lin, W., Li, S., Li, H., Zhou, J., Ni, P., Dong, W., Hu, S., Zeng, C., Zhang, J., Zhang, Y., Li, R., Xu, Z., Li, X., Zheng, H., Cong, L., Lin, L., Yin, J., Geng, J., Li, G., Shi, J., Liu, J., Lv, H., Li, J., Deng, Y., Ran, L., Shi, X., Wang, X., Wu, Q., Li, C., Ren, X., Li, D., Liu, D., Zhang, X., Ji, Z., Zhao, W., Sun, Y., Zhang, Z., Bao, J., Han, Y., Dong, L., Ji, J., Chen, P., Wu, S., Xiao, Y., Bu, D., Tan, J., Yang, L., Ye, C., Xu, J., Zhou, Y., Yu, Y., Zhang, B., Zhuang, S., Wei, H., Liu, B., Lei, M., Yu, H., Li, Y., Xu, H., Wei, S., He, X., Fang, L., Huang, X., Su, Z., Tong, W., Tong, Z., Ye, J., Wang, L., Lei, T., Chen, C., Chen, H., Huang, H., Zhang, F., Li, N., Zhao, C., Huang, Y., Li, L., Xi, Y., Qi, Q., Li, W., Hu, W., Tian, X., Jiao, Y., Liang, X., Jin, J., Gao, L., Zheng, W., Hao, B., Liu, S., Wang, W., Yuan, L., Cao, M., McDermott, J., Samudrala, R., Wong, G.K. and Yang, H. (2005) The Genomes of *Oryza sativa*: A History of Duplications. *PLoS Biol*, 3, e38.

Meiotic Genes and Meiosis in Maize

W. Zacheus Cande, Inna Golubovskaya, C.J. Rachel Wang,
and Lisa Harper

Abstract Meiosis is the specialized cell division required to produce gametes with a haploid chromosome content in all eukaryotes with a sexual cycle. The cellular events that occur in meiosis are evolutionarily conserved, as are many of the proteins associated with meiosis, especially those required for homologous recombination. Maize stands out as one of the premier cytological model organisms for studying meiosis because of its large, well defined chromosomes, the ease in which meiotic stages can be identified cytologically, and its many genetics resources. Powerful forward genetics screens have led to the identification of a large number of maize meiotic mutants although only a few of them have been cloned. In this chapter, we describe the mutant collection, major findings associated with working with these mutants, and the promise of maize cytogenetics for future research.

1 Introduction to Meiosis

1.1 Commitment to Meiosis

In multicellular organisms, the first step to achieving meiosis is to develop the special cells, "meiocytes" or formally, "sporocytes" in plants, that will undergo meiosis. In maize, male meiocytes (microsporocytes) develop in the anthers born on the tassel and give rise to four pollen grains, each containing one vegetative nucleus and two

W.Z. Cande, I. Golubovskaya, C.J.R. Wang, and L. Harper
University of California, Berkeley, Department of Molecular and Cell Biology
zcande@berkeley.edu

L. Harper
USDA-ARS-PGEC, Albany CA USA
ligule@nature.berkeley.edu

J.L. Bennetzen and S. Hake (eds.), *Maize Handbook - Volume II: Genetics and Genomics*,   353
© Springer Science+Business Media LLC 2009

sperm nuclei. Female meiocytes (megasporocytes) develop in the ovule born on the ear, and give rise to the egg. In maize, both male and female meiocytes are visually distinctive because of their larger size and position in the anther or ovule. Developmental fate of the cells to become the meiocytes is under genetic control. In addition to establishing appropriate meiocyte cell fate, a switch from the mitotic cell cycle to a meiotic cell cycle is required. This switch requires installation of the meiotic specific versions of many proteins like histones and cohesins during the S phase immediately preceding the first meiotic division, and establishment of the meiotic cytoskeleton.

1.2 Prophase Stages

The general progression of meiosis itself is highly conserved, and is thus similar in yeasts, animals and higher plants (John 1990; Zickler and Kleckner 1999). Once the meiotic cell fate and the switch to the meiotic cell cycle are established, the cell proceeds through stereotypical stages to produce the gamete. The first meiotic prophase is divided into five stages (see Fig. 1).

In **leptotene**, chromosomes become visible and installation of the axial elements of the synaptonemal complex (SC) is completed. We refer to this distinctive unpaired and unsynapsed prophase chromosome with its axial element as the "leptotene chromosome". Sister chromatids are completely appressed throughout the length of the chromosome, held together by Sister Chromatid Cohesion (SCC)

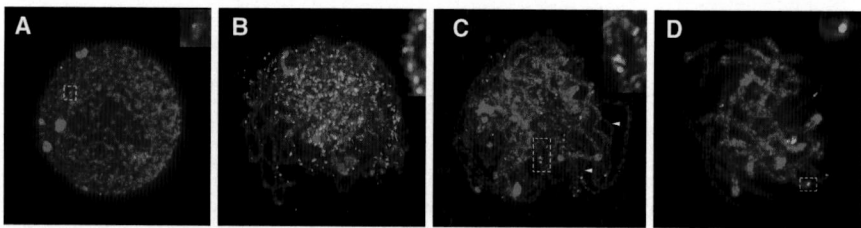

Fig. 1 Maize chromatin structure and the pattern of RAD51 foci distribution during Prophase 1 in wildtype meicytes (A) Leptotene. Few RAD51 foci are seen on newly formed leptotene chromosomes. The conspicuous DAPI-stained chromomeres are heterochromatin. (B) Early zygotene. Massive chromatin remodeling has taken place at the lep-zyg transition and the heterochromatin, especially the knobs, has developed an extended morphology. The maximum numbers of RAD51 foci are seen at this particular stage. Numerous single and small RAD51 foci are present on unsynapsed chromosomes. (Inset B). (C) Late zygotene. Chromosomes are synapsed except for a few unpaired regions (arrowhead). The number of RAD51 foci is greatly decreased in comparison with early zygotene. Many RAD51 foci appear to be dumbbell shaped or doublets. Some foci are elongated (Inset C). (D) Pachytene. Homologous chromosomes are completely paired and synapsed and fewer RAD51 foci are present. These foci in pachytene generally are larger and brighter than those in zygotene and their shapes change as chromosomes pair (Inset D). For more examples of early prophase stages with rotating projections, please see http://jcs.biologists.org/cgi/content/full/117/18/4025/DC1 (Harper, et al. 2004)

protein complexes. The double strand breaks that initiate recombination (probably) occur in leptotene (Pawlowski, Golubovskaya, Timofejeva, Meeley, Sheridan and Cande 2004).

At the **leptotene-zygotene (lep-zyg) transition**, transient remodeling of chromatin architecture occurs; including the elongation of heterochromatin knobs and centromeric heterochromatin, the expansion of euchromatin, and the slight separation of sister chromatids (Dawe, Sedat, Agard and Cande 1994; Carlton, Ananiev and Cande 1998). We speculate that these global and transient changes in chromatin structure are required to initiate or facilitate homologous pairing interactions. Also during the lep-zyg transition, telomeres attach to the nuclear envelope and begin to cluster (Bass, Marshall, Sedat, Agard and Cande 1997; Bass 2003; Harper, Golubovskaya and Cande 2004). This "telomere bouquet" is a highly conserved meiotic event (Dernburg, Sedat, Cande and Bass 1995) that is thought to facilitate many subsequent meiotic prophase processes, such as homologous pairing and synapsis (Golubovskaya, Harper, Pawlowski, Schichnes and Cande 2002; Harper, et al. 2004). More generally, the lep-zyg transition is thought to be a key checkpoint for meiosis progression (Zickler and Kleckner 1998).

During **zygotene**, homologous chromosomes pair and synapse. Homologous pairing requires the initiation of recombination. It is very likely that double strand breaks followed by the homology search and strand invasion actually are the mechanism of homologous pairing. During zygotene, RAD51, a recombination protein, is localized to chromosomes in foci. The number of foci peak in zygotene, reaching a maximum of 500-600 foci (Franklin, McElver, Sunjevaric, Rothstein, Bowen and Cande 1999b). These are likely the sites where homology is being assayed. RAD51 is required for efficient homologous pairing in maize and other organisms (Shinohara, Ogawa and Ogawa 1992; Pawlowski, Golubovskaya and Cande 2003; Li, Chen, Markmann-Mulisch, Timofejeva, Schmelzer, Ma and Reiss 2004). The appearance of RAD51 foci is completely coincident with chromosome pairing (Franklin, et al. 1999b), and mutants that cannot perform recombination also have no homologous pairing (Pawlowski, et al. 2003). Synapsis requires the formation of the tripartite synaptonemal complex (SC) between the two pairing chromosomes. A central element is installed between the two axial elements as the homologous chromosomes "zip up" from the telomeres towards the middle. Axial elements are called lateral elements after they are synapsed. During synapsis, chromosomes can become entangled in "interlocks"; however, the mechanism of their resolution is unknown. Synapsis is a meiosis-specific event, but is completely independent of the homology search. That is, chromosomes that are not homologously paired will still synapse (Pawlowski, et al. 2003; Pawlowski, et al. 2004).

Pachytene is defined as the stage when synapsis is complete and is probably also when recombination is completed. In pachytene, each pair of chromosomes, or the bivalent, appears as one thick strand, and consists of two homologs, each with two sister chromatids, for a total of 4 strands of DNA. Pachytene is the stage when maize chromosomes are often viewed for cytology, as this is the stage when knobs, centromeres and kinetochores, and chromomeres are most visible in aceto-carmine squashes (Dempsey 1994). Maize in particular has beautiful, large (over

100 microns long), and individually distinct pachytene chromosomes. McClintock and many other cytogeneticists used squashes of maize pachytene chromosomes for study.

In **diplotene**, chromosomes further condense as the SC falls apart, releasing homologous chromosomes from each other along their length, except at the sites of chiasmata, the cytological manifestations of crossover recombination events. Chiasma formation is a unique meiotic event, and is absolutely required to hold the homologs together until they separate in the reductional first metaphase to anaphase transition.

During **diakinesis,** the chromosomes condense further, thicken and detach from the nuclear envelope. After diakinesis, the nuclear envelope breaks down, and meiocytes enter metaphase. Sister chromatid cohesion is released along the arms of the sister chromatids of each chromosomes, allowing the resolution of the chiasmata. However, unlike mitotic chromosomes, centromeric cohesion is maintained between sisters, allowing the segregation of the homologous chromosomes, but not sisters, in the reductional division (Meiosis I). Chiasmata are responsible for holding homologous chromosomes together until the onset of anaphase I, and thus allow for the unique meiotic separation of homologous chromosomes. After the first meiotic division, there is a brief interphase with no DNA replication. The second meiotic division (M2) appears cytologically very much like a mitotic division, since cohesion is released at the centromeres allowing sister chromatids to move to opposite poles, in an equational division (Chan and Cande 1998).

2   The Contribution of Maize Meiotic Mutants

In a recent review we have summarized advances in the genetics of meiotic prophase in angiosperms (Hamant, Ma and Cande 2006). Here we discuss our analysis of maize meiotic mutants.

2.1   Classification of Mutants by Phenotype

Maize has excellent forward genetics, and as a result, we now have over 50 meiotic mutants, representing at least 35 genes in maize, mostly found by Inna Golubovskaya and others in our lab (Golubovskaya, Sheridan, Harper and Cande 2003). Most meiotic mutants have been found in forward genetic screens for male sterility. Meiotic mutants usually affect both male and females, and about 10% of all new male sterile mutants are also female sterile. Once a male and female sterile mutant is found, we check the number of bivalents and univalents present in late Prophase I in acetocarmine squashes of meiocytes. Mutants that affect homologous pairing, recombination, synapsis or meiotic chromatin structure usually have defects in their ability to form bivalents. Acetocarmine squashes also allow the examination of all stages of meiosis, and many defects can be recognized by this method. Once a meiotic defect is established, we use a battery of tools to further classify the meiotic mutant. These currently include: terminal transferase dUTP nick end labeling (TUNEL) assay to determine if meiosis-specific DNA breaks are made, RAD51 immunolocalization and counts of bivalents/univalents for reporting on recombination, TEM and immunolocalization of AFD1/REC8 and ASY1/HOP1 for reporting on synapsis, FISH for reporting on pairing and on bouquet formation, and immuno-FISH to analyze the relationship between bouquet formation and axial element installation.

Maize meiotic mutants can be grouped based on their phenotypes: meiotic commitment mutants (including differentiation of meiocytes and the switch to the meiotic cell cycle), desynaptic mutants that affect one or more prophase one events (telomere bouquet, homologous synapsis, homology search, recombination), sister chromatid cohesion and chromosome segregation mutants (including monopolar centromere attachment, meiotic cytoskeleton/spindle, sister chromatid cohesion and condensation mutants), and meiotic exit mutants (see Table 1). Several maize mutants have phenotypes not found yet in any other organism, such as *am1-pra1*, which is arrested at the lep-zyg transition. Because many of the mutants were isolated from *Mutator* active populations, the genes may be tagged. We have cloned several of them (*am1, afd1, phs1, sgo1*). In addition, two Rad51 genes were isolated

Table 1 Maize Meiotic Mutants

Gene/ Location	Mutant alleles	Reference
Differentiation of Meiocytes		
multiple archesporial cells 1 (*mac1*)/ 10S	*mac1*	(Sheridan, Shamrov, Batygina and Golubovskaya 1996; Sheridan, Golubeva, Abrhamova and Golubovskaya 1999)
Switch to Meiotic Cell Cycle		
ameiotic 1 (am1)/ 5S	*am1-1, am1-2, am1-485, am1-489, am1-6, am1-pra1*	(Rhoades 1956; Palmer 1971; Golubovskaya, Grebennikova, Avalkina and Sheridan 1993; Golubovskaya, Avalkina and Sheridan 1997a)
Sister Chromatid Cohesion		
absence of the first division (*afd1*)/ 6.08	*afd1-1, afd1-2, afd1-3, afd1-4*	(Golubovskaya and Mashnenkov 1975; Golubovskaya, Hamant, Timofejeva, Wang, Braun, Meeley and Cande 2006)
shugoshin 1 (sgo1)/ 7.02	*sgo1-1*	(Hamant, Golubovskaya, Meeley, Fiume, Timofejeva, Schleiffer, Nasmyth and Cande 2005)

(continued)

Table 1 (continued)

Gene/ Location	Mutant alleles	Reference
Chromosome Condensation		
Meiotic025 (Mei025)/ 5L	Mei025	(Golubovskaya 1979; Golubovskaya, et al. 2003)
sticky1 (st1)	st1	(Beadle 1937)
elongate1 (el1)/ 8L	el1	(Rhoades and Dempsey 1966; Barrell and Grossniklaus 2005)
Telomere Bouquet		
plural _abnormalities of meiosis_ (pam1)/ 1L	pam1	(Golubovskaya 1977; Golubovskaya, et al. 2002)
Homologous Synapsis		
asynaptic 1 (as1)/ 1S	as1	(Beadle 1930)
desynaptic1 (dy1)	dy1	(Nelson and Clary 1952)
desynaptic 1 (dsy1)	dsy1-1, dsy1-9101	(Golubovskaya and Mashnenkov 1976; Golubovskaya, Grebennikova, Auger and Sheridan 1997b)
desynaptic2 (dsy2)/ 5.03-05	dsy2	(Golubovskaya 1989 ; Franklin, Golubovskaya, Bass and Cande 2003)
desynaptic 9303	dsy9303	(Golubovskaya, et al. 2003)
desynaptic 9305	dsy9305	
desynaptic 9904a	dsy9904a	
desynaptic 9904b	dsy9904b	
desynaptic 9905a	dsy9905a	
desynaptic 9905b	dsy9905b	
desynaptic 9906a	dsy9906a	
desynaptic 9906b	dsy9906b	
mtm99-13[a]	mtm99-13	(Golubovskaya, et al. 2003)
mtm99-14	mtm99-14	
mtm99-25	mtm99-25	
mtm99-30	mtm99-30	
Homology Search		
poor homology synapsis (phs1)[b]/ 9.03	phs1	(Golubovskaya, et al. 2003; Pawlowski, et al. 2004)
desynapticCS[c]	dsyCS	(Staiger and Cande 1990; Golubovskaya, et al. 2003)
segregation II	segII	(Golubovskaya, et al. 2003)
Recombination		
Zmrad51A1/ 3.04	Zmrad51A1	(Franklin, et al. 1999b)
Zmrad51A2/ 7.04	Zmrad51A2	(Li, et al. 2007)
Monopolar Centromere Attachment		
mtm00-10	mtm00-10	NEW
Meiotic Cytoskeleton/Spindle		
divergent 1 (dv1)	dv1	(Clark 1940)
male sterile 43 (ms43)	ms43	(Golubovskaya 1989)
male sterile 28 (ms28)	ms28	
variable 1(va1)/ 7L	va1	(Beadle 1932)
Meiosis Exit		
polymitotic1 (po1)/ 6S	po1, po1-ms6	(Beadle 1929; 1932 ; Liu, Golubovskaya and Cande 1993)

[a]mtm: maize _targeted_ _mutagenesis_.
[b]phs1 was first named as _desynaptic 498_.
[c]desynapticCS was first named as _mutator_ _male_ _sterile_ 25 (mms25).

from a cDNA library specific for early meiotic prophase (Franklin, et al. 1999b) and recently double mutants for these genes were isolated and studied (Li, Harper, Golubovskaya, Wang, Weber, Meeley, McElver, Bowen, Cande and Schnable 2007). From studying these mutants and their wild type proteins, we have learned a lot about meiosis.

2.2   Genetic Control of Entry into Meiosis

Meiosis occurs in cells that have been developmentally targeted to become meiocytes. This process relies on at least one gene, *mac1* (Sheridan, et al. 1996; Sheridan, et al. 1999). While *mac1* has not been cloned in maize, genes defined by very similar mutants have been cloned from rice (Nonomura, Miyoshi, Eiguchi, Suzuki, Miyao, Hirochika and Kurata 2003) and *Arabidopsis* (Canales, Bhatt, Scott and Dickinson 2002; Zhao, Wang, Speal and Ma 2002). The genes encode leucine-rich repeat receptor protein kinases, a component of a signaling cascade probably involved in receiving positional information from surrounding cells to help establish meiotic cell fate.

After this cell fate determination step, the cell cycle is switched from mitotic to meiotic. The *ameiotic1 (am1)* gene is the key in controlling this switch in maize. In most of the *am1* mutant alleles, perfectly developed meiocytes failed to go through meiosis; instead they perform a mitotic division or are arrested in a premeiotic interphase (Golubovskaya, et al. 1993; Golubovskaya, et al. 1997a). Except for the *am1-pra1* allele, the chromatin structure, segregation and cytoskeleton in meiocytes all bear the hallmarks of a mitotic division. In all mutant alleles of *am1*, RAD51 foci are not installed on chromosomes in meiocytes, indicating that meiotic recombination machinery is not functioning. Preliminary expression profiling of several *am1* mutant alleles demonstrates that normal *am1* gene function dramatically alters gene expression immediately before meiosis begins. We have cloned the *am1* gene and it codes for a novel protein with partial similarity to *Arabidopsis* SWITCH1 (in preparation). Little is known about how cell cycle regulation is altered as cells enter meiosis but commitment to meiosis probably occurs at or before meiotic S phase, as unique events occur in meiotic S-phase, such as installation of meiosis-specific histones, condensins and cohesins (Strich 2004). This is likely the stage when AM1 first acts.

2.3   AFD1 and Leptotene Chromosome Structure and Function

Afd1 (absence of first division1) controls meiotic chromosome formation and sister chromatid cohesion (SCC). *Afd1* mutants (*afd1*) bypass the early stages of meiotic chromosome formation, blocking the installation of RAD51 foci, and are epistatic to all other meiotic mutants tested except *am1* (Golubovskaya, et al. 1993).

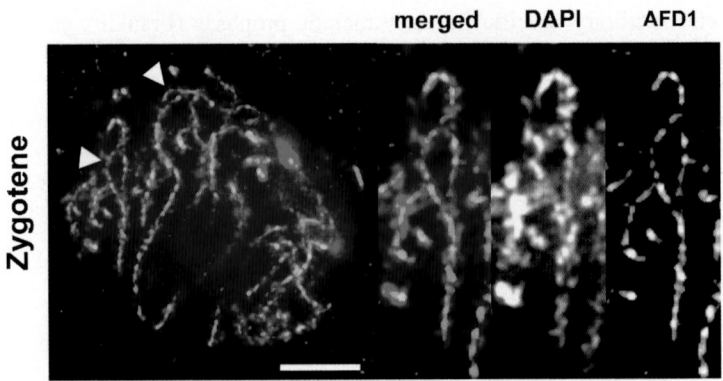

Fig. 2 AFD1 localizes to the axial and lateral elements. Localization of AFD1 in a wild-type nucleus. Chromosomes were stained with DAPI (red) and anti-AFD1 antibody (green) at zygotene, showing an AFD1 signal on both synapsed and unsynapsed chromosome regions. Arrowheads indicate unsynapsed regions. The three panels on the right are magnifications of one unsynapsed region. Individual channels are shown in black and white for clarity. Bar= 5 μm

The *afd1* gene was first identified as a mutant that fails to maintain the centromere cohesion required for the reductional division in meiosis (Golubovskaya, et al. 1975). We cloned *afd1* and showed that it codes for an alpha-kleisin protein, a component of cohesin complex. The cohesin complex is generally composed of four core members: SMC1, SMC3, an alpha-kleisin such as RAD21 or the meiotic specific REC8, and Scc3/STAG3 (Watanabe 2004). While there is one mitotic cohesin complex, several different complexes have been found in meiosis (Revenkova and Jessberger 2005).

Many meiosis-specific variants of cohesion genes are required for the formation of the leptotene chromosome. Maize AFD1/REC8, an ortholog of *Arabidopsis* SYN1 (Bai, Peirson, Dong, Xue and Makaroff 1999), is completely required to build a leptotene chromosome and, in its absence, many subsequent meiotic processes, including bouquet formation, meiotic recombination, homologous chromosome pairing, and synapsis, are deficient or absent. As shown in Fig. 2, AFD1 is associated with the lateral element of the SC in zygotene and pachytene. We established a series of *afd1* alleles with increasing protein level that correlates with an increasing level of axial element (AE) formation and leptotene chromosome architecture (see Fig. 3) as well as preservation of SCC during prophase (Golubovskaya, et al. 2006).

2.4   Role of AFD1 in Coordinating Prophase Events

AFD1 is required for establishing leptotene chromosome structure, bouquet formation, pairing, and synapsis (Watanabe and Nurse 1999; Lee and Orr-Weaver 2001; Pasierbek, Jantsch, Melcher, Schleiffer, Schweizer and Loidl 2001; Cai, Dong,

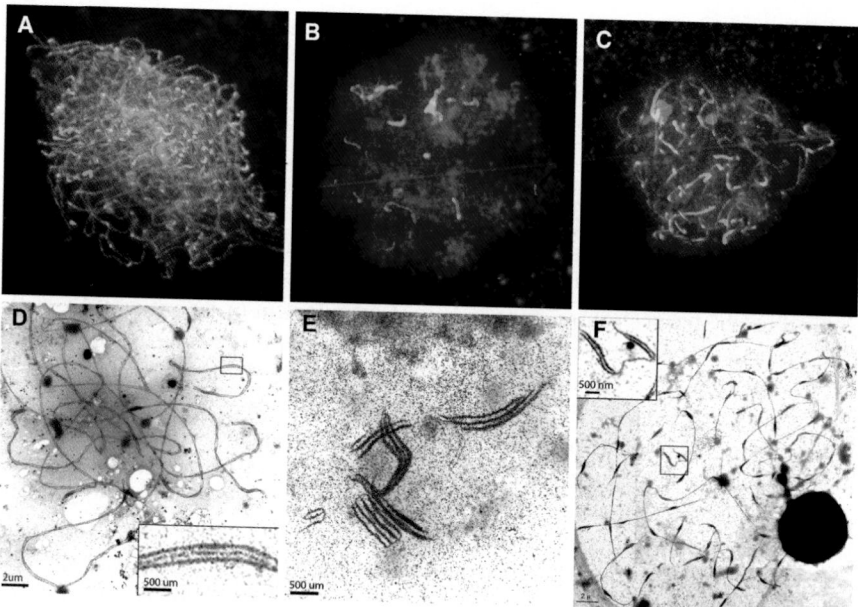

Fig. 3 Maize meiotic chromosome structure in wild type and two *afd1* allele derivatives. Maize meiotic chromosome structure in wild type and two *afd1* allele derivatives shown by immunostaining with antibody against ASY1/HOP1 and TEM of whole-mount spreads of prophase nuclei. Note the similarity of the immunostaining in A-C with the silver staining in D-F. **(A-C)** Immunostaining with an antibody to ASY1/HOP1, a protein associated with axial-lateral elements of the synaptonemal complex (SC). Chromosome stained with DAPI (red), and anti-ASY1/HOP1 (green). **(A)** WT zygotene-pachytene nucleus; well organized chromosome threads and long stretches of ASY1/HOP1 on axial elements are seen. **(B)** *afd1-1* (severe allele) prophase I nucleus; fuzzy chromatin lacking organized chromosome threads. A few short fragments of anti-ASY1/HOP1-stained filaments are present. **(C)** *afd1-4* (mild allele) prophase I nucleus; near- normal early prophase chromosomes associated with elongated filaments containing ASY1/HOP1. **(D-F)** TEM of whole-mount of prophase I nuclei with silver stained axial elements. **(D)** WT pachytene nucleus with completely synapsed homologous chromosomes resulting from the formation of the tripartite SC (a tripartite SC fragment is demonstrated in Inset D; the central element can be seen between the two lateral elements). **(E)** *afd1-1* prophase I nucleus showing a few short filamentous structures. **(F)** *afd1-4* prophase I nucleus showing partial formation of elongated axial elements, however no tripartite SC are formed despite the presence of short synapsed regions (Inset D shows no central element)

Edelmann and Makaroff 2003; Xu, Beasley, Warren, van der Horst and McKay 2005). However, using our series of weaker *afd1* alleles, we dissected these functions and showed more specifically that AFD1 is not required for AE initial recruitment but instead is controlling the extent of AE elongation and their maturation into lateral elements during synapsis. Surprisingly, partial AE elongation was sufficient for leptotene chromosome structure establishment and bouquet formation. In contrast, homologous pairing, synapsis and proper distribution of RAD51 depended on full AE elongation, providing the basis for a model in which AFD1 helps regulate homologous pairing, recombination and synapsis via its role in AE elongation and its impact on the subsequent distribution of the recombination machinery.

2.5   Recombination and RAD51

Analysis of the maize *rad51* double mutant demonstrates that RAD51 functions are required for the precise pairing and synapsis of homologs in maize meiosis (Li, et al. 2007). Moreover, we observed a unique phenotype in *rad51* double mutants in the male that has not been observed in *rad51* mutants of other species: pairing, synapsis and chiasmata between non-homologous chromosomes. This suggests that RAD51 is important for establishing homology. This is similar to what occurs between homeologous chromosomes in the *ph* mutant of wheat (Riley and Chapman 1964; Riley, Chapman, Young and Belfield 1966). However, the low percentage of non-homologous pairing, nearly normal synapsis and close to normal rates of meiotic crossovers from surviving female gametes suggest that other maize RecA homologs such as DMC1 do an imperfect but adequate job to mediate strand invasion in *rad51* double mutants.

2.6   The Link Between Pairing and Recombination

Recombination is initiated by double strand break (DSB) formation generated by the SPO11 protein. The DSBs are resected from 5' to 3' by the MRX complex and plants have a functional homolog of this complex. The single stranded DNA ends subsequently invade the homologous double stranded DNA. This step is catalyzed by RAD51 and DMC1, which are homologs of RecA recombinase. RAD51 functions both in the mitotic and meiotic cell cycle, whereas DMC1 is meiosis-specific in function (Paques and Haber 1999). There is a good correlative evidence to suggest that these early stages of recombination are intimately involved in the homology search (Franklin, et al. 1999b; Pawlowski, et al. 2003); in fact, recombination may be the homology search in plants (Hamant, et al. 2006).

We have extensively studied the relationship between pairing and recombination in 20 meiotic mutants using our battery of tools. The presence of RAD51 foci on chromosomes correlates with the severity of univalents-in 18 of 20 mutants tested (Pawlowski, et al. 2003). The two exceptions to this rule are involved in bouquet formation: *pam1* (Golubovskaya, et al. 2002) and *dsy1* (Bass, Bordoli and Foss 2003), where there are normal numbers of RAD51 foci. In wild-type maize meiocytes, RAD51 forms distinct foci on chromosomes in zygotene, where they reach peak numbers of 500-600 foci, and the numbers of foci are reduced to around 20 in pachytene before they completely disappear (Franklin, et al. 1999b; Pawlowski, et al. 2003). RAD51 foci go through changes in shape and association with the chromosomal axis as chromosomes pair. Since there are only 20 crossovers required (one per arm) to hold bivalents together until anaphase 1, we explored the function of the many additional RAD51 foci. Among the mutants defective in pairing, there are dramatic differences in the number, and shape of Rad51 foci. Mutants with very few RAD51 foci have mostly 20 univalents; mutants with high number of RAD51 foci have significant numbers of bivalents. The majority of the meiotic mutants exhibited numbers of RAD51 foci far greater than the number of foci needed for the formation of the required number of crossovers (Pawlowski, et al. 2003). These data suggest that a full complement of RAD51 foci is essential for proper chromosome pairing.

2.7   PHS1 is Required for Homologous Pairing

From our RAD51 distribution survey, we identified a novel gene, PHS1, involved in loading the recombination machinery onto chromosomes (Pawlowski, et al. 2004). By monitoring homologous pairing using FISH with the 5S rDNA locus, we showed that all chromosomes are synapsed, but none of the 5S rDNA loci were paired in *phs1*. TEM of SCs showed complete, but non-homologus, synapsis. Taken together, pairing in the *phs1* mutant is completely non-homologous, yet synapsis is not impaired. *phs1* mutant meiocytes have about 3 RAD51 foci during zygotene, instead of the normal 500. We proposed that the PHS1 protein is involved in loading RAD51 complexes onto chromosomes and these complexes are completely required for homologous pairing, in agreement with the idea that recombination and pairing are actually the same process. The *phs1* gene was cloned by transposon tagging. It was found to encode a novel protein without significant similarity to any known protein and without any obvious functional features or domains, but putative homologs are present in *Arabidopsis* and other plants (Pawlowski, et al. 2004). Two other maize mutants, *dsyCS* and *segII*, have phenotypes similar to *phs1*.

From this work, we conclude that recombination and homologous pairing are so tightly linked in maize that they cannot be uncoupled, suggesting that the process of recombination is actually part of the homology search. However, synapsis in maize is completely independent of the homology search, recombination or even the formation of the bouquet.

2.8   pam1: The Bouquet Stage and Pairing

In maize, at the lep-zyg transition, there is an abrupt shift in telomere and centromere behavior. Before zygotene, centromeres are constrained to one nuclear hemisphere: afterwards, they have no obvious nuclear organization. Telomeres behave the opposite way: prior to zygotene there is no observable polarity, while afterwards the telomeres are both polarized and clustered on the nuclear envelope (Bass, et al. 1997; Carlton and Cande 2002). In the transition from leptotene to zygotene, this polarization transition can be seen as most telomeres end up in one hemisphere and most centromeres in the other hemisphere. Similar dramatic changes in polarity have been described in other organisms. For example, in fission yeast, centromeres are normally clustered near the spindle pole body in haploid cells, with telomeres loosely grouped at the other end of the nucleus. As cells and nuclei fuse and meiosis begins, the telomeres cluster at the spindle pole body and the centromeres are released (Chikashige, Ding, Imai, Yamamoto, Haraguchi and Hiraoka 1997). The polarization switch in maize is coincident with the transient elongation of both knob heterochromatin and centromeric heterochromatin, consistent with the notion that the large-scale changes in nuclear organization may be mediated by changes at a lower-order chromatin level (Carlton, et al. 2002).

The *pam1* mutant of maize has <u>p</u>lural <u>a</u>bnormalities of <u>m</u>eiosis and a defective bouquet (Golubovskaya, et al. 2002). In *pam1,* telomeres attach normally to the nuclear envelope, and they undergo some initial stages of clustering by making a number of small clumps of telomeres. However, a tight bouquet is not formed. This behavior of chromosomes in *pam1* is similar to rye chromosomes in meiocytes treated with colchicine (Cowan and Cande 2002). Mutant *pam1* meiotic nuclei have aberrant synapsis (monitored by TEM) including nonhomologous synapsis, partner switches and foldbacks, and loss of interlock resolution. There is also a dramatic reduction in homologous pairing. In addition, there is an extreme asynchrony of meiosis within one anther locule. The *pam1* mutant is completely male sterile and almost completely female sterile; however, a few female meiocytes (<1%) do complete normal meiosis. Interestingly, RAD51 foci on zygotene chromosomes looks completely normal, suggesting that the early stages of recombination do not require the bouquet, and that these two processes can be separated. *pam1* is the only mutant that we are aware of in any organism whose telomeres attach normally to the nuclear envelope, but do not cluster. From the similarity of phenotypes of the maize *pam1* and yeast *ndj* mutants, we propose that the bouquet renders meiotic prophase much more efficient and faster. The bouquet is not absolutely required for homologous pairing (as the *pam1* and *ndj* mutants do achieve some homologous pairing), or for synapsis (again, both mutants can synapse), or for completion of meiosis (in both mutants there are always a few survivors). However, the rate and efficiency of all the processes are severely reduced in these bouquet mutants. We can also conclude that the bouquet is required for timely initiation of synapsis, as synapsis is severely delayed in both the *pam1* mutant and the yeast *ndj* mutant.

2.9   Control of Sister Chromatid Cohesion during Meiosis I

In *mtm99-31*, centromeric cohesion is lost precociously at anaphase I. Using a candidate gene approach, we showed that a *Mu* transposon is inserted in the first exon of the *Zmsgo1* gene in this mutant. The *Shugoshin* (SGO) genes have been shown recently to be involved in the protection of REC8 against degradation at the centromeres in fission yeast (Kiburz, Reynolds, Megee, Marston, Lee, Lee, Levine,

Young and Amon 2005; Vaur, Cubizolles, Plane, Genier, Rabitsch, Gregan, Nasmyth, Vanoosthuyse, Hardwick and Javerzat 2005). Using a ZmSGO1-specific antibody, we showed that ZmSGO1 localizes to the centromeres of meiotic cells from leptotene until Metaphase I. (Hamant, et al. 2005).

2.10 A Model of Meiosis

Based on our work and knowledge from other organisms, we have developed a model of early meiotic prophase (Fig. 4). In this diagram, we indicate the molecular events relative to our best estimate of the stage in which these occur. We have placed maize genes at the stage where we think they are likely to first function, based on their published mutant phenotypes (see Table 1 for references). The majority of genes, except *mac1*, *am1* and *afd1*, are defined by failure of proper homologous pairing and synapsis. Hence, they are designated as "desynaptic" (dsy).

During meiotic prophase, several important events occur; the homology search, pairing, synapsis and recombination. We propose, based on work in yeast (Allers and Lichten 2001; Hunter and Kleckner 2001; Bishop and Zickler 2004) and mouse (Moens, Kolas, Tarsounas, Marcon, Cohen and Spyropoulos 2002), that, as pairing and synapsis proceed, most recombinations are resolved as noncrossovers. We think these recombinations are only used for the purpose of the homology search. Subsequently, interference mechanisms insure that only one recombination per arm is resolved as a crossover. These mature into chiasmata, which are required to hold bivalents together until anaphase I.

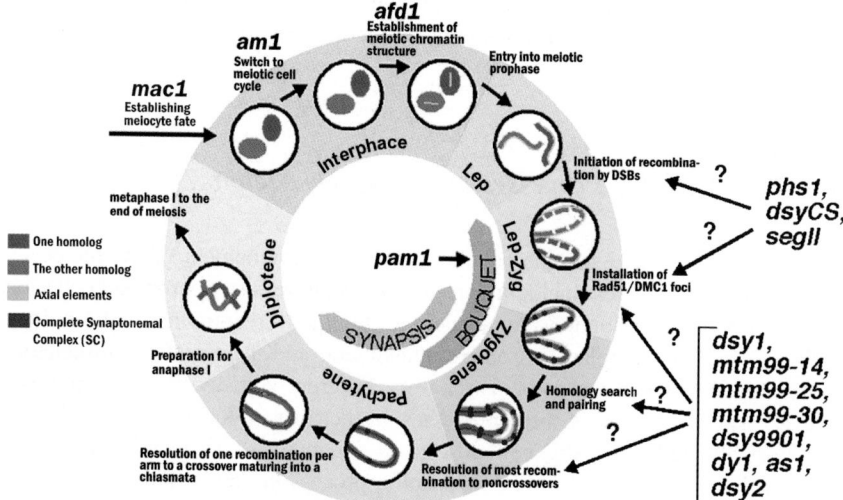

Fig. 4 Stages of meiosis and positioning of mutants in a pathway based on their phenotypes

Phenotypic criteria used to place the *phs1*, *dsyCS*, and *segII* mutants upstream in the pathway relative to other desynaptic mutants include: the lack of installation of the recombination machinery as marked by RAD51 foci, the lack of homologous pairing as monitored by 3D FISH using probes against 5S rDNA, and the extent of non-homologous synapsis as monitored by TEM of silver stained spread chromosomes (Pawlowski, et al. 2004). Analysis of the phenotype of *phs1* suggests that pairing interactions and recombination are in the same pathway and require RAD51 but additionally proves that synapsis is in an independent pathway. Mutants such as *segII* and *dsyCS*, which also have few RAD51 foci and a phenotype similar to *phs1*, may act in concert with *phs1*, whereas other mutants that have more RAD51 foci and variable pairing or recombination may act later. Changes in telomere distribution during bouquet formation provide additional criteria, and were used to identify *pam1*'s involvement in bouquet formation.

3 Cellular and Molecular Processes Required in Meiosis

Meiosis is an ancient, evolutionarily conserved process that requires the intricate coordination and precise timing of a series of unique cellular process. Recombination and independent assortment of chromosomes during this special division generates tremendous diversity in sexual species. While meiotic divisions share aspects with mitosis, several uniquely meiotic events occur such as formation of the telomere bouquet in early prophase, synapsis of homologous chromosomes, and the omission of an S phase between the two meiotic divisions. Recombination is an essential process in meiosis, but it is also required in DNA damage repair; therefore, recombination is not unique to meiosis. In maize, the analysis of meiosis has been very successful at the cytological level for more than 70 years. More recently, the development of molecular techniques has allowed the cloning of several meiotic genes in this organism.

3.1 What Cellular Processes are Uniquely Meiotic?

We can divide meiosis into nine cellular processes: development of the meiocyte, switch to meiotic cell cycle including pre-meiotic S phase, sister chromatid cohesion (SCC), chromosome condensation, meiotic bouquet, recombination/homology search, synapsis, monopolar centromere attachment, and the meiotic cytoskeleton/spindle. Together these make meiosis a very different cell division than mitosis. However, aspects of most of these processes are performed during the somatic cell cycle, possibly with only slight modifications for meiosis. As mentioned earlier, recombination is clearly not a meiosis-specific process in any organism. Although some recombination proteins may be meiosis specific and the rate of recombination is greatly increased in meiosis, recombination proteins are required for DNA damage repair pathways. Their role in DNA damage repair probably pre-dates their role in meiosis (Ramesh, Malik and Logsdon 2005). Recombination is also the most well-studied of these processes.

The two processes that may be the most meiosis specific are synapsis and the homology search, and it is possible that they have the fewest genes in common with somatic processes. The structure of the SC by TEM from many diverse organisms is highly conserved. Some of the proteins required for synapsis have been identified in *S. cerevisiae*, mouse and rat. However, these genes are notorious for their lack of primary sequence conservation, even between mouse and man (Bogdanov, Dadashev and Grishaeva 2003). Proteins involved specifically in the homology search are rare – we know of only two or three so far: PHS1 in maize (Pawlowski, et al. 2004) and Hop2 and Mnd1 in yeast and mouse (Petukhova, Pezza, Vanevski, Ploquin, Masson and Camerini-Otero 2005). This fascinating process is largely undefined except that it involves recombination, and certainly more work is needed to identify more genes involved in the homology search.

4   The Future

4.1   Maize as a Model Organism for Studying Meiosis

Meiosis has been studied in a number of excellent model organisms, and each has uniquely contributed to our overall understanding of this highly conserved process. Meiosis, particularly meiotic recombination, has been studied extensively using molecular genetics in yeast. However, interpreting meiotic mutant phenotypes relies heavily on cytological analysis. The small size of yeast chromosomes and their uncondensed state during most of meiotic prophase mean that yeast is not a suitable organism to answer many important questions about meiotic chromosome structure and behavior. *Drosophila* and *C. elegans* are also excellent organisms for molecular genetic studies, but their meiosis has some unconventional features including SPO11-independent meiotic pairing and synapsis (Dernburg, McDonald, Moulder, Barstead, Dresser and Villeneuve 1998; McKim, Green-Marroquin, Sekelsky, Chin, Steinberg, Khodosh and Hawley 1998). In the mouse, there are excellent reverse genetics tools, but forward genetics to screen for meiotic mutants has been more difficult, and leptotene and zygotene chromosome architecture is diffuse. *Arabidopsis* has excellent reverse genetics tools but there are difficulties in studying chromosome behavior in their small pollen mother cell and forward genetics screens have been less productive than in maize. Maize has strengths that complement these other model systems. Maize is one of the few organisms with a large genome where chromatin structure, homologous pairing and synapsis are amenable to analysis by a combination of cytological, genetic, molecular and biochemical techniques. Maize has been used to analyze the kinetics of chromosome pairing, bouquet formation, axial element formation, and recombination among other meiotic features (Dawe, et al. 1994; Bass, et al. 1997; Franklin and Cande 1998, 1999a; Bass, Riera-Lizarazu, Ananiev, Bordoli, Rines, Phillips, Seda, Agard and Cande 2000; Carlton, Cowan and Cande 2003; Pawlowski, et al. 2003; Pawlowski, et al. 2004) and to identify the lep-zyg transition as a cytologically distinct stage in meiosis (Dawe, et al. 1994).

4.2   Why Study Maize?

It is relatively easy to get lots of meiocytes in a single meiotic stage: It is possible in maize to harvest large amounts of anthers which contain male meiocytes in near-synchronous sub-stages of meiosis. Microsporogenesis (development of the meio-cyte) in maize occurs within the anthers of male florets (flowers) on the large terminal inflorescence (tassel). Each floret contains three anthers, each containing hundreds of meiocytes undergoing meiosis in synchrony. Because the three anthers of a floret are also synchronous with each other, we can stage the meiocytes of one anther (or 1/2 anther) cytologically, and harvest the other two. The anthers and meiocytes are large (1- 3 mm at meiosis and 70-100 microns, respectively), and there are typically thousands of anthers per tassel. We estimate that meiocytes comprise 50% of the mass of each anther. Florets occur in pairs in spikelets, which themselves are paired. When we harvest anthers, we always take primary florets from comparable spikelets. We find these to be the most exactly synchronous, and transcriptome differences have been found between similar-sized upper and lower spikelets (Skibbe & Schnable, personal communication). There is a shallow developmental gradient of meiosis along tassel branches, and there can be 30-42 anthers per tassel branch in exactly the same meiotic stage. From one tassel, hundreds of anthers can be gathered in the same stage within 30-60 minutes. While it is possible to harvest one or two meiotic stages per tassel, usually younger or older tassels are needed to harvest contiguous meiotic stages.

Large amounts of meiocytes make genomics-scale experiments possible: This is a tremendous advantage over other meiosis model systems, especially for genomic and proteomic studies, because biochemical quantities of material in exactly the same sub stage of meiosis can be harvested. In other plant and animal meiosis model systems, the developmental gradient is much steeper, allowing the collection of only small amounts of meiocytes in the same stage per organism (as in mammals (Rossi, Dolci, Sette, Capolunghi, Pellegrini, Loiarro, Di Agostino, Paronetto, Grimaldi, Merico, Martegani and Geremia 2004; Schlecht, Demougin, Koch, Hermida, Wiederkehr, Descombes, Pineau, Jegou and Primig 2004) and *C. elegans* (Reinke, Gil, Ward and Kazmer 2004)), or allowing only mixtures of stages or whole flowers to be collected (for example: Endo, Matsubara, Kokubun, Masuko, Takahata, Tsuchiya, Fukuda, Demura and Watanabe 2002; Endo, Tsuchiya, Saito, Matsubara, Hakozaki, Masuko, Kamada, Higashitani, Takahashi, Fukuda, Demura and Watanabe 2004). In *S. cerevisiae* and *S. pombe*, meiotic stages are taken as time points from a synchronization step and include mixtures of meiotic stages (Primig,

Williams, Winzeler, Tevzadze, Conway, Hwang, Davis and Esposito 2000; Mata, Lyne, Burns and Bahler 2002). *Arabidopsis* is a very good model system for meiosis, but collection of large quantities of anthers is next to impossible. Anther-specific genes have been found in *Brassica oleracea* mRNA, and then homologs have been identified and studied in *Arabidopsis* (Amagai, Ariizumi, Endo, Hatakeyama, Kuwata, Shibata, Toriyama and Watanabe 2003), and a similar strategy is being attempted to study the *Arabidopsis* meiotic proteome (Sanchez-Moran, Mercier, Higgins, Armstrong, Jones and Franklin 2005). While we have learned a great deal from studies in these species and others, in maize we have the opportunity to divide the meiotic stages to a much finer degree, and determine if gene expression is a significant control mechanism from one stage of prophase to the next. Finally, the excellent cytology of maize will allow precise correlation between meiotic stage and transcription profiles. Because maize has large meiotic chromosomes, we have worked out precise criteria for staging meiotic cells based on changes in chromosome morphology, nuclear architecture, chromosome width, and cell appearance; we can even distinguish sub stages of zygotene and pachytene. This will allow us to correlate cytological stages with the transcriptional profile.

What is learned in maize is relevant to other organisms: Meiosis is an extremely evolutionarily conserved process (John 1990). What we learn about meiosis in maize will be relevant to other meiosis model systems, including humans. If fact, we know that the relationships between recombination, synapsis and pairing in maize is very similar to that in yeast and mammals. From an evolutionary prospective, we are learning that there is great value in comparing genes from many species when sequence and functional information are available.

4.3   Advances in Studying Chromosome Architecture

An underlying assumption of work on chromosome function is that the higher level architecture of the meiotic chromosome is a reflection of the underlying DNA sequence and its associated epigenetic marks. Although there has been considerable work studying the function of proteins associated with the axial element, and their organization at the light microscope level using indirect immunofluorescence (Pigozzi and Solari 2003; Higgins, Sanchez-Moran, Armstrong, Jones and Franklin 2005; Ollinger, Alsheimer and Benavente 2005; Tsubouchi and Roeder 2005; Nonomura, Nakano, Eiguchi, Suzuki and Kurata 2006; Osman, Sanchez-Moran, Higgins, Jones and Franklin 2006), an integrative view of meiotic prophase chromosome structure and function has been difficult to come by. EM analysis has revealed that chromatin is organized in loops that project out from the axis (defined by the axial elements, reviewed in Zickler, et al. 1999). Analysis of the looped domain organization of chromatin has been the providence of electron microscopy due to an inability to visualize loops with the resolution available with traditional light microscopy (Zickler, et al. 1999). In the most favorable light microscope preparations, the DNA is organized into chromomeres, with euchromatin (presumably loops with lower orders of folding) as lightly staining regions and heterochromatin as densely staining regions (loops with more compaction). Analysis of spread meiotic chromosomes (reviewed in Zickler, et al. 1999) suggests that the loops are similar in size and spaced evenly along the axial element, thus giving no clue as to how the loop arrangement might be transformed into the complex chromomere patterns observed at the light microscope level. How chromomeres are organized with respect to axial element organization is not known. These are important questions to answer since, as shown by analyzing *afd1* alleles, appropriate chromosome architecture is absolutely required for homologous pairing, recombination and synapsis. In the future, this problem can be tackled using ultrahigh resolution light microscopy, the superb morphology of maize meiotic chromosomes, and the genetic and genomic maize resources to analyze the organization of the meiotic chromosome at multiple levels of resolution, from base pairs to chromomeres. These questions cannot be adequately addressed by viewing the small chromosomes, for example, of yeast. Maize may be the best model organism for these discovery-driven questions pertaining to the functions of chromosome architecture during pairing.

Structured illumination (SI) light microscopy has a resolution less than 100 nm (compared to 250 nm in the XY and 500 nm in Z axis in conventional microscopy) (Gustafsson 2005; Gustafsson, Shao, Carlton, Wang, Golubovskaya, Cande, Agard and Sedat 2008) and can be used to look at changes in chromomere and axial element organization on leptotene, zygotene, and pachytene chromosomes (see Fig. 5). The great advantage of structural illumination over EM is the ability to use FISH and

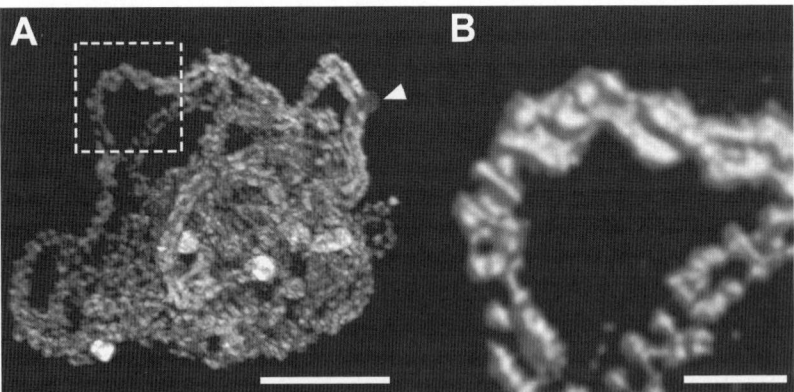

Fig. 5 Maize pachytene chromosomes revealed by structured illumination microscopy (A) A maize pachytene nucleus stained with DAPI and taken with SI light microscopy using the OMX system developed by John Sedat (UCSF), showing the chromosome and centromere (arrowhead) architecture in ultra-high resolution. Bar= 5 μm. (B) A magnified region selected from (A) was processed computationally to highlight surfaces in the 3D rendered image. Note the bulbous projections from the chromosome axis and their resemblance to "loop domains" in EM (as in Zickler, et al. 1999). Bar= 1 μm.

immunolocalizations to easily detect DNA, RNA and proteins on chromosomes. Antibodies against axial components, such as AFD1/REC8 and ASY1/HOP1, can be used to define the prophase chromosome axis.

For the first time, it will be possible to analyze meiotic chromosome architecture at a resolution relevant for mapping higher order chromatin structure and recombination machinery onto axial elements of the prophase chromosome. This research will provide a base line for understanding how chromosome architecture changes during pairing, synapsis and recombination and it will be possible to link these changes to function by analyzing mutants. We may be able to answer questions concerning the complexity of the axial element and how it changes during synapsis. We will know whether underlying axis organization is different in heterochromatic vs. euchromatic regions of the chromosome. Finally, by surveying RAD51 morphology relative to chromosome architecture, we will gain insight into how chromosome architecture creates a favorable environment for homologous pairing and recombination.

References

Allers, T. and Lichten, M. (2001) Differential timing and control of noncrossover and crossover recombination during meiosis. Cell 106, 47–57.

Amagai, M., Ariizumi, T., Endo, M., Hatakeyama, K., Kuwata, C., Shibata, D., Toriyama, K. and Watanabe, M. (2003) Identification of anther-specific genes in a cruciferous model plant, arabidopsis thaliana, by using a combination of arabidopsis macroarray and mrna derived from brassica oleracea. Sexual Plant Reproduction 15, 213–220.

Bai, X., Peirson, B. N., Dong, F., Xue, C. and Makaroff, C. A. (1999) Isolation and characterization of syn1, a rad21-like gene essential for meiosis in arabidopsis. Plant Cell 11, 417–430.

Barrell, P. J. and Grossniklaus, U. (2005) Confocal microscopy of whole ovules for analysis of reproductive development: The elongate1 mutant affects meiosis II. Plant J. 43, 309–320.

Bass, H. W., Marshall, W. F., Sedat, J. W., Agard, D. A. and Cande, W. Z. (1997) Telomeres cluster de novo before the initiation of synapsis: A three-dimensional spatial analysis of telomere positions before and during meiotic prophase. J. Cell Biol. 137, 5–18.

Bass, H. W., Riera-Lizarazu, O., Ananiev, E. V., Bordoli, S. J., Rines, H. W., Phillips, R. L., Seda, J. W., Agard, D. A. and Cande, W. Z. (2000) Evidence for the coincident initiation of homolog pairing and synapsis during the telomere-clustering (bouquet) stage of meiotic prophase. J. Cell Sci. 113, 1033–1042.

Bass, H. W. (2003) Telomere dynamics unique to meiotic prophase: Formation and significance of the bouquet. Cell Mol. Life Sci. 60, 2319–2324.

Bass, H. W., Bordoli, S. J. and Foss, E. M. (2003) The desynaptic (dy) and desynaptic1 (dsy1) mutations in maize (zea mays l) cause distinct telomere-misplacement phenotypes during meiotic prophase. J. Exp. Bot. 54, 39–46.

Beadle, G. W. (1929) A gene for supernumerary mitoses during spore development in zea mays. Science 70, 406–407.

Beadle, G. W. (1930) Genetic and cytological studies of a mendelian asynaptic in zea mays. Cornell Agric. Exp. Sta. Mem. 129, 1–23.

Beadle, G. W. (1932) Genes in maize for pollen sterility. Genetics 17, 413–431.

Beadle, G. W. (1937) Chromosome aberration and gene. Mutation in sticky chromosome plants of zea mays. Cytologia Fujii Jubilee, 43–56.

Bishop, D. K. and Zickler, D. (2004) Early decision; meiotic crossover interference prior to stable strand exchange and synapsis. Cell 117, 9–15.

Bogdanov, Y. F., Dadashev, S. Y. and Grishaeva, T. M. (2003) In silico search for functionally similar proteins involved in meiosis and recombination in evolutionarily distant organisms. In Silico Biol. 3, 173–185.

Cai, X., Dong, F., Edelmann, R. E. and Makaroff, C. A. (2003) The arabidopsis syn1 cohesin protein is required for sister chromatid arm cohesion and homologous chromosome pairing. J. Cell Sci. 116, 2999–3007.

Canales, C., Bhatt, A. M., Scott, R. and Dickinson, H. (2002) Exs, a putative lrr receptor kinase, regulates male germline cell number and tapetal identity and promotes seed development in arabidopsis. Curr. Biol. 12, 1718–1727.

Carlton, P., Ananiev, E. and Cande, W. Z. (1998) Centromere localization in maize meiocytes. Mol. Biol. of the Cell 9, 404A.

Carlton, P. M. and Cande, W. Z. (2002) Telomeres act autonomously in maize to organize the meiotic bouquet from a semipolarized chromosome orientation. J. Cell Biol. 157, 231–242.

Carlton, P. M., Cowan, C. R. and Cande, W. Z. (2003) Directed motion of telomeres in the formation of the meiotic bouquet revealed by time course and simulation analysis. Mol. Biol. Cell 14, 2832–2843.

Chan, A. and Cande, W. Z. (1998) Maize meiotic spindles assemble around chromatin and do not require paired chromosomes. J. Cell Sci. 111, 3507–3515.

Chikashige, Y., Ding, D.-Q., Imai, Y., Yamamoto, M., Haraguchi, T. and Hiraoka, Y. (1997) Meiotic nuclear reorganization: Switching the position of centromeres and telomeres in the fission yeast schizosaccharomyces pombe. EMBO J. 16, 193–200.

Clark, F. J. (1940) Cytogenetic studies of divergent meiotic spindle formation in zea mays Amer. J. Bot. 27, 547–559.

Cowan, C. R. and Cande, W. Z. (2002) Meiotic telomere clustering is inhibited by colchicine but does not require cytoplasmic microtubules. J. Cell Sci. 115, 3747–3756.

Dawe, R. K., Sedat, J. W., Agard, D. A. and Cande, W. Z. (1994) Meiotic chromosome pairing in maize is associated with a novel chromatin organization. Cell 76, 901–912.

Dempsey, E. (1994) Traditional analysis of maize pachytene chromosomes. In: M. Freeling and V. Walbot (Eds.), The maize handbook, Springer, New York, pp. 432–441.

Dernburg, A. F., Sedat, J. W., Cande, W. Z. and Bass, H. W. (1995) Cytology of telomeres. In: E. H. Blackburn and C. W. Greider (Eds.), Telomeres, Cold Spring Harbor Laboratory, Plainview, pp. 295–338.

Dernburg, A. F., McDonald, K., Moulder, G., Barstead, R., Dresser, M. and Villeneuve, A. M. (1998) Meiotic recombination in c. Elegans initiates by a conserved mechanism and is dispensable for homologous chromosome synapsis. Cell 94, 387–398.

Endo, M., Matsubara, H., Kokubun, T., Masuko, H., Takahata, Y., Tsuchiya, T., Fukuda, H., Demura, T. and Watanabe, M. (2002) The advantages of cdna microarray as an effective tool for identification of reproductive organ-specific genes in a model legume, lotus japonicus. FEBS Lett 514, 229–237.

Endo, M., Tsuchiya, T., Saito, H., Matsubara, H., Hakozaki, H., Masuko, H., Kamada, M., Higashitani, A., Takahashi, H., Fukuda, H., Demura, T. and Watanabe, M. (2004) Identification and molecular characterization of novel anther-specific genes in oryza sativa l. By using cdna microarray. Genes Genet. Syst. 79, 213–226.

Franklin, A. E. and Cande, W. Z. (1998) Rad51 distribution is altered in desynaptic2, a maize meiotic mutant that has abnormal chromosome pairing. Mol. Biol. of the Cell 9, 404A.

Franklin, A. E. and Cande, W. Z. (1999a) Nuclear organization and chromosome segregation. Plant Cell 11, 523–534.

Franklin, A. E., McElver, J., Sunjevaric, I., Rothstein, R., Bowen, B. and Cande, W. Z. (1999b) Three-dimensional microscopy of the rad51 recombination protein during meiotic prophase. Plant Cell 11, 809–824.

Franklin, A. E., Golubovskaya, I. N., Bass, H. W. and Cande, W. Z. (2003) Improper chromosome synapsis is associated with elongated rad51 structures in the maize desynaptic2 mutant. Chromosoma 112, 17–25.

Golubovskaya, I., Grebennikova, Z. K., Avalkina, N. A. and Sheridan, W. F. (1993) The role of the ameiotic1 gene in the initiation of meiosis and in subsequent meiotic events in maize. Genetics 135, 1151–1166.

Golubovskaya, I., Avalkina, N. and Sheridan, W. F. (1997a) New insights into the role of the maize ameiotic1 locus. Genetics 147, 1339–1350.

Golubovskaya, I., Sheridan, W., Harper, L. and Cande, W. (2003) Novel meiotic mutants of maize identified from *mu*transposon and ems mutant screens. Maize Genet. Coop. Newsl. 77, 10–13.

Golubovskaya, I. N. and Mashnenkov, A. S. (1975) Genetic control of meiosis. I. Meiotic mutation in corn (*zea mays*) *afd*, causing the elimination of the first meiotic division. Genetika (Russ) 11, 810–816.

Golubovskaya, I. N. and Mashnenkov, A. S. (1976) Genetic control of meiosis: II a desynaptic mutant in maize induced by n-nitroso-n-methylurea. Genetika (Russ) 12, 7–14.

Golubovskaya, I. N. (1979) Genetic control of meiosis. Int Rev Cytol 58, 247–290.

Golubovskaya, I. N. (1989) Meiosis in maize: <u>Mei</u> genes and conception of genetic control of meiosis. Advanced Genetics 26, 149–192.

Golubovskaya, I. N., Grebennikova, Z. K., Auger, D. L. and Sheridan, W. F. (1997b) The maize desynaptic1 mutation disrupts meiotic chromosome synapsis. Developmental Genetics 21, 146–159.

Golubovskaya, I. N., Harper, L. C., Pawlowski, W. P., Schichnes, D. and Cande, W. Z. (2002) The pam1 gene is required for meiotic bouquet formation and efficient homologous synapsis in maize (zea mays l.). Genetics 162, 1979–1993.

Golubovskaya, I. N., Hamant, O., Timofejeva, L., Wang, C. J., Braun, D., Meeley, R. and Cande, W. Z. (2006) Alleles of afd1 dissect rec8 functions during meiotic prophase I. J. Cell Sci. 119, 3306–3315.

Golubovskaya, I. N. and Mashnenkov, A. S. (1977) Multiple disturbances of meiosis in corn are caused by a single recessive mutation pama-a344. Genetika (Russ) 13, 1910–1921.

Gustafsson, M. G. (2005) Nonlinear structured-illumination microscopy: Wide-field fluorescence imaging with theoretically unlimited resolution. Proc. Natl. Acad. Sci. USA 102, 13081–13086.

Gustafsson, M. G. L., Shao, L., Carlton, P. M., Wang, C.-J. R., Golubovskaya, I. N., Cande, W. Z., Agard, D. A. and Sedat, J. W. (2008) Three-dimensional resolution doubling in widefield fluorescence microscopy by structured illumination. Biophysical J. 94:4957–4970.

Hamant, O., Golubovskaya, I., Meeley, R., Fiume, E., Timofejeva, L., Schleiffer, A., Nasmyth, K. and Cande, W. Z. (2005) A rec8-dependent plant shugoshin is required for maintenance of centromeric cohesion during meiosis and has no mitotic functions. Curr. Biol. 15, 948–954.

Hamant, O., Ma, H. and Cande, W. Z. (2006) Genetics of meiotic prophase I in plants. Annu Rev Plant Biol.

Harper, L., Golubovskaya, I. and Cande, W. Z. (2004) A bouquet of chromosomes. J. Cell Sci. 117, 4025–4032.

Higgins, J. D., Sanchez-Moran, E., Armstrong, S. J., Jones, G. H. and Franklin, F. C. (2005) The arabidopsis synaptonemal complex protein zyp1 is required for chromosome synapsis and normal fidelity of crossing over. Genes Dev. 19, 2488–2500.

Hunter, N. and Kleckner, N. (2001) The single-end invasion: An asymmetric intermediate at the double-strand break to double-holliday junction transition of meiotic recombination. Cell 106, 59–70.

Kiburz, B. M., Reynolds, D. B., Megee, P. C., Marston, A. L., Lee, B. H., Lee, T. I., Levine, S. S., Young, R. A. and Amon, A. (2005) The core centromere and sgo1 establish a 50-kb cohesin-protected domain around centromeres during meiosis I. Genes Dev. 19, 3017–3030.

Lee, J. Y. and Orr-Weaver, T. L. (2001) The molecular basis of sister-chromatid cohesion. Annu. Rev. Cell Dev. Biol. 17, 753–777.

Li, J., Harper, L. C., Golubovskaya, I., Wang, C. R., Weber, D., Meeley, R. B., McElver, J., Bowen, B., Cande, W. Z. and Schnable, P. S. (2007) Functional analysis of maize rad51 in meiosis and double-strand break repair. Genetics 176, 1469–1482.

Li, W., Chen, C., Markmann-Mulisch, U., Timofejeva, L., Schmelzer, E., Ma, H. and Reiss, B. (2004) The arabidopsis atrad51 gene is dispensable for vegetative development but required for meiosis. Proc. Natl. Acad. Sci. USA 101, 10596–10601.

Liu, Q., Golubovskaya, I. and Cande, W. Z. (1993) Abnormal cytoskeletal and chromosome distribution in po, ms4 and ms6, mutant alleles of polymitotic that disrupt the cell cycle progression from meiosis to mitosis in maize. J. of Cell Sci. 106, 1169–1178.

Mata, J., Lyne, R., Burns, G. and Bahler, J. (2002) The transcriptional program of meiosis and sporulation in fission yeast. Nat. Genet 32, 143–147.

McKim, K. S., Green-Marroquin, B. L., Sekelsky, J. J., Chin, G., Steinberg, C., Khodosh, R. and Hawley, R. S. (1998) Meiotic synapsis in the absence of recombination. Science 279, 876–878.

Moens, P. B., Kolas, N. K., Tarsounas, M., Marcon, E., Cohen, P. E. and Spyropoulos, B. (2002) The time course and chromosomal localization of recombination-related proteins at meiosis in the mouse are compatible with models that can resolve the early DNA-DNA interactions without reciprocal recombination. J. Cell Sci. 115, 1611–1622.

Nelson, O. E. and Clary, G. B. (1952) Genetic control of semisterility in maize. J. Heredity 43, 205–210.

Nonomura, K., Miyoshi, K., Eiguchi, M., Suzuki, T., Miyao, A., Hirochika, H. and Kurata, N. (2003) The msp1 gene is necessary to restrict the number of cells entering into male and female sporogenesis and to initiate anther wall formation in rice. Plant Cell 15, 1728–1739.

Nonomura, K., Nakano, M., Eiguchi, M., Suzuki, T. and Kurata, N. (2006) Pair2 is essential for homologous chromosome synapsis in rice meiosis I. J. Cell Sci. 119, 217–225.

Ollinger, R., Alsheimer, M. and Benavente, R. (2005) Mammalian protein scp1 forms synaptonemal complex-like structures in the absence of meiotic chromosomes. Mol. Biol. Cell 16, 212–217.

Osman, K., Sanchez-Moran, E., Higgins, J. D., Jones, G. H. and Franklin, F. C. (2006) Chromosome synapsis in arabidopsis: Analysis of the transverse filament protein zyp1 reveals novel functions for the synaptonemal complex. Chromosoma, 1–8.

Palmer, R. G. (1971) Cytological studies of ameiotic and normal maize with reference to premeiotic pairing. Chromosoma 35, 233–246.

Paques, F. and Haber, J. E. (1999) Multiple pathways of recombination induced by double-strand breaks in saccharomyces cerevisiae. Microbiol. and Mol. Biol. Re. 63, 349–404.

Pasierbek, P., Jantsch, M., Melcher, M., Schleiffer, A., Schweizer, D. and Loidl, J. (2001) A caenorhabditis elegans cohesion protein with functions in meiotic chromosome pairing and disjunction. Genes Dev. 15, 1349–1360.

Pawlowski, W. P., Golubovskaya, I. N. and Cande, W. Z. (2003) Altered nuclear distribution of recombination protein rad51 in maize mutants suggests the involvement of rad51 in meiotic homology recognition. Plant Cell 15, 1807–1816.

Pawlowski, W. P., Golubovskaya, I. N., Timofejeva, L., Meeley, R. B., Sheridan, W. F. and Cande, W. Z. (2004) Coordination of meiotic recombination, pairing, and synapsis by phs1. Science 303, 89–92.

Petukhova, G. V., Pezza, R. J., Vanevski, F., Ploquin, M., Masson, J. Y. and Camerini-Otero, R. D. (2005) The hop2 and mnd1 proteins act in concert with rad51 and dmc1 in meiotic recombination. Nat. Struct. Mol. Biol. 12, 449–453.

Pigozzi, M. I. and Solari, A. J. (2003) Differential immunolocalization of a putative rec8p in meiotic autosomes and sex chromosomes of triatomine bugs. Chromosoma 112, 38–47.

Primig, M., Williams, R. M., Winzeler, E. A., Tevzadze, G. G., Conway, A. R., Hwang, S. Y., Davis, R. W. and Esposito, R. E. (2000) The core meiotic transcriptome in budding yeasts. Nat. Genet. 26, 415–423.

Ramesh, M. A., Malik, S. B. and Logsdon, J. M., Jr. (2005) A phylogenomic inventory of meiotic genes; evidence for sex in giardia and an early eukaryotic origin of meiosis. Curr. Biol. 15, 185–191.

Reinke, V., Gil, I. S., Ward, S. and Kazmer, K. (2004) Genome-wide germline-enriched and sex-biased expression profiles in caenorhabditis elegans. Development 131, 311–323.

Revenkova, E. and Jessberger, R. (2005) Keeping sister chromatids together: Cohesins in meiosis. Reproduction 130, 783–790.

Rhoades, M. M. (1956) Genic control of chromosomal behavior. Maize Genet. Coop. Newsl. 30, 38–48.

Rhoades, M. M. and Dempsey, E. (1966) Induction of chromosome doubling at meiosis by the elongate gene in maize. Genetics 54, 505–522.

Riley, R. and Chapman, V. (1964) Cytological determination of the homology of chromosomes of triticum aestivum. Nature 203, 156–158.

Riley, R., Chapman, V., Young, R. M. and Belfield, A. M. (1966) Control of meiotic chromosome pairing by the chromosomes of homoeologous group 5 of triticum aestivum. Nature 212, 1475–1477.

Rossi, P., Dolci, S., Sette, C., Capolunghi, F., Pellegrini, M., Loiarro, M., Di Agostino, S., Paronetto, M. P., Grimaldi, P., Merico, D., Martegani, E. and Geremia, R. (2004) Analysis of the gene expression profile of mouse male meiotic germ cells. Gene Expr Patterns 4, 267–281.

Sanchez-Moran, E., Mercier, R., Higgins, J. D., Armstrong, S. J., Jones, G. H. and Franklin, F. C. (2005) A strategy to investigate the plant meiotic proteome. Cytogenet. Genome Res. 109, 181–189.

Schlecht, U., Demougin, P., Koch, R., Hermida, L., Wiederkehr, C., Descombes, P., Pineau, C., Jegou, B. and Primig, M. (2004) Expression profiling of mammalian male meiosis and gametogenesis identifies novel candidate genes for roles in the regulation of fertility. Mol. Biol. Cell 15, 1031–1043.

Sheridan, W. F., Shamrov, N. A. V. A. I., Batygina, T. B. and Golubovskaya, I. N. (1996) The mac1 gene: Controlling the commitment to the meiotic pathway in maize. Genetics 142, 1009–1020.

Sheridan, W. F., Golubeva, E. A., Abrhamova, L. I. and Golubovskaya, I. N. (1999) The mac1 mutation alters the developmental fate of the hypodermal cells and their cellular progeny in the maize anther. Genetics 153, 933–941.

Shinohara, A., Ogawa, H. and Ogawa, T. (1992) Rad51 protein involved in repair and recombination in s. Cerevisiae is a reca-like protein. Cell 69, 457–470.

Staiger, C. J. and Cande, W. Z. (1990) Microtubule distribution in dv, a maize meiotic mutant defective in the prophase to metaphase transition. Dev. Biol. 138, 231–242.

Strich, R. (2004) Meiotic DNA replication. Curr. Top. Dev. Biol. 61, 29–60.

Tsubouchi, T. and Roeder, G. S. (2005) A synaptonemal complex protein promotes homology-independent centromere coupling. Science 308, 870–873.

Vaur, S., Cubizolles, F., Plane, G., Genier, S., Rabitsch, P. K., Gregan, J., Nasmyth, K., Vanoosthuyse, V., Hardwick, K. G. and Javerzat, J. P. (2005) Control of shugoshin function during fission-yeast meiosis. Curr. Biol. 15, 2263–2270.

Watanabe, Y. and Nurse, P. (1999) Cohesin rec8 is required for reductional chromosome segregation at meiosis. Nature 400, 461–464.

Watanabe, Y. (2004) Modifying sister chromatid cohesion for meiosis. J. Cell Sci. 117, 4017–4023.

Xu, H., Beasley, M. D., Warren, W. D., van der Horst, G. T. and McKay, M. J. (2005) Absence of mouse rec8 cohesin promotes synapsis of sister chromatids in meiosis. Dev. Cell 8, 949–961.

Zhao, D. Z., Wang, G. F., Speal, B. and Ma, H. (2002) The excess microsporocytes1 gene encodes a putative leucine-rich repeat receptor protein kinase that controls somatic and reproductive cell fates in the arabidopsis anther. Genes Dev. 16, 2021-2031.

Zickler, D. and Kleckner, N. (1998) The leptotene-zygotene transition of meiosis. Annu. Rev. Genet. 32, 619-697.

Zickler, D. and Kleckner, N. (1999) Meiotic chromosomes: Integrating structure and function. Annu. Rev. Genet. 33, 603-754.

Homologous Recombination in Maize

Hugo K. Dooner, An-Ping Hsia, and Patrick S. Schnable

Abstract We have divided this chapter into two major sections: somatic and meiotic recombination. Somatic recombination in plants has been mostly monitored with artificial recombination substrates in transgenic systems. Although, in this area, maize has lagged behind other plants that can be more easily transformed, excellent progress has been achieved recently, as detailed in the first section. Specific topics discussed in this section are site-specific and targeted recombination. Research on meiotic recombination, particularly intragenic recombination, has been historically strong in maize relative to other plants, principally because the maize endosperm provides distinct advantages as an experimental unit of observation for recombination studies. It is, at the same time, large enough so that many traits can be scored and small enough so that many kernels can be screened. Many of the genes utilized in meiotic recombinational analyses affect anthocyanin pigmentation in the aleurone layer of the endosperm, as will be evident in the second section. In this section we discuss the distribution of recombination junctions at the genomic, regional, and genic levels, as well as modifiers that affect that distribution. We consider the special case of tandem duplications and gene families as recombination substrates and discuss how recombination has been used as a tool in the genetic analysis of paramutation and disease resistance.

H.K. Dooner
Waksman Institute and Department of Plant Biology, Rutgers University, Piscataway, NJ 08854-8020, USA
dooner@waksman.rutgers.edu

A.-P. Hsia
Department of Agronomy, Iowa State University, Ames, IA 50011-3650, USA
hsia@iastate.edu

P.S. Schnable
Center for Plant Genomics and Department of Agronomy, Iowa State University, Ames, IA 50011-3650, USA
schnable@iastate.edu

J.L. Bennetzen and S. Hake (eds.), *Maize Handbook - Volume II: Genetics and Genomics,*   377
© Springer Science + Business Media LLC 2009

1   Somatic Recombination

1.1   Site-specific Recombination

Site-specific recombination systems have the potential to enable targeted mutagenesis, gene replacement, stable expression of transgenes, and the removal of selectable markers from transgenic plants. Several site-specific recombination systems identified from microorganisms or yeast have been tested in plants, including Cre/*lox* from bacteriophage P1 (Sternberg and Hamilton, 1981), Flp/*FRT* from the 2μ plasmid of *Saccharomyces cerevisiae* (Broach et al., 1982) and R/*RS* from *Zygosaccharomyces rousii* (Araki et al., 1985). These site-specific recombination systems are composed of two elements, the recombinase (Cre, Flp and R) and their respective recognition sequences (*lox*, *FRT* and *RS*). If two recognition sequences are located in the same orientation, DNA between them will be excised by the action of the recombinase. If, however the recognition sequences are located in opposite directions the action of the recombinase will cause an inversion. Recombination catalyzed by the recombinase between a genomic recognition site and an extrachromosomal recognition site will result in the integration of the extrachromosomal DNA molecule into the genome. The functionality of these systems has been demonstrated in tobacco, wheat, rice, Arabidopsis and maize (for a review, see Lyznik et al., 2003).

Several site-specific recombination systems have been tested in maize. The Flp/*FRT* system has been shown to be functional in maize. When Flp and *FRT* were introduced into maize protoplasts (as two plasmids) a reporter gene function was restored following an Flp-catalyzed excision event (Lyznik et al., 1993). This process could be regulated using a heat-shock promoter (Lyznik et al., 1995). It was subsequently shown that Flp could recombine *FRT* sites integrated in the maize genome following transformation with an Flp construct (Lyznik et al., 1996). Similarly, the co-transformation of maize with Cre and a reporter gene flanked by *lox* sites in opposite orientation increases the rate of single copy primary transformants (Srivastava and Ow, 2001). This is explained by the fact that Cre will catalyze the excision of redundant copies of the transgene at an integration site, resulting in single-copy transformants. This approach has the potential to reduce complications caused by multiple transgene copies, such as co-suppression and structural rearrangements (for a review see Matzke and Matzke, 1998).

Site-specific recombination systems have been used to remove selectable markers from transgenic maize lines (for reviews see Puchta, 2003; Darbani et al., 2007). Selectable markers are often used to identify desired transformants but are usually not needed for the expression of desired traits. The subsequent removal from the genome of the selectable markers using site-specific recombination systems will reduce transgene complexity, render the line susceptible to additional rounds of transformation with the same selectable marker, and help to allay public concerns about GMOs. The general strategy is to flank the selectable marker with recombinase recognition sites, allowing the marker to be excised upon the action of the recombinase. Zhang et al. (2003) tested marker removal in maize with the Cre/*lox* system and demonstrated precise and complete removal of marker genes. The first strategy of Zhang et al. was to use genetic crosses to introduce Cre into the genomes of transgenic plants harboring an integrated selectable marker flanked by direct repeats of *lox* sites. Excision occurred early in embryo development and proved stable in progenies. Zhang et al. also tested an auto-excision strategy in which a heat-inducible Cre expression cassette was placed in between the *lox* sites next to the selectable marker on the same construct. Transformed calli were induced by heat treatment to activate Cre, resulting in excision of the selectable marker and the Cre cassette. The auto-excision approach could be applied to vegetable tissues or crops that are vegetatively propagated and does not require a generation for crossing and segregation of the Cre construct. Ream *et al.* (Ream et al., 2005) tested for possible recognition of natural *lox*-like sites in the maize genome using a Cre-expressing line from Zhang et al. (2003). Contrary to previous reports of phenotypic aberrations in Cre-expressing plants (Coppoolse et al., 2003), there were no significant differences in rates of ectopic recombination between Cre+ and Cre− lines as evaluated by pollen abortion rates, frequencies of defective embryos and root-tip karyotypes. It is proposed that this Cre-expressing line has a low level of Cre expression (Zhang et al., 2003), sufficient to catalyze efficient excision but not to catalyze excessive ectopic recombination (Ream et al., 2005).

Site-specific recombination systems also provide the possibility of targeted integration of transgenes into plant genomes. The targeting of transgenes into genomic *lox* sites has been demonstrated in tobacco (Albert et al., 1995), rice (Srivastava et al., 2004; Chawla et al., 2006) Arabidopsis (Vergunst and Hooykaas, 1998) and maize (Kerbach et al., 2005). Kerbach *et al.* (2005) demonstrated targeted integration of a reporter gene into an integrated *lox* site. This experiment was conducted using maize calli expressing Cre via a transient assay. The LY038 line, launched in 2006, is the first commercialized maize line in which a selectable marker was removed via the Cre/*lox* system (Ow, 2007). The selectable marker, *nptII*, flanked by two directly oriented *lox* sites was excised after the desired transformation events were obtained. This approach left behind an integrated *lox* site in LY038 at a locus in the genome that provides stable expression of the transgene, *cordapA*, a lysine-insensitive dihydrodipicolinate synthase. Ow (2007) discussed the use of this *lox* site as a docking site for the introduction of future transgenes to minimize transgene position effects. Ow proposed a gene stacking strategy in which different site-specific recombination recognition sites can be used for each subsequent transgene construct that will be targeted to this specific locus. For example, the first stacking gene will be targeted to this locus via *lox* mediated recombination while introducing the recognition site (A) of a different recombinase. The next stacking gene can be targeted to this locus via site A while introducing recognition site B. The approach may still require screening many events to recover the desired recombinants but it may increase the probability of obtaining stable transgene expression.

Djukanovic et al. (2006) reported that gene conversion was the preferred pathway (vs. site-specific recombination) in maize zygotes generated from fertilization of gametes containing recombinases and recognition sequences, respectively. Although the experimental design allows for gene targeting via either site-specific recombination (Cre/*lox*) or homologous recombination, all recovered recombinants were gene conversion products of homologous recombination. The rate of gene conversion was estimated to be about 1-2% in zygotes containing both the recombinase and the recognition site. Because all analyzed recombination tracks involved the *lox* site, it was postulated that double strand breaks (DSBs) mediated by *lox* have been responsible for this increased rate of recombination (Djukanovic et al., 2006).

Methods have been developed that combine the use of *Ds* transposons and Cre/*lox* for genome manipulation in Arabidopsis (Osborne et al., 1995; Zhang et al., 2003; Woody et al., 2007) and tomato (Coppoolse et al., 2005) and it is conceivable that similar approaches could be applied to maize. For example, a collection of Arabidopsis T-DNA lines, WiscDsLox, became available for mutant screening and potential additional mutagenesis via *Ds* insertion or Cre/*lox* mediated deletion (Woody et al., 2007). The T-DNA construct used to generate these lines contains a *Ds* transposon within the T-DNA borders, a *lox* site outside of the *Ds* sequence and a *lox* site within the *Ds* sequence. The flanking sequences of these T-DNA insertions are available in GenBank for user screening. Other than isolation of T-DNA insertion alleles, the rationale is for users to perform local mutagenesis by activating *Ds* transposition (via the introduction of *Ac*), which will leave behind a *lox* site at the T-DNA integration site and move the other *lox* site to the new *Ds* insertion site. In addition, in the presence of Cre, the two *lox* sites can recombine and result in deletion of the sequences between them, generating deletion mutants that may be useful for functional analyses. However not all lines support the transposition of *Ds* after the introduction of *Ac* and recombination between the "native" and transposed *lox* sites has not been tested (Woody et al., 2007).

1.2 Targeted Recombination

1.2.1 RNA/DNA Oligonucleotides

Chimeric oligonucleotides composed of both RNA and DNA residues have been used to introduce single base changes into episomal and genomic target DNA sequences. The original concept was based on the discovery that the efficiency of RecA-catalyzed strand pairing in *E. coli* increased significantly when RNA was introduced into the targeting molecule to form a double hairpin structure (Kotani et al., 1996). Since then this approach has been shown to generate or correct single base mutations in yeast, mammalian and plant cells (for a review see Igoucheva et al., 2004). Chimeric oligonucleotides were used to create specific point mutations in the acetoactate synthase (ALS, acetohydroxy acid synthase; AHAS in maize) gene(s) to generate transformants that are resistant to herbicides (imidazolinone and sulfonylurea) in maize (Zhu et al., 1999; Zhu et al., 2000), tobacco (Beetham et al., 1999; Kochevenko and Willmitzer, 2003) and rice (Okuzaki and Toriyama, 2004). This approach has not been uniformly successful (Ruiter et al., 2003).

Maize line HiII contains two copies of AHAS108 and five copies of AHAS109 but the DNA sequences of the target site are conserved among these copies (Zhu et al., 2000). Zhu *et al.* (1999) designed two chimeric oligonucleotides to create dominant point mutations in two target sites in AHAS that will confer resistance to imidazolinone and sulfonylurea, respectively. These two oligonucleotides were separately delivered via bombardment and putative mutants were confirmed by sequencing the target sites. Similar experiments were also performed targeting a stably integrated GFP transgene to restore its function. The resulting conversion rates (confirmed conversion event/cells receiving bombardment) for all three targeting experiments are 1-1.5×10^{-4}, which is two to three orders of magnitude higher than the negative control (spontaneous mutation) rate of $10^{-7} - 10^{-8}$, but still substantially lower than observed in mammalian systems. However non-predicted mutations were also discovered. Zhu *et al.* (2000) followed up this research by establishing that the mutant alleles were transmissible with normal segregation. The efficiency and precision of this approach needs to be improved before it can be applied routinely in plants.

1.2.2 Zinc Finger Nucleases (ZFNs)

Gene targeting via homologous recombination (HR) dependent on the endogenous recombination machinery of the host has been widely applied with success in yeast and mice but its efficiency in plants is still not high enough for practical applications (see Puchta, 2002; Hanin and Paszkowski, 2003; Lyznik et al., 2003; Iida and Terada, 2005 for reviews). Various approaches are being studied in plants to increase the rate of HR-mediated gene targeting/repair (for a review see Puchta, 2002) and the introduction of DSBs in the genome has been shown to enhance the frequency of homologous integration at the target site (Puchta et al., 1993). The use of zinc-finger nucleases (ZFN) has been successful in *Xenopus, Drosophila, C. elegans* and human cells and holds promise for gene therapy (for a review see Wu et al., 2007). A ZFN consists of zinc-finger motifs that recognize specific sequences and a non-specific endonuclease, such as FokI (see Durai et al., 2005 for review) that generates DSBs. Typical ZFNs studied so far have three motifs and each motif recognizes a DNA triplet, thus 9-base recognition. Dimerization of the nuclease is required for efficient cleavage of double-strand DNA. The presence of two recognition sites in inverted orientation flanking a 6-base spacer promotes cleavage (Smith et al., 2000; Mani et al., 2005). When a ZFN dimerizes after binding to its properly spaced target sequences in inverted orientation the nuclease will generate a DSB. Recently, two groups reported the use of ZFNs to target specific genome sequences for mutagenesis (Lloyd et al., 2005) and gene modification (Wright et al., 2005) in plants. Lloyd et al. (2005) demonstrated an increased rate of ZFN-mediated mutagenesis in *Arabidopsis*. First, a synthetic 24-base oligo that contained two inverted recognition sequences flanking an *Eco*RI site was transformed into *Arabidopsis*. Transformants were re-transformed with a construct containing a ZFN with three

ZF motifs fused to a heat shock promoter. Seedlings of seven single-locus lines were selected and subjected to heat shock to activate the ZFN. DNA from these seedlings was extracted, amplified via PCR with primers flanking the ZF target sequences, cloned and digested with *Eco*RI. The mutation frequency (loss of the *Eco*RI site) at the ZFN recognition site was estimated to be as high as 8% (43/~550 clones), substantially higher than the control (no heat shock treatment, 0/327 clones). Sequence analysis of these mutant clones revealed mutations ranging from simple deletions of 1-52 bp (78%), simple insertions of 1-4 bp (13%) to deletions/insertions (8%). In addition, 10% of progeny from induced lines transmitted the mutation to the next generation. Wright *et al.* (2005) demonstrated increased rates of homologous recombination (HR) at target sites between an integrated transgene with a ZF target site and donor sequence. A construct containing a defective GUS::NPTII fusion that contained a 600 bp deletion replaced by a ZFN recognition site was introduced into tobacco cells. Protoplasts were isolated from ten such transgenic lines and electroporated with ZFN-encoding DNA and donor DNA that was 5 kb in length and included the 600 bp deleted sequences required to restore reporter gene function. From comparisons to appropriate controls, it was shown that one out of ten recombinant events was generated via HR, a substantial improvement over the one out of 10^6 from previous studies in tobacco. Out of 26 target sites analyzed, five were repaired without other rearrangement. Thus the fidelity of HR is about 20%. Although it remains to be tested if ZFNs will yield similar results in maize, the high rate of targeting from these two studies is encouraging.

2   Meiotic Recombination

2.1.   Distribution of Junctions

2.1.1   Genomic Level

In all eukaryotes studied to date, meiotic recombination is not distributed randomly (reviewed by Mezard, 2006). On a chromosomal level, the first indication that recombination in maize occurs preferentially in some regions and rarely, if at all, in others came from a comparison of genetic and physical maps (reviewed by Harper and Cande, 2000). The large number of translocation breakpoints that had been placed since around 1960 in both the genetic and cytological maps allowed maize geneticists to correlate both maps and draw inferences early on regarding the distribution of recombination along the ten chromosomes. One of the first attempts to collate the data of many geneticists for all 10 chromosomes and represent it in the form of cytogenetic maps was published in the Maize Newsletter 30 years ago (Beckett et al., 1978). A cursory examination of those cytological maps reveals that recombination tends to occur preferentially in distal chromosomal regions and is highly suppressed around the centromeres.

In the past decade, more precise ways to correlate genetic and cytological maps have been developed in maize. One such way is to place genetically mapped sequences directly on chromosomes by fluorescence in situ hybridization (FISH). This technique has been refined to allow the placement of single-copy genes on maize pachytene chromosomes at a remarkably high degree of resolution (Sadder and Weber, 2002; Koumbaris and Bass, 2003; Wang et al., 2006). A recent FISH study with probes from genes located in *9S* (Wang et al., 2006) always resolved two probes located 2 cM (500 kb) apart (*sh1* and *bz1*) and frequently resolved probes less than 100 kb apart (*bz1* and *tac7077*). Nine genetically mapped single-copy genes were simultaneously mapped on chromosome 9 in one FISH experiment. Integration of the genetic linkage map and the new FISH physical map revealed a dramatic reduction of recombination in pericentromeric hetero-chromatic regions and enhanced recombination at the distal end of 9S, thus providing strong support for the distribution of recombination along chromosomes inferred from classical cytogenetics.

The most direct way of determining how recombination junctions are distributed among chromosomes would be to count them. This is possible in maize and a few other organisms, where sites of recombination can be visualized under the EM as recombination nodules or RNs (Stack and Anderson, 2002). RNs are proteinaceous structures 50-200 nm in diameter found in association with the synaptonemal complex (SC) during prophase I of meiosis. The molecular events of meiotic recombination are thought to occur in these bodies. Counting RNs is possible in maize and a few other favorable organisms like tomato because entire synaptonemal complexes can be isolated from microsporocyte protoplasts and examined under the EM. RNs are classified as early (EN) or late (LN), depending on the prophase stage in which they are visible. In addition, ENs and LNs differ in shape, size, and relative numbers. Numerous ENs are observed during leptotene and zygotene and may be involved in homologous synapsis and early events in recombination. LNs, on the other hand, are observed in smaller numbers during pachytene. In maize, the frequency of LNs per unit length of SC is about 8-fold lower than that of ENs (Stack and Anderson, 2002). There is strong correlatory evidence indicating that LNs represent sites of crossing over: their number is similar to the number of chiasmata and, like chiasmata, they display positive interference. Thus, LNs provide extremely high resolution cytological markers for defining the frequency and distribution of crossovers along the length of each chromosome. Unless specified otherwise, the RNs generally discussed in the recombination literature refer to LNs and we will do the same henceforth.

Anderson et al (2003) produced RN maps of all 10 chromosomes by locating 4267 RNs on 2080 SCs at 0.2 μm intervals. All chromosomes exhibited a high frequency of distal RNs and a severe reduction of RNs near the centromeres. Peculiarly, and as yet unexplainably, the total length of the RN recombination map was about twofold shorter than the average maize genetic linkage map, but there was good correspondence between the relative lengths of the individual chromosomes in the two maps. Several possible explanations were considered, particularly the source of the inbred (KYS) used for the RN map, but dismissed experimentally.

The maize RN map was then converted to a map of recombination along the physical length of each of the 10 maize chromosomes by converting the frequency of RNs in each 0.2 μm interval to a cM value, under the assumption that 1 RN represents a crossover event equivalent to 50 cM, and summing along the length of each SC (Anderson et al., 2004). The RN maps (cytological maps of crossover sites) were then used to predict the physical location of the "core" genetic markers (Davis et al., 1999) on each of the 10 maize chromosomes. The predicted location of seven chromosome 9 markers was compared to the observed localization of those markers by in situ hybridization and found to be almost perfectly correlated. Thus, the chromosomal location of any genetically mapped marker can be accurately predicted from the RN map. Recently, using the equations described by Anderson et al. (2004), a web-based tool, the Morgan2McClintock Translator, was developed to automate the prediction of cytological position for any input linkage data (Lawrence et al., 2006).

The ultimate correlation between the physical and genetic (i.e., recombination) maps will come from the maize whole genome sequencing project, currently in progress. Such correlations have been possible in other plants with completely sequenced genomes, like Arabidopsis and rice, and provide a complete picture of how recombination is globally distributed in a genome. A caveat here is that maize has a highly polymorphic genome structure (Wang and Dooner, 2006), so variations in the ratio of physical to genetic distances for very closely linked markers can be expected among different maize lines. This variation will probably average out over longer genetic intervals.

2.1.2   Regional Level

The unevenness of recombination at a local level in maize was first revealed in intragenic recombination studies that compared the genetic and physical lengths of cloned genes (Dooner et al., 1985; Freeling and Bennett, 1985; Wessler and Varagona, 1985). Maize genes, in general, exhibit a much higher amount of recombination per kb than the genome's average of ~0.7 × 10^{-4} cM/kb, i.e., they behave as recombination hotspots. Genes that have been shown to be 50-200 times more recombinogenic than expected include: *a1* (Brown and Sundaressan, 1991; Civardi et al., 1994); *adh1* (Freeling and Bennett, 1985); *B* (Patterson et al., 1995); *bz1* (Dooner et al., 1985; Dooner, 1986); *r* (Eggleston et al., 1995), and *wx* (Wessler and Varagona, 1985). Thus, although genes comprise a small fraction of the maize genome, probably <8% (Messing and Dooner, 2006), they are chromosomal sites where meiotic recombination occurs preferentially. This finding provides support for a prediction of the hypothesis that meiotic recombination in eukaryotes is confined to genes, advanced 30 years ago to account for the discrepancy between the sizes of eukaryotic genomes, which vary greatly, and the lengths of their genetic maps, which vary much less (Thuriaux, 1977).

A second prediction of Thuriaux's hypothesis is that repetitive DNA should not contribute significantly to genetic length. Maize has a large amount of repetitive

DNA, organized mainly as nests of retrotransposons interspersed with genic regions (SanMiguel and Bennetzen, 1998; Messing and Dooner, 2006), so it is a particularly suitable organism for testing this prediction. This has been done at two regions: *bz1* (Fu et al., 2002) and *a1-sh2* (Yao et al., 2002).

In the *bz1* experiment, recombination was examined across homozygous genetic intervals on either side of the *bz1* gene (Fu et al., 2002). This is an important consideration because maize intergenic regions in different haplotypes are polymorphic for the presence or absence of retrotransposons and other large insertions and the insertions themselves are polymorphic in size, composition, and insertion site (Fu and Dooner, 2002; Wang and Dooner, 2006). The genetic intervals were marked by a *Ds* element at *bz1* and an *Ac* element at either the proximal *mkk1* or the distal *stc1* gene. Separating the *bz1* and *mkk1* markers was a 100-kb segment consisting mainly of a 94-kb methylated retrotransposon nest; separating the *bz1* and *stc1* markers was a 6-kb segment consisting mainly of parts of the *bz1* and *stc1* genes. Recombination in the distal side was almost two orders of magnitude higher than in the proximal side, and close to the *bz1* intragenic average, indicating that recombination in the region was confined to the genes and that the retrotransposon cluster contributed little, if any, to genetic length. The authors concluded that the repetitive retrotransposon DNA in maize, which constitutes the bulk of the genome, was most likely recombinationally inert.

In the *a1-sh2* experiment, recombination between the *a1* and *sh2* genes was examined in a heterozygote between two highly divergent haplotypes (Yao et al., 2002). The *a1-sh2* interval measures around 130 kb and contains four genes, *a1*, *yz1*, *x1* and *sh2*, separated by different amounts of polymorphic insertion DNA. Over 100 recombination junctions were located physically relative to SNP and indel polymorphisms and found to be distributed nonuniformly across the interval. The junctions were concentrated within three recombination hotspots, two of which corresponded to genes (*a1* and *yz1*). The third one was single-copy and deemed to be nongenic on the basis of computational and experimental evidence. However, a recent analysis of the sequence revealed it to have high similarity to rice sequences encoding putative *hAT* transposases, suggesting that transposon genes may also serve as recombination hotspots in maize. An unselected recombination junction studied earlier in chromosome 7 was found likewise to fall in single-copy, though apparently nongenic, DNA (Timmermans et al., 1996), but a re-examination of that sequence is not possible, as it is not currently available in the databases. Interestingly, the *x1* gene in the *a1-sh2* interval was not a recombination hotspot, indicating that not all maize genes are hotspots. Recent analysis of the 1-cM *bz1-tac7077* region, which contains five complete genes interspersed with insertions of various kinds, confirms this finding (He and Dooner, 2007). The 80-kb distal portion of the *a1-sh2* interval, which contains large amounts of repetitive DNA, including retrotransposons, and is most likely polymorphic between the parental haplotypes, harbored a single junction, a result consistent with the view that the retrotransposon fraction of the maize genome is not recombinationally active.

The main conclusions to be drawn from the experiments outlined above are that meiotic recombination in maize takes place preferentially in the single-copy component where genes reside and that retrotransposons, which make up most of the repetitive DNA, recombine rarely, if at all. This pattern of recombination most likely evolved as a mechanism to protect the genome from self-destruction at meiosis, when recombination is required for proper chromosomal disjunction. Given the dispersed distribution of retrotransposons in the genome (SanMiguel et al., 1996), ectopic recombination between them would often have serious deleterious consequences. Suppression of recombination in repetitive DNA prevents ectopic exchanges and serves to stabilize the genome.

2.1.3   Gene Level

Many maize genes are recombinational hotspots in the genome, but how is recombination distributed within the genes themselves? In yeast, an organism where all four haploid products of meiosis can be recovered, intragenic recombination is usually detected as gene conversion events, i.e., deviations from a 1:1 segregation of alleles that may or may not be associated with flanking marker exchange (reviewed by Petes et al., 1991; Paques and Haber, 1999). Gene conversion frequencies in yeast often show gradients across the length of the gene, a phenomenon known as conversion polarity: markers located at one end of the gene convert more frequently than markers located centrally or at the other end. The high conversion end of the gene is usually the 5' end, but can also be the 3' end. Polarity gradients are generally accepted to result from DSBs sites located within the high conversion end, where recombination is initiated and from where conversion tracts containing heteroduplex DNA are propagated in a distance-dependent manner. Where the heteroduplex includes sequences derived from two different alleles, a mismatch occurs and subsequent correction of the mismatch results in gene conversion. At a few loci, the high conversion end has been shown to be a region where DSBs are concentrated (Petes, 2001).

In maize, as in other higher plants, only one product of female meiosis is recovered and, although all four products of male meiosis become functional pollen grains, they do not ordinarily stick together and are usually randomized when pollen is shed. The exception to the latter rule is the Arabidopsis *qrt* mutant, which has been useful for the analysis of genome-wide recombination (Copenhaver et al., 2000). Most of the experimental systems used in maize to study intragenic recombination have been designed to screen for the recovery of wild-type individuals following recombination between heteroallelic mutations, i.e., mutations located at different sites within the gene (Figure 1). The location of the recombination junctions in the recombinants is then determined by direct sequencing or restriction analysis. Conclusions derived from these studies vary and it is difficult to draw generalizations, other than that genes tend to behave as recombination hotspots. Complicating matters, most studies have dealt with heteroalleles derived from progenitor alleles differing in multiple sequence polymorphisms, which help in placing recombination junctions, but may affect the outcome of the experiment (Dooner, 2002). As was true of recombination at the regional level, the two systems that have been studied in greatest detail are *bz1* and *a1*.

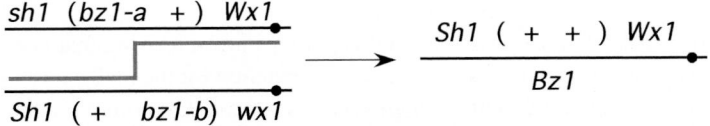

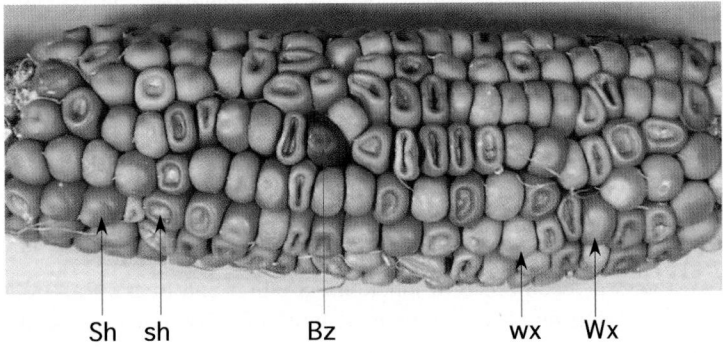

Fig. 1 *Bz* intragenic recombinants (IGRs) occurring at meiosis in plants heterozygous for two heteroallelic *bz1* mutations (*bz1-a* and *bz1-b*) are readily identified as exceptional purple seeds among the bronze testcross progeny, their frequency being a direct measure of the genetic distance between the mutations. The *bz1* locus is flanked by two easily scored endosperm markers, *sh1* and *wx1*, which makes it possible to readily sort Bz IGRs into classes carrying either a crossover (CO) or noncrossover (NCO) arrangement of flanking markers. The ear above was derived from a cross of a *bz1-a/bz1-b* F1 heterozygous female by a triple recessive *sh1 bz-1R wx1* pollen parent. The *bz-1R* reference allele carries a unique internal deletion of *bz1* which serves as pollen marker (Dooner, 1986)

bz1. At *bz1*, intragenic recombination has been examined in heterozygotes between a variety of mutations derived from either the same or different progenitor alleles, i.e., in the absence or presence of heterologies other than the two between which recombination was being measured (Dooner, 1986; Dooner and Kermicle, 1986; Dooner and Martínez-Férez, 1997a; Dooner, 2002). The mutations included point mutations, 1- to 3-kb transposon insertions and 8-bp transposon excisions footprints. Therefore, the differences between *bz1* heteroalleles ranged from just two nucleotides to many nucleotides, small indels, and large insertions. For simplicity, heteroalleles differing at just two positions are referred to as dimorphic and those differing at multiple positions, as polymorphic.

Intragenic recombinants (IGRs) produced by polymorphic pairs of *bz1* alleles are borne almost exclusively on crossover (CO) chromosomes carrying the recombinant arrangement of flanking markers expected from the known physical location of the mutations in the gene (Dooner and Martínez-Férez, 1997a; Dooner, 2002). This outcome differs from that of recombination studies in yeast, but is common in maize studies that have examined recombination between polymorphic heteroalleles (Dooner and Kermicle, 1986; Patterson et al., 1995; Xu et al., 1995). As has been pointed out (Dooner, 2002), heteroalleles used in recombination studies in maize are generally much more polymorphic than those in yeast, where heterologies are often introduced by transformation, so it is perhaps not surprising that none of the current yeast models provides a satisfactory explanation for the maize observations. The distribution of recombination junctions among these IGRs is not random. More junctions than expected by chance occur in regions of uninterrupted homology and fewer junctions in regions with a high density of SNP and indel heterologies (Dooner and Martínez-Férez, 1997a). It appears, therefore, that the density of heterologies affects where meiotic exchanges occur in the *bz1* gene.

A few *Bz* IGRs from polymorphic heterozygotes are borne on noncrossover (NCO) chromosomes carrying parental arrangements of flanking markers. Because chiasma interference in the region is very high (Dooner, 1986), most *Bz* NCO IGRs can be assumed to arise by conversion, rather than double crossing over, and can be used to analyze conversion tracts. In a few cases it has been possible to obtain an estimate of the minimum and maximum lengths of conversion tracts based on the position of the two recombination junctions at either end of the tract relative to flanking polymorphisms. The minimum lengths of three tracts associated with conversion of a *bz1* point mutation near the 3' end of the gene were 965, 983, and 3127 bp and that of two tracts associated with conversion of an excision footprint mutation near the 5' end were 3736 and 3852 bp. One tract associated with conversion of a centrally located excision footprint mutation was shorter, measuring between 109 and 313 bp (Dooner and Martínez-Férez, 1997a; H.K. Dooner and C. Zhan, unpublished results). The average minimum length of these conversion tracts, 2.1 kb, is slightly larger than those of yeast, (1.0 to 2.0 kb: Paques and Haber, 1999). In yeast, multiple heterologies (up to 1%) have been shown to result in longer tracts, which are more likely to be associated with crossing over (Schultes and Szostak, 1990).

The picture given by dimorphic heterozygotes is quite different. First, recombination between dimorphic *bz1* heteroalleles produced both CO and NCO *Bz* IGRs, in ratios apparently dependent on the nature and location of the mutations (Dooner, 2002). As indicated above, NCO IGRs most likely originate by gene conversion. Of course, the absence of additional markers in a dimorphic heteroallelic combination precludes the detection and measurement of a potential conversion tract. Second, cM/kb ratios in dimorphic heterozygotes were consistently higher than in polymorphic heterozygotes, supporting the notion that recombination correlates positively with sequence identity. Third, the cM/kb ratios for intervals defined by five pairs of point mutations located in the central half of the gene were remarkably constant, suggesting that recombination occurred uniformly in that segment of the *bz1* gene (Dooner and Martínez-Férez, 1997a). However, the sample of mutant sites analyzed did not include any at either end of the gene. Preliminary data with additional mutations located close to the 5' and 3' ends of the gene suggest that recombination may be higher at the ends than in the center (Dooner and He, unpublished data). This result is reminiscent of the conversion gradients seen in yeast and other fungi.

A different type of observation made in the above studies relates to the effect of immobile transposable element insertions on recombination (Dooner, 1986; Dooner and Kermicle, 1986; Dooner and Martínez-Férez, 1997a; Dooner, 2002). In dimorphic heterozygotes, recombination between a 1.1-kb *Ds* insertion at the beginning of the *bz1* gene and a 3.3-kb *Ds* insertion in the middle (*bz-m1/bz-m2*) was suppressed fourfold relative to that between the same 1.1-kb *Ds* insertion and an 8-bp excision footprint, which replaced the 3.3-kb *Ds* insertion (*bz-m1/bz-s2*). This observation suggests that transposon insertions, or, at least, a centrally located large transposon, can suppress recombination when the other allele carries an insertion. However, the reciprocal combination of an excision footprint at the beginning of the gene and the centrally located 3.3-kb *Ds* insertion (*bz-s1/bz-m2*) was not examined because of unavailability of a *bz-s1* excision derivative, so it is not possible to determine if the suppressing effect of transposons is position-dependent.

The effect of insertions on recombination in polymorphic heterozygotes is harder to assess because possible insertion effects are confounded by the effects of other sequence heterologies. Dooner and Ralston (1990) examined recombination between two transposon mutations in different progenitor alleles: the 1.1-kb *Ds* insertion in *bz-m1* and a 1.4-kb *Mu1* insertion in the middle of a different progenitor allele (*bz-Mum1* or *bz-Mum4*). Recombination between the two transposons was 2-3 times lower than recombination between the 1.1-kb *Ds* transposon and point mutations located close to *Mu1* and derived from the same *Bz1* progenitor allele (Z. Ma and H.K. Dooner, unpublished characterization of the *bz-Mum* alleles). However, the direct effect on recombination of removing the *Mu1* insertion (by excision) has not been tested. In polymorphic heterozygotes between a *Ds* transposon insertion and a point mutation, Dooner and Martínez-Férez (1997a) observed less recombination than expected (in cM/kb) immediately adjacent to the transposon and concluded that insertions may affect recombination in adjacent regions. Subsequently, it was shown that replacement of the transposon by an excision footprint allele did not alter the overall frequency of *Bz* IGRs (Dooner, 2002), suggesting that the reduced recombination adjacent to the insertion site may have been due to the multiple SNPs and indels that differentiated the two alleles, and not to the presence of the *Ds* insertion in one of them.

a1. Most studies of intragenic recombination at the *a1* locus have involved polymorphic mutations derived from different progenitor alleles. The *a1* gene is very close to *sh2* (130 kb; 0.1 cM), so these studies have not only covered *a1*, but the entire *a1-sh2* region (see earlier section).

Xu et al. (Xu et al., 1995) examined recombination between a 1.4-kb *Mu1* insertion in the 5′ UTR (untranslated region) and a 0.7-kb *rDt* insertion in exon 4 of *a1*. They found that intragenic recombination within a 377-bp segment of the *a1* gene accounted for one-fifth of all recombination events in the *a1-sh2* interval, making the segment a recombination hotspot. This segment, from the middle of exon 1 to the end of exon 2, has been an invariant recombination hotspot in every heterozygote examined to date (Yao et al., 2002; Yandeau-Nelson et al., 2005). What makes this segment a hotspot is not clear at this time. Possible reasons include its proximity to the 5′ end of the gene, a site where recombination frequently initiates in yeast, and its high sequence identity among alleles (Yao et al., 2002). The 1.4-kb *Mu1* insertion in the 5′ UTR was found to reduce recombination rate in the 377-bp hotspot but not to alter its hotspot nature. The majority of recombination junctions mapped to the 377-bp interval in heterozygous combinations of alleles with and without *Mu1*, leading the authors to conclude that the *Mu1* insertion suppressed recombination, but did not alter the distribution of recombination junctions in *a1*.

Two out of the 17 *A1* IGRs isolated by Xu et al. (1995) were borne on NCO chromosomes and most likely arose by conversion of the *rDt* insertion allele. The distal end of those conversion tracts fell in the 377-bp recombination hotspot between the insertions, but the position of the proximal end was not determined. The minimum sizes of those tracts were 1.1 and 1.5 kb. In a subsequent study of the effect of the autonomous *MuDR* element on recombination between the same pair of heteroalleles (discussed below), Yandeau-Nelson et al. (2005) were able to determine the sizes of eight tracts associated with conversion of the 0.7-kb *rDt* insertion. The distal endpoints in six of them mapped to the previously identified 377-bp recombination hotspot, establishing this segment as a hotspot for recombination junctions in both COs and NCOs. Three conversion tracts measured between 621 and 1318 bp and one, between 31 and 683 bp. In four others, the proximal endpoint could not be determined because of a lack of polymorphisms between the parental alleles, but their minimum sizes were 534, 1124, 1320 and 1603 bp. The average minimum conversion tract length at *a1*, 0.9 kb, appears smaller than at *bz1*, but not enough tracts have been analyzed at either locus. The present numbers reveal an overlap in the sizes of the conversion tracts at the two loci and considerable length variation within each locus, particularly at *bz1*, where the six conversion tracts analyzed come from three different sites.

The effect of sequence identity on recombination at the *a1* locus was examined by comparing rates of recombination between dimorphic allelic pairs and polymorphic allelic pairs which defined genetic intervals of roughly the same size and in approximately the same location (Yandeau-Nelson et al., 2006b). A sevenfold higher rate was observed in the dimorphic heterozygotes, most likely from the higher level of sequence identity in the genetic interval.

r. Many *R* "alleles" are complex and carry two or more copies of a gene that encodes a member of the b-HLH family of transcription factors. Functional alleles of *R* regulate where in the plant the anthocyanin biosynthetic pathway is expressed. The main difference between *R* alleles lies in the 5′ end, the region that controls their tissue-specificity. We will limit our discussion in this section to recombination between two simplex alleles, *R-sc* (*Sc*, strong seed color) and *R-p* (*P*, plant color) (Li et al., 2001). These alleles have nonhomologous 5′ flanking sequences and regulate anthocyanin production in nonoverlapping tissues: *Sc* in seed parts (aleurone and scutellum) and *P* in plant parts (coleoptile, roots, and anthers). Li et al. (2001) synthesized heterozygotes between a wild-type *P* and a series of 39 *sc-m* mutants that carried a 2-kb *Ds6* transposon in different locations of the *Sc* gene, from 2.5 kb 5′ of the transcription start site to the end of the coding region, and looked for restoration of *Sc* function by meiotic recombination.

All *sc-m* mutants with *Ds* insertions in a region of the gene stretching from ~400 bp downstream of the transcription start site to the 3′ end produced mainly CO IGRs, as has been found in combinations of polymorphic heteroalleles at other maize loci. The frequency of intragenic recombination in this region is high and the cM/kb ratios fall within the published range for other maize genic regions, confirming that R is a recombination hotspot. In contrast, *sc-m* mutations with *Ds* insertions in 5' regions produced rare NCO *Sc* IGRs. About one-third of such NCO derivatives were identical to the *Sc* progenitor in phenotype and sequence. Surprisingly, the other two-thirds had a novel phenotype that combined the strong seed color of *Sc* with the strong coleoptile color of *P*, plus a strong scutellar node color not seen in either parental allele, yet seen in other naturally occurring *r* alleles.

Molecular characterization of these derivatives revealed them to have arisen by a gene conversion event of the *Ds* insertion site in the *sc-m* mutants involving an exchange between the promoter region of *Sc* and a highly homologous region upstream of *P* that was fragmented by four blocks of nonhomologous DNA. Excluding the nonhomologous blocks, the homologous segments shared by *Sc* and *P* in this region are identical in sequence over 2209 bp of DNA and, in their lack of other heterologies over stretches approaching 1 kb in length, resemble the dimorphic *bz1* insertion heteroalleles that also yield predominantly NCO IGRs.

wx. The *wx* locus occupies a special place in the history of maize genetic research because it was the first one to be recombinationally dissected at a high level of resolution. Oliver Nelson took advantage of the differential staining of Wx and wx pollen grains with an iodine solution to detect recombination between *wx* heteroalleles by the occurrence of Wx pollen grains (Figure 2), and proceeded to construct a fine structure map of the locus using principles of deletion mapping

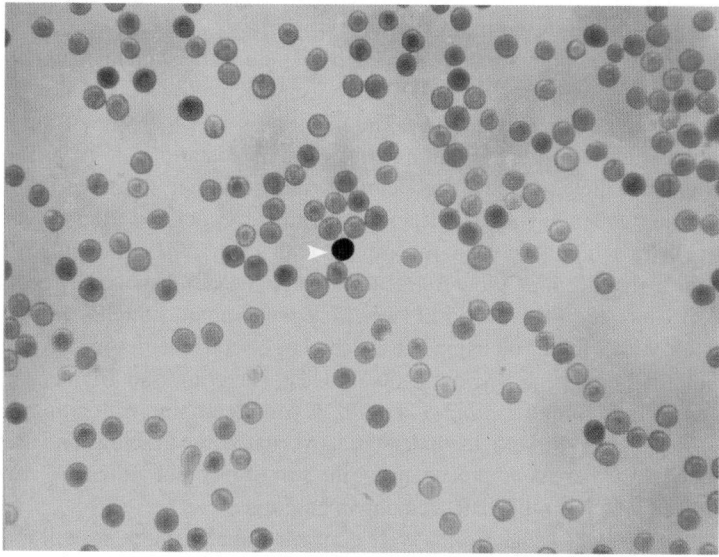

Fig. 2 *Wx1* intragenic recombinants (IGRs) occurring at male meiosis in plants heterozygous for two *wx1* heteroalleles can be identified by staining F1 pollen with an iodine solution. Wx normal pollen stains dark, whereas wx mutant pollen stains light. The arrow points to a dark-staining pollen grain carrying a *Wx1* IGR produced by recombination between the *wx1-C* and *wx1-90* heteroalleles (courtesy of Kitisri Sukhapinda and Peter A. Peterson)

(Nelson, 1962). He also measured intragenic recombination between pairs of *wx* heteroallelic mutations by regular testcrosses to obtain cM estimates of genetic distances. The ratio between genetic and physical distances for the *wx* gene calculated after these same mutations were characterized molecularly (Wessler and Varagona, 1985) was one to two orders of magnitude higher than the average ratio for the maize genome.

Okagaki and Weil (1997) investigated this relationship further. They examined recombination between two *wx* mutations at opposite ends of the gene and found that most *Wx* IGRs were COs, as in most other maize systems. Using a centrally located restriction site polymorphism between the two mutations, they assigned the recombination junctions in these IGRs to either a 1.1- or a 1.5-kb interval. The junctions were distributed in proportion to the physical size of the intervals, providing no evidence for a recombination hotspot within the *wx* gene. The *wx* mutants used in that study were clearly polymorphic in their restriction sites, but the extent of other polymorphisms in the two intervals is not known, as the mutant alleles have not been fully sequenced. Okagaki and Weil also reanalyzed Nelson's earlier recombination data (Nelson, 1968) involving *wx* deletion derivatives, including one that removed a significant part of the promoter, and concluded that, in contrast to yeast, deletion of promoter sequences had little effect on recombination frequencies. Lastly, they determined whether addition of further insertion polymorphisms had any effect on recombination between two insertion mutations and it did not: recombination frequencies between two *Ds* insertion mutations derived from a common progenitor were similar in the presence or absence of another *Ds* element located 0.5 kb upstream of the transcription start site.

adh1. Intragenic recombination at *adh1* can be monitored by staining mature pollen grains for alcohol dehydrogenase (ADH) activity, which allows, as at *wx*, genetic fine structure analysis (Freeling, 1976). The most intriguing outcome of intragenic recombination studies at *adh1* is that a series of point mutations derived from the *Adh1-S* progenitor allele, which do recombine with each other, fail to recombine at any appreciable frequency with a series of point mutations derived from the *Adh1-F* progenitor allele, which can also recombine with each other (Freeling, 1977). Recombination may occur at a low frequency, though, because a naturally occurring potential IGR has been identified in two inbreds (Zeng and Sachs, 1994). The recombination block might be explained by the large sequence dissimilarities both 5′ and 3′ of the *adh1* gene, including its 3′ UTR (Freeling and Bennett, 1985). However, mutants from either *Adh1-S* or *Adh1-F* will recombine with a third allele, an observation that is difficult to understand. As stated by Freeling and Bennett (1985) in a review, "there is no precedent for the sort of recombinational restrictions found at *Adh1*", which, to this date, remain an enigma.

When frequencies of ADH⁺ pollen were converted to cM, under a set of conventional assumptions, *Adh1* was found to be 100 times more recombinogenic than the average maize segment. An interesting observation from the *Adh1* studies was that *Adh1-S* seemed to be considerably more recombinogenic than *Adh1-F*. This conclusion was reached from a comparison of the mean recombination frequency among eight EMS-induced mutants of *Adh1-S* with the mean recombination frequency among the same number of EMS-induced mutants of *Adh1-F*. While trivial explanations, such as differential clustering of mutant sites in the two sets of mutations (which were not sequenced) or genetic background differences cannot be ruled out, this may be a real effect due to the large structural polymorphisms that flank alleles of different origin in maize (Fu and Dooner, 2002; Wang and Dooner, 2006). Similar comparisons with well-characterized series of mutations derived from two (or more) polymorphic progenitor alleles have not been done in maize.

2.2   *Modifiers of Recombination*

It was recognized as early as 1926 that recombination rates in maize are susceptible to genetic control (Stadler, 1926). Two-to-three fold variation in recombination frequencies has been observed in various types of maize populations (Beavis and Grant, 1991; Tulsieram et al., 1992; Fatmi et al., 1993; Williams et al., 1995). These studies and the recent identification of QTLs affecting recombination rates (Esch et al., 2007) demonstrate that variation in recombination rate is under the control of multiple genetic modifiers. These modifiers fall into two general classes. *Cis-* and *trans*-acting modifiers are and are not closely linked, respectively, to the intervals on which they act. The studies described below used experimental designs that could distinguish *cis-* and *trans*-acting modifiers.

2.2.1   Cis-acting Modifiers

A *cis*-acting modifier that increases recombination in the chromosome *9S sh1-bz1* interval has been identified genetically (Timmermans et al., 1997). In another study, recombination rates across several *a1-sh2* intervals derived from teosinte, each having large insertion/deletion polymorphisms (IDPs) as compared to maize, were studied while heterozygous with a common maize-derived *a1-sh2* interval and in the absence of any known polymorphic *trans*-acting modifier (Yao and Schnable, 2005). This analysis identified up to three-fold differences in recombination rates and statistically significant differences in the distributions of recombination break-points across the *a1-sh2* interval that, due to the nature of the experimental design, must be the result of *cis*-acting modifiers.

2.2.2   Trans-acting Modifiers

A *trans*-acting modifier has been identified that affects recombination rates on chromosome 7 and the *c1-sh1* and *bz1-wx1* intervals of chromosome 9 (Timmermans et al., 1997). In another study, the effects of *trans*-acting modifiers on recombination across the ~140-kb/0.1 cM *a1-sh2* interval of chromosome 3L were studied in the absence of polymorphic *cis*-acting factors (Yandeau-Nelson et al., 2006a). Genetic distances across *a1-sh2* varied two-fold among the three analyzed genetic backgrounds. Recombination rates across two genetic intervals on chromosome *1S* were not similarly affected by the *trans*-acting factor(s) that influenced recombination rates across the *a1-sh2* interval, demonstrating that at least some *trans*-acting modifiers do not act globally.

2.2.3 Transposons as Modifiers of Recombination

Although, as discussed above, the large retrotransposon fraction of the maize genome appears to be relatively inert recombinationally (Fu et al., 2002; Yao et al., 2002), other transposons, acting both in *cis* and in *trans*, can both suppress and increase rates of meiotic recombination. Acting in *cis*, heterozygous *Ds* and *Mu1* insertions reduce rates of crossovers (Dooner and Martínez-Férez, 1997a). The presence of active transposons can increase rates of recombination-like losses of duplicated regions (Athma and Peterson, 1991; Lowe et al., 1992; Stinard et al., 1993; Xiao et al., 2000; Xiao and Peterson, 2000). Because these events occur in the absence of meiosis or do not involve the exchange of flanking markers they do not involve meiotic crossovers. *MuDR* (Yandeau-Nelson et al., 2005), but not *Ac* (Dooner and Martínez-Férez, 1997b), autonomous transposons can act in *trans* to increase rates of meiotic crossovers at a locus containing a compatible non-autonomous insertion. Given that *Mu* transposons excise via a process involving the introduction of DSBs (Li et al., 2007a), it is reasonable to conclude that the four-fold increase in crossovers observed at a *Mu* insertion allele in the presence of *MuDR* is a consequence of *MuDR*-induced DSBs stimulating meiotic recombination. *Ac*-induced DSBs are presumably separated temporally or spatially from meiotic recombination (Dooner and Martínez-Férez, 1997b) or *Ac*-induced DSBs are not repaired via the homologous recombination pathway.

MuDR elements can influence the ratios of crossover to non-crossover events. Specifically, CO/NCO ratios are increased more than three-fold in similar crosses involving *MuDR*, acting in *trans* on a *Mu* insertion allele, as compared to those that do not (Yandeau-Nelson et al., 2005).

2.2.4 Meiotic Mutants

Several meiotic mutants reduce recombination rates. The *desynaptic* (*dy*) mutation, which is associated with abnormal telomere localization during the pachytene stage of meiosis, also reduces recombination rates (Ji et al., 1999). Based on this observation it has been hypothesized that telomeres may play a role in the control of meiotic recombination (Bass et al., 2003). Mutations at the *desynaptic2* (*dsy2*) locus reduce recombination by ~70% (Franklin et al., 2003). This is not surprising given that the product of this locus is required for both synapsis of homologous chromosomes and the proper behavior of RAD51 (Franklin et al., 2003). In yeast, the RAD51 protein plays a key role in the repair of DSBs and homologous recombination. Maize plants homozygous for mutant alleles of both copies of *rad51* exhibit cytological disruptions of meiosis in the male that indicate that RAD51 is required for efficient chromosome pairing and its absence results in nonhomologous pairing and synapsis (Li et al., 2007b). Surprisingly, these double mutants do not exhibit reduced rates of crossover during female meiosis.

2.3 Duplications and Tandem Gene Families: a Special Case

It has been estimated that 33% of maize genes are members of tandem arrays of duplicated genes (Messing et al., 2004). Using conservative estimates, approximately 1% of maize genes have a Nearly Identical Paralog (NIP), which is defined as being at least 98% sequence identical (Emrich et al., 2007). About half of NIP families exist as tandem arrays. Members of tandem arrays of duplicated genes can misalign and pair unequally during meiosis. First documented at the *bar* locus of *Drosophila* (Sturtevant, 1925), this has since been observed in many species including at the maize *rp1* (Sudupak et al., 1993; Richter et al., 1995; Sun et al., 2001; Ramakrishna et al., 2002; Mani et al., 2005), *rp3* (Webb et al., 2002), *r1* (Stadler and Neuffer, 1953; Dooner and Kermicle, 1971, 1974; Robbins et al., 1991; Walker et al., 1995), *kn1* (Lowe et al., 1992), 27-kD zein (Das et al., 1991), *p1* (Xiao et al., 2000) and *a1* (Yandeau-Nelson et al., 2006b) loci.

Unequal recombination events can potentially arise via interactions between homologs or sister chromatids. At least at *A1-b*, unequal recombination occurs preferentially between homologs (Yandeau-Nelson et al., 2006b) and duplicated segments pair non-randomly (Yandeau-Nelson et al., 2006b). In addition, the distributions of recombination breakpoints from homolog exchanges differ from those obtained from sister chromatid exchanges.

Given the prevalence of tandem gene arrays in maize and the high rates (10^{-3} and 10^{-4}: Laughnan, 1952; Dooner and Kermicle, 1971) at which tandem arrays participate in unequal recombination, this process likely contributes to the creation of genetic variability (see Section 2.4b), even within homozygous lines.

2.4 Meiotic Recombination as a Tool for Genetic Analysis

2.4.1 Paramutation

Alleles at certain loci (e.g., *R*, *B*, *Pl*, and *P*) show paramutational properties, i.e., the ability to change or effect allelic changes heritably and regularly when combined with other specific alleles. Meiotic recombination has been used to dissect the contribution to paramutation of specific components of paramutagenic alleles at the *B* and *R* loci.

At *B*, a paramutagenic *B'* allele heritably reduces the expression (transcription) of a highly expressed *B-I* allele in *B'/B-I* heterozygotes, changing it into another *B'* allele.

Intragenic recombination with a neutral *B* allele was used to map sequences required for paramutation and expression to the 5′ region of *B′*, between ATG and 150 Kb upstream (Patterson et al., 1995). In those experiments, the *B* locus evidenced several recombinational features seen in other polymorphic heteroallelic systems. Recombination within *B* was much higher than average, all IGRs were COs, and recombination was not evenly distributed within the gene, as most recombination junctions were near the 5′ end of the *B* transcribed region.

In a subsequent experiment, Stam et al (2002b) dissected further the 5′ region of *B'* and were able to localize by fine structure recombination mapping sequences required for *B'* paramutation and allele-specific expression. The 110 kb upstream of the *B'* transcription start site was cloned and sequenced and recombination junctions were identified in 12 IGRs. Sequences required for paramutation mapped to a region 93-106 kb upstream of the transcription start site. The 13-kb region was mostly unique and contained seven tandem repeats of an 853-bp sequence not found elsewhere in the maize genome and present only once in neutral alleles. Sequences required for the tissue specific expression of *B'* were mapped to a distinct region between 8.5 and 49 kb upstream of the transcription start site. The sequences in the 13-kb region required for paramutation were narrowed down further by recombination to a 6-kb segment that included the seven copies of the 853-bp repeat (Stam et al., 2002a). Recombinants with a fewer number of repeats were proportionately less paramutagenic, indicating that the repeats are required for paramutation.

Most studies of paramutation at *R* have analyzed interactions between the paramutagenic *R-st* and the paramutable *R-r:std* or their derivatives. *R-st* is a complex allele, known from early recombinational analysis (Kermicle, 1970), to comprise at least three separable genetic components: *Sc*, conferring strong seed color, *I-R*, a *hAT* transposon in *Sc* (W. Eggleston, pers. comm.), and *Nc*, conferring a near colorless seed phenotype. The *I-R* element is not involved in paramutation because *Sc* excision derivatives of *R-st* are as fully paramutagenic as the parent allele. Interestingly, though, it can recombine with another *I-R* copy located 6 cM away to produce deletions of the intervening segment by unequal crossing (Kermicle, 1984), an observation that suggests that DNA transposable elements, like genes, may be recombinogenic.

Southern blot characterization of the parental *R-st* allele and recombinant derivatives from *R-st* that had arisen by unequal crossing over identified four segments containing at least one *r* coding sequence each in *R-st* and one to three such segments in the recombinant derivatives (Eggleston et al., 1995). The proximal *r* gene, *Sc*, confers strong seed color and the three distal *r* genes (*Nc*s) confer a weak seed color phenotype referred to as near colorless. The level of the near colorless phenotype in the recombinant derivatives was inversely correlated with the number of *r* genes present (Eggleston et al., 1995), whereas their paramutagenic strength was directly correlated with the number of *r* genes (Kermicle, 1984; Kermicle et al., 1995). These observations suggest a repeat-based silencing mechanism in *R-st* paramutagenicity (as in *B'*) and in the regulation of expression of the *Nc* genes, which may be simply cis-suppressed *r* genes in the *R-st* complex.

2.4.2 Organization of the *Rp1* Complex

The *Rp1* locus is a complex of NBS-LRR genes at the tip of *10S* that determines resistance to races of the maize rust fungus (*Puccinia sorghi*). Resolution of its organization has been aided greatly by the analysis of unusual IGRs that arise by both equal and unequal crossing over between different members of the complex.

Hulbert and Bennetzen (1991) used flanking RFLPs to study the genetic fine structure of the locus and the role of recombination in its instability. Susceptible progeny lacking the resistance of either parent were obtained from testcrosses of several *Rp1* heterozygotes. Most susceptible progeny were COs for flanking markers, indicating that they arose by recombination. In most heterozygotes, the flanking marker recombination was unidirectional, allowing the ordering of the resistant alleles relative to each other along the chromosome. However, CO susceptible progeny with both possible recombinant arrangements of flanking markers, in roughly equal numbers, were recovered from one heterozygote, as well as from two homozygotes (Sudupak et al., 1993), recombinational outcomes indicating that unequal crossing over between repeats was the primary mechanism of *Rp1* meiotic instability. The expected reciprocal recombinants with the combined resistance of the two parental types were also obtained (Hulbert et al., 1993). Most interesting, in a screen of novel resistance not present in either parent, four COs were found to be resistant to at least one rust biotype to which the parents were susceptible (Richter et al., 1995), strongly suggesting that recombination within the *Rp1* complex can generate novel resistance genes.

The isolation and molecular characterization of the *Rp1-D* resistant haplotype (Collins et al., 1999) confirmed many of the earlier inferences from genetic experiments. *Rp1-D* is a complex haplotype consisting of nine NBS-LRR paralogs. Analysis of susceptible and double resistant progeny from *Rp1-D/Rp1-J* heterozygotes showed that they had arisen by recombination within the cluster of *Rp1-D* homologous genes in each parent. Likewise, all the susceptible progeny from *Rp1-D* homozygotes had lost at least one of the nine homologs, as expected from an unequal crossing over mode of origin.

Sun et al. (2001) sequenced all nine paralogs in the *Rp1-D* haplotype and were able to order them by determining which ones were either lost or retained in a large set of CO susceptible individuals from *Rp1-D* homozygotes and heterozygotes. They sequenced the recombination junctions in the new chimeric paralog of the susceptible individuals arisen by unequal crossing over in *Rp1-D* homozygotes and determined that they occurred throughout the *Rp1-D* gene, which is located at the distal end of the cluster and is altered by the recombination event with one of the other paralogs. Although four of the eight paralogs recombined with *Rp1-D*, most recombination junctions fell within the same untranscribed paralog. It is clear from the detailed studies of the *Rp1* disease resistance complex summarized above that meiotic recombination has been both an invaluable tool in its analysis as well as a major force in its evolution.

References

Albert, H., Dale, E.C., Lee, E., and Ow, D.W. (1995). Site-specific integration of DNA into wild-type and mutant lox sites placed in the plant genome. Plant J. 7, 649–659.

Anderson, L.K., Salameh, N., Bass, H.W., Harper, L.C., Cande, W.Z., Weber, G., and Stack, S.M. (2004). Integrating genetic linkage maps with pachytene chromosome structure in maize. Genetics 166, 1923–1933.

Anderson, L.K., Doyle, G.G., Brigham, B., Carter, J., Hooker, K.D., Lai, A., Rice, M., and Stack, S.M. (2003). High-resolution crossover maps for each bivalent of Zea mays using recombination nodules. Genetics 165, 849–865.

Araki, H., Jearnpipatkul, A., Tatsumi, H., Sakurai, T., Ushio, K., Muta, T., and Oshima, Y. (1985). Molecular and functional organization of yeast plasmid pSR1. J. Mol. Biol. 182, 191–203.

Athma, P., and Peterson, T. (1991). Ac induces homologous recombination at the maize P locus. Genetics 128, 163–173.

Bass, H.W., Bordoli, S.J., and Foss, E.M. (2003). The desynaptic (dy) and desynaptic1 (dsy1) mutations in maize (Zea mays L) cause distinct telomere-misplacement phenotypes during meiotic prophase. J. Exp. Bot. 54, 39–46.

Beavis, W.D., and Grant, D. (1991). A linkage map based on information from four F2 populations of maize (Zea mays L.). Theor. Appl. Genet. 82.

Beckett, E.B., Burnham, C.R., Coe, E.H., Maguire, M.P., Patterson, E.B., and Phillips, R.L. (1978). Cytogenetic working map. Maize Genet. Newslet. 52, 129–145.

Beetham, P.R., Kipp, P.B., Sawycky, X.L., Arntzen, C.J., and May, G.D. (1999). A tool for functional plant genomics: chimeric RNA/DNA oligonucleotides cause in vivo gene-specific mutations. Proc. Natl. Acad. Sci. USA 96, 8774–8778.

Broach, J.R., Guarascio, V.R., and Jayaram, M. (1982). Recombination within the yeast plasmid 2um circle is site-specific. Cell 29, 227–234.

Brown, J., and Sundaresan, V. (1991). A recombinational hotspot in the maize A1 intragenic region. Theor. Appl. Genet. 81, 185–188.

Chawla, R., Ariza-Nieto, M., Wilson, A.J., Moore, S.K., and Srivastava, V. (2006). Transgene expression produced by biolistic-mediated, site-specific gene integration is consistently inherited by the subsequent generations. Plant Biotechnol. J. 4, 209–218.

Civardi, L., Xia, Y., Edwards, K.J., Schnable, P.S., and Nikolau, B.J. (1994). The relationship between genetic and physical distances in the cloned a1-sh2 interval of the Zea mays L. genome. Proc. Natl. Acad. Sci. USA 91, 8268–8272.

Collins, N., Drake, J., Ayliffe, M., Sun, Q., Ellis, J., Hulbert, S., and Pryor, T. (1999). Molecular characterization of the maize Rp1-D rust resistance haplotype and its mutants. Plant Cell 11, 1365–1376.

Copenhaver, G.P., Keith, K.C., and Preuss, D. (2000). Tetrad analysis in higher plants. A budding technology. Plant Physiol 124, 7–16.

Coppoolse, E.R., de Vroomen, M.J., van Gennip, F., Hersmus, B.J., and van Haaren, M.J. (2005). Size does matter: cre-mediated somatic deletion efficiency depends on the distance between the target lox-sites. Plant Mol. Biol. 58, 687–698.

Coppoolse, E.R., de Vroomen, M.J., Roelofs, D., Smit, J., van Gennip, F., Hersmus, B.J., Nijkamp, H.J., and van Haaren, M.J. (2003). Cre recombinase expression can result in phenotypic aberrations in plants. Plant Mol. Biol. 51, 263–279.

Darbani, B., Eimanifar, A., Stewart, C.N., Jr., and Camargo, W.N. (2007). Methods to produce marker-free transgenic plants. Biotechnol. J. 2, 83–90.

Das, O.P., Poliak, E., Ward, K., and Messing, J. (1991). A new allele of the duplicated 27kD zein locus of maize generated by homologous recombination. Nucleic Acids Res. 19, 3325–3330.

Davis, G.L., McMullen, M.D., Baysdorfer, C., Musket, T., Grant, D., Staebell, M., Xu, G., Polacco, M., Koster, L., Melia-Hancock, S., Houchins, K., Chao, S., and Coe, E.H., Jr. (1999). A maize map standard with sequenced core markers, grass genome reference points and 932 expressed sequence tagged sites (ESTs) in a 1736-locus map. Genetics 152, 1137–1172.

Djukanovic, V., Orczyk, W., Gao, H., Sun, X., Garrett, N., Zhen, S., Gordon-Kamm, W., Barton, J., and Lyznik, L.A. (2006). Gene conversion in transgenic maize plants expressing FLP/FRT and Cre/loxP site-specific recombination systems. Plant Biotechnol. J. 4, 345–357.

Dooner, H.K. (1986). Genetic fine structure of the bronze locus in maize. Genetics 113, 1021–1036.

Dooner, H.K. (2002). Extensive interallelic polymorphisms drive meiotic recombination into a crossover pathway. Plant Cell 14, 1173–1183.

Dooner, H.K., and Kermicle, J.L. (1971). Structure of the R-r tandem duplication in maize. Genetics 67, 437–454.

Dooner, H.K., and Kermicle, J.L. (1974). Reconstitution of the R-r compound allele in maize. Genetics 78, 691–701.

Dooner, H.K., and Kermicle, J.L. (1986). The transposable element Ds affects the pattern of intragenic recombination at the bz and R loci in maize. Genetics 113, 135–143.

Dooner, H.K., and Ralston, E. (1990). Effect of the Mu1 insertion on intragenic recombination at the bz locus in maize. Maydica 35, 333–337.

Dooner, H.K., and Martínez-Férez, I.M. (1997a). Recombination occurs uniformly within the bronze gene, a meiotic recombination hotspot in the maize genome. Plant Cell 9, 1633–1646.

Dooner, H.K., and Martínez-Férez, I.M. (1997b). Germinal excisions of the maize transposon activator do not stimulate meiotic recombination or homology-dependent repair at the bz locus. Genetics 147, 1923–1932.

Dooner, H.K., Weck, E., Adams, S., Ralston, E., Favreau, M., and English, J. (1985). A molecular genetic analysis of insertion mutations in the bronze locus in maize. Mol. Gen. Genet. 200, 240–246.

Durai, S., Mani, M., Kandavelou, K., Wu, J., Porteus, M.H., and Chandrasegaran, S. (2005). Zinc finger nucleases: custom-designed molecular scissors for genome engineering of plant and mammalian cells. Nucleic Acids Res 33, 5978–5990.

Eggleston, W.B., Alleman, M., and Kermicle, J.L. (1995). Molecular organization and germinal instability of R-stippled maize. Genetics 141, 347–360.

Emrich, S.J., Li, L., Wen, T.J., Yandeau-Nelson, M.D., Fu, Y., Guo, L., Chou, H.H., Aluru, S., Ashlock, D.A., and Schnable, P.S. (2007). Nearly identical paralogs: implications for maize (Zea mays L.) genome evolution. Genetics 175, 429–439.

Esch, E., Szymaniak, J.M., Yates, H., Pawlowski, W.P., and Buckler, E.S. (2007). Using crossover breakpoints in recombinant inbred lines to identify quantitative trait loci controlling the global recombination frequency. Genetics, 177, 1851–1858.

Fatmi, A., Poneleit, C.G., and Pfeiffer, T.W. (1993). Variability of recombination frequencies in the Iowa Stiff Stalk Synthetic (Zea mays L.). Theor Appl. Genet. 86, 859–866.

Franklin, A.E., Golubovskaya, I.N., Bass, H.W., and Cande, W.Z. (2003). Improper chromosome synapsis is associated with elongated RAD51 structures in the maize desynaptic2 mutant. Chromosoma 112, 17–25.

Freeling, M. (1976). Intragenic recombination in maize: pollen analysis methods and the effect of parental Adh1 alleles. Genetics 83, 707–719.

Freeling, M. (1977). Allelic variation at the level of intragenic recombination. Genetics 89, 505-509.

Freeling, M., and Bennett, D.C. (1985). Maize Adh1. Annu. Rev. Genet. 19, 297–323.

Fu, H., and Dooner, H.K. (2002). Intraspecific violation of genetic colinearity and its implications in maize. Proc. Natl. Acad. Sci. USA 99, 9573–9578.

Fu, H., Zheng, Z., and Dooner, H.K. (2002). Recombination rates between adjacent genic and retrotransposon regions in maize vary by 2 orders of magnitude. Proc. Natl. Acad. Sci. USA 99, 1082–1087.

Hanin, M., and Paszkowski, J. (2003). Plant genome modification by homologous recombination. Curr. Opin. Plant Biol. 6, 157–162.

Harper, L.C., and Cande, W.Z. (2000). Mapping a new frontier; development of integrated cytogenetic maps in plants. Funct. Integr. Genomics 1, 89–98.

He, L., and Dooner, H.K. (2007). Recombination in a 100-kb genic interval containing Helitrons and retrotransposons. In 49th Annual Maize Genet. Conf. Abstracts (St. Charles, IL), pp. 99.

Hulbert, S.H., and Bennetzen, J.L. (1991). Recombination at the Rp1 locus of maize. Mol. Gen. Genet. 226, 377–382.

Hulbert, S.H., Sudupak, M.A., and Hong, K.S. (1993). Genetic relationships between alleles of the Rp1 rust resistance locus in maize. Mol. Plant-Microbe Int. 6, 387–392.

Igoucheva, O., Alexeev, V., and Yoon, K. (2004). Oligonucleotide-directed mutagenesis and targeted gene correction: a mechanistic point of view. Curr Mol Med 4, 445–463.

Iida, S., and Terada, R. (2005). Modification of endogenous natural genes by gene targeting in rice and other higher plants. Plant Mol. Biol. 59, 205–219.

Ji, Y., Stelly, D.M., De Donato, M., Goodman, M.M., and Williams, C.G. (1999). A candidate recombination modifier gene for Zea mays L. Genetics 151, 821–830.

Kerbach, S., Lorz, H., and Becker, D. (2005). Site-specific recombination in Zea mays. Theor. Appl. Genet. 111, 1608–1616.

Kermicle, J.L. (1970). Somatic and meiotic Instability of R-stippled, an aleurone spotting factor in maize. Genetics 64, 247–258.

Kermicle, J.L. (1984). Recombination between Components of a Mutable Gene System in Maize. Genetics 107, 489–500.

Kermicle, J.L., Eggleston, W.B., and Alleman, M. (1995). Organization of paramutagenicity in R-stippled maize. Genetics 141, 361–372.

Kochevenko, A., and Willmitzer, L. (2003). Chimeric RNA/DNA oligonucleotide-based site-specific modification of the tobacco acetolactate syntase gene. Plant Physiol 132, 174–184.

Kotani, H., Germann, M.W., Andrus, A., Vinayak, R., Mullah, B., and Kmiec, E.B. (1996). RNA facilitates RecA-mediated DNA pairing and strand transfer between molecules bearing limited regions of homology. Mol. Gen. Genet. 250, 626–634.

Koumbaris, G.L., and Bass, H.W. (2003). A new single-locus cytogenetic mapping system for maize (Zea mays L.): overcoming FISH detection limits with marker-selected sorghum (S. propinquum L.) BAC clones. Plant J. 35, 647–659.

Laughnan, J.R. (1952). The action of allelic forms of the gene A in maize. IV. On the compound nature of A and the occurrence and action of Its a derivatives. Genetics 37, 375–395.

Lawrence, C.J., Seigfried, T.E., Bass, H.W., and Anderson, L.K. (2006). Predicting chromosomal locations of genetically mapped loci in maize using the Morgan2McClintock Translator. Genetics 172, 2007–2009.

Li, J., Wen, T.J., and Schnable, P.S. (2007a). The role of RAD51 in the repair of MuDR-induced DSBs in Zea mays L. Genetics, 178, 57–66.

Li, J., Harper, L.C., Golubovskaya, I., Wang, C.R., Weber, D., Meeley, R.B., McElver, J., Bowen, B., Cande, W.Z., and Schnable, P.S. (2007b). Functional analysis of maize RAD51 in meiosis and double-strand break repair. Genetics 176, 1469–1482.

Li, Y., Bernot, J.P., Illingworth, C., Lison, W., Bernot, K.M., Eggleston, W.B., Fogle, K.J., DiPaola, J.E., Kermicle, J., and Alleman, M. (2001). Gene conversion within regulatory sequences generates maize r alleles with altered gene expression. Genetics 159, 1727–1740.

Lloyd, A., Plaisier, C.L., Carroll, D., and Drews, G.N. (2005). Targeted mutagenesis using zinc-finger nucleases in Arabidopsis. Proc Natl Acad Sci USA 102, 2232–2237.

Lowe, B., Mathern, J., and Hake, S. (1992). Active Mutator elements suppress the knotted phenotype and increase recombination at the Kn1-O tandem duplication. Genetics 132, 813–822.

Lyznik, L.A., Rao, K.V., and Hodges, T.K. (1996). FLP-mediated recombination of FRT sites in the maize genome. Nucleic Acids Res. 24, 3784–3789.

Lyznik, L.A., Gordon-Kamm, W.J., and Tao, Y. (2003). Site-specific recombination for genetic engineering in plants. Plant Cell Rep. 21, 925–932.

Lyznik, L.A., Mitchell, J.C., Hirayama, L., and Hodges, T.K. (1993). Activity of yeast FLP recombinase in maize and rice protoplasts. Nucleic Acids Res. 21, 969–975.

Lyznik, L.A., Hirayama, L., Rao, K.V., Abad, A., and Hodges, T.K. (1995). Heat-inducible expression of FLP gene in maize cells. Plant J. 8, 177–186.

Mani, M., Smith, J., Kandavelou, K., Berg, J.M., and Chandrasegaran, S. (2005). Binding of two zinc finger nuclease monomers to two specific sites is required for effective double-strand DNA cleavage. Biochem. Biophys. Res. Commun. 334, 1191–1197.

Matzke, A.J., and Matzke, M.A. (1998). Position effects and epigenetic silencing of plant transgenes. Curr. Opin. Plant Biol. 1, 142–148.

Messing, J., and Dooner, H.K. (2006). Organization and variability of the maize genome. Curr. Opin. Plant Biol. 9, 157–163.

Messing, J., Bharti, A.K., Karlowski, W.M., Gundlach, H., Kim, H.R., Yu, Y., Wei, F., Fuks, G., Soderlund, C.A., Mayer, K.F., and Wing, R.A. (2004). Sequence composition and genome organization of maize. Proc. Natl. Acad. Sci. USA 101, 14349–14354.

Mezard, C. (2006). Meiotic recombination hotspots in plants. Biochem. Soc. Trans. 34, 531-534.

Nelson, O.E. (1962). The waxy locus in maize. I. Intralocus recombination frequency estimates by pollen and by conventional analysis. Genetics 47, 737–742.

Nelson, O.E. (1968). The waxy locus in maize. II. The location of the controlling element alleles. Genetics 60, 507–524.

Okagaki, R.J., and Weil, C.F. (1997). Analysis of recombination sites within the maize waxy locus. Genetics 147, 815–821.

Okuzaki, A., and Toriyama, K. (2004). Chimeric RNA/DNA oligonucleotide-directed gene targeting in rice. Plant Cell Rep. 22, 509–512.

Osborne, B.I., Wirtz, U., and Baker, B. (1995). A system for insertional mutagenesis and chromosomal rearrangement using the Ds transposon and Cre-lox. Plant J. 7, 687–701.

Ow, D.W. (2007). GM maize from site-specific recombination technology, what next? Curr. Opin. Biotechnol. 18, 115–120.

Paques, F., and Haber, J.E. (1999). Multiple pathways of recombination induced by double-strand breaks in Saccharomyces cerevisiae. Microb. Mol. Biol. Rev. 63, 349–404.

Patterson, G.I., Kubo, K.M., Shroyer, T., and Chandler, V.L. (1995). Sequences required for paramutation of the maize b gene map to a region containing the promoter and upstream sequences. Genetics 140, 1389–1406.

Petes, T.D. (2001). Meiotic recombination hot spots and cold spots. Nat Rev Genet 2, 360–369.

Petes, T.D., Malone, R.E., and Symington, L.E. (1991). Recombination in yeast. In The Molecular and Cellular Biology of the Yeast Saccharomyces: Genome Dynamics, Protein Synthesis and Energetics. (Cold Spring Harbor, NY: Cold Spring Harbor Laboratory Press), pp. 407–521.

Puchta, H. (2002). Gene replacement by homologous recombination in plants. Plant Mol. Biol. 48, 173–182.

Puchta, H. (2003). Towards the ideal GMP: homologous recombination and marker gene excision. J. Plant Physiol. 160, 743–754.

Puchta, H., Dujon, B., and Hohn, B. (1993). Homologous recombination in plant cells is enhanced by in vivo induction of double strand breaks into DNA by a site-specific endonuclease. Nucleic Acids Res. 21, 5034–5040.

Ramakrishna, W., Emberton, J., Ogden, M., SanMiguel, P., and Bennetzen, J.L. (2002). Structural analysis of the maize rp1 complex reveals numerous sites and unexpected mechanisms of local rearrangement. Plant Cell 14, 3213–3223.

Ream, T.S., Strobel, J., Roller, B., Auger, D.L., Kato, A., Halbrook, C., Peters, E.M., Theuri, J., Bauer, M.J., Addae, P., Dioh, W., Staub, J.M., Gilbertson, L.A., and Birchler, J.A. (2005). A test for ectopic exchange catalyzed by Cre recombinase in maize. Theor. Appl. Genet. 111, 378–385.

Richter, T.E., Pryor, T.J., Bennetzen, J.L., and Hulbert, S.H. (1995). New rust resistance specificities associated with recombination in the Rp1 complex in maize. Genetics 141, 373–381.

Robbins, T.P., Walker, E.L., Kermicle, J.L., Alleman, M., and Dellaporta, S.L. (1991). Meiotic instability of the R-r complex arising from displaced intragenic exchange and intrachromosomal rearrangement. Genetics 129, 271–283.

Ruiter, R., van den Brande, I., Stals, E., Delaure, S., Cornelissen, M., and D'Halluin, K. (2003). Spontaneous mutation frequency in plants obscures the effect of chimeraplasty. Plant Mol. Biol. 53, 675–689.

Sadder, T., and Weber, G. (2002). Comparison between genetic and physical maps in Zea mays L. of molecular markers linked to resistance against Diatraea spp. Theor. Appl. Genet. 104, 908–915.

SanMiguel, P., and Bennetzen, J.L. (1998). Evidence that a recent increase in maize genome size was caused by the massive amplification of intergene retrotransposons. Ann. Botany 82, 37–44.

SanMiguel, P., Tikhonov, A., Jin, Y.K., Motchoulskaia, N., Zakharov, D., Melake-Berhan, A., Springer, P.S., Edwards, K.J., Lee, M., Avramova, Z., and Bennetzen, J.L. (1996). Nested retrotransposons in the intergenic regions of the maize genome. Science 274, 765–768.

Schultes, N.P., and Szostak, J.W. (1990). Decreasing gradients of gene conversion on both sides of the initiation site for meiotic recombination at the ARG4 locus in yeast. Genetics 126, 813–822.

Smith, J., Bibikova, M., Whitby, F.G., Reddy, A.R., Chandrasegaran, S., and Carroll, D. (2000). Requirements for double-strand cleavage by chimeric restriction enzymes with zinc finger DNA-recognition domains. Nucleic Acids Res. 28, 3361–3369.

Srivastava, V., and Ow, D.W. (2001). Single-copy primary transformants of maize obtained through the co-introduction of a recombinase-expressing construct. Plant Mol. Biol. 46, 561–566.

Srivastava, V., Ariza-Nieto, M., and Wilson, A.J. (2004). Cre-mediated site-specific gene integration for consistent transgene expression in rice. Plant Biotechnol. J. 2, 169–179.

Stack, S.M., and Anderson, L.K. (2002). Crossing over as assessed by late recombination nodules is related to the pattern of synapsis and the distribution of early recombination nodules in maize. Chromosome Res. 10, 329–345.

Stadler, L.J. (1926). The variability of crossing over in maize. Genetics 11, 1–37.

Stadler, L.J., and Neuffer, M.G. (1953). Problems of gene structure. II. Separation of R-r elements (S) and (P) by unequal crossing over. Science 117, 471–472.

Stam, M., Belele, C., Dorweiler, J.E., and Chandler, V.L. (2002a). Differential chromatin structure within a tandem array 100 kb upstream of the maize b1 locus is associated with paramutation. Genes Dev 16, 1906–1918.

Stam, M., Belele, C., Ramakrishna, W., Dorweiler, J.E., Bennetzen, J.L., and Chandler, V.L. (2002b). The regulatory regions required for B' paramutation and expression are located far upstream of the maize b1 transcribed sequences. Genetics 162, 917–930.

Sternberg, N., and Hamilton, D. (1981). Bacteriophage P1 site-specific recombination. I. Recombination between loxP sites. J. Mol Biol 150, 467–486.

Stinard, P.S., Robertson, D.S., and Schnable, P.S. (1993). Genetic isolation, cloning, and analysis of a Mutator-induced, dominant antimorph of the maize amylose extender1 locus. Plant Cell 5, 1555–1566.

Sturtevant, A.H. (1925). The effects of unequal crossing over at the Bar locus in Drosophila. Genetics 10, 117–147.

Sudupak, M.A., Bennetzen, J.L., and Hulbert, S.H. (1993). Unequal exchange and meiotic instability of disease-resistance genes in the Rp1 region of maize. Genetics 133, 119–125.

Sun, Q., Collins, N.C., Ayliffe, M., Smith, S.M., Drake, J., Pryor, T., and Hulbert, S.H. (2001). Recombination between paralogues at the Rp1 rust resistance locus in maize. Genetics 158, 423–438.

Thuriaux, P. (1977). Is recombination confined to structural genes on the eukaryotic genome? Nature 268, 460–462.

Timmermans, M.C., Das, O.P., and Messing, J. (1996). Characterization of a meiotic crossover in maize identified by a restriction fragment length polymorphism-based method. Genetics 143, 1771–1783.

Timmermans, M.C., Das, O.P., Bradeen, J.M., and Messing, J. (1997). Region-specific cis- and trans-acting factors contribute to genetic variability in meiotic recombination in maize. Genetics 146, 1101–1113.

Tulsieram, L., Compton, W.A., Morris, R., Thomas-Compton, M., and Eskridge, K. (1992). Analysis of genetic recombination in maize populations using molecular markers. Theor. Appl. Genet. 84, 65–72.

Vergunst, A.C., and Hooykaas, P.J. (1998). Cre/lox-mediated site-specific integration of Agrobacterium T-DNA in Arabidopsis thaliana by transient expression of cre. Plant Mol. Biol. 38, 393–406.

Walker, E.L., Robbins, T.P., Bureau, T.E., Kermicle, J., and Dellaporta, S.L. (1995). Transposon-mediated chromosomal rearrangements and gene duplications in the formation of the maize R-r complex. EMBO journal 14, 2350–2363.

Wang, C.J., Harper, L., and Cande, W.Z. (2006). High-resolution single-copy gene fluorescence in situ hybridization and its use in the construction of a cytogenetic map of maize chromosome 9. Plant Cell 18, 529–544.

Wang, Q., and Dooner, H.K. (2006). Remarkable variation in maize genome structure inferred from haplotype diversity at the bz locus. Proc. Natl. Acad. Sci. USA 103, 17644-17649.

Webb, C.A., Richter, T.E., Collins, N.C., Nicolas, M., Trick, H.N., Pryor, T., and Hulbert, S.H. (2002). Genetic and molecular characterization of the maize rp3 rust resistance locus. Genetics 162, 381–394.

Wessler, S., and Varagona, R. (1985). Molecular basis of mutations at the waxy locus of maize: correlation with the fine structure genetic map. Proc. Natl. Acad. Sci. USA 82, 4177–4181.

Williams, C.G., Goodman, M.M., and Stuber, C.W. (1995). Comparative recombination distances among Zea mays L. inbreds, wide crosses and interspecific hybrids. Genetics 141, 1573–1581.

Woody, S.T., Austin-Phillips, S., Amasino, R.M., and Krysan, P.J. (2007). The WiscDsLox T-DNA collection: an Arabidopsis community resource generated by using an improved high-throughput T-DNA sequencing pipeline. J. Plant Res 120, 157–165.

Wright, D.A., Townsend, J.A., Winfrey, R.J., Jr., Irwin, P.A., Rajagopal, J., Lonosky, P.M., Hall, B.D., Jondle, M.D., and Voytas, D.F. (2005). High-frequency homologous recombination in plants mediated by zinc-finger nucleases. Plant J. 44, 693–705.

Wu, J., Kandavelou, K., and Chandrasegaran, S. (2007). Custom-designed zinc finger nucleases: What is next? Cell. Mol. Life Sci.1420–682X.

Xiao, Y.L., and Peterson, T. (2000). Intrachromosomal homologous recombination in Arabidopsis induced by a maize transposon. Mol. Gen. Genet. 263, 22–29.

Xiao, Y.L., Li, X., and Peterson, T. (2000). Ac insertion site affects the frequency of transposon-induced homologous recombination at the maize p1 locus. Genetics 156, 2007–2017.

Xu, X., Hsia, A.P., Zhang, L., Nikolau, B.J., and Schnable, P.S. (1995). Meiotic recombination break points resolve at high rates at the 5′ end of a maize coding sequence. Plant Cell 7, 2151–2161.

Yandeau-Nelson, M.D., Nikolau, B.J., and Schnable, P.S. (2006a). Effects of trans-acting genetic modifiers on meiotic recombination across the a1-sh2 interval of maize. Genetics 174, 101–112.

Yandeau-Nelson, M.D., Xia, Y., Li, J., Neuffer, M.G., and Schnable, P.S. (2006b). Unequal sister chromatid and homolog recombination at a tandem duplication of the A1 locus in maize. Genetics 173, 2211–2226.

Yandeau-Nelson, M.D., Zhou, Q., Yao, H., Xu, X., Nikolau, B.J., and Schnable, P.S. (2005). MuDR transposase increases the frequency of meiotic crossovers in the vicinity of a Mu insertion in the maize a1 gene. Genetics 169, 917–929.

Yao, H., and Schnable, P.S. (2005). Cis-effects on meiotic recombination across distinct a1-sh2 intervals in a common Zea genetic background. Genetics 170, 1929–1944.

Yao, H., Zhou, Q., Li, J., Smith, H., Yandeau, M., Nikolau, B.J., and Schnable, P.S. (2002). Molecular characterization of meiotic recombination across the 140-kb multigenic a1-sh2 interval of maize. Proc. Natl. Acad. Sci. USA 99, 6157–6162.

Zeng, Z., and Sachs, M.M. (1994). Intragenic recombination among alleles of the Adh1 gene in maize. Maydica 39, 265–272.

Zhang, W., Subbarao, S., Addae, P., Shen, A., Armstrong, C., Peschke, V., and Gilbertson, L. (2003). Cre/lox-mediated marker gene excision in transgenic maize (Zea mays L.) plants. Theor. Appl. Genet. 107, 1157–1168.

Zhu, T., Mettenburg, K., Peterson, D.J., Tagliani, L., and Baszczynski, C.L. (2000). Engineering herbicide-resistant maize using chimeric RNA/DNA oligonucleotides. Nat Biotechnol. 18, 555–558.

Zhu, T., Peterson, D.J., Tagliani, L., St Clair, G., Baszczynski, C.L., and Bowen, B. (1999). Targeted manipulation of maize genes in vivo using chimeric RNA/DNA oligonucleotides. Proc Natl Acad Sci U S A 96, 8768–8773.

Paramutation: Heritable *in Trans* Effects

Maike Stam and Marieke Louwers

Abstract Paramutation is the heritable transfer of epigenetic information from one allele of a gene to another allele of the same gene. In general, the consequence of this *trans*-communication is a change in gene expression. Paramutation has been observed in plants, fungi and mammals, but is most extensively studied in maize thanks to the long-standing history of maize genetics. For decades, paramutation has been a mystery, but recent progress has shed light on the mechanisms underlying this phenomenon. The identification of MOP1 as an RNA-dependent RNA polymerase shows that RNA plays a crucial role in the *trans*-inactivation process. RNA however appears not the only player in the paramutation process. In this chapter, potential mechanisms will be discussed in light of characteristics that the various paramutation phenomena have in common.

1 Introduction

Gene expression is regulated through DNA sequences located *in cis* and *in trans* and by genetic as well as epigenetic mechanisms. Epigenetic regulation involves heritable changes in gene expression that occur without a change in DNA sequence. Epigenetic mechanisms, such as DNA methylation, histone modifications and the incorporation of histone variants are fundamental for the regulation of eukaryotic gene expression, and thus essential for normal growth and development of multi-cellular organisms. Regulation of gene expression via the binding of proteins to *cis*-acting sequence elements is well studied. Gene regulation *in trans*, i.e. the control of gene expression by sequences on another chromosome, is however less

M. Stam
Swammerdam Institute for Life Sciences, University of Amsterdam, The Netherlands
m.e.stam@uva.nl

M. Louwers
Swammerdam Institute for Life Sciences, University of Amsterdam, The Netherlands
m.l.d/l (.nl) cm l-d louwers@uva.nl
Current address: Crop Design n.v., Technologiepart, 3,9052
zwynaarde, Belgium

J.L. Bennetzen and S. Hake (eds.), *Maize Handbook - Volume II: Genetics and Genomics,*   405
© Springer Science + Business Media LLC 2009

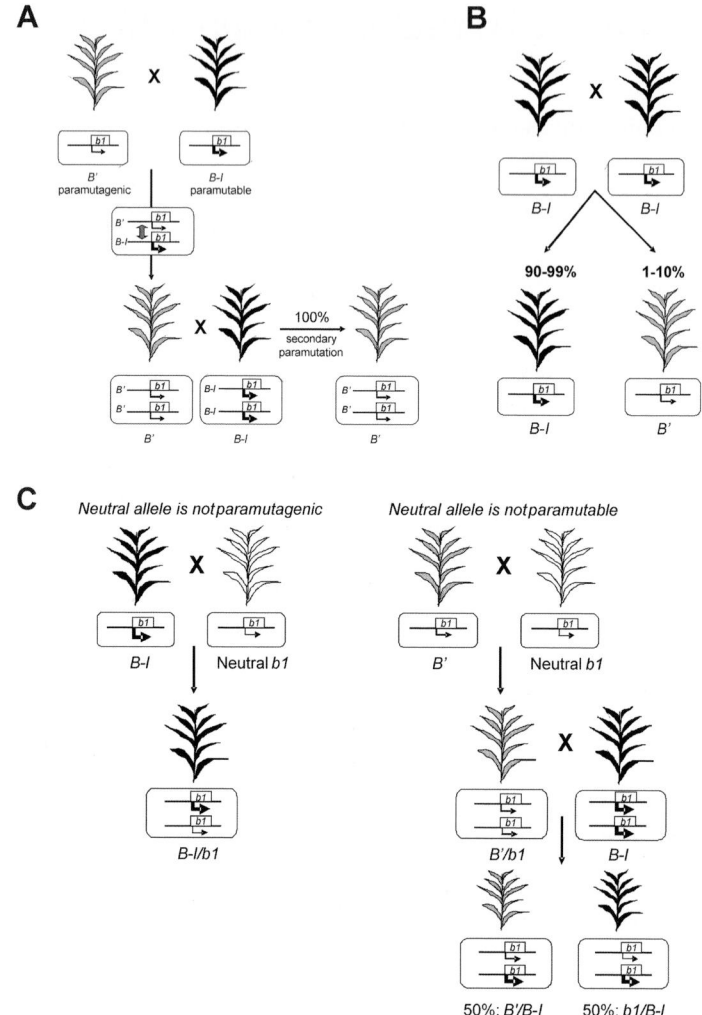

Fig. 1 Paramutation explained by cartoons exemplifying paramutation at the maize *b1* locus. **A.** A maize plant carrying the paramutagenic, inducing *B'* allele (giving rise to a light pigmented plant) is crossed to a plant containing the paramutable, sensitive *B-I* allele (giving rise to a dark pigmented plant). When combined in one nucleus, the paramutagenic *B'* allele *trans*-interacts with the paramutable *B-I* allele, heritably downregulating the expression of the paramutable allele to a low *B'* expression level. As a result, the F1 plants are light pigmented. Often, as is the case for *b1* paramutation, once the paramutable allele is down regulated, it displays secondary paramutation; it has become paramutagenic itself. Consequently, crosses between the F1 and a plant carrying the paramutable *B-I* allele only yield light pigmented progeny. **B.** Spontaneous paramutation. The paramutable *B-I* state can spontaneously change into the paramutagenic *B'* state with a certain frequency. As a result, 1-10% of the progeny of *B-I* plants carries the *B'* allele. **C.** Neutral alleles. Neutral alleles do not participate in paramutation; they are neither paramutagenic nor paramutable. When combined with a paramutable *B-I* allele, a neutral b1 allele does not change the expression level of *B-I*. The inheritance essentially follows the rules of Mendel. When combined with a paramutagenic *B'* allele, a neutral allele does not get paramutated and does not display secondary paramutation. A neutral allele can display any expression level. In this figure, we envisage a neutral allele with a very low expression level.

well examined, but there is increasing evidence that such *trans*-regulatory systems are important in higher eukaryotes, including plants, flies and mammals (Lee et al., 2004; Spilianakis et al., 2005; Bacher et al., 2006; Lee and Wu, 2006; Lomvardas et al., 2006; Xu et al., 2006; Chandler, 2007; Rinn et al., 2007).

Paramutation combines epigenetic and *in trans* regulation of gene expression and is defined as a mitotically and meiotically heritable change in gene expression induced by *trans*-interactions between homologous alleles (Figure 1; Chandler and Stam, 2004; Stam and Scheid, 2005). This change in gene expression is not caused by a change in DNA sequence, but rather a change in DNA methylation (for example see Meyer et al., 1993; Sidorenko and Peterson, 2001; Walker and Panavas, 2001; Rassoulzadegan et al., 2002), and/or chromatin structure (van Blokland et al., 1997; Chandler et al., 2000; Stam et al., 2002) and is therefore an epigenetic phenomenon. Although not known as an epigenetic phenomenon at that time, Alexander Brink chose the term "paramutation" to indicate that the observations were reminiscent of, but not the same as typical genetic mutations (Brink, 1958). Paramutation occurs at a much higher frequency than genetic mutations and is potentially reversible. Paramutation phenomena have been observed in plants, fungi and mammals (van West et al., 1999; Rassoulzadegan et al., 2002; Chandler and Stam, 2004; Rassoulzadegan et al., 2006), but have been studied most extensively in maize. The potential roles of paramutation are a source of speculation, but depending on the phenomenon, it could serve diverse functions such as the defense against foreign DNA, creating a new balance between chromosomes upon polyploidization or hybridization, and providing a heritable mechanism to adapt to environmental changes (reviewed in Chandler and Stam, 2004).

Lately, paramutation has attracted a lot of attention, primarily as a result of two publications (Alleman et al., 2006; Rassoulzadegan et al., 2006). The first paper reported the cloning of a gene affecting paramutation, a crucial step forward in the process of unraveling the mechanisms underlying paramutation. Rassoulzadegan et al. (2002, 2006) showed that paramutation occurs in mice. This indicates that paramutation-like gene regulatory systems are conserved throughout higher eukaryotic evolution, underscoring its importance. In this chapter, we will describe the characteristics that the various paramutation phenomena have in common and discuss potential mechanistic models in the context of these characteristics.

2 Paramutation: Terminology and Common Characteristics

The numerous reported paramutation phenomena share many features, although they also exhibit dissimilarities (extensively described in Louwers et al., 2005). To clarify what paramutation entails, common characteristics of the various paramutation phenomena and definitions associated with paramutation are discussed in this section.

2.1   Alleles Involved in Paramutation

The described paramutation phenomena mainly occur between allelic sequences, but as a result of the use of transgenes an increasing number of cases involve non-allelic homologous sequences. In this review, for simplicity, the term 'alleles' will be used to indicate the affected sequences. There are two types of alleles required for the paramutation process, paramutagenic (inducing) and paramutable (sensitive) alleles. When combined in one nucleus, the two alleles communicate *in trans* and as a result the expression level of the paramutable allele is changed by the paramutagenic allele (illustrated in Figure 1A using paramutation at the maize *b1* locus as an example). The paramutable allele becomes a paramutated allele and is often marked with a prime to indicate that the allele is derived from its paramutable counterpart (for example *B'*).

Most paramutable alleles become paramutagenic upon paramutation; they show secondary paramutation (Figure 1A; e.g. Bateson and Pellew, 1915; Hagemann and Berg, 1978; Meyer et al., 1993; Patterson et al., 1993; Hollick et al., 1995; Rassoulzadegan et al., 2002). In general, the alleles displaying secondary paramutation have the exact same DNA sequence and sequence organization as their paramutagenic counterpart; they are epialleles and their epigenetic state determines whether they are paramutable or paramutagenic. This for instance holds for the alleles involved in *b1* paramutation; upon paramutation, the paramutable *B-I* epiallele changes into the paramutagenic *B'* epiallele (Stam et al., 2002). Some paramutable alleles only become weakly paramutagenic (Brown and Brink, 1960), for others secondary paramutation has not been analyzed or reported (see table in Chandler and Stam, 2004), or does not occur (Park et al., 1996; Hatada et al., 1997; Luff et al., 1999). In all paramutation(-like) systems displaying weak or no secondary paramutation, the paramutable alleles, although possessing sequence homology, have a different sequence organization than their paramutagenic partner (Walker and Panavas, 2001; Kermicle et al., 1995; Panavas et al., 1999). We therefore postulate that those paramutable alleles lack the features required to become paramutagenic.

A number of paramutable alleles display what is called spontaneous paramutation (Figure 1B). Their epigenetic state is inherently unstable and spontaneously changes into the paramutated state at a specific frequency (Bateson and Pellew, 1915; Coe, 1959; Meyer et al., 1993; Hollick et al., 1995). The frequency with which this occurs varies depending on the paramutable allele.

The number of paramutagenic and paramutable alleles reported up to now is low. In fact, most alleles of a gene are neither paramutagenic nor paramutable; in the context of paramutation such alleles are called neutral alleles (Figure 1C). The actual occurrence of alleles involved in paramutation is probably more widespread, but might have gone unnoticed. Most reported paramutation phenomena involve changes in pigment or drug resistance. The discovery of these phenomena was facilitated by the visibility of the affected phenotypes, but also by the fact that downregulation of the expression of pigmentation or antibiotic resistance genes is not lethal to the plant. Paramutation phenomena affecting genes of which the expression is vital to an organism might have been overlooked. It is to be expected that the current availability of powerful genome-wide approaches may help to reveal the existence of many more sequences sensitive to paramutation.

2.2   Three Mechanistic Aspects

Paramutation phenomena are characterized by three important mechanistic aspects: 1) gene expression, 2) paramutagenicity and paramutability, and 3) epigenetic memory. Gene expression concerns the expression level of the alleles involved in paramutation (Figure 2A). In general, the paramutagenic allele displays a low, and the paramutable allele a high expression level. A neutral allele can display any expression level. Paramutagenicity refers to the ability of a paramutagenic allele to paramutate a paramutable allele (Figure 2B); paramutability refers to the ability of a paramutable allele to undergo paramutation. Paramutagenicity and paramutability are not necessarily the same as epigenetic memory. Epigenetic memory represents the epigenetic marks that specify the heritability of a specific epigenetic state, such as the paramutagenic state (Figure 2C). These epigenetic marks are likely to involve DNA modifications and/or a particular chromatin structure (Henderson and Jacobsen, 2007; Martin and Zhang, 2007). Epigenetic memory can be very stable. In such case, the epigenetic state of for example a paramutagenic allele does not readily change into a paramutable state. Multiple generations in the presence of a mutation affecting paramutation are needed to erase the epigenetic marks defining the paramutagenic state. In the case of an unstable epigenetic memory, a paramutagenic state will readily change into a paramutable state in the presence of a mutation affecting paramutation.

Mutations affecting paramutation (Dorweiler et al., 2000; Hollick and Chandler, 2001; Hollick et al., 2005; Hale et al., 2007) have shown that the three aspects can be mechanistically separated from each other. For example, in maize plants homozygous for the *mop1* mutation, the expression level of the paramutagenic *B'* allele is enhanced nearly up to the level displayed by the paramutable *B-I* allele (Dorweiler et al., 2000). In the same plants, the *B'* allele lost its paramutagenicity, but not its epigenetic memory. Upon removal of the *mop1* mutation by crossings, the *B'* allele behaves as *B'* again: it is expressed at a low level and can paramutate *B-I*. This indicates that, although the expression level and paramutagenicity of *B'* were altered in the *mop1* mutant background, the *B'* allele kept its epigenetic memory determining the *B'* epigenetic state. In the presence of another mutation, *rmr1*, the expression level of the paramutagenic maize *Pl'* allele is elevated, but it can still induce paramutation (Hale et al., 2007). Similar observations have been made with the *rmr6-1* mutation (Hollick et al., 2005). The use of mutations affecting paramutation makes it possible to experimentally separate the three aspects of the paramutation process, which is instrumental in unravelling the mechanisms underlying paramutation.

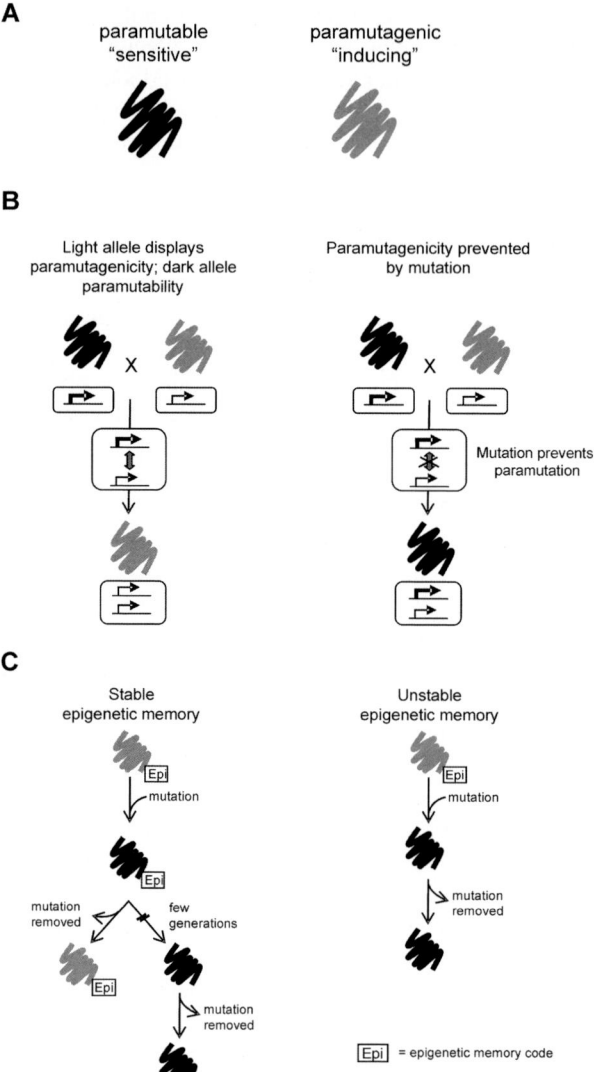

Fig. 2 Mechanistic aspects of paramutation. **A.** Gene expression. Paramutation most often involves a change in gene expression. The paramutable, sensitive allele is usually high expressed (black symbol), such as the maize *B-I* allele. The paramutagenic, inducing allele is mostly low expressed (grey symbol), such as the *B'* allele. **B.** Paramutagenicity and paramutability. Paramutagenicity refers to the ability of a paramutagenic allele to change the expression level of a sensitive allele. Paramutability refers to the ability of a paramutable allele to undergo paramutation. In the left panel the wild-type situation is shown. In the right panel, a mutation preventing paramutation is present in the F1. As a result, the light expressed allele lost its paramutagenicity. **C.** Epigenetic memory. Epigenetic memory represents the epigenetic marks specifying the heritability of a specific epigenetic state, paramutagenic or paramutable. The epigenetic memory can be stable or unstable. In this example, the paramutagenic, low expressed allele becomes high expressed and looses its paramutagenicity in the presence of a muta-tion affecting both aspects. In case of a stable epigenetic memory, once the mutation is removed the

2.3 Transcriptional Gene Silencing

In general, paramutation results in transcriptional gene silencing (e.g. Meyer et al., 1993; Patterson et al., 1993; Hollick et al., 2000; Mittelsten Scheid et al., 2003), but it can also affect other processes, such as transposition (Harrison and Carpenter, 1973; vanHouwelingen et al., 1999) or recombination (Rassoulzadegan et al., 2002). In this review, for simplicity, the term 'gene expression' will be used to indicate the affected process.

2.4 Sequence Features: Repeated Sequences

In most paramutation phenomena repeated sequences are required for the paramutation process (Kermicle et al., 1995; English and Jones, 1998; Luff et al., 1999; van-Houwelingen et al., 1999; Walker and Panavas, 2001; Stam et al., 2002). Repeated sequences are a major trigger for the formation of silenced chromatin (Grewal and Jia, 2007; Zaratiegui et al., 2007). Various transcriptionally downregulated regions in the genome, such as centromeres, telomeres and other heterochromatic regions are packed with repeated sequences. Two types of repeats have been shown to be required for paramutation: inverted repeats (IRs) and direct repeats (DRs).

Some paramutation-like phenomena require sequences in an inverted orientation. Two *dTph1* transposons organized in an IR are necessary for a paramutation-like *trans*-interaction that results in a novel transposition mechanism in petunia plants (vanHouwelingen et al., 1999). Furthermore, all examined paramutable *r1* alleles (16) contain two *r1* genes in an IR organization (Walker and Panavas, 2001).

Other paramutation systems require directly repeated sequences. DRs are needed for *b1* and *r1* paramutation in maize, and *SPT::Ac* paramutation in tobacco (Kermicle et al., 1995; English and Jones, 1998; Stam et al., 2002). With *b1* paramutation, multiple DRs, situated ~100kb upstream of the *b1* coding region, are required for *b1* paramutagenicity and paramutability. *B'* and *B-I* contain seven copies of an 853 bp sequence, while neutral *b1* alleles contain only one. Likewise, although all paramutable *r1* alleles contain IRs, multiple, directly repeated *r1* genes are needed for *r1* paramutagenicity. In both cases, a decrease in copy number of repeats resulted in a decreased paramutagenicity (Kermicle et al., 1995; Panavas et al., 1999; Stam et al., 2002), indicating a crucial role for repeats in paramutation. Whereas the *b1* repeats are relatively small and do not contain coding sequences, the repeated DNA fragments at the *r1* locus are at least 10kb

---

Fig. 2 (continued) allele will revert to its original expression state and is paramutagenic again. The epigenetic memory can however be lost after a few generations in the presence of a mutation affecting paramutation. In that case, once the mutation is removed, the allele will stay in the high expression state, because it cannot 'remember' what the original expression state used to be. In case of an unstable epigenetic memory, already after one generation in the presence of a mutation affecting paramutation, the paramutagenic allele reverts to a high expression state and loses its paramutagenicity

and contain the *r1* coding region and flanking regions. Directly repeated sequences are also observed in other paramutation systems (Sidorenko and Peterson, 2001; Stokes et al., 2002; Mittelsten Scheid et al., 2003; E. Richards, personal communication), and although their role has not yet been reported, we like to speculate that they are somehow involved in the paramutation process. For example, the data reported for *p1* paramutation are consistent with DRs being required for paramutation. The *p1* coding region of the paramutable *P1-rr* allele is flanked on both sides by an almost perfect DR of a 1.2 kb sequence (Athma et al., 1992; Das and Messing, 1994). The neutral *P1-wr* allele (Sidorenko and Peterson, 2001) contains multiple copies of this 1.2 kb sequence, but these copies are separated from each other by several kb (Chopra et al., 1996; Chopra et al., 1998). These data are consistent with the hypothesis that localized repetitiveness of the 1.2 kb sequence is required for *p1* paramutability.

Importantly, merely the presence of a repeat sequence is not sufficient to cause paramutation. The *FWA* gene in *Arabidopsis* contains two small direct repeats at the 5′ end of the gene (one of 38 bp, one of 198 bp), and these repeats can exist in different epigenetic states, resulting in two epialleles, *FWA* and *fwa* (Soppe et al., 2000). At the *FWA* epiallele, gene expression is silenced and the repeats are DNA methylated, at the *fwa* epiallele the gene is active and the repeats are DNA hypomethylated. When combined in one nucleus, the two *FWA* epialleles do however not affect each other's expression level (Soppe et al., 2000; Chan et al., 2006a). The data on *b1* and *r1* paramutation (Kermicle et al., 1995; Panavas et al., 1999; Stam et al., 2002) suggest that the number of *FWA* repeats is too low to cause paramutation. We propose that also the small size of the DRs at the *FWA* locus hampers the occurrence of paramutation. Specific sequence features might in addition be required to allow paramutation. In line with this idea, a transgenic approach, in which various distal *p1* promoter sequences were tested for paramutagenicity, demonstrated that only transgene loci containing the '1.2 kb fragment' were able to induce paramutation of the endogenous *P-rr* allele (Sidorenko and Peterson, 2001). Remarkably, the 1.2 kb sequence is not only required for paramutagenicity, but also acts as an enhancer at the *p1* locus (Sidorenko et al., 1999; Sidorenko et al., 2000). Similarly, the 853 bp repeats required for *b1* paramutation are necessary for *b1* enhancer activity (Stam et al., 2002). This raises an intriguing question for future research: is there a link between enhancer sequences and paramutation?

Repeated sequences might not always be required for paramutation; some systems appear to lack repeated sequences (*A1* and L91 transgenes in plants, *loxP* and recombinant *Ins2* alleles in mouse; Meyer et al., 1993; Duvillie et al., 1998; Qin and von Arnim, 2002; Rassoulzadegan et al., 2002; Qin et al., 2003).

2.5   DNA Methylation and Chromatin Structure

Epigenetic regulation of gene expression is the outcome of an intricate interplay between various epigenetic mechanisms such as DNA methylation, histone modifications and the incorporation of histone variants (Tariq and Paszkowski, 2004; Henikoff and Ahmad, 2005; D'Alessio and Szyf, 2006; Klose and Bird, 2006). Repressed chromatin is usually characterized by DNA hypermethylation and histone modifications such as histone H3 lysine 9 and lysine 27 methylation (H3K9me, H3K27me), while active chromatin is recognized by DNA hypomethylation and histone modifications such as histone H3 lysine 4 methylation and H3 and H4 acetylation.

In most paramutation systems, the transcriptionally silenced paramutagenic allele is DNA hypermethylated, and the transcriptionally active paramutable allele is DNA hypomethylated. When combined in one nucleus, the *trans*-inactivation of the paramutable allele is associated with the acquisition of DNA methylation (Meyer et al., 1993; Eggleston et al., 1995; Forne et al., 1997; Hatada et al., 1997; Walker, 1998; Sidorenko and Peterson, 2001; Walker and Panavas, 2001; Rassoulzadegan et al., 2002; Stam et al., 2002; Mittelsten Scheid et al., 2003). For example, in case of *b1* paramutation, the *B'* repeats, ~100 kb upstream of the *b1* transcription start site, are DNA hypermethylated relative to the *B-I* repeats in all the various tissues and developmental stages examined (R. Bader and M. Stam, unpublished results). When combined in one nucleus, the acquisition of DNA methylation at the *B-I* repeats can already be observed a few days after germination. In the *hygromycin phosphotransferase* (HPT) paramutation system, the acquisition of DNA methylation by the paramutable allele appears a slower process than the accompanying phenotypic change, suggesting that chromatin-based silencing mechanisms precede DNA methylation (Mittelsten Scheid et al., 2003). For a few paramutation systems, no correlation between DNA methylation and paramutation has been observed (English and Jones, 1998; van West et al., 1999; Rassoulzadegan et al., 2006). Importantly, simply a difference in DNA methylation level between two epialleles is not sufficient for paramutation to occur. For example, the hypermethylated, inactivated *Arabidopsis SUPERMAN* allele does not *trans*-inactivate its hypomethylated counterpart (Jacobsen and Meyerowitz, 1997). Similarly, the DNA methylated *FWA* epiallele does not *trans*-inactivate its hypomethylated counterpart (Soppe et al., 2000; Chan et al., 2006a).

What would be the role of DNA methylation in paramutation? In line with the literature, we propose that DNA methylation plays an important role in the heritability of the silenced, paramutagenic epigenetic state (Dieguez et al., 1998; Jones et al., 2001; Martin and Zhang, 2007). This might also explain why paramutation has not yet been described in organisms lacking extensive DNA methylation like *Drosophila, C. elegans, S. pombe* and *S. cerevisiae*.

Epigenetic regulation of gene expression not only involves changes in DNA methylation, but also changes in chromatin structure. For the inactive, paramutagenic maize *b1* and petunia *A1* epialleles it has been shown that they are less accessible to nucleases than their active, paramutable counterparts (vanBlokland et al., 1997; Chandler et al., 2000; Stam et al., 2002). The role of histone modifications in paramutation is the subject of current research. Our recent data indicate that the paramutagenic *B'* and paramutable *B-I* alleles carry a different set of histone modifications at their repeats. In contrast to the DNA methylation, most of these histone modifications are tissue-specifically regulated. They reflect the expression level in the tissue analyzed, rather than the epigenetic state of the alleles (M. Haring and M. Stam, unpublished results). The only exception is the localization of histone H3 lysine 27 dimethylation (H3K27me2) at the repeats. This has been observed in both tissues examined (seedling shoots and husk leaves). We propose that H3K27me2, together with DNA methylation, plays an important role in the epigenetic inheritance of the low *B'* expression state.

2.6   Stability of the Paramutagenic and Paramutable State

The epigenetic states of alleles participating in paramutation can vary between very unstable and extremely stable, and this stability can be influenced amongst others by repetitiveness, zygosity and environmental factors. When thinking about models explaining the various phenomena, the features affecting stability of the various epigenetic states have to be taken into account as well.

The epigenetic state of some paramutable and paramutagenic alleles is very stable (paramutable *P-rr* and *sulf* alleles; paramutagenic *A1*, *b1*, HPT, *p1*, *r1* alleles; Hagemann, 1993; Das and Messing, 1994; Brink and Weyers, 1957; Coe, 1959; Sidorenko and Peterson, 2001; Mittelsten Scheid et al., 2003). Some paramutable and paramutagenic alleles can however spontaneously change into the other epigenetic state (paramutable 'ear rogue', *b1*, *A1*, *Ph-Rh*, and Spr12F *spt* alleles and paramuta-genic *Pl'* allele; Bateson and Pellew, 1915; Coe, 1959; Meyer et al., 1993; Hollick et al., 1995; English and Jones, 1998). Furthermore, a number of alleles also display a series of intermediate epigenetic states, rather than one paramutable and one param-utagenic state (*sulf*, *A1*, *pl1*, and *p1*; Hagemann and Berg, 1978; Hagemann, 1993; Meyer et al., 1993; Hollick et al., 1995; Sidorenko and Peterson, 2001).

The features that determine the stability of a paramutable or paramutagenic state are amongst others repetitiveness (see 2.4), chromosomal location (Hagemann and Berg, 1978; Wisman et al., 1983), environmental conditions and endogenous factors (Meyer et al., 1992), and zygosity (Hagemann and Berg, 1978; Mittelsten Scheid et al., 2003). While some do not, most alleles involved in paramutation contain repeated sequences. The repeats at these alleles vary in size and number. Repeated sequences trigger the formation of silenced chromatin (Grewal and Jia, 2007; Zaratiegui et al., 2007), and we postulate that the size and number of repeats influence allele stability. We expect that the more repeated sequences are present, the more stable the paramutagenic state will be, and the less stable the paramutable state. The paramutable and paramutagenic 7-repeat *B-I* and *B'*, and 5- and 3-repeat recombinant *b1* alleles (Stam et al., 2002) constitute an ideal deletion series to examine the effect of repeat number on allele stability.

Spontaneous paramutation of the petunia *A1* transgene is influenced by environ-mental effects as well as age of the plant (Meyer et al., 1992). Environmental factors like temperature have been shown to influence DNA methylation levels and chromatin structure (Finnegan et al., 2004), and it is highly probable that spontaneous paramutation at the *A1* locus is mediated via DNA methylation and chromatin structure changes as well.

The stability of the epigenetic state can also be affected by the nature or presence of the allele on the homologous chromosome (zygosity). The epigenetic states of the paramutagenic *Pl'*, and paramutable *b1*, *pl1*, *r1*, Spr12F *spt* alleles are less stable when they are homozygous, than when they are heterozygous with a neutral allele or when only one copy of the allele is present (hemizygous; Coe, 1966; Styles and Brink, 1966, 1967; Hollick et al., 1995; English and Jones, 1998; Hollick and Chandler, 1998). The effects of zygosity on the stability of the epigenetic state are more than two-fold and therefore not due to a simple dosage effect of the affected alleles. Remarkably, the reversion of *Pl'* to a higher expression state is only heritable in the presence of a neutral allele, not when the *pl1* allele on the homologous chromosome is lacking. This suggests pairing might be involved in stabilizing the high expression state (Hollick and Chandler, 1998).

The differences in stability of the epigenetic state are reflected in the effect of mutations affecting paramutation. In a homozygous *rmr6* mutant background, *B'*, *Pl'* and *R-st* cannot paramutate their paramutable counterparts and the transcriptional repression displayed by the *B'*, *Pl'* and *R-r'* epigenetic states is lifted (Hollick et al., 2005). The unstable *Pl'* and *R-r'* states can heritably revert to their paramutable states in a homozygous *rmr6* mutant background, while the very stable *B'* state does not. Similarly, the *mop1* mutation prevents establishment of paramutation at *b1*, *pl1* and *r1*, disrupts the maintenance of the low *B'* and *Pl'* expression states, and it can heritably revert *Pl'*, but not *B'*, to its paramutable counterpart (Dorweiler et al., 2000). Except for the stability of their paramutagenic alleles, the *b1* and *pl1* systems are very similar in many respects. Therefore, comparison of the *b1* and *pl1* sequences required for paramutation -once the latter are identified as well- will be very useful in order to determine the features that control the stability of the various epigenetic states.

2.7   Dosage-dependent Paramutation

The frequency and occurrence of paramutation can be influenced by the ploidy level of the organism. *Trans*-inactivation of an active HPT transgene locus occurs in tetraploid, but not in diploid *Arabidopsis* plants (Mittelsten Scheid et al., 2003). In diploid plants, the HPT locus results in resistance to hygromycin. Upon autotetraploidization, transcriptionally active and silenced HPT epialleles were isolated. When combined in a tetraploid background, the silenced epialleles heritably *trans*-inactivated the active epialleles. This paramutation process depends on the tetraploid state. After reduction in ploidy, the silenced epialleles were still stably silenced, but no longer paramutagenic. At the tomato *sulf* locus, ploidy has been shown to affect the frequency of paramutation, rather than the occurrence (Hagemann and Berg, 1978).

A possible reason for differences in paramutation frequency or penetrance in polyploid versus diploid plants might be the more demanding sorting and pairing of multiple homologous chromosomes during meiosis (Weiss and Maluszynska, 2000; Santos et al., 2003). Upon polyploidization, a new balance has to be created between the different chromosomes, which amongst others affects the epigenetic state of certain sequences (Chen, 2007). In autotetraploids for example, although most genes are expressed at a level proportional to the genome copy number, a number of genes show a higher or lower expression level per genome than observed in diploids (Guo et al., 1996; Lee and Chen, 2001).

3 Paramutation Models

The various paramutation phenomena share many features, suggesting they share at least part of the underlying mechanisms. Based on recent progress in the field, two models are proposed to explain paramutation at the molecular level, an RNA model and an RNA-physical interaction model. However, it cannot be excluded that particular paramutation phenomena fit another model better. In this section, the models are described, followed by a discussion of the models using paramutation at the *b1* and other loci as examples. Importantly, neither the production of a paramutagenic RNA nor physical contact would need to occur permanently, provided they last long enough to trigger a heritable change.

3.1 RNA Model

In the RNA model, the *in trans* communication between a paramutagenic and paramutable allele, and the resulting change in gene expression is mediated by a diffusible RNA molecule (Figure 3A). RNAs derived from the paramutagenic allele affect the chromatin structure and transcriptional activity of the paramutable allele. In this RNA model, RNAs are produced from the paramutagenic allele, but not from the paramutable allele.

RNAs, and especially small interfering RNAs (siRNAs), have been implicated in numerous regulatory processes and epigenetic phenomena, and recent evidence indicates they are also involved in paramutation (van West et al., 1999; Alleman et al., 2006; Rassoulzadegan et al., 2006). There are several different siRNA pathways, which display dissimilar outcomes, varying from heterochromatin formation to post-transcriptional gene silencing by RNA decay or translation inhibition (Mello and Conte, 2004; Matzke and Birchler, 2005; Vaucheret, 2006; Grewal and Elgin, 2007; Zaratiegui et al., 2007). Paramutation usually results in transcriptional gene silencing. Thus, RNA-mediated heterochromatin formation models would be most applicable to paramutation. In these models RNAs derived from the target locus are turned into double-stranded RNA by RNA-dependent RNA polymerase and processed by Dicer into siRNAs. The latter contribute to heterochromatin formation via RNA-directed DNA methylation (RdDM) and RNA-directed histone modifications (H3K9 and H3K27 methylation; Ting et al., 2005; Chan et al., 2006b; Liu et al., 2007; Matzke et al., 2007; Zaratiegui et al., 2007). As proposed by Zaratiegui et al.

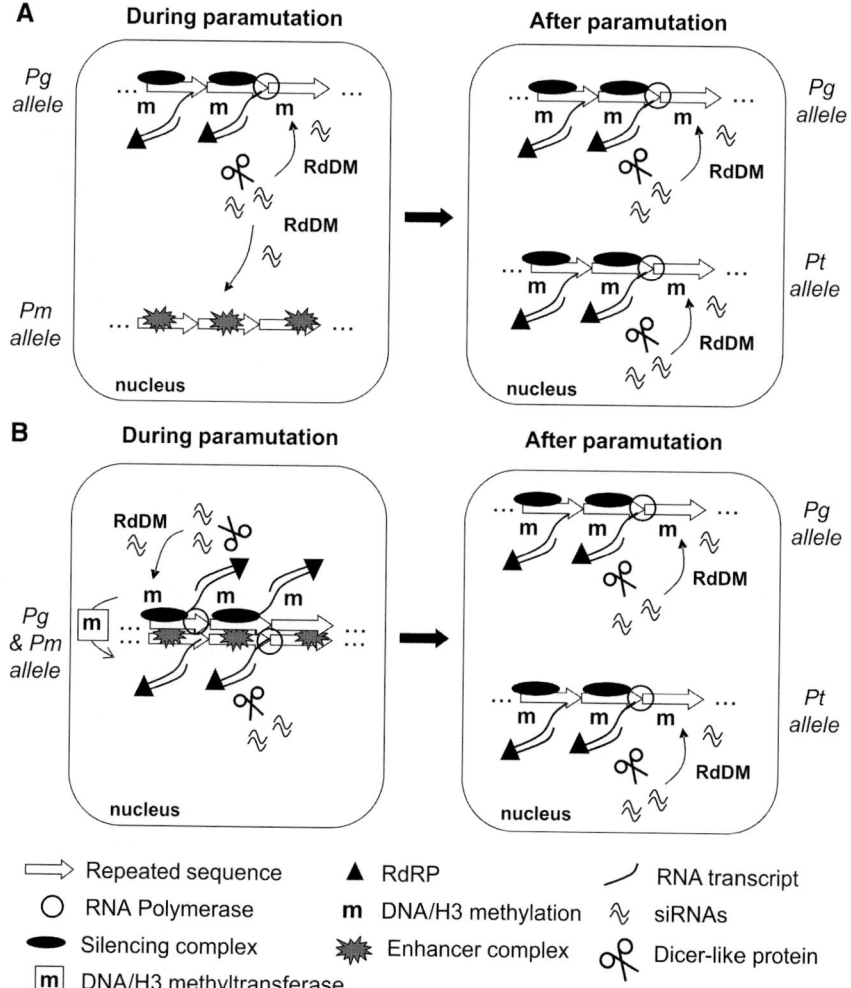

Fig. 3 Models. In this figure, the sequences required for paramutation are depicted as directly repeated sequences (arrows). In the paramutagenic allele they have a repressed chromatin structure, in the paramutable allele they are in an active chromatin state and act as an enhancer. **A.** RNA model. The sequences required for paramutation are transcribed exclusively from the paramutagenic (*Pg*) allele and double-stranded RNA (dsRNA) is produced due to the activity of an RNA-directed RNA polymerase (RdRP). A Dicer-like protein cleaves the dsRNA into siRNAs which then act as diffusible factors and influence the chromatin structure of the paramutable (*Pm*) allele via RNA-dependent DNA methylation (RdDM) and RNA-directed histone modifications. Once the chromatin structure of the paramutable allele is repressed, enhancer complexes can no longer remain associated and silencing complexes bind instead. The allele becomes paramutated (*Pt*) and in this example also paramutagenic. It consequently produces RNA, which is processed into siRNAs. Repeat-derived siRNAs not only affect paramutable alleles, but also reinforce the silent state of the paramutagenic allele. **B.** RNA-physical interaction model. In this model, RNA as well as physical pairing between homologous sequences is required for *in trans* inactivation. Both the paramutable and paramutagenic alleles are transcribed and the RNA processed into siRNAs by a Dicer-like enzyme. However, only the paramutagenic, repressed chromatin state is susceptible to silencing by siRNAs. Physical pairing between the paramutagenic and paramutable allele results in an exchange of proteins that alter the paramutable allele into a paramutated allele. The chromatin state of the paramutated allele is susceptible to the silencing effect of siRNAs.

(2007), the heterochromatinization process might be chromosome-associated, like observed for fission yeast (Grewal and Elgin, 2007). A paramutation-like phenomenon consistent with the RNA model is that reported by Van West et al (1999). This *Phytophtera infestans* system was indicated to involve transcriptional silencing and a sequence-specific, diffusible factor, which is likely to be RNA.

3.2   RNA-Physical Interaction Model

In the RNA-physical interaction model, the *in trans* communication between the paramutagenic and paramutable allele is mediated by RNAs as well as pairing between homologous DNA sequences (Figure 3B; Chandler and Stam, 2004; Stam and Scheid, 2005). In this model, the transfer of the transcriptionally inactive state of the paramutagenic locus onto the paramutable counterpart is dependent on both pairing and specific RNAs, most likely siRNAs. The RNAi machinery has recently been implicated in various aspects of higher-order chromatin organization in the nucleus. For example, the clustering of the heterochromatic telomeres in *S. pombe* (Hall et al., 2003) and the formation of higher-order insulator complexes in Drosophila is disrupted in RNAi mutants (Lei and Corces, 2006). It has therefore been proposed that RNAs act as a 'glue' to promote the clustering of heterochromatic regions into higher-order structures (Grewal and Moazed, 2003). Other precedents of the RNA-physical interaction model are the *trans*-interactions mediated by Polycomb group (PcG) proteins (Grimaud et al., 2006). It has been shown that PcG-like proteins can mediate physical *in trans* interactions (Lavigne et al., 2004) and pairing-dependent silencing in *Drosophila* (Bantignies et al., 2003). Importantly, the RNAi machinery is shown to be required for PcG-mediated silencing, and more specifically for the maintenance of long-range physical interactions (Grimaud et al., 2006). In conclusion, in the RNA-physical interaction model, *trans*-interactions between paramutagenic and paramutable alleles are mediated by both the RNAi machinery and proteins involved in pairing, e.g. PcG-like proteins. Chromosome Conformation Capture (3C; Dekker et al., 2002; Tolhuis et al., 2002) and Fluorescence In Situ Hybridisation (FISH; Grimaud et al., 2006) experiments need to be performed to address the role of physical *in trans* interactions in paramutation.

3.3   Paramutation Models in the Context of Repeated DNA

Repeated sequences, which are required for paramutation in most paramutation phenomena, are compatible with both models proposed. Transcription of repeats, whether in a direct or inverted orientation, can lead to an efficient production of double-stranded RNA (dsRNA), whereby in the case of direct repeats the action of RdRP is required (Martienssen, 2003; Slotkin and Martienssen, 2007; Zaratiegui et al., 2007). After processing of the dsRNAs by Dicer, the resulting siRNAs mediate transcriptional silencing. On the other hand, repeated sequences have been shown to physically pair more often than single copy sequences. In *Arabidopsis*, transgenic *lac* operator arrays associate more often with each other than average euchromatic regions, and the same is observed for an inactive transgenic multicopy HPT locus (Pecinka et al., 2005). Intriguingly, DNA hypermethylation of repeated sequences is shown to increase the frequency of pairing between homologous sequences (Watanabe et al., 2005). In most paramutation systems, paramutagenic alleles have been shown to be hypermethylated (see 2.5), and might therefore display an enhanced pairing frequency, possibly leading to paramutation.

3.4 Paramutation Models in the Context of Recent Data

For long it was only possible to wildly speculate about the mechanisms underlying paramutation. The cloning of the *mop1* mutation and other recent data (Chandler, 2007) has provided a revealing glimpse into the mechanisms involved. *Mediator of paramutation 1* (*mop1*) appeared to encode an RNA-dependent RNA polymerase (RdRP; Alleman et al., 2006; Woodhouse et al., 2006), indicating the involvement of RNA in paramutation. MOP1 is most homologous to the RDR2 protein of *Arabidopsis*; RDR2 is associated with the production of short interfering RNAs (siRNA) that play a role in RNA-dependent DNA methylation of repeats (RdDM; Chan et al., 2004). The *mop1* and *rdr2* mutants do however not behave exactly the same, indicating that MOP1 and RDR2 are functionally divergent to some extent. An *rdr2* mutant for example, exhibits demethylation of the 5S ribosomal DNA (Xie et al., 2004), whereas a *mop1* mutant does not (Dorweiler & Chandler personal communication). Furthermore, in *rdr2* mutants, CNG, and especially CNN DNA methylation is severely reduced (Xie et al., 2004). Such a preference is not obvious when analyzing the DNA methylation pattern at the *b1* repeats in a *mop1* mutant background (R. Bader and M. Stam, unpublished results).

Consistent with the idea that RNA is involved in paramutation, the repeats are transcribed (Alleman et al., 2006) and siRNAs produced (M Arteaga-Vasquez and VL Chandler, personal communication). However, not only the repeats of *B'*, but also those of *B-I* and a neutral *b1* allele are transcribed, at comparable levels (Alleman et al., 2006). In addition, repeat siRNAs are detected from all three alleles (M Arteaga-Vasquez and VL Chandler, personal communication), suggesting that, although RNAs appear required, repeat transcription and siRNAs are not sufficient to establish and/or maintain a paramutagenic state. This finding is reminiscent to results reported by Chan *et al.* (2006), who detected tandem repeat-derived siRNAs from both the silenced and active *FWA* locus. Only the silenced, DNA methylated *FWA* locus could recruit RNA-directed DNA methylation, reinforcing its own silenced state *in cis*. The authors postulate that features of the siRNA-producing locus, such as the DNA methylation level and chromatin structure, determine the susceptibility to siRNA-directed DNA methylation (RdDM). In analogy, we speculate that the chromatin structure of the

B-I hepta-repeat is not susceptible to RdDM, while that of *B'* is. The *B-I* repeats contain a low level of DNA methylation, while those of *B'* are heavily methylated in all tissues analyzed. In addition, the *B'* repeats are H3K27 methylated in both stem and husk tissue (M. Haring and M. Stam, unpublished results). We propose that this tissue-independent DNA and histone methylation makes the *B'* repeats receptive for RdDM and RNA-directed histone modifications, which in turn reinforce the repressed chromatin structure. In this model, the siRNAs derived from the *B'* repeats can contribute to the maintenance of the *B'* repeat DNA methylation pattern and thereby the epigenetic state of the *B'* epiallele, while the repeat siRNAs derived from the *B-I* or neutral *b1* allele cannot.

The previous paragraph mainly discusses the role of siRNAs in the heritable maintenance of the *B'* and *B-I* epigenetic states. How is the *B'* state established? How does the *B'* epiallele change the *B-I* epigenetic state into a *B'* epigenetic state when both epialleles are combined in one nucleus? As discussed above, RNAs are required but not sufficient (Alleman et al., 2006; Chandler, 2007). *In trans* inactivation of *FWA* transgenes requires the *Agrobacterium*-mediated genetic transformation process (Chan et al., 2006a). What are the additional requirements for *b1* paramutation? We propose that pairing, stimulated by the DNA hypermethylated *B'* repeats, is such a requirement (Figure 3B). The physical pairing could allow the transfer of proteins mediating paramutation from the *B'* to the *B-I* allele. The presence of H3K27 methylation at the *B'* repeats suggests that besides RNAs, pairing mediated by PcG-like proteins or other pairing proteins, could be involved. H3K27 methylation has been implicated in both RNAi-dependent heterochromatin formation and PcG protein-mediated chromatin silencing (Peters and Schubeler, 2005; Bantignies and Cavalli, 2006; Weinberg et al., 2006; Liu et al., 2007). Further studies are needed to investigate if this is actually the case. Alternatively, as a result of the different chromatin structure, the *B'* repeat siRNAs obtain specific features that the siRNAs derived from *B-I* and the neutral *b1* allele lack. These features would be required to enable the *B'* siRNAs to affect a paramutable allele *in trans*. Yet another option would be that a specific set of not yet detected *B'* repeat siRNAs are involved in the *trans*-communication leading to a heritable change in gene expression. The presence of these siRNAs, which should only be present in *B'* and not in *B-I* plants, could be limited to a narrow developmental time window, a specific tissue or a set of cells not yet investigated.

The discussion above holds for *b1* paramutation. What about other paramutation systems? Gross and Hollick (2007) examined the occurrence of small RNAs derived from the *pl1* coding region. Similar to the *b1* paramutation system, small RNAs could be detected from all alleles examined, no matter if the allele was paramutagenic, paramutable or neutral. These observations however, do not preclude any models yet. To be able to distinguish between models it is required to examine the properties of the sequences necessary for *pl1* paramutation, which is a subject of current research (Gross and Hollick, 2007). In the *Arabidopsis* HPT tetraploid paramutation system, a change in expression was not observed until the F2 generation, indicating that the paramutable and paramutagenic alleles need to go through meiosis together to allow paramutation.

This suggests the involvement of pairing in paramutation, whereby specific features of pairing during meiosis in a tetraploid system are required. The RNA model seems unlikely in this system, as paramutagenic RNAs are expected to already have an effect in the F1, unless such RNAs are only present late during development, during or close to the meiotic process. Furthermore, this RNA should only be present in tetraploids and not in diploids. Polyploidization is known to provoke changes in gene expression (Chen, 2007). Therefore it is actually possible that paramutagenic RNAs are expressed at different levels or distributed differently in polyploid situations.

4   Concluding Remarks and Future Directions

All paramutation systems have in common that *trans*-interactions between homologous sequences result in heritable changes in epigenetic states. The various systems however also display unique features. Therefore, multiple, slightly different mechanisms might be involved in paramutation. In this chapter we discussed two models, an RNA and an RNA-physical interaction model, but other models might apply as well. To reveal the mechanisms involved, cloning and thorough characterization of the players involved is crucial, also for their effects beyond paramutation. By now, multiple *trans*-acting mutations implicated in various aspects of paramutation have been isolated (Dorweiler et al., 2000; Hollick and Chandler, 2001; Hollick et al., 2005) and excitingly, two of them have been cloned recently (Alleman et al., 2006; Hale et al., 2007) and cloning of others is underway. The complexity and diversity of epigenetic regulation indicates that a view on individual factors is often naïve and shortsighted. It is therefore important that the knowledge on the proteins involved will be combined with the current and future knowledge on their effect on different paramutation phenomena in one and the same organism, such as maize. With such an approach, the common and unique features of the various phenomena can be best utilized and lead to a deeper understanding of the underlying mechanisms.

Paramutation involves changes in DNA methylation and chromatin structure. Although the role of DNA methylation in paramutation is being studied, the role of chromatin structure is still a black box for most paramutation phenomena. Recently, a chromatin immunoprecipitation (ChIP) protocol that is optimized for maize has been published (Haring et al., 2007). This will facilitate the analyses of histone modifications, nucleosome density and other aspects of chromatin structure of paramutagenic and paramutable alleles.

The RNA-physical interaction model discussed in this chapter implicates a role for physical pairing in paramutation. In order to test the involvement of physical interactions we succesfully set up the Chromosome Conformation Capture technique (3C, Dekker et al., 2002; Tolhuis et al., 2002) for maize tissue (Louwers et al., unpublished results). This technique measures physical interactions between chromatin regions and is currently being used to examine the role of long-distance *in cis* interactions in enhancement of *b1* expression (Louwers et al., unpublished results). It will also be used to examine the role of *in trans* interactions in paramutation.

An important aspect for future research is the timing of paramutation. The change from a paramutable into a paramutagenic allele takes place after combining both alleles in one zygote, but does not necessarily occur immediately. Furthermore, the change in epigenetic state most likely involves multiple events such as the change in expression state of the paramutable allele, and the imposition of the epigenetic marks representing the epigenetic memory. Studies on mutations affecting paramutation have shown that these aspects can be separated from each other (Dorweiler et al., 2000; Hollick and Chandler, 2001; Hollick et al., 2005; Hale et al., 2007). In case of *b1* paramutation, the change in expression can be monitored at the seedling stage (Chandler et al., 2000), while the epigenetic memory might only be established in plants carrying ten expanded leaves (Coe, 1966; Ed Coe, personal communication). Besides the timing, the tissue or cell type in which paramutation occurs needs to be determined. Knowledge about the time and place of paramutation is crucial in order to pinpoint the mechanisms involved in paramutation.

At the moment, only a limited number of paramutation phenomena have been reported. Most of those affect a visible phenotype, facilitating their discovery. To test if paramutation is much more common than thus far appreciated, ideally, one would like to study interactions between different alleles of one and the same gene while keeping the genetic background identical. Such an approach is not feasible for a large number of genetic loci. With the increasing numbers of sequenced genomes available, one could however focus on genes and their regulatory sequences showing properties displayed by the known paramutagenic and paramutable alleles, for example directly repeated sequences. The recent discovery that paramutation exists beyond the plant world does suggest that paramutation has more implications than previously anticipated.

References

Alleman, M., Sidorenko, L., McGinnis, K., Seshadri, V., Dorweiler, J.E., White, J., Sikkink, K., and Chandler, V.L. (2006). An RNA-dependent RNA polymerase is required for paramutation in maize. Nature 442, 295–298.

Athma, P., Grotewold, E., and Peterson, T. (1992). Insertional mutagenesis of the maize *P* gene by intragenic transposition of *Ac*. Genetics 131, 199–209.

Bacher, C.P., Guggiari, M., Brors, B., Augui, S., Clerc, P., Avner, P., Eils, R., and Heard, E. (2006). Transient colocalization of X-inactivation centres accompanies the initiation of X inactivation. Nat Cell Biol 8, 293–299.

Bantignies, F., and Cavalli, G. (2006). Cellular memory and dynamic regulation of polycomb group proteins. Curr Opin Cell Biol 18, 275–283.

Bantignies, F., Grimaud, C., Lavrov, S., Gabut, M., and Cavalli, G. (2003). Inheritance of Polycomb-dependent chromosomal interactions in Drosophila. Gene Develop 17, 2406–2420.

Bateson, W., and Pellew, C. (1915). On the genetics of 'rogues' among culinairy peas *(Pisum sativum)*. Journal of Genetics 5, 15–36.

Brink, R., and Weyers, W.H. (1957). Invariable genetic change in maize plants heterozygous for marbled aleurone. Proc. Natl. Acad. Sci. U.S. 43, 1053–1060.

Brink, R.A. (1958). Paramutation at the R locus in maize. Cold Spring Harbor Symp. Quant. Biol. 23, 379–391.

Brown, D., and Brink, R. (1960). Paramutagenic action of paramutant R-r and R-g alleles in maize. Genetics 45, 1313–1316.

Chan, S.W., Zhang, X., Bernatavichute, Y.V., and Jacobsen, S.E. (2006a). Two-step recruitment of RNA-directed DNA methylation to tandem repeats. PLoS Biol 4, e363.

Chan, S.W., Zilberman, D., Xie, Z., Johansen, L.K., Carrington, J.C., and Jacobsen, S.E. (2004). RNA silencing genes control de novo DNA methylation. Science 303, 1336.

Chan, S.W., Henderson, I.R., Zhang, X., Shah, G., Chien, J.S., and Jacobsen, S.E. (2006b). RNAi, DRD1, and histone methylation actively target developmentally important non-CG DNA methylation in arabidopsis. PLoS Genet 2, e83.

Chandler, V.L. (2007). Paramutation: From maize to mice. Cell 128, 641–645.

Chandler, V.L., and Stam, M. (2004). Chromatin conversations: mechanisms and implications of paramutation. Nat Rev Genet 5, 532–544.

Chandler, V.L., Eggleston, W.B., and Dorweiler, J.E. (2000). Paramutation in maize. Plant Mol Biol 43, 121–145.

Chen, Z.J. (2007). Genetic and epigenetic mechanisms for gene expression and phenotypic variation in plant polyploids. Annu Rev Plant Biol 58, 377–406.

Chopra, S., Athma, P., and Peterson, T. (1996). Alleles of the maize P gene with distinct tissue specificities encode Myb-homologous proteins with C-terminal replacements. Plant Cell 8, 1149–1158.

Chopra, S., Athma, P., Li, X.G., and Peterson, T. (1998). A maize Myb homolog is encoded by a multicopy gene complex. Mol Gen Genet 260, 372–380.

Coe, E.H.J. (1959). A regular and continuing conversion-type phenomenon at b locus in maize. Maydica 24, 49–58.

Coe, E.H.J. (1966). The properties, origin and mechanism of conversion-type inheritance at the b locus in maize. Genetics 53, 1035–1063.

D'Alessio, A.C., and Szyf, M. (2006). Epigenetic tete-a-tete: the bilateral relationship between chromatin modifications and DNA methylation. Biochem Cell Biol 84, 463–476.

Das, P., and Messing, J. (1994). Variegated phenotype and developmental methylation changes of a maize allele originating from epimutation. Genetics 136, 1121–1141.

Dekker, J., Rippe, K., Dekker, M., and Kleckner, N. (2002). Capturing chromosome conformation. Science 295, 1306–1311.

Dieguez, M.J., Vaucheret, H., Paszkowski, J., and Mittelsten Scheid, O. (1998). Cytosine methylation at CG and CNG sites is not a prerequisite for the initiation of transcriptional gene silencing in plants, but it is required for its maintenance. Mol Gen Genet 259, 207–215.

Dorweiler, J.E., Carey, C.C., Kubo, K.M., Hollick, J.B., Kermicle, J.L., and Chandler, V.L. (2000). Mediator of paramutation1 is required for establishment and maintenance of paramutation at multiple maize loci. Plant Cell 12, 2101–2118.

Duvillie, B., Bucchini, D., Tang, T., Jami, J., and Paldi, A. (1998). Imprinting at the mouse Ins2 locus: evidence for cis- and trans-allelic interactions. Genomics 47, 52–57.

Eggleston, W.B., Alleman, M., and Kermicle, J.L. (1995). Molecular organization and germinal instability of R-stippled maize. Genetics 141, 347–360.

English, J.J., and Jones, J.D.G. (1998). Epigenetic instability and trans-silencing interactions associated with an SPT:Ac T-DNA locus in tobacco. Genetics 148, 457–469.

Finnegan, E.J., Sheldon, C.C., Jardinaud, F., Peacock, W.J., and Dennis, E.S. (2004). A cluster of Arabidopsis genes with a coordinate response to an environmental stimulus. Curr Biol 14, 911–916.

Forne, T., Oswald, J., Dean, W., Saam, J.R., Bailleul, B., Dandolo, L., Tilghman, S.M., Walter, J., and Reik, W. (1997). Loss of the maternal H19 gene induces changes in Igf2 methylation in both cis and trans. Proc Nat Acad Sci Usa 94, 10243–10248.

Grewal, S.I.S., and Moazed, D. (2003). Heterochromatin and epigenetic control of gene expression. Science 301, 798–802.

Grewal, S.I.S., and Jia, S.T. (2007). Heterochromatin revisited. Nature Reviews Genetics 8, 35–46.

Grewal, S.I.S., and Elgin, S.C.R. (2007). Transcription and RNA interference in the formation of heterochromatin. Nature 447, 399–406.

Grimaud, C., Bantignies, F., Pal-Bhadra, M., Ghana, P., Bhadra, U., and Cavalli, G. (2006). RNAi components are required for nuclear clustering of Polycomb group response elements. Cell 124, 957–971.

Gross, S.M., and Hollick, J.B. (2007). Multiple trans-sensing interactions affect meiotically heritable epigenetic states at the maize pl1 locus. Genetics 176, 829–839.

Guo, M., Davis, D., and Birchler, J.A. (1996). Dosage effects on gene expression in a maize ploidy series. Genetics 142, 1349–1355.

Hagemann, R. (1993). Studies towards a genetic and molecular analysis of paramutation at the *sulfurea* locus of Lycopersicon esculentum Mill. (Lancaster-Basel: Technomic publishing company, Inc.).

Hagemann, R., and Berg, W. (1978). Paramutation at the *sulfurea* locus of *Lycopersicon esculentum* Mill. VII. Determination of the time of occurrence of paramutation by the quantitative evaluation of the variegation. Theoretical and Applied Genetics 53, 113–123.

Hale, C.J., Stonaker, J.L., Gross, S.M., and Hollick, J.B. (2007). A Novel Snf2 Protein Maintains trans-Generational Regulatory States Established by Paramutation in Maize. PLoS Biol 5, e275.

Hall, I.M., Noma, K., and Grewal, S.I.S. (2003). RNA interference machinery regulates chromosome dynamics during mitosis and meiosis in fission yeast. Proc Nat Acad Sci Usa 100, 193–198.

Haring, M., Offermann, S., Danker, T., Horst, I., Peterhaensel, C., and Stam, M. (2007). Chromatin immunoprecipitation: optimization, quantitative analysis and data normalization. Plant Methods 3, 11.

Harrison, B.J., and Carpenter, R. (1973). A comparison of the instabilities at the *Nivea* and *Pallida* loci in *Antirrhinum majus*. Heredity 31, 309–323.

Hatada, I., Nabetani, A., Arai, Y., Ohishi, S., Suzuki, M., Miyabara, S., Nishimune, Y., and Mukai, T. (1997). Aberrant methylation of an imprinted gene U2af1-rs1(SP2) caused by its own transgene. Journal of Biological Chemistry 272, 9120–9122.

Henderson, I.R., and Jacobsen, S.E. (2007). Epigenetic inheritance in plants. Nature 447, 418–424.

Henikoff, S., and Ahmad, K. (2005). Assembly of variant histones into chromatin. Annu Rev Cell Dev Biol 21, 133–153.

Hollick, J.B., and Chandler, V.L. (1998). Epigenetic allelic states of a maize transcriptional regulatory locus exhibit overdominant gene action. Genetics 150, 891–897.

Hollick, J.B., and Chandler, V.L. (2001). Genetic factors required to maintain repression of a paramutagenic maize pl1 allele. Genetics 157, 369–378.

Hollick, J.B., Kermicle, J.L., and Parkinson, S.E. (2005). Rmr6 maintains meiotic inheritance of paramutant states in Zea mays. Genetics 171, 725–740.

Hollick, J.B., Patterson, G.I., Asmundsson, I.M., and Chandler, V.L. (2000). Paramutation alters regulatory control of the maize pl locus. Genetics 154, 1827–1838.

Hollick, J.B., Patterson, G.I., Coe, E.H., Jr., Cone, K.C., and Chandler, V.L. (1995). Allelic interactions heritably alter the activity of a metastable maize pl allele. Genetics 141, 709–719.

Jacobsen, S.E., and Meyerowitz, E.M. (1997). Hypermethylated SUPERMAN epigenetic alleles in Arabidopsis. Science 277, 1100–1103.

Jones, L., Ratcliff, F., and Baulcombe, D.F. (2001). RNA-directed transcriptional gene silencing in plants can be inherited independently of the RNA trigger and requires Met1 for maintenance. Curr Biol 11, 747–757.

Kermicle, J.L., Eggleston, W.B., and Alleman, M. (1995). Organization of paramutagenicity in R-stippled maize. Genetics 141, 361–372.

Klose, R.J., and Bird, A.P. (2006). Genomic DNA methylation: the mark and its mediators. Trends Biochem Sci 31, 89–97.

Lavigne, M., Francis, N.J., King, I.F., and Kingston, R.E. (2004). Propagation of silencing; recruitment and repression of naive chromatin in trans by polycomb repressed chromatin. Mol Cell 13, 415–425.

Lee, A.M., and Wu, C.T. (2006). Enhancer-promoter communication at the yellow gene of Drosophila melanogaster: diverse promoters participate in and regulate trans interactions. Genetics 174, 1867–1880.

Lee, D.W., Seong, K.Y., Pratt, R.J., Baker, K., and Aramayo, R. (2004). Properties of unpaired DNA required for efficient silencing in Neurospora crassa. Genetics 167, 131–150.

Lee, H.S., and Chen, Z.J. (2001). Protein-coding genes are epigenetically regulated in Arabidopsis polyploids. Proc Nat Acad Sci Usa 98, 6753–6758.

Lei, E.P., and Corces, V.G. (2006). RNA interference machinery influences the nuclear organization of a chromatin insulator. Nat Genet 38, 936–941.

Liu, Y., Taverna, S.D., Muratore, T.L., Shabanowitz, J., Hunt, D.F., and Allis, C.D. (2007). RNAi-dependent H3K27 methylation is required for heterochromatin formation and DNA elimination in Tetrahymena. Genes Dev 21, 1530–1545.

Lomvardas, S., Barnea, G., Pisapia, D.J., Mendelsohn, M., Kirkland, J., and Axel, R. (2006). Interchromosomal interactions and olfactory receptor choice. Cell 126, 403–413.

Louwers, M., Haring, M., and Stam, M. (2005). When alleles meet: Paramutation. In Plant Epigenetics, P. Meyer, ed (Oxford, UK: Blackwell Publishing Ltd), pp. 134–173.

Luff, B., Pawlowski, L., and Bender, J. (1999). An inverted repeat triggers cytosine methylation of identical sequences in Arabidopsis. Mol Cell 3, 505–511.

Martienssen, R.A. (2003). Maintenance of heterochromatin by RNA interference of tandem repeats. Nat Genet 35, 213–214.

Martin, C., and Zhang, Y. (2007). Mechanisms of epigenetic inheritance. Curr Opin Cell Biol 19, 266–272.

Matzke, M., Kanno, T., Huettel, B., Daxinger, L., and Matzke, A.J. (2007). Targets of RNA-directed DNA methylation. Curr Opin Plant Biol 10, 512–519.

Matzke, M.A., and Birchler, J.A. (2005). RNAi-mediated pathways in the nucleus. Nat Rev Genet 6, 24–35.

Mello, C.C., and Conte, D. (2004). Revealing the world of RNA interference. Nature 431, 338–342.

Meyer, P., Heidmann, I., and Niedenhof, I. (1993). Differences in DNA-methylation are associated with a paramutation phenomenon in transgenic petunia. Plant J 4, 89–100.

Meyer, P., Linn, F., Heidmann, I., Meyer, H., Niedenhof, I., and Saedler, H. (1992). Endogenous and environmental factors influence 35S promoter methylation of a maize A1 gene construct in transgenic petunia and its colour phenotype. Mol Gen Genet 231, 345–352.

Mittelsten Scheid, O., Afsar, K., and Paszkowski, J. (2003). Formation of stable epialleles and their paramutation-like interaction in tetraploid Arabidopsis thaliana. Nat Genet 34, 450–454.

Panavas, T., Weir, J., and Walker, E.L. (1999). The structure and paramutagenicity of the R-marbled haplotype of Zea mays. Genetics 153, 979–991.

Park, Y.D., Papp, I., Moscone, E.A., Iglesias, V.A., Vaucheret, H., Matzke, A.J.M., and Matzke, M.A. (1996). Gene silencing mediated by promoter homology occurs at the level of transcription and results in meiotically heritable alterations in methylation and gene activity. Plant Journal 9, 183–194.

Patterson, G.I., Thorpe, C.J., and Chandler, V.L. (1993). Paramutation, an allelic interaction, is associated with a stable and heritable reduction of transcription of the maize b regulatory gene. Genetics 135, 881–894.

Pecinka, A., Kato, N., Meister, A., Probst, A.V., Schubert, I., and Lam, E. (2005). Tandem repetitive transgenes and fluorescent chromatin tags alter local interphase chromosome arrangement in Arabidopsis thaliana. J Cell Sci 118, 3751–3758.

Peters, A.H.F.M., and Schubeler, D. (2005). Methylation of histones: playing memory with DNA. Curr Opin Cell Biol 17, 230–238.

Qin, H., and von Arnim, A.G. (2002). Epigenetic history of an Arabidopsis trans-silencer locus and a test for relay of trans-silencing activity. BMC Plant Biol 2, 11.

Qin, H., Dong, Y., and von Arnim, A.G. (2003). Epigenetic interactions between Arabidopsis transgenes: characterization in light of transgene integration sites. Plant Mol Biol 52, 217–231.

Rassoulzadegan, M., Magliano, M., and Cuzin, F. (2002). Transvection effects involving DNA methylation during meiosis in the mouse. Embo J 21, 440–450.

Rassoulzadegan, M., Grandjean, V., Gounon, P., Vincent, S., Gillot, I., and Cuzin, F. (2006). RNA-mediated non-mendelian inheritance of an epigenetic change in the mouse. Nature 441, 469–474.

Rinn, J.L., Kertesz, M., Wang, J.K., Squazzo, S.L., Xu, X., Brugmann, S.A., Goodnough, L.H., Helms, J.A., Farnham, P.J., Segal, E., and Chang, H.Y. (2007). Functional demarcation of active and silent chromatin domains in human HOX loci by noncoding RNAs. Cell 129, 1311–1323.

Santos, J.L., Alfaro, D., Sanchez-Moran, E., Armstrong, S.J., Franklin, F.C., and Jones, G.H. (2003). Partial diploidization of meiosis in autotetraploid Arabidopsis thaliana. Genetics 165, 1533–1540.

Sidorenko, L., Li, X., Tagliani, L., Bowen, B., and Peterson, T. (1999). Characterization of the regulatory elements of the maize P-rr gene by transient expression assays. Plant Mol Biol 39, 11–19.

Sidorenko, L.V., and Peterson, T. (2001). Transgene-induced silencing identifies sequences involved in the establishment of paramutation of the maize *p1* gene. Plant Cell 13, 319–335.

Sidorenko, L.V., Li, X., Cocciolone, S.M., Chopra, S., Tagliani, L., Bowen, B., Daniels, M., and Peterson, T. (2000). Complex structure of a maize *Myb* gene promoter: functional analysis in transgenic plants. Plant J 22, 471–482.

Slotkin, R.K., and Martienssen, R. (2007). Transposable elements and the epigenetic regulation of the genome. Nature Reviews Genetics 8, 272–285.

Soppe, W.J., Jacobsen, S.E., Alonso-Blanco, C., Jackson, J.P., Kakutani, T., Koornneef, M., and Peeters, A.J. (2000). The late flowering phenotype of fwa mutants is caused by gain-of-function epigenetic alleles of a homeodomain gene. Mol Cell 6, 791–802.

Spilianakis, C.G., Lalioti, M.D., Town, T., Lee, G.R., and Flavell, R.A. (2005). Interchromosomal associations between alternatively expressed loci. Nature 435, 637–645.

Stam, M., and Scheid, O.M. (2005). Paramutation: an encounter leaving a lasting impression. Trends Plant Sci 10, 283–290.

Stam, M., Belele, C., Dorweiler, J.E., and Chandler, V.L. (2002). Differential chromatin structure within a tandem array 100 kb upstream of the maize b1 locus is associated with paramutation. Gene Develop 16, 1906–1918.

Stokes, T.L., Kunkel, B.N., and Richards, E.J. (2002). Epigenetic variation in Arabidopsis disease resistance. Gene Develop 16, 171–182.

Styles, E.D., and Brink, R.A. (1966). The metastable nature of paramutable R alleles in maize. I. Heritable enhancement in level of standard *R'* action. Genetics 54, 433–439.

Styles, E.D., and Brink, R.A. (1967). The metastable nature of paramutable *R* alleles in maize. III. Heritable changes in level of *R* action in heterozygotes carrying different paramutable R alleles. Genetics 55, 411–422.

Tariq, M., and Paszkowski, J. (2004). DNA and histone methylation in plants. Trends Genet 20, 244–251.

Ting, A.H., Schuebel, K.E., Herman, J.G., and Baylin, S.B. (2005). Short double-stranded RNA induces transcriptional gene silencing in human cancer cells in the absence of DNA methylation. Nat Genet 37, 906–910.

Tolhuis, B., Palstra, R.J., Splinter, E., Grosveld, F., and deLaat, W. (2002). Looping and interaction between hypersensitive sites in the active beta-globin locus. Mol Cell 10, 1453–1465.

van West, P., Kamoun, S., van 't Klooster, J.W., and Govers, F. (1999). Internuclear gene silencing in Phytophthora infestans. Mol Cell 3, 339–348.

vanBlokland, R., tenLohuis, M., and Meyer, P. (1997). Condensation of chromatin in transcriptional regions of an inactivated plant transgene: evidence for an active role of transcription in gene silencing. Molecular & General Genetics 257, 1–13.

vanHouwelingen, A., Souer, E., Mol, J., and Koes, R. (1999). Epigenetic interactions among three dTph1 transposons in two homologous chromosomes activate a new excision-repair mechanism in petunia. Plant Cell 11, 1319–1336.

Vaucheret, H. (2006). Post-transcriptional small RNA pathways in plants: mechanisms and regulations. Gene Develop 20, 759–771.

Walker, E.L. (1998). Paramutation of the r1 locus of maize is associated with increased cytosine methylation. Genetics 148, 1973–1981.

Walker, E.L., and Panavas, T. (2001). Structural features and methylation patterns associated with paramutation at the r1 locus of Zea mays. Genetics 159, 1201–1215.

Watanabe, K., Pecinka, A., Meister, A., Schubert, I., and Lam, E. (2005). DNA hypomethylation reduces homologous pairing of inserted tandem repeat arrays in somatic nuclei of Arabidopsis thaliana. Plant J 44, 531–540.

Weinberg, M.S., Villeneuve, L.M., Ehsani, A., Amarzguioui, M., Aagaard, L., Chen, Z.X., Riggs, A.D., Rossi, J.J., and Morris, K.V. (2006). The antisense strand of small interfering RNAs directs histone methylation and transcriptional gene silencing in human cells. Rna 12, 256–262.

Weiss, H., and Maluszynska, J. (2000). Chromosomal rearrangement in autotetraploid plants of Arabidopsis thaliana. Hereditas 133, 255–261.

Wisman, S., Ramanna, M.S., and Koornneef, M. (1983). Isolation of a new paramutagenic allele of the sulfurea locus in the tomato cultivar Moneymaker following in vitro culture. Theor. Appl. Genet. 87, 289–294.

Woodhouse, M.R., Freeling, M., and Lisch, D. (2006). Initiation, establishment, and maintenance of heritable MuDR transposon silencing in maize are mediated by distinct factors. PLoS Biol 4, e339.

Xie, Z., Johansen, L.K., Gustafson, A.M., Kasschau, K.D., Lellis, A.D., Zilberman, D., Jacobsen, S.E., and Carrington, J.C. (2004). Genetic and functional diversification of small RNA pathways in plants. PLoS Biol 2, E104.

Xu, N., Tsai, C.L., and Lee, J.T. (2006). Transient homologous chromosome pairing marks the onset of X inactivation. Science 311, 1149–1152.

Zaratiegui, M., Irvine, D.V., and Martienssen, R.A. (2007). Noncoding RNAs and gene silencing. Cell 128, 763–776.

Imprinting in Maize

Nathan M Springer and Jose F Gutierrez-Marcos

Abstract Genomic imprinting in the maize endosperm results in differential expression of maternal and paternal alleles depending on their parental origin. The availability of sequence polymorphisms between different maize inbred lines and the large persistent endosperm of maize collectively provide a unique platform for studying the occurrence and mechanisms of imprinting in plants. Several imprinted genes have been identified in maize by targeted analyses. Genomic screens of allele-specific expression patterns in endosperm tissue have identified additional candidates for imprinting. Imprinted expression in maize is often associated with allele-specific DNA methylation states and it is likely that chromatin modifications are also involved in the establishment and maintenance of imprints.

1 The Phenomenon of Imprinting

Imprinting is an unusual regulatory mechanism that causes a parent-of-origin effect on gene expression and ensures the unique contribution of maternal and paternal genomes to the products of fertilization. In animals, imprinting has only been found in mammals, where it affects gene expression in embryo as well as in placental tissues. In plants, imprinting has only been documented in a few angiosperms, where it appears to be confined to the second product of double fertilization, the endosperm. It is unclear if imprinting also occurs in the embryo.

Genes subject to imprinting exhibit differential allelic expression, without possessing changes to their DNA sequence. Thus, imprinted expression states are under epigenetic control. Further, since active and silent alleles are found in the same nucleus, it is

N.M. Springer
University of Minnesota, Department of Plant Biology
springer@umn.edu

J.F. Gutierrez-Marcos
University of Warwick, Warwick HRI
j.f.gutierrez-marcos@warwick.ac.uk

likely that distinct mechanisms exist which direct the establishment and maintenance of epigenetic differences between parental alleles.

In this chapter, we will discuss examples of imprinting in maize, what is known about the mechanisms of imprinting, and its biological significance.

1.1   Classification of Imprinted Gene Expression

Genomic imprinting causes parent-of-origin effects on gene expression. Depending upon which of the parental alleles are preferentially expressed, imprinting can be described as either maternal imprinting (paternal allele expressed) or paternal imprinting (maternal allele expressed). In plants, these parent-of-origin effects may be manifest in different ways, suggesting the existence of distinct types of imprinting mechanisms (Yadegari et al., 2000; Gutierrez-Marcos et al., 2006). For instance, allele-specific imprinting causes only certain alleles at a particular locus to exhibit an imprinted pattern of expression. On the other hand, gene-specific imprinting is used to describe imprinted expression for all alleles tested at a given locus. Plant imprinted genes have been further classified into two subcategories: binary imprinting or differential imprinting, which describe the relative expression values of parental alleles (Dilkes and Comai, 2004). Binary imprinting results in strict monoallelic expression such that only one of the parental alleles is expressed while the other is silent. Differential imprinting results in bi-allelic expression with one of the parental alleles being expressed at levels lower than expected based on genomic dosage. In addition, imprinted genes have been classified according to the duration of their imprinting; some genes exhibit constitutive imprinted expression throughout endosperm development, whereas other displays a transient pattern of imprinted expression during early stages of endosperm development.

It is also worth noting that the term "imprinting" has been previously used in plants to describe genome-wide or chromosome-wide parent-of-origin effects. It remains unclear if these effects are in fact due to imprinting or to the unequal parental genome dosage contribution in the endosperm (reviewed in Alleman and Doctor, 2000).

2   Identification of Imprinted Genes in Maize

Imprinting was first documented for some alleles of the *R* locus in maize, such as *R-r:standard* (*R-r:std*) (Kermicle, 1970; Kermicle, 1978). *R-r:std* exhibits differential expression in the endosperm depending on whether the allele is transmitted from the male or female parent (reviewed in Kermicle and Alleman, 1990; Alleman and Doctor, 2000). When a dominant *R-r:std* allele is transmitted through the female parent, it will produce a fully colored aleurone. However, when the same *R-r:std* allele is transmitted through the male parent, it will result in a mottled coloration for the aleurone. This differential expression is not caused by the 2:1 maternal to paternal dosage, but by specific parent-of-origin effects on the expression of the *R* locus (Kermicle, 1970). It is worth noting that several other alleles of the *R* locus do not display any evidence for parent-of-origin effects on expression. Therefore, the *R* locus is an example of allele-specific paternal imprinting. The mottled coloration provided by the paternally transmitted *R-std* allele suggests that the imprinting is variable such that some endosperm cells exhibit imprinting while others exhibit bi-allelic expression and that this epigenetic regulation is developmentally regulated. Interestingly, the imprinting of *R* alleles occurs only in the endosperm and does not extend to embryonic tissues, despite being expressed there (Brink et al., 1970).

Further research has since identified other imprinted alleles and loci in maize. These imprinted sequences have been identified based on parent-of-origin dependent phenotypes (similar to *R*), analysis of gene families, homology to *Arabidopsis* imprinted genes or through molecular gene expression screens. We will discuss the set of imprinted genes that have been identified in maize through each of these methods.

2.1 Imprinting Identified by Parent-of-Origin Dependent Phenotypes

There are several examples of genes or loci that condition parent-of-origin dependent phenotypes. These are potentially imprinted loci that have not been verified by molecular testing. The phenotype of the *B-Bolivia* allele is indicative of imprinting while other alleles of *B* do not exhibit parent-of-origin effects (Selinger and Chandler, 2001). The parent-of-origin phenotypes for 10-kDa zein accumulation led to the discovery that the *dzr1* gene, a posttranscriptional regulator of 10-kDa zein accumulation, displays allele-specific imprinting (Chaudhuri and Messing, 1994). In addition, it has been shown that some alleles of *dzr1* show maternal effects that are not caused by dosage effects. Parent-of-origin phenotypes also provide evidence for the potential imprinting of *floury-1* and pH 7.5 *esterase* (Schwartz, 1965), *floury-2* (Di Fonzo et al., 1980) and *Dap* (*defective aleurone pigmentation*) (Gavazzi et al., 1997). For many of these genes, there is limited molecular evidence to conclusively demonstrate their imprinted pattern of expression. An alternative explanation of these parent-of-origin dependent phenotypes is that they may be the result of other parental effects that are distinct from imprinting.

2.2 Imprinting within Maize Gene Families

Several studies in maize have investigated the methylation and expression of parental alleles within gene families. Many of the *alpha-tubulin* genes exhibit differential methylation in endosperm relative to vegetative tissues (Lund et al., 1995a). For several of these *alpha-tubulin* genes, there was evidence for increased levels of expression in the endosperm. Two of the *alpha-tubulin* genes exhibit differential methylation depending upon whether they are transmitted via the male or female parent. There is also evidence for potential imprinting of *zein* gene expression (Bianchi and Viotti, 1988; Lund et al., 1995a). The *zein* genes are specifically expressed in the endosperm and exhibit endosperm-specific demethylation (Bianchi and Viotti, 1988). There is some evidence that the demethylation and expression of some *zein* genes occurs specifically for the maternal allele, suggesting a role for imprinting in regulating *zein* expression. However, other studies did not find evidence for imprinting of specific copies of the *zein* genes (Song and Messing, 2003). The complexity of these gene families makes it difficult to document the specific effect of imprinting at unique loci within the family.

2.3 *Imprinting Amongst Maize Polycomb-Group Genes*

Recently, a series of imprinted maize genes have been identified based on homology to genes that display imprinted expression in *Arabidopsis*. Many of the *Arabidopsis* genes that are known to be imprinted are related to *Drosophila* Polycomb-group genes (Kohler and Makarevich, 2006). There is evidence for a biochemical complex, PRC2, in *Drosophila* and plants that contains four different types of protein. This complex contains four proteins in *Drosophila*, the WD-40 repeat protein *extra sex combs* (*esc*), the SET domain protein *Enhancer of zeste* (*E(z)*), a putative DNA binding protein *Suppressor of variegation 12* (*Su(var)12*), and a nucleosome interacting protein *Nurf55/CAF1* (reviewed in Guitton and Berger, 2005). The *Arabidopsis* genome encodes a single homolog of *esc* (*Fie*), three homologs of *E(z)* (*MEA, CLF* and *SWN*), three homologs of *Su(var)12* (*Vrn2, Fis2, Emf2*) and several *MSI* homologs (reviewed in Kohler and Makarevich, 2006). Phenotypic and allele-specific expression analyses have found evidence for paternal imprinted expression of *MEA* and *FIS2* (Xiao et al., 2003; Jullien et al., 2006). A series of genes homologous to these *Arabidopsis* and *Drosophila* genes were identified within the maize genome (Springer et al., 2002). Subsequent screens for imprinting provided evidence for imprinting of both of the maize *FIE*-like genes, *Fie1* and *Fie2* (Danilevskaya et al., 2003; Gutierrez-Marcos et al., 2003). The *Fie1* gene is only expressed in endosperm tissue and exhibits binary paternal imprinting throughout endosperm development. *Fie2* exhibits expression in other plant tissues and is imprinted during early stages of endosperm development but displays non-imprinted biallelic expression during later stages of endosperm development (Danilevskaya et al., 2003; Gutierrez-Marcos et al., 2003; Gutierrez-Marcos et al., 2006). One of the three E(z)-like genes in maize, *Mez1*, also exhibits imprinted expression through-out endosperm development (Haun et al., 2007). The other two *E(z)*-like genes, *Mez2* and *Mez3,* exhibit non-imprinted bi-allelic expression in the endosperm. Expression analyses of two *Su(var)12*-like genes and three *MSI*-like genes have not revealed evidence for imprinting in the endosperm (Haun WJ, Springer NM, unpublished data). It is quite intriguing that several members of the PRC2-like complex are imprinted in both *Arabidopsis* and maize, but that the specific genes that are imprinted in the two species are different.

2.4   Imprinted Gene Identification through Large-Scale Molecular Screens

Relatively few imprinted genes in plants have been identified based on phenotypic analysis or homology to known imprinted genes in other species. However, it is difficult to estimate the true proportion of genes that are imprinted or the frequency of different types of imprinting in plants based on the examples of imprinting described above. Several groups have performed large-scale unbiased screens of allelic expression patterns in maize endosperm (Guo et al., 2003; Gutierrez-Marcos et al., 2003; Stupar et al., 2007). Each of these screens has revealed additional examples of imprinted gene expression. Guo et al. (2003) utilized a modified differential display approach to profile allelic expression at 10, 14 and 21 days after pollination in maize endosperm tissue. A small proportion of the ~6,500 genes studied exhibited either maternal-like (~5%) or paternal-like (3%) expression patterns. Further analyses of five of the genes that exhibit maternal-like expression suggested that four genes exhibit delayed paternal activation or heterochronic variation between the alleles, while one gene, *Nrp1*, exhibited differential imprinting (Guo et al., 2003). The screen by Gutierrez-Marcos et al. (2003) also used a modified differential-display technique to study allelic variation in endosperm expression levels and found evidence for maternal imprinting at 46 genes, or ~3-5% of the genes analyzed. The imprinted genes included the *Fie* genes, and *Meg1* (Gutierrez-Marcos et al., 2003; Gutierrez-Marcos et al., 2004). In addition, a maternally imprinted gene, *Peg1*, was also identified in this study. Stupar et al. (2007) used microarray expression profiling and a SNP-based allele-specific expression screen to monitor parental effects and imprinting in maize endosperm. This analysis identified 12 examples of paternal-like expression (0.5% of differentially expressed genes) and 120 examples of maternal-like expression (5% of differentially expressed genes). A follow-up SNP-based allele-specific expression profiling identified eight genes with parent-of-origin allelic expression, of which six displayed evidence for a slight paternal bias in endosperm expression and two other exhibited differential maternal imprinting. Collectively, these studies suggest that there are a significant number of imprinted genes in maize endosperm. If we assume that ~2% of the genes expressed in endosperm are imprinted and that there are 20,000 genes expressed in endosperm, we would predict nearly 400 imprinted genes in maize.

The identification and characterization of imprinted genes in maize has revealed several interesting trends. First, imprinting is predominantly observed in endosperm tissue. Molecular analyses for imprinting in embryo tissue have identified few or no imprinted genes (Gutierrez-Marcos et al., 2003; Stupar et al., 2007). Although there are several examples of genes that show delayed paternal activation in embryo tissue, this may not be related to the mechanisms of imprinting (Grimanelli et al., 2005).

Second, paternal imprinting is more common than maternal imprinting. While there are several examples of maternal imprinting in plants, such as maize *Peg1* (Gutierrez-Marcos et al., 2003) and *Arabidopsis PHE1* (Kohler et al., 2005), these are quite rare. Third, most of the genes in the endosperm that are preferentially expressed from the maternal allele do not display imprinting (Guo et al., 2003; Stupar et al., 2007). Therefore, the expression level of genes in the endosperm can be influenced by parent-of-origin effects that are distinct from imprinting such as dosage effects. Fourth, there are multiple types of imprinting observed in the maize endosperm, which suggest that different loci may be imprinted through different mechanisms.

3 Mechanisms of Imprinting

There have been significant advances towards understanding the molecular mechanisms that regulate imprinting in plants. However, there are many open questions that remain to be explained. What are the *cis*-acting elements and *trans*-acting factors required for imprinting? What are the mechanisms that regulate the establishment, maintenance, and interpretation of imprinted marks to discriminate between parental alleles? We will address some of this question in this section and discuss the general requirements for an imprinting mechanism, what is currently known about imprinting marks, and strategies employed to dissect the mechanisms of imprinting in maize.

3.1 Requirements for an Imprinting Mechanism

Imprinting results in the differential expression of the maternal and paternal alleles in the endosperm after fertilization. Since this differential expression is caused without changes to DNA sequence, imprinted expression is likely due to epigenetic differences between parental alleles. Differential expression of two alleles can be caused by allele-specific activation or allele-specific silencing. If the default expression state of a given locus is active expression, then the mechanism of imprinting would be allele-specific silencing. Conversely, allele-specific activation would result in the activation of one allele from a silenced default state. Initially, it was assumed that imprinting in plants was the result of allele-specific silencing. However, studies in *Arabidopsis* (*MEA, FWA* and *FIS2*) and maize (*Fie1*) suggest that imprinting at some loci is the result of specific activation of the maternal alleles. Imprinted expression occurs in the endosperm, where only the maternal alleles are expressed while the paternal alleles remain silent. Incidentally, these genes (with the exception of *MEA*) are also silent in the embryo and other vegetative tissues. By contrast, other imprinted genes in maize, such as *R, Fie2* and *Mez1,* are expressed in other plant tissues and it is probable that their imprinted expression is mediated by an, as yet unknown, allele-specific silencing mechanism. Collectively, these data indicates that both allele-specific activation and allele-specific silencing contribute to imprinted expression in maize.

The process of imprinting can be divided into three distinct phases: establishment, maintenance and interpretation. Maternal and/or paternal alleles can be marked in the sporophyte or in the gametophyte to establish an imprint. Studies in *Arabidopsis* and maize have shown that, for some imprinted loci, the imprint is established in the female gametophyte. The timing of the establishment of the imprint in gametes can be critical for an imprinting effect to take place immediately after fertilization. Following the establishment of the imprint, it is necessary to maintain the epigenetic states between alleles. This requires the stable inheritance of the mark at imprinted sequences over multiple mitotic divisions. Establishment and maintenance of imprints are likely mediated by epigenetic mechanisms such as DNA methylation, histone modification and changes in chromatin structure. Finally, the imprint must be interpreted by transcriptional machinery in order to generate the observed differential expression between parental alleles. These distinct phases of imprinting require concerted action by *cis*-acting elements and *trans*-acting factors.

3.2   Role of DNA Methylation in Imprinting

DNA methylation is frequently associated with epigenetic regulation in higher plants. In plants, methylation is found primarily in symmetric CG and CNG sequence contexts, but can also occur at asymmetric CNN sequence contexts (reference Kaeppler chapter here). As a general rule, DNA methylation in plants is negatively correlated with gene expression. There is significant evidence that DNA methylation plays a key role in the regulation of imprinted gene expression in mammals (Howell et al., 2001). However, the evidence for a direct link between DNA methylation and imprinting is less clear in plants. Genetic studies in *Arabidopsis* have demonstrated that DNA methylation levels are critical to ensure proper endosperm development and seed maturation (Adams et al., 2000; Xiao et al., 2006). A series of genetic studies have demonstrated that DNA methylation is required to establish a default silent state of the *MEA*, *FWA* and *FIS2* loci in *Arabidopsis* (Kinoshita et al., 2004; Jullien et al., 2006; Xiao et al., 2006). Imprinting at these loci is then conditioned by allele-specific demethylation of maternal alleles in the central cell by the DNA glycosylase DEMETER (Choi et al., 2002). However, the small size and transient nature of the endosperm in *Arabidopsis* has made it difficult to perform detailed studies of DNA methylation in this tissue.

Only recently has it been shown in maize that DNA methylation plays an important role in the regulation of specific imprinted sequences in the endosperm. There is evidence for wide-spread hypomethylation of the maternal genome in the maize endosperm. DNA methylation analysis from maize endosperm revealed maternal hypomethylation across large expressed genic regions (Lauria et al., 2004). This global maternal hypomethylation has been postulated to play an important role in the genome-wide maternal expression reportedly occurring during early endosperm development in maize (Grimanelli et al., 2005). A series of experiments have demonstrated a correlation between DNA methylation levels and imprinting. Methylation-sensitive Southern blot analyses have shown an inverse correlation between DNA methylation and gene expression for allele-specific imprinting of *R, alpha-tubulin, dzr1* and *zein* genes (Chaudhuri and Messing, 1994; Lund et al., 1995b; Lund et al., 1995a; Alleman and Doctor, 2000). Similarly, the paternal alleles of gene-specific imprinted *MEG1, FIE1, FIE2* and *MEZ1* genes are also methylated and silenced in the endosperm (Danilevskaya et al., 2003; Gutierrez-Marcos et al., 2004; Gutierrez-Marcos et al., 2006; Haun et al., 2007; Hermon et al., 2007).

Similar to mammals, allele-specific and locus-specific plant imprinted genes exhibit differentially methylated regions (DMRs). The advent of bisulfite sequencing technology has enabled researchers to identify the specific sequences that are targets for differential methylation. For example, *FIE1, FIE2* and *MEZ1* DMRs are localized in 5' upstream regions and are rich in symmetric and asymmetric methylated cytosines (Gutierrez-Marcos et al., 2006; Haun et al., 2007; Hermon et al., 2007). Furthermore, these DMRs were mimicked in transgenic *FIE1* and *FIE2* reporter gene sequences, which were also found to be imprinted, suggesting that these sequences contain *cis*-acting information sufficient for their imprinted regulation (Gutierrez-Marcos et al., 2006).

The detailed analyses of DNA methylation at *FIE1* and *FIE2* provide evidence for the existence of at least two different types of imprinting mechanisms. The *FIE1* DMR is hypermethylated in sperm cells and egg cells, but hypomethylated in central cells, thus establishing the imprinted status of parental alleles prior to fertilization (Gutierrez-Marcos et al., 2006; Hermon et al., 2007). This is very similar to methylation patterns noted for the imprinted *Arabidopsis* loci *MEA, FWA* and *FIS2* (Xiao et al., 2003; Kinoshita et al., 2004; Jullien et al., 2006), suggesting a conserved imprinting mechanism between monocots and dicots. In contrast, bisulfite methylation analysis of the *FIE2* DMR revealed no methylation in sperm cells, eggs or central cells. These data suggest that *de novo* methylation of the paternal *FIE2* allele occurs during early endosperm development (Gutierrez-Marcos et al., 2006), but is subsequently lost at later stages (Hermon et al., 2007). At present, the mechanisms of *de novo* methylation of *FIE2* paternal alleles and its subsequent removal are not well understood.

3.3   *Role of Chromatin Structure in the Regulation of Imprinting*

While DNA methylation has been the main focus of recent research on imprinted genes in plants, much less is known about chromatin structure and regulation of allele-specific silencing at known imprinted loci. Cytogenetic studies of Arabidopsis have revealed that the endosperm possesses a higher-order chromatin organization that is distinct from that of vegetative tissues (Baroux et al., 2007). Dynamic remodeling of chromatin can lead to active or inactive transcriptional states; hence it is likely that chromatin structure of some imprinted parental alleles is important in determining or maintaining their imprinted status. Chromatin conformation and DNA methylation are intimately interrelated. For instance, in *Arabidopsis,* maintenance of CG methylation involves the SWI-SNF chromatin remodeling factor *DECREASE IN DNA METHYLATION1 (DDM1)* (Jeddeloh et al., 1999). Loss of *DDM1* affects imprinting at the *MEDEA* locus (Vielle-Calzada et al., 1999), however it is unclear whether this is due to changes in chromatin structure or to altered DNA methylation state. In addition to *DDM1*, *MEA* itself is involved in its own imprinted regulation and in the regulation of at least one other gene, *PHE1* (Kohler et al., 2003). The MEA protein is capable of binding to its own promoter and direct methylation at histone H3 Lys27 (Gehring et al., 2006; Jullien et al., 2006; Baroux et al., 2007), but it is not known if the other imprinted PcG members, *FIE* and *FIS2*, also behave in this way. The analysis of several *Mez1* alleles with *Mu* insertions in the promoter suggests that the MEZ1 protein is required for efficient silencing of the paternal allele of *Mez1* and *Fie1* in maize (W.J. Haun and N.M. Springer, submitted).

Fortunately, the abundance of maize endosperm compared with that of *Arabidopsis*, provides an opportunity to employ biochemical methods to study chromatin structure at imprinted loci. Interestingly, the loss of imprinting at some loci (e.g. *FIE2* and *MEG1*) coincides with the onset of endoreduplication in the endosperm, which has been linked to the relaxation of chromatin conformation (Zhao and Grafi, 2000). Whether a direct link exists between chromatin conformation and imprinted expression states remains to be determined. Indeed, there is evidence for endosperm-specific enrichment of H3K27me3 at the paternal allele of *Mez1*, *Fie1* and *Nrp1*. In contrast, the maternal alleles for these loci exhibit enrichment for histone acetylation and H3K4me2 (W.J. Haun and N.M. Springer, submitted). The current data provide a descriptive view of chromatin at imprinting loci. The addition of genetic data will help to understand whether these chromatin modifications are required for imprintting. Within the maize community, a large collection of RNAi lines for maize chromatin genes is available (McGinnis et al., 2007), which could provide the ideal genetic resource to further investigate the role of chromatin in imprinted gene regulation.

4   Biological Role of Imprinting

Imprinting is only known to occur in mammals and flowering plants, suggesting that imprinting has evolved recently but independently, as a novel form of gene regula-tion. In mammals, imprinted genes are thought to have arisen via an ancestral mechanism, as they are clustered in discrete chromosomal regions (Walter and Paulsen, 2003). In contrast, imprinting in plants appears to have evolved multiple times in the endosperm of different angiosperm lineages (Haun et al., 2007; Spillane et al., 2007). Research in maize has been instrumental in gathering information about the complex involvement of multiple mechanisms used to imprint different sequences in the endosperm. The occurrence of allele-specific imprinted sequences is reflective of the dynamic nature of the maize genome (Messing and Dooner,

2006). Allele-specific imprinted genes belong to gene families that originated from recent gene duplication events and chromosome rearrangements generated by transposon and retrotransposon activity in some inbred lines (Song and Messing, 2003), and as such became novel targets for epigenetic silencing. Despite being imprinted in the endosperm, these alleles do not appear to play any significant role during endosperm development. On the other hand, locus-specific imprinted genes function in early seed growth regulation, and as such, these genes might have been specifically selected to be under epigenetic control. One of the driving forces responsible for this is thought to be the conflicting interests exerted by maternal and paternal genomes in the allocation of resources from mother to offspring (Haig and Westoby, 1989). Since the endosperm serves a placenta-like role to protect and nourish the growing embryo, it is fitting that imprinting should occur in the endosperm. The maize imprinted gene, *Meg1*, is of particular interest because it is only expressed in the endosperm transfer cells that regulate nutrient trafficking from the sporophytic maternal tissue to the seed (Gutierrez-Marcos et al., 2004), hence supporting evidence for maternal regulation of resource allocation. In addition, the Polycomb Group (PcG) members that comprise the main class of locus-specific imprinted genes, have been shown to be involved in the repression of cell growth and proliferation during early endosperm development in *Arabidopsis*. Further, in *Arabidopsis*, loss of imprinting by mutation in the PcGs or in DNA methyltransferases leads to aberrant endosperm overproliferation and changes in seed size (reviewed in Gehring et al., 2004). Deregulation of the PcG genes in maize may provide clues as to whether these genes have similar roles in endosperm development.

Acknowledgments The authors are grateful to Liliana M. Costa and William Haun for providing valuable discussion and feedback. This work is supported by NSF Grant (DBI-0227310) to N.M.S. and by the Royal Society (RS-78753) to J.F.G-M.

References

Adams, S., Vinkenoog, R., Spielman, M., Dickinson, H.G., and R.J. Scott (2000) Parent-of-origin effects on seed development in Arabidopsis thaliana require DNA methylation. *Development* **127**: 2493–2502.

Alleman, M., and J. Doctor (2000) Genomic imprinting in plants: observations and evolutionary implications. *Plant Mol Biol* **43**: 147–161.

Baroux, C., Pecinka, A., Fuchs, J., Schubert, I., and U. Grossniklaus (2007) The triploid endosperm genome of Arabidopsis adopts a peculiar, parental-dosage-dependent chromatin organization. *Plant Cell* **19**: 1782–1794.

Bianchi, M.W., and A. Viotti (1988) DNA methylation and tissue-specific transcription of the storage protein gene of maize. *Plant Mol. Biol.* **11**: 203–214.

Brink, R.A., Kermicle, J.L., and N.K. Ziebur (1970) Derepression in the Female Gametophyte in Relation to Paramutant R Expression in Maize Endosperms, Embryos, and Seedlings. *Genetics* **66**: 87–96.

Chaudhuri, S., and J. Messing (1994) Allele-specific parental imprinting of dzr1, a posttranscriptional regulator of zein accumulation. *Proc Natl Acad Sci U S A* **91**: 4867–4871.

Choi, Y., Gehring, M., Johnson, L., Hannon, M., Harada, J.J., Goldberg, R.B., Jacobsen, S.E., and R.L. Fischer (2002) DEMETER, a DNA glycosylase domain protein, is required for endosperm gene imprinting and seed viability in arabidopsis. *Cell* **110**: 33–42.

Danilevskaya, O.N., Hermon, P., Hantke, S., Muszynski, M.G., Kollipara, K., and E.V. Ananiev (2003) Duplicated fie genes in maize: expression pattern and imprinting suggest distinct functions. *Plant Cell* **15**: 425–438.

Di Fonzo, N., Fornasari, E., Salamini, F., Reggiani, R., and C. Soave (1980) Interaction of maize mutants *floury-2* and *opaque-7* with *opaque-2* in the synthesis of endosperm proteins. *J. Hered.* **71**: 397–402.

Dilkes, B.P., and L. Comai (2004) A differential dosage hypothesis for parental effects in seed development. *Plant Cell* **16**: 3174–3180.

Gavazzi, G., Dolfini, S., Allegra, D., Castiglioni, P., Todesco, G., and M. Hoxha (1997) Dap (Defective aleurone pigmentation) mutations affect maize aleurone development. *Mol Gen Genet* **256**: 223–230.

Gehring, M., Choi, Y., and R.L. Fischer (2004) Imprinting and seed development. *Plant Cell* **16** Suppl: S203–213.

Gehring, M., Huh, J.H., Hsieh, T.F., Penterman, J., Choi, Y., Harada, J.J., Goldberg, R.B., and R.L. Fischer (2006) DEMETER DNA glycosylase establishes MEDEA polycomb gene self-imprinting by allele-specific demethylation. *Cell* **124**: 495–506.

Grimanelli, D., Perotti, E., Ramirez, J., and O. Leblanc (2005) Timing of the maternal-to-zygotic transition during early seed development in maize. *Plant Cell* **17**: 1061–1072.

Guitton, A.E., and F. Berger (2005) Control of reproduction by Polycomb Group complexes in animals and plants. *Int J Dev Biol* **49**: 707–716.

Guo, M., Rupe, M.A., Danilevskaya, O.N., Yang, X., and Z. Hu (2003) Genome-wide mRNA profiling reveals heterochronic allelic variation and a new imprinted gene in hybrid maize endosperm. *Plant J* **36**: 30–44.

Gutierrez-Marcos, J.F., Pennington, P.D., Costa, L.M., and H.G. Dickinson (2003) Imprinting in the endosperm: a possible role in preventing wide hybridization. *Philos Trans R Soc Lond B Biol Sci* **358**: 1105–1111.

Gutierrez-Marcos, J.F., Costa, L.M., Dal Pra, M., Scholten, S., Kranz, E., Perez, P., and H.G. Dickinson (2006) Epigenetic asymmetry of imprinted genes in plant gametes. *Nat Genet* **38**: 876–878.

Gutierrez-Marcos, J.F., Costa, L.M., Biderre-Petit, C., Khbaya, B., O'Sullivan, D.M., Wormald, M., Perez, P., and H.G. Dickinson (2004) maternally expressed gene1 Is a novel maize endosperm transfer cell-specific gene with a maternal parent-of-origin pattern of expression. *Plant Cell* **16**: 1288–1301.

Haig, D., and M. Westoby (1989) Parent specific gene expression and the triploid endosperm. *Am. Nat.* **134**: 147–155.

Haun, W.J., Laoueille-Duprat, S., O'Connell M, J., Spillane, C., Grossniklaus, U., Phillips, A.R., Kaeppler, S.M., and N.M. Springer (2007) Genomic imprinting, methylation and molecular evolution of maize Enhancer of zeste (Mez) homologs. *Plant J* **49**: 325–337.

Hermon, P., Srilunchang, K.O., Zou, J., Dresselhaus, T., and O.N. Danilevskaya (2007) Activation of the imprinted Polycomb Group Fie1 gene in maize endosperm requires demethylation of the maternal allele. *Plant Mol Biol* **64**: 387–395.

Howell, C.Y., Bestor, T.H., Ding, F., Latham, K.E., Mertineit, C., Trasler, J.M., and J.R. Chaillet (2001) Genomic imprinting disrupted by a maternal effect mutation in the Dnmt1 gene. *Cell* **104**: 829–838.

Jeddeloh, J.A., Stokes, T.L., and E.J. Richards (1999) Maintenance of genomic methylation requires a SWI2/SNF2-like protein. *Nat Genet* **22**: 94–97.

Jullien, P.E., Katz, A., Oliva, M., Ohad, N., and F. Berger (2006) Polycomb group complexes self-regulate imprinting of the Polycomb group gene MEDEA in Arabidopsis. *Curr Biol* **16**: 486–492.

Kermicle, J.L. (1970) Dependence of the R-Mottled Aleurone Phenotype in Maize on Mode of Sexual Transmission. *Genetics* **66**: 69–85.

Kermicle, J.L. (1978) Imprinting of gene action in maize endosperm. In: *Maize breeding and Genetics* (D.B. Walden, ed.) Wiley, New York, pp. 357–371.

Kermicle, J.L., and M. Alleman (1990) Gametic imprinting in maize in relation to the angiosperm life cycle. *Dev Suppl*, 9–14.

Kinoshita, T., Miura, A., Choi, Y., Kinoshita, Y., Cao, X., Jacobsen, S.E., Fischer, R.L., and T.Kakutani (2004) One-way control of FWA imprinting in Arabidopsis endosperm by DNA methylation. *Science* **303**: 521–523.

Kohler, C., and G. Makarevich (2006) Epigenetic mechanisms governing seed development in plants. EMBO Rep **7**: 1223–1227.

Kohler, C., Page, D.R., Gagliardini, V., and U. Grossniklaus (2005) The Arabidopsis thaliana MEDEA Polycomb group protein controls expression of PHERES1 by parental imprinting. *Nat Genet* **37**: 28–30.

Kohler, C., Hennig, L., Spillane, C., Pien, S., Gruissem, W., and U. Grossniklaus (2003) The Polycomb-group protein MEDEA regulates seed development by controlling expression of the MADS-box gene PHERES1. *Genes Dev* **17**: 1540–1553.

Lauria, M., Rupe, M., Guo, M., Kranz, E., Pirona, R., Viotti, A., and G. Lund (2004) Extensive maternal DNA hypomethylation in the endosperm of Zea mays. *Plant Cell* **16**: 510–522.

Lund, G., Messing, J., and A. Viotti (1995a) Endosperm-specific demethylation and activation of specific alleles of alpha-tubulin genes of Zea mays L. *Mol Gen Genet* **246**: 716–722.

Lund, G., Ciceri, P., and A. Viotti (1995b) Maternal-specific demethylation and expression of specific alleles of zein genes in the endosperm of Zea mays L. Plant J **8**: 571–581.

McGinnis, K., Murphy, N., Carlson, A.R., Akula, A., Akula, C., Basinger, H., Carlson, M., Hermanson, P., Kovacevic, N., McGill, M.A., Seshadri, V., Yoyokie, J., Cone, K., Kaeppler, H.F., Kaeppler, S.M., and N.M. Springer (2007) Assessing the efficiency of RNA interference for maize functional genomics. *Plant Physiol* **143**: 1441–1451.

Messing, J., and H.K. Dooner (2006) Organization and variability of the maize genome. *Curr Opin Plant Biol* **9**: 157–163.

Schwartz, D. (1965) Regulation of gene action in maize. In: *Genetics Today* (S.V. Geerst, ed.) Oxford, Pergamon, pp. 131–135.

Selinger, D.A., and V.L. Chandler (2001) B-Bolivia, an allele of the maize b1 gene with variable expression, contains a high copy retrotransposon-related sequence immediately upstream. *Plant Physiol* **125**: 1363–1379.

Song, R., and J. Messing (2003) Gene expression of a gene family in maize based on noncollinear haplotypes. *Proc Natl Acad Sci U S A* **100**: 9055–9060.

Spillane, C., Schmid, K.J., Laoueille-Duprat, S., Pien, S., Escobar-Restrepo, J.M., Baroux, C., Gagliardini, V., Page, D.R., Wolfe, K.H., and U. Grossniklaus (2007) Positive darwinian selection at the imprinted MEDEA locus in plants. *Nature* **448**: 349–352.

Springer, N.M., Danilevskaya, O.N., Hermon, P., Helentjaris, T.G., Phillips, R.L., Kaeppler, H.F., and S.M. Kaeppler (2002) Sequence relationships, conserved domains, and expression patterns for maize homologs of the polycomb group genes E(z), esc, and E(Pc) *Plant Physiol* **128**: 1332–1345.

Stupar, R.M., Hermanson, P.J., and N.M. Springer (2007) Non-additive Expression and Parent-of-origin Effects Identified by Microarray and Allele-specific Expression Profiling of Maize Endosperm. *Plant Physiol*. PMID: 17766400

Vielle-Calzada, J.P., Thomas, J., Spillane, C., Coluccio, A., Hoeppner, M.A., and U. Grossniklaus (1999) Maintenance of genomic imprinting at the Arabidopsis medea locus requires zygotic DDM1 activity. *Genes Dev* **13**: 2971–2982.

Walter, J., and M. Paulsen (2003) The potential role of gene duplications in the evolution of imprinting mechanisms. *Hum Mol Genet* **12** :pec No 2, R215–220.

Xiao, W., Brown, R.C., Lemmon, B.E., Harada, J.J., Goldberg, R.B., and R.L. Fischer (2006) Regulation of seed size by hypomethylation of maternal and paternal genomes. *Plant Physiol* **142**: 1160–1168.

Xiao, W., Gehring, M., Choi, Y., Margossian, L., Pu, H., Harada, J.J., Goldberg, R.B., Pennell, R.I., and R.L. Fischer (2003) Imprinting of the MEA Polycomb gene is controlled by antagonism between MET1 methyltransferase and DME glycosylase. *Dev Cell* **5**: 891–901.

Yadegari, R., Kinoshita, T., Lotan, O., Cohen, G., Katz, A., Choi, Y., Nakashima, K., Harada, J.J., Goldberg, R.B., Fischer, R.L., and N. Ohad (2000) Mutations in the FIE and MEA genes that encode interacting polycomb proteins cause parent-of-origin effects on seed development by distinct mechanisms. *Plant Cell* **12**: 2367–2382.

Zhao, J., and G. Grafi (2000) The high mobility group I/Y protein is hypophosphorylated in endoreduplicating maize endosperm cells and is involved in alleviating histone H1-mediated transcriptional repression. *J Biol Chem* **275**: 27494–27499.

Chromatin, DNA Methylation, RNAi and Epigenetic Regulation

Shawn Kaeppler

Abstract Transcriptional and post-transcriptional control of maize gene expression is proving to be important in many aspects of maize biology. The dramatic growth of research in this area has yielded exciting new discoveries which have enhanced our understanding of gene regulation in maize, and have linked previously diverse phenomena to central underlying mechanisms. In this review, I provide a summary of genes considered to function in chromatin-based transcriptional control of gene expression, in post-transcriptional gene silencing, and which link RNA molecules to heritable states of expression. Information from maize, or most relevant to maize, is cataloged according to gene families and groups. The topic area is too broad to provide a thorough synthesis for any gene family or pathway, but the review should allow the reader an entry point to access most maize information on this topic.

1 Overview

Maize has a rich history of fundamental discoveries in chromatin biology. Research in the areas of paramutation (Hollick et al 1995; Hollick and Chandler 2001; Kermicle et al 1995; Sidorenko and Peterson 2001; Stam et al 2002; Walker 1998; Walker and Panavas 2001), imprinting (Alleman and Doctor 2000; Chadhuri and Messing 1994; Danilevskaya et al 2003; Gutierrez-Marcos et al 2007; Haun et al 2007; Hermon et al 2007), epimutation (Cocciolone and Cone 1993; Das and Messing 1994; Hoekenga et al 2000; Sekhon et al 2007), transgene silencing (McGinnis et al 2006), transposon silencing (Banks et al 1988; Martienssen and Baron 1994; Rudenko et al 2003; Schwartz 1989; Slotkin et al 2003; Woodhouse et al 2006a), and response to stress (Casati et al 2006; Kaeppler et al 2000) is leading to valuable phenonomenological and mechanistic insights. In addition, there is growing appreciation of the role of post-transcriptional gene silencing and RNA-directed

S. Kaeppler
University of Wisconsin-Madison, Department of Agronomy,
smkaeppl@wisc.edu

DNA methylation in establishing and maintaining meiotically heritable chromatin states (Alleman et al 2007; Huettel et al 2007), in interpreting environmental signals and stresses (Sunkar et al 2007; Mallory and Vaucheret 2006), and in modulating the process of development (Chuck et al 2007; Lauter et al 2005; Nogueira et al 2007; Juarez et al 2004; Kidner and Martienssen 2005; Parkinson et al 2007). Mechanisms of transcriptional and post-transcriptional gene silencing in maize are being defined using forward and reverse genetic and genomic analyses. In this chapter, I will provide a gene-based summary of resources and fundamental mechanistic discoveries in maize. The goal is to provide the reader an entry point to access maize information and resources in this diverse and rapidly developing research area.

2 Summary of Chromatin and RNAi Genes

Chromatin is the nucleoprotein complex of DNA and proteins, and chromatin state determines epigenetic variation in transcription (Henderson and Jacobsen 2007). Histones are a core protein component of chromatin, and modifications of histones are important in determining chromatin state. Heterochromatin was originally defined as dark-staining portions of chromatin viewed microscopically and is a densely packaged form of chromatin which is inaccessible to transcription factors. Heterochromatin generally is associated with low or absent levels of transcription, DNA containing high levels of methylated cytosines, and histone modifications including lack of acetylation and specific types of methylation. Euchromatin is light staining when viewed microscopically and is a more transcriptionally active chromatin configuration. Euchromatin is generally associated with active transcription, absence of DNA methylation, and histone acetylation. Functions associated with chromatin formation and conversion include proteins which modify DNA and histones, proteins which facilitate the transition among chromatin states, and proteins that recognize a specific chromatin state and participate in functional complexes.

In plants, RNA-based mechanisms function both to accomplish post-transcriptional gene silencing as well as to provide information that is used to establish and maintain chromatin states. Overlapping pathways produce miRNAs and siRNAs. Post-transcriptional gene silencing (RNAi) is accomplished by proteins that ampify the RNA signal and produce miRNAs and siRNAs that target specific RNAs for degradation (Mallory and Vaucheret 2006; Pressman et al 2007). Small RNAs also link posttranscriptional and transcriptional gene silencing, directing chromatin modification to homologous sequences (Pikaard 2006).

2.1 Genes Curated in ChromDB

A systematic search has been conducted for genes putatively involved in transcriptional and post-transcriptional gene silencing. Maize genes were identified using queries with genes and conserved domains of genes with known function in chromatin and RNA silencing. This information is curated and regularly updated at ChromDB.org. The ChromDB nomenclature for these genes is described in Table 1, and synonyms for specific genes can be found in the database. RNAi stocks have been developed for a number of these genes (McGinnis et al 2006a; McGinnis et al 2007), and stock numbers of the RNA lines as well as additional information for some genes, such as map location, can be accessed in MaizeGDB.

Table 1 Chromatin Genes and Nomenclature. Gene groupings, classes, and nomenclature are consistent with curation at ChromDB.org. Gene designations are shown in capital letters preceding the colon (e.g. HON). According to ChromDB notation, maize genes are numbered using 100 series digits following the gene name (e.g. *Hon101, Hon102*). A short description of each sub-class of gene immediately follows the gene name. When available, a founder gene member is given in parentheses to assist the reader in their search for additional information on classes and subclasses of genes. Genes curated by ChromDB are identified based on homology but, in most cases, function has not yet been proven in maize. Class designations shaded in gray have at least one member represented by an RNAi dominant-negative stock in the Maize Genetics Cooperation Stock Center

Group	Classes
Histones and histone linker proteins	*Linker Histones:* HON: Histone H1; SMH: Single myb histone; HMGA: High Mobility Group family A
	Core Histones and Variants: HTA: Histone H2A; HTB: Histone H2B; HTR: Histone H3; HFO: Histone H4
Nucleosome assembly	*Nucleosome Assembly:* NFA: Nucleosome/chromatin assembly (Nap1); NFE: Nucleosome positioning (Acf1); NFF: Nucleosome/chromatin assembly (Cac1); NFB: Nucleosome/chromatin assembly (Cac2); HIRA: histone chaperone (HIRA); TAFV: TATA binding associated factor 5; SGA: Nucleosome assembly (ASF1); HSC: Histone chaperone for Htz1p/H2A-H2B
RNA polymerase transcription and elongation factors	*PAF Complex Components*—PAFA: PAF1 complex member (Paf1); PAFB: PAF1 complex member (Leo1); PAFC: PAF1 complex member (Ctr9); PAFD: PAF1 complex member (Rtf1); PAFE: PAF1 complex member (Cdc73)
	Other RNA Polymerase II Elongation Factors – GTG: Transcription elongation/nucleosome displacement (Spt4); GTA: Transcription elongation/nucleosome displacement (Spt); GTB: Transcription elongation/nucleosome displacement (Spt6); GTC: Global transcription factor group C (Spt16); GTI: Global transcription factor group I (Spt2); SSRP: FACT [Facilitate Chromatin Transcription] (Pob3 and SSRP); ELF: Transcription elongation factor (Elf1)
Chromatin remodeling complexes	*ATP-dependent nucleosome remodeling:* TUP: Transcription repression (Tup1); CHR: SNF2 super family (Snf2, Ris, Rad26, Ddm1); CHB: SWI/SNF and RISC chromatin remodeling (Swi13, Rsc8); CHC: SWI/SNF and RISC chromatin remodeling (Swp73, Rsc6); CHE: SWI/SNF chromatin remodeling (Snf5); ARP: Actin superfamily; RUVBL: DNA-dependent ATP/helicase (RuvB); SWRCA: SWR complex (SWC4)
	Other Chromatin Remodeling and Associated Proteins: NFC: NURF complex (RBBP4/Caf1); PATPA: Proteosomal ATPases Group A; PATPB: Proteosomal ATPases Group B

(continued)

Table 1 (continued)

Group	Classes
Polycomb group – non SET	EPL: Polycomb group (E(Pc)); VEF: Polycomb Group (*VRN2*, *EMF2*, *FIS2*)
Histone modification	*Acetylation:* HAG: Histone acetyltransferase (GNAT superfamily); HAM: Histone acetyltransferase (MYST family); HAC: Histone acetyltransferase (CREBBP (CBP) family); HAF: Histone acetyltransferase (Taf1)
	Deacetylation: HDA: Histone deacetylase (Rpd3/HDA1 superfamily); SRT: Histone deacetylase (SIR2); HDT: Plant-specific histone deacetylase (*HD2*)
	Methylation: SDG: SET domain histone methyl-transferase (Su(var)3-9, Ez, Trx); PRMT: Protein arginine methyltransferase; DOTS: Nucleosomal histone H3-Lys79 methylase (Dot1)
	Demethylation: HDMA: Histone demethylase (AOF2, LSD1); JMJ: Jumonji domain group
	Phosphorylation: TPK: Threonine phosphorylation-associated kinase; SPK: Serine phosphorylation kinase
	ADP Ribosylation: PARP: Poly (ADP-ribose) polymerase family, member 1
	Ubiquination: HUPA: Histone ubiquination group A (Bre1); HUPB: Histone ubiquination group B (Rad6)
	Sumoylation: SAEO: SUMO activating subunit 1; SAET: SUMO activating subunit 2
	Deimination: PADF: Peptidyl arginine deiminase type IV
	Proline Isomerization: PPI: praline isomerase
	Histone modification-associated and complexes: SWDA: COMPASS complex (SWD1); SWDB: COMPASS complex (SWD2); SWDC: Compass complex (SWD3); HXA: Histone acetyltransferase complex (Ada2); HXB: Histone acetyltransferase complex (Ada3); HMTA: Set1 complex (Ash2, Bre2); YDF: YEATS domain-containing family; HCP: Sin3 complex (SAP18); SNT: Histone deacetylase complex (Sin3)
Modified histone-binding	*Bromodomain-containing:* GTE: Global transcription factor group E (Bdf1, BRD4, BRD2(RING3)); BRD: Diverse bromo-domain containing (polybromo1, RSD, BRD); BRAT: Bromodomain-containing AAA-ATPase; BRWD: Bromodomain-containing WD40 repeat
	Chromodomain-containing: CRD: Diverse group of chromodomain containing proteins (except chromomethylase)
	Other: INGF: Inhibitor of Growth protein group (ING1-5); MRG: MRG-domain containing
DNA modifying	DMT: DNA methyltransferase (Dnmt1, Dnmt3, CMT); DNG: DNA glycosylase
Non-histone DNA binding	HMGB: High mobility group B family; MBD: methyl-binding domain containing; EBP: BAH-PHD domain containing; ARID: ARID/BRIGHT DNA binding domain group 2; DEK: DEK_C domain containing; VIM: Variant in DNA mehtylation (*VIM1*)
Chromosome dynamics	CPC: Condensin complex component (SMC1); CPD: Condensin complex component (Cnd1); CPG: Condensin complex component, non SMC subunit; CPH: Condensin complex component, barren domain containing
RNAi components	AGO: Argonaute gene family; DRB: Double-strand RNA binding group; SGS: Suppressor of gene silencing; NRPDB: RNA polymerase IV small subunit; NRPDA: RNA polymerase IV large subunit; RDR: RNA-dependent RNA polymerase; RHEL: RNA helicase; HEN: HUA enhancer; DCL: Dicer-like group; ERI: Enhanced RNA interference

In the remainder of this chapter, I will highlight specific gene groups and classes for which published information is available on maize genes. The goal of this description is to catalog information specifically related to maize, rather than to provide a synthesis of mechanisms based on research across plants and other organisms.

2.2 Histones and Linker Histones – HMGA, HMGB and SMH Families

HMG proteins are small, abundant components of chromatin (Grasser 2003). HMGA proteins contain AT-hook domains, facilitating the interaction with AT-rich DNA. HMGB proteins contain an HMGB box domain, and bind in a sequence non-specific, but chromatin-dependent, manner to DNA. HMGB proteins promote the assembly of specific chromatin complexes. Grasser *et al*. (2007) review the HMGB family in plants and provide a summary of genes from various species – maize has seven HMGB genes – and information on expression and functionality. This review indicates the existence of a monocot-specific clade of HMG proteins. A theme that I will emphasize at the end of this review is that plants across a number of classes of genes have more family members than found in mammals, insects, yeast, and nematodes. Amplification of genes has occurred subsequent to the divergence of plants and animals, as well as subsequent to the divergence of monocots and dicots. The HMGB family demonstrates both of these themes.

Single-myb histones are a plant-specific class of genes that encode proteins that bind to telomeric DNA *in vitro* and that have unique tripartite structures consisting of a myb domain, a domain similar to H1/H5 histones, and a coiled-coil domain (Marian *et al*. 2003). Maize contains six SMBH genes with pairs of genes in groups which arose prior to the divergence of monocots and dicots (Figure 1). This type of gene duplication is consistent with observations across many chromatin genes, and maize genes in general. Current research is consistent with a role for these proteins in telomere chromatin structure, although the mechanisms of their action remain under study.

2.3 Nucleosome Assembly – NFA and HIRA Families

NFA genes are homologs of yeast and *Drosophila* nucleosome assembly protein 1 (NAP1), a histone chaperone which has been implicated in the deposition of H1 in yeast and H2A/H2B dimers in *Drosophila* (Ito et al 1996). Woodhouse et al. (2006a) reported that RNAi transgenic knockdown stocks individually targeting *Nfa101* and *Nfa104* prevent establishment of *Muk* silencing. These knockdown lines did not reactivate previously silenced *MuDR* elements, indicating their role in the establishment, but not maintenance, of the silenced state.

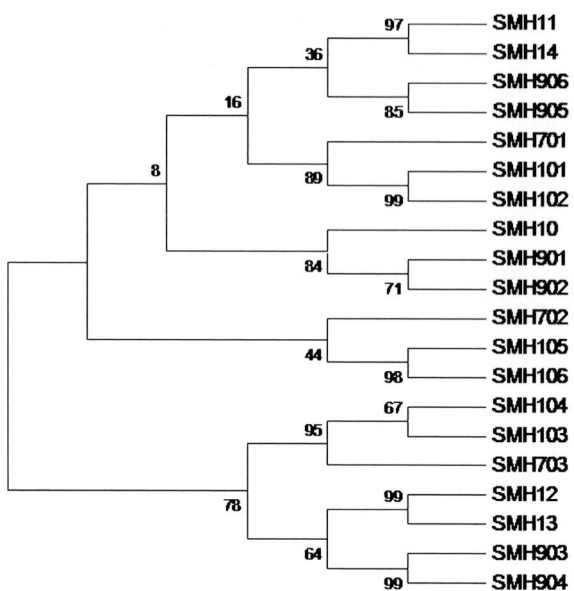

Fig. 1 Single-myb histone (SMH) groups across maize, Arabidopsis, poplar, and rice. The myb domain amino acid sequence of SMH genes available in ChromDB was used to group genes in four species using MEGA2. Maize genes have 100 series numbers following the SMH designation, Arabidopsis numbers are less than 100, rice has 700 series numbers, and poplar has 900 series numbers. Monocot and dicot proteins cluster together within groups, and maize proteins are represented by pairs of similar genes within each group

The HIRA group is founded on yeast genes Hir1p and Hir2p which are histone chaperones that have a role in chromatin-based silencing (Spector et al 1997). Phelps-Durr et al (2005) reported that a maize HIRA protein physically interacts with *Rough sheath2* protein, and that down-regulation of *HIRA* in Arabidopsis resulted in reactivation of *knox* genes in developing leaves. This result supports a role for the Nucleosome Assembly group of genes in maintaining developmental states via participating in silencing of development-specific transcription factors.

2.4   Chromatin Remodeling – CHR and NFC Families

The CHR family is a diverse family of genes in plants. ChromDB lists 48 maize genes in this family, although only a subset are likely to be involved in chromatin remodeling, with others functioning in processes such as recombination and repair. A systematic review of genes in this family is not available, but I will highlight a few key members and their maize homologs. *Ddm1* was first reported in *Arabidopsis* based on a reduced methylation phenotype in individuals homozygous for loss of function (Vongs et al 1993). The maize homologs of *Ddm1* are *Chr101* and *Chr106*. Our attempts to develop RNAi lines containing a construct targeting both *Chr101* and *Chr106* has resulted in an extremely low transformation efficiency and no effectively silenced lines (unpublished data). This observation provides circumstantial evidence that one or both of these maize genes is necessary for cell culture growth and/or plant regeneration. *MOM* was identified in *Arabidopsis* based on the observation that homozygous mutants restored transgene silencing, but that DNA methylation was not decreased in the reactivated transgenes (Amedeo et al 2000). *Chr120* is the maize gene most similar to *MOM*, but phenotypic information on plants mutant for *Chr120* is not available.

A subclass of genes in the CHR group has recently emerged as important for maintenance of heritable chromatin states via a mechanism involving RNA. Hale et al. (2007) discovered that *rmr1* encodes a SNF2 protein that is required for the maintenance of a paramutegenized Pl allele. This subclass of genes is defined by the *Arabidopsis* genes *DRD1* (Kanno et al 2004, 2005) and *CLSY1* (Smith et al 2007) and the maize gene *Rmr1* (Hale et al 2007). Other maize genes in this group include *Chr127* and *Chr156* – no function has been defined for either gene.

Rossi et al. (2001) identified and characterized the maize genes *ZmRbAp1*, *ZmRbAp2*, and *ZmRbAp3*. This is a family of genes that encode WD repeat proteins corresponding to *Nfc101* and *Nfc102* in ChromDB. Rossi et al, (2001) reported high levels of transcription at the initiation of endosperm development, and in shoot apical meristems and leaf primordial, indicating a potential role for these proteins in endosperm formation and plant development.

2.5   Polycomb Group, non-SET

Polycomb group proteins were identified in *Drosophila* for their effect on the polycomb phenotype. Springer *et al.* (2002) documented plant homologs of the polycomb group genes that encode the proteins Enhancer of zeste (*E(z)*), Extra sex combs (*Esc*), and Enhancer of polycomb (*E(Pc)*). Since that publication, the polycomb group gene *Suz(12)* has been characterized and the maize homologs of this gene are *Vefl01* and *Vefl02*. Maize *E(z)* homologs *Mez1*, *Mez2*, and *Mez3* (a.k.a. *Sdg124*, *Sdg125*, and *Sdg126*) contain a SET domain and function as histone methyltransferases, so will be discussed in that context in the next section. *Arabidopsis FIE1* is an *Esc* homolog, and was identified based on its role in seed development (Ohad et al 1999). The inferred function of the polycomb complex in maize is to establish a repressive chromatin state, although specific targets and phenotypes have not yet been demonstrated. Interestingly, maize members of this family, including *Mez1* (a.k.a. *Sdg124*) and *ZmFie1*, are subject to the epigenetic phenomenon of imprinting in the endosperm (Danilevskaya et al 2003; Haun et al 2007; Hermon et al 2007).

3   Histone Modification

Histone modification is important in establishing and maintaining chromatin states. Types of histone modifications and their significance are broadly reviewed by Fuchs *et al.* (2006), although there are many biologically significant subtleties across species that affect the functional interpretation of histone modifications. Histone methylation and acetylation states are the most widely studied chromatin modifications and relevant information on these families will be described below.

3.1   Histone Methylation – SDG Family

Histone methylation is accomplished by proteins containing a SET domain, with gene names based on the presence of this domain in the proteins *Su(var)3-9*, *Enhancer of zeste*, and *Trithorax*. Histone methylation of lysine residues can occur in a mono-, di-, or tri-methyl state. Shi and Dawe (2006) evaluated the cytological distribution of histone methylation across maize chromosomes for modifications at the position H3K4, H3K9, H3K27, and H4K20. They concluded that H3K27me2 marks classical heterochromatin such as knobs and chromomeres, that H3K4me2 is found near chromomeres demarcating the euchromatic space, that H3K9me2 marks euchromatin, and that centromeres and CEPP-C are associated with H3K9me2 and H3K9me3. An intriguing discovery was that H3K27me3 occurs in a small number of euchromatic regions of undefined composition. H4K20 di- and tri-methylation were found to be nearly or completely absent in maize.

   Maize contains over 30 SET-domain containing genes that are likely involved in histone methylation. Reviews by Ng et al. (2007) and Springer et al. (2003) provide useful categorization of these genes and informatic prediction of their likely function. Haun et al. (2007) described expression, evolution, and imprinting of the *Enhancer of zeste* homologs *Sdg124*, *Sdg125*, and *Sdg126*. *Sdg118* is the maize gene most similar gene to Arabidopsis *Kryptonite* and will likely play a similar role. Although SDG genes underlying specific epigenetic phenomena have not yet been reported in the literature, this group of genes will certainly be important, based on the importance of histone methylation in other species.

3.2   Histone Acetylation – HAG, HAM, HAC, HAF Families, and Deacetylation – HDA, SRT, HDT Families

Histone acetylation and deacetylation is a dynamic process central to chromatin structure and remodeling. Information on types and predicted function of histone acetylases and deacetylases, including some maize genes, is provided in Pandey et al. (2002).

Limited information is available on maize histone acetyltransferases. Bhat et al. (2003) cloned histone acetyltransferase *ZmGCN5* (a.k.a. *Hag101*) and demonstrated *in vitro* ability of the protein it encodes to acetylate core histones. The ZmGCN5 protein was shown to interact with the adapter protein encoded by *ZmADA2* and with the bZip factor protein from *ZmO2*. Subsequent research (Bhat et al 2004) demonstrated this interaction *in planta*. Earley et al. (2007) provided information on *in vitro* specificities of histone acetyltransferases in Arabidopsis and the likely progression of the process of histone acetylation. Due to the high level of conservation of histones across organisms, and the generally similar complement of histone acetyltransferases in maize and Arabidopsis, the Earley et al, (2007) results likely provide information that is highly relevant to maize research.

Detailed biochemical studies have defined acetyltransferase and deacetylase activities in maize (Kolle et all 1999, Lopez-Rodas et al 1991). RPD3-type histone deacetylases (HDA class) have been detected in maize embryos (Lechner et al 2000) and mechanisms of regulation have been defined (Pipal et al 2003). The cellular localization, and transcript and protein levels have been characterized for the class members *Hda101*, *Hda102*, and *Hda108*, indicating detectable expression in many tissues, with expression varying with development (Varotto et al 2003). Proteins encoded by this class of histone deacetylases has been shown to interact with ZmRBR1, a regulator of the cell cycle, supporting their role in actively dividing cells (Rossi et al 2003; Varotto et al 2003). A role for this class of histone-deacetylases in the process of development and transcriptional regulation was demonstrated by overexpression and down-regulation of *Hda101* (Rossi et al 2007). Eleven genes are listed in ChromDB as members of the Rpd3 class of histone deacetylases.

The plant specific HD2-type of histone deacetylase was first reported in maize, based on biochemical purification and characterization of the HD2 complex from maize embryos (Brosch et al 1996). An HD2-encoding gene was cloned and the protein characterized as nucleolar (Lusser et al (1997), suggesting a possible function in ribosomal gene function. Four genes are listed in ChromDB as coding for members of the plant-specific HD2 class of histone deacetylases.

4 DNA Methylation

DNA methylation is the molecular mark that has been associated longest with gene silencing. Association of increased methylation with transposon activity provided evidence of a link between modifications outside of the primary sequence and levels of gene expression. DNA methylation was the initial molecular mark for inactive chromatin, and amounts of methylation can vary with the state of gene or transposon activity, through the course of development (e.g. Lauria et al 2004), and in response to stress (e.g. Steward et al 2002). Methylation status has also been used to separate genomic DNA, for example, to produce low-copy-number DNA libraries for sequencing (Emberton et al. 2005, Rabinowicz et al 2005).

4.1 DNA Modification – DMT Family

Plant DNA can be methylated at the 5-position of cytosine residues. Methylation can occur in the symmetric sequence contexts of CG, CAG, CTG, or CCG, and in the asymmetric context in which the next 3′ G is at least two bases removed from the nearest 5′ C.

Maize contains seven DMT genes with proven or likely methyltransferase activity, and one gene (*Dmt106*) with homology to this family but which is missing highly conserved residues (Figure 2; Pavlopoulou and Kossida 2007). Chromomethylase *Dmt102* (a.k.a. *Zmet2*) was shown to condition CNG, but not CG, methylation in maize (Papa et al 2001) consistent with the role of this plant-specific methyltransferase in other species. Makarevitch *et al.* (2007) identified quiescent genes found in specific inbred lines that were reactivated when a *dmt102* mutant allele was introgressed, indicating a role for *Dmt102* in the maintenance of suppression of naturally occurring epialleles. The existence of DRM *de novo* methyltransferases (Cao et al 2000) and Dnmt1-type CG maintenance methyltransferases (Steward et al 2000) in maize has been reported, but their functions in maize have not been proven.

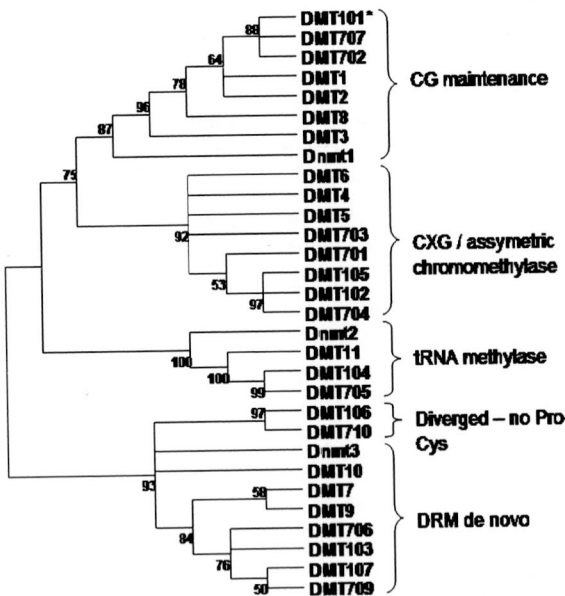

Fig. 2 DNA methyltransferase (DMT) groups across maize, Arabidopsis, and rice. Relationships were determined, based on a conserved portion of the catalytic domain across the three species using MEGA2. Maize genes have 100 series numbers following the SMH designation, Arabidopsis numbers are less than 100, and rice has 700 series numbers. *Dmt101* is represented as a single entity in the tree, but evidence suggests that B73 contains near-identical paralogs of this gene in close proximity in the genome. *Dmt106* is an unusual relative of the DMT family that lacks the Pro-Cys doublet which has been associated with catalytic function. The primary predicted function of each class is based on homology or, in the case of chromomethylases, as proven in maize

Dmt106 is an interesting gene that is transcribed, is apparently capable of producing an intact protein in maize, and is conserved across grasses, yet lacks critical amino acids generally considered to be required for methyltransferase function.

4.2 Methyl-Binding Proteins

DNA methylation in many species is interpreted through interaction with proteins that specifically or preferentially bind methylated DNA. Genes encoding proteins with homology to mammalian methyl-binding proteins have been reported in plants, including maize (Springer and Kaeppler 2005). Studies in Arabidopsis indicate that only some of these proteins preferentially bind methylated DNA whereas others bind DNA of any methylation state equally, or bind only unmethylated DNA (Zemach and Grafi 2006). The MBD family is characteristic of many chromatin proteins in that homology with non-plant proteins is primarily restricted to functional domains, and there is limited homology to non-plant proteins outside those domains. However, sequence alignments of plant proteins often reveal uncharacterized regions of high conservation indicating that chromatin proteins in plants likely operate via unique and mostly uncharacterized signal transduction pathways and chromatin complexes in plants.

The recent discovery that the SRA domain of the *VIM*1 (Woo et al 2007) *and KYP* (Johnson et al 2007) proteins in Arabidopsis preferentially binds methylated DNA provides evidence that previously uncharacterized types of proteins may be as or more impor

5 Role of RNA in Post-Transcriptional and Heritable Gene Silencing – SGS and RDR Families

Evidence for the importance of miRNA silencing in maize development is rapidly accumulating. Juarez et al. (2004) described developmental control of *rolled leaf1* by miRNA166, a pathway that specifies maize leaf polarity. Lauter et al. (2005) discovered that *glossy15*, a gene important in the juvenile to adult transition, is controlled by miRNA172. Recently, the heterochronic mutant *Corngrass1* was found to result from the duplication and thereby overexpression of miRNA156 (Chuck et al 2007). The mechanisms of miRNA production and silencing have been characterized in other species, and maize homologs of these genes have been identified. A report by Nogueira et al. (2007) revealed that *leafbladeless1* (a.k.a. *Sgs101* in ChromDB) is a homolog of Arabidopsis *SGS3*. Mutants of *leafbladeless1* have reduced production of trans-acting siRNA and thus altered the accumulation of miRNA166. An exciting recent discovery provides evidence of a role for RNA in maintaining heritable silencing. Alleman et al. (2006) reported that *mediator of paramutation1* is an RNA-dependent RNA polymerase that is associated with the production of siRNAs. *Mop1* is a gene shown to be required for the establishment and maintenance of paramutation at multiple loci (Dorweiler et al 2000), maintenance of silenced transgenes (McGinnis et al 2006a), maintenance of silenced *MuDR* transposons (Woodhouse et al 2006a, 2006b), and in stabilizing the process of development (Dorweiler et al 2000). Therefore, this discovery provides evidence that the siRNA pathway is important in somatically and meiotically heritable epigenetic silencing across diverse genes and phenotypes.

6 Targeted Gene Silencing Using RNAi

RNAi has been used experimentally to reduce expression of specific maize genes. Segal *et al.* (2003) reported that an inverted repeat transgenic construct was effective in reducing *Opaque2* expression as determined by reduction in target RNA, reduction in 22 kD zein, and an opaque kernel phenotype. Cigan *et al.* (2005) found that an inverted repeat transgene which targeted the *Ms45* promoter was effective in producing a male-sterile phenotype, indicating that dsRNA silencing can be effective in reducing transcription initiation as well as transcript stability. McGinnis et al. (2007) summarized a large functional genomics effort that utilized RNAi to produce mutant stocks for an array of maize chromatin genes. Briefly, RNAi was shown to be effective in reducing target gene expression, and individual constructs could reduce expression of multiple related maize genes. However, there was variability in the degree of silencing within and between constructs, there was difficulty in recovering effectively silenced lines for some constructs likely due to deleterious effects during the transformation process, and there were multiple examples in which expression of the RNAi construct was silenced over sexual generations. The general summary of this article is that RNAi can be an effective tool in maize, but molecular confirmation that the transgene is inducing the desired effect in the relevant tissue is necessary in interpreting the results of studies which use RNAi knockdown stocks.

7 Future Research Priorities

This review highlights exciting discoveries that have been made in maize. Even more so, I think it highlights the incredible amount that remains to be discovered. I will highlight a few areas that I think are important near-term priorities in this area.

1. *Functional characterization of gene families:* An interesting observation, exemplified by the DMT class (Figure 2), is that plants have many more chromatin genes than non-plant counterparts. For example, mammals have 3 known DNA methyltransferases (Dnmt1, Dnmt2a, Dnmt2b) whereas maize has at least 6, including the plant-specific chromomethylase class. This observation holds across many of the gene classes, indicating that plants may use this type of transcriptional control more extensively than mammals. Furthermore, even genes with a characterized function are likely to have additional unreported functions.

As with all genes in maize, functional characterization is a high near-term priority to allow us to better understand basic processes in plants.

2. *Analysis of plant-specific domains*: Domains of chromatin and RNAi proteins that are conserved across diverse species are usually involved in specific binding or catalytic activity. The structure of plant chromatin and RNAi proteins outside these conserved domains usually is dramatically different from their non-plant counterparts, although additional plant-specific regions of conservation can be detected. The presence of plant-specific conserved domains indicates that the pathways and complexes in which plants utilize these proteins are substantially different from those in animals. A largely unexplored area of chromatin biology is the way in which environmental and developmental signals in plants are manifested into heritable chromatin states.

3. *Role of RNA in establishing and maintaining heritable states of expression:* Chromatin-based silencing is a unique mechanism to establish heritable, but reversible, allelic variation. This variation may occur at a frequency substantially more frequent than changes in the primary sequence. The molecular mechanisms controlling the establishment and maintenance of heritable states of expression are just being described. Further research in this exciting area, which links diverse types of heritable silencing phenotypes to common underlying mechanisms, promises to provide valuable mechanistic insight into the processes by which states of expression are established and maintained.

4. *Role of chromatin in endogenous allelic variation:* Examples of alleles subject to epigenetic silencing have been of long-standing interest to the scientific community, but questions have persisted as to whether these examples are rare exceptions or represent a more common basis of allelic variation. The recent study by Markarevitch *et al.* (2007) suggest that epiallelic variation likely is present in many maize lines and is not restricted to unusual genes or alleles. Searches for additional examples of epiallelic variation in mutant stocks for genes beyond *Dmt102* will enhance our understanding of the role of epigenetic modification underlying the vast phenotypic diversity of maize. Since epialleles would not be detected using standard SNP methodology, improved molecular assays for this type of variation among genotypes will be needed to fully explore variation in the maize genome.

Acknowledgments I express my appreciation to Carolyn Napoli for providing the list of chromatin gene groups and subgroups utilized in ChromDB. I apologize to all authors whose articles could not be presented due to space limitations.

References

Alleman, M., and Doctor, J. (2000) Genomic imprinting in plants: observations and evolutionary implications. Plant Mol. Biol. 43,147–161.

Alleman, M., Sidorenko, L., McGinnis, K., Seshadri, V., Dorweiler, J.E., White, J., Sikkink, K., and Chandler, V.L. (2006) An RNA-dependent RNA polymerase is required for paramutation in maize. Nature 442,295–298.

Amedeo, P., Habu, Y., Afsar, K., Mittelsten Scheid, O., and Paszkowski, J. (2000) Disruption of the plant gene MOM releases transcriptional silencing of methylated genes. Nature 405,203–206.

Banks, J.A., Masson, P., and Federoff, N. (1988) Molecular mechanisms in the developmental regulation of the Suppressor-mutator maize transposable element system. Genes Devel. 2,1364–1380.

Bhat, R.A., Borst, J.W., Riehl, M., and Thompson, R.D. (2004) Interaction of maize Opaque-2 and the transcriptional co-activators GCN5 and ADA2, in the modulation of transcriptional activity. Plant Molec. Biol. 55,239–252.

Bhat, R.A., Riehl, M., Santandrea, G., Velasco, R., Slocombe, S., Donn, G., Steinbiss, H-H., Thompson, R.D., and Becker, H-A. (2003) Alteration of GCN5 levels in maize reveals dynamic responses to manipulating histone acetylation. Plant J. 33,455–469.

Brosch, G., Lusser, A., Goralik-Schramel, M., and Loidl, P. (1996) Purification and characterization of a high molecular weight histone deacetylase complex (HD2) of maize embryos. Biochem. 35,15907–15914.

Cao, X., Springer, N.M., Muszynski, M.G., Phillips, R.L., Kaeppler, S.M., and Jacobsen, S.E. (2000) Conserved plant genes with similarity to mammalian *de novo* DNA methyltransferases. Proc. Nat'l. Acad. Sci. USA 97,4979–4984.

Casati, P., Stapleton, A.E., Blum, J.E., Walbot, V. (2006) Genome wide analysis of high-altitude maize and gene knock-down stocks implicates chromatin remodeling proteins in response to UV-B. Plant J. 46,613–627.

Chuck, G., Cigan, A.M., Saeteum, K., and Hake, S. (2007) The heterochronic maize mutant Corngrass1 results from overexpression of a tandem microRNA. Nature Genet. 39,544–549.

Chadhuri, S., and Messing, J. (1994) Allele-specific imprinting of *dzr1*, a post-transcriptional regulator of zein accumulation. Proc. Nat'l. Acad. Sci. USA. 91,4867–4871.

Chuck, G., Cigan, A.M., Saeteurn, K., and Hake, S. (2007) The heterochronic maize mutant *Corngrass1* results from overexpression of a tandem repeat microRNA. Nature Genet. 39,544–549.

Cigan, A.M., Unger-Wallace, E., and Haug-Collet, K. (2005) Transcriptional gene silencing as a tool for uncovering gene function in maize. Plant J. 43,929–940.

Cocciolone, S.M., and Cone, K.C. (1993) *Pl-Bh*, an anthocyanin regulatory gene of maize that leads to variegated pigmentation. Genetics 135,575–588.

Danilevskaya, O.N., Hermon, P., Hantke, S., Muszynski, M.G., Kollipara, K., Ananiev, E.V. (2003) Duplicated fie genes in maize: expression pattern and imprinting suggest distinct functions. Plant Cell 15,425–438.

Das, P. and Messing, J. (1994) Variegated phenotype and developmental methylation changes of a maize allele originating from epimutation. Genetics 136,1121–1141.

Dorweiler, J.E., Carey, C.C., Kubo, K.M., Hollick, J.B., Kermicle, J.L., and Chandler, V.L. (2000) *mediator of paramutation1* is required for establishment and maintenance of paramutation at multiple loci. Plant Cell 12,2101–2118.

Earley, K.W., Shook, M.S., Brower-Toland, B., Hicks, L., Pikaard, C. (2007) *In vitro* specificities of Arabidopsis co-activator histone acetyltransferases: implications for histone hyperacetylation in gene activation. Plant J. 52:615–626.

Emberton, J., J. Ma, Y. Yuan and J.L. Bennetzen (2005) Gene enrichment in maize with hypomethylated partial restriction (HMPR) libraries. *Genome Res.* 15:1441–1446.

Fuchs, J., Demidov, D., Houben, A., and Schubert, I. (2006) Chromosomal histone modification patterns – from conservation to diversity. Trends Plant Sci. 11,199–208.

Grasser, K.D. (2003) Chromatin-associated HMGA and HMGB proteins: versatile co-regulators of DNA-dependent processes. Plant Molec. Biol. 53,281–295.

Grasser, K.D., Launholt, D., and Grasser, M. (2007) High mobility group proteins of the plant HMGB family: dynamic modulators of chromatin. Biochim. Biophys. Acta 1769,346–357.

Gutierrez-Marcos, J.F., Costa, L.M., Pra, M.D., Scholten, S., Kranz, E., Perez, P., and Dickinson, H.G. (2007) Epigenetic asymmetry of imprinted genes in plant gametes. Nat. Genet. 38,876–878.

Hale, C.J., Stonaker, J.L., Gross, S.M., and Hollick, J.B. (2007) A novel SNF2 protein maintains transgenerational regulatory states established by paramutation in maize. PLoS Biology (In Press)

Haun, W.J., Laoueille-Duprat, S., O'Connell, M.J., Spillane, C., Grossniklaus, U., Phillips, A.R., Kaeppler, S.M., and Springer, N.M. (2007) Genomic imprinting, methylation and molecular evolution of maize Enhancer of Zeste (Mez) homologs. Plant J. 49, 325–337.

Henderson, I.R. and Jobsen, S.E. (2007) Epigenetic inheritance in plants. Nature 447, 418–424.

Hermon, P., Srilunchang, K.O., Zou, J., Dresselhaus, T., Danilevskaya, O.N. (2007) Activation of the imprinted Polycomb Group Fie1 gene in maize endosperm requires demethylation of the maternal allele. Plant Mol. Biol. 64, 387–395.

Hoekenga, O.A., Muszynski, M.G., and Cone, K.C. (2000) Developmental patterns of chromatin structure and DNA methylation responsible for epigenetic expression of a maize regulatory gene. Genetics 155,1889–1902.

Hollick, J.B., and Chandler, V.L. (2001) Genetic factors required to maintain repression of a paramutagenic maize pl1 allele. Genetics 157,369–378.

Hollick, J.B., Patterson, G.K., Coe, Jr. E.H., Cone, K.C., and Chandler, V.L. (1995) Allelic interactions heritably influence the activity of a metastable maize pl allele. Genetics 141,709–719.

Huettel, B. Kanno, T., Daxinger, L., Bucher, E., van der Winden, J., Matzke, A.J.M., and Matzke, M. (2007) RNA-directed DNA methylation mediated by DRD1 and PolIVb: A versatile pathway for transcriptional gene silencing in plants. Bioch. Bioph. Acta 1769, 358–374.

Ito, T., Bulger, M., Kobayashi, R., and Kadonaga, J.T. (1996) Drosophila NAP-1 is a core histone chaperone that functions in ATP-facilitated assembly of regularly spaced nucleosomal arrays. Mol. Cell. Biol. 16,3112–3124.

Johnson, L.M., Bostick, M., Zhang, X., Kraft, E., Henderson, I., Callis, J., and Jacobsen, S.E. (2007) The SRA methyl-cytosine-binding domain links DNA and histone methylation. Current Biol. 17,379–384.

Juarez M.T., Kui, J.S., Thomas, J., Heller, B.A., Timmermans, M.C. (2004) microRNA repression of rolled leaf1 specifies maize leaf polarity. Nature 428,84–88.

Kaeppler, S.M., Kaeppler, H.F., and Rhee. Y. (2000) Epigenetic aspects of somaclonal variation in plants. Plant Molec. Biol. 43,179–188.

Kanno, T., Mette, M.F., Kreil, D.P., Aufsatz, W., Matzke, M., and Matzke, A.J.M. (2004) Involvement of putative SNF2 chromatin remodeling protein DRD1 in RNA-directed DNA methylation. Current Biology 14,801–805.

Kanno, T., Aufsatz, W., Jaligot, E., Mette, M.F., Matzke, M., and Matzke, A.J.M. (2005) A SNF2-like protein facilitates dynamic control of DNA methylation. EMBO 6,649–655.

Kermicle, J.L., Eggelston, W.B., and Alleman, M. (1995) Organization of paramutagenecity in R-stippled maize. Genetics 141,361–372.

Kidner, C.A, and Martienssen, R.A. (2005) The developmental role of microRNA in plants. Curr. Opin. Plant Biol. 8, 38–44.

Kolle, D., Brosch, G., Lechner, T., Pipal, A., Helliger, W., Taplick, J., and Loidl, P. (1999) Different types of maize histone deacetylases are distinguished by a highly complex substrate and site specificity. Biochem. 38,6769–6773.

Lauria, M., Rupe, M., Guo, M., Kranz, E., Pirona, R., Viotti, A., and Lund, G. (2004) Extensive DNA hypomethylation in the endosperm of Zea mays. Plant Cell 16,510–522.

Lauter, M., Kampani, A., Carlson, S., Goebel, M., and Moose, S.P. (2005) microRNA 172 down-regulates glossy15 to promote vegetative phase change in maize. Proc. Nat'l. Acad. Sci. 102,9412–9417.

Lechner, T., Lusser, A., Pipal, A., Brosch, G., Loidl, A., Goralik-Schramel, M., Sendra, R., Wegener, S., Walton, J.D., and Loidl, P. (2000) RPD3-type histone deacetylases in maize embryos. Biochem. 39,1683–1692.

Lopez-Rodas, G., Georgieva, E.K., Sendra, R., and Loidl, P. (1991) Histone acetylation in Zea mays I: activities of histone acetyltransferases and histone deacetylases. J. Biol. Chem. 266,18745–18750.

Lusser, A., Brosch, G., Loidl, A., Haas, H., and Loidl, P. (1997) Identification of maize histone deacetylase HD2 as an acidic nucleolar phosphoprotein. Science 277,88–91.

Makarevitch, I., Stupar, R.M., Iniguez, A.L., Haun, W.J., Barbazuk, W.B., Kaeppler, S.M., and Spinger, N.M. (2007) Natural variation for alleles under epigenetic control by the maize chromomethylase zmet 2. Genetics 177,1–12.

Mallory, C.A., and Vaucheret, H. (2006) Functions of microRNAs and related small RNAs in plants. Nat. Genet. 38,S31–S36.

Marian, C.O., Bordoli, S.J., Goltz, M., Santarella, R.A., Jackson, L.P., Danilevskaya, O., Beckstette, M., Meeley, R., and Bass, H.W. (2003) The maize *Single myb histone 1* gene, *Smh1*, belongs to a novel gene family and encodes a protein that binds telomere DNA repeats *in vitro*. Plant Physiol. 133,1336–1350.

Martienssen, R. and Baron, A. (1994) Coordinate suppression of mutations caused by Robertson's mutator transposons in maize. Genetics 136,1157–1170.

McGinnis, K.M., Chandler, V., Cone, K., Kaeppler, H., Kaeppler, S., Kerschen, A., Pikaard, C., Richards, E., Sidorenko, L., Smith, T., Springer, N., and Wulan, T. (2006a) Transgene-induced RNA interference as a tool for plant functional genomics. Meth. Enzymol. 392,1–24.

McGinnis, K.M., Springer, C., Lin, Y., Carey, C.C., and Chandler, V. (2006b) Transcriptionally silenced transgenes in maize are activated by three mutations defective in paramutation. Genetics 173,1627–1647.

McGinnis, K.M., Murphy, N., Carlson, A.R., Akula, A., Akula, C., Basinger, H., Carlson, M., Hermanson, P., Kovacevic, N., McGill, M.A., Seshadri, V., Yoyokie, J., Cone, K., Kaeppler, H.F., Kaeppler, S.M., and Springer, N.M. (2007) Assessing the efficiency of RNA interference for maize functional genomics. Plant Physiol. 143,1441–1451.

Ng, D.W-K., Want, T., Chandrasekharan, M.B., Aramayo, R., Kertbundit, S., and Hall, T.C. (2007) Plant SET domain-containing proteins: structure, function, and regulation. Biochim. Biophys. Acta 1769,316–329.

Nogueira, F.T., Madi, S., Chitwood, D.H., Juarez M.T., and Timmermans, M.C. (2007) Two small regulatory RNAs establish opposing fates of a developmental axis. Genes Devel. 21,750–755.

Ohad, N., Yadegari, R., Margossian, L., Hannon, M., Michaeli, D., Harada, J.J., Goldberg, R.B., Fischer, R.L. (1999) Mutations in RIE, a WD polycomb group gene, allow endosperm development without fertilization. Plant Cell 11,407–416.

Papa, C.M., Springer, N.M., Muszynski, M.G., Meeley, R., and Kaeppler, S.M. (2001) Maize chromomethylase *Zea methyltransferase2* is required for CpNpG methylation. Plant Cell 13,1919–1928.

Pandey, R., Muller, A., Napoli, C.A., Selinger, D.A., Pikaard, C.S., Richards, E.J., Bender, J., Mount, D.W., and Jorgenson, R.A. (2002) Analysis of histone acetyltransferase and histone deacetylase families in Arabidopsis thaliana suggests functional diversification of chromatin modification among multicellular eukaryotes. Nucl. Acids Res. 30,5036–5055.

Parkinson, W.C., Gross, S.M., and Hollick, J.B. (2007) Maize sex determination and abaxial leaf fates are canalized by a factor that maintains repressed epigenetic states. Dev. Biol. 308,462–473.

Pavlopoulou, A. and Kossida, S. (2007) Plant cytosine-5 DNA methyltransferases: Structure, function, and molecular evolution. Genomics 90,530–541.

Phelps-Durr, T.L., Thomas, J., Vahab, P., and Timmermans, M.C.P. (2005) Maize rough sheath 2 and its *Arabidopsis* ortholog ASYMMETRIC LEAVES 1 interact with HIRA, a predicted histone chaperone, to maintain *knox*, gene silencing and determinacy during organogenesis. Plant Cell 17,2886–2898.

Pikaard, C.S. (2006) Cell biology of the *Arabidopsis* nuclear siRNA pathway for RNA-directed chromatin modification. Cold Spring Harbor Symp. Quant. Biol. 71:473–480.

Pipal, A., Goralik-Schramel, M., Lusser, A., Lanzanova, C., Sarg, B., Loidl, A., Lindner, H., Rossi, V., Loidl, P. (2003) Regulation and processing of maize histone deacetylase Hda1 by limited proteolysis. Plant Cell 15,1904–1917.

Pressman, S., Bei, Y., and Carthew, R. (2007). Snapshot: Posttranscriptional gene silencing. Cell 130:570.

Rabinowicz, P.D., Citek, R., Budiman, M.A., Nunberg, A., Bedell, J.A., Lakey, N., O'Shaughnessy, A.L., Nascimento, L.U., McCombie, W.R., and Martienssen, R.A. (2005) Differential methylation of genes and repeats in land plants. Genome Res. 15,1431–1440.

Rossi, V., Locatelli, S., Lanzanova, C., Boniotti, M.B., Varotto, S., Pipal, A., Goralik-Schramel, M., Lusser, A., Gatz, C., Guttierez, C., and Motto, M. (2003) A maize histone deacetylase and retinoblastoma-related protein physically interact and cooperate in repressing gene function. Plant Molec. Biol. 51,401–413.

Rossi, V., Locatelli, S., Varotto, S., Donn, G., Pirona, R., Henderson, D.A., Hartings, H., and Motto, M. (2007) Maize histone deacetylase *hda101* is involved in plant development, gene transcription, and sequence-specific modulation of histone modification of genes and repeats. Plant Cell 19,1145–1162.

Rossi, V., Varotto, S., Locatelli, S., Lanzanova, C., Lauria, M., Zanotti, E., Hartings, H., and Motto, M. (2001) The maize WD-repeat gene *ZmRbAp1* encodes a member of the MSI/RbAp sub-family and is differentially expressed during endosperm development. Mol. Genet. Genomics 265,576–584.

Rudenko, G.N., Ono, A., and Walbot, V. (2003) Initiation of silencing of maize *MuDR/Mu* transposable elements. Plant J. 33,1013–1025.

Schwartz, D. (1989) Gene-controlled cytosine demethylation in the promoter region of the Ac element of maize. Proc. Nat'l. Acad. Sci. USA 86,2789–2793.

Segal, G., Song, R., and Messing, J. (2003) A new opaque variant of maize by a single dominant RNA-interference-inducing transgene. Genetics 165,387–397.

Sekhon, R.S., Peterson, T., and Chopra, S. (2007) Epigenetic modifications of distinct sequences of the *pl* regulatory gene specify tissue-specific expression patterns in maize. Genetics 175,1059–1070.

Shi, J. and Dawe, R.K. (2006) Partitioning of the maize epigenome by the number of methyl groups on histone H3 lysines 9 and 27. Genetics 173,1571–1583.

Sidorenko, L.V. and Peterson, T. (2001) Transgene-induced silencing identifies sequences involved in the establishment of paramutation of the maize *pl* gene. Plant Cell 13,319–335.

Slotkin, R.K., Freeling, M., and Lisch, D. (2003) Mu killer causes the heritable inactivation of the *Mutator* family of elements in *Zea mays*. Genetics 165,781–797.

Smith, L.M., Pontes, O., Searle, I., Yelina, N., Yousafzai, F.K., Herr, A.J., Pikaard, C.S., and Baulcombe, D.C. (2007) An SNF2 protein associated with nuclear RNA silencing and the spread of a silencing signal between cells in *Arabidopsis*. Plant Cell 19:1507–1521.

Spector, M.S., Raff, A., DeSilva, H., Lee, K., and Osley, M.A. (1997) Hir1p and Hir2p function as transcriptional corepressors to regulate histone gene transcription in the *Saccharomyces cereviseae* cell cycle. Mol. Cell. Biol. 17,545–552.

Springer, N.M., Danilevskaya, O.N., Hermon, P., Helentjaris, T.G., Phillips, R.L., Kaeppler, H.F., and Kaeppler, S.M. (2002) Sequence relationships, conserved domains, and expression patterns for maize homologs of the polycomb group genes *E(z), esc,* and *E(Pc)*. Plant Physiol. 128,1332–1345.

Springer, N.M., and Kaeppler, S.M. (2005) Evolutionary divergence of monocot and dicot methyl-CpG-binding domain proteins. Plant Physiol. 138,92–104.

Springer, N.M., Napoli, C.A., Selinger, D.A., Pandey, R., Cone, K.C., Chandler, V.L., Kaeppler, H.F., and Kaeppler, S.M. (2003) Comparative analysis of SET domain proteins in maize and Arabidopsis reveals multiple duplications preceding the divergence of monocots and dicots. Plant Physiol. 132,907–925.

Stam, M., Belele, C., Dorweiler, J.E., and Chandler, V.L. (2002) Differential chromatin structure within a tandem array 100 kb upstream of the maize b1 locus is associated with paramutation. Genes Develop. 16,1906–1912.

Steward, N., Ito, M., Yamaguchi, Y., Koizumi, N., and Sano, H. (2002) Periodic DNA methylation in maize nucleosomes and demethylation by environmental stress. J. Biol. Chem. 277, 37741–37746.

Steward, N., Kusano, T., and Sano, H. (2000) Expression of *ZmMet1*, a gene encoding a DNA methyltransferase from maize, is associated not only with DNA replication in actively proliferating cells, but also with altered DNA methylation status in cold-stressed quiescent cells. Nucl. Acids Res. 28,3250–3259.

Sunkar, R., Viswanathan, C., Zhu, J., and Zhu, J-K. (2007) Small RNAs as big players in plant abiotic stress responses and nutrient deprivation. Trends in Plant Science 12, 301–309.

Varotto, S., Locatelli, S., Canova, S., Pipal, A., Motto, M., and Rossi, V. (2003) Expression profile and cellular localization of maize Rpd3-type histone deacetylases during plant development. Plant Physiol. 133,606–617.

Vongs, A., Kakutani, T., Martienssen, R.A., and Richards, E.J. (1993) *Arabidopsis thaliana* DNA methylation mutants. Science 260, 1926–1928.

Walker, E.L. (1998) Paramutation of the r1 locus of maize is associated with increased cytosine methylation. Genetics 148,1973–1981.

Walker, E.L. and Panavas, T. (2001) Structural features and methylation patterns associated with paramutation at the *r1* locus of *Zea mays*. Genetics 159,1201–1215.

Woo, H.R., Pontes, O., Pikaard, C.S., and Richards, E.J. (2007) VIM1, a methylcytosine-binding protein required for centromeric heterochromatinization. Genes Develop. 21,267–277.

Woodhouse, M.R., Freeling, M., and Lisch, D. (2006a) Initiation, establishment, and maintenance of heritable MuDR transposon silencing in maize are mediated by distinct factors. PLoS Biol. 4,e339.

Woodhouse, M.R., Freeling, M., and Lisch, D. (2006b) The *mop1* (*mediator of paramutation1*) mutant progressively reactivates one of the two genes encoded by the *MuDR* transposon in maize. Genetics 172,579–592.

Zemach, A. and Grafi, G. (2007) Methyl-CpG-binding domain proteins in plants: interpreters of DNA methylation. Trends Plant Sci. 12,80–85.

The B Chromosome of Maize

Wayne Carlson

1 Introduction

Various authors have described the B chromosome in pachytene of meiosis. These include McClintock (1933), Ward (1973a) and Pryor *et al*. (1980). The descriptions vary greatly, depending on the degree of condensation of the bivalent. The distal heterochromatic blocks have been seen as divided into from three to seven subunits. Ward's description probably represents the most commonly seen morphology. He divides the heterochromatin into 4 blocks, with the third most distal being larger than the others. Beckett (1991) labeled these heterochromatic blocks as DH1 to DH4, with DH4 being the most distal. Ward also identified the distal euchromatic tip of the B as being distinct from H4. Fig. 1 shows a pair of Bs in pachytene, while Fig. 2 is a diagram of the pair. Evidence for existence of a small B short arm is genetic, as discussed later.

Randolph (1941) did a comprehensive study of maize B chromosome inheritance. He produced three tables full of data on the progeny of crosses between plants with different B chromosome numbers. Despite Randolph's work, the mode of inheritance of Bs was not clearly understood until Roman (1947, 1948) developed translocations between the B and various A chromosomes. The great advantage of the translocations is that they attach genetic markers to the B, allowing classification of crosses with phenotypic markers rather than by cytological examination. In a typical translocation, two chromosomes are produced, the B-A chromosome and the reciprocal A-B. The B-A chromosome, with the B centromere, shows the same behavior as the standard B chromosome. Roman showed that transmission of the B-A chromosomes through the female was normal but transmission through the male was not. He described an accumulation mechanism, which involves nondisjunction of the B-A (or B) at the

W. Carlson
Department of Biological Sciences
University of Iowa, Iowa City, IA
wrcarlsn@aol.com

J.L. Bennetzen and S. Hake (eds.), *Maize Handbook - Volume II: Genetics and Genomics*,   459
© Springer Science+Business Media LLC 2009

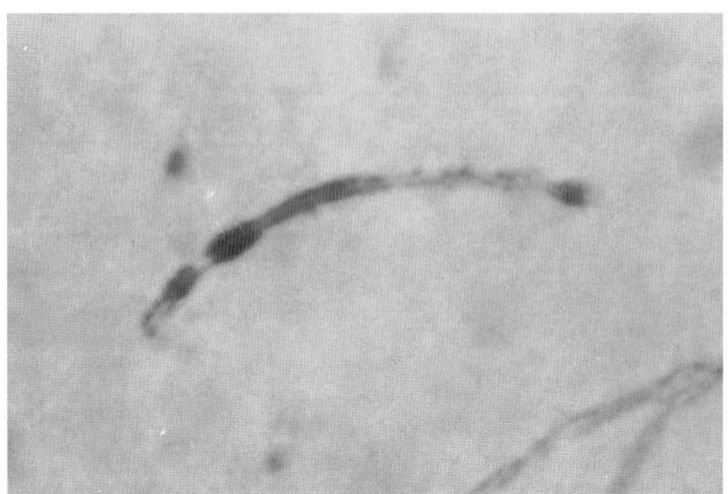

Fig. 1 B chromosome bivalent in pachytene. The light-staining, translucent region at the right tip of the chromosome is the centromere

DH3   C

Fig. 2 Diagram of the B chromosome bivalent in Figure 1. The thin connecting thread between DH3 and DH4 is likely due to stretching of the bivalent during preparation

second pollen mitosis, producing one sperm with two B-A chromosomes and another with zero B-As. Subsequently, the sperm with B-A chromosomes has an advantage in fertilization of the egg.

2   Nomenclature of B-A Translocation Chromosome Types

Inheritance of B-A translocations can result in a large variety of A chromosome constitutions, due to either nondisjunction in the pollen, meiotic segregation or both. Meiotic segregation of the translocations has been discussed by Robertson (1967) and Kindiger et al. (1991). Nondisjunction is reviewed in this chapter.

The most common chromosomes types are classified as 1) balanced, having the complete diploid genome present in proper chromosome dosage 2) hypoploid, having a segmental aneuploidy in which part of one chromosome is in single dosage, and 3) hyperploid, having a segmental aneuploidy in which part of one chromosome is present in triplicate. The balanced diploid can be homozygous, A-B A-B B-A B-A or heterozygous, A A-B B-A. The hypoploid takes the form of A A-B or A-B A-B

B-A. The hyperploid can be A A-B B-A B-A B-A or A-B A-B B-A B-A B-A. Another way of classifying the chromosome types is by referring to the status of the A-B chromosome. This is more specific. The A-B can be heterozygous or homozygous. Among homozygotes, there is the hypoploid homozygote, A-B A-B B-A, the balanced homozygote, A-B A-B B-A B-A and the hyperploid homozygote, A-B A-B B-A B-A B-A. Heterozygotes include the hypoploid heterozygote, A A-B, the balanced heterozygote, A A-B B-A and the hyperploid heterozygote, A A-B B-A B-A.

3   Areas of Study with the B Chromosome

There are at least five types of experiments that utilize the B chromosome. These are: 1) mapping A chromosome genes to chromosome arm, 2) studying the effects of different dosages of A chromosome arms on the organism, 3) analyzing the chromosome type breakage-fusion-bridge cycle, 4) examining various properties of the B chromosome, such as its accumulation mechanisms, affect on crossing over, origin among the A chromosomes, etc., and 5) studying the population dynamics and evolution of the B chromosome. The latter three are discussed below.

4   Analyzing McClintock's Chromosome Type
Breakage-Fusion Bridge Cycle

McClintock (1951) described two types of dicentrics which give rise to breakage-fusion-bridge cycles. There is the chromatid dicentric, involving a connection between the two chromatids of a chromosome. There is also the chromosome dicentric which connects two chromosomes. With the chromatid dicentric cycle, a single bridge is formed in mitotic anaphase, while the chromosome cycle produces double bridges.

Breakage of the bridge in the chromatid cycle sends a broken chromosome to each pole. Later, the broken chromosome undergoes aberrant replication of the telomere-deficient chromosome end, which forms a new chromatid dicentric. This cycle continues at each division in the endosperm, producing variegated phenotypes. The cycle does not ordinarily occur in the plant, due to "healing" of the broken end by addition of telomeric sequences.

The chromosomal cycle does not heal, at least initially, in either the plant or the endosperm. Apparently, the two broken chromosomes, produced at anaphase, fuse with each other before telomeres can be added to stabilize them. The chromosomal cycle is difficult to produce and study because establishing the cycle depends on introducing broken chromosomes from both the sperm and the egg, so that fusion between them can occur.

McClintock utilized modified forms of chromosome 9 to produce both chromatid and chromosomal dicentrics. One of McClintock's specialized forms of chromosome

9 is referred to as duplication 9 (Dp9). It has a duplication of most of the short arm of 9, attached in reverse order to the end of 9S. As a result, foldback pairing and crossing over occur, giving chromatid type dicentrics. Unlike the dicentrics formed by crossing over in inversion heterozygotes, the Dp9 bridge can break to produce non-deficient chromosomes. McClintock, (1939; 1942; 1943) used Dp9 to transmit telomere-deficient chromosomes through both the egg and sperm, establishing a chromosomal dicentric.

It was recognized that the Dp9 system could be adapted for use on the B-9 of TB-9Sb. The duplicated arm of Dp9 was transferred to the B-9 by crossing over. The B-9-Dp9 chromosome can undergo foldback pairing and crossing over at meiosis to give chromatid dicentrics. The chromatid cycle is initiated after bridge formation at anaphase 2 (Fig. 3).

In the male, a broken chromosome arm is transmitted in the first pollen mitosis. DNA replication re-forms a chromatid dicentric. Bridge-breakage sends a broken chromosome to both the tube nucleus and the generative nucleus. The unusual feature is that the chromatid dicentric, present at metaphase of the second pollen mitosis, can undergo nondisjunction, rather than bridge formation. This converts the chromatid cycle to a chromosomal cycle, producing double bridges in plants of the next generation. As a result, formation of the chromosomal cycle is greatly simplified (Carlson 1988).

McClintock showed (1942) that the chromosomal cycle eventually ends, since double bridges are not seen in root tips late in development. The B-9-Dp9 chromosome was used to study the process of healing in the chromosomal cycle (Zheng et al. 1999). The B-9-Dp9 was marked with *Yg2* and the female parent had *yg2 yg2*. Plants with a variegated (yellow and green) phenotype were selected as having the dicentric cycle. In the first sampling of roots from these variegated plants, the great majority (364/410) were confirmed as having the cycle. A group of 137 plants from the group with the chromosomal cycle was selected for further study. A time course study showed that this group declined in the frequency of plants with double bridges from 100% to 7.1% over the ten weeks of sampling. The decline was gradual. Thus, there is no specific time for chromosome healing. Also, various lines of evidence showed that healing can occur at different times in different parts of a plant. Examination of meiosis in plants that had undergone the chromosomal cycle showed a large variety of derivative B-9-type chromosomes, including a group that is very small and referred to as minichromosomes.

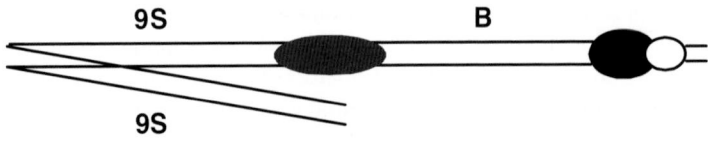

Fig. 3 The B-9-Dp9 chromosome. A crossover between different chromatids will produce a dicentric and an acentric fragment. The latter is lost

Two methods for eliminating double bridges from plants with the B-9-Dp9 chromosome dicentric were considered. One was loss of the dicentric, probably through lagging on the metaphase plate. The other was addition of telomeric sequences to the deficient chromosome ends before they could fuse and re-form the chromosome dicentric. Han *et al.* (2006) demonstrated a third method for eliminating double bridges: centromere inactivation. They examined a collection of 23 B-9 minichromosomes coming from the chromosomal cycle. In 5 cases, they showed, with FISH analysis, that two regions on the minichromosome contained centromeric DNA capable of kinetochore formation. However, immunolabeling with CENH3 (needed for kinetochore activity) demonstrated binding to only one centromere in each case. This is believed to be the first case of centromere inactivation being found in plant chromosomes.

5   Roman's Accumulation Mechanism-Nondisjunction

The rest of this review will cover topics concerning the B chromosome itself, starting with Roman's work. Different sites within the B chromosome have been shown to control nondisjunction at the second pollen mitosis. Each B chromosome region, shown in Fig. 2, will be considered. The order of the chromosome regions discussed is not in terms of chromosome position, but instead convenience of discussion.

5.1   Distal Euchromatic Tip of the B

Roman (1949) found that nondisjunction of the B-4 of TB-4Sa required the presence of the 4-B chromosome. The combination of 4 B-4 in the pollen gave regular disjunction of the B-4, rather than nondisjunction. This discovery began the search for sites on the B that control nondisjunction. The breakpoint in the B of TB-4Sa is at the juncture of the proximal euchromatin and the distal heterochromatin. Any region distal to the breakpoint could be the site required for nondisjunction. Ward (1973b) showed that the 8-B of TB-8La is required for nondisjunction of the B-8. The breakpoint is very distal on the B, so that only the euchromatic tip of the B was transferred to the 8-B. Therefore, one site controlling nondisjunction was localized to a very small region on the B chromosome.

5.2   Proximal Euchromatin

Next, searches for internal sites on the B were initiated by Lin (1978) and Carlson (1973; 1978). Lin constructed numerous B-10L translocations. This allowed him to combine the 10-B from one translocation with the B-10 from another, producing internal deletions of parts of the B chromosome. As long as vital genes for 10L were not deleted, these hybrid translocations could be tested for nondisjunction. Carlson (1973) used a screening method to identify "mutants" of nondisjunction that arose spontaneously in a population of plants carrying TB-9Sb. The method involved selecting a genetic background (W22) that gives an extremely high rate of nondisjunction. In pollen parent crosses with TB-9Sb-W22, progeny that showed a lack of nondisjunction were kept as possible mutants of nondisjunction. The mutants that were found consisted of a) deletions of B chromatin on either the B-9 or 9-B, and b) mutants that lacked any obvious cytological modification. The latter have not been examined.

Lin (1978) found, with his hybrid translocations, that deletion of part of the proximal euchromatin reduced nondisjunction greatly. An exact rate of nondisjunction when this region was deleted could not be determined, due to crossing over between the R-nj marked B-10 of one translocation and the R-scm marked B-10 of the other. However, it was clear that nondisjunction with R-nj (linked to the deletion) was much lower than that of R-scm (complete B present). Therefore, a site controlling nondisjunction was localized to an area within the proximal euchromatin. In addition, the design of the experiment meant that the site in the proximal euchromatin that controls nondisjunction has a *trans* effect on nondisjunction. Carlson (1973, 1978) identified a modified B-9 chromosome that was present in one of the mutant B-9Sb translocations that lack nondisjunction. The B-9 has a deletion of most of the proximal euchromatin and part of the centric heterochromatin. It was demonstrated that this B-9, referred to as 1866, had lost a factor required for nondisjunction. The rate of nondisjunction was minimal (0.4%). By contrast, when standard B chromosomes were present in the same plant as B-9-1866, the rate of nondisjunction rose to 93%. The site, therefore, controls a *trans* function in B nondisjunction. It seems likely that Lin and Carlson have identified the same site, which is located in the proximal euchromatin and is required for nondisjunction.

5.3   B Centromere

Since the centric knob is adjacent to the centromere, the relationship between these two elements is key to understanding the process of stable (0-2) nondisjunction at the second pollen mitosis. Obviously, the B centromere is at the center of the process of nondisjunction. Its behavior must be modified to give migration of the chromatids to one pole. The fact that nondisjunction at the second pollen mitosis can occur at rates as high as 98+% indicates a very efficient process. The question is whether the B centromere actively participates in the process of NDJ or is simply constrained to allow nondisjunction, by nondivision of the centric heterochromatin, as suggested by Pryor et al. (1980). A clue comes from the work of Alfenito and Birchler (1993). They cloned a repetitive sequence, pZmBs (*Zea mays* B specific), that is only found on the B chromosome. This B-specific repeat is the main component of the B centromere. The obvious conclusion is that the centromere is not a passive actor in nondisjunction but is highly modified to help carry out the process.

Han *et al.* (2006) studied centromere inactivation in dicentric B-9 minichromosomes derived from a chromosome type breakage-fusion-bridge cycle, as discussed above. They found a case of transposition of an inactive centromere to the end of 9S, giving a chromosome designated 9Bic-1 (B inactive centromere). The transposed centromere on 9S is inactive, since the dicentric 9Bic-1 chromosome is stable throughout development of plants carrying it. Han et al. (2007) combined the 9Bic-1 chromosome with standard B chromosomes and found that 9S failed to divide properly at the second pollen mitosis, producing chromosome breakage and some nondisjunction. This finding showed that nondivision of the B centromere is separate from its function in kinetochore formation. Similar findings occurred when the inactive B centromere was attached to the tip of 7S.

With the 9Bic-1 chromosome and the 7Bic-1 chromosome, part of the B centromere and the adjacent centric knob are present. Consequently, the results do not distinguish between nondivision of the centromere vs. nondivision of the centric heterochromatin. In another experiment, however, Han et al. (2007) studied a B-9 minichromosome, number 9, that contains the centromere of the B but no detectable amount of centric heterochromatin. Using FISH analysis in the pollen of a plant with minichromosome 9 and one B chromosome, they found cases of nondisjunction by the minichromosome. They concluded that the centromere is the site of nondivision on the B during nondisjunction. However, as noted below, the centric knob may also be a site of nondivision.

5.4   Centric Knob

The heterochromatin adjacent to the centromere in the long arm of the B is related to the maize knobs found on A chromosomes, unlike the distal heterochromatin of the B (see 23.10). Knobs are replicated later than other maize chromatin and the presence of B chromosomes makes their replication occur even later. Pryor et al. (1980) suggested that delayed replication of the B centric knob could account for nondisjunction.

The B centric heterochromatin is the only example in the maize genome of a knob being present adjacent to the centromere. Its location suggests a role in nondisjunction. Rhoades, Dempsey and Ghidoni (1967) and Rhoades and Dempsey (1972) found an inbred line whose behavior supports such a role. They reported finding with the line, referred to as "high loss", that dominant markers in pollen parent crosses were frequently lost. The losses only occurred in high loss plants that contained B chromosomes. The loss of genes was traced to chromosome arms that contained heterochromatic knobs. The B chromosomes caused the knobs to stick together at the second pollen mitosis. The result was dicentric bridges at anaphase. Breakage of the bridges caused the loss of genes. The finding was interpreted as a cross-reaction between the B nondisjunctional system and the knobs of the A chromosomes. Nondivision may normally be confined to the B centric knob. In the high loss background, this specificity is lost and nondivision also occurs with A chromosome knobs.

This interpretation was supported by work with TB-9Sb. An apparent isochromosome of the B-9 was found (Carlson 1970) which was later shown to be a pseudoisochromosome (Carlson and Chou 1981). The chromosome has a complete arm on one side of the centromere, but an arm that lacks the centric knob on the other side, as shown in Fig. 4.

The pseudoisochromosome is unstable and undergoes centric misdivision frequently. As a result, two types of telocentric B-9 chromosomes have been recovered from it. They are referred to as type 1 (no centric heterochromatin) and type 2 (has the centric knob). Four type 1 telocentrics (designated 1852-1855) and two type 2 telocentrics (1856-7) were recovered and tested for nondisjunctional capacity. Three of the type 1 telocentrics (1853-1855) were tested for nondisjunction as 9 9-B B-9 heterozygotes and found to give rates of 1-2% (Carlson, 1973, Table 3). Two of the type 1 telocentrics (1852 and 1854) were tested as homozygotes, giving nondisjunctional rates of approximately 1.3% for 1852 and 4% for 1854. In contrast, both of the type 2 telocentrics gave high rates of nondisjunction. Data for crosses with the type 2 telocentric, 1857, were recorded by Carlson (1973, Table 2). The rate of nondisjunction was 62%. The findings support the idea that the centric heterochromatin is a key factor in nondisjunction.

It was also demonstrated that nondisjunction of the type 1 telocentric could not be restored to normal levels by the addition of standard B chromosomes. With telocentric 1852, the rate of nondisjunction without Bs was 0.8%. In the presence of 3 to 8 standard Bs, the rate rose, but only to 8%. In addition, a sharp rise in variegated endosperm phenotypes was found, indicating instability in the presence of standard Bs (Carlson, 1978). The centric heterochromatin is believed, therefore, to have a necessary *cis* function in nondisjunction.

Three conditions have been found in which the type 1 telocentric can give rise to a significant level of nondisjunction, although well below that of the standard B-9. These are: a) combining the telocentric in the same plant as standard B chromosomes, as noted above, b) conversion of the telocentric to a type 1 isochromosome, and c) testing the telocentric in the hypoploid (9-B 9-B telo B) condition. In all cases, the increase in nondisjunction is accompanied by a considerable instability which is expressed as variegated endosperm phenotypes. Absence of the centric heterochromatin seems to greatly reduce the occurrence of a stable form of nondisjunction (0-2 disjunction) and increases instability, which results from centromeric misdivision (Carlson, 2006).

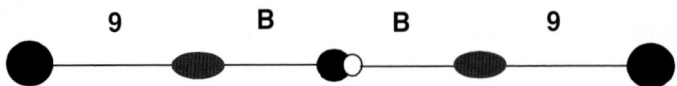

Fig. 4 Diagram of the B-9 pseudoisochromosome

One possible explanation for the various findings with the centromere and the centric knob is that the two elements work together in carrying out 0-2 disjunction. Perhaps, the centric heterochromatin sticks together and protects the nondividing centromere from misdivision, due to chromatid repulsion. This suggestion depends on at least some misdivision occurring at the second pollen mitosis, rather than meiosis (Carlson, 2006).

5.5 B Short Arm

The existence or non-existence of a B short arm has long been a point of contention. McClintock (1933) described the B chromosome as telocentric. Randolph (1941) believed that there is a small short arm. A B-10 chromosome from the translocation TB-10L18 (Lin 1979) and a pseudoisochromosome B-9 from TB-9Sb (Carlson 1970; Carlson and Chou 1981) were identified that have similar types of rearrangement. In both cases, breaks occurred within or near the B centromere with apparent loss of the presumed B short arm.

Lin (1979) found his rearrangement among a large group of B-10L translocations that he constructed. TB-10L18 has a break either in the B short arm or the B centromere. As a result, the B-10 chromosome contains a 10L segment in one arm and almost all of the B chromosome, except the short arm, in the other. Depending on the exact point of breakage, the 10-B may contain a) the B short arm, b) part of the B centromere with the B short arm, or c) part of the centromere only (if there is no short arm). Alfenito and Birchler (1993) found that DNA containing the 10-B chromosome of TB-10L18, but not the B-10, does not show binding to their B centromere-specific sequence. This suggests that there is a B short arm which was transferred to the 10-B, without any centromeric sequences. This is the best evidence for existence of a B short arm.

In terms of a role for the B short arm in nondisjunction, both Lin and Carlson found a lower rate of nondisjunction when the B short arm was absent. For example, a controlled comparison between the standard B-9 and the pseudoisochromosome B-9 gave rates of class 2 nondisjunction (recessive endosperm phenotype) of 66% and 28%, respectively (Carlson and Chou 1981). It appears that the B short arm increases the rate of nondisjunction, but is not required for the event. However, despite the evidence for a role of the B short arm in nondisjunction, there are at least two possible problems with the evidence. First, if the near-terminal location of the B centromere assists in nondisjunction, any attachment of a second arm to the centromere would reduce the rate of nondisjunction. Second, the test of nondisjunctional rate by Lin involved using the 10 B-10 meiotic product. This chromosome combination is duplicated for a portion of 10L. It is possible that the duplication itself reduces the rate of nondisjunction, since background genetic factors are known to affect nondisjunctional rates. The same situation applies to a comparison of the duplicate pseudoisochromosome B-9 to the standard B-9. Therefore, an influence of the B short arm on nondisjunction is possible, but has not been proven.

5.6   Distal Heterochromatin

Lin (1978) used his assortment of B-10L translocations to delete most of the blocks of distal heterochromatin on the B. The B-10 from one translocation, with a breakpoint in a very proximal site on the heterochromatin, was combined with the 10-B from another translocation with a very distal break in the heterochromatin. This combination did not lower the rate of nondisjunction. It appears, therefore, that there is no site in the distal heterochromatin that controls nondisjunction.

6   Roman's Accumulation Mechanism – Preferential Fertilization

When a B chromosome or B-A chromosome undergoes nondisjunction, one sperm receives two B-type chromosomes and the other none. Roman (1948) showed that the sperm with the B-type chromosomes preferentially fertilizes the egg. The process is not 100% effective, but egg-fertilization occurs about two-thirds of the time. There are two possible explanations for the phenomenon. The two sperm might differ, with or without B chromosomes, in the sense that one sperm is designed to fertilize the egg and the other to fertilize the polar cells. If so, the nondisjoining B chromosome is able to migrate to the pole that is destined to fertilize the egg. An alternate explanation is that migration to the poles during nondisjunction is random. The presence of B chromosomes in a sperm confers a selective advantage in egg-fertilization.

It was shown, with TB-9Sb, that the presence of several standard B chromosomes eliminated preferential fertilization of B-9-containing sperm. If the B-9 preferentially migrates to a certain pole, the extra Bs should not affect preferential fertilization. Therefore, the B chromosome confers a selective advantage on sperm in egg fertilization (Carlson, 1969a).

In other studies with TB-9Sb, preferential fertilization was found to fail in crosses of translocation-carrying plants as pollen parents to an inbred *c1 sh1 wx1 gl15* female. The cause was traced to the female tester, which lacked B chromosomes. A gene(s) on the A chromosomes must, therefore, have blocked preferential fertilization (Carlson, 1969a). Chiavarino *et al.* (1998) found a similar effect with the standard B chromosome in a population of maize (see 23.14, below).

Genetic control of preferential fertilization has also been located to a site on the B chromosome. Carlson (2007) studied preferential fertilization by type 1 B-9 chromosomes. These chromosomes lack the centric heterochromatin and possibly some adjacent euchromatin. It was found that type 1 chromosomes are not capable of preferential fertilization. This locates a site controlling preferential fertilization to either the centric heterochromatin or a nearby site in the proximal euchromatin.

7   Beckett's Accumulation Mechanism

Beckett (1982) studied transmission of TB-1La and TB-1Lc when crossed through the male as the translocation heterozygote, 1 1-B B-1. He placed the translocations in a genetic background, inbred line L289, that produces high rates of nondisjunction. Transmission of the 1-B B-1 combination could then be identified by the occurrence of nondisjunction. He found nondisjunction in 56–59% of kernels. The predicted rate of 1-B B-1 transmission is no more than 50%. The heterozygote yields three types of viable meiotic products: 1; 1-B B-1 and 1 B-1. The 1 B-1 is duplicate and selected against. The maximum predicted rate of transmission of 1-B B-1 assumes no survival of the duplicate pollen. Then, 50% of pollen should receive 1-B B-1 and 50% chromosome 1. A value above 50% indicates an advantage in fertilization by B-containing pollen. The accumulation mechanism depends on competition between pollen grains, whereas Roman's system uses competition between the sperm of a pollen grain.

Carlson (1997) looked for the same effect with TB-9Sb. In this case, transmission of 9-B B-9 is seen as the presence of *Wx* (on 9-B) in the kernel. As a result, all cases of 9-B B-9 transmission, with and without nondisjunction, were counted. There was either a small advantage to pollen carrying the translocation or none at all. One possible explanation for the different results could be that background genetic factors influence the ability of B-containing pollen to outcompete pollen without the B chromosome.

8   Localizing B Chromosome in Sperm

The B centromeric sequence, ZmBs, of Alfenito and Birchler (1993) can be used in sperm cells to identify the presence of B chromosomes. Rusche et al. (1997; 2001) used ZmBs to identify pollen in which nondisjunction had occurred, with one sperm binding the sequence and the other not. They also wanted to see whether the sperm with B chromosomes had any preferred location in the pollen grains. They found that location of the sperm with Bs was random, with respect to closeness to the vegetative nucleus or the germination pore.

In another experiment, they compared locations within the sperm for all centromeres (using CentC), knob heterochromatin (using 180 bp sequence) and the B centromere (using ZmBs). They found random locations within sperm cells for the various centromeres and knobs. However, B centromeres were located near the tip of the sperm, in about ¼ of the sperm area. They were located in this area, regardless of whether or not nondisjunction had occurred. The significance of this finding is not known, but could be related to preferential fertilization.

9   Structure of the B Centromere

The B-9 pseudoisochromosome (Fig. 4) is unstable and frequently produces variegated phenotypes in the endosperm. Instability was traced to centric misdivision and the production of unstable, telomere-deficient telocentric chromosomes (Carlson, 2006). These chromosomes replicate incorrectly, causing variegation in the endosperm. The telocentrics are stabilized in the plant through the addition of telomeric sequences (Kaszas et al. 2002).

In addition to telocentrics, the pseudoisochromosome produces new isochromosomes, which either completely lack centric heterochromatin or have it on both sides of the centromere. These are type 1 and type 2 isochromosomes, respectively (Carlson, 2004). Type 1 telocentrics also can give rise to type 1 isochromosomes by misdivision (Carlson and Chou, 1981). This misdivision system became useful in studying the B centromere after the discovery of the B-specific sequence, ZmBs, by Alfenito and Birchler (1993). *In situ* hybridization to a pair of B chromosomes in pachytene showed that ZmBs is mainly present in a region occupied by the centromere. At much lower strength, the sequence binds to a subtelomeric site on the Bs. Localization to the centromere, rather than nearby DNA, was demonstrated by testing various DNA sources, containing different parts of the B chromosome, for binding of the ZmBs sequence.

Kaszas and Birchler (1996) used pZmBs to study centromeres derived from the B-9 misdivision system. The misdivision chromosomes included isochromosomes, telocentrics and ring chromosomes. The number of misdivision events in these chromosomes ranged from 1 to 4. Comparison of the various misdivision centromeres to the centromere of the intact B-9 showed that all misdivision events tested were unique and generally simplified from the progenitor centromere. The centromere of the original isochromosome, with one misdivision event, is considerably reduced in size and complexity compared to that of the B-9. It was clear that the B centromere is a duplicated structure in terms of function and can be rearranged and greatly reduced in size by misdivision, without losing the ability to carry out chromosome division. Jin et al. (2005) identified the functional region of the B centromere, a 700 kb region that binds CENH3, the centromere-specific histone H3 variant. Much of the rest of the approximately 9,000 kb centromere consists of ZmBs repeats. Phelps-Durr and Birchler (2004) used B-9 misdivision to search for the lower limits of centromere size. They found at least partially functional centromeres of lengths less than 200 kb. However, these very small centromeres gave poor transmission from one generation to the next.

10   Uniqueness of B Heterochromatin

There are three classes of heterochromatin on the A chromosomes: 1) knob heterochromatin, 2) centric heterochromatin, and 3) nucleolus organizer heterochromatin. They are cytologically distinct from each other, as reported by several authors.

Knobs are located some distance away from the centromeres of A chromosomes. They are smooth, dark chromosomal sites that are sharply distinct in pachytene of meiosis. Centric heterochromatic regions flank the centromeres and have dark-staining chromomeres. They are diffuse areas of heterochromatin that are not very distinct. The nucleolus organizing region on chromosome 6 contains a very large, smooth and dark area of heterochromatin that often appears similar to knob hetero-chromatin. However, it is attached to the nucleolus and sometimes has a diffuse appearance. (Peacock et al. 1981; Viotti et al. 1985).

The B chromosome contains two unusual classes of heterochromatin. It has a knob, which is similar to the knobs of A chromosomes except that it is adjacent to a centromere. The B chromosome also has large blocks of distal heterochromatin that are distinctly different from knobs. The extended chain of heterochromatic blocks is unlike the usual organization of knobs as single units. In addition, the heterochromatin is more diffuse than knob heterochromatin. (Peacock et al. 1981; Viotti et al. 1985). Also, while knob heterochromatin stains with a C-banding technique, the distal B heterochromatin does not (Ward 1980).

The various types of heterochromatin differ in DNA content. The NOR hetero-chromatin contains most of the rDNA genes and is therefore distinct from other types of heterochromatin which, typically, contain few if any traditional genes (Phillips 1978). Knob heterochromatin contains a 180 bp repeat and/or a 350 bp repeat (Peacock et al. 1981; Ananiev et al. 1998a). Retrotransposons are inserted within the knobs (Ananiev et al. 1998b). Some knobs have mostly the 180 bp repeat, some are mixed for 180 and 350 bp repeats and some contain mostly the 350 bp repeat. The centric knob of the B chromosome gives a strong signal with the 180 bp repeat and a weak one with the 350 bp repeat (Hsu et al. 2003).

Finally, the timing of DNA replication differs between the classes of heterochromatin. Most heterochromatin has a pattern of delayed replication compared to euchromatin. However, NOR DNA replicates at the same time as the euchromatin. The other classes replicate later than euchromatin, with a distinct pattern. Centromeric hetero-chromatin is first to replicate, the distal heterochromatin of the B next, and the knob heterochromatin, including the centric knob of the B, is last (Pryor et al. 1980).

11 Suppression of Meiotic Loss

A collection of TB-9Sb chromosomes, with deletions of different parts of the B chromosome, was assembled to study the control of nondisjunction (Carlson 1978). All these stocks give very little or no B-9 nondisjunction. It was noticed that one of these, the type 1 B-9 chromosome, transmits at a much reduced rate through the female than does the standard B-9 (Carlson, 1986a). The finding prompted the development of controlled comparisons between several different deletion stocks and the standard TB-9Sb (Carlson and Roseman, 1992). The stocks tested included: 1) TB-9Sb-1852, which has the type 1 telocentric B-9; 2) TB-9Sb-1866, which lacks the proximal euchromatin and part of the centric

heterochromatin on the B-9; 3) TB-9Sb-2010, with a terminal deficiency on the 9-B which deletes the euchromatic tip of the B and a small amount of distal heterochromatin; 4) TB-9Sb-14, with a 9-B deletion that includes the distal euchromatic tip of the B and somewhat more than half of the distal heterochromatin; and 5) TB-9Sb-2150, with a terminal 9-B deficiency that includes almost all of its B chromatin, except a short region of distal heterochromatin adjacent to the proximal euchromatin. The tests showed a large amount of meiotic loss for 1852, 1866 and 2150 (45-48% compared to 10-15% for standard TB-9Sb). The rate for TB-9Sb-14 was 30% (compared to 11% for the standard B-9). TB-9Sb-2010 gave 22% loss compared to 21% for standard TB-9Sb. A second test of TB-9Sb-2010 gave 27% loss compared to 22% for the standard TB-9Sb.

It was concluded that there is at least one site on the B-9 that affects meiotic loss. (The 1866 and 1852 B-9 deletions may overlap and could delete the same function controlling meiotic loss). Also, at least one site controlling meiotic loss is located on the 9-B. It appears that the larger the deletion of 9-B distal heterochromatin, the greater the meiotic loss, so several regions could be involved. The distal euchromatic tip of the B does not seem to affect meiotic loss, since TB-9SB-2010 gives meiotic loss rates similar to the standard B-9.

Meiotic loss may be the biggest barrier to establishment of a new B chromosome in a population. When the B chromosome first began spreading through maize populations, its frequency was low and almost all plants containing it would have had only one copy of the chromosome. The likely result would be frequent meiotic loss. As a result, there must have been strong selection for a system that prevents meiotic loss.

The system discussed above most likely operates by producing stable unipolar migration of univalent Bs. It does not prevent all meiotic loss, since the standard B-9 shows a significant amount of loss (Carlson and Roseman, 1992). Further suppression of meiotic loss was provided by the development of nondisjunction at the second pollen mitosis. Roman's accumulation mechanism may have begun its existence as a method for suppressing meiotic loss. Nondisjunction transmits pairs of chromosomes through the pollen, allowing bivalent pairing at meiosis of the next generation. Later, preferential fertilization was added to take advantage of the nondisjunction. It is also possible that univalent migration in meiosis and nondisjunction at the second pollen mitosis have a common origin, in terms of evolution, since the two events are functionally equivalent. Recently, Gonzalez-Sanchez et al. (2003), showed that meiotic loss is also affected by genes on the A chromosomes. (see 23.14, below)

12 Effect of Bs on Crossing Over in A Chromosomes

An influence of the B chromosome on crossing over in A chromosomes has been found in numerous studies, including experiments with the standard B chromosome (Hanson, 1969; Nel, 1973; Rhoades, 1968; Ward, 1976) and with B-A translocations

(Robertson, 1984; Carlson et al, 1993). The most specific effect found with the standard B is the work of Rhoades on transposition 9 (Tp9). This chromosome contains all of chromosome 9 plus a segment of chromosome 3 inserted into 9S. Crossing over between *C1* and *Wx1*, which span the transposed segment, was found to be the same for both N9 N9 and Tp9 Tp9 individuals. Apparently, no crossing over occurred in the transposed segment. However, when Tp9 Tp9 plants containing B chromosomes were tested, a marked increase in crossing over was observed. Tp9 Tp9 sibling plants, with and without Bs, were tested. *C1-Wx1* crossing over was found to be 17.7% without Bs and 40.4% with two Bs. It appears that the B chromosome opened up the transposed segment of chromosome 3 to crossing over.

In regards to B-A translocations, Robertson found (1984), a specific effect on crossing over. Using six different B-A translocations, he found that crossing over in the A A-B bivalents of each was markedly increased compared to that found with a normal bivalent. The effect was found in regions adjacent to the B chromatin and only occurred in pollen parent crosses. Carlson et al. (1993) further examined Robertson's effect using a deletion type TB-9Sb (2150), in which the 9-B chromosome lacks almost all of the distal heterochromatin plus the distal euchromatic tip of the B. While the standard 9-B + 9 bivalent shows a large increase in crossing over through the male, the deletion 9-B + 9 does not. This suggests an effect of the distal half of the B chromosome on crossing over in adjacent regions.

The significance of crossover effects of the B chromosome is not known. It is the only identified function of the B that might be useful to the organism. Perhaps it serves to open up regions, such as the chromosome 3 region of Tp9, to crossing over. The result would be a greater variety of gametes produced for selection, as suggested by Hanson (1969). Another view is that the effect of B chromosomes on crossing over in A chromosomes is a side effect of the real function: promoting recombination between B chromosomes and/or blocking recombination between the B chromosome and its progenitor chromosome(s) in the A set (Carlson, 1994).

13   Origin of the B Chromosome

13.1   *Source of the Progenitor B Chromosome*

The B chromosome may have arisen initially from the A chromosomes (Camacho et al. 2000). Nevertheless, no active genes corresponding to known genes on the A chromosomes have been found on the B (Randolph, 1941). This is not surprising, since the retention of active genes on a supernumerary chromosome could have a harmful, aneuploid effect on the organism. Since Randolph's alleleism tests did not identify the B chromosome source, the search eventually turned to studies of DNA. Chilton and McCarthy (1973) showed that B chromosome DNA is very similar to the DNA of A chromosomes. Subsequent DNA studies have focused on cloned sequences.

The ZmBs sequence of Alfenito and Birchler (1993) is B-specific. It is only found on the B chromosome, primarily as part of the centromere. Page et al. (2001) showed a possible source of this sequence among the A chromosomes. Using a reduced stringency for binding of the sequence, they found, with FISH staining, that two of the A chromosome centromeres fluoresced in root tip metaphases. They identified the chromosome 4 centromere as the binding site by testing various trisomic stocks and looking for fluorescence on three centromeres. The authors concluded that the centromere of chromosome 4 was the likely donor of the B centromere.

A different approach was used by Cheng and Lin (2003). They microdissected B chromosome bivalents from pachytene cells. In this manner, they could clone sequences that are known to be on the B chromosome, but may not be B-specific. They cloned 19 sequences, only one of which was B-specific. The remainder contained sequences found on both the A and B chromosomes. Among the cloned sequences, they identified noncoding sequences for genes known to be present on several different A chromosomes, including 1, 4, 7 and 9. This suggests a complex or multi-step origin of the B chromosome.

It is interesting that Peeters et al. (1985), studying a Himalayan popcorn, found spontaneous fusion of meiotic cells followed by degradation of one of the chromosome sets. During degradation, partial chromosomes sometimes survived and were transmitted along with the normal set. The authors suggested that this may be a rapid process whereby B chromosomes can arise. It seems possible that degradation could lead to parts of several chromosomes joining together, forming a progenitor B chromosome.

13.2   Evolution into a B Chromosome

After the origin of the progenitor B chromosome, its transformation into the modern B chromosome must have required several steps. It can be assumed that Roman's accumulation mechanism was not initially present in the chromosome. Nondisjunction is not a useful function for the A chromosomes and would, instead, be quite harmful. It is possible that Beckett's pollen competition effect (1982) was present in the progenitor chromosome and served immediately as an accumulation mechanism. Otherwise, a selective advantage of the progenitor chromosome would be needed initially, to help maintain it in populations.

Evolution of the progenitor B chromosome into the current chromosome required 1) the loss or heterochromatinization of most functional genes, 2) an effective means of preventing crossing over between the B and its source chromosomes, 3) reduction in the frequency of meiotic loss of univalents, and 4) development of an accumulation mechanism(s). For (1), the elimination of functional genes has a selective advantage in that it avoids harmful aneuploid effects from the duplication of genes. In regards to crossing over (2), it is possible that the B chromosome has functions that promote crossing over between Bs, prevent crossing over between the B and its source A chromosome(s) or both (Carlson, 1994). Meiotic loss (3)

would be of immediate importance in survival of the B chromosome. Therefore, the discovery of a system for suppressing meiotic loss of the univalent B is not surprising (Carlson and Roseman, 1992). Roman's accumulation mechanism (4) may have begun with adaptation of the meiotic loss system to give nondisjunction at the second pollen mitosis (see section: Suppression of Meiotic Loss).

13.3   Relationship to Abnormal Chromosome 10

Another question about B chromosome origin is its possible relationship to the extra chromatin on abnormal chromosome 10. Ting proposed (1958) a common origin for abnormal 10 and the B chromosome. The main elements that abnormal 10 and the B have in common are 1) both are nonessential to the organism, 2) both have accumulation mechanisms which transmit them at a higher than Mendelian frequency, and 3) both have lots of heterochromatin. However, the similarities end there. The B chromosome is cytologically very different from the added chromatin of abnormal 10 (Rhoades 1942). In addition, the great majority of the heterochromatin on the B (the distal heterochromatin) has a different DNA composition (Peacock et al, 1981; Viotti et al, 1985) and staining (banding) properties (Ward, 1980; Aguiar-Perecin, 1985) than the large knob on abnormal 10. Also, the accumulation mechanisms are very different. The B chromosome has a system that operates in the pollen at the second pollen mitosis and at the time of fertilization. Abnormal 10 has an accumulation mechanism that operates in the female during meiosis. Tests for cross reactions between the systems have proven negative (Carlson 1969b; 1986b). It seems clear, therefore, that the extra chromatin on abnormal 10 and the B chromosome arose from different sources.

14   B Chromosome Frequency in Populations

The frequency of the maize B chromosome in populations is affected by at least five factors. These are: 1) the accumulation mechanisms of Roman (1947, 1948) and Beckett (1982), 2) the self-limiting nature of the accumulation mechanisms, 3) the harmful effects of Bs at high frequency (Randolph, 1941; Staub, 1987), 4) meiotic loss of univalent B chromosomes (Carlson and Roseman, 1992; Gonzalez-Sanchez et al. 2003), and 5) the influence of genes on the A chromosomes that affect the accumulation mechanisms and meiotic loss.

One of the factors that needs explanation is the self-limiting nature of the accumulation mechanisms. With Roman's system, preferential fertilization does not occur when several Bs are present in a plant, because both sperm of its pollen grains have B chromosomes and neither has an advantage in egg-fertilization (Carlson, 1969a). A similar result is expected with Beckett's (1982) system. When all pollen grains contain B chromosomes, the system probably does not work.

These limitations of the accumulation mechanisms may take effect before the number of Bs in a population is high enough to cause harm to the organism.

Another element that affects B chromosome frequency and has not been discussed is variation in the genetic background. Rosata et al. (1996) studied a native race of maize from Argentina (Pisingallo), which contains B chromosomes, in order to identify the factors that control frequency of the B. They studied the transmission of B chromosomes from one generation to the next. In one set of crosses, transmission of the B through the pollen parent was studied by making 20 crosses of the type: 0 B x 1 B. The male B transmission rates (B-TR) were determined in the crosses by classifying chromosome numbers in root tips. Nondisjunction was near 100%, so that the progeny had either two Bs or none. The number of progeny examined varied from 21 to 73 per cross. B-TR was found to be highly variable with a range of 0.17 to 0.98. Crosses with a high rate of transmission (0.88-0.98) and crosses with a low rate (0.27-0.28) were chosen to begin a high B-TR line (H) and a low B-TR (L) line. From these lines, plants were selected for crosses of 0 B x 2 B. The transmission rates, after just one generation of selection, were quite different. B-TR for the H line ranged from 0.52-0.80 with a mean of 0.65. In the L line, B-TR varied from 0.34-0.44 with a mean of 0.40.

A similar procedure was used in selecting lines for high and low female transmission of the B. Twenty crosses of 1 B x 0 B were made, with a range of transmission of 0.31-0.58. The average rate was 0.47. In these crosses, almost all the progeny had either 1 B or none. The H line was selected by using crosses with B-TR values of 0.48-0.54. The L line was chosen from crosses with values of 0.31-0.37. In the next generation, crosses were made of 1 B x 0 B and the progeny classified in root tips. The H line gave B-TR rates of 0.28-0.58, with an average of 0.48. The L line gave B-TR rates of 0.30-0.54, with an average of 0.40. The mean values are significantly different between H and L lines, although there is considerable overlap among individual crosses.

Chiavarino et al. (1998) investigated the cause of high and low male transmission rates. They made crosses of 0 B × 2 B both within and between H and L lines. It was shown that transmission for H × H and H x L crosses was 0.71 and 0.70 respectively. Crosses of L x H and L x L both gave values of 0.48. In other words, male transmission of the B was controlled entirely by a gene(s) in the female parent, which lacks B chromosomes. The H line allows preferential fertilization while the L line does not. The finding is similar to the discovery of an inbred tester line of maize that, when crossed as female parent to TB-9Sb, blocks preferential fertilization of the B-9 (Carlson, 1969a). In this case, however, genetic control is segregating in a native race of maize. Chiavarino et al. (2001) found that almost all the selection for H and L lines occurred in the first generation and they suggested that a single gene in the female parent is responsible for controlling male B-TR. Zero B plants from the H and L lines were crossed together, to give a hybrid. When H/L plants were crossed as female to male parents (of either line) containing two B's, an intermediate B-TR was found. This is because the control of preferential fertilization occurs at the haploid level.

Gonzalez-Sanchez et al. (2003) provided further evidence that preferential fertilization is controlled by a single gene, using F_2 data. In addition, they studied

the H and L lines that were selected for female transmission. In order to examine the possibility of meiotic loss as the source of the different female transmission rates, they wanted to look at meiosis. Unfortunately, large scale studies of the female meiosis are not possible. Therefore, they examined microspores for differences in B transmission rates from the male meiosis. Microspores were examined for the presence of the B chromosome by *in situ* hybridization, using a B-specific probe (Alfenito and Birchler, 1993). The H line showed no meiotic loss (50% transmission), while the L line did (0.42-0.43 transmission). Interestingly, meiotic loss is a dominant trait. The hybrids of H x L and L x H all gave reduced rates of B-TR. The source of the B chromosome in the hybrids, whether from the H line or the L line, did not affect the findings. Consequently, it is not polymorphism of the B chromosome that is causing the differences. Genes on the A chromosomes control female transmission, as well as male transmission rates. The meiotic loss trait acts like a simple dominant-recessive system, but whether one or more genes controls meiotic loss is not known.

Gonzalez-Sanchez et al. (2003) discussed the significance of findings with the various H and L lines of the Pisingallo race in B chromosome evolution. They suggested that polymorphism for genes from the A genome, that affect preferential fertilization and meiotic loss, may indicate a host-parasite relationship. The A chromosome genes that affect B chromosome transmission may reflect a response of the organism to the presence of B chromosomes. Camacho et al. (2000) discussed methods of B chromosome evolution in many different species.

15   Future Work

The maize B chromosome has proven very useful in manipulating A chromosome segments, using B-A translocations. However, the future direction of B chromosome research may lie more in analysis of the chromosome itself. For example, a more precise determination of the sources of B chromosome DNA from among the A chromosomes is of interest (23.13.1), since it could show the series of evolutionary steps needed to construct the chromosome. Also, studies of the unique distal B heterochromatin (23.10) may provide insights into heterochromatin formation. A third approach to studying the B chromosome is isolating and sequencing the genes that control nondisjunction and preferential fertilization (23.5 and 23.6). These genes control two important processes: chromosome division and fertilization. Consequently, isolation of these genes could provide insights into basic functions of the organism.

References

Aguiar-Perecin, M. L. R. de. (1985) C-banding in maize. I. Band patterns. Caryologia **38**: 23–30.

Alfenito, M. R. and J. A. Birchler. (1993) Molecular characterization of a maize B chromosome centric sequence. Genetics **135**: 589–597.

Ananiev, E. V., R. L. Phillips and H. W. Rines (1998a) A knob-associated tandem repeat in maize capable of forming fold-back DNA segments: Are chromosome knobs megatransposons? Proc. Natl. Acad. Sci. USA **95**: 10785–10790.

Ananiev, E. V., R. L. Phillips, and H. W. Rines, (1998b) Complex structure of knob DNA on maize chromosome 9: Retrotransposon invasion into heterochromatin. Genetics **149**: 2025–2037.

Beckett, J. B. (1982) An additional mechanism by which B chromosomes are maintained in maize. J. Heredity **73**: 29–34.

Beckett, J. B. (1991) Cytogenetic, genetic and plant breeding applications of B-A translocations in maize. Chapter 25, In: Chromosome Engineering in Plants: Genetics, Breeding, Evolution, Part A. Edited by P. K. Gupta and T. Tsuchiya. Elsevier Science Publishers, Amsterdam

Camacho, J. P. M., T. F. Sharbel and L. W. Beukeboom, (2000) B-chromosome evolution. Phil. Trans. R. Soc. Lond. B **355**: 163–178.

Carlson, W. R. (1969a) Factors affecting preferential fertilization in maize. Genetics **62**: 543–554.

Carlson, W. R. (1969b) A test of homology between the B chromosome of maize and abnormal chromosome 10, involving the control of nondisjunction. Molec. Gen. Genetics **104**: 59–65.

Carlson, W. R. (1970) Nondisjunction and isochromosome formation in the B chromosome of maize. Chromosoma **30**: 356–365.

Carlson, W. R. (1973) A procedure for localizing genetic factors controlling mitotic nondisjunction in the B chromosome of maize. Chromosoma **42**: 127–136.

Carlson, W. R. (1978) Identification of genetic factors controlling centromeric function in maize. Ch. 44 In: Maize Breeding and Genetics, D. B. Walden, editor.

Carlson, W. R. (1986a) The B chromosome of maize. CRC Crit. Rev. Pl. Sci. **3**: 201–226.

Carlson, W. R. (1986b) A further test of homology between the B chromosome and Abnormal 10. Maize Genet. Coop. Newslett. **60**: 68–69.

Carlson, W. R. (1988) The cytogenetics of corn. In: Corn and Corn Improvement, 3rd Edition. G. F. Sprague and J. W. Dudley, editors. Am. Soc. Agronomy, Madison, Wisconsin.

Carlson, W. R. (1994) Crossover effects of B chromosomes may be 'selfish'. Heredity **72**: 636–638.

Carlson, W. R. (1997) Pollen competition effects in crosses of the B-A translocation, TB-9Sb. Maydica **42**: 121–126.

Carlson, W. R. (2004) Detecting centromeric misdivision in crosses of a B-type chromosome in maize. Maydica **49**: 31–36.

Carlson, W. R. (2006) Unstable inheritance of maize B-type chromosomes that lack centric heterochromatin. Genome **49**: 420–431.

Carlson, W. R. (2007) Locating a site on the maize B chromosome that controls preferential fertilization. Genome **50**: 578–587.

Carlson, W. R. and T.-S. Chou (1981) B chromosome nondisjunction in corn: Control by factors near the centromere. Genetics **97**: 379–389.

Carlson, W. R. and R. R. Roseman (1992) A new property of the maize B chromosome. Genetics **131**: 211–223.

Carlson, W. R., R. Roseman and Y.-Z. Zheng (1993) Localizing a region on the B chromosome that influences crossing over. Maydica **38**: 107–113.

Cheng, Y. and B. Y. Lin (2003) Cloning and characterization of maize B chromosome sequences derived from microdissection. Genetics **164**: 299–310.

Chiavarino, A. M., M. Gonzalez-Sanchez, L. Poggio, M. J. Puertas, M. Rosato and P. Rosi (2001) Is maize B chromosome preferential fertilization controlled by a single gene? Heredity **86**:743–748.

Chiavarino, A. M., M. Rosato, P. Rosi, L. Poggio and C. A. Naranjo (1998) Localization of the genes controlling B chromosome transmission rate in maize Zea mays ssp. mays, Poaceae. Am. J. Botany **85**: 1581–1585.

Chilton, M. D. and B. J. McCarthy (1973) DNA from maize with and without B chromosomes: A comparative study. Genetics **74**: 605–614.

Gonzalez-Sanchez, M., E. Gonzalez-Gonzalez, F. Molina, A. M. Chiavarino, M. Rosato and M. J. Puertas (2003) One gene determines maize B chromosome accumulation by preferential fertilisation; another gene(s) determines their meiotic loss. Heredity **90**: 122–129.

Han, F., J. C. Lamb and J. A. Birchler (2006) High frequency of centromere inactivation resulting in stable dicentric chromosomes of maize. PNAS **103**: 3238–3243.

Han, F., J. C. Lamb, W. Yu, Z. Gao and J. A. Birchler (2007) Centromere function and nondisjunction are independent components of the maize B chromosome accumulation mechanism. Plant Cell: 19: 524–533.

Hanson, G. P. (1969) B chromosome-stimulated crossing over in maize. Genetics **63**: 601–609.

Hsu, F. C., C. J. Wang, C. M. Chen, H. Y. Hu and C. C. Chen (2003) Molecular characterization of a family of tandemly repeated DNA sequences, TR-1, in heterochromatic knobs of maize and its relatives. Genetics **164**: 1087–1097.

Jin, W. W., J. C. Lamb, J. M. Vega, R. K. Dawe, J. A. Birchler and J. Jiang (2005) Molecular and functional dissection of the maize B chromosome centromere. Plant Cell **17**: 1412–1423.

Kaszas, E. and J. A. Birchler (1996) Misdivision analysis of centromere structure in maize. EMBO J. **15**: 5246–5255.

Kaszas, E., A. Kato and J. A. Birchler (2002) Cytological and molecular analysis of centromere misdivision in maize. Genome **45**: 759–768.

Kindiger, B., C. Curtis and J. B. Beckett. (1991). Adjacent-II segregation products in B-A translocations of maize. Genome **34**: 595–602.

Lin, B-Y. (1978) Regional control of nondisjunction of the B chromosome in maize. Genetics **90**: 613–627.

Lin, B.-Y. (1979) Two new B-10 translocations involved in the control of nondisjunction of the B chromosome in maize. Genetics **92**: 931–945.

McClintock, B. (1933) The association of non-homologous parts of chromosomes in the mid-prophase of meiosis in Zea mays. Z. Zellforsch. Mikroskop. Anat. **19**: 191–237.

McClintock, B. (1939) The behavior in successive nuclear divisions of a chromosome broken at meiosis. Proc. Natl. Acad. Sci. USA **25**: 405–416.

McClintock, B. (1942) The fusion of broken ends of chromosomes following nuclear fusion. Proc. Natl. Acad. Sci. USA 28: 458–463.

McClintock, B. (1943) Maize Genetics. Carnegie Inst. Wash. Yrbk. **42**: 148–152.

McClintock, B. (1951) Chromosome organization and genic expression. Cold Spring Harbor Symp. Quant. Biol. **16**: 13–47.

Nel, P. M. (1973) The modification of crossing over in maize by extraneous chromosomal elements. Theor. Appl. Genet. **43**: 196–202.

Page, B. T., M. K. Wanous and J. A. Birchler (2001) Characterization of a maize chromosome 4 centromeric sequence: Evidence for an evolutionary relationship with the B chromosome centromere. Genetics **159**: 291–302.

Peacock, W. J., E. S. Dennis, M. M. Rhoades and A. J. Pryor (1981) Highly repeated DNA sequence limited to knob heterochromatin in maize. Proc. Natl. Acad. Sci. USA **78**: 4490–4494.

Peeters, J. P., A. J. F. Griffiths and G. Wilkes (1985) In vivo karyotypic modifications following spontaneous cell fusion in maize (Zea mays L.). Can. J. Genet. Cytol. **27**: 580–585.

Phelps-Durr, T. L. and J. A. Birchler (2004) An asymptotic determination of minimum centromere size for the maize B chromosome. Cytogen. Gen. Res. **106**: 309–313.

Phillips, R. L. (1978) Molecular cytogenetics of the nucleolus organizer region. In: Maize Breeding and Genetics, chapter 43. D. B. Walden, editor. John Wiley and Sons, New York.

Pryor, A., K. Faulkner, M. M. Rhoades and W. J. Peacock (1980). Asynchronous replication of heterochromatin in maize. Proc. Natl. Acad. Sci. USA **77**: 6705–6709.

Randolph, L. F. (1941) Genetic characteristics of the B chromosomes in maize. Genetics 26: 608–631.

Rhoades, M. M. (1942) Preferential segregation in maize. Genetics **27**: 395–407.

Rhoades, M. M. (1968) Studies on the cytological basis of crossing over. In: Replication and Recombination of Genetic Material. Ed. W. J. Peacock, R. D. Brock, pp. 229–241. Canberra: Aust. Acad. Sci.

Rhoades, M. M. and E. Dempsey (1972) On the mechanism of chromatin loss induced by the B chromosome of maize. Genetics **71**: 73–96.

Rhoades, M. M., E. Dempsey and A. Ghidoni (1967) Chromosome elimination in maize induced by supernumerary B chromosomes. Proc. Natl. Acad. Sci. USA **57**: 1626–1632.

Robertson, D. S.(1967) Crossing over and chromosomal segregation involving the B-9 element of the A-B translocation *B-9b* in maize. Genetics **55**: 433–449.

Robertson, D. S. (1984) Different frequency in the recovery of crossover products from male and female gametes of plants hypoploid for B-A translocations in maize. Genetics **107**: 117–130.

Roman, H. (1947) Mitotic nondisjunction in the case of interchanges involving the B-type chromosome in maize. Genetics **32**: 391–409.

Roman, H. (1948) Directed fertilization in maize. Proc. Natl. Acad. Sci. USA 34: 36–42.

Roman, H. (1949) Factors affecting mitotic nondisjunction in maize. Rec. Genet. Soc. Am. **18**: 112.

Rosato, M., A. M. Chiavarino, C. A. Naranjo, M. J. Puertas and L. Poggio (1996) Genetic control of B chromosome transmission rate in *Zea mays* ssp. *mays* (Poaceae). Am. J. Botany **83**: 1107–1112.

Rusche, M. L., H. L. Mogensen, L. Shi, P. Keim, M. Rougier, A. Chaboud and C. Dumas (1997) B chromosome behavior in maize pollen as determined by a molecular probe. Genetics **147**: 1915–1921.

Rusche, M. L., H. L. Mogensen, A. Chaboud, J.-E. Faure, M. Rougier, P. Keim andC. Dumas (2001) B chromosomes of maize (*Zea mays* L.) are positioned nonrandomly within sperm nuclei. Sex. Plant Reprod. **13**: 231–234.

Stark, E. A., I. Connerton, S. T. Bennett, S. R. Barnes, J. S. Parker and J. W. Forster (1996) Molecular analysis of the structure of the maize B-chromosome. Chrom. Res. **4**: 15–23.

Staub, R. W. (1987) Leaf striping correlated with the presence of B chromosomes in maize. J. Hered. **78**: 71–74.

Ting, Y. C. (1958) On the origin of abnormal chromosome 10 in maize (Zea mays L.). Chromosoma **9**: 286–291.

Viotti, A., E. Privitera, E. Sala and N. Pogna (1985) Distribution and clustering of two highly repeated sequences in the A and B chromosomes of maize. Theor. Appl. Genet. **70**: 234–239.

Ward, E. J. (1973a) The heterochromatic B chromosome of maize: The segments affecting recombination. Chromosoma **43**: 177–186.

Ward, E. J. (1973b) Nondisjunction: localization of the controlling site in the maize B chromosome. Genetics **73**: 387–391.

Ward, E. J. (1976) The effect of accessory chromatin on chiasma distribution in maize. Can. J. Genet. Cytol. **18**: 479–484.

Ward, E. J. (1980) Banding patterns in maize mitotic chromosomes. Can. J. Genet. Cytol. **22**: 61–67.

Zheng, Y.-Z., R. Roseman, and W. R. Carlson (1999) Time course study of the chromosome-type breakage-fusion-bridge cycle in maize. Genetics **153**: 1435–1444.

Mitochondria and Chloroplasts

Kathleen J. Newton, David B. Stern, and Susan Gabay-Laughnan

Abstract This chapter will summarize what is currently known about mitochondrial and chloroplast genomes and their expression in maize. For maize mitochondria, most of the studies have concentrated on genome plasticity and its consequences. In contrast, relatively little variation in maize chloroplast genomes has been reported and most of the published work has focused on the control of chloroplast gene expression.

1 Maize Chloroplast Genomes

In stark contrast to the mitochondrial genomes of maize (see below), there is no apparent variation in the overall organization and linear gene order among *Zea* chloroplast DNAs (cpDNAs) isolated from plant tissues. A very few variable sites and indels have been reported (Doebley et al. 1987; Pring and Levings 1978), but large-scale rearrangements have not been observed. Indeed, linear gene order and content is conserved among the grass chloroplast genomes (Saski et al. 2007). Only under conditions that eliminate the requirement for photosynthetic function, i.e., in cultured cells of Black Mexican Sweet corn, have a variety of deletion derivatives of chloroplast genomes been found (Cahoon et al. 2003a, 2003b).

Although restriction enzyme maps and the sequences of specific chloroplast-localized genes were early targets of investigation (especially by Bogorad and colleagues), the complete sequence of only one maize chloroplast genome had been published as of mid-2007 (Maier et al. 1995; updated in Tillich et al. 2001). The 140,384-bp maize chloroplast sequence (GenBank/EMBL accession no. X86563) generated a circular map with a pair of 22,748-bp inverted repeats (IR[A] and IR[B])

K.J. Newton
Division of Biological Sciences, University of Missouri, Columbia, MO 65211, USA

D.B. Stern
Boyce Thompson Institute, Ithaca, NY 14853, USA

S. Gabay-Laughnan
Department of Plant Biology, University of Illinois, Urbana, IL 61801, USA

separating the 12,536-bp small and 82,352-bp large single copy regions (SSC and LSC). A total of 108 genes have been annotated, encoding 74 proteins, 30 transfer RNAs and four ribosomal RNAs (Maier et al. 1995). The number of proteins coded by the cpDNA is only a small fraction of the total chloroplast proteome.

Despite the circular map, it has been argued that the major form of cpDNA *in vivo* is linear, representing replicative forms, at least in part (Bendich 2004; Oldenburg and Bendich 2004). Earlier work on replication models assumed a circular form of the cpDNA, in agreement with the mapping data (Heinhorst and Cannon 1993). Chloroplast DNA is packaged into DNA-protein aggregates called nucleoids, whose maintenance, and perhaps replication, may depend in part on the enzyme sulfite reductase, which is also a DNA-binding protein (Sekine et al. 2007). Chloroplast DNA copy number has been studied using microscopic methods, and postulated to decline in older leaves, and perhaps even have its degradation stimulated by light (Oldenburg et al. 2006; Shaver et al. 2006). This is somewhat counterintuitive, given that an increase in chloroplast mRNA levels is well known to occur upon illumination of leaves (Rodermel and Bogorad 1985), a result recently substantiated through microarray analysis (Cahoon et al. 2007).

2 Chloroplast Transcription

Higher plant chloroplasts typically contain two types of RNA polymerase, one which resembles that of bacteria, and the other which is more phage-like. The eubacterial polymerase core subunits are plastid encoded, hence this is known as plastid-encoded polymerase or PEP. The single-subunit phage-like polymerase is nucleus encoded and thus known as NEP (reviewed by Liere and Börner 2007). However, the core PEP requires sigma factor for its activity, and these are nucleus encoded (see below). Thus, the nuclear genome plays an essential role in all chloroplast transcription.

Chloroplast transcription can be studied *in vitro* following fractionation of stromal proteins, and was shown to prefer supercoiled templates (Zaitlin et al. 1989). The PEP, when highly purified from maize, contains the same subunits as the bacterial enzyme (Hu and Bogorad 1990). Additional proteins have been found in PEP preparations from other plants, but none of these have been shown to be required for transcriptional activity (Pfannschmidt et al. 2000; Suzuki et al. 2004). Sigma factors are encoded by a five-member gene family in maize (Lahiri et al. 1999; Tan and Troxler 1999). Of these, Sig2B has the peculiarity of being targeted both to chloroplasts and mitochondria based on both GFP-based assays and biochemical separations (Beardslee et al. 2002). Interestingly, the corresponding *Arabidopsis* mutant appears to be embryo lethal, consistent with a mitochondrial function (Yao et al. 2003), given that most chloroplast sigma factor mutants have mild phenotypes and all are viable (reviewed in Shiina et al. 2005).

Maize, as apparently is the case in other monocots, possesses duplicate nuclear genes encoding phage-like RNA polymerases. RpoTm is targeted to mitochondria and RpoTp is targeted to chloroplasts (Chang et al. 1999). When present in phage

such as T3 and T7, these types of polymerases require no accessory factors, as is apparently the case for both the chloroplast and mitochondrial NEPs in *Arabidopsis* (Kuhn et al. 2007). These related proteins appear to have overlapping specificities, based on the ability of a mitochondrial promoter to be accurately recognized in a transplastomic context (Bohne et al. 2007).

The chloroplast two-polymerase system would appear to have a hierarchical relationship, given that NEP is necessary to express the PEP core subunits, both for transcribing the *rpo* genes and because NEP is the primary polymerase for transcribing ribosomal protein genes, whose products are required to translate the PEP polypeptides (e.g. Han et al. 1993; Silhavy and Maliga 1998). Indeed, NEP is most abundant in the maize leaf base, and PEP becomes progressively dominant as chloroplasts mature (Cahoon et al. 2004). Based on work in *Arabidopsis*, it has been proposed that the chloroplast tRNA^{Glu} acts as a negative regulator of NEP (Hanaoka et al. 2005), although this has not been studied in maize.

3   Chloroplast RNA Maturation and Degradation

Many chloroplast genes are organized into clusters, which reflects their eubacterial origin. These clusters can generate complex transcription patterns. A classical example is the cluster transcribed from the *psbB* promoter, which includes the downstream genes *psbT*, *psbH*, *petB* and *petD*, the latter two of which contain introns. Using immunprecipitation of polysomes based on nascent peptides, it was shown that all RNA forms are translatable, although more highly processed forms appear to be translated with higher efficiency (Barkan 1988).

The nuclear gene *crp1* is required for processing the *petB-petD* dicstronic RNA, and also for translation of the *petA* message, which encodes cytochrome *f* (Barkan et al. 1994). The CRP1 protein features a pentatricopeptide (PPR) motif (Small and Peeters 2000) and shares some similarity to fungal mitochondrial post-transcriptional regulators. It is stromally located as part of a multicomponent complex (Fisk et al. 1999). Immunoprecipitation of CRP1 revealed its direct interaction with the *psaC* and *petA* mRNAs (Schmitz-Linneweber et al. 2005). An additional nuclear mutant, *atp1*, appears to affect specifically the translation of the ATPase beta subunit (McCormac and Barkan 1999). It is likely that many other gene-specific and pleiotropic regulators remain to be identified, simply based on analogy to accumulating lists of such factors in both plants and the green alga *Chlamydomonas* (reviewed in Herrin and Nickelsen 2004; Leister and Schneider 2003), and the existence of a large mutant collection in maize (Stern et al. 2004).

One of the more thoroughly studied classes of maize mutants affecting chloroplast gene expression is that in which intron splicing is defective. Splicing becomes more efficient as plastids mature in maize, and is very inefficient in certain nonphotosynthetic tissues (Barkan 1989). Mutants in which splicing is totally disrupted for either a single or multiple genes are seedling-lethal, a phenotype which was first discovered in *crs1* and *crs2* (Jenkins et al. 1997). These genes were cloned, and found to include domains

"borrowed" from other proteins, and then were presumably adapted to serve as cpRNA splicing factors (Jenkins and Barkan 2001; Till et al. 2001). Biochemical techniques led to identification of additional splicing factors such as CAF1 and CAF2 (Ostheimer et al. 2003, 2005, 2006; Schmitz-Linneweber et al. 2005), a PPR protein required for *rps12 trans*-splicing (Schmitz-Linneweber et al. 2006), and a protein related to the prokaryotic ribonuclease III (Watkins et al. 2007), which itself is a relative of Dicer-like proteins involved in RNA interference. While chloroplast splicing has not been reconstituted *in vitro* to date, the studies cited above are major steps in elucidating how introns are identified and removed from precursor transcripts.

RNA editing occurs in chloroplasts but to a much lesser extent than in mitochondria. There are hundreds of sites in mitochondrial genes, but usually fewer than 30 for chloroplast genes (reviewed in Shikanai 2006; Takenaka et al. 2007). The first chloroplast editing site was identified in maize (Hoch et al. 1991) and there are now 27-28 reported maize chloroplast editing sites, all of which post-transcriptionally deaminate C residues to U, sometimes in a developmentally regulated manner (Maier et al. 1995; Peeters and Hanson 2002). Maize chloroplast editing has been studied *in vitro* and *in vivo*. It was shown in one study that an *rpoB* editing site was accurately recognized in transplastomic tobacco, attesting to conservation of at least one editing factor (Reed and Hanson 1997). An *in vitro* system was established to study editing *cis* elements (Hayes and Hanson 2007; Hayes et al. 2006), and should lead to further dissection of recognition elements for the presumed editosome. Maize chloroplast editing was reported to be sensitive to heat stress, perhaps due to the transcription rate exceeding the capacity of the editing apparatus (Nakajima and Mulligan 2001). Interestingly, translational capacity, as measured by elongation factor Ef-Tu, is also stimulated by heat stress (Momcilovic and Ristic 2007) although, in maize, chloroplast translation is not required *per se* for RNA editing (Halter et al. 2004).

Polyadenylation occurs in prokaryotes and in organelles, and usually leads to destabilization of transcripts (reviewed in Slomovic et al. 2006). In maize, polyadenylated RNA was identified several decades ago in chloroplasts (Haff and Bogorad 1976), although its function could not be deduced at that time. Polyadenylation also occurs in maize mitochondria (Lupold et al. 1999) and, although it was not readily linked to RNA instability in that study, a contemporaneous investigation in sunflower strongly suggested a role in RNA degradation (Gagliardi and Leaver 1999). Polyadenylated transcripts are substrates for the exonuclease polynucleotide phosphorylase, which generally is encoded by paralogous genes in plants specifying chloroplast and mitochondrial proteins. While the corresponding mutants have not been studied in maize, the *Arabidopsis* mutants stabilize polyadenylated species (Holec et al. 2006; Walter et al. 2002).

4   Maize Mitochondrial Genomes

Multiple mitochondrial genotypes (cytotypes) have been identified within maize–first on the basis of phenotypes and later by mitochondrial DNA (mtDNA) analyses. Cytoplasmic male sterility (CMS) was recognized and analyzed early in the study of maize genetics, and a classification scheme based upon the expression or sup-

pression of male sterility in sets of inbred lines was developed (reviewed by Duvick 1965). Rhoades (1950) accurately predicted that CMS would be associated with mitochondria. The male-sterility determining mitochondrial genotypes fit into three basic groups, CMS-T (Texas), CMS-S (USDA) and CMS-C (Charrua), according to which nuclear *restorer-of-fertility* (*Rf*) gene(s), carried by the different diagnostic inbreds, could counteract the cytoplasmically determined sterility (Beckett 1971). Later, restriction enzyme profiles were shown to be diagnostic for each group (Pring and Levings 1978).

Mitochondrial DNAs isolated from plants that are male fertile in all tested nuclear backgrounds can also be subdivided into different types. The "normal" N genotype, now referred to as NB (Fauron and Casper 1994), is found in many inbred lines, including B37 and B73. A second male-fertile mitochondrial genotype was designated NA by Fauron and Casper (1994) for the A188 inbred line; it is apparently also present in F6, W182BN and Ky21 (McNay et al. 1983; Newton and Walbot 1985). Mapping studies revealed that the NA mitochondrial genome was larger and carried multiple rearrangements relative to NB (Fauron and Casper 1994). A third male-fertile cytoplasmic type, called RU, occurs in some Latin American strains of maize (Weissinger et al. 1982).

Examples of each of five basic types of maize mitochondrial genomes have now been sequenced (Clifton et al. 2004; Allen et al. 2007). The smallest maize mitochondrial genome is CMS-T (535,825 bp) and the largest is CMS-C (739,719 bp). The NA (701,046 bp) mitochondrial genome is 23% larger than NB (569,630 bp). The difference in sizes of the maize mitochondrial genomes is principally accounted for by large duplications (Table 1; see also Allen et al. 2007). All of the genomes mapped as "master circles" except CMS-S, which has long been postulated to exist mostly as a diverse set of linear molecules (Schardl et al. 1984). Again, it should be noted that circular maps do not mean that the genomes exist *in vivo* as large circles (Bendich et al. 1996, Backert et al. 1997).

When the mitochondrial genotypes are compared, it can be seen that the linear gene order is not at all conserved (see Figure 1 of Allen et al. 2007). A minimum of 16 rearrangements must be postulated to account for the differences between the closely related NB and NA mitochondrial genomes (Allen et al. 2007). In contrast to the high amount of rearrangement, the maize mitochondrial genomes have very low rates of nucleotide substitution (Allen et al. 2007). How closely related each genome is to NB can be inferred from the amount of shared sequence as well as numbers of nucleotide substitutions and indels (Table 1). By these criteria, CMS-T and CMS-S are the most divergent from NB (Allen et al. 2007).

Maize has mitochondrial genomes typical of angiosperms (reviewed by Kubo and Newton 2007). All of the maize mitochondrial genomes have the same basic set of 51 functional genes, although the copy number of individual genes and exons varies with genotype, depending on how many occur in repeated segments (Allen et al. 2007). Included in the basic functional set are genes coding for 33 conserved proteins, including subunits of the mitochondrial electron transfer chain (some of the components of complexes, I, II, IV, and V), four proteins involved in the maturation of cytochrome c, and some of the translation machinery (detailed in Allen et al. 2007). In contrast to the chloroplast genome, most of the mitochondrial genes are not organized as recognizable operons, although a few are co-transcribed. Three

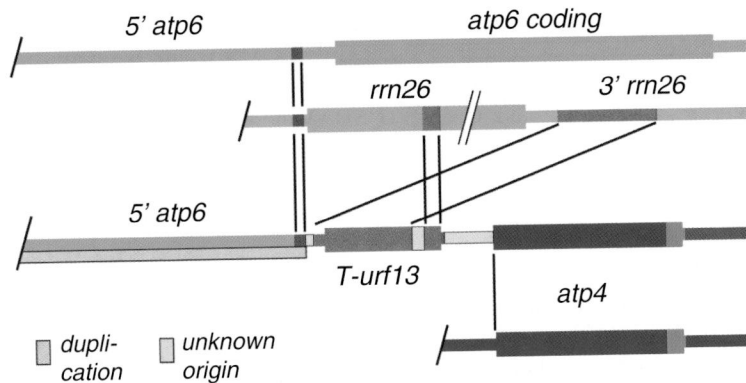

Fig. 1 Multiple recombination events are required to explain the origin of the chimeric 115-codon *T-urf*13 gene in the CMS-T genome. *T-urf*13 itself is composed of parts of the coding and the 3′ non-coding regions of the 26S rRNA gene, as well as some sequences of unknown origin. It is expressed because it has inserted between a strong duplicated promoter from the *atp*6 gene and the required *atp*4 gene, with which it is co-transcribed. There is no other copy of the *atp*4 gene in the CMS-T mitochondrial genome. (Adapted from Dewey et al. 1986, updated with data from Allen et al. 2007.)

ribosomal RNAs and 15 transfer RNAs are included in the basic gene set. The 18S and 5S rRNAs are co-transcribed and the 26S rRNA is independently transcribed (Maloney and Walbot 1990). If rearrangements bring genes into close proximity, they can be co-transcribed as well. Defining motifs specifically required for transcription and translation of mitochondrial genes has been difficult because sequences upstream of coding regions are usually not conserved in plant mitochondrial genomes (reviewed by Hazle and Bonen 2007).

A mitochondrially targeted phage-like RNA polymerase that presumably transcribes the mitochondrial genes has been identified in maize (Chang et al. 1999). Although both chloroplast and mitochondrial RNAs are edited in angiosperms (reviewed in Shikanai 2006; Takenaka et al. 2007), RNA editing is especially extensive for mitochondrial protein-coding genes. The transcription, editing, processing, translation and turnover of plant mitochondrial RNAs have been the subjects of a recent review (Gagliardi and Gualberto 2004). Maize mtRNAs are translated according to the universal code; however, in order to do so, tRNAs needed for the translation of at least five amino acids are imported from the cytosol (see Allen et al. 2007). It is a mystery as to why the large mitochondrial genomes of angiosperms have not retained a complete set of tRNAs.

When the maize NB genome was analyzed (Clifton et al. 2004), nearly three-fourths of its DNA was found to have no sequences in common with anything extant in the databases at that time. Indeed, a comparison of rice and maize mitochondrial genomes showed that functional gene sequences, which are almost identical, are nearly all that they have in common. In contrast, comparisons among maize cytotypes show that the intergenic sequences are highly conserved (Allen et al. 2007). If an intergenic sequence is present, even at a different location, it is highly conserved.

When differences in intergenic regions are seen between genotypes, it is because sequences are simply missing or novel DNA is inserted (Allen et al. 2007). Sequences that originated from the chloroplasts or from mitochondrial plasmids, as well as sequences of unknown origin present in the main mitochondrial genome are particularly variable (Allen et al. 2007). All of the maize mitochondrial genotypes contain plastid DNA at levels ranging from 2.29 to 4.61% of the genome (Table 1). A few tRNAs of originally plastid origin are used by the mitochondrion but the other plastid DNA insertions appear to be dispensable. For example, NB contains a large, 12.6-kb insertion of cpDNA (Stern and Lonsdale 1982; Clifton et al. 2004) that is either reduced or nearly eliminated in the other maize mitochondrial genomes (Allen et al. 2007). Similarly, sequences called R1 and R2, which exist as free linear plasmids in the exotic RU strain of maize (see below), are integrated into the main mitochondrial genome of NB and NA maize. The integrated versions carry deletions at one end of the plasmid (including one copy of the plasmid terminal inverted repeat) that account for their relative stabilities within the genomes. Nonetheless, these plasmid-derived sequences are especially prone to loss over time (Allen et al. 2007), as can be inferred from the differences observed among CMS mitochondrial cytotypes (Table 1). Many insertions of DNA of unknown origin are found within the maize mitochondrial genomes and they also tend to vary among genotypes (Allen et al. 2007).

5   Chimeric ORFs Associated with CMS

In addition to the basic gene set, over a hundred ORFs can be found for each maize mitochondrial genotype (Allen et al. 2007). These ORFs have been defined simply by potential start and stop codons that contain a minimum number of predicted amino acids (70-100) between them. Most of them are not conserved even among maize cytotypes and very few of the predicted ORFs are detectable as stable RNAs (Meyer 2004); therefore, they are unlikely to be translated. However, in a few cases, pieces of known genes are found within ORFs and occasionally, these "chimeric" ORFs are located downstream of a sequence that can act as a promoter. In the cases of CMS-T and CMS-S, novel chimeric ORFs are expressed as stable RNAs and translated into proteins that cause the CMS phenotype (reviewed by Schnable and Wise 1998; Hanson and Bentiolla 2004; Newton et al. 2004; Chase 2007).

In CMS-T, the CMS-associated *T-urf13* is composed of sequences from the 26S ribosomal RNA and its downstream untranslated region, as well as some sequences of unknown origin (Figure 1). It is expressed as an RNA because it is inserted between a strong promoter (a duplicate copy of the *atp6* upstream region) and a required gene, the only copy of *atp4* in the CMS-T genome (Dewey et al. 1986; Allen et al. 2007). This novel gene also has the signals necessary for protein translation, as well as leader sequences for membrane localization. The 13-kD membrane-spanning protein has the remarkable property of being produced at high levels throughout the life cycle and benignly residing in the mitochondrial membrane until, in response

to an as-yet-unknown stimulus in tapetal cells during pollen maturation, it causes changes in the mitochondrial membrane resulting in the premature degeneration of the tapetal cells (Dewey et al. 1987; Wise et al. 1987a; reviewed by Skibbe and Schnable 2005). The detrimental effects of T-URF13 can also be triggered by specific fungal toxins and by the addition of the insecticide methomyl (reviewed by Skibbe and Schnable 2005).

In CMS-S mtDNA, the sterility-associated *orf355-orf77* region occurs within a 4215-bp repeat, designated R. One copy of the repeat was sequenced and shown to be chimeric (Zabala et al. 1997). It includes sequences similar to a portion of the R1 linear plasmid (see below) and rearranged mitochondrial sequences normally located 3′ to the *atp4* gene. Sequences with similarity to the coding and 3′ flanking region of the *atp9* gene are found within *orf77* and a region with no similarity to any known sequences is found within *orf355* (Zabala et al. 1997; Allen et al. 2007). The production of a 1.6-kb transcript from the 4.2-kb repeat, which includes both *orf355* and *orf77*, correlates with the sterility phenotype (Zabala et al. 1997). Pollen of CMS-S plants is normal until late in microspore development, when the pollen grains suddenly abort (Lee et al. 1980). There is no change in other anther tissues.

In CMS-C, no candidate novel chimeric gene has yet been correlated with male sterility (Allen et al. 2007). Dewey *et al.* (1991) reported that rearrangements affected upstream regions of the essential *atp9*, *atp6* and *cox2* genes. We now know that there are two *atp9* genes in the CMS-C mitochondrial genome, one of which has a novel organization (Allen et al. 2007). In addition, the leader sequence of the single *atp6* gene in CMS-C is not shared with those from other maize mitochondrial cytotypes (see Dewey et al. 1991; Allen et al. 2007). It is still unknown whether a rearranged functional gene, presumably either *atp6* or *atp9*, is altered enough in its expression to cause CMS in the C-cytotype. Cytological studies reveal no mitochondrial abnormalities as are found in CMS-T; however, the CMS-C tapetum and microspores disintegrate at an intermediate stage of development (Lee et al. 1979).

6   Mitochondrial DNA Plasmids

Plasmids have been reported in the mitochondria of several higher plants (reviewed by Handa 2007), where they occur in addition to the main mitochondrial genome. In maize mitochondria, both linear and circular plasmids have been found. The high stoichiometry of these molecules relative to the main mitochondrial genome indicates that they replicate autonomously.

Two circular DNA plasmids, 1.4 kb and 1.9 kb in size, have been reported in maize (Kemble and Bedbrook 1980); neither is present in all cytoplasms of maize (Pring and Smith 1985; Smith and Pring 1987). The 1.9-kb minicircle is present in the N, C and T cytoplasms examined, but not in all subtypes of S cytoplasm. The 1.4-kb plasmid is present in some, but not all, N, C and S cytoplasms (Pring and Smith 1985). Both plasmids are transcriptionally active (Smith and Pring 1987) and no

line carries the 1.4-kb plasmid without the 1.9-kb plasmid (Pring and Smith 1985; Smith and Pring 1987). Two additional small, circular DNA plasmids, 1.57 and 1.42 kb in size, have been identified as specific to C cytoplasm (Kemble and Bedbrook 1980). There is no homology between these plasmids and the other DNA plasmids of maize (Kemble and Bedbrook 1980).

While linear DNA plasmids are found relatively frequently in fungal mitochondria (reviewed by Meinhardt et al. 1990, 1997), they are rare in higher plants. Only eight plant species have been found to carry linear plasmids (reviewed by Handa 2007). Interestingly, of the 14 linear plasmids identified thus far, nine occur in *Zea* species (see Table 1 in Handa 2007). These plasmids have terminal inverted repeats (TIRs) and covalently attached proteins at their 5′ ends. Some fungal pathogens of plants contain linear plasmids and these fungi are potentially the source of the linear mtDNA plasmids in maize (reviewed by Handa 2007).

The first linear DNA plasmids to be identified in any eukaryote were the S1 and S2 plasmids specific to the maize S-type cytoplasm (Pring et al. 1977; reviewed by Meinhardt et al. 1990). S1 is 6397 bp (Paillard et al. 1985) and S2 is 5453 bp long (Levings and Sederoff 1983) and there is an ~1150-bp region of sequence homology between them (Kim et al. 1982). S1 and S2 have identical 208-bp TIRs (Braun et al. 1986) and are stabilized by protein(s) bound to their ends (Kemble and Thompson 1982). S1 and S2 are always found to occur together. It is thought that an S1 open reading frame (URF3) encodes a putative DNA polymerase (Kuzmin and Levchenko 1987) while an S2 ORF (URF1) encodes a putative RNA polymerase (Kuzmin et al. 1988; Oeser 1988).

Linear DNA molecules similar to S1 and S2, termed R1 and R2, are found in some races of maize from Latin America (Weissinger et al. 1982, 1983). R2 is very similar to S2. The 7.5-kb R1 plasmid shares 4.9 kb of homology with the S1 plasmid (Weissinger et al. 1982; Levings and Sederoff 1983). Apart from their TIRs, R1 and R2 share no sequence homology (Leving and Sederoff 1983). A model for the origin of S1 from recombination between R1 and R2 has been proposed (Levings and Sederoff 1983).

R-type plasmids have been found in the mitochondria of some teosintes. The D1 and D2 plasmids have been reported in samples of the Mexican source of *Zea diploperennis*. They are apparently the same size as R1 and R2 but are in low copy number (Timothy et al. 1983). The linear DNA molecules M1 (7.5 kb) and M2 (5.4 kb) are present in the Mazoti strain of *Z. luxurians* (Grace et al. 1994). The sequences of the M1 and R1 plasmids are very similar with characteristic small differences in the TIRs (Grace et al. 1994).

A linear 2.3-kb plasmid and its 2.1-kb variant are the most widely distributed maize mitochondrial plasmids (Kemble et al. 1983; Pring & Smith 1985; Leon et al. 1989). The 2.1-kb plasmid has been shown to be a deleted form of the 2.3-kb plasmid (Bedinger et al. 1986). The 2.3-kb plasmid has been sequenced (Leon et al. 1989); it has 170-bp identical TIRs that are associated with stabilizing terminal proteins. Sixteen of the first 17 nucleotides of the 170-bp TIR are identical to the 208-bp TIRs of S1 and S2 (Bedinger et al. 1986). The 2.3/2.1 kb plasmid appears to be maintained because it carries an essential tRNA^{Trp} gene (of plastid origin, Leon et al. 1989) that is the only one present in the within maize mitochondria.

The 2.1-kb plasmid was first found in CMS-T mitochondria and was thought to be specific for that genotype (Kemble et al. 1983). However, the mitochondria of maize inbred lines A188, F6, Ky21and W182BN also contain the 2.1-kb variant of the 2.3-kb plasmid (McNay et al. 1983; Newton and Walbot 1985). We now know that these lines carry NA cytoplasm, which usually has the 2.1-kb variant plasmid. Despite early expectations (e.g. Kemble et al. 1983), there appears to be no combination of maize linear and circular DNA plasmids that is diagnostic for a particular maize cytoplasm (see Table 1 in Pring and Smith 1985).

7   Cytoplasmic Reversions to Fertility

The best-characterized mitochondrial variants in plants are those that result in cytoplasmic male sterility (Figure 2B). Despite claims to the contrary (Lemke et al. 1985, 1988), no *de novo* mutation of normal, fertile cytoplasm to CMS cytoplasm has been observed to arise in a genetically characterized maize strain. Since the mtDNA genomes of CMS strains, especially CMS-S and CMS-T, contain numerous structural differences compared with the NA and NB cytoplasms, it is extremely unlikely that such gross structural differences could arise in a single generation (Lonsdale 1987; Small et al. 1987). Since each type of CMS in maize can be restored to fertility by a specific nuclear restoring allele or alleles (see below), what may appear to be a new CMS mutation could instead be the male sterility trait being uncovered by crosses that remove the restoring allele.

On the other hand, spontaneous reversion to fertility has been observed in field-grown CMS-S plants (Figure 2C), as well as in plants regenerated from tissue cultures that carry S-type male-sterile cytoplasm. CMS-T maize passed through tissue culture has also been observed to generate revertants but no reversion has been known to occur in the field (reviewed by Gabay-Laughnan and Newton 2005). In order to ascertain whether a newly arisen male fertility is due to an alteration in the mtDNA or to a new nuclear restorer mutation, the newly fertile plant is crossed as the pollen parent onto a male-sterile plant with the corresponding progenitor cytoplasm. Because cytoplasmic male sterility is maternally inherited, all progeny of such crosses will be male sterile if the mutation arose in the mtDNA. If the mutation is nuclear, male-fertile plants will be present among the progeny.

Cytoplasmic reversion of CMS-T and CMS-S usually results from rearrangements of the mtDNA generated by aberrant recombination across short repeat regions (reviewed by Gabay-Laughnan and Newton 2005). In all but one of the CMS-T revertants analyzed, the *T-urf13* region is deleted (Dewey et al. 1986, 1987; Wise et al.

Fig. 2 Mutant mitochondrial phenotypes in maize. A. Fertile NB tassel. B. Sterile CMS-C tassel. C. Large revertant sector (lower left) on a CMS-S tassel. D. Rudimentary tassel on a near-homoplasmic mutant NCS2 plant. E. Heteroplasmic NCS4 mutant plant with a severe phenotype. Despite its very small size and abnormal growth pattern, the plant produced both a tassel that shed pollen and a small ear shoot that set seed.

A Fertile NB

B CMS

C CMS-S revert. sector

D NCS2 tassel

E NCS4 plant

1987a; reviewed by Ward 1995). In the exceptional case, a frameshift mutation in the *T-urf13* gene results in the truncation of the protein it encodes (Wise et al. 1987b).

Most CMS-S revertants lose the free S1 and S2 plasmids (Levings et al. 1980; Kemble and Mans 1983; Schardl et al. 1985) and it was initially thought that this alteration was absolutely correlated with reversion to fertility. However, revertants arising in the WF9 nuclear background retain S1 and S2 as free plasmids (Escote et al. 1985; Ishige et al. 1985). We now know that a common feature of all S revertants is a rearrangement of the sterility-associated *orf355/orf77* region and the loss of its 1.6-kb transcript (Zabala et al. 1997).

In field-grown CMS-S plants, cytoplasmic reversion occurring in the ear tissue is observed serendipitously in the next generation as male-fertile plants in progeny where only male-sterile plants were expected. However, reversion in field-grown plants is usually recognized as tassel sectors ranging in size from a single floret to a large sector (Figure 2C); entirely fertile tassels are also observed. How does a researcher recover a mutation for a maternally inherited trait when the mutation is initially observed in the tassel (male reproductive tissue)? Seed from ears carried on plants with new fertility must be planted and tassels on the resulting plants scored in the next generation. Analysis of ears produced on plants with entirely fertile tassels and plants with large tassel sectors has demonstrated that the cytoplasmic revertant can most often be recovered through the female if the main rachis of the tassel is included in the reversion event; that is, when the plant has an entirely fertile tassel or a large tassel sector (Gabay-Laughnan and Laughnan 1990).

The *T-urf13* and *orf355/orf77* regions responsible for CMS-T and CMS-S, respectively, may be viewed as nonessential mitochondrial genes. Thus, revertant plants usually have a male-fertile and otherwise normal phenotype. CMS revertants are stable in subsequent generations (see Laughnan and Gabay-Laughnan 1983), indicating that the revertant plants must be homoplasmic for the rearranged mtDNA.

8   Influence of Nuclear Genes on Cytoplasmic Reversions to Fertility and Plasmid Ratios

Unidentified nuclear genes exist that influence the frequency of CMS reversions (reviewed by Gabay-Laughnan et al. 1995). For example, the inbred line M825 carrying CMS-S cytoplasm is male sterile, but over 10% of plants exhibit some male fertility, usually in the form of tassel sectors (Laughnan et al. 1981). These revertants are predominantly cytoplasmic in nature. That the M825 nucleus is responsible for the reversion events can be confirmed by recurrently crossing different inbred lines with CMS-S mitochondria by M825. The S cytoplasm is stable in the original nuclear background but the converted strain reverts at the same rate as CMS-S M825 (reviewed by Gabay-Laughnan et al. 1995).

Also under nuclear control is the ratio of the S1 and S2 plasmids present in S male-sterile plants (reviewed by Gabay-Laughnan and Newton 2005). While these plasmids occur in equimolar amounts in most CMS-S inbred lines (Pring et al. 1977; Laughnan et al. 1981), in S-M825 the ratio of S1 to S2 is about 5 to 1 (Laughnan

et al. 1981). For CMS-S in the 38-11 line, the ratio of these plasmids is approximately 1 to 3 (Laughnan et al. 1981). In addition, the equimolar plasmid ratios observed in most CMS-S inbred lines can be shifted by the appropriate recurrent backcrosses by M825 or 38-11 (Laughnan et al. 1981).

Whether or not S1 and S2 are retained by CMS-S cytoplasmic revertants is also affected by the inbred nuclear genotype. While most CMS-S inbreds lose S1 and S2 upon cytoplasmic reversion, the plasmids are retained by CMS-S revertants occurring in the WF9 nuclear background (Escote et al. 1985; Ishige et al. 1985).

9 Restorer-of-Fertility Alleles

Restorer-of-fertility (*Rf*) alleles are the most intensively studied of the nuclear genes affecting the expression of mitochondrial genes. Maize restorers do not produce heritable changes in the mtDNA but instead override the mitochondrially encoded male sterility, usually through the modification of CMS-associated transcripts (reviewed by Skibbe and Schnable 2005; Chase 2007). The three major types of CMS in maize (Beckett 1971) are each associated with specific restoring alleles (see Table 3 in Skibbe and Schnable 2005). Restoration of CMS-T is sporophytic and requires a dominant allele from each of two complementary restorers. *Rf1* reduces the amount of *T-urf13* RNA, whereas *Rf2a* is a dominant allele of a gene encoding a mitochondrially targeted aldehyde dehydrogenase, which may serve as a metabolic enhancer of mitochondrial function. Restoration of CMS-C is sporophytic and requires the dominant *Rf4* allele present in many maize inbreds. The dominant alleles of two additional independent loci, *rf5* and *rf6*, have also been reported to be able to restore fertility to CMS-C (reviewed by Skibbe and Schnable 2005).

Restoration of CMS-S is gametophytic (Buchert 1961) and only one restorer, *Rf3*, is required (reviewed by Laughnan and Gabay-Laughnan 1983). It works by disrupting the RNAs from the *orf355-orf77* region of the genome. A number of *restorer-of-fertility* alleles arising by spontaneous mutation at many different nuclear loci have been recovered (Laughnan and Gabay 1973, 1978; reviewed by Gabay-Laughnan et al. 1995). Like *Rf3*, these mutations act as gametophytic restorers-of-fertility by disrupting expression of the S-sterility gene *orf355-orf77*. However, the most common class of these spontaneous restorer mutations, termed *restorer-of-fertility lethal* (*rfl*), results in the loss of function of nuclear genes crucial for the expression of other mitochondrial genes (reviewed by Chase and Gabay-Laughnan 2003, 2004; Wen et al. 2003). This has been shown for the *rfl1* (formerly *RfIII*) mutation that also drastically reduces the expression of the alpha subunit of ATPase (Wen et al. 2003). Late-stage pollen development and pollen germination apparently do not require mitochondrial respiration (reviewed by Tadege and Kuhlemeier 1997; Tadege et al. 1999). Thus, pollen grains with an *rf-lethal* allele and CMS-S cytoplasm can survive and function, but when an *rf-lethal* allele is homozygous in tissues where mitochondrial function is required, the tissues cannot survive and lethality results, typically during kernel development. It should be noted that this class of restorer is also a seed-lethal mutation in N-cytoplasm plants.

10   Abnormal Growth Mutations

Maternally inherited abnormal growth mutations are usually recognized by diagnostic stripes on leaves and/or sectors of aborted kernels on ears. The overall growth of the plant is also affected and mutants may be short, stunted, twisted or otherwise deformed. In maize, such mutants have been named nonchromosomal stripe (NCS; Shumway and Bauman 1967; Coe 1983) and they are correlated with specific mitochondrial DNA (mtDNA) changes (Newton and Coe 1986). The leaf striping phenotypes depend upon the mitochondrial gene affected and may appear as pale green (NCS2), yellow (NCS5, NCS6, NCS7), or striated (NCS3, NCS4) sectors (reviewed by Newton 1995; reviewed by Gabay-Laughnan and Newton 2005).

As is the case in cytoplasmic reversion to fertility, NCS mutations result from abnormal recombination between short repeat regions of mtDNA (reviewed by Gabay-Laughnan and Newton 2005). However, in each of the NCS mutations characterized to date, one of the repeats is contained within an essential gene and the recombination event disrupts the expression of that gene. NCS2 plants (Figure 2D) carry a deletion of the 3′ end of the *nad4* gene (Marienfield and Newton 1994), whereas the NCS5 and NCS6 mutants have lost different amounts of the 5′ end of the *cox2* gene (Lauer et al. 1990; Newton et al. 1990). The NCS3 and NCS4 mutations both involve the gene encoding the *rps3/rpl16* ribosomal protein transcription unit (Hunt and Newton 1991; Newton et al. 1996). Interestingly, in the case of NCS4, which arose in a CMS-S plant, an S2 plasmid aberrantly recombined with the intron of the *rps3* gene, resulting in a new mutant that simultaneously exhibits both abnormal growth and male fertility (Figure 2E).

Since NCS mutations arise in genes whose function is important in kernel development, they are usually lethal. These mutations also have deleterious effects at other stages of plant development (reviewed by Newton 1995; Gabay-Laughnan and Newton 2005). Thus, NCS mutant plants are heteroplasmic for normal and rearranged mtDNAs. The observed leaf sectors represent the sorting out of the mutant mitochondria; these sectors are homoplasmic or nearly so (Newton and Coe 1986; Newton et al. 1989; Baker and Newton 1995).

How does a researcher ensure propagation of maternally inherited mutations observed mainly in vegetative tissue? In NCS mutations, plant height and vigor vary from plant to plant and the extent of leaf striping may vary from leaf to leaf on the same plant (reviewed by Newton 1994). To ensure seed set, the NCS mutation is propagated by pollinating ears borne on the less affected plants. In the next generation, seeds are planted from ears exhibiting sectors of aborted kernels. Phytomers are the basic repeating structural units of a plant. The maize phytomer consists of a leaf, the node to which it is attached, the internode and an axillary meristem (Galinat 1959) that can develop into an ear. By pollinating ears on NCS plants whose phytomers are mostly mutant, a researcher can increase the chances that the NCS mutation will be inherited. It must be noted that, although it is located close to the midrib of the subtending leaf, the ear shoot is clonally related to the internode above it (see Figure 1 in Scanlon 2003).

11 Influence of Nuclear Genes on Generating Mitochondrial Mutations with Abnormal Growth

Most strains of maize have stable mitochondrial genotypes that rarely undergo the rearrangements that lead to NCS mutations. However, in the WF9 nuclear background, such mutations are observed relatively frequently, appearing in 1-2% of plants (Duvick 1965; reviewed by Gabay-Laughnan and Newton 2005). Since the recurrent backcrossing of any nuclear-cytoplasmic combination with pollen from the inbred line WF9 results in an increase in the frequency of NCS mutations, there must be a nuclear allele(s) contributing to the origin of these mutations (reviewed by Gabay-Laughnan and Newton 2005). Once present, each NCS can survive as a stable mitochondrial mutation in a heteroplasmic combination with non-mutant mitochondria. Sorting out of mutant mitochondria during development gives the characteristic striping pattern.

A dramatic and reproducible effect of nuclear genotype on the mitochondrial genome is provided by a naturally occurring "mitochondrial mutator" genotype (Kuzmin et al. 2005). The nuclear genes that cause a dramatic increase in mitochondrial rearrangement mutations originated from an exotic South American popcorn strain, named "P2" (Brown and Duvick 1958). The P2 stock was maintained for many years by D. Duvick at Pioneer HiBred. Multiple mitochondrial lesions in normal and CMS genotypes have been documented as a result of crossing the P2 strain as a male (Kuzmin et al. 2005). Phenotypic defects appear only after a lag time (usually two or three generations) and only in some plants. The effect first shows up as striping on leaves that is suggestive of chloroplast defects, then as sectors of reduced growth and kernel abortion, which strongly suggests that mitochondrial defects are present.

No cpDNA differences have been observed, but *multiple* rearrangement mutations in mtDNA can be seen among the progeny of plants with P2-type striping. Sibling plants can have different phenotypes and mtDNA rearrangements (Kuzmin et al. 2005). In contrast to the NCS mutants, P2 plants show a multiplicity of mtDNA changes. Collectively, the data suggest that the P2 nuclear alleles causing the maternally inherited defects are recessive, that they induce multiple rearrangement mutations, and that the rearrangements amplify and sort out from one another (Kuzmin et al. 2005). In *Arabidopsis*, mutations in the nuclear *chm* locus, now renamed *Msh*1 due to its similarity to a prokaryotic mismatch repair gene (Abdelnoor et al. 2003), give rise to maternally inherited defects and mtDNA rearrangements similar to those caused by P2. Thus, the P2 nuclear genotype represents a natural mutagen for maize mitochondrial genomes (Kuzmin et al. 2005).

12 Future Directions

Now that organelle genome sequences are readily available, many questions remain. Why are plant mitochondrial genomes so plastic in their organization and chloroplast genomes so stable? What nuclear-organellar interactions have contributed

to their divergent evolutionary pathways? In order to answer these types of questions, it is necessary to identify the nuclear genes that control genome stability and transmission. Most of the genes that are involved in the biogenesis and functioning of organelles are nuclear. Maize provides model systems for identifying the nuclear genes involved in the biogenesis of both mitochondria and chloroplasts.

Maize is especially suitable for studying chloroplast biogenesis because embryo-viable but non-photosynthetic, seedling-lethal mutants can be analyzed (Stern et al. 2004). The Photosynthetic Mutant Library (PML) Collection has over 2200 *Mu*-induced photosynthetic mutants, providing a resource for the analysis of nuclear genes required for the expression of chloroplast genes (reviewed by Stern et al. 2004).

A distinctive set of conditions exists in the CMS-S system of maize that could permit the recovery of mutations in nuclear genes essential for mitochondrial gene expression (reviewed by Chase and Gabay-Laughnan 2003, 2004; Wen et al. 2003). The CMS-S sterility-causing locus, *orf355-orf77*, becomes highly expressed in tassel and developing pollen and causes pollen abortion via a cell-death pathway. Meanwhile, the expression of other mitochondrial genes is dispensable because pollen grains survive via ethanolic fermentation (Wen et al. 2003). Because CMS-S restoration is gametophytic, occurring late in pollen development, mutations in nuclear genes that down-regulate mitochondrial gene expression, and thus act to restore fertility to CMS-S pollen grains, can be transmitted and recovered. Exploiting this system could enhance our understanding of the nuclear genes involved in mitochondrial biogenesis (Wen et al. 2003).

A very important area of future research is to increase our understanding of signaling between organelles. When disruptions in mitochondrial or chloroplast function occur, signals are generated and transmitted to the nucleus to alter *nuclear gene expression*. The maize NCS mutations are known to induce expression of nuclear alternative oxidase genes (Karpova et al. 2002), as well as of a specific subset of heat shock proteins (Kuzmin et al. 2004). Altered mitochondrial-nuclear signaling also occurs in maize CMS plants (Hochholdinger et al. 2004) and in *Arabidopsis* plants subjected to electron pathway inhibitors (reviewed by Rhoads and Subbaiah 2007). Pathways involved in plastid-to-nucleus "retrograde regulation" are being studied in *Arabidopsis* and *Chlamydomonas* (reviewed in Nott et al. 2006), but maize should also be an excellent model system for such analyses.

Acknowledgements We thank James Allen for Figure 1 and help organizing Table 1, and Leah Roark for assisting with the preparation of the manuscript. Support for research in the authors' laboratories was from United States Department of Agriculture (for SGL) and the National Science Foundation (DBI-0211935 for DBS and DBI-0110168 for KJN).

References

Abdelnoor RV, Yule R, Elo A, Christensen AC, Meyer-Gauen G, Mackenzie SA (2003) Substoichiometric shifting in the plant mitochondrial genome is influenced by a gene homologous to *MutS*. Proc. Natl. Acad. Sci. USA 100, 5968–5973.

Allen JO, Fauron CM, Minx P, Roark L, Oddiraju S, Lin GN, Meyer L, Sun H, Kim K, Wang C, Du F, Xu D, Gibson M, Cifrese J, Clifton SW, Newton KJ (2007) Comparisons among two fertile and three male-sterile mitochondrial genomes of maize. Genetics 177, 1173–1192.

Backert S, Meissner K, Börner T (1997) Unique features of the mitochondrial rolling circle-plasmid mp1 from the higher plant *Chenopodium album (L.)*. Nucleic Acids Res. 25, 582–589.

Baker RF, Newton KJ (1995) Analysis of defective leaf sectors and aborted kernels in NCS2 mutant maize plants. Maydica 40, 89–98.

Barkan A (1988) Proteins encoded by a complex chloroplast transcription unit are each translated from both monocistronic and polycistronic mRNAs. EMBO J 7, 2637–2644.

Barkan A (1989) Tissue-dependent plastid RNA splicing in maize: transcripts from four plastid genes are predominantly unspliced in leaf meristems and roots. Plant Cell 1, 437–445.

Barkan A, Walker M, Nolasco M, Johnson D (1994) A nuclear mutation in maize blocks the processing and translation of several chloroplast mRNAs and provides evidence for the differential translation of alternative mRNA forms. EMBO J 13, 3170–3181.

Beardslee TA, Roy-Chowdhury S, Jaiswal P, Buhot L, Lerbs-Mache S, Stern DB, Allison LA (2002) A nucleus-encoded maize protein with sigma factor activity accumulates in mitochondria and chloroplasts. Plant J 31, 199–209.

Beckett JB (1971) Classification of male-sterile cytoplasms in maize (*Zea mays L.*). Crop Sci. 11, 724–727.

Bedinger P, de Hostos EL, Leon P, Walbot V (1986) Cloning and characterization of a linear 2.3 kb mitochondrial plasmid of maize. Mol. Gen. Genet. 205, 206–212.

Bendich AJ (1996) Structural analysis of mitochondrial DNA molecules from fungi and plants using moving pictures and pulsed-field gel electrophoresis. J Mol. Biol. 255, 564–588.

Bendich AJ (2004) Circular chloroplast chromosomes: the grand illusion. Plant Cell 16, 1661–1666.

Bohne AV, Ruf S, Börner T, Bock R (2007) Faithful transcription initiation from a mitochondrial promoter in transgenic plastids. Nucleic Acids Res. in press.

Braun CJ, Sisco PH, Sederoff RR, Levings CS III (1986) Characterization of inverted repeats from plasmid-like DNAs and the maize mitochondrial genome. Curr. Genet. 10, 625–630.

Brown WL, Duvick DN (1958) An extreme nuclear-cytoplasmic interaction. Maize Genet. Coop. Newsl. 32, 120–121 (cited with permission).

Buchert JG (1961) The stage of the genome-plasmon interaction in the restoration of fertility to cytoplasmically pollen-sterile maize. Proc. Natl. Acad. Sci. USA 47, 1436–1440.

Cahoon AB, Cunningham KA, Bollenbach TJ, Stern DB (2003a) Maize BMS cultured cell lines survive with massive plastid gene loss. Curr. Genet. 44, 104–113.

Cahoon AB, Cunningham KA, Stern DB (2003b) The plastid *clpP* gene may not be essential for plant cell viability. Plant Cell Physiol. 44, 93–95.

Cahoon AB, Harris FM, Stern DB (2004) Analysis of developing maize plastids reveals two mRNA stability classes correlating with RNA polymerase type. EMBO Rep. 5, 801–806.

Cahoon AB, Takacs EM, Sharpe RM, Stern DB (2007) Nuclear, chloroplast, and mitochondrial transcript abundance along a maize leaf developmental gradient. Plant Mol. Biol. in press.

Chang CC, Sheen J, Bligny M, Niwa Y, Lerbs-Mache S, Stern DB (1999) Functional analysis of two maize cDNAs encoding T7-like RNA polymerases. Plant Cell 11, 911–926.

Chase CD, Gabay-Laughnan S (2003) Exploring mitochondrial-nuclear genome interactions with S male-sterile maize. In: S.G. Pangali (ed) Recent Res. Dev. Genet. Research Signpost, Kerala, pp 31–41.

Chase CD, Gabay-Laughnan S (2004) Cytoplasmic male sterility and fertility restoration by nuclear genes. In: H. Daniell and C.D. Chase (eds) Molecular Biology and Biotechnology of Plant Organelles. Springer, The Netherlands, pp 593–622.

Chase CD (2007) Cytoplasmic male sterility: a window to the world of plant mitochondrial-nuclear interactions. Trends Genet. 23, 81–90.

Clifton SW, Minx P, Fauron CM, Gibson M, Allen JO, Sun H, Thompson M, Barbazuk WB, Kanuganti S, Tayloe C, Meyer L, Wilson RK, Newton KJ (2004) Sequence and comparative analysis of the maize NB mitochondrial genome. Plant Physiol. 136, 3486–3503.

Coe E (1983) Maternally inherited abnormal plant types in maize. Maydica 28, 151–167.

Dewey R, Levings CS III, Timothy DH (1986) Novel recombinations in the maize mitochondrial genome produce a unique transcriptional unit in the Texas male-sterile cytoplasm. Cell 44, 439–449.

Dewey RE, Timothy DH, Levings CS III (1987) A mitochondrial protein associated with cytoplasmic male sterility in the T cytoplasm of maize. Proc. Natl. Acad. Sci. USA 84, 5374–5378.

Dewey RE, Timothy DH, Levings CS III (1991) Chimeric mitochondrial genes expressed in the C male-sterile cytoplasm of maize. Curr. Genet. 20, 475–482.

Doebley JF, Ma DP, Renfroe WT (1987) Insertion/deletion mutations in the *Zea* chloroplast genome. Curr. Genet. 11, 617–624.

Duvick DN (1965) Cytoplasmic pollen sterility in corn. Adv. Genet. 13, 1–56.

Escote LJ, Gabay-Laughnan SJ, Laughnan JR (1985) Cytoplasmic reversion to fertility in *cms-S* maize need not involve loss of linear mitochondrial plasmids. Plasmid 14, 264–267.

Fauron C, Casper M (1994) A second type of normal maize mitochondrial genome: an evolutionary link. Genetics 137, 875–882.

Fisk DG, Walker MB, Barkan A (1999) Molecular cloning of the maize gene *crp1* reveals similarity between regulators of mitochondrial and chloroplast gene expression. EMBO J 18, 2621-2630.

Gabay-Laughnan S., Laughnan JR (1990) Correlation of tassel and ear reversion events in cms-S. Maize Genet. Coop. Newsl. 64, 114–115.

Gabay-Laughnan S, Zabala G, Laughnan JR (1995) S-type cytoplasmic male sterility in maize. In: C.S. Levings, III and I.K. Vasil (eds) Advances in Cellular and Molecular Biology of Plants, Vol. 2: The Molecular Biology of Plant Mitochondria. Kluwer Academic, Dordrecht, The Netherlands, pp 395–432.

Gabay-Laughnan S, Newton KJ (2005) Mitochondrial mutants in maize. Maydica 50, 349–359.

Gagliardi D, Leaver CJ (1999) Polyadenylation accelerates the degradation of the mitochondrial mRNA associated with cytoplasmic male sterility in sunflower. EMBO J 18, 3757–3766.

Gagliardi D, Gualberto JM (2004) Gene expression in higher plant mitochondria. In: H.A. D.A. Day and J.W. Millar (eds) Plant Mitochondria: From Genome to Function. Kluwer Academic Publishers, Dordrecht, The Netherlands, pp 121–142.

Galinat WC (1959) The phytomer in relation to the floral homologies in the American *Maydea*. Bot. Mus. Leaflets, Harvard U. 19, 1–32.

Grace KS, Allen JO, Newton KJ (1994) R-type plasmids in mitochondria from a single source of *Zea luxurians* teosinte. Curr. Genet. 25, 258–264.

Haff LA, Bogorad L (1976) Poly(adenylic acid)-containing RNA from plastids of maize. Biochemistry 15, 4110–4115.

Halter CP, Peeters NM, Hanson MR (2004) RNA editing in ribosome-less plastids of *iojap* maize. Curr. Genet. 45, 331–337.

Han C, Patrie W, Polacco M, Coe EH (1993) Aberrations in plastid transcripts and deficiency of plastid DNA in striped and albino mutants in maize. Planta 191, 552–563.

Hanaoka M, Kanamaru K, Fujiwara M, Takahashi H, Tanaka K (2005) Glutamyl-tRNA mediates a switch in RNA polymerase use during chloroplast biogenesis. EMBO Rep. 6, 545–550.

Handa H (2007) Linear plasmids in plant mitochondria: Peaceful coexistences or malicious invasions? Mitochondrion in press.

Hanson MR, Bentolila S (2004) Interactions of mitochondrial and nuclear genes that affect male gametophyte development. Plant Cell 16 Suppl, S154–169.

Hayes ML, Reed ML, Hegeman CE, Hanson MR (2006) Sequence elements critical for efficient RNA editing of a tobacco chloroplast transcript *in vivo* and *in vitro*. Nucleic Acids Res. 34, 3742–3754.

Hayes ML, Hanson MR (2007) Assay of editing of exogenous RNAs in chloroplast extracts of *Arabidopsis*, maize, pea, and tobacco. Methods Enzymol. 424, 459–482.

Hazle T, Bonen L (2007) Comparative analysis of sequences preceding protein-coding mitochondrial genes in flowering plants. Mol. Biol. Evol. 24, 1101–1112.

Heinhorst S, Cannon GC. (1993) DNA-replication in chloroplasts. J Cell Sci. 104, 1–9.

Herrin DL, Nickelsen J (2004) Chloroplast RNA processing and stability. Photosynth. Res. 82, 301–314.

Hoch B, Maier RM, Appel K, Igloi GL, Kössel H (1991) Editing of a chloroplast mRNA by creation of an initiation codon. Nature 353, 178–180.

Hochholdinger F, Guo L, Schnable PS (2004) Cytoplasmic regulation of the accumulation of nuclear-encoded proteins in the mitochondrial proteome of maize. Plant J 37, 199–208.

Holec S, Lange H, Kuhn K, Alioua M, Börner T, Gagliardi D (2006) Relaxed transcription in *Arabidopsis* mitochondria is counterbalanced by RNA stability control mediated by polyadenylation and polynucleotide phosphorylase. Mol. Cell Biol. 26, 2869–2876.

Hu J, Bogorad L (1990) Maize chloroplast RNA polymerase: the 180-, 120-, and 38-kilodalton polypeptides are encoded in chloroplast genes. Proc. Natl. Acad. Sci. USA 87, 1531–1535.

Hunt MD, Newton KJ (1991) The NCS3 mutation: genetic evidence for the expression of ribosomal protein genes in *Zea mays* mitochondria. EMBO J 10, 1045–1052.

Ishige T, Storey KK, Gengenbach BG (1985) Cytoplasmic fertile revertants possessing S1 and S2 DNAs in S male-sterile maize. Japan J Breed. 35, 285–291.

Jenkins BD, Kulhanek DJ, Barkan A (1997) Nuclear mutations that block group II RNA splicing in maize chloroplasts reveal several intron classes with distinct requirements for splicing factors. Plant Cell 9, 283–296.

Jenkins BD, Barkan A (2001) Recruitment of a peptidyl-tRNA hydrolase as a facilitator of group II intron splicing in chloroplasts. EMBO J 20, 872–879.

Karpova OV, Kuzmin EV, Elthon TE, Newton KJ (2002) Differential expression of alternative oxidase genes in maize mitochondrial mutants. Plant Cell 14, 3271–3284.

Kemble RJ, Bedbrook JR (1980) Low molecular weight circular and linear DNA in mitochondria from normal and male-sterile *Zea mays* cytoplasm. Nature 284, 565-566.

Kemble RJ, Thompson RD (1982) S1 and S2, the linear mitochondrial DNAs present in a male sterile line of maize, possess terminally attached proteins. Nucleic Acids Res. 10, 8181–8190.

Kemble RJ, Gunn RE, Flavell RB (1983) Mitochondrial DNA variation in races of maize indigenous to Mexico. Theor. Appl. Genet. 65, 129–144.

Kemble RJ, Mans RJ (1983) Examination of the mitochondrial genome of revertant progeny from S cms maize with cloned S-1 and S-2 hybridization probes. J Mol. Appl. Genet. 2, 161–171.

Kim BD, Mans RJ, Conde MF, Pring DR, Levings CS III (1982) Physical mapping of homologous segments of mitochondrial episomes from S male-sterile maize. Plasmid 7, 1–14.

Kubo T, Newton KJ (2007) Angiosperm mitochondrial genomes and mutations. Mitochondrion 8, 5–14.

Kuhn K, Bohne AV, Liere K, Weihe A, Börner T (2007) *Arabidopsis* phage-type RNA polymerases: accurate in vitro transcription of organellar genes. Plant Cell 19, 959–971.

Kuzmin EV, Levchenko IV (1987) S1 plasmid from cms-S-maize mitochondria encodes a viral type DNA-polymerase. Nucleic Acids Res. 15, 6758.

Kuzmin EV, Levchenko IV, Zaitseva GN (1988) S2 plasmid from cms-S-maize mitochondria potentially encodes a specific RNA polymerase. Nucleic Acids Res. 16, 4177.

Kuzmin EV, Karpova OV, Elthon TE, Newton KJ (2004) Mitochondrial respiratory deficiencies signal up-regulation of genes for heat shock proteins. J Biol. Chem. 279, 20672–20677.

Kuzmin EV, Duvick DN, Newton KJ (2005) A mitochondrial mutator system in maize. Plant Physiol. 137.

Lahiri SD, Yao J, McCumbers C, Allison LA (1999) Tissue-specific and light-dependent expression within a family of nuclear-encoded sigma-like factors from *Zea mays*. Mol. Cell. Biol. Res. Commun. 1, 14–20.

Lauer M, Knudsen C, Newton KJ, Gabay-Laughnan S, Laughnan JR (1990) A partially deleted mitochondrial cytochrome oxidase gene in the NCS6 abnormal growth mutant of maize. New Biol. 2, 179–186.

Laughnan JR, Gabay SJ (1973) Mutations leading to nuclear restoration of fertility in S male-sterile cytoplasm in maize. Theor. Appl. Genet. 43, 109–116.

Laughnan JR, Gabay SJ (1978) Nuclear and cytoplasmic nuclear gene *Rf3* affects the expression of the mitochondrial mutations to fertility in S male-sterile maize. In: D.B. Walden (ed) Maize Breeding and Genetics. John Wiley & Sons, New York, pp 427–446.

Laughnan JR, Gabay-Laughnan SJ, Carlson JE (1981) Characteristics of cms-S reversion to male fertility in maize. Stadler Symp. 13, 93–114.

Laughnan JR, Gabay-Laughnan S (1983) Cytoplasmic male sterility in maize. Ann. Rev. Genet. 17, 27–48.

Lee S-LJ, Gracen VE, Earle ED (1979) The cytology of pollen abortion in C-cytoplasmic male-sterile corn anthers. Amer. J. Bot. 66, 656–667.

Lee S-LJ, Earle ED, Gracen VE (1980) The cytology of pollen abortion in S cytoplasmic male-sterile corn anthers. Amer. J. Bot. 67, 237–245.

Leister D, Schneider A (2003) From genes to photosynthesis in *Arabidopsis thaliana*. Int. Rev. Cytol. 228, 31–83.

Lemke CA, Gracen VE, Everett HL (1985) A new source of cytoplasmic male sterility in maize induced by the nuclear gene, *iojap*. Theor. Appl. Genet. 71, 481–485.

Lemke CA, Gracen VE, Everett HL (1988) A second source of cytoplasmic male sterility in maize induced by the nuclear gene *Iojap*. J Heredity 79, 459–464.

Leon P, Walbot V, Bedinger P (1989) Molecular analysis of the linear 2.3 kb plasmid of maize mitochondria: apparent capture of tRNA genes. Nucleic Acids Res. 17, 4089–4099.

Levings CS III, Kim BG, Pring DR, Conde MF, Mans RJ, Laughnan JR, Gabay-Laughnan SJ (1980) Cytoplasmic reversion of *cms*-S in maize: association with a transpositional event. Science 209, 1021–1023.

Levings CS, Sederoff RR (1983) Nucleotide sequence of the S-2 mitochondrial DNA from the S cytoplasm of maize. Proc. Natl. Acad. Sci. USA 80, 4055–4059.

Liere K, Börner T (2007) Transcription and transcriptional regulation in plastids. In: R. Bock (ed) Topics in Current Genetics: Cell and Molecular Biology of Plastids. Springer, Berlin/ Heikelberg, pp 121–173.

Lonsdale DM (1987) Cytoplasmic male sterility: a molecular perspective. Plant Physiol. Biocehm. 25, 265–271.

Lupold DS, Caoile AG, Stern DB (1999) Polyadenylation occurs at multiple sites in maize mitochondrial *cox2* mRNA and is independent of editing status. Plant Cell 11, 1565–1578.

Maier R.M, Neckermann K, Igloi GL, Kössel H (1995) Complete sequence of the maize chloroplast genome: gene content, hotspots of divergence and fine tuning of genetic information by transcript editing. J Mol. Biol. 251, 614–628.

Maloney AP, Walbot V (1990) Structural analysis of mature and dicistronic transcripts from the 18 S and 5 S ribosomal RNA genes of maize mitochondria. J Mol. Biol. 213, 633–649.

Marienfeld JR, Newton KJ (1994) The maize NCS2 abnormal growth mutant has a chimeric *nad4-nad7* gene and is associated with reduced complex I function. Genetics 138, 855–863.

McCormac DJ, Barkan A (1999) A nuclear gene in maize required for the translation of the chloroplast *atpB/E* mRNA. Plant Cell 11, 1709–1716.

McNay JW, Pring DR, Lonsdale DM (1983) Polymorphism of mitochondrial DNA 'S' regions among normal cytoplasms of maize. Plant Mol. Biol. 2, 177–187.

Meinhardt F, Kempken F, Kamper J, Esser K (1990) Linear plasmids among eukaryotes: fundamentals and application. Curr. Genet. 17, 89–95.

Meinhardt F, Schaffrath R, Larsen M (1997) Microbial linear plasmids. Appl. Microbial. Biotechnol. 47, 329–336.

Meyer LJ (2004) ORF analysis and tissue-specific differential gene expression in maize mitochondria. M. S. Thesis, University of Missouri, Columbia, MO.

Momcilovic I, Ristic Z (2007) Expression of chloroplast protein synthesis elongation factor, EF-Tu, in two lines of maize with contrasting tolerance to heat stress during early stages of plant development. J Plant Physiol. 164, 90–99.

Nakajima Y, Mulligan RM (2001) Heat stress results in incomplete C-to-U editing of maize chloroplast mRNAs and correlates with changes in chloroplast transcription rate. Curr. Genet. 40, 209–213.

Newton KJ, Walbot V (1985) Molecular analysis of mitochondria from a fertility restorer line of maize. Plant Mol. Biol. 4, 247–252.

Newton KJ, Coe EH (1986) Mitochondrial DNA changes in abnormal growth (nonchromosomal stripe) mutants of maize. Proc. Natl. Acad. Sci. USA 83, 7363–7366.

Newton KJ, Coe EH, Gabay-Laughnan S, Laughnan JR (1989) Abnormal growth phenotypes and mitochondrial mutants in maize. Maydica 34, 291–296.

Newton KJ, Knudsen C, Gabay-Laughnan S, Laughnan JR (1990) An abnormal growth mutant in maize has a defective mitochondrial cytochrome oxidase gene. Plant Cell 2, 107–113.

Newton KJ (1994) Analysis of cytoplasmically inherited mutants. In: M. Freeling and V. Walbot (eds) The Maize Handbook. Springer-Verlag, New York, pp 413–417.

Newton KJ (1995) Aberrant growth phenotypes associated with mitochondrial genome rearrangements in higher plants. In: C.S. Levings, III and I.K. Vasil (eds) The Molecular Biology of Plant Mitochondria. Kluwer Academic, Dordrecht, pp 585–596.

Newton KJ, Mariano JM, Gibson CM, Kuzmin E, Gabay-Laughnan S (1996) Involvement of S2 episomal sequences in the generation of NCS4 deletion mutation in maize mitochondria. Dev. Genet. 19, 277–286.

Newton KJ, Gabay-Laughnan S, DePaepe R (2004) Mitochondrial mutations in plants. In: H.A. D.A. Day and J.W. Millar (eds) Plant Mitochondria: From Genome to Function. Kluwer Academic Publishers, Dordrecht, The Netherlands, pp 121–142.

Nott A, Jung HS, Koussevitzky S, Chory J (2006) Plastid-to-nucleus retrograde signaling. Annu. Rev. Plant Biol. 57, 739–759.

Oeser B (1988) S2 plasmid from *Zea mays* encodes a specific RNA polymerase: an alternative alignment. Nucleic Acids Res. 16, 8729.

Oldenburg DJ, Bendich AJ (2004) Most chloroplast DNA of maize seedlings in linear molecules with defined ends and branched forms. J Mol. Biol. 335, 953–970.

Oldenburg DJ, Rowan BA, Zhao L, Walcher CL, Schleh M, Bendich AJ (2006) Loss or retention of chloroplast DNA in maize seedlings is affected by both light and genotype. Planta 225, 41–55.

Ostheimer GJ, Williams-Carrier R, Belcher S, Osborne E, Gierke J, Barkan A (2003) Group II intron splicing factors derived by diversification of an ancient RNA-binding domain. EMBO J 22, 3919–3929.

Ostheimer GJ, Hadjivassiliou H, Kloer DP, Barkan A, Matthews BW (2005) Structural analysis of the group II intron splicing factor CRS2 yields insights into its protein and RNA interaction surfaces. J Mol. Biol. 345, 51–68.

Ostheimer GJ, Rojas M, Hadjivassiliou H, Barkan A (2006) Formation of the CRS2-CAF2 group II intron splicing complex is mediated by a 22-amino acid motif in the COOH-terminal region of CAF2. J Biol. Chem. 281, 4732–4738.

Paillard M, Sederoff RR, Levings CS (1985) Nucleotide sequence of the S-1 mitochondrial DNA from the S cytoplasm of maize. EMBO J 4, 1125–1128.

Peeters NM, Hanson MR (2002) Transcript abundance supercedes editing efficiency as a factor in developmental variation of chloroplast gene expression. RNA 8, 497–511.

Pfannschmidt T, Ogrzewalla K, Baginsky S, Sickmann A, Meyer HE, Link G (2000) The multisubunit chloroplast RNA polymerase A from mustard (*Sinapis alba L.*). Integration of a prokaryotic core into a larger complex with organelle-specific functions. Eur. J Biochem. 267, 253–261.

Pring DR, Levings CS, Hu WW, Timothy DH (1977) Unique DNA associated with mitochondria in the "S"-type cytoplasm of male-sterile maize. Proc. Natl. Acad. Sci. USA 74, 2904–2908.

Pring DR, Levings CS (1978) Heterogeneity of maize cytoplasmic genomes among male-sterile cytoplasms. Genetics 89, 121–136.

Pring DR, Smith AG (1985) Distribution of minilinear and minicircular mtDNA sequences within Zea. Maize Genet. Coop. Newsl. 59, 49–50 (cited with permission).

Reed ML, Hanson MR (1997) A heterologous maize *rpoB* editing site is recognized by transgenic tobacco chloroplasts. Mol. Cell. Biol. 17, 6948–6952.

Rhoades MM (1950) Gene induced mutation of a heritable cytoplasmic factor producing male sterility in maize. Proc. Natl. Acad. Sci. USA 36, 634–635.

Rhoads DM, Subbaiah CC (2007) Mitochondrial retrograde regulation in plants. Mitochondrion 7, 177–194.

Rodermel SR, Bogorad L (1985) Maize plastid photogenes: mapping and photoregulation of transcript levels during light-induced development. J Cell Biol. 100, 463–476.

Sask C, Lee SB, Fjellheim S, Guda C, Jansen RK, Luo H, Tomkins J, Rognli OA, Daniell H, Clarke JL (2007) Complete chloroplast genome sequences of *Hordeum vulgare, Sorghum bicolor* and *Agrostis stolonifera*, and comparative analyses with other grass genomes. Theor. Appl Genet. 115, 571–590.

Scanlon MJ (2003) The polar auxin transport inhibitor *N*-1-Naphthylphthalamic acid disrupts leaf initiation, KNOX protein regulation, and formation of leaf margins in maize. Plant Physiol. 133, 597–605.

Schardl CL, Lonsdale DM, Pring DR, Rose KR (1984) Linearization of maize mitochondrial chromosomes by recombination with linear episomes. Nature 310, 292–296.

Schardl CL, Pring DR, Lonsdale DM (1985) Mitochondrial DNA rearrangements associated with fertile revertants of S-type male-sterile maize. Cell 43, 361–368.

Schmitz-Linneweber C, Williams-Carrier R, Barkan A (2005) RNA immunoprecipitation and microarray analysis show a chloroplast pentatricopeptide repeat protein to be associated with the 5′ region of mRNAs whose translation it activates. Plant Cell 17, 2791–2804.

Schmitz-Linneweber C, Williams-Carrier RE, Williams-Voelker PM, Kroeger TS, Vichas A, Barkan A (2006) A pentatricopeptide repeat protein facilitates the trans-splicing of the maize chloroplast *rps12* pre-mRNA. Plant Cell 18, 2650–2663.

Schnable PS, Wise RP (1998) The molecular basis of cytoplasmic male sterility and fertility restoration. Trends Plant Sci. 3, 175–180.

Sekine K, Fujiwara M, Nakayama M, Takao T, Hase T, Sato N (2007) DNA binding and partial nucleoid localization of the chloroplast stromal enzyme ferredoxin:sulfite reductase. FEBS J 274, 2054–2069.

Shaver JM, Oldenburg DJ, Bendich AJ (2006) Changes in chloroplast DNA during development in tobacco, *Medicago truncatula*, pea, and maize. Planta 224, 72–82.

Shiina T, Tsunoyama Y, Nakahira Y, Khan MS (2005) Plastid RNA polymerases, promoters, and transcription regulators in higher plants. Int. Rev. Cytol. 244, 1–68.

Shikanai T (2006) RNA editing in plant organelles: machinery, physiological function and evolution. Cell. Mol. Life Sci. 63, 698–708.

Shumway LK, Bauman LF (1967) Nonchromosomal stripe of maize. Genetics 55, 33–38.

Silhavy D, Maliga P (1998) Mapping of promoters for the nucleus-encoded plastid RNA polymerase (NEP) in the *iojap* maize mutant. Curr. Genet. 33, 340–344.

Skibbe DS, Schnable PS (2005) Male sterility in maize. Maydica 50, 367–376.

Slomovic S, Portnoy V, Liveanu V, Schuster G (2006) RNA polyadenylation in prokaryotes and organelles; Different tails tell different tales. Crit. Rev. Plant Sci. 25, 65–77.

Small ID, Isaac PG, Leaver CJ (1987) DNA binding and partial nucleoid localization of the chloroplast stromal enzyme ferredoxin:sulfite reductase. EMBO J 6, 865–869.

Small ID, Peeters N (2000) The PPR motif - a TPR-related motif prevalent in plant organellar proteins. Trends Biochem. Sci. 25, 46–47.

Smith AG, Pring DR (1987) Nucleotide sequence and molecular characterization of a maize mitochondrial plasmid-like DNA. Curr. Genet. 12, 617–623.

Stern DB, Lonsdale DM (1982) Mitochondrial and chloroplast genomes of maize have a 12-kilobase DNA sequence in common. Nature 299, 698–702.

Stern DB, Hanson MR, Barkan A (2004) Genetics and genomics of chloroplast biogenesis: maize as a model system. Trends Plant Sci. 9, 293–301.

Suzuki JY, Ytterberg AJ, Beardslee TA, Allison LA, Wijk KJ, Maliga P (2004) Affinity purification of the tobacco plastid RNA polymerase and *in vitro* reconstitution of the holoenzyme. Plant J 40, 164–172.

Tadege M, Kuhlemeier C (1997) Aerobic fermentation during tobacco pollen development. Plant Mol. Biol. 35, 343–354.

Tadege M, Dupuis II, Kuhlemeier C (1999) Ethanolic fermentation: new functions for an old pathway. Trends Plant Sci. 4, 320–325.

Takenaka M, Verbitskiy D, van der Merwe JA, Zehrmann A, Brennicke A (2007) The process of RNA editing in plant mitochondria. Mitochondrion in press.

Tan S, Troxler RF (1999) Characterization of two chloroplast RNA polymerase sigma factors from *Zea mays*: photoregulation and differential expression. Proc. Natl. Acad. Sci. USA 96, 5316–5321.

Till B, Schmitz-Linneweber C, Williams-Carrier R, Barkan A (2001) CRS1 is a novel group II intron splicing factor that was derived from a domain of ancient origin. RNA 7, 1227–1238.

Tillich M, Schmitz-Linneweber C, Hermann R, Maier RM (2001) The plastid chromosome of maize (*Zea mays*): Update of the complete sequence and transcript editing sites. Maize Genet. Coop. Newsl. 75, 42–44.

Timothy DH, Levings CS III, Hu WW, Goodman MM (1983) Plasmid-like mitochondrial DNAs in diploperenial teosinte. Maydica 28, 139–149.

Walter M, Kilian J, Kudla J (2002) PNPase activity determines the efficiency of mRNA 3′-end processing, the degradation of tRNA and the extent of polyadenylation in chloroplasts. EMBO J 21, 6905–6914.

Ward CG (1995) The Texas male-sterile cytoplasm of maize. In: C.S. Levings, III and I.K. Vasil (eds) The Molecular Biology of Plant Mitochondria. Kluwer Academic, Dordrecht, pp 433–459.

Watkins KP, Kroeger TS, Cooke AM, Williams-Carrier RE, Friso G, Belcher SE, van Wijk KJ, Barkan A (2007) A ribonuclease III domain protein functions in group II intron splicing in maize chloroplasts. Plant Cell 19, 2606–2623.

Weissinger AK, Timothy DH, Levings CS, Hu WW, Goodman MM (1982) Unique plasmid-like mitochondrial DNAs from indigenous maize races of Latin America. Proc. Natl. Acad. Sci. USA 79, 1–5.

Weissinger AK, Timothy DH, Levings CS, Goodman MM (1983) Patterns of mitochondrial DNA variation in indigenous maize races of Latin America. Genetics 104, 365–379.

Wen L, Ruesch KL, Ortega VM, Kamps TL, Gabay-Laughnan S, Chase CD (2003) A nuclear *restorer-of-fertility* mutation disrupts accumulation of mitochondrial ATP synthase subunit α in developing pollen of S male-sterile maize. Genetics 165, 771–779.

Wise RP, Fliss AE, Pring DR, Gengenbach BG (1987a) *Urf*-13-T of T cytoplasm maize mitochondria encodes a 13 KD polypeptide. Plant Mol. Biol. 9, 121–126.

Wise RP, Pring DR, Gengenbach BG (1987b) Mutation to male fertility and toxin insensitivity in Texas (T)-cytoplasm maize is associated with a frameshift in a mitochondrial open reading frame. Proc. Natl. Acad. Sci. USA 84, 2858–2862.

Yao J, Roy-Chowdhury S, Allison LA (2003) AtSig5 is an essential nucleus-encoded *Arabidopsis* σ-like factor. Plant Physiol 132, 739–747.

Zabala G, Gabay-Laughnan S, Laughnan JR (1997) The nuclear gene *Rf3* affects the expression of the mitochondrial chimeric sequence R implicated in S-type male sterility in maize. Genetics 147, 847–860.

Zaitlin D, Hu J, Bogorad L (1989) Binding and transcription of relaxed DNA templates by fractions of maize chloroplast extracts. Proc. Natl. Acad. Sci. USA 86, 876–880.

Part IV
Maize Genetic and Genomic Technologies

Genetic Mapping and Maps

Karen C. Cone and Edward H. Coe

Abstract Early genetic analyses of maize were rooted in genetic mapping, and mapping continues to be an important tool for contemporary maize geneticists. Mapping is extraordinarily easy in maize; consequently many maps have been made. The first genetic map published for maize in 1935 contained 62 loci defined by morphological variants. Current genetic maps contain thousands of loci defined by morphological, biochemical, cytogenetic, and molecular polymorphic variants. These maps serve critically important functions in linking genes to traits, facilitating comparative evolutionary studies, enabling positional cloning, and anchoring the physical map for genome sequencing. Sequencing in turn now makes it possible to derive the map locations of sequenced genes by matching to genomic sequences that have been anchored to the physical map.

1 Definition

A genetic map, or linkage map, is a map of the frequencies of recombination that occur between markers on homologous chromosomes during meiosis. Recombination frequency between two markers is proportional to the distance separating the markers. The greater the frequency of recombination, the greater the distance between two genetic markers; conversely, the smaller the recombination frequency, the closer the markers are to one another. Thus a genetic map is a representation of recombination events and frequencies, rather than a physical map. The genetic and physical order is the same but distances are not. Although the *average* centimorgan

K.C. Cone
Division of Biological Sciences, University of Missouri, Columbia, MO 65211
conek@missouri.edu

E.H. Coe
Plant Genetics Research Unit, USDA-ARS, University of Missouri, Columbia, MO 65211
coee@missouri.edu
Division of Plant Sciences, University of Missouri, Columbia, MO 65211

J.L. Bennetzen and S. Hake (eds.), *Maize Handbook - Volume II: Genetics and Genomics*,   507
© Springer Science+Business Media LLC 2009

is about 180 kb, physical distance is not consistently proportional to recombination frequency for each interval and varies widely along a chromosome (Wei et al., 2007).

2   Utility of Genetic Maps

Genetic maps provide a way to link a genetic region to a trait of interest. Mapping provides a mechanism to track the co-segregation of genetic markers with traits in segregating populations. Such marker tracking can be used in selection (marker-assisted selection) of genes responsible for agronomically important traits and thus serve as an aid in crop improvement (Morgante and Salamini, 2003; Tuberosa and Salvi, 2006).

Genetic maps can be used in comparative studies to understand the processes that led to the evolution and diversification of a species. Between related taxa, comparative mapping can reveal regions of chromosomal synteny or conservation of gene order; and within a taxon, mapping can pinpoint regions of chromosomal duplication derived from ancient polyploidization events (for example, Helentjaris et al., 1988; Moore et al., 1995; Bennetzen and Freeling, 1997; Devos and Gale, 1997; Feuillet and Keller, 2002; Wei et al., 2007).

High-resolution genetic maps are essential tools for positional cloning. Recombinations between markers flanking a cloning target localize the target with increasing precision, as closer mapped markers are incorporated in the analysis. The most tightly linked markers co-segregate with the target. Positional cloning has been used to isolate a number of maize genes in the past couple of years and is likely to see more use in the future (Bortiri et al., 2006a; Bortiri et al., 2006b; Chuck et al., 2007; Salvi et al., 2007).

Genetic maps serve as a foundation for anchoring the physical map. Assemblies of genome fragments are formed into physical contigs (contiguous sequences). The placement of those contigs to chromosomes, and their orientation and order on the chromosomes, are achieved by correlating to genetic maps, which are chromosome-based (Coe et al., 2002; Cone et al., 2002; Wei et al., 2007). A genetic-map skeleton of markers that are matched to the physical map serves as an invaluable aid for genome sequencing and assembly (Messing and Dooner, 2006).

3   Making a Map

The first two basic requirements for genetic mapping are: (a) parents that are polymorphic for measurable traits and detectable markers, and (b) a population segregating for the traits of interest, made by crossing the polymorphic parents. Maize is ideal for genetic mapping, as the vast amount of diversity among maize lines provides a rich source of polymorphisms in traits and markers. The separation of male and female flowers on the plant makes it extremely easy to make controlled crosses, and the large number of progeny kernels from each cross can provide an ample segregating population from a single ear. Moreover, because maize plants can be both outcrossed and self-pollinated, making a mapping population segregating for the trait of interest is as simple as crossing two polymorphic parents and then self-pollinating the F1 to generate an F2, or crossing the F1 to one or both of the two parents to generate a backcross (BC) population.

3.1   Trait and Marker Polymorphisms

The first requirement for genetic mapping is to have parental lines with trait and marker polymorphisms. Maize has tremendous genetic diversity; surveys of diverse collections of maize lines have led to the estimate that the average maize gene contains about 200 nucleotide polymorphisms and 20-30 amino acid polymorphisms (Buckler et al., 2006). At least some of these molecular polymorphisms are likely to underlie diversity in function that is manifest as polymorphisms in trait expression.

Trait Polymorphisms

Hundreds of trait polymorphisms have been mapped in maize. These include: morphological traits with phenotypes explained by alternate alleles of a single gene, such as white/yellow endosperm, colored/colorless aleurone, dwarf/normal plant stature; and quantitative traits involving multiple genes controlling variation in agronomically important characteristics such as productivity; starch, oil, and protein composition of the kernel; and tolerance to biotic and abiotic stresses.

Marker Polymorphisms

Early maize mappers took advantage of isozyme variation and were able to map a large number of isozyme polymorphisms (for example, Wendel et al. [1988]). Nowadays, isozyme markers have been supplanted by DNA markers, which capitalize on the high level of variation in nucleotide sequence across maize lines. The five major types of molecular markers that have been used in maize mapping are restriction fragment length polymorphisms (RFLPs), amplified fragment length polymorphisms (AFLPs), simple sequence repeats (SSRs), insertion-deletion polymorphisms (IDPs), and single nucleotide polymorphisms (SNPs).

RFLP polymorphisms are detected by digesting genomic DNA with restriction enzymes and then detecting the restriction fragments by DNA gel blot hybridization with a radioactive probe made from genomic DNA or cDNA. The high degree of nucleotide sequence polymorphism in maize means that digestion with only four to six restriction enzymes is usually sufficient to detect polymorphism between any two maize lines for any probe tested. This led to widespread use of RFLPs as one of the first molecular marker types for maize (Evola et al., 1986; Helentjaris et al., 1986; Burr et al., 1988; Gardiner et al., 1993). Drawbacks to RFLPs as markers are the labor intensive process involved in preparing the DNA blots and the need to use radioactivity to detect hybridization.

Several types of polymerase chain reaction (PCR)-based markers have been used for mapping genes in maize. Among these are AFLPs and amplified polymorphisms associated with miniature inverted repeat transposable elements (MITEs). AFLPs are detected by digesting genomic DNA with restriction enzymes and then ligating adaptors to the ends of the fragments. Subsets of the restriction fragments can be amplified using primers complementary to the adaptor and the restriction site, and then the fragments are visualized on denaturing acrylamide gels (Vuylsteke et al., 1999). A similar technology was used to develop MITE-associated markers by including a primer complementary to the MITE inverted repeat in the PCR amplification reaction (Casa et al., 2000; Casa et al., 2004).

Probably the most widely used type of PCR-based marker is the SSR. SSRs are tandemly repeated mononucleotide, dinucleotide, trinucleotide or tetranucleotide sequences that are abundant and dispersed across the maize genome (Taramino and Tingey, 1996; Sharopova et al., 2002). SSR polymorphism arises from variation in the number of repeats at a given locus. This variation is detected by PCR using primers that flank the SSR and then fractionating the PCR products by gel electrophoresis to display length differences. The ease and relatively low cost of detection for SSRs makes them an attractive marker type for mapping.

IDPs result from insertions or deletions (InDels). Many IDPs for maize have been developed from InDels in introns or 3′ untranslated regions of transcribed genes (Bi et al., 2006; Fu et al., 2006). IDPs, like SSRs, are easily detected by PCR, using primers that flank the InDel, followed by gel electrophoresis to detect length differences.

SNPs are more abundant than the other types of polymorphisms; on average, between any randomly chosen pair of inbreds, there is one SNP in every 150 bp. Maize SNPs have been developed by comparative sequencing across 14 maize inbreds (Bi et al., 2006). For genotyping with SNPs, alleles can be discriminated by one of two basic approaches – PCR-based primer extension or differential hybridization – and allelic differences can be detected using mass spectrometry, fluorescence, or chemiluminescence methods (reviewed in Kim and Misra, 2007).

3.2 Mapping Populations

There are several types of mapping populations, each with its own advantages. Probably the most versatile population is an F2, as this kind of population can be produced promptly and is easy to analyze; individuals in the population will have one of three possible genotypes (two homozygous and the heterozygous genotypes). In backcross (BC) populations, there are only two possible genotypes (homozygous and heterozygous). Both F2 and BC populations lend themselves well to mapping of one or a few traits, especially if recessive individuals can be analyzed as a pool by bulked segregant analysis (Michelmore et al., 1991; Carson et al., 2004).

Two disadvantages of F2 or BC populations are that phenotypes of individuals in the population can only be scored in a single generation and seed for the population is limited. One way to overcome these difficulties is to self-pollinate the F2 plants to produce a population that can be analyzed as F3 families. Another method is to produce an immortalized F2 (IF2) by chain-pollinating (one male on one sib) and bulking seed within individual F3 families to "fix" the alleles of the F2 parent in the progeny (Gardiner et al., 1993). This produces a larger store of seed, but requires further maintenance of the immortalized population, in which the advanced progenies may be subject to changes by genetic drift.

Two types of populations – doubled haploids and recombinant inbred lines (RILs) – circumvent the problem of limited seed, as both constitute permanent populations; as such, they can be used for assessing phenotypic variation through repeated measures across time and environment (McMullen, 2003). Doubled haploid populations are generated directly from F1 plants. Because they are homozygous, they effectively fix the linkage groups present in the gametes of the F1 and have the same mapping resolution as BC progeny (Snape, 1988). Homozygosity of these populations means that they can be easily maintained by sib- or self- pollinations.

RILs are made by repeatedly self-pollinating single-seed descendants of individuals from an F2 population to produce virtually complete homozygosity for linkage groups originally present in the F2 (Burr et al., 1988; Burr and Burr, 1991). Once homozygosity is attained, RILs can be perpetuated by sib- or self-pollination. The homozygous nature of the lines allows polymorphisms for presence *vs.* absence of a marker to be scored unambiguously. A number of maize RIL populations are publicly available (Maize Genetics Cooperation Stock Center; http://maizecoop.cropsci.uiuc. edu/). This enables multiple researchers to use the RILs; as a result, the genetic information gathered from mapping in RIL populations is cumulative. For codominant markers, the resolution of RILs is essentially equivalent over short intervals to that of F2s or IF2s, because F2 plants contain products from two distinct meioses and RILs accumulate a comparable number of recombination events during their derivation.

The mapping resolution of RILs can be increased by random intermating for one or more generations before the selfing rounds are begun. This strategy was used to create an intermated RIL (IRIL) population derived from crossing B73 and Mo17, self-pollinating the F1 and then randomly mating progeny for four generations before selfing by single-seed descent (Lee et al., 2002). The resulting Intermated B73-Mo17 (IBM) population has a very high mapping resolution, ~ 0.4 centimorgans (1 centimorgan = 1 map unit = 1% recombination). The IBM population (~302 IRILs, conveniently scored as a subset of 286 lines plus the two parent inbreds in three 96-well plates) is publicly available and has been used extensively for genetic mapping by many research groups. The resolution of the full population is sufficient to place on average about one recombination breakpoint within the length of a typical bacterial artificial chromosome (BAC) clone (~140-160 kb). Subsets of 94 lines, equivalent to about 750 tested gametes, can be used for approximate mapping, with the Community IBM Mapping utility (http://www.maizemap.org/CIMDE/cIBMmap.htm).

3.3   Collecting and Analyzing Data to Construct a Map

Once a population segregating traits of interest is obtained, mapping the trait typically involves measuring the phenotype and determining the genotype of each member of the population. Genotyping with the molecular markers used in current mapping is a two-step process. First, DNA samples from the parents of the mapping population are screened for polymorphisms, using markers that span the chromosome(s) of interest. To scan the whole genome, polymorphic markers spaced approximately every 25-30 cM are needed. The second step is to use the polymorphic markers to determine the genotypes for each member in the population, or, in the case of bulked segregant analysis, for the pools of recessive and normal individuals. Cataloguing genotypes for large numbers of individuals and/or markers can be simplified using software specifically designed for collecting and managing mapping data (Sanchez-Villeda et al., 2003).

To construct the map, associations of genotype to phenotype must be derived. For bulked segregant mapping of a simple recessive trait, single-locus associations are made by comparing SSR or RFLP band intensity. Marker alleles linked *in cis* with the recessive trait allele will be overrepresented in the pool of homozygous recessive individuals and underrepresented in the pool of control individuals (Carson et al., 2004). Markers that show evidence of linkage can then be used to determine genotypes for individuals in the pools, and genetic distances can be estimated by calculating recombination frequencies.

For whole-genome mapping, genotype to phenotype correlations require more sophisticated computation. A number of mapping programs are available for mapping traits controlled by single genes, as well as quantitative traits. One of the first mapping packages, still in use, is MAPMAKER/EXP, which constructs genetic linkage maps using data generated from experimental populations (F2, BC and RIL [Lander et al., 1987; Lincoln et al., 1992]). (Note: For closely spaced markers in the IBM population, maps generated as RIL with MAPMAKER present distances approximately 4-fold greater than standard centimorgans [Winkler et al., 2003], an approximation that is useful for comparison with mapping data from other population types.) Other programs allow integration of data from different experiments (JoinMap; Stam, 1993) and adjustments in map distance due to the extra rounds of intermating in IRIL populations (IRILmap; Falque, 2005). Output from MAPMAKER/EXP can feed QTL mapping programs, such as MAPMAKER/QTL (Lincoln et al., 1992) and QTL Cartographer (Basten et al., 1997). Other QTL mapping programs bypass MAPMAKER and generate maps directly from input data (reviewed in Manly and Olson, 1999).

4 Maize Genetic Maps: Past and Present

Mapping in maize has a long history based on cumulative shared information that laid the foundation for modern molecular marker-based maps. The first full genetic maps were a part of the seminal monograph on maize genetics published by Emerson, Beadle, and Fraser in 1935. Data from individual gene-to-gene recombination experiments, made available by cooperating research scientists and collated by G. W. Beadle, were constructed into maps by M. M. Rhoades, which were published in the early issues of the Maize Genetics Cooperation Newsletter (MNL). These maps set the precedent for orientation with the cytologically short-arm end as the starting point, which depended on correlations of cytological data with the genetic data (only chromosome 3 was mis-oriented in the original maps and was corrected soon after). New phenotypically defined genes and new gene-to-gene data were accreted on the 1935 skeleton for the next 70 years. A key innovation in maize mapping came with the division of each chromosome into "bins", which were defined by a set of "core" markers dispersed at regular intervals (Gardiner et al., 1993). The near-immediate impact of defining bins by flanking core markers was that many groups adopted the core markers in their various mapping experiments. As discussed below, when maps contain common markers, making linkages across those maps is possible.

As mapping continued, it soon became clear that genetic maps constructed from different mapping populations could show differences in the order and/or distances between genetic markers (for example, see Sharopova et al., 2002). In addition, some genetic markers proved to be present in some maize lines, but absent from others (Gardiner et al., 1993; Davis et al., 1999). The recent discovery of the Helitron class of transposons has shed light on these anomalies. Helitrons can carry genes or gene segments and can mobilize these gene segments in the genome (Lai et al., 2005; Morgante et al., 2005). As a result, chromosomal segments can exhibit non-colinearity, differing in marker order, distance between markers, or both.

Currently, over 150 maps – many targeted at mapping QTL – derived by over 40 research groups from 130 different mapping populations, are documented in the Maize Genetics and Genomics Database (MaizeGDB). The most current genetic map, IBM2 2005 Neighbors, is one of several maps based on the IBM IRIL population (Coe and Schaeffer, 2005). (A reference guide to the IBM maps is available at MaizeGDB [http://maizegdb.org/neighbors.php].) Fig. 1 and Table 1 highlight the expansion in marker density between the 1935 map and modern maps. The 1935 map contained only a few loci, most defined by morphological traits. By contrast, the IBM maps contain thousands of loci, which include both named genetic loci and loci defined by molecular markers not yet linked to genes. Inspection of a representative genetic interval on chromosome 2 – which includes *lg1, gl2* and *b1* – reveals that, as mapping information accrued and the number of loci found to lie in the *lg1-gl2* and *gl2-b1* intervals increased, the map distances in these intervals did not change from those established in 1942. This observation underscores the incredible accuracy of early mapping efforts.

5   Linking Genetic Maps to Other Maps

5.1   *Genetic to Genetic*

The key to linking genetic maps to one another is the use of common markers for mapping in different populations. If a marker unique to one map is located between common markers on the two maps, its location can be determined by extrapolating from the normalized distance between the common markers. This strategy has been

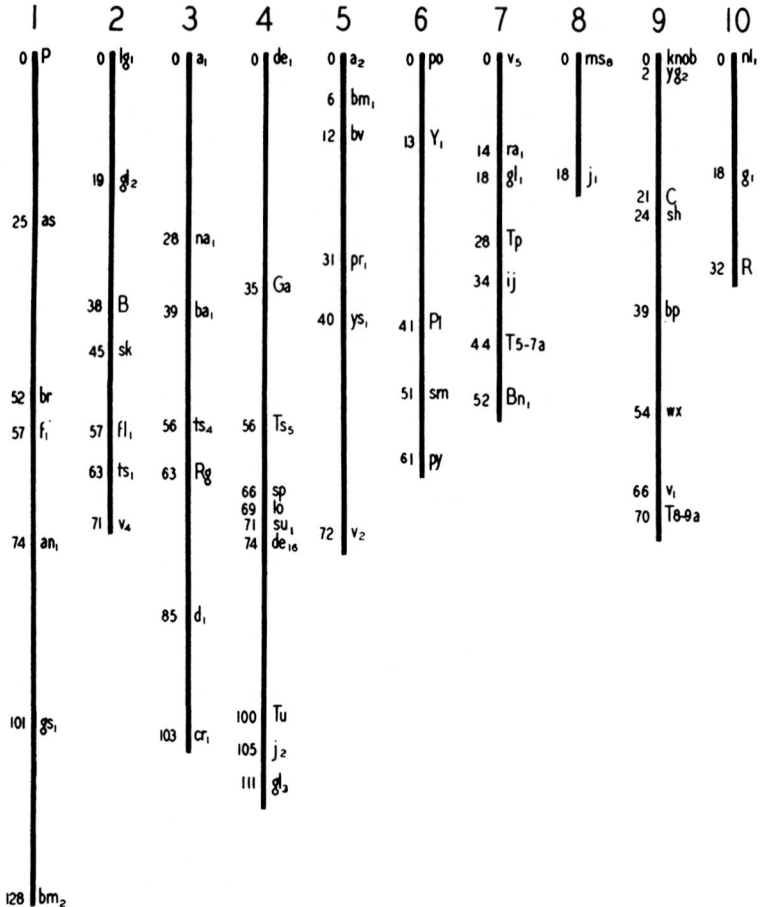

Fig. 1 Diagram of the first maize genetic map. Published in 1935 (Emerson et al., 1935), this map contained 62 loci defined by morphological variants, one cytological feature (the knob on the tip of chromosome 9) and two A-A translocation breakpoints (T5-7a and T8-9a)

applied over the years to generate a number of genetic-to-genetic map linkages, which have been published in various issues of the annual MNL. As marker numbers have increased, however, extrapolation through hand calculation has become extremely laborious. More recently, an algorithm was developed to link genetic maps and applied to great advantage to create the IBM2 Neighbors map series (Coe and Schaeffer, 2005). The most recent version, IBM2 2005 Neighbors, incorporates 14 genetic maps built on the frame of the IBM2. The nearly 35,000 loci represent the locations of markers mapped using either traditional genetic strategies or by placement to BAC clones (see below).

A strategy for linking QTL maps has been developed. This database-enabled approach employs standardized plant ontology terms to categorize phenotypes and

Table 1 Mapping history of a representative gene interval on chromosome 2: *liguleless1* (*lg1*) - *glossy2* (*gl2*) - *booster1* (*b1*)

Year	Map positions (cM) lg1	gl2	b1	Interval lg1-gl2 (cM)	Named loci	gl2-b1 (cM)	Named loci	Loci in map Named	Total	Prepared by (in)
1935	0	19	38	19	0	19	0	62	62	M. Rhoades (Emerson et al., 1935)
1942	11	30	49	19	0	19	0	68	68	M. Rhoades (Rhoades, 1942)
1942	11	30	49	19	0	19	1	86	86	C. Burnham (Hayes and Immer, 1942)
1950	11	30	49	19	0	19	0	89	89	M. Rhoades (Rhoades, 1950)
1955	11	30	49	19	0	19	0	94	94	M. Rhoades (Rhoades, 1955)
1960	11	30	49	19	1	19	1	155	155	M.G. Neuffer (Neuffer, 1960)
1968	11	30	49	19	1	19	1	166	166	M.G. Neuffer (Neuffer et al., 1968)
1975	11	30	49	19	1	19	2	187	187	M.G. Neuffer & E. Coe (Neuffer and Coe, 1975)
1978	11	30	49	19	0	19	1	196	196	E. Coe & M.G. Neuffer (Coe and Neuffer, 1977)
1988	11	30	49	19	0	19	1	393	393	E. Coe et al. (Coe et al., 1988)
1993	11	30	49	19	0	19	5	559	559	E. Coe (Coe, 1993)
1995	30	52	71	22	2	19	7	668	668	E. Coe (Coe et al., 1995)
1997	30	52	71	22	5	19	9	843	843	M.G. Neuffer et al. (Neuffer et al., 1997)
2005	13.7	38	58	24.3	3	20	8	744	2449	E. Coe & M. Schaeffer (Genetic 2005; Coe and Schaeffer, 2005)
2005	13.03	30.6	-	17.57	3	-	-	983	30,387	M. Schaeffer (IBM2 2005 Neighbors Frame; Schaeffer et al. 2006)
2005	13.03	30.6	-	17.57	3	-	-	1,076	33,958	M. Schaeffer (IBM2 2005 Neighbors; Schaeffer et al. 2006)
2007	11.75	30.52	49.26	18.77	6	18.74	4	720	27,628	M. Schaeffer (IBM2 FPC0507; Schaeffer et al., 2006)
2007	11.75	30.52	49.3	18.77	*	18.78	*	1,278	3,210	E. Coe (Genetic 2007, in preparation)

* not yet compiled.

takes advantage of MaizeGDB as the central repository of genotypic and phenotypic data (Schaeffer, 2006). Although QTL results from any given study pertain only to a specific mapping population, compilation of results from multiple studies allows better understanding of the inheritance of that trait.

5.2 Genetic to Cytological

A number of strategies have been developed for linking the genetic and cytological maps. One of the earliest was the use of reciprocal A-A translocations. Translocations were essential in developing the 1935 maps for (a) associating chromosomes to linkage groups, (b) orienting linkage groups, and (c) providing supplemental information on the order of genes (Emerson et al., 1935; Rhoades and McClintock, 1935). Although data from at least 14 translocations were applied, only two were shown on the map (Fig. 1), inasmuch as recombination percentages around translocation breakpoints are reduced and cannot be related reliably to gene-to-gene distances. These 14 translocations were among 89 characterized by Anderson (1935) and used for chromosome placement and mapping in subsequent years. That set has since expanded to 1100 (data from MaizeGDB), a resource that provides up to 2200 breakpoint markers whose cytological coordinates are defined and whose genetic and physical map positions can continue to be useful aids to research.

Two other types of translocations can be used to place mutations or traits to chromosome arm. A set of translocations marked with recessive *waxy1* on chromosome 9 or *sugary1* on chromosome 4 allows association of recessives or dominants to chromosome with as few as 9 or 10 simple F2 progenies (Anderson, 1945). B-A translocations allow placement of recessive traits to narrower chromosomal regions (Roman, 1947; Beckett, 1991). The advantage of this method is that F1 progeny reveal the trait location due to the deficiencies for chromosome arm segments that are produced by these translocations. For both of these translocation methods, additional subsequent mapping with other markers is needed to refine map location.

Oat-maize addition lines offer another way to localize a trait or molecular polymorphism to chromosome arm or segment. Oat-maize addition lines have been made by crossing oat and maize and recovering oat lines that retain one maize chromosome. Using a PCR-based assay, any maize sequence can be localized to chromosome arm by screening for presence or absence of an amplified product in each of the oat-maize lines (Okagaki et al., 2001). Radiation hybrid derivatives of these lines enable more precise localization to specific chromosomal segment (Kynast et al., 2004).

Two other methods have been useful. Fluorescence in situ hybridization (FISH) has been used extensively to localize genes to chromosomes (Koumbaris and Bass, 2003; Kato et al., 2005; Lamb et al., 2007). Recombination nodule maps make it possible to predict the physical positions of genetic markers and to examine the distribution of markers across the maize chromosomes (Anderson et al., 2004; Anderson et al., 2006). A new tool, Morgan2McClintock, integrates recombination nodule and genetic maps to predict the chromosomal distance between genetically mapped markers (Lawrence et al., 2006).

5.3   Genetic to Physical to Genome Sequence

Anchoring of genetic and physical maps with common markers provides (a) association of physical map elements (BACs) with genetic points on the chromosomes, (b) orientation and ordering of physical-element assemblies (contigs), and (c) a framework skeleton for defining a minimum tiling path for sequencing.

Applying genetic map information to aid genome sequencing requires integrating genetic and physical maps. Moreover, the genetic map must be of high enough resolution such that recombination distances separating loci are on the scale of a few BAC lengths. The IBM population was created to meet the need for high resolution (Lee et al., 2002) and served as the foundation for the IBM map constructed by the Maize Mapping Project using data from public and private-sector collaborators.

Concomitant with development of the IBM genetic map, a physical map was constructed using fingerprint contig assembly of BAC clones from three libraries made from the B73 inbred line (Tomkins et al., 2002). Two methods were used to fingerprint the BAC clones. An agarose fingerprinting method resulted in 292,201 fingerprints that were automatically assembled into 4,518 contigs using FPC (Soderlund et al., 1997). High information content fingerprinting generated 350,253 fingerprints that were automatically assembled into 1,500 FPC contigs (Nelson et al., 2005).

A total of 25,908 markers were integrated into the FPC map (Wei et al., 2007). This included 1,902 genetically mapped markers (SSRs, RFLPs, SNPs and InDels) and 24,006 sequence-based markers (ESTs, BAC ends, and 40-bp overlapping oligonucleotide overgo probes). Associating markers to BACs involved three basic strategies. The first was hybridization of BAC libraries arrayed on filters with a suite of probes, including genetically mapped RFLPs (Yim et al., 2002), overgo probes derived from a maize EST unigene set (Gardiner et al., 2004), and overgo probes derived from sequences that had been genetically mapped in maize, sorghum and other grasses (e.g., Draye et al., 2001). The second strategy involved generation of BAC pools by six-dimensional pooling of a portion of one of the BAC libraries–representing six genome equivalents–and screening by PCR with primer pairs derived from single-copy genetically mapped sequences (Yim et al., 2007). The third strategy was sequencing of BAC ends (Messing et al., 2004). After manual editing, the final FPC map contained 721 contigs covering 2,150 Mb (93.5% of the total genome); 421 of the contigs (86.1% of the total genome) are anchored to the genetic map. The integrated map can be accessed at http://www.genome.arizona.edu/fpc/maize.

The FPC map provided the foundation for selecting approximately 19,000 BACs to make up a minimal tiling path for DNA sequencing. Details about the Maize Genome Sequencing Project can be viewed at http://maizesequence.org/overview.html.

6   The Future of Genetic Mapping

Emerging genomic sequence information is paving the way to an improved genetic map. The extraordinary potential of having a sequenced genome will only be realized when targeted traits defined by observation, measurement, or response can be associated with the sequence. Accordingly, advancement of trait analysis requires that markers, annotated functional genes, and the sequence of the genome become tied to trait polymorphisms so that their genetic bases can be determined. These facts call for refined genetic maps, densely populated with markers that are usable in trait-mapping experiments and also placed physically on the genome sequence. Sequencing of the genome has reached the point that over 95% of genes for which there is a sequence can be placed on the genetic-map framework (Coe, personal observation). Applying such mapping *in silico* (sequenced gene to physical map to genetic map), a genetic map is in preparation and will be presented in MaizeGDB. This map uses IBM2 as the framework for genetic positions, has distances adjusted to standard centimorgans, applies the physical map for accretion of further genes, and places other genes on the basis of retrospective data from previous maps.

Finally, it should be noted that strong interest in trait mapping continues, with increasingly diverse materials, and can be expected to produce more mapping populations, genetic placement of a greater and greater range of traits, and higher resolution of trait variations. A deepening resource of information about maize as a research model and as a malleable crop plant will emerge from map-based analyses.

References

Anderson, E. (1935) Chromosomal interchanges in maize. *Genetics* **200**: 70–83.

Anderson, E. (1945) The following tables are compiled for the benefit of those using or wanting to use the sugary and waxy series of translocations for the study of economic or other characters in maize. *MNL* **19**: 5–8.

Anderson, L. K., A. Lai, S. M. Stack, C. Rizzon and B. S. Gaut (2006) Uneven distribution of expressed sequence tag loci on maize pachytene chromosomes. *Genome Res.* **16**: 115–122.

Anderson, L. K., N. Salameh, H. W. Bass, L. C. Harper, W. Z. Cande, G. Weber and S. M. Stack (2004) Integrating genetic linkage maps with pachytene chromosome structure in maize. *Genetics* **166**: 1923–1933.

Basten, C., B. S. Weir and Z.-B. Zeng (1997) QTL Cartographer: A reference manual and tutorial for QTL mapping. http://statgen.ncsu.edu/qtlcart/. Raleigh, N.C., Department of Statistics, North Carolina State University.

Beckett, J. B. (1991) Cytogenetic, genetic and plant breeding applications of B-A translocations in maize. In *Chromosome Engineering in Plants: Genetics Breeding, Evolution.* (P. Gupta and T. Tsuchiya, ed.) Elsevier Science Publishers, New York, pp. 493–529.

Bennetzen, J. L. and M. Freeling (1997) The unified grass genome: Synergy in synteny. *Genome Res.* **7**: 301–306.

Bi, I. V., M. D. McMullen, H. Sanchez-Villeda, S. Schroeder, J. Gardiner, M. Polacco, C. Soderlund, R. Wing, Z. Fang and E. H. Coe (2006) Single nucleotide polymorphisms and insertion-deletions for genetic markers and anchoring the maize fingerprint contig physical map. *Crop Sci.* **46**: 12–21.

Bortiri, E., G. Chuck, E. Vollbrecht, T. Rocheford, R. Martienssen and S. Hake (2006a) *ramosa2* encodes a LATERAL ORGAN BOUNDARY domain protein that determines the fate of stem cells in branch meristems of maize. *Plant Cell* **18**: 574–585.

Bortiri, E., D. Jackson and S. Hake (2006b) Advances in maize genomics: the emergence of positional cloning. *Curr. Op. Plant Biology* **9**: 164–171.

Buckler, E. S., B. S. Gaut and M. D. McMullen (2006) Molecular and functional diversity of maize. *Curr. Op. Plant Biology* **9**: 172–176.

Burr, B. and F. A. Burr (1991) Recombinant inbreds for molecular mapping in maize: theoretical and practical considerations. *Trends Genet.* **7**: 55–60.

Burr, B., F. A. Burr, K. H. Thompson, M. C. Albertson and C. W. Stuber (1988) Gene mapping with recombinant inbreds in maize. *Genetics* **118**: 519–526.

Carson, C., J. Robertson and E. Coe (2004) High-volume mapping of maize mutants with simple sequence repeat markers. *Plant Mol. Biol. Rep.* **22**: 131–143.

Casa, A., C. Brouwer, A. Nagel, L. Wang, Q. Zhang, S. Kresovich and S. Wessler (2000) The MITE family heartbreaker (Hbr): molecular markers in maize. *Proc. Natl. Acad. Sci. USA* **97**: 10083–10089.

Casa, A., A. Nagel and S. Wessler (2004) MITE display. *Methods Mol. Biol.* **260**: 175–188.

Chuck, G., A. Cigan, K. Saeteurn and S. Hake (2007) The heterochronic maize mutant *Corngrass1* results from overexpression of a tandem microRNA. *Nat. Genet.* **39**: 544–549.

Coe, E. (1993) Gene list and working maps. *MNL* **67**.

Coe, E., K. Cone, M. McMullen, S. S. Chen, G. Davis, J. Gardiner, E. Liscum, M. Polacco, A. Paterson, H. Sanchez–Villeda, C. Soderlund and R. Wing (2002) Access to the maize genome: An integrated physical and genetic map. *Plant Physiol.* **128**: 9–12.

Coe, E., D. J. Hancock, S. Kowalewski and M. Schaeffer (1995) Gene list and working maps. *MNL* **69**: 191–256.

Coe, E. and M. Neuffer (1977) The genetics of corn. In: *Corn and Corn Improvement.* (G. Sprague, ed.) American Society of Agronomy, Madison, WI, pp. 111–223.

Coe, E., M. Neuffer and D. A. Hoisington (1988) The genetics of corn. In: *Corn and Corn Improvement.* (G. Sprague and J. Dudley, ed.) American Society of Agronomy, Madison, WI, pp. 81–258.

Coe, E. and M. Schaeffer (2005) Genetic, physical, maps, and database resources for maize. *Maydica* **50**: 285–303.

Cone, K. C., M. D. McMullen, I. V. Bi, G. L. Davis, Y. S. Yim, J. M. Gardiner, M. L. Polacco, H. Sanchez-Villeda, Z. W. Fang, S. G. Schroeder, S. A. Havermann, J. E. Bowers, A. H. Paterson, C. A. Soderlund, F. W. Engler, R. A. Wing and E. H. Coe (2002) Genetic, physical, and informatics resources for maize: on the road to an integrated map. *Plant Physiol.* **130**: 1598–1605.

Davis, G. L., M. D. McMullen, C. Baysdorfer, T. Musket, D. Grant, M. Staebell, G. Xu, M. Polacco, L. Koster, S. Melia-Hancock, K. Houchins, S. Chao and E. H. Coe (1999) A maize map standard with sequenced core markers, grass genome reference points and 932 expressed sequence tagged sites (ESTs) in a 1736-locus map. *Genetics* **152**: 1137–1172.

Devos, K. M. and M. D. Gale (1997) Comparative genetics in the grasses. *Plant Mol. Biol.* **35**: 3–15.

Draye, X., Y. Lin, X. Qian, J. E. Bowers, G. Burow, P. Morell, D. Peterson, G. Presting, S. Ren, R. Wing and A. Paterson (2001) Toward integration of comparative genetic, physical, diversity, and cytomolecular maps for grasses and grains, using the sorghum genome as a foundation. *Plant Physiol.* **125**: 1325–1341.

Emerson, R., G. Beadle and A. Fraser (1935) A summary of linkage studies in maize. *Cornell Univ. Agric. Exp. Stn. Memoir* **180**: 1–83.

Evola, S., F. A. Burr and B. Burr (1986) The suitability of restriction fragment length polymorphisms as genetic markers in maize. *Theor. Appl. Genet.* **71**: 765–771.

Feuillet, C. and B. Keller (2002) Comparative genomics in the grass family: molecular characterization of grass genome structure and evolution. *Ann. Bot.* **89**: 3–10.

Fu, Y., T. J. Wen, Y. I. Ronin, H. D. Chen, L. Guo, D. I. Mester, Y. J. Yang, M. Lee, A. B. Korol, D. A. Ashlock and P. S. Schnable (2006) Genetic dissection of intermated recombinant inbred lines using a new genetic map of maize. *Genetics* **174**: 1671–1683.

Gardiner, J., S. Schroeder, M. L. Polacco, H. Sanchez-Villeda, Z. W. Fang, M. Morgante, T. Landewe, K. Fengler, F. Useche, M. Hanafey, S. Tingey, H. Chou, R. Wing, C. Soderlund and E. H. Coe (2004) Anchoring 9,371 maize expressed sequence tagged unigenes to the bacterial artificial chromosome contig map by two-dimensional overgo hybridization. *Plant Physiol.* **134**: 1317–1326.

Gardiner, J. M., E. H. Coe, S. Melia-Hancock, D. A. Hoisington and S. Chao (1993) Development of a core RFLP map in maize using an immortalized-F2 population. *Genetics* **134**: 917–930.

Hayes, H. and F. Immer (1942). *Methods of Plant Breeding*, McGraw-Hill, New York.

Helentjaris, T., M. Slocum, S. Wright, A. Schaefer and J. Nienhuis (1986) Construction of genetic linkage maps in maize and tomato using restriction fragment length polymorphisms. *Theor. Appl. Genet.* **72**: 761–769.

Helentjaris, T., D. Weber and S. Wright (1988) Identification of the genomic locations of duplicate nucleotide sequences in maize by analysis of restriction fragment length polymorphisms. *Genetics* **118**: 353–363.

Kato, A., J. Vega, F. Han, J. Lamb and J. Birchler (2005) Advances in plant chromosome identification and cytogenetic techniques. *Curr. Op. Plant Biology* **8**: 148–154.

Kim, S. and A. Misra (2007) SNP genotyping: technologies and biomedical applications. *Annu. Rev. Biomed. Eng.* **9**: 289–320.

Koumbaris, G. and H. W. Bass (2003) A new single-locus cytogenetic mapping system for maize (*Zea mays* L.): overcoming FISH detection limits with marker-selected sorghum (*S. propinquum* L.) BAC clones. *Plant J.* **35**: 647–659.

Kynast, R. G., R. J. Okagaki, M. W. Galatowitsch, S. R. Granath, M. S. Jacobs, A. O. Stec, H. W. Rines and R. L. Phillips (2004) Dissecting the maize genome by using chromosome addition and radiation hybrid lines. *Proc. Natl. Acad. Sci. USA* **101**: 9921–9926.

Lai, J. S., Y. B. Li, J. Messing and H. K. Dooner (2005) Gene movement by Helitron transposons contributes to the haplotype variability of maize. *Proc. Natl. Acad. Sci. USA* **102**: 9068–9073.

Lamb, J., T. Danilova, M. Bauer, J. Meyer, J. Holland, M. Jensen and J. Birchler (2007) Single-gene detection and karyotyping using small-target fluorescence *in situ* hybridization on maize somatic chromosomes. *Genetics* **175**: 1047–1058.

Lander, E. S., P. Green, J. Abrahamson, A. Barlow, M. Daley, S. Lincoln and L. Newburg (1987) MAPMAKER: an interactive computer package for constructing primary genetic linkage maps of experimental and natural populations. *Genomics* **1**: 174–181.

Lawrence, C., T. Seigfried, H. Bass and L. K. Anderson (2006) Predicting chromosomal locations of genetically mapped loci in maize using the Morgan2McClintock translator. *Genetics* **172**: 2007–2009.

Lee, M., N. Sharopova, W. D. Beavis, D. Grant, M. Katt, D. Blair and A. Hallauer (2002) Expanding the genetic map of maize with the intermated B73 x Mo17 (IBM) population. *Plant Mol. Biol.* **48**(5): 453–461.

Lincoln, S., M. Daley and E. S. Lander (1992) Mapping genes controlling quantitative traits. http://www.broad.mit.edu/genome_software/other/qtl.html, Whitehead Institute Technical Report. **2007**.

Manly, K. F. and J. M. Olson (1999) Overview of QTL mapping software and introduction to map manager QT. *Mammal. Genome* **10**: 327–334.

McMullen, M. (2003) Quantitative trait locus analysis as a gene discovery tool. In: *Methods in Molecular Biology: Plant Functional Genomics Methods and Protocols*. (E. Grotewold, ed.) Humana Press, Inc., Totowa, NJ, **236**: pp. 141–154.

Messing, J., A. K. Bharti, W. M. Karlowski, H. Gundlach, H. R. Kim, Y. Yu, F. S. Wei, G. Fuks, C. A. Soderlund, K. F. X. Mayer and R. A. Wing (2004) Sequence composition and genome organization of maize. *Proc. Natl. Acad. Sci. USA* **101**: 14349–14354.

Messing, J. and H. K. Dooner (2006) Organization and variability of the maize genome. *Curr. Op. Plant Biology* **9**: 157–163.

Michelmore, R. W., I. Paran and R. V. Kessell (1991) Identification of markers linked to disease-resistance genes by bulked segregant analysis: a rapid method to detect markers in specific genomic regions by using segegrating populations. *Proc. Natl. Acad. Sci. USA* **88**: 9828–9832.

Moore, G., K. M. Devos, Z. Wang and M. D. Gale (1995) Cereal genome evolution - grasses, line up and form a circle. *Curr. Biol.* **5**: 737–739.

Morgante, M., S. Brunner, G. Pea, K. Fengler, A. Zuccolo and A. Rafalski (2005) Gene duplication and exon shuffling by helitron-like transposons generate intraspecies diversity in maize. *Nat. Genet.* **37**: 997–1002.

Morgante, M. and F. Salamini (2003) From plant genomics to breeding practice. *Curr. Op. Biotech.* **14**: 214–219.

Nelson, W. M., A. K. Bharti, E. Butler, F. S. Wei, G. Fuks, H. Kim, R. A. Wing, J. Messing and C. Soderlund (2005) Whole-genome validation of high-information-content fingerprinting. *Plant Physiol.* **139**: 27–38.

Neuffer, M. (1960) Linkage maps of maize chromosomes. *MNL* **40**: 167–172.

Neuffer, M. and E. Coe, Jr (1975) Corn (Maize). In: *Handbook of Genetics.* (R. King, ed.) Plenum Press, New York, **2**: pp. 3–30.

Neuffer, M., L. Jones and M. Zuber (1968). *The Mutants of Maize.* Crop Science Society of America, Madison, WI.

Neuffer, M., E. Coe and S. Wessler (1997). *Mutants of Maize.* Cold Spring Harbor Laboratory, Cold Spring Harbor, NY.

Okagaki, R. J., R. G. Kynast, S. M. Livingston, C. D. Russell, H. W. Rines and R. L. Phillips (2001) Mapping maize sequences to chromosomes using oat-maize chromosome addition materials. *Plant Physiol.* **125**: 1228–1235.

Rhoades, M. (1942) Inasmuch as the writer was assigned chromosome 2 he has from time to time collected additional data on the location of certain genes placed in the map by two-point tests. *MNL* **16**: 4.

Rhoades, M. (1950) Meiosis in maize. *J. Heredity* **41**: 58–67.

Rhoades, M. (1955) The cytogenetics of maize. In: *Corn and Corn Improvement.* (G. Sprague, ed.) Academic Press, New York: pp. 123–220.

Rhoades, M. and B. McClintock (1935) The cytogenetics of maize. *Bot. Rev.* **10**: 292–325.

Roman, H. (1947) Mitotic nondisjunction in the case of interchanges involving the B-type chromosome in maize. *Genetics* **320**: 391–409.

Salvi, S., G. Sponza, M. Morgante, D. Tomes, X. Niu, K. A. Fengler, R. Meeley, E. V. Ananiev, S. Svitashev, E. Bruggemann, B. Li, C. F. Hainey, S. Radovic, G. Zaina, J. A. Rafalski, S. V. Tingey, G. H. Miao, R. L. Phillips and R. Tuberosa (2007) Conserved noncoding genomic sequences associated with a flowering-time quantitative trait locus m maize. *Proc. Natl. Acad. Sci. USA* **104**: 11376–11381.

Sanchez-Villeda, H., S. Schroeder, M. Polacco, M. McMullen, S. Havermann, G. Davis, I. Vroh-Bi, K. Cone, N. Sharopova, Y. Yim, L. Schultz, N. Duru, T. Musket, K. Houchins, Z. Fang, J. Gardiner and E. Coe (2003) Development of an integrated laboratory information management system for the maize mapping project. *Bioinformatics* **19**: 2022–2030.

Schaeffer, M. (2006) Consensus quantitative trait maps in maize: a database strategy. *Maydica* **51**: 357–367.

Schaeffer, M., H. Sanchez-Villeda, M. McMullen and E. Coe (2006) IBM2 2005 Neighbors – 45,000 locus resource for maize. *Plant and Animal Genome Conference Abstracts* **XIV**: 200.

Sharopova, N., M. D. McMullen, L. Schultz, S. Schroeder, H. Sanchez-Villeda, J. Gardiner, D. Bergstrom, K. Houchins, S. Melia-Hancock, T. Musket, N. Duru, M. Polacco, K. Edwards, T. Ruff, J. C. Register, C. Brouwer, R. Thompson, R. Velasco, E. Chin, M. Lee, W. Woodman-Clikeman, M. J. Long, E. Liscum, K. Cone, G. Davis and E. H. Coe (2002) Development and mapping of SSR markers for maize. *Plant Mol. Biol.* **48**: 463–481.

Snape, J. (1988) The detection and estimation of linkage using doubled haploid or single seed descent populations. *Theor. Appl. Genet.* **76**: 125–128.

Soderlund, C., I. Longden and R. Mott (1997) FPC: a system for building contigs from restriction fingerprinted clones. *Comput. Appl. Biosci.* **13**: 523–535.

Stam, P. (1993) Construction of integrated genetic linkage maps by means of a new computer package: JoinMap. *Plant J.* **3**: 739–744.

Taramino, G. and S. Tingey (1996) Simple sequence repeats for germplasm analysis and mapping in maize. *Genome* **39**: 277–287.

Tomkins, J. P., G. Davis, D. Main, Y. Yim, N. Duru, T. Musket, J. L. Goicoechea, D. A. Frisch, E. H. Coe and R. A. Wing (2002) Construction and characterization of a deep-coverage bacterial artificial chromosome library for maize. *Crop Sci.* **42**: 928–933.

Tuberosa, R. and S. Salvi (2006) Genomics-based approaches to improve drought tolerance of crops. *Trends Plant Sci.* **11**: 405–412.

Vuylsteke, M., M. R, R. Antonise, E. Bastiaans, L. Senior, C. Stuber, A. Melchinger, T. Luebberstedt, X. Xia, P. Stam, M. Zabeau and M. Kuiper (1999) Two high-density AFLP linkage maps of *Zea mays* L. : analysis of distribution of AFLP markers. *Theor. Appl. Genet.* **99**: 921–935.

Wei, F., E. Coe, W. Nelson, A. K. Bharti, F. Engler, E. Butler, H. Kim, J. L. Goicoechea, M. Chen, S. Lee, G. Fuks, H. Sanchez-Villeda, S. Schroeder, Z. Fang, M. McMullen, G. Davis, J. E. Bowers, A. H. Paterson, M. Schaeffer, J. Gardiner, K. Cone, J. Messing, C. Soderlund and R. A. Wing (2007) Physical and genetic structure of the maize genome reflects its complex evolutionary history. *PLOS Genetics* **3**: 1254–1263.

Wendel, J. F., M. M. Goodman, C. W. Stuber and J. B. Beckett (1988) New isozyme systems for maize (*Zea mays* L.): aconitate dehydratase, adenylate kinase, NADH dehydrogenase, and shikimate dehydrogenase. *Biochem Genet* **26**: 421–445.

Winkler, C., N. Jensen, M. Cooper, D. Podlich and O. Smith (2003) On the determination of recombination rates in intermated recombinant inbred populations. *Genetics* **164**: 741–745.

Yim, Y., G. Davis, N. Duru, T. Musket, E. Linton, J. Messing, J. McMullen, C. Soderlund, M. Polacco, J. Gardiner and E. Coe (2002) Characterization of three maize bacterial artificial chromosome libraries toward anchoring of the physical map to the genetic map using high-density bacterial artificial chromosome filter hybridization. *Plant Physiol.* **130**: 1686–1696.

Yim, Y. S., P. Moak, H. Sanchez-Villeda, T. A. Musket, P. Close, P. E. Klein, J. E. Mullet, M. D. McMullen, Z. Fang, M. L. Schaeffer, J. M. Gardiner, E. H. Coe and G. L. Davis (2007) A BAC pooling strategy combined with PCR-based screenings in a large, highly repetitive genome enables integration of the maize genetic and physical maps. *BMC Genomics* **8**:47.

Genetic Analyses with Oat-Maize Addition and Radiation Hybrid Lines

Ronald L. Phillips and Howard W. Rines

Abstract Oat-maize addition lines, with individual maize (*Zea mays* L.) chromosomes added to the oat (*Avena sativa* L.) genome via wide hybridization and embryo rescue, simplify the maize genome by 10-fold. Radiation hybrids, derived through gamma irradiation of monosomic addition lines, have less than a complete maize chromosome in an oat genomic background. Maize genes and gene families can be readily located to their respective physical chromosome or chromosome segment by rapid, high-throughput, and straight-forward PCR experiments. Polymorphisms are not required to map a maize DNA fragment with this system. Numerous non-mapping uses of these lines currently underway are described. Seed and/or DNA of the addition lines in various oat and maize backgrounds are available as well as approximately 650 radiation hybrid lines.

1 Uses of Oat-Maize Addition Lines and Radiation Hybrids

An astonishing number of genetic applications are enhanced through the use of oat-maize addition (OMA) and radiation hybrid (RH) lines. Oat-maize addition lines are derived from crosses between oat as the female parent and maize as the male parent (Kynast et al. 2001; Okagaki et al. 2001; Phillips et al. 2003; Kynast et al. 2004; Rines et al. 2005). The progeny with a complete haploid set of oat chromosomes and one or more maize chromosomes are termed oat-maize addition lines (oat-maize

R.L. Phillips
University of Minnesota, Department of Agronomy and Plant Genetics and Microbial and Plant Genomics Institute
phill005@umn.edu

H.W. Rines
USDA-ARS and University of Minnesota, Department of Agronomy and Plant Genetics
rines001@umn.edu

J.L. Bennetzen and S. Hake (eds.), *Maize Handbook - Volume II: Genetics and Genomics*,   523
© Springer Science+Business Media LLC 2009

monosomic addition lines, to be precise). Plants with double the chromosome number, which occurs upon self-pollination, are also referred to as oat-maize addition (OMA) lines, but should be called oat-maize disomic addition lines (Rines, Riera-Lizarazu, and Phillips 1995).

Advantages of using these materials include high resolution mapping (any maize gene can be mapped without the availability of polymorphic forms), the gene's physical chromosome location becomes known, the cloning of small segments of maize chromosomes separated from the remaining portions of the genome is possible, the isolation of additional markers in a small genomic region assists in finding tags for important traits, and oat lines can be screened for useful traits such as disease resistance incorporated into oat from maize.

The principal use today for OMAs and RHs is mapping (Ananiev et al. 1997; Riera-Lizarazu et al. 2000; Kynast et al. 2000; Okagaki et al. 2002; Okagaki et al. 2004). Using OMAs, DNA sequences can be mapped to a specific chromosome in a single experiment involving PCR amplification. No polymorphisms are required since it is a +/- test for each chromosome. The +/- result can be scored across OMAs specific for each of the maize chromosomes in a short period of time. Members of gene families (Zhang et al. 2006; Mica et al. 2006) are easily mapped to their respective chromosomes, again in a single experiment. If the sequence is also amplified in the oat genome, then a polymorphism between oat and maize is required in order to map the maize sequence to chromosome.

An example of mapping maize ESTs and STSs using OMAs was reported in Okagaki et al. (2001). They mapped 72 on chromosome 1; 47 on chromosome 2; 45 on chromosome 3; 42 on chromosome 4; 56 on chromosome 5; 46 on chromosome 6; 43 on chromosome 7; 52 on chromosome 8; 42 on chromosome 9; and 20 on chromosome 10. OMAs have been used to demonstrate the distribution of retroelements among chromosomes (Ananiev et al. 1998a). The mapping of transpositions of transposable elements can be highly facilitated by OMAs (Brutnell and Dooner, personal communication). Additional techniques such as FISH for single-locus mapping (Koumbaris and Bass 2003; Amarillo and Bass 2007) have been successful.

Sets of RHs are derived by irradiating seed of oat-maize monosomic addition lines (Riera-Lizarazu et al. 2000). The RHs range from telosomes (Kynast et al. 2003) to small deletions of the maize chromosome or translocations of segments to an oat chromosome. The telosomes allow the rapid mapping of a DNA sequence to chromosome arm. A diminutive maize chromosome or a translocated fragment from a maize to an oat chromosome allows mapping to the physically missing segment. A series of RHs allows the mapping of a sequence to a small physical region. If the OMA has an associated phenotype, an examination of a RH series may allow placement of the controlling gene to a specific segment of the chromosome by phenotypic analysis of the RHs.

The OMA and RH lines also may be useful for chromosome sorting (Li et al. 2001). The largest maize chromosome is about the size of the smallest oat chromosome. With only one maize chromosome present in an OMA line, sorting of specific maize chromosomes becomes feasible. Broken oat chromosomes creating a fragment matching the maize chromosome in size may result in impure maize chromosome preparations. Such chromosome sorting using OMAs or RHs would facilitate the generation of maize chromosome-specific or subchromosome-specific libraries.

Radiation Hybrid lines have been generated, such as ones for chromosome 2 and 9, that appear to have little more than the centromere region present. In screening plants from irradiated addition line seed, occasionally a line is recovered which retains centromere-specific Cent A or Cent C sequences, but which tests negative for Grande 1, a retroelement widely spread through the maize genome. Cytology confirms that only a small "dot" chromosome is present. These lines may be useful for isolating centromeres or constructing Plant Artificial Chromosomes. OMAs have been used to demonstrate the distribution of retroelements among chromosomes (Ananiev et al. 1998a), and for improved localization of the centromeres in maize chromosome physical and genetic maps (Okagaki et al. 2008).

Meiotic chromosome pairing is another feature that can be analyzed using OMAs. Because the repetitive DNAs of oat and maize are so different, Genomic In Situ Hybridization (GISH) can be readily performed using labeled maize genomic DNA, even without blocking with unlabeled oat DNA. Such technology allows maize chromosome visualization while the 42 oat chromosomes are hardly visible except for the conserved telomere and ribosomal DNA regions. A cell with only added maize chromosomes visible via GISH allows the maize chromosomes to be followed in their meiotic pairing process in oat-maize disomic addition lines (Bass et al. 2000).

Oat-maize addition lines have been used to study many aspects of maize genome structure. One use that increased the efficiency of selecting useful BACs dealt with the identification of chimeric maize BACs. A chimeric BAC may have end sequences from two different chromosomes. Quickly mapping the end sequences to two different homologous chromosomes using a set of OMAs can readily indicate that the BAC is chimeric and may be discarded from the collection avoiding considerable confusion. Improved understanding of knob behavior led to the thought that knobs could be megatransposons (Ananiev et al. 1998b) and also to additional information on retrotransposon invasion into heterochromatin (Ananiev et al.1998c). OMAs have been used to search for missing DNA sequences among maize lines (Okagaki et al. 2006). An evolutionary study on ancestral rice blocks and their related regions in the maize genome also involved OMAs (Odland et al. 2006). B-chromosomes of maize added to the oat genome allowed testing B-chromosome transmission in an alien background (Kynast et al. 2007, 2008). Aspects of centromere structure have been studied with these lines (Jin et al. 2004), and the identification of new centromere-specific elements for analysis of OMAs (Ananiev et al. 1998) contributed to construction of transmissible maize mini-chromosomes (Ananiev et al. 2007).

Although the focus of this chapter is on mapping maize DNA sequences, the OMAs and RHs can be screened for the introgression of useful traits from maize to oat. Since oat is a C3 species and maize is C4, the OMAs and RHs may provide avenues for the introgression of C4 photosynthesis into oat (Kowles et al. 2008). Another trait of interest is disease resistance. For example, crown rust (*Puccinia coronata*) is the most serious disease of oat. Because maize is not infected by crown rust, searching for resistance to crown rust among the various OMAs and RHs conceivably could lead to the transfer of crown rust resistance from maize to oat

(Walch, personal communication). Maize PR (Pathogenesis Related) genes are expressed in certain OMAs (Walch 2007).

2 Development of Oat-Maize Addition Lines

Laurie and Bennett (1986, 1989) brought to our attention, via their research on wheat by maize crosses, the possibility of crossing oat and maize. They found that such crosses resulted in the initiation of an embryo (with little endosperm) that occasionally could be grown into a plantlet by employing a hormone treatment (2,4-D) coupled with embryo rescue procedures. They found that the genomes of wheat and maize indeed formed a hybrid embryo and that the maize chromosomes were eliminated in the early divisions of the embryo. The resulting plants were haploids of wheat. This technique has been utilized to produce doubled haploids of wheat in various wheat breeding programs. Subsequent studies have shown that the maize chromosomes always appear to be completely eliminated resulting in haploid wheat plants. Because doubled haploids also were of interest in oat breeding and genetics, oat by maize crosses were initiated. In contrast to wheat by maize crosses, however, the oat by maize crosses resulted in oat haploid plants (21 chromosomes) only about two-thirds of the time (Riera-Lizarazu et al. 1996). One-third of the rescued embryos retained one or more maize chromosomes in addition to the 21 oat chromosomes. If only one or two maize chromosomes are present, the plants often set seed allowing the possibility of recovering OMAs and subsequent RHs. Interestingly, the haploid oat plants, with or without maize chromosomes, frequently undergo first division restitution (FDR) resulting in gametes with a complete chromosome complement. Therefore, the simple act of self-pollination leads to the recovery of doubled haploids from haploid oat plants, or oat-maize disomic addition lines with 42 oat chromosomes and a pair of maize chromosomes from haploid oat plants with a single maize chromosome.

Plantlets from oat by maize crosses are first screened with the retroelement Grande 1, which is found in 65% of the BACs in a maize BAC library. This implies that Grande 1 will likely be present in any plant carrying maize chromatin. Plants that test positive for Grande 1 are further screened with simple sequence repeat (SSR) markers to determine which maize chromosome is present. SSR markers located in the distal regions of the respective chromosome arms are used to determine if both arms are present. The number of maize chromosomes present can be readily determined cytologically by using GISH (Fig. 1). These techniques enabled the production of a complete set of maize individual chromosome additions to oat (Kynast et al. 2001).

Various genotypes of oat and maize have been used to produce the OMAs. Seneca 60 maize was used by Laurie and Bennett (1986, 1989) due to it being a prolific pollen producer and the ease of growing the strain in the greenhouse. This strain was found to also work well for crossing with oat. Seneca 60 is a hybrid sweet corn variety produced from proprietary inbreds. The hybrid nature of Seneca 60 may be

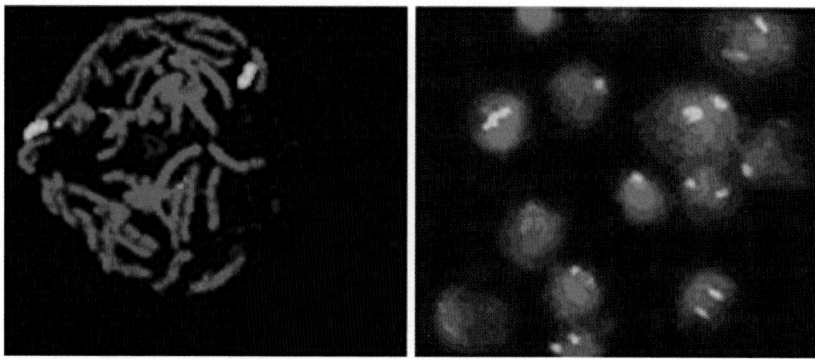

Fig. 1 A pair of maize chromosomes added to oat at the metaphase and interphase stages shown by GISH analysis. (Courtesy of E. Ananiev)

a disadvantage in certain cases since different OMAs for specific chromosomes will each reflect somewhat different genic contents. On the other hand, having the genetic variability among OMAs could be an advantage if one of the inbred parents carries genes deleterious for a specific aspect of development. The OMA for Seneca chromosome 10 may be a specific example. OMA 10 was the most difficult to recover despite it being the smallest maize chromosome. In addition, the OMA for Seneca chromosome 10 never transmitted the whole maize chromosome to progeny but did transmit the short arm. Apparently, there is a gene(s) on the long arm of chromosome 10 that precludes chromosome 10 transmission in this case. Additional OMAs for chromosome 10 were not recovered with Seneca 60; therefore, we could not determine if the putative deleterious gene was on only one perental chromosome of the Seneca 60's chromosome 10 pair. A chromosome 10 with a part of the long arm was transmitted from the B73 OMA chromosome 10 line indicating that the deleterious gene is located distal to the breakpoint. A fertile OMA 10 was recovered in another background (Mo17) that transmits an apparent intact chromosome 10 (Kynast et al. 2005).

Several OMAs have been produced in B73 and Mo17. These are popular inbreds; the hybrid between them has been important in hybrid corn production. In addition, B73 was selected as the maize inbred to be completely sequenced. For these reasons, we have also derived OMA lines in these two maize genetic backgrounds. Currently, self-fertile OMAs are available in B73 background for eight of the ten maize chromosomes, and also for five of the maize chromosomes in Mo17 background. There are no fertile OMAs of chromosomes 3 or 7 in either maize inbred, but with Mo17 as the maize parent, non-fertile haploid OMAs of these chromosomes were recovered, and their DNA isolated.

Multiple chromosome additions also occur from oat by maize crosses. Such lines allow the study of interactions between chromosomes. All maize chromosomes have been identified in OMAs with one or more other maize chromosomes (Fig. 2). At this point in time, whether single or multiple combinations occur at significantly different frequencies among different chromosomes is not known due to the difficulty

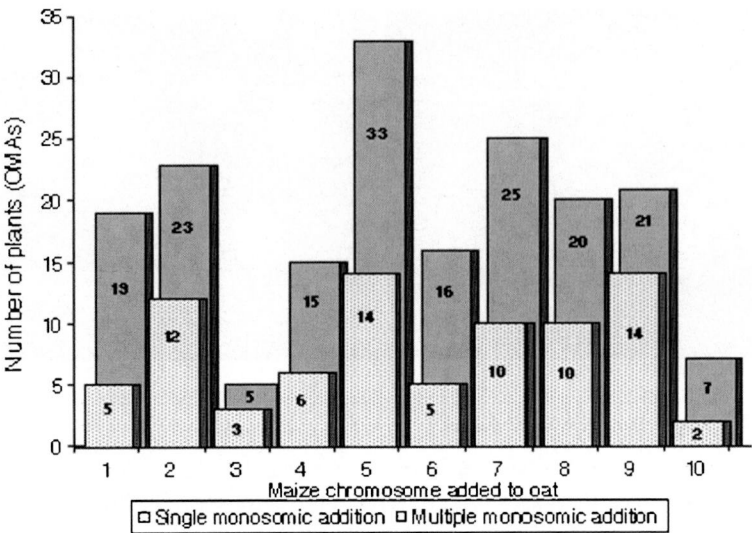

Fig. 2 Recovery of multiple chromosome combinations in OMAs. Using Seneca 60 maize. (Adapted from Fig. 4, Kynast et al. 2001.)

of analyzing for each chromosome in often weak or chimeric plants. In general, the more maize chromosomes present the less vigorous the plant. Only about 10% of the florets of oat fertilized with maize pollen result in embryo formation. Similarly, only about 10% of the embryos were recovered as plants via embryo rescue and only one-third of the plantlets had maize chromatin. Improved embryo rescue procedures increased the frequency of germinating embryos to about 50%, with more embryos testing positive for the presence of maize chromatin. However, many of the additional plantlets recovered contained multiple maize chromosomes and failed to develop to a flowering stage. Even those plants with only one or few maize chromosomes that do flower often do not set seed. Thus, the number of OMAs recovered is low making it difficult to show statistically significant differences for recovery of specific chromosomes and chromosome combinations.

A single maize chromosome in an oat genome background represents only about 2% of the total DNA. Because the plants look like oat with some phenotypic modifications, the amount of maize gene expression or gene silencing is difficult to intuitively assess. Cabral et al. (2007) used the Affymetrix Maize Gene Chip to estimate the amount of gene expression in B73 OMA for chromosome 5 compared to the B73 inbred. They found that about 24% of the chromosome 5 genes expressed in B73 are expressed in OMAs with a B73 chromosome 5. Thus, it appears that a considerable proportion of the maize genes are expressed while many also appear to be silenced or lack the appropriate transacting factors needed for their expression. Cytological examination of OMA 6 cells show that the maize nucleolus organizer does not form a nucleolus, indicating the lack of expression of the maize rDNA. Jin et al. (2004) showed that the histone Cent C in OMA6 is from oat, indicating lack of expression of the gene for maize histone Cent C located on chromosome 6. The application of OMAs and RHs for functional genomics is reviewed in Kynast et al. (2002). The enrichment of maize DNA by representational difference analysis was performed by Chen et al. (1998).

The phenotype associated with OMAs (Table 1) depends on the chromosome and the genetic background of the oat and the maize parents. The most interesting phenotype is the one associated with chromosome 3. OMA3 shows a crooked panicle and abnormal upper leaf ligules. Muehlbauer et al. (2000) showed that abnormal ligules were the result of the ectopic expression of the *Lg3* gene on chromosome 3 of maize. The presence of a single maize chromosome in an oat background may allow the expression of maize genes that are otherwise suppressed by a gene on another chromosome. Further studies along this line of thought would be quite informative relative to control of gene expression. A certain amount of variability in expression may occur among OMAs possessing the same chromosome. For example, five out of six of our first OMAs with chromosome 6 expressed a disease lesion mimic phenotype. This could be due to the ectopic expression of a disease lesion mimic gene on chromosome 6. Although there is such a gene on chromosome 6, many others also exist across the genome. We attempted to locate the gene by evaluating the disease lesion mimic phenotype across a RH series for chromosome 6. Unfortunately, expression of the phenotype was too variable to use that approach to map the gene.

The Seneca 60 OMA lines are available in a variety of oat genetic backgrounds (depending on the specific chromosome), including genotypes GAF-Park, Kanota, Preakness, Starter, Stout, Sun II, and a Minnesota hybrid (MN97201 x MN841801-1). The genetic background appears to have a variety of influences on phenotype, chromosome stability, and transmission (Kynast et al. 2001).

Maintenance of the OMA lines is straightforward since oat is naturally self-pollinating and oat-maize disomic addition lines usually transmit the maize chromosome with high fidelity. However, we would recommend that testing for

Table 1 Oat Maize Addition Line Phenotypes Using Seneca 60 Maize (Adapted from Table 3, Kynast et al. 2001)

Added maize chromosome	Characteristics
1	Erect leaf blade, Photoperiod neutral response, Sectoring among shoots
2	Bluish leaf, Waxy stem
3	Liguleless, Crooked panicle, Outgrowth of aerial axillary buds
4	Lighter green leaf, Small seed, Earlier maturing
5	Branched stem, Sectoring among shoots, Generative instability
6	Disease lesion mimic, NOR amphiplasty
7	Small stature, Instability in some offspring
8	Small stature, Sectoring among roots, Irregular transmission
9	Erratic premature senescence
10	Grassy type, No transmission of long arm

Grande 1 would be expedient to confirm the continued presence of the maize chromosome. Of course, further PCR of appropriate markers or cytological analyses also would be informative to assure that the added maize chromosome has remained intact. The maize chromosome in an oat-maize monosomic addition would segregate as a univalent and confirmation of the progeny genotypes would be required; the maize chromosome would be expected to behave as a univalent and transmit at a low frequency (10%) through the female and perhaps not at all through the male (Riera-Lizarazu et al. 1996). A segment translocated to an oat chromosome in the heterozygous condition would transmit at a much higher frequency, perhaps 50% (Vales et al. 2004). Transmission of RHs will depend on the nature of the chromosome and whether homozygous or heterozygous; usual cytogenetic rules will normally apply.

Optimal plant growth in growth chambers is obtained with 18-20 °C day, 14-16 °C night and a 11-hour light, 13-hour dark photoperiod for 6-8 weeks to enhance vegetative growth followed by a shift to16-hour light, 8-hour dark to promote flowering. Conditions are not as critical if the goal is only vegetative tissue for DNA or RNA extraction although short day lengths maximize vegetative growth. We have no experience growing these materials in the field.

Figures 3 and 4 show the ease of mapping maize DNA sequences using OMAs

(i) In Figures 3A & 3E, cDNA sequences are absent in oat, and map to single maize chromosomes – #7 in 3A, and #8 in 3E.

(ii) In Figures 3B & 3F, cDNA sequences are absent in oat, and map to two maize chromosomes – #6 and #8 in 3B, and #6 and #9 in 3F.

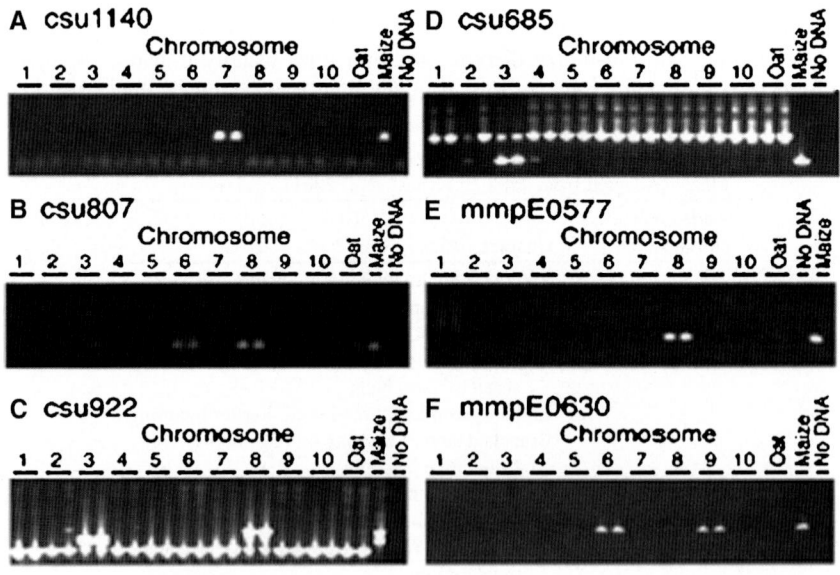

Fig. 3 Mapping maize DNA sequences using OMAs. (Samples run in duplicate.) (See explanation in text.) (Courtesy of R. Okagaki)

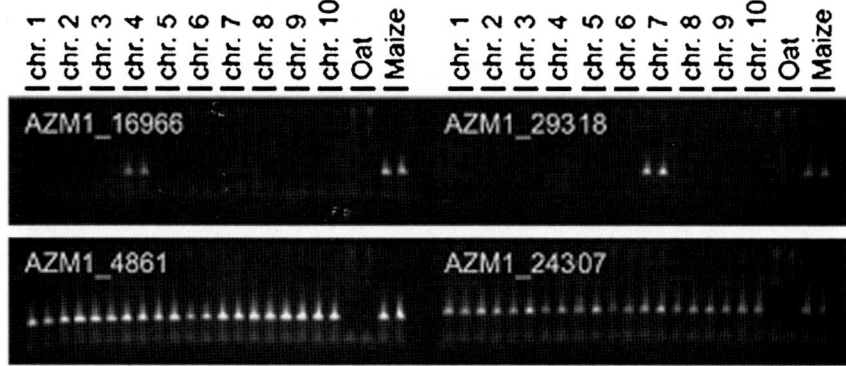

Fig. 4 *Top gel*: genomic sequences map to chromosome 4 (left), and chromosome 7 (right). *Bottom gel*: repetitive sequences map to all maize chromosomes. (Samples run in duplicate.) (Courtesy of R. Okagaki)

(iii) In Figures 3C & 3D, cDNA sequences map to both maize and oat chromosomes, but polymorphisms exist between the species which allow mapping to the appropriate maize chromosomes – #3 and #8 in 3C, and #3 in 3D.

(iv) Figure 4 shows two cases of a genomic sequence mapping to a single maize chromosome, and two where a repetitive sequence maps to all maize chromosomes.

3 Development of Oat-Maize Radiation Hybrid Lines

The concept of radiation hybrids has been well-developed for human gene mapping (Cox et al. 1990). A human cell line is irradiated at high dose and then fused with a Chinese hamster or mouse cell line. Cell lines are then grown which have the rodent genome and a fragment(s) of the human genome. Assuming that the radiation-induced breakage of the human chromosomes is random, the frequency that two sequences on the same chromosome are not both present in the cell line reflects the physical distance between the two sequences. The longer the distance, the greater chance of breakage between the sequences. This approach works well with animals.

Tissue culture of maize leads to considerable somaclonal variation (Lee and Phillips 1987a; Armstrong and Phillips 1988) and cytological disturbances including chromosome breakage (Lee and Phillips 1987b). We decided that greater stability and consistency would occur if we created OMAs by wide hybridization and then irradiated seed of oat-maize monosomic additions. The monosomic condition was used because that precluded the possibility of breakage and reunion of two homologues, as would be possible in disomic additions, which could lead to complicated rearrangements among homologous chromosomes. In addition, the use of monosomic OMAs allowed ready detection of deleted maize chromosome segments whose absence would be masked by the presence of a homologue. Monosomic OMAs were easily produced by crossing the appropriate oat parental genotype by the oat-maize disomic addition line.

We chose to use Cesium 137 as a gamma radiation source to induce chromosome breakage by irradiating monosomic OMA seed. The dosage was adjusted depending on the genetic background in order to achieve a balance between a high chromosome breakage rate and reasonable seed germination percentages. This compromise leads to a lower frequency of chromosome breakage than would be most desirable. Although many useful RH lines can be derived (approximately 650 to date), the frequency and distribution of breaks (Okagaki et al. 2004) has not allowed the use of software programs such as RH Mapper to generate extensive maps. The oat-maize system can be efficiently used for mapping genes, however, in a manner consistent with deletion mapping.

Progeny of plants grown from irradiated monosomic OMA seed are first screened with Grande 1 to determine the presence of maize chromatin. The use of progeny plants allows segregation of induced maize chromosome rearrangements or deletions and ensures that they are meiotically transmissable. Approximately 40 markers are used per maize chromosome to detect the presence or absence of the genomic regions associated with them. This data then defines the RH under investigation. If all markers are present, a plant is considered "normal" (no detectable breaks) and is discarded. The nature of chromosomal changes, such as diminutive maize chromosomes versus translocations of a maize segment to an oat chromosome, can be readily detected by GISH (Riera-Lizarazu et al. 2000). No doubt, oat also undergoes chromosome breakage, however, the inherent buffering of its allohexaploid genome allows recovery of robust plants, even with such disruptions. The RH set used to

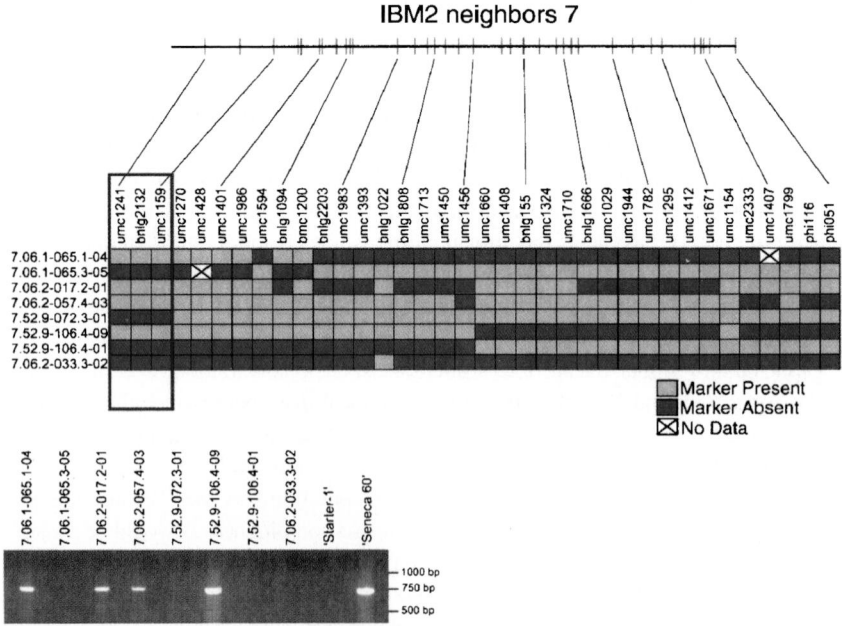

Fig. 5 Mapping of a pathogenesis-related gene (PR1) of maize using a RH series of lines for chromosome 7. The gene is physically located within the region marked by umc1241 to umc1159. (Courtesy of M. Walch)

map a maize sequence (Fig. 5), depends on whether there is information about the location of a sequence and the desired degree of resolution.

Maintenance of RHs depends on the nature of the breakage event. Ideally, the recovery of the homozygous genotype is best since it should breed true. The RH lines produced in our laboratories generally have not been cytologically analyzed, so any increase of seed would require testing molecularly and/or cytologically to confirm the genotype.

4 Availability of OMAs and RHs

The available Oat Maize Addition lines are listed in Table 2.

The available Radiation Hybrid lines are listed on the web site. The various lines are described according to the markers present (+) or absent (-) on the respective chromosome. Some would appear to be telosomes, others are putative translocations, and others are apparently complex chromosomes with various missing segments. No RHs are available for chromosome 8 (Table 3).

Seeds of the OMA lines will be deposited in the USDA-ARS National Center for Genetic Resources, Fort Collins, Colorado. All OMAs and RHs lines are listed at http://agronomy.cfans.umn.edu/Maize_Genomics.html. DNA for most lines is

Table 2 Number of independently derived OMA lines available from various maize chromosome donors.

Maize Chromosome Donor	Oat-Maize Addition Line										
	1	2	3	4	5	6	7	8	9	10	B
Seneca 60		1	11	2	6	3	3	3	2	9	(1)
B73		1	1		3	11	3		1	1	1
Mo17			8	(1)	1	8	3	(1)		2	
A188					1						
bz1-mum9			1					1			
B73 w/Bs from Black Mexican Sweet											2

( ) No seed produced, but limited DNA of the original plant is available.

Table 3 Oat-Maize Chromosome Addition Radiation Hybrid Panels

Maize Chromosome	Number of Lines with Breaks
Chr. 1	67
Chr. 2	81
Chr. 3	32
Chr. 4	72
Chr. 5	71
Chr. 6	120
Chr. 7	22
Chr. 8	0
Chr. 9	203
Chr. 10	17
Total	685

available by contacting the authors of this chapter. Nomenclature for OMA lines follows OMAxy.z where it is an abbreviation for Oat Maize Addition and x indicates the maize chromosome constitution (d for disomic, m for monosomic), y is the number of the maize chromosome added (1 to 10), and z identifies the specific event (Kynast et al. 2001).

5 Introgression of Maize Traits into Oat

C4 photosynthesis confers a more efficient photosynthetic process on certain species, known as C4 species, compared to C3 species. Maize is a C4 species while oat is C3. The ability to add one maize chromosome at a time to oat raises the prospect that OMAs or certain combinations may allow the expression of C4 traits in an oat background. PEPc and PPDK are two key enzymes in the C4 photosynthetic process. The appropriate forms of these genes are known to be located on chromosomes 6 and 9, respectively. Walch (2007) used OMAs and RHs to confirm the chromosome location of these two genes. Kowles et al. (2008) then tested for gene expression via northern blots (Fig. 6) and by enzyme activity assays. It was found that expression of PEPc occurred in OMA 9 lines, expression of PPDK in OMA 6 lines, and expression of both in OMA 6 & 9 double monosomic addition lines (made by crossing the respective disomic OMAs).

CO_2 compensation points were examined with these materials (Kowles et al. 2008) which did not indicate any dramatic changes. Currently, smaller segments of

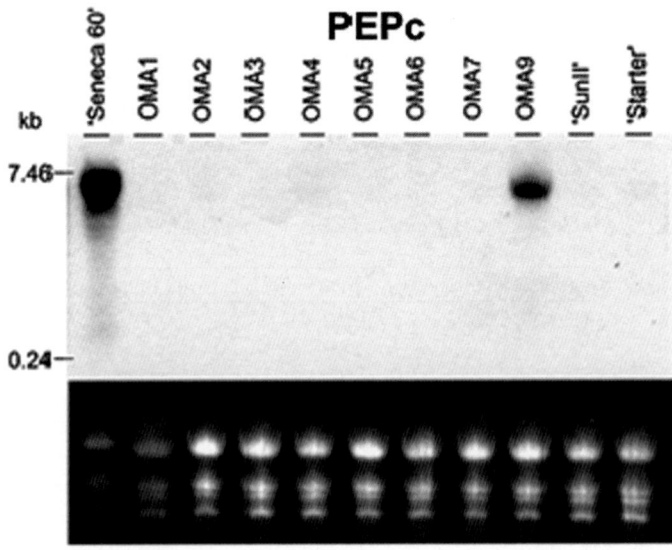

Fig. 6 Expression of C4 gene, PEPc, in OMA 9. (Courtesy of M. Walch)

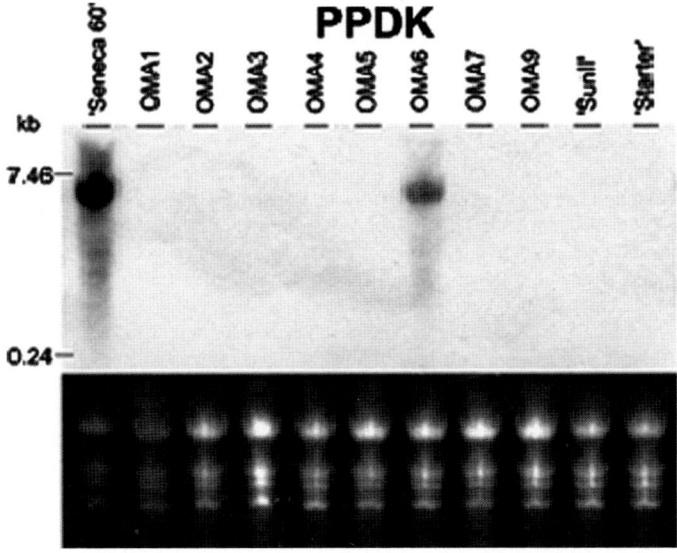

Fig. 7 Expression of C4 gene, PPDK, in OMA 6. (Courtesy of M. Walch)

chromosomes 6 and 9 are being brought together by crossing the appropriate RHs. OMA3 also may be utilized in these studies because the leaf anatomy is modified in that OMA somewhat reminiscent of Kranz anatomy that is often implicated in C4 species.

6   Haploids for Use in Oat Genetics and Breeding

One-half to two-thirds of the plantlets recovered from oat x maize crosses appear to be missing all of the maize chromosomes (Rines and Dahleen 1990). Such plants are considered haploids and show no evidence of the Grande 1 retroelement or other maize sequences tested. Because haploid plants also undergo meiotic restitution by FDR, doubled haploids are recovered in the progeny without the need for colchicine or other doubling treatments; however, monosomics (2n=41) and other aneuploids have been observed in about 25% of the recovered plants (Davis 1992; Rines et al. 1995; Rines et al.1997). Monosomics recovered in this manner have been used to complete a monosomic series for assigning molecular markers and linkage groups to chromosome in oat (Jellen et al. 1997; Fox et al. 2001). Field tests of a limited number of apparent normal doubled haploids indicated their potential value in breeding (Davis 1992). Although these plants would appear to be very useful, the possibility remains that a small segment of maize chromatin could still be present; careful examination is needed of doubled haploid oat lines derived from oat x maize crosses.

Acknowledgments The work reported in this chapter was supported by NSF grants 9872650 and 011134, USDA-NRI /96-35300-3775, University of Minnesota Plant Molecular Genetics Institute, McKnight Presidential Chair in Genomics, Pioneer Hi-Bred Intl. Inc., Quaker Oats Company, Interministerial Commission of Science and Technology of Spain, Midwest Plant Biotechnology Consortium (USDA Prime/Purdue University sub#593-0120-13), the Minnesota Agricultural Experiment Station, and USDA-ARS.

The excellent contributions of the following postdoctoral scientists, graduate students, technicians, and undergraduate students are gratefully acknowledged. The work would not have been accomplished without their dedicated and innovative efforts. The following people are acknowledged: Postdoctoral Scientists Evgueni Ananiev, Ralf Kynast, Ron Okagaki, Oscar Riera-Lizarazu, and Isabel Vales; Graduate Students Carrie Beckenbach, Candida Cabral, Victor Nunez, Silvia Maquieira, Wade Odland, and Matt Walch; Technicians Mark Galatowitsch, Shannon Granath, Paul Huettl, Morrison Jacobs, Suzanne Livingston, Charles Russell, Tina Schmidt, Adrian Stec, Jayanti Suresh, and Herika Zaia; and Undergraduate Students Elizabeth Buescher, Howard Chen, Kim Clafton, Robert Day, Renee Donahue, Andrew Dosdall, Jason Draper, Christian Fluur, Alexis Garreau, Jean Gillian, Rachel Hansen, Peter Johnson, Marit Johnson, Joan LaPorte, Eduardo Lebedenco, Curtis Lindgren, Emily Littrell, Alan Nachtigal, Derek Nelson, Michelle Nilson, Jana Ninkovic, Andy Rausa, Sarah Savage, Tom Secher, Tera Secker, Kevin Spahr, Eric Stevenson, Ruth Swanson, and Kate Thompson.

References

Amarillo FIE, and Bass HW (2007) A transgenomic cytogenetic sorghum (*Sorghum propinquum*) bacterial artificial chromosome fluorescence *in situ* hybridization map of maize (*Zea mays* L.) pachytene chromosome 9, evidence for regions of genome hyperexpansion. Genetics 177:1509–1526

Ananiev EV, Chamberlin MA, Gordon-Kamm WJ, Svitashev S, Wu C (2007) Artificial plant minichromosomes. US Patent Publication 20070271629

Ananiev, EV, Phillips, RL and Rines, HW (1998a) Chromosome-specific molecular organization of maize (*Zea mays* L.) centromeric regions. Proc Natl Acad Sci USA 95:13073–13078

Ananiev EV, Phillips RL, Rines HW (1998b) A knob-associated tandem repeat in maize capable of forming fold-back DNA segments: Are chromosome knobs megatransposons? Proc Natl Acad Sci USA 95:10785–10790

Ananiev EV, Phillips RL, Rines HW (1998c) Complex structure of knob DNA on maize chromosome 9: Retrotransposon invasion into heterochromatin. Genetics 149:2025–2037

Ananiev EV, Riera-Lizarazu O, Rines HW, Phillips RL (1997) Oat-maize chromosome addition lines: A new system for mapping the maize genome. Proc Natl Acad Sci USA 94:3524–3529

Armstrong CL, Phillips RL (1988) Genetic and cytogenetic variation in plants regenerated from organogenic and friable, embryogenic tissue cultures of maize. Crop Sci 28:363–369

Bass HW, Riera-Lizarazu O, Ananiev EV, Bordoli SJ, Rines HW, Phillips RL, Sedat JW, Agard DA, Cande WZ (2000) Evidence for the coincident initiation of homolog pairing and synapsis during the telomere-clustering (bouquet) stage of meiotic prophase. J Cell Sci 113:1033–1042

Cabral C, Springer N, Rines HW, Phillips R (2007) Many maize genes are expressed in an oat background carrying a specific maize chromosome. 49th Annual Maize Genet Conf, St. Charles, IL Abstracts p 112

Chen ZJ, Phillips RL, Rines HW (1998) Maize DNA enrichment by representational difference analysis. Theor Appl Genet 97:337–344

Cox DR, Burmeister M, Price ER, Kim S, Myers RM (1990) Radiation hybrid mapping: A somatic cell genetic method for constructing high-resolution maps of mammalian chromosomes. Science 250:245–250

Davis DW (1992) Characterization of oat haploids and their progeny. University of Minnesota MS Thesis

Fox SL, Jellen EN, Kianian SF, Rines HW, Phillips RL (2001) Assignment of RFLP linkage groups to chromosomes using monosomic F_1 analysis in hexaploid oat. Theor Appl Genet 102:320–326

Jellen EN, Rines HW, Fox SL, Davis DW, Phillips RL, Gill BS (1997) Characterization of "Sun II" oat monosomics through C-banding and identification of eight new "Sun II" monosomics. Theor Appl Genet 95:1190–1195

Jin W, Melo JR, Nagaki K, Talbert PB, Henikoff S, Dawe RK, Jiang J (2004) Maize centromeres: organization and functional adaptation in the genetic background of oat. Plant Cell 16:571–581

Koumbaris GL, Bass HW (2003) A new single-locus cytogenetic mapping system for maize (*Zea mays* L.): Overcoming FISH detection limits with marker-selected sorghum (*S. propinquum* L.) BAC clones. Plant J 35:647–659

Kowles RV, Walch MD, Minnerath JM, Bernacchi CJ, Stec AO, Rines HW, Phillips RL (2008) Expression of C4 photosynthesis enzymes in oat-maize chromosome addition lines. Maydica 53:69–78

Kynast RG, Galatowitxch MW, Hanson L, Huettl PA, Lupke L, Phillips RL, Rines HW (2008) Maternal and paternal transmission to offspring of B-chromosomes of *Zea mays* L. in the alien genetic background of *Avena sativa* L. Maize Genet Coop Newsl 82:23

Kynast RG, Galatowitsch MW, Huettl PA, Phillips RL, Rines HW (2007) Adding B-chromosomes of *Zea mays* L. to the genome of *Avena sativa* L. Maize Genet Coop Newsl 81:6

Kynast RG, Okagaki RJ, Galatowitsch MW, Granath SR, Jacobs MS, Stec AO, Rines HW, Phillips RL (2004) Dissecting the maize genome by using chromosome addition and radiation hybrid lines. Proc Natl Acad Sci USA 101:9921–9926

Kynast RG, Okagaki RJ, Galatowitsch MW, Huettl PA, Jacobs MS, Stec AO, Walch MD, Cabral CB, Odland WE, Rines HW, Phillips RL (2005) Development of the first fertile plants with an addition of the whole maize chromosome 10 to the oat genome. Maize Genet Coop Newsl 79:38–39

Kynast RG, Okagaki RJ, Granath SR, Rines HW, Phillips RL (2003) A novel maize ditelosomic addition to oat cv. Sun II for use in radiation hybrid mapping. Maize Genet Coop Newsl 77:62–63

Kynast RG, Okagaki RG, Odland WE, Russell CD, Livingston SM, Rines HW, Phillips RL (2000) Towards a radiation hybrid map for maize chromosomes. Maize Genet Coop Newsl 74:60–61

Kynast RG, Okagaki RJ, Odland WE, Stec A, Russell CD, Zaia H, Livingston SM, Rines HW, Phillips RL (2001) Oat-maize chromosome manipulation for the physical mapping of maize sequences. Maize Genet Coop Newsl 75:54–55

Kynast RG, Okagaki RJ, Rines HW, Phillips RL (2002) Maize individualized chromosomes and derived radiation hybrid lines and their use in functional genomics. Funct Integr Genomics 2:60–69

Kynast RG, Riera-Lizarazu O, Vales MI, Okagaki RJ, Maquieira S, Chen G, Ananiev EV, Odland WE, Russell CD, Stec AO, Livingston SM, Zaia HA, Rines HW, Phillips RL (2001) A complete set of maize individual chromosome additions to the oat genome. Plant Physiol 125:1216–1227

Laurie DA, Bennett MD (1986) Wheat x maize hybridization. Can J Genet Cytol 28:313–316

Laurie DA, Bennett MD (1989) The timing of chromosome elimination in hexaploid wheat x maize crosses. Genome 32:953–961

Lee M, Phillips RL (1987a) Genetic variants in progeny of regenerated maize plants. Genome 29:834–838

Lee M, Phillips RL (1987b) Genomic rearrangements in maize induced by tissue culture. Genome 29:122–128

Li LJ, Arumuganathan K, Rines HW, Phillips RL, Riera-Lizarazu O, Sandhu D, Zhou Y, Gill KS (2001) Flow cytometric sorting of maize chromosome 9 from an oat-maize chromosome addition line. Theor Appl Genet 102:658–663

Mica E, Gianfranceschi L, Pe ME (2006) Characterization of five microRNA families in maize. J Exp Bot 57:2601–2612

Muehlbauer GJ, Riera-Lizarazu O, Kynast RG, Martin D, Phillips RL, Rines HW (2000) A maize-chromosome 3 addition line of oat exhibits expression of the maize homeobox gene *liguleless 3* and alterations of cell fate. Genome 43:1055–1064

Odland W, Baumgartner A, Phillips R (2006) Ancestral rice blocks define multiple related regions in the maize genome. The Plant Genome 1:S41–S48

Okagaki, RJ, Jacobs MS, Stec AO, Kynast RG, Buescher E, Rines HW, Vales MI, Riera-Lizarazu O, Schneerman M, Doyle G, Friedman KL, Staub RW, Weber DF, Kamps TL, Amarillo IFE, Chase CD, Bass HW, Phillips RL (2008) Maize centromere mapping: A comparison of physical and genetic strategies. J Heredity 95:85–93

Okagaki RJ, Kynast RG, Livingston SM, Russell CD, Rines HW, Phillips, RL (2001) Mapping maize sequences to chromosome using oat-maize chromosome addition materials. Plant Physiol 125:1228–1235

Okagaki RJ, Kynast RG, Odland WE, Stec A, Russell CD, Zaia HA, Rines HW, Phillips RL (2002) A radiation hybrid system for the genetic and physical mapping of the corn genome. Maize Genet Coop Newsl 76:88–89

Okagaki, R.J., Kynast, R.G., Rines, H.W. and Phillips RL (2004) Radiation hybrid mapping. In: Goodman RM (ed) Encyclopedia of plant & crop science. Marcel Dekker Inc, New York, pp 1074–1077

Okagaki RJ, Jacobs M, Stec AO, Kynast RG, Zaia H, Granath SR, Rines HW, Phillips RL (2004) The distribution of chromosome breaks in radiation hybrid lines from chromosome 3. Maize Genet Coop Newsl 78:49–50

Okagaki RJ, Schmidt CM, Stec AO, Rines HW, Phillips RL (2006) How many maize genes will not be in maize inbred B73? Maize Genet Coop Newsl 80:26

Phillips RL, Rines HW, Okagaki RJ, Kynast RG, Donahue R, Odland WE (2003) Genetic and physical mapping of the maize genome through radiation hybrids. Proc Intl Rice Research Conference 2003 Gurdev Khush Symposium. In: Mew TW, Brar DS, Peng S, Dawe D, Hardy B (eds) Rice Science: Innovations and Impact for Livelihood, pp 77–90

Riera-Lizarazu O, Rines HW, Phillips RL (1996) Cytological and molecular characterization of oat x maize partial hybrids. Theor Appl Genet 93:123–135

Riera-Lizarazu O, Vales MI, Ananiev EV, Rines HW, Phillips RL (2000) Production and characterization of maize-chromosome 9 radiation hybrids derived from an oat-maize chromosome addition line. Genetics 156:327–339

Rines HW, Dahleen LS (1990) Haploid oat plants produced by application of maize pollen to emasculated oat florets. Crop Sci 30:1073–1078

Rines HW, Phillips RL, Kynast RG, Okagaki RJ, Odland WE, Stec AO, Jacobs MS, Granath SR. (2005) Maize chromosome additions and radiation hybrids and their use in dissecting the maize genome. In: Tuberosa R, Phillips RL, Gale M (eds) In the wake of the double helix: From the green revolution to the gene revolution. Avenue Media, Bologna, Italy pp 427–441

Rines HW, Riera-Lizarazu O, Nunez VM, Davis DW and Phillips RL (1997) Oat haploids from anther culture and from wide hybridizations. In: SM Jain, K Sopory, RE Veilleux (eds) In Vitro Haploid Production in Higher Plants. Vol 4 Cereals Kluwer Academic Publ, Amsterdam, The Netherlands, pp 205–221

Rines HW, Riera-Lizarazu O, Phillips RL (1995) Disomic maize chromosome-addition oat plants derived from oat x maize crosses. In: Oono K, Takaiwa F (eds) Modification of Gene Expression and Non-Mendelian Inheritance. Natl Inst Agrobiol Res Tsukuba, Japan, pp 235–251

Vales MV, Riera-Lizarazu O, Rines HW, Phillips RL (2004) Transmission of maize chromosome 9 rearrangements in oat-maize radiation hybrids. Genome 47:1202–1210

Walch MD (2007) Expression of maize pathogenesis-related and photosynthetic genes in oat x maize addition lines. University of Minnesota MS thesis

Zhang J, Simmons C, Yalpani N, Crane V, Wilkinson H, Kolomiets M (2006) Genomic analysis of the 12-oxo-phytodienoic acid reductase gene family of *Zea mays*. Plant Mol Biol 59:323–343

Maize Chromosome Tools: Quantitative Changes in Chromatin

David Weber

1 Introduction

This chapter discusses chromosome tools that alter the quantity of chromatin in the maize genome. These tools are monosomes, trisomes, telocentrics, telodisomes, isochromosomes, deficiencies, duplications, and ring chromosomes. The sources of each of these, and some of the ways they can be and have been employed in experimental studies will be discussed. Although maize is considered to be a true diploid, its genome is much more tolerant to quantitative changes of chromatin than the genomes of many other diploids because much of the maize genome is present as duplicated segments.

Evidence that duplicated segments exist in the maize genome has accumulated through the years. Rhoades (1951) pointed out that there were at least 14 cases of duplicate, 2 of triplicate, and 1 of quadruplicate factor inheritance known at the time. Each factor had to be present in the recessive condition for the mutant phenotype to be expressed. From this and other observations, he concluded, "that the architecture of maize contains duplicated regions can hardly be doubted." The presence of bivalents at diakinesis in maize haploids has been interpreted as evidence for interchromosomal homology. Ting (1966) reported that 59.9% of diakinesis cells contained one or more bivalent; however, other studies (Ford 1952; Snope 1967; and Weber and Alexander 1972) reported much lower frequencies of pairs. Exchanges between nonhomologous chromosomes in haploids have identified the locations of presumptive redundant sequences within the maize genome (Weber and Alexander 1972). Helentjaris, Weber, and Wright (1988) found that 62 of 217 cloned maize DNA sequences hybridized with more than one DNA fragment on Southern blots identifying duplicate nucleotide sequences in maize. The genomic locations of the duplicate sequences were determined, and more recent studies by others have identified numerous additional duplicated sequences in maize (summarized

D. Weber
llinois State University
dfweber@ilstu.edu

J.L. Bennetzen and S. Hake (eds.), *Maize Handbook - Volume II: Genetics and Genomics*,   539
© Springer Science+Business Media LLC 2009

in Gaut 2001; Ahn, Anderson, Sorrells, and Tanksley 1993; Moore, Devos, Wang, and Gale 2005; Odland, Baumgarten, Phillips 2006). These and other studies have demonstrated that much of the maize genome is duplicated, and it is becoming increasingly accepted that maize had an allopolyploid ancestry (see chapter in this volume by Messing). However, because five pairs of homoeologous chromosomes are not currently present within the maize genome, the genome has been extensively rearranged subsequent to the apparent allopolyploidisation event.

2 Monosomes

A monosome is a 2n-1 aneuploid cell, tissue, or individual with one chromosome missing from an otherwise diploid organism. Even though the term, monosome, was coined by Blakeslee in 1921 (Blakeslee 1921), the same term has subsequently been used to refer to a single ribosome bound to a mRNA. In this review, the term, monosome, unless otherwise indicated, designates a primary monosome (loss of one of the chromosomes from the normal haploid chromosome complement).

2.1 Sources of Monosomes in Maize

When a monosomic maize plant is crossed with a diploid, no monosomic progeny are recovered. The reason for this is that when a monosomic maize plant undergoes meiosis, haploid (n) and nullisomic (n-1) gametes are produced in equal frequencies. The n-1 gametes invariably abort in maize and other diploids because factors necessary for gametophytic development are located on each of the chromosomes. For this reason, the only viable gametes produced by monosomes are haploid. Therefore, maize monosomes need to be generated each time they are required for studies. Monosomes in diploids are usually produced by post-meiotic events during the development of the gametophyte.

2.1.1 From Untreated Diploids

Although monosomes have appeared spontaneously in the progeny of normal diploids in *Nicotiana alta* (Avery 1929) and *Hyoscyamus niger* (Griesinger 1937), they occurred at an exceedingly low rate. The author is not aware of the spontaneous occurrence of monosomes in maize.

2.1.2 From Treated Diploids

One set of twin monosomic maize plants that also contained a fragment chromosome in part of their meiotic cells was recovered from crosses with X-irradiated pollen (Morgan 1956). Monosomes can be produced by irradiation of pollen because the tube nucleus mediates the metabolism of the pollen grain and loss of a chromosome from the generative nucleus or a sperm (which are metabolically inert) is compatible with survival and functioning of a pollen grain.

2.1.3   From Aneuploids and Polyploids

McClintock (1929a) found a 2n-1 chimeric maize plant in the progeny of a triploid female. The sporocyte of this plant was monosomic for chromosome 10 and the root tip cells contained the diploid number of 20 chromosomes. Fisher and Einset (1940) obtained a monosome for one of the shorter chromosomes (undetermined) in a parthenogenetic diploid of a maize tetraploid. Einset (1943) reported that five monosomes and one monosomic plant plus a fragment chromosome were found among 1916 progeny of trisomic maize plants. Unfortunately, five of the six mono-somic plants were extremely weak and died soon after transplanting to the field. Shaver (1963) reported six plants with chromosome counts in root-tip cells of 19 in over 300 highly maize-like derivatives from hybrids between maize and perennial teosinte. Two died while juvenile, two had germ line counts of 20 chromosomes, one was monosomic for chromosome 6, and one was monosomic for chromosome 9.

Thus, maize monosomes have been found in the progeny of individuals with abnormal chromosome constitutions. It is not known if the abnormal genomic constitution in some way mediated the generation of monosomes in these cases because comparable sample sizes from diploid plants were not analyzed simultaneously.

2.1.4   From Interactions Between Knobs and B Chromosomes

Rhoades, Dempsey, and Ghidoni (1967) found that regions of maize A chromo-somes containing heterochromatic knobs were frequently eliminated at the second microspore division in maize plants containing two or more B chromosomes. This is the same division where B chromosomes undergo non-disjunction. Although the most frequent outcome of this interaction between knobs and B chromosomes was the loss of segments of the chromosome arm with the knob, loss of entire knobbed A chromosomes also occurred, and five monosomic-3 plants were recovered in these studies (Rhoades, Dempsey, and Ghidoni 1967; Rhoades and Dempsey 1972). In addition, loss of the entire knobbed arm occurred to produce a telodisomic plant for 3L (discussed in section 3.1). The authors postulated that heterochromatic knobs of A chromosomes replicate incompletely in the presence of B chromosomes, and the partially replicated knobs then fail to separate at the second microspore division resulting in loss of part or all of the chromosome arm bearing the knob, or even the entire chromosome. They also postulated that heterochromatin of B chromosomes was also affected in the same way, and this is the reason that B chromosomes undergo nondisjunction at this same division.

2.1.5   From the *rl-Xl* Deficiency in Maize

The *rl-Xl* deficiency includes the *Rl* locus on chromosome 10 of maize. It was induced with X-irradiation by L. J. Stadler (unpublished). The dominant allele of the *Rl* locus is necessary for anthocyanin production in the endosperm and other tissues. Kante Satyanarayana (unpublished) at the University of Wisconsin noted that female parents carrying the *rl-Xl* deficiency produced many abnormal progeny, including monosomes.

When *Rl/rl-Xl* female parents are crossed by *rl/rl* male parents, about 55-60% of the kernels produced are colored (with *Rl/Rl/rl* endosperm and *Rl/rl* embryos) and the remaining 40-45% are colorless (with *rl/rl/rl-Xl* endosperm and *rl/rl-Xl* embryos). Thus, the *rl-Xl* deficiency is transmitted with a high efficiency through the female; however, it is not transmitted through the male even though all of the pollen produced by *Rl/rl-Xl* plants is morphologically normal.

Plants germinated from the colored kernels are diploids while those from colorless kernels include 10-18% monosomes and 10-18% trisomes (Weber 1983). Thus, a high rate of non-disjunction is taking place in gametes that contain the *rl-Xl* deficiency but not in gametes that lack the deficiency. Cleary, a factor that is necessary for normal chromosome disjunction is deleted in the *rl-Xl* deficiency. Most of the remaining plants from the colorless kernels are diploid; however, low frequencies of multiply aneuploid plants and plants with terminal deficiencies are also generated (Weber 1969, 1983; Lin 1987).

Nearly all of the monosomes produced by the *rl-Xl* system are monosomic for the entire plant, and monosomes for each of the 10 maize chromosomes have been recovered using this system (Weber 1983; Weber and Schneerman 1989). This is the only series of its type that has been recovered from any higher diploid organism. Each of the monosomic types is distinctively smaller than their diploid siblings; however, they are remarkably vigorous. Most monosomic types are 1-2 m tall at maturity in a genetic background where their diploid siblings are about 2.7 m tall. All monosomic types reach sexual maturity, excellent meiotic samples can be collected from them, and most monosomic types can be crossed. Thus, genetic and cytological analyses are possible with these plants. In addition, doubly and triply monosomic maize plants have been recovered from the *rl-Xl* system, and these were sufficiently vigorous to provide excellent meiotic samples for cytological analysis (Weber 1973).

2.1.5.1   Some Characteristics of the *rl-Xl* System

Some characteristics of the *rl-Xl* deficiency are: it is sub-microscopic (too small to be detected using a light microscope), it is transmitted with a high efficiency through female gametes but not through male gametes, it causes non-disjunction post-meiotically at the second of the three embryo sac divisions (Lin and Coe 1986; Simcox, Shadley, and Weber 1987) and the first male gametophyte division (Zhao and Weber 1988), different monosomic types are produced in very different frequencies, the genetic background influences the activity, and deficiencies are also produced (Weber 1983; Lin 1987).

2.1.5.2   Selecting for Plants that are Monosomic for Specific Chromosomes

To select for monosomes for specific chromosomes, a male parent with a plant-expressed recessive mutation can be crossed onto an *R1/r1-X1* female parent that is homozygous for the dominant allele. F_1 progeny that express the recessive phenotype of this mutation are usually monosomic for the chromosome that carries this mutation; however, a few plants with a deficiency in the chromosome arm that carries the recessive mutation are also recovered. Plants with deficiencies are often taller and more vigorous than the true monosomes for that chromosome. Cytological analysis or analysis with molecular markers in both arms of the presumptive monosomes can be used to distinguish between true monosomes and deficiency-bearing plants.

Monosomes for each of the 10 maize chromosomes have been recovered from the single cross given in Table 1. The genetic markers in the male and female parents and the frequencies that the different monosomes are generated are also given in this table. Mangelsdorf's tester (Mangelsdorf 1948), which has a recessive mutation on each of the 10 chromosomes and is also *r1/r1*, can be crossed as a male parent to *R1/r1-X1* female parents (in the inbred W22). The colorless *r1-X1* deficiency-bearing kernels are selected and planted directly into a field nursery. Five of the mutations in Mangelsdorf's tester (*bm2, lg1, gl1, j1,* and *gl1*) are

Table 1 Genetic markers and frequencies of maize monosomics produced by the *R1/r1-X1* x Mangelsdorf's tester cross

Chromosome	Female parent	Male parent	Marker gene name	Frequency recovered
	(R1/r1-X1)	(r1/r1)		
1	Bm2	bm2	brown midrib- 2	0.03%[a]
2	Lg1	lg1	liguleless	1.24%[a]
3	A1	a1	anthocyaninless	0.21%[b]
4	Su1	su1	sugary endosperm	0.29%[a]
5	Pr1	pr1	red aleurone	0.06%[b]
6	Y1	y1	yellow endosperm	1.89%[a]
7	Gl1	gl1	glossy seedling	1.01%[a]
8	J1	j1	japonica striping	3.46%[a]
9	Wx1	wx1	waxy endosperm	0.49%[a]
10	G1	g1	golden stalk	1.57%[a]

[a] Frequencies for these types are from Weber 1983
[b] Frequencies for these types are from a 1990 planting (Weber, unpub. res.)

expressed in the plant, and plants expressing the recessive phenotypes of these genes can be selected directly as presumptive monosomes for chromosomes 1, 2, 7, 8, and 10 respectively. The other five mutations in Mangelsdorf's tester are expressed in the endosperm of the kernels. Kernels with embryos monosomic for chromosomes with these mutations do not express these mutant phenotypes because loss of a chromosome from the embryo is not accompanied by the loss of the same chromosome from the endosperm. We identify monosomes for these chromosomes in the following way. Semi-sterile (50% or greater pollen abortion) plants of sub-normal stature are identified as possible monosomic plants. These are testcrossed with a line that is *a1, su1, pr1, y1,* and *wx1* (on chromosomes 3, 4, 5, 6, and 9, respectively) and *R1/R1*. All kernels produced by a monosomic-3, -4, -5, -6, or -9 plant express the recessive phenotype of one of the marker mutations in all of the kernels produced corresponding to the marker gene on the monosomic chromosome. Diploids and all other monosomic types will give a 1:1 ratio for that gene. For example, all of the kernels produced by a monosomic-6 plant will have white (*y1/y1/y1*) endosperm whereas all other monosomic types and diploids will produce a 1:1 ratio of white (*y1/y1/y1*) to yellow (*Y1/Y1/y1*) kernels. In addition, monosomes for chromosomes 3, 4, 5, 6, and 9 each have a distinctive plant phenotype that can be recognized (see Weber 1991, 1994).

2.2   Uses of Monosomes

Maize monosomes are powerful experimental tools that have many uses. Examples of how monosomes have been used are given below:

2.2.1   Analysis of Univalent Chromosome Behavior in Monosomes

The behavior of univalent chromosomes is poorly understood. Monosomes in diploid species are ideal for analyzing the behavior of univalent chromosomes because every meiotic cell in a monosome contains a univalent chromosome. Studies of the behavior of univalent chromosomes have been summarized previously (Weber 1983). In addition, maize plants monosomic for two different chromosomes have been recovered, and tests for the possible interaction between non-homologous univalent chromosomes during meiosis have been carried out in these plants (Weber 1973) to test the "distributive pairing" hypothesis proposed by Grell (1962).

2.2.2   Mapping Unplaced Genes to Specific Chromosomes

Monosomes can be used in several different ways to assign unplaced genes to specific chromosomes.

2.2.2.1   Morphological loci

Maize monosomes produced by the *r1-X1* system can be used to identify the chromosomes that carry plant-expressed genes. A pollen parent with a recessive allele of an unplaced gene can be crossed to *R1/r1-X1* female parents that are homozygous for the dominant allele of the gene. Almost all of the F_1 progeny that express the recessive phenotype (pseudodominants) are monosomic for the chromosome that carries this gene. The monosomic chromosome can then be determined by cytological or RFLP analysis, and in some cases, the distinctive plant phenotype of certain monosomic types can reveal the identity of the monosomic chromosome. Simcox and Weber (1985) used this approach to assign the *benzyxanzinless (bx1)* locus to chromosome 4 in maize.

2.2.2.2   Loci With Different Electrophoretic Mobilies

The *r1-X1* system can be used to assign genetic loci with alleles that produce gene products that have different electrophoretic mobilities to specific chromosomes. Maize plants with the *r1-X1* deficiency, an allele of an unplaced gene whose gene product has one electrophoretic mobility, and dominant morphological tester allele(s) can be crossed as a female parent with a male parent that has an allele for a different electrophoretic mobility and a recessive morphological tester allele(s). Monosomes can be identified using the morphological marker(s). The monosomic type that only displays the electrophoretic allele from the male parent is monosomic for the chromosome that carries the unplaced locus. Stout and Phillips (1973) and Weber and Brewbaker (1983) have used this strategy to assign the *H1a* and *px3* genes to chromosomes 1 and 7, respectively. This procedure does not require the cytological identification of the monosomic chromosome.

2.2.2.3   RFLP Loci

Maize monosomes generated with the *r1-X1* system have been especially useful for assigning RFLP loci to specific maize chromosomes. Helentjaris, Weber, and Wright (1986) examined DNAs of *R1/r1-X1* W22 plants and Mangelsdorf's tester using different RFLP probes and restriction enzymes to identify probe-restriction enzyme combinations that revealed a polymorphism between the two parents. Monosomes for the various maize chromosomes were selected from a cross between these two lines, and the monosomic type that only displayed the allele from the male parent is monosomic for the chromosome that carries this RFLP locus. This work identified linkage groups for the maize chromosomes on the maize RFLP map. Duplicated RFLP loci were also mapped using this same approach (Helentjaris, Weber, and Wright (1988). Monosomes generated with the *r1-X1* system were also used to map histone *H3* and *H4* loci in maize (Chaubet, Philipps, Gigot, Guitton, Bouvet, Freyssinet, Schneerman, and Weber 1992). *H3* and *H4* genes are highly duplicated and were found to be located on most, possibly all, of the maize chromosomes.

2.2.3 Altering the Number of Copies of Known Genes

Monosomes can be used to alter the number of copies of previously identified genetic loci.

2.2.3.1 One vs. Two Copies

Monosomes can be compared with diploids to determine the effects of one vs. two copies of previously mapped genetic loci. Philips, Weber, Kleese, and Wang (1974) used this approach to examine the effect of one vs. two copies of the 18S and 28S rRNA genes (which are located at the nuclear organizing region) to determine if gene magnification (Tartof 1971) occurs in maize.

2.2.3.2 Zero vs. One Copies

Monosomic plants produce haploid (n) microspores and microspores that are nullisomic (n-1) for the monosomic chromosome. By comparing these two types of microspores, one compares zero vs. one copy of all genes on a specific chromosome. This approach was used to determine that the 5S rRNA templates are not necessary for nucleolar formation, and to screen other chromosomes in the maize genome for factors that are necessary for nucleolar formation in maize (Weber 1978a).

2.2.4 Exploring the Genome for Gene Dosage Effects

By comparing plants monosomic for a specific chromosome with their diploid siblings, one compares the effects of one vs. two copies of all genes on an entire chromosome. If a gene is present on the monosomic chromosome that expresses dosage effects, the monosomes and diploid will differ for the trait specified by this gene. In this way, the effects of all genetic loci on a specific chromosome are compared without the use of gene mutations.

This approach has been used to screen for genes that affect kernel lipid quantity (Plewa and Weber 1973) and quality (Plewa and Weber 1975), acid-extractable amino acids (the free amino acid pool) (Cook and Weber 1976), intergenic recombination (Weber 1971) and intragenic recombination (Weber 1978b).

3 Telodisomes

A telodisome (= monotelodisome) is a cell, tissue, or individual in which one of the chromosomes in the otherwise diploid genome is a telocentric chromosome. Because each of the 10 chromosomes in the haploid genome of maize can produce two telocentric chromosomes, twenty different telodisomes are possible.

3.1 Sources of Telodisomes

The interaction between heterochromatic knobs of maize A (i.e. standard) chromosomes and B chromosomes discovered by Rhoades et al. (1967, 1971) was discussed in section 2.1.4 of this chapter. A telodisome for the long arm of chromosome 3 (3L) was also recovered from this system (Rhoades et al. 1967, 1972).

Rose and Staub (1990) discovered a centromere cleavage event that produced stable telocentrics for both arms of chromosome 3 (3S and 3L). This occurred in a line of Black Mexican sweet corn containing high numbers of B chromosomes. They also observed (Rose and Staub, personal communication) that the telocentrics underwent non-disjunction at the second pollen grain mitosis when B chromosomes were present to produce pollen grains that lacked either the 3S or 3L telocentric chromosome in a sperm nucleus. This is the same division that the centromeres of maize B chromosomes normally undergo non-disjunction. When such a gamete (containing only 1 of the 2 telocentric chromosomes) is fertilized by normal haploid female gametes, a plant containing one normal homolog plus a telocentric for only one of the two chromosome arms (a telodisome) is formed.

Rose and Staub (unpublished) established a line that was homozygous for the telocentrics for both chromosome arms of chromosome 3 (a double ditelocentric for 3S and 3L). This line was crossed as a male parent by $R1/r1$-$X1$ female parents. The F_1s, (that had a normal chromosome 3, a 3L, and a 3S) were crossed as female parents by $lg2$ $et1$ (on 3L) or $g2$ (on 3S) male parents to see if the $r1$-$X1$ system could cause non-disjunction of telocentric chromosomes. Plants that expressed the $lg2$ and $et1$ or $g2$ phenotypes were recovered, and cytological analysis indicated that they were telodisomes for 3S and 3L, respectively (Weber and Schneerman, unpublished).

The double ditelocentrics for 3S and 3L were also crossed by plants that contained ~10 B chromosomes. The F_1s (which contained ~5 B chromosomes) were crossed as male parents by plants with seedling-expressed mutants on 3L or 3S (listed above). Approximately 1% of the seedlings expressed the marker mutations, and several telodisomes for 3S and 3L were recovered from these plants and analyzed.

3.2 Uses of Telodisomes

Telodisomes can be used to alter the gene dosage in individual chromosome arms. Telodisomics can also be used to determine the arm location of a gene that was previously assigned to a specific chromosome. Chi, Fowler, and Freeling (1994) crossed a plant that carried telocentrics for 3S and 3L as a male parent to female parents that carried a mutation on the short arm ($v19$) or on the long arm ($y3$). F_1 plants were selected that were telodisomic for 3L or 3S. DNAs of the two telodisomic types were probed with the $lg3$ probe, and it was determined that this locus was in 3L. Telodisomics were also used to map the centromeric position on chromosome 3 in maize to a region of 4.2 map units (Muzumdar, Schneerman, and Weber, unpublished; L'Hereux, Muzumdar, Schneerman, and Weber 1993).

4 Trisomes

A trisome is a 2n+1 aneuploid cell, tissue, or individual with an extra chromosome in an otherwise diploid chromosome complement.

4.1 Types of Trisomy

Four classes of trisomes, primary trisomes, secondary trisomes, telosomic trisomes, and tertiary trisomes, occur in diploids, including maize. In diploids, when any of the four classes of trisomes undergoes meiosis, haploid (n) and disomic (n+1) gametes are produced. Both the haploid and disomic gametes produce functional male and female gametophytes. Thus, when any of the four classes of trisomes is crossed (especially as a female parent) with a diploid pollen parent, some of the progeny are trisomic. Thus, the trisomic condition is efficiently transmitted.

In addition, compensating trisomes can occur in allopolyploids where a chromosome is missing from one genome and an extra copy of a homoeologous chromosome from a different genome is present.

4.1.1 Primary Trisomes

Primary trisomes (2n+1) have one complete chromosome in triplicate and the other nine chromosomes in duplicate.

4.1.1.1 Sources of Primary Trisomes

Primary trisomes for each of the 10 maize chromosomes have been isolated and are available from the Maize Genetics Stock Center at the University of Illinois. When a primary trisomic female parent is crossed by a diploid male parent, trisomes are recovered among the progeny; therefore, trisomes can be obtained and efficiently generated for use in experiments. For many trisomic types, kernels from this cross with trisomic embryos are detectably smaller than their diploid sibling kernels (Einset 1943, Fox and Weber 1977). Therefore, one can enrich for trisomes by selecting the smaller kernels from this cross. Also, all of the primary trisomes are smaller and less vigorous than their diploid siblings, and some of them can be recognized by their distinctive phenotypes, as described in Rhoades (1955).

Several different approaches have been used to generate new trisomes. McClintock (1929b) first recovered primary trisomes in maize from n+1 gametes produced by triploids. Beckett (1984) germinated distinctively smaller kernels from the inbred W23. He found that 21% were aneuploid and, of these, 46% were trisomes and 52% had one or more telocentric chromosome or isochromosome. Ghidoni, Pogna, and Villa (1982) recovered trisomes from smaller and deformed kernels from diploid populations. Trisomes are also generated from the *r1-X1* system as previously discussed. Colorless kernels from *R1/r1-X1* silk parents by a Mangelsdorf's tester pollen parent (given in Table 1) could be analyzed cytologically to select trisomes. The trisomes could then be backcrossed with Mangelsdorf's tester to detect trisomes for chromosomes 1, 2, 4, 6, 7, 8, 9, and 10 by their genetic ratios, and also crossed with a line that is *a1, pr1,* and R to select for trisomes for chromosomes 3 and 5 by their genetic ratios. Trisomes for chromosome 8 have been recovered using this experimental approach (Weber and Schneerman 2000).

4.1.1.2 Meiosis and Transmission in Primary Trisomes

In primary trisomes, nearly all of the pachytene cells have a trivalent. Only two of the three homologs of a trivalent are paired at any one point along the length of the chromosome; however, exchanges of pairing partners occur along the length of the trivalent so that all three homologs are involved in synapsis at different points along the chromosome. At diakinesis and metaphase I, trivalents are present in about two-thirds of the cells, and the remaining cells have ten bivalents plus a univalent (Einset 1943). Einset reported that the frequency of trivalents in metaphase I cells was positively correlated with the length of the chromosome, longer chromosomes in general had a higher frequency of trivalents than shorter chromosomes. He also determined the percentage of trisomes in the progeny of trisomic female by diploid male crosses. In general, the transmission of trisomy was higher for longer chromosomes than for shorter chromosomes. Einset (1943) argued that because the number of chiasmata per chromosome is related to the length of the chromosome, longer chromosomes would have a higher probability of being held together by chiasmata and therefore were transmitted more frequently than shorter chromosomes.

If a trisomic plant with the constitution *A1,A1,a1* is testcrossed as a female, a ratio of 5 *A1* to 1 *a1* is expected if half of the ovules are n+1. However, the actual ratios from testcrosses of this type are somewhat less that this predicted ratio because less than half of the ovules are n+1. This is because about a third of the metaphase I cells contain 10 bivalents and 1 univalent, and the univalent chromosome produced is frequently eliminated during meiosis (Einset 1943). Also, if the marker mutation is very far from the centromere, the ratio becomes more random. When the reciprocal cross is made (*a1,a1* female X *A1,A1,a1* male), a ratio of 2 *A1* to 1 *a1* is obtained. The reason for this is that the n pollen grains are almost invariably successful in normal pollinations because they grow more rapidly than n+1 pollen grains (which have an entire chromosome duplicated). However, trisomic progeny can be recovered from this cross if a very sparse amount of pollen is used in the pollination so there is less competition between the n and n+1 pollen.

4.1.1.3 Uses of Trisomes

Primary trisomes are powerful cytogenetic tools. McClintock and Hill (1931) used them to assign genes to six different chromosomes in maize. Rhoades (1935) also used trisomes (and terminal deficiencies) to assign a gene for disease resistance in maize (*Rp1*) to the short arm of chromosome 10. They have also been used in gene dosage studies.

Trisomes have also been compared with diploids to screen entire chromosomes for factors that have dosage effects (for example, Ghidoni 1975; Ward 1978, 1979; Shadley and Weber, 1980 1986). Doubly trisomic plants and singly trisomic plants with a B chromosome have also been examined to test for possible interactions between non-homologous univalent chromosomes during meiosis (Michel and Burnham 1969; Weber 1969) to test the distributive pairing hypothesis by Grell (1962).

4.1.2 Secondary Trisomes and Telosomic Trisomes

Secondary and telosomic trisomes will be considered together because both behave in a similar manner in meiosis and in crosses, and both have similar uses.

In a secondary trisome, an isochromosome (a chromosome that has two identical chromosome arms) is present in addition to the diploid genome. If the gene order in a normal chromosome is a-b-c-d-centromere-e-f-g-h, the gene orders in the two isochromosomes that can be derived from this chromosome are a-b-c-d-centro-mere-d-c-b-a and h-g-f-e-centromere-e-f-g-h. A total of 20 isochromosomes is possible because two different isochromosomes can be generated from each of the 10 chromosomes.

Secondary trisomes have been recovered for several of the 20 chromosome arms in maize including 4S (Schneerman, Weber, Lee, and Doyle 1998), 5S (Rhoades 1933, 1940), 6S and 6L (Maguire 1962), 7L (Muzumdar, Schneerman, Doyle, and Weber (1997), 8S (Yu, Han, Kato, and Birchler 2006), 10L (Emmerling 1958), and unidentified arms (Beckett 1984), where "S" designates the short arm and "L" designates the long arm. In addition, pseudo-isochromosomes (chromosomes with two nearly identical chromosome arms) for several chromosomes were generated by Morris (1955).

In a telosomic trisome, a telocentric chromosome with one arm of an A chromosome is present in addition to the diploid genome. For a chromosome with the gene order of a-b-c-d-centromere-e-f-g-h, the gene order of the two possible telo-centrics would be a-b-c-d-centromere and h-g-f-e-centromere. A total of 20 different telocentric types is possible. Telosomic trisomes have been recovered for 5L (Rhoades 1936, 1940), 6L (Doyle 1974) and unidentified chromosome arms (Beckett 1984).

4.1.2.1 Sources of Secondary Trisomes and Telosomic Trisomes

Rhoades (1933) recovered a secondary trisome for 5L among the selfed progeny of a plant with normal chromosome fives. A plant with an abnormal phenotype was selected, and was found to contain an isochromosome for 5L in addition to the diploid maize genome. He also recovered secondary trisomes for 5L from the progeny of plants telodisomic for 5L (Rhoades 1940). He found that when telodisomes for 5L were crossed as male and female parents with diploids, a respective 0.46% and 0.22% of the progeny were secondary trisomes for 5L (Rhoades 1940). The secondary trisomes and telodisomes could be distinguished from each other because they had strikingly different phenotypes. He suggested that the formation of isochromosomes from telocentric chromosomes occurs by "misdivision" of the centromere. In trisomes, about a third of the diakinesis and metaphase I cells have 10 bivalents and 1 univalent, and the univalent will lack a pairing partner needed for proper disjunction. Normally, the centromere divides longitudinally, but when it is under stress as a lagging univalent at anaphase I, it sometimes divides transversely and the sister chromatids can join at the centromere to produce an isochromosome(s). Alternatively, if the sister chromatids do not join, a telocentric chromosome(s) can be generated. Also, the telocentric univalent in a telocentric trisome can behave in the same way to generate an isochromosome.

The formation of isochromosomes can also be induced by irradiation (Maguire 1962). In addition, chromosomes in which most of the chromosome arms are identical (pseudo-isochromosomes) have been produced by irradiating interchanges involving opposite chromosome arms (Morris 1955).

Deviant genetic ratios from genetically marked trisomes can be used to isolate secondary trisomes and telotrisomes. Doyle (1988) analyzed the progeny of maize primary trisomes and identified "presumptive telocentrics" for several different chromosome arms on the basis of deviant genetic ratios. From these presumptive telocentrics, isochromosomes for 4S (Schneerman et al. 1998) and 7L (Muzumdar *et al.* 1997) and telocentrics for 6L (Doyle 1974) were verified cytologically and genetically. Rhoades (1936, 1940) also isolated a telosomic for 5L from trisomic-5 plants by selecting a plant with an intermediate phenotype between diploids and trisomic-5 plants.

A promising method for producing maize telocentrics and isochromosomes was described by Beckett (1984). He selected smaller kernels from the inbred line, W23, and found that they contained a high frequency of telocentrics, isochromosomes, and also trisomes. Thus, the selection of small kernels may allow recovery of additional telocentrics and isochromosomes.

4.1.2.2   Uses of Secondary Trisomes and Telosomic Trisomes

Both secondary trisomes and telocentric trisomes can be used in a similar way in many experimental approaches. Plants with one allele of a locus on the normal chromosomes and a different allele on the isochromosome or telocentric chromosome can be used to identify the chromosome arm that carries the marker locus. Schneerman et al. (1998) and Weber and Schneerman (unpublished) used this approach to assign the centromeres of chromosomes 4 and 6 to small intervals on the respective chromosomes on the maize RFLP map.

Both secondary trisomes and telocentrics can also be used to alter the dosage of entire chromosomal arms. Both have also been employed to study the behavior of centromeres (Rhoades 1940; Maguire 1962).

4.1.3   Tertiary Trisomes

The extra chromosome in a tertiary trisome is derived from a reciprocal translocation, and is composed of parts of two nonhomologous chromosomes. Translocation heterozygotes can give rise to tertiary trisomes by a 3:1 disjunction of the ring of 4 in meiosis to produce a gamete that contains the 10 normal chromosomes plus a translocated chromosome. When such a gamete is fertilized by a normal haploid gamete a tertiary trisome is produced. Large numbers of different tertiary trisomes are possible. Little cytological work has been carried out with them; however, they have been used to determine the locations of genes on chromosomes (Dempsey and Smirnov 1964).

5   Deficiencies

A deficiency (= deletion) is the loss of a segment of a chromosome. Large deficiencies typically are not transmissible because genetic loci necessary for gametophyte development are present in the deleted segment, such that male and female gametophytes that carry the deletion will abort. Transmissible deficiencies that have been recovered are small, and typically are transmitted more efficiently through female than male gametes. Many transmissible deficiencies can only be transmitted through female gametes (for example, the *r1-X1* deficiency that was discussed previously). Transmissible deficiencies are often too small to be detectable at pachytene (are submicroscopic).

Large deletions are tolerated by the plant. In fact, the loss of an entire chromosome (or even two or three chromosomes) from a diploid maize plant is compatible with the survival of the plant (discussed in section 2.1.5).

5.1   Sources and Uses of Deficiencies

Deficiencies can occur spontaneously, be produced by irradiation, or be produced by various genetic systems. Some of the mechanisms that can produce deficiencies in maize and their uses are described below.

5.1.1   Radiation-Induced Deficiencies

X-ray- and UV-induced deficiencies have been useful in determining the chromosomal locations of various genes in maize. Pollen from plants with a dominant allele can be irradiated and crossed to plants homozygous for the recessive allele. Pachytene analysis of F_1 individuals that express the recessive phenotype identifies the chromosomal segment that contains the locus. Different deficient segments will be present in different F_1 plants with the recessive phenotype; however, all should have a common segment that is deleted, and this segment contains the locus. The cytological locations of several loci determined using this approach are summarized on page 161 in Rhoades (1955).

Stadler and Roman (1948), Neuffer (1957) and Mottinger (1970) each X-ray treated pollen from plants with a dominant allele and pollinated homozygous recessive silk parents to determine if point mutations could be recovered. They each obtained "mutants" that were analyzed using several criteria to determine if each "mutant" was a point mutation or due to a deficiency. They each concluded that all of the mutants recovered were due to deficiencies. These studies describe the types of tests that can be carried out to distinguish between gene mutations and mutant phenotypes produced by a deficiency.

5.1.2   Deficiencies Produced by Genetic Mechanisms

Deficiencies (and duplications) of known chromosomal regions can be generated by several different genetic mechanisms.

First, when a translocation heterozygote undergoes adjacent-1 segregation, deficient (as well as duplicate) gametes are generated. For most translocations, the unbalanced gametes will not survive (abort). However, certain translocations in maize with a breakpoint close to the end of one of the chromosome arms will produce viable duplicate-deficient eggs. Several translocations with this type of behavior are listed in Phillips, Burnham, and Patterson (1971). Also, Patterson (1978) described a system using recessive nuclear male-sterile genes and duplicate-deficient chromosomes generated by adjacent-1 segregation that would be capable of generating large populations of homozygous recessive male-sterile plants that could then be used in a production field for hybrid maize production.

Second, plants can be generated that contain two different translocations with breakpoints close to each other in the same chromosome arms. Meiotic segregation can produce gametes that have a deficiency and a duplication, and if the deleted segment does not contain a factor necessary for gametophyte function, it will be viable (Gopinath and Burnham 1956; Burnham 1978).

Third, translocations between B chromosomes and A chromosomes and compound B-A-A translocations can be used to produce plants that have specific chromosome segments deleted or duplicated.

Fourth, inversions with breakpoints near the ends of a chromosome can also be used to produce deletions and duplications. Rhoades and Dempsey (1953) isolated female-transmissible duplicate-deficient chromosomes from a paracentric inversion heterozygote. These were produced by crossing over within the inversion loop in inversion heterozygotes followed by breakage of the resultant bridge at anaphase I. The duplicate-deficient chromatids recovered contained a deficiency of about 5% of 3L. This region is not necessary for the survival of the female gametophyte.

Fifth, deficiencies and duplications are also produced by mobile genetic elements. Some elements, particularly *Mutator*, have been observed to generate relatively small (several hundred basepairs) deletions during transposon excision. These deletions can be detected by PCR, and provide a useful avenue for the isolation of stable mutants from transposon-insertion alleles (Das and Martienssen, 1995). In addition, certain configurations of maize *Ac/Ds* elements can induce much larger (tens to hundreds of kilobasepairs) deletions and duplications (McClintock 1951; Dooner and Weil 2007; Zhang and Peterson 1999, 2004, 2005). These appear to arise as a consequence of so-called alternative transposition; i.e., transposition reactions involving the termini of different elements (for further details, see the transposable elements chapters in this volume).

Sixth, recombination between tandemly duplicated segments in the maize genome can also produce a deficiency of one of the tandemly duplicated members. This has been demonstrated at the *a1* locus on chromosome 3 (Laughnan1950, 1952) and the *r1* locus (Stadler and Neuffer 1953). Duplications have also been recovered from recombination between tandemly duplicated genomic segments (Dooner and Kermicle 1975).

Seventh, the *r1-X1* system produces deficiencies, as was previously mentioned (Weber 1983; Lin 1987).

Eighth, homozygous deficient plants that have a ring chromosome that includes the deficient segment have been recovered (McClintock 1938, 1941). Various segments of the ring are lost during development, uncovering different segments of the deficiency segment in different portions of the plant. This approach uncovers all of the genetic loci within the ring, and different sectors on the plant have different regions of the deletion uncovered.

6 Duplications

Much of the maize genome is composed of duplicated segments. Evidence for interchromosomal duplications was briefly summarized in the introduction section of this chapter (section 1).

Tandem duplications are also recognized as an important component of the maize genome, and these have been demonstrated for various regions of the maize genome including the *R1* locus (Stadler and Neuffer 1953), the *A1* locus (Laughnan 1952) and the *alcohol dehydrogenase* locus (Schwartz and Endo 1966; Birchler and Schwartz 1979). Laughnan's work at the *A1* locus (Laughnan 1949, 1952) also gave rise to the concept that intrachromosomal recombination involving tandem duplications might occur.

6.1   *Sources and Uses of Duplications*

Some genetic mechanisms that can produce duplications are described below.

The first six mechanisms given in section 4.1 can produce duplications of specific chromosomal segments in addition to deletions, and will not be repeated here. In addition, duplications for specific chromosomal segments can be produced by combining two different B-A translocations (Carlson and Curtis, 1986; Carlson and Roseman 1988). Also, all loci in a ring chromosome in an otherwise diploid genome are duplicated.

Duplications have been primarily used to alter gene dosage in defined genomic regions (Guo and Birchler 1994, 1997). Also, the expression of a desirable gene that expresses gene dosage effects can be increased by duplication of the region that contains the gene of interest (Shadley and Weber 1986).

7 Ring Chromosomes

Ring chromosomes are generated when two breaks occur in a chromosome and the broken ends join to form a closed circle. If a true centromere or centromeric activity is associated with the ring, the ring can be transmitted in mitotic and meiotic divisions.

The first study of ring chromosomes in maize was by McClintock (1932). Most ring chromosomes have been recovered after ionizing radiation (McClintock 1932; Schwartz 1953); however, a ring chromosome was also recovered from untreated material (McClintock 1932). Ring chromosomes can also form spontaneously from certain configurations of B-A translocations (Ghidoni 1973; Carlson 1973).

Ring chromosomes are very unstable and are frequently lost. If a sister-strand exchange occurs between the two chromatids in the ring, a double-sized dicentric ring chromosome will be produced. The dicentric ring breaks or is severed by the cell wall during cytokinesis producing rings that have duplicated or deleted segments, or are lost. Therefore, ring chromosomes change size and genetic content in cell divisions. Ring chromosomes can also give rise to interlocking rings when two exchanges occur within the ring, and these are often lost during cell division. Schwartz (1953) presented evidence that ring chromosomes may undergo sister strand exchange during meiosis.

Because ring chromosomes are very unstable, it is difficult to maintain them, and many of the rings that have been obtained have been lost. Also, exchange between a ring and its homolog produces a single anaphase I bridge without a fragment in anaphase I. These can heal producing derivative chromosomes that have terminal deficiencies or duplications. This procedure was used by Emmerling (1959) and Miles (1970) to generate terminal deletions of abnormal chromosome 10.

Ring chromosomes can be used to alter the copy number of regions in the genome. Diploid plants that have a recessive allele on the normal homologs and also have a ring chromosome with the dominant allele of this gene are useful for certain purposes. Such a plant would be a mosaic with three copies of the gene in tissue where the dominant allele is expressed, and have two copies of the gene in tissue where the dominant allele is not expressed. Additional genetic loci on the ring chromosome would be present in different copy numbers in different tissues in the plant. Also, plants with a ring chromosome that are homozygous deficient for the region contained in the ring have been intensively studied by McClintock (1938). Sectors on the plant that are deficient for each of the genetic factors in the ring are generated in different tissues of the plant, and therefore can be used to detect mutations within the deleted segment of the chromosome.

8 Learning from Changes in Chromatin Content

Thus, even though maize is considered to be a true diploid, its genome is much more tolerant to quantitative changes in chromatin than the genomes of many other diploids. This is probably because much of the maize genome is duplicated due to its allopolyploid ancestry. The rich array of chromosome tools that alter the quantity of chromatin in the maize genome, as well as chromosome tools that alter the structure of the genome (e.g., inversions and translocations [including B-A translocations]) have played an important role in numerous previous studies, and will no doubt continue to make important contributions in future maize genetic studies.

References

Ahn S., Anderson J.A., Sorrells M.E., and Tanksley S.D. (1993) Homeologous relationships of rice, wheat, and maize chromosomes. Mol. Gen. Genet. 241, 483–490.
Avery P. (1929) Chromosome number and morphology in *Nicotiana*. IV. The nature and effects of chromosome irregularities in *N. alta* var Grandifloria. Univ. Calif. Pub. Bot 11, 265–284.
Beckett J.B. (1984) An aneuploid generating system in the maize inbred W23. Genetics 107, s9.
Birchler J.A. and Schwartz D. (1979) Mutational study of the alcohol dehydrogenase-1 FCm duplication in maize. Biochem. Genet. 17, 1173–1180.
Blakeslee A.F. (1921) Types of mutations and their possible significance in evolution. Amer. Nat. 55, 254–267.
Burnham C.R. (1978) Cytogenetics of interchanges. In: D.B. Walden, *Maize Breeding and Genetics,* John Wiley and Sons, New York.
Carlson W. (1973) Instability of the B chromosome. Theor. Appl. Genet. 43, 147–150.
Carlson W. and Curtis C.A. (1986) A new method for producing homozygous duplications in maize. Can. J. Genet. Cytol. 28, 1034–1040.
Carlson W. and Roseman R.R. (1988) Maize Genet. Coop. Newsl. 62, 66–67.
Chaubet N., Philipps F., Gigot C., Guitton V., Bouvet N., Freyssinet G., Schneerman M., and Weber D. (1992) Subfamilies of histone *H3* and *H4* genes are located on most, possibly all of the chromosomes in maize. Theor. Appl. Genet. 84, 555–559.
Chi Y, J. Fowler, and Freeling M. (1994) The *lg3* locus maps to the short arm of chromosome 3. Maize Genet. Coop Newsl. 68, 16.
Cook J.W. and Weber D.F. (1976) Monosomic analysis of the acid extractable amino acids (free amino acid pool) in maize leaves. Maize Genet. Coop. Newsl. 50, 40–42.
Das L, and Martienssen R. (1995) Site-selected transposon mutagenesis at the hcf106 locus in maize. Plant Cell 7: 287–294.

Dempsey E. and Smirnov V. (1964) Cytological location of *gl15*. Maize Genet. Coop. Newsl. 39. 71–73.

Dooner H.K. and Weil C.F. (2007) Give-and-take: interactions between DNA transposons and their host plant genomes. Curr. Opin. Genet. Dev. 17 :486–92.

Dooner, H.K. and Kermicle J.L. (1976) Displaced and tandem duplications in the long arm of chromosome 10 in maize. Genetics 82, 309–322.

Doyle G.G. (1974) Telocentric 6L trisomes and their possible use in the commercial production of corn. Maize Genet. Coop. Newsl. 48, 128–131.

Doyle G.G. (1988) The production of the telocentrics of maize. Maize Genet. Coop. Newsl. 62, 49–50.

Einset J. (1943) Chromosome length in relation to transmission frequencies in maize trisomes. Genetics 28, 349-364.

Emmerling M. (1958) Evidence of non-disjunction of abnormal 10. J. Hered. 49, 203–207.

Emmerling M.H. (1959) Preferential fertilization of structurally modified chromosomes in maize. Genetics 44, 625–645.

Fisher H.E. and Einset J. (1940) Monosomic maize. Maize Genet. Coop. News Lett. 14, 13–14.

Ford L.E. (1952) Some cytogenetic aspects of maize monoploids and monoploid derivatives. Doctoral thesis, Iowa State College, Ames, IA.

Fox W.D. and Weber D.F. (1977) The effects of trisomy on kernel weight. Maize Genet. Coop. Newsl. 51, 6–7.

Gaut B.S. (2001) Patterns of chromosomal duplication in maize and their implications for comparative maps of the grasses. Genome research 11, 55–66.

Ghidoni A. (1975) Effect of trisomy 6 on recombination in chromosome 9 of maize. Genetics 81, 253–262.

Ghidoni A., Pogna N.E., and Villa N. (1982) Spontaneous Aneuploids of Maize (*Zea mays*) in a selected sample. Can J. Genet. Cytol. 24, 705–713.

Gopinath D.M. and Burnham C.R. (1956) A cytogenetic study in maize of deficiency-duplication produced by crossing interchanges involving the same chromosomes. Genetics 41, 382–395.

Griesinger R. (1937) Uber hypo- und hyperdiploide Formen von *Petunia, Hoscyamus, Lamium,* und einige andere Chromosomenzahlungen. Ber. Deut. Bot. Ges. 55, 556–571.

Grell R.F. (1962) A new hypothesis on the nature and sequences of meiotic events in the female of *Drosophila melanogaster*. Proc. Nat. Acad. Sci. (U.S.A.) 48, 165–172.

Guo M. and Birchler J.A. (1994) Trans-acting dosage effects on the expression of model gene systems in maize aneuploids. Science 266, 1999–2002.

Guo M. and Birchler J.A. (1997) Dosage regulation of *Zea mays* homeobox (*ZmHox*) genes and their relationship with the dosage factors of *shrunken 1 (Sh1)* in maize. Dev. Geet. 40, 67–73.

Helentjaris T., Weber D. F., and Wright S. (1986) Use of monosomics to map cloned DNA fragments in maize. Proc. Nat. Acad. Sci. USA 83, 6035–6039.

Helentjaris T., Weber D. F., and Wright S. (1988) Identification of the genomic locations of duplicate nucleotide sequences in maize by analysis of restriction length polymorphisms. Genetics 118, 353–363.

Laughnan J.R. (1950) The action of allelic forms of the gene *A* in maize II. The relation of crossing over to mutation of *Ab*. Proc. Nat. Acad. Sci. (U.S.A.) 35, 167–178.

Laughnan J.R. (1950) The action of allelic forms of the gene *a* in maize. III. Studies on the occurrence of isoquerotin in brown and purple plants and its lack of identity with the brown pigments. Proc. Natl. Acad. Sci. (U.S.A.) 36, 312–318.

Laughnan J.R. (1952) The action of allelic forms of the gene *A* in maize. IV. On the compound nature of *A1-b* and the occurrence and action of it's *A1-b* derivatives. Genetics 37, 375–3905.

L'Heureux T.A., Muzumdar D.A., Schneerman M.C., and Weber D.F. (1997) Haploids and monosomics are produced by maiz plants containing telocentrics for both arms of chromosome 3 and B chromosomes. Maize Genet. Coop Newsl. 71, 66–67.

Lin B-Y. and Coe E. (1986) Monosomy and trisomy induced by the *r-X1* deletion in maize, and associated effects on endosperm development. Can. J. Genet. Cytol. 28, 831–834.

Lin B-Y. (1987) Cytological evidence of terminal deficiencies produced by the *r-X1* deficiency in maize. Genome 29, 718–721.

Maguire M.P. (1962) Pachytene and diakinesis behavior of the isochromosomes 6 in maize. Science 158, 445–446.

Mangelsdorf P.C. (1948) Multiple gene linkage testers. Maize Genet. Coop. Newsl. 22, 22.

McClintock B. (1929a) A 2n-1 chromosomal chimera in maize. J. Hered.20, 218.

McClintock B. (1929b) A cytological and genetical study of triploid maize. Genetics 14, 180–222.

McClintock B. (1938) The production of homozygous deficient with mutant characteristics by means of aberrant mitotic behavior of ring-shaped chromosomes. Genetics 23, 315–376.

McClintock B. (1941) The association of mutants with homozygous deletions in Zea mays. Genetics 26, 542–571.

McClintock B. (1951) Chromosome organization and genic expression. Cold Spring Harbor Symp. on Quant. Biol. 16, 13–47.

McClintock B and Hill H.E. (1931) The cytological identification of the chromosome associated with the R-G Linkage Group in Zea mays. Genetics 16, 175–190.

Michel K. and Burnham C.R. (1969) The behavior of nonhomologous univalents in double trisomics in maize. Genetics 63, 851–864.

Miles J.H. (1970) Influence of modified K10 chromosome on preferential segregation and crossing over in Zea mays. Doctoral thesis, Indian University, Bloomington, IN.

Moore G., Devos K.M., Wang Z., and Gale M.D. (2005) Grasses line up and form a circle. Curr. Biol 5, 737–739.

Morgan, D.T. Jr. (1956) Asynapsis and plasmodial microsporocytes in maize following X-irradiation of pollen. J. Heredity 47, 269–274.

Morris R. (1955) Induced reciprocal translocations involving homologous chromosomes in maize. Am. J. Bot. 42, 546–550.

Mottinger J.P. (1970) The effect of X-rays on the bronze and shrunken loci in maize. Genetics 64, 259–271.

Muzumdar D.A., Schneerman M.C., Doyle G.G., and Weber D.F. (1997) Identification of an isochromosome for the long arm of chromosome 7 in maize. Maize Genet. Coop. Newsl. 71, 67.

Neuffer M.G. (1957) Additional evidence on the effect of X-ray and ultraviolet radiation on mutation in maize. Genetics 42, 273–282.

Odland W., Baumgarten A., Phillips R. (2006). Ancestral rice blocks define multiple related regions in the maize genome. Crop Sci. 46, 41–48.

Patterson E.B. (1978) Properties and uses of duplicate-deficient chromosome complements in maize, In: D.B. Walden, Maize Breeding and Genetics, John Wiley and Sons, New York.

Phillips R.L., Burnham C.R., and Patterson E.B. (1971) Advantages of chromosomal interchanges that generate haplo-viable deficiency-duplications. Crop Sci. 11, 525–528.

Phillips R.L., Weber D.F., Kleese R.A., and Wang S.S. (1974) The nucleolus organizing region of maize (Zea mays L.): Tests for ribosomal gene compensation or magnification. Genetics 77, 285–297.

Plewa M.J. and Weber D.F. (1973) The use of monosomics to detect genes conditioning lipid content in Zea mays L. embryos. Can. J. Genet. Cytol. 15, 313–320.

Plewa M.J. and Weber D.F. (1975) Monosomic analysis of fatty acid composition inembryo lipids in Zea mays L. Genetics 81, 277–286.

Rhoades M.M. (1933) A secondary trisome in maize. Genetics 19, 1031–1038.

Rhoades M.M. (1936) A cytogenetic study of a chromosome fragment in maize. Genetics 21, 491–502.

Rhoades MM (1940) Studies of a Telocentric Chromosome in Maize with Reference to the Stability of its Centromere. Genetics 25, 483–520.

Rhoades M.M. (1951) Duplicate genes in maize. Am Nat 85, 105–110.

Rhoades M.M. (1955) The cytogenetics of maize. In: G.F. Sprague, Corn and Corn Improvement. Academic Press, New York.

Rhoades M.M. and E. Dempsey. (1953) Cytogenetic studies of deficient-duplicate chromosomes derived from inversion heterozygotes in maize. Am J. Bot. 40, 405–424.

Rhoades M.M., Dempsey E., and Ghidoni A. (1967) Chromosome elimination in maize induced by B supernumerary chromosomes. Proc. Nat. Acad. Sci. (U.S.A.) 57, 1626–1632.

Rhoades M.M. and Dempsey E. (1972) On the mechanism of chromatin loss induced by the B chromosome. Genetics 81, 79–88.

Rhoades V.H. (1935) The location of a gene for disease resistance in maize. Proc. Natl. Acad. Sci. (U.S.A.) 21, 243–246.

Rose K.L. and Staub R.W. (1990) Centromere breakage and recovery of both telocentric fragments. Maize Genet. Coop. Newsl. 64, 94–95.

Schneerman M.C., Weber D.F., Lee W., and Doyle G. (1998) RFLP mapping of the centromere of chromosome 4 in maize using isochromosomes for 4S. Theor. Appl. Genet. 96, 361–366.

Schwartz D. (1953) The behavior of an X-ray induced ring chromosome in maize. Am. Nat. 87, 19–28.

Schwartz D. and Endo T. (1966) Alcohol dehydrogenase polymorphism in maize- simple and compound loci. Genetics 53, 709–715.

Shadley J. and Weber D.F. (1980) Identification of a factor in maize that increase embryo fatty acid unsaturation by trisomics and B-A translocational analysis. Can. J. Genet. Cytol. 22, 11–19.

Shadley J. and Weber D.F. (1986) Location of chromosomal regions controlling fatty acid composition of embryo oil in *Zea mays* L. Can. J. Genet. Cytol. 28, 260–265.

Shaver D. (1965) A new maize monosomic. Maize Genet. Coop. Newsl. 39, 24.

Simcox K.D., Shadley J.D., and Weber D.F. (1985) Detection of the time of occurrence of nondisjunction induced by the *r-X1* deficiency in *Zea mays* L. Genome 29, 782–785.

Simcox K.D and Weber D.F. (1985) Location of the *benzyxanzinless (bx1)* locus in maize by monosomic and B-A translocational analyses. Crop Sci. 25, 827–830.

Snope, A.J. (1967) The relationship of abnormal chromosome 10 to B-chromosomes in maize. Chromosoma 21, 243–249.

Stadler L.J. and Neuffer M.G, (1953) Problems of gene structure. II. Separation of the *R1-r* elements (S) and (P) by unequal crossing over. Science 117, 417–472.

Stadler L.J., Roman H. (1948) The effect of X-rays upon mutation of the gene *A* in maize. Genetics 33, 273–303.

Staub R.W. (1990) Centromere fission resulting in short and long arm telosomes of chromosome 3 capable of nondisjunction. Maize Genet. Conf. Abst. 32, T29.

Stout J.T. and Phillips R.L. (1973) Two independently inherited electrophoretic variants of the lysine-rich histones of maize (*Zea mays*). Proc. Natl. Acad. Sci. (U.S.A.) 70. 3043–3047.

Tartof K.D. (1971) Increasing the multiplicity of ribosomal RNA genes in *Drosophila melanogaster*. Science 171, 294–297.

Ting Y.C. (1966) Duplications and meiotic behavior of the chromosomes in haploid maize (*Zea mays* L.). Cytologia 31, 324–329.

Ward E.J. (1978) Influence of trisomy on chiasmata in maize. Maydica 23, 201–208.

Ward E.J. (1979) Chiasma frequency and distribution in a maize family segregating for K10 and trisomy 10. Genetics 92, 223–230.

Weber D.F. (1969) A test of distributive pairing in *Zea mays*. Chromosoma 27, 354–370

Weber D.F. (1971) The use of monosomy to detect genes altering recombination in *Zea mays*. Maize Genet. Coop. Newsl. 45, 32–35.

Weber D.F. (1973) A test of distributive pairing in *Zea mays* utilizing doubly monosomic plants. Theor. Appl. Genet. 43, 167-173.

Weber D.F. (1978a) Nullisomic analysis of nucleolar formation in *Zea mays*. Can. J. Genet. Cytol. 29, 97–100.

Weber D.F. (1978b) Monosomic analysis of intragenic recombination at the waxy locus in maize. Genetics 88, s109–s110.

Weber D.F. (1983) Monosomic analysis in diploid crop plants. In: M.S. Swaminathan, P.K. Gupta, and U. Sinha, (Eds.), *Cytogentics of Crop Plants*, Macmillan, India.

Weber D.F. (1991) Monosomic analsis in maize and other diploid crop plants. In: P.K. Gupta and T. Tsuchiya (Eds.), *Chromosome Engineering in Plants: Genetics, Breeding, and Evolution*. A. Elsevier, Amsterdam.

Weber D.F. (1994) Use of maize monosomics for gene localization and dosage studies. In: M. Freeling and V. Walbot (Eds.), *The Maize Handbook*. Springer-Verlag, New York.

Weber D.F. and Alexander D.E. (1972) Redundant segments in *Zea mays* detected by translocations of monoploid origin. *Chromosoma* **39**, 27–42.

Weber D.F. and Brewbaker J.L. (1983) Location of the *Px* locus in maize to chromosome seven by monosomic analysis. Maize Genet. Coop. Newsl. 57:108.

Weber D.F. and Schneerman M.C. (1989) Identification of monosomic-5 maize plants. Maize Genet. Coop. Newsl. 63, 99.

Weber D. F. and Schneerman M.C. (2000) Use of the *r-X1* deficiency system to recover trisomics for chromosome 8 in maize. Maize 42nd. Genet. Conf. Abst. 38, 37.

Zhang J. and Peterson T. (1999) Genome rearrangements by nonlinear transposons in maize. Genetics 153, 1403–1410.

Zhang J and Peterson T. (2004) Transposition of reversed *Ac* element ends generates chromosome rearrangements in maize. Genetics 167, 1929–1937.

Zhang J and Peterson T. (2005) A segmental deletion series generated by sister-chromatid transposition of *Ac* transposable elements in maize. Genetics 171, 333–344.

Yu W., Han F., Kato A, and Birchler J.A. (2006) Characterization of a maize isochromosome 8S:8S. Genome 49, 200–206.

Zhao Z.-Y. and Weber D.F. (1988) Analysis of nondisjunction induced by the *r-X1* deficiency during microsporogenesis in *Zea mays* L. Genetics 119, 975–980.

Transposon Resources for Forward and Reverse Genetics in Maize

Donald R McCarty and Robert B. Meeley

1 Introduction

The maize geneticists toolkit includes an impressive set of strategies for creating mutations that facilitate identifying genes based on phenotypes (forward genetics) and/or assigning phenotypes to genes identified by sequence (reverse genetics). Key to both forward and reverse genetics strategies are methods for construction and efficient molecular analysis of large, mutagenized maize populations that ideally contain mutations in all genes. Hence, as the technologies for high-throughput phenotype analysis of maize populations advance apace with DNA sequencing and genotyping technologies, the conventional distinction between forward and reverse genetics is likely to blur. Strategies for comprehensive mutagenesis of maize genes include TILLING (Till et al., 2004); RNAi (McGinnis et al., 2007); and transposon insertional mutagenesis, the focus of this chapter. These three approaches have complementary strengths and weaknesses with differences in relative cost per gene, precision, genetic background limitations, scalability, accessibility and relative coverage of the maize genome. While insertional mutagenesis is the most venerable of these technologies, resources based on mutations caused by defined DNA insertions are likely to have an enduring importance in functional genomics for several practical reasons: 1) compared to other types of mutations (e.g. point mutations) insertions are relatively easy to identify and map in the genome using conventional or high-throughput sequencing technologies, 2) large insertions are highly effective in causing significant disruptions of gene function (e.g. null mutations), and 3) the resulting loss-of-function mutations are genetically stable and typically recessive. Recessive, loss of function mutations are an important reference point for functional analysis of a gene.

D.R. McCarty
University of Florida, Gainesville, Florida 32611

R.B. Meeley
Pioneer – A DuPont Company, Johnston, Iowa 50131

J.L. Bennetzen and S. Hake (eds.), *Maize Handbook - Volume II: Genetics and Genomics,*   561
© Springer Science+Business Media LLC 2009

Various DNA elements including random T-DNA insertions introduced by transformation (Alonso et al., 2003), engineered transposons (Muskett et al., 2003; Kolesnik et al., 2004; Raizada et al., 2003), as well as native transposons (Yamazaki et al., 2001) have been employed for large scale insertional mutagenesis of plant genomes. For maize, transposon-based resources are currently favored for several reasons: 1) the relative inefficiency of methods for transformation of maize limits production of large numbers of T-DNA lines; 2) maize is a pre-eminent model for transposon genetics with multiple genetically well-characterized transposon families; and 3) because maize is more easily out-crossed than self-pollinating species such as Arabidopsis and rice, plant populations containing large numbers of independent transpositions are comparatively easy to construct. The so-called "cut and paste" DNA transposons that have been the most favored for genomic resource development in maize include the *Ac/Ds* (Cowperthwaite et al., 2002; Kolkman et al., 2005) and *Robertson's Mutator* (Bensen et al., 1995, May et al., 2003; McCarty et al., 2005) systems. Each of these systems has well-characterized mechanisms enabling genetic control of transposon mobility in the genome. These transposon systems differ in properties that affect their suitability for functional genomics applications including: 1) copy number of active elements in the genome, 2) relative bias for insertion into gene sequences, and 3) propensity for transposition to linked sites.

1.1   High-Copy and Low-Copy Transposon Strategies

In maize, transposon strategies can be classified broadly based on the relative copy number of active elements deployed in the genome. Both low- and high-copy transposons have important applications. *Robertson's Mutator* is well suited to strategies that take advantage of its propensity for achieving and maintaining high-copy numbers in active lines (Alleman and Freeling, 1986; Lisch et al., 1995); whereas, strategies that use *Ac/Ds* and engineered transposon systems (Raizada, 2003) typically focus on analyzing insertions created by one or several identical transposon copies per individual.

The *Ac/Ds* system is most often employed for low-copy strategies that maximally leverage the extraordinary genetics and unique properties of the *Ac/Ds* system. Because *Ac* has a negative dosage effect on transposition rate, copy number can be readily monitored and transposed elements mapped genetically using suitable *Ds* markers (Cowperthwaite et al., 2002; Brutnell and Conrad, 2003; Kolkman et al., 2005). Moveover, the strong bias for *Ac* transposition to linked positions in the genome can be exploited to saturate local regions of the genome or single genes with insertions. A series of lines with strategically placed donor *Ac* elements have been developed that enable targeting of the genome in segments (Kolkman et al., 2005). This focused approach (see Bai et al., 2007) may be particularly effective for mutagenesis of complex loci and genes that occur in tandem duplications. The high germinal reversion frequency of *Ac/Ds* and linked transposition properties combined facilitate generation of new allelic variants from a single insertion allele (Bai et al., 2007). Stabilized (transposition defective) *Ac* elements have been identified that remain capable of activating *Ds* elements in trans (Conrad and Brutnell, 2005). Stable *Ac* in turn can be used to activate germinal reversion of *Ds* alleles facilitating generation and analysis of single transposed *Ds* elements to either linked or unlinked locations. Brutnell and collaborators are using this approach to construct a resource containing 10,000 transposed Ds insertions in hypomethylated regions of the maize genome (visit http://PlantGDB.org/AcDsTagging/). An important feature of *Ac/Ds* resources developed by the Brutnell group is that they are constructed in a homogenous W22 inbred background.

Robertson's Mutator is a particularly efficient mutagen for large scale genomics resources because 1) high *Mu* element copy numbers in active lines and high transposition rate per generation result in relatively high mutation frequencies, 2) *Mu* elements have a strong bias toward insertion into gene sequences in the maize genome (Cresse et al., 1995), 3) *Mu* elements evidently target a very broad spectrum of maize genes indicating that insertion site biases at the sequence level are weak (Cresse et al., 1995; May et al., 2003; Fernandes et al., 2004; Settles et al., 2004; McCarty et al., 2005), 4) *Mu* transposons share a highly conserved 0.2 kbp terminal inverted repeat (TIR) that is very amenable to design of universal PCR primers; and 5) in contrast to *Ac/Ds*, *Mu* elements do not preferentially transpose to linked sites (Lisch et al., 1995), facilitating random coverage of the genome. While both *Ac* and *Mu* show a strong bias for insertion into low-copy regions of the maize genome, *Ac* insertion sites are apparently more likely to be flanked by highly repetitive DNA (Kolkman et al., 2005) compared to *Mu* insertion sites (Cresse et al., 1995; Settles et al., 2004; Fernandes et al., 2004; McCarty et al., 2005) suggesting that *Mu* and *Ac* have different insertion site biases. Taken together, the biological features of *Mu* transposition provide tremendous capability - perhaps greater than any other insertional mutagen - for effective saturation of a large genome within a manageable population size.

1.2   Saturation Transposon Mutagenesis

An ideal functional genomics resource for maize will include sufficient mutations to saturate the genome. In principle, a resource comprised of random insertions that contains an average of 5 loss-of-function mutations per gene will include at least one mutation for greater than 99% of all maize genes. In the case of *Robertson's Mutator*, the total number of insertions required to reach statistical saturation – an average of 5 mutations per gene - is on the order 10^6 insertions. From a forward genetics perspective, the number required for genetic saturation can be derived from the typical single locus forward mutation frequencies ($\sim 10^{-4}$ per locus per individual) observed in active *Mutator* populations (Alleman et al. 1986; Walbot, 2000) and the estimated number transpositions that occur per plant per generation (10-20 insertions; Alleman et al., 1986; McCarty et al., 2005). In other words, the

Table 1 Properties of *Mutator* populations

Resource	Sponsor	Background	Genetic control	Size	Transgenic
TUSC	Pioneer	Heterogenous	No	41,472	No
MTM	Public	Heterogenous	Yes	44,000	No
RescueMu	Public	Heterogenous	No	33,000	Yes
UniformMu	Public	Inbred	Yes	38,000	No

observed forward mutation frequency of 10^{-4} per locus per plant results from a transposition rate of about ~18 transpositions per plant or the creation of 180,000 new germinal insertions in a population of 10,000 plants. On that basis, 5-fold genetic coverage will entail generation of 900,000 independent insertions in a population of 50,000 *Mutator* plants. In fact, several *Mutator* populations of the required scale have been constructed using different genetic backgrounds (Table 1). Hence, the TUSC population alone has been sufficient to obtain mutant alleles for nearly 90% of the genes targeted using the resource. In contrast, while low-copy or single-copy insertion lines have many advantages for genetic, phenotypic and molecular analysis of single gene knockouts, the costs of maintaining and distributing a population of 10^6 single-copy insertion maize lines might be prohibitive.

A downside of using a high-copy transposon such as *Mutator* is that each mutant of interest carries a significant number of background mutations that, if necessary, can be removed by backcrossing to a suitable non-Mutator stock. This downstream burden typically borne by the individual researcher can be mitigated by managing copy number accumulation in the population and constructing the resource in an inbred background (McCarty et al., 2005). Fortunately, recent advances in DNA sequencing technologies have greatly reduced the cost of sequencing *Mu* flanking sequences in large populations of maize, making it feasible to index essentially all insertions in a population. In a fully indexed collection, the user will know the number and locations of background insertions as well as the insertion of interest.

2 Transposon Resources Based on *Robertson's Mutator*

At least four large public or quasi-public transposon resources have been based on *Robertson's Mutator*: TUSC (Meeley and Briggs, 1995), MTM (May et al., 2004), RescueMu (Fernandes et al., 2004), and UniformMu (McCarty et al., 2005). Some salient features of the populations are summarized in Table 1.

2.1 The TUSC Collection

The first reverse-genetics resource developed in maize was named the Trait Utility System for Corn (TUSC), and was created beginning in late 1992 by Pioneer

Hi-Bred Intl. Inc. It took 1.5 years to create a population of 41,472 mutagenized maize plants, and to assemble these into a resource for selected gene disruption in maize, based on serial gene-to-transposon PCR (Meeley and Briggs, 1995; Bensen et al. 1995). Genome-wide mutagenesis was performed using active, high-copynumber Standard *Mu* Lines.

All Standard *Mu* Lines derive from a specific stock first described by Prof. Donald S. Robertson while at Iowa State University. This line tended to a high mutation rate, with mutant phenotypes frequently exhibiting somatic instability; the mutagenic activity appeared dominant, but non-Mendelian (Robertson, 1978; Robertson, 1980). As defined (Lisch et al., 1995), Standard *Mu*-active Lines, contain high copy numbers of *member* elements, often fifty or more, including multiple copies of *MuDR*. These stocks offer practical advantages to handle the size and complexity of the maize genome. By way of simple direct pollinations, it is genetically straightforward to tap the indiscriminate mutagenic potential of high copy numbers of actively transposing *Mu* elements. As discussed above, effective (3-5x) redundant coverage may be expected by fixing 50,000 or fewer mutagenized gametes produced by Standard *Mu* parents. For any selected target sequence, a stepwise PCR regimen can be used against a DNA index created from the population to identify F1 plants that genetically transmit novel *Mu* insertions in the target gene to F2 progeny. This approach capitalizes on the intrinsic property of *Mu* element sequence conservation in their ~220 bp terminal-inverted-repeat (TIR) regions (Bennetzen, 1996). As discussed later, sequence conservation promotes the design of outward-reading PCR primers and robust methods for amplifying and selecting mutations of interest.

While the overall advantages of a high forward mutation rate and anchored PCR amplification have held true for over a dozen years of active screening of the TUSC population, a multi-copy TE family adds extra challenges when attempting to distinguish germinal from somatic insertions and in substantiating a novel link from gene to phenotype. Somatic insertion of *Mutator* elements is rampant in Standard *Mu* lines, and occurs via a cut and paste mechanism that is distinct from the replicative mechanism used during germinal transposition (Rudenko and Walbot, 2001). Such sectors are easy to amplify by PCR, creating potentially high frequencies of false-positive leads for *Mu* insertions in the target gene that can not be expected to transmit to F2 segregants. For TUSC, this limitation was overcome some years ago by extending screens from assays based solely against F1 DNA to include an assay against DNA pools created from F2 progeny. This final step confirms transmission of selected *Mu* alleles and provides reagents for immediate cloning and sequencing of insertion site amplicons. As will be shown later in this chapter, this gain in precision permits a more quantitative evaluation of success rates, and clearly circumvents most early concerns expressed about somatic interference found in TUSC (Walbot, 2000; May et al., 2003).

The TUSC system has been used to rapidly isolate novel mutations for reverse genetics, has assisted many transposon tagging experiments in retrieval of confirmatory mutations, and has even been employed to explore unknown regions of DNA for novel sequence discovery. More importantly, the observations integrated over a dozen years - including over one thousand screening projects - generates a global and highly interpretable view of TE distribution in a complex genome.

2.1.1 Population Construction

The working population for TUSC stands at 41,424 plants. This was constructed in two major phases, called the PV03 (Puerto Vallarta, MX; winter 1993) and BT94 (Johnston, IA; summer, 1994) sub-populations, respectively. In the original crosses, pollen from single Mu-active plants was used to make multiple pollinations to a number of female recipient ears. As an example, the PV03 segment (24,218 F1 plants) was made from approximately 350 inbred x Mu crosses, using some 65 different Standard Mu Lines as donors of mutagenized gametes (Figure 1). A mixture of inbred backgrounds was used as females, a decision based largely on what silks were available at the time. B73 female pedigrees, which amount to approximately 35% of the PV03 segment, are comprised of crosses to 18 Mu donors, averaging a 6% contribution to this portion of the collection. Similar methods were used to construct the BT94 segment, but there is less diversity here since all pollinations were made to a single hybrid female (Pioneer 3394) and relied on a smaller set of selfed Standard Mu Lines with better agronomics. In all cases, pedigree structures were maintained and noted in a relational database for further reference to ensure the independence of selected alleles.

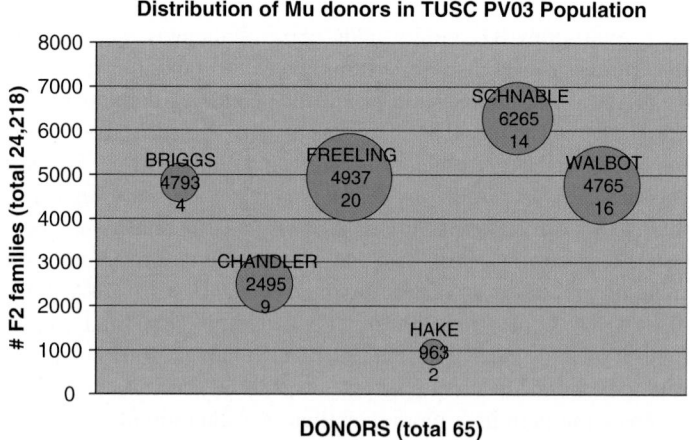

Fig. 1 Sources of the founding Mutator stocks contributed to TUSC. Contributors included Steve Briggs, Vicki Chandler, Mike Freeling, Pat Schnable, Sarah Hake and Virginia Walbot. The lower number represents the number of donor stocks provided from each lab. The circle size, and the upper number represents the total number of F2 families fixed from each lab's stocks in the 1993 PV03 population of 24,218 families. It took approximately 350 independent crosses to create the F1 base for selfing into the PV03 F2 archive

2.1.2 Growouts and Grids

To set up the F1 growouts for self pollination and sampling, between 60 and 80 F1 kernels from each cross were planted in a standard nursery arrangement. At maturity, vigorous efforts were made to self-pollinate all possible plants. Planting large blocks of related F1 material in this fashion does predispose one to encounter pre-existing parental mutations. Such long "pedigree strings" were of limited concern; keeping in mind a greater interest in the new and largely unique germinal transpositions occurring at the time of tassel development in each donor line. Parental insertions are easily spotted in a linear (non-randomized) pooling strategy, and can be discretely managed early in the screening and selection process, as discussed further below. For reverse genetics, parental insertions are of equal value to new, unique insertions where no mutation has been defined previously.

Plant identification and tissue sampling were performed immediately post-pollination. Each selfed plant was tagged using a system of addresses based on the common 96-well microtiter plate. Each address is broken down into Population, Plate, and Coordinate components, respectively. For example, plant no.10,000 corresponds to the address PV03 104 B-04 in this indexing system. Each pedigree is recorded for all plants to ensure the independence of selected mutations.

At harvest, each F2 ear was husked, tagged, dried to proper moisture level, examined, and individually shelled. Harvest and shelling was the first opportunity to examine the extent of the mutagenesis by noting various kernel phenotypes segregating on the ear. Each pedigree record was maintained as a critical element of the inventory. Kernel yields for a given F2 can vary widely, and some barrenness was both expected and observed, but this was a very minor fraction of the intended inventory. Upon harvest, all progeny kernels were stored in envelopes indexed in a series of storage units within a controlled climate facility.

2.1.3 Tissue Sampling and DNA Extraction

Following pollination and tagging, ordered sampling of F1 leaf tissues was conducted. Samples, in the form of leaf discs (0.5 cm dia.), were collected in two replicates. In a first pass through each F1 population, tissue pools were collected from groups of 48 consecutive plants, using their microtiter-based tags as a sampling guide. Six leaf discs were harvested at each plant, using a customized punching tool which drops each disc into a large collection tube. Each pool contains samples from the individuals making up one-half of a microtiter plate array - odd numbered pools contained the 48 samples A1-D12; even numbered pools, E1-H12. Following extraction and resuspension, aliquots of the pool DNAs were arrayed in nine separate 96-well master plates and archived for long term use (Figure 2).

Following the pool sampling, a second independent pass was made to collect individual tissues. Again, six leaf discs were collected per plant in small tubes fitted into megatiter racks. The collection of tissue replicates was both for convenience - it was easily adaptable to our nursery layout - and importance, since it would be

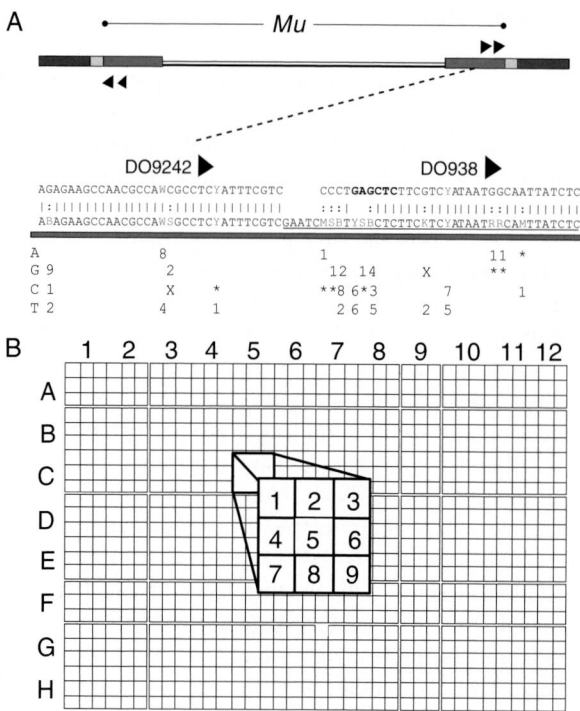

Fig. 2 A) Universal *Mu* PCR primers based on the highly conserved inverted terminal repeat (TIR). B) Pooled DNA samples (48 F1 plants per pool) arrayed in nine 96 well plates to represent the TUSC population

good defense against interference by somatic *Mutator* insertion (discussed below). These replicates permitted independent processing of Pool and Individual samples using high-throughput techniques, and eliminated many of the liquid handling steps otherwise necessary for creation of post-preparation pools. The completed TUSC collection stands ready as a fully processed set of pooled and individual F1 genomic DNAs, indexed to its corresponding F2 progeny seed. This DNA has been in constant use for over 12 years with no appreciable loss of performance in PCR.

2.1.4 *Mu* PCR

A successful resource for knockout screening requires PCR methods for reliable amplification from *Mutator* elements. This was greatly facilitated by the wonderful sequence architecture of the *Mu* element. While each *Mu* element class is distinguished by extremely diverse internal sequences of varying length (Chandler and Hardeman, 1993; Bennetzen, 1996), the unifying definition of a *Mu* element (Lisch et al., 1995) is reflected by common termini. Each member shares a minimal unit of approximately

0.2 kbp of conserved terminal inverted repeat (TIR) sequence, denoting arbitrary left and right borders (Hirschberger et al., 1991, Benito et al., 1994).

Initially, it was predicted that the length and high sequence conservation in the inverted TIRs would confound a PCR by intramolecular pairing into large, stable stem-loops. However, work with a number of primers, and reconstructions with both plasmids and genomic DNA templates from known mutations, produced straightforward PCR conditions for reliable amplification and detection of the target::*Mu* products.

Figure 2A illustrates aspects of TIR orientation and conservation relevant to the *Mu* primers used in TUSC. Much of the early reconstructions (R.B. Meeley, unpublished) and initial screening work (Bensen et al., 1995; Mena et al., 1996) employed DNA oligonucleotide (DO)938 as the MuTIR primer. DO938 is an outward-reading primer, capable of amplifying from either TIR border. This primer contains a single degeneracy and other modifications (Fig. 2A), but one of its biggest drawbacks is that it extends to the terminal -CTC bases of the *Mu* TIR.

An alternative primer, DO9242, has been used almost exclusively since the study by Chuck and cowokers (Chuck et al., 1998). DO9242 is recessed 71 bp into the TIR, and occupies a region of very high sequence identity (Benito et al., 1994, and Fig. 2A), which is further accommodated by two degenerate positions in the primer. This is a long primer (32bp) with a very high T_m, but it has a balanced base composition, and no hairpin stems >3bp. Like many other successful *Mu* TIR primers (Das and Martienssen, 1995; Frey et al., 1998; Gray et al., 1997; Settles et al., 2004), DO9242 permits amplification from left or right TIR borders of all known mobile *Mu* elements, *Mu*1, *Mu*2, *Mu*3, rcy:*Mu*7, *Mu*8, and *MuDR* (Chandler and Hardeman, 1996). The recessed location provides flexibility to detect alleles that might otherwise be missed if nucleotides at the point of insertion are corrupted, and is also useful when secondary nested amplification methods are desired. Importantly, the sequences derived from DO9242 PCR products contain ~39 bp of TIR sequence that is specific to the allele-in-question (Fig 2A). While not unambiguous, similarity searches using this sequence tag are excellent cues as to the identity of the resident *Mu* element at a particular insertion allele. This is useful for subsequent genetic and molecular analyses, including the development of DNA gel-blot, or PCR-based genotyping strategies, and helps advance the annotation and nomenclature of informative mutant alleles. Novel *Mu* elements (*Mu*10-*Mu*12) have been designated using this limited TIR sequence information (Dietrich et al. 2002).

2.1.5   Primer Design and Interpretation of Screening Data

Gene-specific primer (GSP) design is among the most important factors in preparing a reverse genetics screen. The TUSC protocols prefer generally long primers (25-29 nt) with a balanced base composition and GC content near 55%, but the standard parameters of primer design and compatibility are not the main concern. Despite a consistent approach in their derivation, the major limitation to successful primer design is most often rooted in a lack of information about target gene structure.

Obviously, genomic target sequence is preferred to initiate a project. In recent years, the advent of maize sequencing technologies (Yuan et al., 2003; Rabinowicz et al., 1999; Palmer et al., 2003), and genomic contig assemblies, and the pending completion of the maize genome sequence (www.maizesequence.org), have greatly simplified the interpretation of target gene structure and the design of TUSC screening primers. In the early days of TUSC, such detail was seldom available and comparative sequence analysis to rice and Arabidopsis genomes composed much of the primary effort to approximate the locations of putative introns. In cases where the maize target of interest shares a close homologue in Arabidopsis, rice, or other plant genomes, any available annotation or genomic sequence information was used to advance the design of functional screening primers. These were rational approaches that tested and leveraged the conservation of gene structure across species, and were proven very successful in generating large numbers of satisfactory primers. In cases where the databases are of no apparent aid, a cDNA sequence can be broken into arbitrary segments of 300-700 bp, from which a series of forward and reverse primers at regular, and sometimes overlapping, intervals are chosen. Regardless of the strength of their origin, all primers are tested in gene-specific control amplifications (e.g. against B73 gDNA) before declared as "screening-ready". Common tools that facilitate alignments of cDNA to gDNA sequence such as BLAT (Kent, 2002) and MACAW (Schuler et al., 1991) have been indispensable aids in designing good TUSC primers.

The above points are even more important when targeting gene paralogs or members of gene families. In the case of maize, a second copy of a particular gene is nearly a given (Emrich, et al., 2007) since the origins of maize are punctuated by segmental allotetraploidy events (Gaut and Doebley, 1997). Published examples of cases where gene duplications have been specifically targeted in TUSC include the duplicate Zfl1 and Zfl2 (Floricuala /Leafy) paralogs (Bomblies et al., 2003), and duplicates of maize Rad51 genes (Li et al., 2007).

2.1.6 Screening Strategy

Figure 3 summarizes the stepwise process for conducting PCR-based screens in the TUSC program. In practice, sets of working primers are deployed in the quest for *Mu* alleles of targeted genes. In genomic sequence space, primers are spaced about a maximum of 2kb apart, with most genes requiring two to three pairs of working primers, staggered along the gene (by example, primers P1 through P4 in Fig. 3).

PCR reactions and dot blot hybridizations against F1 pools and individual DNAs are performed in replicates. The resulting autoradiographic images are quantitatively scanned and annotated in a relational database. This permits visualizations and comparisons on the cumulative performance of primer sets, producing a thread of data that is used to make quality selections of candidate alleles throughout the process. The most desirable insertion candidates are those producing strong, reproducible signals that may be cross-confirmed by an opposing target primer.

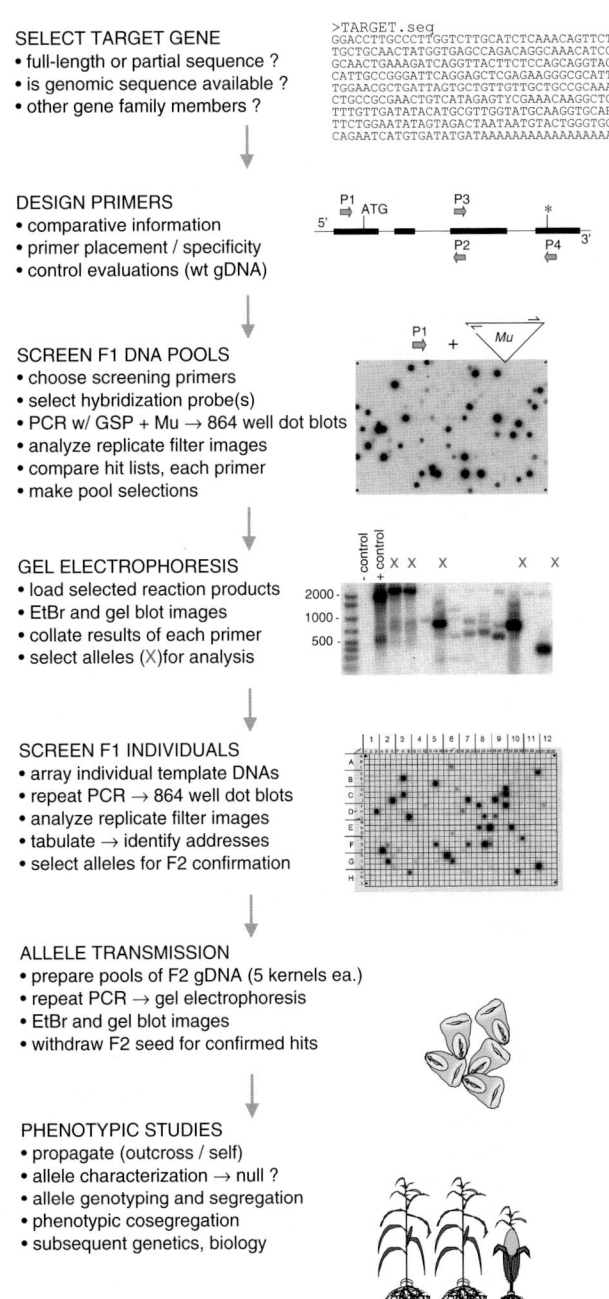

SELECT TARGET GENE
• full-length or partial sequence ?
• is genomic sequence available ?
• other gene family members ?

>TARGET.seq
GGACCTTGCCCTTGGTCTTGCATCTCAAACAGTTCTTGG
TGCTGCAACTATGGTGAGCCAGACAGGCAAACATCCAGG
GCAACTGAAAGATCAGGTTACTTCTCCAGCAGGTACTAC
CATTGCCGGGATTCAGGAGCTCGAGAAGGGCGCATTCCG
TGGAACGCTGATTAGTGCTGTTGTTGCTGCCGCAAAGCG
CTGCCGCGAACTGTCATAGAGTYCGAAACAAGGCTGCAT
TTTGTTGATATACATGCGTTGGTATGCAAGGTGCARATC
TTCTGGAATATAGTAGACTAATAATGTACTGGGTGGCTT
CAGAATCATGTGATATGATAAAAAAAAAAAAAAAAA

DESIGN PRIMERS
• comparative information
• primer placement / specificity
• control evaluations (wt gDNA)

SCREEN F1 DNA POOLS
• choose screening primers
• select hybridization probe(s)
• PCR w/ GSP + Mu → 864 well dot blots
• analyze replicate filter images
• compare hit lists, each primer
• make pool selections

GEL ELECTROPHORESIS
• load selected reaction products
• EtBr and gel blot images
• collate results of each primer
• select alleles (X) for analysis

SCREEN F1 INDIVIDUALS
• array individual template DNAs
• repeat PCR → 864 well dot blots
• analyze replicate filter images
• tabulate → identify addresses
• select alleles for F2 confirmation

ALLELE TRANSMISSION
• prepare pools of F2 gDNA (5 kernels ea.)
• repeat PCR → gel electrophoresis
• EtBr and gel blot images
• withdraw F2 seed for confirmed hits

PHENOTYPIC STUDIES
• propagate (outcross / self)
• allele characterization → null ?
• allele genotyping and segregation
• phenotypic cosegregation
• subsequent genetics, biology

Fig. 3 The TUSC procedure. The seven steps involved in screening the TUSC population for targeted insertions. Addition of the F2 screen enabled confirmation of germinal insertions

Gel analysis of positive pools is used to visualize the complexity and size distribution of target-specific *Mu* products. Here, remnant reaction products from the initial pool screen, or products from an independent pool PCR, can be evaluated for the quality and size diversity of PCR products. Estimations of *Mu* insertion location can be inferred based on PCR product size, relative to the gene model. Both ethidium bromide staining and DNA gel blot hybridization patterns are evaluated, adding new information to the data thread. For a typical gene, project flow can involve the electrophoresis and hybridization of as many as 120 reactions. Naturally, the best cases involve dominant ethidium bromide bands that hybridize to the probe and are cross-confirmed with different target primers. Depending on how much information is known about target gene structure, precise decisions can be made in favor of those candidates with insertions predicted to fall in exons.

Once candidate allele selections are made, the appropriate individual DNA templates are retrieved from storage, and a robotic liquid handler is used to create a 9X array of selected individuals (Fig. 2B). PCR is then repeated with the appropriate primers, followed by another round of dot blot hybridizations, image analysis, and database entry. Again, the cumulative data quality is used to identify the strongest individual hits harboring selected target-specific *Mu* insertions.

A *Mu* insertion profile of each TUSC line will share many insertions in common with its Standard Line parent. Since *Mu* transposes germinally via a replicative mechanism (Alleman and Freeling, 1986), the PCR can be expected to pick up conserved parental insertions. Because large blocks of sibling F1 plants are preserved in consecutive order in our template array, parental alleles produce a regular and conspicuous hybridization pattern that can be interpreted easily. They can either be selected for study, or excluded from further analysis.

2.1.7   Addressing False Positives: Addition of a Confirmatory Test of F2 Progeny

In the first several years of the program, TUSC seed was dispensed to collaborators based on analysis through the F1 individual stage. Many individual families that were PCR-positive at the F1 individual stage failed to produce positive results in F2 segregants. What this indicates is a significant interference by somatic *Mu* insertion events, and this contributed to an apparent high rate of false positives (Walbot, 2000). A point emphasized in Figure 3 is the final step added to the TUSC screening regime that minimizes the detection of false-positives and helps establish clearer measurements of the rate of success of the program. A rapid and final step added to the screen confirms the transmission of candidate target::*Mu* alleles to the F2 generation. DNA pools are prepared from five representative F2 kernels from selected candidates, and the PCR is repeated to provide a completion to the thread of screening data. These F2 reaction products also serve as reagents for cloning and sequence characterization of insertion sites. This effort has permitted a more thorough accounting of insertion rate (Fig. 4) and greatly simplified the downstream work devoted to allele characterization and phenotypic analysis.

2.1.8 Accomplishments and coverage of the TUSC resource

Since the TUSC resource was announced to the maize community in 1995 (Meeley and Briggs, 1995) and made available to academic researchers via collaborative research and materials transfer agreements, over 100 individual investigators have accessed the resource. A review of 438 screening projects that included F2 confirmation tests reveals that reverse genetics screening in the TUSC resource is successful ~90% of the time, and returns an average of greater than three independent insertion alleles per gene (Fig. 4). The estimated overall coverage of 3.3-fold (e.g. 3.3 alleles per target) is in rough agreement with the estimate based on forward mutation frequencies given above.

Examples of how the TUSC resource has helped further the work of maize biologists can be found throughout the recent literature, including key studies in maize architecture/domestication (Gallavotti et al., 2004; Bomblies et al., 2003; Ching et al. 2006), inflorescence development (Mena et al., 1995; Chuck et al., 1998), seed biology (Fu et al. 2002; Lid et al., 2002; Shi et al., 2003; Holding et al, 2007), meiosis (Pawlowski et al., 2004; Golubovskaya et al, 2006; Li et al, 2007), hormone biochemistry (Bensen et al, 1995; Park et al., 2003), and even the genetics of microRNAs (Chuck et al., 2007). TUSC has robust utility, but also has acknowledged limitations. Each has helped define the opportunity for enhanced genetics and technologies that will improve reverse genetics resources for the post genomic era.

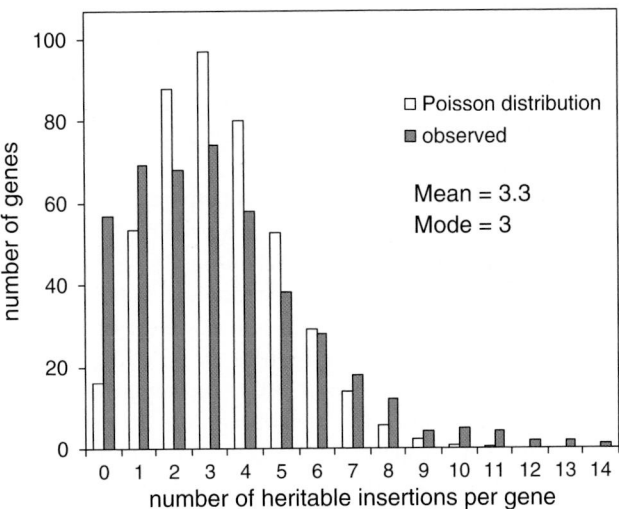

Fig. 4 Frequency of insertion alleles detected in genes targeted using the TUSC resource. The solid bars show the observed distribution of allele number per gene for screens that included F2 confirmation. The open bars show an ideal random distribution with a mean of 3.3 alleles per gene

2.2 Enhanced Mu Populations

The great impact of the TUSC resource on maize research has inspired several efforts in the public sector to develop improved *Mu* populations that are tailored specifically for functional genomics applications.

The RescueMu resource is based on an engineered *Mu* element introduced into an active *Mu* line by transformation (Fernandes et al., 2004). The RescueMu element contains an embedded BlueScript plasmid that enables transposed copies, including up to several kilobases of flanking DNA, to be recovered by plasmid rescue in *E. coli*. In addition to the engineered RescueMu element, the population contains a large number of native *Mu* elements that also contribute to the mutagenic activity. Because only the RescueMu insertions are recovered for molecular analysis, RescueMu has been used effectively as a low-copy transposon tool for purposes of gene discovery enabled by high through-put sequencing of RescueMu insertion sites (Fernandes et al., 2004).

The Maize-Targeted-Mutagenesis (MTM) population was constructed by Martienssen and co-workers (May et al., 2003) as a public resource similar in capability to the TUSC population. MTM lines were generated by crossing highly active *Mu* lines to a stock that carried *Les28*, a dominant marker for *Mu* activity that can be scored in leaves, and a dominant silencing factor similar to the *Mu*-killer described by Slotkin et al. (2003). In this system, silencing of the *les28* marker is used to monitor progressive inactivation of *Mu* during development of F1 plants. To enable reverse genetic screens, DNA samples from a population of 43,776 plants were arrayed into a series of nineteen 48x48 2-dimensional grids. Each grid of 2304 plants is represented by 96 pooled DNA samples such that the population can be screened for gene knockouts by PCR using one or more gene specific primers in combination with a universal *Mu* primer based on the highly conserved *Mu* TIR sequence. In addition, F2 families in the MTM population have been extensively screened for seed and plant phenotypes, providing a resource for forward genetic analysis of diverse plant processes.

The UniformMu resource is designed to incorporate features that facilitate high-throughput molecular and phenotypic analysis of large *Mu* populations. Populations have been developed by backcross introgression of active *Mu* into W22 and B73 inbred backgrounds. The W22 population of 38,000 M2 lines has been screened extensively for seed phenotypes and a subset of seed mutants have been analyzed by MuTAIL sequencing (Settles et al., 2004, 2007; Porch et al., 2006; Suzuki et al., 2006; McCarty et al., 2005). The key genetic properties of the population include: 1) capability for genetic control of germinal and somatic transposition activity, 2) mechanisms for management of transposon copy number and genetic load, and 3) use of a defined inbred genetic background to facilitate phenotypic analysis.

Genetic control of transposon activity in a resource population is important for two reasons. First, as noted above, the presence of large numbers of somatic transposon insertions can complicate molecular analysis and mapping of insertions in the maize genome. Somatic insertions are non-heritable insertions created by transposition events that occur during development of the plant somatic tissues sampled for DNA analysis. In genomic DNA samples from highly active lines, somatic insertions may outnumber the germinal insertions. Based on high throughput sequencing of RescueMu plasmids, Fernandes et al. (2004) estimated that up to 85% of recovered *Mu* flanking sequences may be derived from somatic insertions. Second, once the population is created, genetically stable lines that do not continue to accumulate germinal insertions are highly desirable for purposes of genetic and phenotypic analysis.

Two strategies have been used to enable genetic control of *Mu* activity in large *Mu* populations: 1) epigenetic silencing of *Mu* using a *Mu*-killer stock (May et al., 2004) and 2) genetic control of *Mu* based on Mendelian segregation of *MuDR* (McCarty et al., 2005). Both strategies rely on use of marker genes that report somatic transposition activity. Advantages of the *Mu*-killer mechanism are that inactivation is dominant, independent of *MuDR* copy number, and achievable in a single generation by outcrossing *Mu* lines to a *Mu*-killer stock. A disadvantage is that suppression of somatic *Mu* transposition is apparently incomplete in the F1 (May et al., 2004). The Mendelian approach used in construction of the UniformMu population takes advantage of the backcrossing and marker scheme implemented for management of *Mu* copy number and genetic load (McCarty et al., 2005). A majority of UniformMu lines contained 0, 1, or 2 copies of *MuDR,* enabling selection of stable bronze (non-spotted, *Mu*-inactive) kernels using the *bz1-mum9* marker.

Because transposon insertions accumulate in the genomes of active lines propagated by self-pollination, individual lines may attain very high transposon copy numbers and accrue a correspondingly large and typically variable genetic load. The genetic load and *Mu* copy number can be managed by propagating the population through continuous backcrossing to a non-*Mutator* recurrent parent. In this system, *Mutator* activity and *MuDR* copy number can be monitored using a suitable marker, such as *bz1-mum9* (Chomet, 1994). Moreover, by using an inbred as the recurrent parent, the population is constructed in a highly homogeneous genetic background that is conducive to phenotype analysis for both forward and reverse genetics approaches. McCarty et al. (2005) implemented this strategy in construction of the UniformMu population and showed that it is possible to maintain uniform rates of transposition and forward mutation over at least eight generations, consistent with a steady-state model for transposon mutagenesis (Figure 5). The steady-state model postulates that, for each generation, the average rate of transposition is in approximate equilibrium with the rate that parental transposons are lost due to random segregation in the backcross (e.g. each individual inherits 50% of the insertions that are heterozygous in the *Mu* active parent).

2.1.1   Biases Affecting the Distribution of Mu Insertion Sites in the Genome

Extensive screening and molecular analyses of the large *Mu* populations has provided a wealth of information on the behavior of *Mu* transposon in the maize genome and biases that affect use of the transposon for genomics applications (May et al., 2003; Fernandes et al., 2004; McCarty et al., 2005; Settles et al., 2004, 2007). In addition to the well-known preference of *Mu* for inserting into gene sequences (Cresse et al., 1995),

A Seed mutation frequency

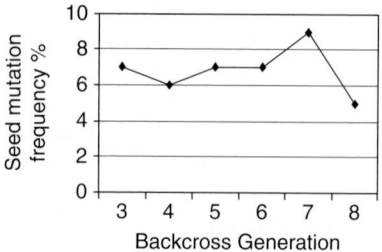

B Steady-state transposon mutagenesis

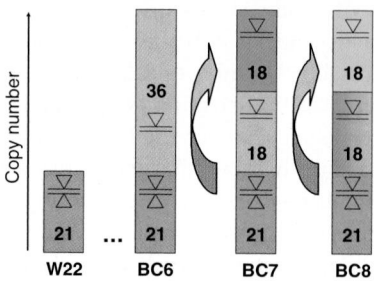

Fig. 5 Transposon mutagenesis in the UniformMu population. **A.** Seed mutation frequencies were consistent through 8 generations of introgression of active *Mu* into a W22 inbred (McCarty et al., 2005). **B.** The stable mutation frequencies and the profile of *Mu* transposons in UniformMu lines are consistent with a steady-state model in which the average germinal transposition rate is balanced by dilution of non-parental insertions through backcrossing. Based on MuTAIL sequencing data, UniformMu lines have an average copy number of ~57 transposons (see McCarty et al., 2005). These include 21 *Mu* elements that are fixed in the W22 inbred, ~18 pedigree-specific parental *Mu* transposons, and ~18 new germinal transpositions that are unique to individuals

two other types of bias in the distribution of *Mu* insertions have been reported: a bias in positions of *Mu* insertions within genes and a bias in the frequency of insertions among genes (the so-called "hotspots" and "cold spots").

Several studies have shown a pronounced bias toward *Mu* insertions near the 5'-end for a wide variety of genes (Dietrich et al., 2002; May et al., 2003). While a 5'bias does not directly affect the likelihood of finding insertions in a gene, it may influence the probability that an insertion will disrupt gene function. For example, insertions that occur near or immediately upstream of the coding sequence may be more susceptible to the phenomenon of *Mu* suppression (Barkan and Martienssen, 1991) than insertions located in interior exons of a gene. *Mu* suppressible mutants have phenotypes that are only expressed if the autonomous *MuDR* element is also present in the genome. In suppressible alleles, epigenetic modification of the *Mu* element in the absence of *MuDR* results in restoration of gene expression to a sufficient level to suppress the mutant phenotype (Barkan and Martienssen, 1991; Martienssen and Baron, 1994). Estimates of the overall frequency of *Mu* suppressible mutant phenotypes vary from 17% of seed phenotypes in the UniformMu population

(McCarty et al., 2005) to greater than 50% of all types of *Mu*-induced mutations screened in the MTM population (May et al., 2003). The apparent disparity between estimates derived from unrelated *Mu* populations is intriguing and worthy of further investigation. McCarty et al. (2005) also reported evidence that the frequency of *Mu* suppression differed among classes of seed mutants.

Evidence for genes that are "hotspots" for *Mu* insertions has been reported in the MTM (May et al., 2004), RescueMu (Fenandez et al., 2004), and TUSC populations (Fu et al., 2002; Fig. 4). An indication that *Mu* insertions are not randomly distributed among diverse maize genes can be seen in the spectrum of mutations identified in the TUSC resource shown in Fig 4. The distribution of insertions per gene is significantly broader than predicted for a simple random process (e.g. compare the observed and Poisson distributions) suggesting that genes vary in their affinity for *Mu* insertions. Deviations from random occur at both the low and high ends of the distribution implying that there may be relative "cold spots" as well as "hotspots" for *Mu* insertions in the maize genome. However, other transmission biases may also inflate the distribution of insertions at the low end. For example, TUSC and other collections (McCarty et al., 2005) have used the *Mu* active parents exclusively as males, creating a potential bias against insertions that are lethal to the haploid microspore or cause dominant male-sterile phenotypes. Intriguingly, thus far the specific genes identified as hotspots are not the same across populations. Whether this is due to limited sampling of maize genes, systematic differences in sampling methods, or the result of real population differences in *Mu* insertion site bias is not yet clear. Thus far, evidence of hotspots has not been detected in the UniformMu population (McCarty et al., 2005). Moreover, the mutants that are over-represented in TUSC and MTM appear to have normal frequencies in UniformMu (D. McCarty, unpublished results) suggesting that genetic background and/or *Mu* copy number per genome may affect the distribution of hotspots.

3   Prospects for Construction of Comprehensive Sequence-Indexed Transposon Resources for Maize

The great utility of comprehensive collections of molecularly characterized insertional mutations for functional analysis of plant genomes is amply demonstrated by the impact T-DNA and transposon resources have had on the Arabidopsis genetics community over the past decade (e.g. Alonso et al., 2003). From the user's perspective, the salient features of these resources are: 1) the precise locations of insertions in the genomes of individual lines have been determined by sequencing of flanking DNA, enabling construction of a searchable sequence-index for each collection; 2) convenient public access to the sequence-index database provided through one or more web portals that link insertion sites to corresponding seed stocks deposited in community stock centers; 3) annotation of mapped insertions is fully integrated with gene annotations and other genome features; and 4) last but not least, the cost of obtaining seed for any desired mutant to the user is nominal.

Several approaches have been taken toward systematic sequencing of genomic DNA flanking transposon insertion sites in maize (Hanley et al., 2000; Settles et al., 2004, 2007; Fernandes et al., 2004). As noted above, a project aimed at the sequencing of up to 10,000 *Ds* insertion sites amplified by inverse PCR is in progress. The *Ac/Ds* database hosted on PlantGDB.org (PlantGDB.org/AcDsTagging/) includes annotations and links to seed stocks containing sequence-indexed transposed *Ds* insertion sites in the W22 inbred.

The RescueMu resource includes sequences from about 54,000 RescueMu plasmids recovered by plasmid rescue from a population of 33,000 plants (Fernandes et al., 2004). The dataset includes at least 3,138 independent germinal insertions. On that basis, the total RescueMu resource is estimated to include at least 8,000 independent germinal RescueMu insertions. High throughput sequence analysis of the native *Mu* insertions in UniformMu lines has been undertaken by adapting and optimizing TAIL PCR methods for amplification of *Mu* flanking sequences (Settles et al., 2004, 2007; McCarty et al., 2005). MuTAIL sequencing has been used to analyze insertions in 148 seed mutants isolated from the population (McCarty et al., 2005). The resulting dataset of 36,000 MuTAIL sequences has identified over 1,700 germinal insertions (McCarty et al., 2005; Settles et al., 2007).

The transposons present in the UniformMu population fall into three classes: 1) a small set of *Mu* insertions that are homozygous in the W22 inbred parent and therefore present in all plants in the population, 2) pedigree-specific parental insertions that have been propagated through one or more generations of backcrossing, and 3) insertions that are unique to individuals. The average number of pedigree-specific parental insertions per individual is roughly equal to the number of insertions that are unique to individual lines (~18 insertions per line). In a given lineage, half of any heterozygous insertions, including the pedigree-specific and line-specific classes as well as new insertions that occur in the male germ line of the *Mu*-active parent, are expected to be lost by segregation each backcross generation. In that case, a steady-state rate of transposition would be maintained if every germinal element transposed in the male germ line about once per generation. This inference is consistent with the analysis of the RescueMu population, where it is estimated that each founder element transposed to a new site once per plant per generation (Fernandes et al., 2004), as well as, the estimate for *Mu*1 transposition rate reported by Alleman and Freeling (1986).

The frequency of pedigree-specific parental insertions in the population will vary, depending on the backcross generation in which they were created. Hence, a small number of *Mu* insertions inherited from the founder *Mutator* lines are widely distributed in the population, whereas more recent transpositions are limited to small pedigrees. Conceivably, the endogenous insertions that are fixed in the W22 inbred contribute to the maintenance of a steady-rate transposition rate by providing a reservoir of donor elements that can be mobilized by *MuDR*. Whether or to what extent the W22 *Mu* insertions contribute to new transposition events is not known. Our preliminary analysis of a similarly constructed UniformMu-B73 population indicates that steady-state mutagenesis can be applied to other inbreds.

In addition to yielding important insights into the properties of *Mu* in large populations, the UniformMu and RescueMu datasets highlight key limitations of conventional high-throughput random sequencing for mapping of *Mu* transposons on a large scale. While random sequencing from representative libraries of *Mu* flanking DNA is a very effective approach for mapping insertions, particularly in high-copy number lines, the number of insertions identified depends on the depth of sequence coverage. Sampling the unique insertion sites to an average depth of 5X coverage – sufficient to identify >90% of insertions in a population of 50,000 plants - would require at least 4,500,000 sequence reads. Importantly, for this estimate, we assume that the highly redundant sequences derived from parental insertions that are widely distributed in the population can be efficiently subtracted from the sequencing libraries at a nominal cost prior to sequencing. In the UniformMu and RescueMu datasets where subtraction was not performed the highly redundant parental insertions account for nearly 50% of the raw sequences. Moreover, because the most efficient strategies for sequencing insertion sites in large numbers of plants typically employ 2-D or higher dimensional pooling schemes similar to the arrays used for gene-specific PCR screens (e.g. Fernandes et al., 2004), both axes would need to be sequenced to circa 5X, increasing the total number of sequence reads by another factor of two (i.e. 9,000,000 total reads). While obtaining the required sequencing throughput using conventional automated Sanger sequencing technology is cost prohibitive, this volume of sequences is well within the range of feasibility for next-generation massively parallel sequencing technologies (e.g. Margulies et al., 2005).

3.1   Application of Massively Parallel DNA Sequencing Technology for Indexing Mu Insertions

We warn the reader that the technologies we will discuss in this section are evolving very rapidly, so some of the specifics mentioned may already be obsolete by the time this chapter appears in print. At present, two leading platforms for massively parallel DNA sequencing are the 454-FLX platform based on pyrosequencing technology (Margulies et al., 2005) and the Illumina 1G platform which employs a sequence-by-synthesis approach (Barski et al., 2007). Very briefly, the 454-FLX platform is currently capable of producing ~500,000 reads with an average length of about 250 bp in a single reaction; whereas, the Illumina platform is capable of generating ~10^6 reads of ~35 bp per reaction (Barski et al., 2007). A third technology, implemented in the SOLiD platform (AppliedBiosystems.com), uses a novel ligation-based sequencing approach that offers high throughput, low cost and read length capabilities roughly comparable to the Illumina instrument. Another important but generally under-appreciated feature of all three technologies is their extra-ordinary sensitivity, such that inputs on the order of just 10^7 template DNA molecules per reaction are sufficient to generate millions of reads. This feature reduces and in some cases eliminates (Eveland et al., 2008) the need for amplification of template DNA upstream of the sequencer.

Although the specifications of these platforms are evolving rapidly, the trade off between raw throughput and read length is likely to be relevant to how these technologies are applied for the foreseeable future. Hence, a comparative strength of 454-FLX is the long read length capability, and the comparative strength of the Illumina platform is a vastly higher throughput of short reads with substantially lower reagent costs. For the application of sequence-indexing of *Mu* insertions, a longer average read length is important for three reasons: 1) longer reads can accommodate DNA "barcode" keys that enable multiplexing of multiple template samples in a single reaction, 2) reads that extend across the junction between the transposon and flanking genomic DNA are more easily validated as representing authentic insertion sites, and 3) reads that include more flanking genomic sequence are more likely to map to a unique location in the maize genome. The latter characteristic is particularly important for resolving closely-related duplicate genes in the maize genome (Emrich et al, 2007). The limitations of short reads (<50 bp) in these respects can be compensated to some extent by the greater depth of sequence coverage, utilization of paired reads, and new methods for assembly of large numbers of short sequence tags. Conversely, techniques that enable synthesis of precisely anchored and oriented template DNA molecules (Eveland et al., 2008) can be used to maximize utilization of the relatively low throughput, but long read capacity of the 454-FLX instrument.

Toward this end, the UniformMu project has developed and tested highly efficient protocols for pyrosequencing of *Mu* insertion sites using the 454-platform. Key features of the protocol are the following.

1) The sequencing template is synthesized from a series of customized adaptor (454-A adaptor) that incorporates a 4-base error-detecting multiplex key (Eveland et al., 2008) that enables up to 64 samples to be combined in a single template library.

2) The A adaptor includes a 22-base *Mu* TIR-specific primer that initiates template synthesis 6 bp upstream of the 3'-end of the TIR. As a result, 6 bp of the TIR located downstream of the primer is included in each sequence read, providing validation of authentic insertion sites in the genome.

3) All sequence reads are precisely anchored and oriented to extend outward from the 3'-end of the TIR and include about 220 bp of flanking sequence [250 bp a verage read length – (4 bp multiplex key + 28 bp of TIR) = 218 bp].

MAGI4_7281

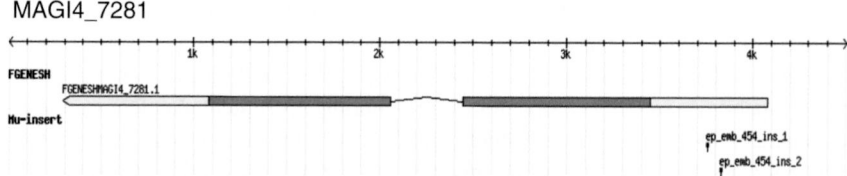

Fig. 6 UniformMu insertion alleles mapped by 454-sequencing. Positions of *Mu* insertions in the 5'-UTR of a predicted PPR domain protein gene annotated in MAGI4_7281 are displayed using the URL import feature of the MAGI GBROWSE server (http://magi.plantgenomics.iastate.edu). The precise insertion sites were identified by oriented 454-sequence reads that are anchored to the 3'-end of the *Mu* TIR (see text). Sequences were mapped by BLASTN analysis of the MAGI4 dataset (Emerich et al., 2004)

In a pilot study, multiplexed template libraries were prepared from 22 pooled DNA samples that form the axes of a 10 X 12 array representing 120 UniformMu lines. The multiplex library was sequenced using 454-FLX (Bao-Cai Tan, Sue Latshaw, D. R. McCarty, unpublished results). Greater than 95% of the reads obtained were validated as correctly oriented *Mu*-flanking DNA sequences. Insertions identified by 454 sequencing have been successfully mapped to individual lines in the array by detecting intersections between the X and Y axis pools. A bioinformatic pipeline has been developed that enables mapping of 454-Mu flanking sequence tags to maize genome sequence assemblies (Fig. 6) as well as automated annotation of *Mu* insertion sites (McCarty et al., 2005).

References

Alleman, M. and M. Freeling (1986) The *Mu* transposable elements of maize: evidence for transposition and copy number regulation during development. *Genetics* 112: 107–19.

Alonso, J. M., A. N. Stepanova, T. J. Leisse, C. J. Kim, H. Chen, P. Shinn, D. K. Stevenson, J. Zimmerman, P. Barajas, R. Cheuk, C. Gadrinab, C. Heller, A. Jeske, E. Koesema, C. C. Meyers, H. Parker, L. Prednis, Y. Ansari, N. Choy, H. Deen, M. Geralt, N. Hazari, E. Hom, M. Karnes, C. Mulholland, R. Ndubaku, I. Schmidt, P. Guzman, L. Aguilar-Henonin, M. Schmid, D. Weigel, D. E. Carter, T. Marchand, E. Risseeuw, D. Brogden, A. Zeko, W. L. Crosby, C. C. Berry and J. R. Ecker (2003) Genome-wide insertional mutagenesis of Arabidopsis thaliana. *Science* 301:653–7.

Bai, L., M. Singh, L. Pitt, M. Sweeney and T. P. Brutnell (2007) Generating novel allelic variation through *Activator* insertional mutagenesis in maize. *Genetics* 175:981–92.

Barkan, A. and R. A. Martienssen (1991) Inactivation of maize transposon-Mu suppresses a mutant phenotype by activating an outward-reading promoter near the end of *Mu1*. *Proc Natl Acad Sci, USA* 88:3502–3506.

Barski, A., S. Cuddapah, K. Cui, T. Y. Roh, D. E. Schones, Z. Wang, G. Wei, I. Chepelev and K. Zhao (2007) High-resolution profiling of histone methylations in the human genome. *Cell* 129:823–37.

Benito, M. I. and V. Walbot (1994) The terminal, inverted repeat sequences of *MuDR* are functionally active promoters in maize cells. *Maydica* 39:255–264.

Bennetzen, J. L. (1996) The *Mutator* transposable element system of maize. *Curr Top Microbiol Immunol* 20:195–229.

Bensen, R. J., G. S. Johal, V. C. Crane, J. T. Tossberg, P. S. Schnable, R. B. Meeley and S. P. Briggs (1995) Cloning and characterization of the maize An1 gene. *Plant Cell* 7:75–84.

Brutnell, T.P. and L. J. Conrad (2003) Transposon tagging using Activator (Ac) in maize. *Methods Mol Biol* 23:157–76.

Chandler, V. L. and K. J. Hardeman (1992) The *Mu* elements of Zea mays. *Adv Genet* 3:77–122.

Chomet, P.S. (1994) Transposon tagging with *Mutator*. In *The Maize Handbook* (Freeling, M. and Walbot, V., eds). New York: Springer Verlag, pp. 243–249.

Chuck, G., R. Meeley and S. Hake (1998) The control of maize spikelet meristem fate by the APETALA2-like gene *indeterminate spikelet 1*. *Genes Dev* 12:1145–1154.

Conrad, L. J. and T. P. Brutnell (2005) Ac-immobilized, a stable source of Activator transposase that mediates sporophytic and gametophytic excision of Dissociation elements in maize. *Genetics* 171:1999–2012.

Cowperthwaite, M., W. Park, Z. Xu, X. Yan, S. C. Maurais and H. K. Dooner (2002) Use of the transposon *Ac* as a gene-searching engine in the maize genome. *Plant Cell* 14:713–26.

Cresse, A. D., S. H. Hulbert, W. E. Brown, J. R. Lucas and J. L. Bennetzen (1995) *Mu1*-related transposable elements of maize preferentially insert into low copy number DNA. *Genetics* 140: 315–24.

Das, L. and R. A. Martienssen (1995) Site-selected transposon mutagenesis at the hcf106 locus in maize. *Plant Cell* 7: 287–94.

Dietrich, C. R., F. Cui, M. L. Packila, J. Li, D. A. Ashlock, B. J. Nikolau and P. S. Schnable (2002) Maize *Mu* transposons are targeted to the 5′ untranslated region of the gl8 gene and sequences flanking *Mu* target-site duplications exhibit nonrandom nucleotide composition throughout the genome. *Genetics* 160: 697–716.

Emrich, S. J., L. Li, T. J. Wen, M. D. Yandeau-Nelson, Y. Fu, L. Guo, H. H. Chou, S. Aluru, D. A. Ashlock and P. S. Schnable (2007) Nearly identical paralogs: implications for maize (*Zea mays* L.) genome evolution. *Genetics* 175:429–39.

Emrich, S. J., S. Aluru, Y. Fu, T. J. Wen, M. Narayanan, L. Guo, D. A. Ashlock and P. S. Schnable (2004) A strategy for assembling the maize (Zea mays L.) genome. *Bioinformatics* 20:140–7.

Eveland, A. L., D. R. McCarty and K. E. Koch (2008) Transcript Profiling by 3′UTR Sequencing Resolves Expression of Gene Families. *Plant Physiol* 146:32–44.

Fernandes, J., Q. Dong, B. Schneider, D. J. Morrow, G. L. Nan, V. Brendel and V. Walbot (2004) Genome-wide mutagenesis of Zea mays L. using RescueMu transposons. *Genome Biol* 5: R82.

Frey, M., C. Stettner and A. Gierl (1998) A general method for gene isolation in tagging approaches: Amplification of insertion mutagenised sites (AIMS). *Plant J* 13:717–721.

Fu, S., R. Meeley and M. J. Scanlon (2002) *Empty pericarp 2* encodes a negative regulator of the heat shock response and is required for maize embryogenesis. *Plant Cell* 14:3119–32.

Gallavotti, A., Q. Zhao, J. Kyozuka, R. B. Meeley, M. K. Ritter, J. F. Doebley, M. E. Pè and R. J. Schmidt (2004) The role of barren stalk1 in the architecture of maize. *Nature* 432:630–5.

Gaut, B. S. and J. F. Doebley (1997) DNA sequence evidence for the segmental allotetraploid origin of maize. *Proc Natl Acad Sci, USA* 94: 6809–6814

Gray, J., P. Close, S. Briggs and G. Johal (1997) A novel suppressor of cell death in plants encoded by the *Lls1* gene of maize. *Cell* 89:25–31

Golubovskaya, I. N., O. Hamant, L. Timofejeva, C. J. Wang, D. Braun, R. Meeley and W. Z. Cande (2006) Alleles of *afd1* dissect REC8 functions during meiotic prophase I. *J Cell Sci* 119:3306–15.

Hanley, S., D. Edwards, D. Stevenson, S. Haines, M. Hegarty, W. Schuch and K. J. Edwards (2000) Identification of transposon-tagged genes by the random sequencing of Mutator-tagged DNA fragments from Zea mays. *Plant J* 23:557–66.

Hershberger, R. J., C. A. Warren and V. Walbot (1991) *Mutator* activity in maize correlates with the presence and expression of the *Mu* transposable element *Mu9*. *Proc Natl Acad Sci USA* 88:10198–202.

Holding, D. R., M. S. Otegui, B. Li, R. B. Meeley, T. Dam, B. G. Hunter, R. Jung and B. A. Larkins (2007) The maize floury1 gene encodes a novel endoplasmic reticulum protein involved in zein protein body formation. *Plant Cell* 19:2569–82.

Kent, W. J. (2002) BLAT–the BLAST-like alignment tool. *Genome Res* 12:656–64.

Kolesnik, T., I. Szeverenyi, D. Bachmann, C. S. Kumar, S. Jiang, R. Ramamoorthy, M. Cai, Z. G. Ma, V. Sundaresan and S. Ramachandran (2004) Establishing an efficient Ac/Ds tagging system in rice: large-scale analysis of Ds flanking sequences. *Plant J* 37:301–14.

Kolkman, J. M., L. J. Conrad, P. R. Farmer, K. Hardeman, K. R. Ahern, P. E. Lewis, R. J. Sawers, S. Lebejko, P. Chomet and T. P. Brutnell (2005) Distribution of *Activator* (*Ac*) throughout the maize genome for use in regional mutagenesis. *Genetics* 169: 981–95.

Li, J., L. C. Harper, I. Golubovskaya, C. R. Wang, D. Weber, R. B. Meeley, J. McElver, B. Bowen, W. Z. Cande and P. S. Schnable (2007) Functional analysis of maize RAD51 in meiosis and double-strand break repair. *Genetics* 176:1469–82.

Lid, S. E., D. Gruis, R. Jung, J. A. Lorentzen, E. Ananiev, M. Chamberlin, X. Niu, R. Meeley, S. Nichols and O. A. Olsen (2002) The *defective kernel 1* (*dek1*) gene required for aleurone cell development in the endosperm of maize grains encodes a membrane protein of the calpain gene superfamily. *Proc Natl Acad Sci USA* 99:5460–5.

Lisch, D., P. Chomet and M. Freeling (1995) Genetic characterization of the *Mutator* system in maize: behavior and regulation of *Mu* transposons in a minimal line. *Genetics* 139:1777–96.

Margulies, M., M. Egholm, W. E. Altman, S. Attiya, J. S. Bader, L. A. Bemben, J. Berka, M. S. Braverman, Y. J. Chen, Z. Chen, S. B. Dewell, L. Du, J. M. Fi-erro, X. V. Gomes, B. C. Godwin, W. He, S. Helgesen, C. H. Ho, G. P. Irzyk, S. C. Jando, M. L. Alenquer, T. P. Jarvie, K. B. Jirage, J. B. Kim, J. R. Knight, J. R. Lanza, J. H. Leamon, S. M. Lefkowitz, M. Lei, J. Li, K. L. Lohman, H. Lu, V. B. Makhijani, K. E. McDade, M. P. McKenna, E. W. Myers, E. Nickerson, J. R. Nobile, R. Plant, B. P. Puc, M. T. Ronan, G. T. Roth, G. J. Sarkis, J. F. Simons, J. W. Simpson, M. Srinivasan, K. R. Tartaro, A. Tomasz, K. A. Vogt, G. A. Volkmer, S. H. Wang, Y. Wang, M. P. Weiner, P. Yu, R. F. Begley and J. M. Rothberg (2005) Genome sequencing in microfabricated high-density picolitre reactors. *Nature* 437:376–80.

Martienssen, R. A. and A. Baron (1994) Coordinate suppression of mutations caused by *Robertson's mutator* transposons in maize. *Genetics.* 136:1157–70.

May, B. P., H. Liu, E. Vollbrecht, L. Senior, P. D. Rabinowicz, D. Roh, X. Pan, L. Stein, M. Freeling, D. Alexander and R. A. Martienssen (2003) Maize-targeted mutagenesis: A knockout resource for maize. *Proc Natl Acad Sci USA* 100:11541–6.

McCarty, D. R., A. M. Settles, M. Suzuki, B. C. Tan, S. Latshaw, T. Porch, K. Robin, J. Baier, W. Avigne, J. Lai, J. Messing, K. E. Koch and L. C. Hannah (2005) Steady-state transposon mutagenesis in inbred maize. *Plant J* 44: 52–61.

McGinnis, K., N. Murphy, A. R. Carlson, A. Akula, C. Akula, H. Basinger, M. Carlson, P. Hermanson, N. Kovacevic, M. A. McGill, V. Seshadri, J. Yoyokie, K. Cone, H. F. Kaeppler, S. M. Kaeppler and N. M. Springer (2007) Assessing the efficiency of RNA interference for maize functional genomics. *Plant Physiol* 143:1441–51.

Meeley, R. and S. Briggs (1995) Reverse genetics for maize. *Maize Genetics News Letter* 69:67–82

Mena, M., B. Ambrose, R. Meeley, S. Briggs, M. F. Yanofsky and R. J. Schmidt (1996) Diversification of C-function activity in maize flower development. *Science* 274:1537–1540.

Muskett, P. R., L. Clissold, A. Marocco, P. S. Springer, R. Martienssen and C. Dean (2003) A resource of mapped dissociation launch pads for targeted insertional mutagenesis in the Arabidopsis genome. *Plant Physiol* 132, 506–16.

Palmer, L. E., P. D. Rabinowicz, A. L. O'Shaughnessy, V. S. Balija, L. U. Nascimento, S. Dike, M. de la Bastide, R. A. Martienssen and W. R. McCombie (2003) Maize genome sequencing by methylation filtration. *Science* 302: 2115–7.

Park, W. J., V. Kriechbaumer, A. Möller, M. Piotrowski, R. B. Meeley, A. Gierl and E. Glawischnig (2003) The Nitrilase ZmNIT2 converts indole-3-acetonitrile to indole-3-acetic acid. *Plant Physiol* 133:794–802.

Pawlowski, W. P., I. N. Golubovskaya, L. Timofejeva, R. B. Meeley, W. F. Sheridan and W. Z. Cande (2004) Coordination of meiotic recombination, pairing, and synapsis by PHS1. *Science* 303:89–92.

Porch, T. G., C. W. Tseung, E. A. Schmelz and A. M. Settles (2006) The maize *Viviparous10/ Viviparous13* locus encodes the Cnx1 gene required for molybdenum cofactor biosynthesis. *Plant J* 45: 250–63.

Rabinowicz, P. D., K. Schutz, N. Dedhia, C. Yordan, L. D. Parnell, L. Stein, W. R. McCombie and R. A. Martienssen (1999) Differential methylation of genes and retrotransposons facilitates shotgun sequencing of the maize genome. *Nat Genet* 23:305–8.

Raizada, M.N. (2003) RescueMu protocols for maize functional genomics. *Methods Mol Biol* 23: 37–58.

Robertson, D. S. (1978) Characterization of a mutator system in maize. *Mutat. Res* 51: 21–28.

Robertson, D. S. (1980) The Timing of *Mu* Activity in Maize. *Genetics* 94:969–978.

Rudenko, G. and V. Walbot (2001) Expression and post-transcriptional regulation of maize transposable element *MuDR* and its derivatives. *Plant Cell* 13:553–570.

Settles, A. M., S. Latshaw and D. R. McCarty (2004) Molecular analysis of high-copy insertion sites in maize. *Nucleic Acids Res* 32: e54.

Settles, A. M., D. R. Holding, B. C. Tan, S. P. Latshaw, J. Liu, M. Suzuki, L. Li, B. A. O'Brien, D. S. Fajardo, E. Wroclawska, C. W. Tseung, J. Lai, Hunter CT, W. T. Avigne, J. Baier, J. Messing, L. C. Hannah, K. E. Koch, P. W. Becraft, B. A. Larkins and D. R. McCarty (2007) Sequence-indexed mutations in maize using the UniformMu transposon-tagging population. *BMC Genomics* 8:116.

Schuler, G. D., S. F. Altschul and D. J. Lipman (1991) A workbench for multiple alignment construction and analysis. *Proteins* 9:180–90.

Slotkin, R. K., M. Freeling and D. Lisch (2003) *Mu* killer causes the heritable inactivation of the *Mutator* family of transposable elements in Zea mays. *Genetics* 165:781–97.

Suzuki, M., A. Mark Settles, C. W. Tseung, Q. B. Li, S. Latshaw, S. Wu, T. G. Porch, E. A. Schmelz, M. G. James and D. R. McCarty (2006) The maize vi*viparous15* locus encodes the molybdopterin synthase small subunit. *Plant J* 45: 264–74.

Till, B. J., S. H. Reynolds, C. Weil, N. Springer, C. Burtner, K. Young, E. Bowers, C. A. Codomo, L. C. Enns, A. R. Odden, E. A. Greene, L. Comai and S. Henikoff (2004) Discovery of induced point mutations in maize genes by TILLING. *BMC Plant Biol* 4:12.

Walbot, V. (2000) Saturation mutagenesis using maize transposons. *Curr Opin Plant Biol* 3:103-7.

Yamazaki, M., H. Tsugawa, A. Miyao, M. Yano, J. Wu, S. Yamamoto, T. Ma-tsumoto, T. Sasaki and H. Hirochika (2001) The rice retrotransposon Tos17 prefers low-copy-number sequences as integration targets. *Mol Genet Genomics* 265:336–44.

Yuan, Y., P. J. SanMiguel and J. L. Bennetzen (2003) High-Cot sequence analysis of the maize genome. *Plant J* 34:249–55.

TILLING and Point Mutation Detection

Clifford Weil and Rita Monde

1   Introduction

Maize reverse genetics is extremely powerful because of the wealth of active transposable elements still residing in the maize genome. In addition to serving as molecular tags for mutated genes, these elements tend to knock out genes into which they insert. Once an initial, reference allele has been identified, it facilitates obtaining an allelic series of mutations within a gene that have a range of phenotypes. Such an allelic series can provide great insight into the function of the gene and its cognate protein. In addition to spontaneous mutations, this allelic series is typically the result of chemical mutagenesis to create point mutations at locations throughout the gene.

Forward, "targeted" mutagenesis has long been the way to do this. Mutagenizing with ethyl methane sulfonate (EMS) and then crossing onto a line homozygous for a reference allele, any mutant progeny that arise in the F1 represent new mutations in the gene of interest. While effective, a new mutagenesis is required for each gene of interest. The capacity to screen any mutagenized population using reverse genetic techniques for any gene of interest is therefore valuable as well.

Targeting Induced Limited Lesions IN Genomes (TILLING) is a reverse genetic method for screening a collection of mutagenized lines for single base changes in any gene [1,2]. These mutagenized lines also serve as a general forward genetic resource, with the advantages of having a clear molecular measure of the mutation density in those lines and reducing the effort necessary for the community to carry out general forward screens. The Maize TILLING Project (MTP) started doing TILLING screens in 2005 using mutant populations developed by our group and several other laboratories.

C. Weil and R. Monde
Agronomy Dept. and Whistler Center for Carbohydrate Research
Purdue University, 915 W. State St.
West Lafayette, IN 47906 USA
cweil@purdue.edu

J.L. Bennetzen and S. Hake (eds.), *Maize Handbook - Volume II: Genetics and Genomics,*   585
© Springer Science + Business Media LLC 2009

2 TILLING Mutagenesis

Treatments with EMS for reverse genetics are typically higher than those for forward genetic screens. The goal is to achieve the maximum mutation density (taken as mutations per kb per 1000 mutant families) that still permits viable, fertile plants, thus minimizing the number of lines needed to identify mutations in any given gene. Because EMS treatment is highly dependent on time and temperature we have tried to develop a consistent and reproducible treatment procedure. As of this writing, our current procedure utilizes 0.09% EMS well-dispersed in paraffin oil at 30°C for 45 minutes, with shaking at 3 minute intervals. We achieve the temperature control for the mutagenesis by using an incubator in a building near the field. Fewer, but more mutagenized seed can be obtained at slightly higher concentrations of EMS. The EMS should be freshly purchased and is best made into a 1% v/v emulsion stock in paraffin oil, kept in a brown glass bottle (EMS is light sensitive) by vigorous stirring on a stir plate. This stock is useable for several weeks but needs to be stirred at high speed for ~30min prior to each use to redisperse the EMS. Dilution to a mutagenesis solution is made in 100ml in a fresh plastic squirt bottle. The stock should be kept stirring vigorously while withdrawing the aliquot for the dilution as the EMS begins to settle out quickly. We typically make the dilution first thing in the morning and then place it in a shaking incubator at 30°C to equilibrate while pollen begins to shed and is collected, typically 1-2 hours. Pollen is collected from plants as soon as it begins to shed and is filtered through a piece of screen over a plastic funnel to remove anthers into a fresh tassel bag. We find that collecting into a plastic tube tends to lead to clumping of the pollen and lower viability. Collection of 20-30 tassels can be done quickly, and then a graduated 50ml plastic tube is used to measure 10ml of pollen; again trying to have the pollen in the plastic tube for as little time as possible. This pollen is immediately put into a fresh tassel bag, transferred to the EMS solution, the bottle capped and shaken to disperse the pollen and the bottle placed in the incubator, but without shaking. Within 2-3 minutes the pollen begins to settle and the bottle is shaken to stir both pollen and EMS. We have noticed that constant shaking at speeds sufficient to keep the pollen and EMS dispersed can decrease the viability of the pollen somewhat and testing this process empirically is encouraged. The bottle of pollen is taken to the field such that the application of the pollen to silks begins at ~45–50min after the start of treatment. The bottle is shaken gently before each application to be sure that the pollen is well-mixed and 0.5-1ml of solution is added to each cut back ear (on which silks have regrown to 2-3cm in length). Ears cut back 18-48h prior to pollination give best results, depending on the inbred and weather conditions.

It is important to note that the M2 and M3 plants that result from this mutagenesis can have a wide range of developmental defects and carry as many as 15,000 different mutations in any one line. While many of these mutations will be silent, 400-500 will be within protein coding exon and backcrossing lines where mutations of interest are identified is strongly advised. At the same time, however, there is analytical power in an allelic series. If, for example, five different mutant lines carrying different mutations in a gene of interest all share a similar mutant phenotype, it is highly unlikely that all five lines share a mutation anywhere else in the genome.

The efficacy of the mutagenesis can be assessed directly in two ways. EMS mutagenesis of pollen in maize allows treatment of pollen in one genotype and crossing of that pollen onto a different genotype. Even though the goal is a generally mutagenized population, a targeted mutagenesis for an easily scored, visible kernel phenotype is a useful tool for measuring the success of the mutagenesis. For example, if *Bz1/Bz1* pollen is mutagenized, using that pollen to fertilize a *bz1* mutant will produce a bronze kernel anytime the *Bz1* gene is mutagenized to inactivity. Note that the female parent in these cases should be detassled if possible to avoid self-contamination and that the bronze kernels in this example are unlikely to carry transmissible *bz1* alleles because mutagenized pollen already has two sperm nuclei, only one of which carries the *bz1* mutation.

The more accurate assessments turn out to be using TILLING itself, screening a new population for the occurrence of mutations in a test set of 10-20 genes across a sample of 384-768 M1 plants and extrapolating.

3   Tissue Collection, DNA Prep and 2-D Pooling

Plants from kernels that result from pollen mutagenesis are heterozygous for any induced mutations they carry. Tissue from these plants is collected by allowing the plants to germinate and then, once the plants have emerged, cutting the top 5 cm of the shoot when the plant is ~10-15 cm tall and placing the furled scutellum and leaves into a 2 ml tube in a 96-well rack. The meristem, still well below the cut, is unharmed and the plants continue to grow normally. Tissue can also be sampled from mature plants (leaf punches or simply tearing strips from healthy leaves), however, we have noticed that the DNA shows some degradation not seen in DNA isolated from young tissue. Both the plant and the tube containing the tissue are labeled with matching barcoded tags. Any fertile plants are selfed to create M2 ears (see below).

The tissue samples are kept at −80°C until they are freeze-dried. Lyophilized tissue is stored at −20°C until DNA preparation. Any of several preparations work well. We use a FastPrep instrument (Q-BIOgene) and a resin-based preparation kit, although paintshaker/CTAB preps are also effective as are 96-well, resin based kits. Using ~2-5mg of the dried, powdered tissue (~ 20mg if using tissue from mature plants) samples yield ~1g of high molecular weight, high quality DNA, more than enough DNA for multiple TILLING screens. The amount of DNA is determined by comparing the DNA sample against known amounts of uncut bacteriophage lambda DNA, then comparing band intensity levels using ethidium staining and a CCD camera. We find that use of MetaPhor agarose (Lonza) greatly improves the quality of the gel images, particularly if mature tissue is the source of the DNA, in that the gDNA runs as a tighter band with less smearing. Once concentrations of the DNA are established, the samples are all brought to 3g/ml.

Individual DNAs are arrayed in 8 X 8 grids in 96 well plates, deliberately leaving 4 columns blank to prevent confusion orienting the plate. Eight-fold pooled samples are prepared by collapsing each row into a single well of one column on a "pool plate" and then collapsing each column of the same 8 X 8 grid into the wells of the adjacent column on the pool plate. Each pool plate this represents six 8 X 8 grids of individuals, each individual is represented twice and has a unique row and column "address". Pool plates are arrayed, still at 3g/ml, and used to make dilutions as needed of 0.5 g/l in TE pH 7.4. The stock plate is relatively stable at 4°C, and best results are obtained if the dilutions are used within a week of being mad.

4   Expansion of Populations for Seed Coverage

While tissue of M1 plants is harvested for TILLING, the goal is to be able to return mutant seed to the community for further study. We therefore self each M1 plant to create M2 ears. Typically ~40-50% of the M1 are self-fertile. Because of the mutational load these M1 plants carry, the M2 ears often have only 40-100 kernels on them. To increase the number of seed with a minimum of disruption to the allele frequency of any induced mutations, we plant families of 40 M2 kernels and then randomly intermate to create a family of M3 ears. Kernels from these M3 ears are then bulked together and samples are distributed to the community by drawing 30-40 random seed from these bulks. For any given mutant allele, the overall distribution of mutant alleles in the M3 seed remains similar, though not identical to the M2. Each M3 family has segregating within it thousands of mutations. While backcrossing of each allele to clean up the genetic background is advisable, if multiple lines carrying different alleles of a given gene all have similar phenotypes it is likely that those mutations are the cause of the observed phenotype because it is unlikely that all of the lines also share mutations in a second gene.

5   TILLING Workflow

5.1   CODDLe Bioinformatics Portal

TILLING initiates with a researcher identifying genomic sequence of interest and a gene model (introns and exons). If the precise gene model is not yet known, a preprocessing utility is available for developing one using protein sequence and genomic DNA. This option can be particularly useful when looking for mutations in genes that have been identified by comparison with other plants, often using TBLASTN. Translation in all reading frames of the maize genomic sequence can often allow identification of regions where the known protein aligns with translation of the maize DNA and canonical splice sites can be identified flanking these regions of alignment to produce working gene models. Using the Codons Optimized for Determining Deleterious Lesions (CODDLe) program (http://www.proweb.org/input/), this sequence can be scanned for regions that are most likely to provide G to A changes that will result in damage or truncation of the gene product. Using the BLOCKS software [3]. CODDLe scans the input sequence for regions of conservation with known sequence motifs and then looks for nonconservative amino acid changes within these conserved regions, nonsense mutations or changes to splice sites (likely to produce framseshifts and premature stop codons). A region of ~1500 bp is selected that has the highest potential for damaging mutations. The user then has the option of accepting this region for analysis or directing CODDLe to analyze another portion of the gene.

Once the sequence to be analyzed has been selected, the program asks whether primers should be chosen to amplify this region. This is done using Primer3, software developed at MIT [4], that has been preset with the PCR parameters used in TILLING (Fig. 1). This step ensures that the primers selected have the best chance of working and eliminates the need to adjust PCR conditions for each new target, thereby improving throughput. A list of primers is returned to the user, who then selects a primer pair and submits the order to MTP via this web-based interface. Once the TILLING request is placed, unlabeled primers are ordered and sent to MTP for prescreening.

5.2   Primer Pre-Screening

Given the ancient tetraploid nature of the maize genome it is important to identify gene-specific primers for TILLING so that polymorphisms between paralogs are not mistaken for induced mutations in a gene of interest. The gene target is amplified from the B73 and W22 inbred lines using TILLING PCR conditions and the amplicon analyzed to ensure that there is only one product and that single primer controls have no product. In addition, the amplicons are sequenced to verify that they are indeed single products and to establish a wild-type reference sequence for both inbreds against which to screen for mutations. If the primers pass these prescreens, the same primers are re-ordered, this time with an infrared fluorescent dye at the 5′ end (IRD700 for the left primer and IRD800 for the right primer). The dyes provide high sensitivity for detection of products with mutations in them (see Fig. 2). If the unlabeled primers do not pass the prescreen, the researcher that submitted the order is notified and allowed to select a new set of primers. MTP works with the researcher to help select the new primers, and >95% of these second rounds of primer selection have been successful.

It is often easier and faster if researchers prescreen primer pairs themselves before placing an order. The PCR protocol (Fig. 1) is a touchdown procedure, and ramp times of the thermal cycler are critical. These conditions are necessary because the dyes themselves appear to interfere somewhat with PCR; thus, a mixture of labeled and unlabeled primers are used in TILLING screens. The prescreening

	Per reaction	Per 100 reactions (made for one 96-well plate)
10 X Ex-Taq buffer	0.49µl	48.7µl
25mM MgCl$_2$	0.58µl	58.1µl
dNTPs (2.5mM for each)	1.18µl	118µl
ddH$_2$O	2.40µl	240µl
TakaraEx-Taq	0.05µl	5.0µl
	4.70µl	470µl
0.9 ng/µl genomic DNA	5.00µl	50.0µl
10µM Left primer	0.15µl	Note: substitute appropriate amount of
10µM Right primer	0.15µl	d d H$_2$O in single and no primer controls
	10.00µl	

The following nine reactions are recommended for each primer pair (TILLING populations are in both B73 and W22 inbreds; if using mutagenized populations of other inbreds they should be inserted here):

B73 Left+Right primers (in duplicate); B73 Left primer only;   B73 Right primer only
W22 Left+Right primers (in duplicate); W22 Left primer only;   W22 Right primer only
TE Left+Right primers;

Thermocycler Profile:

1 cycle of 95 °C 2 minutes

Loop 1: *8 cycles*
    94 °C 20 seconds
    73 °C 30 seconds
        decrease annealing temperature by 1.0 °C per cycle
    Ramp to 72 °C at 0.5 °C /second
    72 °C 1 minute

Loop 2: *45 cycles*
    94 °C 20 seconds
    65 °C 30 seconds
    Ramp to 72 °C at 0.5 °C /second
    72 °C 1 minute

Final extension at 72 °C for 5 minutes

Hold (4 - 10 °C)

Run 2 µl of sample on a 1% agarose gel. Include on this gel 4 µl of Low DNA Mass Ladder (Invitrogen) or an analogous quantitative ladder and a sizing ladder (*e.g.,* 1 Kb Plus ladder (Invitrogen)).

If one band of expected size in the template+Left+Right primer samples, DNA sequence the remaining PCR product directly.

Fig. 1 TILLING Primer Prescreening Protocol

protocol uses the same thermocycler profile and PCR reagent cocktail (though only unlabeled primers). The single primer controls in this prescreen are important because, while not always the case, a primer that generates PCR products on its own often results in a reduced amount of full-length product and an increased amount of non-specific bands, making TILLING gel images difficult to interpret. If a single primer control has one or two low intensity bands, it is generally useable; however, if it generates multiple products and any of these are abundant, it is best to try a different primer.

A robust product is best for successful TILLING. Primers that do not amplify well at this step will be problematic in TILLING because the amount of product will not be sufficient and the gels will be difficult to interpret. Ideally, there should be a single band on an agarose gel of the correct molecular weight for your target. The intensity of 2l of the PCR product should be 120 ng, (taking care not to oversaturate the gel image during quantitation).

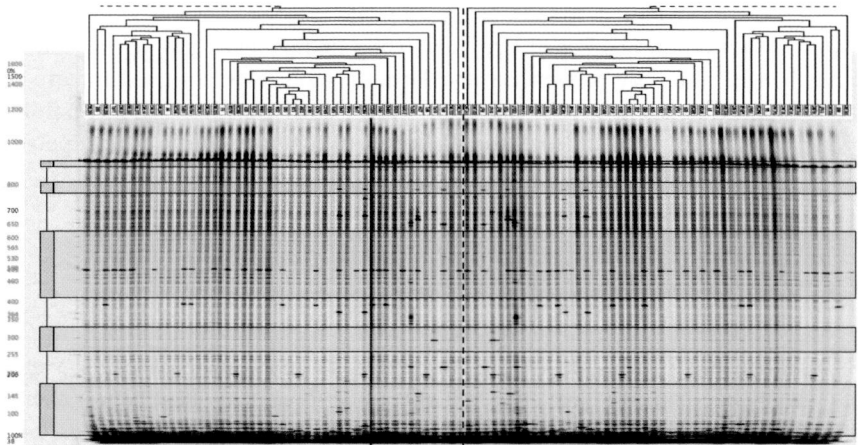

Fig. 2 EcoTILLING Gel Analysis. Size standards (in bp) are shown at far left. Exons (boxes) and introns (lines) of gene model for this target also at left, with extent of exons indicated across the gel. A Phylogenetic tree of the Diversity Lines compared to B73 in each lane is shown at the top of the gel. Each sample is loaded twice for confirmation and the patterns are in mirror image. Only the 700nm wavelength image is shown. Bands are verified by their occurrence in corresponding lanes of the mirror image and the presence of complementary-sized bands in the 800nm channel

5.3 TILLING (mismatch detection)

TILLING works on the premise that point mutations in one copy of a gene that is amplified as part of a pool of nonmutant copies will result in a portion of the amplified DNA having a mismatched base pair where a mutant DNA strand has reannealed with a nonmutant strand. An S1 family nuclease that recognizes the mismatch is used to cleave 3′ to the mismatched bases [5]; this enzyme is typically Cel1, which is easily and inexpensively purified from celery.

In pooled samples where there are no mutations, all amplicons have the same sequence, there are no mismatches to cleave and the entire amplicon is full length with one IRD700 and one IRD800 label at each end. In contrast, if a mutation is present within a pool, mismatched duplexes form between the mutant and the wild type copies and the Cel1 cleaves, producing two molecules of less than full length, complementary in size, one with IRD700 and the other with IRD800 at its 5~ end. These labeled subfragments are distinguished readily in Li-COR, two-color DNA sequencers and positive signals are identified as paired bands (one from a row pool and one from a column pool) that have complementary-sized mates in the opposite channel. The two dimensional readout of the slab gel format has proven valuable for confirming the presence of mutations and minimizing the false positive rate, and works very well in combination with the two-dimensional pooling of the DNA templates described earlier. Mutant individuals are then re-amplified separately and the PCR product sequenced to determine the precise location of the mutation. Bands in the Li-COR slab gels can be estimated to within ~10-20 bp of their actual size, facilitating finding and verifying the correct base change. Capillary sequencers and labeling molecules designed for use in them are more amenable to automating the system, and have been used successfully [6], but often require additional verification work downstream and have poorer resolution of fragment size.

5.4   Analysis and Delivered Readout

As a service, MTP will screen all existing and future mutagenized lines for alleles of a given gene until we have delivered either one truncation allele or two alleles that are predicted to have severe damaging effects on the protein sequence based on the program Sorting Intolerant From Tolerant (SIFT; [7,8]). Any other alleles that are discovered are also returned to the user and, once the delivery algorithm is satisfied, screening is stopped. In the event these alleles are not found, additional mutagenized material is screened and the order remains open for further screening at no additional charge until the delivery algorithm is met. The information is sent electronically but is kept confidential for six months, after which it goes onto a searchable, general TILLING website and to MaizeGDB.

6   TILLING in Nongenic DNA

The initial goal of TILLING has been delivery of mutations within protein coding sequence to help functional characterization of genes. More recently, interest has grown in conserved noncoding DNA, primarily to gain a better understanding of gene regulation. TILLING can be applied to screening noncoding DNA for induced single base changes using the same methodology described above, but starting with a different bioinformatic front end. CODDLe is designed to screen protein coding sequence by examining effects on translated products. It therefore rejects DNA sequence that does not have a gene model associated with it. It is worth noting that changes in conserved noncoding regions may also be important for the function of the adjacent gene; however, databases on such conserved motifs are only just emerging. Just as with standard TILLING, screens of noncoding DNA are carried out as SNP detection among mutagenized inbred lines. All alleles are returned to the user, at present, with no informatic analysis of what the changes might or might not be doing to gene expression, and it is then up to the researcher to decide which alleles are of interest. For example, all the alleles in a putative promoter region can be tested for transcript levels of the associated gene to determine which have an effect. At present, requests to screen noncoding sequence are submitted directly to MTP without going through the usual first steps with CODDLe. In addition, because there is no way at present to evaluate the utility of the alleles found in noncoding sequence, for these requests we will screen a fixed number of lines (~3200).

7 EcoTILLING for Natural Allelic Variation in Gene Targets

In addition to delivering induced alleles of a requested gene, MTP also provides an assessment of the natural allelic diversity available for that gene among the Maize Diversity Lines ([9] and covered elsewhere in this book) and ten recently off-patent inbreds used in commercial breeding. EcoTILLING, originally devised for comparing Arabidopsis ecotypes [10], compares different accessions (in maize, inbred lines) to a reference genome and assays single nucleotide polymorphisms (SNPs). Comparing a representative set of the Diversity Lines in pairwise fashion with the reference B73 genome, one can quickly and easily identify SNPs in the diverse inbreds using TILLING. Because the gene models are known, variation in exons can be differentiated from that in introns and the gene sequenced in the relevant inbred (Fig. 33.2). This method serves as a useful initial screen so that any sequencing can be targeted more effectively.

8 Looking Forward at Reverse Genetics: Targeted Resequencing Using Massively Parallel Methodologies (SequeTILLING)

Over the past two decades, dideoxy (or "Sanger") sequencing has been the predominant sequencing method. Even with extensive automation of the sequencers, Cell-based TILLING has proven to be an efficient and cost-effective way to screen large mutant populations for point mutations in specific gene targets. The alternative, re-sequencing individual genes across thousands of mutant lines, is simply too expensive using Sanger sequencing.

Recently, massively parallel sequencing methods have arisen that have driven the cost of sequencing dramatically downward. Methods such as 454 sequencing [11], four color, sequence-by-synthesis [12] (initially "Solexa sequencing" and now marketed by Illumina) and ligase-based methods [13]) such as that available in the Applied Biosystems SOLiD instrument have also increased the output of single runs of these sequencers to the level of gigabases. The costs per gene are now relatively low (even if the initial outlay for the instrumentation is still high). As a result, resequencing genes across thousands of mutagenized lines has not only become feasible, dozens of genes can be analyzed simultaneously. While resequencing thousands of entire mutagenized genomes is not yet feasible, targeting the resequencing to specific genes is. However, this prospect presents several technological and informatic challenges.

Resequencing mutagenized inbred lines has a distinct advantage over de novo sequencing in that the wild-type sequence serves as a scaffold for sequence assembly. A key feature to using massively parallel methods for sequencing multiple genes across multiple lines is to make this assembly as straightforward as possible. For sequencing instruments that have short read lengths, computer simulations using ~100 kb collections of random maize genes have indicated that, operationally, this means having no repeated sequences of longer than 20 bp within all the genes being analyzed in any given run of the instrument.

Whereas Cell TILLING uses 8-fold pooling of individuals to ensure robust signal quality (thus one mutant allele in 16), massively parallel sequencing instruments allow detection of one allele in 80 and, thus, 40-fold pooling of diploid individuals (E. Cuppen, U. Utrecht, pers. com.). For example, identifying individuals in "2-D pools" by a unique row and column "address", 3200 individuals could be represented by 160 pools developed from two 40 X 40 arrays.

Using available 96-well liquid handling, two genes at a time can be amplified easily from these 160 pools (thus, 320 samples) in one run of a 384-well PCR thermal cycler. Once all the genes for a sequencer run have been amplified, the amplicons made from each individual pool can then be combined, creating 160 samples each containing all the genes being tested.

8.1   Amplicon Shearing

The ABI SOLiD and Ilumina 1G instruments read 25-35 bases at the ends of DNA fragments, thus random shearing is required to ensure each base in the amplicons occurs within 25-35 bases of a fragment end. The Roche 454 instrument is capable of longer reads (up to hundreds of bases), but random shearing is still necessary to ensure that the entire gene is sequenced. Keeping the amplified samples in their microplate format facilitates the workflow and several 96-well sonicating devices are now commercially available to reduce the 2-4 kb amplified genes to ~100-200 bp fragments. These fragment lengths provide better coverage of the sequences and a more effective size for subsequent steps (either emulsion PCR (ePCR) for 454 or SOLiD sequencing or bridge amplification to create sequence clusters (for the Illumina instrument).

8.2   Using Barcode Sequence Tags to Identify Individuals

Once the amplicons are sheared, identifying individuals within the pooled samples will require adding short "barcode sequence tags" unique to each pool to associate a unique row and column address with DNA sequence reads later. A major hurdle in applying the new generation of sequencers to mutation screening has been that they read sequences in short stretches of 25-35 bases at a time. Thus, attaching only one barcode sequence and not a chain of several at each end of a fragment is important. Placing the barcode sequence in a linker adjacent to a site that is cleaved only in the linker and not in the amplified fragment would be one way to solve this problem, leaving only one barcode at each end of the sheared fragments. The two adapters required for whatever sequencing platform is being used would then be ligated onto the barcoded fragment ends, and all the fragments from all the pools sequenced *en masse*.

8.3   Assembly and Analysis

Using the previously determined wild-type sequences as a scaffold, each fragment sequenced can be placed unambiguously on that scaffold and aberrant reads easily identified and removed. Using the scenario described above, at the 5′ end of each read would be the remainder of a restriction site unique to the barcode linker and the 4 bp barcode for each fragment, followed by fragment sequence of varying length depending on the sequence platform used. Mutations would be identified readily as a single base change that occurs repeatedly for an individual (recognizable because it always has either one of two barcodes that correspond to that individual). False positives would be minimized because of the depth of the sequence coverage (typically ~3-5X or greater for each individual allele). Because of the tremendous numbers of reads per instrument run, this methodology also provides the flexibility to screen more genes per run on smaller populations, or smaller numbers of genes in larger populations.

9   Conclusion

Characterizing the function of all the genes in maize will require a spectrum of mutations in each gene, from knockouts to those with partial and even novel functions. Single base changes provide the necessary, wide range of effects, and are particularly useful in conjunction with transposon-based gene discovery and gene knockout methods. The maize community has established efficient ways for creating various types of mutant (and natural) populations as well as propagating and distributing the seed stocks. As the maize genome sequence nears completion, advances in reverse genetics technologies such as TILLING, EcoTILLING and, soon, massively parallel DNA resequencing provide excellent methods for identifying mutations in a wide variety of traits and biological processes., and getting mutant seed carrying them into the hands of the research community.

Acknowledgements We particularly thank Jorja Henikoff of the proWeb group at Fred Hutchinson Cancer Research Center for database and informatics support. MTP has been supported by funding from the National Science Foundation and the United States Dept. of Agriculture.

References

McCallum CM, Comai L, Greene EA, Henikoff S: Targeting induced local lesions IN genomes (TILLING) for plant functional genomics. *Plant Physiol* 2000, 123:439–442.

Till BJ, Colbert T, Tompa R, Enns LC, Codomo CA, Johnson JE, Reynolds SH, Henikoff JG, Greene EA, Steine MN, et al.: High-throughput TILLING for functional genomics. *Methods Mol Biol* 2003, 236:205–220.

Henikoff JG, Pietrokovski S, McCallum CM, Henikoff S: Blocks-based methods for detecting protein homology. *Electrophoresis* 2000, 21:1700–1706.

Rozen S, Skaletsky H: Primer3 on the WWW for general users and for biologist programmers. *Methods Mol Biol* 2000, 132:365–386.

Till BJ, Burtner C, Comai L, Henikoff S: Mismatch cleavage by single-strand specific nucleases. *Nucleic Acids Res* 2004, 32:2632–2641.

Perry JA, Wang TL, Welham TJ, Gardner S, Pike JM, Yoshida S, Parniske M: A TILLING reverse genetics tool and a web-accessible collection of mutants of the legume Lotus japonicus. *Plant Physiol* 2003, 131:866–871.

Ng PC, Henikoff S: Predicting deleterious amino acid substitutions. *Genome Res* 2001, 11:863-874.

Ng PC, Henikoff S: SIFT: Predicting amino acid changes that affect protein function. *Nucleic Acids Res* 2003, 31:3812–3814.

Liu K, Goodman M, Muse S, Smith JS, Buckler E, Doebley J: Genetic structure and diversity among maize inbred lines as inferred from DNA microsatellites. *Genetics* 2003, 165:2117–2128.

Comai L, Young K, Till BJ, Reynolds SH, Greene EA, Codomo CA, Enns LC, Johnson JE, Burtner C, Odden AR, et al.: Efficient discovery of DNA polymorphisms in natural populations by Ecotilling. *Plant J* 2004, 37:778–786.

Margulies M, Egholm M, Altman WE, Attiya S, Bader JS, Bemben LA, Berka J, Braverman MS, Chen YJ, Chen Z, et al.: Genome sequencing in microfabricated high-density picolitre reactors. *Nature* 2005, 437:376–380.

Seo TS, Bai X, Kim DH, Meng Q, Shi S, Ruparel H, Li Z, Turro NJ, Ju J: Four-color DNA sequencing by synthesis on a chip using photocleavable fluorescent nucleotides. *Proc Natl Acad Sci U S A* 2005, 102:5926–5931.

Shendure J, Porreca GJ, Reppas NB, Lin X, McCutcheon JP, Rosenbaum AM, Wang MD, Zhang K, Mitra RD, Church GM: Accurate multiplex polony sequencing of an evolved bacterial genome. *Science* 2005, 309:1728–1732.

Gene Expression Analysis

David S. Skibbe and Virginia Walbot

Abstract A brief history of methods used to elucidate protein function, protein presence, and RNA transcript presence is provided. Gene expression profiling through microarray hybridization, high throughput sequencing, or quantitative reverse transcriptase-polymerase chain reaction methods are reviewed and compared. Proteomics analysis using two dimensional gel electrophoresis followed by protein identification by mass spectrometry is then discussed. Relative costs and prospects for future improvements are also presented.

1 Introduction

To understand how the genome programs plant physiology and development, it is crucial to determine where and when each gene is expressed at the RNA and protein levels and to determine the function(s) of the encoded protein products. The purpose of this Chapter is to review the current methods of gene expression analysis available for maize and to consider methods now in development that will be applicable in the near future.

2 Biochemical Functions of Proteins

Historically, protein functions were more accessible than information about RNA and protein distribution. Many biochemical pathways were worked out before the discovery of messenger RNA! Protein function analysis traditionally required

D.S. Skibbe
Stanford University, Department of Biology
skibbe@stanford.edu

Virginia Walbot
Stanford University, Department of Biology
walbot@stanford.edu

J.L. Bennetzen and S. Hake (eds.), *Maize Handbook - Volume II: Genetics and Genomics,*   597
© Springer Science+Business Media LLC 2009

(partial) protein purification, a feat requiring strong skills in biochemistry, and developing a suitable assay. For these reasons, most proteins with well-defined functions are enzymes with readily assayed activities or structural constituents of readily purified complexes such as ribosomes. Today putative functional assignments are made based on sequence similarity to proteins of known function in other organisms, however, the "gold standard" remains a biochemical demonstration of function. The assignment of dehydrogenase, for example, may be correct, but is insufficient without knowing the substrate(s) metabolized by the enzyme. Therefore, it is important to remember that in most cases sophisticated transcriptome and proteome analyses do not resolve protein functions. Of course, maize genetics has the advantage that many loci are studied because they confer a recognizable phenotype; careful phenotypic analysis can refine where and when the gene product is required and genetic analysis can uncover epistatic interactions with other genes and ultimately connect an unknown gene product to known pathways.

Although we will not discuss protein function analysis in any detail, it is clearly an area in which new breakthroughs of protein purification, protein production with appropriate post-translational modifications in transgenic hosts, availability of antibodies to maize proteins, and scaled-up methods for determining functions are required for there to be rapid progress. Today it is much more facile to quantify RNA abundances and distribution (the transcriptome), followed by the amount and distribution of proteins and post-translationally modified isoforms (the proteome), while identification of precise protein functions remains a laborious procedure. High throughput, massively parallel methods for assessing the transcriptome and proteome have made information about gene expression activity and cellular protein constituents relatively inexpensive and accessible to those with only modest expertise in biochemistry.

3   Transcriptome Profiling

There are multiple steps in the RNA-related components of gene expression: primary transcript synthesis dependent on the rate of transcription, processing (such as 5′ capping, intron splicing, polyadenylation at the 3′ end, and nuclear export), translation (reflecting the proportion of cytoplasmic transcript engaged with ribosomes), and finally mRNA turnover. The easiest aspect of gene expression to quantify is the abundance of poly(A)+ mRNA; these transcripts are primarily found in the cytoplasm and most are assumed to be available for or being actively translated. Nonetheless, it is important to remember that quantifying mRNA abundance does not define transcription rate and may or may not be correlated with the abundance of the encoded protein. Perdurance, the persistence of mRNA beyond the time at which it will be translated, will cause a gene to be considered expressed (and by implication important) at stages after its true time of action.

Maize is a very favorable species for transcriptome profiling because of the large size of the plant body and of the individual organs. It is relatively quick to dissect sufficient tissue from almost any life stage for an experiment. For limited samples such as laser dissected cell types, two stage RNA amplification methods have been and continue to be developed for applications in medicine, and these methods are suitable for any purified RNA sample.

3.1 Hybridization Technologies

RNA blot (northern) hybridization was the traditional method for observing mRNA abundance in isolated total or poly(A)+ RNA samples. This one gene at a time method is accurate but laborious. A major technical advance started with the advent of spotting purified cDNAs onto membranes or glass slides to interrogate them with radioactively or fluorescently labeled cDNA (or cRNA) synthesized from purified RNA. Starting with a platform of 45 cDNAs (Schena et al. 1995), there has been rapid progress in increasing the number and quality of the probe sets displayed. Next came high density spotted cDNA arrays, incorporating up to ~20,000 maize genes, and then oligonucleotide arrays, ranging from 25-mer length from a photolithography-based synthesis system with multiple elements per gene (Affymetrix, current platform for 13,339 genes as described at http://www.affymetrix.com/products/arrays/specific/maize.affx) to synthetic 70-mers spotted onto glass slides (NSF-funded Maize Oligonucleotide Array Project with 58,000 spotted oligos representing about 40,000 genes as described at http://www.maizearray.org/). Already there are platforms to interrogate nearly all maize genes (i.e. 44,000 Agilent array printed 4 x 44K as described on http://www.chem.agilent.com/Scripts/PCol.asp?lPage=494); when the genome is finished and annotated, platforms will be constructed to follow expression of all maize genes.

A key difference among platforms is how quickly new designs can be implemented (Table 1). With Affymetrix arrays, which have become the standard in many genetic communities, a design is finalized and used for one or several years, because the cost of manufacturing the masks for the synthetic reactions is very high. This platform is sensitive to SNPs (even with fifteen 25-mer oligos in the current maize design, see Kirst et al. 2006 for a discussion). As maize is highly polymorphic at many loci and geneticists use a wide variety of inbred and landraces in genotyping, the SNP sensitivity of very short oligo platforms may restrict them to experiments conducted with B73 and a core of inbreds with substantial genomic and cDNA sequence available. New software and more data about maize gene diversity in inbred and exotic lines may permit extracting useful SNP data from Affymetrix chips. The 60-mer and 70-mer oligo platforms based on *in situ* synthesis (Agilent and NimbleGen (soon to be part of Roche)) are very flexible because arrays are manufactured in small batches, and the design software can be updated at each printing. This high flexibility means that new information can be readily incorporated, and custom platforms can be designed to track splicing choices or other features of specific genes for individual users. The downside to this flexibility is that there are likely to be many versions of the array platforms in use, making it

Table 1 Key features of current and possible gene expression interrogation platforms

Source	Content Current/ Possible	Flexibility	Limitations
Affymetrix	13K/whole gene set	Periodic update by company; printing method very expensive	Sensitive to SNP
Agilent	44K/whole gene set	E-mail to update synthesis file	60-mer oligos sensitive to indels but not to SNP
NimbleGen	Custom product/ whole gene set	E-mail to update synthesis file	Sensitivity to SNP and indels depends on oligo length
Maize Oligonucleotide Array	58K/whole gene set	Oligos based on current genome knowledge	70-mer oligos sensitive to indels but not to SNP
Short EST sequencing	200,000 reports of 70-250 bases on the 454; 2-20 million 30 base reads on Solexa/ ABI and others in the future	Technical expertise required for sample preparation but user has ultimate flexibility in sample choice	Not yet a protocol to prepare normalized RNA samples for sequencing; error-prone sequences
ABI short oligoligation to sequence	Very new technology	Technical expertise required for sample preparation but user has ultimate flexibility in sample choice	Very little public data yet available; clever error detection method

imperative that the designs be described precisely in public data repositories. Long oligos synthesized in advance and then spotted into high density arrays were an attractive option just a few years ago. This type of platform suffers from many difficulties in manufacturing that result in poor spot formation, and the platforms lack flexibility because the oligos are used for one to several years and ultimately become based on an outdated genome or cDNA assembly.

3.2 Sequencing for Transcriptome Profiling

As the cost of DNA sequencing continues to decrease dramatically, new technologies have emerged for sample sequencing at very high throughput, with some compromise in quality compared to the best capillary sequencing methods. Quality (phred) scores for DNA sequencing are reported in integers, i.e. phred40, that relates to the error rate, in this case 10^{-4}; phred30 is 10^{-3}, etc. Capillary sequencing has achieved phred40-60 routinely, and with multiple passes over the same sequence, a virtually error free consensus sequence is derived at a high price. For much less effort, deep sequencing is possible using several technologies (454, Solexa) that utilize a sequencing by synthesis strategy. First, RNA is converted into cDNA, and this is used as a template for synthesis of a second strand under controlled conditions. Polymerization of specific bases is monitored by fluorescence in massively parallel reactions located on beads or other substrates with direct capture of data at each cycle. Currently these and other anticipated sequencing technologies yield short read lengths (maximum of about 300 bases for the second generation 454 sequencing technique, http://www.454.com/) to ~20-30 bases for other technologies (see Illumina Solexa, http://www.illumina.com/pages.ilmn?ID=201). Hundreds of thousands to hundreds of millions of short sequences are obtained per run, and these are fitted onto gene models or transcript models to determine which genes have transcripts present.

A limitation with some of these technologies is stuttering (poor synthesis) at homopolymeric runs of the same base and the iterative error associated with mistakes early in a sequencing run (which frameshift the entire string of subsequent bases). 454 sequencing is most prone to this problem, and early adopters have found the Solexa platform to be less prone to this problem. Error detection and elimination strategies are rapidly evolving, i.e. the 2 base code reading developed by ABI as part of their SOLiD technology (http://marketing.appliedbiosystems.com/images/Product/Solid_Knowledge/ABI-5845_SOLID_Product_Spec_Sheet_loresfinal.pdf) which relies on ligation of oligos rather than polymerization. Quality scores in the 1 error in 20 to 40 bases range are the current norm. [Note that the error detection and reporting software is still under development for many applications and may well improve quickly.]

In terms of identifying all transcripts in a complex mixture, there is not yet sufficient data for maize to judge how well the current technologies will perform. Anthers express >20,000 transcripts, for example, with some present in thousands of copies per cell, many present in up to 100 copies, and perhaps the majority present at <10 copies per cell. Without normalization of the original RNA sample, or a subtraction step to remove the most common transcripts, it is unlikely that 200,000 or even 20 million short reads will identify all transcript types with confidence. A particular problem is that 3′ end reads are often employed, and many maize transcripts have multiple 3′ poly(A) addition sites, only a subset of which may be well-defined by high quality EST or full-length cDNA sequencing. Furthermore, the 3′ untranslated region is composed of many simple sequence repeats, including homopolymeric runs that present the greatest difficulty for some of the new sequencing methods. It seems likely that 5′ cap selection and normalization kits will be developed to make these new methods highly effective in describing the transcriptome in the future. In terms of cost, the high throughput sequencing methods are already on a par with microarray profiling and similarly require 3-4 independent biological replicates to attain confidence in the dataset. In late 2007, the cost for these replicates is about $1000 - $2000 by oligo array profiling and about $1500 - $2500 by sequencing once all of the machinery is purchased. The new sequencing technologies are ~4-fold more expensive than the best microarray scanning technologies, however, the sequencers can be used for many other applications. Currently sample preparation is the major expense, and it seems likely that competition among technology companies will reduce this expense through developing kits suitable for smaller scale samples, reducing reagent costs, etc.

3.3 Prospects for Complete Transcriptome Data

Given the push from biomedical applications, such as profiling transcripts in small tumor biopsies, ever higher throughput platforms are highly likely, at decreasing cost, and with better software for error detection and data interpretation. Within 5 years there should be an enormous amount of data about maize organs, tissues, and even isolated cell types. Laser microdissection of cell types or use of cell-based markers (GFP expressed from a cell type specific promoter followed by protoplast formation and cell sorting) are the frontiers in moving transcriptome profiling from the organ level to the unit of function, the cell. These developmental data will complement similarly large datasets analyzing gene expression after environmental perturbations and diseases.

One specific challenge for the maize community is how to engage manufacturers in providing custom products for maize research, given the far smaller number of people working with this species than in medicine. A second challenge is the high initial cost of equipment and the need for special training to operate the equipment efficiently. Centers running fee for service sequencing may be a good option for the maize community.

Equally important considerations are free access to data, data storage, and data display with the maize genome sequence. Federally funded projects require data deposit into GEO or other data warehouses, and there are several websites now providing added value such as cross-species comparison or between tissue or treatment comparisons of data generated by multiple labs, c.f. http://www.plexdb. org/, http://www.maizearray.org/maize_study.shtml). Currently these NSF-funded projects deal with just single platforms. Because maize is so polymorphic and so many different lines are utilized in specific projects, it seems unlikely – and probably unnecessary – for all studies to use precisely the same platform. It is likely that B73 and other lines evaluated in large-scale phenotypic surveys will be analyzed on a single platform, however, custom-built platforms for that specific line may not be suitable for other lines, because of SNP, differential gene content, or different goals (studying intron splicing, for example). Similarly, the high polymorphism of maize will make comparison of short sequence reads among genotypes more challenging than with many other model organisms.

Validation of transcriptome profiling, study design, and appropriate statistical treatment of the data have been contentious issues, and the discussants have often lost sight of the important biology. The purpose of the study is typically to generate a list of genes and processes for more detailed analysis – user confidence in the list and knowledge of the false discovery rate are important in planning genetic analysis, construction of RNAi knockdown plants, biochemical studies, etc. Minimizing false leads while preserving novel findings from the transcriptomics is the primary goal. Now that complex transcriptome experiments are routine, a driving force is reducing cost by reducing redundancy; consequently, multiple independent replicates hybridized in the minimum combinations rather than a reference sample hybridized to each sample have come to dominate the study designs employed. Multiple biological samples are now viewed as crucial, while technical replications have much less value now that platforms are reliable. Data collection at two or more intensity settings can maximize recovery of information from individual array platforms. For the two-color platforms (virtually all except Affymetrix), balanced dye swap designs are now routine, i.e. two biological replicates labeled with red and two labeled with green, to account for difference in dye labeling. Internal "spike-in" controls to a dilution series printed on the matrix now permit direct calculation of the copy number of cDNA or cRNA from the signal intensity. Even a few years ago, these improved features were in the "dream world" – it is likely that in the near future better platforms, data capture, and data handling will be developed.

Returning to the biological questions with large data sets in hand, inevitably brings up the question of testing the validity of specific responses for genes of particular interest or at least for a subset of genes representing different expression levels. A few target genes can be evaluated quantitatively using qRT-PCR, although this method can be scaled up (at some expense) to handle thousands of reactions (Czechowski et al. 2004, and see the next section). An alternative is to hybridize the same samples used on one platform to a second platform, provided the labeling methods are compatible, or to prepare another aliquot of RNA and the probe cocktail from the same or additional biological replicates to apply to a second platform. This platform to platform test can be used to find outliers, which may represent poorly designed probes on one of the platforms, and to determine the relative sensitivity of each platform for the detection of the rare class of mRNAs.

In the future some type of high-throughput *in situ* localization of RNA using PCR-based methods would be highly desirable. Such methods have been described, but none has become an established protocol. Combined with qRT-PCR on extracted samples, both qualitative (where, when) and quantitative (how much transcript) information could become part of the validation effort.

3.4   *Laser Microdissection*

Laser microdissection allows for the enrichment of homogeneous cell populations to the exclusion of neighboring cells (reviewed in Kehr 2003; Schnable et al. 2004; Day et al. 2005; Nelson et al. 2006; Ohtsu et al. 2007b). Specifically, laser microdissection combines the throughput of microarray experiments with the resolution of *in situ* hybridizations. The flexibility of laser microdissection has been demonstrated by a diversity of downstream applications, including cDNA library production, real-time PCR, genomic analyses, proteomic analyses and antibody arrays. Several researchers have taken advantage of the increased signal to noise ratio afforded by laser microdissection to pinpoint differential expression by cell type, for example, in epidermal and vascular tissue (Nakazono et al. 2003), root tissues (Woll et al. 2005; Jiang et al. 2006; Liu et al. 2006; Hochholdinger, this issue), shoot apical meristems (Ohtsu et al. 2007a; Zhang et al. 2007), and to demonstrate the applicability of laser microdissection to a number of tissues (Kerk et al. 2003).

3.5   High-throughput qRT-PCR

Microarray platforms allow for the simultaneous analysis of tens of thousands of transcripts. This platform is often useful when performing global analysis expression analysis when target genes are unknown. In recent years, extensive efforts have been put forth to sequence over 1.1 million maize ESTs (as of 8/31/07). Furthermore, the tissue source used for producing these ESTs is often known. The convergence of readily available sequence data and the decreased costs of oligonucleotides and qRT-PCR reagents allow researchers to perform highly focused experiments on genes of interest regardless of their expression level.

qRT-PCR has been demonstrated as a focused, high throughput tool for gene expression analysis. Czechowski et al. (2004) quantified the accumulation of 1465 *Arabidopsis* transcription factors via qRT-PCR. Transcripts were detected for 87% of the genes, whereas the 22K *Arabidopsis* Affymetrix chip detected only 55% of the transcripts from the same RNA samples. The utility of qRT-PCR was further demonstrated by the ability to detect transcript over six orders of magnitude (0.001 to 100 copies per cell).

3.6   Data Acquisition in Microarray Experiments

DNA microarray experiments can be divided into seven steps: experimental design, array production, RNA isolation and amplification, labeling and hybridization considerations, data acquisition and downstream data analyses. One often-overlooked aspect in maximizing the information extracted from a microarray experiment is at the data acquisition step. For two color microarray experiments, most protocols recommend scanning one time per channel at a setting that minimizes the number of saturated spots. While this approach captures a majority of the information on an individual array, it has the potential to exclude elements on the array from further analysis based solely on signal strength.

The multiple scan approach is a cost-effective method to extract additional information from a microarray experiment at a minimal cost (Skibbe et al. 2006). This method employed a scanning strategy where each array was scanned six times in ascending order of laser strength. The three scans with empirically assigned median intensity values were selected and assigned into "low", "medium" and "high" intensity categories. Then, independent statistical analyses were performed on each group to generate three data sets. Combining the statistically significant datasets of the three intensities increased the detection by 30-40% (vs. a single scan approach) and 10-15% (vs. a double scan approach).

Other approaches have been proposed to increase the number of statistically significant differences, particularly when dealing with saturated spots. The censored regression model utilizes pixel level data of an individual spot to correct for saturation (Dodd et al. 2004). Another model (linear regression method) extends the linear range of signal detection (above the saturated value of 65536) by extrapolating the data from a low intensity scan using a linear regression algorithm (Dudley et al. 2002).

4 Proteomics

Two-dimensional (2-D) gel electrophoresis is an effective method for resolving thousands of proteins on the basis of pI and molecular weight. Traditional 2-D electrophoresis experiments call for the two samples to be run on individual gels, silver stained, imaged, and the resulting images to be computationally merged. Skilled hands can perform these experiments efficiently and effectively, but many factors complicate this experimental setup, including gel-to-gel variation, image merging and detection limits of silver staining. A newer method, 2-D Difference Gel Electrophoresis (2D-DIGE), employing fluorescent CyDyes™, overcomes these limitations (Figure 1).

Briefly, each sample is labeled with either Cy3 or Cy5 DIGE fluor. Proteins from the two mixed samples are then resolved first by pI and second by molecular weight. Then, the samples are visualized by scanning at the respective Cy3 and Cy5 wavelengths. The resulting images are then computationally merged and statistically analyzed. Spots of interest can then be picked, digested with trypsin and sequenced using mass spectrometry. The CyDye™ DIGE fluors have a 125 pg detection limit, and approximately 3000 spots can be routinely resolved using these fluors.

Mass spectrometry analysis of digested proteins often reveals the presence of two or more proteins per spot. Furthermore, the ability to assign peptide fragments hits to an individual protein depends greatly on the number of peptide fragments, the quality of the database, and the organisms populating the database. As with microarray experiments, researchers should use caution when determining matches on arbitrary fold-change and peptide matches per protein. Instead, search algorithms should be set to provide statistical values of confidence. Two of the most commonly used programs that assign statistical cutoffs are MASCOT (Matrix Science, Boston, MA) and Protein Prospector (Chalkley et al. 2005a; Chalkley et al. 2005b).

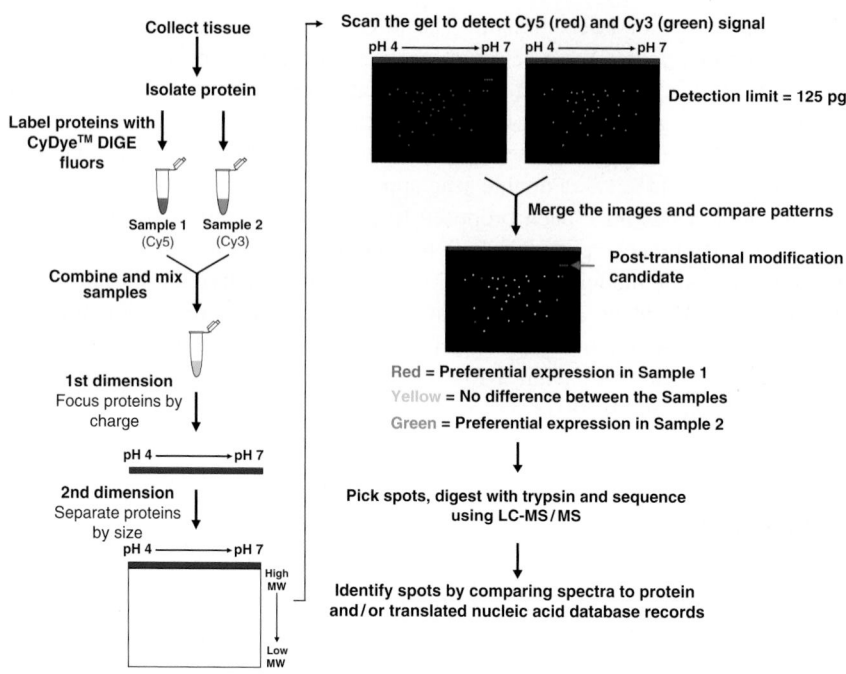

Fig 1 A schematic representation of the 2D-DIGE procedure

Acknowledgments D.S.S. is supported by a NIH Ruth L. Kirschstein post-doctoral fellowship (Grant Number 5F32GM076968). Our continuing contributions to gene expression are supported by NSF Grant 07-01880.

References

Chalkley, R.J., P.R. Baker, K.C. Hansen, K.F. Medzihradszky, N.P. Allen, M. Rexach and A.L. Burlingame (2005) Comprehensive analysis of a multidimensional liquid chromatography mass spectrometry dataset acquired on a quadrupole selecting quadrupole collision cell, time-of-flight mass spectrometer. I. How much of the data is theoretically interpretable by search engines? *Mol. Cell. Proteomics* **4**: 1189–1193.

Chalkley, R.J., P.R. Baker, L. Huang, K.C. Hansen, N.P. Allen, M. Rexach and A.L. Burlingame (2005) Comprehensive analysis of a multidimensional liquid chromatography mass spectrometry dataset acquired on a quadrupole selecting quadrupole collision cell, time-of-flight mass spectrometer. II. New developments in protein prospector allow for reliable and comprehensive automatic analysis of large datasets. *Mol. Cell. Proteomics* **4**: 1194–1204.

Czechowski, T., R.P. Bari, M. Stitt, W.-R. Scheible and M.K. Udvardi (2004) Real-time RT-PCR profiling of over 1400 Arabidopsis transcription factors: unprecedented sensitivity reveals novel root- and shoot-specific genes. *Plant J.* **38**: 366–379.

Day, R.C., U. Grossniklaus and R.C. Macknight, (2005) Be more specific! Laser-assisted microdissection of plant cells. *Trends Plant Sci.* **10**: 397–406.

Dodd, L.E., E.L. Korn, L.M. McShane, G.V.R. Chandramouli and E.Y. Chuang (2004) Correcting log ratios for signal saturation in cDNA microarrays. *Bioinformatics* **20**: 2685–2693.

Dudley, A., J. Aach, M. Steffen and G. Church (2002) Measuring absolute expression with microarrays with a calibrated reference sample and an extended signal intensity range. *Proc. Natl. Acad. Sci. USA* **99**: 7754–7759.

Jiang, K., S. Zhang, S. Lee, G. Tsai, K. Kim, H. Huang, C. Chilcott, T. Zhu and L.J. Feldman (2006) Transcription profile analyses identify genes and pathways central to root cap functions in maize. *Plant Mol. Biol.* **60**: 343–363.

Kehr, J. (2003) Single cell technology. *Curr. Opin. Plant Biol.* **6**: 617–621.

Kerk, N.M., T. Ceserani, S.L. Tausta, I.M. Sussex and T.M. Nelson (2003) Laser capture microdissection of cells from plant tissues. *Plant Physiol.* **132**: 27–35.

Liu, Y., T. Lamkemeyer, A. Jakob, G. Mi, F. Zhang, A. Nordheim and F. Hochholdinger (2006) Comparative proteome analyses of maize (*Zea mays* L.) primary roots prior to lateral root initiation reveal differential protein expression in the lateral root initiation mutant *rum1*. *Proteomics* **6**: 4300–4308.

Nakazono, M., F. Qiu, L.A. Borsuk and P.S. Schnable (2003) Laser–capture microdissection, a tool for the global analysis of gene expression in specific plant cell types: identification of genes expressed differentially in epidermal cells or vascular tissues of maize. *Plant Cell* **3**: 583–596.

Nelson, T., S.L. Tausta, N. Gandotra and T. Liu (2006) Laser microdissection of plant tissue: what you see is what you get. *Annu. Rev. Plant Biol.* **57**: 181–201.

Ohtsu, K., M.B. Smith, S.J. Emrich, L.A. Borsuk, R. Zhou, T. Chen, X. Zhang, M. Timmermans, J. Beck, B. Buckner, D. Janick-Buckner, D. Nettleton, M.J. Scanlon and P.S. Schnable (2007) Global gene expression analysis of the shoot apical meristem of maize (*Zea mays* L.). *Plant J.*, in press.

Ohtsu, K., H. Takahashi, P.S. Schnable and M. Nakazono (2007) Cell type-specific gene expression profiling in plants by using a combination of laser microdissection and high-throughput technologies. *Plant Cell Physiol.* **48**: 3–7.

Schena, M., D. Shalon, R.W. Davis, and P.O. Brown (1995) Quantitative monitoring of gene expression patterns with a complementary DNA microarray. *Science* **270**: 467–470.

Schnable P.S., M. Nakazono, and F. Hochholdinger (2004) Global expression profiling applied to plant development. *Curr. Opin. Plant Biol.* **7**: 50–56.

Skibbe, D.S., X. Wang, X. Zhao, L.A. Borsuk, D. Nettleton and P.S. Schnable (2006) Scanning microarrays at multiple intensities enhances discovery of differentially expressed gene. *Bioinformatics* **22**: 1863–1870.

Woll, K., L.A. Borsuk, H. Stransky, D. Nettleton, P.S. Schnable and F. Hochholdinger (2005) Isolation, characterization and pericycle specific transcriptome analyses of the novel maize (*Zea mays* L.) lateral and seminal root initiation mutant *rum1*. *Plant Physiol.* **139**: 1255–1267.

Zhang, X., S. Medi, L.A. Borsuk, D. Nettleton, B. Buckner, D. Janick-Buckner, J. Beck, M. Timmermans, P.S. Schnable and M.J. Scanlon (2007) Laser microdissection of narrow sheath mutant maize uncovers novel gene expression in the shoot apical meristem. *PLoS Genetics* **3**: 1040–1052.

Maize Transformation

Kan Wang, Bronwyn Frame, Yuji Ishida, and Toshihiko Komari

Abstract Plant genetic transformation technologies have brought fundamental changes to both plant biology laboratory research as well as to modern agricultural field practices. Once a recalcitrant plant for tissue culture and gene delivery, maize is becoming one of the most targeted cereal crops using genetic transformation for both basic and applied purposes. This chapter provides a brief review of the history of maize transformation technology development, but focuses extensively on technical aspects of the methodology, including DNA delivery systems, target tissues and genotypes, selectable markers for transformation, and various issues related to integration and expression of transgenes. Some recent observations and improvements from two maize transformation groups are discussed. It is anticipated that increasing genomics information will assist further enhancement of maize transformation technology leading to more rapid progress in understanding and improvement of this important crop.

1 Introduction

Transformation is an indispensable technology in both applied and basic studies in maize. In 2007, more than 70% of US maize fields were planted with genetically engineered (GE) varieties (http://www.ers.usda.gov/Data/BiotechCrops/), the largest acreage since GE crop commercialization in 1996. Given that the share of transgenic crops in the other two major cereals, rice and wheat, is close to zero, and that the

K. Wang and B. Frame
Iowa State University, Center for Plant Transformation and Department
of Agronomy, Ames, Iowa, USA
kanwang@iastate.edu
bframe@iastate.edu

Y. Ishida, and T. Komari
Japan Tobacco Inc., Plant Innovation Center, Iwata, Shizuoka, Japan
yuji.a.ishida@ims.jti.co.jp
toshihiko.komari@ims.jti.co.jp.

J.L. Bennetzen and S. Hake (eds.), *Maize Handbook - Volume II: Genetics and Genomics*,   609
© Springer Science+Business Media LLC 2009

commercial value of maize seed makes up approximately one half that of the global seed market, the impact of gene transfer technology is more significant for maize than for many other crops. Unlike conventional breeding in which introgression of beneficial traits through hybridization from one variety to another often leads to carry-over of undesired genetic materials, genetic transformation methods allow introduction of a defined DNA segment carrying well characterized genes and regulatory sequences into the plant genome.

For basic research, genetic transformation technology enables scientists to study gene function and regulation by altering or perturbing signaling or metabolic pathways. The scope of transformation experiments covers topics of ever growing complexity such as conducting experiments in over-expression, regulated expression, down-regulation or silencing of exogenous or endogenous genes, analyses of promoters or other regulatory elements, and complementation of mutations with genomic sequences.

Currently, particle bombardment and *Agrobacterium*-mediated methods are the two major options for gene delivery to maize. A number of maize genotypes can be transformed by both methods and these techniques are well developed in maize compared to other major crops such as wheat or soybean. We can proudly state that the basic setup for gene transfer in maize has been accomplished. However, there still are considerable limitations, especially to meet the large demand for transgenic plants in functional genomics. For example, compared to rice, the model cereal crop for genomic research, maize transformation techniques suffer low overall transformation frequency and high genotype dependency. Thus, we still need to achieve a long list of desired improvements.

In this chapter, technology for DNA delivery, maize target tissues and genotypes, selectable markers for transformation, and various issues related to integration of transgenes are discussed, with an emphasis on recent developments.

2 DNA Delivery Technologies

Considerable efforts to develop gene transfer technology in maize have been made since the mid 1980s, and both direct DNA transfer and *Agrobacterium*-mediated methods have been actively studied (Armstrong 1999; Torney et al. 2007). The first successful production of fertile transgenic maize was achieved by particle bombardment in the late 1980s (Gordon-Kamm et al. 1990). Initially, most transgenic maize was produced by direct DNA transfer techniques due to wide-spread skepticism as to whether *Agrobacterium* was suitable for monocotyledonous plant transformation. While some early efforts were made to transform maize using *Agrobacterium*, it was not until 1996 that the first convincing and efficient protocol for *Agrobacterium*-mediated maize transformation was reported (Ishida et al. 1996). Because of the quality of transgenic plants generated by this method (high frequency of single and low copy number transgene integration), *Agrobacterium*-mediated maize transformation is becoming the method of choice in maize research.

2.1 Direct DNA Transfer Methods

Numerous direct DNA transfer methods have been developed to deliver genetic material into plant cells. These methods include particle bombardment, electroporation, polyethylene glycol (PEG) incubation, silica carbide whiskers, microinjection, macroinjection, microlaser, liposome, pollen tube pathway, and electrophoresis (Potrykus and Spangenberg 1995; Armstrong 1999). However, only a few methods have produced transgenic maize plants. Currently, the most efficient and popular direct DNA transfer method is particle bombardment or biolistic gun-mediated transformation. The methods that have been successfully used for generating transgenic maize plants are discussed here.

2.1.1 Electroporation and PEG Methods

Electroporation is a method in which the DNA of interest is introduced into protoplasts by applying an electric pulse to a mixture of the two. The first transgenic maize plants were produced by this method (Rhodes et al. 1988). Protoplasts were isolated from embryogenic suspension cultures that originated from immature zygotic embryos (IEs) of the maize inbred line, A188. After electroporation, protoplasts were placed on plant media allowing callus formation and selection. Calli that survived kanamycin selection was then regenerated into plants. However, none of the transformed plants set seed.

Instead of using an electric pulse, PEG may be added to a mixture of protoplasts and the DNA of interest (PEG method). This chemical is not toxic to plants and has been used in other plant transformation protocols (Armstrong et al. 1990). Indeed, Golovkin et al. (1993) produced fertile transgenic plants of the maize germplasm He/89 using this method.

In spite of this success, methods using protoplasts are inherently difficult in maize. Unlike dicotyledonous plants, protoplasts that regenerate plants cannot be prepared from mesophyll cells in cereals. Regenerable maize protoplasts can be prepared only from embryogenic suspension or callus cultures. Only a limited number of tissue-culture adapted maize genotypes can produce the desired embryogenic cultures. Additionally, prolonged culture leads to loss of capacity for these protoplasts to regenerate plants after transformation and to maintain fertility for seed production. Due to these limitations, transformation techniques based on protoplasts have never been widely used.

To avoid using protoplasts as target material for transformation, electroporation methods were used to transform intact maize tissues. The targeted tissues are briefly treated with an enzyme that digests cell walls, after which DNA is introduced by an electric pulse. Using this approach, maize IEs and Type I callus (compact embryogenic callus, D'Halluin et al. 1992), and suspension cultures and Type II callus (friable embryogenic callus, Laursen et al. 1994; Pescitelli and Sukhapinda 1995) were transformed and transgenic plants were obtained. This method has not been widely adopted mainly because of its low transformation frequency (TF, where transformation frequency is defined as the number of herbicide or antibiotic resistant events recovered per 100 targeted explants).

2.1.2 Particle Bombardment

Using this method, target cells or tissues are bombarded with fine metal particles coated with DNA. Because highly accelerated particles can easily penetrate cell walls, various types of tissues can be targeted. The earliest fertile transgenic maize lines generated by this method were from cell suspension cultures and Type II callus (Fromm et al. 1990; Gordon-Kamm et al. 1990; Walters et al. 1992; Vain et al. 1993). Then, Type I callus, IEs, shoot meristems, organogenic callus from shoot tips and shoot meristem cultures were successfully employed (Wan et al. 1995; Zhong et al. 1996; Zhang et al. 2002; Ahmadabadi et al. 2007). Using this technology, fertile transgenic plants have been efficiently produced in a wide range of maize genotypes and, to date, it has been the major transformation method used for producing genetically engineered commercial varieties of maize (http://www.agbios.com/).

2.1.3 Silicon Carbide Whiskers

Needlelike silicon carbide fibers, which are approximately $20\,\mu m$ in length and referred to as "whiskers", penetrate the cell wall and plasma membrane of target cells to effect DNA uptake and subsequent transformation (Southgate et al. 1998). This technique is relatively simple and fast, and does not require expensive equipment and supplies. In maize, the whisker method was first used to transform non-regenerable Black Mexican Sweet suspension cells (Kaeppler et al. 1992). Production of fertile transgenic maize from suspension cultures and Type II callus have been reported (Frame et al. 1994; Petolino et al. 2000). In addition to low TF, the major limitation of this technology is that it can only deliver DNA to fine cell aggregates such as suspension cultures and friable callus cultures, and, as such, has not been widely employed in maize transformation.

2.2 Agrobacterium-Mediated Transformation

The soil pathogen *Agrobacterium tumefaciens* can genetically transform plant cells with DNA segments (T-DNA) from its tumor-inducing plasmid (Ti plasmid, Gelvin 2003). High frequency transformation protocols based on this naturally occurring machinery for the integration of foreign DNA into plant chromosomes have been continuously reported in many dicotyledonous plants since the early 1980s. Initially, it was assumed that this technology could not be extended to monocotyledonous plants because they are not natural hosts for *A. tumefaciens* (Hernalsteens et al. 1984).

Nevertheless, early attempts to infect maize and other cereals with *A. tumefaciens* laid the ground work for later successes using this biological vector to effectively transform maize.

2.2.1 Early Attempts

Graves and Goldman (1986) infected germinating seedlings of maize with a wild type *Agrobacterium* strain and detected the expression of an *Agrobacterium* opine gene in the inoculated plants two weeks after infection. Grimsely et al. (1987) inoculated apical meristems of maize plants with an *Agrobacterium* strain that carried T-DNA in which maize streak virus DNA had been inserted. The transformed maize displayed symptoms of systemic infection of the virus. Gould et al. (1991) then inoculated apical meristems of maize with *Agrobacterium* and confirmed by Southern blot analysis and the expression of the GUS reporter gene that some seeds from the resulting plants carried the introduced gene. While these early documentations did not offer reproducible results or a workable frequency for maize transformation, they were encouraging. Other reports also included evidence of *Agrobacterium* attachment to maize cell surfaces (Graves et al. 1988) and identification of substances inducing virulence (*vir*) genes by monocots, including maize (Primich-Zachwieja and Minocha 1991).

2.2.2 Development of Efficient Protocols

A highly efficient method of monocot transformation using *Agrobacterium* was finally reported in rice (Hiei et al. 1994), followed by successful reports of transformation in other important cereals such as maize (Ishida et al. 1996), wheat (Cheng et al. 1997), barley (Tingay et al. 1997) and sorghum (Zhao et al. 2000). Types of plant materials used for infection with *Agrobacterium*, choices of vectors and strains of *Agrobacterium*, and optimization of tissue culture techniques are among the key factors that contributed to these achievements.

Ishida et al. (1996) inoculated IEs of the maize inbred A188 with a modified *Agrobacterium* strain that carried a super-binary vector (in which extra copies of *vir* genes assisted DNA transfer) and obtained transgenic maize plants from 5% to 30% of the infected embryos. According to their study, it was necessary to use IEs at a specific stage of development from a specific germplasm of maize, A188. The IEs needed to be obtained from non-stressed healthy plants grown in a well-conditioned greenhouse. In addition, the marker genes, vectors, *Agrobacterium* strains, media composition and concentration of bacteria all had to be optimized. The complexity of multiple factors involved and the narrow ranges of optimal parameters in the process were probably the main reasons why efficient maize transformation by *Agrobacterium* had been difficult.

The benefits observed in rice when using the *Agrobacterium*-mediated transformation method – a relatively high transformation frequency accompanied by a high frequency of single or low copy insertion events with few rearrangements – were also evident in maize. Initially, *Agrobacterium*-mediated gene transfer to maize was not rapidly and broadly adopted because very few genotypes, effectively only the agronomically poor inbred line, A188 and its derivatives, were transformable at a high frequency using the original protocol. Painstaking efforts have been gradually resolving this issue to expand transformable genotypes and to improve the efficiency of transformation (Zhao et al. 2001; Frame et al. 2002; Ishida et al. 2003; Quan et al. 2004; Danilova and Dolgikh 2005; Huang and Wei 2005; Frame et al. 2006a).

2.3   Comparison of Particle Bombardment and Agrobacterium-*Mediated Transformation*

Various methods for transformation of higher plants have been examined extensively in maize and other plants for the last two decades, and it is now quite clear that, for transforming maize, the relevant choice is between the particle bombardment and *Agrobacterium*-mediated methods. The major advantages and disadvantages of the two methods are compared in Table 1.

Particle bombardment is purely a physical process for delivery of DNA into the plant genome. Genes of interest (GOI) can be cloned into popular high copy cloning vectors such as pUC, and pBlueScript. Multiple transgenes can be delivered either by bombarding one plasmid carrying multiple genes or co-bombarding several plasmids that carry the genes separately. To avoid introducing non-target DNA such as vector backbones, the target-only DNA can be purified from the plasmid using restriction enzyme digestion and gel purification prior to the bombardment. A typical DNA fragment used for transformation is less than 15 kb. Genomic DNA fragments as large as 100 kb were used for bombardment to examine the function of a GOI less than 15 kb contained in this fragment (V. Chandler, personal communication). Any maize genotype that can produce Type I or Type II callus responses from IEs with adequate frequency is likely to be transformed using this method. Particle bombardment is also a powerful tool in the analysis of transient expression of transgenes in intact and fully developed tissues. However, relatively large percentages of transgenic plants generated by this method have high copy number integration and complex transgene rearrangement, thus, high incidences of low, unstable or silenced transgene expression were reported (Pawlowski and Somers 1996, 1998; Shou et al. 2004).

Advantages of *Agrobacterium*-mediated transformation include high TF, integration of low numbers of T-DNA copies into chromosomes, and the ability to transfer relatively large T-DNA segments with defined ends and little rearrangement. However, because *Agrobacterium*-mediated transformation intrinsically depends on the complex biological interaction between bacterial and plant cells, specific vectors with components that facilitate the transfer of DNA through this interaction are needed, and manipulation of complicated biological systems is required for optimization of the experimental protocols. Specific binary vectors designed for *Agrobacterium*-mediated transformation must be used for cloning of GOI. One disadvantage of this method

Table 1 Advantages and disadvantages of two transformation methods

Factors	Particle bombardment	*Agrobacterium*-mediated transformation
Average transformation frequencies[*]	5–40%	5–50%
Copy number and rearrangement of transgenes	High (<10% single or low copy[**])	Low (10–80% single or low copy[**])
Biological complication in transformation	None	Complex plant-bacteria interaction
Vector configuration	Requires any high-copy cloning vector such as pUC, pBlueScript, etc.	Requires gene of interest cassette to be cloned into *Agrobacterium* binary vectors
Vector construction	Straightforward	Need to handle large, low-copy plasmids and to transfer plasmids from *Escherichia coli* to *Agrobacterium*
Size of DNA to be transferred	Any size, but DNA larger than 15 kb is less efficient and can be fragmented	Typical sizes of T-DNA are 5–30 kb
Maize genotype requirement	Adequate frequency of Type I or Type II callus response and regenerability	Compatible with *Agrobacterium* interaction, and adequate frequency of Type I or Type II callus response and regenerability
Most successful target tissue for stable transformation	Immature embryos, Type I or Type II callus, suspension cells	Immature embryos, Type I or Type II callus
Transfer of non-target DNA	Target-only DNA can be obtained by DNA restriction digestion and purification, and used to bombard plant cells	Certain percentage of transgenic plants may carry DNA fragments from vector backbone in addition to T-DNA
Transient expression in mature tissues	Well-established for various tissues	Limited success only

[*]Transformation frequency: number of herbicide or antibiotics resistant events recovered per 100 transformed explants. The range of frequencies reasonably expected from experiments using immature embryos of Hi-II or A188 conducted by scientists with varying skills is shown.

[**]Low copy: 1–3 transgene copies.

is that a certain percentage of *Agrobacterium*-derived transgenic plants may carry various DNA fragments from the binary vector backbones (Shou et al. 2004). The strategy to overcome this problem will be discussed under Subheading 5.

Therefore, both methods have their advantages and limitations. There are a very small number of studies directly comparing these two methods in maize (Zhao et al. 1998; Shou et al. 2004). Because these authors prioritized patterns of integration and stable expression of transgenes, they favored *Agrobacterium*-mediated transformation. In general, the decision for choosing a method for maize transformation depends not only on the nature of the project and long term objectives but also on laboratory infrastructure and researcher skills.

2.4 Critical Factors in Particle Bombardment and Agrobacterium-*Mediated Transformation*

No matter which method is chosen, transformation protocols must be optimized for particular experimental requirements. Apparently, more factors need to be considered in *Agrobacterium*-mediated transformation because the system involves complex interactions between bacteria and plant cells. Many of the parameters have been adjusted on the basis of transient expression of reporter genes, which is most often a GUS gene that has an intron in the coding sequence (*intron-gus*, Vancanneyt et al. 1990). The *intron-gus* gene is better expressed in maize cells than the ordinary *gus* gene (Ueki et al. 2004). It is especially important to use the *intron-gus* construct to optimize *Agrobacterium*-mediated transformation experiments. Because only eukaryotic cells can process an intron placed in the coding sequence of a gene, use of the *intron-gus* construct allows one to distinguish whether the GUS foci detected on the infected embryos or callus derives from transformed plant cells or simply from the presence of remnant *Agrobacterium* cells (Ohta et al. 1990).

Stable transformants have never been efficiently obtained under the conditions that yield only limited transient expression. However, it should be noted that such adjustments are less straightforward in maize. It has proved relatively easy to find conditions for high-level transient expression but stable transformants were obtained in only a few instances (Ishida et al. 1996). The main hurdles in transformation are not the delivery of DNA fragments to the plant cells but, rather, the recovery of the cells that have integrated the foreign DNA in their chromosomes via tissue culture systems.

2.4.1 Particle Bombardment

Critical factors for transformation success using particle bombardment have been comprehensively reviewed (Morrish et al. 1993; Potrykus et al. 1998; Kikkert et al. 2005). These factors can be grouped into two categories: physical parameters, which are related to preparation and acceleration of the particles, and biological parameters, which are related to target tissues/cells. Some relevant physical parameters include: the particles types (material, density, size), volume of the particles, particle velocity, the method used to accelerate particles, amount of DNA precipitated onto the surface of the particles, procedures for the precipitation of DNA, degree of vacuum in the sample chamber, and aperture characteristics of the stopping plates.

Three different types of particle bombardment devices based on driving forces and the type of macrocarriers used have been reported. The BioRad PDS-1000/He biolistic gun is a gas (helium) pressure-driven device using plastic film as a macrocarrier (Klein et al. 1987). ACCELL™ is an electrical discharge particle bombardment device using a metalized sheet as a macrocarrier (McCabe and Christou 1993). The Particle Inflow Gun (PIG) is a gas-stream driven device using no macrocarrier (Finer et al. 1992). All three biolistic devices can be used to effectively deliver DNA into cells. However, the BioRad PDS-1000/He is currently the most widely used device in maize transformation, therefore it is the focus of our biolistic gun discussion in this review.

Gold and tungsten particles between 0.6 and 1.2 μm in diameter are effective for transforming maize although, depending on the nature of the targeted explant, choice of particle size can improve TF. For example, reducing the gold particle size used to bombard Type II callus from 1.0 μm to 0.6 μm resulted in significantly higher TF (Frame et al. 2000). Magnetic particles were also shown to be used for plant transformation, in which targeted cells could be magnetically collected after being hit by a particle projectile (Horikawa et al. 1997). In most protocols, DNA is usually applied to the particles by precipitating with 2.5 M $CaCl_2$ and 0.1 M spermidine, even though details of handling may vary. Velocity of the particles, which is controlled by pressure of the gas, and parameters related to sample chambers and stopping plates, vary with tissues targeted.

Two important biological parameters influencing the effectiveness of using particle bombardment to transform maize are osmotic treatments (Vain et al. 1993) and duration of IE pre-culture prior to bombardment (Songstad et al. 1996). To increase TF, subjecting targeted cells to concentrations of 0.35 to 0.4 M sorbitol, mannitol, sorbitol and mannitol, or sucrose before and/or after bombardment have been used (Vain et al. 1993; Dunder et al. 1995). This treatment is assumed to limit cell damage at bombardment by plasmolyzing targeted cells. Secondly, by reducing IE pre-culture duration after dissection and before bombardment from 3 to 2 days, average TF increased from 15±2.4 % to 21±1.8 % (L. Marcell and B. Frame, unpublished results). A robust maize transformation protocol using the BioRad PDS-1000/He gun can be found in Frame et al., 2000, and http://www.agron.iastate.edu/ptf/index.aspx#.

2.4.2 *Agrobacterium*-Mediated Transformation

The *Agrobacterium*-mediated transformation system involves two biological organisms: a disarmed but pathogenic bacterium, and a plant cell. Fine tuning the system requires considerations of both parties to minimize the negative impact of the bacteria on plant cells and maximize the recovery of transformants. Many alterations have been made to the infection and co-cultivation stages with the aim of enhancing interaction between these two organisms. Techniques such as sonication of IEs in a bacterial suspension (Trick and Finer 1997), addition of a surfactant to the bacterial infection culture (Huang and Wei 2005), and vacuum infiltration of embryogenic callus with the bacterial suspension (Danilova and Dolgikh 2005) could be helpful. In addition, using the optimal concentration of bacterial culture (between 0.5 x 10^9 and 1.0 x 10^9 cells per ml) for IE infection (Ishida et al. 1996; Zhao et al. 2001), adjusting co-cultivation media pH to ~ 5.8 (Ishida et al. 1996), addition of 100 μM acetosyringone (Ishida et al. 1996), 5 μM silver nitrate (Zhao et al. 2001), 5 μM $CuSO_4$ (Ishida et al. 2007) or between 2.5 and 3.3 mM L-cysteine (Frame et al. 2002) to the co-cultivation media, and conducting co-cultivation at temperatures between 20 °C and 25 °C (Frame et al. 2002) all improved TF in independent studies.

After co-cultivation, a resting culture in which no selective pressure is applied may be used to increase TF (Zhao et al. 2001). For removal of bacteria from the culture, use of the antibiotic carbenicillin instead of cefotaxime also resulted in higher TF (Zhao et al. 2001; Ishida et al. 2003). MS-based media (Murashige and Skoog 1962) was reported to improve TF of three inbred lines when compared with N6-based media (Chu et al. 1975; Frame et al. 2006a). Addition of 10 µM CuSO$_4$ to medium also improved frequency of regeneration and vigor of regenerated plants (Ishida et al. 2007).

Certain IE treatments prior to infection can also improve TF. Hiei et al. (2006) demonstrated that TF in both maize and rice was improved by treating IE with heat and centrifugation before infection with *Agrobacterium*. This effect was detected both at the level of transient expression and rate of stable transformants recovered. For maize, the optimal conditions were 46°C for 3 min followed by centrifugation at 20,000 x g for 10 min. The effect of heat was greater than that of centrifugation in maize and the combination of the two was the most effective treatment. For example, TF of A188 IEs was increased three-fold by pretreatment with heat and centrifuging from 10% to 30% and transgenic plants were generated from two previously non-transformable genotypes (inbred A634 and a hybrid of A634 x Oh43). Transformation with a less efficient vector (non-superbinary vector) was also enhanced (Hiei et al. 2006).

Several disarmed strains of *Agrobacterium* have been employed in plant transformation and differ in effectiveness. Disarmed strains derived from A281, a hypervirulent L,L-succinamopine type strain (Hood et al. 1986), Ach5, an octopine type strain (Ooms et al. 1982), and C58, a nopaline type strain (Koncz and Schell 1986), are used frequently in cereal transformation. For example, A281 derivative EHA101 (Hood et al. 1986), Ach5 derivative LBA4404 (Hoekema et al. 1983), and C58 derivatives ABI (Koncz and Schell 1986) and C58z707 (Hepburn et al. 1985) have been used successfully in maize transformation in both super-binary and standard binary vector systems (Ishida et al. 1996; Zhao et al. 2001; Frame et al. 2002; Sidorov et al. 2006; Frame et al. 2006a).

Super-binary vectors, which carry *virB* and *virG* genes of A281 (Komari 1990), have been highly effective for maize transformation. It should be noted that super-binary vectors exhibited higher transformation frequency when combined with *Agrobacterium* strain LBA4404 (Hiei et al. 1994).

High frequency *Agrobacterium*-mediated maize transformation protocols using the super-binary vectors can be found in (Ishida et al. 1996; Zhao et al. 2001; Ishida et al. 2007). Protocols using a standard binary vector can be found in (Frame et al. 2002; Frame et al. 2006b).

3 Target Tissues and Genotypes

To date, the most amenable target materials for both biolistic-mediated and *Agrobacterium*-mediated transformation methods have been IEs, IE-derived embryogenic callus and callus derived liquid suspension culture (Armstrong 1999; Hansen and Wright 1999; Torney et al. 2007). Early transformation experiments focused on target tissues such as liquid suspension culture or friable Type II callus (Fromm et al. 1990; Gordon-Kamm et al. 1990; Walters et al. 1992; Frame et al. 2000). Focus on targeting a dedifferentiated cell-state for DNA delivery was aimed at maximizing the chance of producing transgenic plants regenerated from single transformed cells (Torney et al. 2007). While a Type II callus response from targeted IEs facilitates selection and regeneration of transgenic events, this callus type is not readily produced in most inbred germplasms, with some exceptions (Lowe et al. 2006).

The most widely used laboratory genotype, Hi-II (high Type II callus production, Armstrong et al. 1991), is a hybrid line that manifests a Type II embryogenic callus induction frequency (ECIF) of 100%. Because of its inferior agronomic performance and segregating genetic background, however, it is not suitable for some functional genomic studies. As such, efforts have been made to develop IE transformation systems that target inbred lines. Although genotype-specific recalcitrance is common in maize inbred lines using the IE tissue culture system, those that do respond often produce Type I callus instead of the more friable Type II callus typical of Hi-II (Wan et al. 1995; Brettschneider et al. 1997; Frame et al. 2006b) or a mixture of the two (in the case of inbred A188). Like Type II callus, Type I callus produces somatic embryos but is a more compact and differentiated callus phenotype. Transformation of inbred line IEs exhibiting a Type I embryogenic callus response has been achieved using biolistic (Koziel et al. 1993; Brettschneider et al. 1997; Wang et al. 2003) and *Agrobacterium* methods (Ishida et al. 1996; Ishida et al. 2003; Huang and Wei 2005; Frame et al. 2006b).

Additional efforts to identify genotype-independent transformation systems have focused on using the mature seed – in particular the shoot apical meristem (SAM) of germinated seedlings – as an alternative explant to IEs. Heterogenous shoot tip cultures derived from the SAM were targeted for biolistic transformation (Zhong et al. 1992; Zhong et al. 1996; O'Connor-Sánchez et al. 2002; Zhang et al. 2002). Non-chimeric transgenic events were recovered using this approach, but at low frequency. In a recent report, the coleoptilar node region of germinated seedlings of public inbred line H99 was used to produce homogenous, embryogenic Type I callus which was subsequently *Agrobacterium*-transformed to produce stable transgenic events (Sidorov et al. 2006).

It must be emphasized that no matter what types of explants are targeted, using non-stressed healthy maize plants as starting materials is the key to the success of both biolistic and *Agrobacterium*-mediated transformation. Therefore, research efforts and funds invested in greenhouse management and equipment to improve light, temperature, soil and other growth conditions will be well rewarded in the long run.

3.1 IEs and Type II Embryogenic Callus of the Hi-II Genotype

Target materials (IE, Type II callus, and callus-derived suspension cultures) of the Hi-II hybrid genotype used in early transformation successes continue to play important roles in transgenic maize production, although suspension cultures have not been widely used in public laboratories. IEs have been the preferred target material for biolistic (Songstad et al. 1996) and *Agrobacterium*-mediated transformation systems

(Zhao et al. 2001). When cultured on N6 medium with proline, dichlorophenoxyacetic acid (2,4-D) and silver nitrate (Armstrong et al. 1991; Songstad et al. 1991; Songstad et al. 1996), friable, rapidly growing and highly embryogenic callus (Type II) is formed from embryo scutellum cells. The Type II callus produced is suitable for biolistic transformation (Walters et al. 1992; Frame et al. 2000), and can be used as a year-round stock material to alleviate shortages of IEs. *Agrobacterium*-mediated transformation of this callus phenotype (albeit an elite line) has also been reported (Lowe et al. 2006).

Detailed protocols for genetic transformation of Hi-II IEs using *Agrobacterium*-mediated (Zhao et al. 2001; Frame et al. 2002) or biolistic-mediated (Dunder et al. 1995; Songstad et al. 1996; Frame et al. 2000) methods have been published. At the Iowa State University Plant Transformation Facility (ISU PTF), the *Agrobacterium*-mediated transformation frequency (TF) ranges from 1–21% (number of bialaphos resistant callus events per 100 embryos infected) by construct and averaged 7.2% over the 36 constructs (completed during 2007). All these constructs were based on a standard binary vector, pTF101.1 (2x CaMV 35S promoter driving the *bar* gene as selectable marker, Paz et al. 2004), harbored in *Agrobacterium* strain EHA101 (Hood et al. 1986). By comparison, TF using the biolistic-mediated method ranges from 8–41%, averaging 19% over the last 28 constructs bombarded using the published protocol (Frame et al. 2000). Frequency fluctuations may be affected by various experimental factors including Hi-II seed source, ear variation, season, and construct.

In comparing two distinct sources of Hi-II germplasm, both originating from the Maize Genetics Coop (https://maizecoop.cropsci.uiuc.edu), but with one producing seed in Iowa (designated as PTFHT) and the other producing seed in Wisconsin (designated WHT, and kindly provided by H Kaeppler), we found that WHT germplasm transformed at frequencies four times higher than PTFHT germplasm using the biolistic gun. Average TF using our published protocol to co-transform 28 constructs over an 18 month period was 30% for WHT (ranging from 7–57% by construct) and 7% for PTFHT (ranging from 1–13%, L. Marcell and B. Frame, unpublished results).

Interestingly, this difference in TF between the two germplasms is not evident in our *Agrobacterium*-mediated transformation experiments, where fluctuations in TF are more strongly influenced by the construct employed. For example, average TFs for two GOI constructs, both cloned on vector pTF101.1 in EHA101, were 2% and 18% while average TF for the PTFHT and WHT germplasms in these same experiments were 9% and 8%, respectively.

The consistent, differential performance of these two Hi-II germplasms across two different transformation methods is not understood. Callus growth rate from bombarded WHT embryos is higher than that from PTFHT embryos (B. Frame, unpublished results), yet a *germplasm x media* interaction that may account for this difference in growth rate is unlikely given that switching pipeline media did not increase the relative TF of WHT using the *Agrobacterium* transformation method (B. Frame, M. Main, J. Lund, unpublished results). It is possible that there is a *germplasm x antibiotic* interaction that is limiting WHT performance using our *Agrobacterium* method.

Using PTFHT germplasm, TF also varies widely by ear for both transformation technologies. Across three separate *Agrobacterium*-mediated transformation experiments in which embryos from 18 ears were infected using the same pTF101.1-based construct, TF between ears ranged from 0–17% with an average of 5.5%. By comparison, TF between operators and embryo sizes ranged only from 4–6% and 5–6%, respectively (B. Frame, M. Main, J. Lund, unpublished results). These results suggest that if the number of embryos to be infected is fixed, it is expedient to dissect more ears, and fewer embryos per ear.

This range in TF by ear is also consistently observed using our biolistic-mediated transformation method. Embryos from 12 of the above-mentioned 18 ears were also bombarded. Average TF for these ears was 18% but ranged from 0–31% by ear (B. Frame, M. Main, J. Lund, unpublished results). Of interest was whether, for the twelve ears compared across transformation methods, the high transforming ears using the *Agrobacterium* method were also the high transforming ears using the biolistic gun. Results showed that the four lowest transforming ears in *Agrobacterium* experiments (all 2% TF) ranked from lowest (0% TF) to second highest (30% TF) in biolistic gun experiments, suggesting no obvious consistency in ear effect across transformation methods.

To investigate whether transformation rates also fluctuate throughout the year, TF was assessed four times a year for three years in an ongoing study in the ISU PTF. Using only PTFHT germplasm, greenhouse-derived IEs (1.2–2 mm) were harvested 9–12 days post pollination (depending on season) and bombarded (3 days post-dissection) at the equinox (spring and fall) and solstices (winter and summer) using the published protocol (Frame et al. 2000) with fixed parameters as follows: 0.6 μm gold coated with the construct pB184 (maize ubiquitin, *ubi*, promoter driving the *bar* gene, Frame et al. 2000), 650 psi rupture disks, 6 mm gap distance, 6 cm target distance, and 3 days pre-culture. Four to eight ears (240 embryos) were represented in each eight plate bombardment on each date. Average TF for spring and fall were 17±3% and 11±1%, respectively, and for winter and summer were 18±5% and 10±2%, respectively (L. Marcell and B. Frame, unpublished results). These results may reflect lighting constraints dictated by lack of adequate greenhouse cooling – in summer and early fall, we are often forced to turn off greenhouse lights to minimize heat stress on people and plants. In turn, this may affect the vigor, and transformation competency of embryos reaching the lab (Zhao et al. 2001).

3.2   IEs and Type I Callus Culture of Maize Inbreds

Because of inherent genetic disadvantages to using the Hi-II hybrid genotype for transformation, considerable effort has been invested in developing robust methods for transforming maize inbred genotypes. Targeting and transforming IEs that in turn respond to produce Type I callus for selection and regeneration has been used in both biolistic and *Agrobacterium*-mediated transformation systems. Using biolistic methods, stable transformation of inbred lines CG0056 (Koziel et al. 1993), H99, A188 and Pa91 (Brettschneider et al. 1997) and Oh43 (Wang et al. 2003) have been reported. Stable transformation of inbred line A188 IEs using *Agrobacterium*-mediated methods was first reported by Ishida et al. (1996) and proved to be a major breakthrough in transgenic maize production. The relatively high TF achieved, and the high frequency of single or low copy number events obtained in this study, rendered this a useful system for functional genomic studies and for producing transgenic events that were likely to exhibit relatively stable transgene expression in progeny generations (Zhao et al. 1998; Shou et al. 2004).

Subsequently, this method was used to transform several additional proprietary or public inbred lines including H99, W117, Mo17, B104, Ky21 and B114 (Gordon-Kamm et al. 2002; Ishida et al. 2003; Huang and Wei 2005; Frame et al. 2006a). Key to success in these inbred line studies was the identification of parameters that induced targeted IEs to respond with a high ECIF after transformation, as is the case for the Hi-II, Type II maize tissue culture system. Although this is a necessary condition, it is not always a sufficient one for achieving high TF. Frame et al. (2006a) reported that Oh43 demonstrated high ECIF but was transformation incompetent using the *Agrobacterium*-mediated protocol the authors described.

Type I callus from responding embryos can also be targeted for transformation. This approach was used for electroporation (D'Halluin et al. 1992) and biolistic gun transformations (Wan et al. 1995; Wang et al. 2003). In the ISU PTF, we still rely on bombarding Type I callus for those inbred lines with low ECIF or transformation competence using our *Agrobacterium*-mediated method. Using dicamba-based media (Carvalho et al. 1997) modified to contain MS salts and vitamins (Frame et al. 2006a), we are able to biolistically transform Type I callus of inbred lines W22 (1% TF), M37W (1% TF) and W64 (6% TF, B. Frame, unpublished results).

3.3   Mature Seed-Derived Explants for Transformation

While the IE is currently the most reliable target material for maize transformation, its disadvantages (genotype dependent tissue culture response for inbred lines and greenhouse space requirements for donor embryo supply) have fueled many investigations for alternative target explants.

One attractive target tissue is the shoot apical meristem (SAM) in seedlings germinated from mature seeds. It was hypothesized that once germ-line progenitor cells in the SAM were transformed with a selectable marker gene, they would proliferate under selection pressure and subsequently regenerate into shoots without undergoing a proliferative, callus formation stage. However, the small number of predetermined germ-line cells in the meristem presented a major challenge to targeting the SAM directly (Lowe et al. 1995; Bowman and Eshed 2000; Sticklen and Oraby 2005). Only the successful delivery and integration of transgenes into germ-line cells can result in non-chimeric plant that will transmit the transgene to its progeny. To increase the population of transformation and regeneration competent target cells, protocols were established (Zhong et al. 1992, 1996) and subsequently modified (Li et al. 2002; O'Connor-Sánchez et al. 2002; Zhang et al. 2002) to induce and transform (using the biolistic gun) heterogeneous multiple-shoot meristem cultures. Non-chimeric transgenic events were recovered at a low frequency from these experiments, including from the recalcitrant inbred line B73 (Zhang et al. 2002), and two subtropical inbred lines (O'Connor-Sánchez et al. 2002).

Although high frequency transformation of these heterogeneous cultures remains elusive, expectations that a SAM-based tissue culture system may be relatively genotype independent were upheld by Li et al. (2002) and O'Connor-Sánchez et al. (2002), respectively, in which 70% of 45 temperate maize inbred lines tested and all nine tropical and subtropical lines (inbred and hybrid lines) tested produced proliferative, regenerable cultures from SAM explants. Of interest in the literature describing the evolution of maize tissue cultures derived from SAM or explants containing the SAM is early reference to the elastic nature of these cultures (Zhong et al. 1992) and reports that, from these heterogeneous shoot tip cultures, somatic embryogenic callus could be produced through media manipulations of the cytokinin to auxin ratio (Zhong et al. 1992) or the addition of adenine (O'Connor-Sánchez et al. 2002).

By altering growth regular regimes and duration of explant exposure, Sidorov et al. (2006) capitalized on the elastic nature of apical and axillary bud meristems in the coleoptilar node of germinated seedlings to produce homogeneous Type I callus from several proprietary inbred lines and the public inbred genotype, H99. Briefly, coleoptilar node explants were first cultured on a cytokinin/auxin media before being moved to media containing only the auxins 2,4-D and picloram. On this latter media, a vigorous, fast growing Type I embryogenic callus was produced, primarily from axillary bud meristems, and moved into the dark for proliferation. The Type I callus was then stably transformed using *Agrobacterium*-mediated methods at frequencies ranging from 1.5–11%. Furthermore, 60% of the transgenic events recovered were low copy number events (1–2 copies of the transgene). Although genotype independence has yet to be demonstrated using this alternative explant, the work by Sidorov et al. (2006) marks a new direction in maize transformation. In combining the benefits of using *Agrobacterium*-mediated methods over the biolistic-mediated methods already described (Ishida et al. 1996; Zhao et al. 1998; Shou et al. 2004), with those of transforming maize inbred lines using a mature seed explant, this new method overcomes some of the major limitations imposed by dependency on an IE-based transformation system (continuous plant growing that often requires a large greenhouse space) and on the Hi-II hybrid genotype (not fully suitable for genomic and genetic studies).

Recently, a leaf-based regeneration and transformation system was reported in which a relatively high callus induction frequency was achieved from young leaf bases of the maize hybrid Pa91 x H99 (Ahmadabadi et al. 2007). Addition of polyamines such as spermidine to the medium increased callus induction rates and transgenic plants were produced from bombarded, leaf-derived, Type I callus. Significant in this recent report was the author's claim that production of transformation-competent callus is independent of the presence of the SAM in explanted leaves. Furthermore, the authors describe the production and maintenance of embryogenic callus from leaf explants cultured in the light using a recently described growth regulator, phytosulfokine-alpha (Matsubayashi et al. 1997). Light-maintained maize cultures would be particularly useful in chloroplast transformation because effective antibiotic selection to obtain tissue carrying transformed chloroplasts needs to be carried out in the light (Maliga 2004).

4   Selectable Markers for Maize Transformation

Only a small fraction of the cells that are subjected to transformation actually integrate the DNA into the genome and become progenitors of transgenic plants. Therefore, an efficient selection system, which consists of a selective pressure and a selectable marker gene, is required so that transformed cells may preferentially proliferate. Genes that confer resistance to antibiotics or herbicides, such as kanamycin, hygromycin, phosphinothricin (PPT) and glyphosate, are widely used in higher plant research. If the development of herbicide resistant plants is the goal, the trait gene itself could also serve as a selectable marker gene. A selectable marker gene is an intrinsic component of a plant transformation method and the choice of the system is a key factor in successful transformation. In general, a system is deemed effective if transformed and untransformed cells can clearly be differentiated and transformed cells are not irreversibly damaged by the selective pressure. There is no single selection system that works universally well for all plants species. For example, kanamycin resistance has been most frequently employed in the transformation of dicotyledonous plants whereas hygromycin resistance is most effective in rice transformation (Hiei et al. 1994). Several selective agents that have worked reasonably well in maize are listed in Table 2 and discussed below. It is possible that selections better than any of those listed may yet be discovered because more than 15 selectable marker genes that have already been tested in various other plants remain to be examined in maize (Komari et al. 2006). For a comprehensive review of selectable marker genes in transgenic plants, see Miki and McHugh (2004).

4.1   Major Markers in Maize Transformation

The combination of the phosphinothricin acetyltransferase (*bar* or *pat*) genes and the selective agent phosphinothricin (PPT) or its derivatives (bialaphos, PPT+two alanine, or glufosinate, an ammonium salt of PPT) provides a very efficient system for selection in maize and has been effectively used throughout the maize transformation literature since it was first introduced (Fromm et al. 1990; Gordon-Kamm et al. 1990; Ishida et al. 1996; Zhao et al. 2001; Frame et al. 2002, 2006a). Growth of transformed cells is uninhibited whereas untransformed cells are severely inhibited on selective media. Selection of transformed cells is straightforward and deteriorative effects on growth and regeneration of maize cells under the selection pressure is negligible. It is important to note that among the three selective agents used in this system (PPT, bialaphos and glufosinate), bialaphos appears to be the most effective agent for transformation of Hi-II and a number of inbred lines (Dennehey et al. 1994; Wang et al. 2003). Secondly, for selection to work effectively in maize, the *bar* gene should be controlled by a strong promoter.

Transgenic maize can also be effectively recovered using the 5–enolpyruvylshikimate–3–phosphate synthase (EPSPS) selectable marker gene which

Table 2 Selectable markers for maize transformation

Selective pressure	Nature of pressure	Selectable marker gene	Enzyme for protection	Success in maize[*]	Reference
Resistance to herbicides					
Phosphinothricin (PPT) Bialaphos (alanyl PPT) Glufosinate (PPT ammonium) in BASTA®	Inhibition of glutamine synthase	• *bar* from *Streptomyces hygroscopicus* • *pat* from *Streptomyces viridochromogenes*	PPT-N-acetyltransferase (PAT)	+++	Gordon-Kamm et al. 1990 Brettschneider et al. 1997
Glyphosate in Roundup®	Inhibition of 5-enolpyruvylshikimate-3-phosphate synthase (EPSPS)	• *epsps* from maize or *Agrobacterium*	EPSPS insensitive to glyphosate	++	Huang et al. 2004
		• Synthetic *gox*	Glyphosate oxidoreductase	+	Howe et al. 2002
Butafenacil	Inhibition of protoporphyrinogen oxidase (PPO)	*ppo* from *Arabidopsis thaliana*	PPO insensitive to butafenacil	+	Li et al. 2003
Sulfonylurea (SU) Imidazolinone (IZ)	Inhibition of acetolactate synthase (ALS)	*als* from maize or *Arabidopsis thaliana*	ALS insensitive to SU and IZ	+	Fromm et al. 1990
Resistance to other drugs					
Kanamycin G418 Paromomycin	Inhibition of protein synthesis	*nptII* from *Escherichia coli*	Neomycin phosphotransferase	+	D'Halluin et al. 1992
Hygromycin	Inhibition of protein synthesis	*hpt* from *Escherichia coli*	Hygromycin phosphotransferase	++	Walters et al.1992
Methotrexate (MTX)	Inhibition of dihydrofolate reductase (DHFR)	*dhfr* from mouse	DHFR insensitive to MTX	+	Golovkin et al. 1993
Positive selection marker					
Mannose as a sole carbon source	Non-transformed cells cannot assimilate mannose	*pmi* from *Escherichia coli*	Phosphomannose isomerase	++	Negrotto et al. 2000

[*] +++, extremely effective and broadly applicable; ++, effective; +, useful but low in transformation frequency or not extensively studied

confers resistance to glyphosate (Howe et al. 2002). Glyphosate resistance is a deregulated trait used widely in commercialized Roundup Ready™ GE maize varieties in the USA. As such, it can serve as a selectable marker for producing transgenic maize lines destined for commercialization. It should be noted that larger clusters of maize cells may be less sensitive to glyphosate and some non-transgenic calli may be expected to grow using this selection (Y. Ishida, unpublished results). However, selection at the stage of regeneration is more effective and, as also observed in wheat (Zhou et al. 1995; Hu et al. 2003), escapes are rarely regenerated.

The neomycin phosphotransferase II gene (*nptII*) and the antibiotic kanamycin (or its analogs G418 or paromomycin) were employed in early maize transformation attempts (Rhodes et al. 1988; Gould et al. 1991; D'Halluin et al. 1992; Lowe et al. 1995). Although growth of single cells or small clusters of cells is inhibited by these antibiotics, larger clumps of maize cells are much less sensitive. The concentration of antibiotics required in later stages of selection may therefore be high enough to damage selected cells or plants (Y. Ishida, unpublished results). Because of these limitations, the *nptII*/kanamycin selection system is not widely used for maize transformation. In our experience, effective selection pressure for maize was easier to apply using paromomycin than kanamycin (Y. Ishida, unpublished results). For example, we infected IEs of inbred A188 with an *Agrobacterium* strain that carried an *nptII* gene under the control of the *nos* promoter. We modified a standard protocol (Ishida et al. 2007) to include an initial selection pressure of 25 mg/l paromomycin followed by later selection pressure of 50 mg/l paromomycin. Transgenic maize plants were obtained from 7.5% of the infected IEs (Y. Ishida, unpublished results).

The hygromycin phosphotransferase (*hpt*) gene in conjunction with the selection agent, hygromycin, has been used for particle bombardment and *Agrobacterium*-mediated transformation in maize (Walters et al. 1992; Ishida et al. 1996). Growth of untransformed maize cells is suppressed (although not to the degree seen using bialaphos) while transformed cells can be clearly identified on hygromycin containing media. TF using the *hpt* system is typically less than half or one third of that with the *bar* system (Ishida et al. 2007), depending on the genotype used.

Maize may also be transformed using other resistance markers, as listed in Table 2. Although no system has yet been reported that outperforms the *bar*/bialaphos selection system, data related to alternative marker systems are not extensive enough to draw firm conclusions as to their relative efficacy.

Non-antibiotic and non-herbicide selectable marker genes have been developed to overcome concerns about the presence of drug or herbicide resistance genes in commercial transgenic lines. For example, plant cells expressing the phosphomannose isomerase gene (*pmi*) can grow on media with mannose as the sole carbon source. Such markers are referred to as positive selection markers (Joersbo et al. 1998). In maize, the *pmi* gene has been successfully employed with the *Agrobacterium*-mediated transformation method, and direct gene transfer methods such as PEG-mediated and particle bombardment (Evans et al. 1996; Negrotto et al. 2000;

Wang et al. 2000; Wright et al. 2001; Ahmadabadi et al. 2007). Generally, transformation frequency using the *pmi* selection system is as high as with the *bar* selection system. However, distinguishing transformants from this selection system may not be obvious for less experienced researchers due to considerable background growth of untransformed cells on mannose-containing selection media (Y. Ishida, unpublished results).

4.2   Removal of Selectable Markers

A selectable marker gene is indispensable for identification of transformed cells. However, once the transgenic plants are obtained, the selectable marker genes are not only unnecessary but can also be problematic. Concerns have been expressed over the safety of marker genes and unnecessary DNA segments in plants (Miki and McHugh 2004), therefore the demand for marker-free transgenic plants is high. Technically, the presence of a selectable marker gene in transgenic lines is undesired if further transformations are necessary to introduce additional genes. In this case, a different selectable marker gene must be used to facilitate the secondary transformation. A number of studies focused on this issue in many plants, including maize (Depicker et al. 1985; Schocher et al. 1986; Odell et al. 1990; De Block and Debrouwer 1991; Komari et al. 1996; Ebinuma et al. 1997; Daley et al. 1998; McCormac et al. 2001). Marker genes may be removed from transgenic plants or progeny either by genetic segregation following co-transformation or by specific recombination systems.

4.2.1   Co-Transformation

If a gene of interest and a selectable marker gene are placed on separate DNA segments and a plant cell is co-transformed with these segments, there is a high probability that the segments are integrated into different chromosomes and will therefore segregate in later generations. Co-transformation is a classical approach, in which a GOI and a marker gene is purposely placed on separate plasmids and introduced into plant cells using direct transformation methods such as particle bombardment. Because the number of transgene copies tends to be high in direct transformation, and co-transformed segments tend to link together in one locus in these transformants (Zhong et al. 1999), many transformants may need to be screened for low copy transgenics, from which marker-free progeny may be obtained.

For *Agrobacterium*-mediated co-transformation, Komari et al. (1996) designed super-binary vectors that carried two separate T-DNAs. The first T-DNA contained a selectable marker gene, and the second T-DNA carried the GOI. This system was first tested in tobacco and rice and later confirmed to be effective in maize (Miller et al. 2002; Ishida et al. 2004). About 50–90% of initial transformants were found to be co-transformants (carrying both T-DNAs). In these studies, marker-free progeny were obtained from about half of the co-transformants. In another approach, Huang et al.

(2004) took advantage of the fact that the backbone of a standard binary vector could be considered as a second T-DNA delimited by the left border and the right border in the other orientation. Instead of placing both selectable marker gene and GOI in the T-DNA region, they put the selectable marker gene in the backbone and the GOI in the T-DNA (Huang et al. 2004). Transgenic maize lines were successfully generated by this configuration, and marker-free transgenics were subsequently obtained.

4.2.2 Site-Specific Recombination

Recombinases from phages and yeasts, such as Cre, FLP and R, which recombine specific target sites, *loxP*, *FRT* and *RS*, respectively, are powerful tools to remove selection marker genes and other unneeded segments from plants (Ow 2005, 2007). A selectable marker gene may be placed between two directly orientated target sites (e.g., *lox*) in a transformation vector. Once a transformant is obtained, the recombinase (e.g., *cre* gene in this *lox* example) can be introduced into the cell by cross-pollination or additional rounds of transformation. The *lox*-flanked marker gene will be removed by the Cre recombinase and the *cre* gene can be removed by subsequent genetic seg-regation. In maize, Zhang et al. (2003) confirmed that the *nptII* marker gene could be efficiently removed by Cre/*lox* recombination via either crossing or auto-excision. Most recently, this practice has led to regulatory approval in the USA of a commercial corn line, LY308, a high lysine corn targeted for use in the poultry industry (http://www.aphis.usda.gov/brs/aphisdocs2/04_22901p_com.pdf).

5   Improvement of Maize Transformation

The past 20 years have seen tremendous technological advances in maize genetics and genomics research and rapid adoption of GE maize in agriculture. The pace of research has been accelerated by the enhanced efficiency of genetic transformation, a powerful technology that allows us to complement mutants, analyze gene functions, and dissect signaling or metabolic pathways. While we are no longer concerned about whether we can produce transgenic maize, we are now challenged by how we can produce transgenic maize of better quality and greater quantity to meet the increasing demands of genomic research.

5.1   *Expression and Copy Number of Transgenes*

Transgene silencing or unpredictable transgene expression is one of the major concerns in transgenic plant research. Once considered experimental outliers, transgenic events that show gene silencing are now accepted as an important aspect of gene regulation systems in all living organisms. While increased knowledge of the mechanisms for gene silencing is mounting, we are still struggling to take control of this phenomenon and design strategies to minimize its undesired effects in transgenic plants.

It has been widely recognized and also discussed in Subheading 2.3 that if a large number of transgene copies are integrated into the plant genome, there is an increased chance of transgene silencing or suppression of the endogenous genes. Because the *Agrobacterium*-mediated method offers low copy number and simple transgene integration, it is currently the favored method over biolistic gun transformation. In our experience, transformation mediated by *Agrobacterium* in maize has a big advantage over that in rice in terms of copy number of the transgenes. A number of rice and maize (A188) transformation experiments have been carried out using similar super-binary vectors by *bar*, *hpt* and *pmi* selectable markers for years in our laboratory. No matter which marker was employed, the ratio of single copy events was usually 25% to 45% in rice and 50% to 70% in maize (Y. Ishida and Y. Hiei, unpublished results). In these experiments, TF fluctuated to some extent in both rice and maize, but the ratio of single copy events remained quite stable. Although this suggests that overall efficiency of transformation is probably not a reason for this difference, no other explanation has yet been forthcoming

It has also been observed that the type of promoter may affect transgene copy number and the level of silencing of a non-selected transgene (Y. Ishida, unpublished results). Maize inbred line A188 was *Agrobacterium*-transformed with two similar strains that carried identical selectable *bar* gene cassettes but different *gus* gene cassettes (one with CaMV 35S and the other with the *ubi* promoter). Strong GUS expression was detected from almost all of the *ubi* transformants but only 60% of the 35S transformants. Interestingly, Southern analysis revealed that, among these strong expressers, 46% of *ubi* transformants and 74% of 35S transformants, respectively, were single copy events. Among the 40% of weak expressers of 35S transformants, 73% were multiple copy events. One interpretation could be that the 35S promoter might be prone to silencing in multi-copy transformants. Thus, strong 35S-expressers tended to be low-copy transformants.

5.2   Transfer of Vector Backbones

One major drawback of the *Agrobacterium*-mediated method is that high frequencies of vector backbone DNA can be detected in transgenic lines. This has been observed in both monocot and dicot plants (Wenck et al. 1997; De Buck et al. 2000; Shou et al. 2004; Wu et al. 2006). This not-so-precise transgene integration caused by the *Agrobacterium*-mediated transformation process can be a serious issue, especially in development of transgenic events for commercial purposes. Failure of the termination in the process of generation of T-DNA transfer intermediates at the left border of T-DNA appeared to be a major cause of this undesired phenomenon in rice (Kuraya et al. 2004). A method recently examined in rice was to place one, two or three additional copies of the left border sequences in transformation vectors, thereby suppressing the transfer of the segment outside T-DNA to plant cells in a nearly perfect fashion (Kuraya et al. 2004). This approach was recently tested in maize (Y. Ishida, unpublished results). Immature embryos of A188 were infected with *Agrobacterium* strains, in which one, two, three or four repeats of the left border were designed. A *gus* gene was placed outside the T-DNA in these vectors to monitor the carry over of backbone DNA in transformants. Transient expression of GUS in the embryos infected with multiple left border constructs was markedly lower when compared to the embryos infected with the standard one left border construct, suggesting that the strategy used in rice can also be effective in maize.

5.3   Location of Transgene Integration

Another problem, in addition to the copy number issue of transgene integration, is the unpredictability of transgene location in the targeted genome. It is believed that this transgene "position effect" can significantly affect its expression and stability (Matzke and Matzke, 1998; Day et al. 2000; Ow 2005; Kumar et al. 2006). To overcome this limitation, large numbers of independent transformation events are frequently screened to obtain a few that display simple integration structures and satisfactory expression levels. Numerous molecular strategies were also explored. It was found earlier that matrix attachment regions (MAR) could be used to shield the transgene from the genomic environment. For example, if the *gus* gene was flanked by the tobacco RB7 MAR elements, the number of low GUS expression transformants was significantly reduced (Mankin et al. 2003; Halweg et al. 2005).

Using recombinase-based site specific integration to circumvent position effect has been explored extensively (Ow 2005). The strategy is to first establish a founder line in which the construct carrying the recombination sites (e.g., *lox* or *FRT*) is integrated into a suitable chromosome location in the plant that allows adequate gene expression. The existing integrated construct in the founder line can then be used for sequential insertion of transgenes into the same target locus. This approach has been tested in a number of plants (Ow 2002). In tobacco, Cre-directed site-specific integration places a precise single-copy DNA fragment into the target site in about a third of the selected events. In rice, nearly half of the selected events consist of a single precise copy at the target site (Srivastava et al. 2004). These rates are significantly higher than those reported for homology-dependent insertions (Terada et al. 2002). Moreover, half of the precise single copy insertions in tobacco, and nearly all of those in rice, express the transgene within a range that is predictable and reproducible (Day et al. 2000; Srivastava et al. 2004). This indicates that once a suitable target site is found, the plant line can be used for predictable insertion and expression of genes.

The use of zinc finger nucleases (ZFNs) as molecular scissors for gene targeting is gaining increasing attention in both animal and plant systems (Durai et al. 2005; Porteus and Carroll 2005). ZFNs can be used to induce double-stranded breaks (DSBs) in specific DNA sequences, thereby stimulating site-specific homologous recombination in the targeted genomic loci. Successes in plants include Arabidopsis (Lloyd et al. 2005), tobacco (Wright et al. 2005) and maize (Arnold et al. 2007).

While still at its early stage, these studies showed that the use of ZFNs may be promising in targeted plant genome modification.

Recently, two groups reported the development of maize minichromosomes in attempting to overcome the transgene position effect problem. The group of James Birchler (Yu et al. 2006; 2007) used telomere-mediated chromosomal truncation approach to introduce transgenes to an existing chromosome lacking essential genes (such as B chromosomes in maize). They designed a vector carrying Arabidopsis telomere repeats, a selectable marker and a *lox* site-specific recombination site to transform maize IEs using the *Agrobacterium*-mediated method. The addition of transgenic telomeric sequences caused chromosome fragmentation at the site and left the transgenes at the tip of the fractured chromosome. Because the engineered minichromosome provided the *lox* site for site-specific recombination, additional transgenes could then be targeted to the site via the Cre/*lox* recombination system.

The other approach, led by Daphne Preuss, focused on the development of an in vitro-assembled autonomous maize minichromosome (MMC, Carlson et al. 2007). The authors constructed plasmids that carried selectable marker gene *nptII*, fluorescent reporter gene *DsRed*, as well as various segments (7 – 190 kb) of maize genomic DNA containing centromeric repeats and delivered the purified plasmids into maize IEs using the biolistic gun method. They found that 90% of the constructs carrying various centromeric sequences were able to form an autonomous MMC in plant cells. One transgenic line carrying a 19 kb centromeric DNA showed efficient mitotic and meiotic inheritance of MMC through 4 generations.

While these technologies are still at an early stage, they demonstrate the possibility of using chromosome engineering for basic and applied research. These approaches will enable the introduction of several trait genes or genes comprising a complete metabolic pathway on a single DNA fragment with a defined sequence order. More consistent transgene expression may also be expected from these engineered chromosome environments.

5.4 Genome Information for Improvement of Transformation

As maize genetic transformation technology becomes increasingly critical to maize genomic studies, genomic information, conversely, is also becoming important to the improving transformation methods. For example, the tissue culture ability and transformation competency of different maize genotypes may be addressed at the molecular level using various genomic tools. In rice, it was found that ferredoxinnitrite reductase (NiR) was correlated with callus culture regenerability. Overexpressing NiR in a poorly regenerable rice variety increased its regeneration rate significantly (Nishimura et al. 2005). In maize, stimulation of the cell cycle by disruption of the plant retinoblastoma pathway can lead to enhanced transformation frequency (Gordon-Kamm et al. 2002). Efforts in investigating and improving tissue culturability and transformability of maize using marker-assisted breeding and genomic techniques have been reported (Che et al. 2006; Krakowsky et al. 2006;

Lowe et al. 2006). Also, a number of plant genes, including chromatin proteins, have been shown to participate in the T-DNA integration process (Gelvin and Kim 2007).

5.5 Future Goals

By manipulating *Agrobacterium* strains and improving tissue culture conditions, we are now able to transform maize, once considered a plant recalcitrant to transformation using *Agrobacterium* (Hernalsteens et al. 1984). Future improvement of the technology will focus on transformation efficiency (less input for greater output), genotype spectrum (broaden inbred line transformation), integration precision (targeted gene insertion), and delivery size (large DNA fragment transformation for gene stacking or pathway engineering). Increased overall transformation efficiency for a few widely used public inbred genotypes would provide the maize community with immediate, expanded opportunities for genomic research and crop improvement.

Acknowledgements KW and BF thank Marcy Main, Jennie Lund, Lise Marcell and Tina Paque for their technical assistance and discussion and the National Science Foundation (DBI 0110023) for partial support of this work. YI and TK are grateful to Yukoh Hiei for helpful discussions and to Eriko Usami, Maki Noguchi and Kumiko Donovan for technical assistance.

References

Ahmadabadi, M., Ruf, S. and Bock, R. (2007) A leaf-based regeneration and transformation system for maize (*Zea mays* L.). Transgenic Res 16, 437–448.

Armstrong, C. L. (1999) The first decade of maize transformation: a review and future perspective. Maydica 44, 101–109.

Armstrong, C. L., Green, C. E. and Phillips, R. L. (1991) Development and availability of germplasm with high Type II culture formation response. Maize Genet Coop Newsl 65, 92–93.

Armstrong, C. L., Petersen, W. L., Buchholz, W. G., Bowen, B. A. and Sulc, S. L. (1990) Factors affecting PEG-mediated stable transformation of maize protoplasts. Plant Cell Rep 9, 335–339.

Arnold, N., Bauer, T., Collingwood, T., Dekelver, R., Doyon, Y., Gao, Z., McCaskill, D., Miller, J., Mitchell, J., Moehle, E., Rebar, E., Rock, J., Rowland, L., Shukla, V., Simpson, M., Skokut, M., Urnov, F., Worden, S., Yau, K. and Zhang, L. (2007) *Application of designed zinc-finger protein technology in plants*. 2007 Botany & Plant Biology Joint Congress, Chicago, IL, American Society of Plant Biologists. pp. 248–249 (P44015).

Bowman, J. L. and Eshed, Y. (2000) Formation and maintenance of the shoot apical meristem. Trends Plant Sci 5, 110–115.

Brettschneider, R., Becker, D. and Lörz, H. (1997) Efficient transformation of scutellar tissue of immature maize embryos. Theor Appl Genet 94, 737–748.

Carvalho, C. H. S., Bohorova, N., Bordallo, P. N., Abreu, L. L., Valicente, F. H., Bressan, W. and Paiva, E. (1997) Type II callus production and plant regeneration in tropical maize genotypes. Plant Cell Rep 17, 73–76.

Carlson, S. R., Rudgers, G. W., Zieler, H., Mach, J. M., Luo, S., Grunden, E., Krol, C., Copenhaver, G. P., Preuss, D. (2007) Meiotic transmission of an in vitro-assembled autonomous maize minichromosome. PLOS Genetics 3, 1965–1974.

Che, P., Love, T. M., Frame, B. R., Wang, K., Carriquiry, A. L. and Howell, S. H. (2006) Gene expression patterns during somatic embryo development and germination in maize Hi-II callus cultures. Plant Mol Biol 62, 1–14.

Cheng, M., Fry, J. E., Pang, S., Zhou, H., Hironaka, C. M., Duncan, D. R., Conner, T. W. and Wan, Y. (1997) Genetic transformation of wheat mediated by *Agrobacterium tumefaciens*. Plant Physiol 115, 971–980.

Chu, C. C., Wang, C. C., Sun, C. S., Hsu, C., Yin, K. C., Chu, C. Y. and Bi, F. Y. (1975) Establishment of an efficient medium for anther culture of rice through comparative experiments on the nitrogen source. Sci Sinica 18, 659–668.

Daley, M., Knauf, V. C., Summerfelt, K. R. and Turner, J. C. (1998) Co-transformation with one *Agrobacterium tumefaciens* strain containing two binary plasmids as a method for producing marker-free transgenic plants. Plant Cell Rep 17, 489–496.

Danilova, S. A. and Dolgikh, Y. I. (2005) Optimization of Agrobacterium (*Agrobacterium tumefaciens*) transformation of maize embryogenic callus. Russian Journal of Plant Physiology 52, 535–541.

Day, C. D., Lee, E., Kobayashi, J., Holappa, L. D., Albert, H., and Ow, D. W. (2000) Transgene integration into the same chromosome location can produce alleles that express at a predictable level, or alleles that are differentially silenced. Genes Dev 14, 2869–2880.

De Block, M. and Debrouwer, D. (1991) Two T-DNA's co-transformed into *Brassica napus* by a double *Agrobacterium tumefaciens* infection are mainly integrated at the same locus. Theor Appl Genet 82, 257–263.

De Buck, S., De Wilde, C., Van Montagu, M. and Depicker, A. (2000) T-DNA vector backbone sequences are frequently integrated into the genome of transgenic plants obtained by *Agrobacterium*-mediated transformation. Mol Breeding 6, 459–468.

Dennehey, B. K., Peterson, W. L., Ford-Santino, C., Pajeau, M. and Armstrong, C. L. (1994) Comparison of selective agents for use with the selectable marker gene *bar* in maize transformation. Plant Cell Tiss Org 36, 1–7.

Depicker, A., Herman, L., Jacobs, A., Schell, J. and Montagu, M. V. (1985) Frequencies of simultaneous transformation with different T-DNAs and their relevance to the *Agrobacterium*/plant cell interaction. Mol Gen Genet 201, 477–484.

D'Halluin, K., Bonne, E., Bossut, M., De Beuckeleer, M. and Leemans, J. (1992) Transgenic maize plants by tissue electroporation. Plant Cell 4, 1495–1505.

Dunder, E., Dawson, J., Suttie, J. and Pace, G. (1995) Maize transformation by microprojectile bombardment of immature embryos. In: I. Potrykus and G. Spangenberg (Eds.), *Gene Transfer to Plants*. Springer-Verlag, Berlin, pp. 127–138.

Durai, S., Mani, M., Kandavelou, K., Wu, J., Porteus, M. H. and Chandrasegaran, S. (2005) Zinc finger nucleases: custom-designed molecular scissors for genome engineering of plant and mammalian cells. Nucleic Acids Res 33, 5978–5990.

Ebinuma, H., Sugita, K., Matsunaga, E. and Yamakado, M. (1997) Selection of marker-free transgenic plants using the isopentenyl transferase gene. Proc Natl Acad Sci U S A 94, 2117–2121.

Evans, R., Wang, A. S., Hanten, J., Altendorf, P. and Mettler, I. (1996) A positive selection system for maize transformation. In Vitro Cell Dev Biol-Plant 32, 72A (abstract).

Finer, J. J., Vain, P., Jones, M. W. and McMullen, M. D. (1992) Development of the particle inflow gun for DNA delivery to plant cells. Plant Cell Rep 11, 323–328.

Frame, B. R., Drayton, P. R., Bagnall, S. V., Lewnau, C. J., Bullock, W. P., Wilson, H. M., Dunwell, J. M., Thompson, J. A. and Wang, K. (1994) Production of fertile transgenic maize plants by silicon carbide whisker-mediated transformation. Plant J 6, 941–948.

Frame, B. R., McMurray, J. M., Fonger, T. M., Main, M. L., Taylor, K. W., Torney, F. J., Paz, M. M. and Wang, K. (2006a) Improved *Agrobacterium*-mediated transformation of three maize inbred lines using MS salts. Plant Cell Rep 25, 1024–1034.

Frame, B. R., Paque, T. and Wang, K. (2006b). Maize (*Zea mays* L.). In: K. Wang (Eds.), *Agrobacterium Protocols (2nd edition)*. Humana Press Inc., Totowa, NJ, pp. 185–199.

Frame, B. R., Shou, H., Chikwamba, R. K., Zhang, Z., Xiang, C., Fonger, T. M., Pegg, S. E., Li, B., Nettleton, D. S., Pei, D. and Wang, K. (2002) *Agrobacterium tumefaciens*-mediated transformation of maize embryos using a standard binary vector system. Plant Physiol 129, 13–22.

Frame, B. R., Zhang, H., Cocciolone, S. M., Sidorenko, L. V., Dietrich, C. R., Pegg, S. E., Zhen, S., Schnable, P. S. and Wang, K. (2000) Production of transgenic maize from bombarded Type II callus: Effect of gold particle size and callus morphology on transformation efficiency. In Vitro Cellular and Developmental Biology Plant 36, 21–29.

Fromm, M. E., Morrish, F., Armstrong, C., Williams, R., Thomas, J. and Klein, T. M. (1990) Inheritance and expression of chimeric genes in the progeny of transgenic maize plants. Bio/technology 8, 833–839.

Gelvin, S. B. (2003) *Agrobacterium*-mediated plant transformation: the biology behind the "gene-jockeying" tool. Microbiol Mol Biol Rev 67, 16–37.

Gelvin, S. B. and Kim, S. I. (2007) Effect of chromatin upon Agrobacterium T-DNA integration and transgene expression. Biochim Biophys Acta 1769, 410–421.

Golovkin, M. V., Ábrahám, M., Mórocz, S., Bottka, S., Fehér, A. and Dudits, D. (1993) Production of transgenic maize plants by direct DNA uptake into embryogenic protoplasts. Plant Sci 90, 41–52.

Gordon-Kamm, W., Dilkes, B. P., Lowe, K., Hoerster, G., Sun, X., Ross, M., Church, L., Bunde, C., Farrell, J., Hill, P., Maddock, S., Snyder, J., Sykes, L., Li, Z., Woo, Y. M., Bidney, D. and Larkins, B. A. (2002) Stimulation of the cell cycle and maize transformation by disruption of the plant retinoblastoma pathway. Proc Natl Acad Sci U S A 99, 11975–11980.

Gordon-Kamm, W. J., Spencer, T. M., Mangano, M. L., Adams, T. R., Daines, R. J., Start, W. G., O'Brien, J. V., Chambers, S. A., Adams, W. R., Jr., Willetts, N. G., Rice, T. B., Mackey, C. J., Krueger, R. W., Kausch, A. P. and Lemaux, P. G. (1990) Transformation of maize cells and regeneration of fertile transgenic plants. Plant Cell 2, 603–618.

Gould, J., Devey, M., Hasegawa, O., Ulian, E. C., Peterson, G. and Smith, R. H. (1991) Transformation of *Zea mays* L. using *Agrobacterium tumefaciens* and the shoot apex. Plant Physiol 95, 426–434.

Graves, A. C. F. and Goldman, S. L. (1986) The transformation of *Zea mays* seedlings with *Agrobacterium tumefaciens*. Plant Mol Biol 7, 43–50.

Graves, A. E., Goldman, S. L., Banks, S. W. and Graves, A. C. (1988) Scanning electron microscope studies of *Agrobacterium tumefaciens* attachment to *Zea mays*, *Gladiolus* sp., and *Triticum aestivum*. J Bacteriol 170, 2395–2400.

Grimsely, N., Hohn, T., Davies, J. W. and Hohn, B. (1987) *Agrobacterium*-mediated delivery of infectious maize streak virus into maize plants. Nature 325, 177–179.

Halweg, C., Thompson, W. F. and Spiker S. (2005) The rb7 matrix attachment region increases the likelihood and magnitude of transgene expression in tobacco cells: a flow cytometric study. 17, 418–429.

Hansen, G. and Wright, M. S. (1999) Recent advances in the transformation of plants. Trends Plant Sci 4, 226–231.

Hepburn, A. G., White, J., Pearson, L., Maunders, M. J., Clarke, L. E., Prescott, A. G. and Blundy, K. S. (1985) The use of pNJ5000 as an intermediate vector for the genetic manipulation of Agrobacterium Ti-plasmids. J Gen Microbiol 131, 2961–2969.

Hernalsteens, J.-P., Thia-Toong, L., Schell, J. and Montagu, M. V. (1984) An *Agrobacterium*-transformed cell culture from the monocot *Asparagus officinalis*. The EMBO Journal 3, 3039–3041.

Hiei, Y., Ishida, H., Kasaoka, K. and Komari, T. (2006) Improved frequency of transformation in rice and maize by treatment of immature embryos with centrifugation and heat prior to infection with *Agrobacterium tumefaciens*. Pant Cell Tiss Org 87, 233–243.

Hiei, Y., Ohta, S., Komari, T. and Kumashiro, T. (1994) Efficient transformation of rice (*Oryza sativa* L.) mediated by *Agrobacterium* and sequence analysis of the boundaries of the T-DNA. Plant J 6, 271–282.

Hoekema, A., Hirsch, P. R., Hooykaas, P. J. J. and Schilperoort, R. A. (1983) A binary plant vector strategy based on separation of *vir*-and T-region of the *Agrobacterium tumefaciens* Ti-plasmid plant genetics. Nature 303, 179–180.

Hood, E. E., Helmer, G. L., Fraley, R. T. and Chilton, M. D. (1986) The hypervirulence of *Agrobacterium tumefaciens* A281 is encoded in a region of pTiBo542 outside of T-DNA. J Bacteriol 168, 1291–1301.

Horikawa, Y., Yoshizumi, T. and Kakuta, H. (1997) Transformants through pollination of mature maize (*Zea mays* L.) pollen delivered *bar* gene by particle gun. Grassland Science 43, 117–123.

Howe, A. R., Gasser, C. S., Brown, S. C., Padgette, S. R., Hart, J. J., Parker, G. B., Fromm, M. E. and Armstrong, C. L. (2002) Glyphosate as a selective agent for the production of fertile transgenic maize (*Zea may* L.) plants. Mol Breeding 10, 153–164.

Hu, T., Metz, S., Chay, C., Zhou, H. P., Biest, N., Chen, G., Cheng, M., Feng, X., Radionenko, M., Lu, F. and Fry, J. E. (2003) *Agrobacterium*-mediated large-scale transformation of wheat (*Triticum aestivum* L.) using glyphosate selection. Plant Cell Rep 21, 1010–1019.

Huang, S., Gilbertson, L., Adams, T. H., Malloy, K., Reisenbigler, K., Birr, D., Snyder, M., Zhang, Q. and Luethy, M. (2004) Generation of marker-free trangenic maize by regular two-border *Agrobacterium* transformation vectors. Transgenic Res 13, 451–461.

Huang, X. and Wei, Z. (2005) Successful *Agrobacterium*-mediated genetic transformation of maize elite inbred lines. Plant Cell Tiss Org 83, 187–200.

Ishida, Y., Hiei, Y. and Komari, T. (2007) *Agrobacterium*-mediated transformation of maize. Nat Protoc 2, 1614–1621.

Ishida, Y., Murai, N., Kuraya, Y., Ohta, S., Saito, H., Hiei, Y. and Komari, T. (2004) Improved co-transformation of maize with vectors carrying two separate T-DNAs mediated by *Agrobacterium tumefaciens*. Plant Biotechnology 21, 57–63.

Ishida, Y., Saito, H., Hiei, Y. and Komari, T. (2003) Improved protocol for transformation of maize (*Zea mays* L.) mediated by *Agrobacterium tumefaciens*. Plant Biotechnol 20, 57–66.

Ishida, Y., Saito, H., Ohta, S., Hiei, Y., Komari, T. and Kumashiro, T. (1996) High efficiency transformation of maize (*Zea mays* L.) mediated by *Agrobacterium tumefaciens*. Nat Biotechnol 14, 745–750.

Joersbo, M., Donaldson, I., Kreiberg, J., Petersen, S. G., Brunstedt, J. and Okkels, F. T. (1998) Analysis of mannose selection used for transformation of sugar beet. Mol Breeding 4, 111–117.

Kaeppler, H. F., Somers, D. A., Rines, H. W. and Cockburn, A. F. (1992) Silicon carbide fiber-mediated stable transformation of plant cells. Theor Appl Genet 84, 560–566.

Kikkert, J. R., Vidal, J. R. and Reisch, B. I. (2005) Stable transformation of plant cells by particle bombardment/biolistics. In: L. Peña (Eds.), *Transgenic Plants: Methods and Protocols*. Humana Press, Totowa, NJ, pp. 61–78.

Klein, T. M., Wolf, E. D., Wu, R. and Sanford, J. C. (1987) High-velocity microprojectiles for delivering nucleic acids into living cells. Nature 327, 70–73.

Komari, T. (1990) Transformation of cultured cells of *Chenopodium quinoa* by binary vectors that carry a fragment of DNA from the virulence region of pTiBo542. Plant Cell Rep 9, 303–306.

Komari, T., Hiei, Y., Saito, Y., Murai, N. and Kumashiro, T. (1996) Vectors carrying two separate T-DNAs for co-transformation of higher plants mediated by *Agrobacterium tumefaciens* and segregation of transformants free from selection markers. Plant J 10, 165–174.

Komari, T., Takakura, Y., Ueki, J., Kato, N., Ishida, Y. and Hiei, Y. (2006) Binary vectors and super-binary vectors. In: K. Wang (Eds.), *Agrobacterium Protocols*. Humana Press, Totowa, NJ, pp. 15–41.

Koncz, C. and Schell, J. (1986) The promoter of T_L-DNA gene 5 controls the tissue-specific expression of chimaeric genes carried by a novel type of *Agrobacterium* binary vector. Mol Gen Genet 204, 383–396.

Koziel, G. M., Beland, G. L., Bowman, C., Carozzi, N. B., Crenshaw, R., Crossland, L., Dawson, J., Desai, N., Hill, M., Kadwell, S., Launis, K., Lewis, K., Maddox, D., McPherson, K., Meghji, M. R., Merlin, E., Rhodes, R., Warren, G. W., Wright, M. S. and Evola, S. V. (1993) Field performance of elite transgenic maize plants expressing an insecticidal protein derived from *Bacillus thuringiensis*. Bio/Technology 11, 194–200.

Krakowsky, M. D., Lee, M., Garay, L., Woodman-Clikeman, W., Long, M. J., Sharopova, N., Frame, B. and Wang, K. (2006) Quantitative trait loci for callus initiation and totipotency in maize (*Zea mays* L.). Theor Appl Genet 113, 821–830.

Kumar, S., Allen, G. C. and Thompson, W. F. (2006). Gene targeting in plants: fingers on the move. Trends Plant Sci 11, 159–161.

Kuraya, Y., Ohta, S., Fukuda, M., Hiei, Y., Murai, N., Hamada, K., Ueki, J., Imaseki, H. and Komari, T. (2004) Suppression of transfer of non-T-DNA 'vector backbone' sequences by multiple left border repeats in vectors for transformation of higher plants mediated by *Agrobacterium tumefaciens*. Mol Breeding 14, 309–320.

Laursen, C. M., Krzyzek, R. A., Flick, C. E., Anderson, P. C. and Spencer, T. M. (1994) Production of fertile transgenic maize by electroporation of suspension culture cells. Plant Mol Biol 24, 51–61.

Li, W., Masilmany, P., Kasha, K. J. and Pauls, K. P. (2002) Development, tissue culture, and genotypic factors affecting plant regeneration from shoot apical meristems of germinated *Zea mays* L. seedlings. In Vitro Cell Dev Biol-Plant 38, 285–292.

Li, X., Volrath, S. L., Nicholl, D. B. G., Chilcott, C. E., Johnson, M. A., Ward, E. R. and Law, M. D. (2003) Development of protoporphyrinogen oxidase as an efficient selection marker for *Agrobacterium tumefaciens*-mediated transformation of maize. Plant Physiol 133, 736–747.

Lloyd, A., Plaisier, C. L., Carroll, D. and Drews, G. N. (2005) Targeted mutagenesis using zinc-finger nucleases in Arabidopsis. Proc Natl Acad Sci U S A 102, 2232–2237.

Lowe, B., Way, M. M., Kumpf, J. M., Rout, G. R., Warner, D., Johnson, R., Armstrong, C. L., Spencer, M. T. and Chomet, P. S. (2006) Marker assisted breeding for transformability in maize. Mol Breeding 18, 229–239.

Lowe, K., Bowen, B., Hoerster, G., Ross, M., Bond, D., Pierce, D. and Gordon-Kamm, B. (1995) Germline transformation of maize following manipulation of chimeric shoot meristems. Bio/Technology 13, 677–682.

Maliga, P. (2004) Plastid transformation in higher plants. Annu Rev Plant Biol 55, 289–313.

Mankin, S. L., Allen, G. C., Phelan, T., Spiker, S. and Thompson, W. F. (2003) Elevation of transgene expression level by flanking matrix attachment regions (MAR) is promoter dependent: a study of the interactions of six promoters with the RB7 3′ MAR. Transgenic Res 12, 3–12.

Matsubayashi, Y., Takagi, L. and Sakagami, Y. (1997) Phytosulfokine-alpha, a sulfated pentapeptide, stimulates the proliferation of rice cells by means of specific high- and low-affinity binding sites. Proc Natl Acad Sci U S A 94, 13357–13362.

Matzke, A.J., and Matzke, M.A. (1998) Position effects and epigenetic silencing of plant transgenes. Curr Opin Plant Biol 1, 142–148.

McCabe, D. and Christou, P. (1993) Direct DNA transfer using electric discharge particle acceleration (ACCELL™ technology). Plant Cell, Tiss Org 33, 227–236.

McCormac, A. C., Fowler, M. R., Chen, D. F. and Elliott, M. C. (2001) Efficient co-transformation of *Nicotiana tabacum* by two independent T-DNAs, the effect of T-DNA size and implications for genetic separation. Transgenic Res 10, 143–155.

Miki, B. and McHugh, S. (2004) Selectable marker genes in transgenic plants: applications, alternatives and biosafety. J Biotechnol 107, 193–232.

Miller, M., Tagliani, L., Wang, N., Berka, B., Bidney, D. and Zhao, Z. Y. (2002) High efficiency transgene segregation in co-transformed maize plants using an *Agrobacterium tumefaciens* 2 T-DNA binary system. Transgenic Res 11, 381–396.

Morrish, F., Songstad, D. D., Armstrong, C. L. and Fromm, M. (1993) Microprojectile bombardment: A method for the production of transgenic cereal crop plants and the functional analysis of genes. In: A. Hiatt (Eds.), *Transgenic Plants: Fundamentals and Applications*. Marcel Dekker, Inc., New York, pp. 133–171.

Murashige, T. and Skoog, F. (1962) A revised medium for rapid growth and bioassays with tobacco tissue cultures. Physiol Plant 15, 473–497.

Negrotto, D., Jolley, M., Beer, S., Wenck, A. R. and Hansen, G. (2000) The use of phosphomannose-isomerase as a selectable marker to recover transgenic maize plants (*Zea mays* L.) via *Agrobacterium* transformation. Plant Cell Rep 19, 798–803.

Nishimura, A., Ashikari, M., Lin, S., Takashi, T., Angeles, E. R., Yamamoto, T. and Matsuoka, M. (2005) Isolation of a rice regeneration quantitative trait loci gene and its application to transformation systems. Proc Natl Acad Sci U S A 102, 11940–11944.

O'Connor-Sánchez, A., Cabrera-Ponce, J. L., Valdez-Melara, M., Téllez-Rodríguez, P., Pons-Hernández, J. L. and Herrera-Estrella, L. (2002) Transgenic maize plants of tropical and subtropical genotypes obtained from calluses containing organogenic and embryogenic-like structures derived from shoot tips. Plant Cell Rep 21, 302–312.

Odell, J., Caimi, P., Sauer, B. and Russell, S. (1990) Site-directed recombination in the genome of transgenic tobacco. Mol Gen Genet 223, 369–378.

Ohta, S., Mita, S., Hattori, T. and Nakamura, S. (1990) Construction and expression in tobacco of a beta-glucuronidase (GUS) reporter gene containing an intron within the coding sequence. Plant Cell Physiol 31, 805–813.

Ooms, G., Hooykaas, P. J., Van Veen, R. J., Van Beelen, P., Regensburg-Tuink, T. J. and Schilperoort, R. A. (1982) Octopine Ti-plasmid deletion mutants of *Agrobacterium tumefaciens* with emphasis on the right side of the T-region. Plasmid 7, 15–29.

Ow, D. W. (2002) Recombinase-directed plant transformation for the post-genomic era. Plant Mol Biol 48, 183–200.

Ow, D. W. (2005) Transgene management via multiple site-specific recombination systems. In Vitro Cell Dev Biol-Plant, 41, 213–219.

Ow, D. W. (2007) GM maize from site-specific recombination technology, what next? Curr Opin Biotechnol 18, 115–120.

Pawlowski, W. P. and Somers, D. A. (1996) Transgene inheritance in plants genetically engineered by microprojectile bombardment. Mol Biotechnol 6, 17–30.

Pawlowski, W. P. and Somers, D. A. (1998) Transgenic DNA integrated into the oat genome is frequently interspersed by host DNA. Proc Natl Acad Sci U S A 95, 12106–12110.

Paz, M., Shou, H., Guo, Z.-B., Zhang, Z.-Y., Banerjee, A. and Wang, K. (2004) Assessment of conditions affecting *Agrobacterium*-mediated soybean transformation using the cotyledonary node explant. Euphytica 136, 167–179.

Pescitelli, S. M. and Sukhapinda, K. (1995) Stable transformation via electroporation into maize Type II callus and regeneration of fertile transgenic plants. Plant Cell Rep 14, 712–716.

Petolino, J. F., Hopkins, N. L., Kosegi, B. D. and Skokut, M. (2000) Whisker-mediated transformation of embryogenic callus of maize. Plant Cell Rep 19, 781–786.

Porteus, M. H. and Carroll, D. (2005) Gene targeting using zinc finger nucleases. Nat Biotechnol 23, 967–973.

Potrykus, I., Bilang, R., Fütterer, J., Sautter, C., Schrott, M. and Spangenberg, G. (1998) Genetic engineering of crop plants. In: A. Altman (Eds.), *Agricultural Biotechnology*. Marcel Dekker, Inc., New York, pp. 119–159.

Potrykus, I. and Spangenberg, G. (1995) *Gene Transfer to Plants*. Springer-Verlag, Berlin.

Primich-Zachwieja, S. and Minocha, S. C. (1991) Induction of virulence response in *Agrobacterium tumefaciens* by tissue explants of various plant species. Plant Cell Rep 10, 545–549.

Quan, R., Shang, M., Zhang, H., Zhao, Y. and Zhang, J. (2004) Improved chilling tolerance by transformation with *betA* gene for the enhancement of glycinebetaine synthesis in maize. Plant Sci 166, 141–149.

Rhodes, C. A., Pierce, D. A., Mettler, I. J., Mascarenhas, D. and Detmer, J. J. (1988) Genetically transformed maize plants from protoplasts. Science 240, 204–207.

Schocher, R. J., Shillito, R. D., Saul, M. W., Paszkowski, J. and Potrykus, I. (1986) Co-transformation of unlinked foreign genes into plants by direct gene transfer. Bio/Technology 4, 1093–1096.

Shou, H., Frame, B., Whitham, S. and Wang, K. (2004) Assessment of transgenic maize events produced by particle bombardment or *Agrobacterium*-mediated transformation. Mol Breeding 13, 201–208.

Sidorov, V., Gilbertson, L., Addae, P. and Duncan, D. R. (2006) *Agrobacterium*-mediated transformation of seedling-derived maize callus. Plant Cell Rep 25, 320–328.

Songstad, D. D., Armstrong, C. L. and Petersen, W. L. (1991) Silver Nitrate increases Type II callus production from immature embryos of maize inbred B73 and its derivatives. Plant Cell Rep 9, 699–702.

Songstad, D. D., Armstrong, C. L., Petersen, W. L., Hairston, B. and Hinchee, M. A. (1996) Production of transgenic maize plants and progeny by bombardment of Hi-II immature embryos. In Vitro Cell Dev Biol 32, 179–183.

Southgate, E. M., Davey, M. R., Power, J. B. and Westcott, R. J. (1998) A comparison of methods for direct gene transfer into maize (*Zea mays* L.). In Vitro Cell Dev Biol-Plant 34, 218–224.

Srivastava, V., Ariza-Nieto, M. and Wilson, A., J. (2004) Cre-mediated site-specific gene integration for consistent transgene expression in rice. Plant Biotechnol J 2, 169–179.

Sticklen, M. B. and Oraby, H. F. (2005) Shoot apical meristem: a sustainable explant for genetic transformation of cereal crops. In Vitro Cell Dev Biol-Plant 41, 187–200.

Terada, R., Urawa, H., Inagaki, Y., Tsugane, K., and Iida, S. (2002) Efficient gene targeting by homologous recombination in rice. Nat Biotechnol 20: 1030–1034.

Tingay, S., McElroy, D., Kalla, R., Fieg, S. J., Wang, M.-B., Thornton, S. and Brettell, R. (1997) *Agrobacterium tumefaciens*-mediated barley transformation. Plant J 11, 1369–1376.

Torney, F., Frame, B. R. and Wang, K. (2007) Maize. In: E.-C. Pua and M. R. Davey (Eds.), *Transgenic Crops IV*. Springer, Berlin Heidelberg, pp. 73–105.

Trick, H. N. and Finer, J. J. (1997) SAAT: Sonication-assisted *Agrobacterium*-mediated transformation. Transgenic Res 6, 1–8.

Ueki, J., Komari, T. and Imaseki, H. (2004) Enhancement of reporter-gene expression by insertions of two introns in maize and tobacco protoplasts. Plant Biotechnol 21, 15–24.

Vain, P., McMullen, M. D. and Finer, J. J. (1993) Osmotic treatment enhances particle bombardment-mediated transient and stable transformation of maize. Plant Cell Rep 12, 84–88.

Vancanneyt, G., Schmidt, R., O'Connor-Sanchez, A., Willmitzer, L. and Rocha-Sosa, M. (1990) Construction of an intron-containing marker gene: splicing of the intron in transgenic plants and its use in monitoring early events in *Agrobacterium*-mediated plant transformation. Mol Gen Genet 220, 245–250.

Walters, D. A., Vetsch, C. S., Potts, D. E. and Lundquist, R. C. (1992) Transformation and inheritance of a hygromycin phosphotransferase gene in maize plants. Plant Mol Biol 18, 189–200.

Wan, Y., Widholm, J. M. and Lemaux, P. G. (1995) Type I callus as a bombardment target for generating fertile transgenic maize (*Zea mays* L.). Planta 196, 7–14.

Wang, A. S., Evans, R. A., Altendorf, P. R., Hanten, J. A., Doyle, M. C. and Rosichan, J. L. (2000) A mannose selection system for production of fertile transgenic maize plants from protoplasts. Plant Cell Rep 19, 654–660.

Wang, K., Frame, B. R. and Marcell, L. (2003) Maize Genetic Transformation. In: P. K. Jaiwal and R. P. Singh (Eds.), *Plant Genetic Engineering: improvement of food crops*. Sci-Tech Publication, Houston, Texas, USA, pp. 175–217.

Wenck, A., Czako, M., Kanevski, I. and Marton, L. (1997) Frequent collinear long transfer of DNA inclusive of the whole binary vector during *Agrobacterium*-mediated transformation. Plant Mol Biol 34, 913–922.

Wright, D. A., Townsend, J. A., Winfrey, R. J., Jr., Irwin, P. A., Rajagopal, J., Lonosky, P. M., Hall, B. D., Jondle, M. D. and Voytas, D. F. (2005) High-frequency homologous recombination in plants mediated by zinc-finger nucleases. Plant J 44, 693–705.

Wright, M., Dawson, J., Dunder, E., Suttie, J., Reed, J., Kramer, C., Chang, Y., Novitzky, R., Wang, H. and Artim-Moore, L. (2001) Efficient biolistic transformation of maize (*Zea may* L.) and wheat (*Triticum aestivum* L.) using the phosphomannose isomerase gene, *pmi*, as the selectable marker. Plant Cell Rep 20, 429–436.

Wu, H., Sparks, C. A. and Jones, H. D. (2006) Characterisation of T-DNA loci and vector backbone sequences in transgenic wheat produced by *Agrobacterium*-mediated transformation. Mol Breeding 18, 195–208.

Yu, W., Han, F., Gao, Z., Vega, J. M. and Birchler, J. A. (2007) Construction and behavior of engineered minichromosomes in maize. Proc Natl Acad Sci U S A 104, 8924–8929.

Yu, W., Lamb, J. C., Han, F. and Birchler, J. A. (2006) Telomere-mediated chromosomal truncation in maize. Proc Natl Acad Sci U S A 103, 17331–17336.

Zhang, S., Williams-Carrier, R. and Lemaux, P. G. (2002) Transformation of recalcitrant maize elite inbreds using in vitro shoot meristematic cultures induced from germinated seedlings. Plant Cell Rep 21, 263–270.

Zhang, W., Subbarao, S., Addae, P., Shen, A., Armstrong, C., Peschke, V. and Gilbertson, L. (2003) Cre/*lox*-mediated marker gene excision in transgenic maize (*Zea mays* L.) plants. Theor Appl Genet 107, 1157–1168.

Zhao, Z. Y., Cai, T., Tagliani, L., Miller, M., Wang, N., Pang, H., Rudert, M., Schroeder, S., Hondred, D., Seltzer, J. and Pierce, D. (2000) *Agrobacterium*-mediated sorghum transformation. Plant Mol Biol 44, 789–798.

Zhao, Z. Y., Gu, W., Cai, T., Tagliani, L. A., Hondred, D., Bond, D., Krell, S., Rudert, M. L., Bruce, W. B. and Pierce, D. A. (1998) Molecular analysis of T0 plants transformed by Agrobacterium and comparison of *Agrobacterium*-mediated transformation with bombardment transformation in maize. Maize Genet Coop Newsl 72, 34–37.

Zhao, Z.-Y., Gu, W., Cai, T., Tagliani, L. A., Hondred, D., Bond, D., Schroeder, S., Rudert, M. and Pierce, D. A. (2001) High throughput genetic transformation mediated by *Agrobacterium tumefaciens* in maize. Mol Breeding 8, 323–333.

Zhong, G. Y., Peterson, D., Delaney, D. E., Bailey, M., Witcher, D. R., Register, J. C., III, Bond, D., Li, C. P., Marshall, L., Kulisek, E., Ritland, D., Meyer, T., Hood, E. E. and Howard, J. A. (1999) Commercial production of aprotinin in transgenic maize seeds. Mol Breeding 5, 345–356.

Zhong, H., Srinivasan, C. and Sticklen, M. B. (1992) In-vitro morphogenesis of corn (*Zea mays* L.). Planta 187, 483–489.

Zhong, H., Sun, B., Warkentin, D., Zhang, S., Wu, R., Wu, T. and Sticklen, M. B. (1996) The competence of maize shoot meristems for integrative transformation and inherited expression of transgenes. Plant Physiol 110, 1097–1107.

Zhou, H., Arrowsmith, J. W., Fromm, M. E., Hironaka, C. M., Taylor, M. L., Rodriguez, D. J., Pajeau, M., Brown, S. M., Santino, C. G. and Fry, J. E. (1995) Glyphosate-tolerant CP4 and GOX gene as a selectable marker in wheat transformation. Plant Cell Rep 15, 159–163.

Doubled Haploids

Hartwig H. Geiger

Abstract Doubled haploid (DH) maize lines can be produced by *in vitro* and *in vivo* techniques. The *in vitro* approach is focused on anther and microspore culture. However, most maize genotypes proved to be highly recalcitrant. Genetic analyses showed that *in vitro* androgenetic response is under complex multifactorial control. Despite good results with specific genotypes, the technique has not yet become a routine tool in maize research or breeding. In contrast, *in vivo* procedures could be improved considerably and have been widely applied during the last 10 – 15 years. *In vivo* induction of paternal haploids based on the 'indeterminate gametophyte' mutant has become a standard technique for transferring elite seed parent lines into cytoplasms that condition male sterility. Similarly, the induction of maternal haploids by pollination with specific inducer genotypes has become a routine procedure for large-scale DH line production. Both *in vivo* techniques are much less affected by donor genotypes than the *in vitro* procedures.

Progress achieved in the induction of maternal haploids pertains to the induction rate, easily screenable markers for haploid identification, chromosome doubling procedures, and handling of seedlings which survived chromosome doubling. In research, DH lines are mainly being used for mapping purposes and in breeding they are progressively replacing conventional inbred lines. Various DH line-based breeding schemes have been suggested, and computer software has been developed for optimizing the dimensioning of the schemes and for determining the relative merits of alternative breeding strategies. DH lines feature important advantages regarding quantitative genetic, operational, logistic, and economic aspects. The DH-line technology can therefore be considered as one of the most effective tools in modern maize genetics and breeding. The mechanism of *in vivo* induction of gynogenetic haploids is not yet fully understood. Most likely, one of the two inducer sperm cells is not fully functional yet fuses with the egg cell. During subsequent cell divisions,

H.H. Geiger
Universität Hohenheim, Institute of Plant Breeding,
Seed Science, and Population Genetics
70593 Stuttgart, Germany
geigerhh@uni-hohenheim.de

a degeneration process starts and the chromosomes get fragmented and finally are eliminated from the primordial cells leaving only maternal chromosomes. The second sperm cell fuses with the central cell leading to a regular triploid endosperm and a normal-sized functional seed.

1   Introduction

The first haploid maize plant was described by Stadler and Randolph (1929, cited by Randolph 1932). About two decades later, Chase (1947) found haploids at low frequency (about 1 in 1000) in various US Corn-Belt materials. He recognized the great potential of haploids for the genetics and breeding of maize and consequently devoted all of his professional career to promote research and development in this field. Further milestones were the detection of colchicine as a chromosome doubling agent, the successful application of *in vitro* anther and microspore culture techniques (Section 2), and the detection of specific genotypes suited for *in vivo* haploid induction (Section 3). Various studies have shown that both *in vitro* and *in vivo* haploid induction are polygenically controlled characters and QTL (quantitative trait loci) analysis have identified genomic regions affecting haploid induction on almost all chromosomes (Section 4). Today, doubled haploids are widely applied in many fields of maize research and are worldwide used in commercial hybrid maize breeding (Section 5). In research, doubled haploid (DH) genotypes are a valuable tool in structural and functional genomics, proteomics, metabolomics, marker-trait association studies, marker-based gene introgressions, molecular cytogenetics, genetic engineering, etc. In breeding, DH lines allow an increase in the efficacy of selection, reduction in the length of a breeding cycle, simplification of the logistics, and saving time in commercializing a new breeding product.

2   *In vitro* Techniques

Encouraged by the successful application of *in vitro* culture techniques for the production of haploids in many crop plants (Kasha 1972), researchers in many countries tried to establish this approach in maize as well (for reviews see Büter 1997; Pret'ová et al. 2006). However, maize turned out to be an extremely recalcitrant plant species, with only a very few genotypes displaying satisfactory *in vitro* embryogenesis. Petolino and Thompson (1987) analyzed anther culture response in diallele crosses between four US Corn-Belt lines (H99, LH38, Pa91, FR16) known to have haploid regeneration aptitude. Significant differences between the six crosses were observed, ranging from 0.0 to 6.3% anther response (percentage of plated anthers displaying embryo-like structures). Both general and specific combining ability effects were significant. This clearly demonstrated the heritable nature of anther culture responsiveness. The findings were corroborated by Spitkó et al.

(2006) who studied the F_1 generation of a cross between a responsive Chinese line and a recalcitrant Hungarian ('Iodent') line, the first backcross (BC_1) of this cross to the recalcitrant parent, and two selected DH lines obtained from each of the two generations. Both F_1 and BC_1 plants had a higher anther culture ability than the Chinese parent and three of the four selected DH lines ranged between the F_1 resp. BC_1 and the Chinese parent. In addition, these lines displayed acceptable agronomic performance. These results demonstrated that high anther culture ability and agronomic performance can be combined by breeding.

3   *In vivo* Techniques

If maize plants are crossed with specific genotypes, so-called inducers, a certain fraction of kernels possess a haploid rather than a regular diploid F_1 embryo. This phenomenon is called *in vivo* (or *in situ*) haploid induction. Generally, kernels with a haploid embryo have a regular triploid endosperm. Therefore, such kernels display the same germination rate and vigor as those with a diploid embryo (Coe and Sarkar 1964).

Two methods of *in vivo* haploid induction are known in maize leading to paternal (androgenetic) and maternal (gynogenetic) haploids, respectively. In induction crosses aiming at paternal haploids, the inducer is used as female and the donor plant as male parent. Thus the cytoplasm of paternal haploids originates from the inducer but the chromosomes exclusively from the donor plant. For production of maternal haploids, on the other hand, the inducer is used as pollinator, leading to haploids carrying both cytoplasm and chromosomes from the donor. Different inducers are used for induction of paternal and maternal haploids (Sections 8.3.1 and 8.3.2). Both methods of *in vivo* haploid induction are much less dependent on the donor genotype than current *in vitro* techniques (Röber et al. 2005, Spitkó *et al.* 2006).

3.1   Induction of Paternal Haploids

The induction of gynogenetic haploids rests on properties of the mutant 'indeterminate gametophyte' caused by the recessive gene *ig* (Kermicle 1969). Multiple embryological abnormalities have been observed in homozygous *ig* plants. In some embryo sacs not all nuclei divide a third time, leading to various cytological irregularities including egg cells without a nucleus. After fusion with one of the two paternal sperm cells, such an egg cell may develop into a haploid embryo possessing the maternal cytoplasm and only paternal chromosomes. In selected inducer lines, the haploid induction rate ranges from 1 to 2% (Kermicle 1994; Schneerman et al. 2000).

Paternal haploids have gained considerable importance for the creation of cytoplasmic male-sterile (CMS) analogues of seed parent lines in commercial hybrid breeding. For this purpose, induction lines in various CMS-inducing cytoplasms have been developed (Pollacsek 1992; Schneerman et al. 2000). Using these CMS inducer versions, the transfer of new breeding lines into the CMS cytoplasm requires only a single induction cross rather than multiple backcross generations.

3.2 Induction of Maternal Haploids

Chase (1952) reported spontaneous haploid induction rates in US Corn-Belt materials of about 0.1%. This value was far too low for a commercial application of the DH technology, as suggested by the author. A great step forward was the detection of inbred line Stock6 which had a 10 - 20 times higher induction rate (Coe 1959). Stock6 became the "Mother" of all subsequently developed inducers. Considerable progress was reported from groups working in India (Sarkar et al. 1994), Russia and Moldova (Tyrnov and Zavalishina 1984; Chalyk 1994; Shatskaya et al. 1994), France (Lashermes and Beckert 1988; Bordes et al. 1997), and Germany (Deimling et al. 1997; Röber et al. 2005). In the course of time, more effective techniques were developed for rapid identification of haploid embryos or seedlings, for chromosome doubling and for raising and selfing large numbers of doubled haploid plants (generation D_0) in the field. As a consequence, the so-called DH technology has meanwhile become a standard tool in modern maize research and breeding (Seitz 2005; Röber et al. 2005; Presterl et al. 2007).

3.2.1 Induction Rate

According to literature reports, the presently most effective inducer is line RWS developed at the University of Hohenheim, Stuttgart, Germany (Röber *et al.* 2005). It was derived from a cross between the Russian inducer synthetic KEMS (Shatskaya et al. 1994) and the French inducer line WS14 (Lashermes and Beckert 1988) and is adapted to the temperate climate of Central Europe but is effective also in tropical environments (Röber et al. 2005). Averaged across a wide range of donors and environments, it has an induction rate of about 8%. A sister line, RWK-76, developed from the reciprocal cross (WS14 x KEMS) even reached an average induction rate of 9 - 10% (unpubl. data). The same rate was observed for the cross RWS x RWK-76. Although having related parents, this cross is much more vigorous and a better pollen shedder than each of its parents and therefore easier to handle, particularly in adverse environments.

Roux (1995) tested lines Stock6, WS14, and W23*ig* for their ability to induce maternal haploids. Line W23*ig* is an isogenic form of dent line W23 (developed in Wisconsin, USA) except for the *ig* gene that enables the induction of paternal haploids (Section 8.3.1). The non-converted W23 line induces neither paternal nor maternal haploids. In agreement with this, the induction rate of W23*ig* for maternal haploids did not significantly deviate from the frequency of spontaneously occurring haploids. The induction rates were 0.2% for W23*ig*, 2.0% for Stock6, and 7.3% for WS14.

Significant differences between donor genotypes were observed for the induction rate (Roux 1995; Eder and Chalyk 2002; Röber et al. 2005). However, the range of variation was small compared to that reported for anther or microspore culture response (Section 8.2). Environmental conditions also influence the success of *in vivo* haploid induction. Using KEMS and RWS as inducers and a donor genotype marked with the recessive mutant 'liguleless', Röber et al. (2005) obtained an average induction rate of 2.0% in the most adverse and a rate of 16.4% in the most favourable field environment. Optimizing the growing conditions by minimizing biotic and abiotic stress generally raises the induction success (personal communications from various breeders).

3.2.2 Haploid Identification

A key issue in applying the *in vivo* haploid induction approach on a commercial scale is an efficient screening system allowing one to differentiate between kernels or seedlings generated by haploid induction and those resulting from regular fertilization. The most efficient haploid identification marker is the 'red crown' or 'navajo' kernel trait encoded by the dominant mutant allele *R1-nj* of the 'red color' gene *R1*. In the presence of the dominant pigmentation genes *A1* or *A2* and *C2*, *R1-nj* causes deep pigmentation of the aleurone (endosperm tissue) in the crown (top) region of the kernel. In addition, it conditions pigment in the scutellum (embryo tissue). Pigmentation may vary in extent and intensity depending on the genetic background of the particular donor and inducer. In conjunction with additional color genes (*B1*, *Pl1*), *R1-nj* also conditions pigmentation of the coleoptile and root of the seedling.

Nanda and Chase (1966) and Greenblatt and Bock (1967) first used the red crown mutant as a selectable marker in haploid-induction experiments. To be effective, the donor has to have colorless seeds and the inducer needs to be homozygous for *R1-nj* and the aforementioned dominant color genes. A kernel resulting from haploid induction then has a red crown (regular triploid endosperm) and an unpigmented scutellum (haploid maternal embryo), whereas a regular F_1 kernel displays pigmentation of both aleurone and scutellum (Figures 1 and 2). If only the egg cell but not the central cell is fertilized, the kernel has a pigmented (diploid) embryo and a non-pigmented, diploid maternal endosperm and aborts during the early kernel development phase (see "tooth gaps" in Fig. 2). Kernels resulting from (unintended) selfing or outcrossing with other colorless donors don't show any pigmentation. The red crown marker does not work if the donor genome is homozygous for *R1* or for dominant anthocyanin inhibitor genes such as *C1-I*. These genes seldom occur in commercial US Corn-Belt germplasm (Seitz, pers. comm.) but may be a problem in European flint (Röber et al. 2005) or tropical (Belicuas et al. 2007) materials. If *R1-nj* is poorly expressed in the scutellum, a validation of the putative haploids is possible in the early seedling stage by means of the coleoptile and root color gene *Pl*.

An unambiguous distinction between maternal haploid and regular F_1 plants was accomplished by means of a dominant herbicide resistance marker by

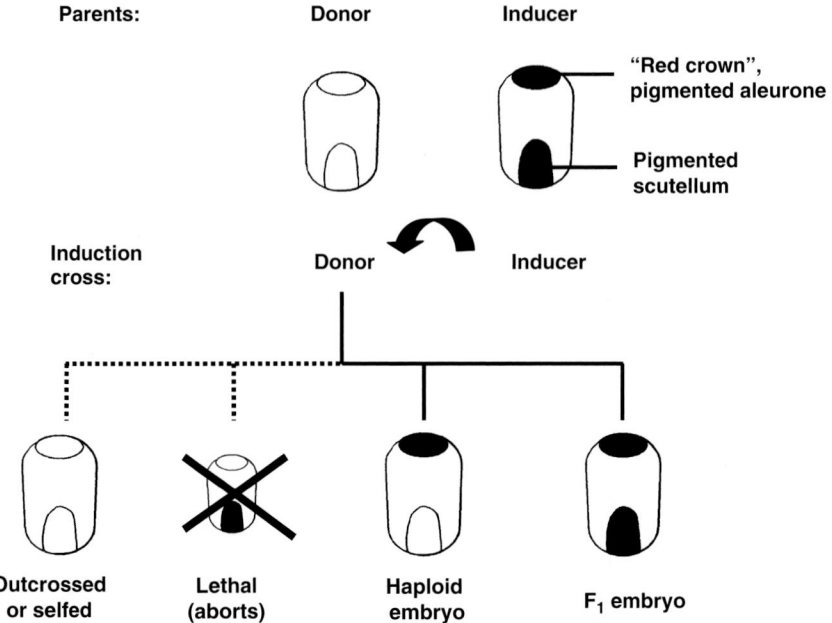

Parents: Donor Inducer

"Red crown", pigmented aleurone

Pigmented scutellum

Induction cross: Donor Inducer

Outcrossed or selfed **Lethal (aborts)** **Haploid embryo** **F₁ embryo**

Fig. 1 Graphical representation of the dominant maize pigmentation marker 'red crown' and its use for identifying kernels with haploid embryo resulting from haploid induction crosses

Geiger et al. (1994). The authors had transferred the phosphinotricin acetyl transferase gene *pat* to the inducer line RWS. Using this transgenic line as inducer, the induced haploid seedlings are sensitive and the F_1 plants resistant. Applying the herbicide to a small leaf region leads to a clear distinction between the two types of seedlings without killing the susceptible haploids. However, for large-scale applications, the method is too laborious.

More recently, Belicuas et al. (2007) showed that microsatellite markers are a reliable and universally applicable haploid identification tool. But the cost of genotyping and the need for adequate high-throughput marker facilities are presently limiting the applicability of this approach.

A cheap and fast solution was suggested by Rotarenko et al. (2007). The authors observed that kernels with a haploid embryo have a significantly lower oil concentration than those with a diploid F_1 embryo. This is due to the reduced size of haploid compared to diploid embryos. Averaged across eight donors, the oil concentrations of the haploid and diploid embryos amounted to 46.7 and 55.6 g kg^{-1}, respectively. The authors did not show single-kernel data which would allow one to judge the precision achievable with this approach. Anyway, because a fast, inexpensive and non-destructive assessment of oil concentration is feasible with single-kernels at high-throughput, the approach deserves further investigation. Inducers with a high oil concentration should be best suited for this approach.

Fig. 2 Ears of a colorless dent single cross pollinated with inducer line RWS featuring the 'red crown' kernel marker (regular triploid F_1 endosperm). Gaps indicate aborted kernels (diploid maternal endosperm) (Photographed by Silvia Koch)

3.2.3 Properties of Haploids

Haploid plants are smaller and less vigorous than corresponding DH or inbred lines (Chase 1952; Chalyk 1994). They are also much more sensitive to any kind of stress. Most haploid plants display a certain degree of female fertility if pollinated by diploids. In a study by Chalyk (1994), 96% of the haploids derived from a synthetic dent population produced ears carrying at least a few kernels. The highest kernel number per ear was 107 and the average 27. Even higher values were obtained by Geiger et al. (2006). The authors analyzed haploid progenies of three elite dent single crosses and all three of them showed some degree of fertility. In the most fertile progeny, the average number of kernels per plant amounted to 80.

On the contrary, most haploids lack male fertility. Early studies of Chase (1952) revealed fertile sections of the tassel in about one percent of the haploid plants studied. In the aforementioned materials, Chalyk (1994) observed some pollen in 3% of the haploids. Results of Zabirova et al. (1993) demonstrate a great influence of the donor genotype on the frequency of pollen-shedding haploids. The authors detected a donor line from which 33% of the induced haploids could successfully be selfed. The line resulted from four cycles of selection for this trait.

In the above reports, none of the authors provide cytological data on the ploidy level of the selfed plants. So they may have been completely haploid or may have undergone spontaneous, possibly sectorial, chromosome doubling during the pre-anthesis phase of development. At Hohenheim, leaves and immature anthers of putative haploids derived from current breeding materials were analyzed by flow cytometry and only haploid ones were bagged for selfing (Geiger and Schönleben, unpubl.). Twenty five out of 390 plants (6.4%) produced between 1 and 11 (average 3.8) seeds per selfed ear. However, three plants (0.8%) yielded 16, 23, and 40 seeds, respectively. Taken together, data indicate that there is promising genetic variation for spontaneous male fertility of haploids in various breeding materials.

3.2.4   Artificial Chromosome Doubling

For decades, artificial chromosome doubling was a serious constraint of haploid induction in maize since, in contrast to most other crop plants, the seedling reaction was highly genotype specific to the usual colchicine treatment. A breakthrough was accomplished by Gayen et al. (1994) who applied the colchicine at the coleoptile stage, 2-3 days after germination. The authors cut off the tip of the coleoptiles and immersed the whole seedling into a 0.06% colchicine solution plus 0.5% DSMO (dimethyl sulfoxide) for 12 hrs at 18°C. Deimling et al. (1997) further increased the efficacy of the method by reducing the roots to 20-30 mm and placing the immersed seedlings in the dark. After the colchicine treatment, the seedlings are carefully washed in water and subsequently grown (during the first days under high humidity) in the greenhouse to the 5- to 6-leaf stage. A few weeks later the plants are transferred to the field. Eder and Chalyk (2002) applied the method to a broad range of donor genotypes and achieved an average doubling rate of 49%. For comparison, the authors tested a colchicine injection method applied in the 3- to 4-leaf stage and reached a doubling rate of only 16%. With both methods, 50 - 60% of the pollen-shedding plants could be selfed.

A gentler method of chromosome doubling was developed by Kato (2002). He treated haploid plants in the flower primordial stage with nitrous oxide gas (NO_2) for 2 days at 600 kPa. Averaged across donor genotypes, 44% of the treated plants produced seed after self-fertilization. However, a very strong influence of the donor genotype on the doubling rate was observed. Furthermore, the method is very laborious and requires special equipment (safe gas chambers) and, therefore, is not easily adaptable to high-throughput applications. On the other hand, the high recovery rate makes it attractive for approaching specific cytogenetic and other scientific problems.

3.2.5   Possible Mechanisms

Principally, two mechanisms leading to maternal haploids have been conceived. (1) One of the two sperm cells provided by the inducer is defective yet able to fuse with the egg cell. During subsequent cell divisions, the inducer chromosomes degenerate and are eliminated stepwise from the primordial cells. The second sperm cell fuses with the central cell and leads to a regular triploid endosperm. (2) One of the two sperm cells is not able to fuse with the egg cell but instead triggers haploid embryogenesis. The second cell fuses with the central cell as under the first hypothesis. At any rate, kernel abortion is expected if the functional sperm cell fuses with the egg cell and the defective one fuses with the central cell or if the central cell remains unfertilized (Fig. 2).

Experimental data in support of the first hypothesis come from the studies of Wedzony et al. (2002). The authors fixed ovaries of selfed RWS plants at intervals during the first 20 days after pollination. Eighteen out of 203 embryos contained micronuclei in every cell of the shoot primordia. Micronuclei varied in number and diameter, displaying the typical characteristics of metabolically inactive chromatin. In some equatorial plates, chromosome fragments were observed. Micronuclei elimination started during the globular state of embryogenesis. These observations are indirectly corroborated by the results of Fischer (2004). The author used microsatellite markers to check for a strictly maternal origin of haploids induced by RWS. Among 624 haploid plants and 309 DH lines (generation D_1), 1.4% of the genotypes possessed one or, rarely, several inducer chromosome segments. Generally, these segments had replaced the homologous maternal segments.

Observations supporting the second hypothesis were reported by Chalyk et al. (2003). The authors found 10 -15% aneuploid microsporocytes in the haploid induction lines MHI and M471H. From this, the authors concluded that part of the viable pollen of these inducers may be aneuploid and will result in sperm cells which only fuse with the central cell.

Another abnormality of MHI was described by Rotarenko and Eder (2003). The authors detected a three times higher heterofertilization rate in crosses with MHI compared with regular inbred lines. This correlated with a higher frequency of haploids induced by a single controlled pollination, compared to open-pollination under topcross conditions. Similar observations have not been made with other inducers.

Taking all of this information together, the mechanism of haploid induction is still not fully understood. Most likely, several reproductive abnormalities are involved and different inducers may vary in this regard. At any rate, researchers should keep in mind that maternal haploids might possess small fractions of the inducer genome.

4   Genetics of Haploid Induction

Both, *in vitro* and *in vivo* haploid induction are under polygenic control. Segregating generations derived from crosses between parents contrasting in haploid induction ability revealed continuous variation for various induction-associated traits (Lashermes and Beckert 1988; Deimling et al. 1997; Röber et al. 2005).

In a QTL study, Cowen *et al.* (1992) analyzed 98 F_3 lines derived from a cross between two contrasting US Corn-Belt lines (B73, recalcitrant; 139/39-05, responsive) for anther culture response, taking the number of embryo-like structures per 100 plated anthers as response criterion. Seventy five RFLP clones were used to genotype 98 F_3 lines. Fifty seven percent of the phenotypic variance among F_3 lines could be explained by the joint effects of two major and two minor QTL. The two major QTL reside on chromosomes 3 and 9. At both loci the alleles for responsiveness are recessive and show strong complementary epistasis. Results indicate that marker-assisted introgression of QTL for responsiveness shows promise for improving the anther culture ability of breeding materials. Further progress can be expected from the rapidly increasing knowledge in functional genomics of microspore embryogenesis in other plant species (for review see Hosp et al. 2007).

Murigneux et al. (1994) conducted QTL analyses in three DH-line populations derived from three crosses between contrasting dent parent lines. In two populations, the DH lines were evaluated *per se* and in the third population as testcrosses. The populations comprised 48, 96, and 95 DH lines, respectively, and were genotyped for at least 100 polymorphic RFLP markers. In each cross, three to four QTL were found for percentage of responding anthers and zero to four for number of embryos per 100 anthers. The QTL are scattered over chromosomes 3, 4, 5, and 7 - 10. QTL positions did not agree between populations in any case. Jointly, the QTL explained only 30 - 40% of the phenotypic variance for percent responding anthers.

Röber (1999) evaluated a population of 211 F_3 plants derived from the cross W23*ig* x Stock6 (the parents of WS14, see Section 8.3.2) for *in vivo* induction of maternal haploids. QTL analysis with 84 polymorphic RFLP markers revealed two QTL located on chromosomes 1 and 2 jointly explaining 17.9% of the phenotypic and 40.7% of the genotypic variance for induction rate. The positive QTL allele on chromosome 1 is dominant and originates from Stock6 (the "high" parent) whereas the one on chromsome 2 is additive and originates from W23*ig* (the "low" parent). Interestingly, no QTL for *in vitro* haploid induction was detected on chromosomes 1 and 2 in the two foregoing anther culture studies.

Because *in vivo* haploid induction could be considered a detrimental trait from the evolutionary point of view, one might expect that the F_1 of a cross between two unrelated inducers would furnish a lower induction rate than the better parent. In an induction experiment with WS14, KEMS, WS14 x KEMS, and KEMS x WS14, Röber et al. (2005) obtained induction rates of 2.0%, 6.9%, 3.9%, and 4.6%, respectively. No significant difference existed between the mid-parent and the mean F_1 value whereas KEMS significantly surpassed all other entries. In another experiment (Geiger, unpublished), RWS, RWK-76, and their cross were compared. In this case, the F_1 reached the same induction rate as the better parent, RWK-76. Thus, neither of the two experiments supports the hypothesis above.

5   Using Doubled Haploids in Breeding

During the last decade, *in vivo* haploid induction has developed into a routinely used tool in hybrid maize breeding. While the induction of paternal haploids has proven to be an efficient tool for converting high combining seed parent lines to isogenic CMS analogues (Pollacsek 1992; Schneerman et al. 2000), the induction of maternal haploids is used for the rapid development of homozygous lines in mainstream breeding and in exploiting genetic resources (Röber et al. 2005). The following three sections will focus on the use of DH lines in recurrent selection (RS) and in hybrid parent line development (LD).

5.1   Quantitative Genetic Aspects

As is well known from quantitative genetics, the gain from selection for performance traits depends on (1) the selection intensity, (2) the heritability coefficient, (3) the genetic correlation between selection criterion and gain criterion, and (4) the genetic standard deviation for the gain criterion (Hallauer and Miranda 1981; Falconer and Mackay 1996). The gain criterion, *i.e.* the criterion for evaluating the selection response, is the GCA (general combining ability) with one or more heterotic group(s). Selection criteria are the performances of the candidate lines *per se* and of their testcrosses.

Strong selection increases the short-term response to selection but reduces the effective population size and thus leads to a steady decline of the genetic variance in the course of medium- to long-term RS and LD programs. To keep this decline within adequate limits, a minimum number of selected lines needs to be saved for starting a new breeding cycle (Gordillo and Geiger 2008a). Since the loss of genetic variance increases with the inbreeding coefficient (F), *i.e.* with the degree of homozygosity of the candidate lines, more lines need to be saved when selecting among DH lines (completely homozygous) compared to early-generation selfed lines (F = 0.5, 0.75, 0.875, ... in generations S_1, S_2, S_3, ..., respectively). Thus, all other things being equal, selection intensity needs to be more restricted when using DH lines.

Contrary to the effective population size, the genetic variance of the selection criterion and consequently the heritability coefficient increase as F increases. This leads to better differentiation among highly homozygous lines *per se* and among their testcrosses (Griffing 1975; Röber et al. 2005; Gordillo and Geiger 2008b). Seitz (2005) compared three sets of S_2 lines and corresponding DH lines evaluated for testcross grain yield with the same testers and in the same environments. On average, the estimated genetic variance among the DH lines was 2.1 times higher than among the S_2 lines. In an analogous experiment with S_3 and DH lines, a 1.6 times higher variance was obtained for the DH lines. These findings were in principal confirmed by Bordes et al. (2007), who compared S_1 lines and single-seed descent (SSD) lines (in S_5) with corresponding DH lines. Combined across locations, the estimated genetic variance for testcross grain yield was 1.6 times higher among the DH lines than among the S_1 lines and 1.2 times higher than among the SSD lines. The greater estimates of the variance between the DH lines compared to the SSD lines ($F = 0.97$) was unexpected and might be attributable to unconscious selection during the selfing phase of the SSD lines.

The genetic correlation between selection criterion and gain criterion also increases with the homozygosity of the candidate lines, *i.e.* the closer the genotypes of the candidate lines agree with those of the homozygous lines to be derived thereof. This again favors the use of DH lines in breeding.

Epistatic gene action may positively or negatively affect hybrid maize performance (Lamkey and Edwards 1999). Selection for positive epistasis is most effective among uniform selection units such as DH or completely inbred lines. Negative epistatic effects are frequently observed in three-way and double-cross hybrids (Melchinger et al. 1986) and are to be expected in progenies of selected lines that are intercrossed to start a new RS or LD cycle. This decline of performance can be explained by a disruption of coadapted gene arrangements accumulated in elite breeding lines. To limit such negative effects, a balance between genetic recombination and fixation of gene arrangements is needed irrespective of whether the breeding method is based on DH or inbred lines.

5.2   Breeding Schemes

Various DH line-based schemes have been suggested for hybrid maize breeding (Gallais 1988; Bordes et al. 2007; Longin et al. 2006 and 2007; Gordillo and Geiger 2008a and b). To ensure long-term breeding progress, the gene pools used for line development need to be continuously improved by RS (Hallauer and Miranda 1981). Most efficiently, this can be accomplished by combining LD and RS in one integrated and comprehensive breeding scheme (Gordillo and Geiger 2008b). However, whereas LD aims at maximizing short-term success, RS is geared towards raising the genetic potential of the breeding population in the long run. This means that, in RS, considerably more lines must be selected as parents for the next cycle than are needed for establishing experimental hybrids in LD. The breeder therefore must decide which weight he or she wants to give to short- and long-term goals when allocating resources.

An example of a breeding scheme using DH lines is given in Figure 3. The dimensioning example refers to a breeding budget of 500,000 EUR. The figures approximately correspond to optimum values as determined by the computer software MBP (version 1.0) developed by Gordillo and Geiger (2008c). In brief, the scheme comprises the following steps:

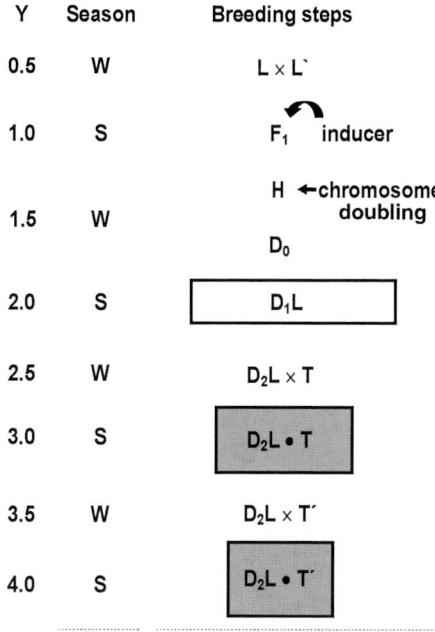

Y	Season	Breeding steps
0.5	W	$L \times L^{\prime}$
1.0	S	F_1 inducer
1.5	W	H ← chromosome doubling
		D_0
2.0	S	D_1L
2.5	W	$D_2L \times T$
3.0	S	$D_2L \cdot T$
3.5	W	$D_2L \times T^{\prime}$
4.0	S	$D_2L \cdot T^{\prime}$

Fig. 3 Scheme for an integrated recurrent selection (RS) and line development (LD) procedure based on doubled haploid (DH) lines. RS cycles are completed after the first stage of testcross evaluation in year 3 and LD cycles after the second stage in year 4. The dimensioning example refers to a breeding budget of 500,000 EUR per cycle and was determined by model calculations applying the optimization software MBP (version 1) (Gordillo and Geiger 2008c) assuming equal weights for RS and LD; W = winter, S = summer, L = line, H = haploid seedling, D_x = DH generation x, T = tester, N_T = number of testers)

- Creating variation by intercrossing 79 selected lines (recombination units).
- *In vivo* haploid induction of maternal haploids in generation F_1.
- Artificial chromosome doubling in the seedling stage, raising and selfing fertile plants (Generation D_0).
- Evaluation of 6400 D_1 lines in single-row observation plots for visually scorable traits at 2 locations.
- Production of testcrosses of 3200 D_2 lines with 1 tester.
- Evaluation of the testcrosses in yield trials at 7 locations.
- Intercrossing the best 79 D_2 lines to commence the next RS cycle and testcrossing of the best 43 lines with 5 testers.
- Evaluation of the testcrosses in yield trials at 15 locations (15 x 5 = 75 plots per line).
- Selection of the 5 best DH lines for production of experimental hybrids.

In the described scheme, the RS cycle extends over six generations and the LD cycle over eight. This takes three and four years, respectively, if sufficient off-season capacities are available. The LD cycle could be reduced to three years by renouncing the second stage of testcross evaluation. The expenses saved in year 4 would then be available for boosting the budget in years 1 - 3. This reduces the expected selection response per cycle but increases the response per year (Longin et al. 2006; Gordillo and Geiger 2008a and b). However, a second stage of testcross selection still seems to be advisable in order to diminish bias due to genotype x year interactions. Moreover, model calculations of Gordillo and Geiger (2008b) revealed that, in the long run, the annual genetic gain from two-stage selection can be further raised if the lines selected for commencing a new cycle are not only taken from the cycle under consideration but also from the first selection stage of the breeding cycle started one year later. This genetic interlinking of yearly staggered breeding cycles allows for a higher selection intensity and accelerates the decline of gametic phase disequilibrium in the respective gene pool.

Longin et al. (2007) investigated the efficiency of early testing for combining ability before *in vivo* haploid induction is applied. The authors compared the two-stage selection scheme in Figure 3 with a combined S_1/DH line scheme. At the first stage, S_1 lines and at the second stage DH lines derived from selected S_1 lines are evaluated for testcross performance. This scheme takes one year more time than the "pure" two-stage DH scheme. The maximum predicted genetic gain was about 10% larger, if computed per cycle, but 3% smaller on an annual basis. To reduce the cycle length of the S_1/DH line scheme, S_1 plants rather than S_1 lines would need to be crossed (as males) to a tester and additionally be used as females in haploid induction. Considerable improvement of the DH technology is needed to obtain a large enough number of DH lines from a single S_1 plant before this accelerated S_1/DH scheme will become applicable.

5.3 Operational, Logistic, and Economic Aspects

Developing maternal DH lines by *in vivo* haploid induction requires specific skills and equipment for (1) large-scale chromosome doubling, (2) raising unregulated seedlings in the greenhouse, (3) transplanting the surviving plantlets to the field, (4) avoiding stress during the growing period, and (5) selfing the adult plants. Under adequate conditions, about one to five DH lines emanate from one donor plant (Eder & Chalyk 2002; Röber et al. 2005). Improvement seems possible for each of the five foregoing steps. However, if several times more DH lines per S_1 donor plant are demanded, an androgenetic approach might, in the long run, be more promising.

Considerable savings are possible in line development by haploid induction compared to inbreeding. No continued selfing is needed for creating homozygous genotypes, and maintenance breeding can be kept to a minimum. The fact that DH lines are genetically fixed units also simplifies the logistics of seed transfer between main and off-season sites, since DH lines need to be shipped only once whereas, in a selfing program, new sublines arise in every segregating generation.

Since DH lines are homozygous and homogeneous from the very first, they fully meet the requirements for obtaining plant variety protection. Thus, DH lines allow the breeder to considerably reduce the time to commercialization which can be considered as the most significant economic advantage of the technology (Seitz 2005; Bordes et al. 2006). Moreover, haploid induction is a very helpful tool in stacking transgenic or other monofactorial traits in potential hybrid parents by marker-assisted backcrossing, since the segregation pattern is much simpler at the haploid than at the diploid level. This makes it feasible to fix a demanded stack in the shortest possible time and with the lowest possible genotyping expenditure.

References

Belicuas PR, Guimarães CT, Paiva LV, Duarte JM, Maluf WR, Paiva E (2007) Androgenetic haploids and SSR markers as tools for the development of tropical maize hybrids. Euphytica 156:95–102

Bordes J, de Vaulx RD, Lapierre A, Pollacsek M (1997) Haplodiploidization of maize (Zea mays L.) through induced gynogenesis assisted by glossy markers and its use in breeding. Agronomie 17:291–297

Bordes J, Charmet G, de Vaulx RD, Pollacsek M, Beckert M, Gallais A (2006) Doubled haploid versus S$_1$-family recurrent selection for testcross performance in a maize population. Theor Appl Genet 112:1063–1072

Bordes J, Charmet G, de Vaulx RD, Lapierre A, Pollacsek M, Beckert M, Gallais A (2007) Doubled-haploid versus single-seed descent and S$_1$-family variation for testcross performance in a maize population. Eyphytica 154:41–51

Büter B, (1997) In vitro haploid production in maize. In: Jain SM, Sopory SK, Veilleux RE (eds.) In vitro haploid production in higher plants. Kluwer Academic Publishers, Dordrecht, Boston, London, pp 37–71

Chalyk ST (1994) Properties of maternal haploid maize plants and potential application to maize breeding. Euphytica 79:13–18

Chalyk S, Baumann A, Daniel G, Eder J (2003) Aneuploidy as a possible cause of haploid-induction in maize. Maize Genet Newsl 77:29–30

Chase SS (1947) Techniques for isolating monoploid maize plants. J Bot 34:582

Chase SS (1952) Monoploids in maize. In: Gowen JW (ed.) Heterosis. Iowa State College Press, Ames, IA, USA, pp 389–399

Coe EH (1959) A line of maize with high haploid frequency. Am Nat 93:381–382

Coe EH, Sarkar KR (1964) The detection of haploids in maize. J Hered 55:231–233

Cowen NM, Johnson CD, Armstrong K, Miller M, Woosley A, Pescitelli S, Skokut M, Belmar S, Petolino JF (1992) Mapping genes conditioning in vitro androgenesis in maize using RFLP analysis. Theor Appl Genet 84:720–724

Deimling S, Röber F, Geiger HH (1997) Methodik und Genetik der in-vivo-Haploideninduktion bei Mais. Vortr Pflanzenzüchtg 38:203–224

Eder J, Chalyk S (2002) In vivo haploid induction in maize. Theor Appl Genet 104:703–708

Falconer DS, Mackay TFC (1996) Introduction to quantitative genetics. 4th ed. Longman Scientific & Technical Ltd. Essex, England

Fischer E (2004) Molekulargenetische Untersuchungen zum Vorkommen paternaler DNA-Übertragung bei der in-vivo-Haploideninduktion bei Mais (Zea mays L.). Ph.D. thesis, University of Hohenheim, Stuttgart, Germany

Gallais A, (1988) A method of line development using doubled haploids: the single doubled descent recurrent selection. Theor Appl Genet 75:330–332

Gayen P, Madan JK, Kumar R, Sarkar KR (1994) Chromosome doubling in haploids through colchicine. Maize Genet Newsl 68:65

Geiger HH, Roux SR, Deimling S (1994) Herbicide resistance as a marker in screening for maternal haploids. Maize Genet Newsl 68:99

Geiger HH, Braun MD, Gordillo GA, Koch S, Jesse J, Krützfeldt BAE (2006) Variation for female fertility among haploid maize lines. Maize Genet Newsl 80:28–29

Gordillo GA, Geiger HH (2008a) Optimization of DH-line based recurrent selection procedures in maize under a restricted annual loss of genetic variance. Euphytica 161:141–154

Gordillo GA, Geiger HH (2008b) Alternative recurrent selection strategies using doubled haploid lines in hybrid maize breeding. Crop Sci 48:911–922

Gordillo GA, Geiger HH (2008c) MBP (version 1.0): A software package to optimize maize breeding procedures based on doubled haploid lines. J Hered 99:227–231

Greenblatt IM, Bock M (1967) A commercially desirable procedure for detection of monoploids in maize. J Hered 58:9–13

Griffing B (1975) Efficiency changes due to use of doubled-haploids in recurrent selection methods. Theor Appl Genet 46:367–386

Hallauer AR, Miranda JB (1981) Quantitative genetics in maize breeding. Iowa State Univ. Press, Ames, IA, USA

Hosp J, de Faria Maraschin S, Touraev A, Boutilier K (2007) Functional genomics of microspore embryogenesis. Euphytica 158:275–285

Kasha KJ (ed.) (1974) Haploids in higher plants. University of Guelph, Canada.

Kato A (2002) Chromosome doubling of haploid maize seedlings using nitrous oxide gas at the flower primordial stage. Plant Breed 121:370–377

Kermicle JL (1969) Androgenesis conditioned by a mutation in maize. Science 166:1422–1424

Kermicle JL (1994) Indeterminate gametophyte (*ig*) - biology and use. In: Freeling M, Walbot V (eds.) The Maize Handbook, Springer-Verlag, New York

Lamkey KR, Edwards JW (1999) Quantitative genetics of heterosis. In: Coors JG, Pandey S (eds.) The genetics and exploitation of heterosis in crops. ASA – CSSA, Madison, WI, pp 31–48

Lashermes P, Beckert M (1988) A genetic control of maternal haploidy in maize (*Zea mays* L.) and selection of haploid inducing lines. Theor Appl Genet 76:405–410

Longin CFH, Utz HF, Reif JC, Schipprack W, Melchinger AE (2006) Hybrid maize breeding with doubled haploids: I. One-stage versus two-stage selection for testcross performance. Theor Appl Genet 112:903–912

Longin CFH, Utz HF, Reif JC, Wegenast T, Schipprack W, Melchinger, AE (2007) Hybrid maize breeding with doubled haploids: III. Efficiency of early testing prior to doubled haploid production in two-stage selection for testcross performance. Theor Appl Genet 115:519–527

Melchinger AE, Geiger HH, Schnell FW (1986) Epistasis in maize (*Zea mays* L.) I. Comparison of single and three-way cross hybrids among early flint and dent inbred lines. Maydica 31:179–192

Murigneux A (1994) Genotypic variation of quantitative trait loci controlling in vitro androgenesis in maize. Genome 37:970–976

Nanda DK, Chase SS (1966) An embryo marker for detecting monoploids of maize (*Zea mays* L.). Crop Sci 6:213–215

Petolino JF, Thompson SA (1987) Genetic analysis of anther culture response in maize, Theor Appl Genet 74:284–286

Pollacsek M (1992) Management of the *ig* gene for haploid induction in maize. Agronomie 12:247–251

Presterl T, Ouzunova M, Schmidt W, Möller EM, Röber FK, Knaak C, Ernst K, Westhoff P, Geiger HH (2007) Quantitative trait loci for early plant vigour of maize grown in chilly environments. Theor Appl Genet 114:1059–1070

Pret'ová A, Obert B, Bartošová Z (2006) Haploid formation in maize, barley, flax, and potato. Protoplasma 228:107–114

Randolph LF (1932) Some effects of high temperature on polyploidy and other variations in maize. Genetics 18:222–229

Röber FK (1999) Fortpflanzungsbiologische und genetische Untersuchungen mit RFLP-Markern zur *in-vivo*-Haploideninduktion bei Mais. Ph.D. thesis, University of Hohenheim, Stuttgart, Germany

Röber FK, Gordillo GA, Geiger HH (2005) *In vivo* haploid induction in maize - Performance of new inducers and significance of doubled haploid lines in hybrid breeding. Maydica 50:275–283

Rotarenco V, Eder J (2003) Possible effects of heterofertilization on the induction of maternal haploids in maize. Maize Genet Newsl 77:30

Rotarenco VA, Kirtoca IH, Jacota AG (2007) Possibility to identify kernels with haploid embryo by oil content. Maize Genet Newsl 81:11 (http://www.agron.missouri.edu/mnl/81/32rotarenco.htm)

Roux SR (1995) Züchterische Untersuchungen zur *in-vivo*-Haploideninduktion bei Mais. Ph.D. thesis, University of Hohenheim, Stuttgart, Germany

Sarkar KR, Pandey A, Gayen P, Mandan JK, Kumar R, Sachan JKS (1994) Stabilization of high haploid inducer lines. Maize Genet Newsl 68:64–65

Schneerman MC, Charbonneau M, Weber DF (2000) A survey of *ig* containing materials. Maize Genet Newsl 74:92–93

Seitz G (2005) The use of doubled haploids in corn breeding. In: Proc. of the 41th Annual Illinois Corn Breeders' School 2005. Urbana-Champaign, Il, USA, p. 1–7

Shatskaya OA, Zabirova ER, Shcherbak VS, Chumak MV (1994) Mass induction of maternal haploids in corn. Maize Genet Newsl 68:51

Spitkó T, Sági L, Pintér J, Marton LC, Barnabás B (2006) Haploid regeneration aptitude of maize (*Zea mays* L.) – Lines of various origin and of their hybrids. Maydica 51:537–542

Tyrnov VS, Zavalishina AN (1984) Inducing high frequency of matroclinal haploids in maize. Dokl Akad Nauk SSSR 276(3):735–738

Weber DF (1986) The production and utilization of monosomic *Zea mays* in cytogenetic studies. In: Reddy GM, Coe EH (eds.) Gene structure and function in higher plants. Oxford and IBH, New Delhi, India, pp 191–204

Wedzony M, Röber F, Geiger HH (2002) Chromosome elimination observed in selfed progenies of maize inducer line RWS. In: VII Intern. Congress on Sexual Plant Reproduction. Maria Curie-Sklodowska University Press, Lublin, Poland, pp 173

Zabirova ER, Shatskaya OA, Shcherbak VS (1993) Line 613/2 as a source of a high frequency of spontaneous diploidization in corn. Maize Genet Newsl 67:67

Databases and Data Mining

Carolyn J. Lawrence and Doreen Ware

Abstract Over the course of the past decade, the breadth of information that is made available through online resources for plant biology has increased astronomically, as have the interconnectedness among databases, online tools, and methods of data acquisition and analysis. For maize researchers, the number of resources available is both impressive and daunting, in many cases leaving them at a loss regarding where to begin. Described here is an historical perspective on the origin of these resources, as well as how they are expected to change and grow in the future. We outline the current types of resources, how they are connected, and methods for data acquisition, analysis, and interpretation. In addition, we offer guidance to assist researchers place data generated by their maize projects into appropriate databases for long-term storage and use.

1 Databases Past and Present

The theory for storing information in relational databases was reported in 1970 by Edgar Codd, who worked for IBM Research (Codd 1970). Subsequently, various methods for storing data relationally were implemented based upon Codd's ideas. Early on, these data resources could only be accessed by direct interaction with the computers that stored the data. However, the creation of ARPANET (the U.S. government's Advanced Research Projects Agency's networking project) in the early 1970s served as the basis for linking various resources together. ARPANET eventually

C.J. Lawrence
USDA-ARS, Corn Insects and Crop Genetics Research Unit and Iowa State University, Departments of Agronomy and Genetics, Development and Cell Biology
carolyn.lawrence@ars.usda.gov

D. Ware
USDA-ARS, U.S. Plant, Soil and Nutrition Research Unit
doreen.ware@ars.usda.gov
Cold Spring Harbor Laboratory
ware@cshl.edu

evolved into the present day Internet, which has brought the utility of databasing to bear on problems ranging from personnel management to shopping online.

Simultaneous with the evolution of database technologies and the creation of the Internet, biologists began to create datasets of ever-increasing size. These datasets included DNA sequence information and molecular biological data, as well as others that were species-specific in nature. A need to store, categorize, and easily access these datasets resulted in the adoption of database technologies by biologists. Coupled with tool-building activities for biological data analysis, the field of bioinformatics was born.

Some of the earliest and most widely utilized publicly accessible biological databases were created to store DNA sequences and to make the sequences accessible via a variety of methods. These include EMBL (the European Molecular Biology Laboratory), DDBJ (the DNA Data Bank of Japan), and GenBank. The first of these, EMBL, which is run by the European Bioinformatics Institute (EBI), began in 1980 (Stoesser et al., 1997), whereas DDBJ (http://www.ddbj.nig.ac.jp/) began work in 1986 (Tateno and Gojobori 1997), and NCBI (the National Center for Biotechnology Information) founded GenBank (http://www.ncbi.nlm.nih.gov/ Genbank/) in 1992 (Benson et al., 1997). All are permanently funded, long-term repositories. To ensure that each of these three equivalent repositories could serve the most comprehensive and up-to-date set of sequences, each agreed to share their data with the other two when all became part of the International Nucleotide Sequence Databases Collaboration.

The plant biology databases AAtDB (An *Arabidopsis thaliana* Database, which later evolved into the *Arabidopsis thaliana* Database, AtDB, then The *Arabidopsis* Information Resource, TAIR; Flanders et al., 1998; Huala et al., 2001) was one of the first plant biological databases to be created. Howard Goodman founded AAtDB in 1991 as a resource to serve information on the model dicot *Arabidopsis thaliana*. Its evolution to become TAIR involved the adoption of AIMS, the *Arabidopsis* Information Management System, which served as the primary stock information and ordering facility of the *Arabidopsis* Biological Resource Center (ABRC). Other plant biological databases that began in 1991 include GrainGenes for the Triticeae (Carollo et al., 2005), RiceGenes for rice (Cartinhour 1997), SoyBase for soy (Grant and Shoemaker, 2007), Dendrome for forest trees (Neale, 2007), and MaizeDB for corn (Polacco and Coe, 1999). MaizeDB, the maize equivalent to AAtDB, was headed by USDA-ARS scientist and past editor of the Maize Newsletter, Ed Coe. MaizeDB served genetics information including (but not limited to) maps, phenotypes, and molecular marker/probe data. The current maize database, MaizeGDB, came into existence when MaizeDB merged with a sequence database called ZmDB (Lawrence et al., 2004). Like AtDB/TAIR, MaizeDB/MaizeGDB also stores data for the Maize Genetics Cooperation–Stock Center (Scholl et al., 2003). More recently, Gramene, a resource for comparative biology among grass species, was established (Ware et al., 2002), and various maize project-specific resources have come on line. GrainGenes, SoyBase, and MaizeGDB operate on permanent funds from the USDA-ARS. Dendrome is permanently funded by the U.S. Forest Service, and the others are not funded long-term.

1.1   Types of Resources

Databases storing genomic information fall into various categories based upon their role within a larger context. A Laboratory Information Management System (LIMS) is the most basic sort of database and interface solution, and can be as simple as a spreadsheet stored on a computer in a particular laboratory. Complex systems where the LIMS is made up of various data pipelines and/or laboratories are generally highly customized and are created to support an individual research group's shared data management needs. Data stored within a LIMS environment represent the group's working information and generally are not made available for use outside of the group that generated the data. In some cases, the LIMS system may eventually be deployed as a public repository, but often with limited support for long-term maintenance. Static Repositories (SRs) are those resources where data (often limited to a single data type) are deposited for long-term storage. The data generally are not changed over time, hence the moniker 'static'. The most well known SR for biological data is GenBank (http://www.ncbi.nlm.nih.gov/Genbank/; Benson et al., 2007), the federally funded resource that stores sequence data for all species. An Automatic Annotation Shop (AAS) harvests data from SRs and runs those data through analysis pipelines to create products that have added value for use by researchers. Of these, JCVI (the J Craig Venter Institute, formerly TIGR, http://www.jcvi.org) is an AAS that provides value-added sequence-based products, including genome assemblies and repeat databases based upon the sequence set stored at GenBank (Chan et al., 2006). Model Organism Databases (MODs), which are generally species-specific, have been created for various plant species, including soybean (SoyBase; http://www.soybase.agron.iastate.edu), *Arabidopsis* (The *Arabidopsis* Information Resource; http://www.arabidopsis.org/), and various other species including *Drosophila*, *C. elegans*, mouse, zebrafish, etc. (Crosby et al., 2007; Bieri et al., 2007; Eppig et al., 2007; Sprague et al., 2006). These databases are built and maintained by teams of information technology specialists and biological curators, and represent highly curated products that recapitulate the biology of a particular species by storing species-specific data types and making available specialized tools for analyzing those data within their specialized biological context. Most MODs store and integrate more than one data type and provide the community with integrated views and specialized tools for analyzing those data within the context of their organism of interest. Clade-Oriented Databases (CODs) store and make accessible those data that can be leveraged by researchers to enable comparative biological analyses, including sequence similarity and genomic synteny information. The CODs are especially important for communities working on groups of species simultaneously, such as potato, tomato, and pepper (SGN; http://www.sgn.cornell.edu; Mueller, 2005). Other CODs include LIS (the Legume Information System; Gonzales et al., 2005) and GrainGenes (for small grains; Carollo et al., 2005).

MaizeGDB (http://www.maizegdb.org/; Lawrence et al., 2007) is the MOD for maize. It is the central repository for all sorts of maize genetics and genomics data, and includes information on maps, loci, gene products, molecular markers, and references, as well as bulletin boards, such as a maize-specific job list and a calendar of upcoming events. MaizeGDB also serves as the clearinghouse for maize nomenclature and supports the activities of the Maize Nomenclature Committee (http://www.maizegdb.org/maize_nomenclature.php) and the Maize Genetics Executive Committee (http://www.maizegdb.org/mgec.php). To best determine how to move MaizeGDB forward to meet the needs of the maize community, a Working Group meets yearly (current membership is listed on the home page at http://www.maizegdb.org), and feedback from researchers who utilize MaizeGDB is solicited, both through the Web interface and in person at meetings, including the Annual Maize Genetics Conference and the International Plant and Animal Genome Conference. Sets of data are taken in over the course of the year by data type (see http://www.maizegdb.org/data_schedule.php), and methods for collaborating with MaizeGDB personnel to incorporate researchers' data into the database are also available online (see http://www.maizegdb.org/data_contribution.php).

Gramene (http://www.gramene.org/; Jaiswal et al., 2006) is the COD that serves maize data alongside information from other grasses to enable cross-species comparisons, including the analysis of synteny information among cereals, which is useful for leveraging data from other grasses to advance maize research.

Other maize resources currently in operation include Panzea (http://www.panzea.org/; Zhao et al., 2006), the Maize WebFPC (http://www.genome.arizona.edu/fpc/maize/; Gardiner et al., 2004), the Maize Genome Sequencing Consortium's genome browser (MaizeSequence.org; http://www.maizesequence.org/), PlantGDB's maize genome browser, which is called ZmGDB (http://www.plantgdb.org/ZmGDB/; Schlueter et al., 2006), the Functional Genomics of Maize Chromatin Consortium database (http://www.chromatin-consortium.org/), and MAGI (Maize Assembled Genomic Island; http://www.magi.plantgenomics.iastate.edu/; Fu et al., 2005). A non-exhaustive list of additional plant-specific databases that are used by maize researchers is shown in Table 1.

Table 1 Online resources utilized by maize researchers that are not maize-specific.

Resource Name	Resource Type	Link	Funding Source(s)
ChromDB	LIMS	http://www.chromdb.org/	NSF
GrainGenes	COD	http://wheat.pw.usda.gov/	USDA-ARS
Gramene	COD	http://www.gramene.org/	NSF, USDA-ARS
GRIN	Static/LIMS	http://www.ars-grin.gov/	USDA-ARS
NCBI (esp. PLANTS)	Static	http://www.ncbi.nlm.nih.gov/ and http://www.ncbi.nlm.nih.gov/genomes/PLANTS/PlantList.html	NIH
PlantGDB	COD/AAS/LIMS	http://www.plantgdb.org/	NSF
PLEXdb	AAS	http://www.plexdb.org/	NSF, USDA-ARS, USDA-CSREES
TAIR	MOD	http://www.arabidopsis.org/	NSF
JCVI	AAS	http://www.jcvi.org/	NSF
UniProt	Static	http://www.pir.uniprot.org/	NIH

1.2   Interconnections among Different Repository Types

Online resources abound for maize. This creates an environment that both assists and stymies researchers. Because many resources are available, it is probable that some available resource stores the data that could help to address a given research question. However, the breadth of resources, coupled with few or no mechanisms to search all resources simultaneously, makes exhaustive searches of available data difficult, if not impossible. Methods that currently are used to interconnect repositories are outlined below.

At the most basic and straightforward level, online resources are interconnected using Web-based hypertext links. Links are stored in a data repository and can provide context-sensitive points of entry into relevant data hosted at another site. Three other methods for interconnecting data among different repositories based on methods and data architectures include the following: data warehousing, data federation, and the use of mediators or portals (reviewed in Lushbough et al., 2008).

Data warehousing represents the least cost-effective method of interconnecting data repositories (Lacroix and Critchlow 2003). In this type of set up, one database duplicates some of the data from another repository. Data federation requires cooperation among all members of the federation (Sheth and Larson 1990). It consists of component databases that are autonomous yet participate in the federation to allow partial and controlled sharing of their data. A mediator is the most flexible approach to data integration. It offers intermediary services that link the data resources and application programs. Mediator approaches integrate information by accessing and retrieving data from multiple resources, abstracting and transforming the retrieved data, integrating the product, and processing the integrated data to return a result (Wiederhold and Genesereth 1997).

In practice, biological databases use all these approaches to integrate data, although data warehousing is probably the most commonly used method of data integration. The use of federation and mediator approaches is growing, due largely to the availability of Web services (technologies that enable one repository to grab information from another resource using defined protocols), ontologies (hierarchical controlled vocabularies that can be used to interconnect related information) and other controlled metadata tags, and the semantic web (a standard for inferring the meanings associated with shared data).

Interconnections via links are often supported by minimal data warehousing. For example, by warehousing GenBank identifiers as well as a protocol for linking to GenBank, any repository could embed linkages from, e.g., molecular markers available at a resource to the marker's GenBank sequence record. A more significant instance of data warehousing is the inclusion of maize molecular markers at Gramene: The marker data were contributed for inclusion in Gramene from MaizeGDB and are currently represented at both repositories.

The NCBI (National Center for Biotechnology Information; Wheeler et al., 2005) is comprised of various databases, including PubMed, GenBank, and various specialized resources, such as PLANTS (http://www.ncbi.nlm.nih.gov/genomes/PLANTS/ PlantList.html). These resources represent a database federation, in that they are each separate databases, but are presented as if they were components of a single repository (see http://www.ncbi.nlm.nih.gov/Database/). Gramene also uses a similar strategy: the genome information, diversity data, pathway information, map, and protein data are housed in separate database structures, but are presented as if they were a single repository within Gramene. In the case of the Genome Browser, Gramene and the maize sequencing project make use of Ensembl (Fernández-Suárez and Schuster, 2007) to store and visualize the data. Currently there are seven sequenced genomes available, 4 monocots, two varieties of rice, maize, sorghum, grape, poplar and sorghum, and 3 dicots, including Arabidopsis. For diversity data, Gramene leverages both the Ensembl and Genomic Diversity and Phenotype Data Model GDPDM to store and distribute diversity data. Currently, Gramene hosts diversity data for maize, rice, and wheat. In the case of pathways, data are stored and displayed using the SRI pathway tools (Hubbard 2002). Gramene supports Web services through Ensembl and GDPC. Ensembl uses the distributed annotation services DAS architecture (Dowel 2001).

A more recent example of implementation of Web services within the plant community is VPIN (the Virtual Plant Information Network; http://vpin.ncgr.org/), which makes use of a mediator approach for data sharing and presentation. VPIN serves data from Gramene, JCVI, and other databases. The underlying technology is SSWAP (Simple Semantic Web Architecture and Protocol), which was developed by D. Gessler and others. Using SSWAP, resources are allowed to define themselves on the Web. Defined documents are available via a non-exclusive discovery server (http://sswap.info) and also can be accessed by third-party servers. Classes of data are deduced based on shared properties, or finding resources based on the type of data they accept (instead of the resource's static categorization). Shown below are some example implementations for each method for creating these interconnections.

MaizeGDB currently uses link integration and data warehousing to enable researchers to get to data of interest stored at sites other than http://www.maizegdb.org/. Linkages from MaizeGDB include (but are not limited to) the following: BioCyc, CerealsDB, Dana-Farber Cancer Institute, Gramene, KEGG, the Maize Genetics Cooperation–Stock Center, MaizeSequence.org, NCBI, PlantGDB, SwissProt/

Steps	Data-Sharing Approach	Example Link(s)
Jump to relevant ZmGI contigs at the Dana-Farber Cancer Insitute from a sequence page at MaizeGDB	Link Integration	http://www.maizegdb.org/cgi-bin/displayseqrecord.cgi?id=BG836376
Find out to which contig at MaizeSequence.org the locus *bnlg1372* belongs via MaizeGDB.	Data Warehousing	http://www.maizegdb.org/cgi-bin/displaylocusrecord.cgi?id=144892
Jump from PubMed's display of Wang and Dooner, 2006 to nucleotide sequences and taxonomy via "Links"	Federation	http://www.ncbi.nlm.nih.gov/sites/entrez?Db=pubmed&Cmd=ShowDetailView&TermToSearch=17101975
Find out (via SSWAP at VPIN) how to create direct linkages to QTL trait symbols at Gramene	Mediation	http://www.sswap.org/ with query "qtl trait symbol"

TREMBL, JCVI, WebFPC, and ZmGDB. Project-specific databases to which MaizeGDB links include ChromDB, the Chromatin Consortium, the Maize TILLING Project site, MAGI, MaGMaP, and others. A more complete list of current maize projects can be accessed at MaizeGDB on the Maize Research Projects page (http://www.maizegdb.org/maizeprojects.php/). Similarly, Gramene uses a combination of linked integration, data warehousing, and Web services methods for obtaining data internally (using a federated approach) and externally, as described above.

MaizeGDB is the hub or focal point for connecting a researcher to relevant resources when initiating a search from a maize-centric perspective. A researcher can, for instance, navigate to GenBank to access the sequences of relevant BACs, or navigate to Gramene to help identify orthologs in rice, wheat, or other grasses. Gramene provides a central location for rice and plant researchers when initiating searches related to rice biology or for cross-species analysis for plants.

2   Data Mining using Currently Available Resources

Databases are only as useful as the information they can provide for given research questions. Below is an example problem that a researcher could solve online using various resources. In addition to access through Web-based displays, in many cases resources offer access to the database through an application programming interface (API) or via wholesale downloads of the database.

2.1   Example Problem 1: Discovering and Developing Molecular Markers for a Genomic Region of Interest Given some Sequence Data

Researcher 1 has created a recessive mutation that disrupts meiotic spindles using Robertson's *Mutator* (*Mu*) (reviewed in Lisch et al., 1995; Lisch, this volume). She also has found that the marker for *bnlg1185* on chromosome 10 identifies a locus within ten centiMorgans (cM) of the mutation responsible for the mutant phenotype. In an effort to narrow down the region before she tries walking to the gene, she plans to check to see which markers might lie closer to the mutation and will find out whether the Maize Genome Sequencing Consortium has sequenced a BAC containing the markers identified.

First she goes to MaizeGDB and uses the search field at the top of any MaizeGDB page (Figure 1A) to search loci for 'bnlg1185'. The locus *bnlg1185* (http://www.maizegdb.org/cgi-bin/displaylocusrecord.cgi?id=144839) is at the top of the results list. On that page, she clicks the link to see the IBM2 2004 Neighbors 10 map. On that page, the locus *bnlg1185* is highlighted in green (Figure 1B). Nearby loci that could be tested to orient the mutation's position relative to *bnlg1185* are *gln1*, which is 20.19 cM proximal to *bnlg1185*, and *csu48*, which is 27.69 cM distal to *bnlg1185*.

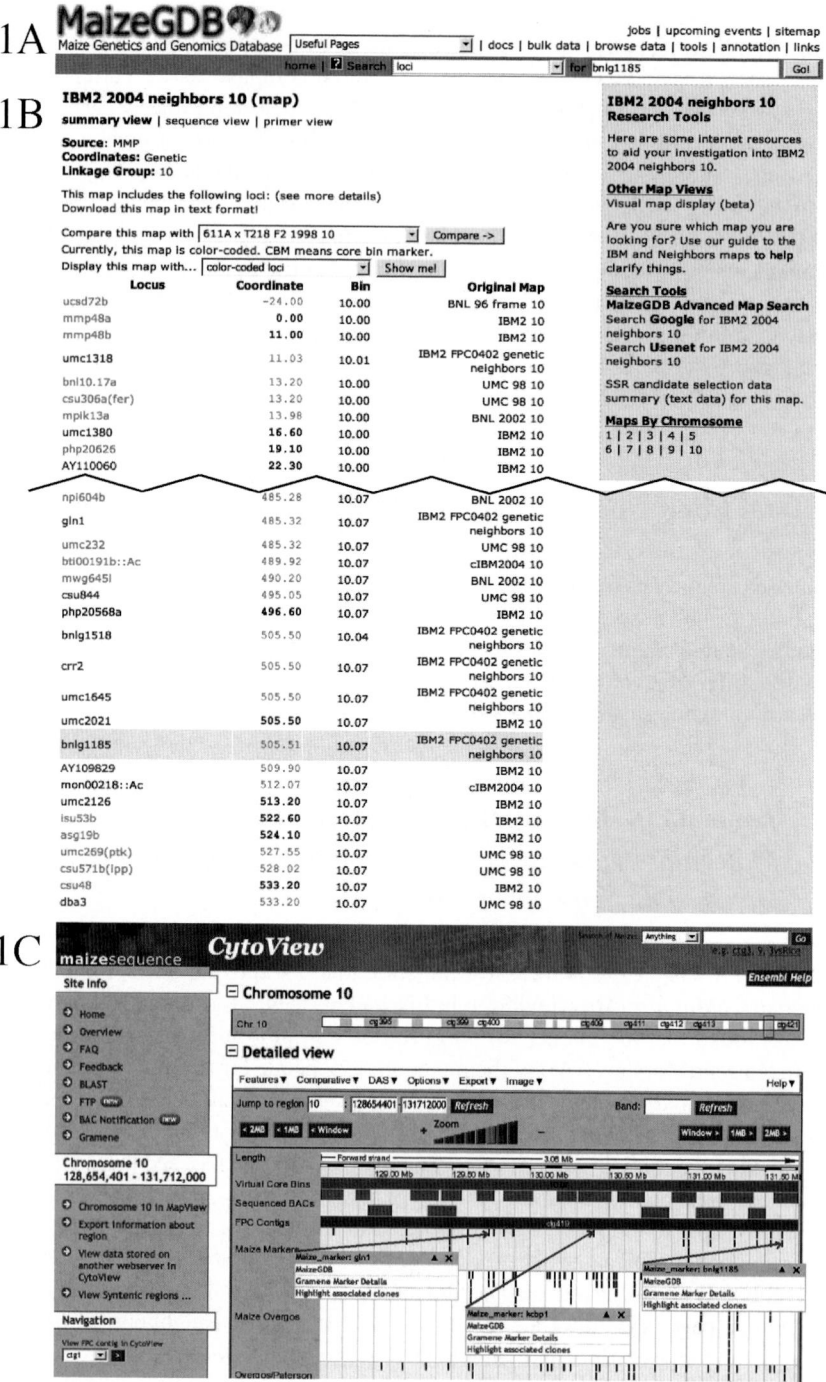

Fig. 1 Screen shots of the MaizeGDB and MaizeSequence.org displays used by Researcher 1. 1A shows the search field at the top of any MaizeGDB page, which can be used to find the locus

Using these two markers, she can orient her mutation via recombination mapping. It turns out that her mutation is approximately 10cM from *bnlg1185* and 40cM from *csu48*. This indicates that the mutation lies between *gln1* and *bnlg1185*.

Next she wants to narrow the interval containing the mutation of interest by selecting markers between *gln1* and *bnlg1185*. She, once again, navigates to MaizeGDB and uses the search field at the top of any page, this time using 'gln1' as a search term. Toward the top of the *gln1* locus page (http://www.maizegdb.org/cgi-bin/displaylocusrecord.cgi?id=61733) is a note that "This locus is part of contig ctg419 at MaizeSequence.org." Similarly, the locus page at MaizeGDB for *bnlg1185* says that the locus is associated with "contig ctg419". Clicking the link to view that contig at MaizeSequence.org, she winds up at the 'CytoView' display of the contig (Figure 1C). Mousing over the row of Maize Markers beneath ctg419, she finds vertical bars indicating the relative locations of *gln1* and *bnlg1185*. Clicking on nearby markers, she sees links to information at MaizeGDB and Gramene, as well as a tool that enables her to highlight associated clones of interest.

Interestingly, on the 'CytoView' display at MaizeSequence.org, a marker labeled 'kcbp1' lies between her markers of interest. When she visits the MaizeGDB locus record for *kcbp1*, she finds that it is, "expressed in all tissues with highest levels in actively dividing cells." Encouraged that this might in fact be (or be very similar to) her gene of interest, she develops primers that are specific for *Mu* as well as primers that are specific to *kcbp1*. If this is the gene where *Mu* inserted, she should be able to get a product with PCR. If not, it's back to narrowing the region by mapping new markers between *bnlg1185* and *gln1*, and subsequently trying out combinations of gene-specific and *Mu*-specific primers with PCR.

2.3 *What to do if no Resources Support Your Workflow Needs*

In instances where the data exist at a particular repository that can be pieced together to answer a particular researcher's questions, but the method by which the data are made available causes the same repetitive set of steps to be carried out many times, the researcher's best course of action is to use links at the repository to contact its personnel, describe the special needs of the project at hand, and to ask for the creation of a customized dataset. Before making contact, it is advisable for the researcher to document exactly how s/he can use use the repository's data to get the necessary set of information, such that the repository's personnel could follow that protocol. Furthermore, it is advisable for the researcher to accompany the

Fig. 1 (continued) *bnlg1185*. From the *bnlg1185* locus page, clicking the link to the IBM2 2004 Neighbors 10 map not only shows the map of interest (1B), but also highlights the location of *bnlg1185*. Using links at the top of the *bnlg1185* and/or *gln1* page, the MaizeSequence.org CytoView page for contig ctg419 is accessed. There (1C), the relative positions of *gln1*, *kcbp1*, and *bnlg1185* are shown

request with a table including at least one example of the desired data. It is often the case that these processes can be automated and will be of use to other researchers in the field. In most instances, a researcher who requests a dataset will receive it promptly, thus allowing valuable time to be spent at the bench, rather than clicking through an interface to collect a set of data.

3 How to get Your Own Project Data into the Mainstream Databases

Researchers often begin to generate data and to store generated data well in advance of knowing whether the data will be useful—after all, it's called research, right? One problem that often arises when preliminary investigations go well and data generation picks up is a tendency to continue to store generated data in an idiosyncratic manner. Datasets that are not formatted in a way that integrates well with existing information stored at the larger data repositories pose a real problem for subsequent integration and results in data quality issues, loss of data, and sometimes a loss of human resources available to the project.

3.1 Choosing a Repository to House Data

The best way for a researcher to find a repository to house project data is to first consider which repositories already hold the types of data to be generated. With that list in hand, the researcher should consider how others might be expected to utilize the data and then contact the repository that best meets the needs of that type of data and analysis. Feedback links at the repository of interest or direct contact with the repository's lead scientist or project manager are two avenues of initial inquiry as to whether the data could be accommodated.

Funding sources affect resource development and maintenance practices as well as data access longevity. For this reason, it is always advisable for a researcher to consider whether the resource most appropriate for storing the data has long-term funding, as well as to inquire about whether and how the research project's funds could support data integration directly. Because allocating funds to the repository may be required, it is important to investigate data warehousing options well in advance of writing proposals to funding agencies.

3.2 Data Types that Help with Resource Integration

Once a repository has been selected for collaboration, it is wise for the researcher to depend upon the personnel who work at that resource for guidance in how to store project data in a manner that will facilitate its incorporation into the data repository. Some data types that are likely to be useful for integration are the following:

Controlled vocabularies	These resource-specific sets of words are assigned to records to enable others to find the data via keyword searches. One example would be assigning the keyword "SSR" to a probe/molecular marker record that is to be included in MaizeGDB. Note that, if the researcher were to fail to inquire in advance how the repository assigns terms, it is possible that the word "microsatellite" would have been assigned, thus causing the new records to be absent from SSR lists generated by the Web interface or other data presentation interfaces.
Ontologies	Ontologies are hierarchically-related controlled vocabularies that are a standard utilized by many resources (i.e., they are not database- or resource-specific). Again, inquiring with the resource into which data will flow is key to finding out which ontologies to use for descriptor assignments during data collection. Ontologies that are likely to be suggested include the Plant Ontologies (PO; Jaiswal et al., 2005) and Gene Ontologies (GO; The GO Consortium, 2000). Including these terms enables repositories to warehouse links to other resources using the annotations. For example, at PO one can find genes for *Arabidopsis*, rice, and maize that affect inflorescences, or parts of inflorescences, and link to individual database records at TAIR, Gramene, and MaizeGDB for detail.
Size and color standards	To help others to know, for example, the size of bands on gels, lengths of floral organs, or the color of a kernel phenotype; and to enable evolving software tools to search images given an example image as a query (Shyu et al., 2007), researchers should take care to collect these sorts of data and to work with the chosen repository well in advance of data submission to define size and color standards for all phenotypic measurements.

4   The Future of Plant Databases and Data Mining

Resources for database creation and development continually diminish. The repositories upon which maize researchers depend are affected by the scarcity of funding, making it difficult to continue to serve researchers at the level to which they have become accustomed. Simultaneously, many researchers create project-specific databases without ensuring future accommodation by a long-term repository, making it difficult to integrate generated data with related information once project personnel have moved on to other things. The good news is that these problems are apparent, and the funding agencies are responding. A new National Science Foundation-funded project to create a plant cyberinfrastructure called the iPlant™ Collaborative (see http://

www.iplantcollaborative.org) has begun, with the intention to create community-based resources that build upon existing resources and also to encourage the development of new technologies and methods of collaboration to better organize resources and their interactions. In addition, a new Project Portal for corn (POPcorn; also funded by the NSF and to be implemented by USDA-ARS personnel) is in the planning stages. The POPcorn resource, which will be ancillary to MaizeGDB, will allow researchers to search all maize projects' data simultaneously from one Web portal and will provide tools to allow dataset upload to MaizeGDB at a research project's close.

Maize researchers are at the cusp of a new era: The sequence of B73 will be available at the end of 2008, and the cost to sequence other inbred lines falls each day. Maize was once a genetics-rich but sequence-poor model for research, but this is changing. The stage is set for a renaissance in maize research where the species' strengths shine: researchers will have access to sequenced genomes, excellent genetics, and unparalleled cytogenetics. These tools will allow a better understanding of metabolism, development, and breeding. With these excellent assets emerging, the need for a well-annotated genome, improved repositories to store diverse data, and improved connections among maize informatics resources becomes paramount. It is anticipated that NCBI will represent the available maize genome sequence and related data as they have for other sequenced species. For MaizeGDB, the database and Web interface will evolve based upon available resources coupled with the community's stated objectives as communicated by the Maize Genetics Executive Committee and the MaizeGDB Working Group. Increased requirements for data handling will emerge and be met, and researchers' ability to utilize all available data will improve as the data stored in various places are shared by increased utilization of federation and mediation approaches, as well as other technologies currently under development.

References

Benson, D.A., Boguski, M.S., Lipman, D.J., and Ostell, J. (1997) GenBank. Nucleic Acids Res. 25(1), 1–6.

Benson, D.A. Karsch-Mizrachi, I., Lipman P., Gelbart, W.M., and the FlyBase Consortium. (2007) FlyBase: genomes by the dozen. Nucleic Acids Res. 35(Database issue), D486–D491.

Bieri, T., D. Blasiar, P. Ozersky, I. Antoshechkin, C. Bastiani, P. Canaran, J. Chan, N. Chen, W.J. Chen, P. Davis, T.J. Fiedler, L. Girard, M. Han, T.W. Harris, R. Kishore, R. Lee, S. McKay, H.M. Muller, C. Nakamura, A. Petcherski, A. Rangarajan, A. Rogers, G. Schindelman, E.M. Schwarz, W. Spooner, M.A. Tuli, K. Van Auken, D. Wang, X. Wang, G. Williams, R. Durbin, L.D. Stein, P.W. Sternberg, and J. Spieth. 2007. WormBase: new content and better access. *Nucleic Acids Res* 35: D506–510.

Carollo, V., Matthews, D.E., Lazo, G.R., Blake, T.K., Hummel, D.D., Lui, N., Hane, D.L., and Anderson, O.D. (2005) GrainGenes 2.0. An improved resource for the small-grains community. Plant Physiol. 139(2), 643–651.

Cartinhour, SW. (1997) Public informatics resources for rice and other grasses. *Plant Mol Biol* 35(1–2),241–251.

Chan, A., Cheung, F., Lee, D., Zheng, L., Whitelaw, D., Pontaroli, A., Sanmiguel, P., Yuan, Y., Bennetzen, J., Barbazuk, W.B., Quackenbush, J., and Rabinowicz, P.D. (2006) The TIGR Maize Database. Nucleic Acids Res. 34, D771–D776.

Codd, E.F. (1970) A relational model of data for large shared data banks. Communications of the ACM 13(6), 377–387.

Dowell, R.D., R.M. Jokerst, A. Day, S.R. Eddy, and L. Stein. 2001. The distributed annotation system. BMC Bioinformatics 2: 7.

Eppig, J.T., Blake, J.A., Bult, C.J., Kadin, J.A., Richardson, J.E., and the Mouse Genome Database Group (2007) The mouse genome database (MGD): new features facilitating a model system. Nucleic Acids Res. 35(Database issue), D630–D637.

Fernández-Suárez, X.M., and Schuster, M.K. (2007) Using the Ensembl genome server to browse genomic sequence data. Curr Protoc Bioinformatics. 1,1.15.

Fu, Y., Emrich, S.J., Guo, L., Wen, T.J., Ashlock, D.A., Aluru, S., and Schnable, P.S. (2005) Quality assessment of maize assembled genomic islands (MAGIs) and large-scale experimental verification of predicted genes. Proc. Natl. Acad. Sci. U.S.A. 102(34), 12282–12287.

Gardiner, J., Schroeder, S., Polacco, M.L., Sanchez-Villeda, H., Fang, Z., Morgante, M., Landewe, T., Fengler, K., Useche, F., Hanafey, M., Tingey, S., Chou, H., Wing, R., Soderlund, C., and Coe, E.H. (2004) Anchoring 93,971 maize expressed sequence tagged unigenes to the bacterial artificial chromosome contig map by two-dimensional overgo hybridization. Plant Physiol. 134,1317–1326.

Gonzales, M.D., Archuleta, E., Farmer, A., Gajendran, K., Grant, D., Shoemaker, R., Beavis, W.D., and Waugh, M.E. (2005) The Legume Information System (LIS): an integrated information resource for comparative legume biology. Nucleic Acids Res. 33(Database issue), D660–D665.

Grant, D. and Shoemaker, R.C. (2007) SoyBase, The USDA-ARS Soybean Genome Database. http://soybase.org.

Huala, E., Dickerman, A.W., Garcia-Hernandez, M., Weems, D., Reiser, L., LaFond, F., Hanley, D., Kiphart, D., Zhuang, M., Huang, W., Mueller, L.A., Bhattacharyya, D., Bhaya, D., Sobral, B.W., Beavis, W., Meinke, D.W., Town, C.D., Somerville, C., and Rhee, S.Y. (2001) The *Arabidopsis* Information Resource (TAIR): a comprehensive database and web-based information retrieval, analysis, and visualization system for a model plant. Nucleic Acids Res. 29(1), 102–5.

Hubbard, T., D. Barker, E. Birney, G. Cameron, Y. Chen, L. Clark, T. Cox, J. Cuff, V. Curwen, T. Down, R. Durbin, E. Eyras, J. Gilbert, M. Hammond, L. Huminiecki, A. Kasprzyk, H. Lehvaslaiho, P. Lijnzaad, C. Melsopp, E. Mongin, R. Pettett, M. Pocock, S. Potter, A. Rust, E. Schmidt, S. Searle, G. Slater, J. Smith, W. Spooner, A. Stabenau, J. Stalker, E. Stupka, A. Ureta-Vidal, I. Vastrik, and M. Clamp. 2002. The Ensembl genome database project. Nucleic Acids *Res* 30: 38–41.

Jaiswal, P., Avraham, S., Ilic, K., Kellogg, E., McCouch, S.R., Pujar, A., Reiser, L., Rhee, S., Sachs, M., Schaeffer, M., et al. (2005) Plant Ontology (PO): a controlled vocabulary of plant structures and growth stages. Comp. Funct. Genomics 6, 388–406.

Jaiswal, P., Ni, J., Yap, I., Ware, D., Spooner, W., Youens-Clark, K., Ren, L., Liang, C., Zhao, W., Ratnapu, K., Faga, B., Canaran, P., Fogleman, M., Hebbard, C., Avraham, S., Schmidt, S., Casstevens, T.M., Buckler, E.S., Stein, L., and McCouch, S. (2006) Gramene: a bird's eye view of cereal genomes. Nucleic Acids Res. 2006 Jan 1;34(Database issue), D717–D723.

Lacroix, Z. and Critchlow, T. (2003) Bioinformatics: Managing Scientific Data. Morgan Kaufmann Publishers, pp. 21–24.

Lawrence, C.J., Dong, Q., Polacco, M.L., Seigfried, T.E., and Brendel, V. (2004) MaizeGDB, the community database for maize genetics and genomics. Nucleic Acids Res. 32(Database issue), D393–D397.

Lawrence, C.J., Schaeffer, M.L., Seigfried, T.E., Campbell, D.A., and Harper, L.C. (2007) MaizeGDB's new data types, resources and activities. Nucleic Acids Res. 35(Database issue), D895–900.

Lisch, D., Chomet, P., and Freeling, M. (1995) Genetic characterization of the *Mutator* system in maize: behavior and regulation of *Mu* transposons in a minimal line. Genetics 139, 1777–1796.

Lushbough, C., Bergman, M.K., Lawrence, C.J., Jennewein, D., and Brendel, V. (2008) BioExtract Server - an integrated workflow-enabling system to access and analyze heterogenous, distributed biomolecular data. IEEE. ACM Transactions on Computational Biology and Bioinformatics. 11

Sept 2008. IEEE computer Society Digital Library. IEEE Computer Society, 10 November 2008 <http://doi.ieeecomputersociety.org/10.1109/TCBB.2008.98.

Mueller, L.A., Solow, T.G., Taylor, N., Skwarecki, B., Buels, R., Binns, J., Lin, C., Wright, M.H., Ahrens, R., Wang, Y., Herbst, E.V., Keyder, E.R., Menda, N., Zamir, D., and Tanksley, S.D. (2005) The SOL Genomics Network: a comparative resource for Solanaceae biology and beyond. Plant Physiol. 138(3), 1310–1317.

Neale, D. (2007) Dendrome, The USDA Forest Service's Forest Tree Genome Database. http://dendrome.ucdavis.edu.

Polacco, M. and Coe, E. (1999) MaizeDB: The maize database. In *Bioinformatics Databases and Systems*, Letovsky, S.I., ed. Kluwer Academic Publishers, Boston.

Schlueter, S.D., Wilkerson, M.D., Dong, Q., and Brendel, V. (2006) xGDB: open-source computational infrastructure for the integrated evaluation and analysis of genome features. Genome Biol. 7(11), R111.

Scholl, R., Sachs, M., and Ware, D. (2003) Maintaining collections of mutants for plant functional genomics. In Grotewold, E., ed. Plant Function Genomics, Totowa, NJ Humana Press Vol. 236, pp. 311–326.

Sheth, A.P. and Larson, J.A. (1990) Federated database systems for managing distributed, heterogeneous, and autonomous databases. ACM Computing Surveys. 22(3), 183–236.

Shyu, C., Green, J.M., Lun, D.P.K., Kazic, T, Schaeffer, M., and Coe, E. (2007) Image analysis for mapping immeasurable phenotypes in maize. IEEE Signal Processing Maga. May, 115–118.

Sprague, J., Bayraktaroglu, L., Clements, D., Conlin, T., Fashena, D., Frazer, K., Haendel, M., Howe, D.G., Mani, P., Ramachandran, S., Schaper, K., Segerdell, E., Song, P., Sprunger, B., Taylor, S., Van Slyke, C.E., and Westerfield, M. (2006) The Zebrafish Information Network: the zebrafish model organism database. Nucleic Acids Res. 34(Database issue), D581–D585.

Stoesser, G., Sterk, P., Tuli, M.A., Stoehr, P.J., and Cameron, G.N. (1997) The EMBL nucleotide sequence database. Nucleic Acids Res. 25(1), 7–14.

Tateno, Y. and Gojobori, T. (1997) DNA Data Bank of Japan in the age of information biology. Nucleic Acids Res. 25(1), 14–17.

The Gene Ontology Consortium (2000) Gene Ontology: tool for the unification of biology. Nature Genet. 25, 25–29.

Wang, Q. and Dooner, H.K. (2006) Remarkable variation in maize genome structure inferred from haplotype diversity at the *bz* locus. Proc. Natl. Acad. Sci. U.S.A. 2006 103(47), 17644–9.

Ware, D., Jaiswal, P., Ni, J., Pan, X., Chang, K., Clark, K., Teytelman, L., Schmidt, S., Zhao, W., Cartinhour, S., McCouch, S., and Stein, L. (2002) Gramene: a resource for comparative grass genomics. Nucleic Acids Res. 30(Database issue), 103–105.

Wheeler, D.L., Barrett, T., Benson, D.A., Bryant, S.H., Canese, K., Church, D.M., DiCuccio, M., Edgar, R., Federhen, S., Helmberg, W., Kenton, D.L., Khovayko, O., Lipman, D.J., Madden, T.L., Maglott, D.R., Ostell, J., Pontius, J.U., Pruitt, K.D., Schuler, G.D., Schriml, L.M., Sequeira, E., Sherry, S.T., Sirotkin, K., Starchenko, G., Suzek, T.O., Tatusov, R., Tatusova, T.A., Wagner, L., and Yaschenko, E. (2005) Database resources of the National Center for Biotechnology Information. Nucleic Acids Res. 33(Database issue), D39–D45.

Wiederhold, G. and Genesereth, M. (1997) The conceptual basis for mediation services. IEEE Expert, 12(5), 38–47.

Zhao, W., Canaran, P., Jurkuta, R., Fulton, T., Glaubitz, J., Buckler, E., Doebley, J., Gaut, B., Goodman, M., Holland, J., Kresovich, S., McMullen, M., Stein, L., and Ware, D. (2006) Panzea: a database and resource for molecular and functional diversity in the maize genome. Nucleic Acids Res. 34(Database issue), D752–D757.

Sequencing Genes and Gene Islands
by Gene Enrichment

Pablo D. Rabinowicz and W. Brad Barbazuk

Abstract Access to the sequence of any gene in a genome greatly accelerates genetics research. Whole genome sequencing is a way to retrieve such information although, for large genomes such as that of maize, it represents a huge effort. Fortunately, a maize genome sequencing project is currently underway but, before this project started, the maize research community benefited from the development of gene enrichment methods that allow selectively cloning and sequencing genes. The application of these methods to maize generated comprehensive gene sequence collections that were extensively used by the community. Once the maize genome project is completed, combination of gene enrichment methods with next-generation sequencing technologies will greatly facilitate genome-wide comparative analysis of different maize inbred lines for functional, population, and evolutionary studies.

1 Introduction

When the human genome project began in the 1990s, sequencing and assembling the 3 gigabase pair (Gbp) human genome represented a colossal enterprise (Lander and Weinberg 2000). The sequences of several smaller genomes were completed in the following years by large collaborative efforts (Fleischmann et al. 1995; Goffeau et al. 1996; The *C. elegans* Sequencing Consortium 1998). These projects allowed the development and improvement of tools and technologies for genomic sequencing and analysis, making it more affordable to attempt the sequencing of increasingly large genomes. Plant genome sequencing started later, and the first plant genome to

P.D. Rabinowicz
Institute for Genome Sciences and Department of Biochemistry and Molecular Biology,
School of Medicine, University of Maryland, Baltimore, MD 21201
prabinowicz@som.umaryland.edu

W.B. Barbazuk
Department of Botany and Zoology, and the Genetics Institute, University of Florida,
Gainesville, FL 32611 bbarbazuk@ufl.edu

J.L. Bennetzen and S. Hake (eds.), *Maize Handbook - Volume II: Genetics and Genomics,*   673
© Springer Science+Business Media LLC 2009

be extensively sequenced was the relatively small (130 Mbp) genome of *Arabidopsis*, which was published in 2000 (The Arabidopsis Genome Initiative 2000). While the *Arabidopsis* project was underway, efforts to sequence larger plant genomes, such as the 400 Mbp rice genome, were set in motion (Sasaki and Burr 2000). However, by the time the *Arabidopsis* genome sequence was published, sequencing Gbp-size plant genomes by existing technologies did not seem realistic, not only due to the insufficient funds available for plant genomics, but also due to the large amount of conserved, highly repetitive sequences characteristic of large plant genomes that would make genome assembly cumbersome. Coincidentally, techniques to selectively clone and sequence plant genes, while avoiding repetitive sequences (gene enrichment), were being developed. Therefore, the notion of sequencing only the genic regions within a plant genome was presented, and the maize research community, gathered in a workshop in Saint Louis, MO in 2001, embraced this idea. A majority of the workshop participants supported a maize genome sequencing approach focused on the low-copy and gene-rich regions of the maize genome anchored to physical and genetic maps (Bennetzen et al. 2001). A "proof of concept" project aimed to sequence the genic fraction of the maize genome (the gene-space) using gene-enrichment techniques started in 2002 and rapidly produced a comprehensive catalog of maize gene tag-sequences for the community to take advantage of, before a whole genome sequence could become a reality. Auspiciously, a maize whole genome sequencing project commenced in 2005, with the goal of producing ordered and oriented, high-quality sequence for all genes, and draft sequences for the intergenic, repetitive fraction of the genome. Here we review different gene-enrichment technologies applied to maize, in the context of whole genome sequencing approaches, and discuss the impact of such technologies in maize and plant genetics and genomics research.

2 Genes and Transposons Structure and Methylation

In order to sequence the maize genome, important challenges must be overcome. Primarily, its large size, estimated in 2.4 Gbp for B73, the inbred line selected for genomic sequencing (Rayburn *et al.* 1993), puts the maize genome among the largest to be sequenced. Furthermore, the vast majority of the maize genome is composed of highly repetitive DNA, most of which is composed, in turn, of a few large families of very conserved retrotransposons (Hake and Walbot 1980; SanMiguel and Bennetzen 1998; Meyers et al. 2001). In general, retrotransposons occupy intergenic regions and are often arranged in groups of nested insertions spanning hundreds of kbp (SanMiguel et al. 1996). Genes, on the other hand, are typically a few kbp-long and are generally separated by repetitive elements. Nevertheless, genes as long as 89 kbp have been reported in maize (Bruggmann et al. 2006), where intronic repetitive elements may occur.

Maize is also an ancient allotetraploid (Gaut and Doebley 1997; Swigonova et al. 2004), and polyploidy generally poses additional challenges for genome sequence assembly due to the presence of highly similar homoeologous genes. In maize, however, a process of gene loss that rendered the maize genome to a partially diploidized state (Ilic et al. 2003; Lai et al. 2004; Langham et al. 2004) may reduce the problem of distinguishing members of pairs of duplicated genes.

The replication of repetitive elements can be deleterious for the host and silencing mechanisms are needed to limit genome damage. DNA methylation is believed to be involved in such mechanisms and, consistently, silent methylated transposons can be reactivated in methylation-defective mutants of *Arabidopsis* (Miura et al. 2001; Singer et al. 2001). DNA methylation in carbon 5 of cytosine (5-methyl cytosine) is found ubiquitously in plants, mainly in symmetrical CpG and CpNpG sequences, but also in asymmetrical sites. Most of the methylated DNA co-localizes with repetitive sequences, although methylation has also been detected in genes by genome-wide tiling microarray analysis of *Arabidopsis*. Nevertheless, the level of methylation observed in genes was lower than that in repeats and pseudogenes, and in many cases methylation was localized towards the 3′ end of the genes (Lippman et al. 2004; Vaughn et al. 2007; Zhang et al. 2006; Zilberman et al. 2007).

3 Genome Sequencing Strategies

3.1 Whole Genome Shotgun

Several plant genomes have been approached using the so-called whole genome shotgun (WGS) strategy in which the ends of multiple random genomic clones are sequenced. Genomic libraries of different insert sizes are used to help prevent cloning biases. Enough sequence data are produced in order to cover the genome several times, which facilitates computational assembly of overlapping sequence reads into contiguous sequences (contigs). The assembly process is achieved by identifying regions of identity between reads that allows aligning and orienting them relative to each other. The redundant reads are then collapsed to produce a consensus sequence. Several assembly algorithms have been developed, and these serve as the foundation for the genome assembly programs currently available (www.phrap.org; Batzoglou et al. 2002; Huang and Madan 1999; Myers et al. 2000; Sutton et al. 1995). Many assembly algorithms are designed for data representing uniform sampling of the genome. However, complex eukaryotic genomes contain repetitive elements and duplicated regions that violate this assumption and affect the performance of the assembler, resulting in gaps or misassemblies (Tang 2007). The assembly software also uses the "mate-pair" reads (sequences from both ends of the same clone) to link contigs into scaffolds, determining their relative order and orientation, and estimating the physical size of the gaps between contigs. The quality of the resulting assembly is mostly influenced by the level of coverage of the genome sequence, the amount of repetitive DNA, and the presence of genomic duplications. If available, genetic markers of known sequence can be aligned to the genome assembly in order to anchor the contigs to the corresponding chromosomes. High quality draft sequences obtained by a WGS approach are generated after 8- to 10-fold sequence coverage, which can be achieved in a relatively short period of time. Nevertheless, assemblies of plant genomes of several hundred Mbp generally result in tens of thousands of contigs.

In the case of maize, the abundance of highly conserved repetitive sequences (Hake and Walbot 1980; SanMiguel and Bennetzen 1998; Meyers et al. 2001), added to its large size, makes a WGS sequencing approach impractical. The low-copy fraction of the genome might be properly assembled but not the conserved repetitive elements, which will likely be clustered together and inappropriately collapsed. A WGS sample covering 1/3 of the maize genome has been released by the Joint Genomics Institute from the US Department of Energy (JGI-DOE; http://www. ncbi.nlm.nih.gov/Traces/trace.cgi?). The analysis of these data will provide accurate estimation of the amount and classes of repetitive elements in the maize genome, and will complement other whole genome sequencing efforts as described in the following section.

3.2   Bacterial Artificial Chromosome-Based Genome Sequencing

Two model plant genomes (*Arabidopsis* and rice) have been sequenced using a bacterial artificial chromosome (BAC)-based strategy. The quality of this sequence is very high because a BAC-based physical map was constructed to aid the selective sequencing of minimally overlapping BAC clones spanning most of the genome, and substantial efforts were subsequently invested in manual curation of the sequence. Clones in a BAC library representing multiple genome equivalents can be clustered together into physical contigs by obtaining a restriction fragment fingerprint of each BAC. Overlapping clones can be identified on the basis that they share restriction patterns (Luo et al. 2003; Marra et al. 1999; Soderlund et al. 2000; Soderlund et al. 1997) and a "tiling path" of contiguous clones can, in turn, be identified to ensure a minimal overlap between BAC clones selected for sequencing. The sequence of each BAC clone is determined by a shotgun sequencing and assembly strategy. This procedure dramatically simplifies the assembly problem faced by the WGS approach, which is particularly relevant in large and repetitive genomes. However, generating a physical map for a large genome is difficult and costly, as is the production of thousands of BAC sub-clone shotgun libraries for sequencing. Another disadvantage of the BAC-based strategy is that some sequences (usually tandem repeats such as those abundant in pericentromeric regions) are unstable in large insert-size libraries and are thus underrepresented. Such cloning biases are minimized in the WGS approach by using multiple libraries with insert sizes ranging between 1.5 to 150 kbp.

The maize genome is currently being sequenced using a BAC-based method. However, because of the difficulty posed by the abundant conserved repeats, the project will target only low-copy sequences for manual curation, so that a high quality sequence of the genic fraction of the genome is delivered, leaving the repetitive (mostly intergenic) content in a draft form. For this purpose, high- and low-copy sequences must be discriminated in the draft BAC sequences to then target the manual finishing efforts to the low-copy regions (http://genome.wustl.edu/genome.cgi?GENOME = Zea%20 mays%20mays%20cv.%20B73; http://www.nsf.gov/awardsearch/showAward. do?AwardNumber=0527192).

The size and repetitive nature of the maize genome conspire to make the accurate assembly of the maize genome a difficult prospect. Nevertheless, some success has been achieved with the ARACHNE sequence assembly suite (Batzoglou *et al.* 2002; Jaffe et al. 2003), which has been modified and tested in limited maize BAC assemblies (Haberer et al. 2005). A technique called retroscaffolding, which orders and orients contigs based on their span of LTR retrotransposon-rich regions of the genome rather than traditional paired-end sequences, may also improve maize genome assembly (Kalyanaraman et al. 2006). Computational methods to facilitate identifying and resolving sequence misassemblies have been developed. One such method was used to identify potential assembly errors due to segmental duplications in the human genome by identifying over-representation of specific sequences within randomly generated data (Bailey et al. 2002). However, its application to highly repetitive genomes such as that of maize may pose significant challenges.

The WGS and BAC-based approaches can be combined to take advantage of the benefits of each method. In some mammalian genome projects, WGS sequence data was added to the BAC sequences to increase the coverage. The WGS data can be generated more rapidly and for less cost than BAC clone sequences and delivered to the community in advance of the BAC sequences (Gibbs et al. 2004; Mouse Genome Sequencing Consortium 2002). In addition to the 0.3X WGS sequence of maize produced by the JGI-DOE, nearly 0.5 million BAC-end sequences and 1 million gene-enriched sequences, representing most of the low-copy fraction of the genome (see below), were available before the maize genome project started. These data will be used to improve BAC sequences in the BAC-based genome project. These additional genomic sequences will allow estimating the fraction of the maize genome that is missing in the BAC-based physical map, and therefore absent in the genome sequence. The low-copy sequences in these data sets can not only enrich the BAC sequences when added to them, but also help identifying genic regions for targeted manual finishing.

3.3   Chromosome-Specific Sequencing

The idea of sequencing BAC clones to improve the assembly of a genome can be applied at the chromosome level. Instead of fragmenting a genome in 100-200 kbp fragments (BAC clones), entire chromosomes can be isolated by flow cytometry and sequenced either by WGS or with a BAC-based approach. This is particularly attractive for evolutionarily recent polyploids, in which very similar copies of genes can be present in the genome, and sequencing isolated chromosomes would overcome this difficulty.

This strategy has serious limitations. One is the difficulty in obtaining sufficient amounts of single-chromosome DNA to construct the necessary libraries. Techniques to uniformly amplify whole genomes can be applied to isolated chromosome preparations to overcome this problem (Dean et al. 2002). Another issue is the feasibility of isolating each chromosome in the genome. In hexaploid wheat, this can be achieved using aneuploid lines that contain only one of the three homoe-ologous chromosomes or chromosome arms, allowing their isolation by flow cytometry. Excellent cytogenetics have been developed for maize (see the chapter by Bass and Birchler in this volume) and a series of oat lines containing maize chromosome additions (see the chapter by Phillips and Rines in this volume) could potentially be used for maize chromosome-specific sorting. This approach has been tested as part of a pilot project by the JGI-DOE for sequencing the maize genome. Given difficulties associated with isolating maize chromosomes, the outcomes of this project remain inconclusive (http://www.maizegdb.org/sequencing_project.php).

4 Gene Enrichment

4.1 EST Sequencing

Random cDNA clones can be sequenced to rapidly obtain information about transcribed regions (Adams et al. 1991). Large numbers of these sequences, called expressed sequence tags (ESTs), have been produced for many species, regardless of the availability of genomic sequence information. EST data provides not only useful coding sequences, but also gene expression pattern information derived from the source tissue for the starting mRNA material. Furthermore, expression levels can be deduced by correlating specific mRNA abundance with the frequency of occurrence of a given cDNA sequence in the EST data set.

For the purpose of gene discovery, the frequency of highly expressed sequences can be reduced by sequencing ESTs from normalized cDNA libraries (Patanjali et al. 1991; Soares et al. 1994). Information regarding the gene expression level is traded for the increased discovery rate of rare cDNA sequences. Due to expression biases, even normalized EST data miss a large proportion of the gene set of the organism under study, resulting in an incomplete collection of gene sequences, regardless of the size of the data set (Barbazuk et al. 2005; Bonaldo et al. 1996).

Although ESTs are single-pass sequences, extended – and often complete – coding sequences can be generated by assembling EST sequences into larger contigs (Childs et al. 2006; Dong et al. 2005; Lee et al. 2005; Wheeler et al. 2006). Furthermore, if genome sequence data is available, ESTs from related species can be aligned to those, using spliced alignment tools, in order to identify genes along with their intron/exon structure.

The approximately 1.3 million maize ESTs existing in GenBank will be a key resource for gene discovery and annotation in the maize genome sequence, and an on-going project aiming to generate 30,000 full-length cDNA sequences (Y. Yu, personal comm.) will be an excellent resource to determine the complete structure of many maize genes.

4.2   Methylation-Sensitive Restriction Digestion Cloning and Sequencing

Due to the differential methylation of plant repetitive elements and genes, the latter can be selected by digesting genomic DNA with cytosine methylation-sensitive restriction enzymes and selecting the smaller fragments. Because repetitive elements are densely methylated, they are resistant to digestion and end up in large fragments, while unmethylated genes are digested to smaller fragments. This approach has been applied to maize using the methylation-sensitive enzyme *Pst*I and selecting 1.5-2.5 kbp DNA fragments in agarose electrophoresis to isolate low-copy sequences for molecular marker development (Burr et al. 1988). Although this method yields a high frequency of low-copy and genic sequences, it only recovers sequences that are flanked by restriction sites spaced appropriately to be included in the size fractionation step.

In order to increase the randomness of this method, partial restriction with multiple methylation-sensitive restriction enzymes (generally frequent-cutters) may be used. This concept has been tested in maize and was called hypomethylated partial restriction (HMPR; Emberton et al. 2005; Figure 1). Because only the ends of HMPR clones are sequenced, enrichment in low-copy sequences is observed due to their proximity to non-methylated restriction sites in the genome. Nevertheless, HMPR clones may include methylated repetitive elements, which can be advantageous, for example, to link two low-copy sequences that are separated by relatively short repetitive regions. Sequencing the ends of large-insert clones (i.e. BACs) generated by digestion of plant genomic DNA with rare-cutter methylation-sensitive restriction enzymes, can allow determining physical linkage between low-copy sequences that are separated by large stretches of repetitive elements. As in HMPR, the end-sequences of such BAC clones are generally enriched in genic and other low-copy sequences. Two variations of this method have also been tested in maize. One of them, called methylation spanning linker libraries (MSLL; Yuan et al. 2002; Figure 1), uses complete restriction enzyme-digestion, so that the resulting sequences are close to the junction between low-copy and repetitive sequences. The second method uses partial restriction enzyme-digestion and it is therefore called methylation spanning partial restriction (MSPR; Yu and Li 2006; Figure 1). MSPR libraries are expected to show a higher level of randomness than MSLL, and Yu and Li constructed a vector that contains cloning sites for methylation-sensitive restriction enzymes to improve cloning efficiency in the case of sticky-end cutters.

Encouraging pilot HMPR studies in maize showed 4- to 14-fold enrichment in genes relative to a library made in a non-methylation sensitive restriction enzyme, while sequencing a small number of MSPR clones resulted in a low frequency of repetitive sequences.

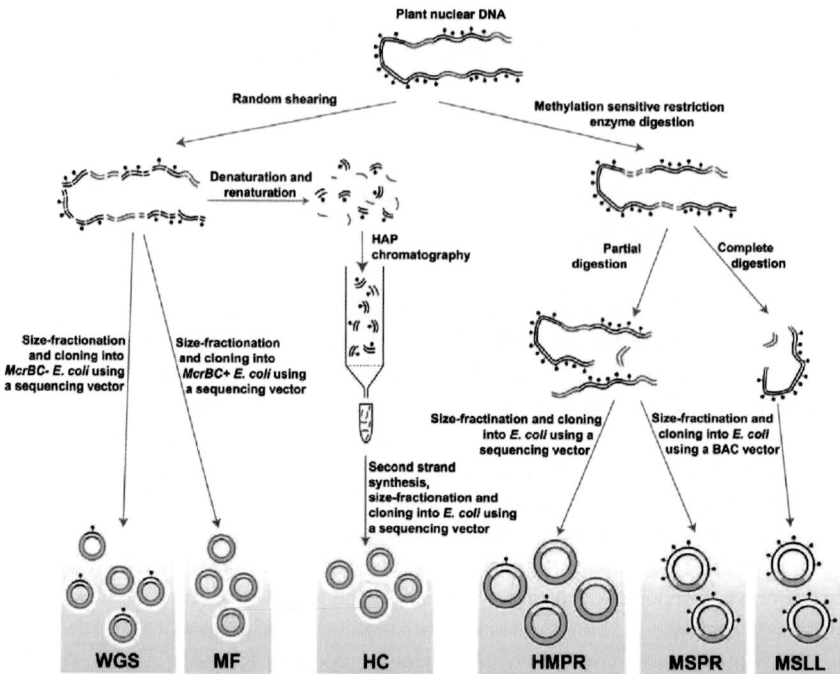

Fig. 1 Schematic representation of genomic DNA-based gene-enrichment sequencing and WGS strategies. Double lines: double-stranded genomic DNA; black dots: methyl groups; circles: plasmid clones; red: genic or low-copy regions; black: repetitive regions; blue: cloning vectors. Reproduced with permission from (Rabinowicz 2007). Copyright 2007 American Chemical Society

4.3 Methylation Filtration

Another way to take advantage of the differential methylation observed in most plant genomes is to make use of the Mcr (modified cytosine restriction) system from *E. coli* (Dila et al. 1990; Raleigh and Wilson 1986). One of the components of this system is the restricion enzyme McrBC, which cuts any DNA that enters the bacterial cell and contains two [G/A]C sites, separated by 40-3,000 bp, in which the cytosine is methylated (Sutherland et al. 1992). Because such methylated sequences are very frequent in plant DNA, McrBC can digest virtually any plant methylated DNA. Therefore, the use of an *McrBC+ E.coli* strain to construct a genomic library results in a selection of hypomethylated sequences, which are enriched in genes and other low-copy sequences. The genomic DNA must be randomly sheared and size fractionated to select fragments smaller than ~3 kbp to allow cloning DNA fragments that contain only low-copy sequences. Chloroplast DNA is mostly non-methylated (Fojtova et al. 2001; McCullough et al. 1992), and must thus be excluded by using nuclear DNA preparations as input for MF library construction.

The selective cloning of plant genomic low-copy sequences using *McrBC+ E. coli* is called methylation filtration (MF; Figure 1) and was initially tested in maize (Rabinowicz et al. 1999). In this pilot experiment, sequences from an MF library were compared to control data from a WGS library, constructed in the same way, except that an *McrBC- E.coli* strain was used as a host. The analysis of several hundred sequences from each library showed that the MF library was 6-fold enriched in gene-like sequences, relative to the WGS control. Later on, two larger MF projects were carried out in maize (Palmer et al. 2003; Whitelaw et al. 2003). For the purpose of gene discovery, a comparison of maize MF and EST sequences concluded that the number of new genes detected in EST data is higher than in MF data when less than approximately 60,000 sequences form each set are analyzed. However, when larger numbers of sequences are included, the gene discovery rate in EST data reaches a plateau much earlier than that of MF data (Palmer et al. 2003). This observation underscores a smaller sampling bias of the genic fraction of the genome in MF than in EST data. Nevertheless, because MF sequences are derived from genomic DNA, MF and EST data complement each other very well, allowing, for example, to deduce intron/exon structure of genes by spliced sequence alignment.

Some repetitive sequences are also recovered in MF libraries, but comparing the sequence composition of MF and WGS repeats revealed that MF repeats are often depleted of CpG and CpNpG motifs, which are the most frequently methylated sequences in plants (Palmer et al. 2003). Some MF repeats do contain those motifs, and they are probably recovered in MF libraries because they are unmethylated – and thus potentially active – transposable elements.

MF has been tested in several other plants and a correlation between the level of gene enrichment and genome size can be observed, particularly among the grasses (Rabinowicz et al. 2005). However, there are some exceptions that may indicate differences in DNA methylation patterns or genome structure. Interestingly, two wheat species showed lower than expected gene enrichment for their large genome size. Gene enrichment is calculated as the ratio between the number of gene-like sequences in the MF dataset versus that in the WGS dataset. In a diploid wheat species, a lower than anticipated level of enrichment is thought to represent an abundance of unmethylated repeats, while a large number of potentially methylated gene-like sequences (allegedly gene fragments or pseudogenes) found in a WGS sample of hexaploid wheat may be responsible for the low gene enrichment observed in this species. MF data from several cultivars of castor bean showed a variability of enrichment in non-methylated sequences among cultivars, suggesting that methylation patterns can be different among closely related plants (Rabinowicz et al., unpublished). Furthermore, *Arabidopsis* genome-wide methylation analyses, uncovered that gene methylation is variable among different ecotypes (Vaughn et al. 2007), while methylation analysis of a number of random maize exons showed that most of them are non-methylated (Rabinowicz et al. 2003). These pieces of evidence suggest that gene methylation in plants varies within and between species.

4.4   High Cot

It has been known for quite some time that, during renaturation of denatured DNA, repetitive sequences re-associate faster than low-copy ones. The product of the concentration of DNA by the re-association time is called C_0t or Cot. Thus, the high Cot (HC; Figure 1) DNA corresponds to low-copy DNA. Only decades later, it was proposed to use Cot analysis to separate and selectively clone HC DNA (Bennetzen et al. 2001). The first examples of the application of Cot analysis to construct gene-enriched plant genomic libraries were reported shortly after in sorghum (Peterson et al. 2002) and maize (Yuan et al. 2003). This technique consists in shearing, heat-denaturing, and slowly re-annealing genomic DNA. Then, the double-stranded (repetitive) DNA is separated from the single-stranded (low-copy) DNA by hydorxyapaptite (HAP) chromatography, due to their differential affinity to HAP.

The second strand of the low-copy DNA is synthesized *in vitro*. The fragments ends are then repaired and the low-copy DNA can be cloned in a sequencing vector. A large-scale HC sequencing project in maize has been carried out in parallel to the MF sequencing project, and the comparison of the two methods showed that MF captures slightly more gene-like sequences, but also more repetitive sequences, than HC (Whitelaw et al. 2003; Yuan et al. 2003). On the other hand, HC shows a larger number of anonymous sequences, which could be intronic and regulatory sequences, as well as novel, low-copy transposons.

4.5   Combination of Different Gene-Enrichment Techniques

The availability of comparable sequence sets of HC and MF data allowed an accurate comparison of both techniques. In order to extend the length of the low-copy sequences and to reduce redundancy, gene-enriched sequences can be assembled, although algorithms that can handle such non-uniform data must by used (Emrich et al. 2004; Lee et al. 2005; Whitelaw et al. 2003). EST clustering tools assemble non-uniform samples and two groups have independently used these tools to provide high quality assemblies. One method used a clustering tool designed to take advantage of parallel processing (Emrich et al. 2004) after identifying potential repetitive sequences based on their frequency of occurrence (Kalyanaraman et al. 2006). The assembled sequences thus produced are called "maize assembled genomic islands" (MAGI). Another method relied on the TGICL clustering utilities (Pertea et al. 2003) and TIGR Assembler (Sutton et al. 1995) after masking out known repetitive elements (Whitelaw et al. 2003).

After separately assembling the MF and HC reads, respectively, the total span of assemblies and singletons was 150 Mbp for MF and 190 Mbp for HC, while the combined assembly of HC and MF reads yielded 300 Mbp (Chan et al., unpublished). This illustrates that the sampling space of these methods only partially overlap and thus, the techniques are complementary. The extent to which these combined gene-enrichment strategies can identify genes has been tested using a set of 78 full-length cDNA sequences from GenBank. The MF and HC sequences combined could be aligned with 95% of these cDNA sequences and they covered 75% of the cDNA nucleotides with an average coverage of 2X (Springer et al. 2004).

Two studies analyzed gene models from maize sequences from complete BAC clones. One of them used finished sequences from several maize BAC clones deposited in GenBank (Barbazuk et al., unpublished). This analysis yielded comparable conclusions to those obtained when using full-length cDNA as reference sequence data. The second study used gene models from two physical contigs of BAC clones whose draft sequence was thoroughly annotated for genes (Bruggmann et al. 2006). Because of the substantial length of the sequence under analysis, it was possible to identify large genes with very large introns that contain repeats. Because MF and HC sequences largely exclude repeats, they did not cover such large genes completely, and the nucleotide coverage resulted lower than in the full-length cDNA study (65%). It is worth noting that this number may be an underestimate if sequences annotated as genes in the analyzed BAC contigs were actually transposable element genes or acquired gene fragments mis-identified as true organismal genes, as it has often been observed in genome annotation projects (Bennetzen et al. 2004).

Limitations of gene-enrichment techniques include a lack of anchoring of the sequence data to the genome, and the absence of relative order and orientation of the sequence contigs, once they are assembled. While these limitations are shared by EST data, MF and HC sequences provide information on 5′ and 3′ flanking sequences of coding regions that cannot be obtained from ESTs. By looking at the gene models predicted in all assembled maize gene-enriched reads over 3 kbp, the average length of flanking sequences were 1.5 and 1.7 kbp, for 5′ and 3′ regions, respectively (Chan et al., unpublished).

It has been proposed that the gene-enrichment methods described so far, combined with traditional genome sequencing techniques, including WGS and BAC-based sequencing, can deliver a high quality draft sequence of a large plant genome such as that of maize, with higher sequence quality in the low-copy regions than in intergenic repeats (Rabinowicz and Bennetzen 2006). The on-going effort to sequence the maize genome will certainly benefit from the availability of gene-enrichment data.

4.6   Transposon Insertion Site Sequencing

Class II (or DNA) transposable elements have been extensively used in maize for transposon-tagging mutagenesis because of their preference to insert in low-copy sequences (Hanley et al. 2000). Therefore, isolating and sequencing the regions flanking the transposon insertions can be used as a gene-enrichment sequencing technique. Several transposon mutagenesis systems have been developed for maize (Bai et al. 2007; May et al. 2003; McCarty et al. 2005) using either *Mutator* or *Ac/ Ds* transposable elements, and a DNA sequence-based index of *Mutator* transposon insertions is under construction (http://www.nsf.gov/awardsearch/showAward.

do?AwardNumber=0703273). Using techniques to amplify flanking sequences, thereby taking advantage of the conserved end-sequence of the transposon (Hanley et al. 2000), a collection of insertion site sequences, mostly enriched in low-copy sequences, can be recovered. A *Mutator*-based system to facilitate sequencing of these insertion site sequences has been developed. It is based on an engineered transposon called *RescueMu* that contains a linearized cloning vector that can be recovered from the plant genome (Raizada et al. 2001). The *RescueMu* element is introduced in the plant genome by transformation and, after transposition, plasmids containing insertion-site sequences can be isolated in the cloning vector. Although this system identifies gene sequences at high frequency, there is a substantial degree of sequence redundancy, probably due to the recovery of insertion sequences that were already present in the parental lines used during the induction of the mutagenesis. Furthermore, it has been shown that transposon insertion is not random, and uneven representation of genes is expected (May *et al.* 2003; Raizada *et al.* 2001).

5 Far-reaching Applications of Gene-Enriched Sequences and Future Prospects

The maize gene-enrichment project data contributed to the development of key genomic resources well in advance of a whole genome sequence. These include a maize gene microarray, in which several thousand gene probes that were not present among the EST data available at the time were derived from the maize gene-enriched sequences (Gardiner *et al.* 2005), and a dense genetic map (Fu et al. 2006) containing single nucleotide and insertion/deletion polymorphisms derived from gene-enriched and other genomic survey sequences in GenBank. In addition, The comprehensive maize gene-enriched sequence collection allowed the estimation of the amount of mitochondrial sequences present in the maize nuclear genome (Clifton et al. 2004) and global single nucleotide polymorphism discovery (Barbazuk et al. 2007). This sequence collection has not only enabled genome-scale investigations, but it has also aided "gene by gene" research as well. For example, many gene family studies and map-based cloning projects have been facilitated by the availability of gene-rich sequences (Bortiri et al. 2006; Colasanti et al. 2006; Jin et al. 2007; Springer and Kaeppler 2005).

The next-generation sequencing technologies can deliver extremely high throughput and depth of coverage at a fraction of the cost of traditional sequencing (Bentley 2006; Margulies et al. 2005). Sequence read lengths are typically shorter than traditional Sanger sequencing, but some platforms are increasing read lengths and re-sequencing complete genomes or selected regions can be very useful when the data are aligned to a reference genome. Improved assembly algorithms for short reads are also being developed (Jeck et al. 2007; Warren et al. 2007). These methods are not based on plasmid cloning in *E. coli*. Rather, they clonally amplify sheared DNA, which is then sequenced by pyrosequencing, dideoxyterminators, or labeled-oligonucleotide sequential ligation. Therefore, any DNA can be used as input, such as cDNA derived from gene transcipts (Bainbridge et al. 2006; Barbazuk et al. 2007; Emrich *et al.* 2007) or small RNAs (Berezikov et al. 2006; Henderson et al. 2006), transcription factor binding site sequences (Johnson et al. 2007), DNA from immunoprecipitated chromatin (Barski et al. 2007; Mikkelsen et al. 2007), etc. Therefore, MF and HC techniques can be adapted for sequencing with these new technologies. For MF, genomic DNA could be digested in vitro with purified McrBC enzyme and the undigested fraction eluted from agarose gels. In the same way, HC DNA can be used as input for ultra high throughput sequencing. Hence, sequencing the gene-space of additional maize inbred lines could be afforded by combining gene-enrichment with the new sequencing technologies, using the B73 genome sequence as a reference.

Although the large size of plant genomes is a major obstacle for sequencing, the experience gained from the application of gene-enrichment techniques to maize will allow the discovery of genes and other functional elements, as well as the extensive genome characterization of new plant varieties and species with a relatively modest investment.

References

Adams, M.D., Kelley, J.M., Gocayne, J.D., Dubnick, M., Polymeropoulos, M.H., Xiao, H., Merril, C.R., Wu, A., Olde, B., Moreno, R.F. et al. (1991) Complementary DNA sequencing: expressed sequence tags and human genome project. *Science* **252:** 1651–1656.

Bai, L., Singh, M., Pitt, L., Sweeney, M., and Brutnell, T.P. (2007) Generating novel allelic variation through activator insertional mutagenesis in maize. *Genetics* **175:** 981–992.

Bailey, J.A., Gu, Z., Clark, R.A., Reinert, K., Samonte, R.V., Schwartz, S., Adams, M.D., Myers, E.W., Li, P.W., and Eichler, E.E. (2002) Recent segmental duplications in the human genome. *Science* **297:** 1003–1007.

Bainbridge, M.N., Warren, R.L., Hirst, M., Romanuik, T., Zeng, T., Go, A., Delaney, A., Griffith, M., Hickenbotham, M., Magrini, V. et al. (2006) Analysis of the prostate cancer cell line LNCaP transcriptome using a sequencing-by-synthesis approach. *BMC Genomics* **7:** 246.

Barbazuk, W.B., Bedell, J.A., and Rabinowicz, P.D. (2005) Reduced representation sequencing: a success in maize and a promise for other plant genomes. *Bioessays* **27:** 839–848.

Barbazuk, W.B., Emrich, S.J., Chen, H.D., Li, L., and Schnable, P.S. (2007) SNP discovery via 454 transcriptome sequencing. *Plant J* **51:** 910–918.

Barski, A., Cuddapah, S., Cui, K., Roh, T.Y., Schones, D.E., Wang, Z., Wei, G., Chepelev, I., and Zhao, K. (2007) High-resolution profiling of histone methylations in the human genome. *Cell* **129:** 823–837.

Batzoglou, S., Jaffe, D.B., Stanley, K., Butler, J., Gnerre, S., Mauceli, E., Berger, B., Mesirov, J.P., and Lander, E.S. (2002) ARACHNE: a whole-genome shotgun assembler. *Genome Res.* **12:** 177–189.

Bennetzen, J.L., Chandler, V.L., and Schnable, P. (2001) National Science Foundation-sponsored workshop report. Maize genome sequencing project. *Plant Physiol.* **127:** 1572–1578.

Bennetzen, J. L., C. Coleman, J. Ma, R. Liu and W. Ramakrishna (2004) Consistent over-estimation of gene number in complex plant genomes. *Curr. Opin. Plant Biol.* **7:** 732–736.

Bentley, D.R. (2006) Whole-genome re-sequencing. *Curr Opin Genet Dev* **16:** 545–552.

Berezikov, E., Thuemmler, F., van Laake, L.W., Kondova, I., Bontrop, R., Cuppen, E., and Plasterk, R.H. (2006) Diversity of microRNAs in human and chimpanzee brain. *Nat Genet* **38:** 1375–1377.

Bonaldo, M.F., Lennon, G., and Soares, M.B. (1996) Normalization and subtraction: two approaches to facilitate gene discovery. *Genome Res.* **6:** 791–806.

Bortiri, E., Jackson, D., and Hake, S. (2006) Advances in maize genomics: the emergence of positional cloning. *Curr Opin Plant Biol* **9:** 164–171.

Bruggmann, R., Bharti, A.K., Gundlach, H., Lai, J., Young, S., Pontaroli, A.C., Wei, F., Haberer, G., Fuks, G., Du, C. et al. (2006) Uneven chromosome contraction and expansion in the maize genome. *Genome Res.* **16:** 1241–1251.

Burr, B., Burr, F.A., Thompson, K.H., Albertson, M.C., and Stuber, C.W. (1988) Gene mapping with recombinant inbreds in maize. *Genetics* **118:** 519–526.

Childs, K.L., Hamilton, J.P., Zhu, W., Ly, E., Cheung, F., Wu, H., Rabinowicz, P.D., Town, C.D., Buell, C.R., and Chan, A.P. (2006) The TIGR Plant Transcript Assemblies database. *Nucleic Acids Res.*

Clifton, S.W., Minx, P., Fauron, C.M., Gibson, M., Allen, J.O., Sun, H., Thompson, M., Barbazuk, W.B., Kanuganti, S., Tayloe, C. et al. (2004) Sequence and comparative analysis of the maize NB mitochondrial genome. *Plant Physiol* **136:** 3486–3503.

Colasanti, J., Tremblay, R., Wong, A.Y., Coneva, V., Kozaki, A., and Mable, B.K. (2006) The maize INDETERMINATE1 flowering time regulator defines a highly conserved zinc finger protein family in higher plants. *BMC Genomics* **7:** 158.

Dean, F.B., Hosono, S., Fang, L., Wu, X., Faruqi, A.F., Bray-Ward, P., Sun, Z., Zong, Q., Du, Y., Du, J. et al. (2002) Comprehensive human genome amplification using multiple displacement amplification. *Proc Natl Acad Sci U S A* **99:** 5261–5266.

Dila, D., Sutherland, E., Moran, L., Slatko, B., and Raleigh, E.A. (1990) Genetic and sequence organization of the mcrBC locus of Escherichia coli K-12. *J. Bacteriol.* **172:** 4888–4900.

Dong, Q., Lawrence, C.J., Schlueter, S.D., Wilkerson, M.D., Kurtz, S., Lushbough, C., and Brendel, V. (2005) Comparative plant genomics resources at PlantGDB. *Plant Physiol.* **139:** 610–618.

Emberton, J., Ma, J., Yuan, Y., SanMiguel, P., and Bennetzen, J.L. (2005) Gene enrichment in maize with hypomethylated partial restriction (HMPR) libraries. *Genome Res.* **15:** 1441–1446.

Emrich, S.J., Aluru, S., Fu, Y., Wen, T.J., Narayanan, M., Guo, L., Ashlock, D.A., and Schnable, P.S. (2004) A strategy for assembling the maize (Zea mays L.) genome. *Bioinformatics* **20:** 140–147.

Emrich, S.J., Barbazuk, W.B., Li, L., and Schnable, P.S. (2007) Gene discovery and annotation using LCM-454 transcriptome sequencing. *Genome Res.* **17:** 69–73.

Fleischmann, R.D., Adams, M.D., White, O., Clayton, R.A., Kirkness, E.F., Kerlavage, A.R., Bult, C.J., Tomb, J.F., Dougherty, B.A., Merrick, J.M. et al. (1995) Whole-genome random sequencing and assembly of Haemophilus influenzae Rd. *Science* **269:** 496–512.

Fojtova, M., Kovarik, A., and Matyasek, R. (2001) Cytosine methylation of plastid genome in higher plants. Fact or artefact? *Plant Sci* **160:** 585–593.

Fu, Y., Wen, T.J., Ronin, Y.I., Chen, H.D., Guo, L., Mester, D.I., Yang, Y., Lee, M., Korol, A.B., Ashlock, D.A. et al. (2006) Genetic dissection of intermated recombinant inbred lines using a new genetic map of maize. *Genetics* **174:** 1671–1683.

Gardiner, J.M., Buell, C.R., Elumalai, R., Galbraith, D.W., Henderson, D.A., Iniguez, A.L., Kaeppler, S.M., Kim, J.J., Liu, J., Smith, A. et al. (2005) Design, production, and utilization of long oligonucleotide microarrays for expression analysis in maize. *Maydica* **50:** 425–435.

Gaut, B.S. and Doebley, J.F. (1997) DNA sequence evidence for the segmental allotetraploid origin of maize. *Proc Natl Acad Sci U S A* **94:** 6809–6814.

Gibbs, R.A. Weinstock, G.M. Metzker, M.L. Muzny, D.M. Sodergren, E.J. Scherer, S. Scott, G. Steffen, D. Worley, K.C. Burch, P.E. et al. (2004) Genome sequence of the Brown Norway rat yields insights into mammalian evolution. *Nature* **428:** 493–521.

Goffeau, A., Barrell, B.G., Bussey, H., Davis, R.W., Dujon, B., Feldmann, H., Galibert, F., Hoheisel, J.D., Jacq, C., Johnston, M. et al. (1996) Life with 6000 genes. *Science* **274:** 546–567.

Haberer, G., Young, S., Bharti, A.K., Gundlach, H., Raymond, C., Fuks, G., Butler, E., Wing, R.A., Rounsley, S., Birren, B. et al. (2005) Structure and architecture of the maize genome. *Plant Physiol* **139:** 1612–1624.

Hake, S. and Walbot, V. (1980) The genome of Zea mays, its organization and homology to related grasses. *Chromosoma* **79:** 251–270.

Hanley, S., Edwards, D., Stevenson, D., Haines, S., Hegarty, M., Schuch, W., and Edwards, K.J. (2000) Identification of transposon-tagged genes by the random sequencing of Mutator-tagged DNA fragments from Zea mays. *Plant J.* **23:** 557–566.

Henderson, I.R., Zhang, X., Lu, C., Johnson, L., Meyers, B.C., Green, P.J., and Jacobsen, S.E. (2006) Dissecting Arabidopsis thaliana DICER function in small RNA processing, gene silencing and DNA methylation patterning. *Nat Genet* **38**: 721–725.

Huang, X. and Madan, A. (1999) CAP3: A DNA sequence assembly program. *Genome Res.* **9**: 868–877.

Ilic, K., SanMiguel, P.J., and Bennetzen, J.L. (2003) A complex history of rearrangement in an orthologous region of the maize, sorghum, and rice genomes. *Proc. Natl. Acad. Sci. U. S. A.* **100**: 12265–12270.

Jaffe, D.B., Butler, J., Gnerre, S., Mauceli, E., Lindblad-Toh, K., Mesirov, J.P., Zody, M.C., and Lander, E.S. (2003) Whole-genome sequence assembly for mammalian genomes: Arachne 2. *Genome Res* **13**: 91–96.

Jeck, W.R., Reinhardt, J.A., Baltrus, D.A., Hickenbotham, M.T., Magrini, V., Mardis, E.R., Dangl, J.L., and Jones, C.D. (2007) Extending assembly of short DNA sequences to handle error. *Bioinformatics*.

Jin, Y., Wang, M., Fu, J., Xuan, N., Zhu, Y., Lian, Y., Jia, Z., Zheng, J., and Wang, G. (2007) Phylogenetic and expression analysis of ZnF-AN1 genes in plants. *Genomics* **90**: 265-275.

Johnson, D.S., Mortazavi, A., Myers, R.M., and Wold, B. (2007) Genome-wide mapping of in vivo protein-DNA interactions. *Science* **316**: 1497–1502.

Kalyanaraman, A., Aluru, S., and Schnable, P.S. (2006) Turning repeats to advantage: scaffolding genomic contigs using LTR retrotransposons. *Comput Syst Bioinformatics Conf*: 167–178.

Lai, J., Ma, J., Swigonova, Z., Ramakrishna, W., Linton, E., Llaca, V., Tanyolac, B., Park, Y.J., Jeong, O.Y., Bennetzen, J.L. et al. (2004) Gene loss and movement in the maize genome. *Genome Res.* **14**: 1924–1931.

Lander, E.S. and Weinberg, R.A. (2000) Genomics: journey to the center of biology. *Science* **287**: 1777–1782.

Langham, R.J., Walsh, J., Dunn, M., Ko, C., Goff, S.A., and Freeling, M. (2004) Genomic duplication, fractionation and the origin of regulatory novelty. *Genetics* **166**: 935–945.

Lee, Y., Tsai, J., Sunkara, S., Karamycheva, S., Pertea, G., Sultana, R., Antonescu, V., Chan, A., Cheung, F., and Quackenbush, J. (2005) The TIGR Gene Indices: clustering and assembling EST and known genes and integration with eukaryotic genomes. *Nucleic Acids Res.* **33**: D71–74.

Lippman, Z., Gendrel, A.V., Black, M., Vaughn, M.W., Dedhia, N., McCombie, W.R., Lavine, K., Mittal, V., May, B., Kasschau, K.D. et al. (2004) Role of transposable elements in heterochromatin and epigenetic control. *Nature* **430**: 471–476.

Luo, M.C., Thomas, C., You, F.M., Hsiao, J., Ouyang, S., Buell, C.R., Malandro, M., McGuire, P.E., Anderson, O.D., and Dvorak, J. (2003) High-throughput fingerprinting of bacterial artificial chromosomes using the snapshot labeling kit and sizing of restriction fragments by capillary electrophoresis. *Genomics* **82**: 378–389.

Margulies, M., Egholm, M., Altman, W.E., Attiya, S., Bader, J.S., Bemben, L.A., Berka, J., Braverman, M.S., Chen, Y.J., Chen, Z. et al. (2005) Genome sequencing in microfabricated high-density picolitre reactors. *Nature* **437**: 376–380.

Marra, M., Kucaba, T., Sekhon, M., Hillier, L., Martienssen, R., Chinwalla, A., Crockett, J., Fedele, J., Grover, H., Gund, C. et al. (1999) zA map for sequence analysis of the Arabidopsis thaliana genome. *Nat. Genet.* **22**: 265–270.

May, B.P., Liu, H., Vollbrecht, E., Senior, L., Rabinowicz, P.D., Roh, D., Pan, X., Stein, L., Freeling, M., Alexander, D. et al. (2003) Maize-targeted mutagenesis: A knockout resource for maize. *Proc. Natl. Acad. Sci. U. S. A.* **100**: 11541–11546.

McCarty, D.R., Settles, A.M., Suzuki, M., Tan, B.C., Latshaw, S., Porch, T., Robin, K., Baier, J., Avigne, W., Lai, J. et al. (2005) Steady-state transposon mutagenesis in inbred maize. *Plant J.* **44**: 52–61.

McCullough, A.J., Kangasjarvi, J., Gengenbach, B.G., and Jones, R.J. (1992) Plastid DNA in Developing Maize Endosperm : Genome Structure, Methylation, and Transcript Accumulation Patterns. *Plant Physiol* **100**: 958–964.

Meyers, B.C., Tingey, S.V., and Morgante, M. (2001) Abundance, distribution, and transcriptional activity of repetitive elements in the maize genome. *Genome Res.* **11**: 1660–1676.

Mikkelsen, T.S., Ku, M., Jaffe, D.B., Issac, B., Lieberman, E., Giannoukos, G., Alvarez, P., Brockman, W., Kim, T.K., Koche, R.P. et al. (2007) Genome-wide maps of chromatin state in pluripotent and lineage-committed cells. *Nature* **448:** 553–560.

Miura, A., Yonebayashi, S., Watanabe, K., Toyama, T., Shimada, H., and Kakutani, T. (2001) Mobilization of transposons by a mutation abolishing full DNA methylation in Arabidopsis. *Nature* **411:** 212–214.

Mouse Genome Sequencing Consortium. (2002) Initial sequencing and comparative analysis of the mouse genome. *Nature* **420:** 520–562.

Myers, E.W., Sutton, G.G., Delcher, A.L., Dew, I.M., Fasulo, D.P., Flanigan, M.J., Kravitz, S.A., Mobarry, C.M., Reinert, K.H., Remington, K.A. et al. (2000) A whole-genome assembly of Drosophila. *Science* **287:** 2196–2204.

Palmer, L.E., Rabinowicz, P.D., O'Shaughnessy, A.L., Balija, V.S., Nascimento, L.U., Dike, S., de la Bastide, M., Martienssen, R.A., and McCombie, W.R. (2003) Maize genome sequencing by methylation filtration. *Science* **302:** 2115–2117.

Patanjali, S.R., Parimoo, S., and Weissman, S.M. (1991) Construction of a uniform-abundance (normalized) cDNA library. *Proc. Natl. Acad. Sci. U. S. A.* **88:** 1943–1947.

Pertea, G., Huang, X., Liang, F., Antonescu, V., Sultana, R., Karamycheva, S., Lee, Y., White, J., Cheung, F., Parvizi, B. et al. (2003) TIGR Gene Indices clustering tools (TGICL): a software system for fast clustering of large EST datasets. *Bioinformatics* **19:** 651–652.

Peterson, D.G., Schulze, S.R., Sciara, E.B., Lee, S.A., Bowers, J.E., Nagel, A., Jiang, N., Tibbitts, D.C., Wessler, S.R., and Paterson, A.H. (2002) Integration of Cot analysis, DNA cloning, and high-throughput sequencing facilitates genome characterization and gene discovery. *Genome Res.* **12:** 795–807.

Rabinowicz, P.D. (2007) Plant genomic sequencing using gene-enriched libraries. *Chem Rev* **107:** 3377–3390.

Rabinowicz, P.D. and Bennetzen, J.L. (2006) The maize genome as a model for efficient sequence analysis of large plant genomes. *Curr. Opin. Plant Biol.* **9:** 149–156.

Rabinowicz, P.D., Citek, R., Budiman, M.A., Nunberg, A., Bedell, J.A., Lakey, N., O'Shaughnessy, A.L., Nascimento, L.U., McCombie, W.R., and Martienssen, R.A. (2005) Differential methylation of genes and repeats in land plants. *Genome Res.* **15:** 1431–1440.

Rabinowicz, P.D., Palmer, L.E., May, B.P., Hemann, M.T., Lowe, S.W., McCombie, W.R., and Martienssen, R.A. (2003) Genes and transposons are differentially methylated in plants, but not in mammals. *Genome Res.* **13:** 2658–2664.

Rabinowicz, P.D., Schutz, K., Dedhia, N., Yordan, C., Parnell, L.D., Stein, L., McCombie, W.R., and Martienssen, R.A. (1999) Differential methylation of genes and retrotransposons facilitates shotgun sequencing of the maize genome. *Nat. Genet.* **23:** 305–308.

Raizada, M.N., Nan, G.L., and Walbot, V. (2001) Somatic and germinal mobility of the RescueMu transposon in transgenic maize. *Plant Cell* **13:** 1587–1608.

Raleigh, E.A. and Wilson, G. (1986) Escherichia coli K-12 restricts DNA containing 5-methylcytosine. *Proc. Natl. Acad. Sci. U. S. A.* **83:** 9070–9074.

Rayburn, A.L., Biradar, D.P., Bullock, D.G., and McMurphy, L.M. (1993) Nuclear DNA content in F1 hybrids of maize. *Heredity* **70:** 294–300.

SanMiguel, P., and J. L. Bennetzen (1998) Evidence that a recent increase in maize genome size was caused by the massive amplification of intergene retrotransposons. *Annals Bot.* **82:** 37–44.

SanMiguel, P., Tikhonov, A., Jin, Y.K., Motchoulskaia, N., Zakharov, D., Melake-Berhan, A., Springer, P.S., Edwards, K.J., Lee, M., Avramova, Z. et al. (1996) Nested retrotransposons in the intergenic regions of the maize genome. *Science* **274:** 765–768.

Sasaki, T. and Burr, B. (2000) International Rice Genome Sequencing Project: the effort to completely sequence the rice genome. *Curr. Opin. Plant Biol.* **3:** 138-141.

Singer, T., Yordan, C., and Martienssen, R.A. (2001) Robertson's Mutator transposons in A. thaliana are regulated by the chromatin-remodeling gene Decrease in DNA Methylation (DDM1). *Genes Dev.* **15:** 591–602.

Soares, M.B., Bonaldo, M.F., Jelene, P., Su, L., Lawton, L., and Efstratiadis, A. (1994) Construction and characterization of a normalized cDNA library. *Proc. Natl. Acad. Sci. U. S. A.* **91:** 9228–9232.

Soderlund, C., Humphray, S., Dunham, A., and French, L. (2000) Contigs built with fingerprints, markers, and FPC V4.7. *Genome Res.* **10:** 1772–1787.

Soderlund, C., Longden, I., and Mott, R. (1997) FPC: a system for building contigs from restriction fingerprinted clones. *Comput. Appl. Biosci.* **13:** 523–535.

Springer, N.M. and Kaeppler, S.M. (2005) Evolutionary divergence of monocot and dicot methyl-CpG-binding domain proteins. *Plant Physiol* **138:** 92–104.

Springer, N.M., Xu, X., and Barbazuk, W.B. (2004) Utility of different gene enrichment approaches toward identifying and sequencing the maize gene space. *Plant Physiol.* **136:** 3023–3033.

Sutherland, E., Coe, L., and Raleigh, E.A. (1992) McrBC: a multisubunit GTP-dependent restriction endonuclease. *J. Mol. Biol.* **225:** 327–348.

Sutton, G., White, O., Adams, M., and Kerlavage, A.R. (1995) TIGR Assembler: a new tool for assembling large shotgun sequencing projects. *Genome Sci. Tech.* **1:** 9–19.

Swigonova, Z., Lai, J., Ma, J., Ramakrishna, W., Llaca, V., Bennetzen, J.L., and Messing, J. (2004) Close split of sorghum and maize genome progenitors. *Genome Res* **14:** 1916–1923.

Tang, H. (2007) Genome assembly, rearrangement, and repeats. *Chem Rev* **107:** 3391–3406.

The Arabidopsis Genome Initiative. (2000) Analysis of the genome sequence of the flowering plant Arabidopsis thaliana. *Nature* **408:** 796–815.

The C. elegans Sequencing Consortium. (1998) Genome sequence of the nematode C. elegans: a platform for investigating biology. *Science* **282:** 2012–2018.

Vaughn, M.W., Tanurd Ic, M., Lippman, Z., Jiang, H., Carrasquillo, R., Rabinowicz, P.D., Dedhia, N., McCombie, W.R., Agier, N., Bulski, A. et al. (2007) Epigenetic Natural Variation in Arabidopsis thaliana. *PLoS Biol* **5:** e174.

Warren, R.L., Sutton, G.G., Jones, S.J., and Holt, R.A. (2007) Assembling millions of short DNA sequences using SSAKE. *Bioinformatics* **23:** 500–501.

Wheeler, D.L., Barrett, T., Benson, D.A., Bryant, S.H., Canese, K., Chetvernin, V., Church, D.M., Dicuccio, M., Edgar, R., Federhen, S. et al. (2006) Database resources of the National Center for Biotechnology Information. *Nucleic Acids Res.*

Whitelaw, C.A., Barbazuk, W.B., Pertea, G., Chan, A.P., Cheung, F., Lee, Y., Zheng, L., van Heeringen, S., Karamycheva, S., Bennetzen, J.L. et al. (2003) Enrichment of gene-coding sequences in maize by genome filtration. *Science* **302:** 2118–2120.

Yu, C. and Li, Z. (2006) Construction of methylation-sensitive partial restriction bacterial artificial chromosome libraries in maize. *Anal. Biochem.* **359:** 141–143.

Yuan, Y., SanMiguel, P.J., and Bennetzen, J.L. (2002) Methylation-Spanning Linker Libraries Link Gene-Rich Regions and Identify Epigenetic Boundaries in Zea mays. *Genome Res.* **12:** 1345–1349.

Yuan, Y., SanMiguel, P.J., and Bennetzen, J.L. (2003) High-Cot sequence analysis of the maize genome. *Plant J.* **34:** 249–255.

Zhang, X., Yazaki, J., Sundaresan, A., Cokus, S., Chan, S.W., Chen, H., Henderson, I.R., Shinn, P., Pellegrini, M., Jacobsen, S.E. et al. (2006) Genome-wide high-resolution mapping and functional analysis of DNA methylation in arabidopsis. *Cell* **126:** 1189–1201.

Zilberman, D., Gehring, M., Tran, R.K., Ballinger, T., and Henikoff, S. (2007) Genome-wide analysis of Arabidopsis thaliana DNA methylation uncovers an interdependence between methylation and transcription. *Nat Genet* **39:** 61–69.

Part V
Genes and Gene Families

Maize Transcription Factors

Erich Grotewold and John Gray

Abstract The availability of the complete sequence of the maize genome is permitting rapid advancement in our understanding of transcription factors and the mechanisms by which they control gene expression. The emerging challenge is to provide information on regulatory proteins, their target genes and the regulatory motifs in which they participate in a way that allows the integration with similar resources being generated in other plants, including other grasses. The most important aspects of maize transcription factors, recommendations for annotation and tools to investigate their function are discussed here.

1 Introduction

Control of gene expression is central to all cellular processes. In eukaryotes, regulation of transcription provides one of the most frequent mechanisms by which gene expression is controlled. This is often performed by the tethering of a particular type of proteins, the transcription factors (TFs), to discrete 3-8 base pairs-long DNA-sequence elements. These elements are distributed across the promoter or other regulatory regions of the gene. These elements, which function in *cis* with respect to the gene that they regulate and are hence known as *cis*-regulatory elements, are often arranged into regulatory modules, each module responsible for a fraction of the overall regulatory output of the gene (Davidson 2001). TFs are organized into hierarchical gene regulatory networks in which one regulatory protein, often in cooperation with others, positively or negatively regulates the expression of another TF. This establishes a variety of regulatory motifs, which, when assembled into

E. Grotewold
The Ohio State University, Department of Plant Cellular & Molecular Biology,
grotewold.1@osu.edu

J. Gray
University of Toledo, Department of Biology
jgray5@UTNet.UToledo.Edu

J.L. Bennetzen and S. Hake (eds.), *Maize Handbook - Volume II: Genetics and Genomics,*   693
© Springer Science+Business Media LLC 2009

regulatory modules, provide the free-scale architecture that characterizes gene regulatory networks (Babu, Luscombe, Aravind, Gerstein and Teichmann 2004; Yu and Gerstein 2006). The advent of genome information for a number of plant species and the development of high-throughput genomic and network visualization tools is permitting assembly of the first pieces in the complex puzzle underlying plant gene regulatory networks. The picture that is likely to emerge, based on findings in *Saccharomyces cerevisiae* and *Escherichia coli* (Yu et al. 2006), is that plant regulatory networks have a pyramid-shaped structure with discrete hierarchical levels. Master TFs are situated at the top of the pyramid and they directly control a few more centrally located TFs. At the bottom of the pyramid are the regulatory proteins that control structural proteins and enzymes responsible for carrying out most cellular processes. The TFs located at the base of the pyramid, however, do not control the expression of any other TFs.

2   Transcription Factors

A significant fraction of a plant genome is devoted to the control of gene expression. TFs, defined for the purpose of this study as proteins containing a domain that can bind DNA in a sequence-specific fashion, comprise 7% or more of all the plant genes (J. L. Riechmann, Heard, Martin, Reuber, Jiang, Keddie, Adam, Pineda, Ratcliffe, Samaha, Creelman, Pilgrim, Broun, Zhang, Ghandehari, Sherman and Yu 2000b; J.L. Riechmann and Ratcliffe 2000a). Indeed, about 1,770 TFs have so far been identified in Arabidopsis, according to AGRIS (Palaniswamy, James, Sun, Lamb, Davuluri and Grotewold 2006). The Rice Transcription Factor Database (RiceTFDB) includes 2,031 proteins involved in transcriptional regulation, many of them corresponding to TFs, as defined above (Riano-Pachon, Ruzicic, Dreyer and Mueller-Roeber 2007). Given that maize has undergone entire genome duplication in the recent past (Gaut and Doebley 1997), it is possible that the maize genome encodes 4,000 TFs or more.

2.1   TFs are Organized Into Families

TFs are classified into families, based on the presence of conserved DNA-recognition domains. Different authors utilize slightly different classifications, thus it is difficult to compare from one study to another the exact number of families. We adopted the family organization that we utilized for Arabidopsis TFs (Davuluri, Sun, Palaniswamy, Matthews, Molina, Kurtz and Grotewold 2003), which corresponds to an expansion of the classification of Riechmann et al. (J. L. Riechmann et al. 2000b). According to this, plant TFs can be classified into 50-60 discrete families (Table I). While the relative number of members in each family might be different between monocots and dicots, as specific TF families are likely to have undergone more recent amplifications than others, for example the R2R3-MYB TFs in grasses (Dias, Braun, McMullen and Grotewold 2003), it is unlikely that TF families will be identified that are restricted to either monocots or dicots (Shiu, Shih and Li 2005). Below, we provide a brief survey of some of the larger TF families that occur in plants and some that have been characterized in maize. We have omitted from Table I those TF families from rice that are not yet well characterized or correspond to subfamilies of those already listed. The omitted rice families (name and number of members) are AUX/IAA (29), BES1 (6), DRP (6) DRT (6), HMG (9), LIM (6), LUG (6), MRF (2), PRF2-Like (2), PLATZ (12), Pesudo ARR-B (5), RWP-RK (12), S1Fa-like (2), SET (31), Sigma70-like (6), SNF2 (36), SRS (5), TAZ (5), ULT (2), and ZIM (16).

2.1.1   Heat Shock (HSF) Family

A wide range of organisms respond to elevated temperatures through the synthesis of heat shock proteins. The expression of these genes is regulated by an ancient group of TFs conserved between animals and plants, the HSFs. While yeast has only one HSF, animals typically have many and plants contain about 20 of these genes. Members of this HSF family contain a DNA-binding component near the N-terminus of the first nuclear localization region, called the HSF domain (PFAM PF00447).

The cDNA or genomic sequences for at least 22 maize HSFs were identified though mining of the PlantGDB database (Fu, Rogowsky, Nover and Scanlon 2006). These were named by homology to their rice homologs, and 16 of these had conserved gene structure (intron locations) across species. The remaining family members have as yet an incomplete genomic sequence. Plant HSFs contain HR-A and HR-B coiled-coil domains that interact with cytoplasmic regulators such as Heat Shock Binding Protein 2 (HSBP2). Variation in flanking non-coiled regions allows for conserved proteins to evolve specificity in this interaction. Based on the distance and conservation between these domains, plant HSFs are divided into three classes, HSF A, B and C. In maize there are 12, 7, and 3 HSF A, B and C genes, respectively (Fu et al. 2006).

2.1.2   MYB Family

MYB factors represent a heterogeneous group of proteins that is ubiquitous in eukaryotes, most notably in plants, and which contain 1-4+ MYB repeats. Accordingly, MYB proteins are usually classified according to the number of repeats that conform to MYB domains. Most MYB proteins from vertebrates consist of three imperfect repeats (R1, R2, and R3), and 3R-MYB proteins are also found in the plants (Braun and Grotewold 1999), where they form a small gene family involved in cell cycle progression (Ito 2005; Ito, Araki, Matsunaga, Itoh, Nishihama, Machida, Doonan and Watanabe 2001). But the large majority of plant MYB proteins correspond to the R2R3-MYB family, characterized by the presence of two MYB repeats, R2 and R3. The R2R3-MYB family is large, with ~130 members in *Arabidopsis* (Stracke, Werber and Weisshaar 2001) and at least twice that number in maize (Rabinowicz, Braun, Wolfe, Bowen and Grotewold 1999). R2R3-MYB genes were proposed to have derived from an ancestral 3R-MYB precursor by the loss of R1 (Dias et al. 2003). The amplification of the R2R3-MYB family occurred 450-200 million years ago (MYA), likely after plants invaded the land (Rabinowicz et al. 1999). In addition, in the grasses, specific sub-groups of *R2R3-MYB* genes appear to be still undergoing amplification (Rabinowicz et al. 1999), and this has been linked to the diversification of plant metabolic pathways (Grotewold 2005). A few plant MYB factors, such as the *Arabidopsis* CAPRICE (CPC) (Wada, Tachibana, Shimura and Okada 1997) and TRIPTYCHON (TRY) proteins (Hulskamp, Misera and Jurgens 1994), contain only one MYB repeat, most likely derived from R2R3-MYB proteins by loss of R2 (Tominaga, Iwata, Okada and Wada 2007). The origin of other single-MYB repeat proteins, such as the *Arabidopsis* CCA1 and LHY1 proteins involved in circadian regulation (Alabadi, Oyama, Yonovsky, Harmon, Mas and Kay 2001; Z.-Y. Wang, Kenigsbuch, Sun, Harel, Ong and Tobin 1997), is less clear.

2.1.3   MADS Family

The MADS domain (MCM1, AGAMOUS, DEFICIENS, and SRF [Serum Response Factor]) is a conserved DNA-binding/dimerization region present in a variety of TFs from different kingdoms. MADS box genes represent a large multigene family in vascular plants (e.g., at least 64 loci in rice). In angiosperms, many of the genes of the MADS family are involved in different steps of flower development, most notably in the determination of floral meristem and organ identity (e.g., AGAMOUS and DEFICIENS). The roles that MADS box genes play, however, are not restricted to controlling the development of plant reproductive structures (J. L. Riechmann and Meyerowitz 1997). A conserved core domain of around 90 amino acids is sufficient for DNA-binding, dimerization and interaction with accessory factors (PFAM PF00319) activities. Within this core is located the DNA-binding region, designated the MADS box.

2.1.4   bHLH Family

The basic helix-loop-helix (bHLH) family of proteins is a group of functionally diverse TFs found in both plants and animals (at least 144 loci in rice). These proteins evolved early in eukaryotes before the split of animals and plants, but appear to function in plant-specific or animal-specific processes. In animals, bHLH proteins are involved in the regulation of a wide variety of essential developmental processes. In contrast, bHLH proteins have not been extensively studied in plants. Those that have been characterized participate in functions that include anthocyanin biosynthesis, phytochrome signaling, globulin expression, fruit dehiscence, and carpel and epidermal development. These TFs are characterized by a highly conserved bHLH domain (PFAM PF00010) that mediates DNA-binding and specific dimerization by the conversion of inactive monomers to trans-activating dimers at appropriate stages of development.

2.1.5 WRKY Family

Members of the WRKY TF superfamily have been identified from a wide range of higher plants and have been linked to regulating pathogen elicited responses (Ross, Liu and Shen 2007; Tian, Lu, Peng and Fang 2006). There are an estimated 100 WRKY loci in rice with roles linked to disease resistance, responses to salicylic and jasmonic acid, seed development and germination mediated by gibberellins, developmental processes including senescence, and responses to abiotic stresses and abscisic acid (Ross et al. 2007). These TFs exhibit an invariant N-terminal sequence WRKYGOR and a $CX_{4-5}Cx_{22-23}HxH$ zinc binding domain, and they regulate expression of target genes that contain W-box elements ($TTTGAC^C/_T$) in their promoter regions (Tian et al. 2006). The three-dimensional structures of two *Arabidopsis* members of the WRKY family have been determined and provide insights into the DNA mechanism of these plant-specific transcription factors (see section 2.2 below).

2.1.6 AP2-EREBP Family

The AP2 (APETALA2)/EREBP (Ethylene Responsive Element Binding Protein) TF family includes many developmentally and physiologically important TFs (Aharoni, Dixit, Jetter, Thoenes, van Arkela and Pereira 2004; Kizis and Pages 2002; Ohto, Fischer, Goldberg, Nakamura and Harada 2005; Zhu, Hoque, Dennis and Upadhyaya 2003). There are an estimated 164 AP2/EREBP loci in rice, making it the largest TF gene family in that species. AP2/EREBP genes are divided into two subfamilies: AP2 genes with two AP2 domains and EREBP genes with a single AP2/ERF (Ethylene Responsive element binding Factor) domain. Interestingly, the expression of AP2-like genes is regulated by the microRNA miR172, and the target site of miR172 is significantly conserved in gymnosperm AP2 homologs, suggesting that regulatory mechanisms of these TFs using microRNA have been conserved over the three hundred million years since the divergence of gymnosperm and flowering plant lineages (Shigyo, Hasebe and Ito 2006). At least six members of this family have been studied in maize (Table II).

2.1.7 GLK (G2-like) Family

This family was first defined in plants by isolation of the maize *Golden2* (*G2* or *Bundle Sheath Defective 1*) gene (Hall, Rossini, Cribb and Langdale 1998). G2 is necessary for chloroplast differentiation underlying the development of bundle

Table 1 TF family frequencies in *Arabidopsis* and rice, and TF identifying features.

TF Family	A.t.	O.s	Plant Specific	DNA-Binding Domain (DBD) and Interaction Domains (ID)
ABI3VP1	11	52	Yes	(Abscisic Acid Insensitive1, Viviparous1TF family) B3 type DBD that binds to CAACG
ALFIN-like	7	9	Yes	C4 and H/C3 zinc fingers that bind GNGGTG or GTGGNG concensus targets
AP2-EREBP	138	164	Yes	(Apetala2 and ethylene-responsive element binding proteins) AP2 domain that binds CAACA
ARF	24	25	Yes	N-term DBD that binds auxin response elements (AuxREs), C-term protein-protein ID
ARID	7	5	No	(A-T Rich Interaction Domain) HTH motif type DBD that bind AT-rich sequences
ARR-B	15	8	Yes	(Arabidopsis Response Regulators Type B) B(GARP-like)-motif that binds GAT sequence
AtRKD/NLP	14	3	No	NIN-like proteins with RWPxRK DBD
BBR/BPC	7	4	Yes	putative zinc finger DBD at C-term binds to (GA/TC)n -dinucleotide repeat enhancer
bHLH	161	144	No	large superfamily with basic helix-loop-helix DBD, can form homo- or heterodimers
bZIP	73	85	No	basic DBD plus leucine zipper ID, several subfamilies
BZR	6	nd	Yes	(Brassinosteroid signalling) Novel N-term DBD that binds CGTG(T/C)G
C2C2-CO-like	30	17	Yes	B-box C_(2)-C_(2) zn finger DBD. Named after CONSTANS-like genes
C2C2-Dof	36	30	Yes	DBD with one zinc C_(2)-C_(2) finger
C2C2-Gata	30	27	No	C_(2)-C_(2) zn finger DBD that binds (A/T)GATA(A/G)
C2C2-YABBY	6	7	Yes	Member of High Mobility group (HMG) -superfamily. Specifies abaxial polarity in leaves
C2H2	211	102	No	Classical xCx_(1-5)Cx_3#x_5#x_2[H/C]x_{(3-6)}[H/C] zn finger DBD
C3H	165	66	No	Cx_8Cx_5Cx_3H type zn finger DBD. Some are RNA binding.
CAMTA	6	6	No	130 aa DBD associated with Calmodulin-binding transcription activator
CCAAT-DR1	2	1	No	Bifunctional (Activator or repressor) DBD that binds CCAAT
CCAAT-HAP2	10	11	No	Bifunctional (Activator or repressor) NF-YA DBD that binds CCAAT
CCAAT-HAP3	10	12	No	Bifunctional (Activator or repressor) NF-YB DBD that binds CCAAT
CCAAT-HAP5	13	21	No	Bifunctional (Activator or repressor) NF-YB DBD that binds CCAAT
CPP	8	11	No	CxCx_4Cx_3YCxCx_6Cx_3CxCx_2C type CXC DBD
CSD	nd	2	No	Cold Shock Domain DBD and a zinc knuckle Cx2Cx4Hx4C
E2F-DP	8	9	No	Winged-helix DBD of Elongation factor 2A that binds symmetric c/gGCGCg/c
EIL	6	7	Yes	(Ethylene Insensitive-Like) novel pentahelical fold in DBD, mediates ethylene response
G2-like	40	46	Yes	(Golden2-like subfamily of GARP TFs)
GEBP	16	6	Yes	(Glabrous1 Enhancer Binding Protein) uncharacterized DBD, C-term Leucine zipper
GRAS	33	54	Yes	(GAI, RGA, SCR) family Conserved leucine repeat I and II,VHIID, PFYRE, and SAW motifs.
GRF	9	9	Yes	(Growth-Regulating Factor1) GRF DBD (containing QLQ and WRC motifs)

Family	A.t.	O.s.		Description
HOMEOBOX	102	91	No	Helix-turn-helix DBD, Second helix binds to DNA, first helix stabilizes structure
HRT	3	1	No	Three unusual Zn finger motifs ($Cx_{8-9}Cx_{7}Cx_{10}Cx_{2}H$)
HSF	21	25	No	(Heat shock Transcription Factor) HSF DBD between 1 C-termand 3 N-term leucine zippers
JUMONJI	5	15	No	Classical C2H2 and finger and eight residue C5HC2 zn finger.
MADS	111	64	No	(MCM1, AGAMOUS, DEFICIENS, and SRF) DBD of 2 antiparalell helices
MYB	133	121	No	Highly conserved 3 repeat N terminal DBD that binds YAAC(G/T)G
MYB-related	75	81	No	Exhibit more divergent MYB domain present either singly or as a repeat
NAC	96	123	Yes	(No Apical Meristem) 160 residue N-term NAC DBD and dimerization domain
Orphan	2	140	N/A	TFs that exhibit unique, hybrid, or distantly related DBDs to other families e.g. LFY
PHD	11	49	No	C4HC3 homeodomain zinc-finger-like (similar but distinct from C3HC4 type RING finger)
RAV	11	nd	Yes	(Related to ABI3/VP1) Contains AP2 and B3 DBDs, recognizes CAACA and CACCTG mtifs
REM	21	nd	Yes	Several repeats of a B3-related domain. Functions in vegetative and floral meristems.
SBP	16	19	Yes	(SQUAMOSA-PROMOTER BINDING PROTEIN) SBP-box DBD with C and H residues
TCP	26	22	Yes	(TB1, CYC and PCFs) non-canonical basic-Helix-Loop-Helix (bHLP) DBD
TRIHELIX	29	19	Yes	One or two amino-terminal trihelix motifs
TUB	10	14	No	(Tubby, TF linked to obesity in mice) Conserved C-term DBD with conserved C residue
VOZ-9	0	2	Yes	Conserved VOZ DBD(Domain B) binds to specific palindromic sequence, GCGTNx_{7}ACGC
WHIRLY	3	N/A	Yes	Novel TFs implicated to regulation of systemic acquired resistance (SAR)
WRKY	72	97	No	Conserved WRKYGQK sequence followed by a C2H2 or C2HC zinc finger motif
ZF-HD	15	15	No	Conserved 54 residue N-term and a ZF-HD class homeobox domain toward the C terminus

A.t. *Arabidopsis thaliana*, O.s. *Oryza sativa*, nd not determined, N/A Not applicable

sheath cells in maize. Other *G2*-like genes in maize and rice defined a G2-like family (GLK) in plants. These TFs poses a conserved HLH DNA-binding motif and a motif referred to as the GLK/C-terminal box (GCT box). The former shares a low level of sequence similarity with a class of eukaryotic DNA-binding domains called TEA, whereas the latter appears to be unique to plants (Cribb, Hall and Langdale 2001; Rossini, Cribb, Martin and Langdale 2001). In angiosperms, GLK factors regulate the development of at least three chloroplast types. Conservation of GLK-mediated regulation of chloroplast development in moss defines this as one of the most ancient conserved regulatory mechanisms in the plant kingdom (Yasumura, Moylan and Langdale 2005). A more recent study, in which the *Arabidopsis GLK1* gene was over-expressed, demonstrated that it regulates a variety of genes linked to pathogen response and detoxification, and thus it may be useful for providing disease resistance in crop plants (Savitch, Subramaniam, Allard and Singh 2007).

2.2   Structures of Plant Transcription Factors

For TFs that are conserved across kingdoms, the three dimensional structure is also likely to be conserved, and useful predictions may be made in regards to critical DNA-binding residues. A good example is provided by the MYB domains present in one of the largest classes of plant TFs. The structure of the R2R3-MYB region of the animal c-MYB protein, which shares about 50% identity to plant R2R3-MYB domains, has been solved (Ogata, Morikawa, Nakamura, Hojo, Yoshimura, Zhang, Aimoto, Ametani, Hirata, Sarai, Ishii and Nishiura 1995; Ogata, Morikawa, Nakamura, Sekikawa, Inoue, Kanai., Sarai., Ishii and Nishimura 1994). Each MYB repeat adopts a helix-turn-helix fold with the third α-helix in each MYB repeat making base-pair contacts. However, despite the sequence identity between plant and animal MYB domains, significant structural differences exist. These differences are best highlighted by the presence of just one Cys residue in animal MYB domains, but two proximal and highly conserved Cys residues in most plant R2R3-MYB domains, which can form an intra-molecular disulfide bond (S-S) under non-reducing conditions (Heine, Hernandez and Grotewold 2004). Recently, the structure of the *RADIALIS (RAD)* protein from *Antirrhinum majus*, has been solved to 1.9 Å (Stevenson, Burton, Costa, Nath, Dixon, Coen and Lawson 2006). Consistent with RAD being part of a distinct group of single MYB repeat proteins that includes proteins from tomato and *Arabidopsis* (Barg, Sobolev, Eilon, Gur, Chmelnitsky, Shabtai, Grotewold and Salts 2005; Corley, Carpenter, Copsey and Coen 2005), the structure of RAD has significant difference from canonical MYB repeats and a notably longer third α-helix (Stevenson et al. 2006).

In addition to the MYB (3R or R2R3) proteins, there are a number of TFs that contain MYB-related motifs (often classified as MYB-related factors, and further sub-classified with different names). For example, the GARP subfamily exhibits a signature 60 amino acid known as the B-motif that resembles the classical MYB repeat. The structure of the B-motif from the *Arabidopsis* ARR10 TF, which is a response regulator involved in His-to-Asp phosphorelay signal transduction systems, was determined by NMR (Hosoda, Imamura, Katoh, Hatta, Tachiki, Yamada, Mizuno and Yamazaki 2002b). The B-motif consists of a helix-turn-helix motif (Fig 1D), and its mechanism of interaction with DNA appears to be similar to that of homeodomain proteins (Hosoda, Imamura, Katoh, Hatta, Tachiki, Yamada, Mizuno and Yamazaki 2002a). In addition, the B-motif contains a nuclear localization signal, making this a multifunctional domain.

While MYBs exemplify a group of regulatory proteins that is likely to have originated prior to the split between plants and animals (Braun et al. 1999; Lipsick 1996), several TF families appear to be unique to the plants (Shiu et al. 2005). Examples include the WRKY, NAM and TCP families, together accounting for more than 10% of the total plant TFs so far described. As of October 2007, out of ~46,000 structures available at PDB (http://www.pdb.org/), only ~14 correspond to plant DNA-binding proteins, indicating a need to increase the structural analysis of plant TFs, particularly those unique to this kingdom.

Another example is provided by the structure of p24, the single-stranded DNA-binding subunit of the plant defense TF PBF-2, which has been solved to 2.3 Å resolution. This protein belongs to a novel family of ubiquitous plant-specific factors known as the WHIRLY family because of their quaternary structure. PBF-2 is composed of four p24 molecules that interact with DNA through a helix-loop-helix motif. This interaction produces a central pore, with β-strands radiating outwards, resulting in a whirligig appearance to the quaternary structure (Fig. 1i). The non-crystallographic C4 symmetry arrangement of p24 subunits is novel for ssDNA-binding proteins and may explain the binding specificity of PBF-2. This structural arrangement also supports the role of PBF-2 in binding melted promoter regions to modulate gene expression (Desveaux, Subramaniam, Despres, Mess, Levesque, Fobert, Dangl and Brisson 2004).

NMR and crystal structures of the WRKY4 and WRKY1 (Fig 1b) proteins from *Arabidopsis* have been determined (Duan, Nan, Liang, Mao, Lu, Li, Wei, Lai, Li and Su 2007; Yamasaki, Kigawa, Inoue, Tateno, Yamasaki, Yabuki, Aoki, Seki, Matsuda, Tomo, Hayami, Terada, Shirouzu, Tanaka, Seki, Shinozaki and Yokoyama 2005b). The novel WRKY4 structure consists of a four-stranded β-sheet, with the zinc-binding pocket located at one end of the β-sheet. The WRKYGQK residues correspond to the most N-terminal β-strand, kinked in the middle of the sequence by the Gly residue, which enables extensive hydrophobic interactions involving the Trp residue and contributes to the structural stability of the β-sheet (Yamasaki et al. 2005b). The WRKY1 crystal structure and site directed mutagenesis studies revealed a five-pleated β-sheet structure and that the DNA-binding residues are located in the β2 and β3 strands. These data were used to develop a model of how the WRKY domain interacts with the W-box, a model that has yet confirmed by protein-DNA complex structural studies (Duan et al. 2007). This approach, however, will guide studies on the specificity of this plant TF family and serves as a model strategy for examining the specificities of other plant TFs.

To our knowledge, the structure of only one plant TF bound to its target DNA-sequence has so far been determined, using heteronuclear multidimensional

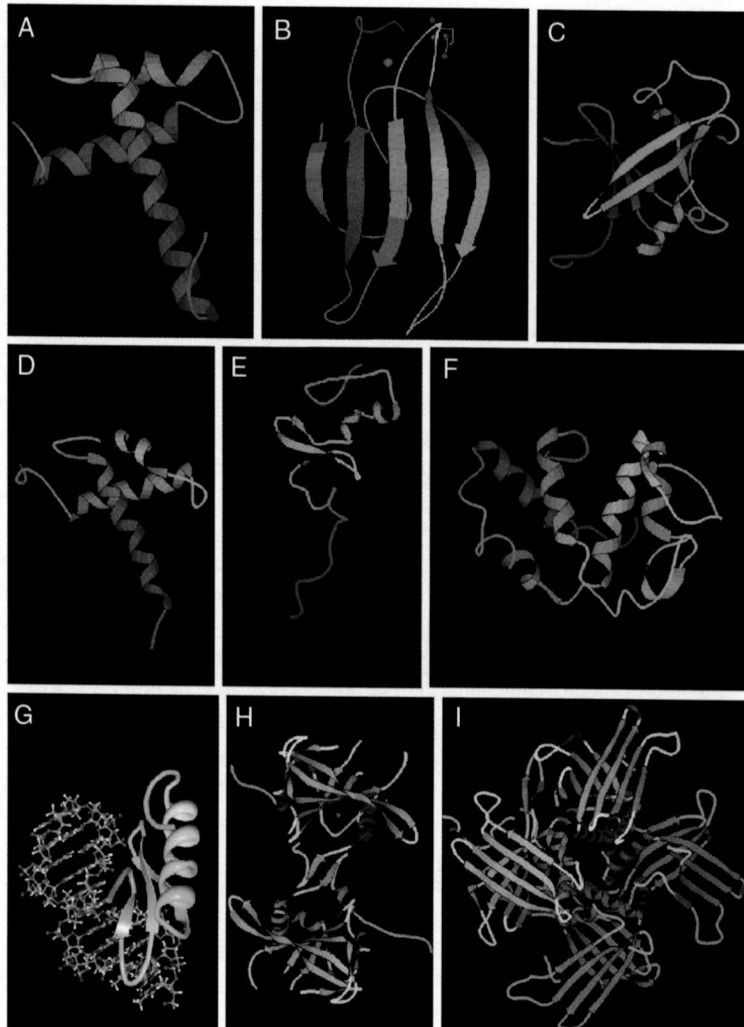

Fig. 1 Crystallographic and NMR Structures of Plant TFs: A: MYB domain of the RAD TF from *Antirrhinum majus* (PDB 2CJJ), B: C-terminal WRKY1 DNA-binding domain from *A. thaliana* (PDB 2AYD), C: B3 DNA-binding domain of the Cold-Responsive TF RAV1 from *A. thaliana* (PDB 1WID), D: B motif of type-B response regulator (ARRs) ARR10-B from *A. thaliana* (PDB 1IRZ), E: DNA-binding domain of the Squamosa Promoter Binding Protein-like4 from *A. thaliana* (PDB 1UL4), F: The major DNA-binding domain of the Ethylene Insensitive3 (EIN3)–like protein3 (EIL3) from *A. thaliana* (PDB 1WIJ), G: GCC-Box binding domain of ATERF1 from *A. thaliana* (PDB 1GCC), H: DNA-binding NAC domain of ANAC (abscisic-acid-responsive NAC) from *A. thaliana* (PDB 1UT7), I: PBF-2 ssDNA binding WHIRLY TF from *Solanum tuberosum* (PDB 1WJ2)

NMR (Allen, Yamasaki, Ohme-Takagi, Tateno and Suzuki 1998). The *Arabidopsis* ERF1 protein belongs to the family of ethylene-responsive element binding proteins (EREBPs) that bind to the consensus nucleotide sequence, AGCCGCC, known as the GCC-box, which has been identified in the promoter region of pathogenesis-related genes (Ohme-Takagi and Shinshi 1995). The GCC-box binding domain (GBD) of AtERF1 was shown to consist of a three-stranded anti-parallel β-sheet and an α-helix packed approximately parallel to the β-sheet (Fig 1g) (Allen et al. 1998). Arg and Trp residues in the β-sheet are identified to contact eight consecutive base pairs in the major groove as well as the sugar phosphate backbones. The target DNA bends slightly at the central CG base pair, thereby allowing the DNA to follow the curvature of the β-sheet. Such studies provide clear insights into the residues that are needed for target binding and help define the regions of TFs that are available for interaction with other regulatory proteins.

Another plant-specific TF that exhibits a classical helix-turn-helix motif is the DNA-binding NAC domain of *Arabidopsis* ANAC (abscisic-acid-responsive NAC, NAM/ATAF1,2/CUC2). The NAC TF family has been implicated in a wide range of plant responses, including developmental processes, among them the formation of the shoot apical meristem, floral organs and lateral shoots, as well as in plant hormonal control and defense (Ernst, Olsen, Larsen and Lo Leggio 2004; Olsen, Ernst, Leggio and Skriver 2005; Olsen, Ernst, Lo Leggio, Johansson, Larsen and Skriver 2004). The crystal structure reveals that the NAC domain consists mainly of a twisted β-sheet surrounded by a few helical elements, and that the protein operates as a functional dimer (Fig 1h).

The structure of a number of other plant TFs has been solved using NMR. A variety of plant-specific TFs, including factors involved in auxin-regulated and abscisic acid-regulated transcription share a DNA-binding domain known as B3. The B3 domain of the *Arabidopsis* cold-responsive transcription factor RAV1 consists of a seven-stranded open β-barrel and two α-helices located at the ends of the barrel (Fig 1c). This TF exhibits a significant structural similarity to the *Eco*RII restriction enzyme and it is likely that the DNA-binding residues are similarly located between these proteins (Yamasaki, Kigawa, Inoue, Tateno, Yamasaki, Yabuki, Aoki, Seki, Matsuda, Tomo, Hayami, Terada, Shirouzu, Osanai, Tanaka, Seki, Shinozaki and Yokoyama 2004). Plants are distinct from animals in utilizing the gas ethylene as a hormone. A number of genes such as Ethylene-Insensitive3 (EIN3) and EIN3-like (EIL) proteins are essential TFs in the transduction of this signal. These TFs bind to the promoters of downstream genes and function in a variety of stress responses. The unique DNA-binding domains of these proteins were revealed by the structure of an EIN3/EIl member from *Arabidopsis* (Yamasaki, Kigawa, Inoue, Yamasaki, Yabuki, Aoki, Seki, Matsuda, Tomo, Terada, Shirouzu, Tanaka, Seki, Shinozaki and Yokoyama 2005a). The EIL DBD structure consists of five α-helices, furnishing a novel fold dissimilar not previously found in other DNA-binding proteins (Fig 1f). By chemical-shift perturbation analyses, a region including the site mutated in the *ein3-3* allele was suggested to be involved in DNA binding (Yamasaki et al. 2005a).

The SBP TF family is another new family of plant TFs that was defined by the isolation of the *Antirrhinum majus* floral meristem identity gene SQUAMOSA (Klein, Saedler and Huijser 1996). This gene family appears to have evolved early in plants as evidenced by the finding of an SBP clade in moss (Riese, Höhmann,

Saedler, Münster and Huijser 2007). The NMR structures of two SBP DNA-binding domains from *Arabidopsis* reveals unusual zinc finger domains comprised of 8 Cys or His residues Cys3HisCys2HisCys or Cys6HisCys, Fig 1e (Yamasaki, Kigawa, Inoue, Tateno, Yamasaki, Yabuki, Aoki, Seki, Matsuda, Nunokawa, Ishizuka, Terada, Shirouzu, Osanai, Tanaka, Seki, Shinozaki and Yokoyama 2004).

While these examples clearly indicate that plant TFs will exhibit a variety of unique DNA-binding domains, to date not a single structure for a monocot TF has been reported. Although it is not anticipated that monocots will exhibit unique TFs that are absent from dicots, there may be subfamilies whose structures will be required to help elucidate the specificities of regulatory networks. The advent of genomic and full-length cDNA resources for rice and maize should help facilitate the attempts to over-express TFs for structural studies.

2.3 Adopting a Uniform Nomenclature for Naming Maize Transcription Factors

The naming of TFs (and other maize gene families) has been complicated by the use of GenBank accession numbers or EST numbers that often end up assigning different attributes to, for example, allelic versions of the same TF. To overcome this, a uniform nomenclature for TFs was developed that mirrors similar efforts in other plants. As part of this initiative, maize TFs would be named by a species identifier (Zm for maize) followed by a short descriptor of the family (e.g., bHLH for basic helix-loop-helix, HD for homeodomain) and a number starting at "1" (e.g. ZmbHLH1). The recently developed GRASSIUS (www.grassius.org) server provides, among other databases, a comprehensive collection of TFs from various grasses, including maize. This community resource provides guidelines on how to name newly identified maize TFs, guidelines that we encourage the community to follow as much as possible in an effort to be able to position the maize "TFome" in the context of other plants.

2.4 A Sample of Characterized Maize Transcription Factors

A survey of the literature reveals that, to date, approximately 70 maize TFs have been characterized in some detail (Table 2). We provide here a sample of some TFs that have contributed significantly to regulatory studies in maize.

2.4.1 C1

C1 (ZmMYB1), encoded by the *COLOR1* (*C1*) gene, was the first plant MYB factor to be identified (Paz-Ares, Ghosal, Weinland, Peterson and Saedler 1987).

It corresponds to a typical R2R3-MYB transcriptional activator of anthocyanin biosynthetic genes. It binds DNA with the consensus $AC/_ACT/_AAC/_AC$, yet with an affinity significantly lower than other MYB factors (Sainz, Grotewold and Chandler 1997a). C1 also contains an acidic transcriptional activation domain located at the C-terminus (Sainz, Goff and Chandler 1997b). *C1* is primarily expressed in the aleurone, and *PL1* (ZmMYB2), a duplicate of *C1*, is responsible for plant body pigmentation (Cone, Cocciolone, Burr and Burr 1993). The MYB domains of C1/PL1 physically interact with the N-terminal region of the bHLH factors R/B (Goff, Cone and Chandler 1992), interactions that are essential for C1/PL1 activity. The specificity of the interaction between R/B and C1/PL1 is furnished by a discrete set of 4-6 solvent-exposed MYB residues (Grotewold, Sainz, Tagliani, Hernandez, Bowen and Chandler 2000).

2.4.2 GLOSSY15

GLOSSY15 encodes a member of a small subclass of AP2-domain proteins that contain two tandem repeats of the conserved AP2 DNA-binding protein, preceded by a putative serine-rich nuclear localization signal and separated by a short linker sequence (Moose and Sisco 1996). Biochemical analyses of the closely-related *Arabidopsis* AINTEGUMENTA protein suggests both AP2 domains may independently bind DNA (Krizek 2003). Like its close paralog *INDETERMINATE SPIKELET1*, *GLOSSY15* mRNA accumulation is negatively regulated by miRNA172 (Lauter, Kampani, Carlson, Goebel and Moose 2005). Analysis of *glossy15* mutations demonstrates that it functions in the regulation of juvenile versus adult leaf identity via the activation or repression of genes associated with epicuticular wax production, macrohair initiation, and the composition of epidermal cell walls (Evans, Passas and Poethig 1994; Moose et al. 1996). Over-expression of GLOSSY15 in transgenic plants prolongs the expression of juvenile leaf identity and delays flowering.

2.4.2 KN1

KN1 (ZmHD1), encoded by the *KNOTTED1* gene (*Kn1*), is a member of the TALE (Three Amino acid Loop Extension) sub-group of homeodomain proteins. *KN1* was originally cloned from a dominant *Ds*-induced mutation that resulted in the formation of leaf "knots" as a consequence of the transposition of proximal leaf cell identities to the leaf blade (Hake, Vollbrecht and Freeling 1989). KN1 recognizes the consensus $TGACAGG/_CT$, and the binding affinity to this site is significantly increased through the interaction with KIP (KN1-Interacting Protein), another TALE HD (H. M. Smith, Boschke and Hake 2002). KN1 is required for the functions of the shoot apical meristem (SAM) and, accordingly, KN1 expression is restricted to undifferentiated SAM cells (L. G. Smith, Green, Veit and Hake 1992), mediated in part by ROUGH

SHEET2 (RS2), encoding a MYB factor with some unusual characteristics (Timmermans, Hudson, Becraft and Nelson 1999).

2.4.3 P1

P1 (ZmMYB3), encoded by the *PERICARP COLOR1* gene (*P1*), is responsible for the regulation of the phlobaphene pigments in several maize floral tissues (Styles and Ceska 1989). Indeed, *P1* alleles are named by suffixes that correspond to the phlobaphene pigmentation present in pericarp and cob glumes. For example, *P1-rr* specifies red pericarp and red cobs and *P1-wr* specifies white (colorless) pericarp and red cob. Sometimes, a third suffix it utilized to represent the accumulation of brown phenolic compounds in silk tissues. P1 corresponds to a typical R2R3-MYB (Grotewold, Drummond, Bowen and Peterson 1994) and belongs to a R2R3-MYB sub-family (the P-to-A clade) that has expanded recently during the radiation of the grasses (Dias et al. 2003; Rabinowicz et al. 1999). P1 regulates a subset of the anthocyanin biosynthetic genes, including *C2* and *A1*, by binding to *cis*-regulatory elements in their promoters that can be classified into high-affinity and low-affinity P1-binding sites, and which fit the P1 DNA-binding consensus $CC^{T/}{}_{A}ACC$ (Grotewold et al. 1994; Sainz et al. 1997a). The *in vitro* DNA-binding activity of P1 is REDOX regulated (Williams and Grotewold 1997) by the formation of an intramolecular S-S bond between two proximal MYB Cys residues (Heine et al. 2004). The *P-wr* allele is formed by a head-to-tail tandem repeat of the *P1* transcribed region (Chopra, Athma, Li and Peterson 1998), resulting in epigenetic modifications that alter *P1* expression (Cocciolone, Chopra, Flint-Garcia, McMullen and Peterson 2001; Sekhon, Peterson and Chopra 2007), and which are affected by the *Ufo1* genetic factor (Chopra, Cocciolone, Bushman, Sangar, McMullen and Peterson 2003). Some maize lines express a second *P* gene, *P2*, closely linked to *P1*, which is primarily responsible for the accumulation of *C*-glycosyl flavones and other phenolics in silks (P. Zhang, Chopra and Peterson 2000).

2.4.4 R1

R1 (ZmbHLH1), encoded by the *RED COLOR1* gene (*R1*), is a member of the R/B family of bHLH anthocyanin regulators (Ludwig and Wessler 1990). R/B genes are complex in structure, and are characterized by a diversity of alleles that furnish anthocyanin pigmentation from a few cell types to pretty much the entire maize plant (Coe and Neuffer 1988). Despite containing a conserved bHLH motif, no direct binding of R/B to DNA has yet been reported. However, R is recruited to the promoter of the *A1* gene through its interaction with C1, as recently shown in ChIP experiments (Hernandez, Feller, Morohashi, Frame and Grotewold 2007). For the regulation of anthocyanin biosynthesis, the role of R appears to be to serve as a platform for the docking of various proteins, which include the WD40 protein

PAC1 (Carey, Strahle, Selinger and Chandler 2004) and RIF1, the latter providing a functional link of R with chromatin functions (Hernandez et al. 2007).

2.4.5 O2

O2 (ZmbZIP1), encoded by the *OPAQUE2* gene (*O2*), is a member of the bZIP family of TFs (Schmidt, Burr, Aukerman and Burr 1990; Schmidt, Burr and Burr 1987). Among the several genes regulated by O2 is a subset of the zein 22 kD endosperm storage proteins which are controlled by the tethering of O2 to the TCCACGTAGA sequence present in the promoters of many of these genes (Schmidt et al. 1990; Sturaro and Viotti 2001). The expression of *O2* in subaleurone layers of the endosperm appears to be controlled by the circadian clock, with peaks of mRNA accumulation at midday and lowest accumulation at midnight (Ciceri, Locatelli, Genga, Viotti and Schmidt 1999).

3 Resources for Investigating Maize TF Function

One aspect of understanding TF function involves identifying the regulatory motifs in which a TF participates. Several experimental and computational approaches are available to determine and predict protein-DNA interactions, respectively. Experimental approaches involve investigating the formation of protein-DNA complexes, for example by electrophoretic mobility shift assays (EMSA), or by exploring the specific DNA sequence recognized by a TF on a given fragment of DNA using chemical or nuclease footprinting techniques. These techniques, however, involve *in vitro* protein-DNA interactions and their application depends on the availability of a DNA fragment containing the regulatory sequences.

Two approaches are commonly used to identify and/or validate the direct *in vivo* targets of a TF. The first one involves expressing a fusion of the TF to GR (GR corresponds to the hormone-binding domain of the glucocorticoid receptor) and identifying the mRNAs induced/repressed in the presence of the GR ligand (dexamethasone, DEX), in the presence of an inhibitor of translation (*e.g.*, cycloheximide, CHX). This method has been extensively used in plants (Morohashi, Zhao, Yang, Read, Lloyd, Lamb and Grotewold 2007; Sablowski and Meyerowitz 1998; Shin, Choi, Yi, Yang, Cho, Kim, Lee, Paek, Kim and Song 2002; Spelt, Quattrocchio, Mol and Koes 2002; D. Wang, Amornsiripanitch and Dong 2006; Wellmer, Alves-Ferreira, Dubois, Riechmann and Meyerowitz 2006). The second approach involves identifying the DNA sequences that a TF binds *in vivo*, using chromatin immunoprecipitation (ChIP) assays (Wells and Farnham 2002). ChIP not only provides a tool to identify the *in vivo* location of TFs on the DNA, but also complements many of the weaknesses of EMSA and footprinting, unless *in vivo* footprinting is performed (Paul and Ferl 1991, 1994), which remains a technically challenging technique. For ChIP, intact tissues or cells are treated with a cross-linking agent that covalently links the protein with the DNA. The chromatin is then sheared (using enzymatic or mechanical methods) and the covalently-linked protein-DNA complex is enriched by immunoprecipitation (IP) using the specific antibodies to the proteins (Morohashi, Xie and Grotewold 2008).

ChIP-chip (a.k.a., ChIP-on-chip or genome-wide location analysis) involves the hybridization of a microarray representing a fraction or the entire genome space with the DNA resulting from the ChIP experiment. ChIP-chip has been applied to plants, either to investigate the location of specific histone modifications [e.g., (X. Zhang, Clarenz, Cokus, Bernatavichute, Pellegrini, Goodrich and Jacobsen 2007)), or towards the identification of TF direct targets (Gao, Li, Strickland, Hua, Zhao, Chen, Qu and Deng 2004; Lee, He, Stolc, Lee, Figueroa, Gao, Tongprasit, Zhao, Lee and Deng 2007). It is unlikely, however, that a tiling array will become soon available for maize, but the combination of ChIP with high-throughput sequencing methods (Mardis 2007; Robertson, Hirst, Bainbridge, Bilenky, Zhao, Zeng, Euskirchen, Bernier, Varhol, Delaney, Thiessen, Griffith, He, Marra, Snyder and Jones 2007) provides an attractive alternative for the identification of direct targets for maize TFs.

4 Conclusions

As a group, TFs constitute a significant portion of the protein-encoding genes. Yet, elucidating their functions has remained challenging, particularly for those that control processes that do not result in immediate phenotypic changes. The availability of a complete genome sequence, together with similar resources soon to be available for related grasses, furnishes unique opportunities to enter the field of comparative regulatory genomics, or comparing how TF function has evolved over relatively short evolutionary times.

Acknowledgments EG is supported by grants from the National Science Foundation (DBI-0701405 & MCB-0437318), the U.S. Dept. of Agriculture CSREES (NRI Grant 2007-35318-17805), the Dept. of Energy (DE-FG02-07ER15881) and JG by a grant from the National Science Foundation (DBI-0701405). A fraction of the salaries and research support in the EG and JG labs are also provided by state funds appropriated to the Ohio Plant Biotechnology Consortium through The Ohio State University, Ohio Agricultural Research and Development Center.

References

Aharoni, A., Dixit, S., Jetter, R., Thoenes, E., van Arkela, G., and Pereira, A. (2004). The SHINE clade of AP2 domain transcription factors activates wax biosynthesis, alters cuticle properties, and confers drought tolerance when overexpressed in *Arabidopsis*. Plant Cell 16, 2463–2480.
Alabadi, D., Oyama, T., Yonovsky, M.J., Harmon, F.G., Mas, P., and Kay, S.A. (2001). Reciprocal regulation between *TOC1* and *LHYICCA1* within the *Arabidopsis* circadian clock. Science 293, 880–883.

Allen, M.D., Yamasaki, K., Ohme-Takagi, M., Tateno, M., and Suzuki, M. (1998). A novel mode of DNA recognition by a beta-sheet revealed by the solution structure of the GCC-box binding domain in complex with DNA. Embo J 17, 5484–5496.

Babu, M.M., Luscombe, N.M., Aravind, L., Gerstein, M., and Teichmann, S.A. (2004). Structure and evolution of transcriptional regulatory networks. Curr Opin Struct Biol 14, 283–291.

Barg, R., Sobolev, I., Eilon, T., Gur, A., Chmelnitsky, I., Shabtai, S., Grotewold, E., and Salts, Y. (2005). The tomato early fruit specific gene *Lefsm1* defines a novel class of plant-specific SANT/MYB domain proteins. Planta 221, 197–211.

Braun, E.L., and Grotewold, E. (1999). Newly discovered plant *c-myb*-like genes rewrite the evolution of the plant *myb* gene family. Plant Physiol. 121, 21–24.

Carey, C.C., Strahle, J.T., Selinger, D.A., and Chandler, V.L. (2004). Mutations in the pale aleurone color1 regulatory gene of the *Zea mays* anthocyanin pathway have distinct phenotypes relative to the functionally similar TRANSPARENT TESTA GLABRA1 gene in Arabidopsis thaliana. Plant Cell 16, 450–464.

Chopra, S., Athma, P., Li, X.G., and Peterson, T. (1998). A maize Myb homolog is encoded by a multicopy gene complex. Mol Gen Genet 260, 372–380.

Chopra, S., Cocciolone, S.M., Bushman, S., Sangar, V., McMullen, M.D., and Peterson, T. (2003). The maize Unstable factor for orange1 is a dominant epigenetic modifier of a tissue specifically silent allele of pericarp color1. Genetics 163, 1135–1146.

Ciceri, P., Locatelli, F., Genga, A., Viotti, A., and Schmidt, R.J. (1999). The activity of the maize Opaque2 transcriptional activator is regulated diurnally. Plant Physiol 121, 1321–1328.

Cocciolone, S.M., Chopra, S., Flint-Garcia, S.A., McMullen, M.D., and Peterson, T. (2001). Tissue-specific patterns of a maize Myb transcription factor are epigenetically regulated. Plant J 27, 467–478.

Coe, E.H., and Neuffer, M.G. (1988). The genetics of corn. In G.F. Sprague and J.W. Dudley, eds *Corn and Corn Improvement*, Madison, WI: American Society of Agronomy, pp. 81–258.

Cone, K.C., Cocciolone, S.M., Burr, F.A., and Burr, B. (1993). Maize anthocyanin regulatory gene *pl* is a duplicate of *c1* that functions in the plant. Plant Cell 5, 1795–1805.

Corley, S.B., Carpenter, R., Copsey, L., and Coen, E. (2005). Floral asymmetry involves an interplay between TCP and MYB transcription factors in *Antirrhinum*. Proc Natl Acad Sci U S A 102, 5068–5073.

Cribb, L., Hall, L.N., and Langdale, J.A. (2001). Four mutant alleles elucidate the role of the G2 protein in the development of C4 and C3 photosynthesizing maize tissues. Genetics 159, 787–797.

Davidson, E.H. (2001). Genomic regulatory systems. (San Diego, California: Academic Press).

Davuluri, R.V., Sun, H., Palaniswamy, S.K., Matthews, N., Molina, C., Kurtz, M., and Grotewold, E. (2003). AGRIS: Arabidopsis gene regulatory information server, an information resource of Arabidopsis cis-regulatory elements and transcription factors. BMC Bioinformatics 4, 25.

Desveaux, D., Subramaniam, R., Despres, C., Mess, J.N., Levesque, C., Fobert, P.R., Dangl, J.L., and Brisson, N. (2004). A "Whirly" transcription factor is required for salicylic acid-dependent disease resistance in *Arabidopsis*. Dev Cell 6, 229–240.

Dias, A.P., Braun, E.L., McMullen, M.D., and Grotewold, E. (2003). Recently duplicated maize *R2R3 Myb* genes provide evidence for distinct mechanisms of evolutionary divergence after duplication. Plant Physiol. 131, 610–620.

Duan, M.R., Nan, J., Liang, Y.H., Mao, P., Lu, L., Li, L., Wei, C., Lai, L., Li, Y., and Su, X.D. (2007). DNA binding mechanism revealed by high resolution crystal structure of *Arabidopsis thaliana* WRKY1 protein. Nucleic Acids Res 35, 1145–1154.

Ernst, H.A., Olsen, A.N., Larsen, S., and Lo Leggio, L. (2004). Structure of the conserved domain of ANAC, a member of the NAC family of transcription factors. EMBO Rep 5, 297–303.

Evans, M.M., Passas, H.J., and Poethig, R.S. (1994). Heterochronic effects of *glossy15* mutations on epidermal cell identity in maize. Development 120, 1971–1981.

Fu, S., Rogowsky, P., Nover, L., and Scanlon, M.J. (2006). The maize heat shock factor-binding protein paralogs EMP2 and HSBP2 interact non-redundantly with specific heat shock factors. Planta 224, 42–52.

Gao, Y., Li, J., Strickland, E., Hua, S., Zhao, H., Chen, Z., Qu, L., and Deng, X.W. (2004). An *Arabidopsis* promoter microarray and its initial usage in the identification of HY5 binding targets in vitro. Plant Mol Biol 54, 683–699.

Gaut, B.S., and Doebley, J.F. (1997). DNA sequence evidence for the segmental allotetraploid origin of maize. Proc. Natl. Acad. Sci. 94, 6809–6814.

Goff, S.A., Cone, K.C., and Chandler, V.L. (1992). Functional analysis of the transcriptional activator encoded by the maize *B* gene: evidence for a direct functional interaction between two classes of regulatory proteins. Genes Dev. 6, 864–875.

Grotewold, E. (2005). Plant metabolic diversity: A regulatory perspective. Trends Plant Sci. 10, 57–62.

Grotewold, E., Drummond, B.J., Bowen, B., and Peterson, T. (1994). The myb-homologous *P* gene controls phlobaphene pigmentation in maize floral organs by directly activating a flavonoid biosynthetic gene subset. Cell 76, 543–553.

Grotewold, E., Sainz, M.B., Tagliani, L., Hernandez, J.M., Bowen, B., and Chandler, V.L. (2000). Identification of the residues in the Myb domain of maize C1 that specify the interaction with the bHLH cofactor R. Proc Natl Acad Sci U S A 97, 13579–13584.

Hake, S., Vollbrecht, E., and Freeling, M. (1989). Cloning *Knotted*, the dominant morphological mutant in maize using *Ds2* as a transposon tag. EMBO J 8, 15–22.

Hall, L.N., Rossini, L., Cribb, L., and Langdale, J.A. (1998). GOLDEN2: A novel transcriptional regulator of cellular differentiation in the maize leaf. Plant Cell 10, 925–936.

Heine, G.F., Hernandez, M.J., and Grotewold, E. (2004). Two cysteines in plant R2R3 MYB domains participate in REDOX-dependent DNA binding. J Biol Chem 279, 37878–37885.

Hernandez, J.M., Feller, A., Morohashi, K., Frame, K., and Grotewold, E. (2007). The bHLH domain of maize R links transcriptional regulation and histone modifications by recruitment of an EMSY-related factor. Proc Natl Acad Sci USA In Press.

Hosoda, K., Imamura, A., Katoh, E., Hatta, T., Tachiki, M., Yamada, H., Mizuno, T., and Yamazaki, T. (2002a). Molecular structure of the GARP family of plant Myb-related DNA binding motifs of the *Arabidopsis* response regulators. Plant Cell 14, 2015–2029.

Hulskamp, M., Misera, S., and Jurgens, G. (1994). Genetic dissection of trichome cell development in *Arabidopsis*. Cell 76, 555–566.

Ito, M. (2005). Conservation and diversification of three-repeat Myb transcription factors in plants. J Plant Res 118, 61–69.

Ito, M., Araki, S., Matsunaga, S., Itoh, T., Nishihama, R., Machida, Y., Doonan, J.H., and Watanabe, A. (2001). G2/M-phase-specific transcription during the plant cell cycle is mediated by c-Myb-like transcription factors. Plant Cell 13, 1891–1905.

Kizis, D., and Pages, M. (2002). Maize DRE-binding proteins DBF1 and DBF2 are involved in rab17 regulation through the drought-responsive element in an ABA-dependent pathway. Plant J 30, 679–689.

Klein, J., Saedler, H., and Huijser, P. (1996). A new family of DNA binding proteins includes putative transcriptional regulators of the *Antirrhinum majus* floral meristem identity gene SQUAMOSA. Mol. Gen. Genet. 250, 7–16.

Krizek, B.A. (2003). AINTEGUMENTA utilizes a mode of DNA recognition distinct from that used by proteins containing a single AP2 domain. Nucleic Acids Res 31, 1859–1868.

Lauter, N., Kampani, A., Carlson, S., Goebel, M., and Moose, S.P. (2005). microRNA172 down-regulates *glossy15* to promote vegetative phase change in maize. Proc Natl Acad Sci U S A 102, 9412–9417.

Lee, J., He, K., Stolc, V., Lee, H., Figueroa, P., Gao, Y., Tongprasit, W., Zhao, H., Lee, I., and Deng, X.W. (2007). Analysis of Transcription Factor HY5 Genomic Binding Sites Revealed Its Hierarchical Role in Light Regulation of Development. Plant Cell.

Lipsick, J.S. (1996). One billion years of Myb. Oncogene 13, 223–235.

Ludwig, S.E., and Wessler, S.R. (1990). Maize R gene family: Tissue-specific helix-loop-helix proteins. Cell 62, 849–851.

Mardis, E.R. (2007). ChIP-seq: welcome to the new frontier. Nat Methods 4, 613–614.

Moose, S.P., and Sisco, P.H. (1996). *Glossy15*, an APETALA2-like gene from maize that regulates leaf epidermal cell identity. Genes Dev 10, 3018–3027.

Morohashi, K., Xie, Z., and Grotewold, E. (2009). Gene-specific and genome-wide ChIP approaches to study plant transcriptional networks. Methods Mol Biol In Press.

Morohashi, K., Zhao, M., Yang, M., Read, B., Lloyd, A., Lamb, R., and Grotewold, E. (2007). Participation of the *Arabidopsis* bHLH factor GL3 in trichome initiation regulatory events. Plant Physiol. 145, 736–746.

Ogata, K., Morikawa, S., Nakamura, H., Sekikawa, A., Inoue, T., Kanai., H., Sarai., A., Ishii, S., and Nishimura, Y. (1994). Solution structure of a specific DNA complex of the Myb DNA-binding domain with cooperative recognition helices. Cell 79, 639–648.

Ogata, K., Morikawa, S., Nakamura, H., Hojo, H., Yoshimura, S., Zhang, R., Aimoto, S., Ametani, Y., Hirata, Z., Sarai, A., Ishii, S., and Nishiura, Y. (1995). Comparison of the free and CNA-complexed forms of the DNA-binding domain from c-Myb. Struct. Biol. 2, 309–319.

Ohme-Takagi, M., and Shinshi, H. (1995). Ethylene-inducible DNA binding proteins that interact with an ethylene-responsive element. Plant Cell 7, 173–182.

Ohto, M.-a., Fischer, R.L., Goldberg, R.B., Nakamura, K., and Harada, J.J. (2005). Control of seed mass by APETALA2. Proc Nat Acad Sci USA 102, 3123–3128.

Olsen, A.N., Ernst, H.A., Leggio, L.L., and Skriver, K. (2005). NAC transcription factors: structurally distinct, functionally diverse. Trends Plant Sci 10, 79–87.

Olsen, A.N., Ernst, H.A., Lo Leggio, L., Johansson, E., Larsen, S., and Skriver, K. (2004). Preliminary crystallographic analysis of the NAC domain of ANAC, a member of the plant-specific NAC transcription factor family. Acta Crystallogr D Biol Crystallogr 60, 112–115.

Palaniswamy, K., James, S., Sun, H., Lamb, R., Davuluri, R.V., and Grotewold, E. (2006). AGRIS and AtRegNet: A platform to link cis-regulatory elements and transcription factors into regulatory networks. Plant Phyisiol. 140, 818–829.

Paul, A.-L., and Ferl, R.J. (1991). *in vivo* footprinting reveals unique cis-elements and different modes of hypoxic induction in maize *Adh1* and *Adh2*. Plant Cell 3, 159–168.

Paul, A.-L., and Ferl, R.J. (1994). *In vivo* footprinting identifies and activating element of the maize *Adh2* promoter specific for root and vascular tissues. Plant J. 5, 523–533.

Paz-Ares, J., Ghosal, D., Weinland, U., Peterson, P.A., and Saedler, H. (1987). The regulatory *c1* locus of *Zea mays* encodes a protein with homology to *myb* proto-oncogene products and with structural sililarities to transcriptional activators. EMBO J. 6, 3553–3558.

Rabinowicz, P.D., Braun, E.L., Wolfe, A.D., Bowen, B., and Grotewold, E. (1999). Maize *R2R3 Myb* genes: Sequence analysis reveals amplification in higher plants. Genetics 153, 427–444.

Riano-Pachon, D.M., Ruzicic, S., Dreyer, I., and Mueller-Roeber, B. (2007). PlnTFDB: an integrative plant transcription factor database. BMC Bioinformatics 8, 42.

Riechmann, J.L., and Meyerowitz, E.M. (1997). MADS domain proteins in plant development. Biol Chem 378, 1079–1101.

Riechmann, J.L., and Ratcliffe, O.J. (2000a). A genomic perspective on plant transcription factors. Curr Op Plant Biol 3, 423–434.

Riechmann, J.L., Heard, J., Martin, G., Reuber, L., Jiang, C., Keddie, J., Adam, L., Pineda, O., Ratcliffe, O.J., Samaha, R.R., Creelman, R., Pilgrim, M., Broun, P., Zhang, J.Z., Ghandehari, D., Sherman, B.K., and Yu, G. (2000b). *Arabidopsis* transcription factors: genome-wide comparative analysis among eukaryotes. Science 290, 2105–2110.

Riese, M., Höhmann, S., Saedler, H., Münster, T., and Huijser, P. (2007). Comparative analysis of the SBP-box gene families in *P. patens* and seed plants. Gene. 2007 401, 28–37.

Robertson, G., Hirst, M., Bainbridge, M., Bilenky, M., Zhao, Y., Zeng, T., Euskirchen, G., Bernier, B., Varhol, R., Delaney, A., Thiessen, N., Griffith, O.L., He, A., Marra, M., Snyder, M., and Jones, S. (2007). Genome-wide profiles of STAT1 DNA association using chromatin immunoprecipitation and massively parallel sequencing. Nat Methods 4, 651–657.

Ross, C.A., Liu, Y., and Shen, Q.J. (2007). The WRKY gene family in rice (*Oryza sativa*). Journal of Integrative Plant Biology 49, 827–842.

Rossini, L., Cribb, L., Martin, D.J., and Langdale, J.A. (2001). The maize *Golden2* gene defines a novel class of transcriptional regulators in plants. Plant Cell 13, 1231–1244.

Sablowski, R.W.M., and Meyerowitz, E.M. (1998). A Homolog of NO APICAL MERISTEM is an immediate target of the floral homeotic genes APETALA3/PISTILLATA. Cell 92, 93–103.

Sainz, M.B., Grotewold, E., and Chandler, V.L. (1997a). Evidence for direct activation of an anthocyanin promoter by the maize C1 protein and comparison of DNA binding by related Myb domain proteins. Plant Cell 9, 611–625.

Sainz, M.B., Goff, S.A., and Chandler, V.L. (1997b). Extensive mutagenesis of a transcriptional activation domain identifies single hydrophobic and acidic amino acids important for activation *in vivo*. Mol. Cell. Biol. 17, 115–122.

Savitch, L.V., Subramaniam, R., Allard, G.C., and Singh, J. (2007). The GLK1 'regulon' encodes disease defense related proteins and confers resistance to Fusarium graminearum in *Arabidopsis*. Biochemical and Biophysical Research Communications 359, 234–238.

Schmidt, R.J., Burr, F.A., and Burr, B. (1987). Transposon tagging and molecular analysis of the maize regulatory locus *opaque-2*. Science 238, 960–963.

Schmidt, R.J., Burr, F.A., Aukerman, M.J., and Burr, B. (1990). Maize regulatory gene *opaque-2* encodes a protein with a "leucine-zipper" motif that binds to zein DNA. Proc Natl Acad Sci U S A 87, 46–50.

Sekhon, R.S., Peterson, T., and Chopra, S. (2007). Epigenetic modifications of distinct sequences of the *p1* regulatory gene specify tissue-specific expression patterns in maize. Genetics 175, 1059–1070.

Shigyo, M., Hasebe, M., and Ito, M. (2006). Molecular evolution of the AP2 subfamily. Gene (Amsterdam) 366, 256–265.

Shin, B., Choi, G., Yi, H., Yang, S., Cho, I., Kim, J., Lee, S., Paek, N.C., Kim, J.H., and Song, P.S. (2002). AtMYB21, a gene encoding a flower-specific transcription factor, is regulated by COP1. Plant J 30, 23–32.

Shiu, S.H., Shih, M.C., and Li, W.H. (2005). Transcription factor families have much higher expansion rates in plants than in animals. Plant Physiol 139, 18–26.

Smith, H.M., Boschke, I., and Hake, S. (2002). Selective interaction of plant homeodomain proteins mediates high DNA- binding affinity. Proc Natl Acad Sci U S A 99, 9579–9584.

Smith, L.G., Green, B., Veit, B., and Hake, S. (1992). A dominant mutation in the maize homeobox gene, *Knotted-1*, causes its ectopic expression in leaf cells with altered fates. Development 116, 21–30.

Spelt, C., Quattrocchio, F., Mol, J., and Koes, R. (2002). ANTHOCYANIN1 of petunia controls pigment synthesis, vacuolar pH, and seed coat development by genetically distinct mechanisms. Plant Cell 14, 2121–2135.

Stevenson, C.E., Burton, N., Costa, M.M., Nath, U., Dixon, R.A., Coen, E.S., and Lawson, D.M. (2006). Crystal structure of the MYB domain of the RAD transcription factor from *Antirrhinum majus*. Proteins 65, 1041–1045.

Stracke, R., Werber, M., and Weisshaar, B. (2001). The R2R3 MYB gene family in *Arabidopsis thaliana*. Curr Opin Plant Biol 4, 447–456.

Sturaro, M., and Viotti, A. (2001). Methylation of the Opaque2 box in zein genes is parent-dependent and affects O2 DNA binding activity *in vitro*. Plant Mol Biol 46, 549–560.

Styles, E.D., and Ceska, O. (1989). Pericarp flavonoids in genetic strains of *Zea mays*. Maydica 34, 227–237.

Tian, Y., Lu, X.-Y., Peng, L.-S., and Fang, J. (2006). The structure and function of plant WRKY transcription factors. Yichuan 28, 1607–1612.

Timmermans, M.C., Hudson, A., Becraft, P.W., and Nelson, T. (1999). ROUGH SHEATH2: a Myb protein that represses knox homeobox genes in maize lateral organ primordia. Science 284, 151–153.

Tominaga, R., Iwata, M., Okada, K., and Wada, T. (2007). Functional Analysis of the Epidermal-Specific MYB Genes CAPRICE and WEREWOLF in *Arabidopsis*. Plant Cell 19, 2264–2277.

Wada, T., Tachibana, T., Shimura, Y., and Okada, K. (1997). Epidermal cell differentiation in *Arabidopsis* determined by a *Myb* homolog, CPC. Science 277, 1113–1116.

Wang, D., Amornsiripanitch, N., and Dong, X. (2006). A genomic approach to identify regulatory nodes in the transcriptional network of systemic acquired resistance in plants. PLoS Pathog 2, e123.

Wang, Z.-Y., Kenigsbuch, D., Sun, L., Harel, E., Ong, M.S., and Tobin, E.M. (1997). A Myb-Related transcription factor is involved in the phytochrome regulation of an *Arabidopsis Lhcb* gene. Plant Cell 9, 491–507.

Wellmer, F., Alves-Ferreira, M., Dubois, A., Riechmann, J.L., and Meyerowitz, E.M. (2006). Genome-wide analysis of gene expression during early *Arabidopsis* flower development. PLoS Genet 2, e117.

Wells, J., and Farnham, P.J. (2002). Characterizing transcription factor bidning sites using formaldehyde crosslinking and immunoprecipitation. Methods 26, 48–56.

Williams, C.E., and Grotewold, E. (1997). Differences between plant and animal Myb domains are fundamental for DNA-binding, and chimeric Myb domains have novel DNA-binding specificities. J. Biol. Chem. 272, 563–571.

Yamasaki, K., Kigawa, T., Inoue, M., Yamasaki, T., Yabuki, T., Aoki, M., Seki, E., Matsuda, T., Tomo, Y., Terada, T., Shirouzu, M., Tanaka, A., Seki, M., Shinozaki, K., and Yokoyama, S. (2005a). Solution structure of the major DNA-binding domain of *Arabidopsis thaliana* ethylene-insensitive3-like3. J Mol Biol 348, 253–264.

Yamasaki, K., Kigawa, T., Inoue, M., Tateno, M., Yamasaki, T., Yabuki, T., Aoki, M., Seki, E., Matsuda, T., Tomo, Y., Hayami, N., Terada, T., Shirouzu, M., Tanaka, A., Seki, M., Shinozaki, K., and Yokoyama, S. (2005b). Solution structure of an *Arabidopsis* WRKY DNA binding domain. Plant Cell 17, 944–956.

Yamasaki, K., Kigawa, T., Inoue, M., Tateno, M., Yamasaki, T., Yabuki, T., Aoki, M., Seki, E., Matsuda, T., Tomo, Y., Hayami, N., Terada, T., Shirouzu, M., Osanai, T., Tanaka, A., Seki, M., Shinozaki, K., and Yokoyama, S. (2004). Solution structure of the B3 DNA binding domain of the *Arabidopsis* cold-responsive transcription factor RAV1. Plant Cell 16, 3448–3459.

Yamasaki, K., Kigawa, T., Inoue, M., Tateno, M., Yamasaki, T., Yabuki, T., Aoki, M., Seki, E., Matsuda, T., Nunokawa, E., Ishizuka, Y., Terada, T., Shirouzu, M., Osanai, T., Tanaka, A., Seki, M., Shinozaki, K., and Yokoyama, S. (2004). A novel zinc-binding motif revealed by solution structures of DNA-binding domains of *Arabidopsis* SBP-family transcription factors. J Mol Biol 337, 49–63.

Yasumura, Y., Moylan, E.C., and Langdale, J.A. (2005). A conserved transcription factor mediates nuclear control of organelle biogenesis in anciently diverged land plants. Plant Cell 17, 1894-1907.

Yu, H., and Gerstein, M. (2006). Genomic analysis of the hierarchical structure of regulatory networks. Proc Natl Acad Sci U S A 103, 14724–14731.

Zhang, P., Chopra, S., and Peterson, T. (2000). A segmental gene duplication generated differentially expressed myb-homologous genes in maize. Plant Cell 12, 2311–2322.

Zhang, X., Clarenz, O., Cokus, S., Bernatavichute, Y.V., Pellegrini, M., Goodrich, J., and Jacobsen, S.E. (2007). Whole-genome analysis of histone H3 lysine 27 trimethylation in *Arabidopsis*. PLoS Biol 5, e129.

Zhu, Q.-H.-H., Hoque, M.S., Dennis, E.S., and Upadhyaya, N.M. (2003). Ds tagging of *branched floretless 1* (BFL1) that mediates the transition from spikelet to floret meristem in rice (*Oryza sativa* L). BMC Plant Biology 3.

The Genetics and Biochemistry of Maize Zein Storage Proteins

Rebecca S. Boston and Brian A. Larkins

Abstract Zeins, the most abundant proteins in the maize seed, have been studied for over 100 years. We have learned a great deal about the structure, synthesis and genetic regulation of these proteins, making them a valuable model system to study storage protein synthesis. In this review, we explore longstanding questions and controversies regarding zeins in light of new information from genomic sequencing, improved protein detection methods and molecular characterization of mutants.

1 Introduction

Research on the genetics and biochemistry of the major seed storage proteins in the maize kernel, which are conventionally known as "zeins", figured prominently in the early days of plant molecular biology. When this field began to emerge in the early 1970's, it primarily featured studies on plant viruses and viroids, the chloroplast and mitochondrial genomes and the crown gall disease caused by *Agrobacterium tumefaciens* (for an overview see Leaver 1980). Following a report that polyribosomes synthesizing zein proteins could be isolated from developing maize endosperm (Larkins and Dalby 1975; Burr and Burr 1976), it became clear that seed storage proteins could become a model system for studying molecular genetics in plants, much in the way hemoglobins were for humans. Subsequently, a number of laboratories began to investigate the organization and expression of genes encoding the seed storage proteins of the major cereal and legume species.

R.S. Boston
North Carolina State University, Department of Plant Biology
boston@ncsu.edu

B.A. Larkins
University of Nebraska, Associate Vice Chancellor for Research
blarkins2@unl.edu

J.L. Bennetzen and S. Hake (eds.), *Maize Handbook - Volume II: Genetics and Genomics*,   715
© Springer Science+Business Media LLC 2009

Among the laboratories that figured prominently in the study of maize storage protein genes were those of the following individuals: Brian Larkins at Purdue University; Benjamin and Frances Burr at Brookhaven National Laboratory; Francesco Salamini at the Istituto Sperimentale per la Cerealicoltura in Bergamo, Italy; Carlo Soave and Angelo Viotti at the University of Milan in Italy; Gunter Feix at Freiburg University in Germany; Irwin Rubenstein at the University of Minnesota, Joachim Messing at the University of Minnesota and later the Waksman Institute at Rutgers University, and Jaume Palau and Pere Puigdomènech at the Institute of Biology in Barcelona, Spain.

The period of gene discovery was followed by efforts to identify and characterize DNA elements and protein factors regulating zein gene expression, and to determine the molecular basis for mutant phenotypes associated with changes in zeins. As high-throughput analysis tools became available zeins once again gained prominence as models for analyzing expression within large gene clusters and for fine-scale comparison across different inbred backgrounds. Space does not allow us to review all of the research and publications that make up the zein story; here we will highlight some of the key findings to date and prospects for future investigation.

2 Fractionation and Properties of Zein Proteins

The maize kernel accumulates large amounts of nitrogen in the form of storage proteins, the largest proportion of which are the prolamins known as zeins. Besides their importance for establishing the seedling, zein storage proteins are a valuable source of amino acids for livestock, and as a result the amino acid composition of zeins has an important nutritional consequence, especially for monogastric animals such as humans and certain livestock. Zeins are rich in proline and glutamine (hence the name prolamin), but they are essentially devoid of lysine and tryptophan, both essential amino acids. Consequently, a great deal of research has been devoted to finding ways of increasing the level of these amino acids in corn. Aside from their valuable nutritional properties, the structure and solubility of zein proteins are also important, because they confer important functional characteristics to flours made from corn. These properties affect food processing and manufacturing, and they also make zein proteins useful for manufacturing a variety of industrial products. Therefore, it is not surprising that the structure and synthesis of zein proteins has been of interest for many years.

The zein fraction is composed of several different types of prolamin proteins that are soluble in aqueous-alcohol solutions, e.g. 60% isopropanol, 70% ethanol, or 95% methanol. Because zeins are deposited as insoluble accretions that are surrounded by proteins cross-linked by disulfide bonds, they dissolve slowly in alcoholic solutions and are insoluble in water. Their extraction is accelerated by increasing the temperature of the alcohol and by including a disulfide reducing agent, but it is also affected by the particle size of the endosperm flour. These factors influence the recovery of the different types of zein proteins, and they are responsible for the complex and often confusing nomenclature that evolved to describe these proteins.

One of the first reports describing "zeine" was published in 1821 by J. Gorham, and the unique properties of this protein fraction attracted the interest of a number of other investigators (Lawton 2002). T.B. Osborne, the father of seed storage protein research (Osborne 1908), developed methods for extracting zein and other prolamin proteins from cereal grains based on their hydrophobic properties. The classical Osborne extraction procedure, which is still widely used today, involves extracting proteins from endosperm with water, 5% saline, 70% alcohol, and 5% NaOH, and this sequentially removes the albumin, globulin, prolamin and glutelin fractions, respectively (Osborne 1897). Originally, two types of zeins were distinguished: "α-zein", which is soluble in 95% aqueous alcohol or 85% isopropanol, and "β-zein", which is soluble in 60% aqueous ethanol (McKinney 1958). Later, more efficient methods for extracting proteins from maize flour and separating zein from the other solubility classes of proteins were described by Paulis et al. (1969) and Landry and Moureaux (1970). With these procedures, following the removal of albumins and globulins, zein was recovered in several steps: the first used aqueous alcohol alone and the second used aqueous alcohol plus a reducing agent, such as β-mercaptoethanol. This yielded what was classified as zein, and a second fraction identified as gluteliln-1 (Landry and Moureaux 1970), alcohol-soluble reduced glutelin (Paulis and Wall 1971), zein-2 (Sodek and Wilson 1971) and "zein-like" protein (Misra et al. 1975) by different research groups. Proteins in these two solubility classes had similar amino acid compositions, but there were distinctive differences, suggesting they contained different polypeptides. The resolution of their relationship was eventually made clear with the application of SDS-PAGE, which showed differences in polypeptide composition.

Asim Esen developed a new technique for extracting and identifying maize prolamin proteins (Esen 1986). His procedure was based on the differential solubility of zeins in the presence of reducing agent. The total zein fraction, which could be efficiently recovered after extraction with 60% isopropanol containing 2% β-mercaptoethanol, was divided into three sub-fractions (SF): SF1, which was soluble in 60% isopropanol alone, principally contained polypeptides of 20- and 24-kDa and appeared to correspond to the α-zein described by McKinney (1958); SF2 contained polypeptides of 17- to 18-kDa that required 60% isopropanol containing 2% β-mercaptoethanol for solubility; SF3 predominantly contained a 27-kDa polypeptide that was soluble in aqueous buffer, providing reducing agent was present. Esen proposed that SF1, SF2 and SF3 be designated as α-, β-, and γ-zeins (Esen 1987). Later, Kirihara et al. (1988) reported a methionine-rich 10-kDa protein in the zein fraction, and Swarup et al. (1995) and Chui and Falco (1995) described a related protein of 18-kDa. These proteins are designated δ-zeins as an extension of Esen's nomenclature (Larkins et al. 1989). The designations of α-, β-, γ- and δ-zeins not only distinguished four structurally distinctive types of proteins in the alcoholic extracts of maize endosperm, but they also fit with the order in which the genes encoding these proteins were isolated.

3   Zein Genes

3.1   Complexity of Zein Coding Sequences

Among the initial challenges of describing the zein fraction was determining how many genes encode these proteins. Separation of zeins by SDS-PAGE revealed polypeptides of different sizes (Lee et al. 1976; Gianazza et al. 1977), but it was unclear whether each protein band was encoded by one or more genes. In vitro translation of zein mRNAs produced major products that co-migrated with the 22- and 19-kDa α-zeins (Larkins and Dalby 1975; Burr and Burr 1976), and later these were shown to be precursor proteins containing signal peptides (Larkins and Hurkman 1978; Larkins et al. 1979). The interpretation of these results was that there were two major zein proteins of 22- and 19-kDa and potentially several minor ones. However, studies in the Italian laboratories showed that when zeins were analyzed by isoelectric focusing (IEF), the 22- and 19-kDa bands could be separated into multiple polypeptides with different charges (Righetti et al. 1977). Furthermore, individual IEF protein bands were shown to be associated with genetic loci on different maize chromosomes (Soave et al. 1978; Valentini et al. 1979; Soave et al. 1981). Clearly, zein genes were more complex than first thought! Subsequent investigations with two-dimensional gel separation (IEF followed by SDS-PAGE) demonstrated that the major zein bands, i.e. the 22- and 19-kDa zeins, were composed of several polypeptides and these could also be obtained from in vitro translation products of zein mRNAs (Viotti et al 1978). Viotti et al. (1979) and Pedersen et al. (1980) attempted to estimate the complexity of zein coding sequences and their frequency in the genome using nucleic acid hybridization of zein cDNAs to endosperm mRNA and genomic DNA. Although the approaches used by these investigators were similar, the interpretations of their results were not: Viotti concluded there were 15 different (non-cross hybridizing) zein coding sequences present in 10 copies, while Pedersen concluded there were approximately four distinct mRNA sequences that contained internal sequence repeats, possibly represented by 2-5 gene copies in the genome. So, the answers ranged from 10-20 genes to 100-150 genes!

3.2   Characterization of the Zein Gene Space

Resolution of questions pertaining to the number of zein genes and how they are regulated required gene cloning, a technology that was still in its infancy in the late 1970s, at least for plants. The first report of zein cDNA clones came from the Feix lab (Weinand et al. 1979) and was followed shortly thereafter by publications by the Rubenstein (Park et al. 1980) and Messing labs (Geraghty et al. 1981). Zein cDNAs provided probes to screen maize genomic libraries for related sequences which showed that, unlike many eukaryotic genes, zein coding sequences lack introns (Wienand et al. 1981; Hu et al. 1982; Pedersen et al.

1982; Spena et al. 1983). Using Southern blots of genomic DNA, researchers were able to obtain a more precise estimate of the number of zein genes and evaluate their relationships. These studies demonstrated that α-zeins are encoded by small multi-gene families, many of which lacked full coding regions and were assumed to be pseudogenes (Hagen and Rubenstein 1981; Burr et al. 1982; Geraghty et al. 1982). The zein genes showed unexpected heterogeneity outside of the coding region, with some genes being clustered tightly enough to be contained within a single genomic clone (Thompson et al. 1992) while others were adjacent to highly abundant repetitive sequences (Kriz et al. 1987). In the absence of enzymatic activity, however, the pseudogene designation was more of a catch-all term than an operationally useful definition.

Much of the confusion surrounding the zein gene organization has recently been resolved by large-scale sequencing and analysis of the zein-containing gene space and comparative analysis with colinear regions of rice and sorghum (see chapter by Messing, this volume and citations within). In addition to improving our understanding of the zein gene space, this work also revealed surprising haplotype variation among maize inbreds in the organization of these genes. For example, the 22-kDa zeins in subfamily z1C are found in two loci, one containing only z1C2 (Song et al. 2001). The second is composed of gene copies oriented head to tail with 22 in the inbred line BSSS53 and 15 in the B73 inbred (Song and Messing 2003). In both lines, the intergenic space contains a number of different retrotransposons that expand the zein cluster on the chromosome and explain previous observations of highly repetitive DNA sequences in zein genomic clones. Surprisingly, within the z1C loci, over half of the genes lack an intact coding region such that expression is limited to the single z1C2 gene in both inbreds, along with five z1C1 genes in BSSS53 and six in B73 (Song and Messing 2003).

The 19-kDa zein sequences are the most abundant with B73 subfamilies having ~50 copies split across z1A (25 members), z1B (20 members) and z1D (5 members; Song and Messing 2002). Like the genes of the z1C subfamily, about half of the 19-kDa α-zein genes encode premature stop codons or truncations that disrupt the predicted coding region.

In contrast to the multi-gene α-zein family, the β-, γ- and δ-zeins are encoded by genes present in only one or two copies (Pedersen et al. 1986; Das and Messing 1987; Kirihara et al. 1988). The copy number of the 27-kDa γ-zein varies from 1 to 2 depending on inbred background (Das and Messing 1987). The 10- and 18-kDa δ-zeins, and the 15-, 16- and 50-kDa γ zeins apparently are encoded by single genes.

The zein gene families were shown to exist in related grass species with similar complexity to that observed in maize. Thus, they probably arose by gene duplication before the divergence of maize, teosinte and Trypsacum but the factors that led to the amplification of α-zein genes, but not β-, γ-, or δ-zein genes, are unknown (Wilson and Larkins 1984). An initial analysis of the effects of polyploidization, paralogous and orthologous zein gene amplification is presented by Messing elsewhere in this volume. Undoubtedly, these genes will serve as excellent models to test hypotheses of evolutionary events as even more genomic sequence data become available.

3.3   Expression of Zein Genes

Efforts to characterize the organization and arrangement of zein genes have been accompanied by studies of zein gene expression. Early on, a detailed RNA profiling analysis revealed the unexpected result that only a few of the many zein genes were highly expressed (Woo et al. 2001). These authors estimated that of the ~250,000 endosperm ESTs surveyed from the B73 inbred, seven 19-kDa zein genes accounted for ~50% of the zein EST's with four z1C genes and the γ and δ-zein ESTs being 15%, 32% and 3%, respectively. This estimate of expressed zein genes was adjusted upward by Song and Messing (2003) whose integration of genomic sequence and expression data revealed additional expressed genes as well as a more refined distinction of multiple genes being represented within the highly expressed unigene sets identified in the earlier study. Using random sequencing of pooled cDNA clones from an RT-PCR library, these authors found six expressed z1C genes with relative expression levels of the identified genes varying by ~2-fold. A striking feature of the z1C subfamily expression was revealed by a comparison of BSSS53 and B73 inbred lines. Whereas each line had at least six expressed zein genes, only three of these were common to both inbreds. Expression levels from the common genes were neither the highest nor the lowest of the complement expressed in either inbred, nor was there consistency across these genes in expression patterns in reciprocal hybrids. This complexity of gene expression at the mRNA level is also consistent with the large line to line variation observed by Consoli and Damerval in a proteomics analysis of zeins (2001). As larger scale comparison across different inbreds becomes more affordable, it should be feasible to dissect the biological implications of these differences and determine the functional significance of the various zein genes.

3.4   Regulatory Mechanisms Affecting Zein Gene Expression

Zein mRNAs and proteins accumulate specifically in endosperm tissue of developing seeds. While one might envision coordination being conferred by a common means of transcriptional regulation, this has not proven to be the case. The first indication of multiple zein regulatory mechanisms came with the discovery of several high lysine mutants (e.g. *o2*, *fl2*, *o7*, *De*-B30*, *Mc*) which had qualitative and quantitative changes in the complement of zeins. In some mutants multiple classes of zeins were affected while in others members of a single zein subfamily were preferentially reduced (e.g. *o2*, *o7*, *De*-B30*; Soave and Salamini 1984).

   At first, the differences were attributed to different regulatory factors affecting subsets of zein genes. This idea was supported by the cloning of the *O2* gene and subsequent identification of it as a transcription factor of the bZIP class (Schmidt et al. 1987, 1990; Motto et al. 1988; Hartings et al. 1989). A second transcription factor, prolamin box binding factor (PBF), was also a promising candidate (Ueda et al. 1994; Vincente-Carbajosa et al. 1997). PBF was identified in a directed search for an endosperm Dof class transcription factor after sequence alignments revealed a conserved cis-acting element. This element was located upstream of all zein subclasses and was capable of interacting with proteins from endosperm nuclear extracts (Maier et al. 1987; Vincente-Carbajosa et al. 1997; Wang et al. 1998; Wang and Messing, 1998). A third putative transcription factor, OHP1, is expressed in endosperm and is able to dimerize with O2 (Pysh et al. 1993). In contrast to O2 and PBF, however, a role for this factor in regulating zein gene expression has not been found.

Both O2 and PBF are synthesized in endosperm and can be detected 1-2 days prior to the appearance of zein mRNAs. This expression pattern thus fits well with these proteins having a regulatory role in the spatial and temporal expression of zeins but other regulatory mechanisms must also be involved. One level of added complexity is related to nitrogen supply. Balconi et al, (1993) assayed RNA and protein for the z1C and z1A,B and D subfamilies in cultured endosperm and found a positive correlation of zein accumulation and nitrogen content in the media. Mutation of nucleotides within the O2 and PBF binding domains decreased promoter activity under high nitrogen as judged by transient expression assays with a GUS fusion protein. No differences were seen, however, between bombardment of WT and o2 endosperm. Precedent for such metabolic control has been well-documented for the GCN4 transcription factor in yeast (Hinnebusch and Natarajan 2002). However, while the nutritional responsiveness seen for zein synthesis has similarities with nutritional regulation of GCN4 in yeast, the O2 independence of the N-responsiveness would seem to rule out an O2 role as a GCN4 like activator in endosperm.

Unexpectedly, several of the other mutants with altered zein levels did not have primary defects in transcription factors. Instead mutations were mapped to sites that lay within the zein genes themselves. These mutants are discussed later in this chapter.

The tremendous diversity in expression patterns raises questions of how many subtle and perhaps not so subtle effects on zein gene expression remain to be discovered. Given the haplotype diversity in gene expression, and the emergence of high through-put transcriptome analysis tools such as 454 sequencing, future work will likely include additional profiles including searches to identify new transcriptional regulators (Emrich et al. 2007). Other longstanding questions include a bias against C or G in the third position of the multi-copy α-zein genes along with maternal imprinting effects that cause an inverse correlation between CG methylation and expression (Lund et al. 1995, 2003). Diurnal changes in the amount of O2 transcripts and differences in transcription of the 22-kDa α-zeins in o2 lines with different mutant alleles were observed by Ciceri et al. (1999, 2000). While these effects are likely to have important physiological implications, they have not yet been fully characterized.

4 Structure and Characteristics of Zein Proteins

Sequencing of zein cDNA clones confirmed that the predominant α-zein proteins contain a high proportion of glutamine and proline and the hydrophobic amino acids, alanine and leucine. They are essentially devoid of lysine and tryptophan, both of which are essential amino acids. The structure of α-zeins is largely defined by a series of tandemly repeated peptides of 20 amino acids: there are 9 repeats in the 19-kDa α-zeins and 10 in the 22-kDa α-zeins (Geraghty et al. 1981; 1982; Pedersen et al. 1982; Marks and Larkins 1982; Spena et al. 1982). Each repeat is flanked by clusters of glutamine residues and appears to have an α-helical conformation, based on circular dichroism and a solution conformational analysis (Argos et al. 1982; Tatham et al 1993). The position of polar amino acids in the repeats is conserved, and it was proposed that this is important for establishing the conformation of the proteins, and/or the formation of intermolecular interactions that influence the aggregation of α-zeins in the protein body (Garrat et al. 1993).

The γ-zeins, which comprise the second most abundant type of maize prolamin, were among the last group of genes to be characterized (Prat et al. 1985, 1987; Pedersen et al. 1986). The γ-zeins are cysteine-rich, disulfide-linked proteins. The 27-kDa γ-zein (a.k.a. glutelin II and reduced soluble protein, see above) is the most abundant protein in this group, followed by the 16-kDa γ-zein. Like the α-zeins, both of these proteins are essentially devoid of lysine and tryptophan. A larger version of these proteins, the 50-kDa γ-zein (Woo et al. 2001), features a longer N-terminus, but shares the conserved cysteine-rich core. The protein originally designated as β-zein (Pedersen et al. 1986) was shown to have a high degree of homology with the γ-zeins and is now termed 15-kDa γ-zein (Woo et al. 2001). All the γ-zeins have six highly conserved cysteine-rich domains. The 27-kDa and 50-kDa γ-zeins contain a block of tandem amino acid repeats at the N-terminus that are missing in the 15- and 16-kDa γ-zeins. This region of the 27-kDa γ-zein contains the hexapeptide, PPPVHL, while the extension of the 50-kDa γ-zein N-terminus has polyglutamine repeats.

Although the two sulfur-rich δ-zeins differ markedly in size, their structures are highly conserved (Kirihara et al. 1988; Swarup et al. 1995; Chui and Falco 1995). The 18-kDa protein differs from the 10-kDa protein by a highly repetitive, methionine-rich 53 amino acid insertion in the center of the protein. Little is known about the structure of these proteins, although they are clearly hydrophobic and occur in the center of the core of the protein body, along with the α-zeins (Esen and Stetler 1992).

Dramatic changes in the zein profile as well as the complement of non-zein endosperm proteins occur when zein synthesis is perturbed (Azevedo et al. 2003). For example, the decrease in the 22-kDa α-zeins in o2 mutants leads to a compensatory increase in many non-zeins (Habben et al. 1993). Unlike the zeins, which lack lysines, the non-zein protein fraction contains a more balanced amino acid profile. In fact, this characteristic led to the designation of o2 and fl2 mutants as "high-lysine" mutants when they were characterized by Oliver Nelson's group in the 1960's (Nelson et al. 1965; Mertz et al. 1964). Although it was relatively straightforward to identify the most abundant compensating non-zeins in o2 mutants, it was not easy to dissect the effect of altering zeins from the loss of other transcriptional targets of the O2 protein.

Hunter et al. surveyed >1400 maize genes represented on Affymetrix arrays for expression in WT, o2, and seven other endosperm mutants (2002). The resulting profiles revealed a few common features but in general, comparison across the mutants emphasized the pleiotropy of the patterns. Segal et al. (2003) took a different approach and used an RNAi construct to silence expression for all members of the 22-kDa zein gene subclass. Using an antisense strategy Huang et al. (2004) targeted 19-kDa zein genes and obtained a substantial reduction in this class of proteins. In both cases, transgenic seeds had increased levels of lysine along with an opaque phenotype. Thus, using RNAi to target specific zeins appears to be a tractable solution not only for manipulating zein levels but also for removing some of the other pleiotropic effects caused by loss of O2.

5 Protein Bodies as Novel Organelles

Control of zein gene expression by O2, PBF, nitrogen supply and probably developmental cues certainly contributes to the complexity of zein synthesis and accumulation. An additional layer of complexity, however, is imposed by the intensive demands placed on the cellular secretory and protein folding pathways as zeins accumulate. All zeins have NH_2-terminal signal sequences that direct co-translational import of the polypeptides into the rough ER where the newly translocated zeins form an insoluble matrix within membrane-bound organelles that bud directly from the ER and remain studded with membrane-bound ribosomes. Formation of these organelles, termed protein bodies, presents an unusual challenge for the endosperm secretory pathway because the large number of highly hydrophobic amino acids that make the zeins insoluble in water also make the proteins particularly prone to aggregation. Folding and assembly of the zeins within the protein bodies leads to an ordered arrangement of zeins that become part of the crystalline-like vitreous endosperm as the seed matures (Lending et al. 1988).

The localization of zeins in cytoplasmic inclusions or "protein granules" was described by Duvick in 1961. Since then, a number of experiments have helped to characterize these unusual organelles and the cellular factors that are involved in their biogenesis. An early hypothesis of protein body formation was based on Anfinsen's theory that the three-dimensional structure of a protein was determined by its primary amino acid sequence (1973). The corollary for zeins would be that they aggregated as deposits in the ER due to their insolubility in an aqueous environment. It now seems that while this idea was overly simplistic, zein-zein interactions do have a critical role in protein body formation. Some of the best evidence for this has come from mutant analysis as well as heterologous expression systems. Geli et al. localized the ER retention domain of γ-zein to the PPPVHL hexapeptide repeat region by surveying deletion constructs in transgenic Arabidopsis (1994). Coleman et al. (1996), Bagga et al. (1995, 1997) and Bellucci et al. (2000) expressed zein genes from different subclasses in transgenic tobacco. Neither α- nor δ-zeins accumulated in the absence of β- or γ-zeins. In contrast, expression of β- or γ-zeins alone or co-expression with α- and δ-zeins allowed the zeins to accumulate in the ER as protein accretions. More direct demonstrations of the zein-zein interactions among the proteins of the γ and α subclasses and the β- and δ-zeins were made in two-hybrid systems from yeast and E. coli (Kim et al. 2002; Randall et al. 2005).

These zein interaction data agree well with ultrastructural images of protein bodies (Lending et al. 1988). Immunostaining of normal protein bodies for specific zein sub-classes shows a concentric arrangement of γ-zeins at the periphery of a 22-kDa α-zein layer surrounding an internal core of 19-kDa α- and 10-kDa δ-zeins (Lending and Larkins 1989; Holding et al. 2007). Protein bodies with normal amounts of γ- and δ zein but lower amounts of α-zein (such as those from *o2* mutants) are smaller than those with a full complement of zeins but the overall morphology is normal (Geetha et al. 1991).

A very different scenario is found with opacity mutants that interfere with the primary structure of zeins. *Floury-2 (fl2)*, *Mucronate (Mc)* and *Defective endosperm B-30 (De*-B30)* all have severe morphological changes in protein body structure yet they alter overall zein levels less than *o2* (Wolf et al. 1969; Soave and Salamini 1984; Lending and Larkins 1992; Zhang and Boston 1992). All three mutants have a coordinated increase in several non-zein proteins coincident with the onset of zein synthesis. At first, these additional proteins were thought to be transcriptional regulatory factors for zeins (Galante et al. 1983). Cloning of the corresponding genes and characterization of the proteins, however, identified them as ER molecular chaperones that are induced by misfolded proteins as part of the cellular unfolded protein response (UPR; Marocco et al. 1991, Fontes et al. 1991; Boston et al. 1991). This discovery led to the hypothesis that zein proteins themselves were triggering the UPR through either misfolding of individual zeins or improper zein-zein interactions that exposed non-native protein domains in the protein body. We now know that the primary defects in the *fl2*, *Mc* and *De*-B30* mutants are simple changes. *De*-B30* and *fl2* have point mutations in the 19- and 22-kDa α-zein, respectively, that block cleavage of the signal sequence after import into the ER/protein body (Coleman et al. 1995; Gillikin et al. 1997; Kim et al. 2004). The *Mc* mutant has a small deletion in the cystine-rich 16-kDa-γ-zein that causes a frame shift in the COOH-terminal third of the protein and a net loss of one cystine residue (Kim et al. 2006).

In addition to harboring elevated molecular chaperones, these mutants also have up-regulated ER associated degradation (ERAD) machinery (Kirst et al. 2005), increases in IRE1, the central protein kinase signal transducer that activates the UPR (J. Gillikin and R. Boston, unpublished results), and a higher level of phosphoinositol content in protein body membranes (Shank et al. 2001). These hallmarks of the unfolded protein response as well as the strong phenotypes seen in the three mutants would be unlikely if protein bodies simply represented terminal storage organs that fill with zeins and remain inert until needed during germination. Instead, they remain as dynamic parts of the protein secretory pathway capable of activating the signal transduction and quality control processes needed to maintain ER homeostasis, and in particular, the continued synthesis and deposition of zeins within the protein secretory system.

While the zein-zein interactions clearly influence protein body assembly and morphology, non-zein proteins also play a role. In the *fl1* mutant, protein bodies look normal, but the endosperm is opaque and immunostaining shows a shift in 22-kDa α-zeins from the periphery to the core of the protein body. The Fl1 protein has several membrane spanning domains and interacts with both 22-kDa and 19-kDa α-zeins in yeast two-hybrid assays yet the distribution of only the 22-kDa α-zeins is affected in *fl1* mutants. Unlike other opaque mutants that affect zeins, no reduction is seen in the amount of zein in *fl1* nor does the altered arrangement of zeins induce an unfolded protein response (Holding et al. 2007).

Having endosperm mutants has facilitated investigation of protein body assembly and function, yet many questions remain unanswered. "What is the initial step in protein body formation?" Based on immunofluorescence studies, zein mRNAs were suggested to share targeting signals that would bring them to subdomains of the ER prior to translocation (Washida et al. 2004). Other experiments, however, showed a random distribution of zein mRNAs (Kim et al. 2002), leaving this an open question. We do not yet understand the determining factors that limit protein body size or whether the zeins themselves are curvature generating proteins (Zimmerberg and Kozlov 2006). Expression of γ-zeins alone leads to formation of small protein bodies in transgenic tobacco, protein bodies in *o2* mutants are smaller than those in normal maize lines, and *De**-*B30* and *fl2* mutants have highly invaginated protein bodies that appear to lack the capacity to separate into individual vesicles. Clearly protein body formation is an ordered process but what are the organizing factors? In the past, the abundance of zeins has complicated analyses of other endosperm proteins by creating a signal/noise problem that was difficult to overcome. As new technologies allow masking of zeins and detection of low abundance molecules, it should be feasible to gain a better understanding of the protein interactions and cellular processes involved in the formation of these unusual organelles and the packing of zeins within them.

Acknowledgements We thank J. Messing and J. Gillikin for helpful suggestions. Support for this work was provided in part by Department of Energy grants # DE-FG02-00ER150065 (R.S.B) and DE-96ER20242 (B.A.L) and USDA grant # CSREES2004-00918 (B.A.L.).

References

Anfinsen, C.B. (1973) Principles that govern the folding of protein chains. Science 181: 223-230.

Argos, P., K. Pedersen, M.D. Marks, and B.A. Larkins (1982) A structural model for maize zein proteins. J. Biol. Chem. 257: 9984–9990.

Azevedo, R.A., C. Damerval, J. Landry, P.J. Lea, C.M. Bellato, L.W. Meinhardt, M. Le Guilloux, S. Delhaye, A.A. Toro, S.A. Gaziola, and B.D. Berdejo (2003) Regulation of maize lysine metabolism and endosperm protein synthesis by opaque and floury mutations. Eur J Biochem. 270:4898–908.

Bagga, S., H. Adams, J.D. Kemp, and C. Sengupta-Gopalan (1995) Accumulation of 15-kilodalton zein in novel protein bodies in transgenic tobacco. Plant Physiol. 107: 13–23.

Bagga, S., H.P. Adams, F.D. Rodriguez, J.D. Kemp, and C. Sengupta-Gopalan (1997) Coexpression of the maize delta-zein and beta-zein genes results in stable accumulation of delta-zein in endoplasmic reticulum-derived protein bodies formed by beta-zein. Plant Cell 9: 1683–1696.

Balconi, C., E. Rizzi, M. Motto, F. Salamini, and R.D. Thompson (1993) The accumulation of zein polypeptides and zein mRNA in cultured endosperms of maize is modulated by nitrogen supply. The Plant J. 3: 325–334.

Bellucci, M., A. Alpini, F. Paolocci, L. Cong, and S. Arcioni (2000) Accumulation of maize γ-zein and γ-zein:KDEL to high levels in tobacco leaves and differential increase of BiP synthesis in transformants. Theor Appl Genet. 101:796–804

Boston, R.S., E.B. Fontes, B.B. Shank, and R.L. Wrobel (1991) Increased expression of the maize immunoglobulin binding protein homolog b-70 in three zein regulatory mutants. Plant Cell 3: 497–505.

Burr, B. and F.A. Burr (1976) Zein synthesis in maize endosperm by polyribosomes attached to protein bodies. Proc. Natl. Acad. Sci. U. S. A. 73: 515–519.

Burr, B., F.A. Burr, T.P. St. John, M. Thomas, and R.W. Davis (1982) Zein storage protein gene family of maize. An assessment of heterogeneity with cloned messenger RNA sequences. J. Mol. Biol. 154: 33–49.

Chui, C.F. and S.C. Falco (1995) A new methionine-rich seed storage protein from maize. Plant Physiol. 107: 291.

Ciceri, P., S. Castelli, M. Lauria, B. Lazzari, A. Genga, L. Bernard, M. Sturaro, and A. Viotti (2000) Specific combinations of zein genes and genetic backgrounds influence the transcription of the heavy-chain zein genes in maize *opaque-2* endosperms. Plant Physiol. 124: 451–460.

Ciceri, P., F. Locatelli, A. Genga, A. Viotti, and R.J. Schmidt (1999) The activity of the maize Opaque2 transcriptional activator is regulated diurnally. Plant Physiol. 121: 1321–1328.

Coleman, C.E., E.M. Herman, K. Takasaki, and B.A. Larkins (1996) The maize γ-zein sequesters α-zein and stabilizes its accumulation in protein bodies of transgenic tobacco endosperm. Plant Cell 8: 2335–2345.

Coleman, C.E., M.A. Lopes, J.W. Gillikin, R.S. Boston, and B.A. Larkins (1995) A defective signal peptide in the maize high-lysine mutant *floury 2*. Proc. Natl. Acad. Sci. U. S. A. 92: 6828–6831.

Consoli, L. and C. Damerval (2001) Quantification of individual zein isoforms resolved by two-dimensional electrophoresis: Genetic variability in 45 maize inbred lines. Electrophoresis 22:2983–2989.

Das, O.P. and J.W. Messing (1987) Allelic variation and differential expression at the 27-kilodalton zein locus in maize. Mol. Cell Biol. 7: 4490–4497.

Duvick, D.N. (1961) Protein granules of maize endosperm cells. Cereal Chem. 38: 374–385.

Emrich, S.J., W.B. Barbazuk, L. Li, and P.S. Schnable (2007) Gene discovery and annotation using LCM-454 transcriptome sequencing. Genome Res. 17: 69–73.

Esen, A. (1986) Separation of alcohol-soluble proteins (zeins) from maize into three fractions by differential solubility. Plant Physiol. 80: 623–627.

Esen, A. (1987) A proposed nomenclature for the alcohol-soluble proteins (zeins) of maize (*Zea mays*-L). J. Cereal Sci. 5: 117–128.

Esen, A. and D.A. Stetler (1992) Immunocytochemical localization of the delta-zein in protein bodies of maize endosperm cells. Amer. J. Bot. 79: 243–248.

Fontes, E.B., B.B. Shank, R.L. Wrobel, S.P. Moose, G.R. OBrian, E.T. Wurtzel, and R.S. Boston (1991) Characterization of an immunoglobulin binding protein homolog in the maize *floury-2* endosperm mutant. Plant Cell 3: 483–496.

Galante, E., A. Vitale, L. Manzocchi, C. Soave, and F. Salamini (1983) Genetic control of a membrane component and zein deposition in maize endosperm. Mol. Gen. Genet. 192: 316–321.

Garratt, R., G. Oliva, I. Caracelli, A. Leite, and P. Arruda (1993) Studies of the zein-like alpha-prolamins based on an analysis of amino acid sequences: implications for their evolution and three-dimensional structure. Proteins: Structure, Function and Genetics 15: 88–99.

Geetha, K.B., C.R. Lending, M.A. Lopes, J.C. Wallace, and B.A. Larkins (1991) *opaque-2* modifiers increase γ-zein synthesis and alter its spatial distribution in maize endosperm. Plant Cell 3: 1207–1219.

Geli, M.I., M. Torrent, and D. Ludevid (1994) Two structural domains mediate two sequential events in γ-zein targeting: protein endoplasmic reticulum retention and protein body formation. Plant Cell 6: 1911–1922.

Geraghty, D., Peifer, M.A., Rubenstein, I. and Messing, J. (1981). The primary structure of a plant storage protein: Zein. Nucl. Acids Res. 9: 5163–5174.

Geraghty, D.E., J. Messing, and I. Rubenstein (1982) Sequence analysis and comparison of cDNAs of the zein multigene family. EMBO J. 1: 1329–1335.

Gianazza, E., V. Viglienghi, P.G. Righetti, F. Salamini, and C. Soave (1977) Amino acid composition of zein molecular components. Phytochem. 16: 315–317.

Gillikin, J.W., F. Zhang, C.E. Coleman, H.W. Bass, B.A. Larkins, and R.S. Boston (1997) A defective signal peptide tethers the *floury-2* zein to the endoplasmic reticulum membrane. Plant Physiol. 114: 345–352.

Habben, J.E., A.W. Kirleis, and B.A. Larkins (1993) The origin of lysine-containing proteins in *opaque-2* maize endosperm. Plant Mol. Biol. 23: 825–838.

Hagen, G. and I. Rubenstein (1981) Complex organization of zein genes in maize. Gene 13: 239–249.

Hartings, H., M. Maddaloni, N. Lazzaroni, N. Di Fonzo, M. Motto, F. Salamini, and R. Thompson (1989) The *O2* gene which regulates zein deposition in maize endosperm encodes a protein with structural homologies to transcriptional activators. EMBO J. 8: 2795–2801.

Hinnebusch, A.G. and K. Natarajan (2002) Gcn4p, a master regulator of gene expression, is controlled at multiple levels by diverse signals of starvation and stress. Eukaryot. Cell 1: 22-32.

Holding, D.R., M.S. Otegui, B. Li, R.B. Meeley, T. Dam, B.G. Hunter, R. Jung, and B.A. Larkins (2007) The maize *Floury1* gene encodes a novel endoplasmic reticulum protein involved in zein protein body formation. Plant Cell 19: 2569–2582.

Hu, N.T., M.A. Peifer, G. Heidecker, J. Messing, and I. Rubenstein (1982) Primary structure of a genomic zein sequence in maize. EMBO J. 1: 1337–1342.

Huang, S., W.R. Adams, Q. Zhou, K.P. Malloy, D.A. Voyles, J. Anthony, A.L. Kriz, and M.H. Luethy (2004) Improving nutritional quality of maize proteins by expressing sense and antisense zein genes. J. Agric. Food Chem. 52: 1958–1964.

Hunter, B.G., M.K. Beatty, G.W. Singletary, B.R. Hamaker, B.P. Dilkes, B.A. Larkins, and R. Jung (2002) Maize opaque endosperm mutations create extensive changes in patterns of gene expression. Plant Cell 14: 2591–2612.

Kim, C.S., B.C. Gibbon, J.W. Gillikin, B.A. Larkins, R.S. Boston, and R. Jung (2006) The maize *Mucronate* mutation is a deletion in the 16-kDa γ-zein gene that induces the unfolded protein response. The Plant J. 48: 440–451.

Kim, C.S., B.G. Hunter, J. Kraft, R.S. Boston, S. Yans, R. Jung, and B.A. Larkins (2004) A defective signal peptide in a 19-kD alpha-zein protein causes the unfolded protein response and an opaque endosperm phenotype in the maize *De*-B30* mutant. Plant Physiol. 134: 380–387.

Kim, C.S., Y.-m. Woo, A.M. Clore, R.J. Burnett, N.P. Carneiro, and B.A. Larkins (2002) Zein protein interactions, rather than the asymmetric distribution of zein mRNAs on endoplasmic reticulum membranes, influence protein body formation in maize endosperm. Plant Cell 14: 655–672.

Kirihara, J.A., J.B. Petri, and J. Messing (1988) Isolation and sequence of a gene encoding a methionine-rich 10-kDa zein protein from maize. Gene 71: 359–370.

Kirst, M.E., D.J. Meyer, B.C. Gibbon, R. Jung, and R.S. Boston (2005) Identification and characterization of endoplasmic reticulum-associated degradation proteins differentially affected by endoplasmic reticulum stress. Plant Physiol. 138: 218–231.

Kriz, A.L., R.S. Boston, and B.A. Larkins (1987) Structural and transcriptional analysis of DNA sequences flanking genes that encode 19 kilodalton zeins. Mol. Gen. Genet. 207: 90–98.

Landry, J. and T. Moureaux (1970) Héterogénéite des glutélines du grain de mais: extraction sélective et composition en acidés aminés des trois fractions isolées. Bull. Soc. Chim. Biol. 52: 1021–1037.

Larkins, B.A. and A. Dalby (1975) *In vitro* synthesis of zein-like protein by maize polyribosomes. Biochem. Biophys. Res. Commun. 66: 1048–1054.

Larkins, B.A. and W.J. Hurkman (1978) Synthesis and deposition of zein in protein bodies of maize endosperm. Plant Physiol. 62: 256–263.

Larkins, B.A., K. Pedersen, A.K. Handa, W.J. Hurkman, and L.D. Smith (1979) Synthesis and processing of maize storage proteins in *Xenopus laevis* oocytes. Proc. Natl. Acad. Sci. U. S. A. 76: 6448–6452.

Larkins, B.A., J.C. Wallace, G. Galili, C.R. Lending, and E. Kawata (1989) Structural analysis and modification of maize storage proteins. Dev. Indust. Micro. 30: 203–209.

Lawton, J.W. (2002) Zein: A history of processing and use. Cereal Chem. 79: 1–18.

Leaver, C.J. (1980) Genome Organization and Expression in Plants. Plenum Press, New York.

Lee, K.H., R.A. Jones, A. Dalby, and C.Y. Tsai (1976) Genetic regulation of storage protein content in maize endosperm. Biochem. Genet. 14: 641–650.

Lending, C.R., A.L. Kriz, B.A. Larkins, and C.E. Bracker (1988) Structure of maize protein bodies and immunocytochemical localization of zeins. Protoplasma 143: 51–62.

Lending, C.R. and B.A. Larkins (1989) Changes in the zein composition of protein bodies during maize endosperm development. Plant Cell 1: 1011–1023.

Lending, C.R. and B.A. Larkins (1992) Effect of the *floury-2* locus on protein body formation during maize endosperm development. Protoplasma 171: 123–133.

Lund, G., P. Ciceri, and A. Viotti (1995) Maternal-specific demethylation and expression of specific alleles of zein genes in the endosperm of *Zea mays* L. The Plant J. 8: 571–581.

Lund, G., M. Lauria, P. Guldberg, and S. Zaina (2003) Duplication-dependent CG suppression of the seed storage protein genes of maize. Genetics 165: 835–848.

Maier, U.G., J.W. Brown, C. Tologcyzki, and G. Feix (1987) Binding of a nuclear factor to a consensus sequence in the 5′ flanking region of zein genes from maize. EMBO J. 6: 17–22.

Marks, M.D. and B.A. Larkins (1982) Analysis of sequence microheterogeneity among zein messenger RNAs. J. Biol. Chem. 257: 9976–9983.

Marocco, A., A. Santucci, S. Cerioli, M. Motto, N. Di Fonzo, R. Thompson, and F. Salamini (1991) Three high-lysine mutations control the level of ATP-binding HSP70-like proteins in the maize endosperm. Plant Cell 3: 507–515.

McKinney, L.L. (1958) Zein. In: The Encyclopedia of Chemistry (G.L. Clark, ed.). Reinhold, N.Y., pp. 319–320.

Mertz, E.T., L.S. Bates, and O.E. Nelson (1964) Mutant gene that changes protein composition and increases lysine content of maize endosperm. Science 145: 279–280.

Misra, P.S., E.T. Mertz, and D.V. Glover (1975) Characteristics of proteins in single and double endosperm mutants of maize. In: High Quality Protein Maize (Bauman, L. F., E. T. Mertz, A. Caballo, and E. W. Sprague, eds.) Dowden, Hutchinson and Ross, Inc., Stroudsburg, PA, pp. 291–305.

Motto, M., M. Maddaloni, G. Ponziani, M. Brembilla, R. Marotta, N. Fonzo, C. Soave, R. Thompson, and F. Salamini (1988) Molecular cloning of the *o2-m5* allele of Zea mays using transposon marking. Mol. Gen. Genet. 212: 488–494.

Nelson, O.E., E.T. Mertz, and L.S. Bates (1965) Second mutant gene affecting the amino acid pattern of maize endosperm proteins. Science 150: 1469–1470.

Osborne, T.B. (1897) Amount and properties of the proteins of the maize kernel. J. Amer. Chem. Soc. 19: 525–532.

Osborne, T.B. (1908) Our present knowledge of plant proteins. Science 28: 417–427.

Park, W.D., E.D. Lewis, and I. Rubenstein (1980) Heterogeneity of zein mRNA and protein in maize. Plant Physiol. 65: 98–106.

Paulis, J.W., C. James, and J.S. Wall (1969) Comparison of glutelin proteins in normal and high-lysine corn. J. Agric. Food Chem. 17: 1301–1305.

Paulis, J.W. and J.S. Wall (1971) Fractionation and properties of alkylated-reduced corn glutelin proteins. Biochim. Biophys. Acta. 251: 57–69.

Pedersen, K., P. Argos, S.V. Naravana, and B.A. Larkins (1986) Sequence analysis and characterization of a maize gene encoding a high-sulfur zein protein of Mr 15,000. J. Biol. Chem. 261: 6279–6284.

Pedersen, K., K.S. Bloom, J.N. Anderson, D.V. Glover, and B.A. Larkins (1980) Analysis of the complexity and frequency of zein genes in the maize genome. Biochemistry 19: 1644–1650.

Pedersen, K., J. Devereux, D.R. Wilson, E. Sheldon, and B.A. Larkins (1982) Cloning and sequence analysis reveal structural variation among related zein genes in maize. Cell 29: 1015–1026.

Prat, S., J. Cortadas, P. Puigdomenech, and J. Palau (1985) Nucleic acid (cDNA) and amino acid sequences of the maize endosperm protein glutelin-2. Nucleic Acids Res. 13: 1493–1504.

Prat, S., L. Perez-Grau, and P. Puigdomenech (1987) Multiple variability in the sequence of a family of maize endosperm proteins. Gene 52: 41–49.

Pysh, L.D., M.J. Aukerman, and R.J. Schmidt (1993) OHP1: a maize basic domain/leucine zipper protein that interacts with Opaque2. Plant Cell 5: 227–236.

Randall, J.J., D.W. Sutton, S.F. Hanson, and J.D. Kemp (2005) BiP and zein binding domains within the delta zein protein. Planta 221: 656–666.

Righetti, P.G., E. Gianazza, A. Viotti, and C. Soave (1977) Heterogeneity of storage proteins in maize. Planta 136: 115–123.

Schmidt, R.J., F.A. Burr, M.J. Aukerman, and B. Burr (1990) Maize regulatory gene opaque-2 encodes a protein with a "leucine-zipper" motif that binds to zein DNA. Proc. Natl. Acad. Sci. U. S. A. 87: 46–50.

Schmidt, R.J., F.A. Burr, and B. Burr (1987) Transposon tagging and molecular analysis of the maize regulatory locus *opaque-2*. Science 238: 960–963.

Segal, G., R. Song, and J. Messing (2003) A new opaque variant of maize by a single dominant RNA-interference-inducing transgene. Genetics 165: 387–397.

Shank, K.J., P. Su, I. Brglez, W.F. Boss, R.E. Dewey, and R.S. Boston (2001) Induction of lipid metabolic enzymes during the endoplasmic reticulum stress response in plants. Plant Physiol. 126: 267–277.

Soave, C., R. Reggiani, N. Di Fonzo, and F. Salamini (1981) Clustering of genes for 20 kd zein subunits in the short arm of maize chromosome 7. Genetics 97: 363–377.

Soave, C. and F. Salamini (1984) Organization and regulation of zein in maize endosperm. Phil. Trans. R. Soc. Lond. B. 304: 341–347.

Soave, C., N. Surman, A. Viotti, and F. Salamini (1978) Linkage relationships between regulatory and structural gene loci involved in zein synthesis in maize. Theor. Appl. Genet. 52: 263–267.

Sodek, L. and C.M. Wilson (1971) Amino acid composition of proteins isolated from normal, opaque-2 and floury-2 corn endosperms by a modified Osborne procedure. J. Agric. Food Chem. 19: 1144–1150.

Song, R., V. Llaca, E. Linton, and J. Messing (2001) Sequence, regulation, and evolution of the maize 22-kD alpha zein gene family. Genome Res. 11: 1817–1825.

Song, R. and J. Messing (2003) Gene expression of a gene family in maize based on noncollinear haplotypes. Proc. Natl. Acad. Sci. U. S. A. 100: 9055–9060.

Spena, A., A. Viotti, and V. Pirrotta (1983) Two adjacent genomic zein sequences: structure, organization and tissue-specific restriction pattern. J. Mol. Biol. 169:799–811.

Spena, A., A. Viotti, and V. Pirrotta (1982) A homologous repetitive block structure underlies the heterogeneity of heavy and light chain zein genes. EMBO J. 1: 1589–1594.

Swarup, S., Timmermans, M.C.P., Chaudhuri, S., and Messing, J. (1995). Determinants of the high-methionine trait in wild and exotic germplasm may have escaped selection during early cultivation of maize. The Plant J. 8: 359–368.

Tatham, A.S., J.M. Field, V.J. Morris, K.J. I'Anson, L. Cardle, M.J. Dufton, and P.R. Shewry (1993) Solution conformational analysis of the alpha-zein proteins of maize. J. Biol. Chem. 268: 26253–26259.

Thompson, G.A., D.R. Siemieniak, L.C. Sieu, J.L. Slightom, and B.A. Larkins (1992) Sequence analysis of linked maize 22 kDa alpha-zein genes. Plant Mol. Biol. 18: 827–833.

Ueda, T., Wang, Z., Pham, N., and Messing, J. (1994). Identification of a transcriptional activator-binding element in the 27-kilodalton zein promoter, the -300 element. Mol. Cell. Biol. 14: 4350–4359.

Valentini, G., C. Soave, and E. Ottaviano (1979) Chromosomal location of zein genes in *Zea mays*. Heredity 42: 33–46.

Vicente-Carbajosa, J., S.P. Moose, R.L. Parsons, and R.J. Schmidt (1997) A maize zinc-finger protein binds the prolamin box in zein gene promoters and interacts with the basic leucine zipper transcriptional activator Opaque2. Proc. Natl. Acad. Sci. U. S. A. 94: 7685–7690.

Viotti, A., E. Sala, P. Alberi, and C. Soave (1978) Heterogeneity of zein synthesized in vitro. Plant Sci. Lett. 13.

Viotti, A., E. Sala, R. Marotta, P. Alberi, C. Balducci, and C. Soave (1979) Genes and mRNAs coding for zein polypeptides in *Zea mays*. Eur. J. Biochem. 102: 211–222.

Wang, Z., Ueda, T., and Messing, J. (1998). Characterization of the maize prolamin box-binding factor-1 (PBF-1) and its role in developmental regulation of the zein multigene family. Gene 223: 321–332.

Wang, Z., and Messing, J. (1998). Modulation of gene expression by DNA-protein and protein-protein interactions in the promoter region of the zein multigene family. Gene 223: 333–345.

Washida, H., A. Sugino, J. Messing, A. Esen, and T.W. Okita (2004) Asymmetric localization of seed storage protein RNAs to distinct subdomains of the endoplasmic reticulum in developing maize endosperm cells. Plant Cell Physiol. 45: 1830–1837.

Weinand, U., C. Bruschke, and G. Feix (1979) Cloning of double stranded DNAs derived from polysomal mRNA of maize endosperm: isolation and characterisation of zein clones. Nucleic Acids Res. 6: 2707–2715.

Wienand, U., P. Langridge, and G. Feix (1981) Isolation and characterization of a genomic sequence of maize coding for a zein gene. Mol. Gen. Genet. 182: 440–444.

Wilson, D.R. and B.A. Larkins (1984) Zein gene organization in maize and related grasses. J. Mol. Evol. 20: 330–340.

Wolf, M.J., U. Khoo, and H.L. Seckinger (1969) Distribution and subcellular structure of endosperm protein in varieties of ordinary and high-lysine maize. Cereal Chem. 46: 253–263.

Woo, Y.M., D.W. Hu, B.A. Larkins, and R. Jung (2001) Genomics analysis of genes expressed in maize endosperm identifies novel seed proteins and clarifies patterns of zein gene expression. Plant Cell 13: 2297–2317.

Zhang, F. and R.S. Boston (1992) Increases in binding protein (BiP) accompany changes in protein body morphology in three high-lysine mutants of maize. Protoplasma 171: 142–152.

Zimmerberg, J. and M.M. Kozlov (2006) How proteins produce cellular membrane curvature. Nat. Rev. Mol. Cell Biol. 7: 9–19.

The Cytochrome P450 Superfamily of Monooxygenases

Alfons Gierl

Abstract Cytochrome P450 monooxygenases (P450) are encoded by a superfamily of genes that is ubiquitously present in bacteria, animals and plants. Plants have many different P450s and use them for biosynthesis and for detoxification. Plant P450s function in primary and secondary metabolism and are involved in biosynthesis of hormons and signalling molecules.

1 Introduction

Nature has evolved several ways to utilize molecular oxygen to functionalize organic molecules. Flavin, metalloporphyrin, non-heme iron, and copper have been recruited as cofactors for oxygenases. The heme-containing cytochromes P450 (P450) are monooxygenases that catalyze the incorporation of one oxygen atom of atmospheric dioxygen into a substrate with the simultaneous reduction of the other oxygen atom to water. These enzymes have been named for the spectral properties of the heme-containing red pigment, which, in their reduced carbon-monoxide bound form, display a typical absorption band at 450 nm.

   P450 genes are not classified according to the specific reaction that is catalyzed, but on the basis of structural homology. They are identified by the root symbol *CYP* followed by a number designating the family (proteins with more than 40% sequence identity), a letter designating a subfamily (more than 55% identity) and a number identifying the individual gene within the subfamily, for example, *CYP71C2*. So-called clans represent groups of homologous families that can be derived from a common progenitor. A detailed compilation of the more than 5000 P450 genes that have been cloned to date and more information on nomenclature can be obtained from http://drnelson.utmem.edu/cytochromeP450.html.

A. Gierl
Technische Universität München, Plant Science Department
gierl@wzw.tum.de

2 Enzymatic Reaction and Structural Features

Cytochrome P450s catalyze the following reaction:

$$RH + O_2 + NAD(P)H + H^+ \rightarrow ROH + H_2O + NAD(P)^+$$

P450s attack non-activated hydrocarbons regio and stereo specifically and catalyze diverse reactions such as hydroxylation, N-, O-, S-dealkylation, decarboxylation, desaturation, epoxydation, ring expansion, and C-C cleavage (Bernhardt 2006). A detailed description of the catalytic chemistry can be found in a recent review (Denisov et al. 2005) and will not be discussed here.

Two main P450 classes have been defined that differ with respect to the redox system involved. Bacterial, mitochondrial and plastidal P450s (class I) are soluble or membrane associated proteins that typically require both a FAD-containing reductase and an iron sulfur containing ferredoxin. No plant P450s have yet been undoubtedly localized to mitochondria. The allenoxide synthase (AOS, Laudert et al. 1996), kaurene oxidase in gibberillic acid biosynthesis (Helliwell et al. 2001) and two P450s involved in carotenoid biosynthesis (Kim and DellaPenna 2006; Tian et al. 2004) represent examples for plastidal enzymes that are localized in different compartments of the organelle. In plants, estimated on the basis of targeting sequences, 5% of all P450s are estimated to fall into class I (Schuler et al. 2006).

The class II P450s are the most common eukaryotic enzymes that obtain electrons from NADPH via a FAD and FMN-containing P450 reductase. Typically these P450s and the P450 reductase are dissociated and anchored in the outer face of the endoplasmic reticulum (Figure 1). There are only one or a few P450 reductases encoded in the genome of a plant. A P450 reductase can transfer the redox equivalents from NADPH to many if not all the different P450 isozymes of an organism. Besides these two main classes, minor P450 classes have been defined (for review see Hannemann et al. 2007).

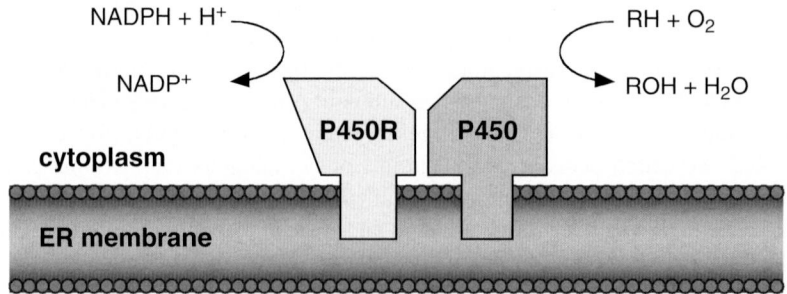

Fig. 1 Class II P450 enzymes. This most common eukaryotic cytochrome P450 system is localized in the endoplasmatic reticulum (ER). The system comprises two independent integral membrane proteins: the cytochrome P450 monooxygenase (P450) and the NADPH-cytochrome P450 reductase (P450R). P450s accept electrons from the FAD and FMN containing P450 reductase. One P450 reductase can transfer the redox equivalents from NADPH to many different P450s

3   Evolution of the Plant P450 Superfamily

The cytochrome P450s have been found in all kingdoms of life and represent one of the largest and oldest gene superfamilies. In animals and plants there are two main functional roles for these enzymes. One is the detoxification of xenobiotics in order to protect the organism. The second functional role is in the biosynthesis of hormones, vitamins and some other metabolites. In plants, P450s are also frequently involved in the biosynthesis of a large number of secondary metabolites that often function as defense toxins against herbivores and pathogens or as flower pigments and fragrance for pollinators. This additional function explains the fact that plants have three to five times as many P450 genes as animals (Nelson 2006).

In part, plant and animal P450s are related by co-evolution. As plants and animals diverged more than a billion years ago, the first herbivores developed. Plants started to produce secondary metabolites in order to combat the herbivores and animals evolved new P450s to detoxify these substances. This co-evolution most likely increased when the animals arrived on land and encountered new terrestrial plant species. In this phase of plant evolution several new P450 clans evolved, one of which is the CYP71 clan (see below) that predominantly encodes enzymes for secondary metabolism. A striking example for plant/animal co-evolution is represented by the corn earworm *Helicoverpa zea*. Four P450s are induced in this herbivore by the plant defense signals jasmonate and salicylate, which in the plant induce the synthesis of toxins after herbivory. This mechanism assures the activation of detoxifying P450s in the herbivore before or concomitant with the induction of plant defense (Li et al., 2002).

The evolutionary analysis of P450 genes was strongly promoted by the advent of complete genome sequences. The genome of *Arabidopsis thaliana* contains 245 P450 genes (Schuler et al. 2006), while the number of P450 genes in rice is approximately 350 (Nelson et al. 2004). Thus the number of P450 genes comprises roughly 1% of the genes in a plant genome.

An elegant phylogenetic analysis was recently provided by comparing P450 genes from the green algae *Chlamydomonas reinhardtii*, the moss *Phycomitrellla patens*, *A. thaliana* and Populus (Nelson 2006). In this way four ancient P450 families have been identified that predate the emergence of land plants, e.g. CYP51, a sterol demethylase that exists in most eukaryotes. With one exception, all remaining nine P450 clans of higher plants are represented by the 71 P450 genes in moss. Therefore a burst of P450 innovations must have occurred after emergence on land and arrival at the moss stage. During that period an "explosion" of new P450 genes can be seen that generated five new clans, amongst them CYP71 that comprises nearly one half of all modern angiosperm P450 genes. Most of the CYP71 enzymes are integrated in secondary metabolism. A comparison of the rice and Arabidopsis P450 genes (Nelson et al. 2004) revealed that most of the known plant P450 families existed before the monocot-dicot divergence that occurred approximately 200 million years ago.

4   Function of 450s in Plants and in Maize

The functional analysis of plant P450s began with the characterization of cinnamate 4-hydroxylase (C4H) from Jerusalem artichoke (Benveniste et al. 1977). This enzyme catalyzes a paramount reaction in the phenylpropanoid pathway, the conversion of cinnamic acid to coumaric acid and controls the flux into lignins, flavonoids and anthocyanins. As indicated above, P450 enzymes function in the detoxification of xenobiotics and in biosynthetic pathways. Biosynthetic P450s play a predominant role in secondary metabolism and hence individual members might often occur only in certain phylogenetic classes. However, several biosynthetic P450 are also essential for primary metabolism catalyzing reactions common to all plants, including maize (Table 1). These P450s catalyze often rate limiting steps and function in the core phenylpropanoid pathway, in lignin biosynthesis, in the modification of fatty acids and the synthesis or degradation of plant hormones including gibberillins, brassinolides, auxins, abscisic acid, and signaling molecules such as jasmonate and salicylic acid (Table 1).

Table 1 Examples of functionally defined P450s common in all plants

CYP Subfamily	Pathway	Enzymatic Activity	References
51A2	Sterols/steroids	Obtusifoliol 14α-demethylase	Kushiro et al. 2001
72B1	Degradation of brassinosteroids	26-hydroxylase	Neff et al. 1999
72C1	Degradation of brassinosteroids	exact substrate not known	Nakamura et al. 2005
			Turk et al. 2005
73A5	Phenylpropanoid	Cinnamate 4-hydroxylase	Mizutani et al. 1993
74A1	Oxylipin	Allene oxide synthase	Laudert et al. 1996
74 B2	Oxylipin	Hydroperoxide lyase	Bate et al. 1998
85 A1	Brassinolide	multiple C6-oxidase	Bishop et al. 1999
85 A2	Brassinolide	multiple C6-oxidase	Shimada et al. 2003
86A1	Fatty acids	ω-hydroxylase for C12 to C18 fatty acids	Benveniste et al. 1998
88A3	Gibberillin	*ent*-kaureoic acid oxidase	Helliwell et al. 2001
88A4	Gibberillin	*ent*-kaureoic acid oxidase	Helliwell et al. 2001
90A1	Brassinolide	23α-hydroxylase for 6-oxo-cathasterone	Szekeres et al. 1996
97A3	Carotenoid	b-ring oxidase	Kim and DellaPenna 2006
98A3	Phenylpropanopid	3′-hydroxylase (C3′H)	Schoch et al. 2001
701 A3	Gibberellin	*ent*-kaurene oxidase	Helliwell et al. 1998
707A1	Degradation of abscisic acid	8′-hydroxylase	Saito et al. 2004
710A1	Sterol	C-22 desaturase for b-sitosterol	Morikawa et al. 2006
734A1	Degradation of brassinosteroids	26-hydroxylase	Turk et al. 2003
735A2	Cytokinin	*trans*-hydroxylase	Takei et al. 2004

Relatively few P450 enzymes have been characterized in maize to a level where not only the gene sequence was identified, but also the exact biochemical function was established. In contrast, approximately 40 different P450s have been functionally dissected in *A. thaliana*. The analysis of the whole P450 gene family was greatly stimulated by the complete genome annotation of this model species.

For maize, three examples comprising gibberellin biosynthesis, the induction of P450s upon treatment with safeners and bacterial toxins, and the function in benzoxazinone biosynthesis are summarized below, to illustrate ongoing P450 research in maize.

The benzoxazinoid DIBOA and its methoxy derivative DIMBOA are secondary metabolites that function as natural pesticides and serve as important factors in resistance to microbial diseases and insects, and as allelo-chemicals (Sicker et al. 2000). These toxins are predominantly expressed in seedlings and during juvenile stages of plant development. In maize, a series of five genes is sufficient to encode the enzymes to synthesize DIBOA (Frey et al. 1997). The first enzyme in this pathway, BX1, catalyzes the formation of indole (see *Insect resistance*, McMullen et al., Volume 1) that is converted to DIBOA by the consecutive action of four different P450 enzymes (Figure 2).

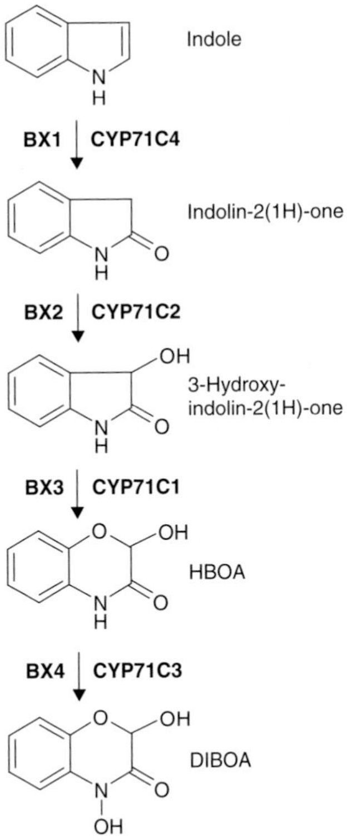

Fig. 2 DIBOA biosynthesis in maize. Indole is synthesized in the plastid by BX1. The four P450 enzymes involved in benzoxazinone biosynthesis have been termed BX2-BX5 (Frey et al. 1997). They are members of the CYP71C subfamily and share an overall amino acid identity of 45%-65%. The stepwise conversion of indole to 2,4-dihydroxy-2*H*-1,4-benzoxazin-3(4*H*)-one (DIBOA) includes the unusual ring expansion to 2-hydroxy-2*H*-1,4-benzoxazin-3(4*H*)-one (HBOA).

The four P450 genes involved in DIBOA biosynthesis have been termed *Bx2-Bx5*. They are members of the CYP71C subfamily of plant cytochrome P450 genes and share an overall amino acid identity of 45-65%. The stepwise conversion of indole to DIBOA includes C- and N-hydroxylations and one ring expansion. The fact that four reactions are catalyzed by P450 enzymes in one biosynthetic pathway is rare in plants and only exceeded by the eight P450 enzymes that are required for biosynthesis of the anticancer drug taxol in *Taxus* (Croteau et al. 2006). Genetic mapping of the *Bx* genes in maize had the surprising result that these genes are clustered on the short arm of chromosome 4, indicating that the P450 genes evolved by tandem gene duplication. The cluster of P450 genes is linked to the *Bx1* gene and to the remaining DIMBOA biosynthetic genes. BX2-BX5 are localized in the endoplasmatic reticulum, but it is not known whether they form an ordered complex of sequential enzymes that function as a metabolon (Srere 1985) channeling the substrates from one P450 to the next.

Plant P450s are generally involved in the protection against animals, insects and microbes and their expression can be induced by bacterial elicitors and xenobiotics, such as safeners, and by wounding (Persans et al. 2001). Several maize P450 genes that are induced by chemicals, such as naphthalic anhydride, triasulfuron and phenobarbital, and bacterial elicitors derived from *Erwinia stuartii* and *Acidovorax avenae* were identified in this study. These P450 transcripts have distinct expression patterns and are specifically regulated by the chemical inducers. The corresponding genes belong to the CYP71C, CYP72A and CYP92A subfamilies. Interestingly, the *Bx4* and *Bx5* genes, involved in DIBOA biosynthesis, are induced in response to wounding and to treatment with naphthalic anhydride and triasulfuron (Persans et al. 2001).

Gibberellins (GAs) are phytohormones required for normal growth and development in higher plants. The *Dwarf3* locus of maize encodes an *ent*-kaurenoic acid oxidase (Helliwell et al. 2001) catalyzing the three steps in GA biosynthesis from *ent*-kaurenoic acid to GA_{12}. The *Dwarf3* gene was isolated by transposon-tagging (Winkler and Helentjaris 1995) using Robertson's Mutator, and it was demonstrated that this locus encodes the CYP88A gene.

5   Biotechnological Application

The molecular isolation of the first plant P450 gene (Bozak et al. 1990) has opened the gate to exploit the potential of the large reaction spectrum of these enzymes that is obviously related to the versatile capacity of plants to synthesize a highly diverse collection of natural products. P450s efficiently catalyze a variety of regio specific and stereo specific and often "delicate" bio-transformations at physiological conditions.

P450s catalyze most oxidation reactions in secondary metabolism and are involved in detoxification of xenobiotics. Therefore these enzymes are targets for metabolic engineering to improve the biosynthesis of drugs and plant resistance against insects and microbes and to modify herbicide tolerance.

The diterpenoid taxol is an important antimitotic drug that is effective against a variety of carcinomas, sarcomas and melanomas. Its complex biosynthesis (Croteau et al., 2006) requires 19 enzymatic steps including the sequential action of eight P450-mediated oxygenations. Taxol is isolated from the bark of Pacific yew (*Taxus brevifolia*). Due to the complex structure it is not possible to improve the availability of this potent drug by total chemical synthesis. This situation has prompted the isolation of biosynthetic intermediates that are more abundant and genetic engineering of plant cell cultures (Ketchum et al., 2007) as well as yeast cells (Dejon et al. 2006).

Cyanogenic glucosides are potent defense compounds in response to herbivores resulting in the release of toxic hydrogen cyanide. The cyanogenic glucoside dhurrin is synthesized in *Sorghum bicolor*. Three consecutive P450 catalyzed reactions convert tyrosine to dhurrin. The potential of metabolic engineering of a defense pathway was demonstrated by implementation of the entire pathway in *A. thaliana* (Tattersall et al. 2001) that naturally does not contain cyanogenic glucosides. The dhurrin containing *A. thaliana* gained resistance against an insect herbivore specific for cruciferous plants.

In addition to the two examples described above, several other biotechnological approaches have been described (for review see Morant et al., 2003). The production of natural products whose biosynthesis involve P450s can be scaled up by heterologous expression in bacteria, yeast or insect cells (for review see Duan and Schuler 2006). Efficient expression of these membrane-localized proteins requires optimization of the redox environment and the protein expression level in the respective system used.

6   Conclusion

The analysis of the evolution and the functional characterization of P450 genes drastically benefited from complete genome annotation. A comparative genomic analysis of *Arabidopsis* and rice revealed several P450 families that are unique for each species (Nelson et al., 2004). Therefore, the ongoing genome sequencing projects will stimulate the functional analysis of the P450 super family and provide more possibilities for biotechnological applications.

References

Bate NJ, S. Sivasankar, C. Moxon, J.M. Riley, J.E. Thompson and S.J. Rothstein (1998) Molecular characterization of an Arabidopsis gene encoding hydroperoxide lyase, a cytochrome P-450 that is wound inducible. *Plant Physiol.* **117**:1393–400.

Benveniste I., J.-P. Salün and F. Durst (1977) Wound-induced cinnamic acid hydroxylase in Jerusalem artichoke tuber. *Phytochem* **16**:69–73.

Benveniste I., N. Tijet, F. Adas, G. Philipps, J.P. Salaun and F Durst (1998) CYP86A1 from Arabidopsis thaliana encodes a cytochrome P450-dependent fatty acid omega-hydroxylase. *Biochem Biophys Res Commun.* **243**:688–93.

Bernhardt R. (2006) Cytochromes P450 as versatile biocatalysts. *J Biotechnol.* **124**:128–145.

Bishop G.J., T. Nomura, T. Yokota, K. Harrison, T. Noguchi, S. Fujioka, S. Takatsuto, J.D. Jones and Y. Kamiya (1999) The tomato DWARF enzyme catalyzes C-6 oxidation in brassinosteroid biosynthesis. *Proc Natl Acad Sci USA.* **96**:1761–1766.

Bozak K.R., H. Yu, R. Sirevåg and R.E. Christoffersen (1990) Sequence analysis of ripening-related cytochrome P-450 cDNAs from avocado fruit. *Proc Natl Acad Sci U S A.* **87**:3904–3908.

Croteau R., R.E.B. Ketchum, R.M. Long, R. Kaspera and M.R. Wildung (2006) Taxol biosynthesis and molecular genetics. *Phytochem Rev* **5**:75–97

Dejong J.M., Y. Liu, A.P. Bollon, R.M. Long, S. Jennewein, D. Williams and R.B. Croteau RB. (2006) Genetic engineering of taxol biosynthetic genes in Saccharomyces cerevisiae. *Biotechnol Bioeng.* **93**:212–224.

Denisov I.G., T.M. Makris, S.G. Sligar and I. Schlichting (2005) Structure and chemistry of cytochrome P450. *Chem Rev.* **105**:2253–2277.

Duan H. and M.A. Schuler (2006) Heterologous expression and strategies for encapsulation of membrane-localized plant P450s. *Phytochem Rev* **5**:507–523.

Frey M., P. Chomet, E. Glawischnig, C. Stettner, S. Grün, A. Winklmair, W. Eisenreich, A. Bacher, R.B. Meeley, S.P. Briggs, K. Simcox and A. Gierl (1997) Analysis of a chemical plant defense mechanism in grasses. *Science* **277**:696–699.

Hannemann F., A. Bichet, K.M. Ewen and R. Bernhardt R. (2007) Cytochrome P450 systems–biological variations of electron transport chains. *Biochim Biophys Acta.* **1770**:330–344.

Helliwell C.A., C.C. Sheldon, M.R. Olive, A.R. Walker, J.A. Zeevaart, W.J. Peacock and E.S. Dennis (1998) Cloning of the Arabidopsis ent-kaurene oxidase gene GA3. Proc Natl Acad Sci U S A. **95**:9019–9024.

Helliwell C.A., P.M. Chandler, A. Poole, E.S. Dennis and W.J. Peacock (2001) The CYP88A cytochrome P450, ent-kaurenoic acid oxidase, catalyzes three steps of the gibberellin biosynthesis pathway. Proc Natl Acad Sci U S A **98**:2065–2070.

Ketchum R.E., L. Wherland and R.B. Croteau (2007) Stable transformation and long-term maintenance of transgenic Taxus cell suspension cultures. *Plant Cell Rep.* **26**:1025–1033.

Kim J. and D. DellaPenna (2006) Defining the primary route for lutein synthesis in plants: the role of Arabidopsis carotenoid beta-ring hydroxylase CYP97A3. Proc Natl Acad Sci U S A. **103**:3474–3479.

Laudert D., U. Pfannschmidt, F. Lottspeich, H. Hollander-Czytko and E.W. Weiler. Cloning, molecular and functional characterization of Arabidopsis thaliana allene oxide synthase (CYP 74), the first enzyme of the octadecanoid pathway to jasmonates. *Plant Mol Biol.* **31**:323–335.

Li X., M.A. Schuler and M.R. Berenbaum MR. (2002) Jasmonate and salicylate induce expression of herbivore cytochrome P450 genes. *Nature* **419**:712–5.

Mizutani M, D. Ohta and R. Sato (1993) Purification and characterization of a cytochrome P450 (truns-cinnamic acid 4-hydroxylase) from etiolated mung bean seedlings. *Plant Cell Physiol* **34**:481–488.

Morant M., S. Bak, B.L.Moller and D. Werck-Reichhart (2003) Plant cytochromes P450: tools for pharmacology, plant protection and phytoremediation. *Curr Opin Biotechnol.* **14**:151–162.

Morikawa T.,M. Mizutani and D. Ohta (2006) Cytochrome P450 subfamily CYP710A genes encode sterol C-22 desaturase in plants. Biochem Soc Trans. **34**:1202–1205.

Nelson D.R. (2004) Cytochrome P450 nomenclature. *Methods Mol Biol.* **320**:1–10.

Nelson D.R., M.A. Schuler, S.M. Paquette, D. Werck-Reichhart and S. Bak (2004) Comparative genomics of rice and Arabidopsis. Analysis of 727 cytochrome P450 genes and pseudogenes from a monocot and a dicot. *Plant Physiol.* **135**:756–772.

Nelson D.R. (2006) Plant cytochrome P450s from moss to poplar. *Phytochem Rev* **5**:193–204.

Persans M.W., J. Wang, M.A. Schuler (2001) Characterization of maize cytochrome P450 monooxygenases induced in response to safeners and bacterial pathogens. *Plant Physiol.* **125**:1126–38.

Saito S., N. Hirai, C. Matsumoto, H. Ohigashi, D. Ohta, K. Sakata and M. Mizutani M. (2004) Arabidopsis CYP707As encode (+)-abscisic acid 8′-hydroxylase, a key enzyme in the oxidative catabolism of abscisic acid. *Plant Physiol.* **134**:1439–1449.

Schoch G., S. Goepfert, M. Morant, A. Hehn, D. Meyer, P. Ullmann and D. Werck-Reichhart (2001) CYP98A3 from Arabidopsis thaliana is a 3′-hydroxylase of phenolic esters, a missing link in the phenylpropanoid pathway. *J Biol Chem.* **276**:36566–36574.

Schuler, M.A., H. Duan, M. Bilgin and S. Ali (2006) Arabidopsis cytochrome P450s through the looking glass: a window on plant biochemistry. *Phytochem Rev* **5**:205–237.

Sicker D, M. Frey, M. Schulzand A. Gierl (2000) Role of natural benzoxazinones in the survival strategy of plants. *Int Rev Cytol.* **198**:319–46.

Shimada Y., H. Goda, A. Nakamura, S. Takatsuto, S. Fujioka and S. Yoshida (2003) Organ-specific expression of brassinosteroid-biosynthetic genes and distribution of endogenous brassinosteroids in Arabidopsis. *Plant Physiol.* **131**:287–97.

Srere P.A. (1987) Complexes of sequential metabolic enzymes. *Annu Rev Biochem* **56**:89–124

Szekeres M., K. Nemeth, Z. Koncz-Kalman, J. Mathur, A. Kauschmann, T. Altmann, G.P. Redei, F. Nagy, J. Schell and C. Koncz C. (1996) Brassinosteroids rescue the deficiency of CYP90, a cytochrome P450, controlling cell elongation and de-etiolation in Arabidopsis. *Cell* **85**:171–82.

Takei K., T. Yamaya and H. Sakakibara H. (2004) Arabidopsis CYP735A1 and CYP735A2 encode cytokinin hydroxylases that catalyze the biosynthesis of trans-Zeatin. *J Biol Chem.* **279**:41866–41872.

Tattersall D.B., S. Bak, P.R. Jones, C.E. Olsen, J.K. Nielsen, M.L. Hansen, P.B. Hoj and B.L. Moller (2001) Resistance to an herbivore through engineered cyanogenic glucoside synthesis. *Science* **293**:1826–1828.

Tian L., V. Musetti, J. Kim, M. Magallanes-Lundback and D. DellaPenna (2004) The Arabidopsis LUT1 locus encodes a member of the cytochrome P 450 family that is required for carotenoid epsilon-ring hydroxylation activity. *Proc Natl Acad Sci U S A.* **101**:402–7.

Turk E.M., S. Fujioka, H. Seto, Y. Shimada, S. Takatsuto, S. Yoshida, M.A. Denzel, Q.I. Torres and M.M. Neff MM. (2003) CYP72B1 inactivates brassinosteroid hormones: an intersection between photomorphogenesis and plant steroid signal transduction. *Plant Physiol.* **133**:1643–1653.

Turk E.M., S. Fujioka, H. Seto, Y. Shimada, S. Takatsuto, S. Yoshida, H. Wang, Q.I. Torres, J.M. Ward, G. Murthy, J. Zhang, J.C. Walker and M.M. Neff (2005) BAS1 and SOB7 act redundantly to modulate Arabidopsis photomorphogenesis via unique brassinosteroid inactivation mechanisms. *Plant J.* **42**:23–34.

Winkler RG and T Helentjaris (1995) The maize Dwarf3 gene encodes a cytochrome P450-mediated early step in Gibberellin biosynthesis. *Plant Cell* **7**:1307–1317.

Cell wall Biosynthetic Genes of Maize and their Potential for Bioenergy Production

Wilfred Vermerris

Abstract The maize cell wall is a complex composite of cellulose, hemicellulosic polysaccharides pectin, proteins, and lignin, and representative of the unique Type II cell wall architecture common among the Poales. The genetic control of cell wall biosynthesis in maize is being actively pursued, using a combination of comparative genomics, forward and reverse genetics, and expression profiling. This has revealed the existence of many multi-gene families with individual members that are differentially expressed. While the precise function of most of the individual genes is yet to be established, it is clear that the existence of multi-gene families enables the synthesis of cell walls that are tailored to the specific needs of individual tissues throughout the life of the plant. A better understanding of cell wall biosynthesis will be of great value for the development of dedicated bioenergy crops used for the production of cellulosic ethanol, especially if cell wall traits are combined with morphological variants with increased biomass production.

1 Introduction

A distinctive feature of plant cells is the presence of a rigid cell wall surrounding the cell membrane. The cell wall is instrumental in providing cell shape, given that removal of the cell wall with the use of hydrolytic enzymes will turn any plant cell into a globular protoplast. Aside from providing structure and mechanical strength, the cell wall enables cell-to-cell contact and communication, protects the cell from biotic and abiotic stress, and mediates the transport of water and minerals.

The cell wall is a complex matrix formed by several different polymers that interact to form a functional structure. The main constituents of plant cell walls are cellulose, hemicellulosic polysaccharides, pectin, lignin, and proteins. The proteins can have a structural, enzymatic or signaling function. The variation in cell shape and function

W. Vermerris
University of Florida Genetics Institute and Agronomy department, Gainesville, FL, Department
of Agricultural & Biological Engineering and Laboratory of Renewable Resources Engineering,
Purdue University, West Lafayette, IN, USA

is dictated by the relative proportion of these polymers, and the way they interact with each other, which in turn depends on the species and the developmental stage.

All cells have a primary cell wall that is formed after cell division. When two cells divide, the newly formed cells are initially separated by a pectin-rich layer that forms around the phragmoplast. As other cell wall polymers are deposited, the pectin layer becomes the middle lamella.

Certain specialized cells, notably those in the xylem and sclerenchyma, contain a secondary cell wall that is deposited in between the membrane and the primary wall. The secondary wall is typically rich in lignin and other phenolic compounds, such as hydroxycinnamic acids. Cells that contain a secondary cell wall often undergo programmed cell death, during which the cell content is purged (lysis) and a hollow tube is formed. The secondary wall is then exposed to the lumen side of this newly formed tube.

Detailed background information on various aspects of cell wall biogenesis can be found in Carpita (1996), Carpita and McCann (2000), Ralph et al. (2004a), Yong et al. (2005), and Vermerris and Nicholson (2006). This chapter will focus on recent developments regarding the genetic control of cell wall biosynthesis in maize and on the implications this has on the production of cellulosic ethanol as an alternative transportation fuel.

2   Cell Wall Constituents

2.1   *Cellulose*

Cellulose is present in both primary and secondary cell walls where it serves as the predominant structural polymer. It generally makes up 15–30% of the primary cell wall and 50–60% of the secondary cell wall. As a consequence, cellulose is the most abundant biopolymer on Earth.

Cellulose is a polymer of β-(1,4)-linked D-glucopyranose molecules. Since the two neighboring glucosyl residues are rotated 180° relative to each other, the repeat unit of cellulose is the dimer of D-glucose, cellobiose (structure **A** in Fig. 1).

In plants cellulose is present in microfibrils, which consist of bundles of on average thirty-six β-(1-4)-linked D-glucans that are bound together through hydrogen bonds. Each D-glucan chain is several thousand glucose units long and therefore limited in length, but because the chains do not all start and end at the same place, the length of the microfibril that these chains form can be considerable – in the order of several hundred micrometers. The diameter of cellulose microfibrils is between 5 and 12 nm wide. Within the microfibril the glucan chains run in parallel direction. More detailed information on the structure of cellulose can be found in Carpita and McCann (2000), Zugenmaier (2001), and Ding and Himmel (2006).

In plant cell walls the site of cellulose synthesis is generally a terminal complex (TC) that has a rosette structure with six-fold symmetry (Mueller and Brown,

1980). Each of the six subunits of a TC consists of six cellulose synthases, which are membrane-bound enzymes that use the nucleotide sugar UDP-glucose as a substrate and that add a glucosyl residue to the growing D-glucan chain. This is why most cellulose microfibrils in plants consist of thirty-six D-glucan chains (Saxena and Brown, 2005; Somerville, 2006).

Plant *cellulose synthase* genes were first identified in an expressed sequence tag (EST) collection derived from developing cotton bolls (Pear et al., 1996). The identification was based on sequence homology with the then recently obtained *cellulose synthase* gene from the bacterium *Acetobacter xylinum* (Saxena et al., 1990). Homology searches revealed the existence of rice and Arabidopsis ESTs encoding putative cellulose synthases (Pear et al., 1996; Cutler and Somerville, 1997). The first genetic proof demonstrating the function of a plant *cellulose synthase* gene came from studies by Arioli et al. (1998) in Arabidopsis. Plant cellulose synthases contain four highly conserved U-motifs likely to be involved in substrate binding, an amino-terminal zinc finger domain, a hypervariable (HVR) domain and a plant-specific domain (Pear et al., 1996; Delmer, 1999). The HVR domain is nowadays more accurately referred to as the class-specific region (Vergara and Carpita, 2001).

Since the initial cloning of the cotton and Arabidopsis *cellulose synthase* genes, homologs from many other species have been identified, resulting in a major research effort aimed at elucidating the mechanism and control of cellulose biosynthesis, summarized in a number of excellent reviews by Delmer (1999), Carpita et al. (2001), Perrin (2001), Doblin et al. (2002), Brown (2004), Saxena and Brown (2005) and Somerville (2006).

With the availability of whole-genome sequences and large gene expression databases it became clear that all plant species contain many genes with homology to *cellulose synthase* (Holland et al., 2000; Richmond and Somerville, 2000). These genes can be classified in gene families based on gene structure and the presence of conserved motifs. The genes encoding actual cellulose synthases are referred to as *CesA* genes, whereas genes with a lower level of homology to *CesA* genes are referred to as *cellulose synthase-like* (*Csl*) genes. The proteins encoded by the *Csl* genes generally lack the zinc finger domain and/or plant-specific sequences (Dhugga, 2001). Nine *Csl* gene families – *CslA* through *CslH* – have been identified based on sequence homology (Richmond and Somerville, 2000; 2001; Hazen et al., 2002), although not all species contain all nine families. Members of some of the *Csl* gene families have been shown to be involved in the synthesis of the backbone of the non-cellulosic polysaccharides: (gluco)mannan in guar (Dhugga et al., 2004) and Arabidopsis (*AtCslA9*; Liepman et al., 2005) or xyloglucan (*AtCslC4*; Cocuron et al., 2007) in Arabidopsis. The *CslD3* gene is involved in root tip growth in Arabidopsis (Wang et al., 2001, Favery et al. 2001) and rice (Kim et al., 2007). The functions of the corresponding maize orthologs have yet to be established, but it is plausible that maize and rice orthologs will have similar if not identical functions.

Molecular genetic analyses in Arabidopsis revealed that the six subunits of the cellulose synthase complex represent three different proteins, and that these three proteins differed between the primary and secondary wall. In the primary wall the subunits of the cellulose synthase complex are CesA1, CesA3, and CesA6 (Arioli et al., 1998; Fagard et al., 2000; Burn et al., 2002), whereas in the secondary wall these subunits are CesA4, CesA7 and CesA8 (Taylor et al., 1999; 2000; 2003). The role of the other four *CesA* genes remains to be fully elucidated, but there is some evidence for redundancy (Burn et al., 2002) or tissue-specific expression (Hamann et al., 2004).

The maize *cellulose synthase* (*ZmCesA*) genes have been described in detail by Holland et al. (2000), Dhugga (2001) and Appenzeller et al. (2004). All three studies were based on EST collections generated at Pioneer Hi-Bred (Johnston, IA). Holland et al. (2000) reported nine maize *CesA* genes. Three additional *CesA* genes, *ZmCesA10–12*, were identified by Appenzeller et al. (2004) in an EST library from the transition zone of the elongating internode, where the rate of secondary cell wall formation is known to increase. Phylogenetic analyses performed with the maize and Arabidopsis *CesA* sequences indicated that *ZmCesA10–12* cluster with Arabidopsis *CesA* genes known to be involved in secondary cell wall synthesis. The nine other *ZmCesA* genes can be grouped into three clusters: *ZmCesA1-3*, *ZmCesA4*, *5* and *9* and *ZmCesA6-8*. Based on the homology with and established function of the Arabidopsis *CesA* genes, all three clusters are likely involved in primary cell wall synthesis. There is, however, more sequence diversity between the maize and Arabidopsis *CesA* orthologs involved in primary cell wall synthesis than between the *CesA* orthologs involved in secondary cell wall synthesis. Furthermore, *ZmCesA12* is more closely related to *ZmCesA6-8* than to *ZmCesA10* and *11*. This led Appenzeller et al. (2004) to hypothesize that ZmCesA6–8 are involved in the later stages of primary wall formation. Gene expression data indicated that *ZmCesA1* and *3–8* were expressed at different levels in the majority of tissues, whereas *ZmCesA9* is expressed only at low levels. The expression of *ZmCesA2* was restricted to the roots, kernels and silks, with very low expression levels in the latter tissue. The *ZmCesA5* gene is the most highly expressed *CesA* gene in the developing endosperm. *ZmCesA10-12* were expressed at high levels in the stalk, a tissue rich in secondary cell wall. These data are consistent with the functions hypothesized based on sequence homology with the Arabidopsis *CesA* genes.

2.2   *Hemicellulosic Polysaccharides*

The term hemicellulose refers to a diverse class of polysaccharides that differ from cellulose in that they are noncrystalline and hydrolysable in acid or alkaline solutions. Hemicellulosic polysaccharides are tethered to the cellulose microfibrils through hydrogen bonds, and thus provide a mechanism to lock the cellulose microfibrils in place (Carpita, 1996; Somerville et al., 2004). Unlike the cell-membrane localized biosynthesis of cellulose, hemicellulosic polysaccharides are synthesized in the Golgi complex and secreted in vesicles (Carpita and McCann, 2000).

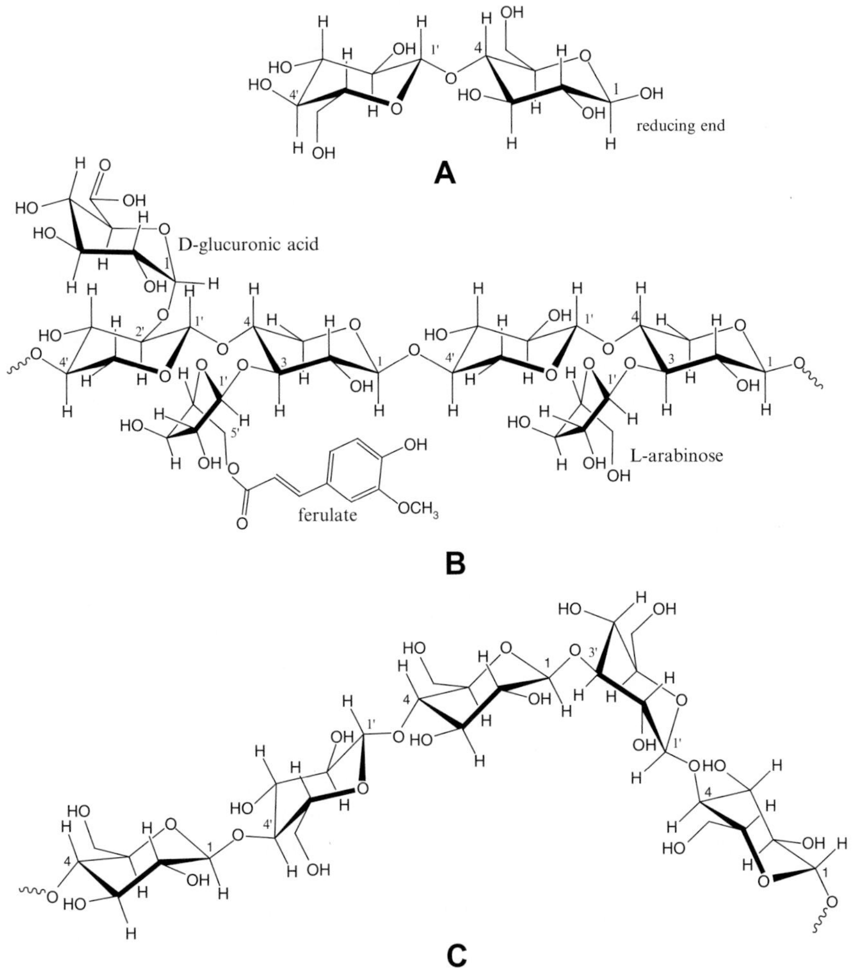

Fig. 1 Structural fragments of maize cell wall polysaccharides. Cellobiose (**A**) is the repeat unit of cellulose. GAX (**B**), the main hemicellulosic polysaccharide consists of a xylan backbone with arabinose, glucuronic acid and ferulic acid substitutions. Mixed-linkage β-glucans (**C**) consist of β-(1,4)-linked cellotriose and cellotetraose units that are connected *via* β-(1,3)-linkages

Hemicellulose composition in maize and other species in the order of Poales is quite unique. Most dicots and noncommelinoid monocots have glucan- or mannan-based polymers as their main cross-linking hemicellulosic polysaccharide and contain a considerable amount of pectin. This is referred to as a Type I wall (Carpita and Gibeaut, 1993). The Type II wall of maize and other grasses contains glucuronoarabinoxylan (GAX) as the predominant hemicellulosic polysaccharide. Aside from tethering cellulose microfibrils, this polymer is thought to substitute pectin to some extent, by controlling pore size and charge of the wall (Carpita,

1996). Mixed-linkage β-glucans represent another hemicellulosic polysaccharide that is unique to the Poales and that is abundant in the primary cell wall of maize.

2.2.1 Glucuronoarabinoxylans

Glucuronoarabinoxylans (GAXs) contain a xylan-based backbone with α-L-arabinose (Ara) and α-D-glucuronic acid (α-D-GlcA) substitutions. GAXs represent the predominant hemicellulosic polysaccharide in commelinoid monocots, where they make up 20-30% of the cell wall. In species belonging to that order, the Ara residues are linked to the xylose residue on the O-3 position. The α-D-GlcA residue is attached to the O-2 position of the xylan backbone. The Ara unit can be further substituted with ferulic acid. This compound is esterified to the O-5 position (see Fig. 1 structure **B**). The feruloylated arabinose residues are spaced approximately 50 xylose residues apart (Carpita and McCann, 2000). The degree of substitution of GAXs also varies as a function of plant development. The highest degree of substitution in maize coleoptiles was observed during maximal growth (Carpita and Whittern, 1986), whereas removal of arabinosyl residues occurred as cell elongation ceased (Carpita and Gibeaut, 1993). The lower degree of substitution is thought to allow the formation of hydrogen bonds between GAX and cellulose, resulting in a more rigid cell wall. The feruloyl substitution on the arabinosyl residues allows the formation of cross-linking diferulate bridges and oxidative coupling with lignin (Iiyama et al., 1990; 1994). In fact, the ferulate substituents have been proposed to act as nucleation sites for lignin in the cell walls of grasses (Hatfield et al., 1998). Myton and Fry (1994) showed that in cell cultures of tall fescue grass (*Festuca arundinacea*), the feruloyl substitution takes place in the protoplast, as opposed to in the cell wall. It is likely that this is also the case in maize.

The biosynthesis of GAX in maize is under investigation. It is likely that the enzymes responsible for the synthesis of the xylan backbone are encoded by *Csl* genes, and that different *Csl* genes will be expressed in different tissues and/or at different developmental stages. The enzymes responsible for adding the Ara substitutions are likely glycosyltransferases (GTs). This class of enzymes is encoded by several large gene families. The function of several GTs has been determined in Arabidopsis (*e.g.* Faik et al., 2002) and this is expected to help identify orthologs in maize.

2.2.2 Mixed-Linkage Beta-Glucans

Mixed-linkage β-glucans are transiently present in the primary wall during cell expansion. They are hydrolyzed when cell growth is complete (Carpita and Gibeaut, 1993). Mixed linkage β-glucans consist of D-glucose residues with no further substitution. The building blocks are cellotriose and cellotetraose units in a ratio of 2:1. The glucose residues within the cellotriose and cellotetraose units are linked by β-(1,4)-linkages, but the linkages between the cellotriose and cellotetraose units are β-(1,3)-linked (see Fig. 1, structure **C**). This latter linkage creates turns in the chain, so that the spatial structure of the polymer looks like a corkscrew. The corkscrew is interrupted every 50 residues by an oligomer of four or more β-(1,4)-linked D-glucose residues (Carpita and McCann, 2000). The rice *CslF* gene family encodes enzymes that catalyze the biosynthesis of mixed-linkage β-glucans (Burton et al., 2006). Based on the evolutionary relationship with rice, this is likely to be the case in maize as well.

2.3   Pectins

Pectins are branched, hydrated polymers rich in D-galacturonic acid. They affect the pore size of the wall (porosity), the charge (and hence the ion binding capacity, and the ability to bind charged cell wall proteins), as well as the pH of the wall. The middle lamella formed after cell division is largely made of pectins. Pectins are thought to influence cell-to-cell contact (Carpita and McCann, 2000). As stated earlier, the maize cell wall is low in pectin content relative to most dicots.

There are three types of pectic polymers. *Homogalacturonan* (HGA) are long polymers (up to 100 nm) of α-(1-4)-linked D-galacturonic acid. A portion of the carboxyl residues is methyl esterified, which affects the properties of the polymer in terms of its charge and ion-binding capacity, and hence the porosity and viscosity. *Rhamnogalacturonan I* (RG I) is a contorted, rod-like molecule made of a repeating disaccharide -2)-α-D-rhamnose-(1-4)-α-D-galacturonic acid-(1-. A portion of the carboxyl groups are acetylated. There are additional substitutions possible with arabinan, galactan, and arabinogalactan side chains. *Rhamnogalacturonan II* (RG II) and xylogalacturonan (XGA) are modified HGA molecules with complex structures. RGII can dimerize around boron residues (Carpita, 1996).

The biosynthesis of pectins is complex and not yet understood very well, especially in maize. Recent advances (with a focus on dicots) were reviewed by Scheller et al. (2007). The backbone of HGA is likely controlled by one or more CSL enzymes, and the substitution of the backbone is likely controlled by the same transferases that are used for the biosynthesis of the hemicellulosic polysaccharides, or by transferases that are similar but encoded by a set of genes specific for pectin biosynthesis.

2.4   Lignin and Hydroxycinnamic Acids

The phenolic polymer lignin is formed *via* oxidative coupling of monolignols. Intermediates from the monolignol biosynthetic pathway serve as precursors for flavonoids, hydroxycinnamic acids, and generally one or more order- or species-specific classes of phenolic compounds. A detailed description of the biosynthesis of these different phenolic compounds is provided by Vermerris and Nicholson (2006). Figure 2 shows a schematic representation of the monolignol biosynthetic pathway with the numbered compounds listed below. This pathway is based primarily on experimental data from Arabidopsis (reviewed by Humphreys and Chapple, 2002). Monolignol biosynthesis in maize is believed to largely follow the same route, although there are some maize-specific variations.

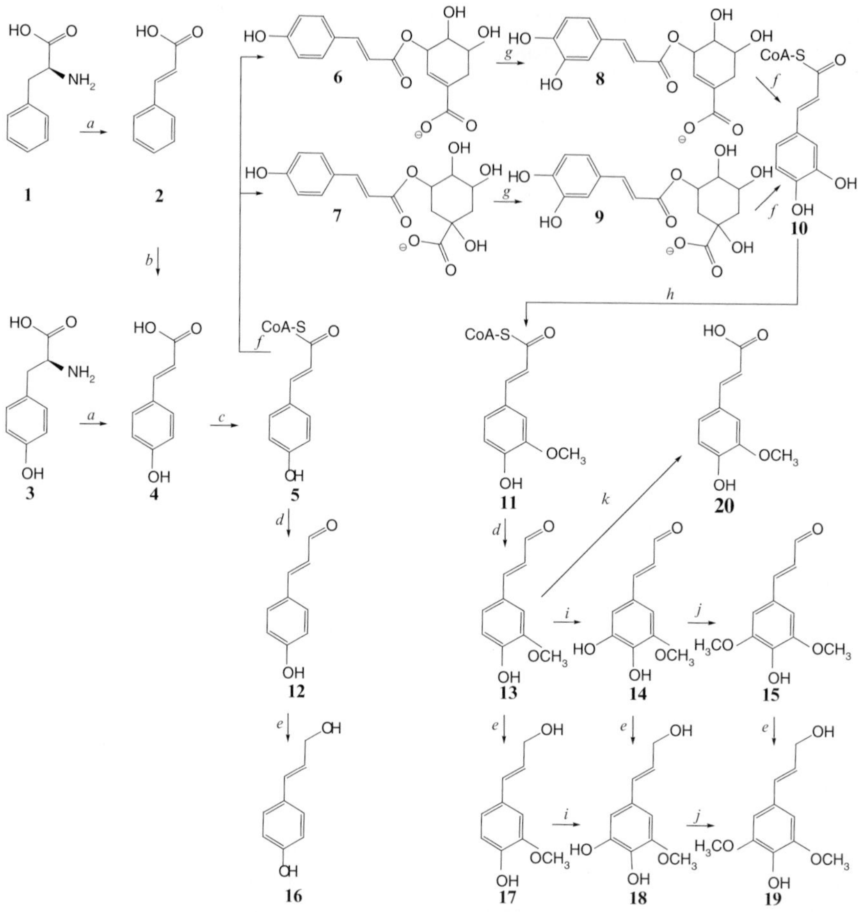

Fig. 2 Monolignol biosynthetic pathway. Compound names are listed in the text. The letters refer to the following enzymes [followed by their abbreviation]: **(a)** phenylalanine/tyrosine ammonia lyase [PAL/TAL], **(b)** cinnamate 4-hydroxylase [C4H], **(c)** 4-cinnamate CoA ligase [4CL], **(d)** cinnamoyl-CoA reductase [CCR], **(e)** cinnamyl alcohol dehydrogenase [CAD] **(f)** hydroxycinnamoyl-CoA:shikimate/quinate hydroxy-cinnamoyl transferase [HCT], **(g)** p-coumaroyl-CoA 3′-hydroxylase [C3′H], **(h)** caffeoyl-CoA O-methyltransferase [CCoAOMT], **(i)** coniferyl aldehyde/coniferyl alcohol 5-hydroxylase [F5H (C5H)], **(j)** coniferaldehyde/coniferyl alcohol O-methyltransferase [COMT], and *(k)* coniferaldehyde dehydrogenase

The amino acid L–phenylalanine (**1**) is commonly regarded as the precursor of monolignols. Deamination by phenylalanine ammonia lyase (PAL) leads to cinnamic acid (**2**). Roesler et al. (1997) showed that maize (and grasses in general) can bypass C4H because maize PAL also recognizes L–tyrosine (**3**) as a substrate, explaining the tyrosine ammonia lyase activity observed in maize extracts (Morrison and Buxton, 1993).

Subsequent hydroxylation of cinnamic acid by cinnamic acid 4-hydroxylase (C4H) results in the formation of p-coumaric acid (**4**), an abundant hydroxycinnamic acid in the maize cell wall. The enzyme 4-coumarate:CoA ligase (4CL) converts *p*-coumaric acid to *p*-coumaroyl-CoA (**5**), which can subsequently undergo two types of modifications: reduction of the carboxyl group on the propane side-chain to an alcohol, and substitution of the phenyl ring. This ultimately leads to the monolignols *p*-coumaryl alcohol (**16**), coniferyl alcohol (**17**) and sinapyl alcohol (**19**), with the latter two being the most common.

A unique feature of sinapyl alcohol in maize and other grasses is the presence of *p*-coumarate esters on the γ-carbon (Ralph et al., 1994). Experimental evidence suggests that the *p*-coumaroylation of sinapyl alcohol occurs prior to their incorporation into lignin (Lu and Ralph., 1999), and that presence of the ester enhances (or allows) the incorporation of sinapyl alcohol into the growing lignin polymer in maize cell walls (Ralph et al., 2004b). The gene encoding the acyl transferase responsible for this reaction has not yet been identified.

The monolignols are synthesized inside the cell and need to be exported to the cell wall where lignin is deposited. Given their toxicity, monolignol transport is likely to involve glucoside intermediates (Lim et al. 2001) or vesicles. The precise mechanism has not been elucidated and may be different in different species. In the cell wall the monolignols are converted to monolignol radicals through the action of peroxidases and/or laccases (Vermerris and Nicholson, 2006). Expression analysis of three known maize peroxidase genes, referred to as *ZmPox1–3*, indicated that ZmPOX2 was likely the predominant peroxidase (de Obeso et al., 2003). Even though *ZmPox3* was expressed at low levels in lignifying tissues (de Obeso et al., 2003), Guillet-Claude et al. (2004) localized this gene to a quantitative trait locus (QTL) associated with lignin content and silage digestibility. The importance of this gene was further substantiated by the correlation between the presence of a *pox3* mutant allele and high cell wall digestibility. The mutant allele carried a miniature inverted repeat transposable element (MITE) in exon 2 which resulted in a truncated protein.

Lignin residues derived from *p*-coumaryl, coniferyl and sinapyl alcohols are referred to as *p*-hydroxyphenyl (H), guaiacyl (G) and syringyl (S) residues, respectively. In the tissue of normal plants H-units make up only a small proportion (<5%) of the lignin. Lignin is formed through the end-wise addition of monolignol radicals to the growing polymer and is thus under chemical as opposed to biological control (Hatfield and Vermerris, 2001). Alternative models involving protein-mediated polymerization of lignin have been proposed (Davin and Lewis, 2000, 2005a,b) but are so far not supported by chemical nor genetic evidence (Ralph et al., 2004a).

The biosynthesis of ferulic acid (**20**) was originally thought to be catalyzed by the enzymes *p*-coumaric acid acid 3-hydroxylase (C3H) and caffeic acid *O*-methyltransferase (COMT) as part of monolignol biosynthesis. This was, however, contradicted by the biochemical and genetic evidence supporting 3-hydroxylation at the level of the shikimate or D-quinate ester of *p*-coumaroyl-CoA (**6** and **7**, respectively; Schoch et al., 2001; Franke et al., 2002a,b). Instead, in Arabidopsis ferulic acid is synthesized from coniferaldehyde (**13**) by coniferaldehyde dehydrogenase encoded by the *REDUCED EPIDERMAL FLUORESCENCE1* (*REF1*) gene (Nair et al., 2004). This route is likely to exist in maize based on the fact that extracts from maize leaves displayed coniferaldehyde dehydrogenase activity (Nair et al. 2004).

2.5   Cell Wall Proteins

The maize cell wall contains structural and non-structural proteins. The structural proteins in the maize cell wall can be classified as hydroxyproline-rich glycoproteins (HRGPs), glycine-rich proteins (GRPs), proline-rich proteins (PRPs), and arabinogalactan proteins (AGPs). For comprehensive reviews, see Carpita (1996) and Cassab (1998).

Non-structural proteins include a large variety of enzymes that are involved in biosynthesis or rearrangement of cell wall polymers, such as peroxidases, hydrolases and transferases. Zhu et al. (2006) used a proteomics approach to catalog water-soluble and lightly ionically-bound cell wall proteins in the primary root elongation zone of maize and identified several Type II wall-specific proteins.

Expansins form an interesting class of non-structural cell wall proteins. These 25–28 kDa glycoproteins facilitate cell expansion during growth by loosening the cell wall (Cosgrove and Li, 1993), have been implicated in the drought response of maize seedlings (Wu et al., 1996), and appear to be involved in abscission, and the response to submergence, light, and stress (Cosgrove et al., 2002). Expansins can be classified as α- and β-expansins. The α-expansins (EXPAs) are grass pollen-specific and are likely to facilitate pollen tube penetration by loosening the cell walls in the style (Cosgrove et al., 1997). The β-expansins (EXPBs) are overall similar in sequence but contain N-linked glycosylation motifs absent in the EXPAs. Most of the β-expansins are present in vegetative tissues, but there is a subgroup of pollen-expressed EXPBs that are referred to as group-1 grass pollen allergens and that are responsible for hay fever and seasonal asthma (Cosgrove et al., 2002).

The expansin-mediated mechanism of cell wall loosening is uncertain, but does *not* appear to involve enzymatic action. Based on the crystal structure of the maize EXPB1 protein, Yennawar et al. (2006) proposed a model in which expansins weaken the non-covalent adhesion between cellulose and arabinoxylans. This relaxes the structure of the cell wall and allows expansion. Subsequent reassociation of the glycan chains after expansion restores cell wall strength. The expansins are encoded by large multigene families in maize. Comparison to Arabidopsis, maize and rice contain approximately the same number of *EXPA* genes, but considerably more *EXPB* genes. This is likely a reflection of the different composition of the cell wall in grasses (Cosgrove et al., 2002).

3   Maize Cell Wall Databases

The progress in genomics has enabled the development of several useful resources to access information on maize genes related to cell wall biosynthesis.

MAIZEWALL (www.polebio.scsv.ups-tlse.fr/MAIZEWALL) is a sequence database and expression profiling resource described by Guillaumie et al. (2007a). This database was generated by identifying maize orthologs to *Zinnia elegans* genes known to be involved in secondary wall formation. An additional set of maize genes was identified in a private sequence database based on homology to known cell wall-related genes from other plant species that were identified using a keyword search. This resulted in a set of 735 genes. Gene expression data for 651 of these genes over the course of plant development is provided.

The Cell Wall Genomics site (cellwall.genomics.purdue.edu) classifies cell wall biogenesis in six distinct stages and displays Arabidopsis, maize and rice genes involved in each of these stages. The more than 1,200 maize cell wall genes were identified in the ZmGI database at The Institute for Genomic Research (now housed at the Dana Farber Cancer Institute; accessible at the web site: compbio.dfci.harvard. edu/tgi/plant.html). This database, containing sequence fragments from expressed maize genes, was searched with sequences from cell wall-related Arabidopsis and rice genes. A benefit of this approach is that it allows incorporation of un-annotated gene sequences. The dendrograms displayed on the web site were generated based on deduced amino acid sequences.

4 Cell Wall Mutants of Maize

The identification of genes involved in cell wall biosynthesis has been aided by the existence of a vast collection of cell wall mutants in Arabidopsis. In contrast, there have traditionally only been a limited number of maize cell wall mutants, despite the long history of maize as a genetic model. This is likely due to a lack of conven-ient screening protocols combined with the large size of maize. Unless coleoptiles or seedlings can be used for screening, maize requires adequate field space, often at locations away from analytical equipment. This also restricts screening efforts to the normal growing season(s).

The historically known maize cell wall mutants – the *brown midrib* and *brittle stalk* mutants – are, however, of considerable interest and value both from a biological a agronomic perspective. With the development of genomics tools, additional novel cell wall mutants have been obtained. The sections that follow will provide an overview of what is known about the classic cell wall mutants of maize, both in terms of their chemical composition and the underlying genes, and recent efforts to generate additional cell wall mutants in maize.

4.1 The brown midrib Mutants

The *brown midrib1* (*bm1*), *bm2*, *bm3* and *bm4* mutants are naturally occurring mutants that have brown vascular tissue in their leaves. The *bm1* mutant was discovered by Eyster (1926) and mapped by Jorgensen (1931). The *bm2* mutant was reported by Burnham and Brink (1932), the *bm3* mutant by Emerson et al. (1935) and the *bm4* mutant by Burnham (1947). These genes map to bins 5.04, 1.11, 4.05, and 9.7, respectively.

There are several reports detailing the changes in chemical composition of the cell wall as a result of these mutations. The *bm1* mutant was shown to accumulate an aldehyde in its cell wall (Kuc and Nelson, 1964; Kuc et al., 1968). This aldehyde was later shown to be coniferaldehyde, an intermediate from the monolignol biosynthetic pathway (**13** in Fig. 2), based on the increased staining observed after incubation of vascular tissue with acid phloroglucinol (Wiesner's reagent) (Provan et al. 1997; Halpin et al., 1998; Vermerris et al., 2002a), as well as compositional data (Provan et al., 1997; Marita et al., 2003; Barrière et al., 2004). Vermerris et al. (2002a) showed that a *bm1* heterozygote in a near-isogenic background has a cell wall composition that is distinct from the corresponding wild-type and homozygous mutant. This dosage effect was also apparent in the flowering dynamics, where the heterozygote flowered earlier than the homozygotes.

Halpin et al. (1998) showed that the *bm1* mutation reduces the expression of at least one of the genes encoding the monolignol biosynthetic enzyme cinnamyl alcohol dehydrogenase (CAD), which explains the enhanced incorporation of coniferaldehyde in the lignin. While Halpin et al. (1998) mapped this *CAD* gene near the *bm1* locus, they did not provide genetic proof showing that *bm1* is a mutation in this *CAD* gene. The *CAD* gene isolated by Halpin et al. (1998) is now referred to as *ZmCAD2* based on its homology with the rice *OsCAD2* gene (Tobias and Chow, 2005). Gene expression studies showed that the expression of four *CAD(-like)* genes, including *ZmCAD2*, was reduced in seedlings of a near-isogenic *bm1* mutant (Guillaumie et al. 2007b). In addition, a relatively large number of other phenylpropanoid- and monolignol-biosynthetic genes were down-regulated, including *PAL*, two *4CLs*, *HCT2*, two *CCoAOMTs*, five *laccase* and two *peroxidase* genes. The *bm1* mutation also resulted in the reduced expression of several regulatory genes, including two *MYB* genes and one *LIM* gene. Of particular interest was the finding that an ortholog of an Arabidopsis *ARGONAUTE* (*AGO*) gene was under-expressed in *bm1* seedlings. AGO proteins are involved in microRNA-mediated post-transcriptional gene silencing. The maize *AGO* gene whose expression was most reduced in *bm1* has a rice ortholog 1 Mb upstream of the *OsCAD2* gene, making this a prime candidate gene for *Bm1*.

Detailed analyses of dissected vascular tissue from leaf tissue of a near-isogenic *bm2* mutant using pyrolysis-mass spectrometry revealed a reduction in both G- and S-residues in the lignin. Furthermore, the developmentally-coordinated lignin gradient observed in wild-type tissue was absent in the *bm2* mutant (Vermerris and Boon, 2001). This effect appears to be restricted to leaf tissue, since the vascular tissue in the *bm2* stem is not brown. This is also consistent with lack of major compositional changes based on chemical analyses using NMR spectroscopy (Marita et al., 2003).

Gene expression analyses revealed reduced expression of several phenylpropanoid biosynthetic genes, including two *4CLs*, *ZmPox2*, *chorismate mutase*, and *chalcone isomerase1*. Several regulatory genes were also under-expressed in *bm2* seedlings (Guillaumie et al., 2007b), but this expression analysis did not produce any obvious candidate. The sequence of the *Bm2* gene obtained via transposon-tagging has not yet provided clear clues about its function (Vermerris et al., unpublished data).

The *Bm3* gene is so far the only *Brown midrib* gene with an established function. The *bm3* mutant was shown to contain a defective *caffeic acid O-methyltransferase* (*COMT*) gene (Vignols et al., 1995). The *bm3-ref* allele contains a B5 retrotransposon insertion in the 5′ end of exon 2, which results in a truncated mRNA. Two additional alleles have been identified: the *bm3-2* allele lacks the 3′ end of the gene (Vignols et al., 1995), whereas the *bm3-3* allele lacks two large, non-contiguous segments of the intron and exon 2 (Morrow et al., 1997). The *bm3* mutations result in reduced COMT activity, which explains the reduction in S-residues and enhanced incorporation of 5-hydroxyconiferyl alcohol (**18** in Fig. 2) observed in the lignin of *bm3* (Lapierre et al., 1988; Provan et al., 1997; Suzuki et al., 1997; Marita et al. 2003). Based on expression analyses in young seedlings, the S-residues that are still present are likely the result of the coordinated up-regulation of two *O-methyltransferase* genes, two *cytochrome P450* genes and the *S-adenosyl methionine synthase3* (*SAMS3*) gene that are normally *not* involved in monolignol biosynthesis (Guillaumie et al., 2007b). Shi et al. (2006) used microarray and suppression subtractive hybridization (SSH) experiments to determine changes in gene expression resulting from the *bm3* mutation. Compared to Guillaumie et al. (2007b), this study included a much larger set of genes, different developmental stages, different tissues, and different genetic backgrounds. Shi et al. (2006) reported that the genetic background influenced which genes were affected as a result of the *bm3* mutation, but that a *PAL*, one *4CL* and two *4CL-like*, and a *CAD* gene were down-regulated in *bm3* plants regardless of the genetic background. SSH analysis revealed down-regulation of three *CCoAOMT* genes in one of the *bm3* mutant lines.

Chemical analyses of the *bm4* mutant indicated reductions in both G- and S-residues, resembling the changes observed in the *bm2* mutant (Barrière et al., 2004). The *bm4* mutation also affects primarily the leaves, explaining why NMR studies of the lignin of *bm4* stems do not reveal major differences relative to the wild-type control (Marita et al., 2003). Guillaumie et al. (2007b) reported over-expression of *PAL*, *4CL1*, *HCT1*, and *CAD* genes in *bm4* seedlings. Interestingly, the *SAMS3*, one of the *OMT* and one of the *cytochrome P450* genes up-regulated in *bm3* was also up-regulated in *bm4*. The *ZmPox2* gene and an *aldehyde dehydrogenase* gene were under-expressed. The genetic basis of the *bm4* mutation is currently not known, but the gene has recently been transposon-tagged and cloning efforts have begun (Vermerris et al., unpublished results).

The gene expression analyses by Shi et al. (2006) and Guillaumie et al. (2007b) revealed the complex regulation of lignin biosynthesis, whereby perturbation of one gene results in modified expression of both structural and regulatory genes.

Recently additional and potentially novel *bm* mutants have become available through the Maize Genetics Stock Center (MGSC) at the University of Illinois and can be requested through the Center's website. A number of these mutant alleles – *bm**-PI228174, -PI251930, -PI262480, and -COOP – have been introgressed to inbred line A619 to enable quantitative analyses on the impact of these mutations on cell wall composition and plant growth and development (Vermerris, unpublished results). They can also be compared more accurately to the traditional four *bm* mutants in this same background (Marita et al., 2003).

4.2   *The* **brittle stalk2** *Mutant*

The *brittle stalk2* (*bk2*) mutant has a stalk that easily breaks (snaps) under mechanical pressure. In many cases modest wind is sufficient to break the stalk. At the moment *bk2* is the only *bk* mutant available, although based on the name, *bk1* must have existed. The *bk* phenotype manifests itself in all tissues, but only after approximately four weeks after planting.

The *Bk2* gene was recently independently cloned by Ching et al. (2006) and by Sindhu et al. (2007) and shown to be orthologous to the Arabidopsis *COBRA-LIKE4* (*COBL4*) gene. The *bk2-ref* allele contains a 1-kb transposon-like element named KITE (Ching et al., 2006). The *AtCOBL4* gene belongs to the twelve-member *AtCOB* gene family whose founding member was identified by Benfey et al. (1993) based on a mutant root phenotype that looked like the head of the snake with the same name. This phenotype was shown to be the result of abnormal anisotropic cell expansion during root development. Furthermore, the *cob* root contained fewer and improperly oriented cellulose microfibrils (Benfey et al., 1993; Roudier et al., 2005). The *COB* gene encodes a glycosyl-phosphatidyl-inositol (GPI)-anchored membrane protein. These proteins are synthesized in the Golgi complex and secreted in the cell wall, where they are attached to a phospholipid molecule at the C-terminal ω-attachment site. Additional characteristic features include a conserved cysteine-rich domain, an N-terminal secretion signal sequence, and a predicted cellulose binding site (Roudier et al., 2002). The precise function of these proteins remains, however, uncertain.

Detailed characterization of the *bk2* phenotype by Ching et al. (2006) and Sindhu et al. (2007) resulted in different conclusions. Ching et al. (2006) reported that disruption of Bk2 function interferes with cellulose deposition in the secondary cell wall of fibers, based on the non-uniform thickness of sclerenchyma walls, cell wall compositional analyses, and the co-expression of *Bk2* with the *ZmCesA10-12* genes. Sindhu et al. (2007) reported a decrease in the number of vascular bundles in the outer periphery of the stalk. They also observed thinner sclerenchyma walls. Based on the increased lignin content in developing and mature *bk2* stems, the lack of a strict correlation between relative cellulose content and brittleness, and the high expression levels of *Bk2* during early development, they proposed that Bk2 plays a role in the coordination between lignin and cellulose deposition, thus ensuring organ flexibility during growth.

A genome-wide analysis of rice and maize revealed the existence of nine and eight *Cob-like* (*Cobl*) genes, respectively (Li et al., 2003; Brady et al., 2007). The rice *Cobl* genes included the *Brittle culm1* gene (Li et al., 2003), which is the closest ortholog of *Bk2*. The maize and rice *Cobl* genes are referred to as *ZmBk2L* and *OsBC1L* genes, respectively. The lack of effect on plant architecture of the *bk2* mutation in maize could indicate a fundamental difference between maize and Arabidopisis with respect to the mechanism of cellulose deposition. Alternatively, a mutation in the maize *ZmBk2L3* gene – functionally the most closely related to *AtCOB* – could result in a change in architecture (Brady et al., 2007), analogous to the structural changes in the xylem resulting from a mutation in the *ZmBk2* ortholog *AtCOBL4* (Brown et al., 2005).

4.3 Novel Cell Wall Mutants of Maize

The project '*Identification and Characterization of Cell Wall Mutants in Maize and Arabidopsis using Novel Spectroscopies*' (funded by the U. S. National Science Foundation) was aimed at gaining a better understanding of the genetic control of cell wall biogenesis. The review by Yong et al. (2005) and the project web site (cellwall.genomics.purdue.edu) contain a detailed description of the different aspects of this project.

The identification of cell wall-related genes in maize benefited from the availability of the Uniform*Mu* population (see Chapter by D. McCarty), which enabled the use of both forward and reverse genetics approaches. The reverse genetics experiments were aimed at identifying phenotypes associated with mutations in cell wall-related genes. Maize genes were selected based on their known function in Arabidopsis, and DNA sequences for the corresponding maize genes were obtained from publicly available databases. This has resulted in a collection of over 100 mutants covering several different classes of genes involved in the biogenesis of several cell wall constituents (K. Koch and D. McCarty, unpublished results).

The forward genetics screen was performed on segregating F_2 families of the Uniform*Mu* population using near-infrared reflectance spectroscopy (NIRS) as a screening tool (Yong et al., 2005; Vermerris et al., 2007). NIRS is a vibrational spectroscopic technique that confers information on the chemical composition of the sample. For the analysis of mature tissues NIRS is preferred over mid-range Fourier transform infrared (FTIR) spectroscopy because it is more sensitive to aromatic compounds that are abundant in the secondary wall. For identification of genes affecting primary cell wall composition and architecture, FTIR spectroscopy is a more suitable technique. This analysis can be performed on isolated walls from coleoptiles grown in the dark for 7 days (Sené et al., 1994; Yong et al., 2005). While the acquisition of NIR spectra is quick and straightforward, the analysis generally requires the use of calibration models that correlate spectral data with compositional data obtained *via* more traditional analytical methods (*cf.* Jung et al., 1998; Brinkmann et al., 2002).

In the case of the screen for maize secondary cell wall mutants, the emphasis was on identifying so-called *spectrotypes*, spectral phenotypes that allowed mutants to be distinguished from the W22 wild-type control. A total of 2,100 F_2 families were screened using NIRS. The large number of leaf samples required the development of a screening method that took into consideration environmental and physiological variation in the field (soil quality, moisture and nutrient availability, time of sample collection), as well as variation resulting from the fact that different individuals were acquiring spectra (Vermerris et al., 2007). Putative spectral mutants were first tested for a heritable phenotype by screening the progeny obtained after self-pollinating the putative mutants. Thirty-nine spectral mutants were identified and confirmed. Interestingly, the vast majority of these mutants (85%) did not have a visible phenotype and could not be distinguished from W22 wild-type plants based on anatomical or architectural differences. Observed visible changes among some of these mutants included variation in height and flowering time, generalizing observations on maize and sorghum *brown midrib* mutants (Vermerris and McIntyre, 1999; Vermerris et al., 2002a,b). All 39 spectral mutants are available through the Maize Genetics Stock Center, and further details, including the spectral data files and phenotypic information, can be obtained through the project web site (cellwall.genomics.purdue.edu/families/7.html).

Confirmed mutants were further analyzed using additional analytical techniques, specifically pyrolysis-molecular beam-mass spectrometry (Py-MBMS). During analytical pyrolysis of cell wall samples a volatile pyrolysate is generated as a result of thermal degradation under anoxic conditions (Evans et al., 1987). The pyrolysate contains break-down products from the major cell wall constituents that can then be identified and quantified (Boon, 1989; Ralph and Hatfield, 1991; Fontaine et al., 2003).

The analysis of Py-MBMS data confirmed cell wall compositional changes in a number of the spectral mutants, and provided more detailed information on cell wall composition. This included changes in the ratio of lignin to carbohydrates or changes in lignin subunit composition. Some spectral mutants did not appear to display a change in cell wall composition based on the Py-MBMS data. This could mean that these mutants were not truly cell wall mutants, but for example mutants in which mineral content or composition, or epicuticular wax composition were changed. An alternative explanation is that these mutants reflect changes in cell wall architecture that are detectable with NIRS but not with Py-MBMS. Additional analyses on a number of mutants are in progress, as are efforts to clone the genes underlying the mutations. The cloning is achieved using the PCR-based *Mu*TAIL method (Settles et al., 2004).

Py-MBMS could be used as an alternative method to identify cell wall mutants, as has been shown for loblolly pine (Sewell et al., 2002). This has recently become much more feasible with the development of autosamplers. This method, as mentioned, cannot readily identify variation in cell wall architecture. The same is true for the analysis of cell walls by gas chromatography (GC) or high-performance liquid chromatography (HPLC) methods. These two methods would require acid or alkaline hydrolysis, typically followed by a derivatization to increase mobility on the column. The Arabidopsis *mur* mutants were identified as part of a GC-based screen of neutral sugar derivatives, so-called alditol acetates, obtained after hydrolysis of hemicellulosic polysaccharides (Reiter et al., 1993). These methods have in common with the NIRS screening described above that variation resulting from the size of the field needs to be taken into account much more so than when samples are collected from plants growing under controlled conditions.

5 Bioenergy Production from Plants

5.1 Cell Walls as Feedstocks for Renewable Energy

Global demand for energy is expected to continue to grow during the next several decades as a result of population growth and an increase in the standard of living in primarily India, China and Latin-America. The majority of energy currently comes from fossil energy sources, specifically coal for the generation of electricity and petroleum for the generation of liquid transportation fuels. The projected global energy demand, the realization that fossil fuels – especially petroleum – are finite resources, and concerns about energy dependence on countries that are not known for their political stability, have prompted an interest in alternative fuels. Recent reports on global climate change resulting from increases in atmospheric carbon dioxide (CO_2) levels have provided further incentives to explore alternative energy sources (Adger et al., 2007).

The main sources of alternative transportation fuels are bio-diesel and ethanol. Bio-diesel is produced *via trans*-esterification of fats and oils, most commonly from oil crops such as soybean (*Glycine max*), oil palm (*Elaeis guineensis*), or canola (oil seed rape; *Brassica napus*). Ethanol is currently produced on a large scale from sugar cane (*Saccharum* spp.) in Brazil (19 billion liter in 2007), and from maize starch in the United States (24.5 billion liter in 2007). While the sugars from sugar cane juice can be readily fermented to ethanol by microorganisms, starch needs to be hydrolyzed first to yield simple sugars that can subsequently be converted to ethanol. Yet the current volume of ethanol produced in the U.S. represents only a small fraction of the volume of gasoline consumed in the U.S. every year and needs to increase drastically in order for ethanol to contribute significantly as an alternative fuel.

The U.S. government has set as a target that 30% of the transportation fuel used in the U.S. (based on the volume consumed in 2004) has to come from renewable resources by the year 2030 ('30 × 30'; U.S. DOE, 2006), which translates to 165 billion liter. In the European Union the target is to substitute 10% of the traditional fuels with biofuels by 2020. In order to safeguard global food and feed production, such an increase in ethanol production is *only* feasible if the fermentable sugars are produced from sugar crops such as sugar cane, sweet sorghum (*Sorghum bicolor*) or sweet potatoes (*Ipomoea batatas*), or from lignocellulosic biomass. Ethanol produced from lignocellulosic biomass is produced by fermentation of sugars obtained from hydrolysis of cell wall polysaccharides – cellulose and hemicellulose – through hydrolysis with (hemi)cellulose-degrading enzymes. This process is referred to as enzymatic saccharification and the product is referred to as cellulosic ethanol. The U.S. Department of Agriculture calculated that 1.3 billion dry tons of biomass per year are needed to meet the '30 × 30' goal, with biomass derived from forest trees and dedicated bioenergy crops such as switchgrass (*Panicum virgatum*) and miscanthus (*Miscanthus xgiganteus*), as well as and agricultural residues such as maize and sorghum stover and wheat straw (USDA-DOE, 2005).

In order for cellulosic ethanol to become a reality the overall production efficiency needs to be improved (Ragauskas et al., 2006). This includes implementation of efficient pretreatment methods such as acid- or base-catalyzed hydrolysis, steam explosion, ammonia fiber explosion, and liquid hot water pretreatment (Yang and Wyman, 2004; Mosier et al., 2005a,b). Pretreatment results in an overall loosening of the cell wall structure, solubilization of hemicellulose in the cell walls of the stover, and reduction in the crystallinity of the cellulose, while minimizing the formation of degradation products that could interfere with the micro-organisms used during fermentation. The development of more efficient and cheaper cellulolytic enzymes (Escovar-Kousen et al., 2004) and the development of recombinant microorganisms that can co-ferment both hexose and pentose sugars (Ho et al., 1998) will also help increase the overall yield of ethanol.

5.2   The Role of Maize in Bioenergy Production

It is important to realize that the feedstocks themselves can also be improved through plant breeding and biotechnological approaches. Since efforts to improve maize, sorghum and small grains have historically focused largely on maximizing grain yield, the development of crops to be used as feedstocks for bioenergy will require an entirely different set of selection criteria. Biomass needs to be produced in large volumes and is bulky, especially when wet, but needs to be transported (most likely by truck) to a processing plant where it is heavily processed. Since the processing cost represents a considerable cost of the final product (McAloon et al., 2000) biomass needs to be as inexpensive as possible. This means developing bioenergy crops that produce maximum biomass yield with only minimal input requirements (water, fertilizer, pesticides, fungicides) and that have the optimal cell wall composition for bioprocessing. In practical terms this translates to selection for tolerance to abiotic and biotic stresses, large plant size, a shift towards a longer vegetative growth phase to allow maximum biomass production, and incorporation of traits that impact (secondary) cell wall composition (Vermerris et al., 2007).

Maize deserves attention as a source of lignocellulosic biomass. With currently over 90 million acres (36 million hectare) of maize grown in the U.S. alone, it is an abundant and inexpensive feedstock. Stover is likely to remain a major source of biomass during the next 5–10 years as the anticipated increase in ethanol production drives the demand for maize starch in the U.S. Furthermore, the germplasm resources for dedicated bioenergy crops are currently still limited, as is the infrastructure (planting- and harvesting equipment, storage and transportation solutions) for these crops. Finally, the genetic resources available in maize – a large collection of mutants, forward and reverse genetics tools, rich sequence information, synteny with rice and sorghum, diverse germplasm sources including exotic accessions, and extensive experience with maize breeding – make maize particularly suitable as a model species for the development of improved, dedicated C4 bioenergy crops.

Modification of cell wall composition is a promising strategy to enhance ethanol production from maize stover. This is evident from experiments with the (classic) *brown midrib* mutants. Vermerris et al. (2007) reported that hydrolysis of unpretreated stover from the maize mutants *bm1* and *bm3* in an inbred A619 background resulted in 144 ± 21 and 152 ± 20 mg glucose per gram stover (dry weight), respectively. This represents a 40-50% increase relative to stover from the wild-type control, which yielded 98 ± 9 mg/g. In contrast, glucose yields obtained after hydrolysis of *bm2* and *bm4* stover did not differ from the wild type. Given that all four mutants contain less lignin (Marita et al., 2003; Barrière et al., 2004), the change in lignin subunit composition appears to be responsible for the improved yield of fermentable sugars from *bm1* and *bm3* stover. Multivariate statistical models based on a large collection of stover samples with varying cell wall composition revealed that a high G/S ratio in the lignin improves the yield of fermentable sugars (Vermerris et al., unpublished results).

Stalk lodging, reduced grain yields, and increased susceptibility to pests and diseases are often-expressed concerns about the *bm* mutants in the context of large-scale production (Pedersen et al., 2005). As part of a 'proof-of-concept' breeding effort aimed at developing maize lines that combined good agronomic performance and high yields of fermentable sugars on a per-gram-dry-weight basis, Vermerris et al. (2007) reported promising results on three inbred lines derived from a cross between inbred lines A619 and B52 that produced twice the yield of fermentable sugars while showing excellent late-season standability, and considerably better tolerance to Fall Armyworm (*Spodoptera frugiperda*) infestation than the standard research inbred lines.

There are obvious similarities between the digestibility of forage and silage by microorganisms in the rumen of animals and the enzymatic saccharification of stover for ethanol production. It is therefore likely that maize hybrids optimized for forage quality with parameters such as digestibile neutral detergent fiber (dNDF) make good candidates for the production of ethanol. Indeed, Weimer et al. (2005) showed a high degree of correlation between the *in-vitro* gas production from incubation of stover with rumen fluid and the ethanol yield following enzymatic saccharification. Production of maize stover is probably economically most feasible when it is a by-product from grain production. This is a result of the combination of high input requirements and the fact that not all of the stover can be collected in order to prevent soil erosion and soil depletion (Wilhelm et al., 2004). Even if all the stover, with a typical yield of 8-9 Mg/ha of dry stover (Frey et al., 2004; Shinners et al., 2007) could be collected, perennial dedicated bioenergy crops such as switchgrass with a yield potential between 9.5 and 23 Mg/ha (McLaughlin and Kszos, 2005) may look more appealing.

Aside from introducing traits that enhance biotic and abiotic stress tolerance, which will improve biomass and grain yield in general, and using cell wall mutants that result in higher yields of fermentable sugars, specific mutations may be of interest to enhance the yield of fermentable sugars from maize stover by increasing overall biomass production. These mutations include the *indeterminate* (*id1*), *Leafy1*, *grassy tiller1* (*gt1*) and the *perennialism1* (*pe1*) mutations, as well as novel

mutations that result in similar phenotypes. The *id1* mutation results in an indeterminate growth pattern whereby the transition from the vegetative to the reproductive
phase is delayed (Colasanti et al., 1998; Colasanti and Sundaresan, 2000; Kozaki et al.,
2004; Colasanti et al., 2006; see also Chapter 3, Volume 1). The dominant *Leafy1*
trait was discovered in 1971 and results in extra nodes and leaves above the ear, low
ear placement, highly lignified stalks, early maturity, and high yield potential
(Shaver, 1983), although its genetic basis has not been elucidated. Note that this
trait is not related to the Arabidopsis *LEAFY* (*LFY*) gene involved in the regulation
of flowering time (Weigel et al., 1992). The maize *Lfy1* trait has been used in forage
breeding programs and in breeding programs aimed at developing cultivars for
regions with short growing seasons (Modarres et al., 1997a,b; Dijak et al., 1999).
The *gt1* mutant produces many long basal branches that can vary in size and
number depending on the genetic background (Tracy and Everett, 1982). The *gt1*
allele also tends to improve regrowth after early-season cuttings by producing multiple
stalks and thus extra biomass. The *pe1* mutation delays tassel formation and,
depending on the environmental conitions and genetic background, may cause
failure of ear formation or replacement of the earbranch by a vegetative branch
(Shaver, 1967). The potential negative impact of *pe1* on ear formation would have
to be controlled in order for this mutation to be of agronomic value.

Given the large number of maize mutants and the apparent tolerance of maize to
mutations that have a major impact on plant architecture, mutations that alter cell
wall composition and mutations that alter biomass production can complement a
breeding program that is focused on good agronomic performance. Successful
inbred lines and hybrids from such a program are likely to be attractive sources
of lignocellulosic biomass. Furthermore, mutations in maize identified as part of
high-throughput forward and reverse genetics screens will be able to advance the
development of dedicated C4 bioenergy crops. In conclusion, there is a bright
future for maize as a bioenergy (model) species.

Acknwledgements The data from the Cell Wall Genomics project presented in this chapter were
generated with funding from the U.S. National Science Foundation Plant Genome Research
Program (DBI-0217552) in collaboration with Nick Carpita (PI), Maureen McCann, Bryan
Penning, Karen Koch, Don McCarty, Steven Thomas and Mark Davis. The saccharification data
were generated in collaboration with Michael Ladisch and Nathan Mosier, with funding from The
Consortium for Plant Biotechnology Research, Inc. (CPBR), U.S. Department of Energy (DOE)
Prime Agreement no. DEFG36-02GO12026. This support does not constitute and endorsement by
DOE or CPBR of the views expressed in this publication. Additional financial support from Dow
AgroSciences, Purdue University, and the University of Florida is gratefully acknowledged. I am
grateful for technical support from William Foster, Javier Campos and Randi Wheeler.

References

Adger, N., Aggarwal, P., Agrawla, S., Alcamo, J., Allali, A. et al. (2007) A report accepted by
Working Group II of the International Panel on Climate Change. http://www.ipcc-wg2.org.
Appenzeller, L., Doblin, M., Barreiro, R., Wang, H., Niu, X., Kollipara, K., Carrigan, L., Tomes, D.,
Chapman, M., and K. S. Dhugga, (2004) Cellulose synthesis in maize: isolation and expression
analysis of the cellulose synthase (*CesA*) gene family. *Cellulose* **11**: 287–299.

Arioli, T., Peng, L., Betzner, A. S., Burn, J., Wittke, W., Herth, W., Camilleri, C., Höfte, H., Plazinski, J., Birch, R., Cork, A., Glover, J., Redmond, J., and R. E. Williamson, (1998) Molecular analysis of cellulose biosynthesis in *Arabidopsis. Science* **279**: 717–720.

Barrière, Y., Ralph, J., Méchin, V., Guillaumie, S., Grabber, J. H., Argillier, O., Chabbert, B., and C. Lapierre, (2004) Genetic and molecular baisis of grass cell wall biosynthesis and degradability. II. Lessons from brown midrib mutants, *C. R. Biologies* **327**: 847–860.

Benfey, P. N., Linstead, P. J., Roberts, K., Schiefelbein, J. W., Hauser, M. T., and R. A. Aeschbacher, (1993), Root development in Arabidopsis: four mutants with dramatically altered root morphogenesis. *Development* **119**: 57–70.

Boon, J.J. (1989) An introduction to pyrolysis mass spectrometry of lignocellulosic material: case studies of barley straw, corn stem and *Agropyron*. In Physico-chemical characterization of plant residues for industrial and feed use (A. Chesson and E. R. Ørskov, eds.) Elsevier Applied Science, London, pp. 25–49.

Brady, S. M., Song, S., Dhugga, K. S., Rafalski, J. A., and P. N. Benfey, (2007) Combining expression and comparative evolutionary analysis. The *COBRA* gene family. *Plant Physiol.* **143**: 172–187.

Brinkmann, K., Blaschke, L., and A. Polle, (2002) Comparison of different methods for lignin determination as a basis for calibration of near-infrared reflectance spectroscopy and implications of lignoproteins. *J. Chem. Ecol.* **28**: 2483–2501.

Brown, D. M., Zeef, L. A. H., Ellis, J., Goodacre, R., and S. R. Turner (2005) Identification of novel genes in Arabidopsis involved in secondary cell wall formation using expression profiling and reverse genetics. *Plant Cell* **17**: 2281–2295.

Brown, Jr., R. M. (2004) Cellulose structure and biosynthesis: What is in store for the 21st century? *J. Polymer Sci.* **42**: 487–495.

Burn, J. E., Hocart, C. H., Birch, R. J., Cork, A. C., and Williamson, R. E. (2002) Functional analysis of the cellulose synthase genes *CesA1, CesA2,* and *CesA3* in Arabidopsis. *Plant Physiol.* **129**: 797–807.

Burnham, C. R. (1947) Untitled. *Maize Genet. Coop. News Lett.* **21**: 36–37.

Burnham, C. R., and R. A. Brink (1932) Linkage relations of a second brown midrib gene (*bm2*) in maize, *J. Am. Soc. Agron.* **24**: 960–963.

Burton, R. A., Wilson, S. M., Hrmova, M., Harvey, A. J., Shirley, N. J., Medhurst, A., Stone, B. A., Newbigin, E. J., Bacic, A., and G. B. Fincher (2006) Cellulose synthase-like *CslF* genes mediate the synthesis of cell wall (1,3;1,4)-β-D-glucans. *Science* **311**: 1940–1942.

Carpita, N. C. (1996) Structure and biogenesis of the cell walls of grasses. *Annu. Rev. Plant Physiol. Plant Mol. Biol.* **47**: 445–476.

Carpita, N. C., and D. M. Gibeaut, (1993) Structural models of the primary cell walls in flowering plants: consistency of molecular structure with the physical properties of the wall during growth. *Plant J.* **3**: 1–30.

Carpita, N. C. and M. C. McCann (2000) The cell wall. In: *Biochemistry and Molecular Biology of Plants* (B. B. Buchanan, W. Gruissem and R. L. Jones, eds.), J. Wiley and Sons, Somerset, NJ, pp. 52–108.

Carpita, N. C, Tierney, M., and M. Campbell (2001) Molecular biology of the plant cell wall: searching for the genes that define structure, architecture and dynamics. *Plant Mol. Biol.* **47**: 1–5.

Carpita, N. C., and D. Whittern (1986) A highly substituted glucuronoarabinoxylan from developing maize coleoptiles. *Carbohydr. Res.* **146**: 129–140.

Cassab, G. I. (1998) Plant cell wall proteins. *Annu. Rev. Plant Physiol. Plant Mol. Biol.* **49**: 281–309.

Ching, A., Dhugga, K. S., Appenzeller, L., Meeley, R., Bourett, T. M., Howard, R. J., and A. Rafalski (2006) *Brittle stalk2* encodes a putative glycosyl-phosphatidyl-inositol anchored protein that affects mechanical strength of maize tissues by altering the composition and structure of secondary cell walls. *Planta* **224**: 1174–1184.

Cocuron, J. C., Lerouxel, O., Drakakai, G., Alonso, A. P., Liepman, A. H., Keegstra, K., Raikhel, N., and C. G. Wilkerson, (2007) A gene from the *cellulose synthase-like C* family encodes a β-1,4 glucan synthase. *Proc. Natl. Acad. Sci. USA* **104**: 8550–8555.

Colasanti, J., and V. Sundaresan (2000) 'Florigen' enters the molecular age: long-distance signals that cause the plant to flower. *Trends Biochem. Sci.* **25**: 236–240.

Colasanti, J., Tremblay, R., Wong, A. Y., Coneva, V., Kozaki, A., and B. K. Mable (2006) The maize *INDETERMINATE1* flowering time regulator defines a highly conserved zinc finger protein family in higher plants. *BMC Genomics* **7**:1–17.

Colasanti, J., Yuan, Z., and V. Sundaresan, (1998) The *indeterminate* gene encodes a zinc finger protein and regulates a leaf-generated signal required for the transition to flowering in maize. *Cell* **93**: 593–603.

Cosgrove, D. J., Bedinger, P., and D. M. Durachko (1997) Group 1 allergens of grass pollen as cell wall-loosening agents. *Proc. Natl. Acad. Sci. USA* **94**: 6559–6564.

Cosgrove, D. J., Li, L. C., Cho, H. T., Hoffmann-Benning, S., Moore, R. C., and D.Blecker (2002) The growing world of expansins. *Plant Cell Physiol.* **43**: 1436–14444.

Cosgrove, D. J., and Z. C. Li, (1993) Role of expansin in cell enlargement of oat coleoptiles. *Plant Physiol.* **103**: 1321–1328.

Cutler S., and C. R. Somerville (1997) Cellulose synthase: cloning *in silico. Curr. Biol.* **7**: R108–R111.

Davin, L. B., and N. G. Lewis (2000) Dirigent proteins and dirigent sites explain the mystery of specificity of radical precursor coupling in lignan and lignin biosynthesis. *Plant Physiol.* **123**: 453–461.

Davin, L. B. and N. G. Lewis (2005a) Dirigent phenoxy radical coupling: advances and challenges. *Curr. Opin. Biotechnol.* **16**: 398–406.

Davin, B. L., and N. G. Lewis (2005b) Lignin primary structures and dirigent sites. *Curr. Opin. Biotechnol.* **16**: 407–415.

Delmer, D. P. (1999) Cellulose biosynthesis: Exciting times for a difficult field of study. *Ann. Rev. Plant Physiol. Plant Mol. Biol.* **50**: 245–276.

de Obeso, M., Caparro-Ruiz, D., Vignols, F., Puigdomènech, P., and J. Rigau (2003) Characterisation of maize peroxidases having differential patterns of mRNA accumulation in relation to lignifying tissues. *Gene* **309**: 23–33.

Dhugga, K. S. (2001) Building the wall: genes and enzyme complexes for polysaccharide synthases. *Curr. Opin. Plan.t Biol.* **4**: 488–493.

Dhugga, K. S., Barreiro, R., Whitten, B., Stecca K., Hazerbroek, J., Randhwa, G.S., Dolan, M., Kinney, A.J., Tomes, D., Nichols, S., and P. Anderson (2004) Guar seed β–mannan synthase is a member of the cellulose synthase super gene family. *Science* **303**: 363–366.

Dijak, M., Modarres, A. M., Hamilton, R. I., Dwyer, L. M., Stewart, D. W., Mather, D. E., and D. L. Smith (1999) Leafy reduced-stature maize hybrids for short-season environments. *Crop Sci.* **39**: 1106–1110.

Ding, S. Y. and M. E. Himmel (2006) The maize primary cell wall microfibril: A new model derived from direct visualization. *J. Agric. Food Chem.* **54**: 597–606.

Doblin, S., Kurek, I., Jacob-Wilk, D. and D. P. Delmer (2002) Cellulose biosynthesis in plants: from genes to rosettes. *Plant Cell Physiol.* **43**: 1407–1420.

Emerson, R. A., Beadle, G. W., and A. C. F raser (1935) A summary of linkage studies in maize. *Cornell Univ. Agric. Exp. Stn. Memoir* **180**: 1–83.

Escovar-Kousen, J.M., D. Wilson, and D. Irwin (2004) Integration of computer modeling and initial studies of site-directed mutagenesis to improve cellulose activity on Cel9A from *Thermobifida fusca. Appl. Biochem. Biotechnol.* **113-116**: 287–297.

Evans, R. J., and T. A. Milne, (1987) Molecular characterization of the pyrolysis of biomass. 1. Fundamentals. *Energy & Fuels* **1**: 123–137.

Eyster, W. H. (1926) Chromosome VIII in maize. *Science* **64**: 22.

Fagard, M., Desnos, T., Desprez, T., Goubet, F., Refregier, G., Mouille, G., McCann, M., Rayon, C., Vernhettes, S., and H. Höfte, (2000) *PROCUSTE1* encodes a cellulose synthase required for normal cell elongation specifically in roots and dark-grown hypocotyls of Arabidopsis. *Plant Cell* **12**: 2409–2423.

Faik, A., Price, N.J., Raikhel, N.V., and K. Keegstra. (2002) An Arabidopsis gene encoding an α-xylosyltransferase involved in xyloglucan biosynthesis. *Proc. Natl. Acad. Sci. USA* **99**: 7797–7802.

Favery, B., Ryan, E., Foreman, J., Linstead, P., Boudonck, K., Steer, M., Shaw, P., and L. Dolan (2001) *KOJAK* encodes a cellulose synthase-like protein required for root haircell morphogenesis in Arabidopsis. *Genes Dev.* **15**: 79–89.

Fontaine, A.-S., Bout, S., Barrière, Y., and W. Vermerris (2003) Variation in cell wall composition among forage maize inbred lines and its impact on digestibility, Analysis of neutral detergent fiber composition by pyrolysis-gas chromatography-mass spectrometry. *J. Agric. Food Chem.* **51**: 8080–8087.

Franke, R., Hemm, M. R., Denault, J. W., Ruegger, M. O., Humphreys, J. M., and C. Chapple (2002a) Changes in the secondary metabolism and deposition of an unusual lignin in the *ref8* mutant of Arabidopsis. *Plant J.* **30**: 47–59.

Franke, R., Humphreys, J. M., Hemm, M. R., Denault, J. W., Ruegger, M. O., Cusumano, J. C., and C. Chapple (2002b) The Arabidopsis *REF8* gene encodes the 3-hydroxylase of phenylpropanoid metabolism. *Plant J.* **30**: 33–45.

Frey, T., Coors, J. G., Shaver, R. D., Lauer, J. G., Eilert, D. T., and P. J. Flannery (2004) Selection for silage quality in the Wisconsin quality synthetic population and related maize populations. *Crop Sci.* **44**: 1200–1208.

Guillaumie, S., San-Clemente, H., Deswarte, C., Martinez, Y., Lapierre, C., Murgneux, A., Barrière, Y., Pichon, M., and D. Goffner (2007a) MAIZEWALL. Database and developmental gene expression profiling of cell wall biosynthesis and assembly in maize. *Plant Physiol.* **143**: 339–363.

Guillaumie, S., Pichon, M., Martinant, J.P., Bosio, M., Goffner, D., and Y. Barrière (2007b) Differential expression of phenylpropanoid and related genes in brown-midrib *bm1, bm2, bm3*, and *bm4* young near-isogenic maize plants. *Planta* **226**: 235–250.

Guillet-Claude, C., Birolleau-Touchard, C., Manicacci, D., Rogowsky, P. M., Rigau, J., Murigneux, A., Martinant, J. P., and Y. Barrière (2004) Nucleotide diversity of the *ZmPox3* maize peroxidase gene: relationships between a MITE insertion in exon 2 and variation in forage maize digestibility. *BMC Genetics* **5**: 1–11.

Halpin, C., Holt, K., Chojecki, J., Oliver, D., Chabbert, B., Monties, B., Edwards, K., Barakate, A., and G. A. Foxon (1998) *Brown-midrib* maize (*bm1*) - a mutation affecting the cinnamyl alcohol dehydrogenase gene. *Plant J.* **14**: 545–553.

Hamann, T., Osborne, E., Youngs, H.L., Misson, J., Nussaume, L., and C. Somerville (2004) Global expression analysis of *CESA* and *CSL* genes in Arabidopsis. *Cellulose* **11**: 279–286.

Hatfield, R., and W. Vermerris (2001) Lignin formation in plants: the dilemma of linkage specificity. *Plant Physiol.* **126**: 1351–1357.

Hatfield, R. D., Ralph, J., and J. H. Grabber (1998) Cell wall cross-linking by ferulates and diferulates in grasses. *J. Sci. Food Agric.* **79**: 403–407.

Hazen, S. P., Scott-Craig, J.S., and J.D. Walton (2002) Cellulose synthase-like genes of rice. *Plant Physiol.* **128**: 336–340.

Ho, N.W.Y., Chen, Z., and A.P. Brainard (1998) Genetically engineered *Saccharomyces* yeast capable of effective cofermentation of glucose and xylose. *Appl. Environ. Microbiol.* **64**: 1852–1859.

Holland, N., Holland, D., Helentjaris, T., Dhugga, K. S., Xoconostle-Cazares, B., and D. P. Delmer, (2000) A comparative analysis of the plant *cellulose synthase* (*CesA*) gene family. *Plant Physiol.* **123**: 1313–1323.

Humphreys, J. M., and C. Chapple (2002) Rewriting the lignin road map. *Curr. Opin. Plant Biol.* **5**: 224–229.

Iiyama, K., Lam, T. B. T., and B. A. Stone (1990) Phenolic acid bridges between polysaccharides and lignin in wheat internodes. *Phytochem.* **29**: 733–737.

Iiyama, K., Lam, T. B., and B. A. Stone (1994) Covalent cross-links in the cell wall. *Plant Physiol.* **104**: 315–320.

Jorgenson, L. R. (1931) Brown midrib and its linkage relations. *J. Am. Soc. Agr.* **23** : 549–557.

Jung, H.G., D.R. Mertens, D.R. Buxton (1998) Forage quality variation among maize inbreds: in vitro fiber digestion kinetics and prediction with NIRS. *Crop Sci.* **38**: 205–210.

Kim, C. M., Park, S. H., Il J. B., Park, S. H., Piao, H. L., Eun, M. Y., Dolan, L., and C. D. Han (2007) *OsCSLD1*, a *cellulose synthase-like D1* gene, is required for root hair morphogenesis in rice. *Plant Physiol.* **143**: 1220–1230.

Kozaki, A., Kake, S., and J. Colasanti (2004) The maize ID1 flowering time regulator is a zinc finger protein with novel DNA binding properties. *Nucl. Acid Res.* **32**: 1710–1720.

Kuc, J., and O. Nelson (1964) The abnormal lignins produced by the *brown-midrib* mutants of maize, I. The *brown-midrib-1* mutant. *Arch. Biochem. Biophys.* **105**: 103–113.

Kuc, J., Nelson, O. E., and P. Flanagan (1968) Degradation of abnormal lignins in the brown-midrib mutants and double mutants of maize. *Phytochem.* **7**: 1435–1436.

Lapierre, C., Monties, B., and Rolando, C. (1988) Mise en évidence d' un nouveau type d' unité constitutive dans les lignines d' un mutant de maïs bm3. *C. R. Acad. Sci. Paris, Ser. III* **307**: 723–728.

Li, L. C., Bedinger, P. A., Volk, C., Jones, D., and D. J.Cosgrove (2003) Purification and characterization of four β-expansins (Zea m 1 isoforms) from maize pollen. *Plant Physiol.* **132**: 2073–2085.

Liepman, A. H., Wilkerson, C. G., and K. Keegstra (2005) Expression of cellulose synthase-like (*Csl*) genes in insect cells reveals that *CslA* family members encode mannan synthases. *Proc. Natl. Acad. Sci. USA* **102**: 2221–2226.

Lim, E.-.K, Li, Y., Parr, A., Jackson, R., Ashford, D. A., and D. J. Bowles (2001) Identification of glucosyltransferase genes involved in sinapate metabolism and lignin synthesis in Arabidopsis, *J. Biol. Chem.* **276**: 4344–4349.

Lu, F. and J. Ralph (1999) Detection and determination of *p*-coumaroylated units in lignins. *J. Agric. Food Chem* **47**: 1988–1992.

Marita, J., Vermerris, W., Ralph, J., and R. D. Hatfield (2003) Variations in the cell wall composition of maize *brown midrib* mutants. *J. Agric. Food Chem.* **51**: 1313–1321.

McAloon, A., Taylor, F., Yee, W., Ibsen, K., and Wooley, R. (2000) Determining the cost of producing ethanol from corn starch and lignocellulosic feedstock. National Renewable Energy Laboratory, Golden, CO. pp. 44. (www.nrel.gov/docs/fy01osti/28893.pdf)

McLaughlin, S. B. and L. A. Kszos, (2005) Development of switchgrass (*Panicum virgatum*) as a bioenergy feedstock in the United States. *Biomass Bioenergy* **28**: 515–535.

Modarres, A. M., Hamilton, R. I., Dwyer, L. M., Stewart, D. W., Dijak, M., and D. L. Smith (1997a) Leafy reduced-stature maize for short-season environments: Yield and yield components of inbred lines. *Euphytica* **97**: 129–138.

Modarres, A. M., Hamilton, R. I., Dwyer, L. M., Stewart, D. W., Mather, D. E., Dijak, M., and D. L. Smith (1997b) Leafy reduced-stature maize for short-season environments: morphological aspects of inbred lines. *Euphytica* **96**: 301–309.

Morrison, T. A., and D. R. Buxton (1993) Activity of phenylalanine ammonia-lyase, tyrosine ammonia-lyase, and cinnamyl alcohol dehydrogenase in the maize stalk. *Crop Sci.* **33:** 1264–1268.

Morrow, S.L., Mascia, P., Self, K.A., and M. Altschuler (1997). Molecular characterization of a brown midrib3 deletion mutation in maize. *Mol. Breeding* **3**: 351–357.

Mosier, N., R. Hendrickson, N. Ho, M. Sedlak, and M.R. Ladisch (2005a) Optimization of pH controlled liquid hot water pretreatment of corn stover. *Biores. Technol.* **96**: 1986–1993.

Mosier, N., C. Wyman, B. Dale, R. Elander, Y.Y.Lee, M. Holtzapple, and M.R. Ladisch (2005b) Features of promising technologies for pretreatment of lignocellulosic biomass. *Biores. Technol.* **96**: 673–686.

Mueller, S. C. and M. Brown, Jr. (1980) Evidence for an intramembrane component associated with a cellulose microfibril-synthesizing complex in higher plants. *J. Cell Biol.* **84**: 315–326.

Myton, K. E., and S. C. Fry, (1994) Intraprotoplasmic feruoylation of arabinoxylans in *Festuca arundinacea* cell cultures. *Planta* **193**: 326–330.

Nair, R. B., Bastress, K. L., Ruegger, M. O., Denault, J. W., and C. Chapple (2004) The *Arabidopsis thaliana REDUCED EPIDERMAL FLUORESCENCE1* gene encodes an aldehyde dehydrogenase involved in ferulic acid and sinapic acid biosynthesis. *Plant Cell* **16**: 544–554.

Pear, J. R., Kawagoe, Y., Schreckengost, W. E., Delmer, D. P., and D. M. Stalker (1996) Higher plants contain homologs of the bacterial *celA* genes encoding the catalytic subunit of cellulose synthase. *Proc. Natl. Acad. Sci. USA* **93:** 12637–12642.

Pedersen, J. F., Vogel, K. P., and D. L. Funnell (2005) Impact of reduced lignin on plant fitness. *Crop Sci.* **45:** 812–819.

Perrin, R. M. (2001) Cellulose: How many cellulose synthases to make a plant? *Curr. Biol.* **11:** R213–R216.

Provan, G. J., Scobbie, L., and A. Chesson (1997) Characterisation of lignin from CAD and OMT deficient *Bm* mutants of maize. *J. Sci. Food Agric.* **73:** 133–142.

Ragauskas, A. J., Williams, C. K., Davison, B. H., Britovsek, G., Cairney, J., Eckert, C. A., Frederick Jr., W. J., Hallett, J. P., Leak, D. J., Liotta, C. L., Mielenz, J. R., Murphy, R., Templer, R., and T. Tschaplinski (2006) The path forward for biofuels and biomaterials. *Science* **311:** 484–489.

Ralph, J., Bunzel, M., Marita, J. M., Hatfield, R. D., Lu, F., Kim, H., Schatz, P. F., Grabber, J. H., and H. Steinhart (2004b) Peroxidase-dependent cross-linking reactions of *p*-hydroxycinnamates in plant cell walls. *Phytochem. Rev.* **3:** 79–96.

Ralph, J., and R. D. Hatfield, (1991) Pyrolysis-GC-MS characterization of forage materials. *J. Agr. Food Chem.* **39:** 1426–1437.

Ralph, J., Hatfield, R. D., Quideau, S., Helm, R. F., Grabber, J. H., and H.-J. G. Jung (1994) Pathway of *p*-coumaric acid incorporation into maize lignin as revealed by NMR. *J. Am. Chem. Soc.* **116:** 9448–9456.

Ralph, J., Lundquist, K., Brunow, G., Lu, F., Kim, H., Schatz, P. F., Marita, J. M., Hatfield, R. D., Ralph, S. A., Christensen, J. H., and W. Boerjan (2004a) Lignins: natural polymers from oxidative coupling of 4-hydroxyphenylpropanoids. *Phytochem.* **3:** 29–60.

Reiter W.-D., Chapple, C. C. S., and C. R. Somerville (1993) Altered growth and cell walls in a fucose-deficient mutant of Arabidopsis. *Science* **261:** 1032–1035.

Richmond, T. A. and C. R. Somerville (2000) The cellulose synthase superfamily. *Plant Physiol.* **124:** 495–498.

Richmond T. A. and C. R. Somerville (2001). Integrative approaches to determining *Csl* function. *Plant Physiol.* **47:** 131–143.

Roesler, J., Krekel, F., Amrhein, N., and Schmid, J. (1997) Maize phenylalanine ammonia-lyase has tyrosine ammonia-lyase activity, *Plant Physiol.* **113:** 175–179.

Roudier, F., Fernandez, A. G., Fujita, M., Himmelspach, R., Borner, G. H. H., Schindelman, G., Song, S., Baskin, T. I., Dupree, P., Wasteneys, G. O. et al. (2005) COBRA, an Arabidopsis extracellular glycosyl-phosphatidyl inositolanchored protein, specifically controls highly anisotropic expansion through its involvement in cellulose microfibril orientation. *Plant Cell* **17:** 1749–1763.

Roudier, F., Schindelman, G., DeSalle, R., and P. N. Benfey (2002) The COBRA family of putative GPI-anchored proteins in Arabidopsis. A new fellowship in expansion. *Plant Physiol.* **130:** 538–548.

Saxena, I. M. and R. M. Brown, Jr. (2005) Cellulose biosynthesis: Current views and evolving concepts. *Ann. Bot.* **96:** 9–21.

Saxena I. M., Lin F. C., and R. M. Brown, Jr. (1990) Cloning and sequencing of the cellulose synthase catalytic sub-unit gene of *Acetobacter xylinum*. *Plant Mol. Biol.* **15:** 673–683.

Scheller, H. V., Jensen, J. K., Sørensen, S. Ø., Harholt, J., and N. Geshi (2007) Biosynthesis of pectin. *Physiol. Plant.* **129:** 283–295.

Schoch, G., Goepfert, S., Morant, M., Hehn, A., Meyer, D., Ullmann, P., and D. Werck-Reichhart (2001) CYP98A3 from *Arabidopsis thaliana* is a 3′-hydroxylase of phenolic esters, a missing link in the phenylpropanoid pathway. *J. Biol. Chem.* **276:** 36566–36574.

Séné, C. F. B., McCann, M., Wilson, R. H., and R. Grinter (1994) Fourier-transform Raman and Fourier-transform infrared spectroscopy. An investigation of five higher plant cell walls and their components. *Plant Physiol.* **106:** 1623–1631.

Settles, A. M., Latshaw, S., and McCarty, D. R. (2004) Molecular analysis of high-copy insertion sites in maize. *Nucl. Acids Res.* **32:** e54.

Sewell, M. M., Davis, M. F., Tuskan, G. A., Wheeler, N. C., Elam, C. C., Bassoni, D. L., and D. B. Neale (2002) Identification of QTLs influencing wood property traits in loblolly pine (*Pinus taeda* L.). *Theor. Appl. Genet.* **104:** 214–222.

Shaver, D. L. (1983) Genetics and breeding of maize with extra leaves above the ear. *Proc. Annu. Corn Sorghum Res. Conf.* **38:** 161–180.

Shaver, D. L. (1967) Perennial maize. *J. Heredity* **58**: 270–273.

Shi, C., Koch, G., Ouzunova, M., Wenzel, G., Zein, I., and T. Lübberstedt (2006) Comparison of maize *brown-midrib* isogenic lines by cellular UV-microspectrophotometry and comparative transcript profiling. *Plant Mol. Biol.* **62**: 697–714.

Shinners, K. J., Binversie, B. N., Muck, R. E., and P. J. Weimer (2007) Comparison of wet and dry corn stover harvest and storage. *Biomass Bioenergy* **31**, 211–221.

Sindhu, A., Langewisch, T., Olek, A., Multani, D. S., McCann, M. C., Vermerris, W., Carpita, N. C., and G. Johal (2007) Maize *Brittle stalk2* encodes a COBRA-like protein expressed in early organ development but required for tissue flexibility at maturity. *Plant Physiol.* (in review).

Somerville, C. R. (2006) Cellulose synthesis in higher plants. *Annu. Rev. Cell Dev. Biol.* **22**: 53–78.

Somerville, C., Bauer, S., Brininstool, G., Facette, M., Hamann, T., Milne, J., Osborne, E., Parezdez, A., Persson, S., Raab, T., Vorwerk, S., and H. Youngs (2004) Towards a systems approach to understanding plant cell walls. *Science* **306**: 2206–2211.

Suzuki, S., Lam, T. B. T., and K. Iiyama (1997) 5-Hydroxyguaiacyl nuclei as aromatic constituents of native lignin. *Phytochem.* **46**: 695–700.

Taylor, N. G., Howells, R. M., Huttly, A. K., Vickers, K., and Turner, S. R. (2003) Interactions among three distinct CesA proteins essential for cellulose synthesis. *Proc. Natl. Acad. Sci. USA* **100**: 1450–1455.

Taylor N. G., Laurie, S., and S. R. Turner (2000) Multiple cellulose synthesis catalytic subunits are required for cellulose synthesis in Arabidopsis. *Plant Cell* **12**: 2529–2539.

Taylor, N. G., Scheible, W. R., Cutler, S., Somerville, C. R., and S. R. Turner, (1999) The *irregular xylem3* locus of Arabidopsis encodes a cellulose synthase required for secondary cell wall synthesis. *Plant Cell* **11**: 769–779.

Tobias C. M., and E. K. Chow (2005) Structure of the *cinnamyl alcohol dehydrogenase* gene family in rice and promoter activity of a member associated with lignification. *Planta* **220**: 678–688.

Tracy, W. F. and H. L. Everett (1982) Variable penetrance and expressivity of grassy tillers, gt. *Maize Genet. Coop. News Lett.* **56**: 77–78.

USDA and U.S. DOE (2005) Biomass as feedstocks for a bioenergy and bioproducts industry: the technical feasibility of producing a billion-ton annual supply. (www.osti.gov/bridge).

U.S. DOE. (2006) Breaking the biological barriers to cellulosic ethanol: a joint research agenda, DOE/SC-0095, U.S. Department of Energy Office of Science and Office of Efficiency and Renewable Energy. (www.doegenomestolife.org/biofuels/).

Vergara, C. E. and N. C. Carpita (2001) β-D-Glycan synthases and the *CesA* gene family: lessons to be learned from the mixed-linkage $(1{\rightarrow}3),(1{\rightarrow}4)$ β-D-glucan synthase. *Plant Mol. Biol.* **47**: 145–160.

Vermerris, W., and J. J. Boon (2001) Tissue-specific patterns of lignification are disturbed in the *brown midrib2* mutant of maize (*Zea mays* L.). *J. Agric. Food Chem.* **49**: 721–728.

Vermerris, W., and L. M. McIntyre (1999) Time to flowering in *brown midrib* mutants of maize: an alternative approach to the analysis of developmental traits. *Heredity* **83**: 171–178.

Vermerris, W. and Nicholson, R. L. (2006) Phenolic Compound Biochemistry. Springer, New York, 276 pp.

Vermerris, W., Saballos, A., Ejeta, G., Mosier, N. S., Ladisch, M. R., and N. C. Carpita, (2007) Molecular breeding to enhance ethanol production from corn and sorghum stover. *Crop Sci.* **47**: S142–S153.

Vermerris, W., Thompson, K. J., and L. M. McIntyre (2002a) The maize *Brown midrib1* locus affects cell wall composition and plant development in a dose-dependent manner. *Heredity* **88**: 450–457.

Vermerris, W., Thompson, K. J., McIntyre, L. M., and J. D. Axtell (2002b) Evidence for an evolutionary conserved interaction between cell wall biogenesis and plant development in maize and sorghum. *BMC Evolutionary Biology* **2** http://www.biomedcentral.com/1471-2148/2/2.

Vignols, F., Rigau, J., Torres, M. A., Capellades, M., and P. Puigdomènech (1995) The *brown-midrib3* (*bm3*) mutation in maize occurs in the gene encoding caffeic acid *O*-methyl transferase. *Plant Cell* **7**: 407–416.

Wang, X., Cnops, G., Vanderhaeghen, R., Block, S. D., Van Montagu, M., and M. Van Lijsebettens (2001) *AtCSLD3*, a cellulose synthase-like gene important for root hair growth in Arabidopsis. *Plant Physiol.* **126**: 575–586.

Weigel, D., Alvarez, J., Smyth, D. R., Yanofsky, M. F., and E. M. Meyerowitz, (1992) *LEAFY* controls floral meristem identity in Arabidopsis. *Cell* **69**: 843–859.

Weimer, P. J., Dien, B. S., Springer, T. L., and K. P. Vogel, (2005) In vitro gas production as a surrogate measure of the fermentability of cellulosic biomass to ethanol. *Appl. Microbiol. Biotechnol.* **67**: 52–58.

Wilhelm, W. W., Johnson, J. M. F., Hatfield, J. L., Voorhees, W. B., and D. R. Linden (2004) Crop and soil productivity response to corn residue removal: A literature review. *Agronomy J.* **96**: 1–17.

Wu, Y., Sharp, R. E., Durachko, D. M., and D. J. Cosgrove (1996) Growth maintenance of the maize primary root at low water potentials involves increases in cell-wall extension properties, expansin activity, and wall susceptibility to expansins. *Plant Physiol.* **111**: 765–772.

Yang, B., and C. E. Wyman (2004) Effect of xylan and lignin removal by batch and flowthrough pretreatment on the enzymatic digestibility of corn stover cellulose. *Biotech. Bioeng.* **86**: 88–95.

Yennawar, N. H., Li, L. C., Dudzinski, D. M., Tabuchi, A., and D. J. Cosgrove (2006) Crystal structure and activities of EXPB1 (Zea m 1), a β-expansin and group-1 pollen allergen from maize. *Proc. Natl. Acad. Sci. USA* **103**: 14664–14671.

Yong, W., Link, B., O'Malley, R., Tewari, J., Hunter, C. T., Lu, C. A., Li, X., Bleecker, A. B., Koch, K. E., McCann, M. C., McCarty, D. R., Staiger, C., Thomas, S. R., Vermerris, W., and N. C. Carpita (2005) Genomics of plant cell wall biogenesis. *Planta* **221**: 747–751.

Zhu, J., Chen, S., Alvarez, S., Asirvatham, V. S., Schachtman, D. P., Wu, Y., and R. E. Sharp, (2006). Cell wall proteome in the maize primary root elongation zone. I. Extraction and identification of water-soluble and lightly ionically bound proteins. *Plant Physiol.* **140**: 313-325.

Zugenmaier, P. (2001) Conformation and packing of various crystalline cellulose fibers. *Prog. Polym. Sci.* 26: 1341–1417.

Part VI
Future Prospects

The Future of Maize

Jeffrey L. Bennetzen

Abstract In the near future, maize will continue to expand and diversify as a research model, as an industrial resource and as a crop for feed and fuel. The generation of the first maize genome sequence, followed by great improvements in genome sequencing technology, will allow the exceptional genetic diversity of maize to be described in multiple sequenced genomes. Maize will become a premier plant system for association genetics, and reverse genetic tools will continue to improve to a point where the genetic basis of phenotypic variation can be comprehensively defined. As a model for gene function in a complex genomic environment, maize will be without peer, leading to a uniquely deep understanding of the dynamic relationship between chromosome packaging, chromatin structure, epigenetics and the evolution of genetic regulatory circuits. As a crop, maize production will continue to consume more acreage worldwide, with mixed benefits and problems. The relentless narrowing of the commercial maize gene pool and its increased use in borderline environments will enhance the potential for both local and worldwide crop failures as the environment changes and when new pathogen races jet through a susceptible global germplasm. The recent preciptious shift to maize grain as an ethanol source is a mistake driven by politics and profit. One hopes that wisdom will prevail in the near future. Moreover, it is vital that the current maize-to-ethanol boom will not obscure the long-term value of maize as an industrial feedstock. Maize already has many uses, from adhesives to plastics, and food scientists will continue to expand this cornucopia. The great diversity, ease of genetic study, and talented research community in maize will ensure its continued place in the first line of model systems for plant biology.

J.L. Bennetzen
Department of Genetics, University of Georgia
maize@uga.edu

J.L. Bennetzen and S. Hake (eds.), *Maize Handbook - Volume II: Genetics and Genomics*,   771
© Springer Science+Business Media LLC 2009

1   Introduction

This chapter will focus on my opinions of what maize has to offer for the next few years of plant science research, and how this will impact the use of maize around the world. I will attempt to briefly capture the spirit and value of past and current maize research, thereby indicating the route to and foundation of the current status of maize as a model organism for biological study. The current roadblocks to further advances will be described, and the ever-expanding potential of maize as a model system and crop will be discussed.

2   The Last Century of Maize Research and Development

At the turn of the 20th century, maize was a very productive crop that dominated agriculture in much of North, Central and South America. The exceptional yield derived from a maize field, an outcome of its great pool of genetic diversity and the corn breeding skills of Native Americans over the several thousand years since its domestication, had begun to lead to its adoption by farmers worldwide as a food and feed crop. Commercialization of a seed industry was well underway (see the chapter by Troyer in this volume). With the rediscovery of Mendel's work at the dawn of the 20th century, maize rapidly became the model system for understanding the genetics of plants (see the chapter by Coe in this volume).

The special characteristics of maize that made it a powerful genetic research system were the ease of forward genetics due to the spatial separation of the male and female gametophyte, a great pool of natural genetic diversity and "naked eye" polymorphisms (Neuffer et al. 1997) that were easy-to-follow phenotypes, and large chromosomes with distinct structures and polymorphism that were available for cytological investigation. All of these strengths continue to contribute to the value of maize as a model for plant biology to this day, with the additional advantages of the development of a comprehensive set of modern research tools and an exceptionally rich and broad community of research scientists. In addition, maize has now been recognized as the most powerful model for understanding the types of complex genomes found in most angiosperms, and for the study of processes like C4 photosynthesis that are not found in the other great plant model, *Arabidopsis thaliana*.

At the level of basic research, maize has been unmatched in plants for the discovery of novel genetic phenomena, from the correlation of physical and genetic exchange in recombination, the origin of ring chromosomes, and the nature of the nucleolar organizer, to the existence and properties of telomeres and transposable elements, all first described by McClintock (Creighton and McClintock 1931; McClintock 1931; 1934; 1941; 1948; see the chapter by Kass and Chomet in this volume). The existence of epigenetics, including paramutation and imprinting (Brink 1958; Kermicle 1970; see the chapters by Springer and Gutierrez-Marcos and by Stam and Louwers in this volume), intragenic recombination in plants (Nelson 1962), the exceptional instability of plant disease resistance genes (Bennetzen et al. 1988), the nature of angiosperm genome organization (SanMiguel et al. 1996), and the existence of cis-acting regulatory elements in plants that can be far-removed from their structural genes (Stam et al. 2002; Clark et al. 2004) were all first discovered in maize. Because of the power of maize genetics, key discoveries in development, cell biology, physiology and stress resistance were also first made in maize, including the relationship between dwarfing mutations and gibberellic acid (Phinney 1956), the identification of the basis of cytoplasmic male sterility (cms) in plants (Dewey et al. 1986) and the first cloning of a nuclear restorer of cms (Cui et al. 1996), the discovery that plants have homeobox genes and that they are developmental regulatory factors (Vollbrecht et al. 1991), the first cloning and description of a plant disease resistance gene (Johal and Briggs 1992), and the discovery that organ shape in plants is independent of cell shape (Smith et al. 1996).

Along with this great expansion in our understanding of maize genetics and biology, the breeding and agronomics of maize production were rapidly producing commercial maize with higher yields, broader adaptability and specialized uses (see the chapters by Troyer, by Johnson and McCuddin and by Lee and Tracy in this volume). The discovery of heterosis and the subsequent use of uniform inbreds to produce dependable hybrids has configured maize improvement into a uniquely productive and sustainable activity. The amazingly consistent year-to-year improvements in maize germplasm over the last 60 years have not been matched by any crop at any time in recorded history (Lee and Tollenaar 2007).

3   Potential Problems, and Solutions, for the Improvement of Maize as a Model and as a Crop

The three biggest technical limitations to maize research and improvement at this time are the absence of a complete and ordered sequence of the maize gene space, robust and inexpensive transgenic production, and an incomplete set of genetic/genomic tools. None of these problems are intrinsic or particularly challenging: instead, they represent an absence of commitment to rectifying these deficiencies by the public and/or corporate maize research communities. For example, transformation and the production of transgenics in a variety of maize genetic backgrounds is apparently not a problem for several industrial labs, where proprietary protocols have been developed. Now that a serious commitment to maize transformation research has been made in the public sector (see the chapter by Wang et al. in this volume), it is likely that this limitation will cease to be significant within the next few years.

Most of the maize genome has been sequenced from inbred B73 (http://maizesequence.org/index.html), generating data that are now in the finishing and annotation stages. One initial limitation to molecular genetics in maize was the somewhat large size of its genome, compared to other proposed model plant species, but these limitations mostly disappear once the genome has been sequenced. All other issues being equal, map-based gene isolation, pathway description, transcriptome characterization, proteomics, and reverse genetic analysis of gene function, for instance, will all proceed as easily in a sequenced large genome as in a sequenced small genome. The 'next generation' sequencing of multiple maize genomes will also be a major activity in the near future, and the higher cost of sequencing so much repetitive DNA in this ~2400Mb genome will be offset by the great genetic, physiological and morphological diversity that is present within the species.

The absence of the full set of genetic tools in maize is particularly striking, given that many of them were first proposed and initiated in this organism. However, projects to generate an indexed set of transposon-tagged genes and high throughput expression analysis tools are underway (see the chapters by McCarty and Meeley and by Skibbe and Walbot in this volume). One hopes that activation tagging, virus-induced gene silencing, and the few remaining other deficiencies in the maize genomics/genetics toolkit will also be rectified within the next few years.

One special opportunity and potential problem for maize research, compared to non-crops like Arabidopsis, is the possible direct economic value of any discoveries that may be made. For this reason, resources developed by industry have not always been made fully available to the academic community. Of course, for-profit corporations must protect their intellectual property from competitors, so the complete sharing of information with the public sector is not always in their best interest. The appropriate response to this absence of complete overlap in intellectual property approaches, and often research goals, between the public and private sectors is to cooperate when possible and to act independently when not. In the past, problems have arisen when the public sector pauses in its research while waiting for the release of a private sector resource that may, or may not, become available. For instance, this phenomenon dramatically stifled maize gene expression analyses and EST-based gene discovery in the 1990s. In the future, the public sector must comprehend that it will sometimes need to finance academic research that has already been done but not released by companies, with the understanding that duplication/competition are not necessarily bad and that the release of a public sector resource often pries free a similar private sector resource. On the positive side, corporations often serve as tremendous partners for cooperative research on maize and can also help educate appropriate target audiences regarding the need for public support of important plant science research.

One negative aspect of modern crop improvement, exemplified very well in maize, has been the narrowing of germplasm due to the short-term pressures for continuous improvement of crop varieties. This had led to a preponderance of elite-by-elite crossing that utilizes only a tiny percent of the genetic diversity present in maize. Although this approach continues to produce inbreds and hybrids with improved yields, the breeding/testing cost per percent gain has risen by orders of magnitude over the last 60 years. Moreover, the genetic narrowing of the germplasm creates the potential for a worldwide catastrophe in maize production. For instance, when new pathogen races arise, and they always do, the lack of variability for disease resistance traits could lead to a worldwide epidemic that would dwarf the more than 700 million bushels of corn lost with cms-T maize in the US in 1970 (Tatum 1971). One solution to this problem would be a concerted effort to broaden the genetic diversity present in maize breeding populations. This activity has been undertaken by the relatively tiny breeding community in the US public sector for maize (Pollak and Salhuana 2001), in a cooperative project with several maize breeding companies. The small size of the public sector group, dwarfed by the hundreds of commercial plant breeders that they have trained, has largely erased their ability to generate elite hybrids that can compete with the high quality, and heavily marketed, hybrids from the major seed companies. However, removed from the constraints for immediate production of superior germplasm that is ready for release, this group has found a unique niche in the movement of exotic maize germplasm into populations that show enough commercial potential that they can be employed by the private sector (e.g., Balint-Kurti et al. 2006).

Like its relatively large genome, the longish generation time of maize has been argued to be a significant factor limiting its use as a model for plant biology. Two generations per year is standard for maize geneticists, and they have a tradition of making crosses in advance of their guaranteed need. That is, because one can generate thousands of F1 or self progeny in one or two minutes of field time, one tends to make some additional crosses on the off-chance that they may be needed and to also duplicate one's analysis. An excellent example of the kind of comprehensive genetic resources that can be generated for maize with relatively little effort, and would be daunting for organisms like Arabidopsis and rice that are more challenging to outcross, is the generation of nested association mapping populations (Buckler et al. 2008). The greenhouse or field space needed to screen thousands of progeny is a greater problem than generation time for maize, but it is partly offset by the ease of detecting rare phenotypes in these large plants. Moreover, as gene analysis shifts somewhat from a forward genetic to reverse genetic mode, field space and generation time both become less significant, while the advantages of a large and vigorous organism for phenotype detection are enhanced.

4 The Maize Grain-to-Ethanol Boom

One severe, and unwarranted, current problem for maize research has been the recent canonization of ethanol from maize grain as a solution to US bioenergy needs. In the short term, maize researchers are being distracted from important issues of maize study and improvement while, in the long term, this policy will come back to haunt maize scientists as somehow abetting in this political/corporate nonsense.

Independent studies indicate that ethanol from maize grain is not a good source of liquid fuel because it uses as much or almost as much energy in the generation of ethanol as is produced (Farrell et al. 2006), particularly when one takes into consideration the energy consumption that is needed to generate the portion of US incomes that are taxed to subsidize maize and maize-to-ethanol conversion. Hence, using maize grain to generate ethanol will greatly increase carbon emissions

(Searchinger et al. 2008), will not significantly decrease US dependence on foreign oil, and already costs US taxpayers billions of dollars in subsidies that could be better used to encourage energy conservation or support research on sustainable alternative energy production. Moreover, movement of maize acreage from food and feed production into use as a feedstock for ethanol has been a contributing factor in the increased cost of food worldwide. Hence, a program foisted on the US taxpayer by politicians hungry for "corn state" votes is rapidly enriching a few farmers and agrichemical/seed corporations at the expense of consumers worldwide. This is especially tragic for the poorest of the poor in the developing world, where higher food costs will increase malnutrition and famine.

It is to the credit of the academic research community, particularly the maize community, that it has not supported this precipitous shift to maize grain as a source for biofuels. Plant scientists need to continue to describe the scientific and economic studies that show the lack of wisdom in current policy. Will biofuels, perhaps including those derived from maize, become a useful part of the world's efforts to decrease carbon emissions and use more renewable energy? Almost certainly, although the relative long-term potential of biofuels compared to other alternative energy sources is not yet clear, and the use of perennial woody species and grasses seems a better feedstock source. It is possible that stover from maize or other annual crops might contribute to the future energy mix, if lignocellulosic conversion of plant biomass becomes a significantly more efficient and less expensive process (Ragauskas et al. 2006). Given its excellent genetics, modern molecular toolkit and close relatedness to candidate perennial bioenergy crops like switchgrass, however, it is clear that maize will be an excellent organism for the study of biomass production traits (Lawrence and Walbot 2007; see the chapter by Vermerris in this volume).

5 Where Next for Maize Research and Improvement

Although this type of evaluation should be an ongoing process, the passage through any great transition provides a unique opportunity to assess where you are and determine where you should go next. The upcoming release of the maize genome sequence provides just such a transition. There are other excellent plant research systems, especially Arabidopsis, where a large and enthusiastic research community has an unmatched set of genetic and genomic tools to enable the next generation of research. So, there is no reason to believe that all types of plant science research need to be pursued at their greatest depths in maize. It seems wiser, in fact essential, that maize researchers focus on those types of questions and approaches that are particularly feasible, and will be particularly informative, in maize.

Maize has unmatched genetic diversity that contributes to dramatic phenotypic variability for traits ranging from the purely esoteric to the commercially vital. Hence, maize deserves to become the pre-eminent plant for association genetics and quantitative trait analyses. The populations, analytical approaches, markers and molecular scoring tools all exist right now (Yu and Buckler 2006; Buckler et al. 2008), and all that is lacking is the serious phenotyping and data analyses that need to be pursued.

The complexity of the maize genome, greater than that of any other sequenced genome, certainly provides an ongoing challenge to analysis, but it also provides a unique opportunity to understand the genome of a fairly average angiosperm. The epigenetic interactions (or lack thereof) between transposable elements, protein-encoding genes, gene fragments, and RNA genes will all be writ large in this complex arrangement of euchromatin and heterochromatin that, in reality, will not be found to be a binary on/off pair of chromatin states but rather a great number of chromatin hues (Bennetzen 2000). The exceptional cytogenetics in maize will allow discovery of the regulated behavior of chromosomes not only in condensation, recombination and segregation, but also in the interphase nucleus (see the chapters by Dawe and by Cande et al. in this volume).

Any developmental or physiological process that can be better characterized in maize than in other model plants, like C4 photosynthesis or the restoration of cytoplasmic male sterility, will now be pursued with a comprehensive genetic and molecular toolkit. The general colinearity of gene order across the grasses (Bennetzen and Freeling 1997) means that discoveries in maize can be genetically related to identical or similar processes in other grass species, thereby empowering research across this entire family of important plants.

From both academic and agricultural perspectives, research into heterosis and abiotic stress tolerance deserve high priority, and can now be approached with the systematic use of a full range of molecular, genetic, biochemical, physiological, computational and genomic tools. With the changing needs of agriculture and the environment, it will also be important to pursue research into optimizing maize quality and yield under the lowest possible input regimens.

Because it produces so much biomass in a small area, maize can be an excellent 'green factory' for the production of pharmaceuticals and other high value products (Ohlrogge and Chrispeels 2003). Fermentation facilities and chemical plants are very expensive to operate, so 'in planta' production can succeed when methods for compound purification are developed and germplasm is engineered with transgenes that have no chance of escape into the wild.

In short, research on any biological question shared with other organisms, or unique to plants, can be pursued in maize. The chosen targets for further study in this tractable species will differ at least somewhat across disciplines, between the public and private sectors, over time, and in different parts of the world. The current trend in the US and Europe toward multi-investigator, highly coordinated, mega-projects (Check 2004; Petsko 2006; 2007) will tend to inhibit creativity and serendipitous breakthroughs, while pushing forward the expected discoveries at a faster rate. One hopes that a balance will be attained that allows the majority of the maize research community to pursue at least some discovery research. Maize now has no major technical limitations to its study, so the only restriction on future progress will be a lack of creativity and commitment or an excess of politics and caution.

References

Balint-Kurti, P.J., M. Blanco, M. Millard, S. Duvick, J.B. Holland, M.J. Clements, R.N. Holley, M.L. Carson and M.M. Goodman (2006) Registration of 20 GEM maize breeding germplasm lines adapted to the southern USA. *Crop Sci.* **46:** 996–998.

Bennetzen, J.L. (2000) The many hues of plant heterochromatin. *Genome Biol.* **1:** 107.1–107.4.

Bennetzen, J.L. and M. Freeling (1997) The unified grass genome: synergy in synteny. *Genome Res.* **7:** 301–307.

Bennetzen, J.L., M.-M. Qin, S. Ingels and A.H. Ellingboe (1988) Allele-specific and *Mutator*-associated instability at the *Rp1* disease resistance locus of maize. *Nature* **332:** 369-370.

Brink, R.A. (1958) Paramutation at the *R* locus in maize. *Cold Spring Harbor Symp. Quant. Biol.* 23: 379–391.

Buckler, E.S., IV, J. Yu, J.B. Holland and M.D. McMullen (2008) Genome-wide complex trait dissection through nested association mapping. *Genetics* **178:** 539–551.

Check, E. (2004) David versus Goliath. *Nature* **432:** 546–548.

Clark, R.M., E. Linton, J. Messing and J.F. Doebley (2004) Pattern of diversity in the genomic region near the maize domestication gene *tb1*. *Proc. Natl. Acad. Sci. USA* **101:** 700–707.

Creighton, H.B. and B. McClintock (1931) A correlation of cytological and genetical crossing-over in *Zea mays*. *Proc. Natl. Acad. Sci. USA* **17:** 492–497.

Cui, X., R.P. Wise and P.S. Schnable (1996) The *rf2* nuclear restorer gene of male-sterile T-cytoplasm maize. *Science* **272:** 1334–1336.

Dewey, R.E., C.S. Levings III and D.H. Timothy (1986) Novel recombinations in the maize mitochondrial genome produce a unique transcriptional unit in the Texas male-sterile cytoplasm. *Cell* **44:** 439–449.

Farrell, A.E., R.J. Pleven, B.T. Turner, A.D. Jones, M. O'Hare, and D.M. Kammen (2006) Ethanol can contribute to energy and environmental goals. *Science* **311:** 506–508.

Johal, G.S. and S.P. Briggs (1992) Reductase activity encoded by the *HM1* disease resistance gene in maize. *Science* **258:** 985–987.

Kermicle, J. L. (1970) Dependence of the R-mottled aleurone phenotype in maize on mode of sexual transmission. *Genetics* **66:** 69–85.

Lawrence, C.J. and V. Walbot (2007) Translational genomics for bioenergy production from fuelstock grasses: Maize as a model species. *Plant Cell* **19:** 2091–2094.

Lee, E.A. and M. Tollenaar (2007) Physiological basis of successful breeding strategies for maize grain yield. *Crop Sci.* **47:** S202–S215.

McClintock, B. (1931) Cytological observations of deficiencies involving known genes, translocations, and an inversion in *Zea mays*. *Missouri Agric. Exp. Station Res. Bull.* **163:** 1–30.

McClintock, B. (1934) The relation of a particular chromosomal element to the development of nucleoli in *Zea mays*. *Zeitschrift fur Zellforschung und Microskopische Anatomie* **21:** 294–328.

McClintock, B. (1941) The stability of broken ends of chromosomes in *Zea mays*. *Genetics* **26:** 234–282.

McClintock, B. (1948) Mutable loci in maize. *Carnegie Inst. Wash. Yearbook* **47:** 155–169.

Nelson, O.E. (1962) The *waxy* locus in maize. I. Intralocus recombination frequency estimates by pollen and by conventional analysis. *Genetics* **47:** 737–742.

Neuffer, M.G., E.H. Coe and S.R. Wessler (1997) *Mutants of Maize*. Cold Spring Harbor Press, New York.

Ohlrogge, J. and M.J. Chrispeels (2003) Plants as chemical and pharmaceutical factories. In M.J. Chrispeels and D.E. Sadava (eds.) *Plants, Genes, and Crop Biotechnology*, Jones and Bartlett Publ., Sudbury, MA, pp. 500–527.

Petsko, G.A. (2006) The system is broken. *Genome Biol.* **7:** 105.

Petsko, G.A. (2007) An idea whose time has gone. *Genome Biol.* **8:** 107.

Phinney, B.O. (1956). Growth response of single-gene dwarf mutants in maize to gibberellic acid. *Proc. Natl. Acad. Sci. USA* **42:** 185–189.

Pollak, L.M. and W. Salhuana (2001) The germplasm enhancement of maize (GEM) project: Private and public sector collaboration. In H.D. Cooper, C. Spillane and T. Hodgkin (eds.) *Broadening the Genetic Base of Crop Production*. CABI Publ., Wallingford, Oxon, United Kingdom, pp. 319–329.

Ragauskas, A.J., C.K. Williams, B.H. Davison, G. Britovsek, J. Cairney, C.A. Eckert, W.J. Frederick, Jr., J.P. Hallett, D.J. Leaks, C.L. Liotta, J.R. Mielenz, R. Murphy, R. Templer and T. Tschaplinski (2006) The path forward for biofuels and biomaterials. *Science* **311:** 484–489.

SanMiguel, P., A. Tikhonov, Y.-K. Jin, N. Motchoulskaia, D. Zakharov, A. Melake-Berhan, P.S. Springer, K.J. Edwards, M. Lee, Z. Avramova and J.L. Bennetzen (1996) Nested retrotransposons in the intergenic regions of the maize genome. *Science* **274:** 765–768.

Searchinger, T., R. Heimlich, R. A. Houghton, F. Dong, A. Elobeid, J. Fabiosa, S. Tokgoz, D. Hayes, and T.-H. Yu (2008) Use of U.S. croplands for biofuels increases greenhouse gases through emissions from land-use change. *Science* **319:** 1238–1240.

Smith, L.G., S. Hake and A.W. Sylvester (1996) The *tangled-1* mutation alters cell division orientations throughout maize leaf development without altering leaf shape. *Development* **122:** 481–489.

Stam, M., C. Belele, W. Ramakrishna, J. Dorweiler, J.L. Bennetzen and V.L. Chandler (2002) The regulatory regions required for *B'* paramutation and expression are located far upstream of the maize *b1* transcribed sequences. *Genetics* **162:** 917–930.

Tatum, L.A. (1971) The southern corn leaf blight epidemic. *Science* **171:** 1113–1116.

Vollbrecht, E., B. Veit, N. Sinha and S. Hake (1991) The developmental gene *Knotted-1* is a member of a maize homeobox gene family. *Nature* **350:** 241–243.

Yu, J. and E.S. Buckler (2006) Genetic association mapping and genome organization of maize. *Curr. Opin. Biotech.* **17:** 155–160.

Index

Printed in the United States of America